Extreme Events in Geospace

Extreme Events in Geospace
Origins, Predictability, and Consequences

Edited by

Natalia Buzulukova

Elsevier
Radarweg 29, PO Box 211, 1000 AE Amsterdam, Netherlands
The Boulevard, Langford Lane, Kidlington, Oxford OX5 1GB, United Kingdom
50 Hampshire Street, 5th Floor, Cambridge, MA 02139, United States

© 2018 Elsevier Inc. All rights reserved.

No part of this publication may be reproduced or transmitted in any form or by any means, electronic or mechanical, including photocopying, recording, or any information storage and retrieval system, without permission in writing from the publisher. Details on how to seek permission, further information about the Publisher's permissions policies and our arrangements with organizations such as the Copyright Clearance Center and the Copyright Licensing Agency, can be found at our website: www.elsevier.com/permissions.

This book and the individual contributions contained in it are protected under copyright by the Publisher (other than as may be noted herein).

Notices
Knowledge and best practice in this field are constantly changing. As new research and experience broaden our understanding, changes in research methods, professional practices, or medical treatment may become necessary.

Practitioners and researchers must always rely on their own experience and knowledge in evaluating and using any information, methods, compounds, or experiments described herein. In using such information or methods they should be mindful of their own safety and the safety of others, including parties for whom they have a professional responsibility.

To the fullest extent of the law, neither the Publisher nor the authors, contributors, or editors, assume any liability for any injury and/or damage to persons or property as a matter of products liability, negligence or otherwise, or from any use or operation of any methods, products, instructions, or ideas contained in the material herein.

Library of Congress Cataloging-in-Publication Data
A catalog record for this book is available from the Library of Congress

British Library Cataloguing-in-Publication Data
A catalogue record for this book is available from the British Library

ISBN: 978-0-12-812700-1

For information on all Elsevier publications
visit our website at https://www.elsevier.com/books-and-journals

Publisher: Candice Janco
Acquisition Editor: Marisa LaFleur
Editorial Project Manager: Carly Demetre
Production Project Manager: Anitha Sivaraj
Cover Designer: Mark Rogers

Cover Credit:
1st image in the banner – Active Regions Galore. Credit: Solar Dynamics Observatory, NASA
2nd image in the banner – Stunning Aurora from Space. Credit: NASA
6th (last) image in the banner – 3D model of Earth's magnetosphere and ring current. Image source: NASA GSFC/Natalia Buzulukova

Typeset by SPi Global, India

Contents

Author Biography's xvii
Foreword xxxv
Acronyms xxxvii
Introduction xlix

PART 1 OVERVIEW OF IMPACTS AND EFFECTS

CHAPTER 1 Linking Space Weather Science to Impacts—The View From the Earth 3
  1 Introduction 3
  2 Space Weather Environments at Earth 4
  3 Geomagnetically Induced Currents—The Impacts of Natural Geoelectric Fields 8
  4 Space Weather Impacts on the Upper Atmosphere 16
     4.1 Overview of the Upper Atmosphere 16
     4.2 Trans-Ionospheric Radio Propagation 17
     4.3 Atmospheric Drag 22
  5 Atmospheric Radiation Environment 25
  6 Satellite Plasma Environments 27
  7 Looking to the Future: How May Space Weather Risks Evolve? 29
     References 31

PART 2 SOLAR ORIGINS AND STATISTICS OF EXTREMES

CHAPTER 2 Extreme Solar Eruptions and their Space Weather Consequences 37
  1 Introduction 37
  2 Overview of Extreme Events 38
  3 Estimates of Extreme Events 42
     3.1 CME Speeds 42
     3.2 Distribution Functions for CME Speeds and Kinetic Energies 43
     3.3 Flare Size Distribution 44
     3.4 Active Regions and Their Magnetic Fields 46
  4 Consequences of Solar Eruptions 51
     4.1 SEP Events 51
     4.2 SEP Fluences 52
     4.3 Large Geomagnetic Storms 54
  5 Summary and Conclusions 58
     Acknowledgments 59
     References 59
     Further Reading 63

CHAPTER 3 Solar Flare Forecasting: Present Methods and Challenges ... 65

1 Introduction: Solar Flares and Societal Impacts ... 66
  1.1 Flare Forecasting History ... 68
2 Present Approaches ... 70
  2.1 Two Basic Paradigms ... 70
  2.2 Event Definitions ... 71
  2.3 Parametrizations ... 72
  2.4 The Statistical Classifiers ... 74
  2.5 Self-Organized Criticality and Related ... 75
  2.6 Operational Versus Research ... 76
  2.7 Evaluation ... 76
  2.8 The Role of Numerical Models ... 78
3 Present Status ... 81
  3.1 How Good? ... 81
  3.2 Why Not So Good? ... 83
  3.3 Extreme Solar Flares ... 84
4 Future ... 86
  4.1 Outlook for the Extreme Extremes ... 87
  4.2 Pertinent Question: Are Forecasts Useful? ... 87
  4.3 Avenues for Improvement ... 88
5 Summary and Recommendations ... 89
Acknowledgments ... 90
References ... 90

CHAPTER 4 Geoeffectiveness of Solar and Interplanetary Structures and Generation of Strong Geomagnetic Storms ... 99

1 Introduction ... 99
2 Methods ... 100
3 Results ... 101
4 Discussion ... 106
5 Conclusions ... 109
Acknowledgments ... 110
References ... 110

CHAPTER 5 Statistics of Extreme Space Weather Events ... 115

1 Introduction ... 115
2 Methodologies ... 117
  2.1 Datasets ... 117
  2.2 Statistical Modeling ... 119

3 Results.. 122
    3.1 Assessing the Validity of the Time Stationarity Assumption 126
    3.2 Analysis of *Dxt* and *Dcx* 127
    3.3 Extreme Space Weather Events in the Ionosphere: The *AE* Index 127
    3.4 Extreme Space Weather Events in the Heliosphere: Energetic Protons 129
4 Discussion 130
5 Future Studies 135
6 Conclusions 136
    Acknowledgments 136
    References 137

CHAPTER 6   Data-Driven Modeling of Extreme Space Weather 139
    1 Introduction 139
    2 Data-Driven Modeling of Space Weather 141
    3 Predictability of Extreme Space Weather 146
    4 Conclusion 149
        Acknowledgments 150
        References 150
        Further Reading 153

PART 3   GEOMAGNETIC STORMS AND GEOMAGNETICALLY INDUCED CURRENTS

CHAPTER 7   Supergeomagnetic Storms: Past, Present, and Future 157
    1 Historical Background 157
    2 Present Knowledge About Geomagnetic Storms 159
        2.1 Interplanetary Causes of Intense Magnetic Storms 159
        2.2 Magnetic Storms: Categories and Types 161
        2.3 Some Important Characteristics of Magnetic Storms 161
    3 Supermagnetic Storms 164
        3.1 Past Supermagnetic Storms 166
        3.2 Supermagnetic Storms: Present (Space-Age Era) 169
        3.3 Supermagnetic Storms: In Future 171
    4 Nowcasting and Short-Term Forecasting of Supermagnetic Storm 173
    5 Conclusions 174
        Acknowledgments 175
        Glossary 175
        References 177
        Further Reading 184

CHAPTER 8 An Overview of Science Challenges Pertaining to Our Understanding of Extreme Geomagnetically Induced Currents 187
1 Introduction 187
   1.1 Geomagnetic Storms at Earth 188
   1.2 Basic Theory of GICs 188
2 Impact on Ground Systems 189
   2.1 Electrical Power Systems 190
   2.2 Oil and Gas Pipelines 192
   2.3 Other Systems 193
3 U.S. Federal Actions Relating to GICs 193
4 Key Science Challenges 194
   4.1 Extreme Drivers 195
   4.2 Modeling Extremes 197
   4.3 Defining Extremes 199
5 Concluding Remarks 201
Acknowledgments 201
References 202

CHAPTER 9 Extreme-Event Geoelectric Hazard Maps 209
1 Introduction 209
2 Defining the Hazard 210
3 Direct Geoelectric Monitoring 211
4 Induction in a Conducting Earth 211
5 Magnetic Observatory Data 212
6 Geomagnetic Waveform Time Series 212
7 Observatory Magnetic Hazard Functions 215
8 Global Magnetic Hazard Functions 216
9 Magnetotelluric Impedances 217
10 Geological Interpretations 219
11 Geoelectric Hazard Maps 221
12 Discussion 225
Acknowledgments 225
References 225

CHAPTER 10 Geomagnetic Storms: First-Principles Models for Extreme Geospace Environment 231
1 Introduction 232
2 Overview of First-Principles Magnetospheric Models 234

3 Modeling of Extreme and Intense Geomagnetic Storms 236
    3.1 Modeling of Carrington-Type Events 238
    3.2 Modeling of Radiation Belt Response for Extreme Storms 239
4 An Example: Geomagnetic Storm of June 22–23, 2015 241
    4.1 The Model Run Setup 241
    4.2 The Results for Pressure and Current Distribution 242
    4.3 Comparison With Ground-Based Magnetometers and Satellite Data 244
5 Role of the Ring Current Plasma in Generation of $d\text{B}/dt$ and Variability of Electric Fields and FACs at Low Latitudes 248
6 Challenges and Future Directions 250
7 Conclusions 251
    Acknowledgments 252
    References 252

CHAPTER 11 Empirical Modeling of Extreme Events: Storm-Time Geomagnetic Field, Electric Current, and Pressure Distributions 259

1 Introduction 259
2 Recent Advances in Empirical Geomagnetic Field Modeling 261
    2.1 Model Structure 261
    2.2 Data Binning 263
    2.3 Model Database 264
3 March 2015 Storm 264
4 Bastille Day Storm 270
5 Conclusion 275
    Acknowledgments 275
    References 276

PART 4 PLASMA AND RADIATION ENVIRONMENT

CHAPTER 12 Extreme Space Weather Events: A GOES Perspective 283

1 Introduction 283
2 GOES Extreme Space Weather Events 286
3 GOES Extreme Events, Cases 1–12 301
4 Concluding Remarks 324
    Acknowledgments 324
    References 325

CHAPTER 13 Near-Earth Radiation Environment for Extreme Solar and Geomagnetic Conditions 349

1 Introduction 349
2 Solar Energetic Particles in Space 350
  2.1 Concept of Extreme SEP Events 350
  2.2 Largest SEP Events and Distribution Function 351
  2.3 Solar Cosmic Rays: Penetration Boundary and Changes of Cutoff Rigidities 354
3 GCR Modulation Over the Solar Cycles 354
4 The Inner Proton Radiation Belt Variations Over Solar Cycles 359
5 Radiation Environment for Crewed Orbital Stations 361
6 Conclusions 367
  Acknowledgments 368
  References 368
  Further Reading 372

CHAPTER 14 Magnetospheric "Killer" Relativistic Electron Dropouts (REDs) and Repopulation: A Cyclical Process 373

1 Introduction 374
2 Solar Wind/Interplanetary Driving and Geomagnetic Characteristics: A Schematic 375
3 Relativistic Electron Dropout and Acceleration: An Example 377
4 Solar Cycle Phase Dependence of Electron Acceleration 379
5 Maximum Energy-Level Dependence of Electron Acceleration 381
6 HILDCAA Duration Dependence of Electron Acceleration 384
7 Are CIR Storms Important? 385
8 Relativistic Electron Variation During ICME Magnetic Storms 387
  8.1 Fast Shock, Sheath, and First Magnetic Storm 387
  8.2 Magnetic Cloud (MC) and Second and Third Storms 389
  8.3 HSS and Storm Recovery Phase 389
  8.4 Relativistic Electron Flux Variability During the Complex Interplanetary Event: Shock Effects 389
  8.5 Electron Acceleration 390
9 Conclusions 390
  References 391

CHAPTER 15 Extreme Space Weather Spacecraft Surface Charging and Arcing Effects 401

1 Limits of Discussion 402
2 Cause of Arcing 402
3 Effects of Arcing 402

    4 Physics of Charging in Geosynchronous Earth Orbit 404
    5 The Spacecraft Charging Equation 405
    6 What Are the Worst Spacecraft Charging Events? 411
    7 Limits on Spacecraft Charging Events 412
    8 The Galaxy 15 Failure 415
    9 Conclusion 416
       References 416

CHAPTER 16 Deep Dielectric Charging and Spacecraft Anomalies 419

    1 Introduction 419
    2 What Is Deep Dielectric Charging? 420
       2.1 The Roles of Ions 422
    3 Space Environments 423
       3.1 Trapped Radiation 423
       3.2 Temporal Variation of the Radiation Belts 423
       3.3 Relative Role of Electrons and Ions 424
    4 Deep Dielectric Charging and Discharging 426
    5 Dependence from Geomagnetic Indices: Dst and Kp Index 428
    6 Delay Time 428
    7 Discharge Event Parameters 429
    8 Spacecraft Design Guidelines 429
    9 Conclusion 431
       Acknowledgment 431
       References 431

CHAPTER 17 Solar Particle Events and Human Deep Space Exploration: Measurements and Considerations 433

    1 Radiation in Space and Health Risks for Astronauts 433
    2 GCR vs SPEs: Different Approaches for Risk Mitigation 435
    3 SPEs as Measured in a Space Habitat (International Space Station) ... 437
    4 The ALTEA Detector Onboard the ISS 438
    5 Results From SPE Measurements in the ISS 438
       5.1 The December 13, 2006, SPE 438
       5.2 The March 7, 2012, SPE 441
       5.3 The May 17, 2012, SPE 444
    6 Final Remarks 446
       6.1 Forecasting at 1 AU 446
       6.2 Forecasting at Space Habitat 447
       6.3 Radiation on Mars 447
       6.4 Countermeasures 448
    7 Conclusions 449
       Acknowledgments 449
       References 449

CHAPTER 18 Characterizing the Variation in Atmospheric Radiation at Aviation Altitudes 453

1 Radiation Sources and Their Effects on Aviation 453
2 Status of Models 456
3 Status of Measurements 457
4 Status of Monitoring for Extreme Conditions 458
5 Classification of Aviation-Relevant Extreme Space Weather Radiation Events 460
6 Example of an Extreme Event 463
7 Conclusion 466
   Acknowledgments 466
   References 467

CHAPTER 19 High-Energy Transient Luminous Atmospheric Phenomena: The Potential Danger for Suborbital Flights 473

1 Introduction 473
2 Phenomenology of TLEs 474
3 Experimental Data on TLE from UVRIR Detector on Board Moscow State University Satellites 477
   3.1 TLE Types Measured by UVRIR Detector 478
   3.2 TLE Distribution Over Photon Numbers 480
   3.3 Series of TLE 480
4 Discussion 483
   4.1 Overview of TLE Models and Relation to Other Space Weather Phenomena 483
   4.2 TLE Energy Deposition 485
   4.3 TLEs as a Radiation Hazard 486
5 Results and Conclusions 487
   References 488
   Further Reading 490

PART 5 IONOSPHERIC/THERMOSPHERIC EFFECTS AND IMPACTS

CHAPTER 20 Ionosphere and Thermosphere Responses to Extreme Geomagnetic Storms 493

1 Historical Background 493
2 Electric Fields and the Creation of Large TEC Increases 495
   2.1 Equatorial Plasma Irregularities and Scintillation 499
3 The Role of Ion-Neutral Coupling 501
   3.1 Buoyancy (Gravity) Waves 503
4 Extreme Nighttime Responses Following the Storm Main Phase (Florida Effect) 504

5 Conclusions and Future Outlook 506
Acknowledgments 508
References 508
Further Reading 511

CHAPTER 21 How Might the Thermosphere and Ionosphere React to an Extreme Space Weather Event? 513
1 Introduction 513
2 Effects of Solar EUV and UV Radiation 517
3 Effect of an Extreme Solar Flare 519
4 Effects of an Extreme CME Driving a Geomagnetic Storm 520
    4.1 Defining the Drivers 521
    4.2 Neutral Atmosphere Response to an Extreme Geomagnetic Storm 525
    4.3 Ionospheric Response to an Extreme Geomagnetic Storm 529
5 Summary and Conclusions 532
Acknowledgments 533
References 534
Further Reading 539

CHAPTER 22 The Effect of Solar Radio Bursts on GNSS Signals 541
1 Introduction 541
    1.1 The Solar Radio Burst (SRB) 541
    1.2 The Global Navigation Satellite System (GNSS) 543
2 Review the Effect of SRBs on GNSS Signals 544
    2.1 Reduction of Signal-to-Noise Ratio (SNR) 545
    2.2 Signal Loss of Lock (LOL) 545
    2.3 Decrease of Positioning Precision 546
    2.4 Effect on Space-Based GNSS 546
    2.5 Threshold Value of SRBs Affecting GNSS 547
3 Extreme SRB Case on December 6, 2006 547
4 Discussions 551
5 Conclusions 552
Acknowledgments 552
References 552

CHAPTER 23 Extreme Ionospheric Storms and Their Effects on GPS Systems 555
1 Introduction 555
2 Global Positioning System 557

 3 The Ionosphere 558
  3.1 Total Electron Content 559
  3.2 Low-Latitude Scintillation 560
  3.3 High-Latitude Scintillation 561
 4 Ionospheric Structures Evident in TEC Data 565
  4.1 Storm Enhanced Densities, Patches and Blobs 565
  4.2 Traveling Ionospheric Disturbances 569
 5 Event Studies for Large Ionospheric Storms 572
  5.1 October 2003 574
  5.2 November 2003 574
  5.3 November 2004 575
 6 System Effects of Ionospheric Storms 575
 7 Discussion 578
  Acknowledgments 580
  References 580
  Further Reading 586

CHAPTER 24 Recent Geoeffective Space Weather Events and Technological System Impacts 587
 1 Introduction 588
 2 Recent Events: Overview 588
 3 Solar Origins of Activity 590
 4 Geospace Response 591
  4.1 Energetic Particles and Magnetic Field Observations at GEO 592
  4.2 Geosynchronous Magnetopause Crossings 593
  4.3 Radiation Environment at GEO 593
  4.4 Radiation Environment at LEO 594
 5 Ionospheric Effects 595
 6 System Impacts 601
  6.1 Technological System Impacts 601
  6.2 Aviation Navigation System Impacts 602
 7 Summary 605
  Acknowledgments 605
  References 605
  Further Reading 609

CHAPTER 25 Extreme Space Weather in Time: Effects on Earth 611
 1 Introduction 611
 2 Space Weather Events From the Current Sun 612
 3 Space Weather Events From the Young Sun 613
  3.1 Solar Superflares and CMEs 614

3.2 The Young Sun's Wind 616
4 3D MHD Model of Super-CME Interaction With the Early Earth 617
  4.1 Effects of CMEs on the Magnetosphere of the Early Earth 618
  4.2 Effects of XUV Flux on Atmospheric Escape From the Young Earth 622
5 Space Weather as a Factor of Habitability 624
6 Conclusions 628
  Acknowledgments 629
  References 629
  Further Reading 632

PART 6 DEALING WITH THE SPACE WEATHER

CHAPTER 26 Dealing With Space Weather: The Canadian Experience 635
1 Introduction 635
2 High-Frequency Radio Communications 637
3 Satellites 639
4 Ground Systems 642
5 Surveying and Navigation 646
6 Concluding Remarks 648
  Acknowledgments 652
  References 652
  Further Reading 655

CHAPTER 27 Space Weather: What are Policymakers Seeking? 657
1 Introduction 657
2 Key Concepts 659
  2.1 Space Weather as a Natural Hazard 659
  2.2 Policy Responses to Natural Hazards 662
  2.3 The Importance of Science 664
3 How to Assess Extreme Risks 666
4 What Knowledge is Needed in the Future 671
  4.1 The Need for Data 671
  4.2 The Need to Learn From Meteorology 672
  4.3 The Need for Better Science and Better Models 674
  References 678
  Further Reading 682

CHAPTER 28 Extreme Space Weather and Emergency Management 683
1 Why Emergency Managers Care 684
2 Understanding the Risks of Extreme Solar Events 684
  2.1 The Hazard 684

    3 What Emergency Managers Need From Researchers and Engineers 688
        3.1 Plain Language 688
        3.2 Expert Analysis Available on Demand.. 688
        3.3 Sharable Information 689
        3.4 Improved Forecast Products.. 690
    4 Conclusion 699
        References 699
        Further Reading.. 700

CHAPTER 29 The Social and Economic Impacts of Moderate and Severe Space Weather.. 701

    1 Introduction 701
    2 Approach... 703
    3 Results... 705
        3.1 Electric Power 705
        3.2 Aviation 707
        3.3 Satellites.. 708
        3.4 GNSS Users.. 708
    4 Next Steps and Concluding Remarks 709
        References 709
        Further Reading.. 710

CHAPTER 30 Severe Space Weather Events in the Australian Context 711

    1 Introduction and Concept Development 711
    2 The Nature of Severe Events and the Regional Context 712
    3 The Severe Event Service 715
    4 Policy Background 716
    5 Stakeholder Technology Groups... 716
    6 Conclusion 717
        References 718

CHAPTER 31 Extreme Space Weather Research in Japan.. 719

    1 Overview and History of Operational Space Weather Forecast 719
    2 Action to Telecommunications and Satellite Positioning 720
    3 Action to Aviation... 721
    4 Action to Satellite Saving 723
    5 Action to GIC... 724
    6 Introduction of PSTEP 725
        References 725

Index 727

Author Biography's

Jeffery Adkins is an economist with the I.M. Systems Group (IMSG) working for NOAA's Office of the Chief Economist. A key focus of Jeff's work is the economic valuation of government products and services. Other interests include the economics of hazard resilience (e.g., meteorological and space weather events) and measurement of the U.S. ocean-based economy (through NOAA's Economics: National Ocean Watch project). Jeff has worked as an economist for the federal government since 1979.

Vladimir Airapetian is a senior astrophysicist at NASA's Goddard Space Flight Center and a research professor at George Mason University and the American University. Dr. Airapetian obtained a bachelor of science with a major in physics, his master of science in theoretical physics from the Yerevan State University in 1982, and his PhD in theoretical astrophysics from Byurakan Astrophysical Observatory, Armenia, in 1990. Prior to joining NASA, Dr. Airapetian was a postdoctoral researcher at Los Alamos National Laboratory under the supervision of Dr. S. Colgate, and the National Solar Observatory. In 1995, he joined the Astrophysics Science Division at NASA GSFC and conducted theoretical astrophysics research of coronal activity of young active stars as well as magnetohydrodynamic modeling of massive winds from cool giant and supergiant stars. He later extended his research to model active processes on the Sun, including MHD models of the solar wind, solar streamers, and three-dimensional reconstructions of global solar corona.

Raichelle Aniceto is a PhD candidate at MIT Space Systems Laboratory, Department of Aeronautics and Astronautics.

Irfan Azeem is the Chief Engineer at ASTRA. He has over 15 years of space physics and aeronomy research experience in academic and private sector settings. His research interests include optical and radar remote sensing of the upper atmosphere, GPS technology and its application for environmental monitoring, ionospheric physics, middle atmospheric dynamics, and CubeSat technology for space weather research. He holds a B.Eng. (Hons) in Electronics Engineering from Hull University, United Kingdom, and M.S. and Ph.D. degrees in Electrical Engineering and Space Sciences from the University of Michigan, Ann Arbor.

Daniel N. Baker, director of the Laboratory for Atmospheric and Space Physics and Distinguished Professor of Planetary and Space Physics, Moog-Broad Reach Chair of Space Sciences, University of Colorado, Boulder. Baker obtained his PhD degree with James A. Van Allen and subsequently held positions at Caltech, Los Alamos National Lab, and NASA's Goddard Space Flight Center. He has published ~700 papers in refereed literature and edited eight books on topics in space physics.

Graham Barnes received his Bachelor of Science degree from Yale University in Mathematics and Physics, and his Masters of Science and PhD in Physics and Astronomy from Cornell University. He is a senior research scientist at NorthWest Research Associates (NWRA). Dr. Barnes is an expert in solar magnetic field data analysis and statistical analysis. He developed linear and nonparametric discriminant analysis code for use in solar flare forecasting, and led the statistical and implementation work for the NWRA DAFFS project. Dr. Barnes has served as principal investigator on numerous scientific research programs, including research focusing on the magnetic topology and energetics of flare-productive solar active regions.

Drs Leka and Barnes have led NWRA's hosting of two international workshops on the comparison and validation of flare forecasting algorithms, and are involved in ongoing related efforts with numerous international partners.

Paul A. Bedrosian received BSc degrees in Physics and Chemistry from the University of Minnesota and a Doctorate in physics from the University of Washington. Upon graduation, he was an Alexander von Humboldt fellow at the GeoForschungsZentrum in Potsdam, Germany, and a Mendenhall fellow at the U.S. Geological Survey in Denver, Colorado, where is he is a research geophysicist. Paul's research is centered upon solid-earth geophysics, particularly the application of electromagnetic methods to a range of topics including mineral and water resources, framework tectonics, and natural hazards.

Victor Benghin was born in 1951 in Moscow, U.S.S.R. He graduated from the Engineering Physics Institute (Technical University) Moscow in 1974 and began employment at the Institute for Biomedical Problems where he remains. He received his PhD in 1985. Since 2005 he has held two jobs at SINP MSU.

Francesco Berrilli is associate professor of solar physics and space climate at University of Rome Tor Vergata and a member of the Lincean Academy. His current research topics include experimental astrophysics with an emphasis on image analysis, astronomical optics, interferometry, and space- and ground-based instruments. Also, solar activity and dynamics with an emphasis on photospheric pattern formation and dynamics, solar activity, solar UV irradiance and stratospheric ozone, space weather and space climate, and star-planet interaction. He is the supervisor or cosupervisor of 16 PhD theses and has about 200 entries in the bibliographic database SAO/NASA Astrophysics Data System.

David H. Boteler has extensive experience in engineering and geophysics, including work on multidisciplinary projects in the Arctic and Antarctic. He spent 2 years running the ionospheric program at Halley Bay, Antarctica, and also initiated a study of the ionospheric conditions that caused blackout of radio communications. In the Arctic, Dr. Boteler was project manager for an interdisciplinary study of the performance of an experimental ice-breaking cargo ship, the M.V. Arctic. Since 1990 he has been a research scientist with Natural Resources Canada, where he specializes in space weather and its effects on technological systems. He has organized a number of studies of geomagnetic effects on power systems and pipelines involving industrial partners from Canada and Scandinavia. From 2000 to 2010, he was director of the International Space Environment Service, the coordinating body for space weather forecast centers around the world.

Pontus C. Brandt received his PhD from the Swedish Institute of Space Physics and Umea University, Sweden in 1999. Dr. Brandt started at the JHU/APL, in 2000, where he is presently a principal staff scientist. His areas of scientific expertise and interests are planetary magnetospheric dynamics, with emphasis on the inner magnetosphere, and Heliospheric interactions with the local interstellar medium.

Natalia Buzulukova obtained her PhD from Moscow Space Research Institute of Russian Academy of Sciences in the field of Applied Physics and Mathematics. From 2003 to 2007, she held a postdoctoral position in Moscow Space Research Institute (IKI). In 2005, she received a fellowship for Government Support of Young Russian Scientists of the Russian Federation. In May 2007, she was appointed to the NASA Postdoctoral Program (NPP) and moved to NASA GSFC. At NASA GSFC, she began to work with models of the Earth ring current/plasmasphere and data analysis. Dr. Buzulukova now is a research scientist in University of Maryland/NASA GSFC. Dr. Buzulukova's scientific interests cover a broad range of topics including magnetohydrodynamic modeling of the Earth's magnetosphere, kinetic Particle-In-Cell simulations of the geomagnetic tail, bounce-averaged models of the Earth's ring current, modeling of the Earth's plasmasphere, modeling of ENA emissions, and modeling of the Earth's radiation belts with application to different space weather phenomena.

Kerri Cahoy is an associate professor of aeronautics and astronautics at MIT and leads the MIT Space Telecommunications, Astronomy, and Radiation (STAR) Lab. She currently works on nanosatellite laser communication systems and weather sensors. She also works on space telescope missions to directly image exoplanets, and is developing a 6U CubeSat to test MEMS deformable mirror technology for high contrast coronagraph wavefront control systems. Her group also works on understanding the relationship between space weather and spacecraft anomalies, and does radiation testing of commercial CubeSat components. Previously, She worked on spacecraft radio systems for space weather and planetary atmospheric sensing, nanosatellites. She worked on the MIT Gravity Recovery and Interior Laboratory lunar mission team at NASA Goddard Space Flight Center. She has used spacecraft radio systems to study the atmospheres and ionospheres of solar system planets.

Ashley Carlton received her Bachelor of Science degrees in physics and mathematics from Wake Forest University in 2011, her Master of Science in aeronautical and astronautical engineering from MIT in 2016, and is currently a PhD candidate at MIT in aeronautical and astronautical engineering. Carlton's undergraduate research studied the morphology and evolution of supernova remnants and the timing of pulsars. In between undergraduate and graduate school, Carlton worked as a science operations team mission planner for the Chandra X-ray Observatory. Carlton's scientific interests include high-energy physics, natural space environments, and modeling effects on spacecraft.

Mihail V. Codrescu is a physicist at the Space Weather Prediction Center in Boulder, Colorado. His doctoral research consisted of using the NCAR three-dimensional, time-dependent thermospheric model (TGCM) to study the importance of an interactive ionosphere in thermosphere ionosphere system simulations. His postdoctoral research consisted of working with the University College of London (UCL) coupled thermosphere-ionosphere model. His recent interest has been in understanding and quantifying the response of the upper atmosphere to auroral precipitation and to the magnetosphere electric field. He has become involved with data assimilation techniques and the application of metrics to quantify the ability to specify and forecast the upper atmosphere.

Kyle Copeland received his bachelor of science degree from the University of Wisconsin-Oshkosh, his master of science in physics (atomic, molecular, and chemical physics emphasis) from the University of Oklahoma and his PhD from the Royal Military College of Canada. For the past 20 years he has worked for the U.S. Federal Aviation Administration's Office of Aerospace Medicine as a researcher at the FAA's Civil Aerospace Medical Institute in Oklahoma City, investigating ionizing radiation in aerospace environments. He currently holds the position of team coordinator of the Numerical Sciences Research Team. Kyle's scientific interests include cosmic radiation environmental modeling and dosimetry.

Geoff Crowley is the Founder and Chief Scientist at Atmospheric & Space Technology Research Associates (ASTRA) LLC, a research and technology development company. Dr. Crowley received his B. Sc. (Hons) in Physics from Durham University, United Kingdom, and a PhD in Ionospheric Physics from the University of Leicester, United Kingdom. Prior to founding ASTRA, Dr. Crowley worked in four large research organizations: NCAR, AFRL (as a contractor via U. Lowell), JHU-APL and SwRI, where he performed fundamental research on various areas of space physics and space weather. He has published over 140 scientific articles on his research. His current research interests include the ionospheric and thermospheric responses to high-latitude electrodynamics, as well as applications of GPS signals, TIDs, HF systems, and Cubesats.

Dario Del Moro is an assistant professor of physics at the University of Rome Tor Vergata since 2008.
   He has worked on a number of areas in solar physics, Earth and planetary sciences, and applied optics research, including dynamics of the low solar atmosphere (photosphere and chromosphere), organization of plasma and magnetic field structures in the solar photosphere, CME propagation; techniques for post facto restoring, compression and processing of astronomical images, adaptive optics systems, and optical instrumentation. He has published more than 44 papers in astronomy or applied optics in peer-reviewed journals and more than 60 conference proceedings.

William F. Denig earned a Bachelor of Science degree in physics from Siena College in Loudinville, NY and a Master of Science degree in nuclear physics from Utah State University (USU). Also from USU he received his PhD in space physics before joining the Air Force Research Laboratory (AFRL) in 1983. Within AFRL he held several positions including being the Laboratory's liaison to the Air Force Space Command, Deputy Chief Scientist for the Battlespace Environment Division and Chief of the Space Weather Center of Excellence. In 2005 he joined NOAA as the Director of the World Data Center for Solar-Terrestrial Physics (Boulder) and the Chief of the Solar and Terrestrial Physics (STP) Division. He is currently the program manager for STP within NOAA's NCEI.

Eelco Doornbos is an assistant professor in the Astrodynamics and Space Missions research group at the Faculty of Aerospace Engineering at Delft University of Technology in The Netherlands. Dr. Doornbos is an expert on the modeling of aerodynamic drag and radiation pressure on satellites for orbit determination and data processing. He has worked extensively on the retrieval of data on the density and wind in the Earth's thermosphere from tracking and acceleration measurements of low orbiting satellites. His current work is focused on improving the thermosphere data processing of the European Space Agency's GOCE and Swarm missions, and on investigating the resulting data in comparison with related data and models.

John Emmert is a research physicist in the Space Science Division of the U.S. Naval Research Laboratory. His research focuses on the climate and weather of Earth's thermosphere and their effects on satellite orbits. He has developed a four-decade database of thermospheric density via assimilation of orbit data on thousands of space objects. He is co-developer of the HWM07 and HWM14 empirical models of atmospheric horizontal winds and of the NRLMSIS 2.0 empirical model of atmospheric temperature, density, and composition. He served on a National Research Council Committee in 2011–12 to assess the U.S. Air Force's astrodynamics standards.

Mariangel Fedrizzi is a research scientist in the Cooperative Institute for Research in Environmental Sciences (CIRES), University of Colorado at Boulder. Her research focuses on the thermosphere-ionosphere system dynamics on different spatial and temporal scales. She has been performing extensive validation of the Coupled Thermosphere Ionosphere Plasmasphere Electrodynamics (CTIPe) physics-based model using a whole range of ground- and space-based observations. CTIPe is used as a tool for short-term forecast at the NOAA Space Weather Prediction Center, and it runs operationally at the Community Coordinated Modeling Center (CCMC) in support of space weather needs of the International Space Station and other NASA missions.

Dale C. Ferguson is the Lead for Spacecraft Charging Science and Technology at the U.S. Air Force Research Laboratory. Dr. Ferguson has over 34 years of experience in spacecraft charging with NASA and the U.S. Air Force. He has a PhD from the University of Arizona, and has taught at New York University, Louisiana State University, Southeast Missouri State University, and Baldwin-Wallace University. Research appointments have been at the Max-Planck Institute for Radioastronomy, Arecibo Observatory, Case Western Reserve University, the NASA Glenn Research Center, the NASA Marshall Space Flight Center, and the Air Force Research Laboratory. He has over 250 scientific publications or presentations to his credit, is the author or co-author of several U.S. and international spacecraft charging standards, and has received the NASA Exceptional Achievement Medal, the Steven V. Szabo Award, and the Guenter Loeser Lectureship Award. Dr. Ferguson is an Associate Fellow of the American Institute of Aeronautics and Astronautics (AIAA), and was the Chair of the AIAA Atmospheric and Space Environments Technical Committee.

Mei-Ching Fok is a research space scientist in the Geospace Physics Laboratory at the NASA Goddard Space Flight Center. Dr. Fok's main research interests are studies of the radiation belts and ring current during geomagnetic active periods, understanding the mechanisms responsible for their intensification and decay by numerical modeling and data analysis. She has developed a complex model, named as the Comprehensive Inner Magnetosphere-Ionosphere (CIMI) Model, that can compute and predict the energetic plasma fluxes in the radiation belts-ring current region. Dr. Fok was a participating scientist of the IMAGE mission and is the Project Scientist of the TWINS mission. She is heavily involved in neutral atom imaging. The modeling tools she developed have contributed greatly in interpreting images from the neutral atom imagers on IMAGE and TWINS. Dr. Fok is also involved in the Van Allen Probes mission. Her CIMI model is running in real time to predict energetic ion and electron fluxes that Van Allen Probes satellites would observe.

Kevin F. Forbes of Catholic University specializes in using time-series econometric methods to address issues ranging from electricity load forecasting, the impacts of space weather on the power grid, and the resolution of the operational challenges posed by the integration of wind and solar energies into the power grid. In terms of space weather, he is near completion of a manuscript titled, "Space Weather,

Tropospheric Weather, Network Effects, and Electricity Flows: Evidence from Ontario Canada and New York State."

Dominic Fuller-Rowell received a BA in computer science from the University of Colorado, Boulder in 2015. Before joining NOAA's Space Weather Prediction Center in 2015 as an associate scientist, he spent 5 years at NOAA's NCEI STP as a student employee. His professional activities have included optimizing ionospheric data assimilation schemes, processing and disseminating real-time and historical GNSS datasets, ionospheric model validation, and implementing forward models of spectral airglow measurements.

Tim Fuller-Rowell is a fellow of the Cooperative Institute for Research in Environmental Sciences (CIRES), University of Colorado at Boulder and a senior scientist at NOAA Space Weather Prediction Center. He is head of the Atmosphere Ionosphere Modeling Group and primary developer of the Coupled Thermosphere Ionosphere Plasmasphere Electrodynamics Model (CTIPe). He was co-principal investigator on the AFOSR-MURI on Neutral Atmospheric Density Interdisciplinary Research, an effort to improve understanding of the upper atmosphere affecting satellite drag and orbit prediction. He is lead on the Space Weather Action Plan Benchmark on upper atmosphere expansion, an effort to quantify the impact of extreme space weather events on orbit prediction and collision avoidance.

Gali Garipov graduated from the Ryazan State Radio Engineering University, engineer-physicist, MSc 1973. His research interests: ultra high energy cosmic rays, extensive air showers, the study of the night airglow, and transient luminous events from satellites and mountain altitudes. He successfully worked in several space projects, including the experiments "Tatiana," "Tatiana-2," "Vulkan-Compass-2," "Chibis-M," "Vernov" and in mountain experiments designed to study the coupling of cosmic radiation and lightning discharges at the mountains of Aragats, Armenia, and Tian-Shan, Kazakhstan. He has published more than 150 papers in refereed journals and books.

Matina Gkioulidou received her BS in Physics from the Aristotle University of Thessaloniki (2006), her MS (2008), and PhD (2012) in space physics from the University of California, Los Angeles (UCLA). She has been senior professional staff physicist at the JHU/APL since 2014. Her areas of expertise are plasma transport and energization in the near-Earth magnetosphere, ring current composition and dynamics, and magnetosphere—ionosphere coupling.

Alex Glocer received a BA with high honors in Physics and Mathematics from Dartmouth College in June 2002, and PhD in Space and Planetary Physics from the Department of Atmospheric, Oceanic, and Space Sciences at the University of Michigan in 2008. In 2008 he was awarded an NPP fellowship at NASA's Goddard Space Flight Center in Greenbelt, Maryland. Dr. Glocer became a NASA employee a year and a half later as a civil servant. During his graduate work, Dr. Glocer developed a polar wind model for studying ionospheric outflow. He also helped to create the multifluid MHD version of the BATS-R-US magnetosphere model, and added new polar wind and radiation belt components to the Space Weather Modeling Framework (SWMF) to enable global studies. For this work he was awarded the Baldwin Award thesis prize for the best thesis in space science and astrophysics at the University of Michigan. Dr. Glocer has continued his work on modeling the space environment, in collaboration with other researchers, since arriving at Goddard Space Flight Center. In 2014 he was awarded the NASA Early Career Achievement Medal.

Nat Gopalswamy is an astrophysicist with the Solar Physics Laboratory, Heliophysics Division of NASA's Goddard Space Flight Center. He is an internationally recognized expert in coronal mass ejections and their space weather consequences, with a deep interest in understanding how the solar variability impacts Earth. He has over 30 years of experience in solar-terrestrial research, working on projects such as SOHO, Wind, STEREO, and SDO. He is also a solar radio astronomer working on thermal and nonthermal radio emission from the Sun using data the Clark Lake Radioheliograph, the Very Large Array, and the Nobeyama Radioheliograph. He has authored or co-authored more than 400 scientific articles and has edited nine books. He has received numerous awards and honors including the 2013 NASA Leadership Medal. He is currently the president of ICSU's Scientific Committee on Solar Terrestrial Physics (SCOSTEP) and the executive director of the International Space Weather Initiative (ISWI). He is a fellow of the American Geophysical Union.

Alexander Grigoriev is a senior researcher at SINP MSU. He graduated from the physics department of MSU in 2003 and received his PhD in 2006. He is the author of 33 scientific publications. His research interests: near-space physics, lightning physics, and neutrino physics. He particularly studies the generation of gamma rays and neutrons in the atmosphere during thunderstorms and estimates the related radiation hazard.

Rajkumar Hajra received his Bachelor of Science, Master of Science in physics, and PhD in space science from Calcutta University, Calcutta, India. Currently, he is a postdoctoral researcher at LPC2E-CNRS, Orléans, France. Rajkumar's scientific interest is interplanetary space weather involving near-Earth plasma, comets, and planets.

Mike Hapgood is head of Space Weather at RAL Space, the space department of the U.K. Science and Technology Facilities Council, and is based at STFC's Rutherford Appleton Laboratory in Oxfordshire.He is also a visiting professor in the physics department at Lancaster University, working with the Space and Planetary Physics group. Mike is an internationally recognized expert in space weather with a deep interest in understanding how the science links to practical impacts. He has led a number of space weather studies funded by the European Space Agency (ESA) and is a past chair of ESA's Space Weather Working Team. More recently he has acted as an advisor to the U.K. government on the risks posed by space weather, and he chairs the U.K.'s Space Environment Impacts Expert Group. Mike is also an editor for American Geophysical Union's Space Weather journal, a leading peer-reviewed journal for the subject with a worldwide audience. Mike has also has also had a long involvement with the Royal Astronomical Society, with past positions including secretary (1998–2008) and vice-president (2008–10).

Lianhuan Hu received his PhD degree from IGGCAS in 2015 and is currently an engineer of IGGCAS.

Mamoru Ishii received his Bachelor of Science, Master of Science and Doctor of Science from Kyoto University. After graduation he visited as a guest researcher at the Max-Planck Institut fuer Aeronomie, and joined the Communications Research Laboratory (former NICT). He worked at the Geophysical

Institute at the University of Alaska Fairbanks as a guest researcher. His present position is the director of the Space Environment Laboratory, National Institute of Information and Communications Technology. Ishii's scientific interests are space weather and data science.

Vladimir Kalegaev DSc is chair of the MSU Space Monitoring Data Center (SMDC) and head of the laboratory of space research of the Skobeltsyn Institute of Nuclear Physics of Moscow State University. He is active in space physics and global electrodynamics of the Earth's magnetosphere as well as spacecraft data analysis and interpretation. The main results regarding the dynamics of the large-scale magnetospheric current systems and particle fluxes during geomagnetic storms were presented in many scientific meetings. Vladimir Kalegaev has published more than 80 papers in refereed journals and monographs. He was vice-chair of COSPAR PRBEM in 2006–14. During 2012–14 he was the Russian representative in the COSPAR-ILWS working group on the Space Weather Roadmap.

Suk-Bin Kang received his PhD in Physics from the Korea Advanced Institute of Science and Technology (2015). Since 2015 he is a NASA Postdoctoral fellow at the NASA Goddard Space Flight Center (NASA/GSFC). His scientific interests are transport, energization, and loss of particles in Earth's inner magnetosphere (ring current and radiation belt models).

W. Kent Tobiska is the president and chief scientist of Space Environment Technologies (SET), director of the Utah State University Space Weather Center (SWC), and president of Q-up, LLC. His career spans work at the NOAA Space Environment Laboratory, UC Berkeley Space Sciences Laboratory, Jet Propulsion Laboratory, Northrop Grumman, SET, SWC, and Q-up. He has been a USAF, NOAA, and NASA principal investigator (PI). He has served as the COSPAR C1 Subcommission (Thermosphere & Ionosphere) Chair, the COSPAR International Reference Atmosphere (CIRA) Task Force Chair, and as a session organizer for multiple COSPAR scientific sessions. He serves as lead U.S. delegate to the International Standards Organization (ISO) for the space environment and developed the ISO solar irradiance as well as Earth atmosphere density standards. He is the AIAA Atmospheric and Space Environment Technical Committee (ASETC) Committee on Standards (CoS) chair. He has been an active participant on the American Meteorological Society (AMS) annual Space Weather Symposium organizing committee, the Research-to-Operations Working Group for the National Research Council Decadal Survey, the NASA Heliophysics Division Science Advisory Committee, and the NASA Living With a Star Steering Committee. Dr. Tobiska is an associate fellow of the AIAA and a member of AGU, COSPAR, AMS, and ISO TC20/SC14 U.S. Technical Advisory Group. He has authored/coauthored more than 200 peer-review scientific papers as well as 10 books and major technical publications.

Boris Khrenov graduated from the physics department of Lomonosov Moscow State University in 1954 and became a researcher at SINP. At present, he is one of the leading scientists in SINP. He was a candidate of physics and mathematics, 1962; doctor of scienses in physics and mathematics, 1987. He won the Lomonosov MSU reward in 1987. He is the author of more than 200 publications on cosmic rays. He has lectured in Great Britain, Poland, the United States and South Korea. He is a participant of MSU EAS experiments, MSU satellite experiments on transient events in the atmosphere, and their application in the search for for EAS of ultrahigh energies from satellites.

Pavel Klimov graduated from the physics department of Lomonosov Moscow State University in 2006. He defended his PhD thesis in 2009 and started scientific work at the Skobeltsyn Institute of Nuclear Physics of Moscow State University (SINP MSU) as a junior researcher. From 2015 until now, he has been the head of the ultra high-energy cosmic ray laboratory in the space science department of SINP MSU. His research interests: ultra high-energy cosmic rays (UHECR) and energetic transient atmospheric phenomena. He is a developer of scientific equipment for a number of satellite missions. He is a PI of the TUS experiment onboard the Lomonosov satellite launched in 2016 and a member of the JEM-EUSO collaboration. He is currently occupied with the TUS data analyses, development of the UV telescope (mini-EUSO), and the KLYPVE mission for UHECR measurements from the ISS.

Colin Komar completed a BS in Physics from Illinois Wesleyan University (2008) and an MS (2010) and PhD (2015) in Physics from West Virginia University. Dr. Komar is now a Research Associate in the Geospace Physics Laboratory at NASA's Goddard Space Flight Center under a cooperative agreement with the Catholic University of America. Dr. Komar's research interests include solar-wind-magnetosphere coupling, inner magnetosphere dynamics, model development, ionospheric outflow, and basic plasma physics.

Haje Korth received his Masters in Physics and Doctor of Natural Sciences from the Technical University of Braunschweig, Germany. After completing graduate studies at the Los Alamos National Laboratory, he went to the JHU/APL, where is a principal staff scientist. He has served as deputy project scientist for the MESSENGER mission to Mercury and is presently Deputy Project Scientist for the Europa Clipper mission. His areas of scientific expertise are planetary magnetospheric dynamics and modeling and magnetosphere-to-ionosphere coupling.

Nikolay Kuznetsov was born in 1944. In 1968 he graduated from the physics department of Lomonosov Moscow State University and became a researcher at SINP. He is a candidate of Physical and Mathematical Sciences since 1978. Currently, PhD N. Kuznetsov is the author of over 200 publications on the accumulation of radiation effects in semiconductors, development of models of radiation environment in space, and predictions of absorbed dose and single-event effects in microelectronics at the impact of cosmic radiation.

Shu T. Lai received the PhD degree in physics from Brandeis University in 1971 and the Certificate of Special Studies in administration and management from Harvard University in 1986. He has authored many papers and two books entitled *Fundamentals of Spacecraft Charging* (Princeton, NJ, United States: Princeton University Press, 2011) and *Spacecraft Charging* (editor) (Reston, VA, United States: AIAA Press, 2011). He was the chair of the AIAA Atmospheric and Space Environments Technical Committee from 2003 to 2005. He is a Fellow of the Institute of Electrical and Electronics Engineers (IEEE), Institute of Physics (IOP), and the Royal Astronomical Society (RAS), and an Associate Fellow of the American Institute of Aeronautics and Astronautics (AIAA). He is a senior editor of the IEEE Transactions on Plasma Science.

Gurbax S. Lakhina graduated from Panjab University, Chandigarh, and got his Masters and PhD degrees in physics from the Indian Institute of Technology, Delhi. He is a former Director of the Indian

Institute of Geomagnetism in Mumbai from 1998 to April 2004. After his retirement, he became CSIR emeritus scientist, and then INSA senior scientist. Currently, he is an NASI-senior scientist Platinum Jubilee fellow at the Indian Institute of Geomagnetism. Prof. Lakhina's main expertise is in the area of nonlinear waves in space plasmas, solar wind interaction with magnetospheres, magnetic storms, and space weather.

Guan Le received her BS in Physics from University of Science and Technology of China and PhD in Space Physics from the University of California, Los Angeles. She is currently a scientist and Senior Fellow at NASA Goddard Space Flight Center.Her research interests includesolar wind-magnetosphere-ionosphere coupling, magnetospheric and ionospheric current systems, ULF waves, and the Earth's magnetopause and boundary layers. Her research activities primarily involve the analysis and interpretation of spacecraft data as they relate to these research areas.

K.D. Leka received her Bachelor of Science degree from Yale University in Astronomy and Physics, and her Masters of Science and PhD in Astronomy from the University of Hawaii. She is presently a senior research scientist at NorthWest Research Associates (NWRA) and a designated foreign professor at Nagoya University. Dr. Leka is an expert in the acquisition, interpretation, and analysis of solar vector magnetic field data, and has worked with a wide range of other solar observational data. She was a pioneer for designing studies of solar phenomena that employ the statistical evaluation of samples from both target and control data. Dr. Leka has long been fascinated with what makes the Sun flare when it does; she has served as principal investigator on numerous projects, including the development and deployment of the NWRA Discriminant Analysis Flare Forecasting System (DAFFS).

Irina G. Lodkina is a research scientist in the Solar Wind Laboratory in the Space Plasma Department of the Space Research Institute (IKI), Russian Academy of Science (RAS), in Moscow, Russia. She graduated from Gubkin Russian State University of Oil and Gas (National Research University) and Lomonosov Moscow State University.

Whitney Lohmeyer received her PhD and Masters from MIT in aeronautics and astronautics in 2015 and 2013, respectively, and her Bachelors of Science from North Carolina State University in aerospace engineering in 2011. Since 2013, she has been working as a systems engineer at OneWeb, a company working to deploy a constellation of global low earth orbit communications satellites. Prior to joining the OneWeb team in September 2013, she spent time at NASA and Inmarsat.

Paul T.M. Loto'aniu received a PhD in space plasma physics from the University of Newcastle, Callaghan NSW Australia. Dr. Loto'aniu has held postdoctoral positions at UCLA and the University of Alberta. Currently, Dr. Loto'aniu is the Magnetometer Instrument Scientist for the NOAA-NASA GOES-R Spacecraft Series and the NOAA-NCEI Instrument Scientist on the DSCOVR mission. His research interests include magnetospheric plasma waves and radiation belts.

Jeffrey J. Love received BSc degrees in Physics and Applied Mathematics from the University of California, Berkeley, and a Doctorate in Geophysics from Harvard University. Jeffrey was a Leverhulme Trust postdoctoral researcher at the University of Leeds, England, and a Châteaubriand postdoctoral researcher at the Centre des Faibles Radioactivités, CEA-CNRS, France. He briefly worked at the Institute of Geophysics and Planetary Physics, Scripps Institution of Oceanography, San Diego, California. Since 2001, Jeffrey has worked as a research geophysicist in the Geomagnetism Program of the U.S. Geological Survey in Denver, Colorado. Jeffrey's work has covered a diversity of geomagnetic subjects: dynamo theory, paleomagnetic statistics, solar-terrestrial interaction, historical geophysics, and geomagnetic hazards.

Mark H. MacAlester is a telecommunications specialist in the Disaster Emergency Communications Division, Federal Emergency Management Agency, U.S. Department of Homeland Security. He received his bachelor of science in emergency management and his master of science in management/emergency management specialization from the University of Maryland. Mark currently supports several federal space weather planning and characterization efforts.

Anthony J. Mannucci is supervisor of the Ionospheric and Atmospheric Remote Sensing Group at NASA's Jet Propulsion Laboratory, a principal member of the technical staff, and a senior research scientist. Dr. Mannucci's research interests include ionospheric responses to the prompt phase of large geomagnetic storms, solar wind-ionosphere coupling focused on solar coronal mass ejections and coronal holes, the response of the ionosphere to solar flares, fundamental physics of magnetosphere-ionosphere coupling, and space weather forecasting.

Richard Marshall is the SWS specialist on geomagnetism, GICs, and affected technologies (power grids, pipelines, aeromagnetic surveys).

Steven Martin received his MS (1990) and PhD (1996) in Astronomy and Astrophysics from the University of Chicago. He is currently a Senior Scientific Software Developer with ADNET Systems, Inc. on contract to NASA/Goddard Space Flight Center.

Daniel Matthiae received his diploma in physics from the Albert-Ludwigs-Universität in Freiburg and his PhD from the Christian-Albrechts-Universität, Kiel. He has studied ground level enhancements and their impact on the radiation exposure at aviation altitudes since he was a PhD student. Dr. Matthiae works as a scientist at the German Aerospace Center (DLR) in the field of radiation protection in space and aviation with focus on modeling and numerical simulations. He is the principal developer of the PANDOCA model for the calculation of the radiation exposure at aviation altitudes.

Matthias M. Meier is the team leader of the Radiation Protection in Aviation Group at the Institute of Aerospace Medicine of the German Aerospace Center (DLR). He graduated in physics as well as in industrial engineering and management and received his PhD from the Goethe-University Frankfurt am Main. Dr. Meier has contributed to the understanding of the interactions of swift heavy ions with matter, which is of great importance for the dosimetry in space. When radiation protection of aircrew due to their occupational exposure to cosmic radiation was regulated by European legislation, he moved closer to Earth again and became a consultant in radiation protection to a number of airlines. He is involved with the investigation of space weather events on radiation exposure at aviation altitudes. Dr. Meier is an associate member of the European Radiation Dosimetry Group (EURADOS) and an associate editor of the Journal of Space Weather and Space Climate (JSWSC). He has more than 25 years of experience in radiation protection and has been working as a scientist at the German Aerospace Center since 1994.

Viacheslav Merkin received his BS (1997) and MS (1999) degrees from Moscow Institute of Physics and Technology, and PhD in Physics from the University of Maryland, College Park, United States, in 2004. Presently he is a senior staff scientist at the JHU/APL. His scientific interests lie in theory and modeling of various space plasma environments, including large-scale magnetohydrodynamic simulations of the Earth's magnetosphere and the solar wind.

Joseph Minow received his BA in chemistry and biology from Western State College (now Western State Colorado University) in Colorado in 1981, an MS in physics from University of Denver in 1987, and a PhD in physics from the University of Alaska Fairbanks in 1997. Minow's interests include characterizing space environments for space system design and operations, analysis and modeling of space plasma and ionizing radiation environments and their effects on space systems, investigation of spacecraft surface and bulk charging phenomenon, and investigation of on-orbit anomalies. He has contributed over 160 conference presentations, journal publications, and technical reports on topics related to the space environment and their effects on space systems. He was awarded the Silver Snoopy Award in 2002 for contributions to an International Space Station spacecraft charging investigation and received numerous NASA group achievement awards throughout his career. He joined NASA's Engineering and Safety Center in 2015, where he serves as the Technical Fellow for Space Environments.

Leonty Miroshnichenko graduated from the physics department of Lomonosov Moscow State University in 1961 (speciality: physics of cosmic rays). In 1967 he was awarded a PhD at SINP MSU; in 1991, he was awarded a doctorate degree (DSc) in IZMIRAN. His research interests: are solar cosmic rays, solar flares, radiation hazards in space and solar-terrestrial relations.

Livio Narici is associate professor of applied physics at the University of Rome Tor Vergata. He is/has been principal investigator of many space projects on ionizing radiation and its influence on humans, both in the ISS and on ground. He is an expert in measurements and analysis of space ionizing radiation, risk assessment, shielding methods, interactions between ionizing radiation, and visual systems. He is/has been a member of several international committees [e.g., ESA TT *New Developments in Space Radiation Biology and Dosimetry*, ESA TT *Ground-based facilities and models for space radiation research*, EU-THESEUS *Radiation Dosimetry* (Chair)]. He is chair of the *Radiation for human exploration* expert group of the Italian Space Agency. He is the author of more than 160 papers in international journals.

David Neudegg is the principal space and radio scientist for the Space Weather Service, (formerly the Ionospheric Prediction Service) in the Australian Bureau of Meteorology. He has worked since 1987 on space weather, management, ionosphere, magnetosphere, HF communications and radar, space plasma waves, spacecraft operations, Antarctic field operations, software development, VHF radar, and atmospheric turbulence at SWS/IPS, Rutherford-Appleton Lab (United Kingdom), University of Leicester (United Kingdom), University of Newcastle (Australia), Australian Antarctic Division, Andrew Corporation (United States-Australia), and University of Adelaide (Australia).

Chigomezyo M. Ngwira is a research associate in the Department of Physics at the Catholic University of America (CUA). He received his PhD in experimental Space Physics from Rhodes University, Grahamstown, South Africa, in April 2012. Dr. Ngwira moved to Washington, DC, to take up a postdoctoral appointment at CUA in March 2012. At the same time, he has also been engaged as a science collaborator with the CCMC at NASA Goddard Space Flight Center in Greenbelt, MD, USA.

Dr. Ngwira's research focuses on space weather, particularly on the response of the near-Earth space environment and on ground effects. He has spent a bit of time on examining the density of the ionosphere, a layer of electrically charged particles that can distort and impede radio signals. At CUA/NASA-GSFC, Dr. Ngwira has worked on projects focused on understanding the influence of extreme space weather on GIC, and on the development of GIC forecasting tools. In 2014, he was recognized by the American Geophysical Union for making "significant contributions in the application and use of the Earth and space sciences to solve societal problems" and received the AGU Science for Solutions Award.

Nadezhda S. Nikolaeva is a senior research scientist in the Solar Wind Laboratory of Space Plasma Department of the Space Research Institute (IKI), Russian Academy of Science (RAS), in Moscow, Russia. She graduated from Lomonosov Moscow State University.

Rikho Nymmik is a leading research worker at SINP MSU. Born on February 26, 1936, in Tallinn (Estonia), he graduated from the physics department of Moscow State University in 1960. In 1968 he was awarded a PhD; in 1998 he was awarded a doctorate degree (DSc). He has published over 260 scientific papers. He was awarded the title of "Honored Scientist of Moscow University" and order "The White Star" of the Estonian Republic.

Terrance Onsager is a physicist at the NOAA SWPC. Onsager is the liaison and coordinator for international space weather activities at SWPC and a Working Group co-coordinator of Goal 6 of the U.S. Space Weather Operations, Research, and Mitigation effort. His research has focused on various topics of solar-terrestrial physics and on directing scientific knowledge toward the growing need for space weather services.

Mikhail Panasyuk graduated from the physics department of Lomonosov Moscow State University in 1969. His scientific activity began in the Skobeltsyn Institute of Nuclear Physics of MSU in 1969 during his postgraduate courses. Now he has positions as director of SINP, head of the space science department and space physics chair of the physics department of MSU. The most important scientific results Prof. Panasyuk has obtained in the field of astrophysics are cosmic rays ("Astroparticle physics") and space radiation studies. He coordinated a number of experiments onboard spacecraft and ISS on radiation studies in near-Earth space. In 2014, the space project "Vernov" was started under his supervision, intent on the studies of the interrelation of the physical processes in near-Earth space and in the upper layers of the Earth's atmosphere. His latest experiment onboard the "Lomonosov" satellite (launched in 2016) is aimed at studying extreme physical processes both in nearby and distant space—ultra high energy cosmic rays, gamma ray bursts, transient microbursts in ultraviolet in the Earth's atmosphere, and precipitations of energetic particles from radiation belts.

Marcin Pilinski is a research associate at the Laboratory for Atmospheric and Space Physics (LASP), University of Colorado at Boulder. His primary research focus is currently the investigation of thermospheric and ionospheric physics at Mars using data from the MAVEN spacecraft. Dr. Pilinski is also an expert in the area of satellite drag including specification of the atmospheric state that influences aerodynamic forces on satellites, and the gas-surface interactions that cause those forces. He is a co-investigator and lead designer on a project to design a new assimilative framework of the Earth's thermosphere for satellite drag specification using both global circulation as well as empirical models. Dr. Pilinski is also the author of the Semi-Empirical Satellite Accommodation Model (SESAM) used to specify gas-surface interaction parameters for the calculation of aerodynamic forces on spacecraft.

Helen Popova is a senior researcher at SINP MSU. She received her PhD in 2011. She is active in space physics, dynamo theory, and computer simulation. The main results regarding the dynamics of the large-scale solar magnetic field were presented in many scientific meetings. Popova has published 29 papers in refereed journals.

Antti A. Pulkkinen is currently director of the Space Weather Research Center (SWRC) operated at the NASA Goddard Space Flight Center. Dr. Pulkkinen received his PhD in theoretical physics from the University of Helsinki, Finland, in 2003. Subsequently he joined the nonlinear dynamics group at NASA Goddard Space Flight Center (GSFC) to carry out his postdoctoral research 2004–2006. Dr. Pulkkinen's PhD and postdoctoral research involved studies on both ground effects of space weather and complex nonlinear dynamics of the magnetosphere-ionosphere system. From 2011 to 2013 Dr. Pulkkinen worked as an associate director of the Institute for Astrophysics and Computational Sciences and as an associate professor at the Catholic University of America (CUA). At CUA Dr. Pulkkinen launched a new Space Sciences and Space Weather program crafted to educate the next generation of space weather scientists and operators.

Dr. Pulkkinen has been leading numerous space weather-related projects where scientists have worked in close collaboration with end users. In many of these projects his work has involved general empirical and first-principles modeling of space weather and investigations of effects on man-made systems in space and on the ground. Recently Dr. Pulkkinen has been leading the development of space weather forecasting activities at Goddard. The new SWRC activity provides space weather services to NASA's robotic missions.

Joachim (Jimmy) Raeder is a professor of physics at the University of New Hampshire with a joint appointment in the UNH Space Science Center. He received his PhD from the Universität zu Köln (Germany) in geophysics and applied mathematics. He is best known as the original developer of the Open Geospace General Circulation Model (OpenGGCM), which is a coupled numerical model of Earth's magnetosphere, ionosphere, and thermosphere that is open for use by the scientific community at NASA's Community Coordinated Modeling Center. Dr. Raeder is also a co-investigator on NASA's THEMIS mission, and he has served on numerous NASA, NSF, and NRC panels, including a term as chair of the NSF Geospace Environment Modeling (GEM) program steering committee. Dr. Raeder's research focuses on many aspects of plasma processes in the Earth's magnetosphere, numerical modeling, and space weather.

Robert J. Redmon received a DBS in electrical engineering and computational mathematics from the University of California, Riverside in 1998, a MS in electrical engineering from the University of Notre Dame in 2000 and a PhD in aerospace engineering sciences from the University of Colorado, Boulder in 2012. He joined NOAA's National Centers for Environmental Information (NCEI) Solar and Terrestrial Physics Group as a physical scientist in 2003. His professional activities have included developing radar systems for ionospheric sounding; facilitating information exchange through the World Data Service; studying auroral processes, and magnetic fields at LEO and GEO; calibrating space-borne magnetometers; and investigating satellite anomalies. Dr. Redmon received the Goodrich Aerostructures engineering scholarship (1993–98), and held an Arthur J. Schmitt Leadership Fellowship (1998–2000).

Pete Riley received his BSc from the University of Wales, in Wales, his MSc from the University of Sussex in England, and his PhD from Rice University, in the United States. Currently he is a senior scientist at Predictive Science Inc. (PSI) in San Diego, California, studying physical processes occurring in the solar corona and inner heliosphere.

Alessandro Rizzo is a post coc at the University of Rome Tor Vergata, and, since 2015 has worked in the human space exploration field with ALTEA (Anomalous Long Term Effects on Astronauts) and LIDAL (Light Ion Detector for ALTEA, for which he is system manager) experiments, as an expert on particle detectors and nuclear physics experimental techniques. He has worked on strong interactions at low energies in mesic atoms at Laboratori Nazionali di Frascati (LNF-INFN), on the realization and characterization of particle detectors such as silicon detectors, and gaseous and scintillation detectors for multiple purposes. His PhD thesis features one of the most precise results on kaonic hydrogen (a mesic atom) radiative transitions. Since 2012 he is a member of the international collaboration CLAS at the Thomas Jefferson National Laboratory (JLAB-USA), participating in CLAS and HPS experiments. Since 2016 he is radiation officer by Italian law. He has more than 80 works in international referred journals.

William S. Schreiner received his PhD degree in aerospace engineering sciences from the University of Colorado at Boulder in 1993. He is the director of the UCAR COSMIC Program Office. He is currently working on the design, development, and operation of a COSMIC-2 mission that is scheduled to launch soon.

A. Surjalal Sharma earned his B.Sc. and M. Sc., both in Physics, at the University of Delhi, Delhi, and his Ph.D. in Physics at the Physical Research Laboratory—Gujarat University, Ahmedabad. After graduation he held visiting positions at the Max Planck Institute for Plasma Physics, Garching, Ruhr University, Bochum and UKAEA Culham Laboratory, Abingdon before a research associateship at the Cornell University, Ithaca. He joined the faculty at the Institute for Plasma Research, Gandhinagar, in 1983 and in 1987 moved to the University of Maryland, where he is currently a Principal Research Scientist. Dr. Sharma is a past President of the AGU Nonlinear Geophysics Focus Group (2006–10). His many awards for outstanding scholarship include AGU Lorenz Lecture Award (2009), UMD Distinguished Research Scientist Prize (2011-Inaugural) and K. R. Ramanathan Lecture and Medal of Indian Geophysical Union (2013). With publications in plasma theory, numerical simulations, complex systems and data science, his research has addressed a wide range of topics including space weather prediction, plasma processes in current sheets, nonequilibrium systems, planetary atmospheres, extreme events, and data-driven modeling Dr. Sharma is an elected Fellow of the American Physical Society.

Mikhail I. Sitnov received his MS and PhD in Theoretical and Mathematical Physics from the Lomonosov Moscow State University, U.S.S.R (1986). Presently he is a principal staff scientist at the Johns Hopkins University Applied Physics Laboratory (JHU/APL). His scientific interests are space plasma physics (empirical geomagnetic field models, kinetic theory, and simulations of the magnetotail), nonlinear dynamics, and data analysis.

Grant K. Stephens received his BS in Physics and Astronomy from the Pennsylvania State University and MS in applied physics from the Johns Hopkins University. He has been a scientific programmer at the JHU/APL since 2010 where he has worked on programs such as the Van Allen Probes mission. His primary scientific interest has been advancing empirical geomagnetic field models.

Graham Steward is the manager of the Australian Space Forecast Center within the Space Weather Service. ASFC is an ICSU-ISES approved Regional Warning Center. The four authors on this paper comprise the Severe Event Service Team supporting ASFC.

Wenjie Sun is an engineer of IGGCAS.

Eric Sutton is a research physicist in the Drivers and Impacts section of the Air Force Research Laboratory (AFRL). He specializes in upper atmospheric and ionospheric data analysis and modeling. He has developed a database of neutral density measurements from the CHAMP and GRACE satellite accelerometer missions, and applying this to the characterization of the upper atmosphere's response to solar and geomagnetic disturbances. With AFRL, his research has focused on modeling the upper atmosphere and ionosphere, with an emphasis on improving the operational satellite drag specification and forecast capabilities of the Air Force. In this capacity, he has led and participated in a number of efforts to improve physics-based thermosphere models, data assimilative methods, and techniques for forecasting solar and geophysical conditions.

Susan Taylor is a senior associate at Abt Associates. She has more than 10 years experience conducting interdisciplinary research associated with the interplay of environmental engineering, natural resources, and social sciences, with an emphasis in resilience. Taylor conducts analytical and numerical modeling to emerging areas of research (e.g., space weather impacts, quantifying resilience, build-design practices of natural infrastructure), as well as rigorous programmatic evaluations.

Michael Terkildsen is the SWS specialist on ionospheric TEC variability, monitoring of TEC using GNSS/GPS, aviation GNSS, and severe event effects on GNSS.

Nikolai A. Tsyganenko received his MS (1970) and PhD (1973) in geophysics from the University of Leningrad, U.S.S.R. After graduation he worked at the Institute of Physics of the above University, the NASA Goddard Space Flight Center, Hughes/Raytheon STX, and University Space Research Association. Since 2014 he has been an associate professor at the University of St. Petersburg. Nikolai's scientific interests are focused on space weather and the data-based modeling of the Earth's magnetosphere. He is a recipient of the Julius Bartels Medal from the European Geosciences Union (2013).

Bruce T. Tsurutani received his Bachelor of Arts and PhD in physics from the University of California at Berkeley. After graduation he went to the Jet Propulsion Laboratory, California Institute of Technology, Pasadena, California, United States, where he is presently a senior research scientist and principal scientist. Bruce is a past president of the AGU Space Physics and Aeronomy Section and an AGU Fleming Medalist and Fellow. Bruce's scientific interests are space weather (from Sun to the atmosphere) and nonlinear plasma waves.

Aleksandr Y. Ukhorskiy received his PhD from the University of Maryland, College Park, in 2003 and then started to work at the JHU/APL, where he is presently a principal staff scientist. His areas of scientific expertise and interests are planetary magnetospheric dynamics, with emphasis on the inner magnetosphere and dynamics of energetic particles.

Rodney Viereck is a physicist at the NOAA Space Weather Prediction Center in Boulder, Colorado, where he is the head of the Applied Research Section. He is also the director of the NOAA Space Weather Prediction Testbed, which focuses on the development of applications and models for operational support of space weather forecasters. His research focuses on solar EUV and X-ray irradiance and the impact of solar variability on the terrestrial atmosphere. He is the instrument scientist for the NOAA Geostationary Operational Environmental Satellite (GOES) solar X-ray and EUV irradiance sensors. He is lead on the Space Weather Action Plan Benchmark on ionospheric variability, an effort to quantify the impact of extreme space weather events on communication and navigation systems.

Weixing Wan graduated from Wuhan University, China, in 1982 and received his M.S. and PhD degrees from the Wuhan Institute of Physics, CAS, in 1984 and 1990, respectively. He was elected as the academician of CAS in 2011. He is currently the director of the Key Laboratory of Earth and Planetary Physics at IGGCAS.

Daniel Weimer is a research professor of space science with the Virginia Tech Center for Space Science and Engineering Research, and resident at the National Institute of Aerospace in Hampton, Virginia. Dr. Weimer's empirical models of ionospheric electric potentials and field-aligned current are extensively used by the space physics community. Current research involves an empirical model for predicting geomagnetic variations at the surface of the Earth, and developing a technique to predict changes in the temperature and number density of the thermosphere resulting from auroral heating, to better predict satellite drag perturbations in low Earth orbit.

Daniel Wilkinson joined NOAA in 1979, where he participated in processing and distributing various solar-terrestrial data sets including the Space Environment Data from the GOES satellites. He received a BS in Physics from Metropolitan Sate University, Denver, in 1978.

Stacey Worman, an associate/scientist at Abt Associates, is an interdisciplinary researcher with a background that includes formal training in the geosciences, engineering, economics, and chemistry. She holds a PhD in Earth and ocean sciences from Duke University and an MS in environmental engineering and a BS in chemistry and economics, both from Vanderbilt University. She has more than 10 years experience building original, reduced complexity models and collecting and analyzing environmental datasets. She is awed and thankful for the enthusiasm, expertise, time, and support of such a diverse array of industry stakeholders on this space weather economics project.

Limei Yan received his PhD degree from Peking University in 2016 and is currently working as a post doc in IGGCAS.

Yuri I. Yermolaev is a Head of Solar Wind Laboratory in Space Plasma Department of Space Research Institute (IKI), Russian Academy of Science (RAS), Moscow, Russia. He graduated from

Moscow Institute of Physics and Technology (MIPT), Russia in 1978. He received his PhD degree in experimental physics from IKI, in 1989, and his habilitation (Doctor degree) in space physics in 2003. He participated in experiments on satellites of Prognoz series, and in space projects Intershock and Interball. His research interests include: solar wind large-scale structures, in particular interplanetary coronal mass ejections and compression regions ahead them, corotating interaction regions, and long-term variations; solar wind fine-scale structures, in particular turbulence, discontinues, and shocks; magnetosphere and its structure and boundaries, and geomagnetic storms. He is member of Scientific Discipline Representatives of SCOSTEP, member of the Solar-Terrestrial Physics Section of Space Council of RAS, and member of editorial boards of Geomagnetism and Aeronomy, and Solar-Terrestrial Physics journals.

Michael Yu. Yermolaev is a research scientist at the Space Research Institute (IKI), Russian Academy of Science (RAS), in Moscow, Russia. He graduated from Tsiolkovsky Moscow State Aviation Technology Institute (Russian State Technological University).

Xinan Yue received his PhD degree in space physics from the Graduate School of the Chinese Academy of Sciences (CAS). For 6 years, he worked as a project scientist in the University Corporation for Atmospheric Research and is now working as a research professor in the Institute of Geology and Geophysics, CAS (IGGCAS). He is serving as an associate editor of the Space Weather Journal associated with AGU.

Boris Yushkov was born in 1952. He graduated from the physics department of the Lomonosov Moscow State University in 1975 and, since that time, has worked at SINP MSU. Now he is a senior researcher of the space science department. He received his PhD in physics from Moscow State University in 1992.

Foreword

In roughly the last decade, the concept of "space weather" has gone from something known mostly to space researchers to a topic of concern to policy specialists, public officials, and many in the general public. In fact, space weather and its potential effects on modern technological systems have joined such natural hazards as hurricanes, floods, earthquakes, and droughts as matters of immense public policy concern.

Space weather—like most natural hazards—has many facets or components to it. To spacecraft engineers and designers, space weather may be largely issues related to intense charged-particle radiation in a given part of near-Earth space. To operators of the bulk power grid, space weather threats may be largely focused on powerful geomagnetic storms produced in the Earth's vicinity by solar outbursts called coronal mass ejections. In yet other contexts, space weather may have much more to do with changes in Earth's ionosphere and upper atmosphere, including drag exerted on orbiting satellites during intense solar storms. Space weather is all these things—and much more. Like the familiar tropospheric weather that we all experience every day in our terrestrial habitats, space weather can range from mild and relatively benign to dangerously disturbed and powerfully disruptive.

While mild and moderate space weather takes an inexorable toll over time on space systems, it is often the most extreme space weather that garners the most attention. Both policymakers and members of the lay public often ask the question, "How bad can it get?" This book seeks to answer that question by looking at "extreme" space weather aspects. The book presents review chapters by many leading authorities on the largest known solar forcing events. The consequences of such solar storms in the Earth's atmosphere, ionosphere, and magnetospheric environs are also covered. There are assessments of how bad space weather has been in the past as well as speculation about how severe space weather events might be in the future.

It is timely to ask these questions about extreme events. Maybe with space weather—more than for any other natural hazard—we as a society are rapidly making ourselves more vulnerable to severe consequences by our increasingly complex and highly interconnected technological systems. An extreme solar storm today could conceivably impact hundreds of the more than 1400 operating Earth satellites. Such an extreme storm could cause a collapse of the electrical grid on continental-sized spatial scales. The effects on navigation, communications, and remote sensing of our home planet could be devastating to our social functioning. Thus, it is imperative that the scientific community address the question of just how severe events might become. At least as importantly, it is crucial that policymakers and public safety officials be informed by the facts on what might happen during extreme conditions.

It is said that, "Climate is what we expect—weather is what we get." We have adapted our modern technological society to conditions relatively near the space climatic average state. We must at least think seriously about how we deal with extreme departures from the average. This book helps substantially in this endeavor.

Daniel N. Baker
Boulder, Colorado—29 June 2017

Acronyms

1D	one dimensional
3D	three dimensional
aa	geomagnetic index derived from an antipodal pair of magnetic observatories in England and Australia. Complete from 1868 to present.
AC	alternating current
ACARS	Aircraft Communications Addressing and Reporting System
ACE	Advanced Composition Explorer; NASA science mission
AD	anno domini
ADEOS	Advanced Earth Observing Satellite
AE	auroral electrojet index
AE8/AP8	NASA radiation belt electron, proton flux models
AE9/AP9/SPM	NASA radiation belt electron, proton flux and space plasma specification model
AEMO	Australian Energy Market Operator
AFRL	Air Force Research Laboratory
AFWA	U.S. Air Force Weather Agency
AGU	American Geophysical Union
AGW	acoustic gravity wave
AIDA	Arecibo Initiative in Dynamics of the Atmosphere (campaign)
AIR	air radiation model
AL	auroral electrojet index-lower
ALTEA	Anomalous Long Term Effects in Astronauts (ISS detector)
AMAX	ascending and maximum phase of solar cycle
AMIE	Assimilative Mapping of Ionospheric Electrodynamics
AMPERE	Active Magnetosphere and Planetary Electrodynamics Response Experiment
Anomalies	events of abnormal spacecraft operations; failures
Ap	linear geomagnetic index derived from a network of magnetic observatories in Europe, North America and Australia. Complete from 1932 to present.
ApSS	Appleman skill score
AR	active regions (Sun)
ARMAS	Automated Radiation Measurements for Aerospace Safety
ASCA	Advanced Satellite for Cosmology and Astrophysics (satellite)
ASI	Italian Space Agency (Agenzia Spaziale Italiana)
ASTRA	Atmospheric & Space Technology Research Associates, LLC
AU	astronomical unit
AVIDOS	Aviation dose model
BATS-R-US	Block-Adaptive-Tree Solar-Wind Roe-Type Upwind Scheme MHD code
BIS	U.K. Department of Business, Innovation and Skills (since 2016 part of Department for Business, Energy & Industrial Strategy)
BOU	Boulder, Colorado, magnetic observatory
BRW	Barrow, Alaska, magnetic observatory
BSL	Stennis Space Center magnetic observatory
BSS	Brier skill score
C/A	coarse acquisition
C/N0, C/No	carrier to noise ratio (signal amplitude)

CANMOS	Canadian Magnetic Observatory System
CARI	Civil Aeromedical Research Institute
CASES	connected autonomous space environment sensor
CAWSES	Climate and Weather of the Sun-Earth System
CCDF	complementary cumulative distribution function
CCG	Canadian Coast Guard
CCMC	Community Coordinated Modeling Center (NASA)
CDAAC	COSMIC Data Analysis and Archive Center
CDAWeb	Coordinated Data Analysis Web (NASA)
CEDAR	Coupling, Energetics and Dynamics of the Atmospheric Regions
CHAIN	Canadian High Arctic Ionospheric Network
CHAMP	CHAllenging Minisatellite Payload (mission)
CI	confidence interval
CI	coronal index
CIMI	Comprehensive Inner Magnetosphere-Ionosphere model of the ring current plasma
CIR	corotating interaction region
CMD	central meridian distance
CME	coronal mass ejection
COMM-ISAC	Information Sharing and Analysis Center for Telecommunications
COMPTON-GRO	Compton Gamma Ray Observatory
CONUS	Contiguous United States
CORONAS	Complex ORbital Observations Near-Earth of Activity of the Sun-Photon (Satellite)
CORS	continuously operating reference station
COSMIC	Constellation Observing System for Meteorology, Ionosphere and Climate
COSPAR	COmmittee on SPAce Research
COSPIN	COsmic Ray and Solar Particle INvestigation
COTS	Commercial-Off-The-Shelf
CP	cathodic protection
CPCP	Cross polar cap potential
CRAND	cosmic ray albedo neutron decay
CRaTER	Cosmic Ray Telescope for the Effects of Radiation
CRCM	Comprehensive Ring Current Model of the ring current plasma
CRL	Communications Research Laboratory
CRRES	Combined Release and Radiation Effects Satellite
CSWFC	Canadian Space Weather Forecast Centre
DA	discriminant analysis
DAU	Data acquisition unit
DC	direct (electric) current
Dcx	disturbance storm time index correcting for seasonally varying errors
DFA	detrended fluctuation analysis
DGPS	Differential GPS
DHS	Department of Homeland Security
DICE	dynamic ionosphere Cubesat experiment
DLR	Deutsches Zentrum für Luft- und Raumfahrt (German Aerospace Center)

DMIN	descending and minimum phase of solar cycle
DMSP	Defense Meteorological Satellite Program
DNA	deoxyribonucleic acid
DoD	Department of Defense
DOE	Department of Energy
DOSTEL	DOSimetry TELescope
DOY	day of year
DRAP	D Region Absorption Prediction
DRB	Defence Research Board
DRTE	Defence Research Telecommunications Establishment
DSCOVR	Deep Space Climate Observatory; NOAA space weather mission
Dst	disturbance storm-time index, 1 h resolution
Dxt	reconstructed disturbance storm time index
EAS	extensive air shower
EC	European Commission
EDP	electron density profile
EEJ	equatorial electrojet current
EGNOS	European Geostationary Navigation Overlay Service
EGRET	Energetic Gamma Ray Experiment Telescope
EIA	Equatorial Ionization Anomaly
EISCAT	European Incoherent SCATter (radar)
EIT	extreme-ultraviolet imaging telescope
ELF	extremely low frequency (radiowaves)
EM	electromagnetic
EMFISIS	Electric and Magnetic Field Instrument Suit and Integrated Science (instrument onboard NASA Van Allen Probes mission)
EMIC	electromagnetic ion cyclotron
EMMREM	Earth-Moon-Mars Radiation Environment Module
EMP	electromagnetic pulse
ENA	energetic neutral atoms
EPB	equatorial plasma bubble
EPCARD	European Program Package for the Calculation of Aviation Route Doses
EPEAD	energetic proton, electron and alpha detector
ERB	Earth's radiation belts
E-region	region of Earth's ionosphere between 90 and 150 km altitude
ESA	European Space Agency
ESP	energetic storm particles
ESRAS	Enhanced Solar Radiation Alert System
ETI	Electrodynamics Thermosphere Ionosphere (challenge)
EUV	extreme ultaviolet
eV	electron-volt (unit of energy)
EVA	Extra vehicular activity
EXIS	extreme ultraviolet and X-ray irradiance sensors
EXPACS	EXcel-based Program for Calculating Atmospheric Cosmic ray Spectrum
FA	fluctuation analysis
FAA	U.S. Federal Aviation Administration

FAC	field-aligned current
FCC	Fort Churchill magnetic observatory
FD	Forbush Decrease
FDOSCalc	Flight Dose Calculator
FEMA	U.S. Federal Emergency Management Agency
FERC	U.S. Federal Energy Regulatory Commission
FL	flight level
FL	Florida
foF2	F2 ionospheric layer critical radio frequency (MHz)
FOV	field of view
FRD	Fredericksburg, Virginia, magnetic observatory
FREE	Flight Route Effective Dose Estimation
F-region	region of Earth's ionosphere between 150 and 800 km altitude
FU	flux unit
GAGAN	GPS Aided GEO Augmented Navigation
Galileo	European GNSS system
GAST-D	GBAS Approach Service Type D
GBAS	ground based augmentation system
GBM	gamma-ray burst monitor
GCRs	galactic cosmic rays
GEO	Geostationary Earth Orbit
GEO	Geosynchronous Earth Orbit
GEONET	Japanese GPS Earth Observation Network System
GeV	giga-electron-volt (unit of energy)
GHz	gigahertz (unit of frequency)
GICs	geomagnetically induced currents
GIE	geomagnetically induced electric field
GIM	global ionospheric mapping or maps
GITM	Global Ionosphere Thermosphere Model
GLE	ground level enhancement
GLONASS	Global Navigation Satellite System
GloTEC	Global Total Electron Content
GMC	Geostationary Magnetopause Crossing
GMD	geomagnetic disturbance
GMDTF	geomagnetic disturbance task force
GNSS	Global Navigation Satellite System
GOES	Geostationary Operational Environmental Satellite system
GONG	Global Oscillations Network Group
GPS	global positioning system, U.S. GNSS system
GRL	Gamma Ray Line
GSE	geocentric solar ecliptic
GSF	Global Shipping Forum
GSFC	NASA Goddard Space Flight Center
GSM	geocentric solar magnetospheric
GUMICS	Grand Unified Magnetosphere-Ionosphere Coupling Simulation MHD code
GW	gravity wave
Gy	gray (radiation absorbed dose units)

HB	Herringbone
HCS	heliospheric current sheet
HEIDI	Hot Electron and Ion Drift Integrator model of the ring current plasma
HEO	highly elliptical orbit
H-events	nonstorm time HILDCAA
HF	high-frequency (3–30 MHz) radio
HILDCAA	high-intensity long-duration continuous AE activity
HILF	high impact, low frequency
HMI	hazardously misleading information
HMI	helioseismic and magnetic imager
HOPE	the Helium Oxygen Proton Electron plasma spectrometer
HPL	horizontal protection level
HPS	heliospheric plasma sheet
HSS	high-speed solar wind stream
HV	high-voltage
Hz	hertz (unit of frequency)
HZE	High (atomic number) Z Energy (particles)
HZETRN	High charge (Z) and Energy TRaNsport code
IARC	International Agency for Research on Cancer
ICAO	International Civil Aviation Organization
ICME	Interplanetary coronal mass ejection
ICRC	International Cosmic Ray Conference
ICRP	International Commission on Radiological Protection
ICRU	International Commission on Radiation Units and Measurements
ICSU	International Council of Scientific Unions
IDM	internal discharge monitor
IE	Ionospheric Electrodynamics
IEC	Integrated electron content
IESD	internal electrostatic discharge
IFB	Interfrequency bias
IGS	International GNSS Service
IGY	international geophysical year
ILWS	International Living With the Star program
IMAGE	Imager for Magnetopause-to-Aurora Global Exploration (mission)
IMF	interplanetary magnetic field
IMP-8	Interplanetary Monitoring Platform 8
INTERMAGNET	International Real-time Magnetic Observatory Network
IO	input-output (economic models)
IPS	Ionospheric Prediction Service
IRI	International reference ionosphere
IS	interplanetary shock
ISES	International Space Environment Service
ISGM	ionospheric spatial gradient monitor
ISIS	International Satellites for Ionospheric Studies
ISO	International Organization for Standardization
ISS-b	Ionosphere Sounding Satellite b
ISUAL	Imager of Sprites and Upper Atmospheric Lightning

IUWDS	International Ursigram and World Days Service
JAXA	Japan Aerospace Exploration Agency
K	kelvins (unit of temperature)
KAK	Kakioka magnetic observatory, Japan
KE	Kinetic energy
keV	kilo-electron-volt (unit of energy)
kHz	kilohertz (unit of frequency)
Kp sum	sum of all of the eight 3-h Kp values in the same day
Kp	planetary 3-hour-magnetic activity index in logarithmic scale. Complete from 1932 to present.
KPVT	Kitt Peak vacuum telescope
KREAM	Korean Radiation Exposure Assessment Model for Aviation Route Dose
KS	Kolmogorov-Smirnov goodness-of-fit statistic
L1	Lagrange point 1. The L1 point is 1.5 million km sunward of Earth.
L5	Lagrange point 5. The L5 point is 60 degrees behind the Earth as seen from the Sun.
LANL	Los Alamos National Laboratory
LAPAN	Lembaga Penerbangan dan Antariksa Nasional (Indonesian Space Agency)
LaRC	NASA Langley Research Center
LASCO	Large Angle and Spectrometric Coronagraph
L-band	radio frequencies between 1 and 2 GHz
LEO	low Earth orbit
LEP	lightning induced electron precipitation
LET	linear energy transfer
LET	low energy telescope
LFM	Lyon-Fedder-Mobarry MHD code
LN	log-normal
LOL	loss of lock
LPV	localizer performance with vertical guidance
LSTID	large scale traveling ionospheric disturbance
LT	Local time
LWS	Living With a Star (program)
MARECS	MARitime European Communications Satellite
MC	magnetic cloud
MD	Maryland
MDI	Michelson Doppler imager
MEO	medium Earth orbit (GNSS orbits)
MEPED	medium energy proton and electron detector
METI	Ministry of Economy, Trade and Industry of Japan
MeV	mega-electron-volt (unit of energy)
MEXT	Ministry of Education, Culture, Sports and Technology of Japan
MHD	Magnetohydrodynamics
MH, MSH	millionth of a solar hemisphere
MHz	megahertz (unit of frequency)
M-I	magnetosphere-ionosphere
MI	Michigan
MIPAS	Michelson Interferometer for Passive Atmospheric Sounding

MLT	magnetic local time
MMS	Magnetospheric Multiscale mission (NASA)
MN	Minnesota
MPE	magnetic potential energy
MPS-LO, -HI	magnetospheric particle sensor low and high
MSAS	MTSAT Satellite-Based Augmentation System
MSIS	Mass Spectrometer and Incoherent Scatter
MSL	Mars science laboratory
MSM	Magnetospheric Specification Model of the ring current plasma
MSTID	medium scale traveling ionospheric disturbance
mSv	millisievert (unit of ionizing radiation dose)
MT	magnetotelluric
MTSAT	Multi-functional Transport SATellite
MUSCAT	Multi-Utility Spacecraft Charging Analysis Tool
MWD	measurements while drilling
N	nitrogen atom
NAIRAS	Nowcast of Atmospheric Ionizing Radiation for Aviation Safety
NARMAX	nonlinear *auto*regressive *m*oving *a*verage model with e*x*ogenous inputs
NASA	U.S. National Aeronautics and Space Administration
Nascap-2k	U.S. Spacecraft Charging Code
NATEC	North American Total Electron Content
NAVIC	Navigation with Indian Constellation
NC	North Carolina
NCEI	U.S. National Centers for Environmental Information
NEM	National Electricity Market
NERC	North American Electric Reliability Corporation
NGDC	National Geophysical Data Center, now merged into NCEI
NHC	U.S. National Hurricane Center
NICT	National Institute of Information and Communications Technology of Japan
NLFFF	nonlinear force free field
NM	Neutron Monitor
NN	the nearest neighbors
NO	nitric oxide, nitrogen oxide (neutral molecule)
NO$^+$	nitric oxide ion (molecular ion)
NOAA	U.S. National Oceanic and Atmospheric Administration
NPC	National Prediction Center
NPDA	nonparametric discriminant analysis
NRC	National Research Council
NRC	Nuclear Regulatory Commission
NSF	U.S. National Science Foundation
NSO	National Solar Observatory
NSTC	U.S. National Science and Technology Council
nT	nano-tesla (unit of magnetic flux density)
NWP	numerical weather prediction
NWRA	NorthWest Research Associates
NWS	National Weather Service
NWT	Northwest Territories

O	oxygen atom
O$^+$	oxygen ion
OECD	Organisation for Economic Co-operation and Development
OP	OVATION Prime
OpenGGCM	Open Geospace General Circulation Model code
OSTP	Office of Science and Technology Policy
OTH	over the horizon
OVATION	Oval Variation, Assessment, Tracking, Intensity and Online Nowcasting
OVSA	Owens Valley Solar Array
P(Y)	precision (encrypted)
P/S	pipe-to-soil
PAMELA	Payload for Antimatter Matter Exploration and Light-nuclei Astrophysics
PANDOCA	Professional Aviation Dose Calculator
PARMA	PHITS-based Analytical Radiation Model in the Atmosphere
PC5	geomagnetic pulsations with characteristic period 160–500 s
PCA	Polar Cap Absorption (event)
PCA	principal component analysis
PC-AIRE	Predictive Code for AIrCrew Radiation Exposure
PDE	probability density estimate
PFRR	Poker Flat research range
PFU	Particle Flux Unit (1 pfu = 1 p$^+$/cm^2/s/sr)
PHITS	Particle and Heavy Ion Transport code System
PIC	particle-in-cell
PIL	polarity inversion line
PJ	Polarization jet
PL	power-law
PMT	photomultiplier tube
PNT	positioning, navigation, and timing
POD	precise orbit determination
POES	Polar Operational Environmental Satellite
PPEF	prompt penetration electric field
PPP	precise point positioning
PRE	prereversal enhancement
PRN	pseudo random noise
PSTEP	Project for Solar-Terrestrial Environment Prediction
PTHN or PTN	poly thin
PWM	pulse width modulation
QE	quasi electrostatic
QZSS	Quasi-Zenith Satellite System
RaD-X	Radiation Dosimetry Experiment
RAE	U.K. Royal Academy of Engineering
RAM	random-access memory
RAM	Ring current-Atmosphere interactions Model of the ring current plasma
RAM-SCB	RAM model with self-consistent magnetic equilibrium solver
RBE	Radiation Belt Environment model
RBSPICE	Radiation Belt Storm Probes Ion Composition Experiment
RC	Ring current

RCM	Rice Convection Model of the ring current plasma
RED	relativistic electron dropout
REPPU	REProduce Plasma Universe MHD code
RF	radio frequency
RFBR	Russian Foundation for Basic Research
RFC	random forest classifier
RHCP	right hand circular polarized
RIMS	ranging integrity monitoring stations
RMS	radiation monitoring system
RMS	root mean square
RNA	ribonucleic acid
RO	radio occultation
ROC	receiver operating characteristic
ROT, ROTI	Rate of TEC Index
RRL	Radio Research Laboratory
RSTN	Radio Solar Telescope Network
RTIGS	real-time IGS
RTK	real time kinematic
RWC	Regional Warning Center
S1, S2, S3, S4, S5	solar radiation storm level on the space weather scales developed by NOAA (http://www.swpc.noaa.gov/noaa-scales-explanation)
SAA	South Atlantic Anomaly
SAID	subauroral ion drift
SAMI-2	Sami is Another Model of the Ionosphere-2
SAMPEX	Solar Anomalous and Magnetospheric Particle Explorer
SAPS	subauroral polarization stream
satcom	satellite communications
SBAS	satellite-based augmentation system
SBIR	Small Business Innovative Research
SC	solar coordinates
SC	solar cycle
SC	South Carolina
SC	sudden commencement of geomagnetic storms
SC	surface charging
SCATHA	Spacecraft Charging AT High Altitudes, Satellite P78-2
SCINDA	Scintillation Network Decision Aid
SCME	super-CME
SCR	solar cosmic rays
SCW	substorm current wedge
SDO	Solar Dynamics Observatory (NASA)
SDU	Silicon detector unit
SEA	Sudden Enhancement of Atmospherics
SEALION	Southeast Asia Low-latitude Ionospheric Network
SED	storm-enhanced (plasma) density
SEDA-AP	Space Environment Data Acquisition Equipment-Attached Payload
SEE	single event effect
SEM	Space Environment Monitor

SEON	Solar Electro-Optical Network
SEP	solar energetic particle
SES	Sudden Enhancement of Signal Strength
SEU	single event upset
SFU	Solar Flux Unit (sfu = 1×10^{-22} Watts/m^2/s/Hz)
SG storms	gradual geomagnetic storms
SH-events	HILDCAA preceded by magnetic storm
SI/SI$^+$	Sudden Impulse
SID	sudden ionospheric disturbance
SILSO	Sunspot Index and Long-term Solar Observations
SINP MSU	Skobeltsyn Institute of Nuclear Physics of Lomonosov Moscow State University
SIR	stream interaction region
SITEC	Sudden Increase in Total Electron Content
SLOSH	Sea, Lake and Overland Surges from Hurricanes
SLT	solar local time
SM	solar magnetic
SMS	Synchronous Meteorological Satellite
SNR	signal-to-noise ratio
SOC	self-organized criticality
SOHO	Solar and Heliospheric Observatory; ESA/NASA science mission
SOON	Solar Observing Optical Network
SPA	Sudden Phase Anomalies
SPAN	Solar Particle Alert Network
SPE	Solar particle event
SPENVIS	ESA's Space Environment Information System
SPIE	Society of Photographic Instrumentation Engineers
SPM	spectromagnetograph
SPR	Solar Particle Release (time)
SRB	solar radio burst
SS	sunspots
SSC	Sudden Storm Commencement
SSN	Sun Spot Number
SSPA	solid state power amplifiers
STEREO	solar terrestrial relations observatory, NASA science mission
STET	Superthermal Electron Transport (STET) code
STFC	Science and Technology Facilities Council, U.K. research council
SuperMAG	Collaboration of organizations and national agencies that currently operate more than 300 ground based magnetometers
SUVI	Solar Ultraviolet Imager
Sv	Sievert (unit of ionizing radiation dose)
SVC	static VAR compensator
SW	solar wind
SW	space weather
SWAP	space weather action plan
SWF	short-wave fadeout
SWMF	Space Weather Modeling Framework

SWPC	Space Weather Prediction Center
SWS	Space Weather Services
SXI	Solar X-Ray Imager
SYM-H	1 min resolution, symmetric disturbance index derived from *H* component of the magnetic field; close to high-resolution version of Dst index
TEC	total electron content
TECU	total electron content unit
TEPC	tissue equivalent proportional counter
TGF	Terrestrial Gamma Flashes
THEMIS	Time History of Events and Macroscale Interactions during Substorms mission
TID	total ionizing dose
TID	traveling ionospheric disturbance
TIDDBIT	traveling ionospheric disturbance detector built in Texas
TISN	Trusted Information Sharing Network
TLE	transient luminous events
TOFxPH	Time of Flight by Pulse Height
TOI	tongue of ionization
TS07D	Tsyganenko-Sitnov 2007 dynamical model of magnetic field
TSS	true skill statistic
TSS-1R	Tethered Satellite System-1 Reflight
TWINS	Two Wide-angle Imaging Neutral-atom Spectrometers mission
UCAR	University Corporation for Atmospheric Research
UHF	ultra high frequency
U.K.	United Kingdom
ULF	ultra-low frequency
UN-COPUOS	United Nations Committee on the Peaceful Use of Outer Space
UNSCEAR	United Nations Scientific Committee on the Effect of Atomic Radiation
U.S.	United States
USAF	U.S. Air Force
USGS	U.S. Geological Survey
USTEC	U.S. Total Electron Content
UT	Universal time
UTC	Coordinated Universal Time
UV	ultraviolet
UVRIR	ultraviolet and Red Infra-Red
VA Probes, VAP	Van Allen Probes
VA	Virginia
VAR	volts-ampere reactive
VarSITI	Variability of the Sun and Its Terrestrial Impact
VERB	Versatile Electron Radiation Belt model of electron radiation belts
VHF	very high frequency. Radio signal frequencies between 30 and 300 MHz
VLF	Very Low Frequency (radiowaves)
VPL	vertical protection level
W	watt (unit of power)
WAAS	wide area augmentation system
WASAVIES	Warning System for AVIation Exposure to Solar energetic particles
WDC	World Data Center

WFR	waveform receiver
WI	Wisconsin
WMO	World Meteorological Organisation
X-rays	electromagnetic radiation with wavelengths between 0.01 and 10 nm
XRS	X-Ray Sensor
XUV	X-ray and EUV bands

Introduction

In everyday life, most people usually do not experience space weather effects. Hurricanes, floods, snowfalls, and heat waves are more common manifestations of "normal" extreme weather.

The effects of space weather are hidden and sometimes intertwined with other factors. However, specialists who work in the fields of radio communications, GNSS positioning, avionics, and satellite design and operations deal with space weather effects on a regular basis. The extreme effects here could include degradation of GNSS services, rerouting of many commercial flights, or loss of satellites after a geomagnetic storm or a substorm. Another example is an extreme solar radiation storm that could pose serious radiation hazards for astronauts/cosmonauts onboard ISS, or future crewed missions to the moon or Mars. Countermeasures and mitigation techniques could be very costly, but the general public would probably not count these cases as "extremes."

In the case of a truly extreme and rare space weather event, things could turn out differently. An historically extreme geomagnetic storm could induce large electrical currents flowing in the Earth (Geomagnetically induced currents, GICs) that will seriously disable electrical power grids, leaving millions of people without electricity. The threat is not hypothetical: the famous geomagnetic storm in March 1989 left 6 million people in the Quebec province of Canada without electricity for 9 h. A geomagnetic storm with the highest intensity ever was recorded on September 1–2, 1859, and was named after the British astronomer Richard C. Carrington, who observed solar eruption during the event. It has been shown recently for the Carrington event to have an intensity ∼3 times bigger than the 1989 storm event. During the Carrington storm, fires were set on telegraph stations. According to a report by the U.S. National Academy of Sciences (2008), the total economic impact from a similar event happened today could exceed $2 trillion, or 20 times greater than the costs of Hurricane Katrina. A report of the Royal Academy of Engineering (2013) estimates the time needed to repair damages for U.K. power grid system in weeks or months. That extreme space weather event will be obviously noticeable for everyone!

On July 23, 2012, the Earth was lucky enough to miss the extreme solar eruption and the ensuing Interplanetary Coronal Mass Ejection (ICME) that according to estimates would have generated a geomagnetic storm with intensity comparable to that of the Carrington event. Today, such threats from extreme space weather are internationally recognized as being serious, that is, having sufficient probability to happen during the next 10–20 years.

Scientific studies are essential for understanding the origins and predicting the consequences of extreme space weather. Public discussion on the topic of extreme space weather include views ranging widely from "the end of civilization as we know it," to "it can't happen here." We need to develop scientific knowledge to describe what the credible risk is and to give insights into how to mitigate the consequences. This book presents an attempt to describe our current scientific understanding of extreme space weather events in geospace. The chapters are written by leading experts from many different fields of space research. The book collectively links solar eruptive events with space weather effects and impacts on the Earth, presents an account of the available modeling tools and data sets, reviews historical knowledge, discusses national and space policy issues, shows how space weather risks have evolved over time, and outlines the future challenges for the community.

1 Introduction

One of questions raised throughout the chapters of this book is how to define an extreme space event. It is obvious that extreme geomagnetic storms like the Carrington event will bring many adverse effects to the technological systems and infrastructure, and therefore it is important to assess risks associated with the occurrence of extreme geomagnetic storms. But there are multiple examples throughout the book showing that extreme space weather effects could occur during milder solar and geomagnetically active conditions. For example, extreme GICs are often associated with substorm activity. However, occurrence of extreme substorms do not correlate well with occurrence of the extreme storms. Substorms are also related to the phenomenon of satellite surface charging. Occurrence of extreme geomagnetic storm would not be a necessary condition for these adverse effects, that is, GICs or spacecraft surface charging. Other such examples of this complexity can be found in different chapters of the book. One important point that we would like to leave the reader with: the definition of extreme effects and their impacts varies from space weather topic to topic. We need further studies to understand the nature of extremes, as well as the set of conditions where and when the extremes occur.

The content of the book can be briefly described as follows.

Chapter 1 by Dr. M. Hapgood gives an introduction to the field of space weather, outlining major space weather effects and impacts, and linking them to their solar origins. This chapter shows how technological advances changed the risk associated with space weather, and how this risk could change in the future.

Chapter 2 by Dr. N. Gopalswamy describes the solar origins of extreme space weather events. Extreme solar eruptive events and their consequences are considered, including statistics for coronal mass ejections, solar flares, solar energetic particles (SEP) events, and the magnetospheric Dst index. Basic physical parameters of extreme solar eruptions and their consequences are estimated, including expected one-in-100-year and one-in-1000-year event sizes.

Chapter 3 by Dr. K. D. Leka and Dr. G. Barnes review the current status of solar flare forecasting techniques and methods. At present, the field of solar flare forecasting uses the most developed statistical methods and approaches in the field of space weather. The chapter describes forecasting history and definitions, presents basic approaches, and discusses methods for the evaluation of performance. The chapter stresses the need for a better understanding of physics behind flare events, and emphasizes the role of first-principle models.

Chapter 4 by Dr. Yu. Yermolaev et al. presents different ways to use statistical methods for the prediction of extreme space weather. They differentiate between different solar wind structures and their efficiency to produce geomagnetic storms.

Chapter 5 by Dr. P. Riley reviews current statistical approaches for the estimation of extreme space weather, and applies it to calculate occurrence rates for geomagnetic storms, substorms, and SEPs. One of the results is that the likelihood to observe an extreme geomagnetic storm in the next 10 years is ~10%. However, 95% confidence interval lies between 1% and 20%, that is from "quite unlikely" to "very probably." It seems this is a natural limitation of current statistical approaches, because the number of points representing tail distributions for extreme events is small.

Chapter 6 by Dr. S. Sharma presents another approach to predict the probability of extreme space weather from the viewpoint of statistical physics of nonequilibrium systems, an active topic of research in magnetospheric physics. The chapter stresses that geospace exhibits both coherent behavior and multiscale behavior, arising from internal dynamics as well as driven by the turbulent solar wind. The modeling and prediction of space weather, including extreme events, requires consideration of both the global and multiscale aspects.

Chapter 7 by Dr. G. Lakhina and Dr. B. Tsurutani reviews the present knowledge about magnetic superstorms. The chapter presents historical background and introduces storm classification and some important characteristics. Historical data from Colaba and Alibag magnetic observatories are shown with the list of intense and superintense storms starting from September 1847. The chapter also estimates some basic characteristics of Carrington events and discusses the maximum intensity for a future possible supergeomagnetic storm.

Chapter 8 by Dr. C. M. Ngwira and Dr. A. Pulkkinen, and Chapter 9 by Dr. J. Love and Dr. P. Bedrosian describe the current understanding of effects and impacts associated with GICs. These chapters nicely present basic GICs theory, provide illustrative examples, overview model results, and derive hazard maps for extreme events. The chapters stress the importance of additional data sets required for improved risk assessment and general understanding of GIC effects, namely, magnetotelluric surveys and direct measurements of geoelectric fields during geomagnetic storms.

Chapter 10 by Dr. N. Buzulukova et al., and Chapter 11 by Dr. M. Sitnov et al. present two different approaches to modeling the geospace environment for extreme conditions, namely, global first-principles models of the terrestrial magnetosphere, and empirical models of storm-time magnetic fields and pressure distributions. The chapters describe the available modeling tools, review the modeling results for extreme geomagnetic storms, present distributions of currents and pressure for the intense storms, and outline future challenges. Chapter 10 also shows an example of nonextreme geomagnetic storm with intense ground d**B**/dt variations, presenting model results and tracing the source of the variations from the ground to the solar wind and magnetospheric drivers.

Chapter 12 by Dr. W. Denig et al. presents detailed data analysis and signatures of extremes for 12 cases starting from the March 1989 geomagnetic storm and ending with the March 2012 storm. For each case, a review of published literature is also performed that helps to connect observed signatures of extreme events with the magnetospheric responses.

Chapter 13 by Dr. M. Panasyuk et al., Chapter 14 by Dr. R. Hajra and Dr. B. Tsurutani, Chapter 15 by Dr. D. Ferguson, Chapter 16 by Dr. S. Lai et al., and Chapter 17 by Dr. L. Narici et al. describe different forms of radiation in the near-Earth space environment and their solar-cycle dependence, as well as their space weather effects and impacts, including discussion of implications for crewed space missions. The role of nonsolar source for the most energetic population, Galactic Cosmic Rays (GCRs), is emphasized in Chapter 13. Chapter 14 demonstrates that the most energetic magnetospheric radiation belts are observed during the declining phase of the solar cycle, helping to define extremes. Chapters 15 and 16 discuss the effects of radiation on spacecraft surface charging, arcing, and deep dielectric charging. Chapter 16 also offers detailed definition of "extremes" for spacecraft surface-charging effects.

Chapter 18 by Dr. W. K. Tobiska et al. and Chapter 19 by G. Gapirov et al. describe atmospheric effects of radiation environment with application to aviation. Chapter 18 characterizes radiation exposure for space weather events at aviation altitudes and introduces a new dose index (D-index) to be used for aviation radiation alerts. Chapter 19 introduces a new and important space weather phenomenon, high-energy transient luminous atmospheric events, or TLEs, related to thunderstorm activity. The chapter presents discussion of TLEs as a new radiation hazard for suborbital flights and the relationship of TLEs to other space weather phenomena.

Chapter 20 by Dr. A. Mannucci and Dr. B. Tsurutani, Chapter 21 by Dr. Fuller-Rowell et al., Chapter 22 by Dr. X. Yue et al., Chapter 23 by Dr. G. Crowley and Dr. I. Azeem, and Chapter 24 by Dr. R. Redmon et al. present different aspects of thermospheric and ionospheric effects and impacts associated with space weather. Chapter 20 explores the dynamics of total electron content (TEC) during extreme ionospheric storms, and the role of prompt penetration electric fields (PPEFs). The latter can have strong influence on low-altitude satellite drag during intense magnetic storm events. Chapter 21 describes how thermosphere-ionosphere (T-I) coupled system might respond to different types of extreme events, namely extreme enhancement in solar radiation and extreme geomagnetic storms. Chapter 22 discusses intense solar radio bursts (SRBs) and their impact on GNSS systems. Chapter 23 explores the effects of ionospheric storms on GPS/GNSS signals and systems. The chapter describes various types of ionospheric disturbances, effects on communications, navigation, and surveillance, providing illustrative examples for a number of intense and extreme storms. Chapter 24 presents a detailed study of three moderate geomagnetic storms where aviation systems were seriously or partially degraded, and explores the solar origins for these events.

Chapter 25 by Dr. V. Airapetian explores relations between space weather, astrophysics and life sciences. This chapter discusses the result of interactions of extreme CME and SEP events with the magnetosphere, ionosphere, and atmosphere of the early Earth and shows how extreme events might have initiated the production of precursors of life.

Chapter 26 by Dr. D. Boteler describes the Canadian experience of dealing with space weather. As a result of their geographical location, Canada is especially prone to space weather adverse effects. The chapter shows many examples of how space weather affects different technological systems, and everyday life.

Chapter 27 by Dr. M. Hapgood and Chapter 28 by M. MacAlester discuss the ways space weather scientists communicate information with other experts, including engineers, economists, emergency managers, and policy makers. The two chapters present complementary views and help the reader understand how space weather scientists should present the information in order for it to be useful to other experts.

The book concludes with three short technical reports. Chapter 29 by Dr. S. Worman et al. presents first results and ongoing efforts of identifying, describing, and quantifying the social and economic impacts of space weather in the United States. Chapter 30 by Dr. D. Neudegg et al. reports on the current state of extreme space weather monitoring in Australia. The last chapter, Chapter 31 by Dr. M. Ishii, briefly describes the current state of space weather research in Japan.

In conclusion, we note that personal and organizational experience of space weather can hardly provide good insights into the real extremes, because by definition these events occur rarely. With the methods provided by science, it is possible to greatly enhance our knowledge by looking at historical events or collecting information from solar observatories about powerful eruptions that missed the Earth (e.g., the July 23, 2012, event). Collection of extreme events could be also enhanced from observations of solar-like stars, making it possible to derive occurrence rates for extremes with greater accuracy, a point that could be essential for policy makers. These methods rely on continuous systematic measurements of geomagnetic activity on the Earth (starting from the 19th century), modern observations of solar activity and solar wind structures, and finally, recent and future observations of solar-like stars. The science of extreme space weather has proven to be a complex interdisciplinary field connected to many surrounding fields. We hope this book will be helpful for those who are interested in understanding these interconnections, as well as the origins, predictability, and consequences of extreme space events in geospace.

This book project started as a Fall AGU 2015 session, "Origins of Extreme Events in Geospace," where the editor was the primary convener. The book chapters have been peer reviewed by the members of the community (excluding three short technical reports in the end of the book). Dr. B. Tsurutani served as a guest editor for Chapter 10 written by Dr. N. Buzulukova et al. The editor would like to take the opportunity to thank the authors' team, without whose help and support this project would not be possible.

Natalia Buzulukova
NASA GSFC, Heliophysics Science Division, Greenbelt, MD, United States
University of Maryland, Astronomy Department, College Park, MD, United States

PART 1

OVERVIEW OF IMPACTS AND EFFECTS

CHAPTER

LINKING SPACE WEATHER SCIENCE TO IMPACTS—THE VIEW FROM THE EARTH

1

Mike Hapgood

RAL Space, STFC Rutherford Appleton Laboratory, Didcot, United Kingdom;
Space and Planetary Physics Group, Lancaster University, Lancaster, United Kingdom

CHAPTER OUTLINE

1 Introduction	3
2 Space Weather Environments at Earth	4
3 Geomagnetically Induced Currents—The Impacts of Natural Geoelectric Fields	8
4 Space Weather Impacts on the Upper Atmosphere	16
4.1 Overview of the Upper Atmosphere	16
4.2 Trans-Ionospheric Radio Propagation	17
4.3 Atmospheric Drag	22
4.4 Atmospheric Radiation Environment	25
4.5 Satellite Plasma Environments	27
4.6 Looking to the Future: How May Space Weather Risks Evolve?	29
References	31

1 INTRODUCTION

Scientific discussions of space weather and impacts traditionally start with a discussion of the solar sources of space weather. They then trace along the flow of energy and momentum through the solar wind to the Earth, then to terrestrial environments where space weather interacts with technology. This is a very natural approach for anyone with a physics background—follow the energy flow. In this chapter we will take the opposite approach by tracing space weather from its societal impacts on human activities on Earth back to its origin on the Sun. This reverse approach has an important advantage in that it enables us to more clearly identify the solar, heliospheric, and terrestrial phenomena that are crucial to assessing and forecasting the societal impacts of space weather, since it ensures we follow the chains of physical processes that produce significant impacts.

It is also worth noting that some space weather phenomena have sources other than the Sun. Galactic cosmic rays are a population of protons and ions that are accelerated to GeV and above energies by shocks produced by supernovae elsewhere in our galaxy. They pervade the galaxy, including our solar system, but turbulence in the heliospheric magnetic field scatters many of these particles so that they do not reach the inner solar system. Nonetheless, a substantial fraction of cosmic rays do reach Earth, a fraction that varies with solar activity (Chapter 13 of this volume). These cosmic rays produce significant radiation effects on satellites and the Earth, and must be considered in any discussion of space weather impacts. Phenomena in the troposphere also contribute to space weather. The strong electric fields in large thunderclouds can generate a variety of energetic phenomena—not just lightning, but also very bright (equaling the brightest aurora) and very short-lived (milliseconds) luminous phenomena in the stratosphere and mesosphere above the clouds (Chapter 19 of this volume). We need to understand extreme instances of transient luminous events so as to assess whether they pose a risk to suborbital space flights and hypersonic aircraft now being developed. Large thunderclouds and other strong convective activity in the troposphere can also generate atmospheric gravity waves that can transport energy and momentum to the upper atmosphere, where they are thought to drive up to 50% of day-to-day space weather variations in the ionosphere and thermosphere (Liu et al., 2013). However, it is not clear if they play any role in extreme space weather events.

The adverse impacts of space weather at Earth encompass a wide and growing range of technological systems, a range that can seem bewildering at times (see Fig. 1). However, these impacts arise from a limited number of terrestrial environments—more precisely from how space weather modifies those environments in ways that disrupt various technologies.

The next section summarizes those environments and outlines how space weather leads to disruptive conditions. Later sections will explore in detail some of the major technological impacts arising from those environments, and a final section will outline likely changes in those impacts given current trends in technology development. These later sections seek to give the reader insights into many key space weather impacts, but do not aim to be comprehensive of every impact. The latter is almost impossible due to the diversity of space weather phenomena and their impacts.

2 SPACE WEATHER ENVIRONMENTS AT EARTH

1. **The natural geoelectric fields that exist on the surface of the Earth**. These are driven by time-varying magnetic fields, in particular those that arise from electric currents flowing in Earth's ionosphere and magnetosphere and that are strongly modulated by a range of space weather effects from the Sun. These geoelectric fields drive unexpected electric currents into any metallic infrastructures that are electrically connected to the surface of the Earth (Chapter 8 of this volume). That electrical "grounding" is an essential design feature of most infrastructures, ensuring electrical safety and sometimes also electromagnetic compatibility with devices near the infrastructure. Power grids are the key infrastructure in this class, which also includes power and control systems on railways; pipelines such as those used for transmission of oil and gas; and power systems on the fiberoptic cables that run under the oceans (a mainstay of modern Internet and voice communications).
2. **The upper atmosphere above about 90 km, both its neutral and ionized components (thermosphere and ionosphere), up to at least 1000 km**. The dynamics of this region are

2 SPACE WEATHER ENVIRONMENTS AT EARTH

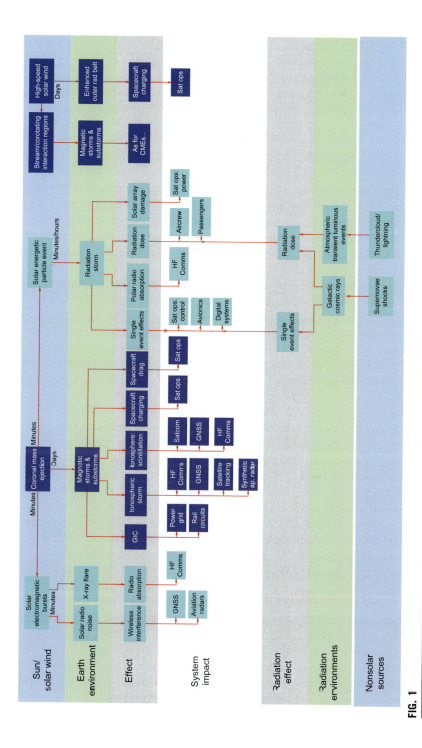

FIG. 1

An overview of space weather processes and impacts, highlighting some of the interrelationships between different effects. Many space weather effects will occur close together in time as they have a common origin in solar phenomena such as coronal mass ejections. This figure outlines the most important associations between space weather effects. The vertical transport of energy and momentum from the troposphere is omitted because of its complexity but may play a role in several system impacts.

dominated by space weather. For example, variations in solar EUV emissions control the heat inputs that drive thermospheric winds and the consequent plasma flows in the ionosphere (King and Kohl, 1965). These emissions also control the ionization rates that create the ionosphere. Variations in the solar wind electric field can sometimes penetrate the magnetosphere and drive unusual plasma flows in the ionosphere, such as the uplift and associated density increases seen early in geomagnetic storms. Energy inputs from solar wind can create intense heating in the polar thermosphere, leading to profound changes in the global circulation of the thermosphere. That heating can also inject molecular species such as nitric oxide (NO) into the thermosphere leading to more rapid loss of ionization via dissociative recombination, and hence the night-time disappearance of the ionosphere that is a distinctive feature of the later phases of geomagnetic storms. In summary, the upper atmosphere is a complex dynamic system comprising neutral and ionized components that are weakly coupled and that both respond strongly to space weather. The neutral component is important technologically because it provides the atmospheric drag that modifies the orbits of satellites passing through this region. Space-weather-induced changes in this drag present a challenge to satellite operators who need accurate orbit predictions to plan satellite operations as well as assess collision risks and re-entry times. In extreme space weather conditions, such as the March 1989 storm, it can lead to loss of knowledge of the positions of thousands of objects in low Earth orbit (LEO) (Air University, 2003). The ionized component is important technologically because of the huge number of systems that use radio signals that are affected by their passage through the ionosphere. These cover frequencies ranging from 50 kHz up to 4 GHz, with lower frequencies (up to 10 or 20 MHz) being reflected and higher frequencies subject to group delay (as the group refractive index of the ionospheric plasma is slightly greater than one) and scintillation (due to plasma instabilities that create fine structures and turbulence in the ionosphere). Frequencies from 1 to 100 MHz can also be subject to absorption. These space weather impacts from the upper atmosphere mostly arise at altitudes below 1000 km. For example, the neutral densities above this altitude are too low to produce significant drag on satellites. But the long path lengths through the tenuous plasma that extends out to 20,000-km altitude (the plasmasphere) can contribute significantly to group delay of GNSS signals (Lunt et al., 1999).

3. **Atmospheric radiation environment** (up to 100 km). Natural radiation includes a contribution from sources in space. There is a slowly changing background of galactic cosmic rays that produces about 8% of the natural radiation background at sea level, but that dominates above 3-km altitude. The atmospheric radiation produced by galactic cosmic rays includes neutrons that can penetrate electronic devices, both on aircraft and on the ground, leading to single-event effects that cause those devices to malfunction or suffer damage. Electronic devices involved in safety-critical applications such as aircraft control are usually designed to mitigate these effects. During major space weather events, solar energetic particles (MeV to GeV protons and ions, energized at shocks ahead of fast coronal mass ejections (CMEs) and at solar flare reconnection sites) can deliver a huge increase in the natural radiation levels for a few hours. The worst observed case was an event in February 1956 that produced a 50-fold increase at the ground (Gold and Palmer, 1956; Marsden et al., 1956), and that was later estimated to have produced a 300-fold increase at aircraft cruising altitudes (Dyer et al., 2007). These huge increases in atmospheric radiation (known as ground level enhancements) can significantly increase single-event effect rates in electronic devices, increasing the risk of malfunctions.
4. **Satellite plasma environments**. The space is full of tenuous plasmas—free electrons, protons, and ions that bombard the surfaces of satellites and can penetrate into them if their energies are

high enough. Satellites in LEO fly through the plasma of the ionosphere, which is relatively cool (say 500–2000 K, giving average particle energies around 0.1 electron-volts (eV) that cause few problems for satellites). But in higher orbits, such as the vitally important geosynchronous orbit (home to many operational satellites for communications, meteorology, and other purposes), the average energy is around 1000 eV. Electrons at these energies can accumulate on the surface of satellites and give the satellite a large potential difference (~1000 V negative) with respect to the local plasma (Chapter 15 of this volume).

Space plasmas are rarely in thermal equilibrium and the energy distributions of electrons, protons, and ions often include strong tails with significant fluxes of particles up to MeV or even GeV (see schematic in Fig. 2). These tails arise from the energization process driven by space weather on the Sun, in Earth's magnetosphere (Horne et al., 2005; Chapter 14 of this volume), and also sometimes Jupiter's magnetosphere. These high-energy particles are a significant threat to satellites. MeV electrons can penetrate satellites, driving the accumulation of electric charge inside dielectric materials such as circuit boards and electrical insulation. This eventually leads to discharges when enough charge has accumulated. Those discharges can generate false signals that trigger anomalous behavior on satellites; even worse, they can cause damage to satellites.

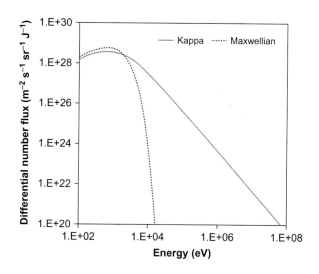

FIG. 2

Most space plasma environments are collisionless, so the velocity distributions of particles such as electrons and protons take the form of a Kappa distribution (Livadiotis and McComas, 2013) as shown by the *red curve*. These distributions have long power law tails at higher energies as well as high fluxes around the average particle energy. This example shows Kappa and Maxwellian distributions for an electron population with average energy ("temperature") of 1 keV and number density of 1×10^6 m^{-3}; and, for the Kappa distribution, with a Kappa value of 2 (corresponding to a power law exponent of −2 in this figure). As you can see, the Kappa distribution has significant particle fluxes at energies many orders of magnitude above the average, while the Maxwellian fluxes *(blue dotted curve)* are limited to a maximum energy about one order of magnitude higher than the average. The differential number flux is expressed in SI units of m^{-2} s^{-1} sr^{-1} J^{-1}; multiply by 1.6x10^{-23} to convert to units of cm^{-2} s^{-1} sr^{-1} eV^{-1}.

> **BOX 1**
> The mainstay of tracking aircraft is primary radar. When an aircraft encounters a radar transmission, a transponder on the aircraft returns a signal to the radar with a coding that identifies the flight. Secondary radar is when the aircraft is passive so the original transmission is scattered by the aircraft and a part of the scattered signal returns to the radar. This weaker signal is important in difficult situations, such as where the aircraft transponder has failed or has been turned off. In such situations interference from solar radio bursts would be a serious concern. Indeed, it would be prudent to quickly check that the interference was natural and not due to human action. This is a good example of how space weather knowledge is important for maintaining the security of our societies. See Knipp et al. (2016) for a broader discussion on how such knowledge has helped to avoid space weather impacts being misinterpreted as hostile actions.

MeV protons and ions can penetrate lightly shielded satellite systems such as solar arrays, causing damage that reduces their performance. At even higher energies, protons and ions can penetrate shielded systems, again causing damage that degrades the performance of electronic devices, for example. These higher energy particles also cause single-event effects inside electronic devices and, as with single-event effects on aircraft, can cause devices to malfunction.

5. **Radio frequency noise.** The Sun is a powerful source of radio noise over a wide range of frequencies that can penetrate Earth's ionosphere (20 MHz to several GHz). Energetic electrons in the corona can generate radio signals at these frequencies due to synchrotron radiation and the cyclotron maser instability. In addition, radio waves can arise from Langmuir waves in the plasma of the corona and solar wind. These solar radio bursts can interfere with radio systems on the dayside on the Earth—most obviously systems that are trying to detect weak man-made signals such as satellite navigation signals from the Global Positioning System (GPS) satellites (Chapter 22 of this volume). Radar systems are also particularly at risk when they are searching for weak signals such as reflections from aircraft (secondary radar—see Box 1).

3 GEOMAGNETICALLY INDUCED CURRENTS—THE IMPACTS OF NATURAL GEOELECTRIC FIELDS

As we noted in Section 1, natural geoelectric fields can drive electric currents to flow through extended conducting infrastructures that are electrically grounded to the Earth. These geomagnetically induced currents (GICs) are intimately related to the electric currents that flow in the solid body of the Earth. The existence of these Earth (or telluric) currents has been known and studied since the development of the electric telegraph in the middle of the 19th century. Indeed, the telegraph was the first technology to be disrupted by GICs arising from space weather (Barlow, 1849).

The existence of Earth currents may seem curious at first because the rocks that make up the Earth's crust are many orders of magnitude less conductive than metals. Common rocks such as granite can be at least a trillion times less conductive than copper. Nonetheless the Earth's crust does have sufficient conductivity to support significant current flow, a feature that is exploited technologically in Earth-return systems where electric power or signals are transmitted from one location to another over a single wire, and the current loop is closed through the body of the Earth. Thus magnetic variations such as those generated by electric currents in the ionosphere can penetrate into the crust and induce electric fields via Faraday's law of magnetic induction. This mechanism is particularly effective because the ionospheric currents vary at low frequencies (a few millihertz) where the skin depth of Earth's crust is large (100s of km). The size of these electric fields will depend not only on the size of the magnetic variations but also on the ability of crustal conductivity to convert this into a large geoelectric field (see Box 2).

As we noted above, these induced electric fields drive electric currents into extended metal infrastructures on the surface such as power grids, pipelines, rail circuits, etc. This occurs because these infrastructures have multiple electrical connections between the infrastructure and the solid body of the Earth, such as at transformers on power grids and rail circuits. Thus there exists an electrical circuit that runs through the infrastructure and closes through the Earth. The GIC that affects these infrastructures arises from the impact of the geoelectric field on this circuit. Thus the Earth connections play a vital role in creating the GIC.

So why do we need those connections? First they play a key role in electrical safety, ensuring that the zero potential of the system matches the electrical potential of the local ground surface. If this were not done, there could arise a significant potential difference between the infrastructure and the local ground. This could be a significant hazard for people working on the infrastructure, possibly also leading to discharges which damage the infrastructure. In addition, ground connections can be important for electromagnetic compatibility. Strongly varying currents in an infrastructure such as the passage of an electric train can generate unwanted signals that interfere with other systems near the infrastructure. Good design of ground connections can reduce the strength of such signals, enabling the infrastructure to be a good neighbor which does not disrupt nearby systems.

BOX 2

Magnetic induction in the Earth's crust is of scientific interest beyond space weather. It is a powerful tool for scientists studying the interior of the Earth. By measuring the natural variations of the geomagnetic and geoelectric fields (a technique known as magneto-telluric (MT) surveying (Simpson and Bahr, 2005)), they can infer information about the conductivity of subsurface layers and thus explore the subsurface structure. This is an important technique for both scientific research and for commercial organizations surveying for oil and minerals. It also provides valuable information for the space weather community in that it can provide direct measurements of the impedance matrix Z that relates magnetic variations (dB/dt) to electric fields (E) as shown in Eq. (1). In this example, we consider that Z is a frequency-dependent two-dimensional matrix that links the electric and magnetic fields in the local surface plane xy. Z varies from place to place on Earth's surface according to the local geology.

$$\begin{pmatrix} Ex \\ Ey \end{pmatrix} = \begin{bmatrix} Zxx(\omega) & Zyx(\omega) \\ Zxy(\omega) & Zyy(\omega) \end{bmatrix} \begin{pmatrix} \dfrac{dBx}{dt} \\ \dfrac{dBy}{dt} \end{pmatrix} \quad (1)$$

As you can see, the values of Z can play an important role in determining where we see extreme geoelectric fields. High values of dB/dt are a significant element in those extremes, but the geoelectric fields will be truly extreme where a large dB/dt combines with large values in the impedance matrix. As a result, the solid-earth geophysics community is an important partner to the wider space weather enterprise because they bring knowledge on these impedances matrices (see Bonner and Schultz, 2017)

Thus it is clear that GIC can enter grounded infrastructures as a result of the geoelectric fields that arise from geomagnetic activity. But what can drive GIC to extreme values and what would be the impact of extreme GIC? Looking first at the drivers:

(1) The topology of the infrastructure is important as this determines the size and distribution of the electric currents produced in response to the pattern of geoelectric fields. This is straightforward if the infrastructure is linear (such as in an isolated pipeline); the current in the segment between adjacent earthing points will be determined by the electric resistance of that segment and geoelectric voltage across the segment, that is, Ohm's Law. But many infrastructures have network topologies, meaning the currents will be determined by Kirchhoff's laws for electrical circuits together with Ohm's law. This leads to a complex pattern of GICs in which some nodes of the network may experience extreme values (Turnbull, 2010). It is important to identify these nodes and consider whether engineering measures are needed to reduce the extreme values or to make that node more robust. For example, in a power grid it would be recommended to equip this node with transformers that have high resilience to GIC and to move less resilient transformers to nodes with lower GIC risks. (Grid transformers have working lifetimes of many decades, so it makes sense to move such assets to optimize long-term resilience.) A key topological factor in considering GIC risks is whether the infrastructure has a tree-like structure or a mesh-like structure, meaning whether segments radiate from one or two central nodes with few interconnections or whether there is a high degree of connectivity between segments of the infrastructure (see Fig. 3). Tree-like topologies can focus GIC while mesh topologies spread these currents, potentially reducing the risk. Thus mesh topologies are considered to be more resilient to GIC. However they can exhibit extreme GIC at the edges of the network as currents entering or exiting the network can concentrate here before being spread by the mesh.

(2) As discussed in Box 2, variations in the conductivity structure of Earth's crust can play an important role in creating extreme GICs. This structure determines the ground impedance that converts geomagnetic variations into geoelectric fields. Roughly speaking this will be higher when the crust

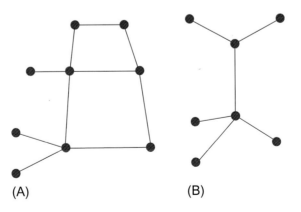

FIG. 3

Simple examples of grid topologies: (A) the network on the left has a mesh topology, while (B) the network on the right has a tree-like topology. The mesh topology enhances general grid resilience by providing multiple paths for many current flows. This is also valuable in the case of space weather as it spreads the GICs.

is dominated by rocks with high electrical resistivity, such as volcanic rocks, and lower when the crust is dominated by rocks with lower resistivity, such as sedimentary rocks containing significant amounts of water (as in natural aquifers such as chalk and limestone). But an accurate assessment of ground impedance requires detailed measurements as noted in Box 2 and is an important element in determining how space weather conditions can drive extreme GICs.

(3) Coastal regions present a special case of the ground conductivity problem. Here there is an abrupt change in the ground impedance, as we have to extend the meaning of this term to include the impedance of seawater (more accurately it is the impedance of Earth's surface, whether land or sea). The "ground impedance" of seawater is much lower than that over land because the electrical conductivity of seawater is thousands of times greater than the rocks underlying the land. Modeling suggests that this abrupt change in conductivity can enhance the geoelectric fields in coastal regions, perhaps only for a few tens of kilometers inland. But this may have important societal impacts because many infrastructures are located in such regions. It should be stressed that this is an area where research is important, as most existing models have considered only the case of simple coastal geometries (Boteler and Pirjola, 2017). We do not know what would be the consequence of high-resolution modeling of fully realistic geometries, for example, reflecting the fractal nature of coastlines (Mandelbrot, 1967). It is also worth noting that these coastal effects can reinforce the edge effect on network topologies as discussed in the previous paragraph. These two effects will inevitably combine in coastal regions and hence make those regions a particular area of space weather vulnerability.

(4) The other important driver of extreme GICs is, of course, geomagnetic activity as reflected in dB/dt, the rate of change of the geomagnetic field. Values of 100s of nT per minute are considered significant and values of 1000s of nT per minute are extreme. There at least three different physical process that can generate such changes, possibly more.

  (a) The most important of these is the ionospheric current systems driven by magnetospheric substorms. These substorm currents are intensifications of the auroral electrojets, the ionospheric currents flowing in the auroral zone as a result of magnetospheric convection driven by momentum coupling from the solar wind (Axford and Hines, 1961). A key part of this convection is a flow of plasma from the nightside to the dayside on magnetic flux tubes that connect to the auroral zone. This flow produces a dayward drift of electrons in the E region ionosphere and hence an eastward current (electrojet) on the evening side of the auroral zone and a westward electrojet on the morning side. When a substorm releases a burst of energy, the electrojets can intensity and become more complex. There is growing evidence (Pulkkinen et al., 2015; Ngwira et al., 2015) that this can include small-scale (500 km) structures containing strong currents that are rapidly changing in strength and direction. These are thought to be a key driver of extreme GICs at high and middle latitudes.

  (b) Another important process is the ring current. This is a torus of electric current that flows westward at altitudes of 10,000–20,000 km above Earth's equator, producing a significant magnetic signature on the ground at low latitudes. The ring current arises from the differential motion of ions and electrons in this region (ions flow westward, electrons eastward) and the injection of additional ions and electrons when a substorm occurs. These additional particles are lost fairly slowly so the ring current will intensify in a series of steps as substorms occur. This can lead to significant dB/dt on the ground at low latitudes and is thought to be a key driver of large GICs at low latitudes (Gaunt and Coetzee, 2007; Thomson et al., 2010).

(c) Our third process is a sudden impulse (sometimes also called a sudden storm commencement). These arise when a large coronal mass ejection hits the Earth—or more accurately the enhanced ram pressure in a CME pushes the magnetopause closer to the Earth. This magnetic compression will propagate toward Earth as an Alfvén wave, taking perhaps a couple of minutes to reach the surface. There it will be seen as a sudden marked increase in the strength of the geomagnetic field. This sudden change may constitute a significant dB/dt and thus could drive large GIC, e.g., in equatorial regions. Recent studies suggest that this effect could be amplified by its interaction with the equatorial electrojet (Carter et al., 2015). Fortunately, this is still a theoretical case as there are, as yet, few vulnerable infrastructures near the magnetic equator. But, as the development of equatorial countries continues, this may become a significant practical issue.

(5) Some authors have suggested that other magnetospheric and ionospheric phenomena could also drive large dB/dt. One example is magnetospheric pulsations, the vibrations of magnetospheric flux tubes driven by Kelvin-Helmholtz instabilities on the flanks of the magnetosphere, and that drive low frequency pulsations in the ground magnetic field (Kappenman, 2005). Another example is magnetic crotchets, spikes in the ground magnetic field following a very intense solar flare. They reflect sharp changes in ionospheric conductivity induced by the X-rays from the flare. These processes require further study.

We now turn to the impact of extreme GICs on grounded infrastructures. How do these currents interfere with normal electrical operation?

In the case of power grids, the primary problem is the disruption of normal transformer operation. GIC will enter a transformer and act as a quasi-DC current, since GIC frequencies of a few millihertz are much lower than the 50 or 60 Hz frequency of power grids. Thus a small GIC can bias the average voltage away from zero and drive the transformer into half-phase saturation, leading to generation of harmonics plus vibration and heating in the transformer. The GIC essentially acts a catalyst so that some of the main power through the transformer is diverted into adverse effects. If large enough, the vibration and heating can cause damage to the transformers while the harmonics in the grid currents can disrupt the operation of other grid devices. If this happens, it is expected that sensors will detect these conditions (many transformers have monitors that can detect gases produced by unexpected heating inside the transformer) and switch off the device before serious damage can occur. In an extreme event, large numbers of switch-offs are expected, probably leading to a cascade in which the whole grid switches off. This is what happened in Quebec in the March 13-14, 1989, storm (Bolduc, 2002) as a number of devices tripped out, starting a cascade that took the Hydro-Québec power grid from nominal operation to complete shutdown in 92 seconds. Fortunately, there was only a small amount of damage; the grid was restarted within 9 hours. Thus this event showed both the spectacular impact of severe space weather, and how good engineering enabled a quick recovery. This blackout was coincident with a very strong substorm that peaked about 6 hours after the onset of the storm (see Fig. 4) and that had a strong footprint over Quebec.

The storm continued for many hours after this, producing many more substorms including two with strong footprints over Northwest Europe, both marked by intense aurora down to 50° latitude. The first substorm occurred some 14 hours after the Quebec substorm and the resulting aurora is shown in Fig. 5. The other occurred a further 4 hours later and the resulting aurora was captured in one of NASA's iconic images from the 1989 storms (see Fig. 6). These images show how the auroral oval will extend

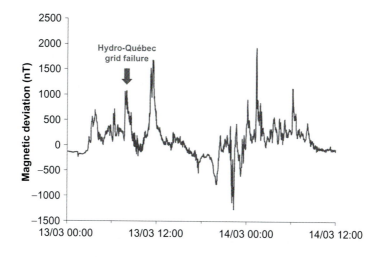

FIG. 4

Deviation (from average) of the horizontal geomagnetic field component at Ottawa, Canada, for 36 hours from 00:00 UTC on March 13, 1989, to 12:00 UTC the following day. This great geomagnetic storm started with a sudden impulse at 01:28 UTC. It was the large substorm around 08:00 UTC that caused the failure of the Hydro-Québec grid.

Data from NOAA/NCEI.

FIG. 5

Intense aurora over Oxfordshire in the south of England during an intense substorm peaking around 21:45 UTC on March 13, 1989.

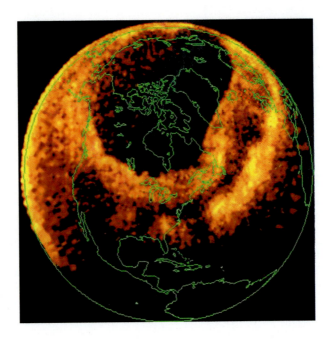

FIG. 6

Wide-angle image of the aurora over the northern hemisphere at 01:51 UTC on March 14, 1989. This is derived from an image taken over the southern hemisphere from the NASA Dynamics Explorer 1 satellite, and mapped to the north by tracing geomagnetic field lines. The right side of the image shows a band of intense aurora located just to the south of the United Kingdom.

From NASA website "Mission to Geospace," page on "Lights Below: The Aurora from Space" https://pwg.gsfc.nasa.gov/istp/outreach/images/Aurora/mar89remap.gif.

to mid-latitudes in an extreme space weather event. In this case the GICs produced by the substorms are thought to be responsible for the tripping out of two U.K. transformers—one at a grid node in Cornwall and the other at a node near Norwich (Erinmez et al., 2002; Boteler, 2012). It is not known at exactly which time the transformers tripped. Fortunately the U.K. grid had sufficient resilience (parallel redundancy of transformers at grid nodes) to handle this, so there was no loss of power on that occasion. But the event showed the potential of space weather to affect the U.K. grid and has led to many improvements, most obviously the gradual deployment of transformers with greater resilience to GIC. Another improvement was the development of procedures to anticipate adverse space weather and to take mitigating actions in response to forecasts, such as making sure all grid lines are on, so as to spread GIC over the system and reduce its impact at any one location.

Power grids in a number of other countries have also experienced significant problems due to space weather, such as the destruction of a power station transformer in New Jersey during the March 1989 storm (Wrubel, 1992), and a short blackout in Malmö during the Halloween storm of 2003 (Pulkkinen et al., 2005). Another major example of the adverse impact of space weather was the gradual loss of several transformers in South Africa following the Halloween and other geomagnetic storms in autumn 2003 (Gaunt, 2014). This demonstrated that ring current variations (and not just substorms) can put disruptive levels of GIC into power grids. The lower latitude of South Africa means that it is rarely exposed to substorms, but is exposed to ring current effects. Analysis of data from transformer sensors suggests that the South African transformers suffered a series of heating events that were correlated in time with a series of geomagnetic events (Thomson and Wild, 2010). This series of events may have gradually degraded the transformers, leading to the rash of failures that significantly reduced the amount of power available via the South African grid.

As a result of our growing understanding of these impacts, power grids around the world are now assessing their vulnerability to space weather, particularly extreme conditions, and adopting a range of measures to mitigate adverse impacts. The examples discussed above, such as more resilient transformers and procedures based on forecasts, are being widely adopted. Other ideas have been considered, such as the use of devices such as capacitors to block GIC from passing through Earth connections. But these have not been widely adopted due to a mixture of extra costs and concerns about the impact of those devices on overall system performance.

We turn now to the impact of GIC on rail systems. This is an area that is still poorly understood, not least due to the greater complexity of the electrical circuits and hence of the interaction of GIC with those circuits. The first direct measurement of GIC in a rail system was reported only in 2016 (Liu et al., 2016) and showed the challenge posed by the interaction of electrical systems for power and signaling through shared earth connections. That measurement showed only a modest GIC, as expected in a modest storm. But there is circumstantial evidence that, in extreme storms, we can expect GIC to disrupt signaling systems. For example, the anomalous changing of signals from green to red has been reported during large geomagnetic storms in 1982 (Wik et al., 2009), 1989 (Eroshenko et al., 2010), and 2003 (Eroshenko et al., 2010). There is also a report of signal system problems during the peak of the severe storm of January 1938 (The Guardian, 1938). These were all "right-side failures" in railway terminology, meaning the system failed to the safe, but disruptive, condition of stopping trains. However, expert analyses (see the report by Krausmann et al., 2015) suggest that a "wrong side failure", that would allow trains to proceed when unsafe, is possible. So this is clearly an area that warrants further research, especially with the ongoing development of high-speed rail systems.

GIC impacts on pipelines have been widely studied (Pirjola, 2005). They are a significant issue because they can interfere with the electrical systems that protect pipelines from corrosion. This protection arises because corrosion of metals is an electrolytic process that can generate a small voltage between the pipe and the surrounding soil. "Cathodic protection" systems apply a small voltage to the pipeline that has the opposite sign to this electrolytic voltage and hence acts to inhibit corrosion. The presence of GIC in a pipeline can bring confusion to this otherwise well-ordered situation. In some cases GIC may oppose the cathodic protection, thus temporarily increasing corrosion rates. More importantly the GIC may confuse the operators of the system, making the system work in unexpected ways and triggering unnecessary work to chase down temporary faults that will go away of their own accord. For example, problems with protection systems in Scotland were reported during the March 1989 storm but operators were advised to just let the storm pass (the author was contacted informally and provided that advice). This is an area with a good understanding of the basic physics but only limited data on actual problems. It would be beneficial to encourage more open reporting of incidents so that the scope of the problem is better understood, so that the industry can be prepared for an extreme space weather event. It would also allow government risk managers to assess whether there is any potential for wider societal or economic impact, such as through an impact on energy distribution. At present, it seems likely that an extreme space weather event will cause consternation in the pipeline industry and a rush for scientific advice.

GIC can also impact the cables that carry telecommunications traffic under the oceans. These are the modern day successors of the undersea cables that carried telegraph signals between the continents from Victorian times well into the 20th century. Those earlier cables sent signals over metal wires and suffered very badly from GIC during severe storms (Lanzerotti, 1979). However, over the last 30 years they have been replaced by fiberoptic cables, providing such a huge increase in bandwidth that they now carry 99% of transoceanic voice and Internet traffic. They are much preferred to the use of satellite links because of the light-time delays inherent in sending signals via satellites in geosynchronous orbit. The change to fiberoptics has eliminated the possibility that GIC can interfere directly with communications signals. But GIC can enter the systems that provide power to optical repeaters along these cables. Excessive power system voltages were reported on fiberoptic cables during the March 1989 storm (Medford et al., 1989) and again during the Bastille Day storm in July 2000 (Lanzerotti et al., 2001). But in both cases the power system proved resilient to these excess voltages. Over the past 15 years no incidents appear to have been reported in the scientific literature, but it is not clear if that reflects work to mitigate the problem or just a lack of reporting. Given the low level of geomagnetic activity in Solar Cycle 24, it is likely that there have been no significant events since 2004, but it is curious that no events were reported during the Halloween storms of October 2003. This is an area that requires assessment. It would be a major societal problem if an extreme event were to knock out a number of these cables.

4 SPACE WEATHER IMPACTS ON THE UPPER ATMOSPHERE
4.1 OVERVIEW OF THE UPPER ATMOSPHERE

Earth's upper atmosphere (90–800 km altitude) is a physically fascinating region. It contains two distinct elements that occupy the same volume of space but that are only weakly coupled, leading to complex physical behavior that scientists have explored over many decades. These two elements are (1) the thermosphere, a neutral component that is collision-dominated and hence behaves as a fluid, and (2) the ionosphere, an ionized component that is collision-dominated below around 150 km, but largely collisionless at higher altitudes and hence behaves as an MHD plasma. Both elements have important space weather impacts: the thermosphere provides the atmospheric drag that satellites experience in low Earth orbit, while the ionosphere is an important factor for the many radio and radar systems that send signals through the upper atmosphere (from ground to satellites, from satellites to ground, and from ground to ground by reflection from the ionosphere). The thermosphere is much denser than the ionosphere; the level of ionization is around one part in a million at 100-km altitude rising to about one part in a thousand at 300-km altitude.

Satellite drag and trans-ionospheric radio propagation are often treated as completely separate space weather issues. They are seemingly unrelated until one understands the underlying science—that the ionosphere and thermosphere are weakly coupled via ion-neutral collisions. These collisions enable exchange of momentum so that strong winds in the thermosphere can drive plasma flows in the ionosphere, and strong plasma flows (such as in the high polar ionosphere directly coupled to the solar wind) can drive winds in the thermosphere. These collisions also enable charge exchange so that neutral molecules such as NO can become ions (i.e., NO^+), thus facilitating the rapid loss of ionization by dissociative recombination (Bates, 1950).

Thus it is convenient, and perhaps valuable, to treat satellite drag and trans-ionospheric radio propagation together so that we can explore where these two different impacts arise from a shared set of physics. In particular, we will emphasize that the density and composition of the thermosphere are key factors controlling the global morphology of the ionosphere. But that density and composition are also key factors in the atmospheric drag experienced by satellites.

4.2 TRANS-IONOSPHERIC RADIO PROPAGATION

Radio waves passing through the ionosphere interact with the plasma that forms this region. As a result the impacts of space weather on trans-ionospheric radio propagation are complex and strongly system-dependent, especially in extreme space weather conditions. For this reason, we first review the key mechanisms through which the ionospheric plasma interacts with radio waves, and then discuss how the complex morphology of the ionosphere arises and is influenced by space weather. Finally we use this science to explore how different trans-ionospheric radio systems are affected by space weather, especially in extreme conditions.

The key mechanisms that affect radio waves propagating through the ionosphere are:

a. Most notably, radio waves below the plasma frequency cannot propagate through the plasma. Thus waves entering the ionosphere at frequencies below the plasma frequency will be reflected, which is an effect that enables terrestrial technologies for communications and radar systems that can work over-the-horizon by bouncing signals off the lower ionosphere. The plasma frequency is the frequency at which the electrons in the plasma naturally oscillate relative to the ions and typically has values between 2 and 20 MHz for conditions in Earth's ionosphere. It varies with the electron number density n thus: $\omega_p = \sqrt{ne^2/(m\epsilon_0)}$, where e is the charge on the electron, m is the effective mass of the electron, and ϵ_0 is the permittivity of free space.

b. Radio waves above the plasma frequency will deliver their signals with a group speed V_g that is slightly less than the speed of light in a vacuum: $V_g = c\sqrt{1 - \omega_p^2/w^2}$ where c is the speed of light and ω is the frequency of the radio wave. Thus signals traveling through the ionosphere are subject to a small delay compared to signal propagation at the speed of light. This group delay is an important factor in some technologies, in particular in satellite navigation that seeks to measure the distance between transmitter and receiver to a precision of a few meters or less (and hence group delays at the 10 ns level are significant).

c. Radio waves can be scattered by irregularities in the density of the ionosphere, such as those produced by a variety of plasma instabilities. The irregularities can act as a diffraction screen producing interference patterns in radio waves that pass through the ionosphere. These patterns will be seen as amplitude and phase scintillation by radio receivers. This scintillation can disrupt signal reception, meaning rapid-phase scintillation may prevent receivers from maintaining a lock on the incoming signal.

d. Radio waves will be absorbed by the ionosphere (rather than reflected or transmitted) if there is a significant density of electrons in a region where the electron-neutral collision frequency is higher than the frequency of the radio wave (i.e., cases such that an electron excited by the radio wave is more likely to lose energy by collisions than by reradiating the wave). This can occur when there is significant ionization at low altitudes ($\sim$90 km) where the electron-neutral collision frequency is high.

The first three of these mechanisms are determined by the global morphology of the ionosphere, especially the so-called F-region between 150 and 500 km in altitude, which contains the bulk of ionosphere plasma. Its morphology arises from a complex interplay of three main factors:

- Production of plasma. Ionospheric plasma is mainly produced by the action of solar EUV on the thermosphere. Thus the production rate varies according to (a) the geometric illumination of the thermosphere, meaning the cosine of the elevation of the Sun, which varies in a very predictable way due to latitude, time of day, and season of the year; (b) the amount of EUV arriving from the Sun, which varies with solar activity, a background showing a marked variation over the solar cycle, plus bursts of extra EUV from active regions and solar flares. Thus production rates vary markedly with the level of space weather on the Sun, but are smoothly distributed across the dayside of the Earth, simply following the elevation of the Sun.
- Transport of plasma. Ionospheric plasma can often be transported over large distances before recombining back to neutral species, so many ionospheric structures arise from plasma flows and not just the local balance of production and loss. One example is the circulatory patterns of often-strong flows in both polar ionospheres, which usually take the form of an antisunward flow over the pole from dayside to nightside driven by momentum from the solar wind (Axford and Hines, 1961), with plasma returning to the dayside by sunward flows at auroral latitudes. This "ionospheric convection" pattern is responsible for many structures in the high latitude ionosphere, including high densities around local noon (as plasma is drawn poleward from lower latitudes) and the occurrence of "polar patches" of relatively cool dense plasma drifting over polar regions (patches of long-lived plasma that have detached from the high density region around local noon; their low temperature distinguishes them from plasma recently produced by electron precipitation, see Carlson (2012)). Another example common at lower latitudes is that strong neutral winds in the thermosphere can transfer momentum to ionospheric plasma. The resulting plasma flows are constrained to follow the geomagnetic field and hence include a vertical component (King and Kohl, 1965). Vertical plasma flows profoundly change loss rates and hence plasma densities; these decrease in the presence of downward flows and increase in the presence of upward flows. Upward flows can also trigger turbulence via the Rayleigh-Taylor instability (Ott, 1978).
- Loss of plasma. Ionospheric plasma is mainly lost by the process of dissociative recombination in which molecular ions such as NO^+ interact with an electron to produce two neutral atoms, N and O in our example. Atomic ions such as O^+, which are common in the ionosphere, must first undergo charge exchange with a neutral molecule, producing a molecular ion. Direct recombination of an atomic ion with an electron, producing a single atom (and a photon), is an inefficient process for loss of plasma; it has a low cross-section as it is hard to conserve momentum and energy in this process. Thus plasma loss rates depend on the composition of the thermosphere and are enhanced when there is a substantial fraction of molecular species as well as the atomic oxygen that is common in the thermosphere. This composition is strongly modulated by space weather in the form of geomagnetic storms. These storms cause intense heating in the polar upper atmosphere (e.g., as energy dissipation from the electric currents and particle precipitation associated with auroral activity). That heating can inject molecular species from lower altitudes into the polar thermosphere, from where they are carried across the global thermosphere by winds that are also driven by this auroral heating.

The interplay of these three factors leads to a complex global morphology that drives many space weather impacts. In particular, strong spatial gradients in plasma densities are a key source of free energy that can drive plasma instabilities.

In contrast our fourth mechanism above, the level of ionospheric absorption, is dominated by local physics. This arises because absorption is a low-altitude effect, so the recombination lifetime is short (seconds or less) and transport plays no part. Ionization at altitudes around 90 km arises from high-energy radiation such as X-rays or high-energy particles. Solar EUV is totally absorbed at higher altitudes by the processes that generate the F-region ionosphere. Thus ionospheric absorption occurs as a result of solar X-ray flares, of increases in the solar X-ray background, from precipitation of solar energetic particles (SEP), and the precipitation of relativistic electrons from Earth's magnetosphere. These different sources determine the spatial and temporal extent of ionospheric absorption. For example, solar flares cause intense but short-lived (a few hours at most) absorption across the dayside of the Earth, while SEPs are directed into polar regions by the geomagnetic field and can cause strong absorption in those regions lasting for several days.

We can now look at how space weather impacts the users of trans-ionospheric radio systems. That range is illustrated in Table 1, where we present some examples of these users and the radio systems they use, together with: (a) the radio frequencies on which those systems operate, (b) the effect of extreme space weather on the radio signals, and (c) the broader impact on the users' activities.

Looking at the table we can see the ultimate impact of extreme space weather, indeed of most space weather, is highly dependent on the system at risk and on how the system is applied to deliver results of value to the user. For example, we have divided the space weather effects on GNSS into four different user categories that highlight this point. (In practice, we could reasonably divide into finer categories, but four will suffice here.) We recognize that loss of signal is an obviously crucial effect for all categories, but that group delay is crucial only for some categories. Looking first at use of GNSS for precise timing we have neglected group delay; this is driven by an assumption that only microsecond timing accuracy is required at present (e.g., consistent with current developments for high-speed financial trading (Aron, 2014; Stafford, 2016)). But when looking at use of GNSS for road-based transport, we assume that an accuracy of at least tens of meters is required (e.g., to quickly identify the correct road/house in densely populated areas) and hence group delay corrections are important. Turning to GNSS use by aviation, we know that an accuracy of a few tens of meters is critical to safety and assume that this level of accuracy will be underpinned by a wide-area augmentation system targeted on aviation users (such as WAAS in the United States and EGNOS in Europe). Thus we have to recognize the challenge that extreme space weather poses for the integrity of wide-area augmentation systems that have to interpolate over large distances. This may be difficult when extreme space weather leads to large spatial gradients and high rates of change. Finally turning to GNSS use in shipping, we assume that accurate group delay corrections will be provided by local augmentation systems around ports (e.g., differential GPS) and that these will be much less susceptible to space weather problems because they are primarily used over short distances.

In summary the impacts of extreme space weather on trans-ionospheric radio propagation are complex, reflecting (a) the range of different interactions between radio waves and ionospheric plasmas, (b) the complex response of the global ionosphere to space weather, and (c) the wide range of users who rely on systems that send radio waves through the ionosphere. It is impossible to summarize these impacts in a simple way, especially in terms of ionospheric and space weather science. The impacts are ultimately determined at radio system level—in particular, how different categories of users exploit trans-ionospheric radio systems and how space weather effects interfere with that exploitation. Thus it is important to pose a range of questions toward users and then explore how those questions link back to the science of space weather. Those questions include:

Table 1 Some Examples of Extreme Space Weather Impacts on Radio Systems

User and Radio System	Frequency	Effect of Extreme Space Weather	Impact of Extreme Space Weather on User's Activities
HF radio communications for civil aviation	2 to 20 MHz	Loss of signal propagation at night-time due to disappearance of ionosphere during severe storm conditions	Night flights over ocean and remote areas (e.g., westbound over North Atlantic) cannot communicate over the horizon using HF systems. Mitigate by using communications technology that does not require presence of ionosphere, e.g., satcom
		Severe distortion of signals due to scintillation	Flights over ocean and remote areas cannot communicate over the horizon using HF systems. Mitigate by using communications technology not affected by scintillation, e.g., on busy routes use chain of VHF links between aircraft as far aircraft with VHF ground contact
		HF radio absorption due to SEP precipitation in polar regions	Flights over high Arctic cannot communicate over the horizon using HF systems. No current mitigation (satcom not yet available)
		HF radio absorption due to strong solar flare	Daytime flights over ocean and remote areas (e.g., eastbound over North Atlantic) cannot communicate over the horizon using HF systems. Mitigate by using communications technology that does not require presence of ionosphere, e.g., satcom
Automatic identification system: telemetry on identity, position, course and speed of ships at sea	~160 MHz	Loss of AIS signals by satellites due to scintillation. AIS uses satellite links when ships are out of sight of land; its VHF frequency is highly vulnerable to scintillation	Loss of wide area information on ship and movements (particularly from busy routes at northern latitudes such as the North Atlantic, North Sea, Northern Sea Route and North Pacific. This information is used by commercial organizations and government bodies. No known mitigation

Table 1 Some Examples of Extreme Space Weather Impacts on Radio Systems—cont'd

User and Radio System	Frequency	Effect of Extreme Space Weather	Impact of Extreme Space Weather on User's Activities
Tracking of satellites in low Earth orbit by ground-based radars	Wide range from 100 MHz to 20 GHz	Ionospheric group delay corrections (important for VHF and UHF systems) inaccurate due to rapid changes and large spatial gradients in TEC	Satellite tracking data less accurate leading to problems in matching observed tracks to known satellites, plus reduced accuracy of orbit predictions, and hence greater uncertainties in collision risk assessments and re-entry forecasts
Synthetic aperture radars for earth observation imaging	300 MHz to 30 GHz	Large variations in TEC along the satellite track add signal phase variations that distort images	Images recorded during extreme events may be of limited use
Use of GNSS by services that precise timing, e.g., communications networks, high-speed financial trading	1.1 to 1.6 GHz	Loss of GNSS signal at mid-latitudes due to scintillation during extreme events. Likely to be intermittent as multiple satellites are available and we need a signal from just one to obtain precise time	Disruption over several days of services at high and mid-latitudes that are dependent on precise time: Readily mitigated by use of holdover clocks
Use of GNSS by road transport delivering services to unfamiliar locations, e.g., emergency services, taxis, home delivery of products, ...	1.1 to 1.6 GHz	Loss of GNSS signal at mid-latitudes due to scintillation during extreme events. Determination of position requires signals from at least three or four satellites, so more at risk than timing applications GNSS services become inaccurate when there are rapid variations in TEC	Service delivery severely delayed and disrupted due lack of knowledge of routes. Particularly critical for emergency services on safety of life work. Mitigate by use of maps where available and where staff are trained. Problems likely to escalate as staff become tired over several days of disruption
Use of GNSS by shipping, especially in confined waters, e.g., entering/exiting ports, close to known obstacles such as rocks and wrecks	1.1 to 1.6 GHz	Loss of GNSS signal at mid-latitudes due to scintillation during extreme events	Increased risk of accident (collisions with obstacles, grounding) with potential for risk to life, the ship and its cargo. Mitigate by use of visual aids and good seamanship
Use of GNSS by aviation	1.1 to 1.6 GHz	Loss of GNSS signal at mid-latitudes due to scintillation during extreme events GNSS augmentation services such as EGNOS and WAAS may become inaccurate when there are rapid variations in TEC. Operators may deprecate use of	Pilots must use alternative navigation systems on board their aircraft, e.g., inertia navigation systems. Aircraft without alternative systems should be grounded

Continued

Table 1 Some Examples of Extreme Space Weather Impacts on Radio Systems—cont'd

User and Radio System	Frequency	Effect of Extreme Space Weather	Impact of Extreme Space Weather on User's Activities
Satellite mobile communications	1 to 2 GHz	augmentation service via integrity flags. Loss of signal at mid-latitudes due to scintillation during extreme events	Loss of communications by users of mobile satcom (satellite phones) including aircraft, ships and ground-based systems in remote areas
Uplink and downlink for control of satellites	2 to 4 GHz	Interruption of satellite contact periods due to scintillation during extreme events, especially if using lower frequencies in this range	Limited ability to monitor satellite status, upload new commands or download data. May need to overschedule satellite contact times to complete planned operation

- Do you have a backup to the radio system at risk from space weather, meaning an alternative communications system, a holdover clock for timing, an alternative navigation system? Absence of an alternative will obviously lead to loss of service during extreme space weather. High-end professional services should usually have backups to ensure resilience.
- Does your backup have a different failure mode in extreme space weather? A backup up with the same failure mode (such as using Galileo as a backup to GPS) is essentially a fake backup. It is fundamental to resilience that backups should have independent failure modes.
- Is your independent backup likely to suffer failure due to some other feature of a severe space weather event, as satcom and high-frequency (HF) communications may both fail at some phase of a severe space weather event? What is the likelihood of multiple failures? Does it warrant provision of a second backup?
- Are your operations teams and end users aware of the risk of interruptions? How would you alert them in the case that an extreme event is forecast/underway? What can they do to prepare for and respond to the event? Would they wish to reschedule critical activities to avoid an extreme event in order to maximize the likelihood of success?

As you will see from this list of questions, any substantive discussion of extreme space weather impacts on trans-ionospheric radio requires extensive interaction with both the engineers responsible for those radio systems and the users of those systems. The biggest risk is a lack of awareness among those engineers and users, the factor that is most likely to cause extensive societal disruption. Thus the most important mitigating action is for space weather experts to raise awareness and to keep doing so amongst new engineers and users.

4.3 ATMOSPHERIC DRAG

As we discussed at the beginning of this section, satellites in LEO experience a drag force as they pass through the thermosphere, the neutral component of the upper atmosphere. This force decreases with altitude, becoming insignificant above about 1000-km altitude. The immediate effect of the drag force is to gradually reduce the altitude of a satellite and increase its orbital speed (drag causes the satellite to lose gravitational potential energy as its altitude decreases, some of this energy is dissipated by the drag and some converted to kinetic energy). These changes, particularly the increase in speed, alter the orbit and hence anyone seeking to forecast the satellite position must keep track of the changing orbit through a combination of drag modeling and tracking of the satellite motion.

There is widespread interest in such forecasts. Most obviously, the operators of any active satellite want to know where it will be in order to plan operations and communications contacts. But there is also wide and growing interest from other bodies such as space agencies and governments in order to anticipate potential for collisions between satellites and for re-entry of large satellites (large enough that debris will survive re-entry and fall to Earth, posing a risk to human activities on the surface or in the air).

These forecasts are sensitive to atmospheric drag, especially at lower altitudes where the drag is large and satellites may be approaching the moment of re-entry. Thus it is important not only to model the atmospheric drag on satellites, but also to anticipate when this drag may become larger or smaller than usual. Larger-than-usual values are particularly important as they may lead to unexpected re-entry. Smaller than usual values will probably delay re-entry but are nonetheless important as they can lead to inaccurate forecasts of satellite position.

Space weather is the major cause of uncertainty in atmospheric drag because it can strongly modulate the density of the thermosphere, as we explored in the discussion of ionospheric effects. In particular, the density will change when geomagnetic activity heats the upper polar atmosphere, causing an upwelling of denser material into the thermosphere while also driving thermospheric winds away from polar regions. The net result is to transport enhanced density worldwide across the thermosphere, changing atmospheric drag and reducing the accuracy of satellite position forecasts. In addition, fluctuations in auroral heating generate atmospheric gravity waves that travel equatorwards; the density fluctuations in these waves add further uncertainties to satellite position forecasts.

These uncertainties will be particularly large in extreme space weather conditions, as intense auroral heating drives major changes in the density and global circulation of the thermosphere. These large uncertainties will make it very difficult to match observed satellite tracks with the forecast tracks of existing satellites, leading to a situation where (a) the location of many LEO satellites is unclear, and, as a direct corollary, (b) the identity of many observed LEO satellite tracks is unknown. It will likely take several days to redetermine which observed track matches which satellite identity. This period of confusion will present a number of challenges:

- Most obvious is the challenge to satellite operations. In particular, operators will need to reschedule a range of activities to match altered satellite locations, such as ground station contacts with satellites, observations of specific areas on Earth, and precautions when passing through regions of high radiation such as the South Atlantic Anomaly. Note that this rescheduling is not a linear time shift. An extreme space weather event will significantly decrease the orbital period of an LEO satellite, so the required time shift (compared to times forecast before the event) will increase with time after the event.
- An extreme event will also challenge work to assess the risk of collisions between satellites. That assessment is critically dependent on accurate knowledge of satellite orbits, not least the orbits of inactive objects such as old satellites and debris from the breakup of satellites. The widespread orbit changes induced by extreme space weather will necessitate a major effort to reassess collision risks as new orbit data becomes available.

- There will be a similar challenge to reassess the risk of re-entry by large satellites, especially those that were approaching re-entry prior to the event. An extreme space weather event is likely to bring forward the re-entry of such objects, perhaps necessitating rapid adoption of plans to deal with the risks posed by large space debris falling to Earth.
- Finally we must recognize that this period of confusion following an extreme space weather event poses challenges that go beyond the technical issues we have so far discussed. Satellites play an important role in the security of countries around the world, so confusion about the location of satellites critical to security can add to any existing concerns between countries, especially if there is preexisting tension. In such situations it is important that decision-makers be aware of when natural phenomena add confusion to a tense situation. Space weather is an important issue here because it can have major technological impacts yet limited impact on human senses. Knipp et al. (2016) have demonstrated how the provision of this awareness was important at the height of the Cold War. It is equally important today.

Finally, it is worth noting that we have a good historical example of this impact of extreme space weather. During the great geomagnetic storm on March 13–14, 1989, U.S. satellite trackers lost awareness of 1500 satellite locations; it took several days to recover that awareness as shown in Fig. 7 (Air University, 2003). Fortunately this event took place in the closing months of the Cold War, a time when international tensions were decreasing markedly.

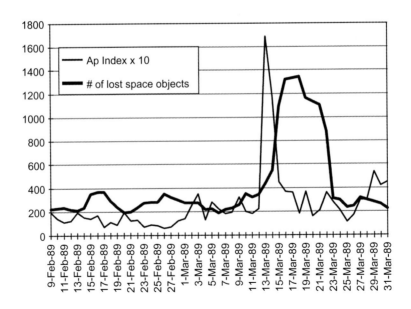

FIG. 7

During the great geomagnetic storm of March 1989, atmospheric drag caused a huge increase in the number of space objects whose location had become unknown, as shown by the peak in the number of lost space objects. This follows the spike in the geomagnetic index Ap caused by the storm.

Courtesy U.S. Air Force, Air University: Space Primer, Chapter 6 on Space Environment. http://www.au.af.mil/au/awc/space/primer/space_environment.pdf.

5 ATMOSPHERIC RADIATION ENVIRONMENT

Aircraft fly in a natural radiation environment that is dominated by radiation from sources in space. This is very different from the situation on Earth's surface where the primary sources are radioactive elements in the Earth's crust, particularly radon gas seeping out of the ground, boosted slightly by radiation from human activities. These terrestrial sources become less important with increasing altitude while radiation from space dominates above an altitude of 3 km.

In quiet space weather conditions, the radiation from space takes the form of a slowly changing background of galactic cosmic rays. Collisions with atmospheric species convert these particles into high-energy neutrons; it is these that we observe at aircraft altitudes. Here they can penetrate electronic devices leading to "single-event effects" in which the interaction of a single particle with material in the device can create extra electrical charge inside the device. The small feature sizes in modern devices means that even a small amount of unexpected charge can cause a malfunction (e.g., bit flip, latch-up) or even permanent damage to the fine structure of the device. Thus SEEs are a major issue for aircraft control systems, and it is important to build resilience into the design of these systems.

A small proportion of the cosmic ray neutrons reach the surface of the Earth, where they contribute about 8% of the natural radiation environment. It is thought that, as with aircraft, these cosmic ray neutrons can penetrate electronic devices and cause single-event effects in ground-based systems. However, there has so far been limited reporting of such effects outside a few specialist areas. One very public example came when a voting machine in Schaerbeek, Belgium, added 4096 votes to the tally of one candidate, thus giving more votes than was possible (PourEVA, 2003). Also, manufacturers of electronic devices have been clear that designers of systems that utilize their devices should consider these radiation effects as a significant source of error; therefore designers must consider whether and how to mitigate those errors (Wood and Caustin, 2006). Nonetheless it is unclear how well this risk is considered in the broader growing use of digital systems.

During an extreme space weather event, the radiation from cosmic rays can be strongly enhanced by a burst of solar energetic particles (SEPs, high energy ions energized by processes in the Sun's corona, e.g., the shocks ahead of fast CMEs and the reconnection events driving solar flares) lasting from a few hours to several days. If the burst of SEPs contains significant fluxes of ions at energies above 400 MeV, these will penetrate Earth's atmosphere, again producing high-energy neutrons by collisions with atmospheric species, increasing the radiation levels at aircraft altitudes and on the ground. These ground level enhancements typically last for a few hours as SEP bursts usually have strong fluxes at energies above 400 MeV only in the early phase of the burst. As noted above, the worst observed case was an event in February 1956 that produced a 50-fold increase at the ground and was estimated to have produced a 300-fold increase at aircraft cruise altitudes. Older atmospheric radiation events have left dateable proxy evidence via C^{14} deposition in tree rings. This is a good proxy since the main natural source of C^{14} is high-energy neutron impacts on N^{14} around aircraft flight altitudes. Recent C^{14} studies (Miyake et al., 2012) suggest that even larger events have happened in the past. An atmospheric radiation event in 774 AD may have had neutron fluxes an order of magnitude higher than February 1956.

These huge increases in atmospheric radiation will significantly increase single-event effect rates in electronic devices, increasing the risk of malfunctions. As noted above, electronic devices embedded in critical systems such as aircraft control are designed to mitigate the risks from SEE. For example, SEEs can be mitigated by parallel redundancy, using three instances of the same control circuit (perhaps on the same chip) so that a malfunction in one circuit is outvoted by the other two working normally.

This is an excellent way to provide system resilience when a system is subject to low levels of stress (e.g., occasional SEEs) but will fail when high levels of stress (e.g., high SEE rates) cause simultaneous malfunctions in parallel control circuits. Thus there is a risk that high SEE rates during an extreme space weather event will overwhelm mitigation and cause malfunctions even in critical systems such as those on aircraft, in ground transport systems, and in other key infrastructures (electric power, gas, water). It is therefore important to carry out high-level tests of the electronic systems that embed devices at risk from SEE, such as through enhanced neutron bombardment of real systems (Hambling, 2014; STFC, 2017).

We should also consider that digital systems are now ubiquitous within many everyday technologies used at home, work, and play. These will likely have less resilience against SEEs since they are easily reset at an individual level. But there is a risk that an extreme space weather event leading to a strong GLE will cause simultaneous widespread disruption of these everyday technologies, potentially causing widespread worry over countries exposed to an extreme event. This psychological impact from extreme space weather has only had limited study; it needs more study to reflect the societal importance of psychological impacts in governmental risk assessments (Cabinet Office, 2015).

Thus it is clear that one of the major risks associated with extreme space weather is widespread disruption of electronic systems due to sharply enhanced atmospheric radiation levels. These enhanced levels will probably only last a few hours (as in the February 1956 event, see Gold and Palmer, 1956, and Marsden et al., 1956), but during that time there will be greatly enhanced SEE rates in digital devices on aircraft and in ground systems, leading to disruption of those systems and an unexpectedly high workload for operations teams. Electronics in unhardened consumer devices may also be disrupted on a wide scale, triggering major concerns if people are not aware that this can happen and that it will clear in a few hours (i.e., that the best policy may be just to let it pass, and to avoid hurried actions that could create further risks).

Finally we should also recognize that the enhanced atmospheric radiation levels at aircraft altitudes may lead to increased radiation doses for crew and passengers on aircraft flying at cruise altitudes during the event, particularly on long flights. In normal space weather conditions, the radiation dose from a single long-haul flight is small, falling within the general principle of keeping such doses as low as reasonably achievable (CDC, 2015). During severe space weather conditions, the radiation dose on such a flight may increase substantially. That increase will likely not reach a level that would pose any immediate health risks, but certainly may reach a level that warrants expert assessment (e.g., to consider whether exposed persons should take action to manage their future exposure to radiation, particularly if working a job that involves radiation exposure). Flights during such severe conditions are best avoided so there is a need for measures to achieve that safely, perhaps by delaying takeoffs once an event has started or by diverting aircraft where these have good communications with ground control. However, there is a significant likelihood that aircraft in flight over oceans or remote land areas may lose communications during a severe space weather event, meaning that ionospheric absorption may black out HF radio and ionospheric scintillation may block satcom. In this case, a diversion could put a flight in danger since it would be unable to obtain information about risks on the diverted route or to advise its position to other aircraft and ground control along the new route. In the event of a loss of communications, it is safer for a flight to continue on its preplanned flight paths so that ground control can anticipate when it will come into range of VHF radio systems. Thus we anticipate that an extreme space weather event will further disrupt aviation through the need to manage radiation doses acquired by aircrew and passengers—with flights delayed or diverted where this can be done safely, and with careful assessment and followup of radiation exposure where not. This is driving research to develop better ways of monitoring and modeling radiation exposure to resolve the many uncertainties in current assessments and measurements.

6 SATELLITE PLASMA ENVIRONMENTS

Satellites are by their nature exposed to space weather, particularly through their interactions with the plasma environments that pervade all space environments. As a result, satellites are constantly bombarded by the charged particles (electrons, protons, and ions) that form the plasma and that can penetrate into satellites if the particle energies are high enough. These can disrupt and damage satellite systems through a number of effects including radiation damage and electrical discharges, and also single-event effects (very similar to those caused by atmospheric radiation). The risks posed by these various effects are recognized by the satellite industry through design standards that enable the construction of satellites with high resilience to these effects as well as through good practice that enables satellite operators to plan for and manage these effects when they do occur.

They are particularly likely to occur during an extreme space weather event. To see why this is so, we must consider that the plasma environment experienced by a satellite varies markedly depending on the orbit of the satellite as well as the current state of space weather. Of particular concern are the higher orbits: (a) geosynchronous orbit where there are hundreds of operational satellites providing vital services such as communications and meteorological observations; and (b) "middle Earth orbit" around 20,000 km altitude which is home to GNSS satellites such as the GPS and Galileo constellations. Polar LEOs are also a concern when they pass through the auroral zone. In all these locations satellites are bombarded by electrons with average energies around 10,000 eV, as discussed above; these can accumulate on the surface of satellites and give the satellite a large potential difference ($\sim$10,000 V negative) with respect to the local plasma, and potential differences of 1000 V between different parts of the satellite. The overall charging of the satellite, often referred to as frame charging, is not usually a problem, but differential charging between parts of the satellite can be a serious problem. (The one exception to this is scientific measurements of low energy ($<$1000 eV) particles; these will be severely perturbed by frame charging and hence such measurements may be supported by devices to control frame charging (Torkar et al., 2005) and by activities to estimate the satellite potential in order to correct particle measurements (Lybekk et al., 2012).)

Thus differential charging is the key challenge to satellite builders. They can deal with this charging by ensuring that the potential is spread evenly over the surface of the satellite, that all surfaces have the same potential so there is no risk of electrical discharges (Chapter 15 of this volume). This was a particular problem in the early days of spaceflight when the charging was not understood, and discharges did disrupt and damage satellites. Today it can be mitigated by ensuring that all surfaces are electrically conductive, such as with a thin metal coating on dielectric materials, and electrically grounded to the body of the spacecraft. Unfortunately, transparent conductors are sometimes a problem as there are few materials with this property, and grounding and testing every single solar array coverslide can be problematic. For other spacecraft surfaces, multilayer insulation, the gold foil covering many satellites, plays a key role in this grounding as well as providing thermal protection. Thus we have some good answers to the problems caused by differential charging, but need to maintain awareness that these problems will be minimized only by following appropriate standards for mitigation of differential charging. It is important that contracts for the design, construction, and verification of new satellites include requirements to follow those standards.

Also, as discussed above, space plasmas are rarely in thermal equilibrium and particle energy distributions often include strong tails with significant fluxes of particles up to MeV or even GeV. These tails arise from energization processes driven by space weather.

For example, high-speed solar wind flowing past the Earth can stimulate the growth of low-frequency plasma waves in Earth's magnetosphere; these, in turn, can energize electrons to energies around 1 MeV (Horne et al., 2005; Chapter 14 of this volume). Electrons at these high energies can penetrate deep into satellites and deposit electric charge inside the materials that make up the satellite. The charge will drain away where those materials are electrically conducting, but will accumulate where they are highly dielectric. The high-energy electron fluxes produced by high-speed solar winds will often persist for days, leading to the accumulation of significant amounts of electrical charge deep inside dielectric materials such as circuit boards and electrical insulation. This "deep dielectric charging" will create strong electric fields that eventually cause an electrical breakdown of the dielectric (see an excellent benchtop demonstration at https://youtu.be/eCz7BL74D4Y). The resulting discharge can generate false signals that trigger anomalous behavior on a satellite, requiring operator intervention to restore nominal behavior. This is a common occurrence, especially in the declining phase of the solar cycle when the Earth is often exposed to high-speed solar wind streams. So, satellite operators are experienced in preparing for and dealing with such anomalies and using spare capacity on other satellites to minimize disruption of the services they deliver (another example of how parallel redundancy provides resilience). However, in extreme space weather conditions, this would be much more challenging. We would expect very high anomaly rates, possibly overwhelming the ability of operators to switch services between satellites, and leading to significant disruption of satellite-based services. We would also expect that a number of satellites will be lost or badly degraded during this process, in some cases because the electrical discharges will be so intense as to damage satellite systems, and in others because the anomalous behavior will leave satellite systems in an unusable state (though many of these cases may eventually be recovered through patient work by operators, perhaps over many months).

Thus deep dielectric charging is a much greater challenge to satellites than surface charging, especially in extreme space weather conditions. It is important to understand that these are two very different charging processes and that there are good engineering methods to mitigate surface charging, while the anomalies caused by deep dielectric charging are best addressed by operator actions and, where sufficient mass can be flown on the satellite, by shielding of sensitive components. These two charging processes are sometimes confused in public discussions, leading to a mistaken belief that good electrical grounding can mitigate deep dielectric charging. It is important to counter this and to encourage satellite operators to be ready for the disruption that may arise from deep dielectric charging during an extreme event.

Another important aspect of particle energization in plasmas is the production of protons and ions at energies of MeV and even GeV, through energization at plasma shocks such as those ahead of fast CMEs and in reconnection regions such as those that also generate the energetic electrons that produce EUV and X-ray solar flares. These are the sources of solar energetic particle bursts. While only the highest SEP energies (>400 MeV) can enhance atmospheric radiation, particles at lower energies down to about 5 MeV can reach geosynchronous orbit and impact the many satellites in that orbit. In addition satellites on interplanetary missions, including those to the Moon, will be exposed to the full range of SEPs. The lower energy SEPs can penetrate lightly shielded satellite systems such as solar arrays and cause damage that will reduce their performance. This is a well-known effect and solar arrays are deliberately designed to overperform, so that their likely decline over a period of 10–15 years will still deliver enough power to run the satellite. Thus the effect of an extreme space weather event will be to age the solar arrays and erode power margins, bringing forward the end of life for the satellite. For most satellites, there will be no immediate loss of that service, rather that operators will need to consider replacing it sooner than previously expected. Hence one interesting consequence of an extreme space weather event may well be to ramp up orders for new satellites for several years following the event.

Now turning to SEPs at higher energies, these can penetrate shielded systems, again causing radiation damage that will degrade performance of, for example, electronic devices. This is typical of the aging of satellite systems and will eventually lead to loss of the satellite. But this is generally a slow process and satellite operators are adept at monitoring how system performance gradually changes as well as at devising procedures to make the best of any remaining performance and thus get the most out of the time and money invested in a satellite. Thus, as with solar arrays, this aspect of extreme space weather will age the satellite and bring forward the time when it must be replaced.

But these higher energy particles also have a much more immediate and dangerous impact. As with particle impacts on electronic devices in aircraft, these particles can cause single-event effects inside electronic devices and hence cause satellite systems to malfunction. Satellite systems are generally designed to have high resilience to SEEs given that they have an obvious and high exposure to high-energy particle radiation, and satellite operators are highly experienced in dealing with those SEEs that do cause problems. Thus in normal space weather conditions, SEEs are just everyday business for satellite operators. But, in extreme conditions, we should expect much higher rates of SEEs that can disrupt satellite applications and put a very high workload on satellite operators. This is likely to have a large short-term impact on everyone who uses services that ultimately rely on satellite applications—which is pretty much everyone in modern society. But once the extreme event is over, we should expect a rapid recovery as experienced satellite operators bring their satellites back into normal operation.

Higher-energy particles can also impact satellites in LEOs. The SEPs that reach geosynchronous orbit can also reach the polar ionosphere as that region is magnetically coupled to the outer magnetosphere, and hence those SEPs will impact on satellites in polar LEOs. These SEPs are complemented by the high-energy protons and ions that are trapped in Earth's geomagnetic field, forming the inner radiation belt. The asymmetries of the current-day geomagnetic field mean that these particles interact with the atmosphere in the South Atlantic off the coast of South America. Thus LEO satellites flying over this area encounter a region of intense radiation, which we term the South Atlantic Anomaly (SAA). Thus LEO satellites can encounter the same problems of radiation damage and single-event effects as those in geosynchronous orbit, and hence they need to be built and operated to the same standards. They are equally likely to suffer major disruption during an extreme space weather event.

7 LOOKING TO THE FUTURE: HOW MAY SPACE WEATHER RISKS EVOLVE?

Finally, and most importantly, we need to reflect that space weather impacts, and the associated risks, evolve as technology develops. This stands out clearly from the history of space weather. The first technological impacts of space weather were those on telegraph and telephone systems and were well established by the beginning of the 20th century, as shown by newspaper reports of the disruption caused by major events. But those impacts have faded from our concerns since the 1980s as copper wires in landline communications systems have been replaced by fiberoptics and hence these systems are no longer subject to interference from GIC in communications lines.

But this gain for resilience against space weather has been more than balanced by the deployment of new technologies that are susceptible to space weather. The most obvious case of this increasing susceptibility is the use of satellite precision timing and navigation systems such as GPS and Galileo. The underlying GNSS technology is particularly vulnerable to space weather effects, due to group delay and scintillation effects in the ionosphere, exacerbated by the relative weak signals that reach GNSS receivers. (A 500 W transmitter on a GNSS spacecraft at 20,000 km altitude delivers around 10^{-15} W of signal into a receiver with 1 cm^{-2} antenna cross-section.) These ionospheric effects are fundamental to the way that GNSS works and are well mitigated in normal space weather conditions, but as discussed above, mitigation is likely to fail during an extreme space weather event.

This vulnerability of GNSS to space weather has been widely addressed in some critical areas, most notably in its use for aircraft navigation. This is an area with a very strong safety culture so it is not surprising to see substantial efforts to address key issues such as the mitigation of group delay (such as via augmentation systems) and the vital role of alternative navigation methods (such as inertial systems; instrument landing systems) as backups against GNSS signal loss. The need for alternatives to vulnerable GNSS timing also seems to have gained traction in the financial services area where accurate time signals are fundamental to high-speed trading (Wilson, 2017). This is not surprising given the amount of money at risk in this area and the willingness to invest in technologies that sustain high-speed trading.

But it is less clear that the vulnerability of GNSS to space weather is appreciated in other areas. A major area for concern is shipping, where significant work has been done to develop an alternative modern navigation method as a backup against loss of GNSS by jamming or space weather (Johnson et al., 2007). But its implementation has now largely been abandoned, leaving much shipping almost entirely reliant on GNSS. It is hard to see this situation as other than a major risk. An extreme space weather event seems very likely to cause major problems with shipping; their safe passage will become heavily dependent on traditional skills of seamanship such as relying on visual observations to guide ships through difficult waters and the use of magnetic compasses or sextants to determine direction and location.

Another area of major concern is the widespread use of GNSS in low-level applications that do not appear to have a high level of individual risk, such as the use of GNSS to provide route guidance to car drivers. These applications generally have a low level of resilience and often rely on the driver comparing the GNSS route guidance to what they can see through the windshield. Unfortunately, there is evidence that some drivers fail to do this, following erroneous guidance down routes that are too narrow for their vehicles (The Telegraph, 2016) or that are dead ends (Kington, 2007). An extreme space weather event will inevitably increase the amount of erroneous guidance because of increased position errors. In addition it is expected that loss of GNSS signal will cause some drivers to become lost, especially when traveling to unfamiliar locations. Thus it is likely that an extreme space weather event could cause extensive problems on the roads, with a large economic impact because of the high number of people and cars involved.

Looking to the future, it seems very likely that GNSS will be embedded in new road transport systems such as driverless cars and road tolling. Thus it is vital to promoting awareness of extreme space weather as these new technologies are developed and to explore how they will respond to such conditions. How will they handle position errors? How will they handle sudden loss of signal? It is unclear whether these issues are being considered in the risk analyses for these new transport systems. It is essential that they are.

Finally we should recognize that some technological developments may allow us to retire some long-standing space weather risks. The obvious example is long-distance communications for aircraft on routes over oceans and remote regions. These communications are currently performed by a mix of L-band satcom and HF radio, both vulnerable to disruption by space weather. But there is a gradual trend to move aircraft communications to satcom at much higher frequencies (above 4 GHz), in order to provide more bandwidth, such as for onboard Internet services (Inmarsat, 2017). This will also eliminate the risk of disruption to trans-ionospheric propagation since those higher frequencies are not affected by space weather. Thus it seems likely that this is an area where space weather impacts will decline over the coming decades. But given the life cycle of passenger aircraft, it may well be 30 years or longer before the risk is fully retired.

REFERENCES

Air University, 2003. Space environment. In: Space Primer, pp. 6–13 (Chapter 6). http://space.au.af.mil/primer/space_environment.pdf.

Aron, J., 2014. Atomic time lord to battle sneaky high-speed trades. New Scientist, 16 April 2014. https://www.newscientist.com/article/dn25423-atomic-time-lord-to-battle-sneaky-high-speed-trades/ (Accessed 8 May 2017).

Axford, W.I., Hines, C.O., 1961. A unifying theory of high-latitude geophysical phenomena and geomagnetic storms. Can. J. Phys. 39, 1433–1464. https://doi.org/10.1139/p61-172.

Barlow, W.H., 1849. On the spontaneous electrical currents observed in wires of the electric telegraph. Philos. Trans. R. Soc. Lond. 139, 61–72. https://doi.org/10.1098/rstl.1849.0006.

Bates, D.R., 1950. Dissociative recombination. Phys. Rev. 78, 492–493. https://doi.org/10.1103/PhysRev.78.492.

Bolduc, L., 2002. GIC observations and studies in the Hydro-Québec power system. J. Atmos. Solar-Terr. Phys. 64, 1793–1802. https://doi.org/10.1016/S1364-6826(02)00128-1.

Bonner IV, L.R., Schultz, A., 2017. Rapid prediction of electric fields associated with geomagnetically induced currents in the presence of three-dimensional ground structure: projection of remote magnetic observatory data through magnetotelluric impedance tensors. Space Weather 15, 204–227. https://doi.org/10.1002/2016SW001535.

Boteler, D.H., 2012. An examination of the causes and consequences of the March 13, 1989 magnetic storm. In: Proceedings of the International Emergency Management Society Conference, October. Space Weather and Challenges for Modern Society, Oslo, Norway, pp. 22–24.

Boteler, D.H., Pirjola, R.J., 2017. Modeling geomagnetically induced currents. Space Weather 15, 258–276. https://doi.org/10.1002/2016SW001499.

Cabinet Office, 2015. National risk register of civil emergencies. https://www.gov.uk/government/collections/national-risk-register-of-civil-emergencies (Accessed 8 March 2017).

Carlson, H.C., 2012. Sharpening our thinking about polar cap ionospheric patch morphology, research, and mitigation techniques. Radio Sci. 47. https://doi.org/10.1029/2011RS004946, RS0L21.

Carter, B.A., Yizengaw, E., Pradipta, R., Halford, A.J., Norman, R., Zhang, K., 2015. Interplanetary shocks and the resulting geomagnetically induced currents at the equator. Geophys. Res. Lett. 42, 6554–6559. https://doi.org/10.1002/2015GL065060.

CDC (Centers for Disease Control and Prevention), 2015. ALARA—as low as reasonably achievable. https://www.cdc.gov/nceh/radiation/alara.html (Accessed 24 April 2017).

Dyer, C.S., Lei, F., Hands, A., Truscott, P., 2007. Solar particle events in the QinetiQ atmospheric radiation model. IEEE Trans. Nucl. Sci. 54 (4), 1071–1075. https://doi.org/10.1109/TNS.2007.893537.

Erinmez, I.A., Kappenman, J.G., Radasky, W.A., 2002. Management of the geomagnetically induced current risks on the national grid company's electric power transmission system. J. Atmos. Solar-Terr. Phys. 64, 743–756. https://doi.org/10.1016/S1364-6826(02)00036-6.

Eroshenko, E.A., Belov, A.V., Boteler, D., Gaidash, S.P., Lobkov, S.L., Pirjola, R., Trichtchenko, L., 2010. Effects of strong geomagnetic storms on Northern railways in Russia. Adv. Space Res. 46, 1102–1110. https://doi.org/10.1016/j.asr.2010.05.017.

Gaunt, C.T., 2014. Reducing uncertainty—responses for electricity utilities to severe solar storms. J. Space Weather Space Clim. 4, A01. https://doi.org/10.1051/swsc/2013058.

Gaunt, C.T., Coetzee, G., 2007. Transformer failures in regions incorrectly considered to have low GIC-ris. In: Proceedings of the IEEE Powertech Conference, July 2007, Lausanne, Switzerland. https://doi.org/10.1109/PCT.2007.4538419.

Gold, T., Palmer, D.R., 1956. The solar outburst, 23 February 1956—observations by the Royal Greenwich Observatory. J. Atmos. Terr. Phys. 8, 287–290.

Hambling, D., 2014. Burnout. New Scientist 223, 42–45. https://doi.org/10.1016/S0262-4079(14)61863-7.

Horne, R.B., et al., 2005. Wave acceleration of electrons in the Van Allen radiation belts. Nature 437, 227–230. https://doi.org/10.1038/nature03939.

Inmarsat, 2017. Global Xpress: first high-speed broadband network to span the world. http://www.inmarsat.com/service/global-xpress/ (Accessed 31 March 2017).

Johnson, G.W., Swaszek, P.F., Hartnett, R.J., Shalaev, R., Wiggins, M., 2007. An evaluation of eLoran as a backup to GPS. In: 2007 IEEE Conference on Technologies for Homeland Security, Woburn, MA. pp. 95–100. https://doi.org/10.1109/THS.2007.370027.

Kappenman, J.G., 2005. An overview of the impulsive geomagnetic field disturbances and power grid impacts associated with the violent Sun-Earth connection events of 29–31 October 2003 and a comparative evaluation with other contemporary storms. Space Weather 3, S08C01. https://doi.org/10.1029/2004SW000128.

King, J.W., Kohl, H., 1965. Upper atmospheric winds and ionospheric drifts caused by neutral air pressure gradients. Nature 206, 699–701. https://doi.org/10.1038/206699a0.

Kington, M., 2007. Sat Nav can take you to the most interesting places. The Independent, Thursday 13 September 2007. http://tinyurl.com/k4naqk8 (Accessed 31 March 2017).

Knipp, D.J., et al., 2016. The May 1967 great storm and radio disruption event: extreme space weather and extraordinary responses. Space Weather 14, 614–633. https://doi.org/10.1002/2016SW001423.

Krausmann, E., Andersson, E., Russel, T., William, M., 2015. Space weather and rail: findings and outlook. EU Joint Research Centre report 98155. https://doi.org/10.2788/211456.

Lanzerotti, L.J., 1979. Geomagnetic influences on man-made systems. J. Atmos. Terr. Phys. 41, 787–796. https://doi.org/10.1016/0021-9169(79)90125-9.

Lanzerotti, L.J., Medford, L.V., Maclennan, C.G., Kraus, J.S., Kappenman, J., Radasky, W., 2001. Trans-Atlantic geopotentials during the July 2000 solar event and geomagnetic storm. Solar Phys. 204, 351–359. https://doi.org/10.1023/A:1014289410205.

Liu, H.-L., Yudin, V.A., Roble, R.G., 2013. Day-to-day ionospheric variability due to lower atmosphere perturbations. Geophys. Res. Lett. 40, 665–670. https://doi.org/10.1002/grl.50125.

Liu, L., Ge, X., Zong, W., Zhou, Y., Liu, M., 2016. Analysis of the monitoring data of geomagnetic storm interference in the electrification system of a high-speed railway. Space Weather 14, 754–763. https://doi.org/10.1002/2016SW001411.

Livadiotis, G., McComas, D.J., 2013. Understanding kappa distributions: a toolbox for space science and astrophysics. Space Sci. Rev. 175, 183–214. https://doi.org/10.1007/s11214-013-9982-9.

Lunt, N., Kersley, L., Bailey, G.J., 1999. The influence of the protonosphere on GPS observations: model simulations. Radio Sci. 34 (3), 725–732. https://doi.org/10.1029/1999RS900002.

Lybekk, B., Pedersen, A., Haaland, S., Svenes, K., Fazakerley, A.N., Masson, A., Taylor, M.G.G.T., Trotignon, J.-G., 2012. Solar cycle variations of the cluster spacecraft potential and its use for electron density estimations. J. Geophys. Res. 117. A01217, https://doi.org/10.1029/2011JA016969.

Mandelbrot, B., 1967. How long is the coast of Britain? Statistical self-similarity and fractional dimension. Science 156, 636–638. https://doi.org/10.1126/science.156.3775.636.

Marsden, P.L., Berry, J.W., Fieldhouse, P., Wilson, J.G., 1956. Variation of cosmic-ray nucleon intensity during the disturbance of 23 February 1956. J. Atmos. Terr. Phys. 8, 278–281.

Medford, L.V., Lanzerotti, L.J., Kraus, J.S., Maclennan, C.G., 1989. Transatlantic earth potential variations during the March 1989 magnetic storms. Geophys. Res. Lett. 16, 10. https://doi.org/10.1029/GL016i010p01145.

Miyake, F., Nagaya, K., Masuda, K., Nakamura, T., 2012. A signature of cosmic-ray increase in AD 774–775 from tree rings in Japan. Nature 486, 240–242. https://doi.org/10.1038/nature11123.

Ngwira, C.M., Pulkkinen, A.A., Bernabeu, E., Eichner, J., Viljanen, A., Crowley, G., 2015. Characteristics of extreme geoelectric fields and their possible causes: localized peak enhancements. Geophys. Res. Lett. 42, 6916–6921. https://doi.org/10.1002/2015GL065061.

Ott, E., 1978. Theory of Rayleigh-Taylor bubbles in the equatorial ionosphere. J. Geophys. Res. 83 (A5), 2066–2070. https://doi.org/10.1029/JA083iA05p02066.

Pirjola, R., 2005. Effects of space weather on high-latitude ground systems. Adv. Space Res. 36, 2231–2240. https://doi.org/10.1016/j.asr.2003.04.074.

PourEVA, 2003. Le Ministre DEWAEL reconnait la faillibilité du vote électronique grâce à un rayon cosmique complice! http://www.poureva.be/article.php3?id_article=36 (Accessed 8 May 2017).

Pulkkinen, A., Lindahl, S., Viljanen, A., Pirjola, R., 2005. Geomagnetic storm of 29–31 October 2003: geomagnetically induced currents and their relation to problems in the Swedish high-voltage power transmission system. Space Weather 3, S08C03. https://doi.org/10.1029/2004SW000123.

Pulkkinen, A., Bernabeu, E., Eichner, J., Viljanen, A., Ngwira, C., 2015. Regional-scale high-latitude extremse geoelectric fields pertaining to geomagnetically induced currents. Earth Planets Space 67, 93. https://doi.org/10.1186/s40623-015-0255-6.

Simpson, F., Bahr, K., 2005. Practical Magnetotellurics. Cambridge University Press, Cambridge. 10.1017/CBO9780511614095.

Stafford, P., 2016. FT explainer: keeping up with high-frequency traders. Financial Times, 5 Sep 2016. https://www.ft.com/content/66fa2bbc-6528-11e6-8310-ecf0bddad227 (Accessed 8 May 2017).

STFC, 2017. ChipIR: instrument for rapid testing of effects of high energy neutrons. http://www.isis.stfc.ac.uk/instruments/chipir/chipir8471.html (Accessed 31 March 2017).

The Guardian, 1938. Article entitled "Radio telephones out of action, The Aurora Borealis, Railway Signals upset", 27 January 1938, p. 14.

The Telegraph, 2016. Article entitled: Sat-nav revolution: new digital map will stop lorries getting stuck under bridges and on narrow lanes. 19 October 2016. http://tinyurl.com/h2ypb3w (Accessed 31 March 2017).

Thomson, A., Wild, J., 2010. When the lights go out…. Astron. Geophys. 51, 5.23–5.24. https://doi.org/10.1111/j.1468-4004.2010.51523.x.

Thomson, A.W.P., Gaunt, C.T., Cilliers, P., Wild, J.A., Opperman, B., McKinnell, L.-A., Kotze, P., Ngwira, C.M., Lotz, S.I., 2010. Present day challenges in understanding the geomagnetic hazard to national power grids. Adv. Space Res. 45, 1182–1190. https://doi.org/10.1016/j.asr.2009.11.023.

Torkar, K., Fehringer, M., Escoubet, C.P., André, M., Pedersen, A., Svenes, K.R., Décréau, P.M.E., 2005. Analysis of cluster spacecraft potential during active control. Adv. Space Res. 36, 1922–1927. https://doi.org/10.1016/j.asr.2005.01.110.

Turnbull, K., 2010. Modelling GIC in the UK. Astron. Geophys. 51, 5.25–5.26. https://doi.org/10.1111/j.1468-4004.2010.51525.x.

Wik, M., Pirjola, R., Lundstedt, H., Viljanen, A., Wintoft, P., Pulkkinen, A., 2009. Space weather events in July 1982 and October 2003 and the effects of geomagnetically induced currents on Swedish technical systems. Ann. Geophys. 27, 1775–1787. https://doi.org/10.5194/angeo-27-1775-2009.

Wilson, H., 2017. London's atomic bombshell for hackers. The Times, 9 May 2017. https://www.thetimes.co.uk/edition/business/londons-atomic-bombshell-for-hackers-gbst7gv2g (Accessed 14 May 2017).

Wood, J., Caustin, E., 2006. Timely testing avoids cosmic ray damage to critical auto electronics. http://www.eetimes.com/document.asp?doc_id=1272752 (Accessed 28 April 2017).

Wrubel, J.N., 1992. Monitoring program protects transformers from geomagnetic effects. IEEE Comput. Appl. Power 5, 10–14. https://doi.org/10.1109/67.111465.

PART 2

SOLAR ORIGINS AND STATISTICS OF EXTREMES

EXTREME SOLAR ERUPTIONS AND THEIR SPACE WEATHER CONSEQUENCES

2

Nat Gopalswamy
NASA Goddard Space Flight Center, Greenbelt, MD, United States

CHAPTER OUTLINE

1 Introduction 37
2 Overview of Extreme Events 38
3 Estimates of Extreme Events 42
   3.1 CME Speeds 42
   3.2 Distribution Functions for CME Speeds and Kinetic Energies 43
   3.3 Flare Size Distribution 44
   3.4 Active Regions and Their Magnetic Fields 46
4 Consequences of Solar Eruptions 51
   4.1 SEP Events 51
   4.2 SEP Fluences 52
   4.3 Large Geomagnetic Storms 54
5 Summary and Conclusions 58
Acknowledgments 59
References 59
Further Reading 63

1 INTRODUCTION

Human society experienced the impact of extreme solar eruptions that occurred on October 28 and 29 in 2003, known as the Halloween 2003 storms. Soon after the occurrence of the associated solar flares and coronal mass ejections (CMEs) at the Sun, people were expecting a severe impact on Earth's space environment and took appropriate actions to safeguard technological systems in space and on the ground. The high magnetic field in the CMEs indeed interacted with Earth's magnetic field and produced two super intense geomagnetic storms. Both CMEs were driving strong shocks that accelerated coronal particles to GeV energies. The shocks arrived at Earth in fewer than 19 h. The consequences were severe: in Malmoe, a southern city in Sweden, about 50,000 people experienced a blackout when the transformer oil heated up by 10°C. About 59% of the reporting spacecraft and about 18% of the onboard instrument groups were affected by these events. In order to protect Earth-orbiting spacecraft from particle radiation, they were put into safe mode (Webb and Allen, 2004). The high-energy particles from the CMEs penetrated Earth's atmosphere, causing significant depletion of stratospheric ozone. The ionospheric total electron content over the U.S. mainland increased tenfold during the period of October 30–31, 2003. Significant enhancement of the density in the magnetosphere also coincided with the arrival of the CMEs at Earth. In addition to the Earth's space environment, the impact of the CMEs was felt throughout the heliosphere, all the way to the termination shock. The detection of the impact was possible because there were space missions located near Mars (Mars Odyssey), Jupiter (Ulysses), and Saturn (Cassini) as well as at the outer edge of the solar system (Voyager 1 and 2). The MARIE instrument on board the Mars Odyssey mission was completely damaged by the energetic particles from these CMEs. The widespread impact of these Halloween events has been documented in about seventy articles published during 2004–05 (see Gopalswamy et al., 2005a for the list of the articles). The solar active region from which the CMEs originated was also very large and had the potential to launch energetic CMEs.

Fig. 1 shows the source active region (10486) with sunspots and its complex magnetic structure as observed by the Magnetic and Doppler Imager (MDI) on board the Solar and Heliospheric Observatory (SOHO). The region produced two of the Halloween events that are of historical importance. The first eruption on October 28, 2003 was seen bright in EUV wavelengths and had the soft X-ray flare size of X17. The CME was a symmetric halo as seen by the Large Angle and Spectrometric Coronagraph (LASCO) on board SOHO. The October 28 and 29, 2003, eruptions were responsible for intense SEP events that had ground level enhancements (GLEs) numbered GLE 65 and GLE 66, respectively. The CMEs had speeds exceeding 2000 km s^{-1} and produced super magnetic storms (Dst < -200 nT) when they arrived at Earth. The Halloween solar eruptions thus turned out to be extreme events both in terms of their origin at the Sun and their consequences in the heliosphere. The two events were observed extremely well by many different instruments from space and ground and the knowledge on space weather events helped us to take appropriate actions to limit the impact when possible. The Sun must have produced such events many times during its long history of 4.5 billion years, but the occurrence now has high significance because human society has become increasingly dependent on technology that can be affected by solar eruptions. It is of interest to know the origin of the extreme events and how big an impact they can cause.

An overview of extreme events on the Sun and their heliospheric consequences is provided in Section 2. Extreme event sizes are estimated in Section 3 for CMEs, flares, and source active regions, assuming the extreme events to be located on the tails of various cumulative distributions. Section 4 considers the heliospheric response of solar eruptions in the form of SEP events and geomagnetic storms. The chapter is summarized in Section 5.

2 OVERVIEW OF EXTREME EVENTS

The definition of an extreme event is not very concrete, but can be thought of as an event on the tail of a distribution. An extreme event can also be thought of as an occurrence that has unique characteristics in its origin and/or in its consequences. For example, a CME that has an extreme speed

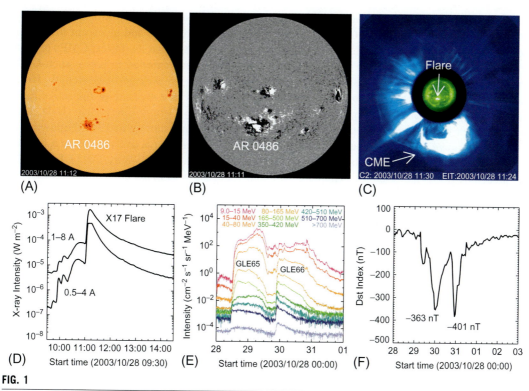

FIG. 1

The solar source and space weather consequences of the October 28, 2003 CME. (A) A continuum image of the Sun from SOHO/MDI showing the sunspot region 10486 (Sunspots appear dark because they are ~2000 K cooler than the surrounding photosphere at ~6000 K.) (B) The sunspot region as seen in a SOHO/MDI magnetogram (*White* is positive and *black* is negative magnetic field region.) (C) A SOHO/LASCO white-light image with superposed SOHO/EIT image showing the flare brightening from the active region 10486 (The dark disk is the occulting disk.) (D) GOES soft X-ray light curve showing the X17 flare in two energy channels (1–8 and 0.5–4 Å); (E) GOES proton intensity in various channels, including the >700 MeV channel indicative of ground level enhancement (GLE) associated with the eruption (GLE65) in (C), as well as the next one (GLE66); (F) Dst index from World Data Center, Kyoto showing the superstorms with Dst = −363 nT associated with the eruption in (C) and Dst = −401 nT associated with the next eruption on October 29, 2003. This figure illustrates the chain of events from the Sun to Earth's magnetosphere considered throughout this paper: active regions, flares, CMEs, SEP events, and geomagnetic storms.

can be considered an extreme event if such an occurrence is extremely rare. Among the thousands of CMEs observed by the Solar and Heliospheric Observatory (SOHO) from 1996 to 2015, only a couple have speeds exceeding 3000 km s^{-1} Therefore, one can consider a CME with speed exceeding 3000 km s^{-1} as an extreme event. But how high can the CME speed get? To answer this question, one has to consider the energy source of CMEs and how that energy is converted to CME kinetic energy. It has been established that CMEs can only be powered by the magnetic energy in closed magnetic field regions on the Sun (see e.g., Forbes, 2000). There are two types of closed field regions that are known to produce CMEs: sunspot regions (active regions) and quiescent filament regions (see e.g., Gopalswamy et al., 2010). Observations have shown that the fastest CMEs originate from active regions because they possess the high magnetic energy needed to power such CMEs. The magnetic energy of an active region depends on its size and the average field strength. Historically, there is a long record of sunspot area, which can be taken as a measure of the active region area. The magnetic fields in sunspots were discovered in 1908 by George Ellery Hale (Hale, 1908) and have been recorded since then with routine field measurements starting in 1915. Following the work of Mackay et al. (1997), one can compute the potential energy in active regions as a measure of the maximum free energy available to power eruptions (see e.g., Gopalswamy, 2011). Essentially, this procedure traces the origin of extreme CMEs to the extremeness in the source region of CMEs, although additional considerations such as the conversion efficiency from the magnetic energy to CME kinetic energy play a role.

Another manifestation of a solar eruption is the flare, which is primarily identified with the sudden increase in electromagnetic emission from the Sun at various wavelengths. The flare phenomenon was originally discovered in white light by Carrington (1859) and Hodgson (1859) and has been extensively observed in the H-alpha line since the beginning of the 20th century. The most common method of flare detection at present is in soft X-rays, and the flare size is indicated by the intensity expressed in units of $W\,m^{-2}$ in the 1–8 Å channel. Flares of size $10^{-4}\,W\,m^{-2}$ are classified as X-class. The largest flare ever observed in the space age had an intensity of X28 or $2.8 \times 10^{-3}\,W\,m^{-2}$ observed on November 4, 2003, from the same active region 10486 a few days after the eruptions described in Fig. 1. The flare was accompanied by a fast ($\sim$2700 km s^{-1}) CME with a kinetic energy of $\sim 6 \times 10^{32}$ erg (Gopalswamy et al., 2005b).

The primary consequences of CMEs are large SEP events and geomagnetic storms, both of which are sources of severe space weather (see e.g., Gopalswamy, 2009). Corotating interaction regions can also cause geomagnetic storms that are more frequent but less severe compared to CMEs (see e.g., Borovsky and Denton, 2006). We do not consider them here. The particles in large SEP events are accelerated at the CME-driven shock, while geomagnetic storms depend on the CME speed and its magnetic content. Each of these space weather events has a chain of effects on Earth's magnetosphere, ionosphere, atmosphere, and even on the ground. In addition, SEPs pose a radiation hazard to astronauts and adversely affect space technology in the near-Earth as well as interplanetary space. It must be noted that SEPs are accelerated also at the flare site, which are responsible for a different type of electromagnetic emission when they propagate toward, and interact with, the solar surface. However, their contribution to the observed SEPs in space is not fully understood, but is usually small compared to that from CME-driven shocks (see e.g., Reames, 2015). Some studies suggest that flares are the dominant sources of high-energy SEPs observed in the interplanetary medium (see e.g., Dierckxsens et al., 2015; Grechnev et al., 2015; Trottet et al., 2015). Cliver (2016) points out that the conclusion is not supported if all the SEP events are included in the correlative analyses. Particles are energized by other mechanisms throughout the heliosphere, providing seed particles to the shock acceleration process (see e.g., Mason et al., 2013; Zank et al., 2014).

The electromagnetic emission from solar flares generally causes excess ionization in the ionosphere, thereby changing the ionospheric conductivity. For example, solar-flare X-rays cause sudden ionospheric disturbances that can affect radio communications. Intense radio bursts are produced by energetic electrons accelerated during flares. If the frequencies of the radio bursts are close to those of GPS and radar signals, the bursts can drown the signals out (see e.g., Kintner et al., 2009).

The extreme space weather consequences thus depend on extreme CME properties. In the case of SEP events, one can think of very strong shocks, which ultimately result from very high CME speeds. Geomagnetic storms also depend on CME speeds as they arrive at Earth's magnetosphere, but they also require intense southward magnetic field in the CME and/or in the shock sheath (e.g., Wu and Lepping, 2002; Gopalswamy et al., 2008, 2015a). The storm strength (as measured by say, the Dst index) can also depend on solar wind density, but the effect is not significant for extreme storms we are interested in (e.g., Weigel, 2010). High-speed shocks that arrive at Earth in less than a day are known as fast-transit events (Cliver et al., 1990; Gopalswamy et al., 2005b). These shocks are considered to be extreme events because they can cause high levels of energetic storm particles (ESPs) at Earth and compress the magnetosphere observed as sudden impulse or sudden commencement (SC) of geomagnetic storms (Araki, 2014). Such shocks are also very strong near the Sun and are highly likely to accelerate SEPs to very high energies. The resulting SEP spectrum is expected to be hard, leading to high-energy particles that affect the Earth's ionosphere and atmosphere.

The consequences at Earth become extreme only under certain conditions because Earth presents only a small cross-section to solar events. This is illustrated in Fig. 2 as the distribution of solar

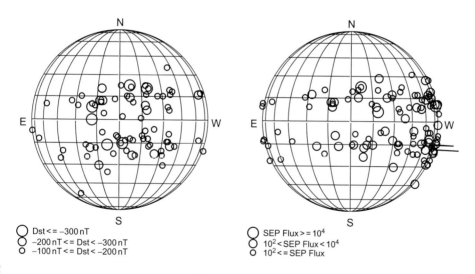

FIG. 2

Solar sources of CMEs causing intense geomagnetic storms (Dst ≤ -100 nT) (left) and large SEP events (intensity >10 pfu in the >10 MeV channel; pfu is the particle flux unit defined as 1 pfu = 1 particle per (cm^2 s sr)) (right) during 1996–2016. The size of the circle indicates the intensity of the event as noted on the plots. The latitude and longitude grids are 15° apart. No correction was made for the solar B0 angle, the heliographic latitude of the central point on the solar disk.

From Gopalswamy, N., 2010. Corona mass ejections: a summary of recent results. In: Proceedings of the 20th National Solar Physics Meeting, held 31 May–4 June, 2010 in Papradno, Slovakia, pp. 108–130.

sources of CMEs that caused intense geomagnetic storms (Dst ≤ -100 nT) and large SEP events (>10 MeV proton intensity >10 pfu). The size of the circles denotes the intensity of events. The most intense geomagnetic storms are associated with CMEs originating very close to the solar disk center (no Dst ≤ -300 nT events beyond a central meridian distance (CMD) of $\sim 20°$). Beyond a CMD of $\sim 30°$, we see only the weaker storms. CMEs originating from close to the disk center head directly toward Earth and deliver a head-on blow to Earth's magnetosphere. This fact was established long ago by Hale (1931) and Newton (1943). The source regions of CMEs producing SEP events have a different distribution: the most intense SEP events generally originate from the western hemisphere of the Sun. At CMD >30° in the eastern hemisphere, SEP events are less frequent and weak (peak >10 MeV intensity <100 pfu). The reason is the heliospheric magnetic structure, which takes the form of an Archimedean spiral (Parker Spiral) along which accelerated particles propagate. Magnetic field lines originating from the western hemisphere of the Sun are connected to Earth and hence particles can be detected at Earth. Thus, an extremely fast CME from the east limb may not produce an extreme space weather event at Earth. However, planets or spacecraft located above the east limb could be affected by such CMEs. There is no such source restriction for solar flares: electromagnetic emissions from flares reach Earth so long as they occur on the front side of the Sun.

We are interested in extreme events both in their origins at the Sun and their consequences and space weather effects. At the Sun, we are interested in the size and magnetic field strength in active regions as well as the amount of energy that can be stored and released in them. The immediate consequences of the energy release are flares and CMEs. CMEs drive shocks that accelerate SEPs from near the Sun and into the heliosphere; they also cause sudden commencement when arriving at Earth. CMEs cause severe geomagnetic storms when they have appropriate field orientation, field strength, and speed. We use cumulative distributions of these events, fit a function to the tail of the distributions, and estimate the size of a one-in-100- and one-in-1000-year events. Traditionally the power-law distribution has been extensively used (e.g., Nita et al., 2002; Song et al., 2012; Riley, 2012), which can lead to overestimates for some types of events. Other distributions such as a lognormal distribution have been found to better represent the data and provide better confidence intervals for extreme-event estimation (Love et al., 2015). Here we use both a power law (e.g., Clauset et al., 2009; Aschwanden et al., 2016) and a version of the Weibull distribution (Weibull, 1951). Our main purpose is to extend the tail to smaller probability regimes without worrying about the theoretical basis of the distributions. Such an approach seems to be consistent with some of the historical extreme events, but may not be unique. It should also be made clear that inferring events on the tail of distributions assumes that the same physics is involved in the inferred parametric regime.

3 ESTIMATES OF EXTREME EVENTS
3.1 CME SPEEDS

Because CMEs are the most energetic phenomena relevant to space weather, we start with the extreme CME events. One of the basic attributes of CMEs is their speed in the coronagraph field of view (FOV). CMEs start from zero speed during eruption, attain a peak speed and then tend to slow down. Observations close to the Sun that were occasionally available in the early phase of SOHO mission

(Gopalswamy and Thompson, 2000; Zhang et al., 2001; Cliver et al., 2004), the STEREO mission (see e.g., Gopalswamy et al., 2009a; Bein et al., 2011), and the ground based Mauna Loa Coronameter (St. Cyr et al., 2015; Gopalswamy et al., 2012) have shown that CMEs attain a peak acceleration ranging from a fraction to several km s^{-2} near the Sun (typically at heliocentric distances <3 Rs). Once the acceleration ceases, CMEs move with constant speed or slowly decelerate due to the drag force exerted by the ambient medium. The average speed in the coronagraph FOV (2–32 Rs) ranges from ~100 to >3000 km s^{-1}. CMEs causing space weather typically have higher speeds.

3.2 DISTRIBUTION FUNCTIONS FOR CME SPEEDS AND KINETIC ENERGIES

Fig. 3 shows the cumulative distribution of CME speeds measured in the FOV of SOHO's Large Angle and Spectrometric Coronagraph (LASCO). The average speeds of several CME populations responsible for energetic phenomena are noted on the plot. The CMEs are related to: metric type II radio bursts (m2) due to shocks in the corona at heliocentric distances <2.5 Rs; magnetic clouds (MC), which are the inter planetary CMEs (ICMEs) with flux rope structure; ICMEs lacking flux rope structure and hence named as ejecta (EJ); interplanetary shocks (S) detected in the solar wind; geomagnetic storms (GM) caused by CME magnetic field or shock sheath; halo CMEs (Halo) that appear to surround the

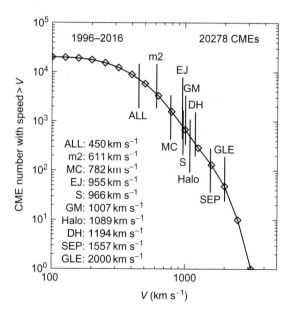

FIG. 3

Cumulative distribution of CME speed (V) from SOHO/LASCO. The CME speeds from https://cdaw.gsfc.nasa.gov have been measured in the sky plane, and no corrections have been applied. The average speeds of CME populations responsible for various coronal and interplanetary phenomena are marked on the plot. The November 10, 2004, CME at 02:26 UT had the highest speed of 3387 km s^{-1}.

From Gopalswamy, N., 2016. History and development of coronal mass ejections as a key player in solar terrestrial relationship, Geosci. Lett. 3, 8–26.

occulting disk of the coronagraph and propagating Earthward or anti-Earthward; decameter-hectometric (DH) type II bursts indicating electron acceleration by CME-driven shocks in the interplanetary medium; SEP events caused by CME-driven shocks; ground level enhancement (GLE) in SEP events indicating the acceleration of GeV particles. The average speed of every one of these populations is significantly greater than the average speed of the general population (450 km s^{-1}). It must be noted that MC, EJ, GM, and Halo are related to the internal structure of CMEs in the solar wind, while the remaining are all related to the shock-driving capability of CMEs. All these CME populations are generally related. What is remarkable about the cumulative distribution is that there are not many CMEs with speeds exceeding ~3000 km s^{-1}. Gopalswamy et al. (2010) attributed the lack of CMEs to speeds >3000 km s^{-1} to the free energy that can be stored in solar active regions and the fractionation of the released energy in the form of CMEs.

The fastest CME in Fig. 3 occurred on November 10, 2004, at 02:26 UT. The average speed in the coronagraph FOV was 3387 km s^{-1}. One might wonder if the high speed of the CME was because of the preceding CMEs that sweep out the ambient material, presenting a low-density (and hence low-drag) medium to the succeeding CME. But the drag depends not only on density, but also on the CME surface area and the square of the excess speed of the CME over the ambient medium. When a low-density medium is created, a CME propagating through such a medium expands and hence acquires a greater area that increases the drag. Similarly, the high speed also increases the drag. So, the net effect may not be a significant decrease in drag. In fact, the November 10, 2004, CME was observed to slow down within the coronagraph FOV (https://cdaw.gsfc.nasa.gov/CME_list/UNIVERSAL/2004_11/htpng/20041110.022605.p302s.htp.html), even though there was a preceding CME from the same source region nine hours before. Therefore, the initial high speed is likely to be due to the propelling force (solar source property) rather than the drag force (ambient medium property).

Fig. 4 shows the cumulative distributions of speeds and kinetic energies of CMEs observed over the past two decades. We have used a power law and the Weibull functions to fit the data points. Clearly, the power law is applicable only over a very limited range of speeds and kinetic energies. On the other hand, the Weibull function fits much better over the entire range, although it has more free parameters. The steep drop in the number of events at high speeds and kinetic energies seems to be real because the current cadence of LASCO is high enough that energetic CMEs are not missed. An event on the tail of the Weibull distribution in Fig. 4 may occur once in 100 years with a speed of 3800 km s^{-1} and once in 1000 years with a speed of ~4700 km s^{-1}. We refer to these events as once-in-100-year and once-in-1000-year events to denote the event size expected once in 100 years and once in 1000 years, respectively. Hereafter we refer to these events as 100-year and 1000-year events for simplicity. From Fig. 4, we can infer that the 100-year and 1000-year kinetic energies as 4.4×10^{33} and 9.8×10^{33} erg, respectively. It must be noted that these kinetic energies are only a few times greater than the highest reported values. We shall return to the reason for these limiting values later.

3.3 FLARE SIZE DISTRIBUTION

Solar flares typically accompany CMEs, but many also occur without CMEs. CMEless flares are confined typically and have an upper limit to their sizes: ~X2.0. About 10% of X-class flares are known to lack CMEs (Gopalswamy et al., 2009b). Here we consider the cumulative distribution of all the flares

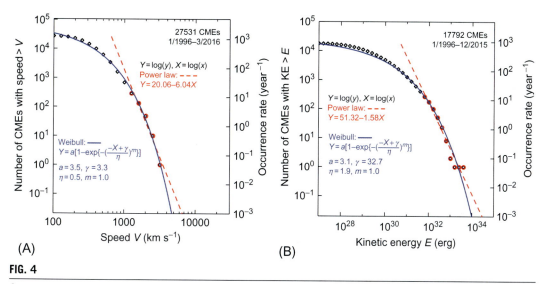

FIG. 4

Cumulative distribution of CME speeds (A) and kinetic energies (B) from SOHO/LASCO catalog (https://cdaw.gsfcs.nasa.gov) for the period 1996–2016. Power-law (e.g., Clauset et al., 2009) and Weibull (Weibull, 1951) fits to the data points are shown. The speed and kinetic energy data points are obtained by binning the original data into 5 data points per decade. The November 10, 2004, CME at 02:26 UT has the highest speed of 3387 km s^{-1} and the September 9, 2005, CME at 19:48 UT has the highest kinetic energy of 4.20×10^{33} erg.

that have been recorded by various GOES satellites since 1969 in the 1–8 Å energy band (see Fig. 5). The distribution shows a break around the X2 level (2×10^{-4} W m^{-2}). According to the Weibull distribution, the 100-year and 1000-year event sizes are X43.9 and X101 respectively. The power law distribution yields similar flare sizes: X42 and X115. The 100-year size is similar to the estimated size of the November 4, 2003, soft X-ray flare recorded in the 1–8 Å energy band by the GOES satellites (Woods et al., 2004; Thomson et al., 2004; Brodrick et al., 2005). The data point corresponding to the largest flare size (X28) in Fig. 5 represents this event. It must be noted that the GOES X-ray sensor saturated at a level of X17.4 for about 12 min during this event, so the X28 value was an initial estimate. Brodrick et al. (2005) concluded that the flare size should be in the range X34–X48, with a mean value of X40. The corrected data point is close to the fitted lines corresponding to the Weibull and power-law functions. Based on solar flare effects on the ionosphere, it has been concluded that the September 1, 1859, flare should have been at least as strong as the November 4, 2003, flare. The flare size estimate for the Carrington flare is in the range X42–X48, with a nominal value of X45 (see Cliver and Dietrich, 2013 and references therein). It is remarkable that the Weibull distribution provides an estimate consistent with several independent estimates of the peak values of the November 4, 2003, flare and the Carrington flare.

The size of the 1000-year flare (X101) is only a factor of ~2 larger than the Carrington flare. The bolometric energy corresponding to an X100 flare is 10^{33} erg (see e.g., Benz, 2008). Flares with bolometric energies $>10^{33}$ erg are considered super flares (Schaefer et al., 2000; Maehara et al., 2012;

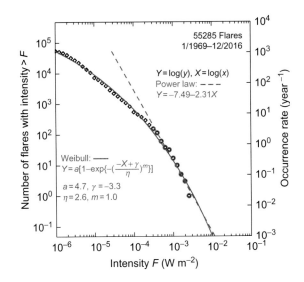

FIG. 5

Cumulative distribution of flare sizes between 1969 and 2016. Weibull and power-law fits to the data points are shown. The November 4, 2003, flare at 19:29 UT has the highest intensity of 2.8×10^{-3} W m^{-2} (X28). The flare data are from https://www.ngdc.noaa.gov/stp/solar/solarflares.html.

Shibata et al., 2013). Thus, the tail of the flare-size distribution suggests that super flares can occur on the Sun once in a millennium. A 10^{34} erg flare can occur on the Sun only once in 125,000 years, too infrequent compared to the once-in-800-years occurrence suggested by Shibata et al. (2013).

3.4 ACTIVE REGIONS AND THEIR MAGNETIC FIELDS

One of the important parameters related to the origin of solar eruptions is the sunspot area, which has been known for a long time. We refer to the sunspot group area as the active region area. We use the whole sunspot area, which includes the penumbra, not just the umbra. Fig. 6 shows the cumulative distribution of the active region area A for the period 1874–2016. The cumulative number decreases slowly until the area reaches $\sim$1000 msh (millionths of solar hemisphere) and then decreases rapidly. The Y-axis on the right-hand side gives the occurrence rate N per year. The rapidly declining part of the distribution fits to a power law, N [year^{-1}] $= 4.68 \times 10^{11} A^{-3.55}$. The maximum observed area was $\sim$5000 msh. Such large-area active regions were observed only twice over the 143-year period used in the distribution (both of these active regions occurred in solar cycle 18). All the data points can also be fit to the Weibull's function, which agrees with the power law at high A. Note that we modified the Weibull function by introducing an additional scale factor "a." According to the power law, a 100-year active region has an area of $\sim$7000 msh and will be considered an extreme event. The Weibull function gives a slightly lower area for the 100-year active region: $\sim$5900 msh. The observation that superflares tend to occur in solar-like stars with large spot areas (more than order of magnitude larger than the

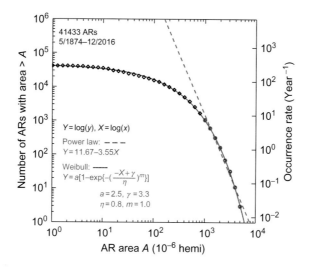

FIG. 6

Cumulative number of active region (sunspot group) areas A from 1874 to 2016. A is expressed in microhemispheres (also known as millionths of solar hemisphere, msh; 1 msh $= 3.07 \times 10^{16}$ cm^2). On the right-hand side Y-axis, the occurrence rate per year (number of active regions in each bin divided by the data interval of 143.5 years). Sunspot group areas are derived from daily photographic images of the Sun recorded at the Royal Greenwich Observatory (ftp://ftp.ngdc.noaa.gov/STP/SOLAR_DATA/SUNSPOT_REGIONS/Greenwich/) for the period 1874–1976. The area data have been extended beyond 1976 by the Solar Observing Optical Network (SOON, https://www.ngdc.noaa.gov/stp/space-weather/solar-data/solar-imagery/photosphere/sunspot-drawings/soon/).

largest sunspot areas) suggest that the same physical process is responsible for the formation of active regions in the Sun and other solar-like stars (Maehara et al., 2017).

While the active region area has been measured systematically since the late 1800s, the measurement of sunspot magnetic fields started only around 1915. There have been a number of investigations in the 20th century that found a good correlation between the sunspot area and the maximum field strength in the umbra (Livingston et al., 2006 and references therein). These investigations also found that the number of regions with field strengths >5000 G is exceedingly low. Livingston et al. (2006) compiled sunspot field measurements of 12,804 active regions in the interval 1917–2004. The cumulative distribution shows only five active regions with sunspot field strengths >5000 G, one of them being 6100 G. These authors also noted that the distribution was a steep power law with an index of −9.5. The relation between sunspot field strength B (G) and the active region area A (msh) has been found to be of the form (Ringnes and Jensen, 1960; Nagovitsyn et al., 2017):

$$B = p \log A + q \qquad (1)$$

where p and q are coefficients that seem to vary between solar cycles and B is in units of 100 G. Ringnes and Jensen (1960) reported the highest correlation ($r = 0.92$ for 43 sunspots) between B and $\log A$, with $p = 23.3$ and $q = -27.0$ during 1945–48. Nagovitsyn et al. (2017) confirmed this B–A relationship for cycles 23 and 24, although they found only q varied between the two cycles. Norton et al. (2013) did not find a significant variation in the umbral field strength when they considered a 3-year period in the rising phase of cycle 24. Schad (2014) found a nonlinear relationship between magnetic field strength and umbral size, over a short period (6 years). Note that both Norton et al. (2013) and Schad (2014) considered umbral areas, not the whole area including the penumbra.

Given the good correlation between the sunspot area and sunspot B, we see that the sharp decline in the number of events with B is consistent with the rapid drop in the number of events with large active region area. Thus, an active region with $A \sim 6000$ msh is expected to have a B of ~ 6100 G, very similar to the extreme case reported by Livingston et al. (2006). It must be noted that the peak field strength is found in the umbra of sunspots, but not throughout the active region area. Nevertheless, we can consider a hypothetical active region with an area of ~ 6000 msh and a peak field strength of 6100 G as an extreme case of source active region that will be used for further discussion. The maximum possible magnetic potential energy (MPE) can be computed as $(B^2/8\pi)A^{1.5}$, where B is the magnetic field strength of the active region that has a sunspot area A. For $B = 6100$ G and $A = 6000$ msh, we get MPE $= 3.7 \times 10^{36}$ erg.

The sunspot magnetic field is thought to emerge from the toroidal field located at the base of the convection zone in the solar interior (see e.g., Basu, 2016). Based on helioseismic techniques, Basu (1997) and Antia et al. (2000) have estimated an upper limit of 3×10^5 G for the field strength at the base of the convection zone. The field strength measured on the surface is about two orders of magnitude smaller than the one at the base of the convection zone. The limit of the field strength in the solar interior ultimately seems to be the physical reason for the size and field strength in solar active regions that determine the free energy available to power eruptions.

Fig. 7 shows scatter plots of the MPE of a large number of active regions with maximum speed and maximum kinetic energy of CMEs originating from the active regions. The active regions were selected based on the fact that they were responsible for one or more of the following: (i) a large SEP event, (ii) a magnetic cloud, and (iii) a major geomagnetic storm. The active region area was computed as the area covered by at least 10% of the peak unsigned magnetic field strength in the active region as observed in a magnetogram from SOHO/MDI or SDO/HMI when the region was close to the central meridian. The MPE is computed as $(<B>^2/8\pi)A^{1.5}$, where $<B>$ is the unsigned average field strength within A (Gopalswamy et al., 2010). Note that the active region area used here is different from the sunspot area used in Fig. 6, which is typically smaller by a factor <10. On the other hand, we use the average B, instead of the peak B used for sunspots. In identifying the maximum CME speed in an active region, we listed all the CMEs from the active region and selected the one with the highest speed to use in the scatter plot. The kinetic energy of the fastest CME from a given active region is taken from the SOHO/LASCO CME catalog. The MPE was computed at the time of the central meridian passage of an active region, and not at the time of the CME with the maximum speed.

Although the scatter is large, the maximum CME speed is significantly correlated with the MPE in the source active region (Fig. 7A), as was shown earlier in Gopalswamy et al. (2010). The regression line, when extrapolated to the maximum possible MPE (3.7×10^{36} erg), gives a speed of ~ 3600 km s^{-1}. This speed is not too different from the highest observed speed by SOHO/LASCO (see Fig. 3). Recall that the Weibull distribution gives this speed for a 100-year event (see Fig. 4). A straight-line fit to the top speeds in the plot corresponding to various MPEs would put the highest speed attainable by a CME as ~ 6700 km s^{-1} in an active region with a potential energy of $\sim 3.7 \times 10^{36}$ erg. This speed is greater than that of the 1000-year event (~ 4700 km s^{-1}) indicated by the Weibull distribution and similar to the speed indicated by the power-law distribution (~ 6500 km s^{-1}).

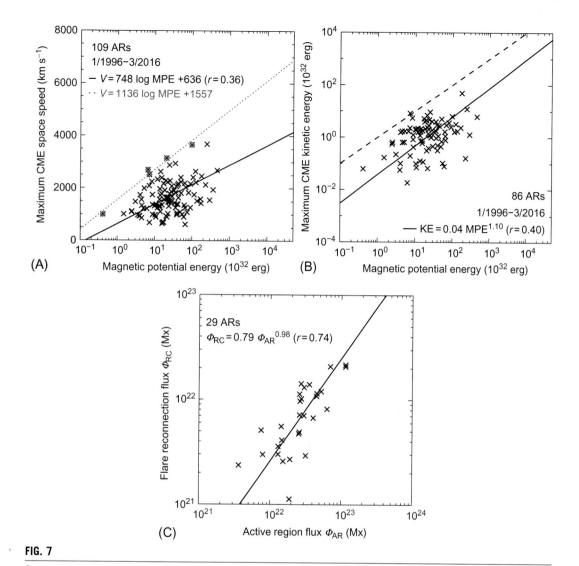

FIG. 7

Scatter plot of the magnetic potential energy (MPE) of active regions with maximum speed (A) and maximum kinetic energy (KE) of CMEs from the active regions (B). Only CMEs with speeds >500 km s^{-1} are included. The regression lines are shown in *solid black*. In the speed plot, the *red line* fits the top 5 data points. In the kinetic energy plot, the dashed line represents equal energies. The correlation coefficients $r=0.36$ and $r=0.40$ are significant despite the large scatter because the corresponding Pearson's critical values are 0.316 and 0.349, respectively, for a significance level of 99.95%. Note that we used deprojected speeds (V_{sp}) as opposed to sky-plane speeds (V_{sky}) used in Figs. 3 and 4. Speeds of full halo CMEs and partial halos are deprojected using a cone model or the empirical formula $V_{sp}=1.10V_{sky}+156$ km s^{-1} (Gopalswamy et al., 2015b) when CMD <60°. For CMEs with CMD >60° a simple geometrical deprojection used. (C) A scatter plot between the reconnected (RC) flux during an eruption and the total flux in the source active region indicates a good correlation ($r=0.74$ with a Pearson critical correlation coefficient of 0.579 at 99.95% confidence level for 29 active regions). RC fluxes from 28 ARs are from Gopalswamy et al. (2017b); for one event, the RC flux was computed using SDO's HMI and AIA data.

(A) From Gopalswamy, N., Akiyama, S., Yashiro, S., Mäkelä, P., 2010. Coronal mass ejections from sunspot and non-sunspot regions. In: Hasan, S.S., Rutten, R.J. (Eds.), Magnetic Coupling Between the Interior and Atmosphere of the Sun. Springer, Berlin, 289 pp.

The maximum CME kinetic energy is also significantly correlated with MPE (see Fig. 7B). Not all CMEs had mass estimates, so the number of CMEs in the kinetic energy plot is smaller than that in the speed plot. The regression line (KE=0.04 MPE$^{1.10}$) gives a kinetic energy of $\sim 4.2 \times 10^{35}$ erg corresponding to the highest possible MPE. This corresponds to an energy-conversion efficiency of $\sim 11\%$. At lower levels of the potential energy, the conversion efficiency is <10%. In the kinetic energy scatter plot, we have also shown the equal-energies line (100% efficiency). We do see a couple of data points that are close to the equal-energies line, but this may be due to the overestimate of the kinetic energies stemming from the uncertainties in mass and speed estimates.

Another way of looking at the energy conversion efficiency is to compare the reconnected (RC) flux during an eruption with the total active region flux. The total active region flux (<B>A) uses the same average magnetic field and area used in computing MPE (<B>^{2}/8π)A$^{1.5}$. The RC flux Φ_{RC} is computed as half the photospheric flux within the area under the posteruption arcade (Gopalswamy et al., 2017a). For a hypothetical region with the largest observed area (6100 msh) and the highest observed field strength (6000 G), the AR flux Φ_{AR} is $\sim 1.12 \times 10^{24}$ Mx. Substituting this value into the regression line ($\Phi_{RC}=0.79\,\Phi_{AR}^{0.98}$), we get $\Phi_{RC} \sim 2.9 \times 10^{23}$ Mx, suggesting that about 26% of the AR flux gets reconnected in the eruption. Gopalswamy et al. (2017b) also reported an empirical relation between Φ_{RC} (in units of 10^{21} Mx) and the CME kinetic energy (in units of 10^{21} erg):

$$KE = 0.19(\Phi_{RC})^{1.87} \tag{2}$$

For $\Phi_{RC}=2.9 \times 10^{23}$ Mx, this relation gives KE=7.7×10^{34} erg. This value is smaller by a factor of 5.5 than that (4.2×10^{35} erg) derived from the scatter plot in Fig. 7B. This is understandable because the KE in Fig. 7B is the maximum value for a given active region, while the one in Eq. (2) has no such constraint; it is simply computed for each eruption considered. The power law function ($Y=51.32-1.58X$ where Y is the log of the occurrence rate per year and X is the KE) in Fig. 4 shows that KE=7.7×10^{34} erg gives an occurrence rate of 1.58×10^{-4} per year; a CME with such KE will occur only once in ~ 6300 years.

In the above discussion, we tacitly assumed that the free energy in active regions is released in the form of CME kinetic energy (eruptive flares). However, there may be no energy going into mass motion in the cases of confined flares. About 10% of X-class flares are known to be confined and the maximum size of a confined flare is $\sim$X1.2 (Gopalswamy et al., 2009b). During solar cycle 24, a huge active region rotated from the east to the west limb of the Sun producing many major X-ray flares including an X3.1 flare. Although there were some narrow CMEs temporally coincided with a couple of the X-class flares, there was no CME associated with most of the X-class flares, including an X2 flare. The active region was NOAA 12192 with an area even larger than that of AR 10486 that resulted in the extreme space weather events shown in Fig. 1. Even the change in the active region area was similar in the two regions (see Fig. 8). The MPE of AR 12192 (2.9×10^{34} erg) was higher than that of AR10486 (1.55×10^{34} erg) by a factor of almost 2, but none of it went into mass motion. For such high magnetic potential energy, one would expect a CME with speed exceeding 3000 km s^{-1} from the correlation plot in Fig. 7B. Based on the investigation of the magnetic environment of AR 12192, it was concluded that the overlying field in the corona was so strong that it did not allow any mass to escape (Thalmann et al., 2015, and references therein). On the contrary, AR 10486 did not have the strong overlying field and had some connectivity to another active region nearby (AR 10484). Thus AR 12192 represents an extreme case in not producing any mass motion.

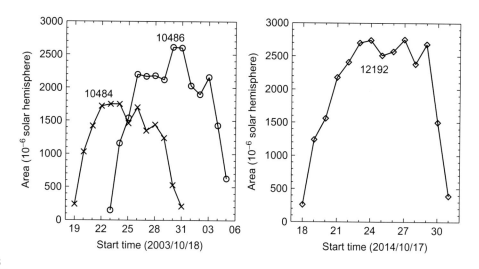

FIG. 8

Observed time variation of the areas of two large active regions from (left) solar cycle 23 (October 2003, AR 10486) and (right) solar cycle 24 (October 2014, AR 12192). In the left plot, a nearby active region (AR 10484) with overlapping disk passage is also shown. The two active regions are at the extreme ends of eruptive behavior. AR data are from NOAA (http://www.swpc.noaa.gov/products/solar-region-summary).

4 CONSEQUENCES OF SOLAR ERUPTIONS

The two primary consequences of CMEs are the SEP events and geomagnetic storms, of which the latter is specific to Earth. SEP events are relevant to any location in the heliosphere. In this section, we consider the distributions of large SEP events (>10 MeV peak intensity ≥10 pfu) and intense geomagnetic storms (Dst ≤ −100 nT). For SEP events, we also consider omnidirectional fluences in the >10 and >30 MeV integral channels. We also discuss the tail of the distributions and how some of the historical events are located on the tails.

4.1 SEP EVENTS

Fig. 9 shows the cumulative distribution of 261 large SEP events from 1976 to 2016 as reported by NOAA (https://umbra.nascom.nasa.gov/SEP/). All SEP events whose >10 MeV proton intensity exceeded anywhere during the event duration are included in the plot. This means, the largest events are energetic storm particle (ESP) events caused when the shock passes by the detector (see e.g., Cohen, 2006). The largest event has a size of $\sim 4.3 \times 10^4$ pfu, which is an ESP event that occurred on March 23, 1991 (Shea and Smart, 1993). The backside event of July 23, 2013, had a peak intensity of $\sim 4.4 \times 10^4$ pfu (Gopalswamy et al., 2016; Mewaldt et al., 2013a,b), but it was a small event at Earth (~ 12 pfu). The Weibull fit can be extrapolated to obtain the size of 100-year and 1000-year events as 2.04×10^5 and 1.02×10^6 pfu, respectively. The power-law fit gives even bigger sizes: 3.03×10^5 and 3.96×10^6 pfu. It must be noted that neither the power-law nor the Weibull fits pass through the last data point. If the

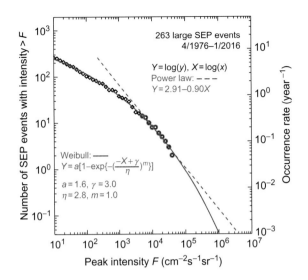

FIG. 9

Cumulative distribution of large SEP events from 1976 to 2016 as reported by NOAA (also listed at NASA's Solar Data Analysis Center, https://umbra.nascom.nasa.gov/SEP/). Both Weibull and power-law fits are shown. The power-law fitted only to the last 10 data points, whereas all data points are used in the case of Weibull distribution. The March 23, 1991, SEP event has the highest peak intensity of 4.3×10^4 cm^{-2} s^{-1} sr^{-1}.

largest measured value is correct, the extrapolated values may be overestimates. We can use the Ellison and Ramaty (1985) or Band et al. (1993) functions to force the fits pass through the last data point. The Band function gives the size of 100-year and 1000-year events as 9.51×10^4 and 3.15×10^5 pfu, respectively. The Ellison-Ramaty function gives slightly lower values: 8.52×10^4 and 1.57×10^5 pfu. We conclude that the maximum size of the 100-year event is $\sim 10^5$ pfu, while the size of the 1000-year can be an order of magnitude larger than this value.

4.2 SEP FLUENCES

Fig. 10 shows the >10 and >30 MeV fluences of 216 large SEP events detected by GOES since 1987. We have shown the Weibull and power-law fits to the occurrence rates. As in the intensity plot, the fitted curves do not pass through the last data point. Ellison-Ramaty (ER) and Band functions can be forced to pass through the last data point. The resulting 100-year and 1000-year fluences are compared in Table 1. The 100-year, >10 MeV fluence values are: 5.11×10^{10} p cm^{-2} (Weibull), 2.43×10^{10} p cm^{-2} (Ellison-Ramaty), and 2.48×10^{10} p cm^{-2} (Band). These values differ only by a factor of ~ 2. The 1000-year, >10 MeV fluence values are: 14.3×10^{10} p cm^{-2} (Weibull), 3.83×10^{10} p cm^{-2} (Ellison-Ramaty), and 4.94×10^{10} p cm^{-2} (Band). The Ellison-Ramaty and Band values are closer to each other, but the Weibull values are higher by a factor of 3–4. The 100-year fluence values for >30 MeV are: 1.58×10^{10} p cm^{-2} (Weibull), 0.63×10^{10} p cm^{-2} (Ellison-Ramaty), and 0.67×10^{10} p cm^{-2} (Band), while the 1000-year fluence values are: 5.09×10^{10} p cm^{-2}

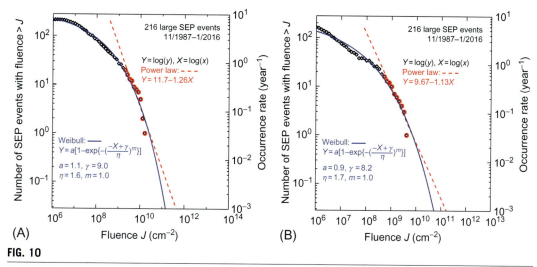

FIG. 10
Cumulative distribution of the omnidirectional SEP fluence in the >10 MeV (A) and >30 MeV (B) ranges. Weibull and power-law fits are shown on the plots. The July 14, 2000, SEP event had the highest fluence of 1.65×10^{10} cm^{-2} (>10 MeV) and 4.31×10^{9} cm^{-2} (>30 MeV). All fluences were computed from time profiles of NOAA's GOES data.

Table 1 Integral Fluence Values for Different Models in Units of 10^{10} p cm^{-2}

Model	100-Year		1000-Year	
	>10 MeV	>30 MeV	>10 MeV	>30 MeV
Weibull	5.11	1.58	14.3	5.09
Power-law	7.08	2.12	43.7	16.3
Ellison-Ramaty	2.43	0.63	3.83	1.02
Band	2.48	0.67	4.94	1.52

(Weibull), 1.02×10^{10} p cm^{-2} (Ellison-Ramaty), and 1.52×10^{10} p cm^{-2} (Band). The Ellison-Ramaty and Band values are consistently close to each other, while the Weibull values are larger by a factor of 3–5. The power law fits yield higher values in all cases, by about an order of magnitude.

Based on SEP event identification made from nitrate deposits in polar ice, Shea et al. (2006) compiled the >30 MeV fluences of events that occurred over the past ~450 years. They concluded from the frequency distribution of these events that the occurrence of >30 MeV fluence exceeding 0.6×10^{10} p cm^{-2} is very rare. However, Wolff et al. (2012) have questioned the statistics on the basis of their finding that most of the nitrate spikes in Greenland ice cores correspond to biomass burning plumes originating in North America. They were also not able to find a nitrate signal even for the Carrington event. In fact, in a simulation study, Duderstadt et al. (2016) concluded that an SEP event large enough and hard enough to produce a nitrate signal in Greenland ice core would not have occurred throughout the Holocene. This conclusion is consistent with the >30 MeV, 100-year fluences obtained in this study. The estimated largest, >30 MeV fluence of 0.6×10^{10} p cm^{-2} was also reported by Webber et al. (2007) for the November 12, 1960, GLE event. Cliver and Dietrich (2013) also estimated the >30 MeV integral fluence to be in the range $(0.5–0.7) \times 10^{10}$ p cm^{-2} for a few GLE events (July 1959, November 1960, and August 1972). Their highest estimate was for the Carrington event: 1.1×10^{10} p cm^{-2} similar to our 100-year fluence from the Weibull distribution. Cliver and Dietrich (2013) noted that the Carrington event is a composite event due to multiple eruptions that happened in quick succession.

Extending the historical data over longer periods became possible with the discovery of two possible SEP events in tree rings. Measurements of ^{14}C in Japanese cedar trees revealed significant increases in the carbon content during two periods: AD 774–775 (Miyake et al., 2012) and AD 992–993 (Miyake et al., 2013). The authors concluded that the two events must be of similar origin. The two events were also identified in Antarctic and Arctic ice cores as enhancements in cosmogenic isotopes such as ^{10}Be and ^{36}Cl (Mekhaldi et al., 2015). There has been considerable debate on the origin of these events (Melott and Thomas, 2012; Usoskin et al., 2013; Hambaryan and Neuhäuser, 2013; Pavlov et al., 2013; Cliver et al., 2014), but the idea that these are due to SEP events seems to be gaining acceptance (Mekhaldi et al., 2015; Usoskin, 2017). In particular, Mekhaldi et al. (2015) provided arguments against cometary and gamma ray burst sources. They also confirmed that an SEP event with a hard spectrum above 100 MeV is needed to cause these enhancements, as suggested by Usoskin et al. (2013). For the present discussion, we take the AD 774/5 and AD 992/3 signals to be consequences of SEP events and compare them with the fluences we obtained in Table 1.

Fig. 11 shows the estimated fluence spectra of the AD 774/5 and AD 992/3 events from Mekhaldi et al. (2015) obtained by scaling the hard spectrum of the January 20, 2005, GLE event. Also shown for comparison are the hard spectrum of the February 23, 1956, GLE, the soft spectrum of the August 4, 1972, GLE, and a recent event on July 23, 2012, which most likely accelerated particles to GeV energies. Superposed on these plots are the 100-year and 1000-year fluences obtained from Fig. 10 using Weibull, Ellison-Ramaty, and Band functions. Clearly, the 100-year and 1000-year fluences are consistent with those of the AD 774/5 and AD 992/3 events. In particular, the 1000-year fluences in the >10 and >30 MeV ranges cover the AD 774/5 and AD 992/3 events with the two-point slope consistent with that of the known SEP events. This comparison also supports the possibility that the AD 774/5 and AD 992/3 events are indeed consequences of SEP events.

4.3 LARGE GEOMAGNETIC STORMS

Geomagnetic disturbances have been recognized since the 1600s, and the term geomagnetic storm was introduced by von Humboldt in the 1800s (see Howard, 2006 for a review). The link to the Sun was recognized by Sabine (1852) as a synchronous variation of sunspot number and geomagnetic activity. The Carrington flare occurred a few years later and was associated with a geomagnetic storm of historical proportions. Fortunately, there were extensive observations of the storm from magnetometers and global aurora (e.g., Tsurutani et al., 2003). This remains a historical extreme event against which many storms are compared (Cliver and Dietrich, 2013). The connection between solar eruptions and geomagnetic storms was established in the early 20th century, including the fact that the storm-causing

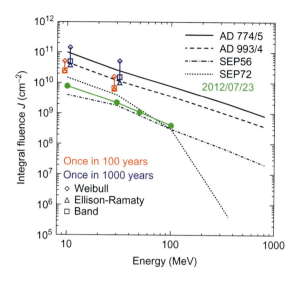

FIG. 11

The 100-year and 1000-year data points derived from the cumulative distributions are superposed on the spectra of the AD 774 and AD 993 particle events obtained by Mekhaldi et al. (2015). Estimates of 100-year and 1000-year event sizes from Weibull, Ellison-Ramaty, and Band functions are shown using different symbols. The data points are shifted slightly to the left (100-year) and right (1000-year) of $X=10$ MeV and $X=30$ MeV to distinguish them. The spectra of the February 23, 1956 (SEP56), and August 4, 1972 (SEP72), solar proton events are also shown from Mekhaldi et al. (2015), who used the reevaluated spectra from Webber et al. (2007). Also shown is the spectrum of the July 23, 2012, extreme event from Gopalswamy et al. (2016).

eruptions occurred close to the disk center of the Sun, and an average delay of ∼1 day was noted between the flare occurrence and the onset of great geomagnetic storms (Hale, 1931; Newton, 1943).

Now we know that the magnetic field in CMEs and in the sheath ahead of shock-driving CMEs is responsible for intense geomagnetic storms (Wilson, 1987; Gonzalez and Tsurutani, 1987). In particular, the strength of the CME magnetic field component oriented in the direction opposite that of Earth's horizontal magnetic field Bz is critically important in causing intense storms along with the speed (V) with which the CME magnetic field impinges on the magnetosphere (see e.g., Wu and Lepping, 2002; Gopalswamy et al., 2008). The following empirical relation reasonably represents how the storm strength (Dst) is determined by V and Bz (Gopalswamy, 2010):

$$Dst = -0.01V|Bz| - 32 nT \qquad (3)$$

The Dst index has been compiled since 1957 and has identified many modern super magnetic storms. Fig. 12 shows a plot of the Dst index as a function of time along with the sunspot number. There are only five super storms that had Dst index <-400 nT. The March 13, 1989, storm is the largest since 1957 with a Dst of -589 nT. The storm was associated with a series of solar eruptions between March 10 and 12. The primary storm started with a sudden commencement at 07:47 UT on March 13. The storm has been attributed to a solar eruption, which occurred at 00:16 UT on March 12, 1989, from N28E09. Details on the solar source (Zhang, 1995) and the interplanetary conditions (Nagatsuma

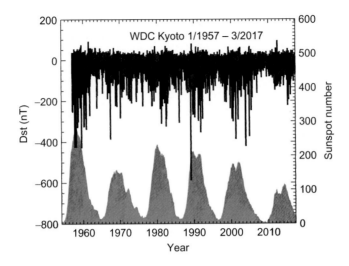

FIG. 12

A plot of the Dst index available at the World Data Center (WDC) in Kyoto, Japan from 1957. The large negative excursions below −100 nT are major storms. The sunspot number is shown at the bottom (*gray*) for reference. The largest storm occurred on March 13, 1989.

et al., 2015) have been reported before. Even though the storm was an extreme event, the flare itself was of moderate size (M7.3). There was no CME data for this event, but from the transit time of 31.5 h one can infer that the CME had a transit speed of $\sim$1300 km s^{-1}. Direct solar wind measurements were also not available, so one has to infer the speed based on empirical relations (Cliver et al., 1990; Belov et al., 2008) as $\sim$960 km s^{-1}. From Eq. (3) with Dst $= -589$ nT, we see that $VBz = 6.2 \times 10^4$ nT km s^{-1}. For $V = 960$ km s^{-1}, we get Bz $= -65$ nT. Nagatsuma et al. (2015) estimated a Bz of ~ -50 nT.

Fig. 13 shows the cumulative distribution of intense geomagnetic storms from the Dst data made available on line at the Kyoto World Data Center. As in many other distributions, the power law fit seems to overestimate the 100-year and 1000-year events. The Weibull distribution fits all the data points. According to the Weibull distribution, a 100-year event has a size of −603 nT, consistent with the March 1989 event; a 1000-year event has a size of −845 nT, consistent with some estimates of the Carrington storm, which occurred about 157 years ago. Although the Dst equivalent of the Carrington storm was estimated as −1760 nT from the geomagnetic record at the Colaba Observatory in India (Tsurutani et al., 2003), many authors have argued for a downward revision. The main arguments are: (i) the Dst index is an hourly average, and (ii) ionospheric/auroral currents might have contributed to the initial sharp spike recorded at the Colaba observatory (see Cliver and Dietrich, 2013 for details). Applying hourly averages to the Colaba data, Siscoe et al. (2006) arrived at a Dst index of −850 nT, similar to the 1000-year event from the Weibull distribution. Recently, Gonzalez et al. (2011) reanalyzed the Colaba data and arrived at a Dst equivalent of $\sim$1160 nT. It must be noted that these estimates are also approximate because the Dst index is actually an average over several equatorial magnetic observatories (see http://wdc.kugi.kyoto-u.ac.jp/dstdir/dst2/onDstindex.html). Cliver and Dietrich suggest a Dst of −900 nT as a nominal value for the Carrington event.

4 CONSEQUENCES OF SOLAR ERUPTIONS

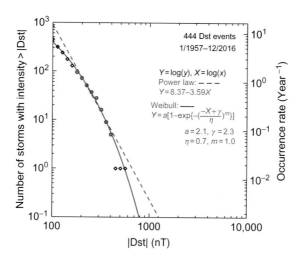

FIG. 13

Cumulative distribution of intense geomagnetic storms (Dst ≤ -100 nT) and their yearly rates using Dst data from 1957 available at the World Data Center, Kyoto. Weibull and Power-law fits to the distribution are shown. The largest storm ($|Dst| = -589$ nT) occurred on 1989 March 14 at 02:00 UT.

The shock transit time of 2381 km s^{-1} when used in Cliver et al. (1990) empirical relation,

$$V = 0.775 Vt - 40 \text{ kms}^{-1} \qquad (4)$$

gives a shock speed at Earth of $\sim$1800 km s^{-1}. The CME speed near the Sun is expected to be $\sim$3000 km s^{-1}. This speed is twice that of the March 13, 1989, event, and hence a doubling of the Dst index is not unexpected. We can also get the 1-AU shock speed of the Carrington event from the average acceleration (a) reported in Gopalswamy et al. (2001):

$$a = -0.0054(u - 406) \text{ ms}^{-2} \qquad (5)$$

where u is the CME speed near the Sun. A lower limit to a can be obtained by replacing the initial speed by the transit speed in Eq. (5), yielding $a = -10.7$ ms^{-2}. From the transit speed and the deceleration, we get a 1-AU speed of $\sim$2044 km s^{-1}, only 13% larger than the speed from Eq. (4). The initial CME speed can then be estimated as 2700 km s^{-1}. Such a speed is well within the observed range of CMEs (see Fig. 3). For Dst $= -900$ nT and $V = 2044$ km s^{-1}, Bz can be estimated as ~ -46 nT. For Dst $= -1160$ nT, only a Bz of -58 nT is needed. These estimates are reasonable if the storm was caused by the shock sheath. If the storm was due to the ICME, one has to allow for the possibility of an ICME speed (V_{ICME}) lower than the shock speed. Using the gas dynamic strong shock limit,

$$V = V_{ICME}(1 + \gamma)/2 \qquad (6)$$

where γ is the adiabatic index. For $\gamma = 5/3$ and $V = 2044$ km s^{-1}, Eq. (6) gives $V_{ICME} = 1533$ km s^{-1}. In this case, Dst $= -900$ nT and -1160 nT would require a Bz of -61 and -78 nT, respectively. These numbers are consistent with a recent backside CME on July 23, 2012, that had Bz ~ -52 nT, $V \sim 2000$ km s^{-1}, and $V_{ICME} \sim 1560$ km s^{-1} at 1-AU (Gopalswamy et al., 2016). The storm strength has been estimated to be similar to that of the Carrington event (Baker et al., 2013; Russell et al., 2013; Mewaldt et al., 2013a,b; Liu et al., 2014a,b; Gopalswamy et al., 2015a).

The requirement of $Bz = -78$ nT to get a Dst value of -1160 nT is not unlikely. Gopalswamy et al. (2017b) obtained an empirical relationship between the peak total magnetic field strength (Bt) and ICME speed:

$$Bt = 0.06 V_{ICME} - 13.58 \text{ nT} \quad (7)$$

For $V_{ICME} = 2000$ km s^{-1}, Eq. (7) can be extrapolated to give $Bt = 106$ nT. From the compilation of Bz and Bt for cycle-23 magnetic clouds in Gopalswamy et al. (2015a), we can see that the magnitude of Bz ~ 0.74 Bt, thus yielding Bz ~ -78 nT for Bt $= 106$ nT. Thus, we conclude that the Carrington storm can be explained by a very fast ICME with high magnetic content and the Dst estimate is consistent with a 1000-year storm.

5 SUMMARY AND CONCLUSIONS

This chapter considered properties of extreme solar eruptions and their consequences assuming that they are located on the tail of their cumulative distributions. In particular, we estimated the sizes of 100-year and 1000-year events. In many cases, these sizes are consistent with known historical events. The Weibull function was used as the baseline function in extrapolating the distributions to estimate the 100-year and 1000-year event sizes. Power-law distributions were also used, but generally they appear to yield overestimates. In some cases, we also used Ellison-Ramaty and Band functions in obtaining conservative estimates of 100-year and 1000-year events. The power laws can be fit only to a subset of the data points and their selection is somewhat subjective. In some cases, even the Weibull distribution may lead to overestimates, but not by as large an extent. Table 2 provides a summary of the 100-year and 1000-year event sizes as a measure of the extremeness of the phenomena considered. The range of values for a given entry between the power law and Weibull distributions give an idea of the uncertainties involved in the event size estimations.

Table 2 Expected 100-Year and 1000-Year Event Sizes Estimated From the Tail of Observed Distributions Fitted to Various Functions

	100-Year		1000-Year	
	Weibull	Power Law	Weibull	Power Law
AR Area (msh)	5780	7090	8200	13,600
CME Speed (km s^{-1})	3800	4484	4670	6564
CME KE (10^{33} erg)	4.40	6.85	9.76	29.5
Flare Size (X1.0 = 10^{-4} W m^{-2})	X43.9	X42.4	X101	X115
Bolometric Flare Energy (10^{32} erg)	4.39	4.24	10.1	11.5
SEP Intensity (10^5 pfu)	2.04	3.03	10.2	39.6
>10 MeV SEP Fluence (10^{10} cm^{-2})	5.11	7.07	14.3	43.7
>30 MeV SEP Fluence (10^{10} cm^{-2})	1.58	2.12	5.09	16.30
Dst (nT)	−603	−774	−845	−1470

We also considered solar active regions as the physical origin of eruptive events and considered the maximum amount of free energy available for powering the eruptions. The limit to the free energy can be traced to the size and magnetic content of active regions. The free energy in an active region is generally not exhausted in a single eruption, so the maximum flare size or the CME kinetic energy is limited by a conversion efficiency, which is not fully understood. Two decades of SOHO observations have shown that there are not many CMEs with speeds exceeding 3000 km s^{-1}. The tail of the Weibull distribution suggests that a 1000-year CME will have a speed of only 4700 km s^{-1}. A 1000-year CME is expected have a kinetic energy of $\sim 10^{34}$ erg. Similarly, a 1000-year flare will have a size of $\sim$X100; the corresponding bolometric flare energy of 10^{33} erg is consistent with the known fact that the CME kinetic energy is typically ten times the flare energy.

The consequences of eruptive events we considered are SEP events and geomagnetic storms. We estimate the >30 MeV fluence of a 1000-year event is in the range $(1-5) \times 10^{10}$ p cm^{-2}, which is consistent with the historical extreme event such as the Carrington event, the AD 774/75 event, the AD 994/95 event, and the recent backside event of July 23, 2012. The Carrington event also serves as the benchmark geomagnetic storm. The tail of the Weibull distribution gives the Dst index of a 1000-year event as -845 nT, which is consistent with the revised estimates of the Carrington storm size. The power law tail gives a larger storm magnitude consistent with higher estimates for the Carrington event, although we think the power law overestimates the event sizes.

ACKNOWLEDGMENTS

I thank P. Mäkelä, S. Akiyama, and S. Yashiro for help with the figures. I thank D. F. Webb and E. W. Cliver for their comments and suggestions that improved the presentation of the material. This work was supported by NASA's Heliophysics Guest Investigator program.

REFERENCES

Antia, H.M., Chitre, S.M., Thompson, M.J., 2000. The Sun's acoustic asphericity and magnetic fields in the solar convection zone. Astron. Astrophys. 360, 335.

Araki, T., 2014. Historically largest geomagnetic sudden commencement (SC) since 1868. Earth Planets Space 66, 164.

Aschwanden, M., et al., 2016. 25 years of self-organized criticality: solar and astrophysics. Space Sci. Rev. 198, 47.

Baker, D.N., Li, X., Pulkkinen, A., Ngwira, C.M., Mays, L.L., Galvin, A.B., Simunac, K.D.C., 2013. A major solar eruptive event in July 2012: defining extreme space weather scenarios. Space Weather 11, 585–591. https://doi.org/10.1002/swe.20097.

Band, D., Matteson, J., Ford, L., et al., 1993. BATSE observations of gamma-ray burst spectra. I—spectral diversity. Astrophys. J. 413, 281.

Basu, S., 1997. Seismology of the base of the solar convection zone. Mon. Not. R. Astron. Soc 288, 572–584.

Basu, S., 2016. Global seismology of the sun. Living Rev. Sol. Phys. 13, 2.

Bein, B.M., Berkebile-Stoiser, S., Veronig, A.M., et al., 2011. Impulsive acceleration of coronal mass ejections. I. Statistics and coronal mass ejection source region characteristics. Astrophys. J. 738, 191.

Belov, A.V., Eroshenko, E.A., Oleneva, V.A., Yanke, V.G., 2008. Connection of Forbush effects to the X-ray flares. J. Atmos. Sol. Terr. Phys. 70, 342. https://doi.org/10.1016/j.jastp.2007.08.021. 200.

Benz, A.O., 2008. Flare observations. Living Rev. Sol. Phys. 5, 1.

Borovsky, J.E., Denton, M.H., 2006. Differences between CME-driven storms and CIR-driven storms. J. Geophys. Res. 111. A07S08, https://doi.org/10.1029/2005JA011447.

Brodrick, D., Tingay, S., Wieringa, M., 2005. X-ray magnitude of the 4 November 2003 solar flare inferred from the ionospheric attenuation of the galactic radio background. J. Geophys. Res. 110. A09S36, https://doi.org/10.1029/2004JA010960.

Carrington, R.C., 1859. Description of a singular appearance in the Sun on September 1. Mon. Not. R. Astron. Soc. 20, 13.

Clauset, A., Shalizi, C.R., Newman, M.E.J., 2009. Power-Law distributions in empirical data. SIAM Rev. 51, 661.

Cliver, E.W., 2016. Flare vs. shock acceleration of high-energy protons in solar energetic particle events. Astrophys. J. 832, 128.

Cliver, E.W., Feynman, J., Garrett, H.B., 1990. An estimate of the maximum speed of the solar wind, 1938–1989. J. Geophys. Res. 95, 17103

Cliver, E.W., Nitta, N.V., Thompson, B.J., Zhang, J., 2004. Coronal Shocks of November 1997 Revisited: The Cme Type II Timing Problem. Sol. Phys. 225, 105.

Cliver, E.W., Dietrich, W.F., 2013. The 1859 space weather event revisited: limits of extreme activity. J. Space Weather Space Clim. 3, A31.

Cliver, E.W., Tylka, A.J., Dietrich, W.F., Ling, A.G., 2014. On a solar origin for the cosmogenic nuclide event of 775 AD. Astrophys. J. 781, 32. https://doi.org/10.1088/0004-637X/781/1/32.

Cohen, C.M.S., 2006. Observations of energetic storm particles: an overview. In: Gopalswamy, N., Mewaldt, R.A., Torsti, J. (Eds.), Solar Eruptions and Energetic Particles.In: Geophysical Monograph 165, pp. 275–282.

Dierckxsens, M., Tziotziou, K., Dalla, S., 2015. Relationship between solar energetic particles and properties of flares and CMEs: statistical analysis of solar cycle 23 events. Sol. Phys. 290, 841.

Duderstadt, K.A., Dibb, J.E., Schwadron, N.A., Spence, H.E., Solomon, S.C., Yudin, V.A., Jackman, C.H., Randall, C.E., 2016. Nitrate ion spikes in ice cores not suitable as proxies for solar proton events. J. Geophys. Res. D 121, 2994–3016.

Ellison, D.C., Ramaty, R., 1985. Shock acceleration of electrons and ions in solar flares. Astrophys. J. 298, 400.

Forbes, T.G., 2000. A review on the genesis of coronal mass ejections. J. Geophys. Res. 95, 23153

Gonzalez, W.D., Tsurutani, B.T., 1987. Criteria of interplanetary parameters causing intense magnetic storms (Dst < 100 nT). Planet. Space Sci. 35, 1101–1109.

Gonzalez, W.D., Echer, E., Clu'a de Gonzalez, A.L., Tsurutani, B.T., Lakhina, G.S., 2011. Extreme geomagnetic storms recent Gleissberg cycles and space era-super intense storms. J. Atmos. Sol. Terr. Phys. 73, 1447–1453.

Gopalswamy, N., 2009. Coronal mass ejections and space weather. In: Tsuda, T., Fujii, R., Shibata, K., Geller, M.A. (Eds.), Climate and Weather of the Sun-Earth System (CAWSES) Selected Papers from the 2007 Kyoto Symposium. Terrapub, Tokyo, pp. 77–120.

Gopalswamy, N., 2010. The CME link to geomagnetic storms. In: Kosovichev, A.G., Andrei, A.H., Rozelot, J.-P. (Eds.), Solar and Stellar Variability: Impact on Earth and Planets. Proc. International Astronomical Union, IAU Symposium. In: vol. 264. Cambridge Univ. Press, Cambridge, pp. 326–335.

Gopalswamy, N., 2011. Coronal mass ejections and their heliospheric consequences. In: Choudhuri, A.R., Banerjee, D. (Eds.), ASI Conference Series, vol. 2. pp. 241–258.

Gopalswamy, N., Thompson, B.J., 2000. Early life of coronal mass ejections. J. Atmos. Sol. Terr. Phys. 62, 1457–1469.

Gopalswamy, N., Lara, A., Kaiser, M.L., Bougeret, J.-L., 2001. Near-Sun and near-Earth manifestations of solar eruptions. J. Geophys. Res. 106, 25261

Gopalswamy, N., Barbieri, L., Cliver, E.W., Lu, G., Plunkett, S.P., Skoug, R.M., 2005a. Introduction to violent Sun-Earth connection events of October-November 2003. J. Geophys. Res. 110. A09S00, https://doi.org/10.1029/2005JA011268.

Gopalswamy, N., Yashiro, S., Liu, Y., Michalek, G., Vourlidas, A., Kaiser, M.L., Howard, R.A., 2005b. Coronal mass ejections and other extreme characteristics of the 2003 October-November solar eruptions. J. Geophys. Res. 110. A09S15, https://doi.org/10.1029/2004JA010958.

Gopalswamy, N., Akiyama, S., Yashiro, S., Michalek, G., Lepping RP, R., 2008. Solar sources and geospace consequences of interplanetary magnetic clouds observed during solar cycle 23. J. Atmos. Sol. Terr. Phys. 70, 245.

Gopalswamy, N., Thompson, W.T., Davila, J.M., Kaiser, M.L., Yashiro, S., Mäkelä, P., Michalek, G., Bougeret, J.-L., Howard, R.A., 2009a. Relation Between Type II Bursts and CMEs Inferred from STEREO Observations. Sol. Phys. 259, 227.

Gopalswamy, N., Akiyama, S., Yashiro, S., 2009b. Major solar flares without coronal mass ejections, Universal Heliophysical Processes. In: Proceedings of the International Astronomical Union, IAU Symposium, Volume 257, pp. 283–286.

Gopalswamy, N., Akiyama, S., Yashiro, S., Mäkelä, P., 2010. Coronal mass ejections from sunspot and non-sunspot regions. In: Hasan, S.S., Rutten, R.J. (Eds.), Magnetic Coupling Between the Interior and Atmosphere of the Sun. Springer, Berlin. 289 pp.

Gopalswamy, N., Xie, H., Yashiro, S., Akiyama, S., Mäkelä, P., Usoskin, I.G., 2012. Properties of Ground Level Enhancement Events and the Associated Solar Eruptions During Solar Cycle 23. Space Sci. Rev. 171, 23.

Gopalswamy, N., Yashiro, S., Xie, H., Akiyama, S., Mäkelä, P., 2015a. Properties and Geoeffectiveness of Magnetic Clouds during Solar Cycles 23 and 24. J. Geophys. Res. Space Phys. 120, 9221–9245.

Gopalswamy, N., Xie, H., Akiyama, S., Mäkelä, P., Yashiro, S., Michalek, G., 2015b. The peculiar behavior of halo coronal mass ejections in solar cycle 24. Astrophys. J. 804, L23–L28.

Gopalswamy, N., Yashiro, S., Thakur, N., Mäkelä, P., Xie, H., Akiyama, S., 2016. The 2012 July 23 backside eruption: an extreme energetic particle event? Astrophys. J. 833, 216–235.

Gopalswamy, N., Yashiro, S., Akiyama, S., Xie, H., 2017a. Estimation of Reconnection Flux Using Post-eruption Arcades and Its Relevance to Magnetic Clouds at 1 AU. Sol. Phys. 292, 65.

Gopalswamy, N., Akiyama, S., Yashiro, S., Xie, H., 2017b. Coronal flux ropes and their interplanetary counterparts. J. Atmos. Sol. Terr. Phys. https://doi.org/10.1016/j.jastp.2017.06.004.

Grechnev, V.V., Kiselev, V.I., Meshalkina, N.S., Chertok, I.M., 2015. Relations between microwave bursts and near-Earth high-energy proton enhancements and their origin. Sol. Phys. 290, 2827.

Hale, GE, 1908. On the probable existence of a magnetic field in Sun-spots. Contributions from the Mount Wilson Observatory/Carnegie Institution of Washington, vol. 30, pp. 1–29.

Hale, G.E., 1931. The spectrohelioscope and its work. Part III. Solar eruptions and their apparent terrestrial effects. Astrophys. J. 73, 379–412.

Hambaryan, V., Neuhäuser, R., 2013. A Galactic short gamma-ray burst as cause for the 14C peak in AD 774/5. Mon. Not. R. Astron. Soc. 430, 32–36.

Hodgson, R., 1859. On a curious appearance seen in the Sun. Mon. Not. R. Astron. Soc. 20, 15–16.

Howard, R.A., 2006. A historical perspective on coronal mass ejections. In: Gopalswamy, N., Mewaldt, R., Torsti, R. (Eds.), Solar Eruptions and Energetic Particles. Geophysical Monograph Series, vol 165. American Geophysical Union, Washington, DC, pp. 7–12.

Kintner, P.M., O'Hanlon, B., Gary, D.E., Kintner, P.M.S., 2009. Global Positioning System and solar radio burst forensics. Radio Sci. 44. RS0A08, https://doi.org/10.1029/2008RS004039.

Liu, Y., et al., 2014a. Mysterious abrupt carbon-14 increase in coral contributed by a comet. Sci. Rep. 4, 3728.

Liu, Y.D., Luhmann, J.G., et al., 2014b. Observations of an extreme storm in interplanetary space caused by successive coronal mass ejections. Nat. Commun. 5, 3841. https://doi.org/10.1038/ncomms4481.

Livingston, W.C., Harvey, J.W., Malanushenko, O.V., Webster, L., 2006. Sunspots with the strongest magnetic fields. Sol. Phys. 239, 41. https://doi.org/10.1007/s11207-006-0265-4.

Love, J.J., Rigler, E.J., Pulkkinen, A., Riley, P., 2015. On the lognormality of historical magnetic storm intensity statistics: implications for extreme-event probabilities. Geophys. Res. Lett. 42, 6544.

Mackay, D.H., Gaizauskas, V., Rickard, G.J., Priest, E.R., 1997. Force-free and potential models of a filament channel in which a filament forms. Astrophys. J. 486, 534.

Maehara, H., Shibayama, T., Notsu, S., et al., 2012. Superflares on solar-type stars. Nature 485, 478.

Maehara, H., et al., 2017. Starspot activity and superflares on solar-type stars. Publ. Astron. Soc. Jpn. 69, 41.

Mason, G.M., Desai, M.I., Mewaldt, R.A., Cohen, C.M.S., 2013. Particle acceleration in the heliosphere. In: Centenary Symposium 2012: Discovery Of Cosmic Rays.AIP Conference Proceedings, vol. 1516. pp. 117–120.

Mekhaldi, F., et al., 2015. Multiradionuclide evidence for the solar origin of the cosmic-ray events of AD 774/5 and 993/4. Nat. Commun. 6, 8611. https://doi.org/10.1038/ncomms9611.

Melott, A.L., Thomas, B.C., 2012. Causes of an AD 774-775 14C increase. Nature 491, E1–E2.

Mewaldt, R.A., Russell, C.T., Cohen, C.M.S., et al., 2013a. A 360° view of solar energetic particle events, including one extreme event. In: Proc. ICRC, Merida, Mexico, 1186.

Mewaldt, R.A., Cohen, C.M.S., Mason, G.M., von Rosenvinge, T.T., Leske, R.A., Luhmann, J.G., Odstrcil, D., Vourlidas, A., 2013b. Solar energetic particles and their variability from the Sun and beyond. AIP Conf. Proc. 1539, 116.

Miyake, F., Nagaya, K., Masuda, K., Nakamura, T., 2012. A signature of cosmic-ray increase in AD 774-775 from tree rings in Japan. Nature 486, 240–242.

Miyake, F., Masuda, K., Nakamura, T., 2013. Another rapid event in the carbon-14 content of tree rings. Nat. Commun. 4, 1748.

Nagatsuma, T., Kataoka, R., Kunitake, M., 2015. Estimating the solar wind conditions during an extreme geomagnetic storm: a case study of the event that occurred on March 13-14, 1989. Earth Planet Space 67, 78.

Nagovitsyn, Y.A., Pevtsov, A.A., Osipova, A.A., 2017. Long-term variations in sunspot magnetic field—area relation. Astron. Nachr. 338, 26–34.

Newton, H.W., 1943. Solar flares and magnetic storms. Mon. Not. R. Astron. Soc. 103, 244.

Nita, G., Gary, D.E., Lanzerotti, L.E., Thompson, D.J., 2002. The peak flux distribution of solar radio bursts. Astrophys. J. 570, 423.

Norton, A., Jones, E.H., Liu, Y., 2013. How do the magnetic field strengths and intensities of sunspots vary over the solar cycle? J. Phys. Conf. Ser. 440012038.

Pavlov, A., et al., 2013. AD 775 pulse of cosmogenic radionuclides production as imprint of a Galactic gamma-ray burst. Mon. Not. R. Astron. Soc. 435, 2878–2884.

Reames, D.V., 2015. What are the sources of solar energetic particles? element abundances and source plasma temperatures. Space Sci. Rev. 194, 303.

Riley, P., 2012. On the probability of occurrence of extreme space weather events. Space Weather. 1002012.

Ringnes, T.S., Jensen, E., 1960. On the relation between magnetic fields and areas of sunspots in the interval 1917–56. Astrophys. Norvegica 7, 99.

Russell, C.T., et al., 2013. The very unusual interplanetary coronal mass ejection of 2012 July 23: a blast wave mediated by solar energetic particles. Astrophys. J. 770, 38.

Sabine, E., 1852. On periodical laws discoverable in the mean effects of the larger magnetic disturbances. Philos. Trans. R. Soc. Lond. 142, 103.

Schad, T.A., 2014. On the collective magnetic field strength and vector structure of dark umbral cores measured by the hinode spectropolarimeter. Sol. Phys. 289, 1477.

Schaefer, B.E., King, J.R., Deliyannis, C.P., 2000. Superflares on ordinary solar-type stars. Astrophys. J. 529, 1026.

Shea, M.A., Smart, D.F., 1993. March 1991 Solar Terrestrial phenomena and related technological consequences. In: Proc. ICRC, Tsukuba city, Japan, vol. 3. p. 739.

Shea, M.A., Smart, D.F., McCracken, K.G., Dreschhoff, G.A.M., Spence, H.E., 2006. Solar proton events for 450 years: The Carrington event in perspective. Adv. Space Res. 38, 232–238.

Shibata, K., et al., 2013. Can superflares occur on our sun? Publ. Astron. Soc. Jpn. 65, 49.

Siscoe, G., Crooker, N.U., Clauer, C.R., 2006. Dst of the Carrington storm of 1859. Adv. Space Res. 38, 173–179.

Song, Q., Huang, G., Tan, B., 2012. Frequency dependence of the power-law index of solar radio bursts. Astrophys. J. 750, 160.

St. Cyr, O.C., Flint, Q.A., Xie, H., Webb, D.F., Burkepile, J.T., Lecinski, A.R., Quirk, C., Stanger, A.L., 2015. MLSO Mark III K-Coronameter Observations of the CME Rate from 1989–1996. Sol. Phys. 290, 2951.

Thalmann, J.K., Su, Y., Temmer, M., Veronig, A.M., 2015. The Confined X-class Flares of Solar Active Region 2192. Astrophys. J. Lett. 801, L23.

Thomson, N.R., Rodger, C.J., Dowden, R.L., 2004. Ionosphere gives size of greatest solar flare. Geophys. Res. Lett. 31. L06803, https://doi.org/10.1029/2003GL019345.

Trottet, G., Samwel, S., Klein, K.-L., Dudok de Wit, T., Miteva, R., 2015. Statistical evidence for contributions of flares and coronal mass ejections to major solar energetic particle events. Sol. Phys. 290, 819.

Tsurutani, B.T., Gonzalez, W.D., Lakhina, G.S., Alex, S., 2003. The extreme magnetic storm of 1–2 September 1859. J. Geophys. Res. 108, https://doi.org/10.1029/2002JA009504. 12–68, SSH 1-1.

Usoskin, I.G., 2017. A history of solar activity over millennia. Living Rev. Sol. Phys. 14, 3.

Usoskin, I.G., Kromer, B., Ludlow, F., Beer, J., Friedrich, M., Kovaltsov, G.A., Solanki, S.K., Wacker, L., 2013. The AD775 cosmic event revisited: the Sun is to blame. Astron. Astrophys. 552, L3.

Webb, D., Allen, J., 2004. Spacecraft and ground anomalies related to the October-November 2003 solar activity. Space Weather 2, 3008.

Webber, W., Higbie, P., McCracken, K., 2007. Production of the cosmogenic isotopes 3H, 7Be, 10Be, and 36Cl in the Earth's atmosphere by solar and galactic cosmic rays. J. Geophys. Res. 112. A10106.

Weibull, W.A., 1951. A statistical distribution of wide applicability. J. Appl. Mech. 293–297 ASME.

Weigel, R.S., 2010. Solar wind density influence on geomagnetic storm intensity. J. Geophys. Res. 115. A09201. https://doi.org/10.1029/2009JA015062.

Wilson, R.M., 1987. Geomagnetic response to magnetic clouds. Planet. Space Sci. 35, 329–335.

Woods, T.N., Eparvier, F.G., Fontenla, J., Harder, J., Kopp, G., McClintock, W.E., Rottman, G., Smiley, B., Snow, M., 2004. Solar irradiance variability during the October 2003 solar storm period. Geophys. Res. Lett. 31. L10802, https://doi.org/10.1029/2004GL019571.

Wolff, E.W., Bigler, M., Curran, M.A.J., Dibb, J.E., Frey, M.M., Legrand, M., McConnell, J.R., 2012. The Carrington event not observed in most ice core nitrate records. Geophys. Res. Lett. 39, L08503.

Wu, C.-C., Lepping, R.P., 2002. Effect of solar wind velocity on magnetic cloud-associated magnetic storm intensity. J. Geophys. Res. 107, 1346

Zank, G.P., Hunana, P., Mostafavi, P., le Roux, J.A., Webb, G.M., Khabarova, O., Cummings, A.C., Stone, E.C., Decker, R.B., 2014. Particle acceleration and reconnection in the solar wind. AIP Conference Proceedings. 1720 (1). 070011.

Zhang, H., 1995. Formation of magnetic shear and an electric current system in an emerging flux region. Astron. Astrophys. 304, 541.

Zhang, J., Kundu, M.R., White, S.M., Dere, K.P., Newmark, J.S., 2001. On the temporal relationship between coronal mass ejections and flares. Astrophys. J. 561, 396.

FURTHER READING

Gopalswamy, N., 2010. Corona mass ejections: a summary of recent results. In: Proceedings of the 20th National Solar Physics Meeting, held 31 May—4 June, 2010 in Papradno, Slovakia, pp. 108–130.

Gopalswamy, N., 2016. History and development of coronal mass ejections as a key player in solar terrestrial relationship. Geosci. Lett. 3, 8–26.

Shea, M.A., Smart, D.F., 2012. Space weather and the ground-level solar proton events of the 23rd solar cycle. Space Sci. Rev. 171, 161.

CHAPTER 3

SOLAR FLARE FORECASTING: PRESENT METHODS AND CHALLENGES

K.D. Leka, Graham Barnes
NorthWest Research Associates, Boulder, CO, United States

CHAPTER OUTLINE

1 Introduction: Solar Flares and Societal Impacts 66
   1.1 Flare Forecasting History 68
2 Present Approaches 70
   2.1 Two Basic Paradigms 70
   2.2 Event Definitions 71
   2.3 Parametrizations 72
   2.4 The Statistical Classifiers 74
   2.5 Self-Organized Criticality and Related 75
   2.6 Operational Versus Research 76
   2.7 Evaluation 76
   2.8 The Role of Numerical Models 78
3 Present Status 81
   3.1 How Good? 81
   3.2 Why Not So Good? 83
   3.3 Extreme Solar Flares 84
4 Future 86
   4.1 Outlook for the Extreme Extremes 87
   4.2 Pertinent Question: Are Forecasts Useful? 87
   4.3 Avenues for Improvement 88
5 Summary and Recommendations 89
Acknowledgments 90
References 90

1 INTRODUCTION: SOLAR FLARES AND SOCIETAL IMPACTS

Today's society relies upon technological systems that are vulnerable to the effects of flares from the Sun (Fisher and Jones, 2007; Baker, 2008; Cannon, 2013; National Science and Technology Council, 2015). Solar flares are abrupt, powerful events that can produce orders-of-magnitude increases across the electromagnetic spectrum, especially at the shorter-wavelength higher-energy regimes. Solar flares can be accompanied by an increase in the flux of high-energy particles with energies over 100 MeV, and coronal mass ejections (CMEs) that send charged plasma into the heliosphere (Fig. 1; Benz, 2017), causing radiation surges, fluctuations in the geomagnetic field, and induced localized charging and ground currents.

Potentially impacted industries range from those that are part of daily life today (mobile phone transmission interruptions, high-frequency communications, aircraft rerouting, and power-grid performance and robustness), to those with slightly more esoteric yet still important consequences (large-scale farm-machine operations, deep-water drilling operations, long-range and polar-route airline crew safety) and to future technologies that will permeate life and the economy (e.g., the NextGen air traffic control system, self-driving cars). Humans in the space environment are particularly vulnerable to the radiation environment from sudden energetic-particle events that often accompany solar flares (Fry et al., 2010); this includes high-altitude military aircraft.

The solar flare-associated ionizing radiation itself has minimal impact on modern communications and precision positioning systems such as the Global Navigation Satellite System (GNSS) and those based on GHz-range frequencies. Still, the moderate-MHz-frequency communication systems that are important for military and remote-location aviation are susceptible to transmission errors and blackouts as the Earth's ionosphere reacts to a sudden influx of X-rays. Additionally, flare-associated radio bursts and the X-ray emission itself are of concern for military threat detection. However, one of the crucial drivers of predicting solar flares is their association with quickly arriving high-energy protons and alpha particles, which put many systems at risk (see NOAA/Space Weather Prediction Center, 2011 for a short synopsis).

Compared with other solar phenomena that impact the heliosphere and the Earth in particular, solar flares are easier to forecast in some ways and more difficult in others. A solar flare is an event that is easily discernible across a broad spectrum of wavelengths, especially for the larger flares. Much of the emission used to define a flare is directed over 4π steradian, again making detection readily available across the solar disk, and (at some wavelengths and for some emission mechanisms) even beyond the solar limb. Smaller flares may indeed be lost in background noise, especially during highly active periods and especially if one considers the full disk emission (see Fig. 1, e.g., during March 6, 2012); still, the detection of solar flares is generally more definitive and objective than, for example, CMEs or solar energetic particle (SEP) events where faint emission, high directionality, and/or unknown connectivity to an originating source can severely impede unambiguous detection.

Solar flares are believed to be the results of magnetic reconnection in the solar atmosphere, which allows energy stored in the magnetic field to suddenly release through thermal and nonthermal heating, acceleration of mass, and emission across the full electromagnetic spectrum (Shibata and Magara, 2011 and references therein). Magnetic reconnection is believed to be the only mechanism that can accomplish such a feat, and estimates suggest that 10%–20% of the excess energy stored in the coronal magnetic fields of complex sunspot groups is released during a large flare event. But understanding these general principles does not a forecast make.

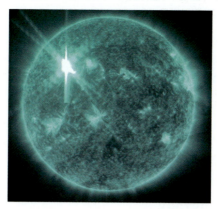
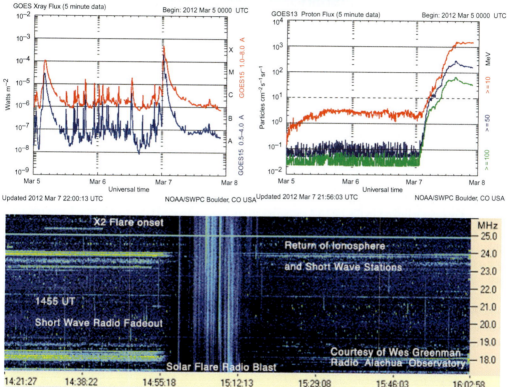

FIG. 1

The sudden increase in ionizing radiation from solar flares routinely disrupts the stability of the Earth's upper atmosphere; they are often associated with CMEs and SEP events. *Top*: A large flare occurred on March 7, 2012, at 00:02 UT from NOAA Active Region No. 11429, saturating the 131 Å channel of the Atmospheric Imaging Assembly on the Solar Dynamics Observatory. *Middle left*: The sun-as-a-star GOES soft-X-ray traces indicating that this flare had a peak output of 5.4×10^{-4} W/m² ("X5.4") over its duration, and was in fact followed by an X1.3 1 hour later. *Middle right*: Although the low-energy proton flux was already elevated prior to the March 7, 2012, flares, the further large increase, especially of the higher-energy protons, led to "strong" radiation and geomagnetic storm conditions, seen here also from the GOES detectors. *Bottom*: Temporal plot of high-frequency (18–25 MHz) radio signals showing the effects of a (different) strong solar flare, including blackout effects lasting many hours at lower frequencies, and a radio blast from the solar event itself.

Top: Data from SDO/AIA. Middle: Images courtesy of NOAA/Space Weather Prediction Center. Bottom: Source as attributed on the figure, reproduced from the archives of SpaceWeather.com, with permission of Dr. Tony Phillips (curator).

The challenges of forecasting these events stem from three things: (1) we do not understand the details of the reconnection initiation, (2) models of solar flares and of some of the presumably relevant phenomena are very simple whereas the observed host sunspot groups are very complex, and (3) because of the nature of this remote-sensing science, it is impossible to observe all the relevant physical quantities everywhere all the time. In effect, solar flare forecasting research is a mechanism to back engineer understanding from the successes and failures of the forecasting itself.

1.1 FLARE FORECASTING HISTORY

Solar flares have historically been classified according to characteristics of emission in the Hydrogen Balmer α (Hα) line, soft X-rays, and radio emissions. As the response of flare-related thermal heating in the solar chromosphere is readily visible in the accessible Hα spectral line, flares have been described in terms of their extent, brightness, impulsiveness, and spatial morphology in Hα line-center emission for almost 80 years. Some standardization from the International Astronomical Union in 1964 (revised in 1975) influenced the reporting from the U.S. Air Force (USAF) Solar Optical Observing Network (SOON) and the National Aeronautics and Space Administration (NASA) Solar Particle Alert Network (SPAN) (Poppe and Jorden, 2006; National Environmental Satellite, Data, and Information Service, 2017). Soft X-ray solar flare reports began with the *Solrad* satellite in 1968, and continued with the launch of the National Oceanic and Atmospheric Administration (NOAA) Geostationary Operational Environmental Satellite (GOES) fleet in 1975 through the present. The standard for solar flare classification started to include characteristics of emission in two pass-bands: 0.5–4 and 1–8 Å. The timing of solar flares was now linked in the NOAA catalogs to sudden increased emission in the latter pass-band, and the peak emission observed in that pass-band has become a standard for categorizing a flare's size (Table 1). The "C," "M," and "X" classifications were the original flux levels defined, leaving room for more sensitive detectors and the now-identified "B" and "A" classes, although these are rarely considered for forecasting. The "X" classifier was supposed to be overtaken by "Y" and "Z" as necessary, but instead the "X" designation has been simply extended beyond X9.9 to X17, etc. (Poppe and Jorden, 2006). Working from Table 1, a group of sunspots' (or active region (AR)'s) history regarding its flare productivity is often described using a "flare index," comprising the summation of peak intensities over a specified time period, following Abramenko (2005b):

$$\text{FI} = \sum (\text{index}_C + 10 \times \text{index}_M + 100 \times \text{index}_X) \qquad (1)$$

where, for example, index$_C$ $\equiv$ 2.3 indicates a maximum flux of 2.3 $\times$ 10^{-6} W/m^2 as reported by the NOAA GOES 1–8 Å detectors (Table 1), similarly for the other indices, including the larger-flare indices

Table 1 Flare Size Nomenclature

Class	Peak 1–8 Å Flux (W/m^2)
A	$10^{-8} \leq \mathcal{F} < 10^{-7}$
B	$10^{-7} \leq \mathcal{F} < 10^{-6}$
C	$10^{-6} \leq \mathcal{F} < 10^{-5}$
M	$10^{-5} \leq \mathcal{F} < 10^{-4}$
X	$\mathcal{F} \geq 10^{-4}$

of "M" and "X." The start, peak, and end times for any particular event are specifically defined by NOAA with regard to the nature of the rise of the emission, its peak magnitude, and its decay, for their catalogs (https://www.ngdc.noaa.gov/stp/solar/solarflares.html); other catalogs may use different definitions.

Radio observations of solar flares became routine in the United States with development of the USAF Radio Solar Telescope Network (RSTN); it had become clear that radio noise storms and metric type II radio bursts had solar origins, caused by coronal shock waves or CMEs often associated with solar flares. Monitoring and understanding radio emissions were crucial during those early decades in the context of space exploration (communications interference) and the Cold War (Knipp et al., 2016).

Forecasting for solar flare events also became more routine in the mid-1960s, as the distinguishing characteristics of sunspots and sunspot groups became associated with flare frequency (McIntosh, 1990; Crown, 2012, and references therein). White-light morphological descriptions and categorizations (e.g., the McIntosh classification system, an extension of the Zurich system) were actually preceded by magnetic-based descriptions (the Mt. Wilson schema, Hale et al., 1919); both are routinely invoked today (Fig. 2 and Table 2). Solar cycles 15–20, but especially the large solar cycle 19, provided the systematic observing capabilities and high number of events to establish statistical relationships between solar flare probabilities and certain sunspot group characteristics.

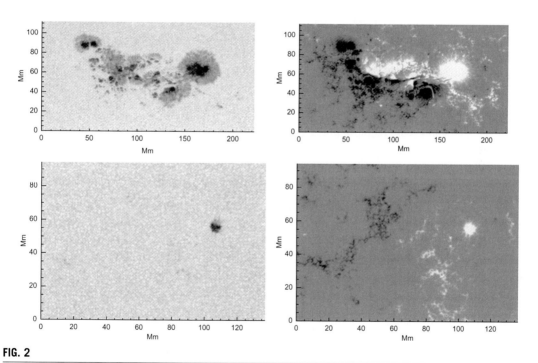

FIG. 2

Left: The photospheric continuum intensity, and *Right*: Radial component of the photospheric magnetic field (*black/white* is negative/positive) for two examples for "extremes" of solar active region characteristics. *Top*: The very complex, very large group of sunspots designated NOAA Active Region No. 11429 that produced many large flares (see Fig. 1). *Bottom*: A very small, very simple group of sunspots, which was almost completely flare-quiet (NOAA Active Region No. 11438). Note scales in megameters (Mm) (see Table 2).

Data from SDO/HMI.

Table 2 Active Region Characteristics for Figs. 2 and 3

NOAA Region Number	Date (at 21:48 TAI)	Size (μH)[a]	Magnetic Class[a]	McIntosh Class (Modified Zurich)[a]	Flare Index[b] Before	Flare Index[b] After
11429	March 8, 2012	950	$\beta\gamma\delta$	Ekc	922.42	243.77
11430	March 8, 2012	180	β	Dao	134.7	13.1
11433	March 16, 2012	100	β	Dso	3.96	0.55
11438	March 25, 2012	20	α	Hsx	1.01	2.54

[a] As listed for 2400 UT in the NOAA Solar Region Summary.
[b] See Eq. (1). Before/after refers to the date in column 2.

2 PRESENT APPROACHES
2.1 TWO BASIC PARADIGMS

The typical approach to forecasting solar flares, whether for operational forecasting or as a research endeavor, generally follows a standard approach. First, observational data of the solar atmosphere are used to characterize the visible magnetic regions, or sunspot groups; typically the data comprise white-light images or magnetic field information, but can extend to $H\alpha$, radio, or other wavelengths. Those characterizations are then used by a statistical classifier by which the flaring likelihood of the observed characteristics is evaluated in the context of historical events. The parameters that describe each region are "physics inspired" (see Section 2.3), meaning that they seek to capture what is believed to indicate an active region's energy storage and propensity for flare activity. As an example, the $\mathcal{R}$ parameter of Schrijver (2007) computes the amount of magnetic flux near those magnetic polarity inversion lines (PILs), which display particularly strong spatial gradients in the field (c.f. Fig. 2: NOAA Active Region No. 11429 displays regions of strong gradients where the magnetic polarity changes); this was argued to be a proxy for detecting current-carrying emerging magnetic flux. Many of the parametrizations are extensions of very early discoveries that solar flares are preferentially associated with solar active regions that deviate significantly from a potential magnetic field configuration (e.g., Severnyi, 1965; Hagyard et al., 1984). Reconnection scenarios do require a reservoir of magnetic energy and a topology conducive to the required instabilities (Priest and Forbes, 2002), and thus a consistent picture of the necessary conditions for flares emerges.

A second approach used by a few methods (primarily for research) prescribes a forecasted flaring rate according to a region's (or the full Sun's) recent flaring rate, by way of a historically determined or otherwise known and described power law (Wheatland, 2001, 2004). A distantly related approach relies upon the idea that solar flares are characterized by a model of self-organized criticality (SOC) (Lu and Hamilton, 1991), and thus will never be deterministically predicted (see review in Aschwanden et al., 2016). Recently this has been extended to a full prediction scheme for larger solar flares (Strugarek and Charbonneau, 2014) based on an avalanche model.

2.2 EVENT DEFINITIONS

A key part of any forecast (or any science) is the definition of terms; in forecasting that becomes crucial as an event definition: what is defined as an event and what is defined as a nonevent. The majority of research investigations currently focus on event definitions that follow operational data products: an event achieves a peak Soft X-ray flux greater than a minimum threshold as defined by the GOES 1-minute averaged Soft X-ray 1–8 Å emission (generally M1.0 or X1.0; see Table 1) at some time within the prescribed time interval (known as the validity period, commonly 24 hours). Operational forecasts usually include additional products such as a 48-hour forecast validity window, or 24-hour validity window that begins after an additional 24 hours (i.e., after a 24-hour latency). Of note, the question being asked is what is the probability of at least one event matching the definition occurring within that time window? This is a different question than *when* would an event occur, or if an event *of a given size* will occur.

In this "forecasting" framework, only knowledge of the Sun up to a certain point in time is considered available, and no knowledge of when/if an event actually occurred is foreshadowed. Research using this paradigm (Leka and Barnes, 2007; Bloomfield et al., 2012; Falconer et al., 2012) most closely matches the needs of an operational forecast. Superposed epoch analysis, on the other hand, asks, "What occurred N hours before an event with this Y particular set of characteristics?" (Leka and Barnes, 2003b; Barnes and Leka, 2006; Mason and Hoeksema, 2010; Reinard et al., 2010; Bobra and Couvidat, 2015). The latter may be a more interesting scientific question with the potential for discovering distinct preevent characteristics. However, this approach can be confounding from an operational forecasting point of view, since operationally one does not have such prescient information. Very relevant research is also conducted outside any forecasting framework whatsoever by simply associating a research-targeted study with an overall flaring rate used simply as an activity indicator (e.g., Welsch et al., 2009; Komm et al., 2011).

Event definitions can have broad ramifications for the direction of research and the usefulness of results. For example, an oft-cited goal is to provide forecasts of extreme events (X10.0 or larger) but such events are so rare that it is difficult to perform well-defined parameter-space distribution estimates. As such, the ability to generate and judge any such predictions suffers (c.f. Section 2.7). On the other hand, the performance is generally higher for event definitions that allow for larger sample sizes (such as C1.0 as a lower threshold) but these smaller events are arguably less impactful for the heliosphere, and hence of less interest (and thus the research may be deemed irrelevant).

Solar flare forecasts and forecast research then is, in part, only as good as the event definition specification. Regarding such options as forecasting flares with or without a CME, or working toward forecasting which flares will produce SEPs—these questions require well-defined event definitions and event lists against which the predictions can be validated. That is, without large databases documenting the relationship of flares with and without CMEs (or SEPs), research into these nuances will be limited. In some cases (e.g., Barnes, 2007), proxies such as the flag for observed high-velocity Hα plasma are used for CMEs, such as appear in NOAA databases. In other cases (Falconer et al., 2008, 2011; Bobra and Ilonidis, 2016), a large endeavor has created specialized event lists for the forecasting research in question. It should be noted, of course, that the NOAA flare catalog is not without bias and missed events (Wheatland, 2001). Either way, any inherent bias (by way of observer, instrumental, viewing angle influences) can reduce the merit of the forecast research.

2.3 PARAMETRIZATIONS

The vast majority of present flare forecast algorithms, both operational and in the research phase, use observed characteristics of solar active regions and historical flaring rates to estimate future flaring probabilities. The earliest such approaches rely on parametrizing the white-light properties of active regions' constituent sunspots, including longitudinal extent, size, and morphology of the largest spot, and the compactness or looseness of the sunspots' positions relative to each other (McIntosh, 1990; Bornmann and Shaw, 1994, see Figs. 2 and 3, Table 2). Observed sunspot group characteristics are then related to historical flaring rates and used as the basis for forecasts (Gallagher et al., 2002; Ahmed et al., 2013). Of note, as pointed out by Bornmann and Shaw (1994), none of the McIntosh classes directly describe the complexity of a sunspot group. Also of note, the NOAA flare forecasts are based on historical-rate lookup tables, but augmented by forecaster (human) expertise (Crown, 2012).

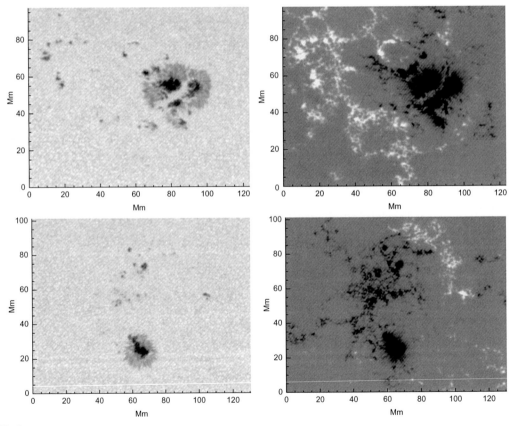

FIG. 3

Same format as Fig. 2, but for two of the more difficult "maybe" scenarios. *Top*: NOAA Active Region No. 11430, which produced an X-class flare a few hours prior to the image shown but was then very quiet for days. *Bottom*: NOAA Active Region No. 11433, whose largest event was a small C-class flare. See Table 2.

Data from SDO/HMI.

2 PRESENT APPROACHES

The size and complexity of sunspot groups as described by the McIntosh (Modified Zurich) classification and the magnetic Mt. Wilson classification are standard benchmarks. The physical characteristics that they describe can be put on a quantitative basis by means of solar photospheric magnetic field data. The line-of-sight component of the magnetic field provides the largest database of active region magnetic information, including decades of daily full-disk observations from the ground-based Solar Tower Telescope at Mt. Wilson (Ulrich et al., 2002), the NASA/National Solar Observatory (NSO) spectromagnetograph (SPM) at the NSO/Kitt Peak Vacuum Telescope (KPVT) (Jones et al., 1992) and the Global Oscillations Network Group (GONG) (Hill et al., 2003). These data are supplemented by the long-running Michelson Doppler Imager (MDI) (Scherrer et al., 1995) on the Solar and Heliospheric Observatory (SoHO) mission (Domingo et al., 1995) and now the Helioseismic and Magnetic Imager (HMI) (Schou et al., 2012; Hoeksema et al., 2014) on the Solar Dynamics Observatory (SDO) mission (Pesnell, 2008). With these data, a sunspot's or region's size can be expressed in terms of total magnetic flux; the qualitative compactness and complexity indices are described by the spatial gradients of the fields and fractal-based indices (Smith et al., 1996; Abramenko, 2005a; McAteer et al., 2005; Leka and Barnes, 2003a,b, 2007; Georgoulis, 2012; Boucheron et al., 2015). Rudimentary models of the associated coronal field from this boundary can quantify its complexity (Barnes and Leka, 2006; Georgoulis and Rust, 2007; Barnes and Leka, 2008). Some algorithms do not parametrize any physics-inspired characteristic but instead characterize the sun through wavelet transforms or similar (e.g., Muranushi et al., 2015). There are difficulties in extending a database by combining these data resources, however, due to different instruments' designs, capabilities, and processing methods. There is also the significant shortcoming of observing only one component of the magnetic vector, as provided by the longest-running magnetic field data sources (Falconer et al., 2011; Leka et al., 2017a).

With larger databases of vector magnetic field data available, more statistical investigations regarding the state of flaring active region magnetic fields have been possible (see Leka and Barnes, 2007, for an early example). Even with increased noise associated with the additional observed components, there is an inherent increase in the amount of information available. Of note, however, is that the vectors are measured essentially at a single height across the photosphere and as such, only an incomplete view of this boundary condition is routinely available. Popular parametrizations concentrate on the degree and extent of strongly sheared horizontal fields (especially near a magnetic PIL), estimates of total energy storage (through measures of magnetic shear and vertical current density), the (related) helicity or magnetic twist (especially as compared to a potential field that would satisfy the same normal-component of the boundary), and various weightings of the same (Fig. 4, left; Leka and Barnes, 2003a; Bobra and Couvidat, 2015; Nishizuka et al., 2017, and references therein). It can be argued that the analysis developed for the line-of-sight data would likely benefit from reduced bias by using the radial component of the field (derivable from the vector field data most directly; although see Leka et al., 2017a; Falconer et al., 2016). In theory, the ability to differentiate between flare-prone and flare-quiet regions should improve with the use of vector field data over solely the line-of-sight component (Barnes et al., 2007; Leka et al., 2017b), although again, early studies concluded that even against parameters derived from vector magnetic field data, persistence (the prior flaring history of a particular active region) performed best (Smith et al., 1996).

Research on nonmagnetic photospheric indications of high flare activity has tested the hypothesis that plasma flows would either contribute to or otherwise react to the forces creating the highly nonpotential magnetic field configurations; if causal, the responsible flow patterns may be detected earlier

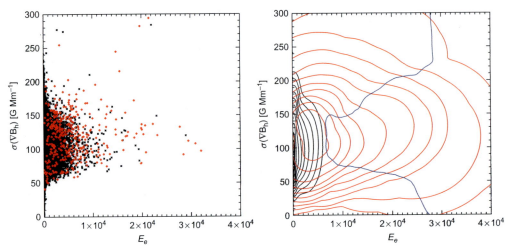

FIG. 4

Example of parameter distributions and NPDA (Barnes et al., 2007; Leka et al., 2017b), for M1.0+, 24-hour validity period, 0-hour latency event definition. *Left*: the parameters shown are (or are substantively similar to) two of the best performance used in numerous forecasting algorithms: a proxy for the magnetic free energy E_e and the standard deviation of the distribution of horizontal gradient of the horizontal component of the magnetic field, derived using SDO/HMI vector magnetic field data from May 2010 through August 2016. Scatter plot shows samples from event (*red*: 434 points) and nonevent (*black*: 28,185 points). *Right*: probability density estimates (*red/black* following the event/nonevent categories) and nonparametric discriminant boundary (at the 50% probability level, *blue line*, where the probability estimates for an event and a nonevent are equal.) for the same. The PDEs are calculated using a nonparametric adaptive kernel, to best represent simultaneous regions with high density and low density of samples.

than magnetic patterns. Surface flows (Welsch et al., 2009) showed that a proxy for the Poynting flux was well correlated with a cumulative flare index, but not significantly more so than magnetic-derived quantities. Helioseismology can be used to infer subsurface flows; when used to evaluate flare-activity indicators, studies have generally concluded that while there are helioseismic signatures associated with flare-productive active regions, they are generally no more informative regarding flare productivity than the most basic magnetic field and size measures (Reinard et al., 2010; Komm et al., 2011; Braun, 2016).

2.4 THE STATISTICAL CLASSIFIERS

Parameters describe the state of the solar atmosphere in myriad ways, but to connect that to forecasting a flare likelihood requires a statistical classifier. That is, a method to relate a historical database of parameters to flaring activity is invoked, such that new data (and its new parameters) can be evaluated as to whether they are likely to produce a flare of the sort prescribed by the event definition in use.

One of the earliest approaches assumes that the timing of any particular flare is a random process, and invokes Poisson statistics (Bornmann and Shaw, 1994; Gallagher et al., 2002) to calculate the probability that the region with a particular value of a parameter will produce one or more flares meeting the event definition within a particular time period. For white-light observations, for example, this can be used to predict the likelihood of a region having a particular McIntosh or modified Zurich class producing at least one M1.0+ in the following 24 hours.

Other statistical methods have been investigated over the last few decades. Neural-network and machine learning algorithms are in vogue, the first implementation being many decades ago (Bornmann and Shaw, 1994). Machine learning is, of course, a term encompassing many algorithms from support vector machines to neural network, random-forest or decision-tree classifiers, nearest-neighbor classifiers, etc. (e.g., Breiman, 2001; Bishop, 2006).

Two somewhat different approaches are to take a simple correlation between flaring rates and the parameter of choice (Falconer et al., 2011, 2014), do this with multiple variables simultaneously (Bornmann and Shaw, 1994), or to fit a nonlinear function to a similar relation (common is the sigma function, Cui et al., 2006, 2007; Georgoulis and Rust, 2007). Alternatively, discriminant analysis (DA) has been employed to "differentiate" or best separate flare-prone from flare-quiet regions (see Fig. 4, right; Neidig et al., 1981; Leka and Barnes, 2003b, 2007; Barnes et al., 2007; Leka et al., 2017b).

The latter is, in fact, what machine-learning algorithms strive for as well—the best separation between the "event" and "nonevent" examples given those two populations' distributions in parameter space. In Fig. 4 (left) we demonstrate the difficulty of achieving this separation: as pointed out in Figs. 2 and 3, those regions least likely to flare and most likely to flare are fairly well separable in parameter space. However, even with fairly large sample sizes and some of the best-performing magnetic characteristics, there is substantial overlap in parameter space between the two populations of active regions.

There are many ways to measure this separation between populations (see Section 2.7), and different algorithms can be trained to optimize on different measures or metrics. The choice of which measure to optimize can result in quite different forecasts. Of course we also do not know the actual populations, only samples thereof. This becomes key with regard to larger or more rare events (see below).

2.5 SELF-ORGANIZED CRITICALITY AND RELATED

A completely different class of forecasting approaches is based on knowledge of the distribution of flare magnitudes, namely that the probability of a flare of size S occurring follows a power law with $f(S) \propto S^{-\gamma}$. The size is typically taken to be the peak flux of the flare, but could be the total energy released instead (see Section 4.3). Power laws like this naturally arise out of the concept of SOC (e.g., Aschwanden et al., 2016). The power-law behavior can be used for forecasting flares directly, without trying to understand the underlying processes (Wheatland, 2004, 2005; Aschwanden and Freeland, 2012), or by viewing flares in the context of SOC, and particularly avalanche models (Lu and Hamilton, 1991; Vlahos et al., 1995; Georgoulis and Vlahos, 1996, 1998; Strugarek and Charbonneau, 2014; Strugarek et al., 2014).

Both approaches are typically most useful for large events. In the approach of Wheatland (2004, 2005), the smaller events are used to derive the power-law index that goes into the forecast, then the probability of a flare occurring within a specified time interval is determined by assuming that flares occur as a Poisson process in time. In the SOC-like approach, a cellular automaton is used to model the build-up and release of energy. The details of how the energy is added and released differ between implementations, but in general the result is a range of flare sizes over which the distribution follows a power law, but with different behavior for small flares, such as a steeper power law, or an excess of large events above what would be expected from a power law (Gilchrist et al., 2012; Riley, 2012). An advantage to the SOC-like approach is that it can forecast not just the energy release but also the timing of a flare (Strugarek and Charbonneau, 2014), but as yet it does not account for specific active region properties.

2.6 OPERATIONAL VERSUS RESEARCH

Numerous approaches following various of the basic tenets described above are being used for research—into the physics of flares, the expected characteristics of superflares, or to improve forecasting methods. However, there are only a few operational systems or operationally ready codes presently running. Of note are the differences between these two regimes, which we highlight here and discuss further regarding forecast evaluation (Section 2.7).

An operational code first and foremost provides a forecast *no matter what*. Data outages, a Sun without sunspots, a Sun with a plethora of sunspots, a Sun with sunspots all near the limb: a forecast must be provided, and without any knowledge of the future, only the past. A forecast is likely required on a particular schedule, for a particular event definition using thresholds and characteristics that are defined ahead of time. These may, for example, follow the NOAA Space Weather Prediction Center (SWPC) definitions (predicting whether a flare of a minimum threshold based on peak Soft X-ray flux will occur sometime in 0–24, 24–48, or 48–72 hours) or may follow some other definition regarding event type and timing (Barnes et al., 2016).

A research code, on the other hand, can focus on establishing correlations between particular phenomena and flare occurrences or on investigating possible improvements in forecasting, but it will not be restricted by operational requirements. Simple differences such as how to handle missing data or long compute times can be less consequential for research. Important issues regarding transition to operations include: bias and preselection (focusing on such a restricted analysis opportunity as to be inapplicable to operational efforts the majority of the time when forecasts are required), reliance on data unavailable in an operational context (even in a scientist's dreams), and equating superposed-epoch analysis (requiring knowledge of upcoming flare occurrences) with forecasting requirements (Section 2.2).

2.7 EVALUATION

Forecast verification relies on statistical metrics for evaluation, for space weather as with terrestrial weather forecasts. NOAA has published meteorology-based metrics for their `M1.0+` and `X1.0` flare forecasts from 1986 to 2013, as can be viewed at http://www.swpc.noaa.gov/content/forecast-verification (Crown, 2012, see also Murray et al. (2017) for a recent validation study of operational forecasts from the U.K. MetOffice), and descriptions of early objective methods usually cited relevant metrics (Neidig et al., 1981). It is a fairly recent development that research-level forecasting efforts have incorporated statistical metrics for evaluations. The early uses of meteorology-based metrics for solar physics research stemmed from the goal of how to approach a remote-sensing science as an experimental design: include both dependent variables and "control" samples, and statistically evaluate how well these samples of different populations could be distinguished (Leka and Barnes, 2003b; Barnes and Leka, 2006). As statistical methods and sample sizes grew, the research realm of flare forecasting is embracing—if still learning—the power and nuances involved (Barnes et al., 2007, 2016; Barnes and Leka, 2008; Bloomfield et al., 2012; Devos et al., 2014; Raboonik et al., 2017).

The use of meteorology-derived metrics is appropriate in the context of rare events. When the occurrence of an event is rare, always forecasting that it will not occur leads to a very high success rate without any knowledge of the event in question. "Skill-score" metrics to measure improvements over a fiducial forecast were developed following the "Findley Affair," highlighting the futility of quoting very high success rates when speaking of extremely rare events (in that case, tornadoes; Murphy, 1996). In general, any skill score value above 0 indicates improvement over the fiducial, with a perfect forecast giving a value of 1.

Skill scores take the form

$$\text{Skill} = \frac{A_{\text{forecast}} - A_{\text{reference}}}{A_{\text{perfect}} - A_{\text{reference}}} \qquad (2)$$

where A_{forecast} is the accuracy of the method under consideration, $A_{\text{reference}}$ is the expected accuracy of a reference method, and A_{perfect} is the accuracy of a perfect forecast (i.e., the entire sample is forecast correctly). Any measure of how well the forecasts correspond to the observed outcome can be used as a measure of the accuracy. Different choices for the reference forecast and for the measure of accuracy lead to different skill scores. For example, using the "rate correct" as a measure of accuracy, and the climatological event rate as a reference forecast results in the Appleman skill score (ApSS), while using the mean square error of probabilistic forecasts, again with the climatological event rate as a reference forecast, results in the Brier skill score (BSS).

Each skill score has different strengths and weaknesses, and answers a slightly different question (Woodcock, 1976; Jolliffe and Stephenson, 2003). A very readable summary appears at http://www.cawcr.gov.au/projects/verification/#What_makes_a_forecast_good. For example, as pointed out by Bloomfield et al. (2012), the true skill statistic (TSS) (also known as the Hanssen & Kuiper Skill Statistic, H&KSS) is not sensitive to differences in sample-size inequality, unlike almost all other skill scores, and as such could ostensibly be used to compare results from methods that were not necessarily tested on a common dataset. This presumes, however, that the same underlying populations were sampled, which may not be the case, sometimes in subtle ways.

Most forecast methods generate the probability of an event occurring rather than a categorical yes/no forecast. Probabilistic forecasts can be evaluated by way of a reliability plot and the receiver operating characteristic (ROC) curve (Fig. 5). A reliability plot compares the forecast probabilities to the observed event rate in bins of forecast probability. A perfectly reliable forecast would have points only along the $x = y$ line. However, perfect reliability is not enough to guarantee perfect forecasts; a forecast method should *also* resolve the events into high and low forecast probabilities, so the reliability plot for a perfect forecast method would only have points in the highest and lowest probability bins, and those points would lie on the $x = y$ line. The example shown in Fig. 5, left, has good reliability (within the error bars) for small to moderate forecast probabilities, but a tendency to overpredict at high forecast probabilities (points lie to the right of the $x = y$ line). However, it lacks the ability to resolve many of the events, so there are very few points with high forecast probability. This lack of resolution is typical of most flare forecasting methods.

Probabilistic forecasts can also be evaluated relative to a categorical outcome by introducing a threshold, $\mathcal{P}$, above which a forecast is considered to be a "yes." An ROC curve shows how the probability of detection and the false alarm rate change when $\mathcal{P}$ is varied. A perfect forecasting method

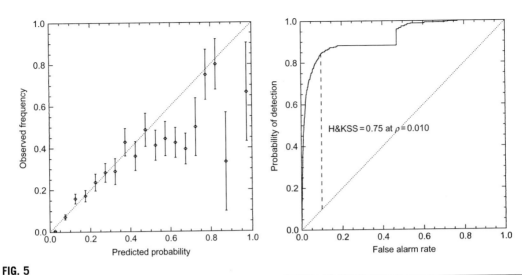

FIG. 5

Evaluation of forecast performance for the method shown in Fig. 4. *Left*: a reliability plot, in which the observed frequency of events is shown as a function of the forecast probability. This example has good reliability (points lie along the $x = y$ line), but has relatively few high forecast probability events. The uncertainties are estimated from the sample size in each bin. *Right*: an ROC plot, showing the probability of detection versus the false alarm rate as a function of the threshold probability. The maximum H&KSS occurs for a low threshold of $\mathcal{P} = 0.010$.

would have an ROC curve consisting of two line segments, one from (0,0) to (0,1) and the second from (0,1) to (1,1). That is, there would be a point with 100% probability of detection and no false alarms. The H&KSS is given by the distance between the ROC curve and the $x = y$ line.

Varying the value of $\mathcal{P}$ will change the value of a skill score for a set of probability forecasts. The value of $\mathcal{P}$ that maximizes one skill score will not, in general, maximize other skill scores. For example, for a reliable forecast method, the ApSS will be maximized for $\mathcal{P} \approx 0.5$, while the H&KSS will be maximized when $\mathcal{P}$ is set to approximately the event rate (see Fig. 5, right). However, when a forecast method has been optimized for one skill score, its performance even on the same dataset should not be compared to that of another method that has been optimized on a different skill score without at least taking into account the value of $\mathcal{P}$ used for each method (see discussion in Barnes et al., 2016).

2.8 THE ROLE OF NUMERICAL MODELS

Numerical models are crucial for exploring what we believe to be the physics of energy build-up, storage, release, and impacts for solar energetic events (e.g., Janvier et al., 2015; Priest and Longcope, 2017; Stepanov and Zaitsev, 2016, as recent examples), including magnetic reconnection initiation and progression. Models are also crucial to understanding the energy redistribution that occurs from the magnetic reconnection into the radiative, kinetic, and accelerative output that is observed (Hudson et al., 2012; Oka et al., 2015; see also Fletcher et al., 2011; Benz, 2017, and references therein). Early (and some contemporary) "models" describe scenarios of instabilities and magnetic configurations leading to magnetic reconnection (Kopp and Pneuman, 1976; Priest, 2016); see McKenzie (2002) including a fun synopsis reproduced at http://solarmuri.ssl.berkeley.edu/~hhudson/cartoons/thepages/McKenzie_CSHKP.html, which is often little more than cartoons (Hudson, 2012). However, observations have been able to test and sometimes uniquely confirm particular characteristics of such models, including reconnection-involved magnetic flux (Qiu, 2009) and the effects of postflare magnetic restructuring (Russell et al., 2016). In the context of forecasting flares, insights are sought from models to understand the necessary and sufficient conditions for a flare to occur.

Modeling efforts take two basic approaches: static and time-dependent. Both have taken full advantage of recent improvements in computer speed and numerical methods. The former generally involves using observational data for boundaries, for example, photospheric vector magnetic field maps, in order to establish estimates of available energy for or describe the magnetic topology associated with particular solar energetic events. By far the most common static models for the coronal magnetic field are nonlinear force free field (NLFFF) extrapolations (e.g., these three on NOAA Active Region #11158: Sun et al., 2012; Inoue et al., 2013; Zhao et al., 2014). In a few cases, nonforce free models, such as the minimum dissipation method (Hu et al., 2008), have also been applied.

The latter refers to 3-D magnetohydrodynamic (MHD) time-dependent models used to investigate mechanisms of instability onset that transpire to initiate an event (e.g., Linton et al., 2001; Aulanier et al., 2010; Kusano et al., 2012; Hassanin and Kliem, 2016; Russell et al., 2016, see Fig. 6). These simulations generally rely on a simple bipolar magnetic configuration embedded within an overlying magnetic arcade, driven by prescribed plasma flow or a modeled emerging magnetic flux rope. In these MHD simulations, the details of magnetic reconnection are generally relegated to numerical resistivity (except for particle-in-cell (PIC) codes, which treat the reconnection explicitly, see e.g., Drake et al., 2006; Knizhnik et al., 2011; Baumann et al., 2013; Zharkova and Siversky, 2015). Data-driven assimilative models are still in their infancy (e.g., Vincent et al., 2012; Wu et al., 2016).

There are also hybrids such as flux constrained equilibrium models (Longcope, 2001), in which the properties of a static model at a given time are determined in part by the evolution of the boundary. Unlike MHD simulations, these models have been applied to a number of specific flare events (e.g., Longcope et al., 2007; des Jardins et al., 2009; Tarr and Longcope, 2012; Tarr et al., 2013).

Two recent examples of the synergy between modeling efforts and observational data analysis in the context of flare prediction relate to the topics of a proposed triggering mechanism and of confined flares. In the former, MHD models explored the ability of an emerging bipole to trigger an eruption due to particular orientations of both it and the overlying field into which it emerges (Kusano et al., 2012), and the implications for subsequent magnetic reconnection. This provided a very testable result, which has been recently explored in case studies (Bamba et al., 2017), and is presently being tested with a statistically significant sample (Bamba et al., 2018). In the latter case, NOAA Active Region No. 12192 was an enigma in that it produced a number of large flares but arguably no CMEs. Although it has been long known that most flares do not produce CMEs, and the differences between eruptive and confined flares have long been studied, NOAA Active Region No. 12192 was special in the number of *large* flares that remained without dramatic, significant coronal mass ejecta. As such, this region has been the subject of modeling efforts, especially examining the region's coronal topology and evolution (Liu et al., 2016; Inoue et al., 2016; Jiang et al., 2016) for differences from CME-productive large active regions (see also Hassanin and Kliem, 2016; Harra et al., 2016).

Models can be pushed beyond their appropriate application. NLFFF modeling of the coronal magnetic field has been recently developed into very accessible tools. However, relying upon free-energy estimates and topological characteristics from NLFFF models without caveat ignores what have been

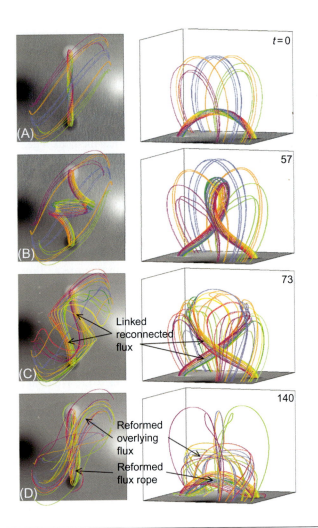

FIG. 6

Initial and early boundary and selected magnetic field line snapshots from a model of magnetic flux emergence and reconnection. The model incorporates an emerging twisted flux rope into an overlying external field; the cited study is a parameter investigation to understand the roles of the relevant field components' characteristics to the susceptibility of the system to torus and kink instabilities (and subsequent eruptive or confined energetic events), and the reconnection scenarios that ensue. For context, compare the boundary of the model (left frames) to the observed magnetic boundary of a flare-productive active region in Fig. 2, top. Panels (A)–(D) show the evolution of the flux system at the timesteps indicated.

Courtesy of B. Kliem, from Hassanin, A., Kliem, B., 2016. Helical kink instability in a confined solar eruption. Astrophys. J. 832, 106. https://doi.org/10.3847/0004-637X/832/2/106.

shown to be serious limitations and uncertainties in these approaches (Schrijver et al., 2006, 2008; Metcalf et al., 2008; DeRosa et al., 2009, 2015). Time-dependent MHD models are crucial for identifying instabilities to which the solar plasma in a flare-ready state may be susceptible (e.g., Titov and Démoulin, 1999; Kliem and Török, 2006). However, the relevant structures are smooth, without realistic plasma properties, and very simple compared to what is observed. Even with data-assimilative techniques, there is an inherent shortfall of information on the inputs to the system (from below the visible surface), precluding true event forecasting from numerical models based on observational data, at least given present capabilities.

Numerical models do provide avenues to explore extreme solar flare events, that is, the realms of parameter space for which there are no observations. Thus far the modeling of proposed "superflares" has focused on relatively simple MHD scale-free approaches (Aulanier et al., 2013; Shibata et al., 2013), to gain some estimates and expectations of, for example, the sizes of superflare host active regions and the estimates of available free magnetic energy. What has not yet been examined, however, are details of nonlinearity in the plasma behavior in these realms, or the context in which such superactive regions might form. If details of magnetic topology so strongly influence the production of CMEs in the context of extremely large field strengths and total flux, then simple scaling arguments on active region size will fail to capture the potential of our Sun to produce extreme solar flare events.

In summary: modeling is crucial as guidance for observational investigations, and observational results (especially those based on large samples) should influence the modeling efforts. Numerical modeling may help describe under which scenarios supersized energetic events will occur, but is presently of limited value as a predictive tool.

3 PRESENT STATUS

How well are we forecasting solar flares? The standard benchmark is of course the NOAA flare forecasts, whose published metrics point to an answer of both "fairly well" and "not very well" depending on the metric. As new objective operational systems and research programs are developed, it is surprisingly difficult to evaluate the improvements made.

3.1 HOW GOOD?

The NOAA flare forecast records for solar cycle 23 have been evaluated using a variety of metrics (and extended online through 2013), and as such do provide a benchmark against which new methods are often compared. As summarized in Crown (2012), BSSs are positive but small ($\ll 0.5$), smaller for larger-threshold event definitions, and, notably, higher when only very complex active regions are considered. The probability of detection and probabilities of false detections are well separated (implying H&KSS scores in the 0.5 range, for thresholds of 0.25, 0.25, and 0.15 for the C1.0+, M1.0+, and X1.0+ thresholds, respectively). Also of note is the documented improvement (especially in the false alarm rates) when human subjective intervention is allowed to alter the default forecasts based on lookup-table probabilities.

An early comparison of a forecasting method using parameters derived from vector magnetic field data (but a much smaller dataset than NOAA) demonstrated how useful this approach could be (Barnes et al., 2007). BSSs were definitely positive for even the largest threshold, yet still $\ll 0.5$. Of note when compared to NOAA forecasts were higher probabilities for event data and lower probabilities for non-event data (better resolution), leading to overall better skill scores (although as noted in that paper, direct comparisons must be made with care, see below).

Following Bloomfield et al. (2012), it has become popular to report the H&KSS due to its insensitivity to sample size ratios. The Poisson statistics-based forecasts highlighted in that report achieved reasonable H&KSS scores (e.g., 0.54 for M1.0+, 24-hour forecasts); but most importantly, the authors explored the variation of the H&KSS with varying the probability threshold $\mathcal{P}$ used.

Very impressive H&KSS thus have been reported recently (quoted here as a summary, without error bars or details): 0.65 M1.0 and 0.69 X1.0+ (Song et al., 2009), 0.82 M1.0+ (Bobra and Couvidat, 2015), 0.52, 0.85 for M1.0+, X1.0+ (Muranushi et al., 2015), 0.50 X1.0+ (Bloomfield et al., 2016), 0.91 for both M1.0+, X1.0+ (Nishizuka et al., 2017). Many authors include full contingency table elements from which metrics can be independently calculated, but only a few include the $\mathcal{P}$ that was used to construct said tables. More complete lists comparing the performance metrics between different studies have been published (Barnes et al., 2007; Barnes and Leka, 2008; Song et al., 2009; Bloomfield et al., 2012, 2016; Bobra and Couvidat, 2015). One can see the broad spread of H&KSS as well as other metrics between the methods in these compilations, lending the question of whether one can in fact simply choose the method reporting the highest H&KSS and thus have the best forecasts.

The answer is "no," for a few reasons.

First and most importantly, regarding these compilations, one should also note the differences in event definition, validity intervals, sample sizes, time interval considered, and consequentially inferred climatological differences between the methods listed. It is due to these differences that direct comparisons of this sort, simply drawing from published results, simply should not be made. Addressing the concern of how to perform direct comparisons was the impetus behind the workshops held by the Flare Forecasting Comparison group in 2009 and 2013 (co-led by NorthWest Research Associates (NWRA), and involving many of the authors cited here). The results from the first workshop (Barnes et al., 2016) investigated what was required to provide meaningful intercomparisons. The point of the study was by no means to compete between methods, but to start to understand what *general* approaches were succeeding where others were not. A significant effort was spent discussing bias, which can be subtly imposed in the context of event definitions, target time intervals, and data restrictions, which can thus give advantage or create disadvantage to any particular method.

Beyond the result that no method was performing particularly well overall, it was demonstrated that a high H&KSS is insufficient to guarantee a good forecasting performance in all circumstances. A single H&KSS (or any single metric) summarizes a method's performance only in a given situation. The authors of Barnes et al. (2016) agreed that for categorical forecasts, the ApSS was intuitive and informative against the climatology; an ROC plot provided a map of the behavior often summarized by a single H&KSS number, but it provided more information regarding said behavior as a function of the probability threshold used (which must be included when reporting the ApSS, as well). The BSS (or, related, the Brier Score and its components) summarizes the behavior of a probabilistic forecast, and is a good companion to a reliability plot. The substantial group of international participants who contributed to Barnes et al. (2016) agreed that this selection of metrics well summarizes the performance of forecasting methods, to the extent that a user should be able to understand relevant performance given their specific requirements.

Second, solar flare forecasting must handle extreme class imbalance problems; except for the case of small flares (in the case of active-region-targeted forecasts) or full-disk forecasts for smaller events, there is almost always a significant imbalance in the sample sizes between the "event" and "nonevent" populations. This, in fact, well reflects the populations: solar flares are statistically rare events, especially the larger events. How a method handles this inequality, both in training and forecast-mode validation reporting, can significantly alter the reported results. Optimal H&KSS scores are generally obtained when the probability threshold for forecasting a "yes" versus a "no" is close to the event rate. Other metrics may or may not behave the same way.

Third and most subtly, some research methods produce forecasts more frequently than their validity period is long (e.g., a forecast is produced every hour for the probability of an event occurring in the next 24-hour interval; Muranushi et al., 2015; Nishizuka et al., 2017). Reporting multiple "hits" and "misses" in this case can be problematic—the incidents are not independent, which is a basic assumption for the classifiers generally used. More subtle may be a bias in the performance due to slow evolution of the Sun relative to the sampling (allowing for the training to effectively use the testing set), leading to an overestimate of the success.

The Community Coordinated Modeling Center (CCMC) at NASA's Goddard Space Flight Center (GSFC) established a "real-time forecasting scoreboard" in 2015 (see https://ccmc.gsfc.nasa.gov/challenges/flare.php), in order to evaluate near real-time flare forecasting methods ostensibly under consistent circumstances, and provide direct comparisons between methods. While this tool may provide a good test of near real-time forecasting success, unfortunately, it will be well into the next solar activity cycle before sufficient numbers of flares have occurred for meaningful comparisons.

As such, the question "which method is best?" is not well posed. It is not yet clear whether recent methods have bestowed significant improvement to the field. There are answerable (but not yet answered) simple questions—such as "How much improvement is there by using magnetic measures than only white light images?" and "How much improvement is gained using vector instead of just the line-of-sight magnetic field data?"—that may be productive to quantitatively investigate. The only direct comparisons thus far have shown no truly impressive performances, demonstrated the requirements for informative method comparisons, and put forth a lot of unanswered questions.

3.2 WHY NOT SO GOOD?

Most approaches to forecasting first parametrize the properties of the source region, then apply a statistical or machine learning technique to issue a forecast based on the parameter values, as described earlier. The present limited success of these approaches could be due to a lack of information in the parameters, the inability of the classifier to make use of the information in the parameters to successfully forecast flares, or a combination of the two. (Of note, Crown (2012) noted a marked improvement from human forecasters, as well.) To test which is the major factor, multiple classifiers can be applied to the same set of parameters to see how much difference the classifier makes.

One such study by Barnes et al. (2017) applied nonparametric discriminant analysis (NPDA) (e.g., Silverman, 1986) and a random forest classifier (RFC) (e.g., Breiman, 2001) to the same set of parameters characterizing the photospheric magnetic field properties of active regions. For an event definition of at least one flare of size C1.0 or above within 24 hours, the accuracy of the two methods was consistent to within the error bars. This supports the conclusion that it is not the statistical or machine learning method that is limiting our ability to forecast flares. In addition, this study looked at how the number of variables (in the case of NPDA) and the number of trees in the forest (in the case of the RFC) affected the accuracy. In both cases, there was little additional gain in accuracy beyond one variable/tree.

In a similar attempt to find improvement between statistical classifiers, Nishizuka et al. (2017) compared support vector machine, k-nearest neighbor, and extremely randomized tree algorithms on a very large dataset and a large number of parameters derived from magnetic, ultraviolet (UV), and Soft X-ray data. Within the stated uncertainties for most event definitions, the three were essentially equivalent in their performance.

The lack of improvement when variables are combined likely means that there are strong correlations among the parameters, so that little independent information is gained by combining them. This was discussed in Leka and Barnes (2003b) albeit with a small sample size, but even Bornmann and Shaw (1994) mentioned the correlation of variables within the McIntosh classification system. Methods considering or computing numerous parameters usually engage in feature selection, "weeding," importance ranking or some other way of removing parameters that do not empirically help the forecasting results (Ahmed et al., 2013; Bobra and Couvidat, 2015). It is not always clear whether lack of weeding hinders performance, and it may contribute to statistical flukes of high performance.

The impact of correlated variables was also presented in Barnes et al. (2016) in which ways to parametrize the photospheric field were compared. As might be expected, different ways to characterize the strong gradient PIL (e.g., the $\mathcal{R}$ parameter proposed by Schrijver (2007) and the WL_{SG2} parameter proposed by Falconer et al., 2008) showed a strong correlation. What was more surprising was that even parameters that were attempting to characterize completely different properties of the active region were also correlated. For example, a measure of the coronal connectivity, B_{eff}, proposed by Georgoulis and Rust (2007) was almost as strongly correlated with the WL_{SG2} as was the $\mathcal{R}$ parameter (Fig. 7).

The larger solar flares are inherently very rare events, and this impacts two aspects of forecasting and related research. First, with regard to metrics reporting, bias must be recognized and the use of appropriate metrics for very unbalanced population sizes ensured. Second, with regard to training or statistical characterization of events, small samples mean that the parameter distributions are difficult to estimate. This is reflected in, for example, the large error bars in reliability plots for event definitions with small samples (Barnes et al., 2016). When an algorithm is optimizing for success using multiple parameters simultaneously, the sample size required for statistically meaningful results becomes very large, very fast. Statistical anomalies caused by outliers in sparse areas of parameter space can be mitigated through cross-validation and jackknife methods.

3.3 EXTREME SOLAR FLARES

The recent National Space Weather Strategy report (National Science and Technology Council, 2015) focused much attention on 100-year or 1000-year "extreme" events. Although the proposed benchmarks are defined in terms of ionospheric impacts (radio blackouts), for peak Soft X-ray solar flare events this translates to roughly X100 or greater. In the context of the Sun as a star, a "Super Flare" or "Extreme Solar Flare" is generally considered one with a total output energy $>10^{33}$ erg (Maehara et al., 2012; Wu et al., 2015; Aulanier et al., 2013); flares producing energies of up to $\asymp 10^{36}$ erg have been recently observed on solar-type stars (Schaefer et al., 2000; Maehara et al., 2012, 2015; Shibayama et al., 2013).

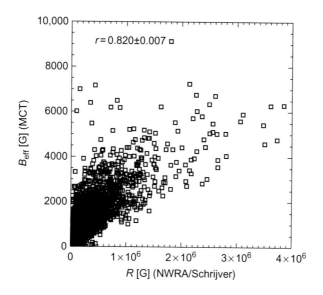

FIG. 7

Scatter plot of parameters characterizing different physical characterizations of solar active regions, yet which are remarkably correlated with each other: x-axis: $\mathcal{R}$ proposed by Schrijver (2007) which measures the amount of magnetic flux near strong-gradient magnetic neutral lines, and y-axis: the B_{eff} measure of the coronal magnetic connectivity proposed by Georgoulis and Rust (2007). Although these measures are based on different physical quantities, they are as strongly correlated as often found with different implementations of the same parameter.

Based on data used in Barnes, G., Leka, K.D., Schrijver, C.J., Colak, T., Qahwaji, R., Ashamari, O.W., Yuan, Y., Zhang, J., McAteer, R.T.J., Bloomfield, D.S., Higgins, P.A., Gallagher, P.T., Falconer, D.A., Georgoulis, M.K., Wheatland, M.S., Balch, C., Dunn, T., Wagner, E.L., 2016. A comparison of flare forecasting methods. I: results from the "All-Clear" workshop. Astrophys. J. 829, 89. https://doi.org/10.3847/0004-637X/829/2/89.

To produce extreme solar flares requires a reservoir of sufficient energy, presumably hosted by the magnetic field in an active region larger and more complex by all measures than any seen in the last few cycles, including the 2003 Halloween storm source regions (NOAA Active Regions No. 10486 and No. 10488). Estimates of the free energy available for flares in any particular active region rarely deviate substantially from $\asymp 10\%$ of the total magnetic energy ($\int |B|^2/8\pi\, dV$), and flares release some portion of that in any given event (it has never been observed that all available free energy is subsumed during an energetic event, nor arguably a substantial fraction; Emslie et al., 2012). Still, solar observations and the historical record can help provide estimates and limits for solar extreme flares.

Recent models have examined the prospects of the Sun producing flares with energies significantly greater than 10^{33} erg. The approaches have examined both the energetics of a model eruption and whether it is realistic to expect the present solar dynamo to be capable of producing a region garnering the $\geq 10^{24}$ Mx required. With the former approach, Aulanier et al. (2013) employed a simple boundary condition (see Section 2.8) with a nondimensionalized MHD model of a shearing field and subsequent loss of equilibrium and eruption, scaled up through estimates of field strengths and region sizes. The conclusion was that a maximum flare energy existed that was not significantly higher than the largest flares thus observed. The question of whether the solar dynamo could even produce an active region with the magnetic flux required to host such an event was studied in Shibata et al. (2013), concluding that it was possible on millennial timescales.

However, there are strong assumptions implied, including that all physics scales essentially linearly with size. Larger sunspots are different than small with respect to temperature, density, field strengths, and complexity of the coronal field topology. It is entirely possible that additional regimes come into play upon the presence of even cooler temperatures (the strong magnetic field suppressing heat transport) and hotter coronal plasma (from heating due to the stronger and more pervasive currents expected to be present). It is increasingly clear that magnetic topology—including the influence of nearby active regions—plays a major role in flare productivity and accompanying CME and energetic particle production (e.g., Nindos and Andrews, 2004; Liu et al., 2016). Some regions still surprise, such as, NOAA Active Region No. 12192, with its high flare productivity and low CME productivity. But regions such as NOAA Active Region No. 11402 should not be overlooked: a medium-sized region of medium complexity that produced three significant (including one X-class) flares.

The present models do not consider scenarios such as the following: instead of an extreme region having access to a standard estimate of 10% of the total magnetic energy, a large (but by no means extreme) region instead for some reason has access to release 90% of the energy in its reservoir in a single event. One contributing factor to the 2003 Halloween geomagnetic storms was the proximity of at least three other active regions, two of those being extremely complex and flare productive; super flares may be the result of not one super active region but multiple regions and their mutual interactions.

There is contradictory evidence regarding whether the Sun has produced Extreme Solar Flares. Historical records of low-latitude auroral sightings serve as proxies for solar activity (Hayakawa et al., 2017), although the energetic relationship between geomagnetic storms and solar *flares* (as opposed to the characteristics of the impacting CME) is by no means straightforward or causal. Spikes in carbon-14 in tree rings or nitrates indicate an influx of energetic particle flux, which may be solar in origin or may be in the form of galactic cosmic rays (Riley, 2012; Miyake et al., 2013); again, the evidence possibly indicates large geomagnetic storms that indirectly implicate large solar flares, but the relation is not straightforward.

Are extreme host regions seen on other Sun-like stars? Possibly, as recent evidence from not just intensity modulation but chromospheric activity indicates (Notsu et al., 2015). If such activity has been found in numerous slowly rotating G-dwarf stars, can it be so unexpected for our Sun? Is it a fluke occurrence during an otherwise normal activity cycle, or does such activity occur as part of an extremely high cycle (antithetical to the recent low-activity cycle)?

The motivation for considering models and solar analogs is that both can help us explore extreme events on the Sun that have not been observed with modern instrumentation. These endeavors may also allow us to explore the physics that occur outside the regime for which good data exist in order to understand and estimate the possibility of extreme solar flares.

4 FUTURE

Successful flare forecasting is a needed tool to protect global infrastructure and communications, as well as provide a "starting point" for forecasting other solar-originating phenomena such as energetic particle events and CMEs. Successful flare forecasting can also be seen as an ultimate test of understanding the underlying physical processes. Limits of success should be illuminating, although interpreting the reasons behind perfect success can be a challenge.

4.1 OUTLOOK FOR THE EXTREME EXTREMES

Of note, few of the solar-observing instruments from Cycle 23 (including the Halloween storm regions) continue to acquire data during the present cycle. The distributions of relevant parameters (and thus flare probabilities) using today's data are not easily applicable to the historical regions without significant calibration work. Long-term observations can thus be more valuable to statistical-based probabilistic forecasts than the next generation of high-resolution scientific experiments. How do we forecast anything beyond "Well, this region is bigger and more complex than anything we've seen before, therefore it has a high probability of doing something very large, sometime in the near future?" Is that satisfactory?

At the same time, moderate events are more frequent, much more difficult to predict with the overlap of parameter space (Fig. 4), and yet can be very impactful in the aggregate (e.g., Mitchell and Schrijver, 2012). Due to the overlap in parameter space, these cases (c.f. Fig. 3) are the most challenging and the strongest test of a forecasting system and even of our basic physical understanding of the underlying phenomena.

The statistical empirical approaches also present very strong constraints on how well any once-in-a-millennium flare could be forecast. For several different definitions of event, Riley (2012) assumes a power-law distribution to infer the probability of an extreme event occurring. However, for the X-ray flux associated with flares, there are indications that for the largest events recorded there are deviations from a power-law distribution, making any estimate of the probability of an extreme event highly suspect; this is especially true if sparse data are treated without maximum likelihood estimators (Bauke, 2007; D'Huys et al., 2016). The SOC-type models also suggest deviations from a power law at large-flare values (Strugarek and Charbonneau, 2014), and even for smaller events the power laws do not always appear to hold (Gilchrist et al., 2012). Indeed, Riley (2012) concludes, "We argued that the flare data [...] deviates significantly from the power-law relationship that holds for lower peak rates, and, thus, cannot be reliably extrapolated beyond observed events."

The outlook for understanding and predicting extreme solar flares thus depends on a more complete understanding of all the factors that contribute to the chain of physics that derives and stores magnetic energy and which then suddenly releases some portion of it as a solar flare. This will only be achieved by the proposed triumvirate of more realistic modeling efforts, more critical and thorough observational analysis, and deeper understanding of solar-analog behavior.

4.2 PERTINENT QUESTION: ARE FORECASTS USEFUL?

Are the present event definitions useful? While they provide good benchmarks against a standard (the historical NOAA forecasts), this single-agency-defined "one size fits none" has no basis in, for example, the relevant time scales or physics of active region evolution. Impacts on the ionosphere can be sensitive to the spectral hardness and duration of enhanced X-ray emission, not just the peak 1–8 Å flux. The physical difference between a `C9.9` and an `M1.0` flare is likely very small, yet if predicting for M1+, the former would have counted as a "correct negative" (e.g., Fig. 4), whereas the latter would count as a "hit"; this does not avail the forecasting method to improved physical understanding. It is also not yet clear what flare or active-region characteristics are uniquely indicative of an impending CME or SEP event (Laurenza et al., 2009; Sun et al., 2015; Kahler et al., 2015).

Presently, the "daily" forecast is the standard: a validity period based on terrestrial convenience rather than prevailing customer need or physics of the flare-initiation process. Such forecasts are useful for situational awareness, but do not lend themselves to actionable items in general (with the exception that a dependable "all-clear" forecast can be an actionable item, especially from the standpoint of human spaceflight; Fry et al., 2010). Is it possible that forecasts for a shorter period are measurably more useful (Al-Ghraibah et al., 2015)? Should peak 1–8 Å flux guide the research, or should other event categorizations be considered? What should direct the research?

4.3 AVENUES FOR IMPROVEMENT
4.3.1 Research into flare trigger mechanisms

Is there a distinct trigger mechanism? Can we answer "why *now*, and not then?" Are there proposed trigger mechanisms that can be ruled out? Or is there simply no such unique trigger available? To identify any unique change in the solar atmosphere that moves a metastable state into an unstable one requires not just modeling effort and high-resolution data analysis (e.g., Kusano et al., 2012; Bamba et al., 2013), but complementary modeling and observational work ruling out counter-examples. The challenge of identifying necessary versus sufficient conditions should not be underestimated.

4.3.2 Addressing correlated variables

To separate flare-prone from flare-quiet regions on any pertinent timescale, the characteristics of regions must be separable in multidimensional parameter space. Hence, the parameters in question need to be complementary but not correlated. As discussed regarding Fig. 7, even parameters that measure ostensibly different phenomena in the solar atmosphere can be highly correlated, and thus do not contribute additional information. Magnetic parameters can be chosen such that the populations are fairly well separated for the bulk of "flare quiet" active regions from the regions that are extremely flare productive (Fig. 4). The largest challenge is in fact separating the populations (or improving the resolution of the forecasts) for regions that are medium-sized, medium-complex, evolving at a medium rate, and hence medium flare-productive—yet still heliospherically impactful.

Productive research would investigate sources of information such that the differences between active regions that otherwise occupy completely overlapping parts of parameter space could be distinguished.

4.3.3 New event definitions

As mentioned earlier, the standard definitions are based on historical instrumentation and convenient timescales. Forecasts may improve (at least for active regions) if the forecasting timing is based on the region's evolutionary epoch (emerging, decaying), or on relevant plasma timescales. It is also possible that the current parameters characterizing the photospheric magnetic field distribution are simply not relevant for peak 1–8 Å flux, but may be more physically relevant to impending total energy release or spectral hardness; using these quantities to define an event may improve the forecast metrics.

4.3.4 Statistical approaches

Many statistical algorithms have been applied to this problem, yet there has not been convincing evidence that significant improvement can be made by using one classifier over another (Barnes et al., 2017; Nishizuka et al., 2017). There is still a wide discipline of feature identification and data mining yet to be exploited, including more autonomous expert systems. Yet for all possible advances, the fundamental limitations remain at the fore: flares are statistically rare events observed with limited remotely sensed data.

4.3.5 New data sources

The information available from the photosphere is limited. Research is needed to identify unique information that will reflect different preevent behavior (whether long-term or very short-term). Some preliminary work has looked at subsurface flows inferred from local helioseismic techniques (Reinard et al., 2010; Komm et al., 2011; Braun, 2016), and the chromosphere in Far-UV (Nishizuka et al., 2017), with mixed results: the former may have a small amount of independent information but is difficult to separate from the influence of the surface magnetic field; the latter appears to confirm the general relation between the presence of strong surface magnetic fields, their complexity and topology, and overall chromospheric and coronal brightness (e.g., Vaiana et al., 1973; van Driel-Gesztelyi et al., 1994; Fisher et al., 1998).

5 SUMMARY AND RECOMMENDATIONS

There are presently significant resources invested in developing improved forecasting methods for solar flares, but bluntly, the majority focus on implementing slightly more automated and quantitative versions of fairly old ideas. As described in Sawyer et al. (1986), even using multivariable statistical prediction algorithms was being tested and evaluated using quantitative skill scores more than 30 years ago. Since these studies, the routine use of magnetic field data and, in particular, vector magnetic field data is the most significant advance, coupled with the increase in computing power and algorithm improvement to make multivariable optimization over large datasets tenable. But these advances have not provided the definitive increase in performance that is sought. Above we demonstrate that improvement should not necessarily be expected from a new statistical classifier. Instead, progress will most likely come from new data sources that provide additional discriminating information, more emphasis on identifying unique trigger scenarios (even if there are multiple such scenarios), and perhaps more physically relevant forecasting timescales. And, of course, more events to study.

The desire, of course, is to not only predict when and how large a flare will occur, but whether it will or will not produce a geoeffective CME or a dangerous energetic particle event. Progress on the former is hampered by the difficulty of establishing an objective database of CME origination regions. Progress on the latter question is hampered not only by an extremely small sample size but a lack of understanding of SEP production physics.

Related, the statistical empirical approaches also present very strong constraints on how well any once-in-a-millennium flare could be forecast. Having no data with which to characterize distribution functions means that different techniques must be applied for any justifiable extrapolation. Yet would the millennial event originate from a "monster" region (extreme in its size and magnetic complexity)?

Or would it be caused by a very large region that somehow utilized a higher fraction of its stored energy than is typical, or otherwise behaved not as we expected?

Whether a new algorithm idea or a new operational method, we urge colleagues to report on results in this research discipline thoroughly. That includes but is not limited to: data requirements and limitations, event definitions, sample sizes for each event definition, optimization specifics for the statistical approach, and probability thresholds used for categorical forecasts and metrics reporting. We suggest providing a combination of metrics, including (but not necessarily limited to) Brier scores and ApSSs, reliability plots, and ROC plots with thresholds indicated, especially that threshold that results in the highest H& KSS.

In summary, the capability of daily probabilistic flare forecasting has not improved substantially in recent decades, and there is still little understood with regard to the physics of flare triggers. In the context of extreme solar flares, it will probably be straightforward to identify a potential source for such an event, that is, an extreme active region. However, it is still beyond our capability to forecast *when* a truly Extreme Solar Flare would occur.

ACKNOWLEDGMENTS

This review is dedicated to Dr. Patrick McIntosh, a mentor, educator, and lasting influence for Dr. Leka and the solar physics community. The authors thank Jim Secan (NWRA) and the referees for very useful suggestions. This material is based upon work supported by the National Science Foundation under Grant No. 1630454, by the NOAA/SBIR program under contracts WC-133R-13-CN-0079 and WC-133R-14-CN-0103, by the NASA/LWS contract NNH09CE72C. Support for writing this manuscript is also acknowledged from the Institute for Space-Earth Environmental Research, Nagoya University, and the NWRA Growth Fund.

REFERENCES

Abramenko, V.I., 2005a. Multifractal analysis of solar magnetograms. Sol. Phys. 228, 29–42. https://doi.org/10.1007/s11207-005-3525-9.

Abramenko, V.I., 2005b. Relationship between magnetic power spectrum and flare productivity in solar active regions. Astrophys. J. 629, 1141–1149. https://doi.org/10.1086/431732.

Ahmed, O.W., Qahwaji, R., Colak, T., Higgins, P.A., Gallagher, P.T., Bloomfield, D.S., 2013. Solar flare prediction using advanced feature extraction, machine learning, and feature selection. Sol. Phys. 283, 157–175. https://doi.org/10.1007/s11207-011-9896-1.

Al-Ghraibah, A., Boucheron, L.E., McAteer, R.T.J., 2015. An automated classification approach to ranking photospheric proxies of magnetic energy build-up. Astron. Astrophys. 579, A64. https://doi.org/10.1051/0004-6361/201525978.

Aschwanden, M.J., Freeland, S.L., 2012. Automated solar flare statistics in soft X-rays over 37 years of GOES observations: the invariance of self-organized criticality during three solar cycles. Astrophys. J. 754, 112. https://doi.org/10.1088/0004-637X/754/2/112.

Aschwanden, M.J., Crosby, N.B., Dimitropoulou, M., Georgoulis, M.K., Hergarten, S., McAteer, J., Milovanov, A.V., Mineshige, S., Morales, L., Nishizuka, N., Pruessner, G., Sanchez, R., Sharma, A.S.,

Strugarek, A., Uritsky, V., 2016. 25 Years of self-organized criticality: solar and astrophysics. Space Sci. Rev. 198, 47–166. https://doi.org/10.1007/s11214-014-0054-6.

Aulanier, G., Török, T., Démoulin, P., DeLuca, E.E., 2010. Formation of torus-unstable flux ropes and electric currents in erupting sigmoids. Astrophys. J. 708, 314–333. https://doi.org/10.1088/0004-637X/708/1/314.

Aulanier, G., Démoulin, P., Schrijver, C.J., Janvier, M., Pariat, E., Schmieder, B., 2013. The standard flare model in three dimensions. II. Upper limit on solar flare energy. Astron. Astrophys. 549, A66. https://doi.org/10.1051/0004-6361/201220406.

Baker, D., 2008. Severe Space Weather Events—Understanding Societal and Economic Impacts: A Workshop Report. The National Academies Press, Washington, DC. ISBN 0-309-12770-X. http://www.nap.edu/openbook.php?record_id=12507.

Bamba, Y., Kusano, K., Yamamoto, T.T., Okamoto, T.J., 2013. Study on the triggering process of solar flares based on Hinode/SOT observations. Astrophys. J. 778, 48. https://doi.org/10.1088/0004-637X/778/1/48.

Bamba, Y., Inoue, S., Kusano, K., Shiota, D., 2017. Triggering process of the X1.0 three-ribbon flare in the great active region NOAA 12192. Astrophys. J. 838, 134. https://doi.org/10.3847/1538-4357/aa6682.

Bamba, Y., Leka, K.D., Barnes, G., Kusano, K., 2018. Photospheric magnetic field properties of flaring vs. flare-quiet active regions. V: on the triggering by emerging flux. Astrophys. J. (in press).

Barnes, G., 2007. On the relationship between coronal magnetic null points and solar eruptive events. Astrophys. J. 670, L53–L56. https://doi.org/10.1086/524107.

Barnes, G., Leka, K.D., 2006. Photospheric magnetic field properties of flaring vs. flare-quiet active regions III: magnetic charge topology models. Astrophys. J. 646, 1303–1318. https://doi.org/10.1086/504960.

Barnes, G., Leka, K.D., 2008. Evaluating the performance of solar flare forecasting methods. Astrophys. J. 688, L107–L110. https://doi.org/10.1086/595550.

Barnes, G., Leka, K.D., Schumer, E.A., Della-Rose, D.J., 2007. Probabilistic forecasting of solar flares from vector magnetogram data. Space Weather 5, 9002. https://doi.org/10.1029/2007SW000317.

Barnes, G., Leka, K.D., Schrijver, C.J., Colak, T., Qahwaji, R., Ashamari, O.W., Yuan, Y., Zhang, J., McAteer, R.T.J., Bloomfield, D.S., Higgins, P.A., Gallagher, P.T., Falconer, D.A., Georgoulis, M.K., Wheatland, M.S., Balch, C., Dunn, T., Wagner, E.L., 2016. A comparison of flare forecasting methods. I: results from the "All-Clear" workshop. Astrophys. J. 829, 89. https://doi.org/10.3847/0004-637X/829/2/89.

Barnes, G., Schanche, N., Leka, K.D., Aggarwal, A., Reeves, K., 2017. A comparison of classifiers for solar energetic events. In: Brescia, M. (Ed.), IAU SymposiumIn: Astroinformatics, vol. 325.

Bauke, H., 2007. Parameter estimation for power-law distributions by maximum likelihood methods. Eur. Phys. J. B 58, 167–173. https://doi.org/10.1140/epjb/e2007-00219-y.

Baumann, G., Haugbølle, T., Nordlund, Å., 2013. Kinetic modeling of particle acceleration in a solar null-point reconnection region. Astrophys. J. 771, 93. https://doi.org/10.1088/0004-637X/771/2/93.

Benz, A.O., 2017. Flare observations. Living Rev. Sol. Phys. 14, 2. https://doi.org/10.1007/s41116-016-0004-3.

Bishop, C.M., 2006. Pattern Recognition and Machine Learning (Information Science and Statistics). Springer-Verlag, Dordrecht.

Bloomfield, D.S., Higgins, P.A., McAteer, R.T.J., Gallagher, P.T., 2012. Toward reliable benchmarking of solar flare forecasting methods. Astrophys. J. 747, L41. https://doi.org/10.1088/2041-8205.

Bloomfield, D.S., Gallagher, P.T., Marquette, W.H., Milligan, R.O., Canfield, R.C., 2016. Performance of major flare watches from the max millennium program (2001–2010). Sol. Phys. 291, 411–427. https://doi.org/10.1007/s11207-015-0833-6.

Bobra, M.G., Couvidat, S., 2015. Solar flare prediction using SDO/HMI vector magnetic field data with a machine-learning algorithm. Astrophys. J. 798, 135. https://doi.org/10.1088/0004-637X/798/2/135.

Bobra, M.G., Ilonidis, S., 2016. Predicting coronal mass ejections using machine learning methods. Astrophys. J. 821, 127. https://doi.org/10.3847/0004-637X/821/2/127.

Bornmann, P.L., Shaw, D., 1994. Flare rates and the McIntosh active-region classifications. Sol. Phys. 150, 127–146. https://doi.org/10.1007/BF00712882.

Boucheron, L.E., Al-Ghraibah, A., McAteer, R.T.J., 2015. Prediction of solar flare size and time-to-flare using support vector machine regression. Astrophys. J. 812, 51. https://doi.org/10.1088/0004-637X/812/1/51.

Braun, D.C., 2016. A helioseismic survey of near-surface flows around active regions and their association with flares. Astrophys. J. 819, 106. https://doi.org/10.3847/0004-637X/819/2/106.

Breiman, L., 2001. Random forests. Mach. Learn. 45, 5–32. https://doi.org/10.1023/A:1010933404324.

Cannon, P., 2013. Extreme Space Weather: Impacts on Engineered Systems and Infrastructure. Royal Academy of Engineering, Prince Philip House, London ISBN 1-903496-95-0.

Crown, M.D., 2012. Validation of the NOAA Space Weather Prediction Center's solar flare forecasting look-up table and forecaster-issued probabilities. Space Weather 10, S06006. https://doi.org/10.1029/2011SW000760.

Cui, Y., Li, R., Zhang, L., He, Y., Wang, H., 2006. Correlation between solar flare productivity and photospheric magnetic field properties. 1. Maximum horizontal gradient, length of neutral line, number of singular points. Sol. Phys. 237, 45–59. https://doi.org/10.1007/s11207-006-0077-6.

Cui, Y., Li, R., Wang, H., He, H., 2007. Correlation between solar flare productivity and photospheric magnetic field properties. II. Magnetic gradient and magnetic shear. Sol. Phys. 242, 1–8. https://doi.org/10.1007/s11207-007-0369-5.

DeRosa, M.L., Schrijver, C.J., Barnes, G., Leka, K.D., Lites, B.W., Aschwanden, M.J., Amari, T., Canou, A., McTiernan, J.M., Régnier, S., Thalmann, J.K., Valori, G., Wheatland, M.S., Wiegelmann, T., Cheung, M.C.M., Conlon, P.A., Fuhrmann, M., Inhester, B., Tadesse, T., 2009. A critical assessment of non-linear force-free field modeling of the solar corona for active region 10953. Astrophys. J. 696, 1780–1791. https://doi.org/10.1088/0004-637X.

DeRosa, M.L., Wheatland, M.S., Leka, K.D., Barnes, G., Amari, T., Canou, A., Gilchrist, S.A., Thalmann, J.K., Valori, G., Wiegelmann, T., Schrijver, C.J., Malanushenko, A., Sun, X., Régnier, S., 2015. The influence of spatial resolution on nonlinear force-free modeling. Astrophys. J. 811, 107. https://doi.org/10.1088/0004-637X/811/2/107.

des Jardins, A., Canfield, R., Longcope, D., McLinden, E., Dillman, A., 2009. Signatures of magnetic stress prior to three solar flares observed by Rhessi. Astrophys. J. 693, 886–893. https://doi.org/10.1088/0004-637X/693/1/886.

Devos, A., Verbeeck, C., Robbrecht, E., 2014. Verification of space weather forecasting at the regional warning center in Belgium. J. Space Weather Space Clim. 27 (27), A29. https://doi.org/10.1051/swsc/2014025.

D'Huys, E., Berghmans, D., Seaton, D.B., Poedts, S., 2016. The effect of limited sample sizes on the accuracy of the estimated scaling parameter for power-law-distributed solar data. Sol. Phys. 291, 1561–1576. https://doi.org/10.1007/s11207-016-0910-5.

Domingo, V., Fleck, B., Poland, A.I., 1995. The SOHO mission: an overview. Sol. Phys. 162, 1–37. https://doi.org/10.1007/BF00733425.

Drake, J.F., Swisdak, M., Che, H., Shay, M.A., 2006. Electron acceleration from contracting magnetic islands during reconnection. Nature 443, 553–556. https://doi.org/10.1038/nature05116.

Emslie, A.G., Dennis, B.R., Shih, A.Y., Chamberlin, P.C., Mewaldt, R.A., Moore, C.S., Share, G.H., Vourlidas, A., Welsch, B.T., 2012. Global energetics of thirty-eight large solar eruptive events. Astrophys. J. 759, 71. https://doi.org/10.1088/0004-637X/759/1/71.

Falconer, D.A., Moore, R.L., Gary, G.A., 2008. Magnetogram measures of total nonpotentiality for prediction of solar coronal mass ejections from active regions of any degree of magnetic complexity. Astrophys. J. 689, 1433–1442. https://doi.org/10.1086/591045.

Falconer, D., Barghouty, A.F., Khazanov, I., Moore, R., 2011. A tool for empirical forecasting of major flares, coronal mass ejections, and solar particle events from a proxy of active-region free magnetic energy. Space Weather 9, S04003. https://doi.org/10.1029/2009SW000537.

Falconer, D.A., Moore, R.L., Barghouty, A.F., Khazanov, I., 2012. Prior flaring as a complement to free magnetic energy for forecasting solar eruptions. Astrophys. J. 757, 32. https://doi.org/10.1088/0004-637X.

Falconer, D.A., Moore, R.L., Barghouty, A.F., Khazanov, I., 2014. MAG4 versus alternative techniques for forecasting active region flare productivity. Space Weather 12, 306–317. https://doi.org/10.1002/2013SW001024.

Falconer, D.A., Tiwari, S.K., Moore, R.L., Khazanov, I., 2016. A new method to quantify and reduce the net projection error in whole-solar-active-region parameters measured from vector magnetograms. Astrophys. J. 833, L31. https://doi.org/10.3847/2041-8213/833/2/L31.

Fisher, G., Jones, B. (Eds.), 2007. Integrating Space Weather Operations and Forecasts Into Aviation Operations: Policy Workshop Report. American Meterological Society.

Fisher, G.H., Longcope, D.W., Metcalf, T.R., Pevtsov, A.A., 1998. Coronal heating in active regions as a function of global magnetic variables. Astrophys. J. 508, 885–898. https://doi.org/10.1086/306435.

Fletcher, L., Dennis, B.R., Hudson, H.S., Krucker, S., Phillips, K., Veronig, A., Battaglia, M., Bone, L., Caspi, A., Chen, Q., Gallagher, P., Grigis, P.T., Ji, H., Liu, W., Milligan, R.O., Temmer, M., 2011. An observational overview of solar flares. Space Sci. Rev. 159, 19–106. https://doi.org/10.1007/s11214-010-9701-8.

Fry, D.J., Zapp, N., Biesecker, D., Leka, K.D., Barnes, G., de Koning, C.A. (Eds.), 2010. Proceedings of the First All-Clear Forecasting Workshop. Lyndon B. Johnson Space Center. National Aeronautics and Space Administration, Houston, TX.

Gallagher, P., Moon, Y.J., Wang, H., 2002. Active-region monitoring and flare forecasting. Sol. Phys. 209, 171–183.

Georgoulis, M.K., 2012. Are solar active regions with major flares more fractal, multifractal, or turbulent than others? Sol. Phys. 276, 161–181. https://doi.org/10.1007/s11207-010-9705-2.

Georgoulis, M.K., Rust, D.M., 2007. Quantitative forecasting of major solar flares. Astrophys. J. 661, L109–L112. https://doi.org/10.1086/518718.

Georgoulis, M.K., Vlahos, L., 1996. Coronal heating by nanoflares and the variability of the occurrence frequency in solar flares. Astrophys. J. 469, L135. https://doi.org/10.1086/310283.

Georgoulis, M.K., Vlahos, L., 1998. Variability of the occurrence frequency of solar flares and the statistical flare. Astron. Astrophys. 336, 721–734.

Gilchrist, S.A., Wheatland, M.S., Leka, K.D., 2012. The free energy of NOAA solar active region AR 11029. Sol. Phys. 276, 133–160. https://doi.org/10.1007/s11207-011-9878-3.

Hagyard, M.J., Smith, J.B.J., Teuber, D., West, E.A., 1984. A quantitative study relating observed shear in photospheric magnetic fields to repeated flaring. Sol. Phys. 91, 115–126.

Hale, G.E., Ellerman, F., Nicholson, S.B., Joy, A.H., 1919. The magnetic polarity of Sun-spots. Astrophys. J. 49, 153. https://doi.org/10.1086/142452.

Harra, L.K., Schrijver, C.J., Janvier, M., Toriumi, S., Hudson, H., Matthews, S., Woods, M.M., Hara, H., Guedel, M., Kowalski, A., Osten, R., Kusano, K., Lueftinger, T., 2016. The characteristics of solar X-class flares and CMEs: a paradigm for stellar superflares and eruptions? Sol. Phys. 291, 1761–1782. https://doi.org/10.1007/s11207-016-0923-0.

Hassanin, A., Kliem, B., 2016. Helical kink instability in a confined solar eruption. Astrophys. J. 832, 106. https://doi.org/10.3847/0004-637X/832/2/106.

Hayakawa, H., Tamazawa, H., Uchiyama, Y., Ebihara, Y., Miyahara, H., Kosaka, S., Iwahashi, K., Isobe, H., 2017. Historical auroras in the 1990s: evidence of great magnetic storms. Sol. Phys. 292, 12. https://doi.org/10.1007/s11207-016-1039-2.

Hill, F., Bolding, J., Toner, C., Corbard, T., Wampler, S., Goodrich, B., Goodrich, J., Eliason, P., Hanna, K.D., 2003. The GONG++ data processing pipeline. In: Sawaya-Lacoste, H. (Ed.), GONG+ 2002. Local and Global Helioseismology: The Present and Future, ESA Special Publication, vol. 517. pp. 295–298.

Hoeksema, J.T., Liu, Y., Hayashi, K., Sun, X., Schou, J., Couvidat, S., Norton, A., Bobra, M., Centeno, R., Leka, K.D., Barnes, G., Turmon, M., 2014. The helioseismic and magnetic imager (HMI) vector magnetic field pipeline: overview and performance. Sol. Phys. 289, 3483–3530. https://doi.org/10.1007/s11207-014-0516-8.

Hu, Q., Dasgupta, B., Choudhary, D.P., Büchner, J., 2008. A practical approach to coronal magnetic field extrapolation based on the principle of minimum dissipation rate. Astrophys. J. 679, 848–853. https://doi.org/10.1086/587639.

Hudson, H.S., 2012. Grand archive of flare and CME cartoons. Available from: http://solarmuri.ssl.berkeley.edu/hhudson/cartoons/.

Hudson, H.S., Fletcher, L., Fisher, G.H., Abbett, W.P., Russell, A., 2012. Momentum distribution in solar flare processes. Sol. Phys. 277, 77–88. https://doi.org/10.1007/s11207-011-9836-0.

Inoue, S., Hayashi, K., Shiota, D., Magara, T., Choe, G.S., 2013. Magnetic structure producing X- and M-class solar flares in solar active region 11158. Astrophys. J. 770, 79. https://doi.org/10.1088/0004-637X/770/1/79.

Inoue, S., Hayashi, K., Kusano, K., 2016. Structure and stability of magnetic fields in solar active region 12192 based on the nonlinear force-free field modeling. Astrophys. J. 818, 168. https://doi.org/10.3847/0004-637X/818/2/168.

Janvier, M., Aulanier, G., Démoulin, P., 2015. From coronal observations to MHD simulations, the building blocks for 3D models of solar flares (invited review). Sol. Phys. 290, 3425–3456. https://doi.org/10.1007/s11207-015-0710-3.

Jiang, C., Wu, S.T., Yurchyshyn, V., Wang, H., Feng, X., Hu, Q., 2016. How did a major confined flare occur in super solar active region 12192? Astrophys. J. 828, 62. https://doi.org/10.3847/0004-637X/828/1/62.

Jolliffe, I.T., Stephenson, D., 2003. Forecast Verification: A Practioner's Guide in Atmospheric Science. Wiley, New York.

Jones, H.P., Duvall Jr., T.L., Harvey, J.W., Mahaffey, C.T., Schwitters, J.D., Simmons, J.E., 1992. The NASA/NSO spectromagnetograph. Sol. Phys. 139, 211–232. https://doi.org/10.1007/BF00159149.

Kahler, S.W., Ling, A., White, S.M., 2015. Forecasting SEP events with same active region prior flares. Space Weather 13 (2), 116–123. https://doi.org/10.1002/2014SW001099.

Kliem, B., Török, T., 2006. Torus instability. Phys. Rev. Lett. 96 (25), 255002. https://doi.org/10.1103/PhysRevLett.96.255002.

Knipp, D.J., Ramsay, A.C., Beard, E.D., Boright, A.L., Cade, W.B., Hewins, I.M., McFadden, R.H., Denig, W.F., Kilcommons, L.M., Shea, M.A., Smart, D.F., 2016. The May 1967 great storm and radio disruption event: extreme space weather and extraordinary responses. Space Weather 14, 614–633. https://doi.org/10.1002/2016SW001423.

Knizhnik, K., Swisdak, M., Drake, J.F., 2011. The acceleration of ions in solar flares during magnetic reconnection. Astrophys. J. 743, L35. https://doi.org/10.1088/2041-8205/743/2/L35.

Komm, R., Ferguson, R., Hill, F., Barnes, G., Leka, K.D., 2011. Subsurface vorticity of flaring versus flare-quiet active regions. Sol. Phys. 268, 389–406. https://doi.org/10.1007/s11207-010-9552-1.

Kopp, R.A., Pneuman, G.W., 1976. Magnetic reconnection in the corona and the loop prominence phenomenon. Sol. Phys. 50, 85–98. https://doi.org/10.1007/BF00206193.

Kusano, K., Bamba, Y., Yamamoto, T.T., Iida, Y., Toriumi, S., Asai, A., 2012. Magnetic field structures triggering solar flares and coronal mass ejections. Astrophys. J. 760, 31. https://doi.org/10.1088/0004-637X/760/1/31.

Laurenza, M., Cliver, E.W., Hewitt, J., Storini, M., Ling, A.G., Balch, C.C., Kaiser, M.L., 2009. A technique for short-term warning of solar energetic particle events based on flare location, flare size, and evidence of particle escape. Space Weather 7, S04008. https://doi.org/10.1029/2007SW000379.

Leka, K.D., Barnes, G., 2003a. Photospheric magnetic field properties of flaring vs. flare-quiet active regions I: data, general analysis approach, and sample results. Astrophys. J. 595, 1277–1295.

Leka, K.D., Barnes, G., 2003b. Photospheric magnetic field properties of flaring vs. flare-quiet active regions II: discriminant analysis. Astrophys. J. 595, 1296–1306.

Leka, K.D., Barnes, G., 2007. Photospheric magnetic field properties of flaring vs. flare-quiet active regions. IV: a statistically significant sample. Astrophys. J. 656, 1173–1186. https://doi.org/10.1086/510282.

Leka, K.D., Barnes, G., Wagner, E.L., 2017a. Evaluating (and improving) estimates of the solar radial magnetic field component from line-of-sight magnetograms. Sol. Phys. 292, 36. https://doi.org/10.1007/s11207-017-1057-8.

Leka, K.D., Barnes, G., Wagner, E.L., 2017b. The NWRA classification infrastructure: description and extension to the discriminant analysis flare forecasting system (DAFFS). Journal of Space Weather and Space Climate. submitted for publication.

Linton, M.G., Dahlburg, R.B., Antiochos, S.K., 2001. Reconnection of twisted flux tubes as a function of contact angle. Astrophys. J. 553, 905–921. https://doi.org/10.1086/320974.

Liu, L., Wang, Y., Wang, J., Shen, C., Ye, P., Liu, R., Chen, J., Zhang, Q., Wang, S., 2016. Why is a flare-rich active region CME-poor? Astrophys. J. 826, 119. https://doi.org/10.3847/0004-637X/826/2/119.

Longcope, D.W., 2001. Separator current sheets: generic features in minimum-energy magnetic fields subject to flux constraints. Phys. Plasmas 8, 5277–5290. https://doi.org/10.1063/1.1418431.

Longcope, D.W., Beveridge, C., Qiu, J., Ravindra, B., Barnes, G., Dasso, S., 2007. Modeling and measuring the flux reconnected and ejected by the two-ribbon flare/CME event on 7 November 2004. Sol. Phys. 244, 45–73. https://doi.org/10.1007/s11207-007-0330-7.

Lu, E.T., Hamilton, R.J., 1991. Avalanches and the distribution of solar flares. Astrophys. J. 380, L89–L92. https://doi.org/10.1086/186180.

Maehara, H., Shibayama, T., Notsu, S., Notsu, Y., Nagao, T., Kusaba, S., Honda, S., Nogami, D., Shibata, K., 2012. Superflares on solar-type stars. Nature 485, 478–481. https://doi.org/10.1038/nature11063.

Maehara, H., Shibayama, T., Notsu, Y., Notsu, S., Honda, S., Nogami, D., Shibata, K., 2015. Statistical properties of superflares on solar-type stars based on 1-min cadence data. Earth Planets Space 67, 59. https://doi.org/10.1186/s40623-015-0217-z.

Mason, J.P., Hoeksema, J.T., 2010. Testing automated solar flare forecasting with 13 years of Michelson Doppler Imager magnetograms. Astrophys. J. 723, 634–640. https://doi.org/10.1088/0004-637X/723/1/634.

McAteer, R.T.J., Gallagher, P.T., Ireland, J., 2005. Statistics of active region complexity: a large-scale fractal dimension survey. Astrophys. J. 631, 628–635. https://doi.org/10.1086/432412.

McIntosh, P.S., 1990. The classification of sunspot groups. Sol. Phys. 125, 251–267.

McKenzie, D.E., 2002. Signatures of reconnection in eruptive flares [invited]. In: Martens, P.C.H., Cauffman, D. (Eds.), Multi-Wavelength Observations of Coronal Structure and Dynamics, p. 155.

Metcalf, T.R., DeRosa, M.L., Schrijver, C.J., Barnes, G., van Ballegooijen, A.A., Wiegelmann, T., Wheatland, M.S., Valori, G., McTtiernan, J.M., 2008. Nonlinear force-free modeling of coronal magnetic fields. II. Modeling a filament arcade and simulated chromospheric and photospheric vector fields. Sol. Phys. 247, 269–299. https://doi.org/10.1007/s11207-007-9110-7.

Mitchell, S., Schrijver, K., 2012. Solar-flare induced disturbances in the U.S. Electric Grid and Their Economic Impact.

Miyake, F., Masuda, K., Nakamura, T., 2013. Another rapid event in the carbon-14 content of tree rings. Nat. Commun. 4, 1748. https://doi.org/10.1038/ncomms2783.

Muranushi, T., Shibayama, T., Muranushi, Y.H., Isobe, H., Nemoto, S., Komazaki, K., Shibata, K., 2015. UFCORIN: a fully automated predictor of solar flares in GOES X-ray flux. Space Weather J.

Murphy, A.H., 1996. The Finley affair: a signal event in the history of forecast verification. Weather Forecasting 11, 3–20.

Murray, S.A., Bingham, S., Sharpe, M., Jackson, D.R., 2017. Flare forecasting at the Met Office Space Weather Operations Centre. Space Weather 15, 577–588. https://doi.org/10.1002/2016SW001579.

National Environmental Satellite, Data, and Information Service, 2017. Solar-terrestrial physics. Available from: https://www.ngdc.noaa.gov/stp/.

National Science and Technology Council, 2015. National Space Weather Strategy. https://obamawhitehouse.archives.gov/sites/default/files/microsites/ostp/final_nationalspaceweatherstrategy_20151028.pdf.

Neidig, D.F., Wiborg, P.H., Seagraves, P.H., Hirman, J.W., Flowers, W.E., 1981. An objective method for forecasting solar flares. Air Force Geophysics Lab.

Nindos, A., Andrews, M.D., 2004. The association of big flares and coronal mass ejections: what is the role of magnetic helicity? Astrophys. J. 616, L175–L178. https://doi.org/10.1086/426861.

Nishizuka, N., Sugiura, K., Kubo, Y., Den, M., Watari, S., Ishii, M., 2017. Solar flare prediction model with three machine-learning algorithms using ultraviolet brightening and vector magnetograms. Astrophys. J. 835, 156. https://doi.org/10.3847/1538-4357/835/2/156.

NOAA/Space Weather Prediction Center, 2011. NOAA Space Weather Scales. http://www.swpc.noaa.gov/noaa-scales-explanation.

Notsu, Y., Honda, S., Maehara, H., Notsu, S., Shibayama, T., Nogami, D., Shibata, K., 2015. High dispersion spectroscopy of solar-type superflare stars. II. Stellar rotation, starspots, and chromospheric activities. Publ. Astron. Soc. Jpn 67, 33. https://doi.org/10.1093/pasj/psv002.

Oka, M., Krucker, S., Hudson, H.S., Saint-Hilaire, P., 2015. Electron energy partition in the above-the-looptop solar hard X-ray sources. Astrophys. J. 799, 129. https://doi.org/10.1088/0004-637X/799/2/129.

Pesnell, W., 2008. The solar dynamics observatory: your eye on the Sun. In: 37th COSPAR Scientific Assembly, COSPAR, Plenary Meeting, vol. 37. p. 2412.

Poppe, B.B., Jorden, K.P., 2006. Sentinels of the Sun: Forecasting Space Weather. Johnson Books, Boulder, CO, USA.

Priest, E., 2016. MHD structures in three-dimensional reconnection. Magn. Reconnect. 427, 101. https://doi.org/10.1007/978-3-319-26432-5_3.

Priest, E.R., Forbes, T.G., 2002. The magnetic nature of solar flares. Astron. Astrophys. Rev. 10, 313–377. https://doi.org/10.1007/s001590100013.

Priest, E.R., Longcope, D.W., 2017. Flux-rope twist in eruptive flares and CMEs: due to zipper and main-phase reconnection. Sol. Phys. 292, 25. https://doi.org/10.1007/s11207-016-1049-0.

Qiu, J., 2009. Observational analysis of magnetic reconnection sequence. Astrophys. J. 692, 1110–1124. https://doi.org/10.1088/0004-637X/692/2/1110.

Raboonik, A., Safari, H., Alipour, N., Wheatland, M.S., 2017. Prediction of solar flares using unique signatures of magnetic field images. Astrophys. J. 834, 11. https://doi.org/10.3847/1538-4357/834/1/11.

Reinard, A.A., Henthorn, J., Komm, R., Hill, F., 2010. Evidence that temporal changes in solar subsurface helicity precede active region flaring. Astrophys. J. 710, L121–L125. https://doi.org/10.1088/2041-8205.

Riley, P., 2012. On the probability of occurrence of extreme space weather events. Space Weather. 10 (2). https://doi.org/10.1029/2011SW000734.

Russell, A.J.B., Mooney, M.K., Leake, J.E., Hudson, H.S., 2016. Sunquake generation by coronal magnetic restructuring. Astrophys. J. 831, 42. https://doi.org/10.3847/0004-637X/831/1/42.

Sawyer, C., Warwick, J.W., Dennett, J.T., 1986. Solar Flare Prediction. Colorado Assoc. Univ. Press, Boulder, CO.

Schaefer, B.E., King, J.R., Deliyannis, C.P., 2000. Superflares on ordinary solar-type stars. Astrophys. J. 529, 1026–1030. https://doi.org/10.1086/308325.

Scherrer, P.H., Bogart, R.S., Bush, R.I., Hoeksema, J.T., Kosovichev, A.G., Schou, J., Rosenberg, W., Springer, L., Tarbell, T.D., Title, A., Wolfson, C.J., Zayer, I., MDI Engineering Team, 1995. The solar oscillations investigation–Michelson Doppler Imager. Sol. Phys. 162, 129–188. https://doi.org/10.1007/BF00733429.

Schou, J., Scherrer, P.H., Bush, R.I., Wachter, R., Couvidat, S., Rabello-Soares, M.C., Bogart, R.S., Hoeksema, J.T., Liu, Y., Duvall, T.L., Akin, D.J., Allard, B.A., Miles, J.W., Rairden, R., Shine, R.A., Tarbell, T.D., Title, A.M., Wolfson, C.J., Elmore, D.F., Norton, A.A., Tomczyk, S., 2012. Design and ground calibration of the helioseismic and magnetic imager (HMI) instrument on the solar dynamics observatory (SDO). Sol. Phys. 275, 229–259. https://doi.org/10.1007/s11207-011-9842-2.

Schrijver, C.J., 2007. A characteristic magnetic field pattern associated with all major solar flares and its use in flare forecasting. Astrophys. J. 655, L117–L120. https://doi.org/10.1086/511857.

Schrijver, C.J., DeRosa, M.L., Metcalf, T.R., Liu, Y., McTiernan, J., Régnier, S., Valori, G., Wheatland, M.S., Wiegelmann, T., 2006. Nonlinear force-free modeling of coronal magnetic fields. Part I: a quantitative comparison of methods. Sol. Phys. 235, 161–190. https://doi.org/10.1007/s11207-006-0068-7.

Schrijver, C.J., DeRosa, M.L., Metcalf, T., Barnes, G., Lites, B., Tarbell, T., McTiernan, J., Valori, G., Wiegelmann, T., Wheatland, M.S., Amari, T., Aulanier, G., Démoulin, P., Fuhrmann, M., Kusano, K., Régnier, S., Thalmann, J.K., 2008. Nonlinear force-free field modeling of a solar active region around the time of a major flare and coronal mass ejection. Astrophys. J. 675, 1637–1644. https://doi.org/10.1086/527413.

Severnyi, A.B., 1965. The nature of solar magnetic fields (the fine structure of the field). Soviet Astron. 9, 171.

REFERENCES

Shibata, K., Magara, T., 2011. Solar flares: magnetohydrodynamic processes. Living Rev. Sol. Phys. 8, 6. https://doi.org/10.12942/lrsp-2011-6.

Shibata, K., Isobe, H., Hillier, A., Choudhuri, A.R., Maehara, H., Ishii, T.T., Shibayama, T., Notsu, S., Notsu, Y., Nagao, T., Honda, S., Nogami, D., 2013. Can superflares occur on our Sun? Publ. Astron. Soc. Jpn 65, 49. https://doi.org/10.1093/pasj/65.3.49.

Shibayama, T., Maehara, H., Notsu, S., Notsu, Y., Nagao, T., Honda, S., Ishii, T.T., Nogami, D., Shibata, K., 2013. Superflares on solar-type stars observed with Kepler. I. Statistical properties of superflares. Astrophys. J. Supp. Ser. 209, 5. https://doi.org/10.1088/0067-0049/209/1/5.

Silverman, B.W., 1986. Density Estimation for Statistics and Data Analysis. Chapman and Hall, London.

Smith Jr., J.B., Neidig, D.F., Wiborg, P.H., West, E.A., Hagyard, M.J., Adams, M., Seagraves, P.H., 1996. An objective test of magnetic shear as a flare predictor. In: Solar Drivers of Interplanetary and Terrestrial Disturbance, ASP Conference Ser, vol. 95. pp. 54–65.

Song, H., Tan, C., Jing, J., Wang, H., Yurchyshyn, V., Abramenko, V., 2009. Statistical assessment of photospheric magnetic features in imminent solar flare predictions. Sol. Phys. 254, 101–125. https://doi.org/10.1007/s11207-008-9288-3.

Stepanov, A.V., Zaitsev, V.V., 2016. The challenges of the models of solar flares. Geomagn. Aeron. 56, 952–971. https://doi.org/10.1134/S001679321608020X.

Strugarek, A., Charbonneau, P., 2014. Predictive capabilities of avalanche models for solar flares. Sol. Phys. 289, 4137–4150. https://doi.org/10.1007/s11207-014-0570-2.

Strugarek, A., Charbonneau, P., Joseph, R., Pirot, D., 2014. Deterministically driven avalanche models of solar flares. Sol. Phys. 289, 2993–3015. https://doi.org/10.1007/s11207-014-0509-7.

Sun, X., Hoeksema, J.T., Liu, Y., Wiegelmann, T., Hayashi, K., Chen, Q., Thalmann, J., 2012. Evolution of magnetic field and energy in a major eruptive active region based on SDO/HMI observation. Astrophys. J. 748, 77. https://doi.org/10.1088/0004-637X/748/2/77.

Sun, X., Bobra, M.G., Hoeksema, J.T., Liu, Y., Li, Y., Shen, C., Couvidat, S., Norton, A.A., Fisher, G.H., 2015. Why is the great solar active region 12192 flare-rich but CME-poor? Astrophys. J. 804, L28. https://doi.org/10.1088/2041-8205/804/2/L28.

Tarr, L., Longcope, D., 2012. Calculating energy storage due to topological changes in emerging active region NOAA AR11112. Astrophys. J. 749, 64. https://doi.org/10.1088/0004-637X/749/1/64.

Tarr, L., Longcope, D., Millhouse, M., 2013. Calculating separate magnetic free energy estimates for active regions producing multiple flares: NOAA AR11158. Astrophys. J. 770, 4. https://doi.org/10.1088/0004-637X/770/1/4.

Titov, V.S., Démoulin, P., 1999. Basic topology of twisted magnetic configurations in solar flares. Astron. Astrophys. 351, 707–720.

Ulrich, R.K., Evans, S., Boyden, J.E., Webster, L., 2002. Mount Wilson synoptic magnetic fields: improved instrumentation, calibration, and analysis applied to the 2000 July 14 flare and to the evolution of the dipole field. Astrophys. J. Supp. Ser. 139, 259–279. https://doi.org/10.1086/337948.

Vaiana, G.S., Krieger, A.S., Timothy, A.F., 1973. Identification and analysis of structures in the corona from X-ray photography. Sol. Phys. 32, 81–116. https://doi.org/10.1007/BF00152731.

van Driel-Gesztelyi, L., Hoffman, A., Démoulin, P., Schmieder, B., Csepura, G., 1994. Relationship between electric currents, photospheric motions, chromospheric activity, and magnetic field topology. Sol. Phys. 149, 309–330.

Vincent, A., Charbonneau, P., Dubé, C., 2012. Numerical simulation of a solar active region. I: Bastille Day Flare. Sol. Phys. 278, 367–391. https://doi.org/10.1007/s11207-012-9953-4.

Vlahos, L., Georgoulis, M., Kluiving, R., Paschos, P., 1995. The statistical flare. Astron. Astrophys. 299, 897.

Welsch, B.T., Li, Y., Schuck, P.W., Fisher, G.H., 2009. What is the relationship between photospheric flow fields and solar flares? Astrophys. J. 705, 821–843. https://doi.org/10.1088/0004-637X/705/1/821.

Wheatland, M.S., 2001. Rates of flaring in individual active regions. Sol. Phys. 203, 87–106. https://doi.org/10.1023/A:1012749706764.

Wheatland, M.S., 2004. A Bayesian approach to solar flare prediction. Astrophys. J. 609, 1134–1139. https://doi.org/10.1086/421261.

Wheatland, M.S., 2005. A statistical solar flare forecast method. Space Weather 3, S07003. https://doi.org/10.1029/2004SW000131.

Woodcock, F., 1976. The evaluation of yes/no forecasts for scientific and administrative purposes. Mon. Weather Rev. 104, 1209–1214. https://doi.org/10.1175/1520-0493(1976)104.

Wu, C.J., Ip, W.H., Huang, L.C., 2015. A study of variability in the frequency distributions of the superflares of G-type stars observed by the Kepler mission. Astrophys. J. 798, 92. https://doi.org/10.1088/0004-637X/798/2/92.

Wu, S.T., Zhou, Y., Jiang, C., Feng, X., Wu, C.C., Hu, Q., 2016. A data-constrained three-dimensional magnetohydrodynamic simulation model for a coronal mass ejection initiation. J. Geophys. Res. 121, 1009–1023. https://doi.org/10.1002/2015JA021615.

Zhao, J., Li, H., Pariat, E., Schmieder, B., Guo, Y., Wiegelmann, T., 2014. Temporal evolution of the magnetic topology of the NOAA active region 11158. Astrophys. J. 787, 88. https://doi.org/10.1088/0004-637X/787/1/88.

Zharkova, V.V., Siversky, T., 2015. Particle acceleration in 3D single current sheets formed in the solar corona and heliosphere: PIC approach. J. Phys. Conf. Ser. 642, 012032. https://doi.org/10.1088/1742-6596/642/1/012032.

CHAPTER 4

GEOEFFECTIVENESS OF SOLAR AND INTERPLANETARY STRUCTURES AND GENERATION OF STRONG GEOMAGNETIC STORMS

Yuri I. Yermolaev, Irina G. Lodkina, Nadezhda S. Nikolaeva[1], Michael Yu. Yermolaev

Space Research Institute, Moscow, Russia

CHAPTER OUTLINE

1 Introduction	99
2 Methods	100
3 Results	101
4 Discussion	106
5 Conclusions	109
Acknowledgments	110
References	110

1 INTRODUCTION

One of the manifestations of extreme events in the magnetosphere of the Earth is a magnetic storm, which is generally shown by the intensification of the ring current, the depression of the geomagnetic field, and the reduction of the Dst index (the horizontal H component of the magnetic field measured at four low-latitude ground stations) (Sugiura, 1964; Sugiura and Kamei, 1991; Gonzatez et al., 1996; Daglis et al., 1999, and also Chapter 7 of this volume). As has been discovered by direct space experiments in the early 1970s, magnetic storms are generated only when the interplanetary magnetic field (IMF) has the southward component (Dungey, 1961; Fairfield and Cahill, 1966; Rostoker and Falthammar, 1967;

---
[1]Deceased

Russell et al., 1974; Burton et al., 1975; Akasofu, 1981). As IMF in quasistationary SW types is oriented approximately parallel to the solar equator, only disturbed SW types can contain an IMF component perpendicular to the ecliptic plane and induce the magnetic storm on Earth. Such disturbed SW types are: (1, 2) ICMEs, including magnetic clouds (MC) and non-MC Ejecta; (3) Sheath, a compression region before the fast leading edge of ICME; and (4) CIR, a compression region before the fast SW stream from the coronal hole, as well as a combination of these disturbed types, especially Sheath+ICME (see reviews and recent papers, for instance, by Tsurutani and Gonzalez (1997), Gonzalez et al. (1999), Huttunen and Koskinen (2004), Yermolaev and Yermolaev (2006), Borovsky and Denton (2006), Tsurutani et al. (2006), Pulkkinen et al. (2007a,b), Zhang et al. (2007), Yermolaev et al. (2010a,b, 2012, 2013, 2015), Yermolaev and Yermolaev (2010), Gonzalez et al. (2011), Kilpua et al. (2015), and references therein). The study and prediction of magnetic storms include analysis of all chains of physical processes from the Sun to the geomagnetosphere. The aim of this paper is to investigate solar and interplanetary sources of magnetic storms, including conditions resulting in extreme magnetic storms.

Recently we studied distributions and waiting times T of a generation of extreme magnetic storms for a total storm dataset over the period 1963–2012 as well as for storms induced by different interplanetary drivers over the period 1976–2000 (Yermolaev et al., 2013). We assumed that tails of distributions have a power law shape (approximations in log-log space are used in papers by Love et al. (2015) and Riley and Love (2017); for the effects of distribution shape please see also Chapter 5 of this volume). We calculated tails of distributions in the area of moderate and strong storms where statistics of events are sufficiently high, and then extrapolated the distributions to the area of extreme magnetic storms where statistics are poor. Such an approach allowed us to obtain two estimations of waiting times: (1) on the basis of the total storm dataset, we found that the most probable estimations of waiting times T for extreme magnetic storms with minimum Dst < −500, −1000, and −1700 nT are ∼20, ∼250, and ∼1500 years (with accuracy of factors ∼1.5, ∼2, and ∼3), respectively; and (2) on the basis of storms induced by the most geoeffective driver MC, we found lower limits of waiting times for extreme storms T_{MC} (−1000) ∼120 and T_{MC} (−1700) ∼500 years. In the present work we increased the analyzed intervals: 1963–2015 and 1976–2005. An increase in statistics allows us to investigate the resistance of earlier obtained results to the change of dataset volume and reliability of the conclusions.

2 METHODS

For analysis we use plasma parameters and interplanetary magnetic field data from the OMNI dataset (http://omniweb.gsfc.nasa.gov, King and Papitashvili, 2004), data on the Dst index for the period 1963–2015 (see http://wdc.kugi.kyoto-u.ac.jp/index.html), and our catalog of large-scale solar wind phenomena covering 1976–2005 (see ftp://www.iki.rssi.ru/pub/omni, Yermolaev et al., 2009) obtained with OMNI data. The technique for identification of solar wind phenomena and determining the connection between magnetic storms and their interplanetary drivers is described in detail in papers (Yermolaev et al., 2009, 2010a).

To calculate the average temporal profile of parameters in the phenomena with different durations, we use the double method of superposed epoch analysis. The time between points of interval was rescaled (proportionally increased/decreased) in such a way that respective beginnings and ends for all intervals of a selected data type coincide (Yermolaev et al., 2010b, 2015). Recently we calculated the average temporal profiles of several interplanetary and magnetospheric parameters for eight usual sequences of phenomena: (1) SW/CIR/SW, (2) SW/IS/CIR/SW, (3) SW/Ejecta/SW, (4) SW/Sheath/Ejecta/SW, (5) SW/IS/Sheath/Ejecta/SW, (6) SW/MC/SW, (7) SW/Sheath/MC/SW, and (8) SW/IS/Sheath/MC/SW (where SW is undisturbed solar wind, IS means interplanetary shock) (Yermolaev et al., 2015) and these results are used in this work.

To study the behavior of storms as a function of storm intensity, we calculated two types of frequency distributions of events based on observations: standard and integral. Standard distribution is derived in the form $dN = F(x)dx$, where dN is the number of events recorded with the parameter x of interest between x and $x+dx$, and $F(x)$ is frequency distributions (see recent reviews and papers by Koons (2001), Tsubouchi and Omura (2007), Crosby (2011), Gorobets and Messerotti (2012), Riley (2012), Love (2012), Yermolaev et al. (2013), Love et al. (2015), Riley and Love (2017), and references therein). In addition to the standard (differential) distribution function, we calculated the integral distribution function $F^*(x)$ which presents the number of events recorded between x and infinity. It should be noted that the integral distribution function $F^*(x)$ has higher statistics in the semiinfinity intervals (between x and infinity) and obtained results have higher statistical significance. For the storms with large intensities, the number of storms is low, and the integral distribution function of magnetic storms F^* has better statistics and accuracy than the standard distribution function F. It allows us to estimate the probability and waiting time for the extreme magnetic storms. The integral probability is calculated using the formula $P_j(Dst0) = K_j(Dst0)/N_j$, where N_j is the total number of solar wind events of type j observed during the period 1976–2005; $K_j(Dst0)$ is the number of solar wind events of type j that induced magnetic storms with minimum Dst < Dst0. The waiting time is the average period between consecutive observations of magnetic storms with intensity Dst_{min} < Dst0. The waiting time for data without a solar wind driver selection $T(Dst0)$ is calculated as $T(Dst0) = t_1/F^*(Dst0)$, where $t_1 = 53$ years is the duration of the period of observation, 1963–2015, and with driver selection $T_j(Dst0) = \left[P_j(Dst0)N_j/(t_2^* r)\right]^{-1}$, where $t_2 = 30$ years (duration of observations from 1976 to 2005) and $r = 0.70$ (ratio of durations of data existence to total duration of observation).

3 RESULTS

Fig. 1 gives an overview of the solar and magnetospheric data for the 40-year interval between 1976 and 2015, including 1635–2160 Carrington rotations of the Sun (~27 days) and four maxima of 21–24 solar cycles. The left panel shows the time variations of solar X-ray flares of strong M (6313 events) and extreme X (490 events) classes on the visible side of the Sun; the right panel presents the moderate ($-100 < Dst_{min} < -50$ nT; 762 events) and strong ($Dst_{min} < -100$ nT; 273 events) magnetic storms. The ratio of the number of flares to the number of storms is about seven. After the exclusion of flares located far 45° from the Sun-Earth line, the ratio is about three. There are cases when flares of C class are considered as a possible solar source of magnetosphere disturbances. Thus, the number of flares is significantly higher than the number of magnetic storms. This is the main reason for a high number of false alarms for storm predictions made on the basis of solar flare observations.

Flare-storm relation had been suggested after the Carrington event in 1859 when, after a strong solar flare, the strongest magnetic storm was recorded on Earth (Carrington, 1859; Tsurutani et al., 2003). Fig. 1 shows that solar-magnetosphere relationships are more complicated. Long-term values of both parameters (i,e., averaged over several Carrington rotations) demonstrate a good correlation but it is difficult to find a point-to-point correspondence between data on the left and right panels. It means that

102 CHAPTER 4 GEOEFFECTIVENESS OF SOLAR AND INTERPLANETARY STRUCTURES

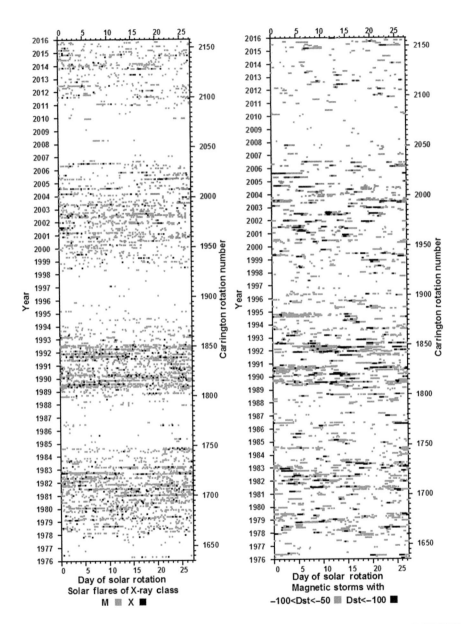

FIG. 1

Time variations in solar X-ray flares (on the left) and magnetic storms (on the right) for the period 1976–2015.

there is not a clear functional relation between them but they can have a common source on the Sun. On one hand, it is well known that flares and CMEs are different physical processes and geomagnetic activity is induced by CMEs/ICMEs but not by flares (Gosling, 1993). On the other hand, the largest solar flares are associated with CMEs and the fraction of solar flares increases from 20% for flares of

3 RESULTS

classes between C3.0 and C9.0 to 100% for the strong (greater of X3.0) solar flares (Yashiro et al., 2005). So, there may be an indirect connection between extreme solar flares and geomagnetic storms.

Analysis of magnetospheric data like in Fig. 1 shows the existence of magnetic storms that are not preceded by any solar activity (including solar flares) and that demonstrate a 27-day period in storm occurrence (i.e., these storms are recurrent). As is well known, the sources of such storms are coronal holes that can live several months and emit fast streams of SW. The fast SW streams can interact with preceding slow SW streams from the coronal streamer belt and form the geoeffective compression region CIR with a southward component of IMF (e.g., Borovsky and Denton, 2010). So, there are two solar sources (CMEs and coronal holes) resulting in four geoeffective SW types (MC, Ejecta, Sheath, and CIR).

Fig. 2 shows the temporal profiles of measured and pressure-corrected (Burton et al., 1975) Dst and Dst* indices in the most typical sequences of these nonstationary types of solar wind: (1) SW/CIR/SW, (2) SW/IS/CIR/SW, (3) SW/Ejecta/SW, (4) SW/Sheath/Ejecta/SW, (5) SW/IS/Sheath/Ejecta/SW, (6) SW/MC/SW, (7) SW/Sheath/MC/SW, and (8) SW/IS/Sheath/MC/SW (where SW is undisturbed

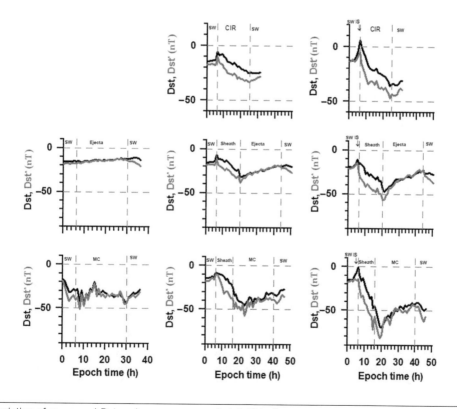

FIG. 2

Time variation of measured Dst and pressure-corrected Dst* indices averaged by the double method of superposed epoch analysis (Yermolaev et al., 2009) in eight potentially geoeffective sequences of SW events: (1) CIR, (2) IS/CIR, (3) Ejecta, (4) Sheath/Ejecta, (5) IS/Sheath/Ejecta, (6) MC, (7) Sheath/MC, and (8) IS/Sheath/MC. *Dashed vertical lines* show boundaries of SW types which are indicated in the upper parts of each panel.

solar wind, and IS means interplanetary shock) (Yermolaev et al., 2015) obtained by the method of double-superposed epoch analysis (Yermolaev et al., 2010b) for all (independent of storm generation) events over the period 1976–2000. The time variation of the Dst index is close to the inverted distribution (histogram) of a number of storms (minima of Dst index) in the corresponding sequences of nonstationary types of solar wind: the higher number of Dst minima in the time bin and the lower value of the average Dst index in this bin (Yermolaev et al., 2017). Such behavior of distribution of storms can be explained by the following reasons: (1) the magnitude of the magnetic field is higher in Sheath that comes before Ejecta in comparison with the field in Ejecta, and the magnitude of the magnetic field in Sheath before MC is close to the magnitude of the field in MC (Yermolaev et al., 2015); (2) the level of fluctuations of parameters in Sheath is higher than in ICME; and (3) the pressure in Sheath is higher than in ICME (differences between Dst and Dst* indices in CIR and Sheath in Fig. 2 are explained by higher pressure in compression regions in comparison with ICME).

Many studies investigated the so-called "CME-induced" storms as an independent type of storm (e.g., Borovsky and Denton 2006; Turner et al., 2009; Keesee et al., 2013). The second and third rows of Fig. 2 show that the contribution of Sheath in storm generation can be significant for "CME-induced" storms: (1) the maxima of storms (minima of Dst index) are observed in the end of Sheath and the beginning of ICME, and (2) the fraction of time of the storm main phase associated with Sheath is 33%, 40%, 52%, and 57%, respectively, in sequences SW/IS/Sheath/MC/SW, SW/Sheath/MC/SW, SW/IS/Sheath/Ejecta/SW, and SW/Sheath/Ejecta/SW. So, Sheath-induced storms should be identified and analyzed separately.

It is important to note that CMEs can interact with other CMEs and SW transient streams if they have different speeds (see, for instance, papers by Gopalswamy et al. (2004), Mishra et al. (2015), Mostl et al. (2015), and references therein). Also, interacting (multiple) ICMEs can be sources of strong magnetic storms with $Dst < -100$ nT (Farrugia et al., 2006a,b; Xie et al., 2006), because the magnitude and components of IMF in Sheath and ICME can increase due to compression in such complex events. A number of papers showed that several extreme magnetic storms (see, for instance, events on March 31, 2001, minimum Dst value of -387 nT; April 11–13, 2001, $Dst_{min} = -271$ nT (Wang et al., 2003); October 28–30, 2003, $Dst_{min} = -363$ nT (Veselovsky et al., 2004; Skoug et al., 2004); November 20, 2003, $Dst_{min} = -472$ nT (Ermolaev et al., 2005); November 8–10, 2004, $Dst_{min} = -373$ nT (Yermolaev et al., 2005)) have been generated by multiple interacting magnetic clouds. Because of the small number of such events, we do not mark out these events in a separate class, and we analyze them together with usual Sheaths and ICMEs in noninteracting ICMEs.

As we indicated in Section 1, here we make storm analyses similar to one made in our previous paper (Yermolaev et al., 2013). But in contrast with that previous work, we increase the number of all storms from 1963–2012 to 1963–2015 (~6%) and selected storms from 1976–2000 to 1976–2005 (~20%). Tables 1 and 2 present the number of storms (N) and values of standard (F) and integral (F^*) distributions in corresponding intervals of magnetic storms of different intensity ($|Dst_{min}|$). The presentation of this data both for intervals analyzed in our previous paper and for wider intervals analyzed in this study allows us to estimate the changes connected with the increase in intervals of observation. Fig. 3 shows distributions F and F^* for magnetic storms. Standard and integral distribution functions F and F^* have similar shapes: there are the flat tops at small values of $|Dst|$ and the power law tails at large values of $|Dst|$. The curves for the total period and its part are close to each other in agreement with our published results, and this fact allows us to suggest that the statistical characteristics of the parameters are similar for the total (1963–2015) and shorter (1976–2005) periods of observation. These data show that both distributions F and F^* have power law tails. We can approximate the tail of F^* in the $|Dst0|$ range of

Table 1 Distribution Functions F of Magnetic Storms Occurrence for 1963–2012/1963–2015 and 1976–2000/1976–2005 Intervals

|Dst| Window	1963–2012/1963–2015		1976–2000/1976–2005	
	N	F	N	F
50–75	545/607	0.4735/0.4895	319/412	0.4288/0.4598
75–100	290/311	0.2520/0.2508	211/228	0.2836/0.2545
100–125	135/136	0.1173/0.1097	92/106	0.1237/0.1183
125–150	73/75	0.0634/0.0605	49/63	0.0659/0.0703
150–175	40/40	0.0348/0.0323	30/30	0.0403/0.0335
175–200	19/19	0.0165/0.0153	11/14	0.0148/0.0156

Table 2 Integral Distribution Functions F^* of Magnetic Storm Occurrences for 1963–2012/1963–2015 and 1976–2000/1976–2005 Intervals

|Dst| Window	1963–2012/1963–2015		1976–2000/1976–2005	
	N	F^*	N	F^*
≥ 50	1151/1240	1.0000/1.0000	744/896	1.0000/1.0000
≥ 70	691/716	0.6003/0.5774	477/541	0.6411/0.6038
≥ 100	316/322	0.2745/0.2597	214/256	0.2876/0.2857
≥ 150	108/110	0.0938/0.0887	73/87	0.0981/0.0971
≥ 200	49/52	0.0426/0.0419	32/43	0.0430/0.0480
≥ 250	22/24	0.0191/0.0194	13/22	0.0175/0.0246
≥ 300	11/11	0.0096/0.0089	–/10	–/0.0112
≥ 350	7/7	0.0061/0.0056	–/6	–/0.0067

(150–350 nT) by a power law with index $\gamma(F^*) = -3.4 \pm 0.3$ with sufficient statistics (see Table 2). The power law approximation of the standard distribution function F has index $\gamma(F) = -4.4 \pm 0.5$, i.e., $\gamma(F^*) - \gamma(F) = 1$ in good agreement with the definition of F^* and F functions.

In contrast to Tables 1 and 2 and Fig. 3, where we present the data for magnetic storms independently of their interplanetary sources, Table 3 and Fig. 4 show data for various types of solar wind: CIR, green crosses; Sheath, blue triangles; MCs, red squares; Ejecta, yellow open squares; and ICMEs (ME = MC+Ejecta), pink circles. In agreement with our previous results, the highest probability is observed for the magnetic clouds. As the |Dst0| index increases from 50 to 200 nT, the probability falls by almost one order of magnitude. The probability for other types is less by 3–4 times at |Dst0| = 50 and 5–15 times at 200 nT, (i.e., the probability for the magnetic clouds decreases more slowly with increasing |Dst0| index than for other drivers). The only essential difference from the previous results is the higher number (and probability) of the strong storms with |Dst| > 150 nT generated by compression regions for Sheath cases. The analysis shows that the most part of such events in additional time interval are observed during well-known extreme events of October-November of 2003 and 2004.

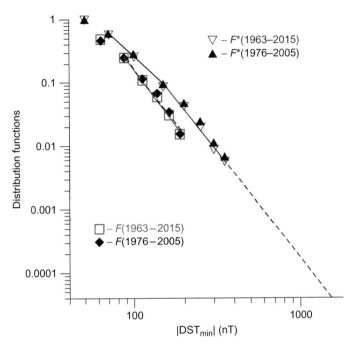

FIG. 3

Standard *F* and integral *F** distributions functions of magnetic storm occurrences for the periods 1963–2015 and 1976–2005.

4 DISCUSSION

Data from Figs. 3 and 4 allow us to calculate the waiting time for the magnetic storms with intensity higher than some fixed value. We calculate the waiting times for three levels of storms: with values of Dst < −500 nT (close to the extreme storm on March 13, 1989), and < −1000 nT and < −1700 nT (the low and high estimations of the Carrington storm on September 1, 1859 (Siscoe et al., 2006; Tsurutani et al., 2003), respectively). The integral distribution F^* in Fig. 3 has a power law tail that does not change with an increase in the number of storms (the observation interval). The waiting times estimated on the basis of F^* for the total set of storms are $T(-500$ nT$) \sim 24$ years, $T(-1000$ nT$) \sim 250$ years, and $T(-1700$ nT$) \sim 1500$ years (with error 10%, 40%, and 100%), respectively. Statistics of the sets for storms induced by various interplanetary drivers are not good enough to make reliable approximations of power law tails. To take into account a possibility that the probability decreases faster than the power law (linear approximation in log-log space (thin lines) in Fig. 5), we also make a square approximation in log-log space (thick lines). The lowest waiting times (e.g., the highest probability) are observed for MCs (T_{MC}). The most probable estimations of waiting times for the extreme magnetic storms with minimum Dst ≤ −500, −1000, and −1700 nT are ~ 20, ~ 120, and ~ 500 years (with the accuracy of factors ~ 1.5, ~ 2, and ~ 3), respectively. It is necessary to note that although the statistics for separate interplanetary drivers are poorer and errors are very large, the differences between $T_{MC}(-1000$ nT$)$

Table 3 Probabilities P_j Generation of Magnetic Storms With $|Dst| \geq 50, 70, 100, 150,$ and 200 nT for Different Types of Interplanetary Drivers for 1976–2000/1976–2005 Intervals

Types of Drivers	N_j	1976–2000/1976–2005									
		$|Dst| \geq 50$		$|Dst| \geq 70$		$|Dst| \geq 100$		$|Dst| \geq 150$		$|Dst| \geq 150$	
		K_j	P_j	K_j	P_j	K_j	P_j	K_j	P_j	K_j	P_j
CIR	718/836	120/142	0.167/0.170	66/79	0.092/0.095	24/31	0.033/0.037	6/6	0.084/0.007	5/5	0.007/0.006
Sheath	642/749	93/127	0.145/0.170	72/95	0.112/0.127	37/50	0.058/0.067	16/23	0.026/0.031	7/13	0.011/0.017
MC	101/132	57/76	0.564/0.576	44/63	0.436/0.477	28/42	0.227/0.319	14/19	0.139/0.144	7/9	0.069/0.068
Ejecta	1127/1258	139/162	0.123/0.129	89/98	0.079/0.078	45/48	0.040/0.039	12/12	0.011/0.009	3/4	0.003/0.003
ME	1228/1390	196/238	0.159/0.171	133/161	0.108/0.116	73/91	0.059/0.066	26/30	0.021/0.022	10/13	0.008/0.009

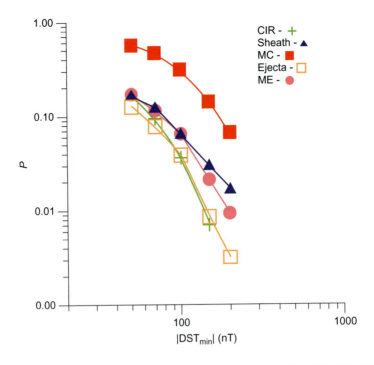

FIG. 4

Dependence of integral probability $P(|Dst| > Dst0)$ of the storm occurrence with $|Dst| > Dst0$ for different types of solar wind.

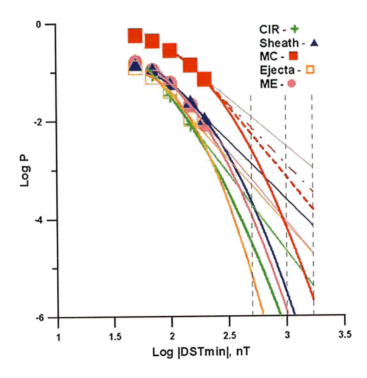

FIG. 5

Approximations of the integral probability $P(|Dst| > Dst0)$ dependence of storm occurrence with $Dst < Dst_{min}$ for different types of solar wind. *Vertical dashed lines* mark storms with $Dst < -500$ nT, -1000 nT, and -1700 nT.

~120, $T_{MC}(-1700\text{ nT}) \sim 500$ and $T(-1000\text{ nT}) \sim 250$, $T(-1700\text{ nT}) \sim 1500$ years are within the limits of the errors. We believe that estimations $T(-1000\text{ nT}) \sim 250$ years and $T(-1700\text{ nT}) \sim 1500$ years are more realistic and estimations $T_{MC}(-1000\text{ nT}) \sim 120$ years and $T_{MC}(-1700\text{ nT}) \sim 500$ years are lower limits of waiting times for extreme storms. It should be noted that the addition of the extreme storms of October-November 2003 and 2004 (storms generated by multiple interacting ICMEs with Sheaths) do not significantly change our previous results for MC distribution but do change results for Sheath distribution; the probability for Sheath decreases more slowly with the increasing |Dst0| index than for other drivers and for Sheath in our previous study. We believe that this is the result of the joint consideration of various cases: lonely and interacting Sheath/MC structures. Cases of Sheath and MC in interacting structures should be selected in separate classes.

It should be noted that there are two different cases of estimations of waiting times. For extreme magnetic storms with a minimum $Dst \leq -500\text{ nT}$, the waiting times $T(-500\text{ nT}) \sim 24$ years and $T_{MC}(-500\text{ nT}) \sim 20$ years are less than the durations of used data sets 53 and 30 years. However, for storms with a minimum $Dst \leq -1000\text{ nT}$ and $Dst \leq -1700\text{ nT}$ the waiting times >100 years are significantly larger than the durations of data sets. Therefore, the accuracy of estimates in the second case is very low because the assumption of stationarity of the Sun at these scales may be violated (e.g., Fig. 1 shows that the solar and magnetospheric activities in the current 24 solar cycle are lower than in previous cycles). For discussion of the validity of the time stationary assumption, please see Chapter 5 of this volume.

5 CONCLUSIONS

We study experimental data for solar and interplanetary sources of geomagnetic storms. Using OMNI data from 1976 to 2005, we classified different types of solar wind events (MC, Ejecta, CIR, and Sheath), identified magnetic storms corresponding to these types of interplanetary drivers, and calculated standard and integral distributions (probabilities) for the generation of magnetic storms. From these results, we make the following conclusions.

1. The main solar sources of magnetic storms are CMEs and coronal holes. These two classes of events define geoeffective interplanetary types of solar wind: ICMEs including MC and Ejecta types, and compression regions including Sheath before fast ICME and CIR before high-speed streams from coronal holes. It is necessary to take into account that (1) so-called "CME-induced" storms are the results of complex phenomena including two geoeffective parts: Sheath and ICME; and (2) the majority of extreme storms are generated by interacting Sheath/ICME structures and these interacting structures should be studied separately.
2. Standard and integral occurrence distribution functions F and F^* at large values of |Dst| have the power law tails with indices $\gamma(F) \sim -4.4 \pm 0.5$ and $\gamma(F^*) \sim -3.4 \pm 0.3$, respectively. These values of indices are close to results obtained in the previous works by Tsubouchi and Omura (2007) ($\gamma(F) \sim -4.9$) and Riley (2012) ($\gamma(F^*) \sim -3.2$), but our results satisfy the requirement $\gamma(F^*) - \gamma(F) = 1$.
3. The integral probabilities of CIR-, Ejecta-, and Sheath-induced storms with $Dst \leq -50\text{ nT}$ are less by 3-4 times than for MC. For extreme storms with $Dst \leq -200\text{ nT}$, their integral probabilities are less by a factor of 5-15. Contribution of Sheath-induced storms for interacting Sheath/ICME should be additionally analyzed.

4. When approximating the data in the Dst region from -150 down to -300 nT and then extrapolating these approximating functions to extreme storms with Dst ≤ -500 nT, -1000 nT, and -1700 nT, we could estimate two waiting times: (1) the most probable estimations of waiting times for extreme magnetic storms $T(-500 \text{ nT}) \sim 24$ years, $T(-1000 \text{ nT}) \sim 250$ years, and $T(-1700 \text{ nT}) \sim 1500$ years (with error 10%, 40%, and 100%) were obtained for total storm set, and (2) the lower limits of waiting times for extreme magnetic storms $T_{MC}(-500 \text{ nT}) \sim 20$ years, $T_{MC}(-1000 \text{ nT}) \sim 120$ years, and $T_{MC}(-1700 \text{ nT}) \sim 500$ years (with accuracy of factors ~ 1.5, ~ 2, and ~ 3), respectively, were estimated for storms generated by MC.

ACKNOWLEDGMENTS

The authors are grateful for the opportunity to use data from the OMNI database (http://omniweb.gsfc.nasa.gov) and Dst index (http://wdc.kugi.kyoto-u.ac.jp/). This work was supported by the Russian Foundation for Basic Research, project 16–02–00125, and the Program 7 of Presidium of the Russian Academy of Sciences.

REFERENCES

Akasofu, S.-I., 1981. Energy coupling between the solar wind and the magnetosphere. Space Sci. Rev. 28, 121–190.

Borovsky, J.E., Denton, M.H., 2006. Differences between CME-driven storms and CIR-driven storms. J. Geophys. Res. 111. A07S08. https://doi.org/10.1029/2005JA011447.

Borovsky, J.E., Denton, M.H., 2010. Solar wind turbulence and shear: a superposed-epoch analysis of corotating interaction regions at 1 AU. J. Geophys. Res. 115, A10101. https://doi.org/10.1029/2009JA014966.

Burton, R.K., McPherron, R.L., Russell, C.T., 1975. An empirical relationship between interplanetary conditions and Dst. J. Geophys. Res. 80, 4204–4214.

Carrington, R.C., 1859. Description of a singular appearance seen in the Sun on September 1. Mon. Not. R. Astron. Soc. XX, 13.

Crosby, N.B., 2011. Frequency distributions: from the sun to the earth. Nonlinear Process. Geophys. 18, 791–805. https://doi.org/10.5194/npg-18-791-2011.

Daglis, I.A., Thorne, R.M., Baumjohann, W., Orsini, S., 1999. The terrestrial ring current: origin, formation, and decay. Rev. Geophys. 37 (4), 407–438. https://doi.org/10.1029/1999RG900009.

Dungey, J.W., 1961. Interplanetary magnetic field and the auroral zones. Phys. Rev. Lett. 6, 47–48.

Ermolaev, Y.I., et al., 2005. Solar and heliospheric disturbances that resulted in the strongest magnetic storm of November 20, 2003. Geomagn. Aeron. 45 (1), 23–50 (in Russian) (Geomagn. Aeron., Engl. Transl., 45 (1), 20–46).

Fairfield, D.H., Cahill Jr., L.J., 1966. The transition region magnetic field and polar magnetic disturbances. J. Geophys. Res. 71, 155–169.

Farrugia, C.J., Matsui, H., Kucharek, H., Jordanova, V.K., Torbert, R.B., Ogilvie, K.W., Berdichevsky, D.B., Smith, C., Skoug, R., 2006a. Survey of intense Sun-Earth connection events (1995–2003). Adv. Space Res. 38 (3), 498–502.

Farrugia, C.J., Jordanova, V.K., Thomsen, M.F., Lu, G., Cowley, S.W.H., Ogilvie, K.W., 2006b. A two-ejecta event associated with a two-step geomagnetic storm. J. Geophys. Res. 111.A11104. https://doi.org/10.1029/2006JA011893.

Gonzalez, W.D., Tsurutani, B.T., Clúa de Gonzalez, A.L., 1999. Interplanetary origin of geomagnetic storms. Space Sci. Rev. 88, 529–562.

Gonzalez, W.D., Echer, E., Tsurutani, B.T., Clúa de Gonzalez, A.L., Dal Lago, A., 2011. Interplanetary origin of intense, superintense and extreme geomagnetic storms. Space Sci. Rev. 158, 69–89. https://doi.org/10.1007/s11214-010-9715-2.

Gonzatez, W.D., Joselyn, J.A., Kamide, Y., Kroehl, H.W., Rostoker, G., Tsurutani, B.T., Vasyliunas, V.M., 1996. What is a magnetic storm? J. Geophys. Res. 99, 5771–5792.

Gopalswamy, N., Yashiro, S., Krucker, S., Stenborg, G., Howard, R.A., 2004. Intensity variation of large solar energetic particle events associated with coronal mass ejections. J. Geophys. Res. (Space Phys.) 109, A12105.

Gorobets, A., Messerotti, M., 2012. Solar flare occurrence rate and waiting time statistics. Sol. Phys. 281, 651–667. https://doi.org/10.1007/s11207-012-0121-7.

Gosling, J.T., 1993. The solar flare myth. J. Geophys. Res. 98 (A11), 18937–18949. https://doi.org/10.1029/93JA01896.

Huttunen, K.E.J., Koskinen, H.E.J., 2004. Importance of postshock streams and sheath region as drivers of intense magnetospheric storms and high-latitude activity. Ann. Geophys. 22, 1729–1738. https://doi.org/10.5194/angeo-22-1729-2004.

Keesee, A.M., Elfritz, J.G., Fok, M.-C., McComas, D.J., Scime, E.E., 2013. Superposed epoch analyses of ion temperatures during CME- and CIR/HSS-driven storms. J. Atmos. Sol. Terr. Phys. 115, 67–78. https://doi.org/10.1016/j.jastp.2013.08.009.

Kilpua, E.K.J., Hietala, H., Turner, D.L., Koskinen, H.E.J., Pulkkinen, T.I., Rodriguez, J.V., Reeves, G.D., Claudepierre, S.G., Spence, H.E., 2015. Unraveling the drivers of the storm time radiation belt response. Geophys. Res. Lett. 42, 3076–3084. https://doi.org/10.1002/2015GL063542.

King, J.H., Papitashvili, N.E., 2004. Solar wind spatial scales in comparisons of hourly wind and ACE plasma and magnetic field data. J. Geophys. Res. 110 (A2), A02209. https://doi.org/10.1029/2004JA010804.

Koons, H.C., 2001. Statistical analysis of extreme values in space science. J. Geophys. Res. 106 (A6), 10,915–10,921.

Love, J.J., 2012. Credible occurrence probabilities for extreme geophysical events: earthquakes, volcanic eruptions, magnetic storms. Geophys. Res. Lett. 39, L10301. https://doi.org/10.1029/2012GL051431.

Love, J.J., Rigler, E.J., Pulkkinen, A., Riley, P., 2015. On the lognormality of historical magnetic storm intensity statistics: implications for extreme-event probabilities. Geophys. Res. Lett. 42 (16), 6544–6553. https://doi.org/10.1002/2015GL064842.

Mishra, W., Srivastava, N., Chakrabarty, D., 2015. Evolution and consequences of interacting CMEs of 9–10 November 2012 using STEREO/SECCHI and in situ observations. Solar Phys. 290, 527.

Mostl, C., Rollett, T., Frahm, R.A., Liu, Y.D., Long, D.M., Colaninno, R.C., Reiss, M.A., Temmer, M., Farrugia, C.J., Posner, A., Dumbovic, M., Janvier, M., Demoulin, P., Boakes, P., Devos, A., Kraaikamp, E., Mays, M.L., Vrsnak, B., 2015. Strong coronal channelling and interplanetary evolution of a solar storm up to Earth and Mars. Nat. Commun. 6, 7135.

Pulkkinen, T.I., Partamies, N., Huttunen, K.E.J., Reeves, G.D., Koskinen, H.E.J., 2007a. Differences in geomagnetic storms driven by magnetic clouds and ICME sheath regions. Geophys. Res. Lett. 34, L02105. https://doi.org/10.1029/2006GL027775.

Pulkkinen, T.I., Partamies, N., McPherron, R.L., Henderson, M., Reeves, G.D., Thomsen, M.F., Singer, H.J., 2007b. Comparative statistical analysis of storm time activations and sawtooth events. J. Geophys. Res. 112.A01205. https://doi.org/10.1029/2006JA012024.

Riley, P., 2012. On the probability of occurrence of extreme space weather events. Adv. Space Res. 10, S02012. https://doi.org/10.1029/2011SW000734.

Riley, P., Love, J.J., 2017. Extreme geomagnetic storms: probabilistic forecasts and their uncertainties. Space Weather 15 (1), 53–64. https://doi.org/10.1002/2016SW001470.

Rostoker, G., Falthammar, C.-G., 1967. Relationship between changes in the interplanetary magnetic field and variations in the magnetic field at the earth's surface. J. Geophys. Res. 72, 5853–5863.

Russell, C.T., McPherron, R.L., Burton, R.K., 1974. On the cause of magnetic storms. J. Geophys. Res. 79, 1105–1109.
Siscoe, G.L., Crooker, N.U., Clauer, C.R., 2006. Dst of the Carrington storm of 1859. Adv. Space Res. 38, 173–179.
Skoug, R.M., Gosling, J.T., Steinberg, J.T., McComas, D.J., Smith, C.W., Ness, N.F., Hu, Q., Burlaga, L.F., 2004. Extremely high speed solar wind: 29–30 October 2003. J. Geophys. Res. 109, A09102. https://doi.org/10.1029/2004JA010494.
Sugiura, M., 1964. Hourly values of equatorial Dst for the IGY. Ann. Int. Geophys. Year 35, 9–45.
Sugiura, M., Kamei, T., 1991. Equatorial Dst index 1957–1986. IAGA Bull. vol. 40.
Tsubouchi, K., Omura, Y., 2007. Long-term occurrence probabilities of intense geomagnetic storm events. Adv. Space Res. 5, S12003. https://doi.org/10.1029/2007SW000329.
Tsurutani, B.T., Gonzalez, W.D., 1997. The interplanetary causes of magnetic storms: a review. In: Tsurutani, B.T. et al., (Ed.), In: Magnetic Storms, Geophys. Monogr. Ser., vol. 98. AGU, Washington, DC, pp. 77–89.
Tsurutani, B.T., Gonzalez, W.D., Lakhina, G.S., Alex, S., 2003. The extreme magnetic storm of 1–2 September 1859. J. Geophys. Res. 108 (A7), 1268. https://doi.org/10.1029/2002JA009504.
Tsurutani, B.T., et al., 2006. Corotating solar wind streams and recurrent geomagnetic activity: a review. J. Geophys. Res. 111, A07S01. https://doi.org/10.1029/2005JA011273.
Turner, N.E., Cramer, W.D., Earles, S.K., Emery, B.A., 2009. Geoefficiency and energy partitioning in CIR-driven and CME-driven storms. J. Atmos. Sol. Terr. Phys. 71, 1023–1031.
Veselovsky, I.S., et al., 2004. Solar and heliospheric phenomena in October November 2003: causes and effects. Kosm. Issled. 42 (5), 453–508 (in Russian) (Cosmic Res., Engl. Transl., 42 (5), 435–488).
Wang, Y.M., Ye, P.Z., Wang, S., 2003. Multiple magnetic clouds: several examples during March–April 2001. J. Geophys. Res. 108 (A10), 1370. https://doi.org/10.1029/2003JA009850.
Xie, H., Gopalswamy, N., Manoharan, P.K., Lara, A., Yashiro, S., Lepri, S.T., 2006. Long-lived geomagnetic storms and coronal mass ejections. J. Geophys. Res. 111, A01103. https://doi.org/10.1029/2005JA011287.
Yashiro, S., Gopalswamy, N., Akiyama, S., Michalek, G., Howard, R.A., 2005. J. Geophys. Res. 11, A12S05. https://doi.org/10.1029/2005JA011151.
Yermolaev, Y.I., Yermolaev, M.Y., 2006. Statistic study on the geomagnetic storm effectiveness of solar and interplanetary events. Adv. Space Res. 37 (6), 1175–1181.
Yermolaev, Y.I., Yermolaev, M.Y., 2010. Solar and interplanetary sources of geomagnetic storms: space weather aspects. Izv. Russ. Acad. Sci. Atmos. Oceanic Phys. Engl. Transl. 46 (7), 799–819.
Yermolaev, Y.I., et al., 2005. A year later: solar, heliospheric, and magnetospheric disturbances in November 2004. Geomagn. Aeron. 45 (6), 723–763 (in Russian) (Geomagn. Aeron., Engl. Transl., 45 (6), 681–719).
Yermolaev, Yu.I., et al., 2009. Catalog of large-scale solar wind phenomena during 1976–2000. Kosm. Issled. 47 (2), 99–113 (Cosmic Res., Engl. Transl., pp. 81–94).
Yermolaev, Yu.I., Nikolaeva, N.S., Lodkina, I.G., Yermolaev, M.Yu., 2010a. Relative occurrence rate and geoeffectiveness of large-scale types of the solar wind. Kosm. Issled. 48 (1), 3–32 (Cosmic Res., Engl. Transl., 2010, 48 (1), 1–30).
Yermolaev, Yu.I., Nikolaeva, N.S., Lodkina, I.G., Yermolaev, M.Yu., 2010b. Specific interplanetary conditions for CIR-, Sheath-, and ICME-induced geomagnetic storms obtained by double superposed epoch analysis. Ann. Geophys. 28, 2177–2186.
Yermolaev, Y.I., Nikolaeva, N.S., Lodkina, I.G., Yermolaev, M.Y., 2012. Geoeffectiveness and efficiency of CIR, sheath, and ICME in generation of magnetic storms. J. Geophys. Res. 117, A00L07. https://doi.org/10.1029/2011JA017139.
Yermolaev, Y.I., Lodkina, I.G., Nikolaeva, N.S., Yermolaev, M.Y., 2013. Occurrence rate of extreme magnetic storms. J. Geophys. Res. Space Phys.. 118https://doi.org/10.1002/jgra.50467.
Yermolaev, Yu.I., Lodkina, I.G., Nikolaeva, N.S., Yermolaev, M.Yu., 2015. Dynamics of large-scale solar wind streams obtained by the double superposed epoch analysis. J. Geophys. Res. Space Phys.. 120https://doi.org/10.1002/2015JA021274.

Yermolaev, Y.I., Lodkina, I.G., Nikolaeva, N.S., Yermolaev, M.Y., Ryazantseva, M.O., 2017. Some problems of identification of large-scale solar wind types and their role in the physics of the magnetosphere. Cosm. Res. 55 (3), 178–189.

Zhang, J., et al., 2007. Solar and interplanetary sources of major geomagnetic storms (Dst < -100 nT) during 1996–2005. J. Geophys. Res. 112, A10102. https://doi.org/10.1029/2007JA012321.

CHAPTER 5

STATISTICS OF EXTREME SPACE WEATHER EVENTS

Pete Riley
Predictive Science Inc., San Diego, CA, United States

CHAPTER OUTLINE

1 Introduction 115
2 Methodologies 117
    2.1 Datasets 117
    2.2 Statistical Modeling 119
3 Results 122
    3.1 Assessing the Validity of the Time Stationarity Assumption 126
    3.2 Analysis of *Dxt* and *Dcx* 127
    3.3 Extreme Space Weather Events in the Ionosphere: The *AE* Index 127
    3.4 Extreme Space Weather Events in the Heliosphere: Energetic Protons 129
4 Discussion 130
5 Future Studies 135
6 Conclusions 136
Acknowledgments 136
References 137

1 INTRODUCTION

Space weather refers to the state and future conditions of the space environment surrounding the Earth, including within the Earth's magnetosphere. As a discipline, it is predominantly operational in nature. Extreme space weather refers to those conditions that are so far removed from the norm that they are rare. Unfortunately, defining what makes a space weather event, or space weather conditions extreme, is difficult. An event, for example, can be extreme with respect to one parameter (say, the Dst index) but not another (say, coronal mass ejection (CME) speed). However, some events, such as those occurring on September 1, 1859 (e.g., Riley, 2012) and July 23, 2012 (e.g., Riley et al., 2016), are undoubtedly extreme.

The Carrington event is perhaps the quintessential space weather event and has been studied in great detail, at least to the extent possible because of the limited data associated with it (e.g., Riley, 2012). Similarly, the July 23, 2012, event, which was observed by the STEREO A spacecraft, located at 1 AU from the Sun but away from the Sun-Earth line, has significantly more solar and heliospheric data but no geomagnetic data because the ICME did not hit the Earth (Russell et al., 2013; Baker et al., 2013; Liu et al., 2014). While there are a number of similarities between the two, there are also important distinctions. However, both can serve as proxies for a generic "extreme" event. And, while much can be learned from studying specific events, broad extrapolations may lead to inaccurate inferences. In this chapter we summarize our current knowledge with respect to the statistics of extreme space weather events, and, in particular, in estimating the probability of occurrence over the next decade.

A number of studies have considered the properties of extreme solar events. Koons (2001), for example, applied extreme value statistics to show that the extreme events observed within several space weather datasets, including the magnetic index Sp, are well fit by extreme value models. Tsubouchi and Omura (2007) assumed a power-law distribution for magnetic storms to infer that an event as large as the March 1989 storm ($|Dst| > 280$ nT) would occur every 60 years or so. Finally, it should be noted that quasipower-law behavior is not limited to space physics, but can be found in a wide range of natural hazards, such as earthquakes, volcanic eruptions, floods, wildfires, and landslides, to name just a few (Sachs et al., 2012).

Riley (2012) first considered the likelihood of a Carrington-like event occurring on a decadal timescale. They showed that even using simple "time-to-event" analysis, this likely was ~9%. Using more sophisticated statistical tools, they suggested that this number ranged from a few percent up to 12%, depending on the parameter under study (i.e., how one defined an "extreme" event). They considered a number of solar, heliospheric, and magnetospheric indices or parameters. Although solar flares (which have often been connected with power-law distributions) would seem to be an ideal parameter, Riley (2012) showed that this was a questionable inference, and that these data were not well suited for further analysis. Similarly, CME speed appeared to show a broken power-law distribution, with a knee just above 2000 km/s. The Dst index generally showed what was interpreted to be a power-law distribution, as did solar proton events (SPEs), as evidenced by >30 MeV proton fluences.

Love (2012) independently also considered the probability of rare geomagnetic storms, but focused on an estimate of the uncertainties associated with them, which were only indirectly addressed by Riley (2012). He found that the 95% confidence intervals for a storm exceeding −589 nT lay between 3.4% and 38.6%, that is, from quite unlikely to very probable. Thus, a crucial part of the analysis that we must attempt to refine is not so much the forecast estimate but a reduction in the confidence intervals.

Love et al. (2015) questioned the basic assumption that the data, and the Dst index in particular, could be adequately described by a power-law distribution. By including more moderate storms in their analysis, they argued that a log-normal distribution was a better fit to the data than a pure power-law distribution. Surprisingly, this difference only modified the likelihood modestly.

Most recently, Riley and Love (2017) have incorporated and generalized the concepts from the previous studies to produce the most robust probabilistic forecasts and their uncertainties. This is the approach we follow in the analysis below. They showed that the probability of another Carrington-like event within the next 10 years is 10.3% with confidence interval [0.9, 18.7].

Riley and Love (2017) also considered how the forecast might vary within a solar cycle. In some sense, this admits that the assumption of time stationarity breaks down. However, there are not a sufficient number of data points in any of the datasets to look at limited portions within a single cycle (during which one can assume stationarity). To address this, they extracted data ±2.5 years around each solar minimum and each solar maximum and combined them into two new datasets, one representing solar minimum-like and the other solar maximum-like conditions. This was possible because of the assumption that the events are independent of one another. They found that the forecasts differed drastically between quasiminimum and maximum conditions. In particular, they estimated that the likelihood of another event as large, or larger than 850 nT, is 1.4% during solar minimum conditions and 28% for solar maximum conditions.

In this chapter, we review the basic techniques for estimating the likelihood of an extreme space weather event, focusing both on the estimate itself as well as the uncertainties associated with it, which, we believe, is equally important. We add to previous studies by estimating the probability of an extreme event based on the auroral electrojet AE index, and extend our previous work investigating the Dst index to also include the Dst proxy parameters Dxt and Dcx, which extend significantly further back in time. Additionally, we refine our previous analysis of SPEs by applying a more rigorous approach.

2 METHODOLOGIES

In this section, we summarize the primary datasets used in our analysis as well as the statistical modeling approach used to make the probabilistic forecasts of extreme events.

2.1 DATASETS

Predicting an extreme space weather event first requires that we define what one is. This in turn means that we must choose parameters that describe events that are distributed in some measure of severity. As a concrete example, we could look at the speeds of CMEs, which vary from a few hundred kilometers per second to almost 4000 km s^{-1}. These speeds are not distributed normally, but have a power-law or log-normal distribution (Riley and Love, 2017). We can arbitrarily define "extreme" events as those that have speeds, $v > 3000$ km s^{-1}.

For continuous datasets, on the other hand, we must define "events" within the data stream. The Dst index, for example, is a continuous time series. Magnetic storms are identified as intervals of depressed Dst over an extended period of time. We can capture the storm by identifying the minimum value within the event.

Previously, Riley (2012) studied a broad range of datasets, including solar flare data, CME speeds, the Dst index, and SPEs as determined from nitrate data, discussing the relative merits and limitations of each. In this chapter, we focus on the Dst index and two proxy measures for Dst, Dxt, and Dcx; the auroral ejectrojet index, AE; and a measure of SPEs (specifically, all >30 MeV proton events with fluences exceeding 10^9 pr cm^{-2}). These add to the data previously analyzed but, also for data already analyzed, extend the methodologies applied. This is obviously not an exhaustive list. We could, for example, consider CME mass and energy, IMF Bz intervals, and the equatorward edge of the diffuse aurora, to name a few. Analysis of these, however, should probably be driven by need. Kp is an obvious index that would benefit from such an analysis; however, because it only takes on a limited number of values, it is not clear how to apply these techniques to such a parameter.

To complement the *Dst* index, we also consider the *Dxt* and *Dcx* indices (Karinen and Mursula, 2006). These indices are attempts to correct defects in the *Dst* index, and, additionally extend the duration of the index back in time to 1932. *Dxt* is considered to be a reconstructed Dst index. It correlates highly with *Dst* (0.987 for hourly values) and, additionally, corrects errors present in the original *Dst* index (Karinen and Mursula, 2005). *Dcx* has been proposed as a corrected version of the *Dst* index to account for seasonally varying quiet-time levels that can raise |*Dst*| by as much as 44 nT for an individual storm (Karinen and Mursula, 2006). The corrections appear robust in that they improve the index's correlation with sunspots and other geomagnetic indices. For our purposes, the main advantage lies in: (1) the extended duration of the dataset (80.1 years), allowing us to test the robustness of our forecasts over this extended period; and (2) the impact of changes in the peak values for severe storms. For example, the two notable storms of March 23, 1940 ($Dcx/Dxt_{min} = -355/-360$ nT) and September 18, 1941 ($Dcx/Dxt_{min} = -404/-417$ nT) are contained within the datasets, but not within the *Dst* dataset. Of course, this comes with an important caveat that a storm value of -850 nT in *Dst* might not be the same as -850 nT in *Dxt* or *Dcx*. Additionally, it is worth emphasizing that these data may contain artifacts not present in the more comprehensive and well-studied *Dst* dataset. For example, during the September 18, 1941 storm, the light trace at the Honolulu observatory dropped off the edge of the photographic paper for 10 hours, suggesting that the measurement "saturated." However, a data gap is not noted in the *Dcx* data, nor is it readily apparent from an inspection of the time series. The errors introduced by such "saturation" of the dataset remain to be analyzed, estimated, and, if possible, mitigated. Nevertheless, with these caveats in mind, we believe the analysis of these complementary datasets is justified.

Additionally, we also investigate the properties of the *AE* index. The *AE* index was designed to provide a global estimate of auroral zone magnetic activity, driven by ionospheric currents both below and within the auroral oval (Davis and Sugiura, 1966). In essence, it is the amount by which the horizontal magnetic field around the auroral oval deviates from quiet-time values. In practice, calculating the *AE* index is a relatively complex process involving measurements of 10–13 observatories situated along the auroral zone, the normalization of these data to base values that are computed from quiet days, and the computation of two intermediary indices, known as AU and AL. Thus, while recognizing that the index is not a simple fiduciary for some simple physical variable, it does provide a measure for the overall activity of the electrojets, which, in turn, demonstrates its usefulness as a measure of substorm activity. The analysis of *AE* complements that of *Dst*, which relies on measurement stations located at low latitudes and, hence, minimizing the effects of the auroral zone.

There are several caveats that should be kept in mind when using the *AE* index (Mandea and Korte, 2010). First, diurnal variations might be present due to the limited latitudinal extent and uneven longitudinal distribution of the *AE* stations used to derive the index. Second, when the number of available stations falls below 12, there is the possibility that ejectrojet enhancement will be missed. Third, although the index is primarily driven by ionospheric electrojet current, magnetospheric currents (e.g., the equatorial ring current) may sometimes contribute to AE, complicating its interpretation.

Finally, we apply our techniques to a dataset derived from nitrate records in polar ice cores, which, at least to some, are believed to capture SPEs. It stretches back more than 400 years, and thus represents the only dataset within which the actual Carrington event is embedded. In principle, they are a measure of the flux of a population of highly energized particles, accelerated either by the flare or CME-driven shock associated with an extreme event. However, they are not without controversy or caveats. First, while the nitrate spikes were, until recently, believed by space physicists to be a record of large, historical space weather events (McCracken et al., 2001), ice-core chemists are skeptical (Wolff et al., 2008). They posit that no viable mechanism exists by which SPEs could be imprinted within the ice, suggesting instead that high concentrations of sea salt provide a simpler and more consistent explanation for the deposition of aerosol nitrates. Additionally, recent work by space physicists also casts doubt on their solar origin (Duderstadt et al., 2016). Second, there are only 70 events spanning the 450 years for which we have data. The largest event in the dataset, with a fluence of 18.8×10^9 cm^{-1}, occurred in 1859. That is, the largest event in the last 400 years was the Carrington event. More importantly, however, with such a limited number of events, the statistics of the fit and the resulting probability estimates will be more prone to error. In spite of these limitations, and under the assumption that they are capturing extreme solar phenomena, we can estimate their likelihood of occurrence.

2.2 STATISTICAL MODELING

Here, we review a method for estimating the likelihood of another Carrington-like event by assuming that the events are distributed in severity in a way that can be described by a continuous curve (e.g., exponential, log-normal, or power-law). However, several other approaches could—at least superficially—be used, including event trees, similarity judgments, and time to event. However, for the reasons outlined by Riley (2012), none of these approaches are amenable for studying extreme space weather phenomena. Thus, we focus on an extrapolation technique that assumes we can continue the curve of well-observed events out into the region of frequency-severity space for which there are few or even no observed events.

Riley (2012) showed that a range of space weather phenomena can—at least qualitatively—be described by a quasipower-law distribution. To generalize this, and test whether another distribution is more appropriate, we consider the following three types of distributions: power-law (PL), log-normal (LN), and exponential (E). To this, we can add so-called "cut-off" distributions where the data dramatically drop off at some point.

First, a set of events, x, obeys a power-law (or Pareto) distribution if the probability of occurrence, $p(x)$, can be written:

$$p(x) = C_1 x^{-\alpha} \tag{1}$$

where the exponent, α, is a fixed parameter, and C_1 is estimated from the location at which the curve intercepts the y-axis. Similarly, a set of events, x, is said to follow a log-normal distribution if the probability of occurrence, $p(x)$, obeys:

$$p(x) = \frac{C_2}{x} e^{-(\ln x - \mu)^2 / 2\sigma^2} \tag{2}$$

where μ and σ are parameters that must be fit based on the observations, and C_2 is another constant. Finally, a set of events, x, is said to follow an exponential distribution if:

$$p(x) = C_3 e^{-\lambda x} \tag{3}$$

where λ is a free parameter whose value is estimated based on a best fit to the measurements, and C_3 is another constant. This triplet of distributions, we believe, reasonably encompasses the relevant phase space with power-laws falling off least quickly, exponentials falling off most rapidly, and log normals generally lying between the two.

2.2.1 Estimating the best-fit parameters to a model

Riley (2012) described a technique for estimating the likelihood of a space weather event for power-law distributions, based on earlier work by McMorrow (2009). Following this, we define the complementary cumulative distribution function (CCDF), $P(x)$, as the probability of an event of magnitude equal to or greater than some critical value x_{crit}:

$$P(x \geq x_{crit}) = \int_{x_{crit}}^{\infty} p(x')dx' \tag{4}$$

which, for a finite dataset, simplifies to:

$$P(x \geq x_{crit}) = \frac{C}{\alpha - 1} x_{crit}^{-\alpha+1} \tag{5}$$

Here, the CCDF also obeys a power law with a lower exponent ($\alpha - 1$). CCDFs offer a number of benefits over the original power-law distributions: (1) they circumvent issues associated with noisy tails; (2) the slope can be computed using the maximum likelihood estimate (MLE):

$$\alpha - 1 = N \left[\sum_{i=1}^{N} \ln \frac{x_i}{x_{min}} \right]^{-1} \tag{6}$$

where x_i are the measured values of x, N is the number of events in the dataset, and x_{min} is some appropriate minimum value of x, below which the power-law relationship breaks down (Newman, 2005); and (3) the CCDF naturally generates the probability of occurrence of some event of a particular strength or greater, not the probability of an event of size x.

Using Eq. (5) we can estimate the number of events as large as or larger than x_{crit} during the period covered by the dataset, E:

$$E(x \geq x_{crit}) = NP(x \geq x_{crit}) \tag{7}$$

where N is the total number of events within the dataset.

Finally, again under the assumption that the events happen independently, we can employ the Poisson distribution to derive the probability of one or more events greater than x_{crit} occurring during some time Δt:

$$P(x \geq x_{crit}, t = \Delta t) = 1 - e^{-N \frac{\Delta t}{\tau} P(x \geq x_{crit})} \tag{8}$$

where τ is the total time span of the dataset. Eqs. (5), (6), (8) thus provide a robust technique for calculating the probability that an event of severity exceeding x_{crit} will occur some time within the next Δt years.

Similar expressions can be written for log-normal and exponential distributions, and Eq. (8) can be used to estimate probabilities based on these distributions.

2.2.2 Identifying the tail in the distribution

It is unlikely that natural phenomena display tail-like behavior throughout their entire distribution. At the lowest frequencies, saturation effects likely dominate. Similarly, at the highest frequencies, a cut-off must be anticipated at some (even remote) point, based on, say, physical constraints (e.g., maximum possible available energy). Thus, we need to identify a lower limit in severity, above which we can reasonably argue that a tail-like distribution exists.

Riley et al. (2012) and Love et al. (2015) identified the minimum values in severity (x_{min}) manually, and, arguably, somewhat subjectively. In particular, Riley et al. (2012) chose $x_{min} = 120$ nT while Love et al. (2015) used $x_{min} = 63$ nT as a lower bound for the $|Dst|$ index. Here, we use an approach for optimizing the value of x_{min} based on minimizing the Kolmogorov-Smirnov (KS) goodness-of-fit statistic between the model and the data (Kolmogorov, 1933; Smirnov, 1948; Press, 2007). Essentially, the KS statistic aims to balance the inclusion of tail data to improve sample statistics while at the same time omitting low-severity data that may not reflect the true nature of the tail.

2.2.3 Nonparametric bootstrapping

Following Efron and Tibshirani (1994), we apply a technique known as nonparametric bootstrapping to estimate the confidence intervals of our predictions. The observed data is randomly sampled and new pseudo-datasets are constructed from the events drawn (and replaced). Intuitively, this process can be understood as follows: You place all the data into a bag and randomly pick one out. You record the value, replace the datapoint back into the bag and choose another. To each of these, one of the three distribution profiles (PL, LN, or E) is fit. The bootstrap approach is straightforward to apply and generally provides reasonable estimates of standard errors and confidence intervals when the sample size is large.

Once a sufficiently large number of pseudo-datasets (i.e., bootstraps) have been computed and fit, the variability within these profiles can be used to define, say, 95% (i.e., between 2.5% and 97.5%) confidence intervals.

2.2.4 Model comparison

Using the techniques outlined thus far, we can use the computed bootstrapped fits to test whether a PL distribution is plausible by computing a p-value. We define the null hypothesis (H_0) to be that the power law adequately describes the data, and the alternative hypothesis (H_1), that some other distribution is better. Thus, if $p > 0.1$, say, then the difference between the data and the model can be attributed to statistical fluctuations and we cannot reject H_0. On the other hand, if p is small, say, <0.1, then the PL model is not a plausible fit to the data. It should be emphasized that merely because the p-value is large, this does not guarantee that the PL model is correct; for that we must apply a model comparison test.

Vuong's test is one such model comparison test that relies on a likelihood ratio test for selecting one model over another (Vuong, 1989). Specifically, it uses the Kullback-Leibler divergence, which is a measure of the difference between two probability distributions, say, A and B (Kullback and Leibler, 1951; Joyce, 2011). The criteria estimates the information gained or lost when model B is used to approximate model A. Alternatively it can be thought of as a metric that measures the distance between A and B.

In our case, we compute Vuong's test statistic, R_V, which compares two models under the hypothesis that both classes of distribution are equally far from the true distribution. If true, the log-likelihood ratio would have a mean value of zero. R_V moves toward $\pm\infty$ if one model is substantially better than the other. Additionally, one-sided and two-sided p-values can be computed to estimate the significance of the R_V statistic. The one-sided approach tests the null hypothesis (H_0) that both classes of distributions are equally far from the true distribution, against the alternative hypothesis (H_1) that model A is closer to the true distribution. The two-sided version tests the null hypothesis (H_0) that both classes of distributions are equally far from the true distribution, against the alternative (H_1), that one of the distributions is closer. In both cases, we reject H_0 if $p < p_{crit}$, where in this case, we conservatively chose $p_{crit} = 0.05$.

For a complementary approach, including estimates of the probabilities for magnetic storms produced by different drivers (CIRs, sheath, MC, and CME), please see Chapter 4 in this volume.

3 RESULTS

In this section we apply the previously described techniques to *Dst*, *Dxt*, *Dcx*, *AE*, and SPE >30 MeV fluences.

Beginning with the $|Dst|$ index, in Fig. 1, we show a time series of severe ($|Dst| > 100$ nT) geomagnetic storms. We note several points. First, only one event approached 600 nT, and, moreover, only five events exceeded 400 nT. Second, the storms appear to cluster on the timescale of ~11 years, in effect mimicking but trailing the sunspot cycle. Third, bimodal peaks can be seen at and after solar maximum, matching CME rates (Riley et al., 2006). Fourth, there is a tendency for the strongest storms to become stronger from 1965 to 2007. In particular, while the largest five storms around 1970 were between 200 and 300 nT, the five most intense storms around 2005 were between 350 and 450 nT. On the other hand, the most recent decade shows a relative dearth of events, and a particular lack of intense storms. In summary then, there appear to be both periodic and secular variations in the time series.

Fig. 2 summarizes the probability estimates using the three possible distributions. The panel on the left (A) shows the CCDF, which, as discussed earlier, is the probability that an event as large or larger than some critical value will occur during a unit time interval. The open circles show all the events. The advantage of using the CCDF rather than the underlying $p(x)$ is self-evident: The data are not binned in x but rather summed so that the number of data points used to construct each open circle is the sum of all the data points to the right of itself (e.g., Riley et al., 2012). The points are well represented by a straight line at least up to ~280 nT. Beyond this, with the exception of the most severe storm, they appear to "fall off" this trajectory.

The colored curves show a selection of fits to the bootstraps. Specifically, for each of 1000 bootstrapped pseudo-datasets, a PL, LN, and E distribution was fit. Of these, 100 randomly chosen ones are displayed. The general conclusion, at least visually, is that: (1) the PL profiles capture the lower severity measurements but overestimate the likelihood of the most severe events; (2) the LN profiles underestimate the low-severity events but capture the trends at higher severity; and (3) the E profiles overestimate the low-severity frequency and underestimate the high-severity frequency of events.

Fig. 2B summarizes the likelihood of observing an event as severe or more severe as the most severe event observed (i.e., $Dst < -589$ nT) during the entire span of the data (~57 years). For a PL/LN/E distribution, the median probabilities are: 0.95, 0.63, and 0.13, respectively.

Using Eq. (8), we can estimate the probability of such an event occurring over the next decade to be 20.3/3.0/0.02% for a power-law, log-normal, or exponential distribution (see also Table 1). Moreover, we can use the bootstrap results to estimate confidence intervals in these predictions. For the power-law distribution, for example, our best estimate is 20.3% for 95%CI [12.5,30.2]. Table 1 also shows the forecast when only data from 1964 through the present is included in the analysis. In this case, estimates drops by almost a factor of two. Additionally, and not shown, if we require that an event exceeds a threshold of -1700 nT, a value closer to that suggested by Tsurutani et al. (2003) for the Carrington

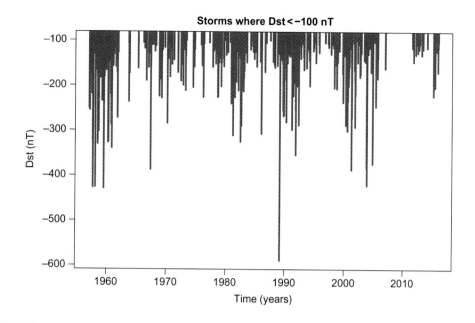

FIG. 1

Magnetic storms, defined by events where |Dst| exceeds 100 T, are shown as a function of time. Individual storms are identified as contiguous intervals where |Dst| exceeded 100 nT. The data were obtained from NASA's COHOWEB.

Based on Riley, P., Love, J.J., 2017. Extreme geomagnetic storms: probabilistic forecasts and their uncertainties. Space Weather 15, 53–64.

event, the probability of occurrence over the next decade decreases to 1.5%. This is quite comparable to the result obtained by Yermolaev et al. (2013), who estimated that the likelihood of a storm with $|Dst| > 1760$ nT was not higher than one event every 500 years (or, a probability of ~2% per decade).

Fig. 3 summarizes the main statistical parameters for the bootstrap fits. For each parameter, the cumulative mean and 25% and 75% quantiles are shown as a function of iteration (i.e., the number of bootstraps). Thus, as the number of bootstraps is increased, our estimate for the different parameters improves. The best-fit values are: $x_{min} \sim 123$ nT and $\alpha \sim 3.72$, and the number of points used to construct the tail statistics is ~250.

Using the computed bootstrap fits, we can also test the hypothesis of whether the power-law distribution is plausible. The *p*-value for this is shown in Fig. 3D. Unlike the more usual approach for interpreting *p*-values, this one is set up such that a value <0.05 provides strong evidence against the power-law hypothesis. On the other hand, values above 0.05, or, more conservatively, above 0.1 would suggest that a power-law distribution is plausible. Thus, the value determined for the *Dst* index, ~0.2, suggests that we cannot rule out the power-law as the underlying distribution. On the other hand, this result does not mean that this is the correct distribution. For that, we need to apply Vuong's test and make direct comparisons amongst viable models.

In Table 2 we compare Vuong's test statistic (R_V) for each of three comparisons: Power-law vs. log-normal; power-law vs. exponential; and log-normal vs. exponential. Additionally, we provide *p*-values for both the one-sided and two-sided tests. Larger positive values of R_V provide support

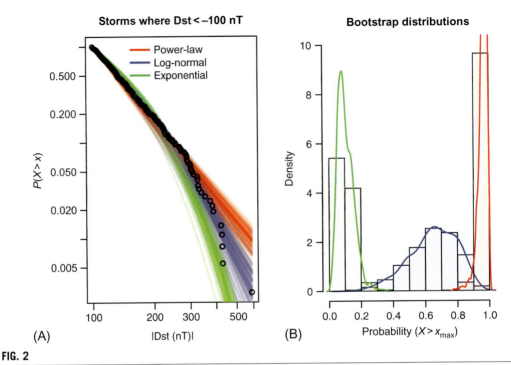

FIG. 2

(A) Complementary cumulative distribution function (CCDF) for the geomagnetic storms shown in Fig. 1. Bootstrap fits for the three distributions (power-law, log-normal, and exponential) are superimposed. (B) Histogram and density plots showing the probability of an event as large as or larger than the largest event in the dataset (−589 nT) over the duration of the dataset (~49 years). The density curve colors follow the convention given in the legend within panel (A).

Table 1 Best Estimates and Confidence Intervals for *Dst*

Distribution	Median (%)	2.5% (%)	97.5% (%)
Power-law (1964–2016)	10.3	0.9	18.7
Power-law (1957–2016)	20.3	12.5	30.2
Log-normal	3.0	0.6	9.0
Exponential	0.02	0.004	0.08

for the first model over the second model. Thus, we can infer from these results that the log-normal distribution is slightly favored against the power-law, and much more so against the exponential distribution. Similarly, the power-law is favored over the exponential. However, for these results to be statistically significant, we require p-values < 0.05. In almost all cases, the calculated p-values exceed this, and, in most cases, substantially so. Only the two-sided log-normal vs. exponential p-value is less than this threshold (0.02), which would allow us to firmly discount the exponential distribution.

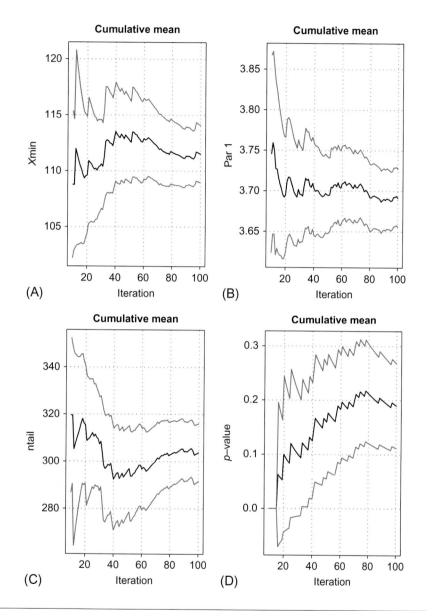

FIG. 3

Summary of statistical parameters for power-law bootstrap fit to the *Dst* dataset as a function iteration (i.e., the number of bootstraps). (A)–(D) The cumulative means of (A) x_{min}, (B) α (Par1), (C) n_{tail}, and (D) the *p*-value.

Table 2 Vuong's Test Statistics for the *Dst* Index, Where P-L, L-N, and Exp. Refer to Power-Law, Log-Normal, and Exponential Distributions

Statistic	P-L vs. L-N	P-L vs. Exp.	L-N vs. Exp.
R_V	−0.488	1.08	2.29
One-sided	0.313	0.86	0.989
Two-sided	0.626	0.279	0.0219

Table 3 Best Estimates and Confidence Intervals for the *Dst* Index for Each Solar Cycle From 1957 Through Early 2016

Cycle	Interval	Power-Law	Log-Normal
19	57–64	65.02 [27.54, 90.67]	16.45 [1.05, 46.45]
20	64–76	0.15 [1 × 10^{-4}, 4.76]	2.5 × 10^{-12} [0, 1.17]
21	76–86	14.34 [3.36, 37.50]	0.14 [1 × 10^{-4}, 3.74]
22	86–96	0.044 [4.2 × 10^{-11}, 4.33]	1.8 × 10^{-7} [0, 0.1]
23	96–08	12.82 [3.89, 30.27]	3.21 [0, 14.0]
24	08–16	0.049 [2.4 × 10^{-9}, 4.43]	2 × 10^{-7} [0, 0.1]

3.1 ASSESSING THE VALIDITY OF THE TIME STATIONARITY ASSUMPTION

The approach adopted here has relied on the assumption that the data are time stationary. As we have discussed, however, there are both cyclic and secular variations in space weather phenomena. To better understand the impact of this variability, we have repeated our analysis for each of five epochs: Solar cycles 19, 20, 21, 22, and 23/24, which cover the time period from 1957 through 2012. Breaking the data into these five intervals necessarily increases the uncertainties associated with any predictions we make. Nevertheless, it may, in principle, provide some information about the intrinsic variability from one cycle to another. Because the number of events following the end of cycle 23 was so small, we grouped cycle 24 events with cycle 23.

Table 3 summarizes the probabilities estimated using both the power-law and log-normal distributions, using the same analysis as described above. We note the following points. First, there is considerable variability from one cycle to the next, suggesting that either (1) time stationarity is not a reasonable approximation; or, and more likely, (2) the limited sample size for a single decade is not large enough to compute a meaningful estimate of the probability. Due to the significant scatter, we cannot discern any obvious secular trend in these probabilities. On the other hand, with the exception of the 57–64 interval, an estimate of ~4% per decade is contained within all confidence bounds. Additionally, when estimates are made for the intervals 1957–2016 and 1964–2016 (Table 1), the former produces significantly higher forecasts. Thus, we suggest that the interval from 1957 to 1964 was indeed associated with a significantly larger probability for an extreme event.

3.2 ANALYSIS OF *Dxt* AND *Dcx*

Fig. 4 summarizes the CCDFs for *Dxt* and *Dcx*. To a large degree, the profiles are quite similar to those produced using the original *Dst* record. When computing the 10-year forecasts for an event as large or larger than −850 nT using *Dxt* and *Dcx*, we find values of 17.7% and 15.9%, respectively, with confidence bounds that are comparable to those computed for the *Dst* index.

3.3 EXTREME SPACE WEATHER EVENTS IN THE IONOSPHERE: THE *AE* INDEX

Fig. 5 shows all *AE* "events" for which the index exceeded 1000 nT. We used the same procedure to identify "events" as with the *Dst* index. In particular, we identified all contiguous intervals for which the *AE* index exceeded 200 nT, then for each of these located the peak value and timing of that peak. We then discarded those events that did not exceed 1000 nT. Over the ∼58-year time span (from 1957 to 2012), there were 1165 events matching this criteria. Solar cycle modulation and any possible secular trends are much less pronounced in this dataset, although an argument can be made that the last eight years have been unusually quiet. The most severe *AE* "event" occurred on March 23, 1991, reaching a value of 3195 nT.

Fig. 6A shows the CCDF for all *AE* "events" that were larger than 1000. Bootstrap solutions for each of the three distributions are shown again in red (PL), blue (LN), and green (SE). The PL bootstraps are, at least subjectively, consistent with the observed tail: Arguably only two of the most severe events are not enveloped within the bootstrap spread. Similarly, at least visually, the LN distributions capture the fall off reasonably well. Only the E distributions fail to capture the profile correctly, dropping off too rapidly to be able to account for the largest events. It could be argued that x_{min} was not set correctly for these data. However, as we have discussed above, by relying on the KS statistic we have removed the subjectivity of fitting to less (the more severe) data in order to include more data and improve the statistics of the fit. Only with some extrinsic knowledge, such as there being an established break in the data, say, produced by different physical mechanisms, would we be justified in manually setting x_{min}. In summary, the data appear to trace the border between the PL and LN distributions.

Fig. 6B compares the probability distributions for predicting an event as large as or larger than, the largest event within the dataset (3195 nT). Focusing on the PL and LN distributions, we note the significant difference in estimates: A median probability of 0.67 or 0.34 depending on whether one chooses a PL or LN distribution, respectively. Table 4 summarizes these estimates and also provides 95% confidence intervals on these predictions.

The best-fit model parameters based on the bootstraps gave: $x_{min} \sim$ 1165 nT and $\alpha \sim$ 7.16. With a *p*-value of ∼0.45, we cannot reject the hypothesis that the data are well described by a power law. Additionally (not shown), we remark on a positive correlation between x_{min} and α: As x_{min} increases, so does α. This suggests that the slope of the curve is becoming increasingly steeper at higher values of AE, perhaps indicating that the underlying distribution is not best approximated by a straight line (i.e., constant α) but by a log-normal distribution, at least for the most severe events.

We also estimated Vuong's statistics associated with these fits. Again, the log-normal distribution was slightly favored over the power-law, and more strongly over the exponential. Similarly, the power-law was favored over the exponential. None, however, were statistically significant. Thus, based on statistical arguments alone, we cannot reject any.

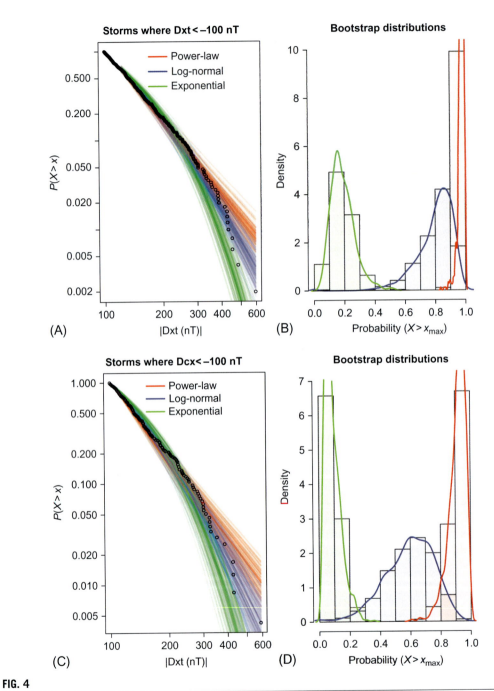

FIG. 4

As Fig. 2 but for *Dxt* (A and B) and *Dcx* (C and D).

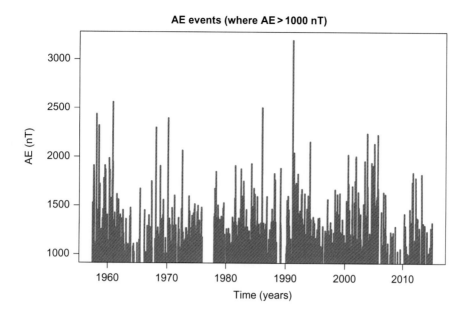

FIG. 5

Large auroral "substorms," defined by events where the *AE* index exceeds 1000 nT, are shown as a function of time. Individual storms are identified as contiguous intervals where *AE* exceeded 1000 nT.

Data from NASA's COHOWEB.

3.4 EXTREME SPACE WEATHER EVENTS IN THE HELIOSPHERE: ENERGETIC PROTONS

Fig. 7 shows all >30 MeV proton events with fluences exceeding 10^9 pr cm^{-2} as a function of time between 1562 and 1944. Although it is not possible to show rigorously because of the limited sample size, there is no obvious trend in the distribution of event sizes or temporal clustering to suggest that the time series is obviously nontime stationary. Of note is that the Carrington event is substantially larger than the other events in the dataset, with the second largest event producing a fluence of only 59% of the value of the 1859 event.

Fig. 8 shows the CCDF as a function of size and the probability density functions assuming power-law, log-normal, or exponential distributions. Specifically, the probability of observing an event as large as or larger than, the largest event observed in the dataset over a period as long as the interval over which the data were collected. Because the largest event contained within the nitrate record is presumed to be the 1859 event, this provides a more direct (albeit less reliable) estimate of the Carrington event. Thus, over the 382-year time span, the median probability of observing the Carrington event was 0.67/0.015/0.033 for the PL/LN/E distributions, respectively (Table 5).

The bootstrap fits, however, underscore the degree of uncertainty with these estimates. Based on KS statistics, only ~35 points were retained for the analysis. It is worth noting that the determination of x_{min} can be defended visually here. The shape of the curve appears to take three distinct slopes below ~4×10^{-9} cm^{-3}. Visually, at least, the log-normal and exponential distribution, which are overlaid

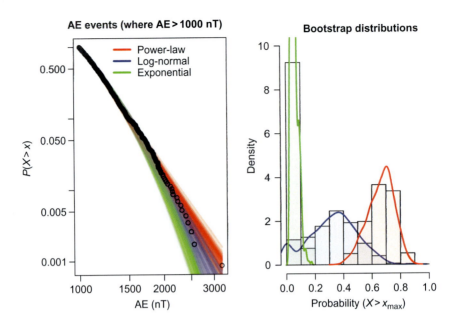

FIG. 6

As Fig. 1 for the auroral "substorms" shown in Fig. 5. Bootstrap fits for the three distributions (power-law, log-normal, and exponential) are superimposed. (*Right*) Histogram and density plots showing the probability of an event as large as or larger than, the largest event in the dataset (3295 nT) over the duration of the dataset (~51 years).

Table 4 Best Estimates and Confidence Intervals for AE

Distribution	Median (%)	2.5% (%)	97.5% (%)
Power-law	17.66	10.28	25.89
Log-normal	6.88	0.00	16.08
Exponential	1.08	0.40	2.08

upon one another, appear to capture the profile best, with the exception of the most severe (Carrington) event, which falls directly in the middle of the PL bootstrap fits. Application of Vuong's test this time slightly favors the log-normal over the exponential, and to a smaller extent the log-normal over the power-law. In turn, the power-law distribution is ever-so-slightly favored over the exponential distribution. Again, however, the comparisons are not statistically significant, using either the one-sided or two-sided p-values.

4 DISCUSSION

Although the analysis presented here has incorporated more statistical methodologies than our previous studies (Riley et al., 2012; Love et al., 2015), somewhat paradoxically, our analysis produced results that are even less constrained. Because we cannot be confident of the underlying distribution, and,

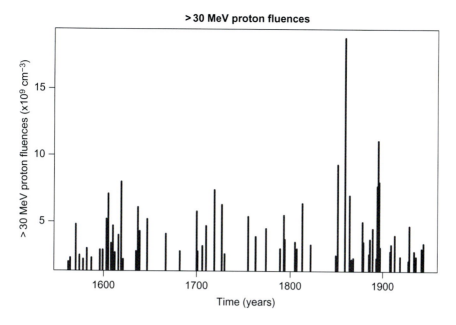

FIG. 7

Large SPEs, as defined by their inferred >30 MeV proton fluence exceeding 10^9 cm^{-3}, are shown as a function of time.

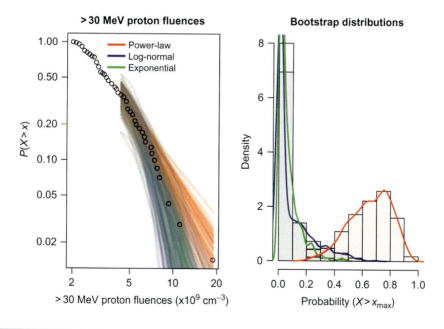

FIG. 8

As Fig. 1 but for the energetic SPEs shown in Fig. 7. Bootstrap fits for the three distributions (power-law, log-normal, and exponential) are superimposed. (*Right*) Histogram and density plots showing the probability of an event as large as or larger than the largest event in the dataset (18.8×10^9 cm^{-3}) over the duration of the dataset (~382 years).

Table 5 Best Estimates and Confidence Intervals for SPEs

Distribution	Median (%)	2.5% (%)	97.5% (%)
Power-law	2.85	1.06	5.67
Log-normal	0.04	1.5×10^{-9}	1.61
Exponential	0.086	0.0022	0.78

indeed, it appears that the data lie firmly between the power-law and log-normal distributions, our basic estimate could be as low as 3.0% or as high as 20.3%. While averaging these two values (11.65%) would yield a value that was remarkably similar to our two previous estimates (11% and 12%) (Riley et al., 2012; Love et al., 2015), in reality, it cannot be both. Statistical inference, however, can only be used up to a point. Beyond this we must use our best judgment, incorporating all relevant information to arrive at an informed estimate with credible confidence intervals. To do this, however, we must first address several important assumptions.

In our analysis, we assumed that the data are time stationary. At both ends of the temporal spectrum, we have shown that this approximation breaks down. The solar cycle modulates many solar parameters on the time scale of a decade or so (e.g., Riley et al., 2000). In particular, the largest 2% of geomagnetic storms (the so-called "super storms"), typically occur just after solar activity maxima (Bell et al., 1997). Thus, the dataset used to make forecasts should be at least this long, and any predictions made must necessarily be solar-cycle averaged estimates. On the other hand, we showed that forecasts based on data from solar minimum intervals might be as low as 1.4%/decade, but as high as 28% during solar maximum. It is interesting to note that this statistical result does not support the anecdotal view held by many in the scientific community that the most extreme storms tend to occur at or near solar minimum (based on the two extreme events of 1859 and 2012). It might, however, be possible to incorporate this knowledge into the prediction. For example, we might anticipate that the probability of an extreme event might peak a year or two after solar maximum, in conjunction with the peak in the rate of CMEs.

Similarly, over longer timescales, there is ample evidence for nonstationarity (e.g., Riley et al., 2012). In particular, Lockwood et al. (2009) inferred that there was a 10% likelihood that the Sun will fall into another grand minimum configuration over the next 40 years or so. If further substantiated, this information could be convolved with the current predictions for an extreme event on the premise that such phenomena would be significantly less likely to occur during Maunder-minimum-like conditions (Riley et al., 2015). Although speculative due to the absence of significant sunspots and, hence, active regions, there would be far fewer CMEs that could drive geoeffective space weather. Indeed, if the nitrate records in fact provide estimates of large SPEs, the Maunder minimum period lasting from ~1645 to ~1710 was the most quiescent period during the entire ~400-year interval. As a rough approximation, we can compute the frequency of space weather events during the Maunder minimum interval and compare that with a 65-year window centered on 1900, say, and use that ratio to adjust our forecasts based on the Dst index. There were seven events between 1645 and 1710 (1647, 1667, 1682, 1700, 1701, 1706, and 1710), and 12 events between 1867.5 and 1930.5 (1889, 1893, 1894, 1895, 1896, 1897, 1908, 1909, 1913, 1919, 1928, and 1929), suggesting that the current forecasts might need to be lowered by 58% if we are entering a grand solar minimum.

Although our analysis of decadal subsets of the Dst index did not yield any systematic trend in forecasts for Carrington events, constructing two long-window datasets (1957–2008 and 1964 to early

2016) suggests that the latter window is associated with a substantially lower forecast. Although the two intervals are roughly comparable (~51 years), the active latter half of the 1950s/early 1960s was replaced with the unusually quiet period surrounding and following the 2008/2009 solar minimum (also known as the "Eddy" minimum). This is further supported by the analysis of the *Dxt* dataset. When divided equally, estimates for an event that exceeds 850 nT are: 26.1%/decade for the period 1932–1973, and 12.3%/decade for 1974 to early 2016. If this captures the overall trend in solar activity into the future, we would anticipate that the future rate of occurrence of extreme events will be notably less than that estimated from the full (60 years) dataset. Obviously, if such conditions do ensue, probabilistic forecasts for extreme events may decrease; however, not necessarily to society's benefit. During periods of very low activity, for example, radiation from galactic cosmic rays (GCRs) will be higher, posing larger risks for passengers and airline crew as well as avionics. And, while SEP events themselves may decrease, the consequences from the ones that are produced may be more severe (e.g., Barnard et al., 2011).

The second major assumption addressed in this study is whether the data are best represented by a PL, LN, or E distribution. We inferred that the LN and PL distributions were consistent with the *Dst*, *Dxt*, *Dcx*, and *AE* datasets. For SPEs, either an L-N or E distribution is consistent with the data. The statistical results described and interpreted here, however, provide no guidance on the underlying causes for observing such distributions. In fact, where statistical summaries are ambiguous, we can reasonably resort to any available theories that might tend to favor one distribution over another. Several studies have alluded to the idea that substorms in particular can be described by self-organized criticality (SOC) (Angelopoulos et al., 1999; Klimas et al., 2000), which provides a natural explanation for the presence of power-law distributions for geomagnetic indices. On the other hand, we must temper this with the caveat that SOC has been invoked, and in fact was developed precisely because the measurements are observed to follow a power-law. By extension, we could posit that, in analogy with the Abelian Sandpile model (Bak et al., 1987), the magnetotail becomes progressively loaded until some specific threshold is reached, then reconnects (presumably in the form of multiple injections/substorms) and produces the observed $|Dst|$ maxima. However, this is undoubtedly a simplistic interpretation of a considerably more complex system. Love et al. (2015) has argued that perhaps the act of combining smaller storms, which do not apparently follow a power-law distribution, with larger storms that do may result in a distribution that is better approximated by a log-normal distribution. Indeed, this may be the case as they included storms down to values exceeding $|Dst| > 63$ nT. When KS statistics are used to identify $x_{min} (= 122$ nT), our results are consistent with a power-law distribution (but not inconsistent with a log-normal distribution).

A related but distinct assumption about the distribution is that it extends into a region of severity that we have observed rarely, if at all. Clearly, this assumption must fail at both extremes of the severity spectrum. In the low-severity portion of the spectrum, the curve usually flattens because smaller events are less easily identified or measured. At the high-severity portion of the spectrum, several factors may be important. First, "small-number" statistics may produce a curve profile that veers away from what would otherwise be a straight line. However, it is worth considering that even if the fluctuations at the extreme of the tail are random, we would expect a bias toward the undersampled region of this phase space, because the errors would not be expected to be distributed normally in log-log space. Second, in any finite-sized system, and particularly when the power-law distribution falls off more rapidly than higher frequency rates would suggest, the cut-off may be a real physical limitation. In practice, there must be a cut-off at some point. The key issue is whether that cut-off is near to, or far from, the critical event under consideration. If the latter, we do not have to modify our analysis. However, if the former, we must account for the fact that events larger than the cut-off cannot contribute to our integrated estimate of events as large as, or larger than, some threshold. If we do not account for this, our estimates will be inflated.

For the *Dst* index, we can inquire what possible limits there might be. The absolute limit for $|Dst|$ is approximately 31,000 nT, which represents the complete cancelation of the Earth's magnetic field at the equator. Vasyliūnas (2011) has argued that the limiting value is considerably lower: ~2500 nT. To arrive at this estimate, he set the plasma pressure equal to the magnetic pressure of the dipole field at the equator of each flux tube. This suggests a strong earthward gradient of the plasma pressure, which, through the relation $\mathbf{J}_\perp \sim (\mathbf{B} \times \nabla)\mathbf{B}/B^2$, implies a strong westward current through the magnetosphere. Vasyliunas used the Dessler-Parker-Sckopke relationship (Dessler and Parker, 1959; Sckopke, 1966) to arrive at the 2500-nT limit as a physical cut-off for *Dst*. Adopting this value would only marginally reduce the forecasts for a power-law distribution. In particular, using the full range of data from the Kyoto Observatory, the probability of an event as large or larger than 2500 nT is ~1.3%. Thus, our estimate of 20.3% would only be reduced to ~19.0%.

An interesting but as yet unexplored possibility is that if we could: (1) provide firm limits to the cut-off, and (2) argue for a log-normal distribution, this would allow us to fix the right-most portion of the curve, which, in turn, would allow us to better constrain the fit to the data and, hence, provide more accurate forecasts.

Our current forecast for an extreme event, where $Dst < -850$ nT, is 20.3%, assuming the data are distributed according to a power law, which is larger than two earlier estimates of ~12% (Riley et al., 2012) and ~11% (Love et al., 2015), although certainly within the overlapping confidence intervals. The disagreement between the current value and that in Riley et al. (2012) is due primarily to the addition of data from 1957 to 1963, which was a period of relatively high solar activity, and disproportionately added more high-severity storms to the data being analyzed. The disagreement with Love et al. (2015) is due to their use of (1) a lower, hand-picked value for x_{min} and (2) use of a log-normal distribution. In particular, the incorporation of the low-severity events strongly influenced the fit of the log-normal curve producing higher forecast estimates than would have been produced with a larger value of x_{min}. Thus, it is worth reemphasizing just how sensitively these results depend on the dataset under study as well as the techniques used to analyze them, and, as was pointed out by Love (2012), it is that it is not only the forecast estimates that must be communicated, but also the uncertainties and assumptions that accompany them.

Estimating the likelihood for future extreme space weather events can be of considerable value to decision makers. However, effectively communicating this information can be difficult. Probabilistic estimates with associated uncertainties can be phrased in any number of ways. Based on these results, for example, the likelihood of another extreme event on the scale of the Carrington or July 23 event over the next 12 months is only 2.3%. On the other hand, that same event has a 90% probability over the next 100 years. More importantly, while our study of extreme space weather events is important in its own right, it is perhaps the relative risk of a Carrington event as compared with, say, another earthquake on the scale of the 1906 San Francisco event or Hurricane Katrina, that is of more value to policy makers. Current 30-year probability estimates of an earthquake in California as large as or larger than, magnitude 8 are 4% (Field and Milner, 2008). Our estimate for an extreme space weather event on the scale of or larger than the "Carrington" event over the next decade is ~20%. The 30-year estimate is 50%/decade (PL distribution) or 10% (LN distribution), which are 12 and 2.5 times larger than the earthquake forecast. But even these comparisons can be misleading because it is the consequences of each disaster that society cares more about.

We should distinguish between natural disasters from natural hazards. The former refers to the actual event, such as a geomagnetic storm, flood, earthquake, or avalanche. It is the interaction between the disaster and society that makes a disaster a natural hazard (e.g., Blackwell, 2014). Practically speaking, the distinction is clear: estimating the damages caused by a disaster is more important than accurately describing the characteristics of the disaster itself. This is the information that both government agencies and insurance actuaries are primarily concerned with. Since there are often scaling laws between observables (e.g., magnitude earthquakes) and related physical parameters (e.g., earthquake rupture areas), we can infer the probability distributions of one from the other. Thus, we suggest that probabilistic forecasts can be significantly more informative when coupled not only with the magnitude of the related damages, but also with the likely costs associated with mitigating those damages. And, as noted earlier, these should be compared across multiple catastrophes allowing policy makers to allocate limited resources most effectively.

5 FUTURE STUDIES

A number of opportunities exist to build upon this work. First, constraining the uncertainties in the forecasts must be a core objective. Without tighter constraints, it is difficult for policy makers to make informed decisions about how to spend limited resources. But, to constrain the forecasts further is difficult. The spread in the bootstraps represents a baseline noise that cannot be removed with a limited number of events. Additionally, we cannot definitively establish which distribution best fits the data. If it were a log-normal distribution, the 10-year likelihood is a mere 4%. If it is a power-law distribution, it is 2.6 times higher (10.3%). Perhaps more sophisticated statistical techniques can be applied to delineate between the two? A further uncertainty concerns the assumption of time stationarity. Since the unusually quiet solar minimum of 2008/2009, the Sun has apparently entered a more quiescent state. Should this continue in the future, probabilistic forecasts based on space-era activity, which may have been anomalously high, would lead to erroneously large estimates for the 10-year forecasts. Instead, if the Sun were to maintain more solar-minimum-like activity, then the forecasts would be reduced by almost an order of magnitude (1.4%). Should we be entering into a more extreme solar wind state, such as a Maunder minimum (Riley et al., 2015), even this may be an overestimate. Thus, an interesting refinement to the techniques described here would be to develop estimates based on historical data, but which include a factor encapsulating what we think is a good estimate of solar activity over the next decade.

Our analysis has followed a purely statistical approach. However, physics-based modeling could be incorporated into the analysis. If, for example, firm upper limits for the size of extreme events could be found, that would modify the forecast estimates. As noted earlier, Vasyliūnas (2011) argued that the Dst index cannot exceed ~2500 nT. If so, this would reduce our estimate of 10.3% to 9%. If the physical limit were even lower, that would have a more substantial effect.

Another area that is ripe for analysis lies on the "consequences" side of the Sun-to-Earth chain. Can we use these statistical tools to estimate the probability of an electric field in the Earth's lithosphere exceeding some value? Or the current in the power grid? Or cosmic ionizing radiation within aircraft? In part, a lack of available data makes these questions difficult to address.

Statistical models can only answer carefully chosen questions in a limited way. We gain little in the way of a deeper understanding from such endeavors. Instead, they simply highlight a problem worthy of more careful investigation. This is where physics-based models can help. What are the solar wind parameters required to produced at Dst value of -1600 nT? What phenomena at the Sun and in the solar wind are required to generate such conditions? Knowledge of the underlying physical processes that produce extreme space weather events should ultimately result in the most accurate and useful forecasting tools. What made the July 23, 2012 extreme event so unique? Would we be able to recognize these or similar conditions in the future? And would we be able to extrapolate what we observe at the Sun to what would likely impact the Earth in the days to follow? This itself is a two-pronged question. First, given surface observations of an eruption, can we forecast the effects that will likely ensue at Earth? Second, and much more difficult, given a particular set of observations, can we reliably predict that there will be an eruption, and, if so, what the properties of that eruption will be at Earth?

6 CONCLUSIONS

In this chapter, we have described and applied a general technique for assessing both the likelihood and uncertainties of an extreme space weather event on the scale of or larger than, the Carrington event of 1859. In addition to the previously assumed power-law distribution, we also considered both log-normal and exponential distributions as alternatives for explaining the observed distribution in severity. Using the Dst index as the fiduciary measurement for defining a space weather event, we inferred that the probability of another event within the next decade exceeding 850 nT was 20.3% for 95% CI [12.5, 30.1] for a PL distribution but only 3.0% for 95% CI [0.6, 9.0] for an LN distribution. Increasing the threshold to 1700 nT reduced the estimate to 3.6% for 95% CI [1.7, 6.6].

Our studies thus far have not established whether a power-law or log-normal distribution best fit the data. We found that both are, within statistical uncertainties, consistent with the data. We also sought to establish tighter limits on our forecasts for the probability of another extreme event within the next decade; however, we found that depending on which datasets, intervals, and distributions were used to make the estimates, the results varied substantially.

In conclusion, under the assumptions that: (1) a PL distribution best represents the data; (2) the PL distribution is likely an upper limit to the behavior of the tail; (3) we are entering a period of lower solar activity; (4) a good definition of the Dst index for an extreme event is that $Dst < -850$ nT, we conclude that the best estimate for the probability of such an event over the next decade is approximately 10% for 95% CI [1, 20].

ACKNOWLEDGMENTS

The authors gratefully acknowledge the support of NSF's FESD and NASA's LWS program, under which this work was performed. The Dst index was obtained from the Kyoto World Data Center (http://wdc.kugi.kyoto-u.ac.jp/dstdir/), while the Dxt and Dcx indices were provided by the Dcx server of the University of Oulu, Finland (http://dcx.oulu.fi).

REFERENCES

Angelopoulos, V., Mukai, T., Kokubun, S., 1999. Evidence for intermittency in Earth's plasma sheet and implications for self-organized criticality. Phys. Plasmas (1994–Present) 6 (11), 4161–4168.

Bak, P., Tang, C., Wiesenfeld, K., 1987. Self-organized criticality: an explanation of the 1/f noise. Phys. Rev. Lett. 59 (4), 381.

Baker, D.N., Li, X., Pulkkinen, A., Ngwira, C.M., Mays, M.L., Galvin, A.B., Simunac, K.D.C., 2013. A major solar eruptive event in July 2012: defining extreme space weather scenarios. Space Weather 11 (10), 585–591.

Barnard, L., Lockwood, M., Hapgood, M.A., Owens, M.J., Davis, C.J., Steinhilber, F., 2011. Predicting space climate change. Geophys. Res. Lett. 381, L16103. https://doi.org/10.1029/2011GL048489.

Bell, J.T., Gussenhoven, M.S., Mullen, E.G., 1997. Super storms. J. Geophys. Res. 102, 14189–14198. https://doi.org/10.1029/96JA03759.

Blackwell, C., 2014. Power law or lognormal? Distribution of normalized hurricane damages in the United States, 1900–2005. Nat. Hazards Rev. 16 (3) 04014024.

Davis, T.N., Sugiura, M., 1966. Auroral electrojet activity index AE and its universal time variations. J. Geophys. Res. 71 (3), 785–801.

Dessler, A.J., Parker, E.N., 1959. Hydromagnetic theory of geomagnetic storms. J. Geophys. Res. 64 (12), 2239–2252.

Duderstadt, K.A., Dibb, J.E., Schwadron, N.A., Spence, H.E., Solomon, S.C., Yudin, V.A., Jackman, C.H., Randall, C.E., 2016. Nitrate ions spikes in ice cores are not suitable proxies for solar proton events. J. Geophys. Res. 121 (6), 2994–3016.

Efron, B., Tibshirani, R.J., 1994. An Introduction to the Bootstrap. CRC Press, Boca Raton, FL.

Field, E.H., Milner, K.R., 2008. Forecasting California's Earthquakes: What Can We Expect in the Next 30 Years? No. 2008–3027. U.S. Geological Survey.

Joyce, J.M., 2011. Kullback-Leibler divergence. International Encyclopedia of Statistical Science, Springer, pp. 720–722.

Karinen, A., Mursula, K., 2005. A new reconstruction of the Dst index for 1932–2002. Ann. Geophys. 23 (2), 475–485.

Karinen, A., Mursula, K., 2006. Correcting the *dst* index: consequences for absolute level and correlations. J. Geophys. Res. 111 (A8), 1–8.

Klimas, A.J., Valdivia, J.A., Vassiliadis, D., Baker, D.N., Hesse, M., Takalo, J., 2000. Self-organized criticality in the substorm phenomenon and its relation to localized reconnection in the magnetospheric plasma sheet. J. Geophys. Res. 105 (A8), 18765–18780.

Kolmogorov, A.N., 1933. Sulla determinazione empirica delle leggi di probabilita. Giorn. Ist. Ital. Attuari 4, 1–11.

Koons, H., 2001. Statistical analysis of extreme values in space science. J. Geophys. Res. 106, 10915–10921.

Kullback, S., Leibler, R.A., 1951. On information and sufficiency. Ann. Math. Statist. 22 (1), 79–86.

Liu, Y.D., Luhmann, J.G., Kajdič, P., Kilpua, E.K.J., Lugaz, N., Nitta, N.V., Möstl, C., Lavraud, B., Bale, S.D., Farrugia, C.J., et al., 2014. Observations of an extreme storm in interplanetary space caused by successive coronal mass ejections. Nat. Commun. 5, 3481–3489.

Lockwood, M., Rouillard, A.P., Finch, I.D., 2009. The rise and fall of open solar flux during the current grand solar maximum. Astrophys. J. 700, 937–944. https://doi.org/10.1088/0004-637X/700/2/937.

Love, J.J., 2012. Credible occurrence probabilities for extreme geophysical events: earthquakes, volcanic eruptions, magnetic storms. Geophys. Res. Lett. 39. https://doi.org/10.1029/2012GL051431. L10301.

Love, J.J., Rigler, E.J., Pulkkinen, A., Riley, P., 2015. On the lognormality of historical magnetic storm intensity statistics: implications for extreme-event probabilities. Geophys. Res. Lett. 42 (16), 6544–6553.

Mandea, M., Korte, M., 2010. Geomagnetic Observations and Models. IAGA Special Sopron Book Series, Springer. https://books.google.com/books?id=DOMEIl7hlxsC. ISBN 9789048198580.

McCracken, K.G., Dreschhoff, G.A.M., Zeller, E.J., Smart, D.F., Shea, M.A., 2001. Solar cosmic ray events for the period 1561–1994: 1. Identification in polar ice, 1561–1950. J. Geophys. Res. 106, 21585–21598.

McMorrow, D., 2009. Rare events. JASON, The MITRE Corporation. http://www.dtic.mil/cgi-bin/GetTRDoc?AD=ADA510224&Location=U2&doc=GetTRDoc.pdf.

Newman, M., 2005. Power laws, pareto distributions and zipf's law. Contemp. Phys. 46, 323–351.

Press, W.H., 2007. Numerical Recipes: The Art of Scientific Computing, third ed. Cambridge University Press, Cambridge.

Riley, P., 2012. On the probability of occurrence of extreme space weather events. Space Weather 10 (null), S02012.

Riley, P., Love, J.J., 2017. Extreme geomagnetic storms: probabilistic forecasts and their uncertainties. Space Weather 15, 53–64.

Riley, P., Linker, J.A., Mikic, Z., Lionello, R., 2000. Solar cycle variations and the large-scale structure of the heliosphere: MHD simulations. In: IAU Joint Discussion, vol. 7. http://adsabs.harvard.edu/abs/2000IAUJD..7E.12R.

Riley, P., Schatzman, C., Cane, H.V., Richardson, I.G., Gopalswamy, N., 2006. On the rates of coronal mass ejections: remote solar and in situ observations. Astrophys. J. 647, 648–653. https://doi.org/10.1086/505383.

Riley, P., Lionello, R., Linker, J.A., Mikic, Z., Luhmann, J., Wijaya, J., 2012. Global MHD modeling of the solar corona and inner heliosphere for the whole heliosphere interval. Solar Phys. 274, 361–375. https://doi.org/10.1007/s11207-010-9698-x.

Riley, P., Lionello, R., Linker, J.A., Cliver, E., Balogh, A., Charbonneau, P., Crooker, N., DeRosa, M., Lockwood, M., Owens, M., et al., 2015. Inferring the structure of the solar corona and inner heliosphere during the maunder minimum using global thermodynamic magnetohydrodynamic simulations. Astrophys. J. 802 (2), 105.

Riley, P., Caplan, R.M., Giacalone, J., Lario, D., Liu, Y., 2016. Properties of the fast forward shock driven by the July 23, 2012 extreme coronal mass ejection. Astrophys. J. 819. https://doi.org/10.3847/0004-637X/819/1/57 57.

Russell, C.T., Mewaldt, R.A., Luhmann, J.G., Mason, G.M., von Rosenvinge, T.T., Cohen, C.M.S., Leske, R.A., Gomez-Herrero, R., Klassen, A., Galvin, A.B., Simunac, K.D.C., 2013. The very unusual interplanetary coronal mass ejection of 2012 July 23: a blast wave mediated by solar energetic particles. Astrophys. J. 770. https://doi.org/10.1088/0004-637X/770/1/38. 38.

Sachs, M., Yoder, M., Turcotte, D., Rundle, J., Malamud, B., 2012. Black swans, power laws, and dragon-kings: earthquakes, volcanic eruptions, landslides, wildfires, floods, and SOC models. Eur. Phys. J. 205, 167–182.

Sckopke, N., 1966. A general relation between the energy of trapped particles and the disturbance field near the Earth. J. Geophys. Res. 71 (13), 3125–3130.

Smirnov, N., 1948. Table for estimating the goodness of fit of empirical distributions. Ann. Math. Statist. 19 (2), 279–281.

Tsubouchi, K., Omura, Y., 2007. Long-term occurrence probabilities of intense geomagnetic storm events. Space Weather 51. https://doi.org/10.1029/2007SW000329. S12003.

Tsurutani, B.T., Gonzalez, W.D., Lakhina, G.S., Alex, S., 2003. The extreme magnetic storm of 1–2 September 1859. J. Geophys. Res. 108, 1268. https://doi.org/10.1029/2002JA009504.

Vasyliūnas, V.M., 2011. The largest imaginable magnetic storm. J. Atmos. Sol.-Ter. Phys. 73 (11), 1444–1446.

Vuong, Q.H., 1989. Likelihood ratio tests for model selection and non-nested hypotheses. Econometrica, 307–333.

Wolff, E.W., Jones, A.E., Bauguitte, S.J.B., Salmon, R.A., 2008. The interpretation of spikes and trends in concentration of nitrate in polar ice cores, based on evidence from snow and atmospheric measurements. Atmos. Chem. Phys. 8 (18), 5627–5634. https://doi.org/10.5194/acp-8-5627-2008. http://www.atmos-chem-phys.net/8/5627/2008/.

Yermolaev, Y.I., Lodkina, I.G., Nikolaeva, N.S., Yermolaev, M.Y., 2013. Occurrence rate of extreme magnetic storms. J. Geophys. Res. 118 (8), 4760–4765.

CHAPTER

DATA-DRIVEN MODELING OF EXTREME SPACE WEATHER

6

A. Surjalal Sharma
University of Maryland, College Park, MD, United States

CHAPTER OUTLINE

1 Introduction 139
2 Data-Driven Modeling of Space Weather 141
3 Predictability of Extreme Space Weather 146
4 Conclusion 149
Acknowledgments 150
References 150
Further Reading 153

1 INTRODUCTION

Extreme events are phenomena in which interaction among complex and interdependent processes leads to large deviations from the typical behavior of the system. They can in turn lead to natural hazards and consequent disasters depending on a combination of physical, social, and other factors. The extreme events exhibit inherent variability and nonequilibrium behavior; therefore the development of models with high predictive capability requires the understanding of the underlying fundamental processes and the interactions among them. Due to the wide range of space and time scales of the processes and their interactions, the modeling of extreme events on the basis of first-principle processes alone remains a challenge. The difficulty arises from many considerations, including the nonequilibrium nature of the natural systems, lack of adequate data, and incomplete understanding of the interactions among the physical processes, including the drivers. The first-principles modeling is based on the fundamental processes underlying extreme events, drawing on the advances in the specific discipline. The modeling of variability of complex large-scale natural systems based on observational data is referred to as data-driven modeling, reflecting the strong dependence on data, and is an essential component of data-enabled science (NSF, 2010). The data-driven modeling approach targets the understanding of extreme events from the viewpoint of statistical physics of nonequilibrium systems, which is active research in which the theoretical foundations are under development. Unlike the statistical physics of equilibrium systems, the theoretical foundations of nonequilibrium systems are under development.

Complexity science is widely recognized as studies of organized behavior of systems that are intermediate between perfect order and perfect disorder. Interactions among interdependent components of a system lead to a competition between organized (interaction-dominated) and irregular (fluctuation-dominated) behavior, and consequently to complexity (Sharma et al., 2012). Early studies of complex systems were focused on model dynamical systems with a small number of degrees of freedom, such as the well-known Lorenz attractor, and have provided deeper understanding of the inherent properties. The understanding achieved with such dynamical systems has enabled a new focus on extended systems with many interdependent components that exhibit regular as well as irregular behavior. In such systems, the behavior of the whole system is more than the sum of its parts (Anderson, 1972). Many natural and anthropogenic phenomena exhibit features readily recognized as typical of complex systems. For example, the plasma in geospace is inherently nonlinear and exhibits instabilities, nonlinear behavior, and nonlinear coupling among the unstable modes. While these processes are complex and lead to irregular behavior, on larger scales they exhibit global features. During a magnetospheric substorm, a multitude of plasma processes—from the microscopic to the macroscopic—act together to yield large-scale coherence. The empirical evidence of the episodic or convulsive nature of the magnetosphere (Siscoe, 1991) is now understood in terms of its low-dimensional dynamics (Sharma, 1995). However, the geospace dynamics are not limited to low-dimensional or global behavior and their multiscale features are evident. Thus geospace exhibits both coherence, described by low dimensionality, and multiscale behavior, arising from the internal dynamics as well as driven by the turbulent solar wind. The modeling and prediction of space weather, including the extreme events, require consideration of both the global and multiscale aspects.

Numerical simulation is an essential approach in the study of a wide variety of complex phenomena. For example, plasmas are inherently nonlinear and simultaneous interaction of a large number of degrees of freedom in such nonlinear systems limits the theoretical analysis. The fundamental laws governing plasmas are known but their consequences cannot be worked out because of the complexity, making numerical simulation an essential approach (Dawson, 1983). In this approach a numerical model of the system is constructed, that is based on the fluid or kinetic plasma descriptions. Then numerical experiments are carried out on a computer, allowing the system to evolve from given initial conditions in accordance with the laws used. This approach or paradigm of numerical simulation, often referred to as the third approach after experiment and theory, has provided much of the advances in our understanding of complex systems.

The recent explosion of data in a wide variety of systems has important implications for data-driven modeling in general. Although science has always been data-driven, recently there has been a dramatic change in the amount of data used in research. There are two origins of large and massive data (Big Data) in research. The first is large-scale numerical simulations that are now carried out on computers with tens of petaflop capacity that can generate exabytes of data. As an example, the simulations of medium-range weather forecast models generate many exabytes, and multiple runs are needed for forecasts with the ensemble modeling approach. The second source of massive data is observations, including imaging by spacecraft-borne imagers. The challenges in research using the massive or complex data have stimulated a new approach that is now referred to as data-enabled science (NSF, 2010). This emerging approach in science is often referred to as the fourth paradigm, the first three being experiment, theory, and simulation (Hey et al., 2009). This new approach is not just data exploration and understanding, but is the use of the data to enable discovery and new understanding. In many nonequilibrium systems including plasmas, significant advances have been achieved with this approach, especially in the modeling of multiscale phenomena (Sharma et al., 2005). In fact, the earliest forecasting tools for space weather were based on data-enabled models developed from the data of the solar wind-magnetosphere coupling (Sharma, 1995).

Recent advances in nonlinear dynamics and complexity science provide a new approach to the understanding of extreme events (Sharma et al., 2012) and this paper presents an overview of the data-driven modeling of extreme space weather. It should be noted that many phenomena in geosciences are inherently nonequilibrium and the data-enabled approach has led to important advances in the modeling of such systems (Ruddell et al., 2013).

2 DATA-DRIVEN MODELING OF SPACE WEATHER

Geospace is a large-scale open system described by a large number of variables, while only a small number of these may be monitored for long durations. In such systems the inherent nonlinearity along with the contraction of its phase space due to dissipation leads to low dimensionality and coherent behavior. The dynamics in such systems can be reconstructed from the time series data of a single variable, based on the embedding theorem (Packard et al., 1980; Takens, 1981; Abarbanel et al., 1993; Vassiliadis et al., 1990; Sharma et al., 1993). The early data-driven models of space weather are based on the correlated time series data of input $I(t)$ and output $O(t)$. The input is typically the solar wind convective electric field given by the product VB_z of the solar flow speed V and the north-south component B_z of the interplanetary magnetic field. The output time series is often the geomagnetic indices, such as the auroral electrojet indices (AL or AE) and the disturbance time index Dst (Sharma, 1995). The data-driven models have been used to provide reliable forecasts of magnetospheric substorms (Vassiliadis et al., 1995; Ukhorskiy et al., 2002; Ukhorskiy et al., 2004a; Chen and Sharma, 2006), relativistic electron fluxes in the radiation belt (Ukhorskiy et al., 2004b), and magnetic storms (Valdivia et al., 1996). Also these studies using time series data have shown that the magnetosphere has features that are clearly global, such as during substorms and storms, and in the same time there are stochastic or multiscale features that lead to their variability. A model for the global features can be obtained by the mean field technique of averaging outputs corresponding to similar states of the system in the reconstructed phase space (Ukhorskiy et al., 2002; Ukhorskiy et al., 2004a,b; Sharma et al., 2005). With the mean-field model, longer-term predictions are obtained that can be used conveniently for forecasting space weather as the model parameters do not need to be changed during operational forecasting.

In this approach, the models are obtained from observational data, and have the advantage of being unencumbered by modeling assumptions. In the time delay embedding technique for the reconstruction of the phase or state space, an m-component phase vector X_i is constructed from a time series $x(t)$ as: $X_i = \{x_1(t_i), x_2(t_i), \cdots, x_m(t_i)\}$, where $x_k(t_i) = x(t_i + (k-1)\tau)$ and τ is a time delay. In an input-output model of the solar wind–magnetosphere system, the solar wind convective electric field VB_s is often used as the input and the geomagnetic activity indices AL, AE, and Dst as the output. Thus the input-output vector in the m dimensional state space is constructed as

$$X_i = (I_1(t_i), I_2(t_i), \cdots, I_{M_I}(t_i), O_1(t_i), O_2(t_i), \cdots, O_{M_O}(t_i)),$$

where the number of dimensions of the input (M_I) and of the output (M_O) can be chosen independently but usually $M_I = M_O = m$. The dimension parameter m is chosen so that the reconstructed space embodies the dynamical characteristics of the system. Such input-output models provide a way to analyze the cause-effect relationship and are used in the studies of causal relationships in space weather. The $2m$-dimensional state vector X_i at $t = t_1, t_2, \ldots t_N$, can now be used to construct a suitable trajectory matrix that embodies the dynamics of the system in an orthonormal vector space (Sharma et al., 1993).

The trajectories in the neighborhood of the current state, at time t, can be used to predict its next position by using its neighboring trajectories (Sharma, 1995; Vassiliadis, 2006). Thus, the trajectories in a small neighborhood of the current state determine the next state and the next output O_{n+1} can be described as a nonlinear function of the input I_n and current output O_n as

$$O_{n+1} = F(I_n, O_n)$$

With a given set of nearest neighbors at $t = n$, their next states ($t = n + 1$) are obtained and their average yields the prediction. The terms other than the average or mean field are not modeled readily in a consistent manner and are often ignored in developing a global model. The predictions of the AL index using the mean-field technique in data-driven modeling are shown in Fig. 1 (Ukhorskiy et al., 2002).

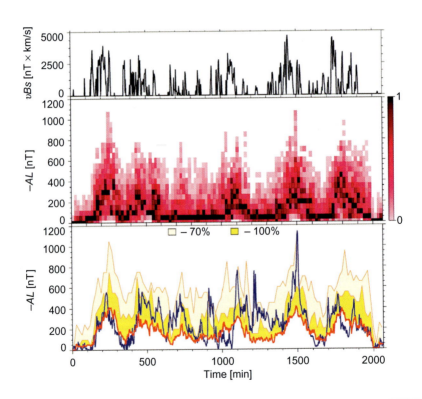

FIG. 1

Prediction of geomagnetic activity by data-driven model using solar wind (*top panel*, VBs) and auroral electrojet index AL (*bottom two panels*). The *middle* panel shows the distribution of past values of AL index for the same solar wind conditions shown in the *top* panel. The bottom panel shows the actual AL (*blue*), predicted AL (*red*) and probabilities of the past actual values exceeding the predictions in 70% (*dark yellow*) and 100% (*light yellow*) of all events (Ukhorskiy et al., 2004a).

These predictions were obtained using the correlated data of the solar wind variable VB_s as the input (*top* panel) and auroral index AL as the output (Bargatze et al., 1985). For each value of the solar wind the corresponding values of the AL index in the database are plotted in Fig. 1 (*middle* panel), with the darker shades showing higher occurrences. These occurrence statistics are the event occurrence statistics and can be used to obtain the probabilities of the AL values. The probabilities of the AL values exceeding the predicted values are obtained using the statistics of the past events, and the two cases of 70% and 100% are shown in Fig. 1 (*bottom* panel).

The data-driven modeling and prediction uses a mean-field model to obtain the predicted value as the mean of the most likely states, namely the nearest neighbors of the current state. The nearest neighbors however can be distributed over a range of values of the input and output variables, in particular for big events, and the model can be improved by assigning weights on the nearest neighbors according to their distance from the current state. A simple weighting of the neighbors by factors inversely proportional to the distances leads to significant improvements in the model and its predictions (Chen and Sharma, 2006). Although the weighted mean-field model captures some of the most abrupt changes, the lack of adequate data of big storms and substorms and their nonequilibrium nature make dynamical prediction of extreme events much less reliable. The distribution of all events, as shown in Fig. 1 (*middle* panel) for AL index, however, provides a measure of the likelihood of extreme events. This approach to quantifying predictability is discussed in detail in the next section.

The global MHD simulations provide the large-scale features of the magnetosphere (Lyon, 2000) and the dynamical features from the simulations have been compared with the data-driven model. For this purpose, the global MHD simulations were carried out for the solar wind conditions used in the data-driven modeling using the Bargatze et al. (1985) data set. The large number of number of substorms in this data set required extensive runs of the global MHD simulations on supercomputers. The simulation data was then used to compute the equivalent of the auroral electrojet index AL and the three leading variables obtained in the same manner as the observational data (Shao et al., 2003). The two results, one from observational data and the other from simulations, yield similar dynamical features and both show phase transition-like behavior (Sharma et al., 2001). It should be noted that the simulations require extensive resources, more than 300 h on supercomputers, and on the other hand the simplified dynamical model derived from data uses a typical workstation.

The continuous monitoring of the solar wind data at the first Lagrange point L1, currently by DSCOVR, ACE, SOHO, and Wind spacecraft, provides rich data to analyze the solar wind driving of the magnetosphere-ionosphere system. Among these spacecraft at L1, ACE has provided extensive data of the solar wind since 1997 and has been used in data-driven modeling for near real-time forecasting of space weather http://spp.astro.umd.edu/gmprediction). Further, the availability of the data from four spacecraft at L1 provides an excellent opportunity for studies of the solar wind conditions driving space weather.

In a study of extreme space weather during the peak of a solar cycle, the time series data from ~60 ground magnetometers in the high-latitude region for the year 2001 (Solar Cycle 23) were used for modeling the geospace during strong driving. Compared to the Bargatze et al. (1985) database, the number of large events in this period was nearly an order of magnitude larger in the solar wind. The corresponding magnetospheric response showed significant increase in the large events, but not as dramatic as in the case of the solar wind, as shown in Fig. 2 (Chen and Sharma, 2006). This result shows the lack of a linear relationship and highlights the understanding that magnetospheric response to solar wind is nonlinear. Although the large values of VBz occurred over a wide range, the large substorms are prominent in a small range around AL ~500 nT.

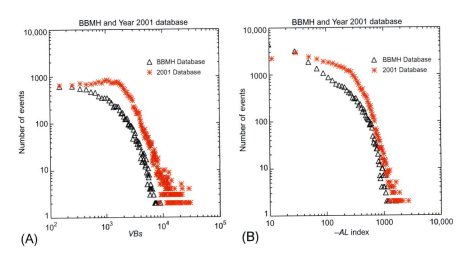

FIG. 2

Statistics of events in Bargatze et al. (1985) and Year 2001 databases. (A) *VBs* (bin size=100); (B) *AL* (bin size=20). The events represent averages over the indicated bin sizes. The differences in the magnetospheric response in the two databases are not as pronounced as in the solar wind variable (Chen and Sharma, 2006).

The forecasting of space weather in terms of the geomagnetic indices, such as *AL* and *Dst*, by data-driven models provides the global features of the geospace disturbances. The spatial structure of geomagnetic activity is, however, essential to the understanding of many phenomena in geospace as well as for regional or local forecasts of space weather. The development of such models has used data from observations by distributed ground-based magnetometers (Valdivia et al., 1999a; Valdivia et al., 1999b) and spacecraft-based imagers (Rosa et al., 1998, 1999).

The data for the period January–June 1979 from 12 ground magnetometers in the auroral region (Kamide et al., 1998) were used to model the spatial structure of space weather. The mutual information function (Abarbanel et al., 1993), which provides the correlation of all orders, was used to obtain the correlations among the stations and the development of substorms in the midnight sector and their spread in longitude were obtained. This result showed the advantage over linear correlation function, which did not yield the observed spatial-temporal development of substorms (Chen et al., 2008). The spatial structure of geomagnetic disturbances in spacecraft images has been analyzed (Uritsky et al., 2006), based on the self-organized criticality (Aschwanden et al., 2016; Sharma et al., 2016). The multiscale nature, seen mostly in the time series data, is seen in the images and provides the spatiotemporal behavior.

The analysis of the time evolution of the spatial structure of space weather from ground-based data requires removal of the diurnal variation, which can be achieved by projecting the data on a coordinate system that is stationary relative to Earth. The time series data of 60 ground-based magnetometers for 2002 were used to develop a spatiotemporal model of geospace disturbances; an example of the predictions is shown in Fig. 3. The upper panel of Fig. 3 shows the solar wind-induced electric field

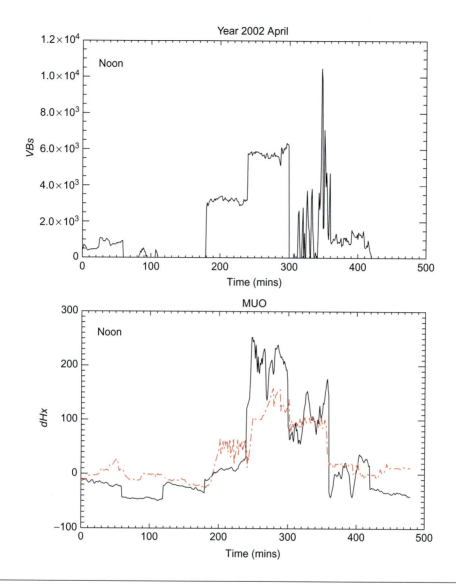

FIG. 3

Dynamical modeling of spatial structure of magnetic field at ground magnetometer station MUO obtained with the spatiotemporal model.

from ACE data and the lower panel shows the time evolution of the horizontal component of the magnetic field (black: actual, *red*: predicted) at magnetometer station MUO (Muonio, Lat. 23.530 degrees north, Long. 68.020 degrees east). This is a typical case of prediction of the local magnetic field variations and in general the predictions track the actual values in some sectors but miss in others (Chen and Sharma, 2017).

3 PREDICTABILITY OF EXTREME SPACE WEATHER

The data-driven modeling provides a direct approach to the use of data to yield the inherent features of the system and prediction of its time evolution. The predictions are based on the evolution of past events (nearest neighbors) in the data that are similar to the current state and are obtained by using techniques designed to capture the leading features of the dynamical trajectories. Among these techniques, the local-linear filter uses a Taylor expansion up to the linear term, and uses the nearest neighbors to obtain the coefficients of the expansion (Sharma, 1995; Vassiliadis, 2006). The variability of the dynamical features, which is usually high in large-scale systems such as space weather, limits the predictive ability of the local-linear filter. The zero-order term represents an average over the nearest neighbors and yields more reliable or consistent predictions, referred to as the mean-field prediction (Ukhorskiy et al., 2002; 2003, 2004a,b; Chen and Sharma, 2006). In the case of extreme events the number of events is small, leading to the inherent difficulty in modeling and prediction. In the distribution of events the extremes lead to the heavy-tail and heavy-tailed distribution functions, such as a Pareto or lognormal are used to model them (Newman, 2005; Clauset et al., 2009). The natural hazards associated with extreme events are rare, but with high impact. New approaches are needed for studies of these low-probability, high-impact events (Sharma et al., 2012). For example, 80% of losses from claims arising from natural hazards come from 20% of events (the 80–20 rule) (Embrechts et al, 1997). Another feature of practical importance is that such distribution functions cannot be fitted reliably to an analytical form and consequently a characterization of the tail region requires other approaches, such as characterization in terms of scaling exponents (Clauset et al., 2009; Setty, 2014; Sharma et al., 2017).

The statistical property of extreme events in the context of the total distribution of events is illustrated in Fig. 4 for geomagnetic activity. The data of the auroral electrojet index AL for the last 30 years yield the distribution function (open circles), which is fitted to a Gaussian distribution, shown by the *green* curve. The large events are negligible in this fit, as expected. The fit to an exponential function, shown by the *red* curve, has significantly more large events than the Gaussian distribution but the big events are largely missing. The tail of the actual distribution function is much more prominent than either of the two functions (Gaussian and exponential), but it cannot be fitted well to an analytic form. Such probability distribution functions (pdf) are common in many natural systems, such as earthquakes, floods, river runoffs, geospace storms, etc. The data on the extreme events are usually limited and the nature of their distribution is not known with high accuracy. However, if the heavy tail of the distribution can be characterized quantitatively, estimates of its contribution to the overall distribution function can be obtained, thus yielding the likelihood or probability of extreme events (Clauset et al., 2009). This ability to yield probabilities without modeling assumptions is an advantage of the data-driven approach to quantifying uncertainty and thus predictability.

The modeling of the extreme events is largely based on their statistical properties and requires a clear separation of the trends in the data, using techniques such as detrended fluctuation analysis (Peng et al., 1994; Kantelhardt et al., 2001; Bryce and Sprague, 2012). The scaling properties of the probability density functions have been studied for many systems in order to characterize extreme events. For example, the studies of the floods of the Nile river using the R/S analysis (Hurst, 1951) led to the well-known Hurst exponent (Feder, 1988; Clauset et al., 2009) and is now widely used in characterizing nonequilibrium phenomena. The scaling exponent reflects the presence of correlations and

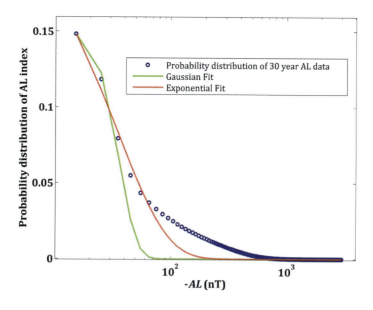

FIG. 4

The distribution function of auroral electrojet index AL (open circles) from the database of the last 30 years. The large events are negligible in the Gaussian fit (*green* curve) and significant in the exponential fit (*red* curve). The extreme events are significant and form the heavy tail of the distribution of events.

in the presence of nonstationarity and periodicity, such as intermittency exhibited by many systems, detrending of the data is widely used in the computation of the scaling exponents from time series data.

The rich database of the auroral electrojet index *AL* provides a suitable data set for the studies of long-range correlations in the magnetosphere and thus for characterizing the extreme events in space weather. In order to analyze the long-range correlations in space weather, the *AL* index data were used (Sharma and Veeramani, 2011). The hourly averaged *AL* for the period 1978–88, which covers a typical 11-year solar cycle and 5-min averaged data for 1978 were used to compare the scaling properties using the detrended fluctuation analysis (DFA). This technique for time series data is accomplished in four steps (Peng et al., 1994; Kantelhardt et al., 2001; Bryce and Sprague, 2012). The first step computes the profile of the time series data x_i as:

$$Y(i) = \sum_{k=1}^{i} (x_k - <x>)$$

The subtraction of the global mean $<x>$ of the dataset however is not essential as the third step, described below, usually removes this and other trends. In the second step, the profile $Y(i)$ is divided into $N_L = N/L$ nonoverlapping segments of length *L*. In order to avoid a loss of data in case N is not a multiple of *L*, the same process is repeated starting from the other end of the data set, thus yielding two N_L segments. In the third step the trends in the data are removed by defining a local trend $q_j(i)$ for each segment *j* by a fitting procedure, such as a least-squares fit. The detrended time series for the segment duration *L* is then defined as:

$$Y_L(i) = Y(i) - q_j(i)$$

In most cases the local trend is usually represented by a polynomial and a quadratic function is widely used. In the fourth step, the variance of each segment $Y_L(i)$ is computed:

$$F_L^2(j) = <Y_j^2(i)> = \frac{1}{L}\sum_{i=1}^{L} Y_L^2((j-1)L+i)$$

This leads to the detrended fluctuation function $F(L)$, which is defined by

$$F^2(L) = \frac{1}{N}\sum_{i=1}^{2N_L} F_L^2$$

In the presence of long-range correlations, the fluctuation function scales as

$$F(L) \approx L^\alpha$$

For uncorrelated or short-range correlated data, such as data with a Gaussian distribution, the exponent $\alpha = 0.5$ and larger values, such as $\alpha > 0.5$, show the presence of long-range correlations.

The detrended fluctuation analysis of the 5-min averaged AL data yields a scaling function $F(L)$ shown in Fig. 5 (Sharma and Veeramani, 2011). Also shown in this figure is the function using the fluctuation analysis (FA) (Peng et al., 1994), which does not remove the trends in the data. The DFA function $F(L)$ yields an exponent $\simeq 0.9$, thus showing long-range correlations. The hourly

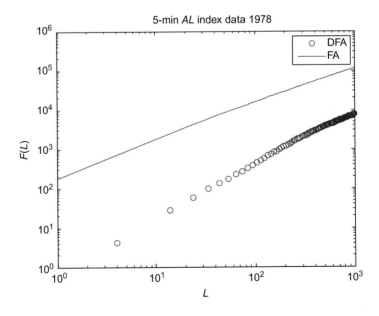

FIG. 5

Detrended fluctuation analysis of 5-min averaged data of AL for 1978. The function F(L) computed using the detrended fluctuation analysis (DFA) yields has a scaling exponent of ~0.9, thus showing the presence of long-range correlations. The fluctuation analysis (FA) yields a similar value.

averaged data covering 11 years yields an exponent of 0.87, and thus the two data sets exhibit similar scaling behavior (Sharma and Veeramani, 2011). The scaling behavior obtained from FA and DFA are similar in the case of the auroral indices, which by definition includes trend removal (Mayaud, 1980). The studies of the times scales in auroral indices (Taka!o and Timonen, 1994) used a structure function that is similar to the FA and showed the break in the scaling obtained earlier from a power law analysis (Tsurutani et al., 1990).

The heavy tail in the distribution of events is often expressed as a power law dependence so that the probability distribution function f of the event size x has a form $f(x) \propto x^{-\gamma}$ for large values of x. The exponent γ is related to the scaling exponents, such as the Hurst exponent, by the so-called Taqqu's theorem (Taqqu et al., 1997). The tail exponent γ (a measure of heavy-tail feature in probability distributions) to the Hurst exponent H (a measure of long-range correlations) are related as $H = (3-\gamma)/2$. For the Hurst exponent value ~ 0.85, that is the cases of AL index and fractional Gaussian noise (Setty and Sharma, 2015), this yields a tail exponent of $\gamma = 1.3$. Another method for estimating the tail exponent is the Hill estimator, which uses the data to construct a convergent series (Hill, 1975). This method however does not yield the tail exponent directly and is known to have inaccuracies connected with data length and crossover features of the exponent (Resnick, 1997). The Hill estimator technique in the case of fractional Gaussian noise yields a tail exponent of 1.29, in agreement with the value of 1.3 obtained using the Hurst exponent and Taqqu's theorem. However the convergence of the exponent in the case of observational data such as the AL index is less clear (Setty, 2014). With a known tail exponent, the power-law tail is well specified and this can be combined with the distribution function of the bulk of the events to obtain the complete distribution function. With the latter the probability of extreme events can then be computed from the distribution function of all events.

4 CONCLUSION

Space weather research is motivated by the quest to understand the effects of solar activity on the near-Earth space environments and the potential impacts on the infrastructure systems and technologies in space and on Earth. The growing importance of extreme space weather events underscores the need to develop the modeling and prediction capabilities for these low-probability, high-impact events. An important strategy in the development of these capabilities is integration of data and models, and data-driven modeling provides such an approach (Sharma et al., 2017). The complexity science is a natural framework for data-driven modeling and has the advantage of being independent of modeling assumptions. Further, the recent explosive growth in massive and complex data (Big Data) from observations and simulations requires development of data-enabled approaches and complexity science provides an important perspective for developing new techniques and methods.

The modeling and prediction of space weather using the reconstruction of phase space from observational data (Sharma, 1995) provided early forecasting tools; this has developed into the characterization of predictability of extreme events. The emphasis in the modeling of extreme events is on quantifying uncertainty in order to specify the likelihood or probability. This requires reliable characterization of the fluctuations in the system, with the large or extreme events included as much as possible. In the distribution of fluctuations, extreme events are essentially the heavy tails of the distribution function. A proper characterization of the fluctuations requires a careful removal of trends in the data and this is achieved with techniques such as detrended fluctuation analysis (Sharma and Veeramani,

2011). This analysis yields the scaling exponent, that is the Hurst exponent, which can be used to specify the nature of the heavy tail in the distribution function, and consequently the probability of extreme events. The data of the geomagnetic indices (*AL* and *Dst*) have been used to obtain the Hurst exponents and they can, in turn, yield the exponents when the heavy tail is represented by a power law. With the details of the heavy tail, the probabilities of events exceeding large values can be obtained. The effectiveness of this procedure, however, is strongly dependent on the availability of large amounts of good data and in practice is subject to constraints arising from limited data lengths and quality. Further, the natural system represented by the data can exhibit additional features such as crossover in the exponent and their implications need to be analyzed with care (Setty, 2014).

The data-driven modeling based on complex systems theory complements other data-enabled techniques such as NARMAX (Wei et al., 2011) and the first-principles-based numerical simulations. An integrated framework for space weather modeling that brings together the different but complementary approaches is very likely the most effective strategy in modeling and prediction of extreme events.

ACKNOWLEDGMENTS

The research at the University of Maryland is supported by NASA and NSF grants.

REFERENCES

Abarbanel, H.D.I., Brown, R., Sidorowich, J.J., Tsimring, L.S., 1993. The analysis of observed chaotic data in physical systems. Rev. Mod. Phys. 65, 1331–1392.

Anderson, P.W., 1972. More is different. Science 177, 393–396.

Aschwanden, M.J., Crosby, N., Dimitropoulou, M., Georgoulis, M.K., Hergarten, S., Jensen, H.J., McAteer, J., Milovanov, A.V., Mineshige, S., Morales, L., Nishizuka, N., Pruessner, G., Sanchez, R., Sharma, A.S., Strugarek, A., Uritsky, V., 2016. 25 Years of self-organized criticality: solar and astrophysics. Space Sci. Rev. 198, 47–166. https://doi.org/10.1007/s11214-014-0054-6.

Bargatze, L.F., Baker, D.N., McPherron, R.L., Hones, E.W., 1985. Magnetospheric impulse response for many levels of geomagnetic activity. J. Geophys. Res. 90, 6387–6394.

Bryce, R.M., Sprague, K.B., 2012. Revisiting detrended fluctuation analysis. Sci. Rep. 2, 315. https://doi.org/10.1038/srep00315.

Chen, J., Sharma, A.S., 2006. Modeling and prediction of the magnetospheric dynamics during intense geospace storms. J. Geophys. Res. 111. A04209. https://doi.org/10.1029/2005JA011359.

Chen, J., Sharma, A.S., 2017. Modeling and prediction of the spatio-temporal structure of space weather. J. Geophys. Res. Submitted.

Chen, J., Sharma, A.S., Edwards, J., Shao, X., Kamide, Y., 2008. Spatio-temporal dynamics of the magnetosphere during geospace storms: mutual information analysis. J. Geophys. Res. 113. A05217. https://doi.org/10.1029/2007JA012310.

Clauset, A., Shalizi, C.R., Newman, M.E.J., 2009. Power-law distributions in empirical data. SIAM Rev. 51, 661–703.

Dawson, J., 1983. Particle simulation of plasmas. Rev. Mod. Phys. 55, 403–447.

Embrechts, P., Kluppelberg, C., Mikosh, T., 1997. Modeling Extreme Events for Insurance and Finance. Springer.

Feder, J., 1988. Fractals. Plenum, New York.

Hey, T., Tansley, S., Tolle, K., 2009. The Fourth Paradigm: Data-Intensive Scientific Discovery. Microsoft Research, Redmond.

Hill, B.M., 1975. A simple general approach to inference about the tail of a distribution. Ann. Stat. 3, 1163–1174.

Hurst, H.E., 1951. Long term storage capacity of reservoirs. Trans. Am. Soc. Civ. Eng. 116, 770–808.

Kamide, Y., Baumjohann, W., Daglis, I.A., Gonzalez, W.D., Grande, M., Joselyn, J.A., McPherron, R.L., Phillips, J.L., Reeves, E.G.D., Rostoker, G., Sharma, A.S., Singer, H.J., Tsurutani, B.T., Vasyliunas, V.M., 1998. Current understanding of magnetic storms: storm-substorm relationships. J. Geophys. Res. 103, 17,705–17,728.

Kantelhardt, J.W., Koscielny-Bunde, E., Rego, H., Havlin, S., Bunde, A., 2001. Detecting long-range correlations with detrended fluctuation analysis. Physica A 295, 441–454.

Lyon, J.G., 2000. The solar wind–magnetosphere–ionosphere system. Science 288, 1987.

Mayaud, P.N., 1980. Derivation, Meaning, and Uses of Geomagnetic Indices. American Geophysical Union, Washington, DC.

Newman, M.E.J., 2005. Power laws Pareto distributions and Zipf's law. Contemp. Phys. 46, 323–351.

NSF. 2010. Data-enabled science in the mathematical and physical sciences, (https://www.nsf.gov/attachments/122032/public/Data_Enabled_Science_Workshop_Report.pdf (last accessed 24.03.2017).

Packard, N.H., Crutchfield, J.P., Farmer, J.D., Shaw, R.S., 1980. Geometry from a time series. Phys. Rev. Lett. 45, 712–716.

Peng, C.K., Buldyrev, S.V., Havlin, S., Goldberger, A.L., 1994. Quantification of scaling exponents and crossover phenomena in nonstationary heartbeat time series. Phys. Rev. E 49, 1685–1689.

Resnick, S.I., 1997. Heavy tail modeling and teletraffic data. Ann. Stat. 25, 1805–1869.

Rosa, R.R., Sharma, A.S., Valdivia, J.A., 1998. Characterization of localized turbulence in plasma extended systems. Physica A 257 (509), 163.

Rosa, R.R., Sharma, A.S., Valdivia, J.A., 1999. Characterization of asymmetric fragmentation patterns in spatially extended systems. Int. J. Mod. Phys. C 10, 147–163.

Ruddell, B.L., Brunsell, N.A., Stoy, P., 2013. Applying information theory in the geosciences to quantify process uncertainty, feedback, scale. Eos. Trans. Am. Geophys. Soc. 94, 56.

Setty, V.A., 2014. Application of Fluctuation Analysis to Characterize Multiscale Nature and Predictability of Complex Systems, Ph. D. Dissertation, University of Maryland, College Park.

Setty, V.A., Sharma, A.S., 2015. Characterizing detrended fluctuation analysis of multifractional Brownian motion. Physica A 419, 698–706.

Shao, X., Sitnov, M.I., Sharma, A.S., Papadopoulos, K., Goodrich, C.C., Guzdar, P.N., Milikh, G.M., Wiltberger, M.J., Lyon, J.G., 2003. Phase transition-like behavior of magnetospheric substorms: global MHD simulation results. J. Geophys. Res. 108, 1037. https://doi.org/10.1029/2001JA009237.

Sharma, A.S., 1995. Assessing the magnetosphere's nonlinear behavior: its dimension is low, its predictability high, (U.S. National Rep. To the IUGG (1991-1994)). Rev. Geophys. 33, 645–650.

Sharma, A.S., Kalnay, E.E., Bonadonna, M., 2017. Predictive capability for extreme space weather events. Eos 98 (7), 9. https://doi.org/10.1029/2017EO071721.

Sharma, A.S., Veeramani, T., 2011. Extreme events and long-range correlations in space weather. Nonlin. Processes Geophys. 18, 719–725. https://doi.org/10.5194/npg-18-719-2011.

Sharma, A.S., Vassiliadis, D.V., Papadopoulos, K., 1993. Reconstruction of low dimensional magnetospheric dynamics by singular spectrum analysis. Geophys. Res. Lett. 20, 335.

Sharma, A.S., Sitnov, M.I., Papadopoulos, K., 2001. Substorm as nonequilibrium transitions of the magnetosphere. J. Atmos. Solar. Terr. Phys. 63, 1399.

Sharma, A.S., Baker, D.N., Borovsky, J., 2005. Nonequilibrium phenomena in the earth's magnetosphere: phase transition, self-organized criticality and turbulence. In: Sharma, A.S., Kaw, P.K. (Eds.), Nonequilibrium Phenomena in Plasmas. Springer, Berlin.

Sharma, A.S., Baker, D.N., Bhattacharyya, A., Bunde, A., Dimri, V.P., Gupta, H.K., Gupta, V.K., Lovejoy, S., Main, I.G., Schertzer, D., von Storch, H., Watkins, N.W., 2012. Complexity and extreme events in geosciences: an overview. In: Sharma, A.S., Dimri, V.P., Bunde, A., Baker, D.N. (Eds.), Complexity and Extreme Events in Geosciences. Geophysical Monograph Series, vol. 196. American Geophysical Union, Washington, DC, pp. 1–16.

Sharma, A.S., Aschwanden, M.J., Crosby, N.B., Klimas, A.J., Milovanov, A.V., Morales, L., Sanchez, R., Uritsky, V., 2016. 25 Years of self-organized criticality: space and laboratory plasmas. Space Sci. Rev. 198, 167–216. https://doi.org/10.1007/s11214-015-0225-0.

Siscoe, G.L., 1991. The magnetosphere: a union of interdependent parts. Eos Trans. Am. Geophys Union 72, 494–495.

Taka!o, J., Timonen, J., 1994. Characteristic time scale of auroral electrojet data. Geophys. Res. Lett. 21 (7), 617–620.

Takens, F., 1981. Detecting strange attractors in "Dynamical Systems and Turbulence, Warwick 1980". In: Lecture Notes in Mathematics, Volume 898. Springer-Verlag. pp. 366–381. ISBN 978-3-540-11171-9.

Taqqu, M.S., Willinger, W., Sherman, R., 1997. Proof of a fundamental result in self-similar traffic modeling. ACM SIGCOMM Comput. Commun. Rev. 27, 5–23.

Tsurutani, B.T., Sugiura, M., Iyemori, T., Goldstein, B.E., Gonzalez, W.D., Akasofu, S.I., Smith, E.J., 1990. The nonlinear response of AE to the IMF Bs driver: a spectral break at 5 hours. Geophys. Res. Lett. 17 (3), 279–282.

Ukhorskiy, A.Y., Sitnov, M.I., Sharma, A.S., Papadopoulos, K., 2002. Global and multiscale aspects of magnetospheric dynamics in local-linear filters. J. Geophys. Res. 107, 1369. https://doi.org/10.1029/2001JA009160.

Ukhorskiy, A.Y., Sitnov, M.I., Sharma, A.S., Papadopoulos, K., 2003. Combining global and multiscale features in the description of solar wind—magnetosphere coupling. Ann. Geophys. 21, 1913.

Ukhorskiy, A.Y., Sitnov, M.I., Sharma, A.S., Anderson, B.J., Ohtani, S., Lui, A.T.Y., 2004a. Data-derived forecasting model for relativistic electron intensity at geosynchronous orbit. Geophys. Res. Lett. 31 (9). L09806. https://doi.org/10.1029/2004GL019616.

Ukhorskiy, A.Y., Sitnov, M.I., Sharma, A.S., Papadopoulos, K., 2004b. Global and multiscale dynamics of the magnetosphere: from modeling to forecasting. Geophys. Res. Lett. 31. L08802. https://doi.org/10.1029/2003GL018932.

Uritsky, V.M., Klimas, A.J., Vassiliadis, D., 2006. Analysis and prediction of high-latitude geomagnetic disturbances based on a self-organized criticality framework. Adv. Space Res. 37, 539–546.

Valdivia, J.A., Sharma, A.S., Papadopoulos, K., 1996. Prediction of magnetic storms using nonlinear models. Geophys. Res. Lett. 23, 2899.

Valdivia, J.A., Vassiliadis, D., Klimas, A.J., Sharma, A.S., Papadopoulos, K., 1999a. Spatio-temporal activity of magnetic storms. J. Geophys. Res. 104 (A6), 12239–12250.

Valdivia, J.A., Vassiliadis, D., Klimas, A.J., Sharma, A.S., 1999b. Modeling the spatial structure of the high latitude magnetic perturbations and the related current systems. Phys. Plasmas 6, 4185–4194.

Vassiliadis, D.V., 2006. Systems theory of Geospace plasma dynamics. Rev. Geophys. 44. https://doi.org/10.1029/2004RG000161.

Vassiliadis, D., Sharma, A.S., Eastman, T.E., Papadopulos, K., 1990. Low dimensional chaos in magnetospheric activity from time-series AE data. Geophys. Res. Lett. 17, 1841–1844.

Vassiliadis, D., Klimas, A.J., Baker, D.N., Roberts, D.A., 1995. A description of the solar wind-magnetosphere coupling based on nonlinear filters. J. Geophys. Res. 100, 3495–3512.

Wei, H.-L., Billings, S.A., Sharma, A.S., Wing, S., Walker, S.N., Boynton, R.J., 2011. Forecasting relativistic electron flux using dynamic multiple regression models. Ann. Geophys. 29, 415–420. https://doi.org/10.5194/angeo-29-415-2011.

FURTHER READING

Boynton, R.J., Balikhin, M.A., Billings, S.A., Sharma, A.S., Amariutei, O.A., 2011. Data derived NARMAX Dst model. Ann. Geophys. 29 (965–971), 2011. https://doi.org/10.5194/angeo-29-965-2011.

Gao, J., Hu, J., Tung, W.W., Cao, Y., Sashar, N., Roychowdhury, V.P., 2006. Assessment of long-range correlations in time series: how to avoid pitfalls. Phys. Rev. E. 73. 016117. https://doi.org/10.1103/PhysRevE.73.016117.

Siscoe, G., 2000. The space-weather enterprise: past, present, and future. J. Atmos. Solar Terr. Phys. 62, 1223–1232.

Sitnov, M.I., Sharma, A.S., Papadopoulos, K., Vassiliadis, D., Valdivia, J.A., Klimas, A.J., 2000. Phase transition-like behavior of the magnetosphere during substorms. J. Geophys. Res. 105, 12955.

Sitnov, M.I., Sharma, A.S., Papadopoulos, K., Vassiliadis, D., 2001. Modeling substorm dynamics of the magnetosphere: from self-organization and self-organized criticality to nonequilibrium phase transitions. Phys. Rev. E. 65. 016116.

PART 3

GEOMAGNETIC STORMS AND GEOMAGNETICALLY INDUCED CURRENTS

CHAPTER 7

SUPERGEOMAGNETIC STORMS: PAST, PRESENT, AND FUTURE

Gurbax S. Lakhina, Bruce T. Tsurutani

Indian Institute of Geomagnetism, Navi Mumbai, India California Institute of Technology, Pasadena, CA, United States

CHAPTER OUTLINE

1 Historical Background 157
2 Present Knowledge About Geomagnetic Storms 159
   2.1 Interplanetary Causes of Intense Magnetic Storms 159
   2.2 Magnetic Storms: Categories and Types 161
   2.3 Some Important Characteristics of Magnetic Storms 161
3 Supermagnetic Storms 164
   3.1 Past Supermagnetic Storms 166
   3.2 Supermagnetic Storms: Present (Space-Age Era) 169
   3.3 Supermagnetic Storms: In Future 171
4 Nowcasting and Short-Term Forecasting of Supermagnetic Storm 173
5 Conclusions 174
Acknowledgments 175
References 177
Further Reading 184

1 HISTORICAL BACKGROUND

A new branch of science, *geomagnetism*, came into existence in AD 1600 after the publication of *De Magnete* by William Gilbert (Gilbert, 1600). Geomagnetism was thought to have great potential for ship navigation at the time. Edmund Halley prepared the first map of the Earth's magnetic field declination by the beginning of the 18th century (Cook, 1998). The daily or diurnal variations of magnetic declination were discovered by George Graham in 1722 (Graham, 1724a,b). In 1741, large magnetic declination perturbations were observed simultaneously in London (by George Graham) and in Uppsala (by Andreas Celsius). Celsius related large magnetic declination perturbations and auroral displays over Uppsala, Sweden (Stern, 2002). The credit for discovering the phenomenon of magnetic storms goes to Alexander von Humboldt of Germany. He was working on a project to record the local magnetic declination in Berlin, each night starting from midnight to morning at intervals of half an hour, from May 1806 until June 1807. On the night of December 21, 1806, von Humboldt observed strong magnetic deflections for six consecutive hours in conjunction with the overhead display of northern lights (aurora borealis). He noticed that when the aurora disappeared at dawn, so did the magnetic fluctuations. Von Humboldt had the acute insight to conclude from these observations that the magnetic disturbances on the ground and the auroras in the polar sky were associated with the same phenomenon. He called this phenomenon "Magnetische Ungewitter" or a magnetic storm (von Humboldt, 1808). Many years later, it was confirmed that such "magnetic storms" were indeed a worldwide phenomena by the observations from the worldwide network of magnetic observatories (Schröder, 1997).

Research on geomagnetic activity and solar activity (sunspot observations) was conducted independently in the beginning of the 19th century. An amateur German astronomer, S. Heinrich Schwabe, began observations of sunspots in 1826. In 1843 Schwabe reported a ~10 year periodic variation of sunspots (Schwabe, 1843). Johann von Lamont reported a ~10 year periodicity in the daily variation of magnetic declination at the Munich Observatory in 1851, but he did not relate it to the sunspot cycle (Lamont, 1867; Schröder, 1997). Edward Sabine, based on the data from the wordwide network of magnetic observatories (Sabine, 1851, 1852), was the first to realize that geomagnetic activity paralleled the then recently discovered sunspot cycle. Thus, a connection between geomagnetic activity and sunspots was established.

On the morning of September 1, 1859, Richard Carrington was observing a large group of sunspots (which we now call an active region, or AR). He was extremely surprised when he saw the sudden appearance of "two brilliant beads of blinding white light" over the sunspots. The intensity of the beads increased with time for a short while and then diminished, and finally the beads disappeared (Carrington, 1859). The whole sequence lasted ~5 min. This is now considered to be the first well-documented observation of a white light (visible) solar flare on record. Richard Carrington was not the only one to record that solar flare, as it was also observed by Richard Hodgson (Hodgson, 1859) from his observatory (also in London). The two observers published simultaneously, giving confirmation of the event. However, the 1859 flare became known as the Carrington flare in recent times. On the very next day, a severe geomagnetic storm was recorded by observatories worldwide, particularly the Kew observatory and the Colaba, Bombay, observatory. Carrington was aware of this fact as he had carefully noted the occurrence of the magnetic storm, but he avoided connecting it with the solar flare. He wrote "one swallow does not make a summer" (Carrington, 1859). William Thomson, later Lord Kelvin, was convinced that there was no connection between solar and geomagnetic activities. In 1863, he showed that the Sun as a magnet was incapable of causing magnetic storms from a direct interaction. During his presidential address to the Royal Society in 1892, he stated, "It seems as if we may also be forced to conclude that the supposed connection between magnetic storms and sunspots is unreal, and that the seeming agreement between periods has been mere coincidence" (Kelvin, 1892). From his exhaustive study of sunspots, Walter Maunder demonstrated a clear latitude drift of sunspots during the sunspot cycle with a well-known butterfly diagram (Maunder, 1904a). He also proved a correlation between geomagnetic disturbances and significant developing sunspots on the surface of the sun (Maunder, 1904b,c, 1905). Whereas Maunder's conclusions about the heliolatitude drift of sunspots were more or less accepted by the astronomical community, his claim about a clear relationship between the sunspots and magnetic storms was bitterly criticized by Chree (1905) and others during the meeting report of the Royal Astronomical Society (1905) (http://adsabs.harvard.edu/abs/1905Obs%E2%80%A6.28%E2%80%A677). Association between large solar flares (arising from active sunspot regions) and magnetic storms was finally established only when sufficient statistics were gathered (Hale, 1931; Chapman and Bartels, 1940; Newton, 1943).

The advent of the space era has provided a tremendous impetus in the study of solar-terrestrial relationships. This has led to an explosion in understanding of geomagnetic storms and their solar and interplanetary causes. As a result, *space weather,* a new branch of space science, has come into existence. Essentially, space weather refers to conditions in the Sun-Earth system that can influence the performance and reliability of space-borne and ground-based technological systems and can endanger human life or health.

Geomagnetic storms are considered to be one of the most important components of space weather. During geomagnetic storms, energy is effectively transferred from the solar wind into the Earth magnetosphere, causing energization of the ring current and radiation belts, intense particle precipitation into the ionosphere, occurrence of large magnetospheric substorms, and formation of giant current loops flowing between magnetosphere and ionosphere. These effects are directly related to many technological impacts, including life-threatening power outages, satellite damage, satellite communication failures, navigational problems, and loss of LEO satellites. During "extreme" geomagnetic storms, all these effects are magnified.

2 PRESENT KNOWLEDGE ABOUT GEOMAGNETIC STORMS

According to the modern definition (Rostoker et al., 1997), a geomagnetic storm is characterized by a "main phase during which the horizontal (H) component of the Earth's low-latitude magnetic fields is significantly depressed over a time span of one to a few hours followed by its recovery, which may extend over several days." It is believed that a geomagnetic storm is caused by the movement of the intensified westward ring current (consisting of $\sim$10–300 keV magnetospheric electrons and ions) closer to Earth. This produces a depression in the H component of the geomagnetic field or the main phase of the geomagnetic storm. The recovery phase of the geomagnetic storm begins when the ring current starts to decay. The physical process of the loss of energetic particles is: charge exchange, Coulomb collisions, wave-particle interactions, and ring current energetic particle convection out of the magnetosphere (magnetopause shadowing) (Kozyra and Liemohn, 2003). The intensity of a geomagnetic storm is measured by the Dst index or by the SYM-H index. Both indices measure the symmetric ring current intensity (Iyemori, 1990). Whereas the Dst is an hourly index expressing the intensity of the ring current, the SYM-H index is the same, but computed at a 1-min time cadence (see Lakhina and Tsurutani, 2016). During magnetic storms, auroral activity intensifies and the region of auroral activity expands. Consequently, the auroras are no longer confined to the auroral oval (60–70 degrees). Auroras during magnetic storm main phases can be observed at subauroral to mid-latitudes regions.

2.1 INTERPLANETARY CAUSES OF INTENSE MAGNETIC STORMS

The main mechanism of energy transfer from the solar wind to the Earth's magnetosphere is magnetic reconnection (Dungey, 1961). Fig. 1 shows a schematic of the magnetic reconnection process. The Sun is shown on the left side and the Earth's magnetosphere on the right side of Fig. 1. The interplanetary magnetic field (IMF) carried by the solar ejecta (ICME, for example) is shown in the middle of Fig. 1. The IMF is directed southward which is opposite to the dayside magnetospheric magnetic field, a

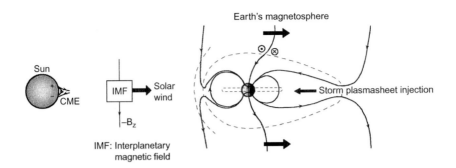

FIG. 1

A schematic of the magnetic reconnection process for the case of oppositely directed IMF and dayside Earth's magnetosphere. The magnetic reconnection subsequently leads to injection of plasma in the nightside magnetosphere.

From Tsurutani, B.T., Gonzalez, W.D., Lakhina, G.S., Alex, S., 2003. The extreme magnetic storm of 1–2 September 1859, J. Geophys. Res. 108(A7), 1268. doi:10.1029/2002JA009504.

situation most favorable for the magnetic reconnection process to operate. As a result of reconnection, the magnetic field lines on the dayside magnetosphere are eroded, and are transported to the nightside magnetotail region. Such accumulation of the magnetic field lines in the nightside magnetotail region in turn drives magnetic reconnection in the nightside magnetotail. This causes near-midnight plasma injection that leads to the excitation of auroras at high-latitude nightside regions. The energetic protons and electrons of the injected plasma drift to the west and to the east, respectively, forming a ring of current around the Earth. This "ring current" is responsible for producing a depression in the H component of the Earth's magnetic field recorded at near-equatorial observatories. It is found that the decrease in the equatorial magnetic field strength is directly proportional to the total energy of the ring current particles and, thus, could be used as a proxy for the energetics of the magnetic storm (Dessler and Parker, 1959; Sckopke, 1966; Carovillano and Siscoe, 1973).

Solar phenomena (flares, coronal holes) and related interplanetary phenomena (coronal mass ejections or ICMEs and co-rotating interaction regions or CIRs), can directly drive geomagnetic storms (Gonzalez et al., 1994; Tsurutani et al., 1995a,b, 2006). The main cause of intense magnetic storms is long duration (hours) southward IMFs (Echer et al., 2008a). The southward IMFs considerably enhance the efficiency of magnetic reconnection, the process that transfers energy from solar wind to the magnetosphere (Tsurutani and Gonzalez, 1997). This causes strong plasma injection from the magnetotail towards the inner magnetosphere leading to ring current intensification.

We point out that a one-to-one relationship between the occurrence of solar flares and ICMEs does not exist. Also there is no strong association between the flare intensity and the speed and magnetic field strength of the ICME. Several authors have reported ICME-related intense (-250 nT $<$ Dst <-100 nT) magnetic storms that are not associated with solar flares (Tsurutani et al., 1988; Tang et al., 1989; Tang and Tsurutani, 1990; Kamide and Kusano, 2015). Further, only the magnetic cloud (MC) portions of ICMEs and not their upstream sheaths can cause magnetic storms with intensities Dst <-250 nT (Tsurutani et al., 1992a,b; Echer et al., 2008b). Tsurutani et al. (1995a, b) have shown that CIRs do not appear capable of causing such intense storms.

However, it is found that *large* solar flares (energies $\sim 10^{24}$–10^{25} J) always occur together with CME releases (Burlaga et al., 1981; Klein and Burlaga, 1982). This is because magnetic reconnection at the Sun is responsible for both phenomena at these intense levels (Shibata et al., 1995; Magara et al., 1995; Benz, 2008; Chen, 2011; Shibata and Magara, 2011). Only such intense CMEs/flares are relevant for the supermagnetic storms (Dst < -500 nT) considered in this paper.

2.2 MAGNETIC STORMS: CATEGORIES AND TYPES

Magnetic storms can be grouped into two categories as shown in Fig. 2. *Sudden impulse* (SI^+) storms are characterized by a sudden increase in the horizontal magnetic field intensity shortly before the main phase (see top panel of Fig. 2). These storms are typically caused by fast ICMEs (a requirement for producing an upstream shock). The impact of the interplanetary shock ahead of the ICME compresses the magnetosphere leading to a sudden increase in magnetic field strength, called a sudden impulse. The period between the SI^+ and the onset of the storm main phase is called the *initial phase*. However, all magnetic storms do not have initial phases. Geomagnetic storms that do not have an SI^+ are called *gradual geomagnetic* (SG) *storms* (bottom panel of Fig. 2).

Each category of storms is further classified into two types depending upon how the main phase is achieved. In single-step or Type 1 storms, the main phase occurs in one step as shown in the top panel of Fig. 3. Here, the ring current is intensified due to the magnetotail injection of energetic particles and decays to a prestorm level in one step. However, in Type 2 or in two-step storms, the main phase undergoes a two-step growth in the ring current, that is, before the ring current has decayed to a significant prestorm level, a new major particle injection occurs, leading to further buildup of ring current and further decrease of Dst. Two-step storms are caused by the compressed southward IMF in the sheath region downstream of the ICME shocks (first main phase) followed by the southward fields of magnetic cloud (second main phase) (Tsurutani and Gonzalez, 1997; Kamide et al., 1998). In general, depending upon the solar and interplanetary conditions, multi-ring current injections can occur that can cause Type 3 (three-step) or even higher step magnetic storms (Richardson and Zhang, 2008).

2.3 SOME IMPORTANT CHARACTERISTICS OF MAGNETIC STORMS

As stated earlier, the intensity of the magnetic storm is measured by the Dst index or SYM-H index at the peak of the main phase. The magnetic storms are called weak when Dst > -50 nT, moderate when $-50 >$ Dst > -100 nT, and intense when Dst < -100 nT (Kamide et al., 1998) and superintense when Dst < -500 nT (Tsurutani et al., 2003; Lakhina et al., 2005, 2012; Lakhina and Tsurutani, 2016).

Taylor et al. (1994) have shown that SI^+ magnetic storms result from interplanetary shocks associated with ICMEs while the gradual storms are caused by fast-slow stream interfaces and CIRs.

Yokoyama and Kamide (1997) conducted a superposed epoch analysis of more than 300 storms and concluded that the southward component of the IMF plays a crucial role in both triggering the main phase and in determining the magnetic storm intensity. Interestingly, the strength of the intense magnetic storm and its main-phase duration were found to be directly proportional to the strength and duration of the IMF B_z, respectively (Vichare et al., 2005; Alex et al., 2006; Rawat et al., 2007, 2010). A high solar wind dynamic pressure, in the presence of a steady southward IMF B_z with a large magnitude, is found to enhance the ring current energy leading to a severe geomagnetic storm (Rawat et al., 2010). Yokoyama and Kamide (1997) also found that the magnetic storm intensity

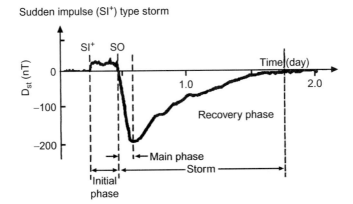

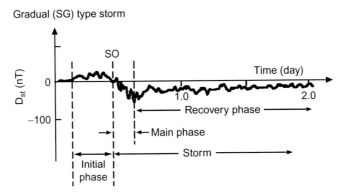

FIG. 2

Schematics of a magnetic storm: sudden impulse (SI⁺) type driven by an ICME (top panel) and gradual (SG) type caused by a CIR (bottom panel). All storms may not have initial phases. The SO stands for storm onset (i.e., onset of the storm main phase).

From Tsurutani, B.T., et al., 2006. Corotating solar wind streams and recurrent geomagnetic activity: a review. J. Geophys. Res. 111, A07S01. doi:10.1029/2005JA011273; Lakhina, G.S., Alex, S., Tsurutani, B.T., Gonzalez, W.D., 2012. Super magnetic storms: hazard to society. In: Sharma, A.S., Bunde, A., Dimri, V.P., Baker, D.N. (Eds.), Extreme Events and Natural Hazards: The Complexity Perspective, Geophys. Mon. Ser., vol. 196. AGU, Washington, p. 267. doi:10.1029/2011GM001073.

depends on the duration of the main phase. The more intense storms have a longer main phase, but this relationship does not seem to apply for superintense magnetic storms.

Generally, ICME-driven storms have higher intensities (intense to superintense storm level) and shorter durations as compared to CIR-driven storms (moderate to intense storm level) intensities and durations. However, CIR-driven storms are associated with hotter plasma sheets and higher fluxes of relativistic magnetospheric electrons. Intense substorms or supersubstorms (Tsurutani et al., 2015; Hajra et al., 2016) are found to be hazardous not only to the spacecraft but also to the performance of the onboard instruments. For example, Galaxy 15 anomalies were caused by electron injection events

2 PRESENT KNOWLEDGE ABOUT GEOMAGNETIC STORMS

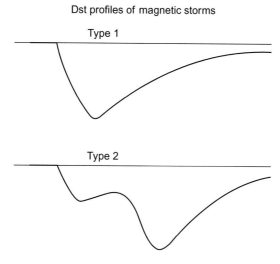

FIG. 3

Schematics of the Dst profile of Type 1 (top panel) and Type 2 (bottom panel) geomagnetic storms.

From Kamide, Y., Yokoyama, N., Gonzalez, W., Tsurutani, B., Daglis, I., Brekke, A., Masuda, S., 1998. Two-step development of geomagnetic storms. J. Geophys. Res. 103(A4), 6917–6921.

during an intense substorm (Allen, 2010). Further, the ionospheric currents due to intense substorms rather than the ring current intensity (Dst) are found to control the severity of geomagnetically induced currents (GICs) that can harm ground-based technologies, such as power grids and long pipe lines, etc. Furthermore, neither the strength of the superintense substorms nor their occurrence rate show any strong association with the intensity of the associated geomagnetic storms (Gonzalez et al., 1999; Borovsky and Denton, 2006; Kataoka and Miyoshi, 2006; Tsurutani et al., 2006, 2015; Richardson et al., 2006; Huttunen et al., 2008; Echer et al., 2008a; Allen, 2010; Yermolaev et al., 2012; Hajra et al., 2016).

Tsurutani et al. (2006) showed that the number of intense (Dst < −100 nT) storms caused by ICMEs follow the solar cycle sunspot number. However, for weak-to-moderate storms caused mainly by CIRs, there is a much smaller solar cycle dependence. The CIR-generated magnetic storms were found to have very long "recovery phases" as compared to those driven by ICMEs. The CIR "recoveries" in the high-speed streams proper can be weeks long (Tsurutani et al., 1995a,b). Further, coincident with intervals of high-speed solar wind streams, relativistic "killer" electrons suddenly appear in the outer magnetosphere during the "recovery" phase of the magnetic storms (Chapter 14 of this volume). Such energetic electrons can pose great danger for Earth-orbiting spacecraft. At present, the exact mechanism for relativistic electron acceleration in the outer magnetosphere is not known. However, there are two popular mechanisms: electron radial diffusion due to ultralow frequency (ULF) waves that break the particle's third adiabatic invariant (Hudson et al., 2000; Su et al., 2015), and energy diffusion by cyclotron-resonant interactions of electrons with a chorus that breaks their first adiabatic invariant (Summers et al., 2004; Hajra et al., 2015). For more details about the mechanisms of appearance and dropout of relativistic electrons in the radiation belts, we refer the reader to Chapter 14 of this volume.

Alex et al. (2005) and Rawat et al. (2006) found that solar energetic particle (SEP) events with high flux levels or a "plateau" after the shock passage produce much more intense storms than the events where the SEP flux levels decrease after the shock passage. Further, SEP events having longer preshock southward IMF B_z duration produced stronger main-phase storms. These results can be used as precursory signature for intense magnetic storms for space weather studies.

Tsurutani et al. (2008) studied the effects of prompt penetration electric fields (PPEFs) on the ionosphere during the great magnetic storm of October 30–31, 2003. They found that PPEFs cause the uplift of equatorial ionosphere leading to a much wider equatorial ionization anomaly in latitude. Further, the total electron content (TEC) increased from the pre-PPEF value of ~50–70 TECU to a peak value of ~270–330 TECU during the peak PPEF period.

Gonzalez et al. (2011) have studied the solar cycle and seasonal distributions of the occurrence of intense geomagnetic storms in the space era. An important result of this study is that intense storms have a dual-type distribution in the solar cycle, one at solar maximum and the second at the descending phase of the cycle, and a seasonal distribution showing the equinoctial peaks and an additional peak in July.

3 SUPERMAGNETIC STORMS

Supermagnetic storms (Dst < -500 nT) are relatively rare. But they are of utmost importance to society because of our ever-increasing dependence on sophisticated technology in space and on the ground. Supermagnetic storms can pose many threats: the safety of astronauts due to harmful radiation, damage to satellites, satellite communication and navigational failures, life-threatening power outages, corrosion of long pipelines due to strong geomagnetically induced currents (GICs), loss of LEO satellites, and disruption of cell phone and general communication usage (NRC Report, 2008; RAE Report, 2013; Lei et al., 2008; Thayer et al., 2008; Mannucci et al., 2005; Tsurutani et al., 2012; Lakhina et al., 2012). It is, therefore, crucial to have knowledge about the causes and occurrence of supermagnetic storms such as Carrington-type events in order to assess their potential for damaging society (Tsurutani et al., 2003; Cliver and Svalgaard, 2004; Vasyliunas, 2011; Lakhina et al., 2012; Hapgood, 2012; Riley, 2012; Cliver and Dietrich, 2013; Cid et al., 2014).

Table 1 gives a partial list of intense and superintense storms based on the magnetic field data from the Colaba and Alibag Observatories (Lakhina et al., 2012). It is noted that within the space era (since 1958), only one true supermagnetic storm has occurred. It happened on March 13–14, 1989, and had an intensity of Dst $= -589$ nT (SYM-H $= -710$ nT). The Canadian Hydro-Quebec power system was damaged due to the intense ionospheric currents during this storm (Allen et al., 1989; Bolduc, 2002). Another very intense magnetic storm, reaching almost the level of a superstorm with Dst < -490 nT, occurred on November 20, 2003. Only these two events can qualify as possible superstorms during the space era. Before the space age, a regularly maintained magnetic observatory network has been in existence for the past ~175 years, and many superintense storms seem to have occurred as seen from Table 1. Therefore, research on historical geomagnetic storms can extend the database of magnetic storms of superintensities (Lakhina et al., 2005, 2012).

Table 1 A List of Intense and Superintense Magnetic Storms From Colaba and Alibag Magnetic Observatories

Sr. No.	Year	Month	Day	ΔH (nT)	Dst (nT)	Remark
1	1847	September	24	471.3		Colaba
2	1847	October	23	534.8		Colaba
3	1848	November	17	404.0		Colaba
4	1857	December	17	306.0		Colaba
5	1859	September	1–2	1722.0		Colaba
6	1859	October	12	984.0		Colaba
7	1872	February	4	1023.0		Colaba
8	1872	October	15	430.0		Colaba
9	1882	April	17	477.0		Colaba
10	1882	November	17	445.0		Colaba
11	1882	November	19	446.0		Colaba
12	1892	February	13	612.0		Colaba
13	1892	August	12	403.0		Colaba
14	1894	July	20	525.0		Colaba
15	1894	August	10	607.0		Colaba
16	1903	October	31	819.0		Colaba
17	1909	September	25	>1500.0		Alibag
18	1921	May	13–16	>700.0		Alibag
19	1928	July	7	779.0		Alibag
20	1935	June	9	452.0		Alibag
21	1938	April	16	532.0		Alibag
22	1944	December	16	424.0		Alibag
23	1957	January	21	420.0	−250	Alibag
24	1957	September	4–5	419.0	−324	Alibag
25	1957	September	13	582.0	−427	Alibag
26	1957	September	29	483.0	−246	Alibag
27	1958	February	11	660.0	−426	Alibag
28	1958	July	8	610.0	−330	Alibag
29	1960	April	1	625.0	−327	Alibag
30	1972	June	18	230.0	−190	Alibag
31	1972	August	9	218.0	−154	Alibag
32	1972	November	1	268.0	−199	Alibag
33	1980	December	19	479.0	−240	Alibag
36	1981	March	5	406.0	−215	Alibag
35	1981	July	25	367.0	−226	Alibag
37	1982	July	13–14	410.0	−325	Alibag
38	1982	September	5–6	434.0	−289	Alibag
39	1986	February	9	342.0	−307	Alibag
40	1989	March	13	Loss	−589	Alibag

Continued

Table 1 A List of Intense and Superintense Magnetic Storms From Colaba and Alibag Magnetic Observatories—cont'd

Sr. No.	Year	Month	Day	ΔH (nT)	Dst (nT)	Remark
41	1989	November	17	425.0	−266	Alibag
42	1991	March	24	Loss	−298	Alibag
43	1992	February	9	225.0	−201	Alibag
44	1992	February	21	304.0	−171	Alibag
45	1992	May	10	503.0	−288	Alibag
46	1998	September	25	300.0	−207	Alibag
47	2000	April	6	384.0	−288	Alibag
48	2000	July	15	407.0	−301	Alibag
49	2001	March	31	480.0	−358	Alibag
50	2001	April	11	332.0	−256	Alibag
51	2001	November	6	359.0	−277	Alibag
52	2001	November	24	455.0	−213	Alibag
53	2003	August	18	254.0	−168	Alibag
54	2003	October	29	441.0	−345	Alibag
55	2003	October	30	506.0	−401	Alibag
56	2003	November	20	749.0	−472	Alibag
57	2004	July	27	342.0	−182	Alibag
58	2004	November	8	459.0	−383	Alibag
59	2005	May	15	352.0	−263	Alibag
60	2005	August	24	457.0	−216	Alibag

From Lakhina, G.S., Alex, S., Tsurutani, B.T., Gonzalez, W.D., 2012 Super magnetic storms: hazard to society. In: Sharma, A.S., Bunde, A., Dimri, V.P., Baker, D.N. (Eds.), Extreme Events and Natural Hazards: The Complexity Perspective, Geophys. Mon. Ser., vol. 196. AGU, Washington, p. 267. doi:10.1029/2011GM001073.

3.1 PAST SUPERMAGNETIC STORMS

One can note from Table 1 that several historical geomagnetic storms might have been of superintensities. The magnetic storms of September 1–2, 1859, (Carrington event) and February 4, 1872, have attracted great attention because of the unusual auroral sighting at low latitudes, electrical shocks and fires caused by arcing from currents induced in telegraph wires in Europe and the United States, and the disruption of telegraph communications at that time. Modern-day knowledge about the interplanetary and solar causes of the magnetic storms, and the older observations of the auroras and other effects attributed to magnetic storms, can be used to analyze historic magnetograms to deduce the possible causes of historic magnetic storms and their intensities. This approach has been applied successfully for the Carrington storm (Tsurutani et al., 2003; Lakhina et al., 2005, 2012; Lakhina and Tsurutani, 2016) and for the February 4, 1872, and other storms where interplanetary data is not available, such as March 13–14, 1989, by Lakhina and Tsurutani (2016). Here we will recapitulate some main points concerning the Carrington storm, which we believe to be the most intense magnetic storm (Dst = −1760 nT) in recorded history.

3 SUPERMAGNETIC STORMS

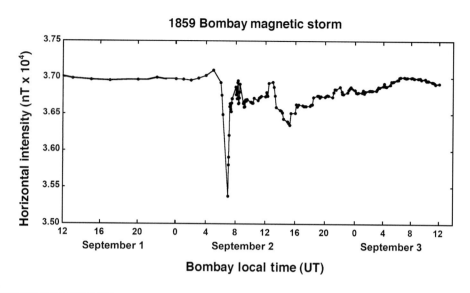

FIG. 4

The Colaba (Bombay) magnetogram for the September 1–2, 1859 magnetic storm.

From Tsurutani, B.T., Gonzalez, W.D., Lakhina, G.S., Alex, S., 2003. The extreme magnetic storm of 1–2 September 1859. J. Geophys. Res. 108(A7), 1268. doi:10.1029/2002JA009504.

Tsurutani et al. (2003) reduced the Colaba Observatory (Mumbai, India) ground magnetometer data of September 1–3, 1859. The auroral reports based on newspapers and personal correspondences with Sydney Chapman (Kimball, 1960; Loomis, 1861), and recently obtained space-age knowledge of interplanetary causes of intense storms were applied to determine the probable causes of this supermagnetic storm event. Here, we will briefly review the main characteristics of this storm (Tsurutani et al., 2003; Lakhina et al., 2005, 2012).

Fig. 4 shows the horizontal component magnetogram of September 1–3, 1859, as deduced from the Colaba Observatory (Mumbai, India) recordings by Tsurutani et al. (2003). The amplitude of the SI^+ preceding the magnetic storm is $\sim+120$ nT. The maximum depression of the H-component at the peak of the storm main phase is $\Delta H \approx -1600$ nT. The duration of the main phase of the storm is ~ 1.5 h. Auroral observations (Kimball, 1960; Loomis, 1861) were used to deduce the position of the plasmapause to be at $L=1.3$. From this information, the magnetospheric convection electric field was determined to be $E_c \sim 20$ mV m^{-1}. Assuming a 10% magnetic reconnection efficiency (Gonzalez et al., 1989), the interplanetary solar wind electric field was estimated to be $E \sim 200$ mV m^{-1}.

Considering the transit time of the ICME from the Sun to the Earth of ~ 17 h and 40 min (Carrington, 1859), the average shock transit speed is found to be $V_{shock}=2380$ km s^{-1}. The relationship, $V_{sw}=0.775\, V_{shock}$, between the solar wind speed, V_{sw}, at 1 AU and the average shock transit speed derived by Cliver et al. (1990), gives solar wind speed $V_{sw} \sim 1850$ km s^{-1} at 1 AU. Then, the relationship, B (nT) $=0.047\, V_{sw}$ (km s^{-1}), between the V_{sw} and magnetic field B of the ejecta (Gonzalez et al., 1998) predicts the magnetic cloud magnetic field magnitude B to be ~ 90 nT at 1 AU. Therefore, the maximum possible interplanetary electric field for this ICME can be $E \sim 160$ mV m^{-1}. Incidentally, this value of the interplanetary electric field compares well with the above estimate based on auroral location and reconnection efficiency ($E \sim 200$ mV m^{-1}). The peak intensity of this superintense storm was estimated to be Dst = -1760 nT by assuming the ring current decay time of 1.5 h (same as the main-phase duration) and using the empirical relation for the evolution of the ring current (Burton et al., 1975). This value of the peak intensity is consistent with the Colaba local measurement of $\Delta H = -1600$ nT. Analyzing the profile of the Colaba magnetogram and by a process of elimination, Tsurutani et al. (2003) deduced that the most likely mechanism for this intense, short-duration superstorm would be a magnetic cloud with intense southward magnetic fields. It has been proposed that the second and third depressions in Dst in Fig. 4 were probably caused by the new ring current injections from the successive ICMEs near the end of the fast-recovery phase of the main (Carrington) storm, thus prolonging the overall "recovery" of the complex storm (Lakhina et al., 2012; Lakhina and Tsurutani, 2016).

Based on computer simulations, Li et al. (2006) have suggested that a high-density plasma plug could reproduce such a short time scale for the Carrington storm event with a very fast recovery after the main phase of the storm (Cid et al., 2014). Although the exact nature of this plasma plug has not been specified, it could be the high plasma density solar filaments (the most sunward part of CMEs) that are shown to play prominent roles in extreme ICME events (Kozyra et al., 2013). More detailed research and simulations are needed to test this hypothesis or to identify another possible explanation.

We point out that there is some disagreement on the minimum value of the Dst = -1760 nT for the Carrington storm in the literature. The earliest estimate for the Dst minimum for this storm was Dst = -2000 nT by Siscoe (1979) which has been revised to Dst = -850 nT by Siscoe et al. (2006). Their revised estimate is based on the assumption that taking hourly averages of the Colaba magnetogram can act as a proxy for Dst. This yielded the maximum H excursion of -859 nT instead of -1600 nT as in the unaveraged magnetogram. Siscoe et al. (2006) claim that the ICME sheath caused ~ -1600 nT depression in H seen in the Colaba magnetogram. Furthermore, some authors have raised the issue that the large drop in H component recorded at Colaba during the Carrington storm could have been caused mainly by field-aligned currents (Akasofu and Kamide, 2005; Cid et al., 2015). In our opinion the above objections are not convincing (Tsurutani et al., 2005; Lakhina and Tsurutani, 2016). Firstly, it should be noted that the hourly average of the H component of the Colaba magnetogram, or any other magnetic observatory, does not represent the true Dst. The standard Dst is based on hourly observations from several magnetic observatories widely distributed in longitude rather than from a single observatory. Secondly, the estimate of minimum Dst for the Carrington magnetic storm derived by Tsurutani et al. (2003) and Lakhina et al. (2005) is based on the calculation of the interplanetary electric field by two independent methods, namely the auroral observations and from the shock transit time from the Sun to the Earth and use of relationships between interplanetary parameters (Cliver et al., 1990; Gonzalez et al., 1998). This value of the interplanetary electric field is used in the Burton relation (Burton et al., 1975) along with a ring current decay time of 1.5 h (taken from the Colaba magnetogram) to derive the minimum Dst = -1760 nT. This value is consistent with the H depression of -1600 nT in the Colaba magnetogram. Nowhere it is implied that the H depression of -1600 nT represents the Dst value of the storm. This is a single-station observation. Incidently, interplanetary sheath fields are unlikely to have caused the extraordinary H component depression of ~ 1600 nT in the Colaba magnetogram as claimed by Siscoe et al. (2006). The shocks ahead of the fast ICME can compress the magnetic field to a maximum value of four times (Kennel et al., 1985) that of the quiet field. Therefore, ICME sheath fields are expected to be ~20–40 nT which are much too low to produce the required interplanetary electric fields that can cause the H depression of ~1600 nT seen in the Colaba magnetogram (Kamide et al., 1998; Tsurutani et al., 2003). Thirdly, it should be noted that the Colaba Observatory is a near-equatorial (10^0) station. Field-aligned current cannot contribute significantly to the magnetic signature at Colaba Observatory as its location is away from the equatorial electrojet influence as well as far away from auroral ionospheric current influences during supermagnetic storms (Tsurutani et al., 2005; Lakhina and Tsurutani, 2016).

3.2 SUPERMAGNETIC STORMS: PRESENT (SPACE-AGE ERA)

As previously mentioned, the March 13–14, 1989, storm is the only event in the space-age era that can qualify as a supermagnetic storm (Allen et al., 1989). Unfortunately, the solar wind data upstream of the Earth's magnetosphere during this event is not available. The ground magnetograms show that this magnetic storm was quite unusual with a long and complex main phase.

Fig. 5 shows Dst for the Hydro-Quebec event. The peak Dst value attained at the end of the main phase was −589 nT. The corresponding maximum SYM-H value was ~−710 nT (Lakhina and Tsurutani, 2016). The storm main phase lasted for ~23 h with an onset at ~0200 UT March 13 and an approximate end at ~0100 UT March 14. The superstorm started with an SI^+ at about 0128 UT on March 13 with a magnitude of about +40 nT. There was a second SI^+ at 0747 UT on the same day with a magnitude of ~+80 nT (Allen et al., 1989). The first main phase of the storm, starting from ~0230 UT to up until ~0900 UT, was probably caused by the sheath magnetic fields, and it produced a Dst of ~−150 nT. During the second main phase lasting from ~1100 to 1200 UT, there was a sharp and smooth drop of Dst to ~−250 nT. This suggests that the second storm main phase was probably caused by a magnetic cloud with a southward B_z component. The third main phase of the storm has multiple sharp Dst excursions, most likely caused by multiple magnetic cloud southward magnetic fields. The Dst attained a peak value of −589 nT at 0100 UT on March 14. The recovery phase of the storm started ~0100 UT on March 14 and lasted till 0600 UT that day.

Fortunately, the interplanetary plasma and magnetic field data and ground magnetic data were available for the Halloween (October 29–30, 2003) and November 20, 2003, near supermagnetic storms. The intense Halloween storms were caused by two fast ICMEs with speeds ~2000 km s^{-1} (Mannucci et al., 2005; Alex et al., 2006). The impact of the first ICME caused a double storm. The first main phase due to southward sheath magnetic field occurred at ~0900 UT on October 29 with peak Dst of about ~−200 nT. The second main phase was caused by the southward magnetic cloud magnetic fields. The peak Dst value of ~−350 nT was attained at ~0125 UT on October 30. The total duration of the storm main phase was ~18 h. However, before the storm "recovery" could be completed, a second ICME impacted the Earth's magnetosphere. The strong southward field within the magnetic cloud caused a new single-step storm with a peak Dst of ~−400 nT at 2315 UT on October 30 (Mannucci et al., 2005). The main phase of this storm lasted for ~5 h.

The near supermagnetic storm of November 20, 2003, was a single-step storm with peak Dst ~ −490 nT (Alex et al., 2006; Mannucci et al., 2008). An intense southward magnetic field in the magnetic cloud portion of an ICME travelling with a speed of about 1100 km s^{-1} was responsible for causing this near superintense storm. The storm main phase lasted for ~8 h.

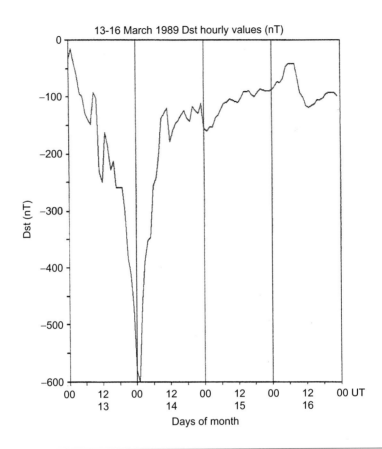

FIG. 5

Dst profile of the March 1989 supermagnetic storm from March 13–16.

From Silbergleit, V.M., Zossi de Artigas, M.M., Manzano, J.R., 1996. Austral electrojet indices derived for the great storm of March 1989. Ann. Geofis. XXXIX(6), 1177–1184.

The study of magnetic storms in the space-age era has clearly shown that the intensity of a magnetic storm depends on both the amplitudes and duration of the southward magnetic fields within the causative ICMEs.

On July 23, 2012, an extremely fast ICME with an initial speed of 2500 ± 500 km s^{-1} directed away from the Earth was observed by STEREO-A. The magnetic cloud of this ICME had an average transit speed of 1910 km s^{-1} with a peak magnetic field strength of 109 nT at 1 AU. Had this extreme ICME been directed toward Earth, it would have produced a superintense magnetic storm with Dst $= -1182$-nT (Baker et al., 2013; Russell et al., 2013; Ngwira et al., 2013a,b). Liu et al. (2014) have suggested that an interaction of two successive CMEs emitted from the Sun produced such a strong magnetic cloud magnetic field.

3.3 SUPERMAGNETIC STORMS: IN FUTURE

Since modern society is becoming increasingly dependent on sophisticated space and ground technologies, future extreme events like the Carrington event can cause much more damage to society. Therefore, it is vital to know whether extreme events like the Carrington storm or even more intense storms can occur in the future. And what is the probability of occurrence of such extreme supermagnetic storms?

3.3.1 Maximum possible intensity of a supermagnetic storm

Tsurutani and Lakhina (2014) have investigated the maximum intensity of a superstorm and other related effects when a possible extreme CME hits the magnetosphere. We shall discuss the main highlights of their model. Tsurutani and Lakhina (2014) made two simple assumptions, (1) they considered the extreme value of CME speeds to be 3000 km s^{-1} near the Sun, and (2) they considered a $\sim$10% decrease in speed (minimum drag on the ICME), or an ICME speed $\sim$2700 km s^{-1} at 1 AU. The first assumption is justified in view of the observations of CMEs by SOHO and other spacecraft missions (Yashiro et al., 2004; Schrijver et al., 2012). The second assumption can be justified if there are multiple CME releases (and multiple flares) from the solar active regions (ARs). In such a case, the associated ICMEs tend to create a low interplanetary drag environment by "cleaning out" the upstream solar wind plasma (Tsurutani et al., 2008, 2014). Tsurutani and Lakhina (2014) employed the Rankine-Hugoniot conservation conditions to derive the shock speed (Tsurutani and Lin, 1985). Considering upstream slow solar wind speed of 350 km s^{-1}, a proton number density of 5×10^6 m^{-3}, and a downstream proton density of 20×10^6 m^{-3}, corresponding to a maximum jump of a factor of 4 in density (Kennel et al., 1985), they estimated the maximum shock speed (in Earth's reference frame) of $V_S = 3480$ km s^{-1} giving a shock transit time from the Sun to the Earth of $\sim$12.0 h (the August 1972 event took 14.6 h and the Carrington event 17.6 h). Such a fast ICME shock will have an Alfvén Mach number of $\sim$63 and magnetosonic Mach number of $\sim$45. Shocks with such high Mach numbers have not been observed to date. The largest magnetosonic Mach number of the shock observed to date is $\sim$28, and that is for the shock associated with the extreme ICME of July 23, 2012 (Riley et al., 2016).

The ram pressure downstream of the ICME shock was calculated to be 244 nPa ($\sim$240 times increase over the upstream pressure). Due to the impingement of this shock on the magnetosphere, the magnetopause will be pushed inward from its quiet time position of $\sim$11.9 R_E to a new subsolar position at $\sim$5.0 R_E from the center of the Earth, where R_E is an Earth radius (6371 km). So far, the lowest magnetopause position detected is at 5.2 R_E for the August 1972 storm. The magnitude of the SI$^+$ resulting from the magneotspheric compress was found to be $\Delta H \approx +234$ nT which exceeds the SI$^+$ amplitude of 202 nT recorded at Kakioka, Japan on March 24, 1991 (Araki et al., 1997).

Fast time variations in the magnetic field, of the order of $dB/dt \sim 30$ nT s^{-1}, produced by the passage of the shock through the magnetosphere, could generate a maximum magnetospheric electric field of the order of 1.9 V m^{-1}. This would produce much stronger new radiation belt fluxes in the magnetosphere, stronger than the one created on March 24, 1991 (composed of $\sim$15 MeV electrons) when an interplanetary shock hit the Earth's magnetosphere (Blake et al., 1992; Li et al., 1993). The electric field amplitude of the March 1991 event was estimated to be $\sim$300 mV m^{-1} (Wygant et al., 1994).

Using an empirical relationship between the speed and magnetic field strength of the ICME at 1 AU (Gonzalez et al., 1998), the magnetic cloud field strength of $\sim$127 nT was estimated by Tsurutani and

Lakhina (2014). This yielded a maximum strength of the interplanetary electric field of ~340 mV m^{-1}, which is nearly twice the estimated value for the Carrington storm (Tsurutani et al., 2003). If we accept the fact that the intensity of all magnetic storms has a linear dependence on the interplanetary electric field (Burton et al., 1975; Echer et al., 2008b), the maximum possible intensity of a superstorm is expected to be twice the intensity of the Carrington storm, that is, Dst ≈ −3500 nT, which incidentally surpasses the maximum possible Dst limit of −2500 nT derived by Vasyliunas (2011). Vasyliunas derived this limit on the maximum Dst by setting the effective plasma pressure equal to the magnetic pressure of the dipole field at the equator of each flux tube, and applying the Dessler-Parker-Sckopke theorem. Incidently, Dst = −31,000 nT represents the absolute limit for the H component depression theoretically (Parker and Stewart, 1967). This could happen when the complete cancelation of the Earth's magnetic field at the equator occurs. Thus, two models independently predict that geomagnetic storms with intensities equal to or greater than that of the Carrington storm event can theoretically occur in the future.

3.3.2 Occurrence probability of carrington-type superstorms

It is not an easy task to provide a definite answer to the question "How often will a Carrington storm occur?" However, many people have made predictions, and we will briefly summarize some of those. This is discussed in more detail in Chapters 4 and 5 of this volume.

Willis et al. (1997) applied extreme value statistics to the daily aa indices from 1844–1993 (14 solar cycles) to find the first, second, and third largest geomagnetic storms per solar cycle. Their prediction, with a 99% probability, is that there will be no storm with $aa > 550$ for the next 100 solar cycles.

Applying extreme value theory to a Dst dataset (1957–2001), Tsubouchi and Omura (2007) predicted an occurrence frequency of a March 1989 storm (Dst = −589 nT) or greater intensity. Tsubouchi and Omura got a value of once in 60 years. For a Carrington-type magnetic storm (Dst = −1760 nT), they obtained a value of once every ~40,000 years.

Assuming that the frequency of occurrence of a storm scales as an inverse power of the intensity of the storm, Riley (2012) predicted that the probability of storms with Dst < −850 nT occurring within the next decade was ~12%. On the other hand, Love (2012) has also predicted a probability for another Carrington-type event in the next 10 years to be ~0.063. Applying lognormal statistics to the Dst time series for the years 1957–2012, Love et al. (2015) predicted that the maximum likelihood for a magnetic storm with intensity exceeding Dst < −850 nT will be about 1.13 times per century, which corresponds to approximately the same frequency as found by Riley (2012). Most recently, Riley and Love (2016) have generalized their approach, and their best estimate for the probability of a supermagnetic storm with Dst < −850 nT occurring within the next decade is ≈10%. Please see Chapter 5 of this volume for a more comprehensive analysis of these and similar methods.

Kataoka (2013) has predicted a probability of 4%–6% for a Carrington-level magnetic storm occurring during solar cycle 24. Yermolaev et al. (2013) have predicted that the occurrence frequency of a Carrington-type storm cannot exceed one event every 500 years (also please see Chapter 4 of this volume).

4 NOWCASTING AND SHORT-TERM FORECASTING OF SUPERMAGNETIC STORM

From the above discussion, it is evident that there is a large variability in the predictions of future superstorm occurrence. It is apparent that it is extremely difficult to predict when a Carrington-level supermagnetic storm will occur. For a reliable prediction of such events, we need to have either full understanding of the physical processes causing extreme ICMEs and magnetic storms or good empirical statistics of the tail of their distributions. At this stage, we do not have complete understanding of the physical processes, or have good data of the tail distributions. Therefore, we feel that at this time it is not possible to estimate the probabilities of occurrence of Carrington-level superstorms with any reasonable level of confidence.

However, in recent times, with the availability of various space missions that provide continuous monitoring of the Sun (the NASA-European Space Agency's Solar and Heliospheric Observatory (SOHO) spacecraft), and the interplanetary medium near 1 AU (e.g., NASA's Advanced Composition Explorer (ACE) spacecraft), there has been much improvement in the capability of predicting magnetic storms or space weather in real-time (nowcasting) and at a short-time scale (short-term forecasting). The nowcasting is essentially based on the use of spacecraft measurements of solar wind parameters at the Sun-Earth libration point 1 (L1 point) (Ogilvie et al., 1978; Tsurutani and Baker, 1979; Temerin and Li, 2002, 2006; Wang et al., 2003; Boynton et al., 2011; Ji et al., 2012). Two main interplanetary parameters used in most of the nowcasting or near real-time prediction schemes are the solar wind speed and the southward component of the IMF (Burton et al., 1975; Wu and Lundstedt, 1996; O'Brien and McPherron, 2000). The main advantage of nowcasting is that it has high levels of accuracy, up to 90%. However, it is limited to advance warnings of only ~0.5 h, which may not be sufficient for preventing the magnetic storm hazards.

Short-term forecasts can predict events with time scales of several hours to several days (Joselyn, 1995; Srivastava and Venkatakrishnan, 2004; Kim et al., 2005). At present, there are not any real short-term forecasting models that are operational. Instead, they are more "proof of concept" of future forecasting systems. The short-term forecasts are based essentially on the properties of CMEs and their associated shock waves, which are thought to be responsible for triggering intense and superintense magnetic storms (Gonzalez and Tsurutani, 1987; Gonzalez et al., 1994; Tsurutani et al., 1995a,b, 2006; Echer et al., 2008a,b). Since the CME parameters observed near the Sun are used, these forecasts can give a lead time of ~2–3 days. Therefore such forecasts are very practical and give sufficient time to prepare for the hazards associated with the intense magnetic storms (Brueckner et al., 1998; Gopalswamy et al., 2000; Srivastava, 2005a,b; Cho et al., 2010; Kim et al., 2014). These forecasts are also relevant for the supermagnetic storms where the lead time may be reduced to a day or less. Two main tasks for the success of the short-term forecasts are to predict the arrival time of a CME at Earth, and the magnitude of the ensuing geomagnetic storm (Gopalswamy et al., 2001; Moon et al., 2002; Cho et al., 2003; McKenna-Lawlor et al., 2006; Kim et al., 2010). Earlier models for short-term forecasts were empirical in nature using the initial speed of the ICME as an input parameter and predicting the ICME arrival at 1 AU (Gopalswamy et al., 2001). Recently, the empirical models based solely on initially observed CME parameters were extended to forecast geomagnetic storm occurrence and intensity (Dst) (Srivastava, 2005b; Kim et al., 2010). More recently, Kim et al. (2014) have developed a two-step forecast model of geomagnetic storms using CME parameters near the

Sun and solar wind conditions near 1 AU. This model's predictions have a higher success rate than the earlier models based on CME parameters alone (Gopalswamy et al., 2001; Moon et al., 2002; Cho et al., 2003; Srivastava, 2005a,b; McKenna-Lawlor et al., 2006; Kim et al., 2010). It should be noted, however, that forecasting the ICME arrival at 1 AU may be difficult when multiple CMEs are launched from an active region (Echer et al., 2009). Echer et al. (2009) surmised that complex interactions in the interplanetary medium between the Sun and Earth complicated their prediction of the arrival of the November 2004 ICMEs.

The field of nowcasting and short-term forecasting of magnetic storms is developing fast due to its application to space weather hazards (Stamper et al., 2004; Sharifi et al., 2006; Khabarova, 2007; Tsagouri et al., 2013; Love et al., 2014; Posner et al., 2014; Schrijver et al., 2015; Savani et al., 2015).

5 CONCLUSIONS

We have given an overview of the status of supermagnetic storm research. Starting from a brief historical background, we first summarized the present knowledge and then discussed its application to the analysis of past (September 1–2, 1859 Carrington event) and present (March 13–14, 1989 event) superstorms where no solar wind data was available. We then discussed the maximum possible intensity of a superstorm as well as the prediction of supermagnetic storms. Lastly, we reviewed the status of nowcasting and short-time forecasting of supermagnetic storms.

We have shown that at the present time, it is not possible to make a precise prediction of when and how often an extreme supermagnetic storm with similar or higher intensity than that of the Carrington event could occur. However, it has been theoretically shown that storms even stronger than the Carrington event can occur (provided there is corresponding extreme CME). Under ideal conditions, an extreme CME having speed of 3000 km s^{-1} near the Sun and a ~10% decrease as a maximum ICME drag during its passage through the slow solar wind plasma from the Sun to 1 AU (Tsurutani and Lakhina, 2014), can excite a supermagnetic storm with intensity reaching at least the saturation value of Dst = −2500 nT predicted by Vasyliunas (2011). The transit time from the Sun to the Earth of an ICME shock will just be ~12.0 h. The maximum amplitude of the interplanetary electric field at Earth due to this ICME will be ~340 mV m^{-1}. Associated storm-time ionospheric electric fields would cause major uplift of the dayside ionosphere with substantially increased ion-neutral drag (Tsurutani et al., 2012). Supermagnetic storms of such intensity could cause havoc to modern society: loss of many low-orbiting satellites due to the additional drag, wide spread failure of telecommunications and Internet networks, failure of global positioning system (GPS) systems, and loss of navigation and power grid networks.

During the past few decades, there has been a lot of activity directed towards understanding Sun-Earth connections and space weather. Several countries—United States, Canada, Mexico, Japan, the United Kingdom, the European Union, Australia, China, South Africa—have developed their own national space weather programs based on the availability of ground-monitoring network and space missions. The ground-monitoring networks include a variety of instruments (such as magnetometers and high frequency (HF) radars) that make measurements of magnetic fields, ULF waves, total electron content using GPS systems, ionospheric irregularities, scintillations, and airglow. Space missions such as ACE, STEREO, and the deep space climate observatory (DSCOVR) are being utilized to collect interplanetary data of solar wind particles. On the other hand, solar monitoring data of flares, CMEs, X-rays and ultraviolet (UV) rays are being collected from SOHO, Hinode, GOES, DSCOVR, Solar Dynamics Observatory (SDO), etc. International programs such as International Living With the Star (ILWS), Climate and

Weather of Sun-Earth System (CAWSES), and Variability of the Sun and Its Terrestrial Impact (VarSITI) have made contributions to space weather research and prediction capabilities. As mentioned above, the nowcasting of magnetic storms has drastically improved. Short-time scale forecasting is also improving rapidly with a better understanding of the physical processes contributing to space weather. The next step is developing, testing, and validating short-time numerical space weather prediction techniques that will be able to predict space weather-related effects with the same level of accuracy as in the field of meteorology (Siscoe, 2007). Some techniques have already been developed to incorporate physics-based models, rather than just empirical models. Ideally, these types of models should cover the region from the Sun to the upper atmosphere of Earth (Tóth et al., 2007; Mannucci et al., 2015). There has been tremendous progress in developing these models in recent years and gradual improvement in prediction accuracy is expected in the coming decade.

ACKNOWLEDGMENTS

Portions of this research were conducted at the Jet Propulsion Laboratory, California Institute of Technology, under contract with NASA. GSL thanks the National Academy of Sciences, India, for support under the NASI-Senior Scientist Platinum Jubilee Fellowship. The authors thank B. I. Panchal for help in preparing Fig. 2.

GLOSSARY

Aurora A natural display of light in the polar sky. It has usually green and red optical lines, sometimes blue colors. Auroral lights are produced by the collision of energetic electrons with atoms and molecules of gases such as oxygen and nitrogen in the upper ionosphere. Auroras occur in both hemispheres in a band of latitudes called the auroral oval. The location of the auroral oval depends on the geomagnetic activity, but it usually extends from 67° to 76° magnetic latitudes. During magnetic storms, the auroral oval expands to both lower and higher magnetic latitudes. Auroras occurring in the Northern hemisphere are called the Aurora Borealis whereas those occurring in the Southern hemisphere are called the Aurora Australis.

Chorus A right-hand, circularly polarized electromagnetic planar whistler mode wave. Chorus is generated near the geomagnetic equatorial plane or in the dayside magnetospheric minimum magnetic field pockets by the loss cone instability excited by anisotropic energetic ($\sim$10–100 keV) electrons. Cyclotron resonant interaction of high-energy electrons with chorus has been proposed as a mechanism for acceleration of electrons to relativistic energies.

Coronal mass ejection (CME) A transient outflow of plasma and magnetic fields from or through the solar corona. CMEs are often, but not always, associated with disappearing solar filaments, erupting prominences, and solar flares. Large-scale closed coronal structures are a common site for CME releases. Magnetic reconnection is believed to be responsible for CMEs. The average mass and energy of the material ejected during CME can be a few times 10^{15} g and 10^{31} erg, respectively. If the magnetic field of the CME is southwardly directed, reconnection between the CME field and the Earth's field can lead to substorms and magnetic storms.

Corotating Interaction Regions (CIRs) Created when a high speed solar wind stream emanating from a coronal hole overtakes a slower (upstream) solar wind stream. The interaction leads to compression of both magnetic fields and plasma. Because coronal holes are often long lasting, the high speed streams and their interaction with slow speed streams appear to "corotate," thus the name CIR. CIRs can give rise to magnetic storms of typically weak to moderate intensity.

Dst index The "disturbance storm time" index is a measure of variations in the horizontal component of geomagnetic field due to the presence of an enhanced equatorial ring current. It is computed from magnetic data from $\sim$4 near-equatorial stations at hourly intervals.

Geomagnetically induced currents (GICs) Induced currents produced by rapid temporal or spatial changes of magnetospheric and ionospheric currents during substorms and magnetic storms. Auroral electrojet currents with intensities of $\sim 10^6$ amperes flowing at $\sim$100 km above the surface of the Earth can cause strong induced currents in power grid lines and pipelines. GICs can corrode long east-west extensions of pipelines and damage high-voltage power transformers.

Interplanetary coronal mass ejection (ICME): A CME is characterized by an outer loop, a dark region and a filament. The most common part of a CME detected at 1 AU from the Sun is the magnetic cloud which corresponds to the dark region of the CME. Since not all parts of a CME are detected at 1 AU, this is referred to as an Interplanetary Coronal Mass Ejection or ICME.

Magnetic declination Denotes the angle on the horizontal plane between magnetic north and geographic north. It is taken as positive when magnetic north is east of geographic north, and negative when it is to the west. Magnetic declination varies with location on the Earth's surface, and it also changes over time.

Magnetic reconnection A plasma process that converts magnetic energy to plasma kinetic energy accompanied by a change in the magnetic field topology. Magnetic reconnection prevents the excessive build-up of magnetic energy in current sheets found in space and astrophysical plasmas. It allows the transfer of magnetic flux and plasma mass between separate magnetic flux regions. Magnetic reconnection is most efficient when plasmas with oppositely directed magnetic fields are brought together. Magnetic reconnection is commonly invoked to explain the energization of plasmas and acceleration of charged particles associated with solar flares, and indirectly with magnetic storms and substorms, etc.

Ring Current Formed due to motion of trapped energetic electrons and ions (energies $\sim$10 to $\sim$300 keV) injected earthward from the plasma sheet in the magnetotail. The presence of magnetic field gradients and curvature forces the electrons and ions to undergo eastward and westward azimuthal drifts near the equatorial plane of the magnetosphere, respectively. These two oppositely directed drifts comprise a westward ring of current known as the Ring Current and is the main signature of a magnetic storm.

Solar flare A sudden release of energy in the solar atmosphere lasting minutes to hours, from which electromagnetic radiation (from EUV to X-ray wavelengths) and energetic charged particles are emitted. Solar flares most commonly occur at complex sunspots called active regions. Solar flares are classified according to their X-ray brightness in the wavelength range 1–8 Å. There are three categories: X-class flares are the most intense with intensities $I > 10^{-4}$ W m^{-2}, M-class flares are medium-sized with $10^{-5} \leq I < 10^{-4}$ W m^{-2}, and C-class flares are small with $10^{-6} \leq I < 10^{-5}$ W m^{-2}. Here, we denotes the peak X-ray burst intensity measured at the Earth. Each category for X-ray flares has nine subdivisions ranging from 1 to 9, e.g., X1 to X9, M1 to M9, and C1 to C9.

Substorms They occur due to injection of energetic (from 100 eV to a few tens keVs) charged particles by an explosive energy release from the near-Earth magnetotail into the nightside magnetosphere. Precipitation of these energetic charged particles into the auroral zone ionosphere produces intense auroral displays. Magnetic reconnection is believed to be responsible for the energy release. A substorm typically lasts from $\sim$30 min to an hour.

Sunspots The dark spots on the Sun's photosphere. Sunspots appear dark as the embedded strong magnetic fields push the hot plasma out, thereby reducing the temperature compared to the surrounding area. Magnetic fields in the sunspot regions keep on evolving and become complex. Such sunspot areas are called active regions and are the potential sites for solar flares. Sunspots are the indicator of solar activity. They undergo a cyclic behavior with a period of $\sim$11 years; this is commonly known as a *solar cycle*.

Symmetric-H (SYM-H) index Same as Dst but computed at a higher resolution of 1 min resolution instead of 1 h used for Dst.

Ultralow frequency (ULF) waves Represent a portion of the radio frequency spectrum from $\sim$1 mHz to 30 Hz. ULF waves are produced by a variety of plasma processes occurring in the magnetosphere and the solar wind.

REFERENCES

Akasofu, S.-I., Kamide, Y., 2005. Comment on "The extreme magnetic storm of 1–2 September 1859" by B. T. Tsurutani, W. D. Gonzalez, G. S. Lakhina, and S. Alex. J. Geophys. Res. 110, https://doi.org/10.1029/2005JA011005 09226.

Alex, S., Pathan, B.M., Lakhina, G.S., 2005. Response of the low-latitude geomagnetic field to the major proton event of November 2001. Adv. Space Res. 36, 2434–2439.

Alex, S., Mukherjee, S., Lakhina, G.S., 2006. Geomagnetic signatures during the intense geomagnetic storms of 29 October and 20 November 2003. J. Atmos. Sol. Terr. Phys. 68, 769–780.

Allen, J., Sauer, H., Frank, L., Reiff, P., 1989. Effects of the March 1989 solar activity. Eos. Trans. AGU 70, 1479.

Allen, J., 2010. The Galaxy 15 anomaly: another satellite in the wrong place at a critical time. Space Weather 8. S06008. https://doi.org/10.1029/2010SW000588.

Araki, T., et al., 1997. Anomalous sudden commencement on March 24, 1991. J. Geophys. Res. 102, 14,075–14,086.

Baker, D.N., Li, X., Pulkkinen, A., Ngwira, C.M., Mays, M.L., Galvin, A.B., Simunac, K.D.C., 2013. A major solar eruptive event in July 2012: defining extreme space weather scenarios. Space Weather 11, 585–591.

Benz, A.O., 2008. Flare observations. Living Rev. Sol. Phys. 5, 1.

Blake, J.B., Kolassinski, W.A., Fillius, R.A., Mullen, E.G., 1992. Injection of electrons and protons with energies of tens of MeV into $L<3$ on March 24, 1991. Geophys. Res. Lett. 19, 821–824.

Bolduc, L., 2002. GIC observations and studies in the Hydro-Quebec system. J. Atmos. Sol. Terr. Phys. 64, 1793–1802.

Borovsky, J., Denton, M.H., 2006. Differences between CME-driven storms and CIR-driven storms. J. Geophys. Res. https://doi.org/10.1029/2005JA011447.

Boynton, R.J., Balikhin, M.A., Billings, S.A., Wei, H.L., Ganushkina, N., 2011. Using the NARMAX OLS-ERR algorithm toobtain the most influential coupling functions that affect the evolution of the magnetosphere. J. Geophys. Res. 116. A05218, https://doi.org/10.1029/2010JA015505.

Brueckner, G.E., Delaboudiniere, J.-P., Howard, R.A., Paswaters, S.E., St. Cyr, O.C., Schwenn, R., Lamy, P., Simnett, G.M., Thompson, B., Wang, D., 1998. Geomagnetic storms caused by coronal mass ejections (CMEs): March 1996 through June 1997. Geophys. Res. Lett. 25, 3019–3022.

Burlaga, L.F., Sittler, E., Mariani, F., Schwenn, R., 1981. Magnetic loop behind an interplanetary shock: Voyager, Helios, and IMP 8 observations. J. Geophys. Res. 86, 6673–6684.

Burton, R.K., McPherron, R.L., Russell, C.T., 1975. An empirical relationship between interplanetary conditions and Dst. J. Geophys. Res. 80, 4204–4214.

Carovillano, R.L., Siscoe, G.L., 1973. Energy and momentum theorems in magnetospheric processes. Rev. Geophys. 11, 289.

Carrington, R.C., 1859. Description of a singular appearance seen in the Sun on September 1, 1859. Mon. Not. R. Astron. Soc. 20, 13–15.

Chapman, S., Bartels, J., 1940. Geomagnetism, vol. 1. Oxford Univ Press, New York. pp. 328–337.

Chen, P.F., 2011. Coronal mass ejections: models and their observational basis. Living Rev. Sol. Phys. 8, 1.

Cho, K.-S., Moon, Y.-J., Dryer, M., Fry, C.D., Park, Y.-D., Kim, K.-S., 2003. A statistical comparison of interplanetary shock and CME propagation models. J. Geophys. Res. 108 (A12), 1445. https://doi.org/10.1029/2003JA010029.

Cho, K.-S., Bong, S.-C., Moon, Y.-J., Dryer, M., Lee, S.-E., Kim, K.-H., 2010. An empirical relationship between coronal mass ejection initial speed and solar wind dynamic pressure. J. Geophys. Res. 115. A10111. https://doi.org/10.1029/2009JA015139.

Chree, C., 1905. Review of Maunder's recent investigations on the cause of magnetic disturbances. J. Geophys. Res. 10, 9–14. https://doi.org/10.1029/TE010i001p00009.

Cid, C., Palacios, J., Saiz, E., Guerrero, A., Cerrato, Y., 2014. On extreme geomagnetic storms. J. Space Weather Space Clim. 4, 28. https://doi.org/10.1051/swsc/2014026.

Cid, C., Saiz, E., Guerrero, A., Palacios, J., Cerrato, Y., 2015. A Carrington-like geomagnetic storm observed in the 21st century. J. Space Weather Space Clim. 5, https://doi.org/10.1051/swsc/2015017. 16.

Cliver, E., Feynman, J., Garrett, H., 1990. An estimate of the maximum speed of the solar wind, 1938–1989. J. Geophys. Res. 95 (A10), 17103–17112.

Cliver, E.W., Dietrich, W.F., 2013. The 1859 space weather event revisited: limits of extreme activity. J. Space Weather Space Clim. 3, 31. https://doi.org/10.1051/swsc/2013053.

Cliver, E.W., Svalgaard, L., 2004. The 1859 solar-terrestrial disturbance and the current limits of extreme space weather activity. Solar Phys. 224, 407–422.

Cook, A., 1998. Edmond Halley: Charting the Heavens and the Seas. Oxford University Press, Oxford. ISBN: 0198500319.

Dessler, A.J., Parker, E.N., 1959. Hydromagnetic theory of magnetic storms. J. Geophys. Res. 64, 2239.

Dungey, J.W., 1961. Interplanetary magnetic field and the auroral zones. Phys. Rev. Lett. 6, 47.

Echer, E., Gonzalez, W.D., Tsurutani, B.T., Gonzalez, A.L.C., 2008a. Interplanetary conditions causing intense geomagnetic storms ($Dst \leq \_100$ nT) during solar cycle 23 (1996–2006). J. Geophys. Res. 113. A05221. https://doi.org/10.1029/2007JA012744.

Echer, E., Gonzalez, W.D., Tsurutani, B.T., 2008b. Interplanetary conditions leading to super intense geomagnetic storms ($Dst \leq -250$ nT) during solar cycle 23. Geophys. Res. Lett. 35, 03–06. https://doi.org/10.1029/2007GL031755.

Echer, E., Tsurutani, B.T., Guarnieri, F.L., 2009. Solar and interplanetary origins of the November 2004 superstorms. Adv. Space Res. 44, 615–620.

Gilbert, W., 1600. De Magnete. Chiswick, London (English translation by P. F. Mottelay, Dover, New York, 1958.).

Gonzalez, W.D., Tsurutani, B.T., 1987. Criteria of interplanetary parameters causing intense magnetic storms ($Dst < -100$ nT). Planet. Space Sci. 35, 1101–1109.

Gonzalez, W.D., Tsurutani, B.T., Gonzalez, A.L.C., Smith, E.J., Tang, F., Akasofu, S.I., 1989. Solar wind-magnetosphere coupling during intense magnetic storms (1978–1979). J. Geophys. Res. 94 (A7), 8835–8851.

Gonzalez, W.D., Tsurutani, B.T., Clúa de Gonzalez, A.L., 1999. Interplanetary origin of geomagnetic storms. Space Sci. Rev. 88, 529–562.

Gonzalez, W., Joselyn, J., Kamide, Y., Kroehl, H., Rostoker, G., Tsurutani, B., Vasyliunas, V., 1994. What is a geomagnetic storm? J. Geophys. Res. 99 (A4), 5771–5792.

Gonzalez, W.D., de Gonzalez, A.L.C., Dal Lago, A., Tsurutani, B.T., Arballo, J.K., Lakhina, G.S., Buti, B., Ho, C.M., Wu, S.-T., 1998. Magnetic cloud field intensities and solar wind velocities. Geophys. Res. Lett. 25 (7), 963–966.

Gonzalez, W.D., Echer, E., Clade Gonzalez, A.L., Tsurutani, B.T., Lakhina, G.S., 2011. Extreme geomagnetic storms, recent Gleissberg cycles and space era super intense storms. J. Atmos. Sol. Terr. Phys. 73 (11–12), 1447–1453. https://doi.org/10.1016/j.jastp.2010.07.023.

Gopalswamy, N., Lara, A., Lepping, R.P., Kaiser, M.L., Berdichevsky, D., St. Cyr, O.C., 2000. Interplanetary acceleration of coronal mass ejections. Geophys. Res. Lett. 27, 145–148.

Gopalswamy, N., Lara, A., Yashiro, S., Kaiser, M., Howard, R.A., 2001. Predicting the 1-AU arrival times of coronal mass ejections. J. Geophys. Res. 106, 29,207–29,217.

Graham, G., 1724a. An account of observations made of the variation of the horizontal needle at London, in the latter part of the year 1722, and beginning of 1723. Philos. Trans. R. Soc. Lond. A 33, 96–107. https://doi.org/10.1098/rstl.1724.0020.

Graham, G., 1724b. Observations of the dipping needle, made at London, in the beginning of the year 1723. Philos. Trans. R. Soc. Lond. A 33, 332–339. https://doi.org/10.1098/rstl.1724.0062.

Hajra, R., Tsurutani, B.T., Echer, E., Gonzalez, W.D., Santolik, O., 2015. Relativistic (E > 0.6, > 2.0 and > 4.0 MeV) electron acceleration at geosynchronous orbit during high-intensity, long-duration, continuous AE activity (HILDCAA) events. Astrophys. J. 799, 39. https://doi.org/10.1088/0004-637X/799/1/39.

Hajra, R., Tsurutani, B.T., Echer, E., Gonzalez, W.D., Gjerloev, J.W., 2016. Supersubstorms (SML ≤ -2500 nT): magnetic storm and solar cycle dependences. J. Geophys. Res. Space Phys. 121, 7805–7816. https://doi.org/10.1002/2015JA021835.

Hale, G.E., 1931. The spectrohelioscope and its work. part III. Solar eruptions and their apparent terrestrial effects. Astrophys. J. 73, 379–412.

Hapgood, M.A., 2012. Prepare for the coming space weather storm. Nature 484, 311–313. https://doi.org/10.1038/484311a.

Hodgson, R., 1859. On a curious appearance seen in the Sun. Mon. Not. R. Astron. Soc. 20, 15–16.

Hudson, M.K., Elkington, S.R., Lyon, J.G., Goodrich, C.C., 2000. Increase in relativistic electron flux in the inner magnetosphere: ULF wave mode structure. Adv. Space Res. 25, 2327–2337.

Huttunen, K.E.J., Kilpua, S.P., Pulkkinen, A., Viljanen, A., Tanskanen, E., 2008. Solar wind drivers of large geomagnetically induced currents during the solar cycle 23. Space Weather 6. S10002. https://doi.org/10.1029/2007SW000374.

Iyemori, T., 1990. Storm-time magnetospheric currents inferred from midlatitude geomagnetic field variations. J. Geomag. Geoelec. 42, 1249–1265.

Ji, E.-Y., Moon, Y.-J., Gopalswamy, N., Lee, D.-H., 2012. Comparison of Dst forecast models for intense geomagnetic storms. J. Geophys. Res. 117. A03209. https://doi.org/10.1029/2011JA016872.

Joselyn, J.A., 1995. Geomagnetic activity forecasting: the state of the art. Rev. Geophys. 33, 383.

Kamide, Y., Yokoyama, N., Gonzalez, W., Tsurutani, B., Daglis, I., Brekke, A., Masuda, S., 1998. Two-step development of geomagnetic storms. J. Geophys. Res. 103 (A4), 6917–6921.

Kamide, Y., Kusano, K., 2015. No major solar flares but the largest geomagnetic storm in the present solar cycle. Space Weather 13, 365–367. https://doi.org/10.1002/2015SW001213.

Kataoka, R., Miyoshi, Y., 2006. Flux enhancement of radiation belt electrons during geomagnetic storms driven by coronal mass ejections and corotating interaction regions. Space Weather 4. S09004. https://doi.org/10.1029/2005SW000211.

Kataoka, R., 2013. Probability of occurrence of extreme magnetic storms. Space Weather 11, 214–218. https://doi.org/10.1002/swe.20044.

Kelvin, W.T., 1892. Address to the Royal Soc. Nov. 30. Proc. R. Soc. Lond. A 52, 302–310.

Kennel, C.F., Edmiston, J.P., Hada, T., 1985. A quarter century of collisionless shock research. In: Collisionless Shocks in the Heliosphere: A Tutorial Review. Geophys. Mon. Ser., vol. 34. AGU, Washington, DC, pp. 1–36.

Khabarova, O.V., 2007. Current problems of magnetic storm prediction and possible ways of their solving. Sun Geosph. 2 (1), 32–37.

Kim, R.-S., Cho, K.-S., Moon, Y.-J., Kim, Y.-H., Yi, Y., Dryer, M., Bong, S.-C., Park, Y.-D., 2005. Forecast evaluation of the coronal mass ejection(CME) geoeffectiveness using halo CMEs from 1997 to 2003. J. Geophys. Res. 110. A11104. https://doi.org/10.1029/2005JA011218.

Kim, R.-S., Cho, K.-S., Moon, Y.-J., Dryer, M., Lee, J., Yi, Y., Kim, K.-H., Wang, H., Park, Y.-D., Kim, Y.-H., 2010. An empirical model for prediction of geomagnetic storms using initially observed CME parameters at the Sun. J. Geophys. Res. 115. A12108. https://doi.org/10.1029/2010JA015322.

Kim, R.-S., Moon, Y.-J., Gopalswamy, N., Park, Y.-D., Kim, Y.-H., 2014. Two-step forecast of geomagnetic storm using coronal mass ejection and solar wind condition. Space Weather 12, 246–256. https://doi.org/10.1002/2014SW001033.

Kimball, D.S., 1960. A Study of the Aurora of 1859. Sci. Rep. 6, UAG-R109Univ. of Alaska, Fairbanks.

Klein, L.W., Burlaga, L.F., 1982. Magnetic clouds at 1 AU. J. Geophys. Res. 87, 613.

Kozyra, J.U., Liemohn, M.W., 2003. Ring current energy input and decay. Space Sci. Rev. 109, 105–131.

Kozyra IV, J.U., Escoubet, W.B.M., Lepri, C.P., Liemohn, M.W., Gonzalez, W.D., Thomsen, M.W., Tsurutani, B.T., 2013. Earth's collision with a solar filament on 21January 2005: overview. J. Geophys. Res. 118, 5967–5978. https://doi.org/10.1002/jgra.50567.

Lakhina, G.S., Tsurutani, B.T., 2016. Geomagnetic storms: historical perspective to modern view. Geosci. Lett. 3, 5. https://doi.org/10.1186/s40562-016-0037-4.

Lakhina, G.S., Alex, S., Tsurutani, B.T., Gonzalez, W.D., 2012. Super magnetic storms: hazard to society. In: Sharma, A.S., Bunde, A., Dimri, V.P., Baker, D.N. (Eds.), Extreme Events and Natural Hazards: The Complexity Perspective. Geophys. Mon. Ser., 196, AGU, Washington. https://doi.org/10.1029/2011GM001073 p. 267.

Lakhina, G.S., Alex, S., Tsurutani, B.T., Gonzalez, W.D., 2005. Research on historical records of geomagnetic storms, in Coronal and Stellar Mass Ejections. In: Dere, K.P., Wang, J., Yan, Y. (Eds.), Proceedings of the 226th Symposium of the International Astronomical Union held in Beijing, China, September 13–17, 2004, Cambridge Univ. Press, Cambridge. UK. pp. 3–15.

Lamont, J., 1867. Handbuch des Magnetismus. Leopold Voss, Leipzig.

Lei, J., Thayer, J.P., Forbes, J.M., Wu, Q., She, C., Wan, W., Wang, W., 2008. Ionosphere response to solar wind high-speed streams. Geophys. Res. Lett. 35, 19105. https://doi.org/10.1029/2008GL035208.

Li, X.L., Roth, I., Temerin, M., Wygant, J.R., Hudson, M.K., Blake, J.B., 1993. Simulation of the prompt energization and transport of radiation belt particles during the March 24, 1991 SSC. Geophys. Res. Lett. 20, 2423–2426.

Li, X., Temerin, M., Tsurutani, B.T., Alex, S., 2006. Modeling of 1–2 September 1859 super magnetic storm. Adv. Space Res. 38, 273–279.

Liu, Y.D., Luhmann, J.G., Kajdic, P., Kilpua, E.K., Lugaz, N., et al., 2014. Observations of an extreme storm in interplanetary space caused by successive coronal mass ejections. Nat. Commun. 5, 3481.

Loomis, E., 1861. On the great auroral exhibition of Aug. 28th to Sept. 4, 1859, and on auroras generally. Am. J. Sci. 82, 318–335.

Love, J.J., 2012. Credible occurrence probabilities for extreme geophysical events: earthquakes, volcanic eruptions, magnetic storms. Geophys. Res. Lett. 39, 10301. https://doi.org/10.1029/2012GL051431.

Love, J.J., Rigler, E.J., Pulkkinen, A., Balch, C.C., 2014. Magnetic storms and induction hazards. Eos. Trans. AGU 95 (48), 445–452.

Love, J.J., Rigler, E.J., Pulkkinen, A., Riley, P., 2015. On the lognormality of historical magnetic storm intensity statistics: implications for extreme-event probabilities. Geophys. Res. Lett. 42, 6544–6553. https://doi.org/10.1002/2015GL064842.

Magara, T., Shibata, K., Yokoyama, T., 1995. Evolution of eruptive flares. I. plasmoid dynamics in eruptive flares. Astrophys. J. 487, 437–446.

Mannucci, A.J., Tsurutani, B.T., Iijima, B.A., Komjathy, A., Saito, A., Gonzalez, W.D., Guarnieri, F.L., Kozyra, J.U., Skoug, R., 2005. Dayside global ionospheric response to the major interplanetary events of October 29–30, 2003 "Halloween Storms" Geophys. Res. Lett. 32, https://doi.org/10.1029/2004GL021467. L12S02.

Mannucci, A.J., Tsurutani, B.T., Abdu, M.A., Gonzalez, W.D., Komjathy, A., Echer, E., Iijima, B.A., Crowley, G., Anderson, D., 2008. Superposed epoch analysis of the dayside ionospheric response to four intense geomagnetic storms. J. Geophys. Res. 113, 00–02. https://doi.org/10.1029/2007JA012732.

Mannucci, A.J., Verkhoglyadova, O.P., Tsurutani, B.T., Meng, X., Pi, X., Wang, C., Rosen, G., Lynch, E., Sharma, S., Ridley, A., Manchester, W., Van Der Holst, B., Echer, E., Hajra, R., 2015. Medium-range thermosphere-ionosphere storm forecasts. Space Weather 13, 125–129. https://doi.org/10.1002/2014SW001125.

Maunder, E.W., 1904a. Note on the distribution of sun-spots in the heliographic latitude, 1874 to 1902. Mon. Not. R. Astron. Soc. 64, 747–761. https://doi.org/10.1093/mnras/64.8.747.

Maunder, E.W., 1904b. Magnetic disturbances, 1882 to 1903, as recorded at the Royal Observatory Greenwich, and their association with sun-spots. Mon. Not. R. Astron. Soc. 65, 2–18. https://doi.org/10.1093/mnras/65.1.2.

Maunder, E.W., 1904c. Demonstration of the solar origin of the magnetic disturbances. Mon. Not. R. Astron. Soc. 65, 18–34. https://doi.org/10.1093/mnras/65.1.18.

Maunder, E.W., 1905. Magnetic disturbances as recorded at the Royal Observatory Greenwich, and their association with sun-spots. Mon. Not. R. Astron. Soc. 65, 538–559. https://doi.org/10.1093/mnras/65.6.538.

McKenna-Lawlor, S.M.P., Dryer, M., Kartalev, M.D., Smith, Z., Fry, C.D., Sun, W., Deehr, C.S., Kecskemety, K., Kudela, K., 2006. Near real-time predictions of the arrival at Earth of flare-related shocks during Solar Cycle 23. J. Geophys. Res. 111. A11103. https://doi.org/10.1029/2005JA011162.

Moon, Y.-J., Dryer, M., Smith, Z., Park, Y.-D., Cho, K.-S., 2002. A revised shock time of arrival (STOA) model for interplanetary shock propagation: STOA-2. Geophys. Res. Lett. 29 (10), 1390. https://doi.org/10.1029/2002GL014865.

Newton, H.W., 1943. Solar flares and magnetic storms. Mon. Not. R. Astron. Soc. 103, 244–257.

National Research Council, 2008. Severe Space Weather Events-Understanding Societal and Economic Impacts. National Academies Press, Washington, DC.

Ngwira, C.M., Pulkkinen, A., Mays, M.L., Kuznetsova, M.M., Galvin, A.B., et al., 2013a. Simulation of the 23 July 2012 extreme space weather event: what if this extremely rare cme was earth directed? Space Weather 11, 671–679.

Ngwira, C.M., Pulkkinen, A., Kuznetsova, M.M., Glocer, A., 2013b. Modeling extreme Carrington-type space weather events using three-dimensional global MHD simulations. J. Geophys. Res. 119, 4456–4474. https://doi.org/10.1002/2013JA019661.

O'Brien, T.P., McPherron, R.L., 2000. An empirical phase space analysis of ring current dynamics: Solar wind control of injection and decay. J. Geophys. Res. 105, 7707–7719.

Ogilvie, K.W., Durney, A., von Rosenvinge, T., 1978. Descriptions of experimental investigations and instruments for the ISEE spacecraft. IEEE Trans. Geosci. Electron. GE-16 (3), 151–153. https://doi.org/10.1109/TGE.1978.294535.

Parker, E.N., Stewart, H.A., 1967. Nonlinear inflation of a magnetic dipole. J. Geophys. Res. 72, 5287–5293.

Posner, A., Hesse, M., St. Cyr, O.C., 2014. The main pillar: assessment of space weather observational asset performance supporting nowcasting, forecasting, and research to operations. Space Weather 12, 257–276. https://doi.org/10.1002/2013SW001007.

Rawat, R., Alex, S., Lakhina, G.S., 2006. Low-latitude geomagnetic signatures during major solar energetic particle events of solar cycle-23. Ann. Geophys. 24, 3569–3583.

Rawat, R., Alex, S., Lakhina, G.S., 2007. Geomagnetic storm characteristics under varied interplanetary conditions. Bull. Astron. Soc. India 35, 499–509.

Rawat, R., Alex, S., Lakhina, G.S., 2010. Storm-time characteristics of intense geomagnetic storms (Dst < - 200 nT) at low-latitudes and associated energetics. J. Atmos. Sol. Terr. Phys. 72, 1364–1371. https://doi.org/10.1016/j.jastp.2010.09.029.

Richardson, I.G., et al., 2006. Major geomagnetic storms (Dst < -100 nT) generated by corotating interaction regions. J. Geophys. Res. 111, https://doi.org/10.1029/2005JA011476. A07S09.

Richardson, I.G., Zhang, J., 2008. Multiple-step geomagnetic storms and their interplanetary drivers. Geophys. Res. Lett. 35, https://doi.org/10.1029/2007GL032025 L06S07.

Riley, P., 2012. On the probability of occurrence of extreme space weather events. Space Weather 10, https://doi.org/10.1029/2011SW000734. 02012.

Riley, P., Caplan, R.M., Giacalone, J., Lario, D., Liu, Y., 2016. Properties of the fast forward shock driven by the 2012 July 23 extreme coronal mass ejection. Astrophys. J. 819, 57. https://doi.org/10.3847/0004-637X/819/1/57. (11 pp.).

Riley, P., Love, J.J., 2016. Extreme geomagnetic storms: probabilistic forecasts and their uncertainties. Space Weather 15. https://doi.org/10.1002/2016SW001470.

Rostoker, G., Friedrich, E., Dobbs, M., 1997. Physics of magnetic storms. In: Tsurutani, B.T., Gonzalez, W.D., Kamide, Y., Arballo, J.K. (Eds.), Magnetic Storms. Geophys. Monogr. Ser., 98, AGU, Washington, pp. 149–160.

Royal Academy of Engineering Report, 2013. Extreme Space Weather Impacts on Engineered Systems and Infrastructure. Royal Academy of Engineering, London.

Royal Astronomical Society, 1905. The Observatory, XXVIII, No. 354.

Russell, C.T., Mewaldt, R.A., Luhmann, J.G., Mason, G.M., von Rosenvinge, T.T., et al., 2013. The very unusual interplanetary coronal mass ejection of 2012July 23: a blast wave mediated by solar energetic particles. Astrophys. J. 770, 38.

Sabine, E., 1851. On periodical laws discoverable in mean effects on the larger magnetic disturbances. Philos. Trans. R. Soc. Lond. 141, 103–129.

Sabine, E., 1852. On periodical laws discoverable in mean effects on the larger magnetic disturbances, ii. Philos. Trans. R. Soc. Lond. A 142, 234–235.

Savani, N.P., Vourlidas, A., Szabo, A., Mays, M.L., Richardson, I.G., Thompson, B.J., Pulkkinen, A., Evans, R., Nieves-Chinchilla, T., 2015. Predicting the magnetic vectors within coronal mass ejections arriving at Earth: 1 Initial architecture. Space Weather 13, 374–385. https://doi.org/10.1002/2015SW001171.

Schrijver, C.J., Beer, J., Baltensperger, U., Cliver, E.W., Güdel, M., et al., 2012. Estimating the frequency of extremely energetic solar events, based on solar, stellar, lunar, and terrestrial records. J. Geophys. Res. 117. 08103. https://doi.org/10.1029/2012JA017706.

Schrijver, C.J., Kauristie, K., Aylward, A.D., Denardini, C.M., Gibson, S.E., Glover, A., Gopalswamy, N., Grande, M., Hapgood, M., Heynderickx, D., Jakowski, N., Kalegaev, V.V., Lapenta, G., Linker, J.A., Liu, S., Mandrini, C.H., Mann, I.R., Nagatsuma, T., Nandy, D., Obara, T., O'Brien, T.P., Onsager, T., Opgenoorth, H.J., Terkildsen, M., Valladares, C.E., Vilmer, N., 2015. Understanding space weather to shield society: a global road map for 2015–2025 commissioned by COSPAR and ILWS. Adv. Space Res. 55, 2745–2807.

Schröder, W., 1997. Some aspects of the earlier history of solar terrestrial physics. Planet. Space Sci. 45, 395–400.

Schwabe, S.H., 1843. Solar observations during 1843. Astron. Nachr. 20 (495), 234–235.

Sckopke, N., 1966. A general relation between the energy of trapped particles and the disturbance field near the Earth. J. Geophys. Res. 71, 3125.

Sharifi, J., Araabi, B.N., Lucas, C., 2006. Multi-step prediction of Dst index using singular spectrum analysis and locally linear neurofuzzy modeling. Earth Planets Space 58, 331–341.

Shibata, K., Masuda, S., Shimojo, M., Hara, H., Yokoyama, T., Tsuneta, S., Kosugi, T., Ogawara, Y., 1995. Hot plasma ejections associated with compact-loop solar flares. Astrophys. J. Lett. 451, 83.

Shibata, K., Magara, T., 2011. Solar flares: magnetohydrodynamic processes. Living Rev. Sol. Phys. 8, 6.

Siscoe, G.L., 1979. A quasi-self-consistent axially symmetric model for the growth of a ring current through earthward motion from a pre-storm configuration. Planet. Space Sci. 27, 285–295.

Siscoe, G., Crooker, N.U., Clauer, C.R., 2006. Dst of the Carrington storm of 1859. Adv. Space Res. 38, 173–179.

Siscoe, G., 2007. Space weather forecasting historically viewed through the lens of meteorology. In: Bothmer, V., Daglis, I. (Eds.), Space Weather, Physics and Effects. Springer, Berlin.

Srivastava, N., Venkatakrishnan, P., 2004. Solar and interplanetary sources of major geomagnetic storms during 1996–2002. J. Geophys. Res. 109. A010103. https://doi.org/10.1029/2003JA010175.

Srivastava, N., 2005a. Predicting the occurrence of super-storms. Ann. Geophys. 23, 2989–2995.

Srivastava, N., 2005b. A logistic regression model for predicting the occurrence of intense geomagnetic storms. Ann. Geophys. 23, 2969–2974.

Stamper, R., Belehaki, A., Buresova, D., Cander, L.R., Kutiev, I., Pietrella, M., Stanislawska, I., Stankov, S., Tsagouri, I., Tulunay, Y.K., Zolesi, B., 2004. Nowcasting, forecasting and warning for ionospheric propagation: tools and methods. Ann. Geophys. 47 (2/3), 957–983.

Stern, D.P., 2002. A millennium of geomagnetism. Rev. Geophys. 40 (3), 1007. https://doi.org/10.1029/2000RG000097.

Su, Z., et al., 2015. Ultra-low-frequency wave-driven diffusion of radiation belt relativistic electrons. Nat. Commun. 6, 10096. https://doi.org/10.1038/ncomms10096(2015).

Summers, D., Ma, C., Meredith, N.P., Horne, R.B., Thorne, R.M., Anderson, R.R., 2004. Modeling outer-zone relativistic electron response to whistler-mode chorus activity during substorms. J. Atmos. Sol. Terr. Phys. 66, 133.

Tang, F., Tsurutani, B.T., Gonzalez, W.D., Akasofu, S.I., Smith, E.J., 1989. Solar sources of interplanetary southward B_z events responsible for major magnetic storms (1978–1979). J. Geophys. Res. 94, 3535–3541.

Tang, F., Tsurutani, B.T., 1990. Reply. J. Geophys. Res.. 95, 10721.

Taylor, J.R., Lester, M., Yeoman, T.K., 1994. A superposed epoch analysis of geomagnetic storms. Ann. Geophys. 12, 612–624.

Temerin, M., Li, X., 2002. A new model for the prediction of Dst on the basis of the solar wind. J. Geophys. Res. 107 (A12), 1472. https://doi.org/10.1029/2001JA007532.

Temerin, M., Li, X., 2006. Dst model for 1995–2002. J. Geophys. Res. 111. A04221. https://doi.org/10.1029/2005JA011257.

Thayer, J.P., Lei, J., Forbes, J.M., Sutton, E.K., Nerem, R.S., 2008. Thermospheric density oscillations due to periodic solar wind high speed streams. J. Geophys. Res. 113. 06307. https://doi.org/10.1029/2008JA013190.

Tóth, G., Zeeuw, D.L., Gombosi, T.I., Manchester, W.B., Ridley, A.J., Sokolov, I.V., Roussev, I.I., 2007. Suntothermosphere simulation of the 28–30 October 2003 storm with the Space Weather Modeling Framework. Space Weather 5, S06003. https://doi.org/10.1029/2006SW000272.

Tsagouri, I., Belehaki, A., Bergeot, N., Cid, C., Delouille, V., et al., 2013. Progress in space weather modeling in an operational environment. J. Space Weather Space Clim. 3, A17.

Tsubouchi, K., Omura, Y., 2007. Long-term occurrence probabilities of intense geomagnetic storm events. Space Weather 5, 12003. https://doi.org/10.1029/2007SW000329.

Tsurutani, B.T., Baker, D.N., 1979. Substorm warnings: an ISEE-3 real time data system. Eos 60, 702.

Tsurutani, B.T., Lin, R.P., 1985. Acceleration of >47 keV ions and >2 keV electrons by interplanetary shocks at 1 AU. J. Geophys. Res. 90, 1.

Tsurutani, B.T., Gonzalez, W.D., Tang, F., Akasofu, S.I., Smith, E.J., 1988. Origin of interplanetary southward magnetic fields responsible for major magnetic storms near solar maximum(1978–1979). J. Geophys. Res. 93 (A8), 8519–8531.

Tsurutani, B.T., Gonzalez, W.D., Tang, F., Lee, Y.T., Okada, M., Park, D., 1992a. Reply to L. J. Lanzerotti: Solar wind RAM pressure corrections and an estimation of the efficiency of viscous interaction. Geophys. Res. Lett. 19((19), 1993–1994.

Tsurutani, B.T., Gonzalez, W.D., Tang, F., Lee, Y.T., 1992b. Great magnetic storms. Geophys. Res. Lett. 19, 73.

Tsurutani, B.T., Gonzalez, W.D., Gonzalez, A.L.C., Tang, F., Arballo, J.K., Okada, M., 1995a. Interplanetary origin of geomagnetic activity in the declining phase of the solar cycle. J. Geophys. Res. 100, 21717–21733.

Tsurutani, B.T., Gonzalez, W.D., Gonzalez, A.L.C., Tang, F., Arballo, J.K., Okada, M., 1995b. Interplanetary origin of geomagnetic activity in the declining phase of the solar cycle. J. Geophys. Res. 100, 21717–21733 A11.

Tsurutani, B.T., Gonzalez, W.D., 1997. The interplanetary causes of magnetic storms: a review. In: Tsurutani, B.T. et al., (Ed.), Magnetic Storms. Geophys. Monogr. Ser., 98, AGU, Washington, DC, pp. 77–89. https://doi.org/10.1029/GM098p0077.

Tsurutani, B.T., Gonzalez, W.D., Lakhina, G.S., Alex, S., 2003. The extreme magnetic storm of 1–2 September 1859. J. Geophys. Res. 108 (A7), 1268. https://doi.org/10.1029/2002JA009504.

Tsurutani, B.T., Gonzalez, W.D., Lakhina, G.S., Alex, S., 2005. Reply to comment by S.-I. Akasofu and Y. Kamide on "The extreme magnetic storm of 1–2 September 1859" J. Geophys. Res.. 110,https://doi.org/10.1029/2005JA011121 09227.

Tsurutani, B.T., et al., 2006. Corotating solar wind streams and recurrent geomagnetic activity: a review. J. Geophys. Res. 111, https://doi.org/10.1029/2005JA011273 A07S01.

Tsurutani, B.T., Verkhoglyadova, O.P., Mannucci, A.J., Saito, A., Araki, T., Yumoto, K., Tsuda, T., Abdu, M.A., Sobral, J.H.A., Gonzalez, W.D., McCreadie, H., Lakina, G.S., Vasyliunas, V.M., 2008. Prompt penetration electric fields(PPEFs) and their ionospheric effects during the great magnetic storm of 30–31 October 2003. J. Geophys. Res. 113. A05311, https://doi.org/10.1029/2007JA012879.

Tsurutani, B.T., Verkhoglyadova, O.P., Manucci, A.J., Lakhina, G.S., Huba, J.D., 2012. Extreme changes in the dayside ionosphere during a Carrington type magnetic storm. J. Space Weather Space Clim. 2, 05. https://doi.org/10.1051/swsc/20122004.

Tsurutani, B.T., Lakhina, G.S., 2014. An extreme coronal mass ejection and consequences for the magnetosphere and Earth. Geophys. Res. Lett. 41. https://doi.org/10.1002/2013GL058825.

Tsurutani, B.T., Echer, E., Shibata, K., Verkhoglyadova, O.P., Mannucci, A.J., Gonzalez, W.D., Kozyra, J.U., Pätzold, M., 2014. The interplanetary causes of geomagnetic activity during the 7–17 March 2012 interval: a CAWSES II overview. J. Space Weather Space Clim. 4, 02. https://doi.org/10.1051/swsc/2013056.

Tsurutani, B.T., Hajra, R., Echer, E., Gjerloev, J.W., 2015. Extremely intense (SML $\leq$ -2500 nT) substorms: isolated events that are externally triggered? Ann. Geophys. Commun. 33, 519–524. https://doi.org/10.5194/angeocom-33-519-2015.

Vasyliunas, V.M., 2011. The largest imaginable magnetic storm. J. Atmos. Sol. Terr. Phys. 73, 1444–1446. https://doi.org/10.1016/j.jastp.2010.05.012.

von Humboldt, A., 1808. Die vollständigsteallerbisherigenbeobachtungenüber den einfluss des nordlichts auf die magnetnadel. Ann. Phys. 29, 425–429.

Vichare, G., Alex, S., Lakhina, G.S., 2005. Some characteristics of intense geomagnetic storms and their energy budget. J. Geophys. Res. 110. A03204. https://doi.org/10.1029/2004JA010418.

Wang, C.B., Chao, J.K., Lin, C.-H., 2003. Influence of the solar wind dynamic pressure on the decay and injection of the ring current. J. Geophys. Res. 108 (A9), 1341. https://doi.org/10.1029/2003JA009851.

Willis, D.M., Stevens, P.R., Crothers, S.R., 1997. Statistics of the largest geomagnetic storms per solar cycle (1844–1993). Ann. Geophys. 15, 719–728.

Wu, J.-G., Lundstedt, H., 1996. Prediction of geomagnetic storms from solar wind data using Elman recurrent neural networks. Geophys. Res. Lett. 23, 319–322.

Wygant, J., Mozer, F., Temerin, M., Blake, J., Maynard, N., Singer, H., Smiddy, M., 1994. Large amplitude electric and magnetic field signatures in the inner magnetosphere during injection of 15 MeV electron drift echoes. Geophys. Res. Lett. 21, 1739–1742.

Yashiro, S., Gopalswamy, N., Michalek, G., St. Cyr, O.C., Plunkett, S.P., Rich, N.B., Howard, R.A., 2004. A catalog of white light coronal mass ejections observed by the SOHO spacecraft. J. Geophys. Res. 109. A07105. https://doi.org/10.1029/2003JA010282.

Yermolaev, Y.I., Nikolaeva, N.S., Lodkina, I.G., Yermolaev, M.Y., 2012. Geoeffectiveness and efficiency of CIR, sheath, and ICME in generation of magnetic storms. J. Geophys. Res. 117. https://doi.org/10.1029/2011JA017139 A00L07.

Yermolaev, Y.I., Lodkina, I.G., Nikolaeva, N.S., Yermolaev, M.Y., 2013. Occurrence rate of extreme magnetic storms. J. Geophys. Res. 118, 4760–4765. https://doi.org/10.1002/jgra.50467.

Yokoyama, N., Kamide, Y., 1997. Statistical nature of geomagneticstorms. J. Geophys. Res. 102, 14215.

FURTHER READING

Ji, E.-Y., Moon, Y.-J., Kim, K.-H., Lee, D.-H., 2010. Statistical comparison of interplanetary conditions causing intense geomagnetic storms (Dst ≤ -100 nT). J. Geophys. Res. 115. A10232. https://doi.org/10.1029/2009JA015112.

Kim, R.-S., Cho, K.-S., Kim, K.-H., Park, Y.-D., Moon, Y.-J., Yi, Y., Lee, J., Wang, H., Song, H., Dryer, M., 2008. CME earthward direction as an important geoeffectiveness indicator. Astrophys. J. 677, 1378.

Silbergleit, V.M., Zossi de Artigas, M.M., Manzano, J.R., 1996. Austral electrojet indices derived for the great storm of March 1989. Ann. Geofis. XXXIX (6), 1177–1184.

Tóth, G., Sokolov, I.V., Gombosi, T.I., Chesney, D.R., Clauer, C.R., Zeeuw, D.L., Hansen, K.C., Kane, K.J., Manchester, W.B., Oehmke, R.C., Powell, K.G., Ridley, A.J., Roussev, I.I., Stout, Q.F., Volberg, O., Wolf, R.A., Sazykin, S., Chan, A., Yu, B., Kóta, J., 2005. Space weather modeling framework: a new tool for the space science community. J. Geophys. Res. 110, A12226.

Tsurutani, B.T., Gonzalez, W.D., 1995. The efficiency of "viscous interaction" between the solar wind and the magnetosphere during intense northward IMF events. Geophys. Res. Lett. 22 (6), 663–666.

Tsurutani, B.T., Kamide, Y., Arballo, J.K., Gonzalez, W.D., Lepping, R.P., 1999. Interplanetary causes of great and super intense magnetic storms. Phys. Chem. Earth 24, 101–105.

Tsurutani, B.T., Verkhoglyadova, O.P., Mannucci, A.J., Araki, T., Sato, A., Tsuda, T., Yumoto, K., 2007. Oxygen ion uplift and satellite drag effects during the 30 October 2003 daytime superfountain event. Ann. Geophys. 25, 569–574.

CHAPTER 8

AN OVERVIEW OF SCIENCE CHALLENGES PERTAINING TO OUR UNDERSTANDING OF EXTREME GEOMAGNETICALLY INDUCED CURRENTS

Chigomezyo M. Ngwira[*,†], Antti A. Pulkkinen[†]

The Catholic University of America, Washington, DC, United States NASA Goddard Space Flight Center, Greenbelt, MD, United States[†]*

CHAPTER OUTLINE

1 Introduction 187
   1.1 Geomagnetic Storms at Earth 188
   1.2 Basic Theory of GICs 188
2 Impact on Ground Systems 189
   2.1 Electrical Power Systems 190
   2.2 Oil and Gas Pipelines 192
   2.3 Other Systems 193
3 U.S. Federal Actions Relating to GICs 193
4 Key Science Challenges 194
   4.1 Extreme Drivers 195
   4.2 Modeling Extremes 197
   4.3 Defining Extremes 199
5 Concluding Remarks 201
Acknowledgments 201
References 202

1 INTRODUCTION

Concern about the vulnerability of man-made technological assets to Earth-directed extreme space weather events has significantly grown in the last two decades. For example, space weather-driven geomagnetically induced currents (GICs) that can disrupt the operation of assets such as power transmission grids, oil and gas pipelines, and telecommunications cables are a serious problem for society today (e.g., Barlow, 1849; Davidson, 1940; Boteler and Jansen van Beek, 1999; Pirjola, 2000; Molinski et al., 2000; Pulkkinen et al., 2001; Eroshenko et al., 2010, and references therein).

The ultimate challenge for the scientific community is to gain a better understanding of extreme storms to ensure that we can more accurately assess and predict the occurrence and impact of extreme events. In this chapter, we present an overview of extreme GICs that highlights some of the most pressing science challenges, known severe impacts within the last 80 years, and recent U.S. federal actions. While the topics in this chapter touch on a wide range of space weather processes, the discussions are focused primarily on GICs. For detailed insight on certain aspects of space weather phenomena, interested readers are referred to the other chapters in this book.

1.1 GEOMAGNETIC STORMS AT EARTH

Large, violent explosions of solar plasma from the Sun's corona, well known as coronal mass ejections (CMEs), are the major source of strong geomagnetic storms at Earth (e.g., Gopalswamy et al., 2005; see also Chapter 7 of this volume). CMEs contain plasma and an embedded solar magnetic field known as the interplanetary magnetic field (IMF). The most intense disturbances are produced under southward IMF conditions when IMF orientation is oppositely directed to the Earth's magnetic field. Under southward IMF conditions, a more efficient coupling pattern is established that in turn stimulates a chain of complex dynamic processes within the magnetosphere–ionosphere (M-I) system driving phenomena such as geomagnetic disturbances (GMDs), auroral displays at high-latitude locations, ionospheric irregularities, and traveling ionospheric disturbances. CMEs are recognized also as the most geoeffective driver of intense GICs (see e.g., Huttunen et al., 2008; Pulkkinen et al., 2010a; Ngwira et al., 2013b).

1.2 BASIC THEORY OF GICS

GICs are essentially the ground manifestation of a complex space weather chain of events initiated by solar eruptions. On a fundamental level, the key aspect of the continuous solar wind and magnetosphere interaction that drives GICs is the variation in near-space electric current systems. Intense time-varying currents cause rapid fluctuation of the geomagnetic field on the ground. The physical principle of the flow of GICs in technological systems is governed by Faraday's law of induction: a changing magnetic field induces currents in conductors. More specifically, Faraday's law

$$\nabla \times \mathbf{E} = -\frac{\partial \mathbf{B}}{\partial t} \qquad (1)$$

relates the temporal variation of the geomagnetic field to the formation of the geoelectric field. The geoelectric field then drives an electric current inside the Earth, according to Ohm's law $\mathbf{J} = \sigma \mathbf{E}$. The induced geoelectric field observed on the surface is independent of the technological system and depends on M-I currents, which in turn are dependent on space weather conditions and on electromagnetic induction that is determined by the Earth's geology (e.g., Pirjola, 1982).

Overall, the GIC problem demands a two-step approach (Pirjola, 2000, 2002). In step one, geophysical aspects involving the estimation of the geoelectric field based on the knowledge of magnetosphere-ionosphere currents and the Earth's geological structure are considered. Step two is the engineering piece in which the system response is modeled based on the determined geoelectric field and detailed information about the particular ground system (e.g., Lehtinen and Pirjola, 1985; Molinski et al., 2000; Pirjola, 2000).

A simple but illustrative model for calculating the geoelectric field assumes a plane wave propagating vertically downward and that the Earth is a uniform (or a layered) half-space with conductivity σ (Cagniard, 1953; Pirjola, 1982). Consider a single frequency ω, then the horizontal geoelectric field E_y component can be expressed in terms of the perpendicular horizontal geomagnetic field component B_x as:

$$E_y(\omega) = -\sqrt{\frac{\omega}{\mu_0 \sigma}} e^{\frac{i\pi}{4}} B_x \qquad (2)$$

where μ_0 is the permeability of free space and the layer of air between the ground and the ionosphere is considered to have zero conductivity so that there is no significant attenuation of external electromagnetic fields. Eq. (1) is the "basic equation of magnetotellurics" because it outlines the basis for deriving the Earth's conductivity by using geoelectric and geomagnetic field measurements recorded at the surface.

The plane wave method is a well-established and widely used method for GIC applications (see e.g., Pirjola, 2002; Viljanen et al., 2006a; Ngwira et al., 2008; Pulkkinen et al., 2015a; da Silva Barbosa et al., 2015). Fig. 1 shows a comparison of the plane wave model geoelectric field and measured field. Measured geoelectric fields were recorded at Browns Ferry station under the NSF-funded EarthScope USArray MT project. The model geoelectric fields were computed from Eq. (2) using recorded geomagnetic field data and a 1-D layered Earth conductivity specific to this particular region, according to Fernberg (2012). Clearly, there is a good match between the modeled and measured E_y component but the E_x component amplitudes are off even though the general trends are well captured. This example highlights the need to improve our modeling abilities.

It is important to note that the plane wave approach has also been applied to cases of extreme geomagnetic storms. For instance, it has been extensively used to study extreme GIC events (Ngwira et al., 2008, 2013b, 2015; Wik et al., 2009; Pulkkinen et al., 2012). More recently, Ngwira et al. (2014) also employed the plane wave method to study surface geoelectric field response based on a hypothetical simulation of an extreme Carrington-type event. The modeled fields captured the global response pattern. Therefore, indications thus far suggest that the approximation can be applied also to extreme cases but further investigations are still required.

In general, information about the geoelectric field that is produced on the ground during a GMD event is obtained using the approach above. Now it is relatively straightforward to calculate the level of GICs flowing through a given node for any ground system. This purely engineering task, requiring a detailed description of the specific system, is beyond the scope of this chapter. For further details regarding the execution of this process, interested readers are encouraged to consult Lehtinen and Pirjola (1985) and/or Viljanen and Pirjola (1994), and references therein.

2 IMPACT ON GROUND SYSTEMS

GICs have the potential to disrupt different technologies that we depend on for economic vitality and national security, including the electric power grid, oil and gas pipelines, and communications

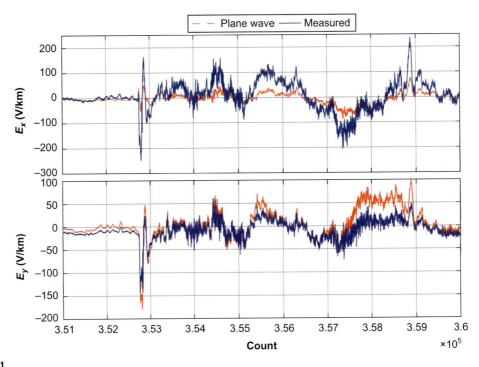

FIG. 1

A comparison of modeled and measured geoelectric fields during a storm event on October 7, 2015. The measured fields were recorded at Browns Ferry USArray station. Geoelectric field measurements courtesy of Anna Kelbert, USGS.

networks. This critical infrastructure makes up a diverse, complex, and interdependent network of systems where a failure of one could likely impact another.

Table 1 provides a list of extreme GIC effects that have taken place over the last 80 years. The source (or reference) of this information is also included. In the rest of this section we briefly describe GIC effects on different ground systems.

2.1 ELECTRICAL POWER SYSTEMS

It is widely acknowledged that a variety of grounded conductors can be disrupted by GICs; however, electric power systems constitute probably the most critical technological infrastructure vulnerable to these deleterious effects due to long transmission lines. For that reason electric power systems are highly singled out by many authors because electric energy cuts across many sectors of modern society and economy.

Large amplitude GICs pose a serious threat to the sustainability, reliability, and availability of electricity. The earliest publicly reported impact of GICs on power systems was recorded following the March 24, 1940 geomagnetic storm effects on North American power grids (Davidson, 1940).

Table 1 Examples of Extreme GIC Effects on Ground Systems Within the Last 80 Years

Date	Associated Effects/Impacts
March 24, 1940	Power circuit disturbances in North America. Earliest publicly reported impact on power grids (Davidson, 1940; McNish, 1940).
February 11, 1958	Blown AC power supply fuses for Finnish telecommunications cable system; Fire damage to telegraph equipment (Nevanlinna et al., 2001; Wik et al., 2009).
November 13, 1960	30 Swedish high-voltage power network line breakers tripped (Wik et al., 2009).
August 4, 1972	Communications cable system outage in mid-western United States (Boteler and Jansen van Beek, 1999).
July 13–14, 1982	Transformers and lines tripped in the Swedish high-voltage power system; Railway traffic light anomalies; Effects on telecommunications (Wik et al., 2009).
March 13–14, 1989	Hydro-Quebec blackout; New York transformer damage; Railway track anomalies (Boteler, 2001b; Bolduc, 2002; Eroshenko et al., 2010).
March 24, 1991	GIC > 100 A measured in Finland; 220 kV lines and a transformer tripped (Viljanen et al., 1999; Wik et al., 2009).
November 9, 1991	220 kV line tripping; Large pipe-to-soil voltages in an oil pipeline (Wik et al., 2009).
April 6, 2000	Largest GIC on Swedish transformer; False railway track occupations/blockages (Wik et al., 2009; Eroshenko et al., 2010).
November 6, 2001	Railway automatic system failure (Eroshenko et al., 2010).
October 29–30, 2003	Malmö blackout; Excessive transformer heating; Triggered emergency procedures at nuclear power plants in Canada and Northeastern United States; Reported damaged to high-voltage transformers in South Africa; Railway automatic system failure (Pulkkinen et al., 2005; Wik et al., 2009; Gaunt and Coetzee, 2007; Eroshenko et al., 2010).
November 20, 2003	Railway automatic system failure (Eroshenko et al., 2010).
November 8, 2004	Transformer GIC exceeding 100 A measured in southern Sweden (Wik et al., 2009).
May 14, 2005	Railway automatic system failure (Eroshenko et al., 2010).

The primary effect of GICs is that they can push power transformers into so-called half-cycle saturation, causing: (1) generation of hot-spots in the windings and/or structural components (e.g., Girgis and Ko, 1991; Lehtinen and Elovaara, 2002; Price, 2002); (2) increased reactive power consumption of the transformer (e.g., Xuzhu et al., 2001; Berge et al., 2011); and (3) injection of even and odd current harmonics into the power system (e.g., Walling and Khan, 1991; Shu et al., 1993; Bernabeu, 2014).

For the electric grid, the most serious concern is related to rarely occurring geomagnetic superstorms. Collapse of the Hydro-Quebec power network grid in Canada during the March 13, 1989 superstorm is the most serious example of the impact space weather-driven GICs can have on power systems. The GICs produced during this superstorm caused a blackout of the entire Hydro-Quebec network on a time scale of just under two minutes (Boteler, 2001b; Bolduc, 2002, and references therein). The power outage that affected millions of people for about 9 hours was the result of widespread transformer saturation. During the same event, a generator step-up power transformer was destroyed in New Jersey, the United States.

More recently, the Halloween storm of October 2003 caused the failure of a high-voltage power transmission system in Malmö, Sweden (e.g., Pulkkinen et al., 2005; Wik et al., 2009). The incident caused a power blackout that lasted for about an hour and left around 50,000 customers without electricity (Wik et al., 2009). In South Africa, at a mid-latitude location that was previously considered much less prone to GIC impacts, possible transformer damages were reported during the same storm of October 2003 (Gaunt and Coetzee, 2007).

The concern of any harmful impacts on power grids has triggered renewed interest in extreme GICs and space weather in general (e.g., Thomson et al., 2011; Pulkkinen et al., 2012; Ngwira et al., 2013b, 2014, 2015; Marshall et al., 2012; Fiori et al., 2014; Pulkkinen et al., 2015a; Schrijver et al., 2015; Carter et al., 2016). As a result, there is growing recognition that the GIC problem is not only a high-latitude phenomenon but affects also mid- to low-latitudes (see e.g., Trivedi et al., 2007; Ngwira et al., 2008; Watari et al., 2009; Marshall et al., 2011; Torta et al., 2012; Carter et al., 2015; Adebesin et al., 2016).

In addition, the very first NASA Living With a Star (LWS) Institute Working Group that targeted the GIC problem was selected competitively as the pilot activity for the new LWS element (Pulkkinen, 2016). The new NASA LWS Institutes program element was launched in 2014. In 2015, the LWS Institute GIC Working Group held two small working group meetings focused on well-defined problems that called for intense, direct interactions between colleagues in interdisciplinary fields. This facilitated the development of a deeper insight of the variety of processes that link solar activity to Earth's environment and the power grid. See Pulkkinen et al. (2017) for more details on the LWS Institute GIC Working Group findings, and Pulkkinen (2016) for the special collection of papers, a product of work by teams of researchers from more than 20 different international organizations.

2.2 OIL AND GAS PIPELINES

Oil and gas pipelines play a key role throughout the world as a mode of transporting energy commodities over long distances from their sources to end-users or consumers. These pipelines are also affected by GICs produced by geomagnetic disturbances whose effects are not immediate but rather may need the cumulative action of several disturbances (Boteler, 2001a; Marshall et al., 2010). Unprotected buried pipelines can be susceptible to corrosion damage (e.g., Seager, 1991; Martin, 1993; Gummow and Eng, 2002). Corrosion is an electrochemical process occurring when a current flows from the pipe into the soil (Gummow and Eng, 2002). To protect the steel pipe from corrosion, pipelines are covered by an insulating coating in conjunction with a cathodic protection system that keeps the pipeline at a negative voltage on the order of 1 V with respect to the ground. This helps to prevent telluric current (or GIC) flows from the pipeline to the soil.

Generally, there isn't much information about GIC impacts on pipelines, but one of the most extended studies of space weather effects on pipelines was conducted on the Finnish natural gas pipeline from August 1998 to May 1999 by Pulkkinen et al. (2001). The project was implemented to: (1) derive a model for computing GICs and pipe-to-soil (P/S) voltages, (2) perform measurements of GICs and P/S voltages in the pipeline, and (3) derive statistically based predictions of GIC occurrences and P/S voltages at different locations in the pipeline network. The study concluded that the largest P/S voltage variations occurred at the ends of pipelines, while the largest GIC flows were in the middle portions of the pipeline network. This conclusion was consistent with earlier published results by Boteler et al. (2000).

2.3 OTHER SYSTEMS

GICs can also impact telecommunication systems (e.g., Nevanlinna et al., 2001; Karsberg et al., 1959). In fact, it is fairly well accepted that the first recorded space weather impact on human-made technology was on telegraph systems (e.g., Barlow, 1849; Prescott, 1860; Cade III, 2013). Today, fiberoptic cables have a decreased probability of problems because they do not carry GIC. On the other hand, the metal wires used in conjunction with those cables to provide the power needed for repeat stations might suffer from GIC-related impact.

For instance, Nevanlinna et al. (2001) report on a case of sudden interruption of two coaxial phone cable systems in the Finnish telecommunications network during the great magnetic storm of February 11, 1958. The problem was a result of blown fuses connected to the AC power feed at the repeater stations. In North America, a major geomagnetic storm on August 4, 1974 is reported to have caused an outage of a communications cable system in the mid-western United States (e.g., Anderson et al., 1974; Boteler and Jansen van Beek, 1999). A detailed analysis of this geomagnetic storm's response performed by Boteler and Jansen van Beek (1999) revealed that the outage was associated with a highly localized ionospheric eastward electrojet source.

On railway networks, there are very few well-documented cases concerning GIC effects (e.g., Wik et al., 2009; Eroshenko et al., 2010, and references therein). One of the earliest reports is a malfunctioning of traffic lights on Swedish railways during the great geomagnetic storm of July 1982 (see Wik et al., 2009, and references therein). A more recent event on the Russian railways was caused by an anomaly in the signal system operation across some divisions within the high-latitude regions, as documented by Eroshenko et al. (2010). The anomaly manifested as false traffic light signals about the occupation of the railways that appeared precisely during the main phases of the strongest part of the 17 geomagnetic storms examined.

3 U.S. FEDERAL ACTIONS RELATING TO GICs

It can be argued that GICs are currently the greatest space weather hazard-related global concern. The concern pertains specifically to extreme storms and extreme GICs that have the potential to cause significant and direct disruption of our daily lives. Also, the awareness about the threat posed by GICs has been evolving very rapidly and, especially in the United States, the elevated awareness has led to a number of policy and regulatory actions.

On the policy side, action has been taken at the White House level in terms of development of the National Space Weather Strategy and National Space Weather Action Plan (SWAP) (National Science and Technology Council, 2015b,a). While GICs are not the only space weather hazard addressed in these policies, the phenomenon does play a central role in them. Importantly, the U.S. federal government is in the process of executing the SWAP and the initial extreme event benchmark studies that are the first step in the plan have been completed. In the SWAP context, "extreme" is defined as a 1-in-100-

year event and a theoretical maximum event characterized using key physical parameters of interest. The initial SWAP geoelectric field work related to GICs was recently reported by Love (2012). It is also worth noting that in the United Kingdom, GICs were recently introduced as a part of the space weather element in the National Risk Registry (Cabinet Office, 2015).

On the U.S. regulatory side, the Federal Energy Regulatory Commission (FERC) recently ordered development of GMD standards that will direct the power transmission industry response to GIC hazards (Federal Energy Regulatory Commission, 2015). In response to both industry concerns about the GIC hazard and the FERC GMD order, the North American Electric Reliability Corporation (NERC) that is the key North American hub for the power transmission industry's coordinated bulk system reliability actions, took the leading role in organizing industry response to the hazard. NERC work has resulted in a number of publications (e.g., North American Electric Reliability Corporation, 2012, 2013, 2016a,b) that guide industry on GIC issues. As a regulatory authority, NERC also drafts the standards in response to the FERC order. The GMD standard pertaining to hazard assessments drafted by NERC includes a 1-in-100-year event benchmark that was tailored for somewhat simplified utility level usage (North American Electric Reliability Corporation, 2016a). The NERC GMD standard was approved by FERC in September 2016, which means that all U.S. utilities operating at or above 200 kV will be addressing the hazard posed by extreme GICs.

Reflecting the actions in North America and the United Kingdom, many nations around the globe have increased interest in characterizing the hazard posed by GICs. Nations such as the United States, the United Kingdom, Canada, Finland, Norway, Sweden, China, Japan, Brazil, Namibia, South Africa, and Australia all have launched hazard-assessment campaigns to understand and mitigate the possible GIC impact on their systems. In all these works, understanding the extremes such as 1-in-100-year events is the central theme. It is thus clear that there is a great demand for improved understanding of extreme GICs. The challenge for scientists is to characterize the spatiotemporal evolution of extreme geoelectric fields and GICs for the past, present, and future (see Pulkkinen et al., 2017, for more discussion on this).

4 KEY SCIENCE CHALLENGES

This section elaborates on some critical science topics that continue to hinder our understanding of extreme events. Here science is perceived as the fundamental Space and Earth sciences research that enriches awareness and improved physics-based modeling of the physical processes relating to GICs. The ability to monitor, model, forecast, and understand extreme space weather events is among the most pressing scientific objectives of our highly technology-dependent society.

Over the past few decades, major advancements in understanding the processes operating within distinct regions of the near-Earth space environment and the interaction between the regions has led to a full-grown insight of the important processes driving GICs (Pulkkinen et al., 2017). However, even though they have a potentially high impact, there is still limited knowledge of extreme events because they occur infrequently. As a result, there are insufficient historical recordings to adequately examine our ability to more accurately model extreme events, from which we can gain better understanding of the physical processes operating in the solar wind-magnetosphere-ionosphere coupled system.

We note that the influence of the Earth's geology on the GIC induction process is not addressed in the present study but a treatment of this topic is offered in Chapter 9 of the same volume and references therein, for the benefit of interested readers.

4.1 EXTREME DRIVERS

Understanding the basic principles responsible for the initiation and development of dynamic M-I currents that create intense GICs on the ground is among the most pressing science issues (e.g., Pulkkinen et al., 2017, and references therein). Two important aspects in this regard are: (1) interplanetary solar wind structures that drive extreme GICs on the ground, and (2) the response of M-I processes under extreme driving.

Solar wind recordings at L1 provide key insight into external driving conditions and are fundamentally important for space weather research (Akasofu, 1981; Borovsky and Funsten, 2003; Echer et al., 2008). Solar wind changes play a central role in controlling the generation and evolution of M-I dynamics (Gonzalez, 1990; Weigel et al., 2002; Ebihara et al., 2005; Tsurutani et al., 2015). While it is generally agreed that CME-driven storms are the most geoeffective source of large GICs (e.g., Huttunen et al., 2008; Adebesin et al., 2016), it is difficult to determine when and if the response of M-I processes during "ordinary" space conditions can be applied to extreme conditions of solar wind (Nagatsuma et al., 2015).

The magnetosphere responds dramatically to major changes in the solar wind. Several studies demonstrate that the response of the M-I coupled system is more complex under extreme conditions (e.g., Chun and Russell, 1997; Ebihara et al., 2005; Ngwira et al., 2014; see also Chapter 20 of this volume). In an extended study of extreme geoelectric fields on the ground, a follow on study of Pulkkinen et al. (2012), Ngwira et al. (2013b) make known that rapid changes in solar wind conditions trigger strong enhancement of the equatorial electrojet (EEJ) current, generating intense geoelectric fields in the EEJ zone. The suggested primary mechanism for this dramatic activity is penetration of high-latitude electric fields. Also, these sudden changes in the solar wind alter the auroral electric current due to related changes in FACs generating intense GICs at high-latitudes, as recently reported by Adebesin et al. (2016). Ebihara et al. (2005) found that apart from moving to much lower latitudes than usual, FACs also developed a complex distribution structure during the superstorm on November 20, 2003. Then, by simulating a Carrington-type event (Carrington, 1859; Tsurutani et al., 2003) using the University of Michigan physics-based Space Weather Modeling Framework (SWMF) and extreme solar wind driving conditions, Ngwira et al. (2014) got similar results of complex FACs in support of earlier findings.

Magnetospheric substorms have long been identified as one of the leading causes of intense high-latitude GICs (e.g., Pulkkinen et al., 2005; Viljanen et al., 2006b; Ngwira et al., 2014, and references therein). Substorms are elemental physical processes of solar wind energy storage and explosive release in Earth's magnetotail region, which encompass basic plasma physics. Closely related to the substorm expansion phase is the enhancement of the westward current across the bulge of expanding aurora that is fed by FACs (McPherron and Chu, 2016). Enhancement of auroral electrojet currents is a direct manifestation of the substorm current wedge (SCW) (e.g., Murphy et al., 2013; McPherron and Chu, 2016, and references therein), which is a central element of the expansion phase associated with magnetic field dipolarization in the magnetotail and intensification of the westward electrojet after auroral substorm onset.

196　CHAPTER 8 AN OVERVIEW OF SCIENCE CHALLENGES PERTAINING TO GICs

Exactly what is driving the development of complex M-I processes and their connection to the development of extreme GICs is still open to debate and calls for further consideration. Thus, besides substorms, other major GIC drivers are also worth considering, including, geomagnetic pulsations (Pulkkinen et al., 2005; Viljanen et al., 1999; Pulkkinen and Kataoka, 2006) and sudden storm commencements (Kappenman, 2003; Pulkkinen et al., 2005; Fiori et al., 2014; Carter et al., 2015).

Now, one of the major characteristics common to most, if not all, extreme events is that peak GIC activity tends to occur in relatively isolated spots/locations (Boteler and Jansen van Beek, 1999; Pulkkinen et al., 2015a; Ngwira et al., 2015, and references therein). It is important that we grasp those fundamental processes that give rise to local-, regional-, and global-scale structure and dynamics in the near-Earth space environment. This is one of the top priorities for space weather research today. Because the solar wind-magnetosphere-ionosphere is a coupled system, one of the issues that makes it difficult to explicate the causes is that several different mechanisms can be contributing at the same time, as Onsager et al. (2004) point out. The challenge for us is to find creative ways (or tools) to separate the different contributions with great accuracy.

The induced geoelectric field is responsible for currents that flow on ground-based systems; thus it is the primary quantity that dictates the magnitude of GICs. Fig. 2 contains an illustration of a highly localized extreme geoelectric field (or GIC event) on November 20, 2003. Ngwira et al. (2015) provide a summary of 12 localized extremes that occurred between the years 1982–2005. In the figure, a very

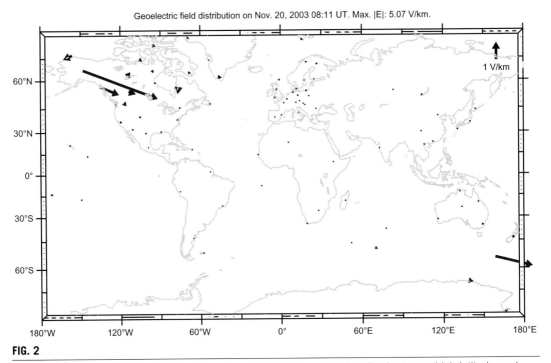

FIG. 2

Calculated global geoelectric fields for November 20, 2003 showing a localized extreme high-latitude peak geoelectric field at the College magnetometer station in Alaska. Note that the peak geoelectric field amplitude indicated on the top of the figure refers to a single station maximum.

intense geoelectric field (>1 V/km) appears at a single site in the Alaskan region. These intense geoelectric fields are seen only at three time instants (08:10, 08:11, and 08:12 UT). Ngwira et al. (2015) further reveal that these localized extreme fields at single sites can be associated with structures that differ greatly from global and regional average fields. However, the physical processes that control the evolution of the localized extremes have not been sufficiently explored.

Obviously, storm-time variability of the geoelectric field is a direct reflection of dynamic processes in the M-I coupled system. Consequently for GIC applications, it is critical that we understand the processes that produce complex current structures in the solar wind, magnetosphere, and ionosphere. The task at hand is to establish: (1) What is the role of external extreme drivers that regulate the state of the whole system? (2) How does the M-I system respond to extreme driving? The ultimate goal is to build a better understanding of the spatiotemporal characteristics of M-I currents under extreme driving. Like most of the major scientific challenges in the geosciences, there is growing recognition that an integrated approach that consolidates multiple disciplines and multiple sets of space-based and ground-based measurements is needed to enrich our knowledge on CMEs and M-I processes that amplify geomagnetic hazards.

4.2 MODELING EXTREMES

A number of GIC system flow analysis techniques exist today (Lehtinen and Pirjola, 1985; Boteler et al., 2000; Pulkkinen et al., 2001; Zhang et al., 2012; Bernabeu, 2013; Overbye et al., 2013; Boteler, 2014; Boteler and Pirjola, 2014, and references therein). A description of the geoelectric field in the near vicinity of the ground system and information about the particular network system parameters must be furnished. For this community, the geoelectric field is the foremost quantity of interest because it is the main interface between the geophysical processes and the engineering application of GIC.

Assessing the geomagnetic hazard to ground systems is often dependent on our ability to provide information to end-users about the expected extremes. Thus, modeling space weather events with sufficient lead time and accuracy is crucial for adopting appropriate operational strategies. Achieving a more reliable forecast framework pertaining to GICs depends on our ability to model ground magnetic field perturbations with high accuracy. Significant progress has been made in the overall modeling framework both from first principles physics-based approaches and from empirical techniques (e.g., Raeder et al., 1996; Weigel et al., 2003; Pulkkinen et al., 2010b, 2013; Tóth et al., 2014; Wiltberger et al., 2015; Wintoft et al., 2015). There has been a growing number of studies that focus on modeling extreme geomagnetic events in general (e.g., Manchester IV et al., 2006; Li et al., 2005; Baker et al., 2013; Ngwira et al., 2014; Liou et al., 2014; Shen et al., 2014; Cid et al., 2014). In addition, other studies specifically focus on GICs (e.g., Ngwira et al., 2013a, 2014; Kelly et al., 2017). As pointed out in Section 4.1, unlocking the mysteries of the M-I system processes under extreme driving conditions is critical for our modeling.

Extreme events in particular push the boundaries of our understanding of the physics of space weather and our ability to model extremes. Interestingly, Ngwira et al. (2014) found that the ionospheric cross polar cap potential (CPCP) values were much higher than normally observed but the transpolar potential relationship between the CPCP and the IMF was fulfilled with saturation occurring at substantially elevated levels. In addition, they also concluded that even under extreme driving, the MHD model was able to reproduce known ground geoelectric field global characteristic distribution (see for instance Figure 4 in Pulkkinen et al., 2012). Furthermore, Fig. 3 depicts the ionospheric electric

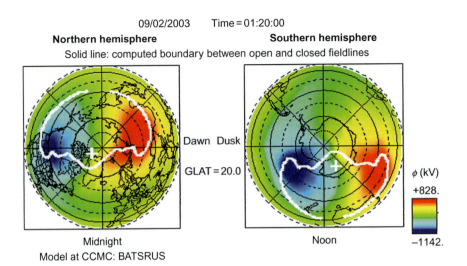

FIG. 3

Ionospheric electrical potential output from MHD simulation of an extreme Carrington-type event by Ngwira et al. (2014). Geomagnetic pole is at the center of each segment and the outer circle depicts 20 degrees geomagnetic latitude. The *solid white trace* shows that the dayside polar cap boundary location is significantly displaced to much lower latitudes under extreme driving conditions.

potentials in the Northern and Southern hemispheres based on simulation results by Ngwira et al. (2014). The solid white trace shows the polar cap boundary location. This figure shows that the polar cap was significantly shifted toward the dayside magnetosphere. The unusual shift of the polar boundaries could expose southern locations to high latitude ionospheric phenomena and thus also expose southern locations to elevated GIC risk.

However, many challenges remain concerning the capabilities of current models, as Welling et al. (2017) have shown. Having reanalyzed results from Pulkkinen et al. (2013) with a goal of better understanding the performance of different models, Welling et al. (2017) found that all models evaluated had a bias toward underprediction, especially during active times. As a result, it is important to keep in mind that current models are still fairly immature and we don't know exactly how valid the results are under extreme driving conditions. Cid et al. (2014) offers an analysis and discussion of some of the key issues related to extreme storms. This has serious implications for future research targeting to improve operational forecast models.

Also, modeling short-term variability and small-scale storm-time features, which are important for GICs, is still a major challenge. Take for example results in Fig. 4 that display geomagnetic field perturbations from observations (red) and model (blue) for the B_x (top) and B_y (middle) components. Bottom panel contains the rate of change of B_x component for the event on March 17, 2015. All data sets were sample at 1-minute resolution. The two stations used in the illustration are Fredericksburg, the United States (left) and Ottawa, Canada (right). The model magnetic field perturbations were obtained from SWMF simulations (Tóth et al., 2012, and reference within) performed at the Community Coordinated Modeling Center (CCMC). The global magnetosphere was represented by BATSRUS

4 KEY SCIENCE CHALLENGES 199

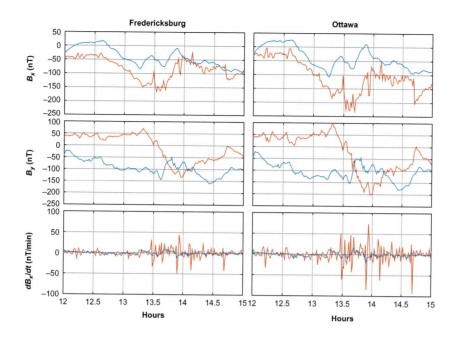

FIG. 4

Example of comparison between observations (*red*) and SWMF model (*blue*) magnetic field perturbations at Fredericksburg and Ottawa for the event on March 17, 2015. Top to bottom are: magnetic field B_x, B_y, and dB_x/dt, respectively. This plot illustrates that current models are able to capture the envelop of the variations but not the rapid changes that are important for GICs. All the data were sampled at 1-minute resolution.

(block-adaptive tree solar wind Roe-type upwind scheme: Powell et al., 1999) coupled to RCM (Rice Convection Model: Toffoletto et al., 2003) using an inner magnetosphere grid resolution of 1/8 R_e with 2 million cells. Evidently, while the model is able to capture the envelope of the variations, the fast changes are missed. It is key to point out that similar results have been noted for other models but only SWMF outputs are shown here for illustration purposes. Nevertheless, this is a serious challenge needing further attention to improve current techniques in order to deliver highly actionable information. However, there are a number of ongoing efforts to further improve these techniques (e.g., Owens et al., 2014; Devos et al., 2014; Pulkkinen et al., 2015b).

4.3 DEFINING EXTREMES

Generally, a wide range of parameters is used to define storm-time intensity, but from a GIC standpoint, it is the geoelectric field extreme that must be defined. This is because the geoelectric field intensity is closely connected to *dB/dt* and not the amplitude of the storm as defined by the Dst or Kp index. The Dst or disturbance storm time index is a measure of geomagnetic activity used to classify the intensity of geomagnetic storms, while Kp is a planetary 3-hour range index for quantifying disturbances in the horizontal geomagnetic field component.

The aforementioned point is clearly emphasized in Fig. 5, which shows from top to bottom: the SYM-H index (high resolution Dst), AE index, horizontal geomagnetic field at Fort Churchill, dB/dt, and calculated geoelectric fields. At points A and B, strongly enhanced geoelectric fields are produced when the SYM-H levels are close to typical quiet-time or prestorm-time values but the dB/dt is high. However, at point C when the SYM-H is at minimum, corresponding to peak of main phase, the geoelectric field is much lower than for case A and B with corresponding lower dB/dt. Therefore, it is critical that we develop GIC specific quantities that better define the extremes.

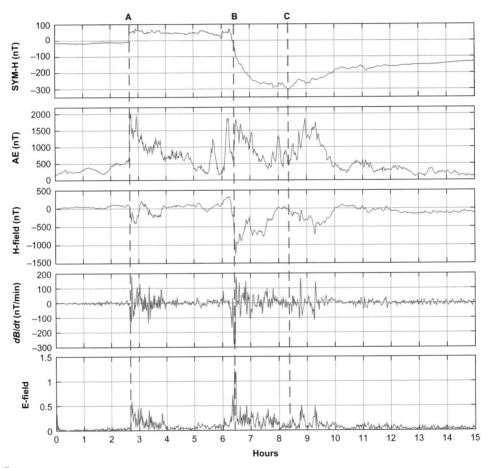

FIG. 5

Plot showing the SYM-H (*top*) and AE (*second*) indices for the geomagnetic storm on May 15, 2005. The horizontal (H) geomagnetic field, dH/dt, and the computed induced geoelectric field at Fort Churchill, Canada are shown in panels three, four, and five, respectively. Points A and B represent times when SYM-H is close to nominal conditions but corresponding geoelectric fields are strongly enhanced, while point C represents an instance when SYM-H is at peak of main phase but corresponding geoelectric fields are low.

However, one of the major obstacles to developing a geoelectric field-based index for predicting intensity of space weather events is that the geoelectric field is rarely measured. There are only a few places on the globe that have continuous geoelectric field recording facilities. Without these measurements, studying the spatiotemporal characteristics or later on developing regional or global geoelectric field indices is hindered. While data from MT campaigns such as the EarthScope USArray MT project (http://ds.iris.edu/spud/emtf) help to alleviate some of our data challenges, long datasets that capture many storms are preferable.

5 CONCLUDING REMARKS

The GIC signal consists of information about the entire space weather chain because GICs are the ground manifestation of the coupled solar wind-magnetosphere-ionosphere dynamic system. The study of GIC events is therefore also key for advancing awareness of the behavior of the whole complex system.

Enriching our knowledge of the solar wind-magnetosphere-ionosphere interaction is one of the top priorities in space weather. Extreme events in particular continue to challenge our understanding of the physics of space weather. Because extreme events occur infrequently but have potentially high impact, this is a major barrier to our understanding of complex processes that produce extreme GICs. There are several ongoing efforts around the world to resolve outstanding challenges we face today. However, there is still a lot more to accomplish as a community in terms of achieving understanding and later on defining extreme GICs in the context of expected impact.

To improve current understanding and modeling abilities, it is very important that we have a variety of space-based and ground-based instruments monitoring conditions from Sun to Earth and a storm's progression from activity on the solar surface to its impact at Earth. Availability of quality geospace measurements of extreme events is critical for testing and interpreting space weather observations. Our limited knowledge about the temporal and spatial complexity of magnetosphere-ionosphere currents is partly because of the difficulty of making simultaneous measurements everywhere. Ideally, all electrodynamic parameters required for modeling would be recorded routinely at high resolution across the globe, but in reality, what is prevailing is a sparse collection of single-point measurements irregularly distributed in time and space. Therefore, acquiring more geospace observations will also help to improve the accuracy of our modeling capabilities.

ACKNOWLEDGMENTS

Results presented in this study rely on ground-based magnetometer data obtained from INTERMAGNET. We thank the national institutes that support magnetic observatories and INTERMAGNET for promoting high standards of magnetic observatory practice. The SYM-H and AE index data are freely available from CDAWeb database hosted by the NASA-GSFC Space Physics Data Facility. The first author is supported by NASA Grant NNG11PL10A 670.135 and NNG11PL10A 670.157 to CUA/IACS.

REFERENCES

Adebesin, B.O., Pulkkinen, A., Ngwira, C.M., 2016. The interplanetary and magnetospheric causes of extreme db/dt at equatorial locations. Geophys. Res. Lett. 43, 11501–11509. https://doi.org/10.1002/2016GL071526.

Akasofu, S.I., 1981. Energy coupling between the solar wind and magnetosphere. Space Sci. Rev. 28, 121.

Anderson, C.W., Lanzerotti, L.J., Maclenna, C.G., 1974. Outage of the L-4 system and the geomagnetic disturbances of August 4, 1972. Bell Syst. Tech. J. 53 (9), 1817–1837.

Baker, D.N., Li, X., Pulkkinen, A., Ngwira, C.M., Mays, M.L., Galvin, A.B., Simunac, K.D.C., 2013. A major solar eruptive event in July 2012: defining extreme space weather scenarios. Space Weather 11, 1–7. https://doi.org/10.1002/swe.20097.

Barlow, W.H., 1849. On the Spontaneous Electrical Currents Observed in the Wires of the Electric Telegraph. Philos. Trans. R. Soc. Lond. 139, 61–72.

Berge, J., Varma, R.K., Marti, L., 2011. Laboratory validation of the relationship between geomagnetically induced current (GIC) and transformer absorbed reactive power. Electrical Power and Energy Conference (EPEC), 2011 IEEE, pp. 491–495.

Bernabeu, E.E., 2013. Modeling geomagnetically induced currents in the Dominion Virginia Power using extreme 100-year geoelectric field scenarios—part 1. IEEE Trans. Power Delivery 28 (1), 516–523.

Bernabeu, E.E., 2014. Single-phase transformer harmonics produced during geomagnetic disturbances: theory, modeling, and monitoring. IEEE Trans. Power Delivery 30 (3), 1323–1330.

Bolduc, L., 2002. GIC observations and studies in the Hydro-Québec power system. J. Atmos. Sol.-Terr. Phys. 64 (16), 1793–1802.

Borovsky, J.E., Funsten, H.O., 2003. Role of solar wind turbulence in the coupling of the solar wind to the Earth's magnetosphere. J. Geophys. Res. 108 (A6), 1246. https://doi.org/10.1029/2002JA009601.

Boteler, D.H., 2001a. Assessment of geomagnetic hazard to power systems in Canada. Nat. Hazards 23, 101–120.

Boteler, D.H., 2001b. Space weather effects on power systems. In: Song, D., Singer, H.J., Siscoe, G.L. (Eds.), Space Weather. AGU Geophysical Monograph 125, pp. 347–352.

Boteler, D.H., 2014. Methodology for simulation of geomagnetically induced currents in power systems. J. Space Weather Space Clim. 4, A21. https://doi.org/10.1051/swsc/2014018.

Boteler, D.H., Jansen van Beek, G., 1999. August 4, 1972 revisited: a new look at the geomagnetic disturbance that caused the L4 cable system outage. Geophys. Res. Lett. 26 (5), 577–580.

Boteler, D.H., Pirjola, R.J., 2014. Comparison of methods for modelling geomagnetically induced currents. Ann. Geophys. 32, 1177–1187. https://doi.org/10.5194/angeo-32-1177-2014.

Boteler, D.H., Pirjola, R., Trichtchenko, L., 2000. On calculating the electric and magnetic fields produced in technological systems at the Earth's surface by a wide electrojet. J. Atmos. Sol.-Terr. Phys. 62 (3), 1311–1315.

Cabinet Office, 2015. National Risk Register of Civil Emergencies. https://www.gov.uk/government/uploads/system/uploads/attachment_data/file/419549/20150331_2015-NRR-WA_Final.pdf.

Cade III, W.B., 2013. The first recorded space weather impact. Space Weather 11, 489. https://doi.org/10.1002/swe.20091.

Cagniard, L., 1953. Basic theory of the magneto-telluric methods of geophysical prospecting. Geophysics 18 (3), 605.

Carrington, R.C., 1859. Description of a singular appearance seen in the Sun on September 1, 1859. Royal Astronomical Society, XX, 13–15.

Carter, B.A., Yizengaw, E., Pradipta, R., Halford, A.J., Zhang, K., 2015. Interplanetary shocks and the resulting geomagnetically induced currents at the equator. Geophys. Res. Lett. 42. https://doi.org/10.1002/2015GL065060.

Carter, B.A., Yizengaw, E., Pradipta, R., Weygand, J.M., Piersanti, M., Pulkkinen, A., Moldwin, M.B., Norman, R., Zhang, K., 2016. Geomagnetically induced currents around the world during the 17 March 2015 storm. J. Geophys. Res. 121, 10496–10507. https://doi.org/10.1002/2016JA023344.

Chun, F.K., Russell, C.T., 1997. Field-aligned currents in the inner magnetosphere: control by geomagnetic activity. J. Geophys. Res. 102 (A2), 226–227.

Cid, C., Palacios, J., Saiz, E., Guerrero, A., Cerrato, Y., 2014. On extreme geomagnetic storms. J. Space Weather Space Clim. 4, A28. https://doi.org/10.1051/swsc/2014026.

da Silva Barbosa, C., Hartmann, G.A., Pinheiro, K.J., 2015. Numerical modeling of geomagnetically induced currents in a Brazilian transmission line. Adv. Space Res. 55, 1168–1179. https://doi.org/10.1016/j.asr.2014.11.008.

Davidson, W.F., 1940. The magnetic storm of March 24, 1940–effects in the power system. Edison Electr. Inst. Bull. 8, 365–366.

Devos, A., Verbeeck, C., Robbrecht, E., 2014. Verification of space weather forecasting at the Regional Warning Center in Belgium. J. Space Weather Space Clim. 4, A29. https://doi.org/10.1051/swsc/2014025.

Ebihara, Y., Fok, M.C., Sazykin, S., Thomsen, M.F., Hairston, M.R., Evans, D.S., Rich, F.J., Ejiri, M., 2005. Ring current and the magnetosphere-ionosphere coupling during the superstorm of 20 November 2003. J. Geophys. Res. 110, A09S22. https://doi.org/10.1029/2004JA010924.

Echer, E., Gonzalez, W.D., Tsurutani, B.T., 2008. Interplanetary conditions leading to superintense geomagnetic storms (Dst ≤ -250 nT) during solar cycle 23. Geophys. Res. Lett. 35, L06S03. https://doi.org/10.1029/2007GL031755.

Eroshenko, E.A., Belov, A.V., Boteler, D., Gaidash, S.P., Lobkov, S.L., Pirjola, R., Trichtchenko, L., 2010. Effects of strong geomagnetic storms on Northern railways in Russia. Adv. Space Res. 46, 1102–1110. https://doi.org/10.1016/j.asr.2010.05.017.

Federal Energy Regulatory Commission, 2015. Reliability Standard for Transmission System Planned Performance for Geomagnetic Disturbance Events, 18 CFR Part 40, Docket No. RM15-11-000.

Fernberg, P., 2012. One-dimensional earth resistivity models for selected areas of continental United States and Alaska. EPRI Technical Update 1026430, Palo Alto, CA.

Fiori, R.A.D., Boteler, D.H., Gillies, D.M., 2014. Assessment of GIC risk due to geomagnetic sudden commencements and identification of the current systems responsible. Space Weather 12, 76–91. https://doi.org/10.1002/2013SW000967.

Gaunt, C.T., Coetzee, G., 2007. Transformer failure in regions incorrectly considered to have low GIC-risks. In: IEEE Power Tech., Conference Paper 445, Lausanne, July, pp. 807–812.

Girgis, R.S., Ko, C.D., 1991. Calculation techniques and results of effects of GIC currents as applied to two large power transformers, Transmission and Distribution Conference, Proceedings of the 1991 IEEE Power Engineering Society, pp. 553–559.

Gonzalez, W., 1990. A unified view of solar wind-magnetosphere coupling functions. Planet. Space Sci. 38, 627–632.

Gopalswamy, N., Yashiro, S., Michalek, G., Xie, H., Lepping, R.P., Howard, R.A., 2005. Solar source of the largest geomagnetic storm of cycle 23. Geophys. Res. Lett. 32, L12S09. https://doi.org/10.1029/2004GL021639.

Gummow, R.A., Eng, P., 2002. GIC effects on pipeline corrosion and corrosion control systems. J. Atmos. Sol.-Terr. Phys. 64 (16), 1755–1764.

Huttunen, K.E.J., Kilpua, S.P., Pulkkinen, A., Viljanen, A., Tanskanen, E., 2008. Solar wind drivers of large geomagnetically induced currents during solar cycle 23. Space Weather 6, S10002. https://doi.org/10.1029/2007SW000374.

Kappenman, J.G., 2003. Storm sudden commencement events and the associated geomagnetically induced current risks to ground-based systems at low-latitude and midlatitude locations. Space Weather 1 (3), 1016. https://doi.org/10.1029/2003SW000009.

Karsberg, A., Swedenborg, G., Wyke, K., 1959. The influences of Earth magnetic currents on telecommunication lines. Tele (English edition), Televerket (Swedish Telecom), 1, No. 1, Stockholm, Sweden, pp. 1–21.

Kelly, G.S., Viljanen, A., Beggan, C.D., Thomson, A.W.P., 2017. Understanding GIC in the UK and French high-voltage transmission systems during severe magnetic storms. Space Weather 15, 99–114. https://doi.org/10.1002/2016SW001469.

Lehtinen, M., Elovaara, J., 2002. GIC occurrences and GIC test for 400 kV system transformer. IEEE Trans. Power Delivery 17, 555–561.

Lehtinen, M., Pirjola, R., 1985. Currents produced in earthed conductor networks by geomagnetically-induced electric field. Ann. Geophys. 3 (4), 479–484.

Li, X., Temerin, M., Tsurusani, B.T., Alex, S., 2005. Modeling of 1–2 September 1859 super magnetic storm. Adv. Space Res. https://doi.org/10.1016/j.asr.2005.06.070.

Liou, K., Wu, C.C., Dryer, M., Wu, S.T., Rich, N., Plunkett, S., Simpson, L., Fry, C.D., Schenk, K., 2014. Global simulation of extremely fast coronal mass ejection on 23 July 2012. J. Atmos. Sol.-Terr. Phys. 121, 32–41.

Love, J.J., 2012. Credible occurrence probabilities for extreme geophysical events: earthquakes, volcanic eruptions, magnetic storms. Geophys. Res. Lett. 39, L10301. https://doi.org/10.1130/G32655.1.

Manchester IV, W.B., Ridley, A.J., Gombosi, T.I., De Zeeuw, D.L., 2006. Modeling the Sun-Earth propagation of a very fast CME. Adv. Space Res. 38, 253–262.

Marshall, R.A., Waters, C.L., Sciffer, M.D., 2010. Spectral analysis of pipe-to-soil potentials with variations of the Earth's magnetic field in the Australian region. Space Weather 8, S05002. https://doi.org/10.1029/2009SW000553.

Marshall, R.A., Smith, E.A., Francis, M.J., Waters, C.L., Sciffer, M.D., 2011. A preliminary risk assessment of the Australian region power network to space weather. Space Weather 9, S10004. https://doi.org/10.1029/2011SW000685.

Marshall, R.A., Gorniak, H., Walt, T.V.D., Waters, C.L., Sciffer, M.D., Miller, M., Dalzell, M., Daly, T., Pouferis, G., Hesse, G., Wilkinson, P., 2012. Observations of geomagnetically induced currents in the Australian power network. Space Weather 11, 1–11. https://doi.org/10.1029/2012SW000849.

Martin, B.A., 1993. Telluric effects on a buried pipeline. Corrosion 49 (4), 349.

McNish, A.G., 1940. The magnetic storm of March 24, 1940. Terr. Magn. Atmos. Electr. 45, 3.

McPherron, R.L., Chu, X., 2016. Relation of the auroral substorm to the substorm current wedge. Geosci. Lett. 3, 12.

Molinski, T.S., Feero, W.E., Damsky, B.L., 2000. Shielding grids from solar storms. IEEE Spectrum 37, 55–60.

Murphy, K.R., Mann, I.R., Rae, I.J., Waters, C.L., Frey, H.U., Kale, A., Singer, H.J., Anderson, B.J., Korth, H., 2013. The detailed spatial structure of field-aligned currents comprising the substorm current wedge. J. Geophys. Res. 118, 7714–7727. https://doi.org/10.1002/2013JA018979.

Nagatsuma, T., Kataoka, R., Kunitake, M., 2015. Estimating the solar wind conditions during an extreme geomagnetic storm: a case study of the event that occurred on March 13–14, 1989. J. Geophys. Res. 67, 78. https://doi.org/10.1186/s40623-015-0249-4.

National Science and Technology Council, 2015. National Space Weather Action Plan, Executive Office of the President (EOP), USA. https://www.whitehouse.gov/sites/default/files/microsites/ostp/final_nationalspace-weatheractionplan_20151028.pdf.

National Science and Technology Council, 2015. National Space Weather Strategy, Executive Office of the President (EOP), USA. Available from https://www.whitehouse.gov/sites/default/files/microsites/ostp/final_nationalspace-weatherstrategy_20151028.pdf.

Nevanlinna, H., Tenhunenb, P., Pirjola, R., Annanpalo, J., Pulkkinena, A., 2001. Breakdown caused by a geomagnetically induced current in the Finnish telesystem in 1958. J. Atmos. Sol.-Terr. Phys. 63, 1099–1103.

Ngwira, C.M., Pulkkinen, A., McKinnell, L.A., Cilliers, P.J., 2008. Improved modeling of geomagnetically induced currents in the South African power network. Space Weather 6, S11004. https://doi.org/10.1029/2008SW000408.

Ngwira, C.M., Pulkkinen, A., Mays, M.L., Kuznetsova, M.M., Galvin, A.B., Simunac, K., Baker, D.N., Li, X., Zheng, Y., Glocer, A., 2013a. Simulation of the 23 July 2012 extreme space weather event: what if this extremely rare CME was Earth-directed? Space Weather 11, 671–679. https://doi.org/10.1002/2013SW000990.

Ngwira, C.M., Pulkkinen, A., Wilder, F.D., Crowley, G., 2013b. Extended study of extreme geoelectric field event scenarios for geomagnetically induced current applications. Space Weather 11, 121–131. https://doi.org/10.1002/swe.20021.

Ngwira, C.M., Pulkkinen, A., Kuznetsova, M.M., Glocer, A., 2014. Modeling extreme "carrington-type" space weather events using three-dimensional MHD code simulations. J. Geophys. Res. 119, 4456–4474. https://doi.org/10.1002/2013JA019661.

Ngwira, C.M., Pulkkinen, A., Bernabeu, E., Eichner, J., Viljanen, A., Crowley, G., 2015. Characteristics of extreme geoelectric fields and their possible causes: localized peak enhancements. Geophys. Res. Lett. 42. https://doi.org/10.1002/2015GL065061.

North American Electric Reliability Corporation, 2012. 2012 Special reliability assessment interim report: effects of geomagnetic disturbances on bulk power systems.

North American Electric Reliability Corporation, 2013. White Paper supporting network applicability of EOP-010-1. http://www.nerc.com/pa/Stand/Pages/Project-2013-03-Geomagnetic-Disturbance-Mitigation.aspx.

North American Electric Reliability Corporation, 2016. Benchmark Geomagnetic Disturbance Event Description, Project 2013-03 GMD Mitigation Standards Drafting Team. http://www.nerc.com/pa/Stand/Pages/Project-2013-03-Geomagnetic-Disturbance-Mitigation.aspx.

North American Electric Reliability Corporation, 2016. Transformer Thermal Impact Assessment White Paper, Project 2013-03 GMD Mitigation Standards Drafting Team. http://www.nerc.com/pa/Stand/Pages/Project-2013-03-Geomagnetic-Disturbance-Mitigation.aspx.

Onsager, T.G., Chan, A.A., Fei, Y., Elkington, S.R., Green, J.C., Singer, H.J., 2004. The radial gradient of relativistic electrons at geosynchronous orbit. J. Geophys. Res. 109, A05221. https://doi.org/10.1029/2003JA010368.

Overbye, T.J., Shetye, K.S., Hutchins, T.R., Qiu, Q., Weber, J.D., 2013. Power grid sensitivity analysis of geomagnetically induced currents. IEEE Trans. Power Syst. 28 (4). https://doi.org/10.1109/TPWRS.2013.2274624.

Owens, M.J., Horbury, T.S., Wicks, R.T., McGregor, S.L., Savani, N.P., Xiong, M., 2014. Ensemble downscaling in coupled solar wind-magnetosphere modeling for space weather forecasting. Space Weather 12, 395–405. https://doi.org/10.1002/2014SW001064.

Pirjola, R., 1982. Electromagnetic induction in the Earth by a plane wave or by fields of line currents harmonic in time and space. Geophysica 18 (1–2), 1–161.

Pirjola, R., 2000. Geomagnetically induced currents during magnetic storms. IEEE Trans. Plasma Sci. 28 (6), 1867–1873.

Pirjola, R., 2002. Review on the calculation of the surface electric and magnetic fields and geomagnetically induced currents in ground based technological systems. Surv. Geophys. 23 (1), 71–90.

Powell, K.G., Roe, P.L., Linde, T.J., Gombosi, T.I., De Zeeuw, D.L., 1999. A solution-adaptive upwind scheme for ideal magnetohydrodynamics. J. Comput. Phys. 154 (2). https://doi.org/10.1006/jcph.1999.6299.

Prescott, G.B., 1860. History, Theory, and Practice of Electric Telegraph. Ticknor and Fields, Boston, MA.

Price, P.R., 2002. Geomagnetically induced current effects on transformers. IEEE Trans. Power Delivery 17, 1002–1008.

Pulkkinen, A., 2016. Introduction to NASA Living With a Star (LWS) Institute GIC working group special collection. Space Weather. https://doi.org/10.1002/2016SW001537.

Pulkkinen, A., Kataoka, R., 2006. S-transform view of geomagnetically induced currents during geomagnetic superstorms. Geophys. Res. Lett. 33, L12108. https://doi.org/10.1029/2006GL025822.

Pulkkinen, A., Viljanen, A., Pajunpää, K., Pirjola, R., 2001. Recordings and occurrence of geomagnetically induced currents in the Finnish natural gas pipeline network. J. Appl. Geophys. 48, 219–231.

Pulkkinen, A., Lindahl, S., Viljanen, A., Pirjola, R., 2005. Geomagnetic storm of 29–31 October: geomagnetically induced currents and their relation to problems in the Swedish high-voltage power transmission system. Space Weather 3, S08C03. https://doi.org/10.1029/2004SW000123.

Pulkkinen, A., Kataoka, R., Watari, S., Ichiki, M., 2010a. Modeling geomagnetically induced currents in Hokkaido, Japan. Adv. Space Res. 46, 1087–1093. https://doi.org/10.1016/j.asr.2010.05.024.

Pulkkinen, A., Rastätter, L., Kuznetsova, M., Hesse, M., Ridley, A., Raeder, J., Singer, H.J., Chulaki, A., 2010b. Systematic evaluation of ground and geostationary magnetic field predictions generated by global magnetohydrodynamic models. J. Geophys. Res. 115, A03206. https://doi.org/10.1029/2009JA014537.

Pulkkinen, A., Bernabeu, E., Eichner, J., Beggan, C., Thomson, A.W.P., 2012. Generation of 100-year geomagnetically induced current scenarios. Space Weather 10, S04003. https://doi.org/10.1029/2011SW000750.

Pulkkinen, A., Rastätter, L., Kuznetsova, M., Singer, H., Balch, C., Weimer, D., Toth, G., Ridley, A., Gombosi, T., Wiltberger, M., Raeder, J., Weigel, R., 2013. Community-wide validation of geospace model ground magnetic field perturbation predictions to support model transition to operations. Space Weather 11 (6), 369–385. https://doi.org/10.1002/swe.20056.

Pulkkinen, A., Bernabeu, E., Eichner, J., Viljanen, A., Ngwira, C.M., 2015. Regional-scale high-latitude extreme geoelectric fields pertaining to geomagnetically induced currents. Earth Planets Space 67, 93. https://doi.org/10.1186/s40623-015-0255-6.

Pulkkinen, A., Mahamood, S., Ngwira, C., Balch, C., Lordan, R., Fugate, D., Jacobs, W., Honkonen, I., 2015. Solar storm GIC forecasting: solar shield extension—development of the end-user forecasting system requirements. Space Weather 13. https://doi.org/10.1002/2015SW001283.

Pulkkinen, A., Bernabeu, E., Thomson, A., Viljanen, A., Pirjola, R., Boteler, D., Eichner, J., Cilliers, P.J., Welling, D., Savani, N.P., Weigel, R.S., Love, J.J., Balch, C., Ngwira, C.M., Crowley, G., Schultz, A., Kataoka, R., Anderson, B., Fugate, D., Simpson, J.J., MacAlester, M., 2017. Geomagnetically induced currents: science, engineering and applications readiness. Space Weather. https://doi.org/10.1002/2016SW001501.

Raeder, J., Berchem, J., Ashour-Abdalla, M., 1996. The importance of small scale processes in global MHD simulations: some numerical experiments. In: Chang, T., Jasperse, J.R. (Eds.), The Physics of Space Plasmas. vol. 14. p. 403.

Schrijver, C.J., Kauristie, K., Aylward, A.D., Denardini, C.M., Gibson, S.E., Glover, A., Gopalswamy, N., Grande, M., Hapgood, M., Heynderickx, D., Jakowski, N., Kalegaev, V.V., Lapenta, G., Linker, J.A., Liu, S., Mandrini, C.H., Mann, I.R., Nagatsuma, T., Nandi, D., Obara, T., O'Brien, T.P., Onsager, T., Opgenoorth, H.J., Terkildsen, M., Valladares, C.E., Vilmer, N., 2015. Understanding space weather to shield society: a global road map for 2015–2025 commissioned by COSPAR and ILWS. Adv. Space Res. 55, 2745–2807.

Seager, W.H., 1991. Adverse telluric effects on the northern pipelines, society of petroleum engineers. International Arctic Technology Conference, Anchorage, AK, SPE22178, May 7, pp. 299–302.

Shen, F., Shen, C., Zhang, J., Hess, P., Wang, Y., Feng, X., Cheng, H., Yang, Y., 2014. Evolution of the 12 July 2012 CME from the Sun to the Earth: data-constrained three-dimensional MHD simulations. J. Geophys. Res. 119, 7128–7141. https://doi.org/10.1002/2014JA020365.

Shu, L., Yilu, L., De La Ree, J., 1993. Harmonics generated from a DC biased transformer. IEEE Trans. Power Delivery 8, 725–731.

Thomson, A.W.P., Dawson, E.B., Reay, S.J., 2011. Quantifying extreme behavior in geomagnetic activity. Space Weather 9, S10001. https://doi.org/10.1029/2011SW000696.

Toffoletto, F., Sazykin, S., Spiro, R., Wolf, R., 2003. Inner magnetospheric modeling with the Rice Convection Model. Space Sci. Rev. 107, 175–196.

Torta, J.M., Serrano, L., Regué, J.R., Sánchez, A.M., Roldán, E., 2012. Geomagnetically induced currents in a power grid of northeastern Spain. Space Weather 10, S06002. https://doi.org/10.1029/2012SW000793.

Tóth, G., van der Holst, B., Sokolov, I.V., De Zeeuw, D.L., Gombosi, T.I., Fang, F., Manchester, W.B., Meng, X., Najib, D., Powell, K.G., Stout, Q.F., Glocer, A., Ma, Y., Opher, M., 2012. Adaptive numerical algorithms in space weather modeling. J. Atmos. Sol.-Terr. Phys. 231, 870–903. https://doi.org/10.1016/j.jastp.2011.02.006.

Tóth, G., Meng, X., Gombosi, T.I., Rastätter, L., 2014. Predicting the time derivative of local magnetic perturbations. J. Geophys. Res. 119, 310–321. https://doi.org/10.1002/2013JA019456.

Trivedi, N.B., Vitorello, I., Kabata, W., Dutra, S.L.G., Padilha, A.L., Bologna, M.S., de Pádua, M.B., Soares, A.P., Luz, G.S., de A. Pinto, F., Pirjola, R., Viljanen, A., 2007. Geomagnetic conjugate observations of large-scale traveling ionospheric disturbances using GPS networks in Japan and Australia. Space Weather 5, S04004. https://doi.org/10.1029/2006SW000282.

Tsurutani, B.T., Gonzalez, W.D., Lakhina, G.S., Alex, S., 2003. The extreme magnetic storm of 1-2 September 1859. J. Geophys. Res. 108 (A7), 1268. https://doi.org/10.1029/2002JA009504.

Tsurutani, B.T., Hajra, R., Echer, E., Gjerloev, J.W., 2015. Extremely intense (SML ≤ -2500 nT) substorms: isolated events that are externally triggered? Ann. Geophys. 33, 519–524. https://doi.org/10.5194/angeocom-33-519-2015.

Viljanen, A., Pirjola, R., 1994. Geomagnetically induced currents in the Finnish high-voltage power system. Surv. Geophys. 15, 383–408.

Viljanen, A., Amm, O., Pirjola, R., 1999. Modeling geomagnetically induced currents during different ionospheric situations. J. Geophys. Res. 104 (A12), 28059–28071.

Viljanen, A., Pulkkinen, A., Pirjola, R., Pajunpää, K., Posio, P., Koistinen, A., 2006a. Recordings of geomagnetically induced currents and a nowcasting service of the Finnish natural gas pipeline. Space Weather 4, S10004. https://doi.org/10.1029/2006SW000234.

Viljanen, A., Tanskanen, E.I., Pulkkinen, A., 2006b. Relation between substorm characteristics and rapid temporal variations of the ground magnetic field. Ann. Geophys. 24, 725–733.

Walling, R.A., Khan, A.N., 1991. Characteristics of transformer exciting-current during geomagnetic disturbances. IEEE Trans. Power Delivery 6, 1707–1714.

Watari, S., Kunitake, M., Kitamura, K., Hori, T., Kikuchi, T., Shiokawa, K., Nishitani, N., Kataoka, R., Kamide, Y., Aso, T., Watanabe, Y., Tsuneta, Y., 2009. Measurements of geomagnetically induced current (GIC) in a power grid in Hokkaido, Japan. Space Weather 7, S03002. https://doi.org/10.1029/2008SW000417.

Weigel, R.S., Vassiliadis, D., Klimas, A.J., 2002. Coupling of the solar wind to the temporal fluctuations in the ground magnetic fields. Geophys. Res. Lett. 29 (19), 1915. https://doi.org/10.1029/2002GL014740.

Weigel, R.S., Klimas, A.J., Vassiliadis, D., 2003. Solar wind coupling to and predictability of the ground magnetic fields and their derivatives. J. Geophys. Res. 108 (A7), 1298. https://doi.org/10.1029/2002JA009627.

Welling, D.T., Anderson, B.J., Crowley, G., Pulkkinen, A.A., Rastätter, L., 2017. Exploring predictive performance: a reanalysis of the geospace model transition challenge. Space Weather 15, 192–203. https://doi.org/10.1002/2016SW001505.

Wik, M., Pirjola, R., Lundstedt, H., Viljanen, A., Wintoft, P., Pulkkinen, A., 2009. Space Weather events in July 1982 and October 2003 and the effects of geomagnetically induced currents on Swedish technical systems. Ann. Geophys. 27, 1775–1787.

Wiltberger, M., Merkin, V., Lyon, J.G., Ohtani, S., 2015. High-resolution global magnetohydrodynamic simulation of bursty bulk flows. J. Geophys. Res. 120, 4555–4566. https://doi.org/10.1002/2015JA021080.

Wintoft, P., Wik, M., Viljanen, A., 2015. Solar wind driven empirical forecast models of the time derivative of the ground magnetic field. J. Space Weather Space Clim. 5, A7. https://doi.org/10.1051/swsc/2015008.

Xuzhu, D., Yilu, L., Kappenman, J.G., 2001. Comparative analysis of exciting current harmonics and reactive power consumption from GIC saturated transformers. In: Power Engineering Society Winter Meeting, 2001 IEEE, vol. 1, pp. 318–322.

Zhang, J.J., Wang, C., Tang, B.B., 2012. Modeling geomagnetically induced electric field and currents by combining a global MHD model with a local one-dimensional method. Space Weather 10, S05005. https://doi.org/10.1029/2012SW000772.

CHAPTER 9

EXTREME-EVENT GEOELECTRIC HAZARD MAPS

Jeffrey J. Love*, Paul A. Bedrosian[†]

U.S. Geological Survey, Geomagnetism Program, Denver, CO, United States *U.S. Geological Survey, Crustal Geophysics and Geochemistry Science Center, Denver, CO, United States[†]*

CHAPTER OUTLINE

1 Introduction	209
2 Defining the Hazard	210
3 Direct Geoelectric Monitoring	211
4 Induction in a Conducting Earth	211
5 Magnetic Observatory Data	212
6 Geomagnetic Waveform Time Series	212
7 Observatory Magnetic Hazard Functions	215
8 Global Magnetic Hazard Functions	216
9 Magnetotelluric Impedances	217
10 Geological Interpretations	219
11 Geoelectric Hazard Maps	221
12 Discussion	225
Acknowledgments	225
References	225

1 INTRODUCTION

Geoelectric fields are induced in the Earth's electrically conducting crust, mantle, and ocean by natural time-dependent geomagnetic field variation generated by dynamic processes in the Earth's surrounding space weather environment. This induction occurs all the time, during both calm and stormy conditions. But during intense magnetic storms and substorms, induced geoelectric fields can drive quasi-direct currents in bulk electric power grids of sufficient strength to interfere with their operation, sometimes causing blackouts and damaging transformers (e.g., Molinski, 2002; Piccinelli and Krausmann, 2014). Perhaps the most dramatic realization of this natural hazard occurred in March 1989 (e.g., Allen et al., 1989), when an intense magnetic storm caused the collapse of the entire

Hydro-Québec electric power grid in Canada (Bolduc, 2002; Béland and Small, 2005). Historically, the Carrington event of 1859 (Tsurutani et al., 2003; Siscoe et al., 2006; Cliver and Dietrich, 2013; Hajra et al., 2016; Lakhina and Tsurutani, 2017) was, by some measures, the most intense magnetic storm ever recorded; it disrupted telegraph systems around the world (e.g., Boteler, 2006). In contrast, a magnetic storm in August 1972 that was, by most measures, of modest global intensity, induced localized geoelectric fields that interrupted the operation of a major telecommunications cable in the Central United States (Anderson et al., 1974). The Halloween magnetic storm of October 2003 caused operational failures in Sweden's grid systems (Pulkkinen et al., 2005). More recently, satellites in orbit around the Sun have recorded powerful coronal mass ejections that (fortunately) were not Earth-directed, but if they had been, they might have produced a superstorm as intense as the Carrington event (Baker et al., 2013; Liu et al., 2014). It has been suggested that especially hazardous geoelectric fields could have been induced (Ngwira et al., 2013).

It is often said that there is nothing new under the Sun. Accepting that, and noting previous occurrences of magnetic superstorms, we might reasonably anticipate the occurrence of another superstorm in the future, even one as intense as the Carrington event. Such an extreme space weather event could cause widespread disruption of numerous types of technological systems (e.g., Lanzerotti, 2001), possibly even causing a continental-scale failure of electric power grids (e.g., Kappenman, 2012). This disruption would carry long-lasting and deleterious consequences for society (e.g., Baker et al., 2008). In support, then, of the investigation of geoelectric hazards that might be realized in the future (e.g., Thomson et al., 2009; Love et al., 2014; Pulkkinen, 2015) and building on previous closely related work (Bedrosian and Love, 2015; Love et al., 2016a; Love et al., 2016b), here we develop and present maps of the extreme-value geoelectric amplitudes for the continental United States. We construct these maps using a data-derived ("empirical") parameterization of induction—convolving a latitude-dependent statistical map of ground-level geomagnetic disturbance, derived from long, historical time series of magnetic observatory data, with Earth-surface impedances, measured at various geographic locations during magnetotelluric surveys.

The U.S. National Science and Technology Council (Jonas and McCarron, 2015; NSTC, 2015, Goal 1.1) and the International Council for Science, Committee on Space Research (Schrijver, 2015) have identified space-weather induction hazard research as a high priority. The results presented here inform both hazard assessments and the development of real-time hazard mapping projects (e.g., Pulkkinen et al., 2003; Rigler et al., 2014). Maps of storm-time geoelectric amplitudes might be used to estimate geomagnetically induced currents in power-grid systems (e.g., Pirjola, 2002; Ngwira et al., 2008; Horton et al., 2012). They also inform government (e.g., Veeramany et al., 2016) and private sector (e.g., Riswadkar and Dobbins, 2010; Benfield, 2013; Lloyd's, 2013) assessments of risks to power-grid infrastructure associated with magnetic storms (e.g., Hapgood, 2011).

2 DEFINING THE HAZARD

A natural "hazard" can be defined, in general terms, as a natural phenomenon or event that might have a negative impact on people, infrastructure, or the environment. Depending on the nature of the hazard, the word is often further specialized to mean the probability that a damaging event will occur within a specified window of time and within a given geographic area (Varnes, 1984, p. 10; Panel on Seismic Hazard Analysis, 1988, p. 94); this is, for example, how the word "hazard" is used in the insurance industry (e.g., Smolka, 2006). Furthermore, one can define "risk" as the product of hazard probability times the "exposure" of an asset to the hazard times the "vulnerability" of the asset should the hazard occur (e.g., Smith and Petley, 2009, Chapter 1). From this, we can understand that risk can be relatively high, even for the rare occurrence of a hazard, if the exposure and vulnerability of the asset are comparatively high.

For an electric power-grid system, geoelectric fields induced in the Earth's conducting interior during magnetic storms are the primary concern (e.g., Boteler, 2001), and with their estimation, geomagnetically induced currents (GIC) in power-grid lines can be calculated (e.g., Horton et al., 2012). The operational interference and infrastructure damage that these GICs can wreak on a power-grid system are issues of exposure and vulnerability that would need to be assessed as part of an engineering study (e.g., NERC, 2013, 2014b). Our focus here is the study of the natural factors: the combination of geomagnetic disturbance and Earth impedance, which determine the threshold magnitude that hazardous geoelectric fields will exceed, on average, once every 100 years in terms of a statistical average over time and across the continental United States.

3 DIRECT GEOELECTRIC MONITORING

In principle, geoelectric fields could be monitored over a given geographic region for long durations of time from numerous fixed-site observatories. In practice, long-term geoelectric monitoring is performed at only a few stations around the world, such as at the Kakioka (KAK) observatory in Japan (Fujii et al., 2015). These data are useful for validating modeling and prediction methods. But due to the localized complexity of the solid-Earth conductivity structure, storm-time geoelectric fields measured at one site can differ significantly from geoelectric fields measured at another just a hundred kilometers away (e.g., McKay and Whaler, 2006; Bedrosian and Love, 2015; Bonner and Schultz, 2017). And this, together with the effort required to maintain geoelectric monitoring systems (e.g., Ferguson, 2012), makes it impractical to maintain a geoelectric monitoring network of sufficient density to evaluate geoelectric hazards on a continental scale. Given this reality, we adopt an alternative method for mapping geoelectric hazards, one using magnetotelluric survey data, historical magnetic observatory data, and a simple parameterization of induction (following Love et al., 2016b).

4 INDUCTION IN A CONDUCTING EARTH

Ground-level geomagnetic disturbance $\mathbf{B}(t, x, y)$ and the induced surface geoelectric field $\mathbf{E}(t, x, y)$ are vector functions of time t and geographic location (x, y). It is convenient to consider their Fourier transformation from the time domain to the frequency domain,

$$\mathcal{F}\{\mathbf{B}(t)\} = \mathbf{B}(\omega) \quad \text{and} \quad \mathcal{F}\{\mathbf{E}(t)\} = \mathbf{E}(\omega), \tag{1}$$

where, for sinusoidal variation with period T, the angular frequency is $\omega = 2\pi/T$. We focus attention on horizontal geomagnetic (north B_x, east B_y) and geoelectric (E_x, E_y) components. In the "low-frequency" limit, and within a realistic range of Earth conductivity, Maxwell's displacement current can be ignored. Under these and other simplifying assumptions (e.g., Simpson and Bahr, 2005, Chapter 2.1), the laws of electromagnetism that govern induction in an electrically conducting medium (e.g., Stratton, 1941, Chapter 5) can be distilled down to a linear equation; for horizontal geomagnetic $\mathbf{B}_h(\omega, x, y)$ and geoelectric $\mathbf{E}_h(\omega, x, y)$ field components,

$$\mathbf{E}_h(\omega,x,y) = \frac{1}{\mu}\mathbf{Z}(\omega,x,y;\sigma(\mathbf{r})) \cdot \mathbf{B}_h(\omega,x,y), \tag{2}$$

where μ is magnetic permeability, assumed to be the free-space constant. The complex impedance tensor $\mathbf{Z}$ has units of ohms (Ω), and the response function $\mathbf{Z}/\mu$ has units of (V/km)/nT. These are functions of variational frequency and geographic location, and they are nonlinearly dependent on the 3D electrical conductivity structure $\sigma(\mathbf{r})$ (or, equivalently, the resistivity $\rho = 1/\sigma$) within the Earth's volume ($\mathbf{r}$ is position vector). Eq. (2) is the basis of our simple parameterization of induction. For now, it is worth emphasizing that field variation is attenuated in the Earth's interior as a "skin effect" that depends on the average of the conductivity of rock beneath the site and within a volume of Earth having a characteristic dimension given by a diffusive skin depth,

$$\delta = \sqrt{\frac{2}{\mu\sigma\omega}}. \tag{3}$$

The conductivity within such a volume of rock is determined by a myriad of properties, including mineral content, the proportion of melt and solid phase, water content, clay content, porosity, and interconnectivity of cracks and grain boundaries (e.g., Yoshino, 2011; Evans, 2012). In detail, how these properties (and conductivity) are distributed within the Earth is reflective of its complex geological history and tectonic structure.

5 MAGNETIC OBSERVATORY DATA

As an ingredient for our estimate of geoelectric hazards, we seek a global description of the statistics of storm-time geomagnetic disturbance. For this, we analyze time series that have been collected at ground-based magnetic observatories (e.g., Love, 2008) since the 1970s and 1980s. We perform a statistical analysis of decades of observatory data collected from the 34 locations shown on the geomagnetic coordinate map in Fig. 1 (Love et al., 2016a, Table 1). A horizontal-component geomagnetic disturbance is recorded at each observatory as discrete, sequential samples that we represent as $B_x(t_i)$ and $B_y(t_i)$, for time-stamp values $t_i, t_{i+1}, t_{i+2}, \ldots$, where $\tau = t_{i+1} - t_i$ is the 1-min (60-s) sampling interval. Sinusoidal signals with periods shorter than 2 min are squelched in the acquisition process by a combination of analog and digital filters. We examine a total of more than 567 million individual 1-min "definitive" data values, for which we have removed a few instances of spikes and artificial noise.

6 GEOMAGNETIC WAVEFORM TIME SERIES

In using Eq. (2) to estimate the induction of geoelectric fields, it is useful to focus analysis on sinusoidal geomagnetic variation $\mathbf{B}_h(\omega, x, y)$ at specific periods (frequencies). Three issues affect our choice of these periods. First, with Fourier transformation, observatory time series recording storm-time

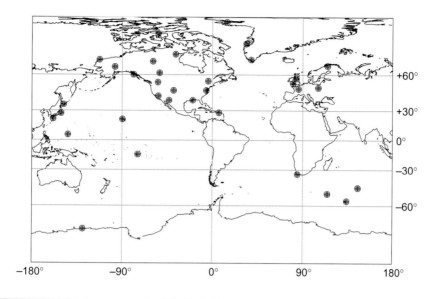

FIG. 1

Geomagnetic coordinate, Miller-projection map of the locations of the observatories for which 1-min data are used in this analysis.

magnetic disturbance can be decomposed into sinusoidal variational constituents having a broad range of periods (or frequencies). Over the course of each storm, the amplitudes of the constituent sinusoids change, with the highest amplitudes for the highest frequencies tending to occur during storm initial and main phases. Second, electric utility companies, in their evaluation of the vulnerability of high-voltage transformers, are concerned with geomagnetic field variation occurring from periods of about 10 s to 1000 s (10^{-1} to 10^{-3} Hz) (e.g., NERC, 2014b). And, third, the 1-min observatory data limit the range of sinusoidal periods that can be practically considered. By Nyquist's theorem (e.g., Priestley, 1981, Chapter 7.1), sinusoidal signals can only be resolved for periods longer than twice the 1-min sampling interval, or 120 s.

Mindful of these factors, we choose to focus on geomagnetic variation at two different sinusoidal periods, $T = 2\pi/\omega = 240$ s (4 min) and 1200 s (20 min), measured, respectively, over windows of length $W = 600$ s (10 min) and 3600 s (1 h). For each of the $\omega = 2\pi/T$ and for both the north B_x and east B_y component time series recorded at each observatory, we perform a minute-by-minute fit of the function

$$\xi(t;\mathbf{a},\omega) = a_0 + a_1 \cdot t + a_s \cdot \sin(\omega t) + a_c \cdot \cos(\omega t). \tag{4}$$

The parameters $\mathbf{a} = \{a_0, a_1, a_s, a_c\}$ are obtained for each minute t_i of each time series using a simple algorithm that minimizes the squared residual difference between ξ and consecutive 1-min data values; so, for example, for the north B_x component, we minimize the running quantity

$$\chi^2(t_i;\mathbf{a},\omega) = \sum_{j=1}^{N}\left[B_x(t_{i+j}) - \xi(t_{i+j};\mathbf{a},\omega)\right]^2, \tag{5}$$

where N is the number of minutes, either 10 or 60, in the two chosen waveform window lengths W; a similar running fit is made for the east B_y component. Here, a_0 is a 10 min running average, and a_1 is the slope of a linear trend in the data. The parameters a_s and a_c are equivalent to sliding-window, short-time Fourier amplitudes (e.g., Boashash, 2016). They can be combined to obtain the amplitude of a square-window "waveform,"

$$b_x(t_i;\omega) = \sqrt{a_s^2 + a_c^2}. \qquad (6)$$

And, again, a similar function applies for the east B_y component. Frequency resolution is limited by the length of the window, $\Delta\omega \simeq 2\pi/W$. From this, we can appreciate that an estimate of a given amplitude b is affected by Fourier harmonics within a frequency band of width $(\Delta\omega)/\omega \simeq T/W$. In Fig. 2A, we show geomagnetic north $B_x(t)$ and east $B_y(t)$ component variation recorded at the USGS, Fredericksburg, Virginia (FRD), observatory during the Halloween storm of October 2003. In Fig. 2B and C, we show the amplitudes $\{b_x, b_y\}$ for each of the chosen waveforms.

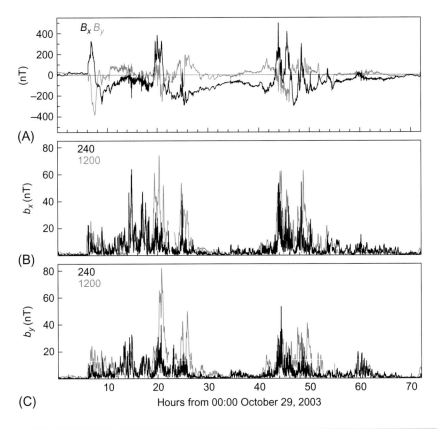

FIG. 2

(A) 1-min, north $B_x(t)$ and east $B_y(t)$ component variation recorded at the Fredericksburg, Virginia (FRD), observatory, (38.20 degrees N, 77.37 degrees W) geographic, (48.62 degrees N, 7.10 degrees W) geomagnetic, during the Halloween storm of October 29–31, 2003. (B and C) The running amplitudes $\{b_x, b_y\}$ of the $T = 2\pi/\omega = 240$-s, $W = 10$-min waveform and the $T = 2\pi/\omega = 1200$-s, $W = 1$-h waveforms.

7 OBSERVATORY MAGNETIC HAZARD FUNCTIONS

Because the time series of waveform amplitudes $\{b_x, b_y\}$ records the continuous evolutionary change in spectral content of the observatory time series, they are autocorrelated—the amplitude at one instance in time t_i is similar to the previous and subsequent values at t_{i-1} and t_{i+1}, for example. But to properly perform a statistical analysis, autocorrelation must be substantially removed (Priestley, 1981, Chapter 5.3.2; von Storch, 1995). We follow Love et al. (2016a, Section 4) and "decluster" the amplitudes using a simple algorithm that gives for each observatory time series the largest waveform amplitudes within one-day windows of time. So, for example, during the Halloween storm of October 2003, Fig. 2B, for 240-s FRD b_x waveforms, only two individual amplitude values are saved after declustering: one of 64.0 nT realized at about hour 15 (after 00:00 of October 29), and one of 54.4 nT realized at about hour 48. The empirical cumulative probabilities Λ for declustered FRD waveform amplitudes, $\{b_x, b_y\}$ for years 1982–2014, are shown as histograms in Fig. 3A and B.

We assume, furthermore, that the declustered waveform amplitudes $\{b_x, b_y\}$ can be modeled (e.g., Love et al., 2015, 2016a) by a lognormal distribution (e.g., Aitchison and Brown, 1957; Crow and Shimizu, 1988) that is truncated at a chosen lower-size threshold. For reference, a log-normal distribution is often realized from the multiplication of numerous independent random variables, each of which is positive and drawn from an arbitrary distribution having a mean and variance. While it might be argued that the amplitude of magnetic disturbance is effectively determined by the multiplication of several physical factors, such a seemingly reasonable expectation should be tempered with recognition that these processes are not likely to act entirely independently. In the end, an assessment of the success or failure of a statistical hypothesis is made on the basis of its consistency with actual data. Conscious of these qualifications, then, a random positive variable b is the realization of a lognormal statistical process if its probability density is

$$\lambda(b|v,\epsilon^2) = \frac{1}{b\sqrt{2\pi\epsilon^2}} \exp\left[-\frac{(\ln b - v)^2}{2\epsilon^2}\right], \tag{7}$$

where v and ϵ^2 are model parameters. The occurrence probability for events with size exceeding b is given by the cumulative

$$\Lambda(b|v,\epsilon^2) = \int_b^\infty \lambda(\xi|v,\epsilon^2)d\xi. \tag{8}$$

We define a rate function,

$$\rho(b|A,v,\epsilon^2), = A \cdot \lambda(b|v,\epsilon^2), \tag{9}$$

where A is a normalizing amplitude such that,

$$A \int_\theta^\infty \lambda(\xi|v,\epsilon^2)d\xi = \frac{N(b_j \geq \theta)}{D}, \tag{10}$$

where N is the number of days such that $b \geq \theta$, and where D is the total duration of the observatory time series. We obtain model parameters $\{A, v, \epsilon^2\}$ by fitting Eq. (9) to the b_j data using the maximum likelihood method (e.g., James, 2006, Chapter 8.3). Fits are shown in Fig. 3A and B, and while these provide a reasonably good representation of the data, all of these have Kolmogorov-Smirnov p-values (e.g., Press et al., 1992) that are greater than 0.67, meaning that there is no significant motivation to reject the hypothesis that the data are, indeed, log-normal distributed. From the fitted functions,

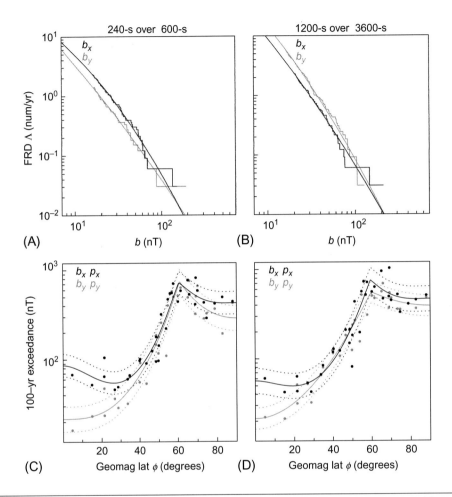

FIG. 3

(A and B) Cumulative Λ for 240-s and 1200-s declustered waveform amplitudes $\{b_x, b_y\}$ derived from data collected at the Fredericksburg, Virginia, magnetic observatory, 1982–2014, and giving the probability of the number of days per year in which the amplitude can be expected to exceed the threshold b. FRD data are shown as *histogram*; log-normal functions are shown as *smooth lines*. (C and D) Once-per-century cumulative exceedances for 240-s and 1200-s waveform amplitudes $\{b_x, b_y\}$ from 34 observatories and functional fits to these values $\{p_x(\phi), p_y(\phi)\}$. *Dotted lines* show (proportional) 1 standard deviation dispersion.

we estimate, as extrapolations, exceedances with an average return rate of once per century. For 240-s FRD b_x, the extrapolation is to 178.8 nT; for b_y, it is to 186.7 nT. For 1200-s FRD b_x, the extrapolation is to 211.2 nT; for b_y, it is to 213.8 nT.

8 GLOBAL MAGNETIC HAZARD FUNCTIONS

On a global scale, geomagnetic disturbance is roughly organized by geomagnetic latitude (e.g., Moe and Nebergall, 1969). In Fig. 3C and D, we plot, as a function of observatory location geomagnetic latitude ϕ, once-per-century exceedance amplitudes $\{b_x, b_y\}$ for 34 observatories, where each amplitude value is taken from an extrapolation from decades of 1-min data like the extrapolations made for FRD in Fig. 3A and B (Love et al., 2016a, 2016b, Figure 3(b)). These results can be compared with those made for Europe (e.g., Thomson et al., 2011, Figure 6), and globally but for one or several magnetic storms (e.g., Pulkkinen et al., 2012, Figure 4(c,d); Ngwira et al., 2013, Figure 2). We furthermore fit a simple function of geomagnetic latitude consisting of a fourth-order polynomial, parameterized by $\{\alpha_0, ..., \alpha_4\}$, plus a term that allows for a kink, parameterized by $\{\beta, \gamma, \psi\}$:

$$p(\phi) = 10^{e(\phi)}, \tag{11}$$

where

$$e(\phi) = \beta \cdot \left|\left(\frac{\phi}{90 \text{ degrees}}\right) - \psi\right|^\gamma + \sum_{k=0}^{4} \alpha_k \cdot \left(\frac{\phi}{90 \text{ degrees}}\right)^k. \tag{12}$$

We fit $e(\phi)$ to the logarithms of the once-per-century amplitudes $\{b_x, b_y\}$ using a least-squares algorithm, subject to derivative constraints that ensure smoothness at the geomagnetic equator and poles. The vertical-axis range in both $\{p_x, p_y\}$ is about an order of magnitude, most of which is seen between the auroral latitude of 60 degrees and the mid-latitude of 40 degrees. Observatory-to-observatory scatter about the fitted $p(\phi)$ curves is noted, and we plot the (proportional) 1 standard deviation [1 SD, lower and upper] given by $10^{\pm s} \cdot p(\phi)$; for both $\{b_x, b_y\}$, the 1-s dispersion is approximately $[0.74, 1.35] \cdot p(\phi)$. The functions $\{p_x, p_y\}$ serve as reference measures of geomagnetic disturbance that we use for estimating geoelectric hazards, and the 1-s dispersion quantifies most of the error in the geoelectric hazard estimates.

It is worth emphasizing that many spatiotemporal details of magnetic disturbance are unique for each storm. Indeed, it is occasionally said, "If you have seen one storm, you have seen one storm." (e.g., Friedel et al., 2002, p. 266). This simple but important point becomes apparent when working with the data used to construct Fig. 3. Geomagnetic data collected over a few decades typically record extreme-value disturbance values that are less than that expected from a statistical extrapolation of a log-normal model for a once-per-century disturbance. But because of localized differences in geomagnetic disturbance that are realized from one storm to another, there are exceptions. So, for example, for data collected at the Boulder, Colorado observatory (BOU) since 1979, 240-s amplitudes $\{b_x, b_y\}$ for the 1989 storm exceeded those that would be expected from a statistical extrapolation to once-per-century values. As another example, since 1978, the largest 240-s amplitudes $\{b_x, b_y\}$ recorded at the Barrow, Alaska (BRW), observatory occurred, respectively, on September 10, 2005, and February 21, 1994, and each during storms of modest global intensity (both amplitudes exceeded once-per-century values).

9 MAGNETOTELLURIC IMPEDANCES

Magnetotellurics is a geophysical method for estimating the electrical conductivity structure of the Earth's solid interior (e.g., Unsworth, 2007). Empirical impedance tensors, $\mathbf{Z}^e(\omega, x, y)$, are derived (e.g., Egbert, 2007b; Chave, 2012) from simultaneous measurement of geomagnetic variation using either induction or fluxgate magnetometers, and geoelectric field variation, obtained by measuring voltage difference between pairs of grounded electrodes (Simpson and Bahr, 2005, Chapter 3;

Ferguson, 2012). With a magnetotelluric survey, impedances are obtained from multiple sites, and these are used to construct models of subsurface electrical conductivity structure (e.g., Egbert, 2007a; Rodi and Mackie, 2012). Since 2006, the National Science Foundation has supported a national-scale magnetotelluric survey in the United States (Schultz, 2010) through the EarthScope program (Williams et al., 2010); so far, large geographic parts of the continental United States have been covered. In a separate, smaller project, the USGS performed a magnetotelluric survey of the Florida peninsula in 2015 consistent with EarthScope protocols. These surveys have been (and are being) accomplished through temporary deployments of magnetotelluric measurement systems at various locations with an approximate 70-km spacing. The EarthScope impedance tensors (Schultz et al., 2006–2018) are well defined across a frequency band of 10^{-4} to 10^{-1} Hz (periods of 10–10,000 s); errors are less than 5%. Empirical impedance tensors have been used to invert for 3D models of Earth conductivity $\sigma(\mathbf{r})$ (e.g., Bedrosian and Feucht, 2014; Meqbel et al., 2014; Yang et al., 2015; Bedrosian, 2016), informing fundamental understanding of North American geology and tectonic evolution. A fringe benefit, not widely anticipated when the EarthScope project was initiated in 2006, is the utility of magnetotelluric data for assessing geoelectric hazards (e.g., Bedrosian and Love, 2015; Love et al., 2016b; Love et al., 2017; Bonner and Schultz, 2017).

A simple examination of the effect of Earth conductivity structure on induction can be made by using empirical magnetotelluric impedance tensors to calculate the geoelectric amplitude that would be generated by a reference geomagnetic signal. For this, consider synthetic geomagnetic variation given by

$$\mathbf{B}_x(t) = \hat{\mathbf{x}} \cdot b_x(\omega) \cdot \sin(\omega t + \phi), \tag{13}$$

where b_x is amplitude and ϕ is phase, and where $\hat{\mathbf{x}}$ is a unit vector in the north direction. In the frequency domain, this north geomagnetic variation gives rise to the geoelectric field

$$\mathbf{E}_h^x(\omega, x, y) = \frac{1}{\mu} \mathbf{Z}^e(\omega, x, y) \cdot \mathbf{B}_x(\omega), \tag{14}$$

where, in general, $\mathbf{E}_h^x$ is a vector field with both north E_x^x and east components E_y^x. The corresponding geoelectric amplitude is given by

$$E_h^x(\omega, x, y) = \frac{1}{\mu} \zeta_x(\omega, x, y) \cdot b_x(\omega), \tag{15}$$

where $\zeta_x = |\mathbf{Z}^e \cdot \hat{\mathbf{x}}|$ is a scalar impedance for induction driven by a north, sinusoidally varying geomagnetic field of frequency ω and unit amplitude, $b_x(\omega) = 1$. Similar definitions can be made for a geoelectric amplitude E_h^y generated by east-oriented, sinusoidally varying geomagnetic field $\mathbf{B}_y$ and impedance $\zeta_y = |\mathbf{Z}^e \cdot \hat{\mathbf{y}}|$. Using the EarthScope and USGS empirical magnetotelluric impedances, in Figs. 4 and 5, we show maps of synthetic geoelectric amplitudes (scalar impedances) that would be induced by spatially uniform north $B_x(\omega)$ and east $B_y(\omega)$ geomagnetic variation having amplitude $b = 1$ nT and varying in time as sinusoids with periods of $T = 240$ s and 1200 s. Site-to-site and broader regional differences in these synthetic amplitudes are due to local differences in impedance that are related to subsurface 3D conductivity structure. From these maps we see that synthetic geoelectric amplitudes in the Northwest are smaller than those in the Central United States and in the Southeast—the median amplitudes differ by about a factor of two.

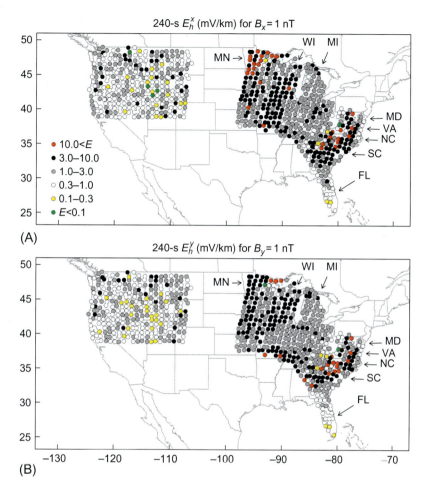

FIG. 4

Maps showing synthetic geoelectric amplitudes (scalar impedances, mV/km) at EarthScope and U.S. Geological Survey sites for (A) north-south and (B) east-west geomagnetic induction with an amplitude of $b(\omega) = 1$ nT and at period $T = 2\pi/\omega = 240$ s.

10 GEOLOGICAL INTERPRETATIONS

The maps of synthetic geoelectric amplitudes (scalar impedances) shown in Figs. 4 and 5 can be interpreted in terms of the geological and tectonic history of North America (e.g., Bally and Palmer, 1989; Whitmeyer and Karlstrom, 2007). The conductivity structure of the Northern Central United States is of particular interest (Yang et al., 2015; Bedrosian, 2016). Consider the southern edge of the Superior craton in Northern Minnesota (MN), where thick Archean lithosphere is juxtaposed against younger, sulfide-bearing sedimentary rocks. Generally speaking, geoelectric field amplitudes are large above the resistive igneous craton and small above conductive sedimentary rocks. These differences are evident in the

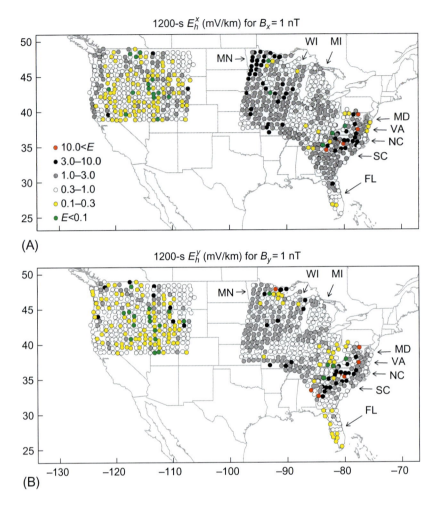

FIG. 5

Maps showing synthetic geoelectric amplitudes (scalar impedances, mV/km) at EarthScope and U.S. Geological Survey sites for (A) north-south and (B) east-west geomagnetic induction with an amplitude of $b(\omega) = 1$ nT and at period $T = 2\pi/\omega = 1200$ s.

synthetic geoelectric amplitudes for this region, which at 240 s and 1200 s correspond to apparent conductivities that span three orders of magnitude: from 10^{-4} S/m to 10^{-1} S/m (Bedrosian and Love, 2015, Figure 6). Furthermore, induced geoelectric field vectors can also be highly polarized for this region, sometimes oriented at odd angles relative to the inducing geomagnetic vector (Bedrosian and Love, 2015). Significant differences in amplitude and polarization from one location to another nearby are indicative of complex 3D subsurface conductivity structure (e.g., Hermance, 2011). In contrast, the Michigan (MI) basin has relatively thick sequences of Phanerozoic sediments resting upon a Proterozoic basement (e.g., Sloss, 1988). For geomagnetic variation at 240 s and 1200 s, impedances throughout this region correspond to conductivities of about 10^{-2} S/m (Bedrosian and Love, 2015, Figure 6), and induced geoelectric vectors at 240 s are of similar amplitude and direction. Subsurface conductivity structure that is nearly 1D depth dependent would give rise to such synthetic geoelectric amplitudes. Along the Florida (FL) Peninsula, the southward decrease in synthetic geoelectric amplitude is seen for both 240 s and 1200 s. This trend parallels a southward increase in the thickness of conductive Mesozoic sediments [e.g., Barnett] with distance away from the Florida-Bermuda fracture zone (e.g., Barnett, 1975). In contrast, higher synthetic amplitudes are seen along the Southern Appalachians in North Carolina (NC) and South Carolina (SC), with a resistive lithospheric keel beneath the Piedmont Belt and a general absence of deep sedimentary cover (e.g., Horton Jr. and Zullo, 1991) corresponding to generally high resistivity (Murphy and Egbert, 2017).

Among the surveyed sites, the largest synthetic geoelectric amplitude for north geomagnetic induction of unit amplitude $b_x = 1$ nT and period 240 s is in Central Virginia (VA): 46.03 mV/km (EarthScope site: VAQ58, 37.38 degrees N, 77.58 degrees W); for east induction, $b_y = 1$ nT and period 240 s, the largest synthetic amplitude is in Northern Minnesota (MN): 51.89 mV/km (MNC37, 47.83 degrees N, 91.99 degrees W). The smallest synthetic geoelectric amplitudes for north and east geomagnetic induction at 240 s are also in Virginia: 0.13 mV/km and 0.09 mV/km (VAQ55, 37.88 degrees N, 79.81 degrees W). These values might be considered anomalous—possibly an indication of localized "static-shift distortion" in the empirical impedances that can be caused by near-surface, small-scale inhomogeneities in subsurface conductivity and from topography (e.g., Groom and Bahr, 1992; Jones, 2012). A variety of techniques can be used to "correct" for static shift (e.g., Sternberg et al., 1988); however, we emphasize that impedance tensors with static shifts are neither "bad data" nor are they an artifact of data processing (e.g., Bonner and Schultz, 2017, Section 2.1). From the standpoint of geoelectric hazards for power-grid systems, the effect of static shift is likely negligible because the geomagnetically induced voltage across a distribution line is obtained by integration

$$V = \oint_G \mathbf{E}_h \cdot d\mathbf{l} \tag{16}$$

over the path of the power grid. Because this path has a characteristic dimension of about 100–1000 km (e.g., Horton et al., 2012, Table II), the voltage that is induced by geomagnetic disturbance on a power grid will effectively average over smaller-scale static shift.

11 GEOELECTRIC HAZARD MAPS

In estimating extreme-value geoelectric fields, it is helpful to consider an idealized (synthetic) halfspace of uniform electrical conductivity. For this, surface impedance (e.g., Simpson and Bahr, 2005, Chapter 2.4) can be represented as a scalar function of frequency ω and conductivity σ,

$$\frac{1}{\mu} Z(\omega, \sigma) = \sqrt{\frac{\omega}{\mu \sigma}}. \tag{17}$$

For a reference geomagnetic amplitude, we can use this to estimate an induced geoelectric amplitude,

$$E(\omega, \sigma) = \sqrt{\frac{\omega}{\mu \sigma}} \cdot b(\omega). \tag{18}$$

Analyses of historical data [Thomson et al., 2011, Figure 6; Love et al., 2016, Figure 4] show that geomagnetic variation at geomagnetic latitudes of $\phi \simeq 50$ degrees can attain amplitudes of $b(\omega) = 250$ nT at $T = 240$ s. At a site with an effective Earth conductivity of $\sigma = 10^{-4}$ S/m (a relatively resistive value that can be realized in some geological settings), the geoelectric field would have an amplitude of about 5 V/km. This is comparable to amplitudes anticipated in other analyses using synthetic Earth conductivity models (e.g., Kappenman, 2003; Ngwira et al., 2013).

More realistic estimates of geoelectric amplitude can be made by using the magnetotelluric impedance tensors $\mathbf{Z}^e$, which record the effects of Earth conductivity structure. Drawing upon the results presented in Figs. 3–5, we can estimate the geoelectric hazard that would be exceeded only once per century in response to an extreme-event magnetic storm, for waveforms $T = 2\pi/\omega = 240$ s (4 min) and 1200 s (20 min), respectively, for windows of length $W = 600$ s (10 min) and 3600 s (1 h),

$$E_e(x,y) \simeq \frac{1}{\mu} \zeta(x,y) \cdot p(\phi(x,y)). \tag{19}$$

In Figs. 6 and 7, we plot the exceedances E_e^x and E_e^y, respectively, for the disturbance functions $\{p_x, p_y\}$. It is important to make a proper interpretation of these maps—they represent point-wise estimates of once-per-century geoelectric amplitude. The maps do not show the geoelectric amplitude that would be expected for a single, once-per-century magnetic storm, or, in fact, any single magnetic storm. Indeed, for any given geographic location, when a once-per-century geoelectric exceedance value is realized for a particular storm, it might not be realized at another geographic location; it is possible that the geoelectric amplitude for the same storm, but at another location, would fall below its once-per-century exceedance value.

Once-per-century geoelectric exceedance amplitudes range over about two orders of magnitude (a factor of 100 or so)—this is the combined result of local site-to-site differences in impedance and the latitude-dependence of geomagnetic disturbance. Where magnetotelluric surveys have been completed in the United States, Minnesota (MN) and Wisconsin (WI) have some of the highest geoelectric hazards. Among the surveyed sites, once-per-century 240-s (1200-s) north geomagnetic variation b_x induces geoelectric amplitudes E_e^x with a median value of 0.34 V/km (0.29 V/km); for once-per-century 240-s (1200-s) east geomagnetic variation, the geoelectric amplitudes E_e^y have a median value of 0.23 V/km (0.16 V/km). These values are much less than the typical 5 V/km value in scenario analyses performed for the North American Electric Reliability Corporation NERC (2014a, pp. 23–24). However, Figs. 6 and 7 show us that storm-time geoelectric amplitudes have a wide and granular distribution with geographic location. At many sites in the United States, once-per-century 240-s and 1200-s geoelectric amplitudes are comparable to the instantaneous 1-min value of 0.4 V/km measured at KAK, during the sudden commencement event of March 24, 1991. Across large parts of the Northern Central United States, once-per-century 240-s geoelectric amplitudes exceed the 2 V/km that Boteler (1994) has inferred was responsible for bringing down the Hydro-Québec electric power grid in Canada in March 1989. Some amplitudes in Northern Minnesota even exceed the once-per-century reference amplitude of 8 V/km anticipated by NERC (2014a, p. 5).

Among the surveyed sites, for once-per-century 240-s (over 10 min) waveforms and for north geomagnetic induction b_x, the largest geoelectric exceedance amplitude $E_e^x = 14.00$ V/km (EarthScope site: MNB36) is in Northern Minnesota, (48.27 degrees N, 92.71 degrees W) geographic, (58.03 degrees N, 27.16 degrees W) geomagnetic. For east geomagnetic induction b_x, the largest 240-s geoelectric amplitude $E_e^y = 23.35$ V/km (MNC37) is also in Northern Minnesota, (47.83 degrees N, 91.99 degrees W) geographic, (57.64 degrees N, 26.16 degrees W) geomagnetic. Just 123 km away, at another site (RED36), (47.19 degrees N, 93.07 degrees W) geographic, (56.92 degrees N,

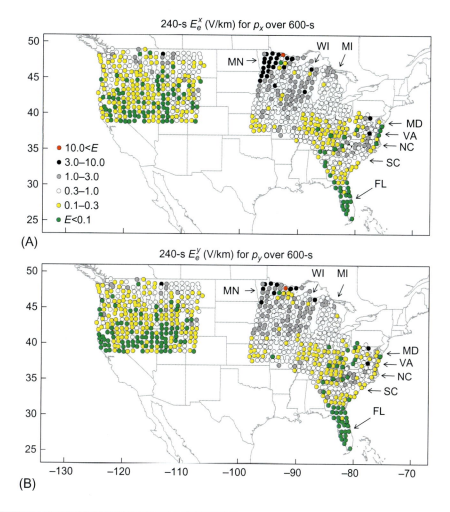

FIG. 6

Maps showing maximum once-per-century geoelectric exceedances (V/km) at EarthScope and U.S. Geological Survey magnetotelluric survey sites for geomagnetic induction at $T = 2\pi/\omega = 240$ s for $W = 10$ min: (A) for north induction E_e^x, (B) for east induction E_e^y.

27.35 degrees W) geomagnetic, once-per-century b_x and b_y induction generates geoelectric amplitudes of only 0.08 and 0.02 V/km. These extreme differences in once-per-century geoelectric field amplitude reflect orders of magnitude differences in the local spatial distribution in the crust and mantle (Bedrosian, 2016). In general, geoelectric exceedance amplitude for once-per-century 1200-s waveforms are smaller than for 240-s waveforms, though, by definition, the 1200-s waveform is of longer duration (1 h) than for 240-s waveform (10 min). For once-per-century 1200-s (over 1 h) waveforms, and for north geomagnetic induction b_x, the largest geoelectric exceedance amplitude $E_e^x = 4.16$ V/km (MNB36) is in Northern Minnesota. For east geomagnetic induction b_x, the largest 1200-s geoelectric

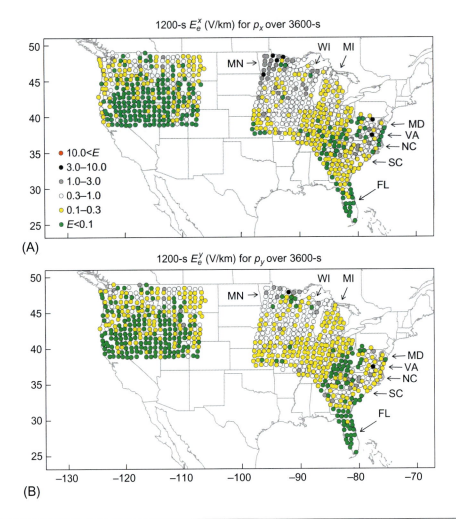

FIG. 7

Maps showing maximum once-per-century geoelectric exceedances (V/km) at EarthScope and U.S. Geological Survey magnetotelluric survey sites for geomagnetic induction at $T = 2\pi/\omega = 1200$ s for $W = 1$ h: (A) for north induction E_e^x, (B) for east induction E_e^y.

amplitude $E_e^y = 7.82$ V/km (MNC37) is also in Northern Minnesota. In each of these cases, as discussed in Section 8, we estimate the statistical 1-s dispersion of individual geoelectric amplitudes as being approximately $[0.73, 1.36] \cdot E_h$. An estimate (Anderson et al., 1974) that geoelectric amplitudes could have reached ~7 V/km at 10^{-2} Hz during the magnetic storm of August 4, 1972, obtained for a synthetic, 1D Earth-conductivity model for Illinois exceeds once-per-century exceedances for that part of the country, though the occurrence of such an amplitude there cannot be simply dismissed on statistical grounds.

12 DISCUSSION

Improvement in the simple statistical parameterization of induction hazards adopted here can be obtained for regions near long-running geomagnetic observatories by direct convolution of multidecade magnetometer time series with local impedance tensors (e.g., Kelbert et al., 2017), and then analyzing the resulting long estimated time series of the geoelectric field using methods like those described in Section 7. Looking beyond the statistical hazard maps developed here, a related follow-on project would be the development of time-series scenario maps for individual magnetic storms—convolving a time-dependent map of ground-level geomagnetic disturbance, derived from ground-based magnetometer data (e.g., Pulkkinen et al., 2003; Rigler et al., 2014), with a map of Earth-surface impedance, derived from a model of Earth conductivity that is, itself, derived from magnetotelluric survey data. Such a project could be further developed into a real-time geoelectric mapping service. Important ongoing magnetotelluric survey work is presently underway in the Northeast United States, a region that encompasses several major metropolitan centers together with electric power-grid infrastructure, all of which are situated on top of complicated geological and tectonic structures and at latitudes where geomagnetic disturbance can be locally intense. The development of geoelectric hazard maps for this part of the United States would be very useful. Additional geomagnetic monitoring and magnetotelluric surveying of Southern Canada would not only lead to improved hazard mapping in the Northern United States, but it would enable risk assessment of integrated North American continental, electric power-grid networks.

ACKNOWLEDGMENTS

We thank S.W. Cuttler, C.A. Finn, J. McCarthy, and J.L. Slate for reviewing a draft manuscript. We thank C.C. Balch, C.E. Black, S. Jonas, A. Kelbert, J.U. Kozyra, A. Pulkkinen, E.J. Rigler, R. Rutledge, A.T. Sabata, and R.M. Waggel for useful conversations. Magnetic observatory data can be obtained from either the Kyoto or Edinburgh World Data Centers or from INTERMAGNET. We thank the national institutes that support magnetic observatories and INTERMAGNET for promoting high standards of observatory practice (http://www.intermagnet.org). Geoelectric data can be obtained from the Kakioka Magnetic Observatory. EarthScope impedance tensors can be obtained from the Data Management Center of the Incorporated Research Institutions for Seismology (http://ds.iris.edu/ds/products/emtf) (Kelbert et al., 2011).

REFERENCES

Aitchison, J., Brown, J.A.C., 1957. The Lognormal Distribution: With Special Reference to Its Uses in Economics. Cambridge University Press, Cambridge, UK, pp. 1–176.

Allen, J., Frank, L., Sauer, H., Reiff, P., 1989. Effects of the March 1989 solar activity. Eos Trans. Am. Geophys. Union 70 (46), 1479–1488. https://doi.org/10.1029/89EO00409.

Anderson, C.W., Lanzerotti, L.J., MacLennan, G., 1974. Outage of the L4 system and the geomagnetic disturbances of 4 August 1972. Bell Labs Tech. J. 53 (9), 1817–1837.

Baker, D.N., et al., 2008. Severe Space Weather Events—Understanding Societal and Economic Impacts. The National Academy Press, Washington, DC, pp. 1–144. https://doi.org/10.17226/12507.

Baker, D.N., Li, X., Pulkkinen, A., Ngwira, C.M., Mays, M.L., Galvin, A.B., Simunac, K.D.C., 2013. A major solar eruptive event in July 2012: defining Extreme Space Weather scenarios. Space Weather 11 (10), 585–591. https://doi.org/10.1002/swe.20097.

Bally, A.W., Palmer, A.R., 1989. The Geology of North America: An Overview. Geological Soc. Am., Boulder, CO, pp. 1–629. https://doi.org/10.1130/DNAG-GNA-A

Barnett, R.S., 1975. Basement structure of Florida and its tectonic implications. Gulf Coast Assoc. Geol. Soc. Trans. 25, 122–142.

Bedrosian, P.A., 2016. Making it and breaking it in the Midwest: continental assembly and rifting from modeling of EarthScope magnetotelluric data. Precambrian Res. 278, 337–361. https://doi.org/10.1016/j.precamres.2016.03.009.

Bedrosian, P.A., Feucht, D.W., 2014. Structure and tectonics of the Northwestern United States from EarthScope USArray magnetotelluric data. Earth Planet. Sci. Lett. 402, 275–289. https://doi.org/10.1016/j.epsl.2013.07.035.

Bedrosian, P.A., Love, J.J., 2015. Mapping geoelectric fields during magnetic storms: synthetic analysis of empirical United States impedances. Geophys. Res. Lett. 42 (23), 10160–10170. https://doi.org/10.1002/2015GL066636.

Béland, J., Small, K., 2005. Space Weather effects on power transmission systems: the cases of Hydro-Québec and Transpower New Zealand Ltd, Daglis, I.A. (Ed.), Effects of Space Weather on Technology Infrastructure. Springer, Dordrecht, pp. 287–299.

Benfield, A., 2013. Geomagnetic Storms. Aon Benfield, Sydney, Australia, pp. 1–12.

Boashash, B., 2016. Heuristic formulation of time-frequency distributions. In: Boashash, B. (Ed.), Time-Frequency Signal Analysis and Processing: A Comprehensive Reference. Academic Press, New York, NY, pp. 65–102.

Bolduc, L., 2002. GIC observations and studies in the Hydro-Québec power system. J. Atmos. Solar-Terr. Phys. 64 (16), 1793–1802. https://doi.org/10.1016/S1364-6826(02)00128-1.

Bonner, L.R., Schultz, A., 2017. Rapid prediction of electric fields associated with geomagnetically induced currents in the presence of three-dimensional ground structure: projection of remote magnetic observatory data through magnetotelluric impedance tensors. Space Weather 15 (1), 204–227. https://doi.org/10.1002/2016SW001535.

Boteler, D.H., 1994. Geomagnetically induced currents: present knowledge and future research. IEEE Trans. Power Delivery 9 (1), 50–58.

Boteler, D.H., 2001. Assessment of geomagnetic hazard to power systems in Canada. Nat. Hazards 23 (2–3), 101–120.

Boteler, D.H., 2006. The super storms of August/September 1859 and their effects on the telegraph system. Adv. Space Res. 38 (2), 159–172.

Chave, A.D., 2012. Estimation of the magnetotelluric response function. In: Chave, A.D., Jones, A.G. (Eds.), The Magnetotelluric Method. Cambridge University Press, Cambridge, UK, pp. 165–218.

Cliver, E.W., Dietrich, W.F., 2013. The 1859 Space Weather event revisited: limits of extreme activity. J. Space Weather Space Clim. 3, A31. https://doi.org/10.1051/swsc/2013053.

Crow, E.L., Shimizu, K., 1988. Lognormal Distributions: Theory and Applications. Marcel Dekker, New York, NY, pp. 1–387.

Egbert, G.D., 2007. Inverse EM modeling. In: Gubbins, D., Herrero-Bervera, E. (Eds.), Encyclopedia of Geomagnetism and Paleomagnetism. Springer, Dordrecht, pp. 219–223.

Egbert, G.D., 2007a. Robust electromagnetic transfer functions estimates. In: Gubbins, D., Herrero-Bervera, E. (Eds.), Encyclopedia of Geomagnetism and Paleomagnetism. Springer, Dordrecht, pp. 866–870.

Evans, R.L., 2012. Earth's electromagnetic environment: 3A. Conductivity of Earth materials. In: Chave, A.D., Jones, A.G. (Eds.), The Magnetotelluric Method. Cambridge University Press, Cambridge, UK, pp. 50–95.

Ferguson, I.J., 2012. Instrumentation and field procedures. In: Chave, A.D., Jones, A.G. (Eds.), The Magnetotelluric Method. Cambridge University Press, Cambridge, UK, pp. 421–479.

Friedel, R.H.W., Reeves, G.D., Obara, T., 2002. Relativistic electron dynamics in the inner magnetosphere: a review. J. Atmos. Solar-Terr. Phys. 64 (2), 265–282. https://doi.org/10.1016/S1364-6826(01)00088-8.

Fujii, I., Ookawa, T., Nagamachi, S., Owada, T., 2015. The characteristics of geoelectric fields at Kakioka, Kanoya, and Memambetsu inferred from voltage measurements during 2000 to 2011. Earth Planets Space 67, 62. https://doi.org/10.1186/s40623-015-0241-z.

Groom, R.W., Bahr, K., 1992. Corrections for near surface effects: decomposition of the magnetotelluric impedance tensor and scaling corrections for regional resistivities: a tutorial. Surv. Geophys. 13 (4–5), 341–379.

Hajra, R., Tsurutani, B.T., Echer, E., Gonzalez, W.D., Gjerloev, J.W., 2016. Supersubstorms (SML < −2500 nT): magnetic storm and solar cycle dependences. J. Geophys. Res. 121 (8), 7805–7816. https://doi.org/10.1002/2015JA021835.

Hapgood, M.A., 2011. Towards a scientific understanding of the risk from extreme space weather. Adv. Space Res. 47 (12), 2059–2072.

Hermance, J.F., 2011. Magnetotelluric interpretation. In: Gupta, H.K. (Ed.), Encyclopedia of Solid Earth Geophysics. Springer-Verlag, Dordrecht, pp. 822–829.

Horton, R., Boteler, D.H., Overbye, T.J., Pirjola, R., Dugan, R.C., 2012. A test case for the calculation of geomagnetically induced currents. IEEE Trans. Power Delivery 27 (4), 2368–2373. https://doi.org/10.1109/TPWRD.2012.2206407.

Horton Jr., J.W., Zullo, V.A., 1991. The Geology of the Carolinas. University of Tennessee Press, Knoxville, TN, pp. 1–424.

James, F., 2006. Statistical Methods in Experimental Physics. World Scientific Publishing, Amsterdam, pp. 1–345.

Jonas, S., McCarron, E.D., 2015. Recent U.S. policy developments addressing the effects of geomagnetically induced currents. Space Weather 13 (11), 730–733. https://doi.org/10.1002/2015SW001310.

Jones, A.G., 2012. Distortion of magnetotelluric data: its identification and removal. In: Chave, A.D., Jones, A.G. (Eds.), The Magnetotelluric Method. Cambridge University Press, Cambridge, UK, pp. 219–302.

Kappenman, J.G., 2003. Storm sudden commencement events and the associated geomagnetically induced current risks to ground-based systems at low-latitude and midlatitude locations. Space Weather 1, 1016. https://doi.org/10.1029/2003SW000009.

Kappenman, J.G., 2012. A perfect storm of planetary proportions. IEEE Spectr. 49, 26–31.

Kelbert, A., Egbert, G.D., Schultz, A., 2011. IRIS DMC Data services products: EMTF the magnetotelluric transfer functions. https://doi.org/10.17611/DP/EMTF.

Kelbert, A., Balch, C.C., Pulkkinen, A., Egbert, G.D., Love, J.J., Rigler, E.J., Fujii, I., 2017. Methodology for time-domain estimation of storm time geoelectric fields using the 3-D magnetotelluric response tensors. Space Weather 15 (7), 874–894. https://doi.org/10.1002/2017SW001594.

Lakhina, G.S., Tsurutani, B.T., 2017. Super geomagnetic storms: past, present and future. In: Buzulukova, N. (Ed.), Extreme Events in Geospace. Springer, Dordrecht, The Netherlands.

Lanzerotti, L.J., 2001. Space Weather effects on technologies. In: Song, P., Singer, H.J., Siscoe, G.L. (Eds.), Space Weather. Am. Geophys. Union, Washington, DC. https://doi.org/10.1029/GM125p0011.

Liu, Y.D., et al., 2014. Observations of an extreme storm in interplanetary space caused by successive coronal mass ejections. Nat. Commun. 5, 3481. https://doi.org/10.1038/ncomms4481.

Lloyd's, 2013. Emerging Risk Report: Solar Storm Risk to the North American Electric Grid. Lloyd's of London, London, UK, pp. 1–22.

Love, J.J., 2008. Magnetic monitoring of Earth and Space. Phys. Today 61 (2), 31–37. https://doi.org/10.1063/1.2883907.

Love, J.J., Rigler, E.J., Pulkkinen, A., Balch, C.C., 2014. Magnetic storms and induction hazards. Eos Trans. Am. Geophys. Union 95 (48), 445–446. https://doi.org/10.1002/2014EO480001.

Love, J.J., Rigler, E.J., Pulkkinen, A., Riley, P., 2015. On the lognormality of historical magnetic storm intensity statistics: implications for extreme-event probabilities. Geophys. Res. Lett. 42 (16), 6544–6553. https://doi.org/10.1002/2015GL064842.

Love, J.J., Coïsson, P., Pulkkinen, A., 2016a. Global statistical maps of extreme-event magnetic observatory 1 min first differences in horizontal intensity. Geophys. Res. Lett. 43 (9), 4126–4135. https://doi.org/10.1002/2016GL068664.

Love, J.J., Pulkkinen, A., Bedrosian, P.A., Jonas, S., Kelbert, A., Rigler, E.J., Finn, C.A., Balch, C.C., Rutledge, R., Waggel, R.M., Sabata, A.T., Kozyra, J.U., Black, C.E., 2016b. Geoelectric hazard maps for the continental United States. Geophys. Res. Lett. 43 (18), 9415–9424. https://doi.org/10.1002/2016GL070469.

Love, J.J., Bedrosian, P.A., Schultz, A., 2017. Down to Earth with an electric hazard from Space. Space Weather 15 (5), 658-662. https://doi.org/10.1002/2017SW001622.

McKay, A.J., Whaler, K.A., 2006. The electric field in northern England and southern Scotland: implications for geomagnetically induced currents. Geophys. J. Int. 167 (2), 613–625. https://doi.org/10.1111/j.1365-246X.2006.03128.x.

Meqbel, N.M., Egbert, G.D., Wannamaker, P.E., Kelbert, A., Schultz, A., 2014. Deep electrical resistivity structure of the northwestern U.S. derived from 3-D inversion of USArray magnetotelluric data. Earth Planet. Sci. Lett. 402, 290–304. https://doi.org/10.1016/j.epsl.2013.12.026.

Moe, K., Nebergall, D., 1969. Variation of geomagnetic disturbance with latitude. J. Geophys. Res. 74 (5), 1305–1307. https://doi.org/10.1029/JA074i005p01305.

Molinski, T.S., 2002. Why utilities respect geomagnetically induced currents. J. Atmos. Solar-Terr. Phys. 64 (16), 1765–1778. https://doi.org/10.1016/S1364-6826(02)00126-8.

Murphy, B.S., Egbert, G.D., 2017. Electrical conductivity structure of southeastern North America: implications for lithospheric architecture and Appalachian topographic rejuvenation. Earth Planet. Sci. Lett. 462, 66–75. https://doi.org/10.1016/j.epsl.2017.01.009.

NERC, 2013. Application Guide: Computing Geomagnetically-Induced Current in the Bulk-Power System. North American Electric Reliability Corporation, Atlanta, GA, pp. 1–39.

NERC, 2014a. Benchmark Geomagnetic Disturbance Event Description. North American Electric Reliability Corporation, Atlanta, GA, pp. 1–26.

NERC, 2014b. Transformer Thermal Impact Assessment: Project 2013-03 (Geomagnetic Disturbance Mitigation). North American Electric Reliability Corporation, Atlanta, GA, pp. 1–16.

Ngwira, C.M., Pulkkinen, A., Leila Mays, M., Kuznetsova, M.M., Galvin, A.B., Simunac, K., Baker, D.N., Li, X., Zheng, Y., Glocer, A., 2013. Simulation of the 23 July 2012 Extreme Space Weather event: what if the extremely rare CME was Earth-directed? Space Weather 11 (12), 671–679. https://doi.org/10.1002/2013SW000990.

Ngwira, C.M., 2013. Simulation of the 23 July 2012 Extreme Space Weather event: what if the extremely rare CME was Earth-directed? Space Weather 11 (12), 671–679. https://doi.org/10.1002/2013SW000990.

Ngwira, C.M., Pulkkinen, A., McKinnell, L.A., Cilliers, P.J., 2008. Improved modeling of geomagnetically induced currents in the South African Power Network. Space Weather 6 (11), S11004. https://doi.org/10.1029/2008SW000408.

Ngwira, C.M., Pulkkinen, A., Wilder, F.D., Crowley, G., 2013. Extended study of extreme geoelectric field event scenarios for geomagnetically induced current applications. Space Weather 11 (3), 121–131. https://doi.org/10.1002/swe.20021.

NSTC, 2015. Executive Office, National Space Weather Strategy. National Science and Technology Council, Washington, DC, pp. 1–13.

Panel on Seismic Hazard Analysis, 1988. Probabilistic Seismic Hazard Analysis. National Academy Press, Washington, DC, pp. 1–97.

Piccinelli, R., Krausmann, E., 2014. Space Weather and Power Grids—A Vulnerability Assessment. European Union, Luxembourg, pp. 1–53.

Pirjola, R., 2002. Review on the calculation of surface electric and magnetic fields and of geomagnetically induced currents in ground-based technological systems. Surv. Geophys. 23 (1), 71–90.

Press, W.H., Teukolsky, S.A., Vetterling, W.T., Flannery, B.P., 1992. Numerical Recipes. Cambridge University Press, Cambridge, UK, pp. 1–963.

Priestley, M.B., 1981. Spectral Analysis and Time Series. Academic Press, London, UK, pp. 1–890.

Pulkkinen, A., 2015. Geomagnetically induced currents modeling and forecasting. Space Weather 13 (11), 734–736. https://doi.org/10.1002/2015SW001316.

Pulkkinen, A., Amm, O., Viljanen, A., BEAR Working Group, 2003. Ionospheric equivalent current distributions determined with the method of spherical elementary current systems. J. Geophys. Res. 108 (A2), 1053–1061. https://doi.org/10.1029/2001JA005085.

Pulkkinen, A., Lindahl, S., Viljanen, A., Pirjola, R., 2005. Geomagnetic storm of 29–31 October 2003: geomagnetically induced currents and their relation to problems in the Swedish high-voltage power transmission system. Space Weather 3 (8), S08C03. https://doi.org/10.1029/2004SW000123.

Pulkkinen, A., Bernabeu, E., Eichner, J., Beggan, C., Thomson, A.W.P., 2012. Generation of 100-year geomagnetically induced current scenarios. Space Weather 10, S04003. https://doi.org/10.1029/2011SW000750.

Rigler, E.J., Pulkkinen, A.A., Balch, C.C., Wiltberger, M.J., 2014. Dynamic geomagnetic hazard maps in space weather operations. Abstract Fall Meeting, SM31A-4178, AGU, San Francisco, CA.

Riswadkar, A.V., Dobbins, B., 2010. Solar Storms: Protecting Your Operations Against the Sun's "Dark Side" Zurich Services Corp., Schaumburg, IL, pp. 1–12.

Rodi, W.L., Mackie, R.L., 2012. The inverse problem. In: Chave, A.D., Jones, A.G. (Eds.), The Magnetotelluric Method. Cambridge University Press, Cambridge, UK, pp. 347–420.

Schrijver, C.J., 2015. Understanding Space Weather to shield society: a global road map for 2015–2025 commissioned by COSPAR and ILWS. Adv. Space Res. 55 (12), 2745–2807. https://doi.org/10.1016/j.asr.2015.03.023.

Schultz, A., 2010. A continental scale magnetotelluric observatory and data discovery resource. Data Sci. J. 8, IGY6–IGY20.

Schultz, A., Egbert, G.D., Kelbert, A., Peery, T., Clote, V., Fry, B., Erofeeva, S., Staff of the National Geoelectromagnetic Facility and Their Contractors, 2006–2018. USArray TA magnetotelluric transfer functions. https://doi.org/10.17611/DP/EMTF/USARRAY/TA.

Simpson, F., Bahr, K., 2005. Practical Magnetotellurics. Cambridge University Press, Cambridge, UK, pp. 1–254.

Siscoe, G.L., Crooker, N.U., Clauer, C.R., 2006. Dst of the Carrington storm of 1859. Adv. Space Res. 38 (2), 173–179.

Sloss, L.L., 1988. Tectonic evolution of the Craton in Phanerozoic time. In: Sloss, L.L. (Ed.), Sedimentary Cover—North American Craton: US, vol. D-2. GeoScienceWorld, Boulder, CO, pp. 25–52.

Smith, K., Petley, D.N., 2009. Environmental Hazards: Assessing Risk and Reducing Disaster. Routledge, New York, NY, pp. 1–382.

Smolka, A., 2006. Natural disasters and the challenge of extreme events: risk management from an insurance perspective. Phil. Trans. R. Soc. Lond. Ser. A 364, 2147–2165.

Sternberg, B.K., Washburne, J.C., Pellerin, L., 1988. Correction for the static shift in magnetotellurics using transient electromagnetic soundings. Geophysics 53 (11), 1459–1468. https://doi.org/10.1190/1.1442426.

Stratton, J.A., 1941. Electromagnetic Theory. McGraw-Hill Book Company, New York, NY, pp. 1–615.

Thomson, A.W.P., McKay, A.J., Viljanen, A., 2009. A review of progress in modelling of induced geoelectric and geomagnetic fields with special regard to induced currents. Acta Geophys. 57 (1), 209–219.

Thomson, A.W.P., Dawson, E.B., Reay, S.J., 2011. Quantifying extreme behavior in geomagnetic activity. Space Weather 9 (10), S10001. https://doi.org/10.1029/2011SW000696.

Tsurutani, B.T., Gonzalez, W.D., Lakhina, G.S., Alex, S., 2003. The extreme magnetic storm of 1–2 September 1859. J. Geophys. Res. 108, 1268. https://doi.org/10.1029/2002JA009504.

Unsworth, M., 2007. Magnetotellurics. In: Gubbins, D., Herrero-Bervera, E. (Eds.), Encyclopedia of Geomagnetism and Paleomagnetism. Springer, Dordrecht, pp. 670–673.

Varnes, D.J., 1984. Landslide Hazard Zonation: A Review of Principles and Practices. United Nations Educational Scientific and Cultural Organization, Paris, France, pp. 1–63.

Veeramany, A., Unwin, S.D., Coles, G.A., Dagle, J.E., Millard, D.W., Yao, J., Glantz, C.S., Gourisetti, S.N.G., 2016. Framework for modeling high-impact, low-frequency power grid events to support risk-informed decisions. Int. J. Disaster Risk Reduct. 18, 125–137. https://doi.org/10.1016/j.ijdrr.2016.06.008.

von Storch, H., 1995. Misuses of statistical analysis in climate research. In: von Storch, H., Navarra, A. (Eds.), Analysis of Climate Variability: Applications and Statistical Techniques. Springer-Verlag, New York, NY, pp. 11–25.

Whitmeyer, S.J., Karlstrom, K.E., 2007. Tectonic model for the proterozoic growth of North America. Geosphere 3 (4), 220–229. https://doi.org/10.1130/GES00055.1.

Williams, M.L., Fischer, K.M., Freymueller, J.T., Tikoff, B., Tréhu, A.M., et al., 2010. Unlocking the secrets of the North American continent: an EarthScope science plan for 2010–2020. EarthScope 1–78.

Yang, B., Egbert, G.D., Kelbert, A., Meqbel, N.M., 2015. Three-dimensional electrical resistivity of the north-central USA from Earthscope long period magnetotelluric data. Earth Planet. Sci. Lett. 422, 87–93. https://doi.org/10.1016/j.epsl.2015.04.006.

Yoshino, T., 2011. Electrical properties of rocks. In: Gupta, H.K. (Ed.), Encyclopedia of Solid Earth Geophysics. Springer-Verlag, Dordrecht, pp. 270–276.

CHAPTER 10

GEOMAGNETIC STORMS: FIRST-PRINCIPLES MODELS FOR EXTREME GEOSPACE ENVIRONMENT

Natalia Buzulukova[*,†], Mei-Ching Fok[*], Alex Glocer[*], Colin Komar[*,‡], Suk-Bin Kang[*], Steven Martin[*,§], Chigomezyo M. Ngwira[*,‡], Guan Le[*]

NASA Goddard Space Flight Center, Greenbelt, MD, United States University of Maryland, College Park, MD, United States[†] The Catholic University of America, Washington, DC, United States[‡] ADNET Systems, Inc., Bethesda, MD, United States[§]

CHAPTER OUTLINE

1 Introduction 232
2 Overview of First-Principles Magnetospheric Models 234
3 Modeling of Extreme and Intense Geomagnetic Storms 236
   3.1 Modeling of Carrington-Type Events 238
   3.2 Modeling of Radiation Belt Response for Extreme Storms 239
4 An Example: Geomagnetic Storm of June 22–23, 2015 241
   4.1 The Model Run Setup 241
   4.2 The Results for Pressure and Current Distribution 242
   4.3 Comparison With Ground-Based Magnetometers and Satellite Data 244
5 Role of the Ring Current Plasma in Generation of dB/dt and Variability of Electric Fields and FACs at Low Latitudes 248
6 Challenges and Future Directions 250
7 Conclusions 251
Acknowledgments 252
References 252

1 INTRODUCTION

A geomagnetic storm is an interval of disturbed near-Earth plasma environment when the energy is effectively pumped from the disturbed solar wind (SW) into the Earth magnetosphere. During geomagnetic storms, energetic particles ($\sim$1–200 keV) form an intense current flowing near the Earth at distances $\sim$2–8 Earth radii (R_E). This current depletes the geomagnetic field, and can be measured at low-latitude stations by deviations from horizontal (H) magnetic field components, resulting in the so-called disturbance storm-time (Dst) index (1 h temporal resolution) or SYM-H index (1 min temporal resolution). These indices are by definition the measure of storm intensity. The storm main phase is the time interval when the Dst index substantially decreases (Gonzalez et al., 1994). It usually corresponds to intervals of strong negative (southward) interplanetary magnetic field (IMF) B_z component. The distribution of ring current plasma is asymmetric during the main phase with respect to local time (azimuthal angle in hours), and becomes symmetric during the recovery phase of storm. When the ring current particles are lost (Kozyra and Liemohn, 2003), Dst will recover to its initial value. Global properties of the ring current and ring current plasma could be inferred from ground-based magnetometers (Grafe, 1999; Weygand and McPherron, 2006; Love and Gannon, 2010), satellite magnetic field measurements (Le et al., 2004; Jorgensen et al., 2004; Tsyganenko and Sitnov, 2007; Sitnov et al., 2008) and from global imaging of the ring current with energetic neutral atom (ENA) measurements (Roelof, 1987; Perez et al., 2000; Fok et al., 2003). (It should be noted that studies of Le et al. (2004) and Jorgensen et al. (2004) do not differentiate between main phase and recovery phase, binning the data in the amplitude of Dst index.) For a description of the basic characteristics of the ring current, we refer the reader to reviews by Gonzalez et al. (1994), Daglis et al. (1999), Ebihara and Ejiri (2003), Kozyra and Liemohn (2003), Ebihara (2016) as well as to Chapter 7 of this volume.

Ring current plasma is the main reservoir of kinetic energy in the magnetosphere and has a significant impact on the near-Earth space environment. The ring current population is normally considered separately from particles of the "radiation belts" with energies from a few hundred keV to MeV, which overlap the same region of space. An important difference between ring current and radiation belt populations is that the former substantially modify the storm-time magnetic field and hence require a self-consistent description. The latter is determined by their source region, as well as by the electromagnetic field, which is not affected by the radiation belt particles. The different names also emphasize different prevailing transport processes and sources/losses defined by particle type and energy. For example, charge-exchange losses are a very important mechanism for ring current ions, but it is absent for radiation belt electrons. Ring current and radiation belt populations are also called "trapped" radiation to distinguish them from solar energetic particles (SEPs) and galactic cosmic ray (GCR) radiation. SEPs and GCR radiation can penetrate the magnetosphere and if they do not have collisions, they exit as well.

Geomagnetic storms are a key component of space weather since many space weather effects intensify during geomagnetic storms, and related impacts could potentially affect Earth-based technological systems. The enhanced global magnetospheric electric field (the "convection electric field") transports the plasmasheet plasma from the magnetotail into the inner magnetosphere and forms the asymmetric ring current, deposits energy into the ionosphere-thermosphere system, causes atmospheric upflow and increases spacecraft drag, rearranges motion of ionospheric plasma, and contributes to the development of ionospheric irregularities. Other examples include the generation of intense and intermittent magnetospheric and ionospheric currents that are responsible for elevated geomagnetically induced currents (GICs) on the Earth's surface, and enhanced fluxes of ring current electrons and radiation belt electrons that are responsible for surface charging and deep dielectric charging of spacecraft. Many of these effects are described in detail elsewhere in this book.

In this Chapter, we concentrate on the magnetospheric portion of geomagnetic storm effects and provide descriptions of currently available 3D global modeling tools designed to reproduce the near-Earth space environment during geomagnetic storms. The outputs of these models could be used as input for ionosphere/thermosphere models (2D maps of ionospheric electric fields and precipitation), GIC models ($d\mathbf{B}/dt$ variations on the Earth's surface), or environment models for spacecraft charging (ring current or radiation belt electron fluxes at spacecraft locations). For example, an ionospheric module (see Fig. 1) calculates magnetic disturbances from magnetospheric and ionospheric currents at any given location on the Earth's surface. $d\mathbf{B}/dt$ variations serve as an input to GICs models that are essentially different forms of Faraday's law (e.g., Eq. 1 from Chapter 1 of this book). We also briefly review previously published results on modeling of intense/extreme geomagnetic storms, and provide an example simulation of the June 22–23, 2015, magnetic storm with minimum Dst = −204 nT. For this event, we show the distribution of pressure and current density in the magnetosphere, provide comparisons with spacecraft measurements, and present a data-model comparison of modeled magnetograms with ground-based observations. We analyze sources of observed $d\mathbf{B}/dt$ variations, try to identify the responsible magnetospheric processes, and emphasize the role of the ring current at low latitudes. We also discuss challenges for the future improvement of two-way coupled models.

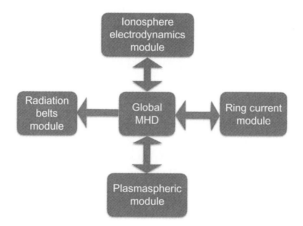

FIG. 1

An example showing coupling between different modules in global coupled model. *Single direction arrows* represent one-way coupling when the information is passed only from the MHD model to the given model. *Bidirectional arrows* represent two-way coupling where information is passed back and forth between the MHD model and a given module. Two-way coupling with an ionospheric module is normally present in all 3D models of magnetosphere. Ring current, radiation belt, and plasmasphere modules are optional. Ring current or plasmasphere modules can also be one-way coupled with an MHD code. In this chapter we simulate a geomagnetic storm with global code including coupling between all five modules, with ring current and ionosphere being two-way coupled and radiation belts and plasmasphere being one-way coupled.

2 OVERVIEW OF FIRST-PRINCIPLES MAGNETOSPHERIC MODELS

Global magnetospheric dynamics during geomagnetic activity are simulated with 3D magnetohydrodynamics (MHD) models of the Earth magnetosphere. The most commonly used codes are (i) the Block-Adaptive-Tree Solar wind Roe-type Upwind Scheme (BATS-R-US) code (Powell et al., 1999) as a part of the space weather modeling framework (SWMF) (Tóth et al., 2005, 2012); (ii) the Lyon-Fedder-Mobarry (LFM) code (Lyon et al., 2004); (iii) the OpenGGCM (Open Geospace General Circulation Model) code (Raeder et al., 2008); (iv) the GUMICS (Grand Unified Magnetosphere—Ionosphere Coupling Simulation) code (Laitinen et al., 2007); and (v) the REPPU (REProduce Plasma Universe) code (Tanaka, 2000; Ebihara et al., 2014). SW input for these models is taken from observations by NASA Global Geospace Science (GGS) Wind, NASA Advanced Composition Explorer (ACE) and launched in 2015 NOAA/NASA Deep Space Climate Observatory (DSCOVR) spacecraft. It is also important to mention many SW-magnetosphere-ionosphere studies rely on OMNI database at CDAWeb (https://cdaweb.sci.gsfc.nasa.gov/index.html/) where WIND and ACE data are available.

Some codes are part of larger frameworks and allow the use of output from first principles models of the 3D inner heliosphere as an input to the magnetospheric codes. It could be tempting to use the outputs from solar models as an input to magnetospheric models, but it is still difficult to produce the required input parameters (e.g., temporal profiles of magnetic field components, density, and SW velocity) with needed accuracy. An example of this can be seen for Halloween storm of October 29, 2003 (minimum Dst ~ -350 nT) with a so-called "Sun-to-thermosphere" model is described in detail by Tóth et al. (2007). The modeled temporal profiles of IMF, density, and velocity significantly differ from upstream SW data. As a result, the modeled Dst profile is also different from the observations. It is concluded that the current model cannot accurately reproduce the magnetic structure of the coronal mass ejection (CME) as seen by ACE, but can reproduce general hydrodynamic characteristics of the CME reasonably well. However, magnetic field profiles, B_z in particular, are of the primary importance for modeling geomagnetic storms, and this type of inputs is currently difficult to obtain. At the same time, Tóth et al. (2007) note the version of model that uses input from ACE data ("magnetosphere-to-thermosphere" model) performs reasonably well in reproducing the Dst profile and cross polar cap potential (CPCP) for the 2003 Halloween storm. We will therefore concentrate on terrestrial magnetospheric modeling with input from ACE/Wind monitors in this chapter. Modeling of SW input for global magnetospheric models from first principles solar models remains a challenge for the future.

A distinctive feature of magnetospheric MHD codes is the coupling with separate modules that represent different plasma domains in the terrestrial magnetosphere. This is dictated by the plasma characteristics in the different domains of the near-Earth space plasma environment (De Zeeuw et al., 2004; Ridley et al., 2004; Tóth et al., 2005; Fok et al., 2006; Zhang et al., 2007; Moore et al., 2008). For example, collisional plasma in the dayside ionospheric F-region has density $\sim 10^6$ cm^{-3} and temperature of a few electron volt, while the ring current plasma has density ~ 1 cm^{-3}, temperatures between 10 and 100 keV, and is essentially collisionless. Even the definition of temperature could be misleading in some regions, for example, at geosynchronous (GEO) orbit, since there is a mixture of plasmas with different characteristics: "cold" ionospheric, ring current, and radiation belt populations.

The MHD module could be coupled with different modules representing ionosphere (and thermosphere for some models), ring current, radiation belts, and plasmasphere as depicted in Fig. 1. The bi-directional arrow emphasizes that information can be passed back and forth between the MHD code and a given module, resulting in so-called "two-way coupling." Likewise, single direction arrows denote so-called "one-way coupling," whereby information is passed only from the MHD magnetospheric module and there is no feedback. Two-way coupling between the magnetosphere and ionosphere is a necessary element for global MHD codes. Two-way coupling between a MHD model and a RC module is nontrivial because it requires matching of the MHD solution with the RC solution in the 3D volume where the modeling region of the ring current model overlaps with MHD modeling region. In this approach it is assumed there is a region where both models are valid, so the boundary conditions for the ring current model are taken from MHD model. At the moment, the question of validity both of MHD models and ring current models in the transition zone has not been studied well and presents a challenge for the future work (see also Section 4.5). The ring current plasma alters the background magnetic field, generates field-aligned currents (FACs), alters the structure of magnetospheric electric fields, produces pressure build-up in the inner magnetosphere, and results in the observable depression of the Dst and SYM-H indices. Therefore, representation of the ring current in the model is critical for an accurate description of geomagnetic storms. The standard MHD approach fails in the Earth's inner magnetosphere, where the ring current flows. The main reasons are (i) the standard adiabatic equation of MHD state becomes inaccurate; (ii) the standard plasma frozen in condition assumed in MHD is violated. Near the Earth (distances inside of $\sim$7–8 R_E at the magnetic equator), the magnetic field strength and its gradients become substantial, giving rise to energy-dependent magnetic drifts. Single fluid approximations are no longer valid, and it is necessary to include RC module (De Zeeuw et al., 2004) with transport equations written in terms of bounce-averaged quantities for hundreds (or even thousands) of "fluids" distributed in energy, pitch-angle, or first and second adiabatic invariants. RC models can also incorporate losses of particles in the loss cone, from charge exchange, and from Coulomb interactions (see review by Ebihara and Ejiri, 2003 and references therein).

Ring current models are initially designed to use empirical models of the electromagnetic fields and plasma boundary conditions, the so-called "stand-alone ring current models." There has been a successful history of different stand-alone RC models in the community: the rice convection model (RCM) (Harel et al., 1981; Toffoletto et al., 2003); the magnetic equilibrium version of RCM, RCM-E (Lemon et al., 2003; Chen et al., 2015); the Magnetospheric Specification Model (MSM) (Wang et al., 2003, 2004); the Comprehensive Ring Current Model (CRCM) (Fok et al., 1993, 1995, 2001), or its successor, the Comprehensive Inner Magnetosphere-Ionosphere model (CIMI) (Fok et al., 2014); the Ring current-Atmosphere interactions Model (RAM) (Jordanova et al., 1994, 2006) and the version with a self-consistent magnetic equilibrium solver RAM-SCB (Zaharia et al., 2006; Zaharia, 2008); and the Hot Electron and Ion Drift Integrator (HEIDI) code (Liemohn et al., 1999). These RC models have been proven to be very useful in describing many observable features in the inner magnetosphere. Numerous simulation studies together with data analyses have been shown to reproduce observed magnetic depression measured by Dst, ring current fluxes, ENA emissions from the ring current, plasmapause changes, and many other effects related to the ring current (Ebihara and Ejiri, 1998; Jordanova et al., 2006; Fok et al., 2001, 2003; Ebihara et al., 2004, 2005; Lemon et al., 2003; Buzulukova et al., 2008, 2010a; Chen et al., 2015).

Realistic descriptions of electromagnetic fields in the simulation domain have been a general challenge for RC models. Since stand-alone RC models use empirical models for all input parameters, simulations are not fully self-consistent. In addition, the simulation domain for RC models is limited in space and does not include magnetotail or the open field line regions. In additional, it is problematic to describe substorm activity (Tsurutani et al., 2015; Hajra et al., 2016) with empirical models of magnetic and electric field, as well as models of plasma density and temperature, required for stand-alone

RC models. (For completeness we note there is ongoing development of substorm magnetic field empirical model, see e.g., Stephens et al. (2017).) Therefore, much effort has been devoted to the development of coupled MHD-RC models. The latest versions of coupled MHD and ring current models have demonstrated their capabilities to qualitatively describe dynamics of the Earth's global and inner magnetosphere with reasonable agreement with ion fluxes observed by Time History of Events and Macroscale Interactions during Substorms (THEMIS) mission, the magnetic field measured at geostationary orbit (GEO) by Geostationary Operational Environmental Satellite system (GOES), and ENA maps from Two Wide-angle Imaging Neutral-atom Spectrometers (TWINS) mission (Tóth et al., 2012; Pembroke et al., 2012; Glocer et al., 2013; Meng et al., 2013; Cramer et al., 2017).

There is also a class of models that use a combination of data assimilation techniques and physics-based models. This approach takes advantages of available datasets from Geospace missions, and uses the data to correct a physics-based solution. Examples include data assimilation of ring current fluxes in order to correct ring current model outputs (Godinez et al., 2016), and data assimilation of low-altitude magnetic perturbations in order to correct MHD solution (Merkin et al., 2016). These models are being actively developed now as an alternative to pure empirical or physics-based models.

3 MODELING OF EXTREME AND INTENSE GEOMAGNETIC STORMS

The developments of extreme magnetic storms have been studied both with coupled ring current models (one-way or two-way) and in stand-alone RC models. Kozyra et al. (1998) examined the ring current buildup and decay during the great storm on February 6–10, 1986 (minimum Dst ~ -300 nT). They simulated the ring current ion distributions with the stand-alone model of Fok et al. (1995), which considered losses from charge exchange and Coulomb collisions along particle drifts. Comparing the calculated total ring current energy content with the Dst index, the model failed to reproduce the initial rapid recovery observed in Dst. On the other hand, intense ion precipitation was observed by LEO NOAA-6 and DMSP satellites during the recovery phase. With the inclusion of ion precipitation losses (additional pitch-angle scattering into the loss cone) estimated by NOAA and DMSP data, the model produces excellent agreement with the observed ring current energy content during the entire recovery phase, indicating that ion precipitation loss could be important during major storms.

Another example is the great storm on August 12, 2000 with minimum Dst ~ -200 nT which has been studied both with the stand-alone CRCM (Fok et al., 2003; Ebihara and Fok, 2004) and with one-way coupled CRCM—BATS-R-US MHD (Buzulukova et al., 2010b). The IMAGE satellite has provided global ENA images of the ring current throughout the storm. The ENA images observed an asymmetric ring current during the main phase and the gradual evolution to a symmetric ring current in the recovery phase. An outstanding feature revealed by the ENA images was that during the storm's main phase, the ENA flux was peaked in the postmidnight sector near dawn rather than around dusk as was thought previously (Ebihara and Ejiri, 2003). This flux enhancement at dawn was measured over a broad energy range, up to 200 keV ions. Brandt et al. (2002) examined observations made by IMAGE high energy neutral atom (HENA) instrument during several storms. They interpreted the shifting of ring current fluxes was a result of distortions in the global electric field caused by the ring current plasma-induced electric field (the so-called eastward-skewed electric field). Ebihara et al. (2004) performed an in-depth study of the postmidnight storm-time enhancement of the ring current by comparing HENA data with simulations from CRCM. They tested several mechanisms and found that a shielding electric field produced by the Region 2 Birkeland (ionospheric field-aligned) currents from ring current plasma was the dominant contributor to the eastward skew to the electric field. Disagreements between ENA flux observations peaking in the postmidnight sector and maps of magnetic disturbances (e.g., Le et al., 2004; Jorgensen et al., 2004; Shi et al., 2006; Love and Gannon, 2010) peaking in the dusk sector has been explained by Buzulukova et al. (2010a). It has been shown the H^+–H charge-exchange cross-sections abruptly decreases in the H^+ energy range 10–100 keV, therefore the part of the ring current plasma that is responsible for the most bright ENA emissions peaks in the postmidnight sector (with energies <100 keV) while the part of the ring current that carries the most pressure and magnetic disturbances (with energies >100 keV) is located in the dusk sector. The aforementioned discrepancy between ENA fluxes and plasma fluxes (and also pressure) holds only for neutral H and H^+ ENA fluxes and is not important for O and O^+ ENA fluxes because corresponding charge-exchange cross-section (for loss of O^+) does not decrease for energies between 10–100 keV. For completeness, we note works of Le et al. (2004) and Jorgensen et al. (2004) do not differentiate between storm main phase and recovery phase. In contrast, the work of Sitnov et al. (2008) demonstrates the existence of a postmidnight current density enhancement in the empirical model of magnetic field (Tsyganenko and Sitnov, 2007) during the main phase of the April 22, 2001 geomagnetic storm. At the same time, Sitnov et al. (2008) report a magnetic field depression located in the dusk sector. It is hard to compare directly the results of different studies because Buzulukova et al. (2010a) do not present ring current density distribution and Sitnov et al. (2008) do not present ring current pressure distribution. The comparison between the empirical model of Tsyganenko and Sitnov (2007) and the first-principles model for recent intense storms will be discussed in Section 4.2.

Another example is the intense storm of November 20, 2003 with the Dst index attaining a minimum value of -422 nT at 21:00 Universal Time (UT). NOAA 17 observed a deep penetration of ring current ions to $L = 2\, R_E$ at the peak of the storm and fast recovery to higher L shells during the early recovery phase. Ebihara et al. (2005) simulated the development of this storm using CRCM and compared results with observations. They found that both the observed and simulated Region 2 field-aligned currents (FAC) showed signatures of multiple current sheets. This complicated FAC distribution could be modeled by the fluctuating convection strength and particle density in the nightside source region during the storm. However, the CRCM did not reproduce the fast ion losses at low L-shells as seen by NOAA 17. Higher geocoronal densities or additional pitch-angle diffusion was needed in order to reproduce the rapid initial recovery. Fok et al. (2011) revisited this extreme event, but with a different simulation tool (the one-way coupled CRCM—LFM MHD model). Instead of using empirical models for the magnetic field and plasma sheet distribution as in Ebihara et al. (2005), the global magnetic configurations from the LFM MHD code were applied and the distribution at the CRCM boundary was established by test-particle tracing of ion trajectories from their source regions. With the elevated O^+/H^+ energy density ratio at the inner plasma sheet during this super storm, Fok et al. (2011) successfully reproduced the recovery of the Dst index. It is well-known that the ring current composition including O^+/H^+ energy density ratio varies with geomagnetic activity, to be $\sim 3/7$ for small/moderate storms, and ~ 2 for intense storms (see Table 1 from Daglis et al., 1999). Since charge-exchange losses are greater for O^+ than for H^+ (different dependence from energy in the critical energy range 10–200 keV), it could be indicative of a different channel for energy losses during intense storms.

Since the successful launch of the four Magnetospheric Multiscale (MMS) spacecraft in March 2015, we have a unique combination of several multispacecraft missions to study the near-Earth space environment, namely, MMS mission, Van Allen Probes mission, THEMIS mission, TWINS mission, Swarm mission, Defense Meteorological Satellite Program (DMSP), GOES, and Polar Operational Environmental Satellites (POES) spacecraft. Together with the measurements of FACs provided by the Active Magnetosphere and Planetary Electrodynamics Response Experiment (AMPERE) and ground-based magnetometer data sets from the INTERMAGNET and SuperMAG projects, there is a unique opportunity to understand the fundamental processes behind geomagnetic activities on different spatial and temporal scales. The two biggest geomagnetic storms of Solar Cycle 24 have minimum Dst ~ -200 nT in March and June of 2015, and both are excellent events for modeling. Le et al. (2016) analyzed the data from Swarm, AMPERE, DMSP, MMS, and GOES observations for March 2015 storm, and found the magnetopause location is closer to Earth than what would have been expected based on the pressure balance. The June 22–23 storm has been studied with the two-way coupled BATS-R-US and RCM global model and compared with observations from MMS, Van Allen Probes, AMPERE, and DMSP spacecraft (Reiff et al., 2016). Two substorms have been identified on June 23 (Nakamura et al., 2016), and the coupled model was able to reproduce B field variations comparable with the data from MMS 1 and Van Allen Probes A s/c. The model also showed that around 03 UT of June 23, the MMS 1 was on very stretched magnetic field lines, close to separatrix between opened and closed field lines. In addition, the coupled model reproduces electric convection field structures as seen by DMSP and magnitude of FACs as seen by AMPERE.

3.1 MODELING OF CARRINGTON-TYPE EVENTS

There are a few attempts to analyze the behavior of the magnetosphere for a Carrington-type geomagnetic storm. Although there were no direct observations of the SW during the Carrington 1859 (Carrington, 1859) magnetic storm, Tsurutani et al. (2003) was able to infer many parameters from ancillary information. From Carrington (1859) statements of the flare time and the time of the concomitant magnetic storm, Tsurutani et al. (2003) assuming standard deceleration of the SW obtained a speed of ~ 1850 km/s at Earth. Using an empirical relationship between ICME speed at 1 AU and the magnetic cloud (MC) magnetic field magnitude (Gonzalez et al., 1994), Tsurutani et al. (2003) derived an MC magnetic field strength of 87 nT. Assuming the MC field was directed entirely southward, Tsurutani et al. derived a storm intensity of Dst $= -1760$ nT.

In Chapter 7 of this volume, Lakhina and Tsurutani assumed a slow SW density of 5 cm^{-3} and a shock compression factor of 4 (Kennel et al., 1985), to derive a sheath density ~ 20 cm^{-3}. This plus an SW speed of ~ 1850 km/s can explain the Carrington SI$^+$ intensity of ~ 110 nT. Li et al. (2006) conjectured that the Carrington event had the following parameters: SW velocity ~ 2000 km/s; B_z in the MC of -80 nT and a large density of ~ 2000 cm^{-3} sunward of the cloud. This is now thought to be the ICME filament (Chapter 7 of this volume). These estimations were made with the model of Dst index proposed by Temerin and Li (2002). The maximal density ~ 2000 cm^{-3} was needed to reproduce the fast recovery observed by Colaba magnetometer. The highest SW density in the records is reported by Burlaga et al. (1998) and Crooker et al. (2000) to be 185 cm^{-3} for the filament structure of January 11, 1997.

Manchester et al. (2006) used a 3D MHD model of the sheath and MC and found the following parameters for 1 AU: SW velocity ~ 1500 km/s, $B_z \sim -200$ nT, and density ~ 800 cm^{-3} at the sunward edge of the compressed MC. Ngwira et al. (2014) used BATS-R-US MHD code coupled with RCM ring current model-to-model Carrington event and study magnetospheric configuration and response of GICs. For the event, the following ad hoc SW parameters were used based on the results of Li et al.

(2006) and Manchester et al. (2006): minimal IMF $B_z \sim -220$ nT; maximal SW velocity ~ -2000 km/s, and maximal plasma density ~ 400 cm^{-3}.

One can estimate maximal value of SW plasma density in the sheath before MC arrival. Here we follow the same argumentation presented in Chapter 7 of this volume. If we assume quiet SW has density ~ 5 cm^{-3}, then according to Kennel et al. (1985), the maximal density jump across ICME shock has a factor of 4. Thus one comes up with a maximal density of about 20 cm^{-3}. If there are multiple CMEs and multiple shocks located close to each other, one might see a factor of 4×4 or $4 \times 4 \times 4$, for two or three shocks going one after another. Therefore one might expect maximal value for the density to be equal ~ 320 cm^{-3}. Three consequent shocks were indeed observed in the past (the case of June 22–23, 2015 [Fig. 2] and also the November 7–8 superstorm, Tsurutani et al., 2008), with maximal sheath SW density ~ 80 cm^{-3}. However, as noted in Chapter 7 of this volume, the Colaba magnetometer record showed only one SI$^+$ (one shock) so their estimations is that the sheath density for Carrington event was ~ 20 cm^{-3}.

It seems extreme values of SW density are not needed to produce a storm of the same intensity as Carrington event (Baker et al., 2013). This work demonstrated that the big ICME of July 23–24, 2015, which missed the Earth, could possibly lead to the extreme, Carrington type geomagnetic storm with minimum Dst ~ -1200 nT, if hit the Earth at right time. Plasma density is one of the important parameters defining location of the magnetopause and shape of the magnetosphere. Magnetospheric plasma density is also one of the inputs for the ring current models. Therefore it is important to understand the changes of SW density during extreme events, in order to be used as an input for the global MHD models or ionosphere-thermospheric models. (For modeling of Carrington-like event with a thermosphere-ionosphere model please see Chapter 21 of this volume.)

3.2 MODELING OF RADIATION BELT RESPONSE FOR EXTREME STORMS

The response of radiation belts to 2003 Halloween superstorm (minimum Dst ~ -400 nT) (Baker et al., 2004) has been extensively studied. For example, modeling the storm with Versatile Electron Radiation Belt (VERB) 3D code (Subbotin et al., 2010) showed fair agreement with the observations (Shprits et al., 2011). Deep injection of MeV electrons during the storm's main phase was reproduced from strong radial diffusion. Furthermore, rapid enhancement of MeV electrons during the storm recovery phase has been predicted from local acceleration due to whistler mode chorus waves. VERB-3D code also simulated the Bastille Day storm (minimum Dst ~ -300 nT) in July 2000 (Kim et al., 2016). The VERB-3D code extrapolated the parameterization of radial diffusion coefficients (Brautigam and Albert, 2000) beyond $Kp > 6$ and reproduced deep penetration of MeV electrons down to $L = 2R_E$. These results showed that plasmaspheric hiss played an important role to form the slot region and split structure of the radiation belts during the recovery phase of the storm.

However, extreme events in the radiation belts present unique challenges for models and predictions. Most radiation belt models solve the Fokker-Plank equation in terms of the first, second, and third adiabatic invariants (Schulz and Lanzerotti, 1974), assuming linear or quasilinear wave-particle interactions. The models thus require diffusion coefficients for each variable and the statistical distribution or a parameterized proxy of plasma waves (Fok et al., 2008; Albert et al., 2009; Subbotin et al., 2011; Tu et al., 2013; Glauert et al., 2014). During extreme events, geomagnetic, and wave activity are far from those statistically derived, which can result in discrepancies between model results and observations.

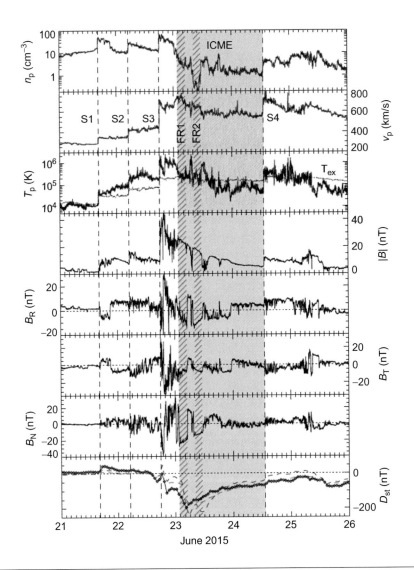

FIG. 2

Solar wind origins for geomagnetic storm of June 22–23, 2015. The *shaded* region is the interval of interplanetary CME. Two *hatched* regions are supposedly flux ropes. There are three shocks (S1, S2, S3) detected before ICME arrival. In the *bottom panel*, Dst index estimations also are shown from equations of Burton et al. (1975) (*blue-dashed line*) and O'Brien and McPherron (2000) (*red-dashed line*).

(From Liu, Y.D., Hu, H., Wang, R., Yang, Z., Zhu, B., Liu, Y.A., Luhmann, J.G., Richardson, J.D., 2015. Plasma and magnetic field characteristics of solar coronal mass ejections in relation to geomagnetic storm intensity and variability. Astrophys. J. Lett. 809, L34.)

4 AN EXAMPLE: GEOMAGNETIC STORM OF JUNE 22–23, 2015

In this section we present the model results of June 22–23, 2015 geomagnetic storm. The simulation was performed using the global BATS-R-US MHD model coupled with the CIMI ring current model to adequately describe the inner magnetosphere. We present comparisons with the magnetic field observations from MMS1, THEMIS-A, and the ground-based observatories. We discuss the global distribution of the currents and pressure, and make connections between SW structures and signatures in $d\mathbf{B}/dt$ variations.

At ~18:40 UT of June 22, 2015, an interplanetary CME (ICME) hit the Earth, resulting in the development of an intense geomagnetic storm. The NOAA Space Weather Prediction Center (http://www.swpc.noaa.gov/) issued a G4 (severe) alert for the storm level at 18:58 UT on June 22. The Dst index reached a minimum value − 204 nT the next day, June 23 around 05:00 UT. This geomagnetic storm together with another similar event on March 17, 2015 (with minimum Dst of − 223 nT) have been two of the biggest geomagnetic storms of the Solar Cycle 24 (as of July, 2017). SW origins for the June storm as observed by WIND are shown in Fig. 2 (Fig. 4 from Liu et al., 2015).

There were a few CME structures detected between 18–21 June, with the bigger driver occurring during the last one, which occurred on June 21, 2015, with detected CME speed of 1300 km/s, and a predicted arrival time at WIND (1.02 AU from the Sun) around 17:00 UT of June 22 (Liu et al., 2015). The corresponding interplanetary CME and its preceding shock hit the Earth on June 22 around 18:40 UT, with just a 1 h delay from the prediction based on the empirical model of Gopalswamy et al. (2000). However, the arrival model is not able to predict observable magnetic field variations, suggestive of the presence of the current sheet and flux rope formation (Liu et al., 2015). As we will show later, these fluctuations ~40 nT translate into a substantial $d\mathbf{B}/dt$ signal on the Earth. In addition, there are also signals from Sudden Impulse (SI$^+$) and signatures from intense substorms detected by ground-based magnetometers, that translate into intense $d\mathbf{B}/dt$ variations. The $d\mathbf{B}/dt$ variations from this storm could be classified as "significant" (values on the order of 10^2 nT/min are considered to be significant and 10^3 nT/min to be extreme, see Chapter 1 of this volume).

In Chapter 7 of this volume, a superstorm is defined as an event where Dst index drops below −500 nT. The value of Dst index could be translated into occurrence rate for magnetic storms with Dst index below some threshold. Chapters 4 and 5 of this volume present the statistical modeling of occurrence rate as a function of Dst index threshold. Therefore, one can define "extreme geomagnetic storm" based on this approach, that is, in terms of Dst threshold or corresponding occurrence rate. However, if the requirement for definition of "extremes" is extreme effects and impacts, then the Dst threshold is not only one factor needed to be considered. Examples are presented through different chapters of the book. The case of the June 22–23 storm is a good example. The storm of this intensity cannot be called "extreme geomagnetic storm" on the basis of occurrence. However, as demonstrated in Chapter 24 of this volume, geomagnetic storms of this size (and weaker) can lead to disruption of navigation systems. While occurrence of extreme storms could be a sufficient condition for many adverse effects to occur, it is not a necessary condition at the same time. It seems necessary and sufficient conditions can vary from field to field, and require additional elaboration.

4.1 THE MODEL RUN SETUP

For this study, we use the anisotropic version of the global MHD BATS-R-US code (Meng et al., 2012). The code is configured to solve the ideal anisotropic MHD equations on a Cartesian grid for the Earth's magnetosphere. BATS-R-US is a part of SWMF (Tóth et al., 2005, 2012), and in this particular configuration, it is two-way coupled CIMI model (Fok et al., 2014). The BATS-R-US model is also coupled in the two-way mode with the Ionospheric Electrodynamics (IE) module (Ahn et al., 1983; Ridley et al., 2004). The IE module solves the equation for the ionospheric potential at the ionosphere surface, with FAC mapped from the inner MHD boundary. The inner boundary is a sphere of 2.5 R_E centered on the Earth. The model resolution in the inner magnetosphere and inner tail region is chosen to be 1/8 R_E, with the total number of cells being $\sim 1.2 \times 10^7$. The 3D computational domain is a box with defined in Geocentric Solar Magnetospheric (GSM) coordinates from -224 to 32 R_E in the X-direction, and from -64 to 64 R_E in both Y- and Z-directions. The grid resolution varies from 1/8 R_E in the inner magnetosphere to 4 R_E near the outer edges of the simulation domain. To be inside the ring current module, MHD field line should satisfy the following two conditions: (i) the line should be closed, and (ii) intersection of the line with the geomagnetic equator (GSM) should be inside 9 R_E. The value 9 R_E is defined empirically, in order to get the majority of the ring current inside the simulation domain. In order to eliminate any numerical effects related to nonzero $\nabla \cdot \mathbf{B}$ at the boundary, the IMF B_x component is taken to be constant and equal to 5 nT.

CIMI is the combination of the CRCM model (Fok and Moore, 1997; Fok et al., 1996, 2001), the Radiation Belt Environment (RBE) model and a cold plasmasphere model. It solves the bounce-averaged Boltzmann equation to calculate non-Maxwellian, anisotropic distribution functions of H^+ and O^+ ions and electrons, within energy range 0.1–300 keV (ions) and 1–3000 keV (electrons). It is essential to use the anisotropic version of the MHD (Meng et al., 2013) to correctly couple the MHD code with CIMI model, since plasma drifts in the inner magnetosphere generate anisotropy, and the CIMI model is anisotropic. For the anisotropic plasma distribution to be consistent with the magnetic field line, anisotropic MHD is required. The reason is that isotropic MHD requires constant pressure along closed field lines for an equilibrium solution, but anisotropic MHD does not. Moreover, the pressure along the field line in CIMI will not be constant. In the extreme case of a nearly equatorially mirroring population, for example, the pressure will be much higher at the equator than elsewhere.

CIMI has capabilities to calculate fluxes of both the ring current and radiation belt populations. As demonstrated in Chapter 15, the dynamics of the electron population with energies 10–40 keV is a key parameter for determining the level of surface dielectric charging, while more energetic ions and electrons are responsible for deep dielectric charging effects (Chapter 16). While we do not address dynamics of electron fluxes in this chapter, two-way coupled models are good candidates to understand, specify, and predict electron fluxes from keV to MeV energies for a given satellite orbit. The model output could be used as input for various models of radiation environment to develop, quantify, and possibly mitigate charging effects.

4.2 THE RESULTS FOR PRESSURE AND CURRENT DISTRIBUTION

Fig. 3 depicts a meridional cut through the global 3D simulation domain showing the results for the total current density at 03:00 UT July 23, 2015 near the storm peak. The main magnetospheric current systems presented here are the asymmetric ring current, tail current, magnetopause current, and FACs (more visible in the regions of converging magnetic field). The gray line is the projection of an additional plane through the approximate geomagnetic equator in the inner magnetosphere, where inner magnetospheric current and pressure attain maximum values. Current density is shown in $\mu A/m^2$, and the *XYZ* axes in the left bottom corner shows the orientation of the GSM coordinate system.

4 AN EXAMPLE: GEOMAGNETIC STORM OF JUNE 22–23, 2015

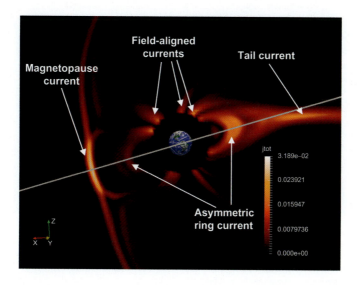

FIG. 3

A meridional cut through the global 3D simulation domain showing the results for the total current density at 03:00 UT July 23, 2015 near the storm peak. The main magnetospheric current systems are presented: the asymmetric ring current, tail current, magnetopause current, and field-aligned currents (more visible in the regions of converging magnetic field). The gray line is the projection of a plane through the approximate geomagnetic equator in the inner magnetosphere. Current density is shown in µA/m^2, and the XY Z axis in the *left bottom corner* shows the orientation of the GSM coordinate system. The inner region with $r < 2.5\ R_E$ is outside the simulation domain.

The inner region with $r < 2.5\ R_E$ is outside the simulation domain. Fig. 4 shows the distribution of total current density **J** (panel A) and pressure P (panel b) through the equatorial plane from Fig. 3. (As the model is anisotropic, pressure is measured as 1/3 of the trace of the pressure tensor.) The pressure peak is located in the dusk sector, in agreement with previous studies. The maximum pressure value is ∼60 nPa, and the pressure peak is located at ∼ 4 R_E. The maximum current density in the ring current region is 0.02 µA/m^2. It is larger than the maximum current density in the tail region, but is smaller than the maximum magnetopause current density. The peak in the current density does not coincide with the plasma pressure peak which is consistent with the equation of magnetostatic equilibrium $\mathbf{J} \times \mathbf{B} = \nabla P$. Both plasma pressure and current density reach their maximum in the dusk sector.

It is interesting to compare the results of our simulations with the results of the empirical Tsyganenko model TS07D (Tsyganenko and Sitnov, 2007) for another storm of similar intensity, namely March 17, 2015 event with minimum Dst of −222 nT (see Chapter 11 of this volume). Namely, we can compare distributions of current density (Fig. 2 from Chapter 11 of this volume) and pressure (Fig. 3 from Chapter 11 of this volume) in the inner magnetosphere with our results. The maximum current density from TS07D varies from ∼0.01 to 0.034 µA/m^2 through the main phase of the March 2015 storm, with global distribution comparable to output from the coupled model. The maximal pressure value from TS07D varies from 116 to 164 nPa. The value of 164 nPa belongs to the interval where the SW and IMF parameters had to be interpolated whereas the actual data was missing. Therefore the

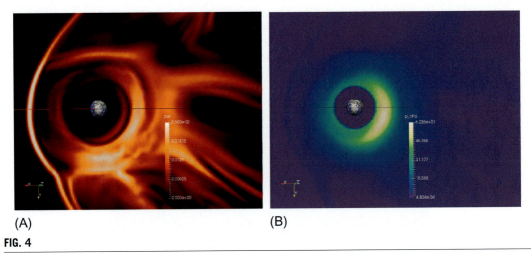

(A) **(B)**

FIG. 4

The distribution of total current density **J** (A) and pressure P (B) through the equatorial plane from Fig. 3. Current density is shown in $\mu A/m^2$, and pressure is in nPa.

value 116 nPa can be used as a reference number. Taking into account that this comparison is done for two different events, the agreement is very reasonable.

4.3 COMPARISON WITH GROUND-BASED MAGNETOMETERS AND SATELLITE DATA

In recent years, the community has been actively trying to understand how accurately global models can describe geomagnetic variations using statistical approaches (Pulkkinen et al., 2013; Glocer et al., 2016). In this study we analyze one event, and identify the responsible magnetospheric processes.

Fig. 5 shows a comparison between observations of geomagnetic observatories (SuperMAG, Gjerloev, 2012) and the model results. In addition to the output from the global coupled model, a second run with the BATS-R-US MHD model without a ring current module was performed for the same event, to highlight differences introduced by the ring current. The green lines show the results for the two-way coupled model, blue lines are MHD-only results, and black lines show data for the northward component B_n of magnetic field variation and corresponding dB_n/dt for three geomagnetic observatories, Fort Churchill (FCC), Barrow (BRW), and Stennis Space Center (BSL). FCC and BRW stations are located at high latitudes, and BSL station chosen to show variations at low latitudes. There are four intervals of interest highlighted in the figure. For the time interval between 18:30 and 21:00 UT on June 22 (Interval 1), there are intense variations in both B_n and dB_n/dt. The data from BSL show signature of SI^+ and follow trends in the Dst index. FCC observatory measured $dB_n/dt > 400$ nT/min in magnitude, the most intense variations for the event under consideration. The intense dB_n/dt are observed not only during SI^+, but also later during the storm's main phase. For the case 2, the BRW observatory observed intense variations in B_n and dB_n/dt between 01:00 and 02:00 UT June 23. Intervals 2 and 3 include signatures from two substorms identified in Nakamura et al. (2016), namely, with onset occurring on June 23 at 03:16 and 05:09 UT. We note the contribution from GICs (induction effects) is not included in the model, therefore limiting direct comparison between the model and observations.

4 AN EXAMPLE: GEOMAGNETIC STORM OF JUNE 22–23, 2015

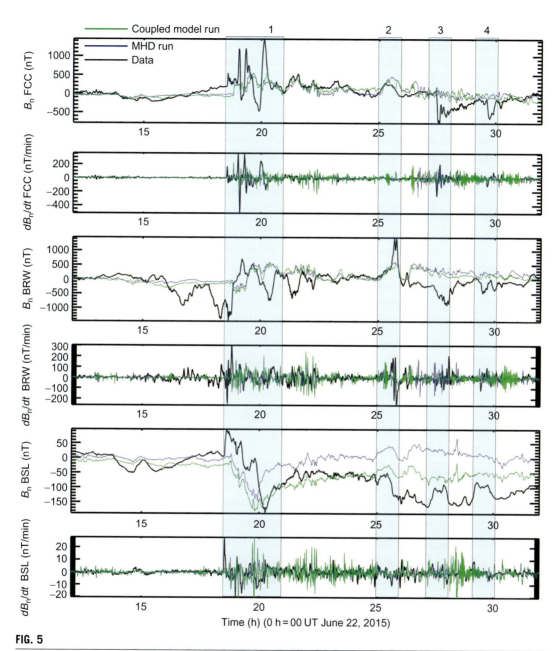

FIG. 5

Comparison between observations (*black*), two-way coupled model (*green*), and MHD (*blue*) for northward component B_n of magnetic field variations and corresponding dB_n/dt for three geomagnetic observatories, Fort Churchill (FCC), Barrow (BRW), and Stennis Space Center (BSL). Corresponding geomagnetic latitude/longitude are 67.78 degrees/328.93 degrees (FCC); 69.72 degrees/246.97 degrees (BRW); and 39.87 degrees/340.06 degrees (BSL). The data are shown for the time interval beginning at June 22, 2015 at 12:00 UT to 08:00 UT on June 23. All data are sampled at 1 min resolution.

(The reader is referred to Chapters 1, 8, and 9 of this volume for the basic GICs theory). The model reproduces general trends of B_n variations, but fails to describe small-scale (substorm-scale) variation. It seems this is a general challenge for global models, as also acknowledged in Chapter 8 of this volume (Fig. 4). Nevertheless, comparison between model results and data leads to a few interesting conclusions. It is generally acknowledged that global models cannot reproduce fast fluctuations of the observed magnetic field (e.g., Chapter 8 of this volume). Our results show that it is partially true, and magnetic field profiles are different between observations and the model, especially for high latitude stations. However, calculated dB_n/dt variations are comparable with observations. Fluctuations scale well with the level of activity: before the start of activity (12:00 to 15:00 UT on June 22), observed and calculated dB_n/dt are low, being comparable during the storm's main phase. Another interesting conclusion follows from comparison of BSL dB_n/dt for two runs with different model configuration. At low latitudes, the run with the ring current module produces higher fluctuations in B_N then the normal MHD run. These results emphasize the role of the ring current for $d\mathbf{B}/dt$ variations for low latitude stations.

Fig. 6 shows the comparison between MMS1 magnetic field and model results for two runs. Data are shown for the time period from June 22, 18:00 UT to June 23, 08:00 UT. Results from two-way coupled model simulation and MHD-only run are shown by solid gray lines and dotted gray lines, correspondingly. Two panels show the results for GSM B_z components and $|\mathbf{B}|$. The four intervals as previously noted in previous figures are also marked here. A sharp increase in magnitude of $\mathbf{B}$ at ~18:40 UT on June 22 points out the shock's arrival at Earth. Both model and data demonstrate substantial fluctuations in the magnetic field after that time. MMS was in the tail, moving toward the inner magnetosphere. The interesting wave-like structures can be seen in the plots of B_z around Intervals 3 and 4, where the B_x component approaches 0 and increases thereafter. There are also intervals when the magnetic field increases, with the field changing configuration from more stretched (small B_z) to more

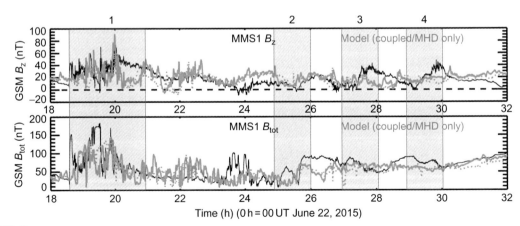

FIG. 6

Comparison between MMS1 observations (*black*), two-way coupled model (*gray solid line*), and MHD (*gray dotted line*) for B_z GSM and $|\mathbf{B}|$. The data are shown for time interval from June 22, 2015 18:00 UT to 08:00 UT on June 23, 2015. MMS1 location for the interval 1 was (GSM, R_E): $X \sim [-8;-5.5]$, $Y \sim 7.5$, $Z \sim [1.5;2]$; for the interval 2: $X \sim -10$, $Y \sim 6$, $Z \sim 1$; for the interval 3: $X \sim -10$, $Y \sim 5$, $Z \sim [0.3;0.6]$; for the interval 4: $X \sim -10$, $Y \sim 4$, $Z \sim [0;-0.3]$.

4 AN EXAMPLE: GEOMAGNETIC STORM OF JUNE 22–23, 2015

dipole-like (increased B_z). Both Intervals 3 and 4 are identified as magnetospheric substorms with onsets happening at 03:16 and 05:09 UT, when B_z is close to 0, with subsequent dipolarizations (Nakamura et al., 2016). The simulations also demonstrate formation of multiple dipolarizations which are smaller in size, although the modeled structures do not coincide exactly with the substorm dipolarizations. It is important to note that the model is able to predict magnetic field profiles with very reasonable accuracy. Since model data are compared along the orbit of MMS1, this comparison demonstrates that the model is doing a good job reproducing the magnetic field structures in the tail.

Fig. 7 displays (from top to bottom): a comparison between THEMIS-A GSM B_z and simulations in the same format as in Fig. 6; modeled total current density along THEMIS-A's orbit; SW B_z from

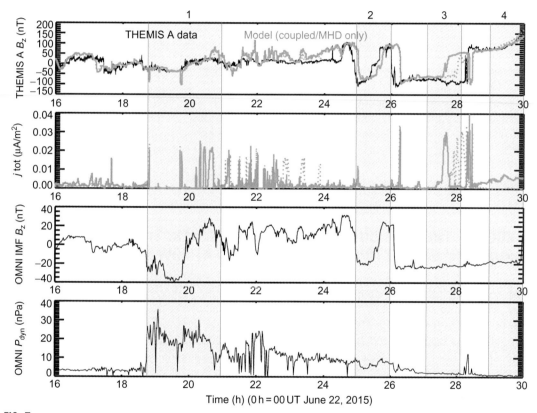

FIG. 7

From *top to bottom*: comparison between THEMIS-A B_z observations (*black*), two-way coupled model (*gray solid line*), and MHD (*gray dotted line*); modeled total current density |**J**| calculated for THEMIS-A orbit with two-way coupled model (*gray solid line*) and MHD (*gray dotted line*); OMNI B_z magnetic field plot; OMNI SW dynamic pressure plot. The data are shown for time interval June 22, 2015 at 16:00 UT to June 23, 06:00 UT. THEMIS A location for the interval 1 was (GSM, R_E): $X \sim [9.5;10]$, $Y \sim [-7;-6]$, $Z \sim [-2.5;-1.5]$; for the interval 2: $X \sim 9$, $Y \sim -3$, $Z \sim [-2.5]$; for the interval 3: $X \sim 8$, $Y \sim [-2;-1]$, $Z \sim -2$; for the interval 4. $X \sim [5,6.6]$, $Y \sim [-0.2;0.7]$, $Z \sim [-1.5]$.

OMNI which is used as an input to the model; SW dynamic pressure, also from OMNI. The agreement between the data and the simulation is very good. THEMIS-A was at the dayside, near the magnetopause where strong current density is calculated in the model, and the magnetic field observed by THEMIS-A is compressed SW magnetic field. Near 04:00 UT on June 23, THEMIS-A enters the inner magnetosphere, and the OMNI B_z profile and THEMIS-A B_z profile begin to diverge, and show the modeled spacecraft crossing the magnetopause current layer. The aforementioned time intervals are again highlighted here. A very interesting observation is a change in B_z from southward to northward for the Interval 2, when BRW station observed $dB_n/dt \sim 200$ nT/min in magnitude. Changes in the SW B_z component are about -40 nT, but near the magnetopause these changes are ~-150 nT for the model and ~-200 nT for the THEMIS-A B_z. Dynamic pressure does not have sharp changes during Interval 2 and stays in the range 5–10 nPa. Therefore, we explain BRW dB_n/dt observations by variation of SW magnetic field. In particular, changes to B_z from southward to northward will cause a rearranging of magnetopause and ionospheric currents. Therefore, it is likely the change of SW IMF orientation that caused the appearance of a strong signal in dB_n/dt for the BRW station. For Interval 1, we observe large variations (an order of magnitude) in dynamic pressure near 21:00 UT, that coincide with the most pronounced variations in dB_n/dt for FCC, ~400 nT/min in amplitude. For Intervals 3 and 4, we observe constant southward B_z, constant dynamic pressure, and conclude the sources of variations in $d\mathbf{B}/dt$ for these intervals are magnetospheric substorms.

Comparison between spacecraft magnetic field observations and the model results gives very reasonable agreement, while magnetic field profiles for ground-based stations differ in the model substantially, especially at high latitudes. This suggests that some important physical mechanism(s) are still missing.

5 ROLE OF THE RING CURRENT PLASMA IN GENERATION OF $d\mathbf{B}/dt$ AND VARIABILITY OF ELECTRIC FIELDS AND FACS AT LOW LATITUDES

To understand why the two-way coupled simulation generates additional variations in dB_n/dt versus those seen in the MHD-only simulation, let us consider the idealized case with a southward to northward and back to southward B_z turning simulated by Buzulukova et al. (2010b) which utilized the one-way coupled BATS-R-US-CRCM model. Fig. 8 shows the dynamics of the ring current pressure distribution as a function of SW B_z and CPCP in one-way coupled model (BATS-R-US-CRCM). The first and third rows show plasma pressure distribution in the equatorial plane, while the second and fourth rows show distribution of Birkeland currents calculated from the pressure distribution, projected into the equatorial plane. Under southward SW B_z, the value of CPCP is ~110 kV and the ring current pressure starts to build up ($t = 0.8$ h and $t = 2.6$ h of the simulation). The enhanced ring current pressure around $t = 2.6$ h generates Region II Birkeland currents and enhanced electric fields in the dusk sector, so-called polarization jet (PJ) or subauroral ion drift (SAID) structures (Galperin et al., 1973; Spiro et al., 1979). The electric field generated by Region II Birkeland currents partially cancels out the outer electric field (shielding effect). Around $t = 2.5$ h, the SW B_z is turned northward and CPCP quickly drops to low values. In the absence of the "outer" electric field source, the asymmetric ring current pressure generates Region II currents and so-called "overshielding" electric field directed opposite to the normal dawn-dusk electric field. Overshielding is a transient phenomenon, and the ring current starts to become symmetric ($t = 2.9$ h and $t = 4.6$ h). When the SW B_z turns southward again, and CPCP

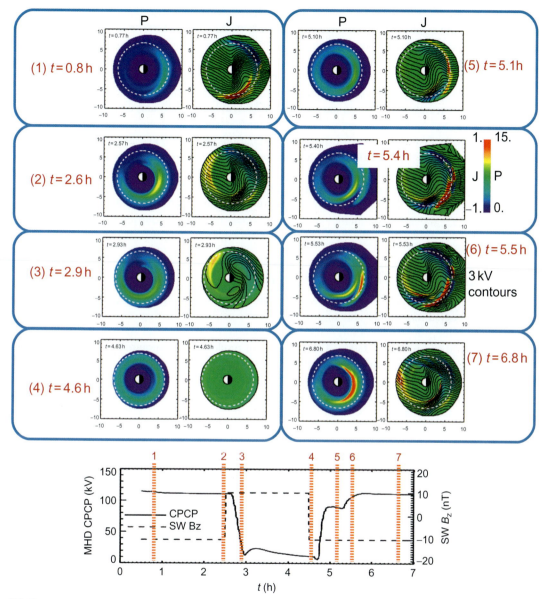

FIG. 8

Dynamics of the ring current pressure distribution, electric fields, and Birkeland currents in the inner magnetosphere during southward-northward-southward IMF B_z turning. The simulations are done with one-way coupled model (BATS-R-US-CRCM). The *first and third columns* are color contour maps of ring current pressure at the equatorial plane, in nPa. The *second and fourth columns* are color contour maps of Birkeland ionospheric currents projected into the equatorial plane, in µA/m², overlapped with electric field equipotentials. The *bottom panel* shows IMF B_z and MHD calculated CPCP.

(Modified from Buzulukova, N., Fok, M.C., Pulkkinen, A., Kuznetsova, M., Moore, T.E., Glocer, A., Brandt, P.C., Tóth, G., Rastätter, L., 2010. Dynamics of ring current and electric fields in the inner magnetosphere during disturbed periods: CRCM-BATS-R-US coupled model. J. Geophys. Res. (Space Phys.) 115, A05210.)

begins to grow, the ring current plasma reorganizes and becomes asymmetric again, with strong Region II Birkeland currents and SAID/PJ structures in the dusk sector ($t = 5.1$ h). The obtained variations in the electric fields are important for understanding of ionosphere-thermosphere response during geomagnetic storms and mechanisms of generation of ionospheric irregularities (Chapters 20 and 23 of this volume). Finally, the model reproduces substorm-like injections ($t = 5.5$ h) with their own structures of pressure and currents distributed throughout the inner magnetosphere. All these changes are reflected in the distribution of plasma pressure and FACs. When SW B_z and CPCP change, the ring current particles start to drift in a newly established configurations of electric (and magnetic) fields, creating changes in FACs that translate into $d\mathbf{B}/dt$ variations at the ground. Therefore, inclusion of the ring current module into the global model increases the intensity of modeled $d\mathbf{B}/dt$ fluctuations. The effect will be more pronounced at the low latitudes where the ring current is the major source of disturbances. This complex pattern of electric fields, pressure distribution and Birkeland currents is obtained with very simple SW variations for southward-northward-southward (-10 nT/10 nT/-10 nT) B_z turning. In the real events, one might expect to see more complex and dynamic picture.

6 CHALLENGES AND FUTURE DIRECTIONS

Coupling of different physical models is nontrivial, and progress is made in incremental steps. In the current state, coupled models are able to reproduce many of the effects that stand-alone ring current models can describe, but provide additional information relating to the global context and include self-consistent electric and magnetic fields. The ring current models also can output particle distribution functions (bounce-averaged) and pitch-angle information. In comparison to stand-alone models, coupled models are a huge step forward allowing the use of nonempirical and self-consistent input parameters. The inclusion of the ring current into global MHD models allows improvements to the solution of the inner magnetosphere and offers ability to model Region 2 Birkeland currents via pressure-driven magnetic fluctuations, therefore improving the description of electromagnetic fields in the magnetosphere. At present, three of the five MHD models (LFM, BATS-R-US, and OpenGGCM) mentioned at the beginning of the chapter have coupled ring current models. The number of published works that use coupled models to understand the dynamics of the global magnetosphere-ionosphere-inner magnetosphere system grow every year.

There are imperfections and limitations of the coupled codes that should be discussed and need to be accounted for the future development. For example, the dependence of the results should be affected by the location of the code's boundary. Because ideal MHD cannot describe gradient-curvature drifts and reproduce the pressure build-up, one can imagine that by moving the boundary of the code closer to the Earth will change boundary conditions for the ring current module and at some point result in the coupled model being less effective. On the other hand, moving the boundary further to the tail could violate the basic assumption of most ring current models: the bounce-averaged approximation. Traditionally the first approach, that is, the exclusion of ring current from the MHD solution, is considered to be acceptable, while the second is not. One also can argue the regions of validity of the kinetic ring current models and MHD models may not overlap at all. In particular MHD models become inaccurate close to the Earth (e.g., inside geosynchronous orbit) because of the energy-dependent drifts. However, at the radial distance where the MHD becomes imprecise, the ring current models could not be appropriate yet because of the missing inertial and substorm effects. These questions are not studied well.

The problem potentially could be solved by two-way coupling of a ring current module with a hybrid code (e.g., Lin et al., 2014) but this type of coupling is not developed so far. At the present moment, the theory of two-way coupled models is developed very poorly, contrary to MHD theory or ring current/radiation belt theory.

Some problems of the coupled models come from MHD solution itself. As it has been showed recently (Gordeev et al., 2017), all MHD codes at the NASA Community Coordinated Modeling Center (CCMC) reproduce substorms differently, even if all the models use the same SW inputs. For coupled models, this means different MHD codes will provide a given model with different inputs, and consequently different results are guaranteed as output. Obvious factors affecting the MHD solution are the grid resolution and numerical scheme employed, other parameters that control the code behavior are numerous. The coupling adds to this list (see e.g., the list of parameters that control the coupling between OpenGGCM and RCM; Cramer et al., 2017). It would be desirable to understand how code parameters control the final solution, in order to separate numerical from physical effects. Answering these questions will require collaboration between different groups of code developers, as well as research funding for code development, and most importantly, a willingness to share information. The community needs to realize that without a solid theoretical background in this direction, it will be hard to move forward and create more reliable codes for space-weather applications.

7 CONCLUSIONS

We have briefly reviewed first-principles global models of the Earth magnetosphere. We have demonstrated global coupled models that combine MHD, ring current, and ionosphere electrodynamics modules have evolved so they can describe complex and interrelated physical processes in Earth's magnetosphere during geomagnetic storms. These processes include: formation of main magnetospheric current systems including intense ring current near the Earth, pressure build-up in the inner magnetosphere, complex pattern of electric fields and plasma drifts in the inner magnetosphere, formation of Region 1 and 2 Birkeland currents, and dynamics of radiation belt population and the ring current ions and electrons. The output from the global models can be used to drive GIC models and spacecraft charge analysis programs. Examples from the published literature on modeling of intense and extreme geomagnetic storms are briefly reviewed. As an illustration, we presented the simulations of a geomagnetic storm of June 22–23, 2015, one of the two larger storms so far for the Solar Cycle 24. The simulations are done with two-way coupled BATS-R-US MHD-CIMI ring current model. Calculated equatorial pressure and total current density distributions are compared with the output from the empirical model of the magnetic field TS07D for the March 2015 storm with similar intensity (see Chapter 11 of this volume). The agreement between the coupled first-principles model and the empirical model is quite reasonable if one takes into account that the comparison is carried out for two different storms. The ground-based magnetometers on June 22–23, 2015 registered intense variations of $dB_n/dt \sim 200-400$ nT/min. The level of variations could be classified as "significant" (values on the order of 10^2 nT/min are considered to be significant and 10^3 nT/min to be extreme, see Chapter 1 of this volume). The following SW sources of variations have been identified: (i) sharp fluctuations of SW dynamic pressure, (ii) fast IMF B_z turning from southward to northward direction, (iii) intervals of stable southward IMF B_z/SW pressure with substorm onsets. We pointed out the model reproduces the global structure of the magnetic field but it is still a challenge to describe localized structures and substorms. The role of ring current plasma in generation of $d\mathbf{B}/dt$

variations at low latitudes is discussed. The storm of this intensity cannot be called "extreme geomagnetic storm" on the basis of occurrence. There is no doubt that a Carrington-like event will lead to serious adverse effects and significantly impact many technological systems. However, as demonstrated in Chapter 24 of this volume, geomagnetic storms of the size Dst ~ -200 nT (and weaker) can lead to serious disruption of navigation systems. We note that classification of extremes could be done both in terms of occurrence rate (Dst index threshold), and in terms of consequences (extreme effects and impacts). From this point of view, the definition of extremes, as well as sets of sufficient and necessary conditions for an extreme event to occur, will vary from field to field.

ACKNOWLEDGMENTS

This work was carried out using the SWMF/BATS-R-US tools developed at the University of Michigan's Center for Space Environment Modeling (CSEM). THEMIS-A magnetic field data were obtained from CDAWeb. OMNI datasets were used as input parameters for the simulations. Results presented in this study rely on ground-based magnetometer data obtained from SuperMAG project (PI Jesper W. Gjerloev). The authors thank the Canadian Magnetic Observatory System (CANMOS) for FCC data and U.S. Geomagnetism Program (Jeffrey J. Love) for BRW and BSL data. The authors acknowledge support from NASA LWS Grant WBS 936723.02.01.09.47. Natalia Buzulukova acknowledges support from the MMS Project at Goddard Space Flight Center. Natalia Buzulukova thanks Bruce Tsurutani and Gurbax Lakhina for the discussion of SW origins for the Carrington event.

REFERENCES

Ahn, B.H., Akasofu, S.I., Robinson, R.M., Kamide, Y., 1983. Electric conductivities, electric fields and auroral particle energy injection rate in the auroral ionosphere and their empirical relations to the horizontal magnetic disturbances. Plan. Space Sci. 31, 641–653. https://doi.org/10.1016/0032-0633(83)90005-3.

Albert, J.M., Meredith, N.P., Horne, R.B., 2009. Three-dimensional diffusion simulation of outer radiation belt electrons during the 9 October 1990 magnetic storm. J. Geophys. Res. (Space Phys.). 114, https://doi.org/10.1029/2009JA014336. A09214.

Baker, D.N., Kanekal, S.G., Li, X., Monk, S.P., Goldstein, J., Burch, J.L., 2004. An extreme distortion of the Van Allen belt arising from the Halloween solar storm in 2003. Nature 432, 878–881. https://doi.org/10.1038/nature03116.

Baker, D.N., Li, X., Pulkkinen, A., Ngwira, C.M., Mays, M.L., Galvin, A.B., Simunac, K.D.C., 2013. A major solar eruptive event in July 2012: defining extreme space weather scenarios. Space Weather 11, 585–591. https://doi.org/10.1002/swe.20097.

Brandt, P.C., Ohtani, S., Mitchell, D.G., Fok, M.C., Roelof, E.C., Demajistre, R., 2002. Global ENA observations of the storm mainphase ring current: implications for skewed electric fields in the inner magnetosphere. Geophys. Res. Lett. 29 (20), 1359. https://doi.org/10.1029/2002GL015160.

Brautigam, D.H., Albert, J.M., 2000. Radial diffusion analysis of outer radiation belt electrons during the October 9, 1990, magnetic storm. J. Geophys. Res. (Space Phys.) 105, 291–310. https://doi.org/10.1029/1999JA900344.

Burlaga, L., Fitzenreiter, R., Lepping, R., Ogilvie, K., Szabo, A., Lazarus, A., Steinberg, J., Gloeckler, G., Howard, R., Michels, D., Farrugia, C., Lin, R.P., Larson, D.E., 1998. A magnetic cloud containing prominence material—January 1997. J. Geophys. Res. (Space Phys.) 103, 277. https://doi.org/10.1029/97JA02768.

Burton, R.K., McPherron, R.L., Russell, C.T., 1975. An empirical relationship between interplanetary conditions and Dst. J. Geophys. Res. 80, 4204–4214. https://doi.org/10.1029/JA080i031p04204.

Buzulukova, N., Fok, M.C., Moore, T.E., Ober, D.M., 2008. Generation of plasmaspheric undulations. Gephys. Res. Let. 35, https://doi.org/10.1029/2008GL034164. L13105.

Buzulukova, N., Fok, M.C., Goldstein, J., Valek, P., McComas, D.J., Brandt, P.C., 2010. Ring current dynamics in moderate and strong storms: comparative analysis of TWINS and IMAGE/HENA data with the comprehensive ring current model. J. Geophys. Res. (Space Phys.). 115, https://doi.org/10.1029/2010JA015292. A12234.

Buzulukova, N., Fok, M.C., Pulkkinen, A., Kuznetsova, M., Moore, T.E., Glocer, A., Brandt, P.C., Tóth, G., Rastätter, L., 2010. Dynamics of ring current and electric fields in the inner magnetosphere during disturbed periods: CRCM-BATS-R-US coupled model. J. Geophys. Res. (Space Phys.) 115, A05210. https://doi.org/10.1029/2009JA014621.

Carrington, R.C., 1859. Description of a singular appearance seen in the Sun on September 1, 1859. Mon. Not. R. Astron. Soc. 20, 13–15. https://doi.org/10.1093/mnras/20.1.13.

Chen, M.W., Lemon, C.L., Guild, T.B., Keesee, A.M., Lui, A., Goldstein, J., Rodriguez, J.V., Anderson, P.C., 2015. Effects of modeled ionospheric conductance and electron loss on self-consistent ring current simulations during the 5–7 April 2010 storm. J. Geophys. Res. (Space Phys.) 120, 5355–5376. https://doi.org/10.1002/2015JA021285.

Cramer, W.D., Raeder, J., Toffoletto, F.R., Gilson, M., Hu, B., 2017. Plasma sheet injections into the inner magnetosphere: two-way coupled OpenGGCM-RCM model results. J. Geophys. Res. (Space Phys.) 122 (5), 5077–5091. https://doi.org/10.1002/2017JA024104.

Crooker, N.U., Shodhan, S., Gosling, J.T., Simmerer, J., Lepping, R.P., Steinberg, J.T., Kahler, S.W., 2000. Density extremes in the solar wind. Geophys. Res. Lett 27, 3769–3772. https://doi.org/10.1029/2000GL003788.

Daglis, I.A., Thorne, R.M., Baumjohann, W., Orsini, S., 1999. The terrestrial ring current: origin, formation, and decay. Rev. Geophys. 37, 407–438. https://doi.org/10.1029/1999RG900009.

De Zeeuw, D.L., Sazykin, S., Wolf, R.A., Gombosi, T.I., Ridley, A.J., Tóth, G., 2004. Coupling of a global MHD code and an inner magnetospheric model: initial results. J. Geophys. Res. (Space Phys.) 109, A12219. https://doi.org/10.1029/2003JA010366.

Ebihara, Y., 2016. Ring current. In: Khazanov, G. (Ed.), Space Weather Fundamentals. CRC Press, Boca Raton, FL, pp. 149–172.

Ebihara, Y., Ejiri, M., 1998. Modeling of solar wind control of the ring current buildup: a case study of the magnetic storms in April 1997. Geophys. Res. Lett. 25, 3751–3754. https://doi.org/10.1029/1998GL900006.

Ebihara, Y., Ejiri, M., 2003. Numerical simulation of the ring current: review. Space Sci. Rev. 105, 377–452. https://doi.org/10.1023/A:1023905607888.

Ebihara, Y., Fok, M.C., 2004. Postmidnight storm-time enhancement of tens-of-keV proton flux. J. Geophys. Res. (Space Phys.) 109, A12209. https://doi.org/10.1029/2004JA010523.

Ebihara, Y., Fok, M.C., Wolf, R.A., Immel, T.J., Moore, T.E., 2004. Influence of ionosphere conductivity on the ring current. J. Geophys. Res. (Space Phys.) 109, A08205. https://doi.org/10.1029/2003JA010351.

Ebihara, Y., Fok, M.C., Sazykin, S., Thomsen, M.F., Hairston, M.R., Evans, D.S., Rich, F.J., Ejiri, M., 2005. Ring current and the magnetosphere-ionosphere coupling during the superstorm of 20 November 2003. J. Geophys. Res. (Space Phys.). 110, https://doi.org/10.1029/2004JA010924. A09S22.

Ebihara, Y., Tanaka, T., Kikuchi, T., 2014. Counter equatorial electrojet and overshielding after substorm onset: global MHD simulation study. J. Geophys. Res. (Space Phys.) 119, 7281–7296. https://doi.org/10.1002/2014JA020065.

Fok, M.C., Moore, T.E., 1997. Ring current modeling in a realistic magnetic field configuration. Geophys. Res. Lett. 24, 1775–1778. https://doi.org/10.1029/97GL01255.

Fok, M.C., Kozyra, J.U., Nagy, A.F., Rasmussen, C.E., Khazanov, G.V., 1993. Decay of equatorial ring current ions and associated aeronomical consequences. J. Gephys. Res. 98, 19. https://doi.org/10.1029/93JA01848.

Fok, M.C., Moore, T.E., Kozyra, J.U., Ho, G.C., Hamilton, D.C., 1995. Three-dimensional ring current decay model. J. Geophys. Res. 100, 9619–9632. https://doi.org/10.1029/94JA03029.

Fok, M.C., Moore, T.E., Greenspan, M.E., 1996. Ring current development during storm main phase. J. Geophys. Res. 101, 15311–15322. https://doi.org/10.1029/96JA01274.

Fok, M., Wolf, R.A., Spiro, R.W., Moore, T.E., 2001. Comprehensive computational model of Earth's ring current. J. Geophys. Res. 106, 8417–8424. https://doi.org/10.1029/2000JA000235.

Fok, M.C., Moore, T.E., Wilson, G.R., Perez, J.D., Zhang, X.X., Brandt, P.C., Mitchell, D.G., Roelof, E.C., Jahn, J.M., Pollock, C.J., Wolf, R.A., 2003. Global ENA image simulations. Space Sci. Rev. 109, 77–103. https://doi.org/10.1023/B:SPAC.0000007514.56380.fd.

Fok, M.C., Moore, T.E., Brandt, P.C., Delcourt, D.C., Slinker, S.P., Fedder, J.A., 2006. Impulsive enhancements of oxygen ions during substorms. J. Geophys. Res. (Space Phys.). 111, https://doi.org/10.1029/2006JA011839. A10222.

Fok, M., Horne, R.B., Meredith, N.P., Glauert, S.A., 2008. Radiation belt environment model: application to space weather nowcasting. J. Geophys. Res. 113, 3. https://doi.org/10.1029/2007JA012558.

Fok, M.C., Moore, T.E., Slinker, S.P., Fedder, J.A., Delcourt, D.C., Nosé, M., Chen, S.H., 2011. Modeling the superstorm in November 2003. J. Geophys. Res. (Space Phys.). 116, https://doi.org/10.1029/2010JA015720. A00J17.

Fok, M.C., Buzulukova, N.Y., Chen, S.H., Glocer, A., Nagai, T., Valek, P., Perez, J.D., 2014. The comprehensive inner magnetosphere-ionosphere model. J. Geophys. Res. (Space Phys.) 119, 7522–7540. https://doi.org/10.1002/2014JA020239.

Galperin, Y.I., Ponomarev, V.N., Zosimova, A.G., 1973. Direct measurements of drift rate of ions in upper atmosphere during a magnetic storm. II. Results of measurements during magnetic storm of November 3, 1967. Cosmic Res. 11, 249.

Gjerloev, J.W., 2012. The SuperMAG data processing technique. J. Geophys. Res. (Space Phys.). 117 (A9) https://doi.org/10.1029/2012JA017683.

Glauert, S.A., Horne, R.B., Meredith, N.P., 2014. Three-dimensional electron radiation belt simulations using the BAS Radiation Belt Model with new diffusion models for chorus, plasmaspheric hiss, and lightning-generated whistlers. J. Geophys. Res. (Space Phys.) 119, 268–289. https://doi.org/10.1002/2013JA019281.

Glocer, A., Fok, M., Meng, X., Toth, G., Buzulukova, N., Chen, S., Lin, K., 2013. CRCM + BATS-R-US two-way coupling. J. Geophys. Res. (Space Phys.). https://doi.org/10.1002/jgra.50221.

Glocer, A., RastŠtter, L., Kuznetsova, M., Pulkkinen, A., Singer, H.J., Balch, C., Weimer, D., Welling, D., Wiltberger, M., Raeder, J., Weigel, R.S., McCollough, J., Wing, S., 2016. Community-wide validation of geospace model local K-index predictions to support model transition to operations. Space Weather 14 (7), 469–480. https://doi.org/10.1002/2016SW001387.

Godinez, H.C., Yu, Y., Lawrence, E., Henderson, M.G., Larsen, B., Jordanova, V.K., 2016. Ring current pressure estimation with RAM-SCB using data assimilation and Van Allen Probe flux data. Geophys. Res. Lett. 43 (23), 11948–11956. https://doi.org/10.1002/2016GL071646.

Gonzalez, W.D., Joselyn, J.A., Kamide, Y., Kroehl, H.W., Rostoker, G., Tsurutani, B.T., Vasyliunas, V.M., 1994. What is a geomagnetic storm? J. Geophys. Res. 99, 5771–5792. https://doi.org/10.1029/93JA02867.

Gopalswamy, N., Lara, A., Lepping, R.P., Kaiser, M.L., Berdichevsky, D., St. Cyr, O.C., 2000. Interplanetary acceleration of coronal mass ejections. Geophys. Res. Lett. 27, 145–148. https://doi.org/10.1029/1999GL003639.

Gordeev, E., Sergeev, V., Tsyganenko, N., Kuznetsova, M., Rastäetter, L., Raeder, J., Tóth, G., Lyon, J., Merkin, V., Wiltberger, M., 2017. The substorm cycle as reproduced by global MHD models. Space Weather 15, 131–149. https://doi.org/10.1002/2016SW001495.

Grafe, A., 1999. Are our ideas about Dst correct? Ann. Geophys. 17, 1–10. https://doi.org/10.1007/s00585-999-0001-0.

Hajra, R., Tsurutani, B.T., Echer, E., Gonzalez, W.D., Gjerloev, J.W., 2016. Supersubstorms (SML ≤ -2500 nT): magnetic storm and solar cycle dependences. J. Geophys. Res. (Space Phys.) 121, 7805–7816. https://doi.org/10.1002/2015JA021835.

Harel, M., Wolf, R.A., Reiff, P.H., Spiro, R.W., Burke, W.J., Rich, F.J., Smiddy, M., 1981. Quantitative simulation of a magnetospheric substorm. I—Model logic and overview. J. Gephys. Res. 86, 2217–2241. https://doi.org/10.1029/JA086iA04p02217.

Jordanova, V.K., Kozyra, J.U., Khazanov, G.V., Nagy, A.F., Rasmussen, C.E., Fok, M.C., 1994. A bounce-averaged kinetic model of the ring current ion population. Gephys. Res. Lett. 21, 2785–2788. https://doi.org/10.1029/94GL02695.

Jordanova, V.K., Miyoshi, Y.S., Zaharia, S., Thomsen, M.F., Reeves, G.D., Evans, D.S., Mouikis, C.G., Fennell, J.F., 2006. Kinetic simulations of ring current evolution during the Geospace Environment Modeling challenge events. J. Geophys. Res. (Space Phys.) 111, A11S10. https://doi.org/10.1029/2006JA011644.

Jorgensen, A.M., Spence, H.E., Hughes, W.J., Singer, H.J., 2004. A statistical study of the global structure of the ring current. J. Geophys. Res. (Space Phys.). 109, https://doi.org/10.1029/2003JA010090. A12204.

Kennel, C.F., Edmiston, J.P., Hada, T., 1985. A quarter century of collisionless shock research. vol. 34. American Geophysical Union Geophysical Monograph Series, Washington, DC, pp. 1–36. https://doi.org/10.1029/GM034p0001.

Kim, K.C., Shprits, Y.Y., Blake, J.B., 2016. Fast injection of the relativistic electrons into the inner zone and the formation of the split-zone structure during the Bastille Day storm in July 2000. J. Geophys. Res. (Space Phys.) 121, 8329–8342. https://doi.org/10.1002/2015JA022072.

Kozyra, J.U., Liemohn, M.W., 2003. Ring current energy input and decay. Space Sci. Rev. 109, 105–131. https://doi.org/10.1023/B:SPAC.0000007516.10433.ad.

Kozyra, J.U., Fok, M.C., Sanchez, E.R., Evans, D.S., Hamilton, D.C., Nagy, A.F., 1998. The role of precipitation losses in producing the rapid early recovery phase of the Great Magnetic Storm of February 1986. J. Geophys. Res. (Space Phys.) 103, 6801–6814. https://doi.org/10.1029/97JA03330.

Laitinen, T.V., Palmroth, M., Pulkkinen, T.I., Janhunen, P., Koskinen, H.E.J., 2007. Continuous reconnection line and pressure-dependent energy conversion on the magnetopause in a global MHD model. J. Geophys. Res. (Space Phys.). 112 (A11) https://doi.org/10.1029/2007JA012352. A11201.

Le, G., Russell, C., Takahashi, K., 2004. Morphology of the ring current derived from magnetic field observations. Ann. Geophys. 22, 1267–1295. https://doi.org/10.5194/angeo-22-1267-2004.

Le, G., LŸhr, H., Anderson, B.J., Strangeway, R.J., Russell, C.T., Singer, H., Slavin, J.A., Zhang, Y., Huang, T., Bromund, K., Chi, P.J., Lu, G., Fischer, D., Kepko, E.L., Leinweber, H.K., Magnes, W., Nakamura, R., Plaschke, F., Park, J., Rauberg, J., Stolle, C., Torbert, R.B., 2016. Magnetopause erosion during the 17 March 2015 magnetic storm: combined field-aligned currents, auroral oval, and magnetopause observations. Geophys. Res. Lett. 43 (6), 2396–2404. https://doi.org/10.1002/2016GL068257.

Lemon, C., Toffoletto, F., Hesse, M., Birn, J., 2003. Computing magnetospheric force equilibria. J. Geophys. Res. (Space Phys.). 108, https://doi.org/10.1029/2002JA009702. 1237.

Li, X., Temerin, M., Tsurutani, B.T., Alex, S., 2006. Modeling of 1–2 September 1859 super magnetic storm. Adv. Space Res. 38, 273–279. https://doi.org/10.1016/j.asr.2005.06.070.

Liemohn, M.W., Kozyra, J.U., Jordanova, V.K., Khazanov, G.V., Thomsen, M.F., Cayton, T.E., 1999. Analysis of early phase ring current recovery mechanisms during geomagnetic storms. Gephys. Res. Lett. 26, 2845–2848. https://doi.org/10.1029/1999GL900611.

Lin, Y., Wang, X.Y., Lu, S., Perez, J.D., Lu, Q., 2014. Investigation of storm time magnetotail and ion injection using three-dimensional global hybrid simulation. J. Geophys. Res. (Space Phys.) 119, 7413–7432. https://doi.org/10.1002/2014JA020005.

Liu, Y.D., Hu, H., Wang, R., Yang, Z., Zhu, B., Liu, Y.A., Luhmann, J.G., Richardson, J.D., 2015. Plasma and magnetic field characteristics of solar coronal mass ejections in relation to geomagnetic storm intensity and variability. Astrophys. J. Lett. 809, https://doi.org/10.1088/2041-8205/809/2/L34. L34.

Love, J.J., Gannon, J.L., 2010. Movie-maps of low-latitude magnetic storm disturbance. Space Weather. 8, https://doi.org/10.1029/2009SW000518. 06001.

Lyon, J.G., Fedder, J.A., Mobarry, C.M., 2004. The Lyon-Fedder-Mobarry (LFM) global MHD magnetospheric simulation code. J. Atmos. Sol.-Terr. Phys. 66, 1333–1350. https://doi.org/10.1016/j.jastp.2004.03.020.

Manchester, D.L., Ridley, A.J., Gombosi, T.I., DeZeeuw, D.L., 2006. Modeling the Sun-to-Earth propagation of a very fast CME. Adv. Space Res. 38, 253–262. https://doi.org/10.1016/j.asr.2005.09.044.

Meng, X., Tóth, G., Liemohn, M.W., Gombosi, T.I., Runov, A., 2012. Pressure anisotropy in global magnetospheric simulations: a magnetohydrodynamics model. J. Geophys. Res. (Space Phys.). 117, https://doi.org/10.1029/2012JA017791. A08216.

Meng, X., Tóth, G., Glocer, A., Fok, M.C., Gombosi, T.I., 2013. Pressure anisotropy in global magnetospheric simulations: coupling with ring current models. J. Geophys. Res. (Space Phys.) 118, 5639–5658. https://doi.org/10.1002/jgra.50539.

Merkin, V.G., Kondrashov, D., Ghil, M., Anderson, B.J., 2016. Data assimilation of low-altitude magnetic perturbations into a global magnetosphere model. Space Weather 14, 165–184. https://doi.org/10.1002/2015SW001330.

Moore, T.E., Fok, M.C., Delcourt, D.C., Slinker, S.P., Fedder, J.A., 2008. Plasma plume circulation and impact in an MHD substorm. J. Geophys. Res. (Space Phys.). 113, https://doi.org/10.1029/2008JA013050. A06219.

Nakamura, R., Sergeev, V.A., Baumjohann, W., Plaschke, F., Magnes, W., Fischer, D., Varsani, A., Schmid, D., Nakamura, T.K.M., Russell, C.T., Strangeway, R.J., Leinweber, H.K., Le, G., Bromund, K.R., Pollock, C.J., Giles, B.L., Dorelli, J.C., Gershman, D.J., Paterson, W., Avanov, L.A., Fuselier, S.A., Genestreti, K., Burch, J.L., Torbert, R.B., Chutter, M., Argall, M.R., Anderson, B.J., Lindqvist, P.A., Marklund, G.T., Khotyaintsev, Y.V., Mauk, B.H., Cohen, I.J., Baker, D.N., Jaynes, A.N., Ergun, R.E., Singer, H.J., Slavin, J.A., Kepko, E.L., Moore, T.E., Lavraud, B., Coffey, V., Saito, Y., 2016. Transient, small-scale field-aligned currents in the plasma sheet boundary layer during storm time substorms. Geophys. Res. Lett. 43, 4841–4849. https://doi.org/10.1002/2016GL068768.

Ngwira, C.M., Pulkkinen, A., Kuznetsova, M.M., Glocer, A., 2014. Modeling extreme "Carrington-type" space weather events using three-dimensional global MHD simulations. J. Geophys. Res. (Space Phys.) 119, 4456–4474. https://doi.org/10.1002/2013JA019661.

O'Brien, T.P., McPherron, R.L., 2000. Forecasting the ring current index Dst in real time. J. Atmos. Sol.-Terr. Phys. 62, 1295–1299. https://doi.org/10.1016/S1364-6826(00)00072-9.

Pembroke, A., Toffoletto, F., Sazykin, S., Wiltberger, M., Lyon, J., Merkin, V., Schmitt, P., 2012. Initial results from a dynamic coupled magnetosphere-ionosphere-ring current model. J. Geophys. Res. (Space Phys.). 117, https://doi.org/10.1029/2011JA016979. A02211.

Perez, J.D., Fok, M.C., Moore, T.E., 2000. Deconvolution of energetic neutral atom images of the Earth's magnetosphere. Space Sci. Rev. 91, 421–436.

Powell, K.G., Roe, P.L., Linde, T.J., Gombosi, T.I., de Zeeuw, D.L., 1999. A solution-adaptive upwind scheme for ideal magnetohydrodynamics. J. Comput. Phys. 154, 284–309. https://doi.org/10.1006/jcph.1999.6299.

Pulkkinen, A., Rastätter, L., Kuznetsova, M., Singer, H., Balch, C., Weimer, D., Toth, G., Ridley, A., Gombosi, T., Wiltberger, M., Raeder, J., Weigel, R., 2013. Community-wide validation of geospace model ground magnetic field perturbation predictions to support model transition to operations. Space Weather 11 (6), 369–385. https://doi.org/10.1002/swe.20056.

Raeder, J., Larson, D., Li, W., Kepko, E.L., Fuller-Rowell, T., 2008. OpenGGCM simulations for the THEMIS mission. Space Sci. Rev. 141, 535–555. https://doi.org/10.1007/s11214-008-9421-5.

Reiff, P.H., Daou, A.G., Sazykin, S.Y., Nakamura, R., Hairston, M.R., Coffey, V., Chandler, M.O., Anderson, B.J., Russell, C.T., Welling, D., Fuselier, S.A., Genestreti, K.J., 2016. Multispacecraft observations and modeling of the 22/23 June 2015 geomagnetic storm. Geophys. Res. Lett. 43 (14), 7311–7318. https://doi.org/10.1002/2016GL069154.

Ridley, A., Gombosi, T., Dezeeuw, D., 2004. Ionospheric control of the magnetosphere: conductance. Ann. Geophys. 22, 567–584. https://doi.org/10.5194/angeo-22-567-2004.

Roelof, E.C., 1987. Energetic neutral atom image of a storm-time ring current. Geophys. Res. Lett. 14, 652–655. https://doi.org/10.1029/GL014i006p00652.

Schulz, M., Lanzerotti, L.J., 1974. Particle diffusion in the radiation belts. In: Physics and Chemistry in Space. Springer, Berlin.

Shi, Y., Zesta, E., Lyons, L.R., Yumoto, K., Kitamura, K., 2006. Statistical study of effect of solar wind dynamic pressure enhancements on dawn-to-dusk ring current asymmetry. J. Geophys. Res. (Space Phys.). 111, https://doi.org/10.1029/2005JA011532. A10216.

Shprits, Y., Subbotin, D., Ni, B., Horne, R., Baker, D., Cruce, P., 2011. Profound change of the near-Earth radiation environment caused by solar superstorms. Space Weather. 9, https://doi.org/10.1029/2011SW000662, S08007.

Sitnov, M.I., Tsyganenko, N.A., Ukhorskiy, A.Y., Brandt, P.C., 2008. Dynamical data-based modeling of the storm-time geomagnetic field with enhanced spatial resolution. J. Geophys. Res. (Space Phys.). 113, https://doi.org/10.1029/2007JA013003. A07218.

Spiro, R.W., Heelis, R.A., Hanson, W.B., 1979. Rapid subauroral ion drifts observed by Atmosphere Explorer C. Geophys. Res. Lett. 6, 657–660. https://doi.org/10.1029/GL006i008p00657.

Stephens, G.K., Sitnov, M.I., Korth, H., Gkioulidou, M., Ukhorskiy, A.Y., 2017. Global empirical model of substorms derived from spaceborne magnetometer data. In: 13th International Conference on Substorms (ICS-13), Available from: http://ics13.unh.edu/cgi-bin/showab.pl?id=steglob, http://ics13.unh.edu/.

Subbotin, D., Shprits, Y., Ni, B., 2010. Three-dimensional VERB radiation belt simulations including mixed diffusion. J. Geophys. Res. (Space Phys.). 115, https://doi.org/10.1029/2009JA015070, A03205.

Subbotin, D.A., Shprits, Y.Y., Ni, B., 2011. Long-term radiation belt simulation with the VERB 3-D code: comparison with CRRES observations. J. Geophys. Res. (Space Phys.). 116, https://doi.org/10.1029/2011JA017019. A12210.

Tanaka, T., 2000. Field-aligned-current systems in the numerically simulated magnetosphere. vol. 118. American Geophysical Union Geophysical Monograph Series, Washington, DC, p. 53. https://doi.org/10.1029/GM118p0053.

Temerin, M., Li, X., 2002. A new model for the prediction of Dst on the basis of the solar wind. J. Geophys. Res. (Space Phys.). 107, https://doi.org/10.1029/2001JA007532. 1472.

Toffoletto, F., Sazykin, S., Spiro, R., Wolf, R., 2003. Inner magnetospheric modeling with the rice convection model. Space Sci. Rev. 107, 175–196. https://doi.org/10.1023/A:1025532008047.

Tóth, G., Sokolov, I.V., Gombosi, T.I., Chesney, D.R., Clauer, C.R., De Zeeuw, D.L., Hansen, K.C., Kane, K.J., Manchester, W.B., Oehmke, R.C., Powell, K.G., Ridley, A.J., Roussev, I.I., Stout, Q.F., Volberg, O., Wolf, R.A., Sazykin, S., Chan, A., Yu, B., Kóta, J., 2005. Space weather modeling framework: a new tool for the space science community. J. Geophys. Res. (Space Phys.) 110, A12226. https://doi.org/10.1029/2005JA011126.

Tóth, G., de Zeeuw, D.L., Gombosi, T.I., Manchester, W.B., Ridley, A.J., Sokolov, I.V., Roussev, I.I., 2007. Sun-to-thermosphere simulation of the 28–30 October 2003 storm with the Space Weather Modeling Framework. Space Weather 5, https://doi.org/10.1029/2006SW000272. 06003.

Tóth, G., van der Holst, B., Sokolov, I.V., de Zeeuw, D.L., Gombosi, T.I., Fang, F., Manchester, W.B., Meng, X., Najib, D., Powell, K.G., Stout, Q.F., Glocer, A., Ma, Y.J., Opher, M., 2012. Adaptive numerical algorithms in space weather modeling. J. Comput. Phys. 231, 870–903. https://doi.org/10.1016/j.jcp.2011.02.006.

Tsurutani, B.T., Gonzalez, W.D., Lakhina, G.S., Alex, S., 2003. The extreme magnetic storm of 1–2 September 1859. J. Geophys. Res. (Space Phys.). 108, https://doi.org/10.1029/2002JA009504. 1268.

Tsurutani, B.T., Echer, E., Guarnieri, F.L., Kozyra, J.U., 2008. CAWSES November 7–8, 2004, superstorm: complex solar and interplanetary features in the post-solar maximum phase. Geophys. Res. Lett. 35, https://doi.org/10.1029/2007GL031473. L06S05.

Tsurutani, B.T., Hajra, R., Echer, E., Gjerloev, J.W., 2015. Extremely intense (SML ≤ -2500 nt) substorms: isolated events that are externally triggered? Ann. Geophys. 33, 519–524. https://doi.org/10.5194/angeo-33 519 2015.

Tsyganenko, N.A., Sitnov, M.I., 2007. Magnetospheric configurations from a high-resolution data-based magnetic field model. J. Geophys. Res. (Space Phys.). 112, https://doi.org/10.1029/2007JA012260. A06225.

Tu, W., Cunningham, G.S., Chen, Y., Henderson, M.G., Camporeale, E., Reeves, G.D., 2013. Modeling radiation belt electron dynamics during GEM challenge intervals with the DREAM3D diffusion model. J. Geophys. Res. (Space Phys.) 118, 6197–6211. https://doi.org/10.1002/jgra.50560.

Wang, C.P., Lyons, L.R., Chen, M.W., Wolf, R.A., Toffoletto, F.R., 2003. Modeling the inner plasma sheet protons and magnetic field under enhanced convection. J. Geophys. Res. (Space Phys.). 108, https://doi.org/10.1029/2002JA009620. 1074.

Wang, C.P., Lyons, L.R., Chen, M.W., Toffoletto, F.R., 2004. Modeling the transition of the inner plasma sheet from weak to enhanced convection. J. Geophys. Res. (Space Phys.). 109, https://doi.org/10.1029/2004JA010591. A12202.

Weygand, J.M., McPherron, R.L., 2006. Dependence of ring current asymmetry on storm phase. J. Geophys. Res. (Space Phys.). 111, https://doi.org/10.1029/2006JA011808. A11221.

Zaharia, S., 2008. Improved Euler potential method for three-dimensional magnetospheric equilibrium. J. Geophys. Res. (Space Phys.). 113, https://doi.org/10.1029/2008JA013325. A08221.

Zaharia, S., Jordanova, V.K., Thomsen, M.F., Reeves, G.D., 2006. Self-consistent modeling of magnetic fields and plasmas in the inner magnetosphere: application to a geomagnetic storm. J. Geophys. Res. (Space Phys.). 111, https://doi.org/10.1029/2006JA011619. A11S14.

Zhang, J., Liemohn, M.W., De Zeeuw, D.L., Borovsky, J.E., Ridley, A.J., Toth, G., Sazykin, S., Thomsen, M.F., Kozyra, J.U., Gombosi, T.I., Wolf, R.A., 2007. Understanding storm-time ring current development through data-model comparisons of a moderate storm. J. Geophys. Res. (Space Phys.) 112, A04208. https://doi.org/10.1029/2006JA011846.

CHAPTER 11

EMPIRICAL MODELING OF EXTREME EVENTS: STORM-TIME GEOMAGNETIC FIELD, ELECTRIC CURRENT, AND PRESSURE DISTRIBUTIONS

Mikhail I. Sitnov*, Grant K. Stephens*, Matina Gkioulidou*, Viacheslav Merkin*, Aleksandr Y. Ukhorskiy*, Haje Korth*, Pontus C. Brandt*, Nikolai A. Tsyganenko[†]

Johns Hopkins University Applied Physics Laboratory, Laurel, MD, United States St. Petersburg State University, St. Petersburg, Russian Federation[†]

CHAPTER OUTLINE

1 Introduction	259
2 Recent Advances in Empirical Geomagnetic Field Modeling	261
2.1 Model Structure	261
2.2 Data Binning	263
2.3 Model Database	264
3 March 2015 Storm	264
4 Bastille Day Storm	270
5 Conclusion	275
Acknowledgments	275
References	276

1 INTRODUCTION

Extreme space weather events, including strong magnetic storms, are inherently difficult for empirical modeling. This is because such modeling usually involves averaging over the whole dataset, as in the case of climatological models, or its subsets containing many similar past events investigated in the past. On the other hand, first-principles models often cannot provide a comprehensive description of the heliosphere or even its separate regions. In particular, global MHD models of the magnetosphere (e.g., Lyon et al., 2004; Powell et al., 1999; Raeder, 2003) do not describe energy-dependent particle drifts, which provide the storm time ring current. At the same time, ring current simulations especially designed to describe the energy-dependent particle drifts (e.g., Fok et al., 1993; Liemohn et al., 2001; Zaharia et al., 2006) depend on external boundary conditions (e.g., at geosynchronous orbit) and on the magnetic field, whose sources are in part outside the model region and thus cannot be described self-consistently. In spite of its strong potential, the comprehensive description of the whole magnetosphere by combined models, coupling ring current and global MHD simulations (e.g., De Zeeuw et al., 2004; Pembroke et al., 2012; Glocer et al., 2013), remains difficult to achieve (see also Buzulukova et al., 2018 in this book for the simulation results of June 2015 even with a two-way coupled model). For instance, the coupled code stability and its ability to generate a consistent ring current may depend on the position of the interface boundary between the codes and other factors, such as the plasma beta parameter (Pembroke et al., 2012). This is likely a manifestation of a fundamental issue with the physical coupling of the inner magnetosphere governed by slow-flow equations and a more distant tail where MHD equations are applicable. There exists a significantly large transition region that is sufficiently close to Earth such that gradient and curvature drifts are important but far enough from Earth that inertial terms in the transport equations cannot be neglected.

A way to attack the problem of modeling and forecasting of the extreme events in the magnetosphere appeared due to a new generation of empirical geomagnetic field models TS07D (Tsyganenko and Sitnov, 2007; Sitnov et al., 2008; Stephens et al., 2016), where the number of ad hoc constraints on the structure of magnetospheric currents had been drastically reduced. Based on millions of points of space-borne magnetometer data, these models provide the ability to reconstruct and nowcast (Sitnov et al., 2012) distributions of the magnetic field with high resolution and high selectivity in the activity level, reflected by such parameters as the $Sym - H$ index and the solar wind electric field. The obtained magnetic field can then be used to calculate the current density vector distributions and to assess, through the integration of the Lorentz force $\mathbf{j} \times \mathbf{B} = \triangle p$, the distributions of the plasma pressure (Sergeev et al., 1994; Stephens et al., 2013). Such distributions can then be compared with global plasma distributions inferred from the energetic neutral atom imaging (ENA) (Brandt et al., 2001; Perez et al., 2016) that are particularly well resolved for strong storms. Such a comparison is particularly useful for extreme events as it helps to adjust empirical models to improve their description of those undersampled events. It is tempting to use the obtained pressure distributions in the first-principles models, and in particular, in global MHD models, where those distributions can be utilized to effectively adjust the equation of state. They can also be used to improve the external boundary conditions for kinetic ring current models, to directly validate those models, or to optimize their coupling with MHD models, for example, to determine the optimum value of the plasma beta parameter at the model interface (Pembroke et al., 2012).

A critical element in the pressure reconstruction is the resolution of the eastward current located at $\sim 3R_E$ (Hoffman and Bracken, 1965; McEntire et al., 1985; Lui et al., 1987), which limits the pressure integral sufficiently close to the Earth. In the past that current was described by ad hoc custom-tailored modules describing the ring current (e.g., Tsyganenko, 2002, and refs. therein). At the same time, in the original versions of the TS07D model (Tsyganenko and Sitnov, 2007; Sitnov et al., 2008) that were free from earlier ad hoc constraints, the eastward current could not be resolved because of the insufficient spatial resolution caused by a small number of the equatorial basis functions and the use of the datasets outside $\sim 4R_E$. The resolution of the eastward current was first demonstrated by Stephens et al. (2016), who used a new version of the TS07D model with an unprecedented number (~300) of equatorial basis functions and a new database, which approximately doubled the original model database, and which included a key subset of data in the radial distance interval 1.5–4 R_E.

In this work we use the next generation of TS07D with the new nowcasting data binning procedure (Sitnov et al., 2012). With the advanced description of the field-aligned currents (Sitnov et al., 2017), we use it to investigate two extreme storm events, the July 2000 Bastille Day storm and the March 2015 event, which remains so far the strongest storm in the Van Allen Probes era. The obtained pressure distributions are compared with similar global ENA distributions retrieved from IMAGE and TWINS missions (Brandt et al., 2001; Perez et al., 2016). We also compare the results for the March 2015 storm with the pressure estimates obtained by integrating proton and oxygen particle measurements that have become possible due to the Van Allen Probes mission (Mauk et al., 2013; Mitchell et al., 2013; Funsten et al., 2013). Finally, we assess the accuracy of the proposed empirical reconstructions by comparing the original global input and state parameters of the magnetosphere used in the reconstruction process with their values averaged over the corresponding statistical bins.

2 RECENT ADVANCES IN EMPIRICAL GEOMAGNETIC FIELD MODELING
2.1 MODEL STRUCTURE

A new generation of empirical geomagnetic field model development was motivated by the dramatic increase in the amount of data over the last decades from single-probe GOES, Geotail, and Polar missions as well as multiprobe missions Cluster, THEMIS, Van Allen Probes, and MMS. These models, known as the TS07D family (Tsyganenko and Sitnov, 2007; Sitnov et al., 2008, 2012; Stephens et al., 2016), drastically differ from earlier models in their spatial structure and their response functions to solar wind and IMF perturbations. These advancements make the new models less dependent on the ad hoc assumptions built in the model structure. They are also extensible, that is, providing higher resolution as long as more datasets become available.

The key distinctive feature of TS07D is the description of the magnetic field of all equatorial currents (tail and ring systems as well as their shielding subsystems at the magnetopause) by a regular expansion into a series of basis functions (Tsyganenko and Sitnov, 2007).

$$\mathbf{B}(\rho,\phi,z) = \sum_{n=1}^{N} a_{0n}^{(s)} \mathbf{B}_{0n}^{(s)} + \sum_{m=1}^{M}\sum_{n=1}^{N} a_{mn}^{(o)} \mathbf{B}_{mn}^{(o)} + \sum_{m=1}^{M}\sum_{n=1}^{N} a_{mn}^{(e)} \mathbf{B}_{mn}^{(e)}, \tag{1}$$

where $\{\rho, \phi, z\}$ is a cylindrical coordinate system with the Z axis normal to the equatorial plane, $\mathbf{B}_{\alpha\beta}^{(\gamma)}$ are the basis functions and $a_{\alpha\beta}^{(\gamma)}$ are scaling coefficients in the expansion. N and M are integer parameters, which determine the number of radial (ρ) and azimuthal (ϕ) expansions, respectively. The set of functions $\mathbf{B}_{\alpha\beta}^{(\gamma)}$ is built on the basis of the general solution of Ampère's equation for an infinitely thin equatorial current layer that can be obtained by the separation of variables method. These functions are naturally split into azimuthally symmetric modes denoted by the index (s) and two families of asymmetric solutions, (o) and (e), having factors $\sin(m\phi)$ and $\cos(m\phi)$ with $m \geq 1$ in their vector-potential representation and corresponding to its odd and even parity components with respect to the plane $y = 0$.

The original solutions of the Ampère's equation are then generalized to assume the finite half-thickness D of the equatorial current sheet.

An example of the basis function corresponding to the first term in Eq. (1) can be expressed through the azimuthal component of the vector potential $(A_\phi)^{(s)}_{0n}(k,\rho,z) = J_1(k_n\rho)\exp\left(-k_n\sqrt{z^2+D^2}\right)$ where J_1 is the Bessel function of the 1st order, $k_n = n/\rho_0$, and ρ_0 is the radial scale, corresponding to the largest wavelength in the radial expansion. In contrast to the other free parameters of the model, such as the scaling coefficients $a^{(\gamma)}_{\alpha\beta}$ and current sheet thickness D, the parameters ρ_0, N, and M are fixed as they determine the adopted resolution of the model.

An advantage of the basis function expansion approach is that by increasing the number of terms M and N in Eq. (1) and the radial scale ρ_0, arbitrary spatial resolution can be achieved bounded only by the necessity of proper magnetometer data coverage to constrain the model. In this study, following (Stephens et al., 2016), we use $\rho_0 = 20R_E$ and $(N, M) = (20, 6)$ resulting in a total of 260 scaling coefficients in the equatorial expansion in Eq. (1). The solar wind dynamic pressure is incorporated into the equatorial current system by replacing the scaling coefficient $a^{(\gamma)}_{\alpha\beta}$ with two groups of coefficients, one of which is a function of the dynamic pressure $(a^{(\gamma)}_{0,\alpha\beta} + a^{(\gamma)}_{1,\alpha\beta}\sqrt{P_{dyn}})$, doubling the number of scaling coefficients.

In principle, a regular expansion approach could be further extended to the whole 3D space as was proposed, in particular, by Andreeva and Tsyganenko (2016) and Tsyganenko and Andreeva (2016). Although that approach may offer a viable alternative to the current TS07D method, it still needs to be properly coupled with a flexible field-aligned current model in order to decently resolve its structure at low altitudes. In this study we follow another trade-off approach in the description of the nonequatorial 3D current structures. This description uses the empirical finding that such 3D currents close to Earth flow largely along the field lines, they are strongly localized in latitude, and they can be roughly split in Region 1 (R1) and Region 2 (R2) sense currents (Iijima and Potemra, 1976). In the original version of TS07D, the contribution of these field-aligned currents (FACs) to the model magnetic field was described following the earlier approach of Tsyganenko (2002) as the field created by two deformed conical current layers corresponding to R1 and R2 currents. The size of each system was an adjustable parameter, while their azimuthal distribution was controlled by the relative contributions of two groups of basis functions with odd and even symmetry due to factors $\sin(k\phi)$ and $\cos(k\phi)$, respectively, where k is a positive integer number. The first group represented the main ("antisymmetric") part of the FAC system, in which the duskside currents had the same magnitude but opposite direction to those at dawn, while the second group had an even (or "symmetric") distribution of currents with respect to the noon-midnight meridian plane. The linear combination of these two groups allows one to model the azimuthal rotation of the original antisymmetric distribution of FACs. The FAC system was described by the 1st and 2nd harmonics ($k = 1, 2$) with the odd ($\sin(k\phi)$) symmetry for R1 currents and the first harmonic ($k = 1$) with both odd and even symmetry for R2 allowing its azimuthal rotation. In the present model, following Sitnov et al. (2017), we use a group of additional FAC modules similar to the original ones but shifted in latitude. The new modules were shown by (Sitnov et al., 2017) to reproduce the characteristic spiral pattern on the nightside (Iijima and Potemra, 1976; Papitashvili et al., 2002; Anderson et al., 2008) associated with the Harang discontinuity (Harang, 1946). This approach increases the flexibility of the FAC current system description with the minimum impact on the model performance and feasibility, which might be problematic in case of fully 3D expansions similar to Eq. (1).

2.2 DATA BINNING

In order to relax the ad hoc assumptions regarding the evolution of the geomagnetic field in response to the solar wind/IMF variations, we use a combination of the dynamical system approach (Packard et al., 1980; Vassiliadis et al., 1995) and a version of the nearest neighbor (NN) data-mining algorithm (Vassiliadis et al., 1995; Mitchell, 1997; Wu et al., 2008, and refs. therein). In this approach (Sitnov et al., 2008, 2012) the magnetosphere is considered as a global system whose state and variations are determined by several system-level parameters: the solar wind electric field vB_z (v is the X-component of the solar wind and B_z is the Z-component of the magnetic field in the GSM coordinate system) and the $Sym - H$ index. Adding the time derivative of the latter parameter to the set of state parameters is suggested both by the dynamical system theory and also by a simple consideration that the time derivative of $Sym - H$ helps differentiate between the main and recovery phases of storms. To exclude the effects of substorms, the global state parameters are averaged over a few hours. In the nowcasting mode adopted in this study (only the past information about the global state of the magnetosphere is used for its magnetic field reconstruction), the key binning parameters are as follows

$$\langle vB_z|(t) \propto \int_{-\Pi/2}^{0} vB_z(t+\tau) \cos\left(\frac{\pi\tau}{\Pi}\right) d\tau, \qquad (2)$$

with $\Pi/2 = 6$ hours. A similar average is applied to the pressure-corrected $Sym - H$ index (Tsyganenko, 1996), while the trends of its evolution are described by the parameter

$$\frac{D\langle Sym - H|}{Dt}(t) \propto \int_{-\Pi/2}^{0} Sym - H(t+\tau) \cos\left(\frac{2\pi\tau}{\Pi}\right) d\tau. \qquad (3)$$

The global parameters $\langle vB_z|$, $\langle Sym - H|$, and $D\langle Sym - H|/Dt$ determine the 3D state space of the magnetosphere on storm scales, where the subsequent NN data mining is performed. At any moment of interest t the magnetosphere is characterized by the query vector in this space $\mathbf{G}(t) = (\langle vB_z|(t), \langle Sym - H|(t), D\langle Sym - H|/Dt(t))$, where all the historical moments from the model database can also be placed. Then K_{NN} nearest neighbors $\mathbf{G}_{NN}^{(i)}$ of the query point can be found based on the condition that their distance is less than a fixed number $R = |\mathbf{G}_{NN}^{(i)} - \mathbf{G}(t)| < R_{K_{NN}}$. A natural and physics-based metric, which determines the distance between two points in this space, can be provided by normalizing all the state and input parameters by their standard deviations:

$$R^2 = \frac{\langle vB_z|^2}{\sigma_{\langle vB_z|}^2} + \frac{\langle Sym - H|^2}{\sigma_{\langle Sym - H|}^2} + \frac{(D\langle Sym - H|/Dt)^2}{\sigma_{D\langle Sym - H|/Dt}^2}. \qquad (4)$$

The obtained K_{NN} points in the database are then used to create a snapshot of the magnetic field at the moment of interest. The number K_{NN} is chosen to have a sufficiently dense distribution of measurements for the reconstruction of the spatial structure of the geomagnetic field. On the other hand, it should be much less than the whole size K_{DB} of the model database to provide a sufficient selectivity of the NN method (in particular, to distinguish between main and recovery phases of storms). Our selection $K_{NN} = 8000$ provides approximately one data point per cubic Earth radius in a cube with the size $20R_E$ equal to the largest mode of the equatorial expansion (1) and the overall cross section scale of the magnetosphere. The size of the model database $K_{DB} \sim 2.5 \cdot 10^6$ (Stephens et al., 2016) gives a promise of high selectivity with $K_{NN}/K_{DB} \approx 0.003$.

2.3 MODEL DATABASE

The spacecraft magnetometer database used for the original TS07D model (Sitnov et al., 2008) was comprised of the Geotail, Polar, Cluster, IMP-8, and the GOES 8, 9, 10, and 12 missions and covered the period from 1995 to 2005. Except for Polar, each of these mission's perigee was $R \gtrsim 4R_E$, which is beyond the location of the eastward ring current ($R \lesssim 3R_E$). Additionally, Polar data within $R < 3.2R_E$ was removed. The magnetometer data was averaged to 15 min, except for Polar where data corresponding to $R < 5R_E$ was averaged to 5 min instead. Furthermore, the four Cluster spacecraft were merged together to form a single Cluster dataset.

In (Stephens et al., 2016) the original TS07D database was expanded by adding seven additional spacecraft: the five spacecraft from the Time History of Events and Macroscale Interactions during Substorms (THEMIS) mission and two from the Van Allen Probes (hereafter VA Probes) mission. Data within $R < 1.5R_E$ were removed for these seven spacecraft. Additionally, the THEMIS data occasionally contained large anomalies in the magnetic field values, so all data with the external field $B_{ext} \geq 200$ nT was also removed. These datasets were averaged to 5 min when $R < 5R_E$ and 15 min when $R \geq 5R_E$.

In this study, the magnetometer database has been modified further. The THEMIS and VA Probes datasets were augmented to include newly available data, and now covers from the start of the missions through December 31, 2015 and March 26, 2016, respectively. The Polar and Cluster data from the original TS07D model were replaced with reprocessed versions, which now span the entire mission lifetime for the Polar mission (1996–2006) and extended the Cluster data through May 2, 2015. Unlike in the original database, each of the four Cluster spacecraft are now treated as separate magnetometer datasets. The Polar and Cluster datasets were filtered to exclude data within $R < 3.2R_E$ and were averaged to a resolution of 5 min when $R < 5R_E$ and 15 min when $R \geq 5R_E$.

3 MARCH 2015 STORM

The March 2015 storm is particularly interesting because it is the strongest storm so far within the VA Probes era, when direct particle measurements of the ring current particles became available, making it possible to get in situ assessments of the plasma pressure. This storm was also monitored by the ENA imagers of the TWINS mission (Perez et al., 2016). Because this event was an interesting example of the strong radiation belt activity, the solar wind and IMF conditions of this CME-driven storm were already discussed in detail in earlier works (e.g., Baker et al., 2016). In Fig. 1 we provide an overview of this storm in terms of the global parameters, the pressure-corrected $Sym - H$ index, the solar wind/IMF input parameters key for the model data mining, and the solar wind dynamical pressure. In Fig. 1D we additionally show the averaged parameter $\langle Sym - H|$, which is directly used in the NN data binning procedure. Vertical dotted lines in this figure indicate the specific moments f_1–f_6, for which we make the empirical reconstructions discussed below.

Fig. 2 provides the equatorial current distributions for all six moments. For the sake of simplicity, the dipole tilt effects are excluded (although they are taken into account in the model fitting procedure). Panels f_1–f_3 show that in the main phase the ring current is strongly asymmetric and is concentrated mainly in the premidnight sector. This is consistent with earlier reconstructions using previous model versions and considering weaker storms (Sitnov et al., 2008; Stephens et al., 2016). Panels f_1–f_3 also show a pronounced eastward current, which includes its connection with the westward current forming the characteristic horse shoe or banana-like pattern (Roelof, 1989; Liemohn et al., 2013) first resolved

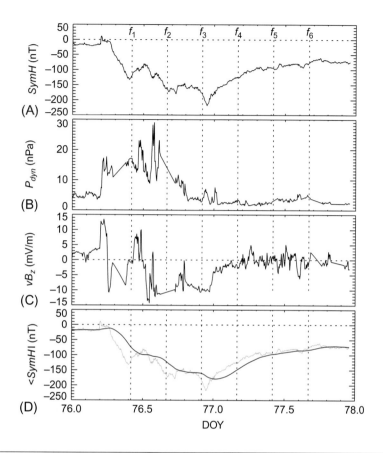

FIG. 1

(A) Evolution of the pressure-corrected $Sym - H$ index during the March 2015 magnetic storm. Panels (B) and (C) show the corresponding solar wind and IMF parameters, the dynamic pressure P_{dyn}, and the electric field parameter vB_z. Panel (D) shows one of the TS07D model binning parameters $\langle Sym - H|$, compared to the original $Sym - H$ index (*grey line*). Zero levels are shown by *horizontal dotted lines*. *Vertical dotted lines* mark the moments f_1–f_6 discussed below in Figs. 2–4.

in the high-resolution modification of TS07D by Stephens et al. (2016). In the recovery phase (panels f_4–f_6), both westward and eastward currents become more azimuthally symmetric and banana currents disappear.

The reconstruction of the plasma pressure is made, assuming the quasistatic balance between the Lorentz force and the gradient of the (presumably) isotropic plasma pressure $\triangle p = \mathbf{j} \times \mathbf{B}$. Following Sergeev et al. (1994) and Stephens et al. (2013), the right-hand side of the force balance equation is integrated in the equatorial plane from the mid-tail boundary $x_0 = -20R_E$ to the selected point (x, y) with $x > x_0$ as follows

$$p(x,y) = p_0 + \int_{x_0}^{x} \mathbf{j} \times \mathbf{B} \cdot d\mathbf{x}. \tag{5}$$

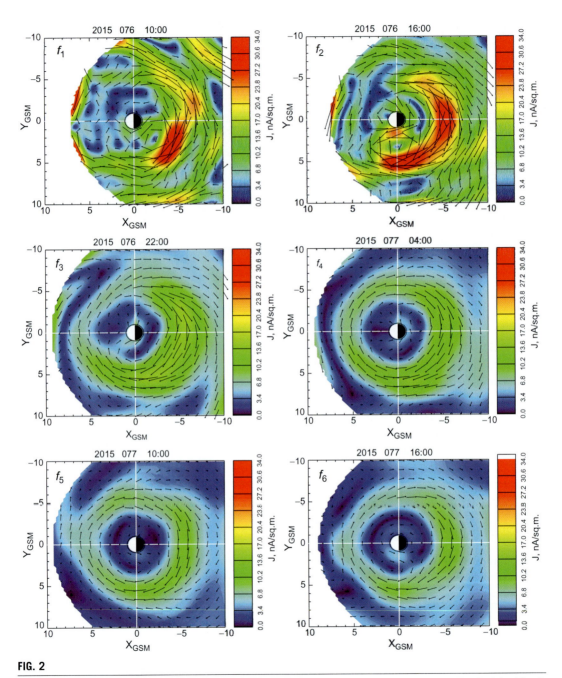

FIG. 2

Equatorial current distributions (using a 0-degree tilt angle) for six moments f_1–f_6 marked in Fig. 1 of the March 2015 magnetic storm. The *arrows* are vectors of the current density whose absolute value is color coded.

The initial value $p_0 = p(x_0, y)$ of the equatorial mid-tail plasma pressure can be estimated using the known magnetic field B_0 in the tail lobes and assuming another force balance across the plasma sheet $p_0 = B_0^2/2\mu_0$. Here, for simplicity and in view of the earlier results (Stephens et al., 2013), we use an ad hoc estimate $p_0 = 1$ nPa. In an ideal case of the perfect isotropic pressure force balance, one would expect to obtain the same plasma pressure independent of the path in the integral (5). Possible violations of this simplified picture were discussed for earlier custom-tailored empirical models in a number of papers (e.g., Zaharia and Cheng, 2003; Tsyganenko, 2010). According to Fig. 9 in (Perez et al., 2016) and Fig. 4 discussed later, the pressure for this storm was indeed quite close to isotropic. To check the consistency of the path integration procedure, here we also calculated the pressure using integration similar to Eq. (5) from the dayside magnetopause to the tail with the same initial value $p_0 = 1$ nPa. Two pressure distributions, p_n and p_d, provided by the integrations from the tail and the dayside magnetopause, respectively, were then sewn together at the terminator line $x = 0$ using the simple matching algorithm $p = \nu p_n + (1 - \nu)p_d$ with $\nu = (1 - \tanh(x/s_0))/2$ and the smoothing scale $s_0 = 1.5R_E$. The resulting distributions shown below in Figs. 3 and 7 suggest that the dayside and nightside pressure reconstruction procedures are consistent with each other. Investigations of other pressure reconstruction methods, including the solution of the elliptic equation $\triangle^2 p = \triangle \cdot (\mathbf{j} \times \mathbf{B})$ (Zaharia and Cheng, 2003), will be investigated elsewhere.

A fundamental problem for earlier pressure reconstructions (e.g., Stephens et al., 2013) was an artificial divergence of the pressure integral (5) near the Earth because of the unresolved eastward current. As is seen from Fig. 2, it is resolved now, and the resulting pressure distributions are presented in Fig. 3. They are compared with the pressure profiles obtained using particle data from the VA Probe B and shown in Fig. 4. The perpendicular and parallel pressure components are obtained using a combination of data from two instruments.

The Radiation Belt Storm Probe Ion Composition Experiment (RBSPICE) instrument (Mitchell et al., 2013) is a time-of-flight (ToF) versus energy measurement system, providing fluxes for hydrogen (10–600 keV), helium (75–600 keV), and oxygen (40 keV to 1 MeV) ions. The Helium Oxygen Proton Electrons (HOPE) instrument (Funsten et al., 2013) is a mass spectrometer measuring fluxes of hydrogen, helium, and oxygen within the energy range 0.001–50 keV. For this particular event, in order to calculate proton pressure we have combined proton intensities from two different RBSPICE measurement techniques: namely the ToF by Energy (TOFxE) product measuring energies ~ 50 keV to 600 keV, and the ToF by Pulse Height (TOFxPH) measuring energies from ~ 10 to 45 keV. In order to calculate the oxygen pressure we have combined HOPE oxygen intensities (1 eV to 50 keV) together with RBSPICE ones (150 keV to 1 MeV). Long-term intercalibrations between the two instruments indicated that the HOPE intensities should be multiplied by a factor of 2 for this event. Then data from two instruments were combined to form composite images of the total plasma pressure covering the whole energy range and both ion species.

In Fig. 3 we also show the VA Probe B orbit and location at the moment of interest mapped to match the equatorial pressure distribution. To perform this mapping, first the original GSM position of the spacecraft is traced to the magnetic field neutral plane prompted by the assumption that pressure is constant along a magnetic field line, which follows from the assumption of an isotropic pressure in force balance with the Lorentz force. However, this neutral plane location includes the dipole tilt deformations built into the model. To compare this position with the equatorial pressure distribution in which the dipole tilt effects have been ignored (because of the north-south symmetry of the magnetic field model the equatorial plane is equivalent to the magnetic field neutral plane), the inverse of the

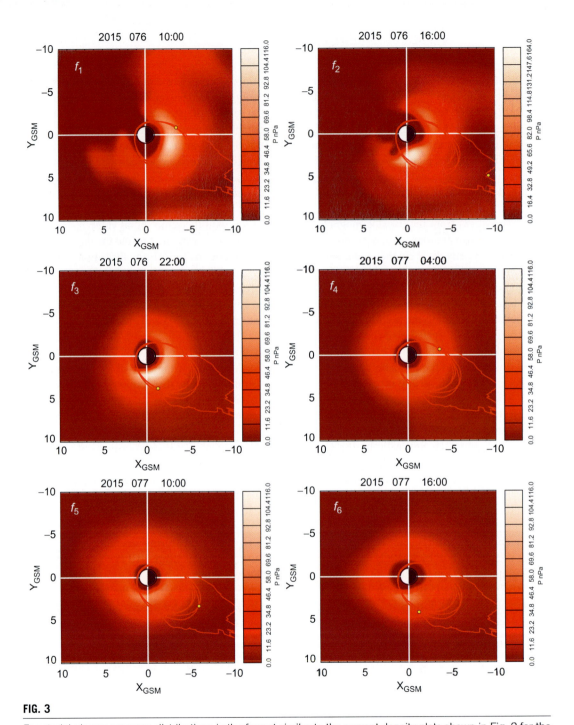

FIG. 3

Equatorial plasma pressure distributions in the format similar to the current density plots shown in Fig. 2 for the March 2015 storm. *Red contours* show the Van Allen Probe B orbit traced to the neutral plane of the current distribution along the model magnetic field lines, with the *yellow circle* showing the location of the probe on the orbit at the moment of interest. Note that in contrast to Figs. 2, 6, and 7, where current and pressure plots have the same maximum values for all panels in each figure, here we made an exception for panel 3f_2 to avoid highlighting that moment with interpolated solar wind and IMF data resulting in anomalously high pressure peak value $p_{max} \approx 164$ nPa.

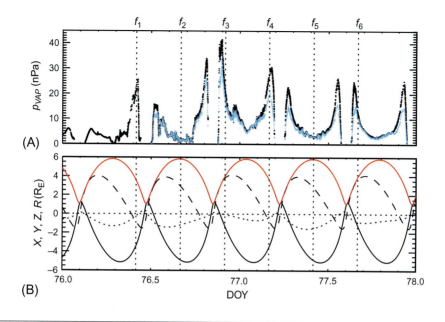

FIG. 4

(A) Evolution of the perpendicular (*black diamonds*) and parallel (*blue diamonds*) plasma pressure during the March 2015 storm according to probe B plasma instrument data of the Van Allen Probes mission. (B) GSM coordinates X (*black solid line*), Y (*dashed*), Z (*dotted*), and radial distance R (*red line*) of the probe B. *Horizontal and vertical solid lines* are similar to those in Fig. 1.

dipole tilt deformations must be applied, that is, inverting Eq. 7 in (Tsyganenko, 1998). In the near-Earth approximation, this transformation reduces to the standard rotation between the SM and GSM coordinate system. For simplicity, this inversion is approximated by rotating the GSM magnetic neutral plane position to the SM coordinate system. The resultant x- and y-coordinates for the whole storm are red contours in Fig. 3 while the moment of interest is a yellow circle.

The obtained pressure distributions reveal both interesting points of consistency and important differences from other sources of data. In particular, consistent with the global ENA imaging results for the main phase of this storm (Perez et al., 2016), the pressure peak locations in panels f_2 and f_3 are close to 19:00 MLT. At the same time, the peaks are closer to the Earth. The radial distance R_p of the pressure peak in panel f_3 is $2.25R_E$, in contrast to $3-4R_E$ range found by Perez et al. (2016) for their 30 keV ENA images. However, our result is close to the VA Probe B measurements (Fig. 4), which show that near the $Sym-H$ minimum (panel f_3) the pressure peak approaches the lower radial bound of the RBSPICE instrument range (e.g. Gkioulidou et al., 2014, Fig. 1) for this event $R=2.5R_E$. The residual difference of $0.25R_E$, together with very steep pressure profiles seen in Fig. 4, may also be of importance as it helps explain the remaining substantial difference between the heights of the pressure peaks: 42 nPa in VA Probes data versus 116 nPa in the TS07D reconstruction. (The even higher pressure peak with $p_p=164$ nPa seen in the panel f_2 is rather an artifact caused by the interpolation of the missing solar wind/IMF data as is seen from Fig. 1.) As is seen from Fig. 3, the difference might be also caused by the fact that spacecraft flew around the pressure peak. Note here that the analysis of the pressure distributions for another VA Probes era storm, the 17 March 2013 event (Menz et al., 2017), revealed strong temporal and local time variations of the plasma pressure. Another possible reason for the residual difference in the pressure peak values might be violation of the isotropic plasma approximation, which becomes particularly substantial near some of the pressure peaks in Fig. 4. At the same time, at the maximum of the parallel pressure ($p_\parallel = 35.63$ at DOY $= 76.9$), the perpendicular pressure $p_\perp$ exceeds $p_\parallel$ by only 16%.

Note that the remaining discrepancy between the plasma pressure distributions calculated from Van Allen Probes particle data and those derived from the empirical geomagnetic field model should not necessarily be attributed to the problem of the latter approach. The procedure of the pressure calculation from particle measurements also has a number of issues, such as the lower radial bound of the RBSPICE instrument discussed above. Another issue, which is particularly important and yet insufficiently investigated, is the contribution of the oxygen species, whose impact increases with the storm strength (e.g. Nosé et al., 2005); it may also be different in different storm phases (e.g. Ohtani et al., 2006). At the same time, the oxygen data from the HOPE instrument are insufficiently investigated, and the specific value of the constant renormalization factor 2 used in this work may require further elaboration.

4 BASTILLE DAY STORM

The July 2000 Bastille Day magnetic storm is another interesting event in a short list of historical supermagnetic storms (Lakhina et al., 2013). Moreover, for this storm an independent source of information on the plasma pressure distributions is available due to the IMAGE mission observations (Brandt et al., 2001). This storm, including its CME driver, was discussed in many papers (e.g., Lepping et al., 2001; Brandt et al., 2001; Raeder et al., 2001; Chen et al., 2003; Tsyganenko et al., 2003). In Fig. 5 we only provide the global state and input parameters key for our empirical model and necessary to provide the frame of reference for the following discussion of the results, including the specific moments f_1–f_6 for which the current and pressure distributions are provided in Figs. 6 and 7.

The current and plasma pressure distributions for this storm have characteristic features similar to those of the March 2015 event. They include the local time asymmetry of the westward and eastward currents seen in panels f_1–f_3 of Fig. 6 (although the banana current loop seen in panel f_2 now extends over 12 hours on the nightside) with their subsequent azimuthal symmetrization in the recovery phase shown by panels f_4–f_6. The strong eastward current near the dayside magnetopause seen in Fig. 4f_3 is likely a part of the magnetopause current system. It was revealed and discussed before in the statistical analysis of the geomagnetic field for different Dst bins (Le et al., 2004, Figs. 7 and 8) and in earlier TS07D current reconstructions (Sitnov et al., 2010, Figs. 7 and 8).

New features include the enhancement of the plasma pressure on the dayside (panels f_1 and f_4 in Fig. 7) in spite of the seemingly opposite current asymmetry, with stronger westward currents on the nightside (similar panels in Fig. 6). It appears due to asymmetry of the eastward currents, which limit the dayside integration of the Lorentz force parameter $\mathbf{j} \times \mathbf{B}$ closer to the Earth, making its accumulation stronger as compared to the nightside, due to the stronger equatorial magnetic field. Another unexpected pressure enhancement on the dawn side in the recovery phase, which is seen in Fig. 7f_5 and f_6, is likely caused by limitations of the pressure reconstruction method discussed above.

It is also interesting that the peak current near the end of the main phase (panel f_3) of this storm, which is stronger than that in the March 2015 event, is located further away from the Earth ($R_p =$

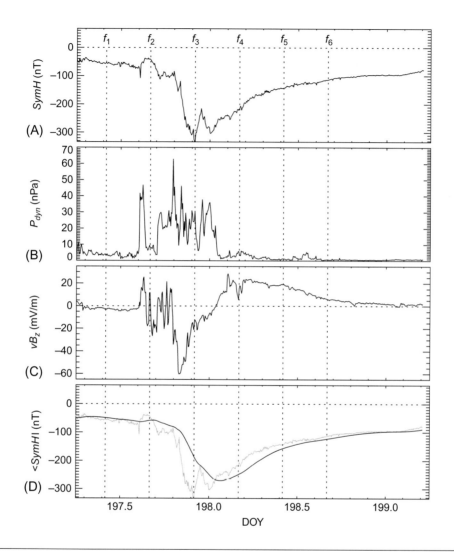

FIG. 5

Evolution of the global state and input parameters of the magnetosphere in the format similar to Fig. 1 during the July 2000 magnetic storm. *Vertical dotted lines* mark the moments $f_1 - f_6$ discussed below in Figs. 6 and 7.

$2.74R_E$, MLT $\approx$ 22:00). Although this feature contradicts the positive correlation between the parameter $1/R_p$ and the Sym-H index value during the storm evolution (similar dependence was reported between the depth of the storm distortion of the inner magnetosphere and the Dst index in Tsyganenko et al., 2003 and Ebihara et al., 2002), it is consistent with the ENA distributions obtained from the IMAGE mission, showing an increase of flux around $L = 3$ (Fig. 3b in Brandt et al., 2001).

The possibility of comparing the TS07D pressure distributions with the ENA emissions is particularly important for this extreme event for the following reasons. Strong storms represent only a very small subset of the model database spanning more than two decades of observations. As a result, in spite

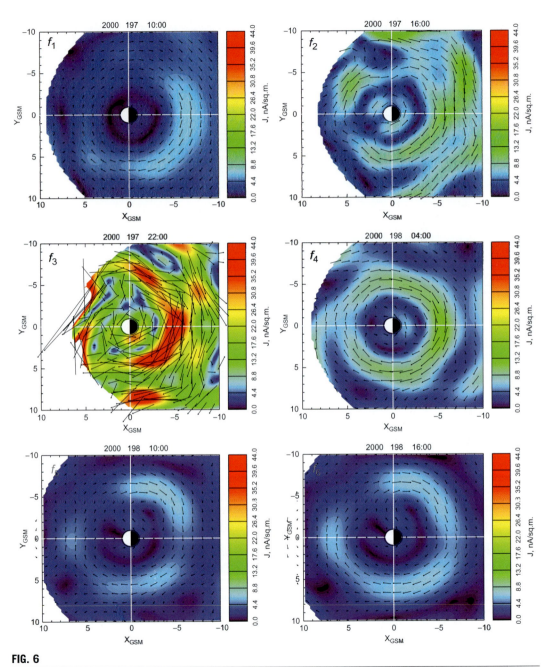

FIG. 6

Equatorial current distributions in the format similar to Fig. 2 for the July 2000 storm.

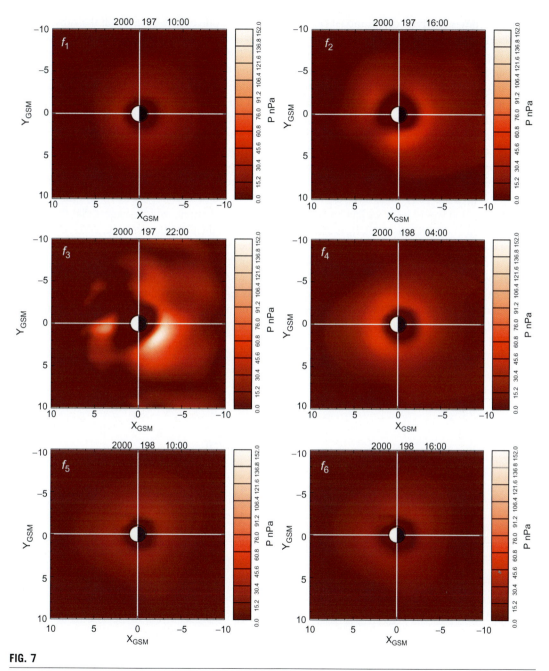

FIG. 7

Equatorial pressure distributions in the format similar to Fig. 3 for the July 2000 storm.

of a very strong selectivity of our model ($K_{NN}/K_{DB} \approx 0.003$), one cannot expect that the subset of the nearest neighbors used for the pressure reconstruction includes mainly storms of comparable strength. This is confirmed by Fig. 8, which presents the original TS07D model binning parameters compared with their values averaged over the corresponding NN subset of K_{NN} points. It shows, in particular, that while for the March 2015 storm its strength might be underestimated by $\approx 30\%$ in terms of the SymH index, it is underestimated by more than a factor of two for the Bastille Day storm (note that the parameter vB_z is underestimated even more, by a factor of five). The plots in Fig. 8A and D might be used to renormalize the current and pressure distributions shown in Figs. 2, 3 and 6, 7, respectively. However, this can be done either by increasing the current strength (by the factor $Sym-H/\langle Sym-H \rangle_{NN}$, where $\langle Sym-H \rangle_{NN}$ is the Sym-H index averaged over the NN subset), or by a similar radial contraction of the current distribution. In the case of the March 2015 storm, such a contraction is hardly possible because the pressure peak is already very close to Earth. In the case of the Bastille Day storm, such a radial contraction would be at variance with the IMAGE observations, which suggest that the present radial location of the pressure peak R_p is already in accord with the data (Brandt et al., 2001, Figs. 3 and 4). As a result, the actual values of the peak pressure in these extreme storms, after the aforementioned Sym-H renormalization, reach ≈ 200 and 300 nPa, respectively, raising new challenges for direct plasma observations in the former case as well as for the present empirical geomagnetic field modeling. Further extension of these results to even stronger storms can be done by combining the renormalization method described above with the extrapolation of the probability distributions of the parameters SymH and R_p, as is done, for example, for the *Dst* index by Riley (2012), Riley and Love (2017), and Chapter 5 of this volume.

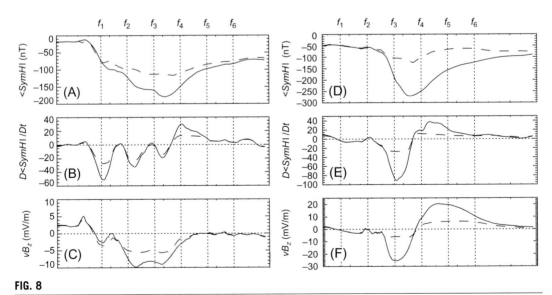

FIG. 8

(A)–(C) The original TS07D model binning parameters $\langle Sym-H \rangle$, $D\langle Sym-H \rangle/Dt$ (arbitrary units), and $\langle vB_z \rangle$ (*solid lines*) and their analogs averaged over $L_{NN} = 8000$ nearest neighbors used in every model output to construct current and pressure distributions (*dashed lines*) for the March 2015 storm. (D)–(F) Similar parameters for the July 2000 storm.

5 CONCLUSION

We presented empirical reconstructions of the magnetic field, electric currents, and force-balanced plasma pressure distributions for two strong storms, the March 2015 and July 2000 events. This included high spatial resolution and high selectivity of the model, which uses only about 0.3% of more than 2 million historical magnetometer records to build a snapshot of the magnetosphere at the moment of interest. The latest version of the new generation empirical models TS07D was used for the reconstruction with 260 basis functions for equatorial currents, flexible field aligned current distributions reproducing the spiral structure of low-altitude current distributions, and the nowcasting binning functions that allow one to use only past information on the global state of the magnetosphere and its solar wind input to build the model. We compared the modeling results with in situ observations of the plasma pressure made with the help of the VA Probe B and with ENA emissions of the storm time plasma showing its pressure distributions and available for both events due to TWINS and IMAGE missions. Our pressure reconstruction became possible due to the resolution of the innermost eastward current whose contribution limits the pressure integral (5) and determines the location and amplitude of the pressure peak. It is shown that a combination of the statistical analysis of the model bins and direct ENA observations of the pressure spatial structure may give realistic pictures of the storm time pressure distributions even for strong storm events. These distributions can then be used to improve the first-principles modeling of these extreme events, and in particular, to adjust the equation of state in global MHD models of the magnetosphere.

Further optimization of the empirical models for reconstruction of the extreme events might be achieved by weighing the nearest neighbors (e.g., Mitchell, 1997). However, standard methods of distance-weighted NNs will likely distort the spatial distribution of the reconstructed magnetic field similar to the effects of the inhomogeneity of the original data distributions (see, for example, Fig. 4 in Tsyganenko and Sitnov, 2007). One can also try to reduce the number of building blocks of the model (e.g., the number of the equatorial current basis functions), which would allow one to reduce the number of points in the binning subset K_{NN} and hence to increase the selectivity. However, this should unavoidably restore the number of ad hoc constraints on the model structure, bringing the empirical modeling back to earlier custom-tailored and event-oriented models (e.g., Tsyganenko, 2002; Kubyshkina et al., 1999). Without such constraints the empirical model with a reduced number of degrees of freedom fails to resolve the eastward current and hence to resolve the pressure peak (Stephens et al., 2016). The good news is that the number of historical data and even the list of strong storms are being further populated, and this progress includes both in situ observations and remote imaging of plasma pressure distributions. Thus, one can expect steady progress in the empirical modeling of magnetic storms and other extreme events leading eventually to their successful forecasting and mitigation.

ACKNOWLEDGMENTS

The authors thank D.G. Mitchell and E.C. Roelof for useful discussions. This work was supported by the NSF grant AGS-1157463 as well as NASA grants NNX15AF53G, NNX16AB78G, and NNX16AB80G. The magnetospheric data was obtained from the public archive at NASA/GSFC Space Physics Data Facility via their CDAWEB ftp site of CDF files (ftp://cdaweb.gsfc.nasa.gov/). Solar wind and IMF data were obtained from the OMNI database (http://omniweb.gsfc.nasa.gov/ow_min.html). The OMNI data files include the *SymH* index that is offered by the World Data Center for Geomagnetism of Kyoto University. The data used to produce figures and analysis in the paper are available upon request.

REFERENCES

Anderson, B.J., Korth, H., Waters, C.L., Green, D.L., Stauning, P., 2008. Statistical Birkeland current distributions from magnetic field observations by the Iridium constellation. Ann. Geophys. 26 (3), 671–687. https://doi.org/10.5194/angeo-26-671-2008.

Andreeva, V.A., Tsyganenko, N.A., 2016. Reconstructing the magnetosphere from data using radial basis functions. J. Geophys. Res. 121 (3), 2249–2263. https://doi.org/10.1002/2015JA022242.

Baker, D.N., Jaynes, A.N., Kanekal, S.G., Foster, J.C., Erickson, P.J., Fennell, J.F., Blake, J.B., Zhao, H., Li, X., Elkington, S.R., Henderson, M.G., Reeves, G.D., Spence, H.E., Kletzing, C.A., Wygant, J.R., 2016. Highly relativistic radiation belt electron acceleration, transport, and loss: large solar storm events of March and June 2015. J. Geophys. Res. 121 (7), 6647–6660. https://doi.org/10.1002/2016JA022502.

Brandt, P.C.S., Mitchell, D.G., Roelof, E.C., Burch, J.L., 2001. Bastille Day storm: global response of the terrestrial ring current. Sol. Phys. 204 (1), 377–386.

Buzulukova, N., Fok, M.-C., Glocer, A., Kang, S.B., Martin, S., Ngwira, C.M., Le, G., 2018. Geomagnetic storms: first-principles models for extreme geospace environment. In: Buzulukova, N. (Ed.), Extreme Events in Geospace. Elsevier, pp. 231–258.

Chen, M.W., Schulz, M., Lu, G., Lyons, L.R., 2003. Quasi-steady drift paths in a model magnetosphere with AMIE electric field: implications for ring current formation. J. Geophys. Res. 108, A5. https://doi.org/10.1029/2002JA009584.

De Zeeuw, D.L., Sazykin, S., Wolf, R.A., Gombosi, T.I., Ridley, A.J., Tóth, G., 2004. Coupling of a global MHD code and an inner magnetospheric model: initial results. J. Geophys. Res. 109, A12. https://doi.org/10.1029/2003JA010366.

Ebihara, Y., Ejiri, M., Nilsson, H., Sandahl, I., Milillo, A., Grande, M., Fennell, J.F., Roeder, J.L., 2002. Statistical distribution of the storm-time proton ring current: POLAR measurements. Geophys. Res. Lett. 29 (20), 30-1–30-4. https://doi.org/10.1029/2002GL015430.

Fok, M.C., Kozyra, J.U., Nagy, A.F., Rasmussen, C.E., Khazanov, G.V., 1993. Decay of equatorial ring current ions and associated aeronomical consequences. J. Geophys. Res. 98 (A11), 19381–19393. ISSN 2156-2202. https://doi.org/10.1029/93JA01848. https://doi.org/10.1029/93JA01848.

Funsten, H.O., Skoug, R.M., Guthrie, A.A., MacDonald, E.A., Baldonado, J.R., Harper, R.W., Henderson, K.C., Kihara, K.H., Lake, J.E., Larsen, B.A., Puckett, A.D., Vigil, V.J., Friedel, R.H., Henderson, M.G., Niehof, J.T., Reeves, G.D., Thomsen, M.F., Hanley, J.J., George, D.E., Jahn, J.M., Cortinas, S., De Los Santos, A., Dunn, G., Edlund, E., Ferris, M., Freeman, M., Maple, M., Nunez, C., Taylor, T., Toczynski, W., Urdiales, C., Spence, H.E., Cravens, J.A., Suther, L.L., Chen, J., 2013. Helium, oxygen, proton, and electron (HOPE) mass spectrometer for the Radiation Belt Storm Probes Mission. Space Sci. Rev. 179 (1), 423–484. https://doi.org/10.1007/s11214-013-9968-7.

Gkioulidou, M., Ukhorskiy, A.Y., Mitchell, D.G., Sotirelis, T., Mauk, B.H., Lanzerotti, L.J., 2014. The role of small-scale ion injections in the buildup of Earth's ring current pressure: Van Allen Probes observations of the 17 March 2013 storm. J. Geophys. Res. https://doi.org/10.1002/2014JA020096.

Glocer, A., Fok, M., Meng, X., Toth, G., Buzulukova, N., Chen, S., Lin, K., 2013. CRCM + BATS-R-US two-way coupling. J. Geophys. Res. 118 (4), 1635–1650. https://doi.org/10.1002/jgra.50221.

Harang, L., 1946. The mean field of disturbance of polar geomagnetic storms. Terr. Magn. Atmos. Electr. 51 (3), 353–380. https://doi.org/10.1029/TE051i003p00353.

Hoffman, R.A., Bracken, P.A., 1965. Magnetic effects of the quiet-time proton belt. J. Geophys. Res. 70, 3541–3556. https://doi.org/10.1029/JZ070i015p03541.

Iijima, T., Potemra, T.A., 1976. The amplitude distribution of field-aligned currents at northern high latitudes observed by Triad. J. Geophys. Res. 81 (13), 2165–2174. https://doi.org/10.1029/JA081i013p02165.

Kubyshkina, M.V., Sergeev, V.A., Pulkkinen, T.I., 1999. Hybrid input algorithm: an event-oriented magnetospheric model. J. Geophys. Res. 104 (A11), 24977–24993. https://doi.org/10.1029/1999JA900222.

Lakhina, G.S., Alex, S., Tsurutani, B.T., Gonzalez, W.D., 2013. Supermagnetic storms: hazard to society. In: Extreme Events and Natural Hazards: The Complexity Perspective. American Geophysical Union, pp. 267–278. https://doi.org/10.1029/2011GM001073.

Le, G., Russell, C., Takahashi, K., 2004. Morphology of the ring current derived from magnetic field observations. Ann. Geophys. 22, 1267–1295. https://doi.org/10.5194/angeo-22-1267-2004.

Lepping, R.P., Berdichevsky, D.B., Burlaga, L.F., Lazarus, A.J., Kasper, J., Desch, M.D., Wu, C.C., Reames, D.V., Singer, H.J., Smith, C.W., Ackerson, K.L., 2001. The Bastille Day magnetic clouds and upstream shocks: near-earth interplanetary observations. Sol. Phys. 204 (1), 285–303. https://doi.org/10.1023/A:1014264327855.

Liemohn, M.W., Kozyra, J.U., Clauer, C.R., Ridley, A.J., 2001. Computational analysis of the near-Earth magnetospheric current system during two-phase decay storms. J. Geophys. Res. 106, 29531–29542. https://doi.org/10.1029/2001JA000045.

Liemohn, M.W., Ganushkina, N.Y., Katus, R.M., de Zeeuw, D.L., Welling, D.T., 2013. The magnetospheric banana current. J. Geophys. Res. 118, 1009–1021. https://doi.org/10.1002/jgra.50153.

Lui, A.T.Y., McEntire, R.W., Krimigis, S.M., 1987. Evolution of the ring current during two geomagnetic storms. J. Geophys. Res. 92, 7459–7470. https://doi.org/10.1029/JA092iA07p07459.

Lyon, J.G., Fedder, J.A., Mobarry, C.M., 2004. The Lyon-Fedder-Mobarry (LFM) global MHD magnetospheric simulation code. J. Atmos. Sol.-Ter. Phys. 66 (1516), 1333–1350. https://doi.org/10.1016/j.jastp.2004.03.020.

Mauk, B.H., Fox, N.J., Kanekal, S.G., Kessel, R.L., Sibeck, D.G., Ukhorskiy, A., 2013. Science objectives and rationale for the Radiation Belt Storm Probes mission. Space Sci. Rev. 179 (1), 3–27. https://doi.org/10.1007/s11214-012-9908-y.

McEntire, R.W., Lui, A.T.Y., Krimigis, S.M., Keath, E.P., 1985. AMPTE/CCE energetic particle composition measurements during the September 4, 1984 magnetic storm. Geophys. Res. Lett. 12 (5), 317–320. https://doi.org/10.1029/GL012i005p00317.

Menz, A.M., Kistler, L.M., Mouikis, C.G., Spence, H.E., Skoug, R.M., Funsten, H.O., Larsen, B.A., Mitchell, D.G., Gkioulidou, M., 2017. The role of convection in the buildup of the ring current pressure during the 17 March 2013 storm. J. Geophys. Res. https://doi.org/10.1002/2016JA023358.

Mitchell, T.M., 1997. Machine Learning, first ed. McGraw-Hill, Inc., New York, NY. ISBN: 0070428077, 9780070428072.

Mitchell, D.G., Lanzerotti, L.J., Kim, C.K., Stokes, M., Ho, G., Cooper, S., Ukhorskiy, A., Manweiler, J.W., Jaskulek, S., Haggerty, D.K., Brandt, P., Sitnov, M., Keika, K., Hayes, J.R., Brown, L.E., Gurnee, R.S., Hutcheson, J.C., Nelson, K.S., Paschalidis, N., Rossano, E., Kerem, S., 2013. Radiation belt storm probes ion composition experiment (RBSPICE). Space Sci. Rev. 179 (1), 263–308. https://doi.org/10.1007/s11214-013-9965-x.

Nosé, M., Taguchi, S., Hosokawa, K., Christon, S.P., McEntire, R.W., Moore, T.E., Collier, M.R., 2005. Overwhelming O+ contribution to the plasma sheet energy density during the October 2003 superstorm: Geotail/EPIC and IMAGE/LENA observations. J. Geophys. Res. 110 (A9). https://doi.org/10.1029/2004JA010930.

Ohtani, S., Brandt, P.C., Singer, H.J., Mitchell, D.G., Roelof, E.C., 2006. Statistical characteristics of hydrogen and oxygen ENA emission from the storm-time ring current. J. Geophys. Res. 111 (A6). https://doi.org/10.1029/2005JA011201.

Packard, N.H., Crutchfield, J.P., Farmer, J.D., Shaw, R.S., 1980. Geometry from a time series. Phys. Rev. Lett. 45, 712–716. https://doi.org/10.1103/PhysRevLett.45.712.

Papitashvili, V.O., Christiansen, F., Neubert, T., 2002. A new model of field-aligned currents derived from high-precision satellite magnetic field data. Geophys. Res. Lett. 29 (14), 28-1–28-4. https://doi.org/10.1029/2001GL014207.

Pembroke, A., Toffoletto, F., Sazykin, S., Wiltberger, M., Lyon, J., Merkin, V., Schmitt, P., 2012. Initial results from a dynamic coupled magnetosphere-ionosphere-ring current model. J. Geophys. Res. 117 (A2). https://doi.org/10.1029/2011JA016979.

Perez, J.D., Goldstein, J., McComas, D.J., Valek, P., Fok, M.C., Hwang, K.J., 2016. Global images of trapped ring current ions during main phase of 17 March 2015 geomagnetic storm as observed by TWINS. J. Geophys. Res. 121 (7), 6509–6525. https://doi.org/10.1002/2016JA022375.

Powell, K.G., Roe, P.L., Linde, T.J., Gombosi, T.I., Zeeuw, D.L.D., 1999. A solution-adaptive upwind scheme for ideal magnetohydrodynamics. J. Comput. Phys. 154 (2), 284–309. https://doi.org/10.1006/jcph.1999.6299.

Raeder, J., 2003. Global magnetohydrodynamics—a tutorial. In: Büchner, J., Dum, C., Scholer, M. (Eds.), Lecture Notes in Physics. In: Space Plasma Simulation, vol. 615. Springer Verlag, Berlin, p. 212.

Raeder, J., Wang, Y.L., Fuller-Rowell, T.J., Singer, H.J., 2001. Global simulation of magnetospheric space weather effects of the Bastille Day storm. Sol. Phys. 204 (1), 323–337. https://doi.org/10.1023/A:1014228230714.

Riley, P., 2012. On the probability of occurrence of extreme space weather events. Space Weather 10 (2). https://doi.org/10.1029/2011SW000734.

Riley, P., Love, J.J., 2017. Extreme geomagnetic storms: probabilistic forecasts and their uncertainties. Space Weather 15 (1), 53–64. https://doi.org/10.1002/2016SW001470.

Roelof, E.C., 1989. Remote sensing of the ring current using energetic neutral atoms. Adv. Space Res. 9 (12), 195–203. https://doi.org/10.1016/0273-1177(89)90329-3.

Sergeev, V.A., Pulkkinen, T.I., Pellinen, T.I., Tsyganenko, N.A., 1994. Hybrid state of the tail magnetic configuration during steady convection events. J. Geophys. Res. 99, 23571. https://doi.org/10.1029/94JA01980.

Sitnov, M.I., Tsyganenko, N.A., Ukhorskiy, A.Y., Brandt, P.C., 2008. Dynamical data-based modeling of the storm-time geomagnetic field with enhanced spatial resolution. J. Geophys. Res. 113, https://doi.org/10.1029/2007JA013003, A07218.

Sitnov, M.I., Tsyganenko, N.A., Ukhorskiy, A.Y., Anderson, B.J., Korth, H., Lui, A.T.Y., Brandt, P.C., 2010. Empirical modeling of a CIR-driven magnetic storm. J. Geophys. Res. 115, https://doi.org/10.1029/2009JA015169, A07231.

Sitnov, M.I., Ukhorskiy, A.Y., Stephens, G.K., 2012. Forecasting of global data-binning parameters for high-resolution empirical geomagnetic field models. Space Weather 10 (9). https://doi.org/10.1029/2012SW000783.

Sitnov, M.I., Stephens, G.K., Ukhorskiy, A.Y., Korth, H., Anderson, B.J., Tsyganenko, N.A., 2017. Empirical reconstruction of the storm-time geomagnetic field using flexible field-aligned current modules. In: Haaland, S., Forsyth, C., Runov, A. (Eds.), AGU Books, AGU/WileyIn: Dawn-Dusk Asymmetry in Planetary Plasma Environments.

Stephens, G.K., Sitnov, M.I., Kissinger, J., Tsyganenko, N.A., McPherron, R.L., Korth, H., Anderson, B.J., 2013. Empirical reconstruction of storm time steady magnetospheric convection events. J. Geophys. Res. 118, 6434–6456. https://doi.org/10.1002/jgra.50592.

Stephens, G.K., Sitnov, M.I., Ukhorskiy, A.Y., Roelof, E.C., Tsyganenko, N.A., Le, G., 2016. Empirical modeling of the storm time innermost magnetosphere using Van Allen Probes and THEMIS data: eastward and banana currents. J. Geophys. Res. 121 (1), 157–170. https://doi.org/10.1002/2015JA021700.

Tsyganenko, N.A., 1996. Effects of the solar wind conditions on the global magnetospheric configurations as deduced from data-based field models. In: Proceedings of the Third International Conference on Substorms (ICS-3), ESA SP-389, Versailles, France, May 12–17, 1996, pp. 181–185.

Tsyganenko, N.A., 1998. Modeling of twisted/warped magnetospheric configurations using the general deformation method. J. Geophys. Res. 103, 23551–23564. https://doi.org/10.1029/98JA02292.

Tsyganenko, N.A., 2002. A model of the near magnetosphere with a dawn-dusk asymmetry. 1. Mathematical structure. J. Geophys. Res. 107, https://doi.org/10.1029/2001JA000219, 1179.

Tsyganenko, N.A., 2010. On the reconstruction of magnetospheric plasma pressure distributions from empirical geomagnetic field models. J. Geophys. Res. 115(A7), https://doi.org/10.1029/2009JA015012.

Tsyganenko, N.A., Andreeva, V.A., 2016. An empirical RBF model of the magnetosphere parameterized by interplanetary and ground-based drivers. J. Geophys. Res. 121 (11), 10786–10802. https://doi.org/10.1002/2016JA023217.

Tsyganenko, N.A., Sitnov, M.I., 2007. Magnetospheric configurations from a high-resolution data-based magnetic field model. J. Geophys. Res. 112, https://doi.org/10.1029/2007JA012260, A06225.

Tsyganenko, N.A., Singer, H.J., Kasper, J.C., 2003. Storm-time distortion of the inner magnetosphere: how severe can it get? J. Geophys. Res. 108 (A5). https://doi.org/10.1029/2002JA009808.

Vassiliadis, D., Klimas, A.J., Baker, D.N., Roberts, D.A., 1995. A description of the solar wind-magnetosphere coupling based on nonlinear filters. J. Geophys. Res. 100 (A3), 3495–3512. https://doi.org/10.1029/94JA02725.

Wu, X., Kumar, V., Ross Quinlan, J., Ghosh, J., Yang, Q., Motoda, H., McLachlan, G.J., Ng, A., Liu, B., Yu, P.S., Zhou, Z.H., Steinbach, M., Hand, D.J., Steinberg, D., 2008. Top 10 algorithms in data mining. Knowl. Inf. Syst. 14 (1), 1–37. https://doi.org/10.1007/s10115-007-0114-2.

Zaharia, S., Cheng, C.Z., 2003. Can an isotropic plasma pressure distribution be in force balance with the T96 model field? J. Geophys. Res. 108 (A11). https://doi.org/10.1029/2002JA009501.

Zaharia, S., Jordanova, V.K., Thomsen, M.F., Reeves, G.D., 2006. Self-consistent modeling of magnetic fields and plasmas in the inner magnetosphere: application to a geomagnetic storm. J. Geophys. Res. 111 (A11). https://doi.org/10.1029/2006JA011619.

PART 4

PLASMA AND RADIATION ENVIRONMENT

CHAPTER

EXTREME SPACE WEATHER EVENTS: A GOES PERSPECTIVE

12

William F. Denig*, Daniel Wilkinson*,[†], Robert J. Redmon*

NOAA, Boulder, CO, United States University of Colorado, Boulder, CO, United States[†]*

CHAPTER OUTLINE

1 Introduction	283
2 GOES Extreme Space Weather Events	286
3 GOES Extreme Events, Cases 1–12	301
4 Concluding Remarks	324
Acknowledgments	324
References	325

KEY POINTS

- NOAA operational space weather data provide a consistent, long-term record of the solar and near-Earth space environments.
- Cases of extreme space weather vary in terms of the solar forcing functions and their terrestrial response.
- NOAA Space Weather Scales provide a useful set of metrics for assessing the severity of space weather.

1 INTRODUCTION

Since the mid-1970s, spacecraft operated by the U.S. National Oceanic and Atmospheric Administration (NOAA) within the Geostationary Operational Environmental Satellite (GOES) program have continuously monitored the sun and the near-Earth space environment at geosynchronous altitudes. The primary functions of the program are to monitor and predict weather events, in their broadest sense, in order to preserve societal health and safety. For example, images of terrestrial weather patterns provided by GOES are used by trained meteorologists to issue alerts and warnings of severe weather such as tornado outbreaks and hurricane landfalls. In a similar vein, GOES serves a space weather mission by providing the environmental products used by NOAA for monitoring and predicting disruptive space weather (Grubb, 1975). The set of solar and space environmental measurements provided by GOES has evolved over time, although the consistency in the types of measurements allows one to present the long-term time series shown in Fig. 1 within the discussion to follow.

284 CHAPTER 12 EXTREME SPACE WEATHER EVENTS: A GOES PERSPECTIVE

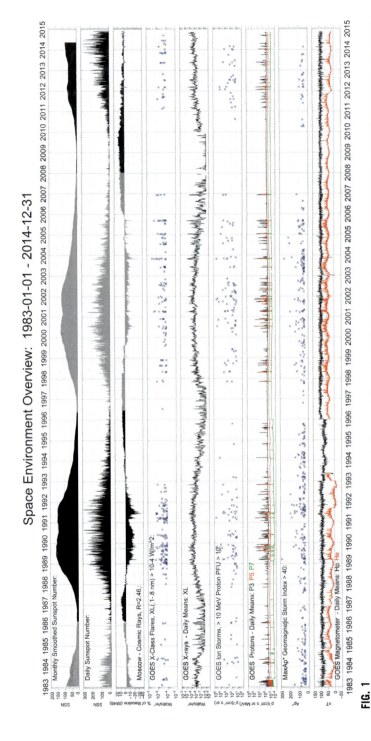

FIG. 1

Overview of the space environment for 1983–2014.

1 INTRODUCTION

Focusing on space weather, the original requirements for GOES were developed to assess the impact of solar flares on radio-wave communications and for detecting the occurrence of solar proton events (SPEs) for astronaut health and safety within NASA. Vestiges of these original requirements remain in the set of Space Weather Scales (see http://www.swpc.noaa.gov/) that NOAA uses to place space weather conditions in context, although these scales and GOES data, in general, have operational utility well beyond the original requirements. In addition and aside from their operational applicability, the GOES space weather products are heavily used by the research community for detailed scientific studies as well as providing the global and historical contexts for other investigations. It is this latter use that is the focus of this chapter. To this end, we have selected 12 cases that are representative of extreme space weather. For each case there exists a body of published work describing in detail the conditions in the solar and space environments that might lead one to judge the case as extreme. It is beyond the scope of this chapter to provide an exhaustive analysis of each case. Rather, our intent is to focus on and describe the GOES measurements for each case and to use the body of published work as supporting information.

The GOES space weather instruments include sensors for measuring solar X-ray irradiance and the in situ space environment, notably energetic charged particles and the local magnetic field. These sensors have been the workhorses for the GOES space weather mission; their derived set of space environmental products provides the focus for this chapter. Over time, the GOES program has introduced new space weather sensors and enhanced capabilities for existing sensors that are not included here. The pertinent capabilities for the GOES sensors for this report are described as follows. The solar X-ray irradiance sensor (XRS) measures total solar emissions within the short (XS) and long (XL) wavelength bands of 0.05–0.4 nm and 0.1–0.8 nm, respectively. The 1-min GOES XL data product is used to provide official X-ray flare classifications where the most intense X-class flares have irradiance values exceeding 10^{-4} watts per square meter; that is, an X1 flare $= 10^{-4}$ W/m^2; X10 $= 10^{-3}$ W/m^2; etc. Data presented here for the in situ energetic particle sensors consist of the integrated proton fluxes above discrete energy thresholds. NOAA space weather operators declare an SPE when the flux of energetic protons above 10 MeV crosses a threshold of 10 particles per square centimeter per second per steradian (10^1 p$^+$/[cm^2 s sr]) or 10 particle flux units (pfu). The local magnetic field is presented in terms of its vector components where, within a geographic sense, the H_e component points earthward, H_p is aligned with the Earth spin axis, and H_n points eastward. We refer to this as the EPN coordinate system in accordance with the conventional "right-hand rule." While GOES typically measures the geomagnetic field within the terrestrial magnetosphere, there are occasions during extreme space weather conditions when the magnetopause near noon is compressed inside geostationary Earth orbit (GEO) due to the pressure of the solar wind exerted on the nose of the magnetopause (Shue et al., 1998). At these times, GOES near local noon measures the interplanetary field within the magnetosheath. A convenient rule of thumb for identifying times when the magnetopause is inside GEO is a strongly negative Hp component of the GOES magnetometer—applicable only when GOES is near local noon. For broader descriptions of the GOES sensors we direct readers to the set of papers included in Society of Instrumentation Engineers (SPIE) volumes #2812 (Washwell, 1996) and #6689 (Fineschi and Viereck, 2007). Present day capabilities for the current GOES-16 space weather sensors can be found at the GOES-R website; (http://www.goes-r.gov/spacesegment/spacecraft.html) and in the literature (Chamberlin et al., 2009; Dichter et al., 2015; Eparvier et al., 2009).

Fig. 1 is a summary plot for 30+ years of GOES space environmental data plus other supporting ancillary data. From the top and working down, the first and second panels are time-series plots of the monthly and daily sunspot numbers (SSNs) that include data from solar cycles 22–24. These data are the original versions (version 1.0) of the SSN available from the World Data Center for Sunspot Index and Long-term Solar Oscillations (SILSO). The clear pattern discerned in the SSN series is the cyclic pattern of increasing and decreasing SSNs for active and quiet levels, respectively, of solar activity. Panel 3 is a record of the cosmic ray background relative to the long-term baseline as measured by the Moscow neutron monitor (NM) that, in a general sense, shows the inverse relationship between the cosmic ray background and solar activity (Forbush, 1958). Moscow data were used here due to the longevity of the record that spanned the GOES measurement period. The cosmic ray cutoff rigidity for Moscow is near 2 GeV/nucleon (Herbst et al., 2013) such that only incoming protons above 2 GeV are effectively measured by the ground-based Moscow NM. Panels 4–5 present data from the GOES XRS: panel 5 is the daily means whereas panel 4 is the set of discrete X-class flares detected by GOES over this sample period. Note that both the average irradiance and frequency of X-class flares increase during more active periods of solar activity near solar maximum. Similarly, the occurrence of GOES SPEs (i.e., ion storms (panel 6) and daily means (panel 7)) for particle fluxes >10 MeV (black), >50 MeV (red), and >100 MeV (green) increase during the more active solar intervals. Panel 8 is a measure of the "geoeffectiveness" of geomagnetic storms according to the Ap* classification (Allen and Wilkinson, 1992) wherein the average number of storms with Ap>40 increases at solar max. Finally, panel 9 is the magnetic field measured by GOES for the Hp (black) and He (red) components; note the scatter in the measurements around solar max due to increased disturbances in the background field due to an increased number of geomagnetic storms and substorms (Gonzalez et al., 1994). The annual oscillation in the field is due to the GOES orbit whereas the abrupt offsets in the magnetic field record are calibration shifts from one operational satellite to the next.

2 GOES EXTREME SPACE WEATHER EVENTS

Table 1 summarizes the set of extreme space weather cases included in this chapter. An immediate observation referring to Fig. 1 is that all events listed in column 1 tended to occur away from the solar minimum intervals near 1986, 1996, and 2008. While this is a not wholly unexpected tendency, it is important to note that disruptive space weather can occur at any time and can cause serious societal consequences (Baker, 2008). Consider, for example, the space weather conditions that resulted in the Galaxy 15 satellite anomaly that occurred in Apr. 2010 (Loto'aniu et al., 2015) during a nominally quiet solar interval. It should also be noted that "extreme" is in the eye of the beholder and that one person's extreme event may have only marginal consequences for another. One can get a sense of this in Table 1 by considering columns 2–4, which provide a space weather assessment in terms of the NOAA Space Weather Scales for geomagnetic storms (G-scale), radio blackouts (R-scale), and solar radiation (S-scale). According to the NOAA website, the scales "were introduced as a way to communicate to the general public the current and future space weather conditions and their possible effects on people and systems" (http://www.swpc.noaa.gov/noaa-scales-explanation). Since the time range of interest for this chapter predates the establishment of the official NOAA Space Weather

Table 1 Listing of extreme space weather events (1989–2012)

Extreme Event Cases 01 to Case 12	G-Scale G1-5	R-Scale R1-5	S-scale S1-5	Kp	X-ray	SEP pfu	Ap*	Dst nT
Mar. 10–15, 1989	G5	R3	S3	9_o	X4.5	1860	285	−589
Sep. 26–Oct. 1, 1989	G0	∼R4	S3	4_o	X9.8	2960	52	−151
Oct. 18–31, 1989	G4	R4	S4	8_+	X13.0	42,200	162	−268
Mar. 22–28, 1991	G5	∼R4	S4	9_o	X9.4	43,500	181	−298
Jun. 1–20, 1991	G5	R4	S3	9_o	X12^a	3000	196	−223
Jul. 14–16, 2000	G5	R3	S4	9_o	X5.7	24,000	192	−301
Nov. 4–8, 2001	G5	R3	S4	9_o	X1.0	>10,000	141	−292
26-Oct.—06-Nov. 2003	G5	R4	S4	9_o	X17.2^b	29,500	252	−383
15-Jan.—24-Jan. 2005	G4	R3	S3	8_o	X7.1	5040	91	−97
06-Sep.—16-Sep. 2005	G4	R4	S3	8_-	X17.0	1880	101	−139
05-Dec.—17-Dec. 2006	G4	∼R3	S3	8_+	X9.0	1980	120	−159
08-Mar.—17-Mar. 2012	G4	R3	S3	8_o	X5.4	6530	90	−131

^{a}Saturated. ^{b}X28 (estimated) flare on 04-Nov-2003 was the largest during the 26-Oct.–06-Nov. 2003 interval although the space weather impacts from this flare were limited (Balch et al., 2004).

Scales, the authors have assigned the scale values in columns 2–4 in Table 1 based on measurement criteria in columns 5–7 wherein the G-scale is derived from the maximum value of the planetary 3-h Kp index (column 5), the R-scale from the maximum GOES X-ray flare classification (column 6), and the S-scale from the maximum GOES SPE flux (column 7) for each extreme event interval. The X-ray classification is consistent with the report of Baker (1970). Our Kp-based classification of geomagnetic storms might differ in detail from the related 1-h Disturbance Storm Time (Dst) and 1-h Ap* geomagnetic indices although extreme geomagnetic storm events typically stress the highest index levels for all geomagnetic indices (Bell et al., 1997). Table 1 includes the additional Ap* and Dst indices, which are also included in the discussions in Section 3. General descriptions of the differences among the Kp, Dst, and Ap* indices are provided by Mayaud (1980) and by Allen (1982).

Figs. 2–13 for the 12 cases listed in Table 1 are presented in a standard format although the duration of the events, in days, varies. The first case considered is the geomagnetic superstorm of Mar. 1989, the data for which is provided in Fig. 2. The format for this figure is discussed in detail here in order to minimize redundant discussions in subsequent cases. Fig. 2 includes time-series space environmental data from the GOES (space) and the cosmic ray record from the Moscow NM (ground). The top panel plots the 1-min averaged X-ray irradiance within the XS and XL bands where, again, the XL band is used for the official NOAA X-ray classification and used here to determine the R-scale value. GOES integrated proton fluxes are plotted in panel 2 for energies >1 MeV, >5 MeV, >10 MeV, >30 MeV, >50 MeV, >60 MeV, and >100 MeV. Within the discussion, other solar-terrestrial data available from various sources are referenced in the acknowledgments. The GOES and ancillary data are useful

288 CHAPTER 12 EXTREME SPACE WEATHER EVENTS: A GOES PERSPECTIVE

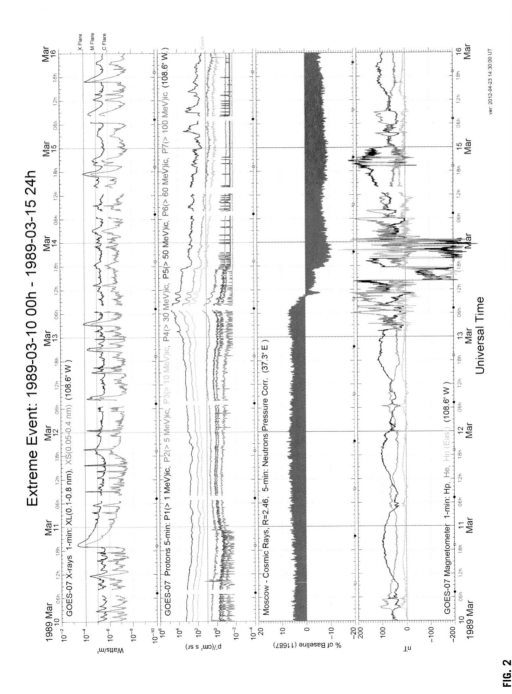

FIG. 2

Case 1: Extreme Event Case for Mar. 10–15, 1989.

2 GOES EXTREME SPACE WEATHER EVENTS 289

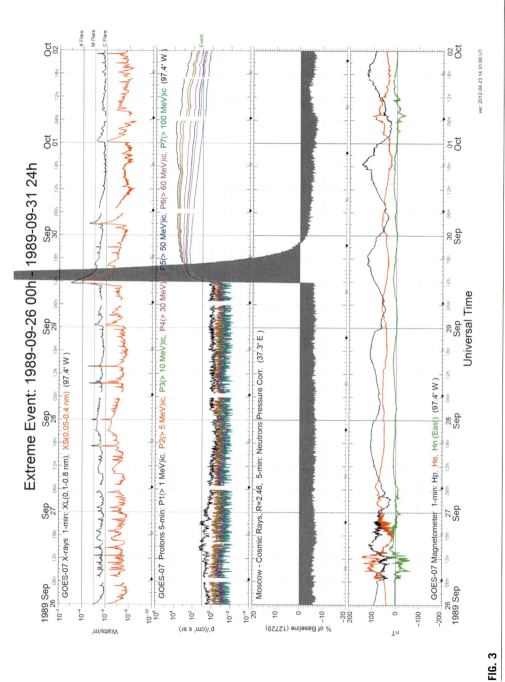

FIG. 3

Case 2: Extreme Event Case for 26-Sep.—01-Oct. 1989.

290 CHAPTER 12 EXTREME SPACE WEATHER EVENTS: A GOES PERSPECTIVE

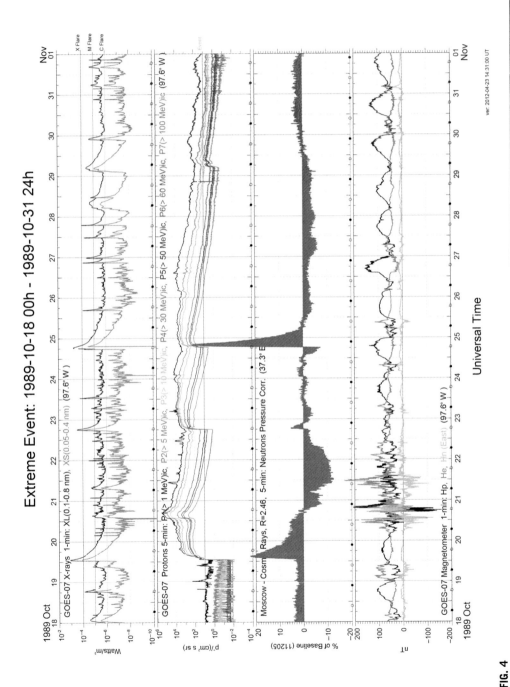

FIG. 4

Case 3: Extreme Event Case for Oct. 18–31, 1989.

2 GOES EXTREME SPACE WEATHER EVENTS   291

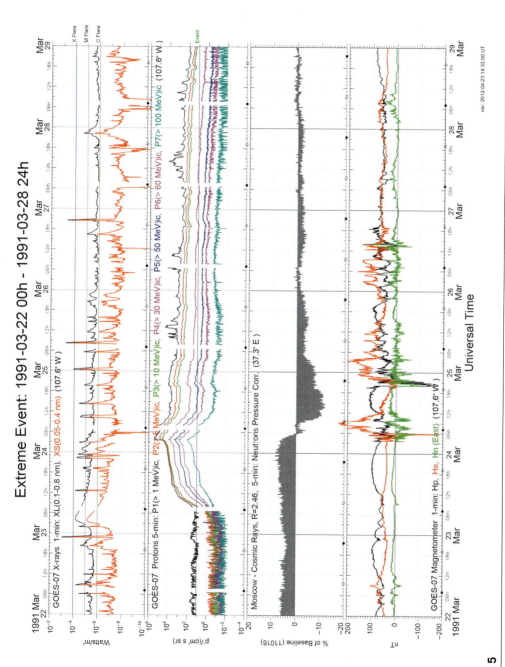

FIG. 5

Case 4: Extreme Event Case for Mar. 22–28, 1991.

292 CHAPTER 12 EXTREME SPACE WEATHER EVENTS: A GOES PERSPECTIVE

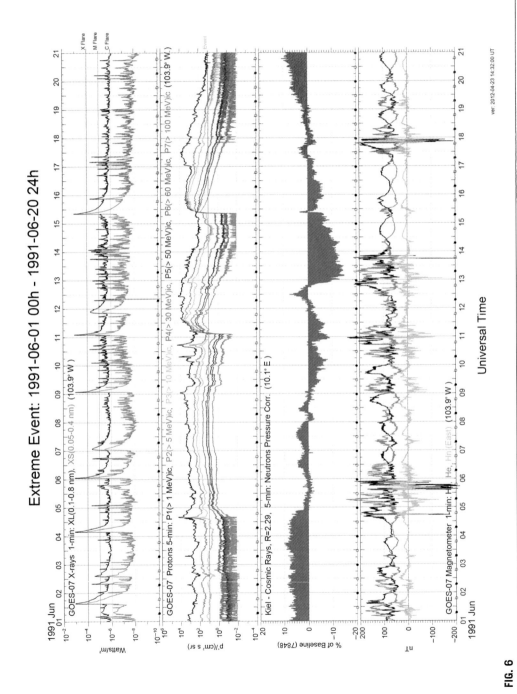

FIG. 6

Case 5: Extreme Event Case for Jun. 1–20, 1991.

2 GOES EXTREME SPACE WEATHER EVENTS   293

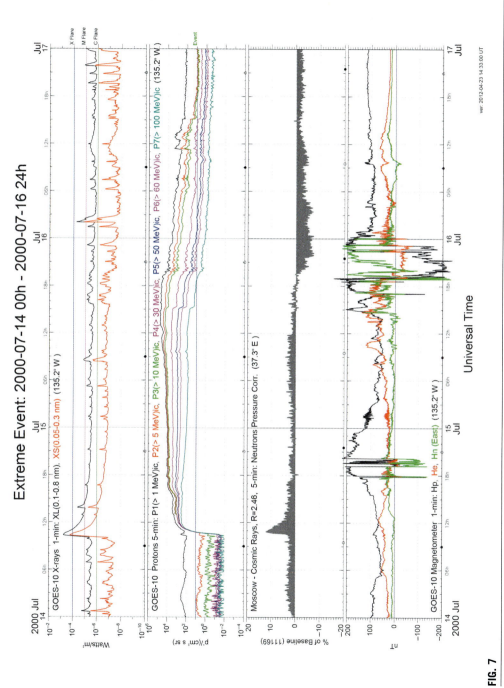

FIG. 7

Case 6: Extreme Event Case for Jul. 14–16, 2000C.

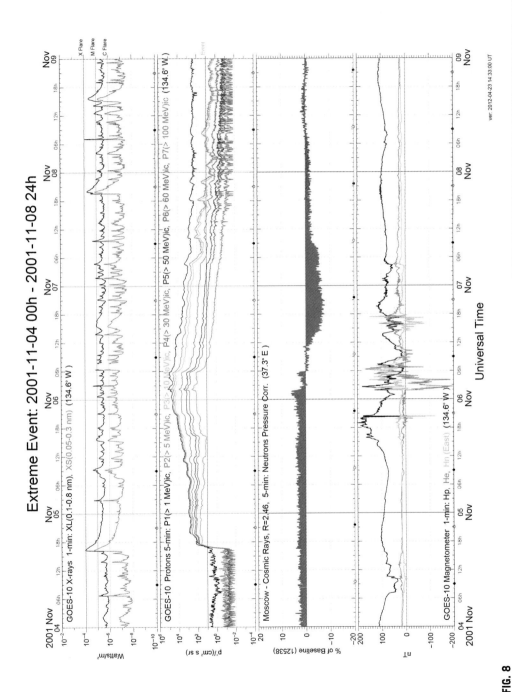

FIG. 8

Case 7: Extreme Event Case for Nov. 4–8, 2001.

2 GOES EXTREME SPACE WEATHER EVENTS 295

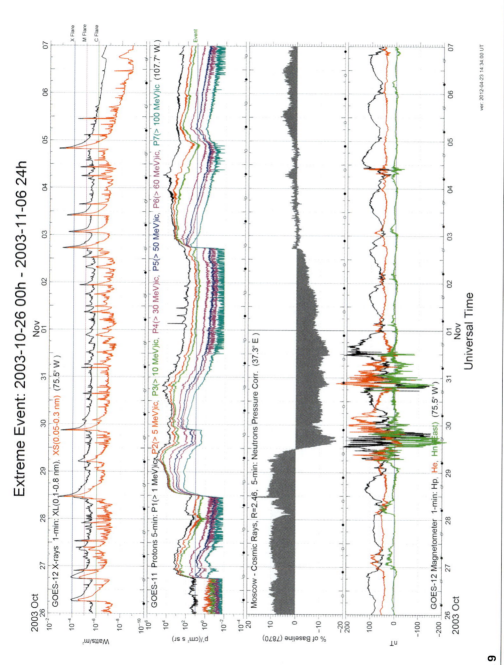

FIG. 9

Case 8: Extreme Event Case for Oct. 26–Nov. 6, 2003.

296 CHAPTER 12 EXTREME SPACE WEATHER EVENTS: A GOES PERSPECTIVE

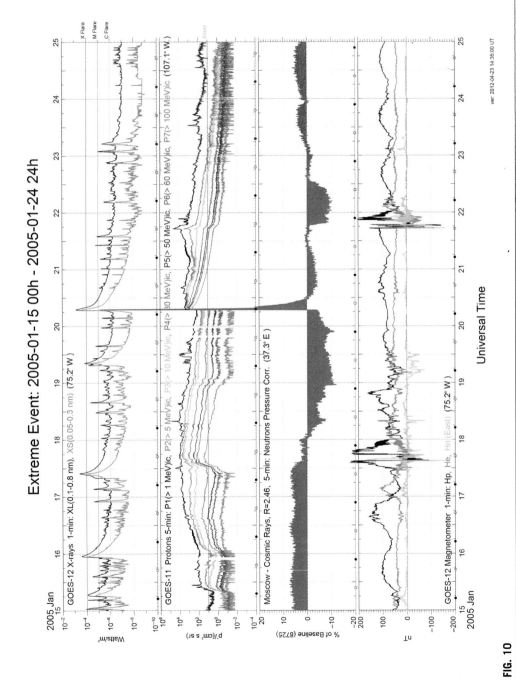

FIG. 10

Case 9: Extreme Event Case for Jan. 15–24, 2005.

2 GOES EXTREME SPACE WEATHER EVENTS 297

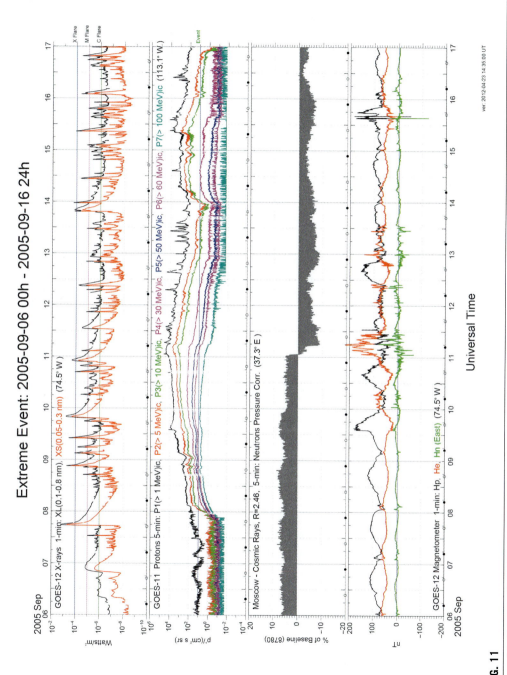

FIG. 11

Case 10: Extreme Event Case for Sep. 6–16, 2005.

298 CHAPTER 12 EXTREME SPACE WEATHER EVENTS: A GOES PERSPECTIVE

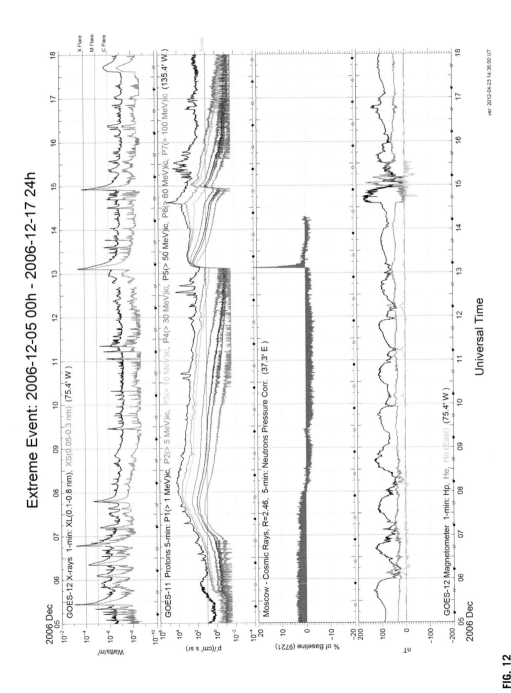

FIG. 12

Case 11: Extreme Event Case for Dec. 5–17, 2006.

2 GOES EXTREME SPACE WEATHER EVENTS 299

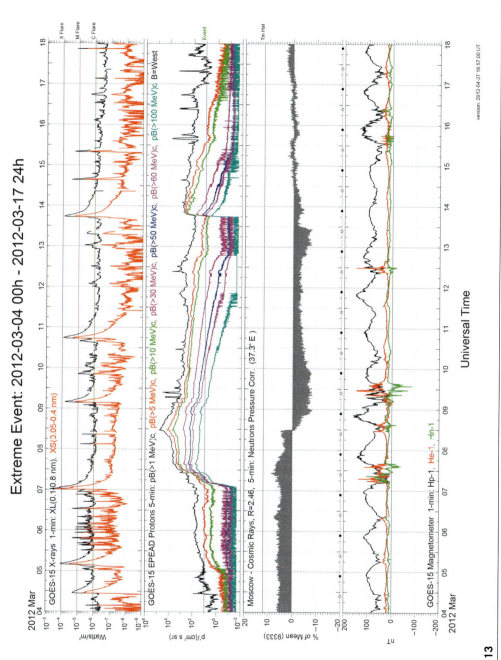

FIG. 13

Case 12: Extreme Event Case for Mar. 4–17, 2012.

to set the extreme event within an environmental context; that is, the occurrence of flares, active region locations, ground-level enhancements (GLEs), and Forbush Decreases (FDs). A GLE refers to a sudden and transient increase, typically minutes, in the cosmic ray response from cascade products, typically neutrons and muons, from highly energetic solar protons interacting with the atmosphere (Gopalswamy et al., 2012; Chapters 1 and 13 of this volume). Throughout this chapter we use the GLE numbering scheme in accordance with Shea and Smart (1992) that is replicated at the World Data Center for Solar-Terrestrial Physics, Moscow, Russia (http://cosrays.izmiran.ru/). An FD is a more persistent decrease, measured on the order of days, in the cosmic ray background due to the scattering of energetic cosmic rays out of interplanetary space (Cane, 2000). As far as we know there is no catalog of FDs. In this chapter we use the convention that an FD is a persistent decrease, of up to several days duration, in the cosmic ray background (Chauhan et al., 2011). GLEs and FDs can be detected by ground-based neutron (Simpson, 2000) and muon (Etchegoyen et al., 2005) detectors. A GLE can be as much as several hundred percent above the cosmic ray background whereas a major FD is a depression in the cosmic ray record with a typical magnitude of order 10% or greater. The top panel of Fig. 2 is a time-series plot of the 1-min averaged ray irradiance in both the XS and XL wavelength bands. Units are in W/m^2. The horizontal lines and key on the right depict the thresholds for C-class ($>10^{-6}$ W/m^2), M-class ($>10^{-5}$ W/m^2), and X-class ($>10^{-4}$ W/m^2) flares, which are derived from the maximum XL irradiance. For the case shown, there is a long-duration X-class flare (X4.5) that occurred late on Mar. 10 and several other flares that barely reached the X-class level over the next 5 days. In accordance with the R-scale, an X4.5 would be classified as a strong (R3) event with significant overall impact. The next panel down plots the energetic protons measured in the geospace. These are 5-min averages of the integrated proton fluxes for energies >1 MeV to >100 MeV at 6 energy thresholds. Units are $p^+/(cm^2/s/sr)$ or, equivalently, pfu. The Solar Radiation S-scale is determined by the greater than 10 MeV (green trace) channel for integrated flux levels which, in this case, exceeded 10^3 pfu for a strong S3 event. The third panel is the cosmic ray flux measured by the Moscow NM. As noted earlier, a GLE within the context of the Moscow measurements would be a positive percentage spike in the measurement for incoming particles of energy ~2 GeV/nucleon and greater; there was no GLE detected by the Moscow NM for this event. The long duration negative deflection in this record is an indication of an FD and possibly that of a Coronal Mass Ejection (CME) in interplanetary space (Cane, 2000). Note that the catalog of CMEs derived from Solar and Heliophysics Observatory (SOHO) satellite observations is only available since 1996, although some earlier observations do exist (Kahler, 1993). Analysis of the detailed transient behavior of the NM record around the time of the percentage decrease is beyond the scope of this work. Finally, the bottom panel is a plot of the GOES magnetic field components within the EPN coordinate system. The large and rapid variations in the magnetic field components on Mar. 13–14, 1989, are indicative of a sizable geomagnetic storm. Again, detailed analysis and assessment of these variations are beyond the scope of this work. Measurements of $Kp=9_o$ (see Table 1) indicate an extreme G5 geomagnetic storm. Supporting measurements of $Ap^*=285$ (no units) justify this event as an extreme case—in fact, an $Ap^*=285$ ranks this event as the third largest since the index was adopted in 1932 and the largest value for the GOES time period (1974 to date). A short-duration (2.6 min) sudden impulse (SI) of large amplitude 60 nT was detected at the start of this storm period at 01:27 UT on Mar. 13 (Araki, 2014). Because the distinction between an SI and a Sudden Storm Commencement (SSC) is the association of the latter with a geomagnetic storm (Burlaga and Ogilvie, 1969), we have chosen to refer only to SIs in this chapter. The related *Dst* level associated with this event was a rare −589 nT response (Riley, 2012).

Within each GOES panel the relevant satellite is identified. For Case 01, measurements from GOES-07 were used in all three GOES panels (X-ray flux, energetic protons, and magnetic field) although in the different cases other GOES satellites were used depending upon their operational availability. The open (closed) dots along the timeline (abscissa) represent when the GOES satellite was at local noon (midnight).

3 GOES EXTREME EVENTS, CASES 1–12

The 12 cases of extreme space weather cover a broad range of conditions including exceptionally intense flares, large SPEs, substantial GLEs, and geomagnetic superstorms. This section includes a brief summary for each case highlighting the most salient features of the event. Case 1 (Mar. 1989) was a warning for modern society that deleterious space weather was a serious matter that should not be ignored. The second case (Sep. 1989) featured the largest GLE (#42) measured at the time since the historic event of February 1956. In Case 3 (Oct. 1989) a series of large X-class flares was associated with a prolonged interval of discrete SPEs that kept the local geospace at a heightened radiation level for the latter half of the month. Case 4 (Mar. 1991) included the Combined Release and Radiation Effects Satellite (CRRES) time frame and the first observation of a rapid injection of energetic charged particles into the "slot" region between the inner and outer van Allen radiation belts. Case 5 (Jun. 1991) featured six powerful X-class flares that were associated with multiple large geomagnetic storms. For case 6 (Bastille Day storm of Jul. 2000) the GOES magnetometer indicated that the magnetopause was pushed inside GEO for at least several hours following the arrival of a CME. The severe radiation storm in Nov. 2001 for Case 7 forced the astronauts aboard the International Space Station (ISS) to seek shelter. In Case 8 (Halloween storms of Oct.–Nov. 2003), the civilian channel for Global Positioning System (GPS) receivers was adversely impacted from a flare-induced solar radio burst (SRB). In Case 9 (Jan. 2005) the large GLE #68 was very prompt, indicative of a magnetically well-connected event and consistent with the GOES SPE record. Case 10 (Sep. 2005) featured multiple X-class flares and a slow-rise SPE. In Case 11 (Dec. 2006) the extreme SRB detected in association with an X6.5 flare was the largest ever recorded, impacting GPS receivers and revealing a potential vulnerability with the Federal Aviation Administration's Wide Array Augmentation System (WAAS). During Case 12 (St. Patrick's Day Storms of Mar. 2012), a Climate and Weather of the Sun-Earth System (CAWSES) study period was under way in which GPS meteorology (Davies and Hartmann, 1997) was extensively used for probing the ionosphere.

CASE 1 (MAR. 10–15, 1989)
Notable features {G5/R3/S3}
- Largest Ap* geomagnetic storm during the GOES era.
- X4.5 flare on Mar. 10 at 19:12 UT from NOAA Active Region 15395 at solar coordinate (SC) location N31E22.
- Expanded auroral zone of particle precipitation down to 40 degrees and measured E-field to 35 degrees.
- Modeled ionospheric response for increased ion temperatures, enhanced NO+ densities with an increased transition height from molecular to atomic ions.

The geomagnetic superstorm of Mar. 13–14, 1989, was a wake-up call to modern society on the hazards posed by extreme space weather (Baker, 2008). While the strength of this storm was likely not as severe as the classic Carrington event of 1859 (Cliver and Dietrich, 2013), the resulting geomagnetic disturbance caused major disruptions to ground and space-based systems (Allen et al., 1989). The most notable impact was the electrical power blackout in Quebec, Canada, that lasted nine hours and affected 6 million customers. Without the intervention of alert power grid operators, it was likely that a cascade outage along the eastern U.S. seaboard would have occurred. A transatlantic telecommunications cable measured a potential difference of ∼700 V (Medford et al., 1989).

As discussed in the last section and as indicated in Fig. 2, the solar disturbance directly associated with this extreme geomagnetic event was a strong X4.5 flare (R3) that first erupted from Active Region 15395 at 18:48 UT on Mar. 10. The peak of the flare was at 19:12 UT with a duration lasting until 21:58 UT, according to the NCEI records. For the remainder of this chapter the timing for flares and other geophysical events will include the start (s), end (e), and peak (p) times parenthetically as {s: 18:48 UT; e: 21:58 UT; p: 19:12 UT} where the Mar. 10 flare was used here as the example. U.S. Air Force (USAF) space weather operators within the Solar Electro-Optical Network (SEON) simultaneously detected this flare in H-alpha of Importance "3" and Brightness "B" (Reid, 1963) from a compact large bipolar group (FKC) type according to the McIntosh (1990) classification. The Mount Wilson magnetic classification for the group was that of a beta-gamma-delta (BGD) configuration for a group containing large, closely separated spots of opposite polarity (Smith and Howard, 1968). In other words, this region was primed and ready for a significant flare eruption.

Allen et al. (1989) noted that for the period of interest here, the near-Earth space environment was in a continued state of heightened space weather as evidenced by the enhanced energetic proton environment shown in panel 2 of Fig. 2. The >10 MeV proton flux peaked at 1860 pfu (S3) near 09:00 UT on Mar. 13, although the maximum is uncertain due to an unfortunate data gap. Overall, during this period multiple X-class flares and numerous M-class flares erupted mostly from Active Region 15395. The largest flare in the sequence was an X15 flare that erupted on Mar. 6, well before the extreme interval of interest here. The cosmic ray record (panel 3) prior to Mar. 13 did not indicate any significant disturbances within interplanetary space suggesting that, perhaps, there were no CMEs or at least none that were geoeffective; recall that this is before the start of the SOHO CME catalog. Also, throughout the entire period there was no striking evidence of any GLE in the Moscow NM record although, as pointed out by Allen et al. (1989), there was an ongoing polar-cap absorption (PCA) event consistent with the enhanced energetic particle environment at GOES.

The Mar. 10 flare had a more significant overall impact on the near-Earth space environment in spite of the fact that the flare occurred on the eastern hemisphere of the sun (SC N35E69) and not in a particularly good geoeffective location (Kahler, 1993). While the postflare energetic particle environment gradually increased above its already heightened preflare level, the gradual increase suggests that this increase may not have been directly related to the flare but rather the result of shock acceleration due to a possible CME structure (Feynman and Hundhausen, 1994). Evidence supporting this conjecture is provided by the start of both an FD and geomagnetic disturbance (panel 4) on Mar. 13. Unraveling the detailed response of the GOES magnetic field is not straightforward. However, the clear and sustained reversal of the north-south aligned Hp magnetic component from 14:00 to 18:00 UT, while GOES-07 was on the dayside, suggests that the terrestrial magnetopause was compressed inside GEO. Allen et al. (1989) indicate that the magnetopause standoff distance may have been compressed to 4.7 Earth radii (R_E), which is about half its normal distance.

Published reports indicate the extent and consequences of this extreme event. Morton and Mathews (1992) showed that the structure of the lower ionosphere at mid-latitudes was significantly disturbed from its normal configuration with the speculation that an imposed E-field was responsible for the unexpected vertical transport of plasma. Evidence supporting this conclusion was provided by Rich and Denig (1992), who showed that the high-latitude electric fields were measured down to 35 degrees magnetic latitude (MLat) by Defense Meteorological Satellite Program (DMSP) spacecraft. These data were presented within context of a greatly expanded auroral zone having measured particle precipitation down to 40 degrees magnetic latitude. Interestingly enough, these authors noted that the measured auroral parameters of precipitation and drift were not exceedingly large during this extreme event but that the expanded equatorial boundary of the auroral zone was responsible for the extreme energy input to the atmosphere. Medford et al. (1989) estimated that the eastward electrojet current was very intense and of order 10^6 amps (delta-H $\sim$ 1000 nT). Reported modeling of the high-latitude ionospheric response to the geomagnetic storm by Sojka et al. (1994) was consistent with increased convection within the polar cap affecting the chemistry of the ionosphere through collisionally induced ion temperature increases.

The Mar. 1989 event was the first significant geomagnetic storm that could be assessed using the advance modeling capabilities and space-based measurements available in the early 1990s (Cid et al., 2014). Despite this, the number of reports describing the conditions leading up to and within the event was not as numerous as the number of publications in subsequent events. However the importance of this extreme event, as noted earlier, was the attention it brought to the field of space weather and to the potential impacts of a devastating space weather event.

CASE 2 (SEP. 26–OCT. 1 1989)
Notable features {G0/~R4/S3}
- Largest cosmic ray GLE measured since Feb. 23, 1956.
- Cosmic ray proton of up to 100 GeV estimated from ground-based NM data.
- Geoeffective X9.8 flare from Active Region 15698 at SC S24W90 @ 11:33 UT.
- GLE time profile suggests bidirectional flows from a backscattered source.

The GOES environmental data for the Sep. 1989 extreme space weather event period are provided in Fig. 3. Quite distinct from Case 1, the most obvious artifact in Fig. 3 is the exceedingly large response of the Moscow NM to GLE #43 on Sep. 29. This GLE was of order several hundred percent above background (Humble et al., 1991) making it possibly the second-largest GLE on record at the time (Cramp et al., 1993; Miroshnichenko et al., 2000; Smart and Shea, 1991) surpassed only by GLE #5 on Feb. 23, 1956, which had a percentage increase of order 4500 (Belov et al., 2005). The GLE #43 on Mar. 29 was also the first to be detected by underground muon detectors (Miroshnichenko et al., 2000). While the rigidity of the Moscow NM is somewhat greater than 2 GeV, published reports indicate that the maximum energy for the incoming solar protons was greater than 19 GeV (Swinson and Shea, 1990), peaking at 30 GeV (Lovell et al., 1998) or possibly 100 GeV (Dvornikov and Sdobnov, 1995). This event is of interest because of its GLE response although there were no notable societal impacts reported for this event period. The Moscow NM record has no clear indication of a discrete FD throughout the interval from Sep. 26–Oct. 1, 1989.

The Sep. 29 GLE was clearly associated with an X9.8 flare {s: 10:47 UT; e: 14:35 UT; p: 11:33 UT} that nearly reached the severe R4 level on the NOAA Space Weather Scales (top panel of Fig. 3). The flare erupted from Active Region 15698, which was located on the presumably geoeffective western limb of the sun near SC S24W90, although the GOES magnetic field record indicates that the event was largely nongeoeffective. When last clearly observed on Sep. 28 at SC S25W89, this group had a much simpler configuration than on earlier days, being that of a unipolar group reaching its final stage of evolution with an A classification, according to McIntosh specification (McIntosh, 1990). Beyond the X-ray measurements, the solar flare was detected in both gamma rays and within the solar radio spectrum (100 kHz–10 GHz) (Bhatnagar et al., 1996). There were some interesting characteristics for the gamma ray line (GRL) measurements at 2.2 MeV, which is the emission from neutral capture by hydrogen (Cliver et al., 1993). Vestrand and Forrest (1993) argue from a geometric perspective that a possible reason for the anomalously strong GRL was that the emissions were from an extended source not limited to a local flare whereas Cliver et al. (1993) alternatively suggest that the source of the GRL was a CME-driven coronal shock. Supporting evidence for the latter mechanism is provided by Kahler (1994) and others (Bhatnagar et al., 1996), who reported an associated CME having an onset at 11:22 UT and a fast ejection speed of 1828 km/s. A herringbone (HB) type II SRB was observed at 11:25–11:56 UT by the former Weisseneu Observatory (Urbarz, 1969), presumably associated with accelerated electrons from the flare blast wave (Cairns and Robinson, 1987; Cane and White, 1989). Closer to Earth, the direct flare effects included a sudden ionospheric disturbance (SID) {s: 10:47 UT; e: 13:00 UT; p: 11:32 UT} of maximum Importance 3 measured by numerous stations across the full range of types, that is, Shortwave Fade (SWF), Sudden Enhancement of Atmospherics (SEA), Sudden Phase Anomalies (SPA), and Sudden Enhancement of Signal Strength (SES) (Lincoln, 1964), having start and peak times in close agreement with those of the X-ray flare. The ionospheric response to the Sep. 29 flare was modeled by Remenets and Beloglazov (1992) wherein the authors note that the response was surprisingly modest in contrast to the intensity of the observed GLE. A weak SI (maximum amplitude of 21 nT), presumably associated with the observed CME, affected Earth at 17:16 UT on Sep. 30, followed by a largely inconsequential magnetic disturbance with a maximum $Kp=4_o$ (G0). The slight geomagnetic disturbance associated with this SI on Sep. 30 is evident in the bottom panel of Fig. 3. The more pronounced geomagnetic disturbance ($Kp=7+$) on Sep. 26 was presumably associated with an earlier solar event, possibly the M3.3 flare that erupted from the unrelated Active Region 15708 on Sep. 25 {s: 23:44 UT, e: 24:06 UT, p: 23:45 UT}. The Sep. 25 solar flare and Sep. 26 geomagnetic storm are not parricularly pertinent to Case 2 and are not considered in the overall ratings in Table 1.

The details of the Sep. 29 GLE (#43) have been discussed by numerous authors, including an extensive retrospective by Miroshnichenko et al. (2000). Fig. 3 indicates the onset of an SPE near 12:00 UT that was more precisely timed with a start at 11:45 UT and peak at 14:55 UT (Sauer, 1993) with a maximum of 2960 pfu (S3). The occurrence of the SPE is nearly simultaneous with the start of GLE #43. A visual inspection of Fig. 3 suggests that the GLE response was prompt although this event was, in fact, a classic gradual GLE (Moraal et al., 2016). Again, the GLE of Sep. 29 was historic in its measured intensity and, to some degree, the energy of the incoming protons. Most experts agree that solar protons can be accelerated to very high energies in CME-driven shocks within either the solar corona or interplanetary space (Kahler, 1994), although other mechanisms may be involved or act in concert with CME shocks (Klein et al., 1999; Ryan, 2000; Moraal et al., 2016). GLE #43 was also interesting in its double-peaked response, which was seen most clearly at higher latitude stations due to the stations' lower cutoff rigidities (Torsti et al., 1991; see also Lovell et al., 1998; Bhatnagar et al., 1996; and Miroshnichenko et al., 2000). This double-peaked response for GLE #43 is not readily apparent in the Moscow NM data presented here although it is apparent in the original records. Various physical mechanisms have been offered for the double peak, including an artifact of a primary and secondary energy release (Bhatnagar et al., 1996) or cosmic ray backscatter from beyond 1 astronomical unit (AU) (Cramp et al., 1993; Lovell et al., 1998).

Given the environmental circumstances surrounding this event period, it is likely that any CME associated with the event was either small or non-Earth directed. The lackluster response of the geomagnetic field and the absence of any clear FD signature would not qualify this event as an example of extreme space weather. Rather, it is the intensity of GLE #43 that makes this event so interesting and worthy of consideration here.

CASE 03 (OCT. 18–31, 1989)
Notable features {G4/R4/S4}
- Unprecedented number of GLEs #43, #44, #45 within a 6-day interval.
- Large (~13%) FD on Oct. 21 associated with X13 flare.
- Enhanced radiation environment with multiple SPEs.
- Satellite solar cells suffered premature radiation degradation.

GOES data for the extreme space weather interval in late Oct. 1989 are provided in Fig. 4. Events during this period included multiple X-class flares, GLEs, SPEs, and related phenomena, all of which could be traced back to the solar activity originating from Active Region 15747. The area of Active Region 15747 throughout this interval was large and of order 1000 MH (millionths of a solar hemisphere) or 0.1% of the solar disk with a complex beta-delta (BD) to BGD magnetic configuration (Smith and Howard, 1968) and compact EKC sunspot group classification (McIntosh, 1990). The three largest X-ray flares were an X13 (R4) on Oct. 19, an X2.9 on Oct. 22, and an X5.7 on Oct. 24, each of which had an associated SPE and GLE. Table 2 lists the flares and associated energetic particle events. The X13 flare was associated with the largest SPE of 42,200 pfu (S4) and GLE #43, although #45 was larger from the perspective of the Moscow NM. Only the Oct. 19 flare had an associated geomagnetic storm of any significance with a maximum $Kp = 8_+$ (G4) on Oct. 20–21 which followed a modest SI (16 nT) at 09:17 UT on Oct. 20. The additional SPE seen in Fig. 4 near 13:00 UT on Oct. 20 (Cane and Richardson, 1995) was not clearly associated with any flare activity or, as argued by Lario and Decker (2002), any CME-driven shock. The SPEs kept the geospace engulfed in high-energy protons and at a sustained radiation storm level for 10-MeV protons from Oct. 19 through nearly the end of the month

Table 2 Case 3: X-class solar flares for Oct. 18–31, 1989

Date 1989	Start UT	End UT	Peak UT	Class	Location SC	SPE Time UT	SPE Flux pfu	GLE Number
Oct. 19	1229	2013	1255	X13.0	S27E10	1310	42,200	#43
Oct. 22	1708	2108	1757	X2.9	S27W31	1755	5330	#44
Oct. 24	1736	2624	1831	X5.7	S30W57	1825	2530	#45

(Torsti et al., 1995). Deep into the atmosphere, this series of SPEs was responsible for significant increases in polar mesospheric NOx (NO and NO$_2$) and for related decreases in polar mesospheric ozone (Jackman et al., 2008), which had a long-term (months) impact on the upper atmosphere. During this interval multiple satellites experienced solar-cell degradation, which resulted in a 5–7 year loss of usable life and may have been responsible for the premature loss of the Insat-1C communications satellite (Odenwald, 2015). While the Moscow NM response for GLE #45 on Oct. 24 was the largest within this interval, it was actually the GLEs on Oct. 19 (GLE #43) and Oct. 22 (GLE #44) that garnered the most interest.

Considering first the solar event of Oct. 19, Shea et al. (1991) reported that the cosmic ray particles detected during the initial phase had an atypical particle anisotropy, suggesting that these particles were the decay product of solar-flare generated relativistic neutrons. Using, in part, data from the Oct. 19 event, Klein et al. (1999) suggested that particle acceleration occurred in the lower to middle corona with the possibility that a CME may have been a contributing factor. This is in contrast to Kahler (1994), who took the position that the most energetic SPEs result from particle acceleration at a single CME shock located at 5–15 solar radii. It is clear from the Moscow GLE #43 signature for Oct. 19–20 that there was one decay rate superimposed upon another, which implies a dual-energization source. The Moscow NM data also indicate that GLE #43 was followed by an interesting FD response suggesting that there may in fact have been two CMEs, one following the other in the geospace. As was noted earlier (bottom panel of Fig. 4), the Oct. 19 flare and CMEs were associated with the G4 geomagnetic storm that occurred on Oct. 20 and persisted for the next 36 hours or so.

The Oct. 22 GLE #44 was not quite noteworthy from the perspective of the Moscow NM, although the event did spur keen interest in the cosmic ray community. Cramp et al. (1997) reported that the energy of the incoming solar protons for this event was ≥ 1 GeV. Given that the cosmic ray rigidity for Moscow is ~ 2 GeV/nucleon, it is possible that the bulk of the energetic protons was not accessible to the detector. Other community interest in GLE #44 concerned its extremely anisotropic onset (Duldig et al., 1993; Cramp et al., 1997; Shea and Smart, 1997). Nemzek et al. (1994) reported that the signature for this event, from the perspective of ground and satellite measurements, was an intense initial spike made up of two individual peaks followed by a longer, slower pulse, which suggested a multisource event as reported by Shea and Smart (1997) (see also Kahler, 1993).

CASE 4 (MAR. 22–28, 1991)
Notable features {G5/~R4/S4}
- New radiation belt formed following deep particle injection event.
- Energetic particles within the slots region pose an unanticipated risk to satellites.
- Largest SPE measured within the GOES interval with a deep FD.

The extreme event of Mar. 1991 is interesting from the standpoint that it was the first observation of trapped radiation within the magnetosphere at a location previously considered to be relatively benign. The typical configuration for the van Allen radiation belts consists of an inner belt of magnetically trapped energetic protons and electrons at equatorial radial distances, or L-shells, from $R_e \sim 1–3$ and an outer belt of energetic electrons at L-shells from $\sim 4–5$. Separating the inner and outer belts is a location commonly referred to as the slot region, due to the paucity of energetic charged particles. Measurements made during the extreme space weather interval of Mar. 1991 showed that, at times, the slot region could be filled with quasitrapped particles that could persist for years (Mullen et al., 1991; Bell et al., 1997). GOES data for the Mar. 22–28, 1991, interval are provided in Fig. 5. During this 7-day period, there were four X-class flares that erupted from Active Region 16555, although it is the X9.4 flare (~R4) on Mar. 22 {s: 22:43 UT, e: 23:17 UT, p: 22:45 UT} that appears to have been the most closely related to the extreme space weather events that ensued. The location (in hydrogen alpha) of this flare was SC S26E28, which is not from a magnetically well-connected location for solar-terrestrial impacts. On Mar. 22, Active Region 16555 had a compact McIntosh classification (McIntosh, 1990) of "FKC" and a flare-ready BGD magnetic configuration (Jaeggli and Norton, 2016; Smith and Howard, 1968). Terekhov et al. (1996) reported a strong neutron capture GRL suggesting that protons during the flare's impulsive phase were accelerated within the flare region to relativistic energies. As indicated in Fig. 5, an SPE began at 08:20 UT on Mar. 23 that peaked at 43,500 pfu (S4) around 04:00 UT on Mar. 24 and maintained high space radiation levels through Mar. 28. The severe SPE was the largest measured during the GOES lifetime, although the SPE associated with the solar-terrestrial events of Aug. 1974 was reported to have been significantly larger, peaking at 110,000 pfu (Kohl et al., 1973). As an aside we note that the 1974 SPE is the only known event that apparently reached an extreme S5 on the NOAA Space Weather Scales. The slow rise in the SPE for Mar. 1991 was most likely due to the poorly geoconnected solar location. The peak of the >10-MeV particle flux was nearly coincident with the occurrence of a large SI (delta B = 103 nT) at 03:41 UT on Mar. 24 (Araki et al., 1997; Araki, 2014) and the start of a geomagnetic storm interval as indicated by the fluctuating GOES magnetic field. The geomagnetic field was greatly disturbed over the next few days with the 3-h Kp peaking at 9_o (G5) several times over the next 32 h. Measurements from the Moscow NM registered a slow rise in the background cosmic ray flux starting just after the X9.4 flare and peaking around the time of the SI on Mar. 24, before diminishing over the next few hours as a major (>10%) FD. The geomagnetic storm and FD were most likely the result of a CME impacting the Earth early on Mar. 24 (Le et al., 2003). An immediate concern for spacecraft manufacturers and operators was that a region of space previously considered relatively void of highly energetic particles could at times present an unexpected and persistent radiation hazard to satellites (Blake et al., 1992a; Gussenhoven et al., 1992). During this interval the GOES-07 satellite suffered a 2–3 year loss of planned lifetime due to solar panel degradation caused by the intense SPE (Shea et al., 1992; Shea and Smart, 1993). Also impacted were radio communications at high latitudes and service disruptions on the Hydro-Quebec power grid (Odenwald, 2015). Another unfortunate consequence of space weather during this interval was the failure of the Maritime European Communications Satellite (MARECS-A) (Bedingfield et al., 1996).

The unexpected filling in of the slot region was first reported by Mullen et al. (1991) from energetic proton measurements obtained on Mar. 24 from CRRES. Blake et al. (1992a,b) then reported, again from CRRES measurements, that nearly simultaneous with the proton injection very highly energetic electrons were also injected into L-shells between 2 and 3. An apparently related consideration for the sudden increase in radiation particles at low L-shells was a bipolar electric field pulse that contributed to the injection and energization of the electrons and ions. Hudson and colleagues (Li et al., 1993; Hudson et al., 1995, 1997, 2004; see also Kress et al., 2007) have modeled the Mar. 24 magnetospheric response to a CME-induced compression and imposed bipolar electic-field pulse. They have shown that conservation of the first adiabatic invariant ($\mu = v_{\perp}^2/2B$) was likely responsible for the energization of incoming solar and resident charged particles that became stably trapped at low L shells. The populations of newly trapped electrons and protons at low L-shells persisted for years following the

Mar. 24 event (Daly et al, 1992; Looper et al., 1994; Hudson et al., 2004). Subsequent to this event it was found that injection and energization of energetic particles into the slot region is not as rare an occurrence as previously thought (Hudson et al., 2004; Kress et al., 2007).

CASE 5 (JUN. 1–20, 1991)
Notable Features {G5/R4/S3}
- Six X-class solar flares erupted during the first 20 days of Jun. 1991.
- Extreme geomagnetic storm occurred Jun. 4–6 with a compressed magnetosphere.
- Radiation storm levels persisted throughout the entire period.
- Lackluster GLE response in the Moscow NM records from the Jun. 11 and Jun. 15 solar flares.

Environmental data for the extreme space weather period that existed during the first 20 days of Jun. 1991 are provided in Fig. 6. Six X-class flares occurred during this interval as Active Region #16659 traversed the solar disk, impacting high-frequency radio communications and exposing astronauts to higher than normal radiation levels (Odenwald, 2015). Five of the six solar flares for this 20-day interval exceeded the capabilities of the GOES-07 XRS at X12 (R4) with the remaining flare being the very respectable X10 on Jun. 9. The relevant characteristics of the flares and Active Region #16659 on the flare days are listed in Table 3. Throughout this entire interval the geospace was very disturbed with the energetic particle background exceeding radiation storm levels from Jun. 4–20. Multiple SPEs and geomagnetic storms occurred during the interval. The three SPEs that occurred on Jun. 4 (224 pfu, R2), Jun. 11 (3000 pfu, R3), and Jun. 15 (1400 pfu, R3), respectively, can be clearly discerned in the second panel of Fig. 6. Also clear in the bottom panel are the major geomagnetic disturbances that existed on Jun. 4–6, Jun. 10–11, Jun. 12–13, and Jun. 17. A maximum $Kp=9_o$ (G5) occurred midday on Jun. 5 at a time when the magnetosphere was compressed inside GEO on the dayside as suggested by a strongly negative Hp component (black curve) of the GOES magnetometer. Nine sudden storm commencements (SSCs) were detected during this interval with the largest being a significant 77-nT event that occurred on Jun. 5 at 09:19 UT. The Moscow cosmic ray data (panel 3 of Fig. 6) also indicated a greatly disturbed geospace with multiple FDs occurring during the interval Jun. 4–19 with a notable >10% decrease occurring on Jun. 13–14. There are reports of numerous GLEs having been detected during this interval (Kovaltsov et al., 1995; Muraki et al., 1991) although the GLE catalog only lists GLE #51

Table 3 Case 5: X-class flares for Jun. 1–20, 1991 associated with Active Region #16659

Date 1991	Start UT	End UT	Peak UT	Class	Location SC	Im/Br	Ha	Station SEON	Config
Jun. 1	1509	1614	1529	X12[a]	N25E90	1F	1FE	RAMY	
Jun. 4	0337	0730		X12[a]	N30E70	3B	3BE	LEAR	BD/DKC
Jun. 6	0054	0212	0112	X12[a]	N33E44	4B	3BE	HOLL	BD/EKC
Jun. 9	0137	0424	0140	X10	N34E04	3B	3BE	PALE	BGD/FKC
Jun. 11	0209	0320	0229	X12[a]	N31W17	3B	3BE	LEAR	BGD/FKC
Jun. 15	0633	1117	0831	X12[a]	N33W69	3B	3BE	SVTO	BD/EKC

[a]GOES-07 XRS Saturation at X12.

and GLE #52, which occurred on Jun. 11 and Jun. 15, respectively. Within the Moscow NM record GLE #52 on Jun. 15 was a rather modest enhancement above background of less than 10% that peaked at around 09:00 UT in association with the threshold-limited X12 flare which peaked at 08:31 UT on the same day. The ground effect of GLE #51 on Jun. 11 cannot be clearly discerned in the Moscow NM records in Fig. 6. However, a more detailed review of the data available from the Cosmic Ray Department of Solar-Terrestrial Division of IZMIRAN does indicate a weak response in association with the X12 flare at 02:29 UT (peak intensity) on Jun. 11. We speculate that the lackluster response in the Moscow NM records for GLEs #51 and #52 suggests that the maximum energy of the incoming protons was close to the cutoff rigidity of ~2 Gev/nucleon, although this suggestion is somewhat at odds with Kockarov et al. (1994). Also, it is noted that there were no direct observations of CMEs, although the presence of classical two-step FDs is a strong indication that the geomagnetic disturbances were associated with CMEs (Cane, 2000) from which shock-accelerated protons might be expected. Published reports of gamma ray emissions (Akimov et al., 1996) support this contention.

Various authors have focused on specific events for the Jun. 1–20, 1991, periods. For example, the extreme geomagnetic storm of Jun. 4–6 has been extensively studied in terms of its near-Earth effects. There were several SIs associated with this magnetic storm with a doublet SI occurring at 14:05 UT {26 nT} and 15:35 UT {42 nT} on Jun. 4 and a separate/stronger SI occurring at 09:19 UT {77 nT} on Jun. 5. While it is risky to do so, it may be reasonable to assume that the Jun. 4 doublet was associated with the solar event on Jun. 1 whereas the separate SI on Jun. 5 was associated with the solar event on Jun. 4. The start of the SPE and minor FD that occurred in the latter half of Jun. 4 suggests that a CME associated with the Jun. 1 solar event was approaching Earth prior to the SI at 14:05 UT and that a faster CME associated with the Jun. 4 solar event reached Earth at 09:19 UT on Jun. 5. This second CME was presumably responsible for the increased solar wind pressure that caused the dayside magnetopause to cross inside of GEO. Burke et al. (1998; 2000) found that the rapid increase in the solar wind dynamic pressure had some interesting consequences, including the penetration of electric fields to subauroral latitudes that led to the development of subauroral ion drifts (SAID) and a highly asymmetric high-latitude potential distribution. Garner (2000) and Garner et al. (2004) modeled the generation mechanisms for subauroral phenomena by the deep penetration of plasma sheet protons (but not electrons) to low L-shells. Kozyra et al. (2002) investigated the cause for changes in the ring current composition during the course of the storm whereas Liemohn et al. (2002) investigated the role of adiabatic energization and de-energization in determining the strength of the ring current. For this storm period, Daglis et al. (2000) suggested that the importance of substorms on the buildup of the ring current was underappreciated. A separate study on the thermospheric response to the Jun. 11–13 magnetic storms (maximum $Kp=8_-$) was conducted by Smith et al. (1994) who examined the role of the IMF B_Y component on momentum coupling between high and middle latitudes.

Quite distinct from the investigations of the geomagnetic storms of Jun. 1991, much of the direct solar interest was on the flares that occurred later in the month. During the X12 solar flare of Jun. 11 gamma rays exceeding 1 GeV were measured by space-borne platforms (Bertsch et al., 1996; Kanbach et al., 1993). The source of these high-energy gamma rays was attributed to nuclear interactions and electron bremsstrahlung although variations in the intensity and spectral shape of the emissions over the course of the flare suggested changes in the flare acceleration mechanisms (Dunphy et al., 1999). For the solar flare of Jun. 15, the detected gamma ray energies were even higher and measured up to 4 GeV (Akimov et al., 1996; Rank et al., 1997). These data were used to calculate the energy distribution of the source population responsible for the emissions (Young et al., 2000). Energetic protons stochastically accelerated within the flare were found to be consistent with the measured gamma-ray spectrum whereas ground-based NM measurements support separate proton acceleration processes occurring within the impulsive and postimpulsive phases of the flare (Kockarov et al., 1994; Lockwood et al., 1993). Comparing the time profiles of the neutral-capture GRL at 2.22 MeV, Rank et al. (2001) found "striking" similarities in the events of Jun. 9, Jun. 11, and Jun. 15 in the extended flare phase, which the authors attribute to a continuous particle acceleration mechanism.

CASE 6 (JUL. 14–16, 2000)—BASTILLE DAY STORM
Notable Features {G5/R3/S4}
- X5.7 flare associated with a massive, high-speed halo CME.
- Shock front of magnetic cloud formed new proton/electron belt at $L \sim 2.5$.
- Prompt ionospheric effects from increased, flare-induced, solar irradiance.
- Long-term variations in atmospheric NOx from energetic particle precipitation.

The Bastille Day Storm refers to the extreme geomagnetic storm ($Kp=9_o$; G5) that occurred on Jul. 15, 2000. Preceding the storm was an X5.7 flare (R3) that erupted on Jul. 14 {s: 10:03 UT; e: 10:43 UT; p: 10:24 UT} at an Earth-effective location of SC N22W07 within Active Region #19077. In H-alpha this flare classification was 3B {Importance: 3; brilliance: B} (Wittman, 2012). Active Region #19077 had a flare-favorable beta-gamma-delta (BGD) configuration in accordance with the Mount Wilson magnetic classification (Smith and Howard, 1968) and a modified Zurich classification of FKI (McIntosh, 1990). Fig. 7 provides a record of this 3-day extreme event for Jul. 14–16, 2000, from a GOES perspective. The solar X-ray record (top panel) clearly shows the occurrence of the X-class flare just after 10:00 UT on Jul. 14. The second panel indicates that the start of a strong SPE (24,000 pfu; S4) which crossed the 10-MeV event threshold of 10 pfu at 10:45 UT with a quick rise due to the favorable flare location. The Moscow NM record in panel 3 includes a modest but long-duration GLE (GLE #59), also in close proximity with the time of the flare. Case 6 was the first extreme space weather event in our set for which routine CME event listings were available from the Large Angle and Spectrometric Coronagraph (LASCO) instrument aboard the Solar and Heliospectric Observatory (SOHO), launched in Dec. 1995. On Jul. 14, 2000, a full-halo CME lifting off the sun was observed at 10:54 UT, having an Earth-directed speed of 1673 km/s and mass of 1.4×10^{16} g. Based on the characteristics of this CME, Wang et al. (2001) and Zank et al. (2001) accurately predicted that the remnants of this CME would be detected by the Voyager 2 spacecraft (at ~ 70 AU) in the Jan. 2001 time frame (Burlaga et al., 2001). Of the two periods of geomagnetic disturbance indicated in panel 4, the disturbance on Jul. 15 was presumably associated with the Jul. 14 flare whereas the disturbance on Jul. 14 may have been associated with an X1.9 flare (not shown, but no observed CME) that occurred on Jul. 12 from the same Active Region #19077. A major SI of 112 nT amplitude was detected at 14:37 UT on Jul. 15, corresponding to the start of the extreme geomagnetic activity. During this storm period, the GOES-10 Hp magnetic component was negative for several hours while GOES-10 was on the dayside, indicating that the magnetopause was compressed inside of GEO at this time. The Bastille Day Storm caused star-tracking navigation systems on several satellites to malfunction and the Advanced Satellite for Cosmology and Astrophysics (ASCA) satellite began to tumble in orbit (Prolss, 2004). An aurora was seen in El Paso, Texas. Power grids and GPS systems were degraded.

The X5.7 solar flare that preceded the Bastille Day geomagnetic storm was not excessively large, although the associated CME was certainly geoeffective. A number of authors have investigated the underlying structure of Active Region #19077 to better understand possible generation mechanisms for the flare and the associated CME (Deng et al., 2001; He and Wang, 2008). In a general sense, the BGD magnetic configuration and FKI classification suggest that the active region was prime for flaring and for CME emergence (Liu and Zhang, 2001) although the magnetic topology of the filament channel provided some additional clues into the magnetic reconnection process (He and Wang, 2008). Relatedly, Kosovichev and Zharkova (2001) found it surprising that rapid (10–15 min) variations in the magnetic field over a large surface area within the lower solar atmosphere were present at flare initiation. Somov et al. (2002) attempted to explain the large-scale structure and dynamics of the flare in terms of three-dimensional reconnection at a magnetic separatrix in the corona.

Closer to Earth, the Jul. 14 flare had an immediate impact on the upper atmosphere (ionosphere and thermosphere) and magnetosphere as a consequence of the CME and resultant geomagnetic storm. Several authors have examined the prompt effects of changes in the extreme ultraviolet (EUV) and X-ray irradiance during the flare radiative phase, finding that the dayside atmosphere responded dramatically on time scales of minutes (Tsurutani et al., 2005) leading to pronounced changes in the ionospheric density and structure (Meier et al., 2002; Huba et al., 2005; Dmitriev et al., 2006). Other immediate effects included GLE #59, which was detected near 10:30 UT on Jul. 14 (Duldig, 2001; Duldig et al., 2003; Wang, 2009; Mishev and Usoskin, 2016a,b). The anisotropic nature of this GLE determined from the NM network was suggestive of a "magnetic bottleneck" for relativistic solar protons, which were reflected back from an earlier CME, which, at the time, was beyond Earth (Bieber et al., 2002). The LASCO catalog indicates that a halo CME (mass unknown) erupted from the sun on Jul. 11 at 13:25 UT with a speed of 1076 km/s, which (ballistically) would place the magnetic cloud just beyond Earth consistent with the suggestion of Bieber et al. (2002). The prompt SPE {24,000 pfu; S4} was one of the largest observed in solar cycle 24, which allowed Lario et al. (2001) to place an upper limit on recent ~50–100 MeV SPE intensities at Earth. When the fast CME {1674 km/s} associated with the Jul. 14 X5.7 flare finally impacted Earth the next day, it affected the global magnetic field topology of the coupled magnetospheric-ionospheric system (Rastatter et al., 2002). The impulsive nature of the SI pushed that dayside magnetopause inside GEO, produced a new long-lived proton belt (Hudson et al., 2004; Kim et al., 2016), and formed an unstable ring current (Tsurutani et al., 2009).

The Bastille Day storm also had major impacts on the ionosphere. Basu et al. (2001; 2005) reported that transient increases in the ring current were due to the subauroral penetration of high-latitude electric fields, which affected the local electron density structure and increased ionospheric scintillation effects. Horvath and Lovell (2010) studied the development of a storm-enhanced plasma density (SED) feature and its relationship to the equatorial ionization anomaly (EIA), concluding that SED development was a complex process that defied a single (simple) explanation. Other ionospheric and atmospheric consequences of the Bastille Day storm were changes to the global total electron content (TEC) patterns (Kil et al., 2003), increases of ionospheric-induced radio wave absorption near the South Atlantic Anomaly (SAA) (Nishino et al., 2006), and long-term changes to the atmospheric constituents, particularly NO_x (NO and $NO2$) (Jackman et al., 2008).

CASE 7 (NOV. 4–8, 2001)
Notable Features {G5/R3/S4}
- SPE injection results in stably trapper particles within the slot region.
- Electron density variability controlled by SPEs with proton energies >50 MeV.
- Preconditioning of the interplanetary medium controlled energetic particle signatures.
- Storm-induced upwelling drastically impacted ionospheric height profiles.

The GOES environmental records for the extreme space weather event of Nov. 4–8, 2001, are provided in Fig. 8. At first blush it is hard to say why this interval might be classified as extreme except for the long-duration SPE seen in panel 2 for Nov. 4–7, 2001. An interesting feature of the SPE was its initial rapid rise at 17:05 UT to ~30 pfu (S1) and subsequent long duration climb to just over 10,000 pfu (S4) (visual inspection) at 03:00 UT on Nov. 6. The SPE was initially seen on Nov. 4 in conjunction with a minor X1 (R3) flare {top panel; s: 16:03 UT/e: 16:57 UT/p: 16:20 UT} located at SC N06W18. Seen in H-alpha, this flare had a 3B classification (Wittman, 2012). The flare originated from Active Region #9684 which, at the time (18:20 UT), had a BGD magnetic configuration (Smith and Howard, 1968) and EKI classification (McIntosh, 1990) with 39 spots, a longitudinal extent of 13 degrees, and an area of about 600 millionths (<0.1%) of the solar disk. Associated with the flare was a CME having an initial estimated speed of 1810 km/s, corresponding roughly to ~1 day transit time to Earth. The geomagnetic storm (panel 4) that followed on Nov. 6 had a maximum $Kp=9_o$ (G5) and a duration of less than 1 day. An SI, signaling the start of the storm interval when the CME presumably hit the Earth, occurred at 01:51 UT on Nov. 6 with an amplitude of 80 nT and a duration of 2.4 min. The solar wind dynamic pressure at the start of this event increased to almost 20 nPa (Marshall et al., 2012). The Moscow NM record (panel 3) for Nov. 4 contains only a hint of the GLE (Event #62) that was reported for this day between 15:00 and 16:00 UT although the FD that apparently resulted from the CME interacting with the cosmic ray background can be seen in the Moscow NM data in conjunction with and following the geomagnetic storm and residual SPE on Nov. 6. The hourly Dst for this storm had a maximum offset of −292 nT near 07:00 UT on Nov. 6. During this disturbed interval astronauts in the ISS were advised to seek shelter and auroras were seen as far south as Texas. Power grid alarms were tripped and one transformer failed in New Zealand (Marshall et al., 2012).

Interest in extreme space weather events continued into the 21st century with a broader set of scientists studying their physical characteristics and societal impacts. Increased satellite drag is a known consequence of strong geomagnetic storms (Allen et al., 1989) and this early November event offered Mishin et al. (2007) the opportunity to examine the prompt thermospheric effects of short and mesoscale perturbations from enhanced electric fields in the coupled ionosphere and thermosphere (Cooper et al., 1995). Precipitating energetic particles can also affect the lower atmosphere, down to 50 km, and disrupt Very Low Frequency (VLF, 3–30 kHz) radio wave propagation (Clilverd et al., 2005). The penetration of polar electric fields to subauroral latitudes during the storm main phase (also the storm of Mar. 31, 2001) was studied by Veenadhari et al. (2010) who found that the resultant ionospheric currents increased the ionospheric conductivity and enhanced the equatorial ionization anomaly (EIA), otherwise known as the Appleton anomaly. The emergence of GPS as an ionospheric probe allowed researchers (Marayama and Ma, 2004) to extract exquisite knowledge regarding the structure of the ionospheric Chapman layers due to upward plasma diffusion, although the role of penetrating electric fields needs to also be considered. The CME that was the cause of the Nov. 6 extreme geomagnetic storm had an abrupt impact on the near-Earth space environment. Blake et al. (2005) (see also Kuznetsov et al., 2003a, 2005a; Rawat et al., 2006) reported that the CME shock energized ions and electrons to 10s of MeV and injected these particles onto low L-shells that subsequently became stably trapped within the nominal slot region (see case #4, Mar. 22–28, 1991). However, other authors (Alex et al., 2005) argued that an interplanetary cloud resulted in multiple injections and that substorm activity may be involved in the overall energy transfer process (Marshall et al., 2012).

The previous paragraph considered the resultant effects from the active sun on Nov. 4 and, at the risk of throwing off the timing flow, we return to the prompt effects of the X1 flare on Nov. 4 and the progression of this space weather event through the geospace. X-ray, and gamma-ray emissions are ubiquitous within solar flares, which Kuznetsov et al. (2003b) have catalogued for the period Aug. 2001–Aug. 2002. Comparing multiple flares with high-energy (>100 MeV) gamma ray emissions, Kurt et al. (2009) reported similar spectral signatures associated with neutral pion generation and decay within the flare. SPEs can be generated within solar flares by magnetic reconnection or by energization processes associated with CMEs in the corona or interplanetary space (Gopalswamy et al., 2012). The Nov. 4 flare, among several other flares, was studied by Nitta et al. (2003), who reported that the energetic particle signatures were consistent with a slowly rising and expanding structure within the lower corona. On the relationship between solar particle release (SPR) times and GLEs, Reames (2009) found some consistency in the interplanetary transition time for magnetically well-connected solar events ($\sim$30 degrees to 90 degrees west helio-longitude) although the Nov. 4 event at SC N06W18 was called out as a puzzling exception.

CASE 8 (OCT. 26–NOV. 6, 2003)—HALLOWEEN STORMS
Notable Features {G5/R4/S4}
- Severe compression/reconfiguration of the van Allen radiation belts.
- Severe SPE on Oct. 20 was the third largest during the GOES lifetime.
- SPE impacts persisted for months in the composition of the middle atmosphere.
- Dual-source Morton wave initiated on the flanks of a coronal CME.
- First event with reported GPS interference due to SRB.

The interval from late October into early Nov. 2003 was a period of extreme space weather during which there were significant socioeconomic impacts (Balch et al., 2004; Wik et al., 2009; Cid et al., 2014). The GOES space environmental data for Oct. 26–Nov. 6 are provided in Fig. 9. Table 4 is a listing of the multiple X-class flares that erupted from the sun from Active Regions #10494, #10486, and #10488 during this interval. A record of the flares is provided in the top panel of Fig. 9. The largest flares in this interval were the severe X17 (R4) flare on Oct. 28 and the extreme >X28 flare (not included in the Space Weather Scales) on Nov. 4, the latter of which saturated the XRS on GOES-12. Our focus herein is the back-to-back X17 and X10 flares of Oct. 28 and Oct. 29, respectively, which had the greatest Earth impacts as indicated by the GOES magnetometer response for Oct. 29–31 (panel 4). The GOES data indicate that there were two distinct geomagnetic storms on Oct. 29 and Oct. 30, which we suggest were most likely related to the Oct. 28 and Oct. 29 flares, respectively. Each storm had a maximum $Kp=9_o$ corresponding to an extreme G5 event. The magnetometer record also indicates that the dayside magnetopause was compressed inside of GEO (strongly negative Hp component) at least intermittently during each storm (Dmitriev et al., 2005). Only the first storm had a clear onset indicated by a significant SI

Table 4 Case 8: X-class solar flares for Oct. 26–Nov. 6, 2003

Date 2003	Start UT	End UT	Peak UT	Class	Location SC	Im/Br	Wilcox	Group AR#	McIntosh
Oct. 26	0557	0733	0654	X1.2	S15E44	3B	BGD	10486	FKC
Oct. 26	1721	1921	1819	X1.2	N02W38	1N	BGD	10484	EKC
Oct. 28	0951	1124	1110	X17.2	S16E08	4B	BGD	10486	FKC
Oct. 29	2037	2101	2049	X10.0	S15W02	2B	BGD	10486	FKC
Nov. 2	1703	1739	1725	X8.3	S14W56	2B	BGD	10486	EKC
Nov. 3	0109	0145	0130	X2.7	N10W83	2B	BGD	10488	FKC
Nov. 3	0943	1019	0955	X3.9	N08W77	2B	BGD	10488	FKC
Nov. 4	1929	2006	1950	X28	S19W33	3B	BGD	10486	EKC

Bolded text identifies flares associated with Active Region #10486.

{$\Delta B = 68$ nT, $\Delta t = 3.4$ min} that occurred at 06:10 UT on Oct. 29. A prior SI {$\Delta B = 8$ nT; $\Delta t = 3.2$ min} that occurred at 02:05 UT on Oct. 28 was probably related to an earlier flare and CME, which are discussed below. Panel 2 is a record of energetic protons measured during this 12-day period indicating a series of successive SPEs. The X17 flare was associated with a severe SPE measuring 29,500 pfu (S4) whereas the subsequent SPE, related to the X10 flare, was a strong S3 event at 3300 pfu although it is unclear how the second event can be clearly discerned from the first. Regardless, the X17-related SPE was one of the larger SPEs on record since the classic Aug. 1972 event (Kohl et al., 1973; Mewaldt et al., 2005a,b) although, within the GOES records, the SPEs that occurred on Oct. 19, 1989, (Case 3) and Mar. 23, 1991, (Case 4) were larger. Also associated with the X17 (Oct. 28) and X10 flares (Oct. 29) were respective halo CMEs. The X17 flare was associated with an ultrafast CME having an estimated speed of 2459 km/s {mass unknown; first observed: 11:30 UT} whereas the speed for the X10-related CME had a very respectable 2029 km/s {mass: 1.7×10^{16} g; first observed: 20:54 UT}. The Moscow NM detected a modest, ~15%, GLE #65 which we associate with the Oct. 28 CME and related flare (Moraal et al., 2005; Vashenyuk et al., 2005a; Kane, 2009). Prior to GLE #65 there was a significant decrease in the cosmic ray background that was possibly related to an X1.2 flare {see Table 4; s: 17:21 UT; e: 19:21 UT; p: 18:19 UT} on Oct. 26 and an associated CME {speed: 1537 km/s; mass: 2×10^{16} g; first observed: 17:54 UT} from Active Region #10484 at a geoeffective location of SC N02W38. Following GLE #65, the Moscow NM intensity decreased from mid-morning on Oct. 29 through noon as the X17.2-related CME presumably reached Earth and disrupted the cosmic ray background. The minimum Dst for the two successive storms was −353 nT at 01:00 on Oct. 29 and −383 nT at 23:00 UT later the same day. A Dst less than −250 nT is considered to be a geomagnetic super storm (Hamilton et al., 1988). Due to the time of year, the extreme geomagnetic disturbances on Oct. 29–31 are often collectively referred to as the Halloween storms. These extreme geomagnetic storms disrupted airline communications and GPS and disabled the Advanced Earth Observing Satellite (ADEOS-II) (Cerruti et al., 2006). The Federal Aviation Administration's WAAS was severely affected and the drillship Global Shipping Forum (GSF) C.R. Luigs encountered significant disruptions in differential GPS (DGPS) positioning (Odenwald, 2015).

Chen et al. (2005) reported that an SRB associated with the X17 flare {s: 09:51 UT; e: 11:24 UT; p: 11:10 UT} on Oct. 28 was responsible for severe signal corruption on GPS receivers using, in this case, codeless/semicodeless tracking. The authors note that the peak solar radio flux between 4000 and 12,000 solar flux units (sfu $= 1 \times 10^{-22}$ Watts/m^2/s/Hz) on the 1410-MHz channel of the Radio Solar Telescope Network (RSTN) was below the accepted interference threshold of 40,000 sfu for GPS interference and well below the 200,000 sfu level predicted for significant GPS impacts (Klobuchar et al., 1999). However, Cerruti et al. (2006) point out that Chen et al.'s finding should not be surprising given the increased sensitivity to the carrier-to-noise ratio for receivers using codeless or semicodeless tracking (see the related discussion in Case 10 for Sep. 6–16, 2005). Rodriguez-Bilbao et al. (2015) also point out that a significant Sudden Increase in Total Electron Content (SITEC) event associated with the same X17 flare may have been a compounding problem for GPS receiver performance.

The October to November storms are perhaps the most extensively studied space weather event to date (Gopalswamy et al., 2005a–c, 2006; Lopez et al., 2004; Plunkett;, 2005). While the effects of this storm were severe in the local geospace, remnants of the related CMEs ejected from the sun during this period were eventually detected at the far reaches of the heliosphere by Voyager 2 some 6 months later when the spacecraft was located at 73 astronomical units (AU) (Burlaga et al., 2005; Richardson et al., 2005; Lario et al., 2005). This event has also been used as the benchmark for astronaut radiation exposure simulations (Kozarev et al., 2010; Mertens et al., 2010; PourArsalan et al., 2010; Schwadron et al., 2010) and for estimating risks to terrestrial power networks from geomagnetically induced currents (Marshall et al., 2012; Matandirotya et al., 2015).

In late October, Active Region #10486 was a complex BGD/FKC region, which Liu and Hayashi (2006) (see also Liu et al., 2007) describe as being "favorable" for launching fast CMEs. In a numerical simulation of an erupting flux rope for the Oct. 28 event, Krall et al. (2006) found that their model nicely reproduced the timing, strength, and orientation of the CME at 1 AU (see also Shiota and Kataoka, 2016). Energetic particles in the geospace can be accelerated by flares and/or at CME shock fronts. With regard to the solar flares of Oct. 28 and Oct. 29, Hurford et al. (2006) contend that the particle acceleration source regions leading to the production of the 2.22 MeV neutron-capture GRL suggest an acceleration process within the flare versus a shock-driven process (see also Veselovsky et al., 2004; Watanabe et al., 2006). This argument is in agreement with Miroshnichenko et al. (2005a,b) where the anisotropic distribution and timing of cosmic ray signatures from the ground NM network were consistent with protons and electrons having been accelerated within the flaring active region (see also Bieber et al., 2005a). This is not to diminish the importance of interplanetary shocks in the acceleration of solar particles (electrons and ions) because it is probable that the higher energy protons (>2 GeV/nucleon for the Moscow NM) responsible for GLE #65 were accelerated at the shock front of the observed CME (Munakata et al., 2005). An interesting observation associated with the X10 flare on Oct. 29 was the occurrence of solar Moreton waves, which, as explained by Muhr et al. (2010), were due to an expanding coronal CME (see also Liu and Hayashi, 2006).

The Halloween storm period also had a dramatic impact on the terrestrial magnetosphere, ionosphere, and lower atmosphere (Mannucci et al., 2005; Miyashita et al., 2005; Liu et al., 2006; Yamauchi et al., 2006; Pawlowski and Ridley, 2009; Shprits et al., 2011). The CME that was associated with the X17 flare on Oct. 28 presumably reached just after 06:00 UT the following day. The compression of the dayside magnetopause and the prompt penetration of the solar wind electric field restructured the van Allen belts and effectively filled the nominal slot region with extremely energetic electrons (10–20 MeV) (Baker et al., 2004; Kress et al., 2007, 2008; Shprits et al., 2006) while depleting the energetic proton flux within the inner belt (Looper et al., 2005). Within the ionosphere, extraordinary density variations were observed in direct response to the flare (Dmitriev et al., 2006; Manju et al., 2009; Tsurutani et al., 2005; Villante and Regi, 2008; Zhang and Xiao, 2005; Osepian et al., 2009; Clilverd et al., 2006) and the geomagnetic storm (Cander and Mihajlovic, 2005; Chi et al., 2005). Other ionospheric features observed during the storm were a rare 6300-angstrom measurement of daytime aurora over Boston (Pallamraju and Chakrabarti, 2005) and a deep penetration of the high-latitude electric field to subauroral latitudes, which increased the severity of radiowave scintillation (Basu et al., 2005) and the related generation of subauroral polarization streams (SAPS) (Foster and Rideout, 2005). Prompt electric fields at the start of the Halloween storm period may have also been responsible for enhanced ionospheric ion outflow (H^+ and O^+) that would tend to mass load the magnetosphere (Harnett et al., 2008). Thermospheric effects from the geomagnetic storms of Oct.–Nov. 2003 included 200–300% increases in the local neutral density at satellite altitudes (Sutton et al., 2005) and an interesting overcooling effect due to increased nitrogen oxide (NO) (Lei et al., 2012). Observed thermospheric recovery times of 6–8 h were not well modeled (Lei et al., 2011). Signification changes in the composition of the lower atmosphere during this period that persisted long after the event were attributed to the severe SPE, including significant ozone losses (Degenstein et al., 2005; Funke et al., 2011; Verronen et al., 2005) and related composition changes (Jackman et al., 2008). As a holistic exercise to understand cause and effect, Toth et al. (2007) used the Space Weather Modeling Framework (SWMF) (Toth et al., 2005) in a sun-to-thermosphere simulation of the Halloween storms, which yielded reasonable results.

CASE 9 (JAN. 15–24, 2005)
Notable Features {G4/R3/S3}
- 5 X-class flares (maximum X7.1) erupted from Active Region #10720.
- Largest GLE measured since the 1956 event.
- Defining example of a two-component GLE with prompt and delayed responses.
- Rapid variations in the flare.

The space environmental conditions for Jan. 15–25, 2005, are provided in Fig. 10. This period included the most intense GLE recorded in almost 50 years. GLE #69 was first detected at 06:49 UT on Jan. 20 in association with an X7.1 (R3) solar flare {s: 06:36 UT; e: 07:26 UT; p: 07:01 UT} that erupted from Active Region #10720 at a geoeffective location of SC N14W61. Other X-class flares detected during this interval were X1.2 and X2.6 flares on Jan. 15, an X3.8 flare on Jan. 17, and an X1.3 flare on Jan. 19, all from the same Active Region #10720, as this active region rotated across the solar disk from SC N14E08 (Jan. 15) to SC N11W81 (Jan. 22). It is apparent from Fig. 10 that there was a series of successive SPEs that had some correlation to the flares, although the correlations were not one-to-one. While the SPE correlation was strong for the X7.1 flare and GLE #69 on Jan. 20, it is interesting to note that only the slow-rise SPE on Jan. 15–16 is included in the NOAA SPE catalog {s: 02:10 UT on Jan. 16}. The obvious reason for this is that once the radiation environment from >10 MeV protons crossed above the event threshold (see indication on the right of Fig. 10) it remained above this threshold until late on 22-Jan. It is also interesting to note that the X2.6 solar flare {s: 22:25 UT; e: 23:31 UT; p: 23:02 UT} and initial rise in the SPE background at ~23:00 UT (visual inspection) on Jan. 15 were likely associated with an exceedingly fast halo CME {speed = 2861 km/s; mass = 1.8×10^{16} g; first observed: 23:07 UT on Jan. 15}. This CME was likely responsible for the geomagnetic disturbance that started on Jan. 16 but transitioned into a full-fledged geomagnetic storm {maximum $Kp=8_-$, G4} on Jan. 18, following a modest 12 nT SI at 07:48 UT on Jan. 17. Data in Fig. 10 also suggest that the multiday evolution of the FD may have been correlated with the X3.8 flare {s: 06:59 UT; e: 10:07 UT; p: 09:52 UT} on Jan. 17 and the initial rise in an SPE at 10:30 UT that increased to 5040 pfu (S3) as the result of an associated halo CME {speed: 2094 km/s; mass: 3.2×10^{16} g; first observed: 09:30 UT on Jan. 17}. However, complicating the association was the massive halo CME ejected at 23:07 UT on Jan. 15, which was previously noted.

Later in the month there was a halo CME (speed: 882 km/s; mass unknown; first observed: 06:54 UT) associated with the X7.1 flare {s: 06:36 UT; e: 07:26 UT; p: 07:01 UT} and GLE #69 {s: 06:49 UT} on Jan. 20, which was apparently responsible for a 41 nT SI at 21:17 UT on Jan. 21 and a severe G4 geomagnetic storm ($Kp=8_o$). A significant (>10% depression) FD was in process on Jan. 21–22. While the SPE on Jan. 20 only reached the level of a strong S3 event for 1000 pfu, it was nevertheless dangerous because it was magnetically well connected to Earth and the radiation particle flux arrived so quickly after the start of the X-ray flare that there was little warning for astronauts on the ISS to take cover (Odenwald, 2015).

The characteristics and nature of GLE #69 provide the rationale for declaring this an extreme space weather event. This GLE was (and remains) possibly the largest event measured since GLE #5 on Feb. 23, 1956 (Meyer et al., 1956; Vashenyuk et al., 2006; Plainaki et al., 2007; Butikofer et al, 2009; Kane, 2009). Measurements of GLE #69 made by neutron monitors at various worldwide locations had count rate increases anywhere from several percent (Miyasaka et al, 2005) up to factors exceeding 50 (Fluckiger et al., 2005; Bieber et al., 2013; Belov et al., 2005). Timing of the GLE relative to the flare was also uncharacteristically prompt (D'Andrea and Poirer, 2005; Zhu et al., 2005) with a quick rise time of 5 min, indicating that the magnetic connection from the Sun to Earth was excellent for this event (Simnett, 2006, 2007; Simnett and Roelof, 2005); recall that the flare erupted from SC N14W61. The energy of the incoming protons was determined to be no less than 20 GeV (Bieber et al., 2005b; Bostanjyan et al., 2007; Chilingarian, 2009). A controversy remains as to whether highly energetic transient events in the geospace for this event were due to acceleration processes occurring at the flare site (Mewaldt et al., 2005b; Grechnev et al., 2008; Labrador et al., 2005; Moraal et al., 2016) and/or at the shock front of the CME (Simnett et al., 2005; Gopalswamy et al., 2005d; 2006; Masson et al., 2009). Numerous authors noted that GLE #69 had the signatures of two, and possibly three, distinct acceleration mechanisms (Bieber et al., 2013; Masson et al., 2009; McCracken et al., 2008; Moraal et al., 2005, 2016; Perez-Peraza et al., 2008; Struminsky, 2005; Timofeev et al., 2005; Vashenyuk et al, 2005b). Subsequently McCracken et al. (2008) proposed that this event be considered as the defining example of a GLE.

The solar and near-Earth space environments also garnered considerable interest for the events of Jan. 15–24, 2005. Arkhangelskaja et al. (2005) examined the spectral characteristics of gamma ray emissions for the X-class flares in this interval showing the not-surprising presence of the positron annihilation line (0.51 MeV), the neutral capture on hydrogen line (2.22 MeV), and other nuclear lines (see also Kuznetsov et al., 2005b). What was surprising were the rapid temporal variations seen in gamma ray flux for the Jan. 20 solar flare, indicating the presence of thin structures within the flaring region (Fletcher et al., 2015). At the other extreme, Manchester et al. (2014) modeled the generation and propagation through space of a flare-related CME, referenced to the Jan. 20 event, suggesting that the atypical structure of the cloud intercepting Earth included dense solar filament materials behind the main CME sheath (see also Foullon et al., 2007). A companion paper by Kozyra et al. (2013, 2014) modeled the magnetospheric response to this CME cloud and, in particular, a northward turning of the IMF that tended to inflate the now-closed magnetosphere and lead to a variety of interesting observational phenomena (McKenna-Lawlor et al., 2010; Saiz et al, 2005; Simnett, 2003; Du et al., 2008). Dmitriev et al. (2014) noted that at the height of the storm, the magnetopause was located inside GEO for several hours as indicated by the negative Hp offset in Fig. 10 at a time when the solar wind dynamic pressure reached an extraordinarily high level of 86 nPa (Liu et al., 2011). The extreme compression of the magnetopause increased the population of high-energy protons within the inner magnetosphere (Dandouras et al., 2009). Li et al. (2011) and Lu et al. (2016) examined the energy input to the atmosphere at high latitudes for the separate geomagnetic storms of Jan. 17 and Jan. 20, respectively. A general consideration of predicting storm-driven thermospheric mass densities for years 2002–2005 has been offered by Liu et al. (2010). Other interesting aspects of the event included a detailed analysis of the FD response on Jan. 17 to a fast-halo CME on Jan. 15, which occurred in conjunction with an X2.6 flare as noted earlier (Plainaki et al., 2007; Jamsen et al., 2007; Chauhan et al., 2011), and the chemical composition changes in the lower atmosphere resulting from the energetic particle precipitation during this interval (Vashenyuk et al., 2005b; Verronen et al., 2007; Osepian et al., 2009).

CASE 10 (SEP. 6–15, 2005)
Notable Features {G4/R4/S3}
- X17 Flare-fourth largest in 15 years.
- SRB degraded performance in coded GPS receivers.
- Slow rise SPE from a nongeoeffective solar location.
- Twofold increase in the E-region density with n_mE greater than n_mF2.

The GOES space environmental data for Sep. 6–16, 2005, are provided in Fig. 11. During this interval numerous X-class flares erupted from Active Region #10808, as listed in Table 5 along with some pertinent attributes. The largest flare in this interval was an X17 (R4), detected at 17:17 UT on Sep. 7 from SC S11E77, that was the fourth-largest flare observed within the previous 15 years (http://www.noaanews.noaa.gov/stories2005/s2499.htm). This flare has historical significance because this was the first direct observation of GPS L1 carrier-to-noise ratio degradation from solar radio noise interference (Cerruti et al., 2006; Sreeja, 2016). Data from the USAF RSTN radio telescope at Sagamore Hill (Massachusetts, United States) near 17:30 UT on Sep. 7 recorded a maximum value of 370,000 sfu on the 15,400-MHz fixed-frequency channel, although, more relevant to GPS, was the measured value at 1415 MHz (GPS L1 band is 1575 MHz) of 160,000 sfu. To place this in perspective, the nominal "threat threshold" (all sources) established for GPS receivers is 40,000 sfu (Klobuchar et al., 1999); that is, a factor of four less than the solar radio intensity measured in association with the X17 flare. Another interesting aspect of this interval was the slow-rise SPE in the second panel, which is also correlated with the X17 flare. The slow rise of this event was most likely due to the nonfavorable geoeffective location of the flare event on the solar disk at SC S11E77. This SPE passed the event threshold (10 pfu) at 02:15 UT on Sep. 8 and had a peak flux of 1880 pfu (S3) at 04:25 UT on Sep. 11. While it is clear that the start of the SPE was well correlated with the X17 flare, it is likely that the flares on Sep. 8–15 also contributed to the overall event profile. The proton flux finally dropped below the event threshold level early on Sep. 16. Despite studies indicating that proton acceleration

Table 5 Case 10: X-class solar flares for Sep. 6–16, 2005

Date 2005	Start UT	End UT	Peak UT	X-ray Class	Location SC	Im/Br	Wilcox	Group AR#	McIntosh
Sep. 7	1717	1803	1740	X17.0	S11E77	3BE	B	10808	DKI
Sep. 8	2052	2117	2106	X5.4	S12E75	2BE	BG	10808	EKC
Sep. 9	0243	0307	0300	X1.1	S12E68	–	BGD	10808	FKC
Sep. 9	0942	1008	0959	X3.6	S11E66	SFE	BGD	10808	FKC
Sep. 9	1913	2036	2004	X6.2	S12E67	–	BGD	10808	FKC
Sep. 10	1634	1651	1643	X1.1	S11E47	2B	BGD	10808	DKC
Sep. 10	2130	2243	2211	X2.1	S13E47	2B	BGD	10808	DKC
Sep. 11	1244	1353	1312	M3.0	–	–	–	–	–
Sep. 13	1919	2057	1927	X1.5	S09E01	SFE	BGD	10808	FKC
Sep. 13	2315	2330	2322	X1.7	S10E04	1BE	BGD	10808	FKC
Sep. 15	0830	0846	0838	X1.1	S12W14	2NE	B	10808	EKC

within the X17 flare was of order 10 GeV (Gonzalez et al, 2015), there was no clear indication in the Moscow cosmic ray record (panel 3) of any GLE on Sep. 7. This may have been due to the fact that Moscow was in the late evening at the time of the X17 flare (17:17 UT on Sep. 7). On the other hand, the Chacaltaya (Bolivia) and Mexico City NMs, which were on the dayside, did detect relativistic solar neutrons from the direct flare site (Watanabe et al., 2007; 2009; Sako et al., 2006, 2008; Valdes-Galicia et al., 2009). The Moscow NM (Panel 3) did record a modest FD early on Sep. 11; this event has been studied by numerous authors (Jamsen et al., 2007; Papaioannou et al., 2009; Diego and Storini, 2009; Chauhan et al., 2011).

Table 6 lists the notable CMEs and the associated flares for Sep. 2005. It is perhaps noteworthy that the CME catalog included no CME that could be clearly associated with the X17 flare on Sep. 7. However, Table 6 includes a massive and fast-halo CME {speed: 2257 km/s; mass: 1.6×10^{17} g; first detected at 19:48 UT} on Sep. 9 that was most likely associated with the same-day X6.2 flare. This CME was likely responsible for the FD (panel 3) that was first detected early on Sep. 11 (Chauhan et al., 2011; Diego and Storini, 2009; Jamsen et al., 2007). Other fast-halo CMEs were observed at 21:52 UT on Sep. 10, 13:01 UT on Sep. 11, and 20:00 UT on Sep. 13, which appeared to be associated with the nearly simultaneous X2.1, M3 {s: 12:44 UT, e: 13:53 UT, p: 13:12 UT} and X1.5 flares,

Table 6 Case 10: Notable CMEs, solar flares and SIs for Sep. 6–16, 2005

CME 2005	K.E. erg	Flare	Class/Location	SI
None listed	–	Sep. 8 @ 20:52 UT	X5.4; SC S12E75	Sep. 9 @ 14:01 UT
Sep. 9 @ 19:48 UT	4.2×10^{33}	Sep. 9 @ 19:13 UT	X6.2; SC S12E67	Sep. 11 @ 01:14 UT
Sep. 10 @ 21:52 UT	7.0×10^{32}	Sep. 10 @ 21:30 UT	X2.1; SC S13E47	Sep. 12 @ 06:22 UT
Sep. 11 @ 13:01 UT	2.3×10^{32}	Sep. 11 @ 12:44 UT	M30; SC S13E42	Not geoeffective
Sep. 13 @ 20:00 UT	3.3×10^{32}	Sep. 13 @ 19:19 UT	X1.5; SC S09E01	Sep. 15 @ 09:04 UT

respectively. The bottom panel of Fig. 11 indicates that, from a GOES perspective, multiple geomagnetic disturbances occurred during this interval of extreme space weather. The maximum Kp for the interval was a $Kp=8_-$ (G4) on Sep. 11. At the risk of making associations based on limited data, we have nevertheless attempted to correlate in Table 6 the detected SIs with the observed flares and CMEs (Gosling, 1993; Hudson, 1995; Reames, 1995; Svestka, 1995). Perhaps not surprising from these associations is that the halo CMEs can be geoeffective regardless of location on the solar disk when associated with the larger flares. But, again, we understand that associations such as these need to be taken with a grain of salt pending more rigorous studies. As an aside, we note that the CMEs from this period apparently contributed to observations of energetic particles at 5 AU (Malandraki et al., 2008).

As has become the case, the extreme space weather events of Sep. 6–16, 2005, generated significant interest within the solar and space physics communities (Gopalswamy et al., 2006). Studies within the solar community for this interval were mostly concerned with the solar magnetic topology of Active Region #10808 (Wang and Liu, 2010) and the related filament eruption (Nagashima et al., 2007) on and around the Sep. 13 X1.5 and X1.7 flares, including precursor signatures (Chifor et al., 2007; Canou et al., 2009) and successive (interconnected) flare eruptions (Li et al., 2007; Liu et al., 2009). In general, X-ray and EUV emissions from flare processes have immediate, mostly short-term impacts on the dayside ionosphere and thermosphere (Qian et al., 2011; Xiong et al., 2011; 2015). In response to the X17 flare on Sep. 7, the E-region density (n_mE) increased twofold and, somewhat surprisingly, exceeded the density of the F-region (n_mF_2) (Xiong et al., 2011). However, the more dramatic near-Earth impacts for this period were associated with the magnetic storm period of Sep. 9–14. Mechanisms responsible for the observed ionospheric positive storm phase response (Pawlowski et al, 2008; Zhu et al., 2016) include the penetration of high-latitude electric fields to lower latitudes (Goncharenko et al., 2007; Klimenko et al., 2011a,b) and neutral wind mechanisms (Lu et al., 2008; Pavlov and Pavlova, 2011) that can result in upward plasma flows into regions having lower recombination rates. A consequence of these storms, as noted by Nagai et al. (2006), was a significant enhancement of relativistic electrons within the slot region (see also, Rodger et al., 2007; 2010). Energetic particle precipitation during this period, whether scattering out of the radiation belts or directly from SPEs, was found to be an important element in the chemical balance of the lower atmosphere at this time (Nozawa et al., 2010; Verronen et al., 2015).

CASE 11 (DEC. 5–17, 2006)
Notable Features {G4/~R4/S3}
- Multiple X-class flares from Active Region #10930.
- Most intense SRB ever recorded with levels exceeding 1,000,000 sfu.
- GPS receivers (dayside) had vertical range errors of up to 60 m.
- One of the largest and last GLEs in solar cycle 23.

Considerations for extreme space weather in early Dec. 2006 involved the largest recorded SRB on Dec. 6, which disrupted GPS positioning (Afrainovich et al. 2006). The GOES space environmental data for Dec. 5–17, 2006, are provided in Fig. 12. During this interval there was a set of four X-class flares, each of which erupted from the same Active Region #10930. Table 7 lists the X-class flares including the relevant timings and locations along with the evolving configurations for Active Region #10930. The largest solar event of the series was a severe X9 (~R4) flare that erupted

Table 7 Case 11: X-class solar flares for Dec. 5–17, 2006

Date 2006	Start UT	End UT	Peak UT	Class	Location SC	Im/Br	Wilcox	Group AR#	McIntosh
Dec. 5	10:18	10:45	10:35	X9.0	S07E68	2NE	A	10930	HHX
Dec. 6	18:29	19:00	18:47	X6.5	S05E64	3BE	BG	10930	EHI
Dec. 13	02:14	02:57	02:40	X3.4	S06W23	4BE	BGD	10930	DKC
Dec. 14	21:07	22:26	22:15	X1.5	S06W46	2BE	BD	10930	DKI

on Dec. 5 from SC S07E68. The magnetic classification (Wilcox) of Active Region #10930 changed from a relatively benign unipolar sunspot group (alpha) on Dec. 5 to a more complex bipolar configuration around the times of the later flares (Smith and Howard, 1968). Similarly, the modified Zurich sunspot classification (McIntosh, 1990) for Active Region #10930 changed from a small unipolar group to a much more complex region during this 13-day event. Given the initial simplicity of the region, it is curious that the largest X9 flare on Dec. 5 occurred while the region was in its formative stage. The flares on Dec. 5 and Dec. 6 erupted from nonfavorable geoeffective locations, from a charged particle perspective, whereas the later flares on Dec. 13–14 were better positioned. To this point, the slow rise in the SPE background for the flares on Dec. 5 and Dec. 6 was indicative of a poorly geoconnected event with the official time of onset for the SPE being 15:55 UT on Dec. 5 when the flux of >10 MeV protons crossed the 10 pfu threshold, as indicated in the second panel of Fig. 12. While the initial rise for the SPE was associated with the larger flare on Dec. 5, it is clear that the rate of rise of the event increased as a result of the second flare (X6.5), reaching a maximum of 1980 pfu (S3) at 19:30 UT on Dec. 7. There was a poorly resolved halo CME {first observed @ 20:12 UT on Dec. 6} which may have been responsible for the modest disturbance in the geomagnetic field on Dec. 7–8 having a maximum Kp of 5_o, indicative of a minor (G1) storm. As is apparent in Fig. 12, a GLE (#70) was detected by the Moscow NM records for Dec. 13 just prior to an unfortunate data outage.

While the solar-geophysical conditions for Dec. 5–6 are, at first blush, not particularly compelling as an example of an interval of extreme space weather, it is the SRB associated with the X6.5 flare on Jun. 6 that provides the rationale to justify this interval as extreme. At 19:32 UT on Jun. 6, the 1415 MHz channel of the USAF RSTN receiver at Palahuam Hawaii hit its maximum threshold of 132,000 sfu. This should be compared to the measured 1,000,000 sfu at 1.4 GHz as reported by Cerruti et al. (2008) from the Owens Valley Solar Array (OVSA). (Note: there is an apparent one-hour shift in the timing of the X6.5 solar flare which has not been resolved). GPS receivers on the sunlit side of the Earth were significantly impacted (Afraimovich et al., 2006; Sreeja, 2016) with estimated horizontal (vertical) range errors of 20 m (60 m) and intermittent loss of lock (Carrano et al., 2009). In studying the degradation of GPS capabilities, Yue et al. (2013) found that solar radio noise as low as ~1807 sfu (referenced to the RSTN 1.4 GHz channel) could have a measureable impact. Note that 1807 sfu is a factor of more than 500 less than the measured solar flux on Dec. 6.

A secondary reason for justifying early Dec. 2006 as an interval of extreme space weather was the solar-terrestrial activity associated with the X3.4 flare and GLE #70 on Dec. 13. The resultant geomagnetic storm of Dec. 14 was selected as an event for the coupling, energetics, and dynamics of the atmospheric regions (CEDAR) electrodynamics thermosphere ionosphere (ETI) challenge (Shim et al., 2012; 2014). A halo CME {speed: 1774 km/s; mass: 4.7×10^{15} g; first observed at 02:54 UT} that was launched in association with the Dec. 13 flare was the most effective, being the largest CME observed since the Halloween storms of 2003 (Liu et al., 2008). On Dec. 14, another flare (X1.2) and an associated halo CME {$v = 1042$ km/s; 7.5×10^{15} g; first observed at 20:30 UT} from the same Active Region #10930 occurred which Fan (2016) explained in terms of a coronal flux rope underlying the first flare current sheet. McKenna-Lawlor et al. (2008) modeled the propagation of these CMEs as distinct structures at Earth that subsequently merged before being observed at Mars. Notable signatures in Fig. 12 are the prompt arrival of solar energetic particles {event start time: 03:10 UT; 698 pfu/S3} and nearly simultaneous occurrence GLE #70 {start: ~03:00 UT} following the X3.4 flare on Dec. 13 and the strong geomagnetic storm that ensued the following day. Butikofer et al. (2009) noted at the time that GLE #70 was one of the largest, and in retrospect the last, GLEs in solar cycle 23 {start date: Oct. 1986; end date: Oct. 2008}. To put things in perspective, Matthia et al. (2009) calculated a 20% increase in the effective dose for passengers on a polar flight for this 5-min event. Numerous other authors have studied various characteristics of GLE #70 (Timashkova et al., 2008; Vashenyuk et al., 2008a,b; Beisembaev et al., 2009; Grigoryev et al., 2009; Mishev and Usoskin, 2016a,b; Plainaki et al., 2009) whereas others have discussed the relatively modest FD that was first observed (Fig. 12) late on Dec. 7, which then persisted until the premature end of the Moscow NM record on Dec. 14 (Angelov et al., 2009; Papailoiu et al., 2009). Measurements within the geospace indicate that accelerated protons were measured up to 600 MeV (Mulligan et al., 2008; Myagkova et al., 2009; von Rosenvinge et al., 2009; Adriani et al., 2011) although the Moscow NM records suggest that the proton energies were much higher. The magnetic cloud from the Dec. 13 CME subsequently impacted the Earth as a 49-nT SI at 14:14 UT on Dec. 14 (Fuquene Magnetic Observatory, Columbia). The geomagnetic storm that ensued over the next 24 hours reached a maximum Kp of 8_+ (G4). The magnetospheric response to the Dec. 13 event was modeled by Dandouras et al. (2009) who found a "very strong magnetospheric compression" although there was no indication in the GOES magnetometer data (Fig. 12) that the magnetopause was pushed inside of GEO in spite of the fact that GOES-12 was near noon. Adriani et al. (2016) noted that changes in the magnetospheric configuration during the Dec.14 storm affected the geomagnetic cutoff latitudes for high-energy ($\geq$MeV) protons. Several authors have reported on the ionospheric and thermospheric responses to the flares and geomagnetic storm in mid-December, noting the role of penetration of high-latitude electric fields on positive (increased ionospheric densities), as well as negative (decreased ionospheric densities) storm effects (Lei et al., 2008; Wang et al., 2010; Kumar et al., 2015; Zhu et al., 2016).

CASE 12 (MAR. 4–17, 2012)—ST. PATRICK'S DAY STORMS
Notable Features {G4/R3/S3}
- Climate and Weather of the Sun-Earth System (CAWSES-II).
- First time FD was measured with neutron scintillation detectors.

The extreme space weather events that occurred in early- to mid-Mar. 2012 are often referred to collectively as the St. Patrick's Day Storms. Interesting enough, Mar. 2013 and Mar. 2015 were also periods of heightened space weather and are also referred to in the same context (Verkhoglyadova et al., 2016) although this chapter only concerns the St. Patrick Day's events for 2012. The space weather activity during this period was responsible for numerous satellites anomalies (Guhathakurta, 2013; Zheng et al., 2013; Horne et al., 2013; Odenwald, 2015) and impacts to GPS positioning

Table 8 Case 12: Flares and CMEs for Mar. 4–17, 2012

Date	Start UT	End UT	Peak UT	Class	Location SC	Group AR	Wilcox	CME UT	Speed (km/s)
Mar. 5	02:30	04:43	04:09	X1.1	N16E54	11429	BD	04:00	1531
Mar. 7	00:02	00:40	00:24	X5.4	N17E27	11429	BD	00:24	2684
Mar. 7	01:05	01:23	01:14	X1.3	N22E12	11430	B	01:30	1825
Mar. 9	03:22	04:18	03:53	M6.3	N18W02[a]	11429[a]	B	04:26	950
Mar. 10	17:15	18:30	17:44	M8.4	N17W27[a]	11429[a]	B	18:00	1296
Mar. 13	17:12	18:25	17:41	M7.9	N19W59	11429	B	17:36	1884

[a]Flare location and active region number were not available from the X-ray flare reports. Values assumed from the solar region reports.

(Filjar et al., 2013). During the interval of Mar. 4–17, 2012, the solar-terrestrial environments were complicated due to multiple solar flares and possibly interacting CMEs. The GOES space environmental data are provided in Fig. 13. Table 8 lists the major (M-class and X-class) flares, which can be seen in the top panel of the figure. Also listed in Table 8 are the CMEs that were observed at the times of the associated flares with the largest flare occurring on Mar. 7 at X5.4 (R3). Note that this activity early on Mar. 7 was actually two flares and associated CMEs, separated by about an hour, which erupted from the separate Active Regions #11429 and #11430, respectively. With regard to the radiation environment, the first flare of the series on Mar. 5 was associated with a noticeable increase in the energetic protons at GEO (panel 2) having an elongated rise time due to the active region's poor magnetic connection to Earth at SC N16E54. The paired flares early on Mar. 7 were also at poorly connected locations, yet the energetic particle environment was sufficient to reach the SPE threshold of 10 pfu at 05:10 UT on Mar. 7 with the flux eventually reaching 6530 pfu (S3) at 11:15 UT on Mar. 8. Neither of the next two M-Class flares (M6.3 @ 03:53 UT on Mar. 9 and M8.4 @ 17:44 UT on Mar. 10) and associated CMEs had any significant impact on the charged particle background, although there were some related transients in the >1 MeV channel. The lack of any significant effect may be due to the rather low speeds for the associated CMEs in Table 8, which presumably limited any shock-induced acceleration of solar protons. The X7.9 flare {p: 17:41 UT} and associated CME {first observed at 17:36 UT} on Mar. 13 were apparently associated with the moderate 469 pfu (S2) SPE at 18:10 UT.

The bottom panel of Fig. 13 indicates that the magnetic field was intermittently disturbed. Multiple SIs were detected throughout the interval; 04:19 UT on Mar. 7 (18 nT), 11:03 UT on Mar. 8 (60 nT), 09:14 UT on Mar. 12 (73 nT), and 13:06 UT on Mar. 15 (36 nT). Speculating on cause and effect, it is interesting to note that the SI on Mar. 8 occurred about the time that the energetic proton fluxes peaked (second panel) and the Moscow NM responded to a significant FD. For this interval, a maximum $Kp=8_o$ (G4) was reached late in the morning of Mar. 9 (bottom panel) with no indication that the magnetopause was pushed inside of GEO at any time from Mar. 4–17.

The Mar. 7–17 2012, interval was a Climate and Weather of the Sun-Earth System (CAWSES-II) special study period (Tsurutani et al., 2012, 2014). As a result of the interest in CAWSES there was widespread interest in using the data from this period to study the ionospheric responses to the immediate effects of the flares and SPEs (Romanova et al., 2014; Lebed et al., 2015; 2017) and the delayed effects related to the geomagnetic storms (Habarulema et al., 2015; 2016; 2017; Habarulema and

Ssessanga, 2017; Belehaki et al., 2017). While it is known that the ionosphere as an element of space weather can impact GPS, it is interesting to note that GPS has now become a tool for ionospheric research (Olawepoa et al., 2015; Prikryl et al., 2015a,b; Tesema et al., 2015). A number of authors also used this interval to characterize the extended FD observed during this interval (see panel 3 of Fig. 13) (Alekseenko et al., 2013; Castillo et al., 2013; Lingri et al., 2013, 2015; Martucci, 2014), although any suggestion that there may have been a GLE associated with the flares and CMEs cannot be verified in the Moscow NM records (Berlov et al., 2015). With regard to the CMEs, Shen et al. (2013, 2014) and others (Liu et al., 2013; Gopalswamy et al., 2014; Wu et al., 2013; Wu et al., 2016; Ding et al., 2016) modeled the propagation of these structures from the Sun to Earth and beyond (Liu et al., 2014). As noted previously, the acceleration of energetic charged particles can occur as a result of processes close to the flare site (Hamidi et al., 2014; Wang et al., 2014; Koga et al., 2015) or in CME shocks (Schmidt and Cairns, 2014; Schmidt et al., 2014; Ding et al., 2016). Mesospheric measurements made during the CAWSES-II interval revealed the impact of energetic charged particle precipitation on the local chemistry (von Clarmann et al., 2013; Verkhoglyadova et al., 2014).

4 CONCLUDING REMARKS

Since 1974 the GOES spacecraft have been NOAA's environmental sentinels in GEO, providing the measurements required for operational space weather. Over their continuing lifetime, these critical satellites have monitored extremes in space weather that have stressed society's technological infrastructure in ways not fully appreciated before. Known technological vulnerabilities are now better understood and, in some cases, have been effectively mitigated. For example, spacecraft design standards include not only space weather climatologies but the magnitudes and probabilities for extreme space weather. The electric power grid, commercial aviation, manned spaceflight, communications, and satellite-based navigation are other aspects of our technology-dependent society that require knowledge of the effects of deleterious space weather. In this chapter we have focused on the continuity of measurements provide by GOES and how this program has been the persistent workhorse for operational space weather. As our understanding of the impacts of space weather has evolved, so too has our knowledge of solar-terrestrial physics. Over the last 40 years that the GOES program has been operational, the role of CME shocks versus solar flare processes in the acceleration of charged particles to extreme energies has been somewhat of an epiphany for solar physicists. The variability of the van Allen radiation belts and their response to interplanetary forcing functions has also been revealing, with the surprising result that the normally quiescent slot region can become a radiation hazard for satellites. While the GOES space environmental data have provided the basis against which extreme space weather can be assessed, the capabilities of the spacecraft have also evolved to address current and future needs. The GOES-R series of spacecraft is NOAA's newest set of operational environmental satellites that will continue to monitor the sun and local geospace through the mid-2030s.

ACKNOWLEDGMENTS

All of the GOES data discussed within this chapter are readily available from the NOAA National Centers for Environmental Information (NCEI). Aside from GOES, other datasets available from NCEI and included herein are: (1) listing of solar particle events affecting Earth also from SWPC, (2) Air Force Solar Electro-Optical Nework data for H-alpha flares, solar region reports and solar radio events, and (3) listing of ground level enhancements (GLE) provided by the Air Force Research Laboratory (courtesy of M.A. Shea and D.F. Smart). Moscow cosmic ray data (CRD) were provided by Cosmic Ray Department of Solar-Terrestrial Division of IZMIRAN. Records of storm sudden commencements (SSC) were acquired from Observatori de l'Ebre whereas values for the magnetic index Kp and the 1-h Disturbance Storm Time (Dst) index were obtained from the German Research Center for Geosciences and World Data Center for Geomagnetism, Kyoto, repectively. The catalog of Coronal Mass Ejections (CME) was acquired from the NASA Goddard Space Flight Center[1].

REFERENCES

Adriani, O., Barbarino, G.C., Bazilevskaya, G.A., Bellotti, R., Boezio, M., Bogomolov, E.A., Bonechi, L., Bongi, M., Bonvicini, V., Borisov, S., Bottai, S., Bruni, A., Cafagna, F., Campana, D., Carbone, R., Carlson, P., Casolino, M., Castellini, G., Consiglio, L., DePascale, M.P., DeSantis, C., DeSimone, N., DiFelice, V., Formato, V., Galper, A.M., Gridhsntseva, L., Gillard, W., Jerse, G., Karelin, A.V., Koldashov, S.V., Krutkov, S.Y., Kvashnin, A.N., Leonov, A., Malakhov, V., Marcelli, L., Mayorov, A.G., Menn, W., Mikhailov, V.V., Mocchiutti, E., Monaco, A., Mori, N., Nikonov, N., Osteria, G., Palma, F., Papini, P., Pearse, M., Picozza, P., Pizzolotto, C., Ricci, M., Ricciarini, S.B., Sarkar, R., Rossetto, L., Simon, M., Sparvola, R., Spillantini, P., Stozhkov, Y.I., Vacci, A., Vannuccini, E., Vasilyev, G., Veronov, S.A., Wu, J., Yurkin, Y.T., Zampa, G., Zamopa, N., Zverev, V.G., 2011. Observations of the 2006 December 13 and 14 Solar Particle Events in the 80 MeV n−1-3 GeV n−1 range from space with the PAMELA detector. Astrophys. J. 742 (2), 11. https://doi.org/10.1088/0004-637X/742/2/102.

Adriani, O., Barbarino, G.C., Bazilevskaya, G.A., Bellotti, R., Boezio, M., Bogomolov, E.A., Bongi, M., Bonvicini, V., Bottai, S., Bruno, A., et al., 2016. PAMELA's measurements of geomagnetic cutoff variations during the 14 December 2006 storm. Space Weather 14, 210–220. https://doi.org/10.1002/2016SW001364.

Afraimovich, E.L., Demyanov, V.V., Smolkov, G.Y., 2006. The total failures of GPS functioning caused by the powerful solar radio burst on December 13, 2006. Earth Planets Space. https://doi.org/10.1186/BF03352940.

Alekseenko, V., Arneodo, F., Bruno, G., Fulgione, W., Gromushkin, D., Shchegolev, O., Stenkin, Y., Stepanov, V., Sulakov, V., 2013. Registration of Forbush decrease 2012/03/08 with a global net of the thermal neutron scintillation *en*-detectors. In: Proc. 23rd European Cosmic Ray Symposium (and 32nd Russian Cosmic Ray Conference), J. Phys.: Conference Series 409, 4 pp. 03-07 July 2012, Moscow, Russia, https://doi.org/10.1088/1742-6596/409/1/012190.

Alex, S., Pathan, B.M., Lakhina, G.S., 2005. Response of the low latitude geomagnetic field to the major proton event of November 2001. Adv. Space Res. 36 (12), 2434–2439. https://doi.org/10.1016/j.asr.2004.01.026.

Allen, J.H., 1982. Some commonly used magnetic activity indices: Their derivation, meaning and use. In: Proceeding of a Workshop on Satellite Drag, NOAA ERL. Space Environment Services Center, 18-19 March 1982, Boulder, CO, pp. 114–134.

Allen, J., Sauer, H., Frank, L., Reiff, P., 1989a. Effects of the March 1989 solar activity. Eos. Trans. AGU 70 (46), 1479–1488. https://doi.org/10.1029/89EO00409.

Allen, J.H., Wilkinson, D.C., 1992. Solar-terrestrial activity affecting systems in space and on Earth. In: Proc. Solar-Terrestrial Predictions – IV, Vol. 1, 75-107, 18-22 May 1992, Ottawa, Canada.

---

[1] This CME catalog is generated and maintained at the CDAW Data Center by NASA and The Catholic University of America in cooperation with the Naval Research Laboratory. SOHO is a project of international cooperation between ESA and NASA.

Akimov, V.V., Ambroz, P., Belov, A.V., Berlicki, A., Chertok, I.M., Karlicky, M., 1996. Evidence for prolonged acceleration based on a detailed analysis of the long-duration solar gamma-ray flare of June 15, 1991. Sol. Phys. 166 (1), 107–134. https://doi.org/10.1007/BF00179358.

Angelov, I., Malamova, E., Stamenov, J., 2009. The Forbush decrease after the GLE on 13 December 2006 detected by the muon telescope at BEO—Moussala. Adv. Space Res. 43 (4), 504–508. https://doi.org/10.1016/j.asr.2008.08.002.

Araki, T., 2014. Historically largest geomagnetic sudden commencement (SC) since 1868. Earth Planets Space 66 (164), 6. https://doi.org/10.1186/s40623-014-0164-0.

Araki, T., Fugitani, S., Emoto, M., Yumoto, K., Shiokawa, S., Ichinose, T., Luehr, H., Orr, O., Milling, D.K., Singer, H., Rostoker, G., Tsunomura, S., Yamada, Y., Liu, C.F., 1997. Anomalous sudden commencement on March 24, 1991. J. Geophys. Res. 102 (A7), 14075–14086. https://doi.org/10.1029/96JA03637.

Arkhangelskaja, I.V., Arkhangelskii, A.I., Glyanenko, A.S., Kotov, Y.D., Kuznetsov, S.N., 2005. The investigation of January 2005 solar flares gamma-emission by Avs-F apparatus data onboard Coronas-F satellite in 0.1-20 Mev energy band. In: Danesy, D., Poedts, S., De Groof, A., Andries, J. (Eds.), The Dynamic Sun: Challenges for Theory and Observations.Proc. 11th European Solar Physics Meeting, ESA SP-600, 11–16 September 2005, Leuven, Belgium, p. 4.

Baker, D.M., 1970. Flare classification based upon X-ray intensity, paper AIAA 70-1370. In: AIAA Observation and Prediction of Solar Activity Conference, 16–18 November 1970, Huntsville, AL, p. 4.

Baker, D.N., 2008. Severe Space Weather—Understanding Societal and Economic Impacts: A Workshop Report. National Academies Press, Washington, DC. ISBN: 978-0-309-12769-1. 144 p. https://doi.org/10.17226/12507.

Baker, D.N., Kanekal, S.G., Li, X., Monk, S.P., Goldstein, J., Burch, J.L., 2004. An extreme distortion of the Van Allen belt arising from the "Halloween" solar storm in 2003. Nature 432, 878–881. https://doi.org/10.1038/nature03116.

Balch, C., et al., 2004. Service Assessment: Intense Space Weather Storms, October 19–November 07, 2003, U.S. Dept. of Commerce, NOAA, 50 p. Available from: https://www.weather.gov/media/publications/assessments/SWstorms_assessment.pdf (Accessed 06 October 2017).

Basu, S., Basu, S., Groves, K.M., Yeh, H.-C., Su, S.-Y., Rich, F.J., Sultan, P.J., Keskinen, M.J., 2001. Response of the equatorial ionosphere in the South Atlantic Region to the Great Magnetic Storm of July 15, 2000. Geophys. Res. Lett. 28, 3577–3580. https://doi.org/10.1029/2001GL013259.

Basu, S., Basu, Su., Groves, K.M., MacKenzie, E., Keskinen, M.J., Rich, F.J., 2005. Near-simultaneous plasma structuring in the midlatitude and equatorial ionosphere during magnetic superstorms. Geophys. Res. Lett. 32. L12S05, https://doi.org/10.1029/2004GL021678.

Bedingfield, K.L., Leach, R.D., Alexander, M.B., 1996. Spacecraft system failures and anomalies attributed to the natural space environment, RP-1390. NASA Marshall Space Flight Center, Huntsville, AL, 43 p.

Beisembaev, R.U., Drobzhev, V.I., Dryn, E.A., Kryakunova, O.N., Nikolaevskiy, N.F., 2009. Solar extreme events on the data of Alma-Ata neutron monitor: identification of ground level enhancements. Adv. Space Res. 43, 509–514. https://doi.org/10.1016/j.asr.2008.10.017.

Belehaki, A., Kutiev, I., Marinov, P., Tsagouri, I., Koutroumbas, K., Elias, P., 2017. Ionospheric electron density perturbations during the 7–10 March 2012 geomagnetic storm period. Adv. Space Res. 59 (4), 1041–1056. https://doi.org/10.1016/j.asr.2016.11.031.

Bell, J.T., Gussenhoven, M.S., Mullen, E.G., 1997. Super storms. J. Geophys. Res. 102 (A7), 14189–14198. https://doi.org/10.1029/96JA03759.

Belov, A.V., Eroshenko, E.A., Mavromichali, H., Plainaki, C., Yanke, V.G., 2005. Ground level enhancement of the solar cosmic rays on January 20, 2005. In: Acharya, B.S., Gupta, S., Jagadeesan, P., Jain, A., Karthikeyan, S., Morris, S., Tonwar, S. (Eds.), Proc. ICRC 29, Vol. 1, 03–10 August 2005, Pune, India, pp. 189–192.

Berlov, A., Eroshenko, E., Kryanunova, O., Nikolayevskiy, N., Malimbayev, A., Tsepakina, I., Yanke, V., 2015. Possible ground level enhancements at the beginning of the maximum of Solar Cycle 24. In: Proc. 24th European Cosmic Ray Symposium, J. Phys.: Conference Series 632, p. 8. https://doi.org/10.1088/1742-6596/632/1/012063.

Bertsch, D.L., Dingus, B.L., Esposito, J.A., Fichtel, C.E., Hartman, R.C., Hunter, S.D., Kanbach, G., Kniffen, D.A., Lin, Y.C., Mattox, J.R., Mayer-Hasselwander, H.A., Michelson, P.F., Montigny, C.V., Nolan, P.L., Schneid, E., Sreekumar, P., Thompson, D.J., 1996. Gamma ray observations of the early June 1991 solar flares. Astron. Astrophys. uppublished manuscript. Available from: https://www.astro.umd.edu/~share/GLAST/references/solarflare/bertsch_pp.pdf (Accessed 06 October 2017).

Bhatnagar, A., Jain, R.M., Burkepile, J.T., Chertok, I.M., Magun, A., Urbarz, H., Zlobec, P., 1996. Transient phenomena in the energetic behind-the-limb solar flare of September 29, 1989. Astrophys. Space Sci. 243, 209–213.

Bieber, J.W., Droge, W., Evenson, P.A., Pyle, R., Ruffolo, D., Pinsook, U., Tooprakai, P., Rujiwarodom, M., Khumlumlert, T., Krucker, S., 2002. Energetic Particle Observations during the 2000 July 14 Solar Event. Astrophys. J. 567 (1), 622–634.

Bieber, J.W., Clem, J., Evenson, P., Pyle, R., Ruffolo, D., Saiz, A., 2005a. Relativistic solar neutrons and protons on 28 October 2003. Geophys. Res. Lett. 32. L03S02, https://doi.org/10.1029/2004GL021492.

Bieber, J., Clem, J., Evenson, P., et al., 2005b. Largest GLE in half a century: neutron monitor observations of the January 20, 2005 event. In: Proc. 29th International Cosmic Ray Conference, vol. 1, 03–10 Aug. 2005, Pune, India, pp. 237–240.

Bieber, J.W., Clem, J., Evenson, P., Pyle, R., Saiz, A., Ruffolo, D., 2013. Giant ground level enhancement of relativistic solar protons on 2005 January 20 Spaceship Earth observations. Astrophys. J. 771 (92), 13. https://doi.org/10.1088/0004-637X/771/2/92.

Blake, J.B., Slocum, P.L., Mazur, J.E., Looper, M.D., Selesnick, R.S., Shiokawa, K., 2005. Geoeffectiveness of shocks in populating the radiation belts. In: Lui, A.T.Y., Kamide, Y., Consolini, G. (Eds.), Multiscale Coupling of Sun-Earth Processes. pp. 125–133. https://doi.org/10.1016/B978-044451881-1/50010-1.

Blake, J.B., Gussenhoven, M.S., Mullen, E.G., Fillius, R.W., 1992a. Identification of an unexpected space radiation hazard. IEEE Trans. Nucl. Sci. 39, 1761–1764. https://doi.org/10.1109/23.211364.

Blake, J.B., Kolasinski, W.A., Fillius, R.W., Mullen, E.G., 1992b. Injection of electrons and protons with energies of tens of MeV into L<3 on 24 March 1994. Geophys. Res. Lett. 19, 821–824. https://doi.org/10.1029/92GL00624.

Bostanjyan, N.Kh., Chilingarian, A.A., Eganov, V.S., Karapetyan, G.G., 2007. On the production of highest energy solar protons at 20 January 2005. Adv. Space Res. 39 (9), 1454–1457. https://doi.org/10.1016/j.asr.2007.03.024.

Burke, W.J., Maynard, N.C., Hagan, M.P., Wolf, R.A., Wilson, G.R., Gentile, L.C., Gussenhoven, M.S., Huang, C.Y., Garner, T.W., Rich, F.J., 1998. Electrodynamics of the inner magnetosphere observed in the dusk sector by CRRES and DMSP during the magnetic storm of June 4–6, 1991. J. Geophys. Res. 103 (A12), 29399–29418. https://doi.org/10.1029/98JA02197.

Burke, W.J., Rubin, A.G., Maynard, N.C., Gentile, L.C., Sultan, P.J., Rich, F.J., deLa Beaujardière, O., Huang, C.Y., Wilson, G.R., 2000. Ionospheric disturbances observed by DMSP at middle to low latitudes during the magnetic storm of June 4–6, 1991. J. Geophys. Res. 105 (A8), 18391–18405. https://doi.org/10.1029/1999JA000188.

Burlaga, L.F., Ogilvie, K.W., 1969. Causes of sudden commencements and sudden impulses. J. Geophys. Res. 74 (11), 2815–2825. https://doi.org/10.1029/JA074i011p02815.

Burlaga, L.F., Ness, N.F., Richardson, J.D., Lepping, R.P., 2001. The Bastille Day Shock and Merged Interaction Region at 63 AU: Voyager 2 observations. Sol. Phys. 204, 399–410. https://doi.org/10.1023/A:1014269926730.

Burlaga, L.F., Ness, N.F., Stone, E.C., McDonald, F.B., Richardson, J.D., 2005. Voyager 2 observations related to the October-November 2003 solar events. Geophys. Res. Lett. 32. L03S05, https://doi.org/10.1029/2004GL021480.

Butikofer, R., Flückiger, E.O., Desorgher, L., Moser, M.R., Pirard, B., 2009. The solar cosmic ray ground-level enhancements on 20 January 2005 and 13 December 2006. Adv. Space Res. 43 (4), 499–503. https://doi.org/10.1016/j.asr.2008.08.001.

Cairns, I.H., Robinson, R.D., 1987. Herringbone bursts associated with type II solar radio emission. Sol. Phys. 3 (2), 365–383. https://doi.org/10.1007/BF00148526.

Cander, L.R., Mihajlovic, S.J., 2005. Ionospheric spatial and temporal variations during the 29–31 October 2003 storm. J. Atmos. Sol. Terr. Phys. 67 (12), 1118–1128. https://doi.org/10.1016/j.jastp.2005.02.020.

Cane, H.V., 2000. Coronal mass ejections and Forbush decreases. Space Sci. Rev. 93, 55–77. https://doi.org/10.1023/A:1026532125747.

Cane, H.V., White, S.M., 1989. On the source conditions for Herringbone structure in type II solar radio bursts. Sol. Phys. 120, 137–144. https://doi.org/10.1007/BF00148539.

Cane, H.V., Richardson, I.G., 1995. Cosmic ray decreases and solar wind disturbances during late October 1989. J. Geophys. Res. 100 (A2), 1755–1762. https://doi.org/10.1029/94JA03073.

Canou, A., Amari, T., Bommier, V., Schmieder, B., Aulanier, G., Li, H., 2009. Evidence for a pre-eruptive twisted rope using the Themis Vector Magnetograph. Astrophys. J. Lett. 693 (1), L27–L30.

Carrano, C.S., Bridgwood, C.T., Groves, K.M., 2009. Impacts of the December 2006 solar radio bursts on the performance of GPS. Radio Sci. 44. RS0A25, https://doi.org/10.1029/2008RS004071.

Castillo, M., Salazar, H., Villasenor, L., 2013. Observation of the March 2012 Forbush decrease with the engineering array of the High Altitude Water Cherenkov Observatory. In: Saa, A. (Ed.), Proc. ICRC 33, 02–09 Jul 2013, Rio De Janeiro, Brazil, p. 4.

Cerruti, A.P., Kintner, P.M., Gary, D.E., Lanzerotti, L.J., de Paula, E.R., Vo, H.B., 2006. Observed solar radio burst effects on GPS/Wide Area Augmentation System carrier-to-noise ratio. Space Weather 4. S10006, https://doi.org/10.1029/2006SW000254.

Cerruti, A.P., Kintner Jr., P.M., Gary, D.E., Mannucci, A.J., Meyer, R.F., Doherty, P., Coste, A.J., 2008. Effect of intense December 2006 solar radio bursts on GPS receivers. Space Weather 6. S10D07, https://doi.org/10.1029/2007SW000375.

Chamberlin, P.C., Woods, T.N., Eparvier, F.G., Jones, A.R., 2009. Next generation X-ray sensor (XRS) for GOES-R satellite series, in Solar Physics and Space Weather Instrumentation III. In: Fineschi, S., Fennelly, J.A. (Eds.), Proc. SPIE 7438, 743802, https://doi.org/10.1117/12.82680.

Chauhan, M.L., Jain, M., Shrivastava, S.K., 2011. Study of large Forbush decrease events of solar cycle. In: Proc. ICRC 32, vol. 10, 11–18 August 2011, Beijing, China, pp. 261–263. https://doi.org/10.7529/ICRC2011/V10/0097.

Chen, Z., Gao, Y., Liu, Z., 2005. Evaluation of solar radio bursts' effect on GPS receiver signal tracking within International GPS Service Network. Radio Sci. 40, 11. https://doi.org/10.1029/2004RS003066. RS3012.

Chi, P.J., Russell, C.T., Foster, J.C., Moldwin, M.B., Engebretson, M.J., Mann, I.R., 2005. Density enhancement in plasmasphere-ionosphere plasma during the 2003 Halloween Superstorm: Observations along the 330th magnetic meridian in North America. Geophys. Res. Lett. 32. L03S07, https://doi.org/10.1029/2004GL021722.

Chilingarian, A., 2009. Statistical study of the detection of solar protons of highest energies at 20 January 2005. Adv. Space Res. 43, 702–707. https://doi.org/10.1016/j.asr.2008.10.005.

Chifor, C., Tripathi, D., Mason, H.E., Dennis, B.R., 2007. X-ray precursors to flares and filament eruptions. Astron. Astrophys. 472, 967–979. https://doi.org/10.1051/0004-6361:20077771.

Cid, C., Palacios, J., Saiz, E., Guerrero, A., Cerrato, Y., 2014. On extreme geomagnetic storms. J. Space Weather Space Clim. 4, A28. https://doi.org/10.1051/swsc/2014026.

Clilverd, M.A., Rodger, C.J., Ulich, T., Seppala, A., Turunen, E., Botman, A., Thomson, N.R., 2005. Modeling a large solar proton event in the outhern polar atmosphere. J. Geophys. Res. 110. A09307, https://doi.org/10.1029/2004JA010922.

Clilverd, M.A., Seppala, A., Rodger, C.J., Thomson, N.R., Verronen, P.T., Turunen, E., Ulich, T., Lichtenberger, J., Steinbach, P., 2006. Modeling polar ionospheric effects during the October–November 2003 solar proton events. Radio Sci. 41. RS2001, https://doi.org/10.1029/2005RS003290.

Cliver, E.W., Dietrich, W.F., 2013. The 1859 space weather event revisited: limits of extreme activity. J. Space Weather Spac. 3, A31. https://doi.org/10.1051/swsc/2013053.

Cliver, E., Kahler, S., Vestrand, W., 1993. On the origin of gamma-ray emission from the behind-the-limb flares on 29 September 1989. In: Leahy, D.A., Hickws, R.B., Venkatesan, D. (Eds.), Proc. ICRC 23, 19-30 July 1993, Alberta, Canada. Vol. 3. pp. 91–94.

Cooper, M.L., Clauer, C.R., Emery, B.A., Richmond, A.D., Winningham, J.D., 1995. A storm time assimilative mapping of ionospheric electrodynamics analysis for the severe geomagnetic storm of November 8–9, 1991. J. Geophys. Res. 100 (A10), 19329–19342. https://doi.org/10.1029/95JA01402.

Cramp, J.L., Duldig, M.L., Fluckiger, E.O., Humble, J.E., 1993. The GLE of 29 September 1989. In: Leahy, D.A., Hickws, R.B., VenkatesanIntern, D. (Eds.), Proc. ICRC 23, 9-30 July 1993, Alberta, Canada. Vol. 3. pp. 47–50.

Cramp, J.L., Duldig, M.L., Flückiger, E.O., Humble, J.E., Shea, M.A., Smart, D.F., 1997. The October 22, 1989, solar cosmic ray enhancement: an analysis of the anisotropy and spectral characteristics. J. Geophys. Res. 102 (A11), 24237–24248. https://doi.org/10.1029/97JA01947.

Daglis, I.A., Kamide, Y., Mouikis, C., Reeves, G.D., Sarris, E.T., Shiokawa, K., Wilken, B., 2000. "Fine structure" of the storm-substorm relationship: Ion injections during Dst decrease. Adv. Space Res. 25 (12), 2369–2372. https://doi.org/10.1016/S0273-1177(99)00525-6.

Daly, E.J., van Leeuwen, F., Evans, H.D.R., Perryman, M.A.C., 1992. Radiation-belt and transient solar-magnetospheric effects on Hipparcos radiation background. IEEE Trans. Nucl. Sci. 41, 2376–2382. https://doi.org/10.1109/23.340590.

Dandouras, I.S., Reme, H., Cao, J., Escoubet, P., 2009. Magnetosphere response to the 2005 and 2006 extreme solar events as observed by the Cluster and Double Star spacecraft. Adv. Space Res. 43 (4), 618–623. https://doi.org/10.1016/j.asr.2008.10.015.

D'Andrea, C., Poirer, J., 2005. Ground level muons coincident with the 20 January 2005 solar flare. Geophys. Res. Lett. 32. L14102, https://doi.org/10.1029/2005GL023336.

Davies, K., Hartmann, G.K., 1997. Studying the ionosphere with the Global Positioning System. Radio Sci. 32 (4), 1695–1703. https://doi.org/10.1029/97RS00451.

Degenstein, D.A., Lloyd, N.D., Bourassa, A.E., Gattinger, R.L., Llewellyn, E.J., 2005. Observations of mesospheric ozone depletion during the October 28, 2003 solar proton event by OSIRIS. Geophys. Res. Lett. 32. L03S11, https://doi.org/10.1029/2004GL021521.

Deng, Y., Wang, J., Yan, Y., Zhang, J., 2001. Evolution of magnetic nonpotentiality in NOAA AR 9077. Sol. Phys. 204 (11), 13–28. https://doi.org/10.1023/A:1014258426134.

Dichter, B.K., Galica, G.E., McGarity, J.O., Tsu, S., Golightly, M.J., Lopate, C., Connell, J.J., 2015. Specification, Design and calibration of the space weather suite of instruments on the NOAA GOES-R program spacecraft. IEEE Trans. Nucl. Sci. 62 (6), 2776–2783. https://doi.org/10.1109/TNS.2015.2477997.

Diego, P., Storini, M., 2009. Modulation signatures on cosmic-ray periodicities before a Forbush decrease. In: Szabelski, J., Giller, M. (Eds.), Proc. ICRC 31, 07–15 July 2009, Lodz, Poland, p. 4.

Ding, L.-G., Cao, X.-X., Wang, Z.W., Le, G.-M., 2016. Large solar energetic particle event that occurred on 2012 March 7 and its VDA analysis. Res. Astron. Astrophys. 16(8)https://doi.org/10.1088/1674-4527/16/8/122.

Dmitriev, A.V., Chao, J.-K., Suvarova, A.V., Ackerson, K., Ishisaka, K., Kasaba, Y., Kojima, H., Matsumoto, H., 2005. Indirect estimation of the solar wind conditions in 29–31 October 2003. J. Geophys. Res. 110. A09S02, https://doi.org/10.1029/2004JA010806.

Dmitriev, A.V., Yeh, H.-C., Chao, J.-K., Veselovsky, I.S., Su, S.-Y., Fu, C.C., 2006. Top-side ionospheric response to extreme solar events. Ann. Geophys. 24, 1469–1477. www.ann-geophys.net/24/1469/2006/.

Dmitriev, A.V., Suvorova, A.V., Chao, J.-K., Wang, C.B., Rastaetter, L., Panasyuk, M.I., Lazutin, L.L., Kovtyukh, A.S., Veselovsky, I.S., Myagkova, I.N., 2014. Anomalous dynamics of the extremely compressed magnetosphere during 21 January 2005 magnetic storm. J. Geophys. Res. 119, 877–896. https://doi.org/10.1002/2013JA019534.

Du, A.M., Tsurutani, B.T., Sun, W., 2008. Anomalous geomagnetic storm of 21–22, A storm main phase during northward IMFs January 2005. J. Geophys. Res. 113. A10214, https://doi.org/10.1029/2008JA013284.

Duldig, M.L., 2001. Fine time resolution analysis of the 14 July 2000 GLE. In: Kampert, K.-H., Heinzelmann, G., Spiering, C. (Eds.), Proc. ICRC 27, 05–15 Aug 2001, Hamburg, Germany, vol. 8. pp. 3363–3366.

Duldig, M.L., Cramp, J.L., Humble, J.E., Bieber, J.W., 1993. The ground-level enhancements of 1989 September 29 and October 22. Publ. Astron. Soc. Aust. 10 (3), 211–217. https://doi.org/10.1017/S1323358000025698.

Duldig, M., Bombardieri, D.J., Humble, J.E., 2003. Fine time resolution analysis of the Bastille Day 2000 GLE. In: Kajita, T., Asaoka, Y., Kawachi, A., Matsubara, Y., Sasaki, M. (Eds.), Proc. ICRC 28, 31-July–07-August 2003, Trukuba, Japan, pp. 3389–3392.

Dunphy, P.P., Chupp, E.L., Bertsch, D.L., Schneid, E.J., Gottesman, S.R., Gottfried, K., 1999. Gamma-rays and neutrons as a probe of flare proton spectra: The solar flare of 11 June 1991. Sol. Phys. 187, 45–57. https://doi.org/10.1023/A:1005143603547.

Dvornikov, V.M., Sdobnov, V.E., 1995. The GLE of 29 September 1989: Time variations of the cosmic-ray rigidity spectrum. In: Iucci, N., Lamanna, E. (Eds.), Proc. ICRC 24, 28 August–08 September 1995, Rome, Italy, vol. 4. pp. 232–235.

Eparvier, F.G., Critsers, D., Jones, A.R., McClintock, W.E., Snow, M., Woods, T.N., 2009. The Extreme Ultraviolet Sensor (EUVS) for GOES-R. In: Fineschi, S., Fennelly, J.A. (Eds.), Proc. SPIE 7438, Solar Physics and Space Weather Instrumentation III, 26 Aug 2009., p. 9. https://doi.org/10.1117/12.826445.

Etchegoyen, A., Bauleo, P., Bertou, X., Bonifazi, C.B., Filevich, A., Medina, M.C., Melo, D.G., Rovero, A.C., Supanitsky, A.D., Tamashiro, A., 2005. Muon-track studies in a water cherenkov detector. Nucl. Instrum. Methods Phys. Res. 545 (3), 602–612. https://doi.org/10.1016/j.nima.2005.02.016.

Fan, Y., 2016. Modeling the Initiation of the 2006 December 13 coronal mass ejection in AR 10930: the structure and dynamics of the erupting flux rope. Astrophys. J. 824 (2), 2. https://doi.org/10.3847/0004-637X/824/2/93.

Feynman, J., Hundhausen, A.J., 1994. Coronal mass ejections and major solar flares: the great active center of March 1989. J. Geophys. Res. 99 (A5), 8451–8464. https://doi.org/10.1029/94JA00202.

Filjar, R., Brcic, D., Kos, S., 2013. Single-frequency horizontal GPS positioning error response to a moderate ionospheric storm over northern adriatic. Advances in Marine Navigation, Marine Navigation and Safety of Sea Transportation. CRC Press, London, UK. ISBN: 9781138001060.

Fineschi, S., Viereck, R.A., 2007. Solar physics and space weather instrumentation II. Proc. SPIE. 6689, https://doi.org/10.1117/12.759511.

Fletcher, L., Cargill, P.J., Antiochos, S.K., Gudiksen, B.V., 2015. Structures in the outer solar atmosphere. Space Sci. Rev. 188 (1), 211–249. https://doi.org/10.1007/s11214-014-0111-1.

Fluckiger, E.O., Butikofer, R.B., Moser, M.R., Desorghe, L., 2005. The cosmic ray ground level enhancement during the Forbush Decrease in January 2005. In: Acharya, B.S., Gupta, S., Jagadeesan, P., Jain, A., Karthikeyan, S., Morris, S., Tonwar, S. (Eds.), Proc. ICRC 29, 03–10 August 2005, Pune, India, vol. 1. pp. 225–228.

Forbush, S.E., 1958. Cosmic-ray intensity variations during two solar cycles. J. Geophys. Res. 63 (4), 651–669. https://doi.org/10.1029/JZ063i004p00651.

Foster, J.C., Rideout, W., 2005. Midlatitude TEC enhancements during the October 2003 superstorm. Geophys. Res. Lett. 32. L12S04, https://doi.org/10.1029/2004GL021719.

Foullon, C., Owen, C.J., Dasso, S., Green, L.M., Dandouras, I., Elliott, H.A., FazakerleyY, A.N., Bogdanova, V., Crooker, N.U., 2007. Multi-spacecraft study of the 21 January 2005 ICME: evidence of current sheet substructure near the periphery of a strongly expanding fast magnetic cloud. Sol. Phys. 244, 139–165. https://doi.org/10.1007/s11207-007-0355-y.

Funke, B., Baumgaertner, A., Calisto, M., Egorova, T., Jackman, C.H., Kieser, J., Krivolutsky, A., López-Puertas, M., Marsh, D.R., Reddmann, T., Rozanov, E., Salmi, S.-M., Sinnhuber, M., Stiller, G.P., Verronen, P.T., Versick, S., von Clarmann, T., Vyushkova, T.Y., Wieters, N., Wissing, J.M., 2011. Composition changes after the "Halloween" solar proton event: the High Energy Particle Precipitation in the Atmosphere (HEPPA) model versus MIPAS data intercomparison study. Atmos. Chem. Phys. 11, 9089–9139. https://doi.org/10.5194/acp-11-9089-2011.

Garner, T.W., 2000. A case study of the June 4–5, 1991 magnetic storm using the Rice Convection Model., PhD dissertation, Rice University, 164 p. http://hdl.handle.net/1911/17966.

Garner, T.W., Wolf, R.A., Spiro, R.W., Burke, W.J., Fejer, B.G., Sazykin, S., Roeder, J.L., Hairston, M.R., 2004. Magnetospheric electric fields and plasma sheet injection to low L-shells during the 4–5 June 1991 magnetic storm: Comparison between the Rice Convection Model and observations. J. Geophys. Res. 109. A02214, https://doi.org/10.1029/2003JA010208.

Goncharenko, L.P., Foster, J.C., Coster, A.J., Huang, C., Aponte, N., Paxton, L.J., 2007. Observations of a positive storm phase on September 10, 2005. J. Atmos. Terr. Phys. 69 (10-11), 1253–1272. https://doi.org/10.1016/j.jastp.2006.09.011.

Gonzalez, W.D., Joselyn, J.A., Kamide, Y., Kroehl, H.W., Rostoker, G., Tsurutani, B.T., Vasyliunas, V.M., 1994. What is a geomagnetic storm? J. Geophys. Res. 99 (A4), 5771–5792. https://doi.org/10.1029/93JA02867.

Gonzalez, L.X., Valdés-Galicia, J.F., Sánchez, F., Muraki, Y., Sako, T., Watanabe, K., Matsubara, Y., Nagai, Y., Shibata, S., Sakai, T., 2015. Re-evaluation of the neutron emission from the solar flare of 2005 September 7, detected by the solar neutron telescope at Sierra Negra. Astrophys. J. 814 (2), 7. https://doi.org/10.1088/0004-637X/814/2/136.

Gopalswamy, N., Barbieri, L., Lu, G., Plunkett, S.P., Skoug, R.M., 2005a. Introduction to the special section: Violent sun-earth connection events of October–November 2003. Geophys. Res. Lett. 32. L03S01, https://doi.org/10.1029/2005GL022348.

Gopalswamy, N., Barbieri, L., Cliver, E.W., Lu, G., Plunkett, S.P., Skoug, R.M., 2005b. Introduction to violent sun-earth connection events of October–November 2003. J. Geophys. Res. 110. A09S00, https://doi.org/10.1029/2005JA011268.

Gopalswamy, N., Yashiro, S., Liu, Y., Michalek, G., Vourlidas, A., Kaiser, M.L., Howard, R.A., 2005c. Coronal mass ejections and other extreme characteristics of the 2003 October–November solar eruptions. J. Geophys. Res. 110. A09S15, https://doi.org/10.1029/2004JA010958.

Gopalswamy, N., Xie, H., Yashiro, S., Usoskin, I., 2005d. Coronal mass ejections and ground level enhancements. In: Acharya, B.S., Gupta, S., Jagadeesan, P., Jain, A., Karthikeyan, S., Morris, S., Tonwar, S. (Eds.), Proc. ICRC 29, 03–10 August 2005, Pune, India, vol. 1. pp. 169–172.

Gopalswamy, N., Yashiro, S., Akiyama, S., 2006. Coronal mass ejections and space weather due to extreme events. In: Gopalswamy, N., Bhattacharyya, A. (Eds.), Proc. International Living With A Star (ILWS) Workshop 2006, Goa, India, pp. 19–24.

Gopalswamy, N., Xie, H., Yashiro, S., Akiyama, S., Makela, P., Usoskin, I.G., 2012. Properties of ground level enhancement events and the associated solar eruptions during solar cycle 23. Space Sci. Rev. 171 (1), 23–60. https://doi.org/10.1007/s11214-012-9890-4.

Gopalswamy, N., Xie, H., Akiyama, S., Makela, P.A., Yahiro, S., 2014. Major solar eruptions and high-energy particle events during solar cycle 24. Earth Planets Space 66 (104), 15. https://doi.org/10.1186/1880-5981-66-104.

Gosling, J.T., 1993. The solar flare myth. J. Geophys. Res. 98 (A11), 18937–18949. https://doi.org/10.1029/93JA01896.

Grechnev, V.V., Kurt, V.G., Chertok, I.M., Uralov, A.M., Nakajima, H., Altyntsev, A.T., Belov, A.V., Yushkov, B.Yu., Kuznetsov, S.N., Kashapova, L.K., Meshalkina, N.S., Prestage, N.P., 2008. An extreme solar event of 20 January 2005: Properties of the flare and the origin of energetic particles. Sol. Phys 252, 149–177. https://doi.org/10.1007/s11207-008-9245-1.

Grigoryev, V.G., Starodubtsev, S.A., Dvornikov, V.M., Sdobnov, V.E., 2009. Estimation of the solar proton spectrum in the GLE70 event. Adv. Space Res. 43 (4), 515–517. https://doi.org/10.1016/j.asr.2008.08.010.

Grubb, R.N., 1975. The SMS/GOES space environment monitor subsystem. NOAA TM ERL SEL- 42, Space Environment Laboratory, Boulder Colorado.

Guhathakurta, M., 2013. Interplanetary space weather: a new paradigm. Eos. Trans. AGU 94 (18), 165–172. http://onlinelibrary.wiley.com/doi/10.1002/2013EO180001/pdf.

Gussenhoven, M.S., Mullen, E.G., Sperry, M., Kerns, K.J., Blake, J.B., 1992. The effect of the March 1991 storm on accumulated dose for selected satellite orbits. CRRES dose models. IEEE Trans. Nucl. Sci. 39, 1765–1772. https://doi.org/10.1109/23.211365.

Habarulema, J.B., Ssessanga, N., 2017. Adapting a climatology model to improve estimation of ionosphere parameters and subsequent validation with radio occultation and ionosonde data. Space Weather 15, 84–98. https://doi.org/10.1002/2016SW001549.

Habarulema, J.B., Katamzi, Z.T., Yizengaw, E., 2015. First observations of poleward large-scale traveling ionospheric disturbances over the African sector during geomagnetic storm conditions. J. Geophys. Res. 120, 6914–6929. https://doi.org/10.1002/2015JA021066.

Habarulema, J.B., Katamzi, Z.T., Yizengaw, E., Yamazaki, Y., Seemala, G., 2016. Simultaneous storm time equatorward and poleward large-scale TIDs on a global scale. Geophys. Res. Lett. 43, 6678–6686. https://doi.org/10.1002/2016GL069740.

Habarulema, J.B., Katamzi, Z.T., Sibanda, P., Matamba, T.M., 2017. Assessing ionospheric response during some strong storms in solar cycle 24 using various data sources. J. Geophys. Res. 122, 1064–1082. https://doi.org/10.1002/2016JA023066.

Hamidi, Z.S., Monstein, C., Shariff, N.N.M., 2014. Radio observation of coronal mass ejections (CMEs) due to flare related phenomenon on 7th March 2012. Intern. Lett. Chem. Phys. Astron. 11 (3), 243–256. https://doi.org/10.18052/www.scipress.com/ILCPA.30.243.

Hamilton, D.C., Gloeckler, G., Ipavich, F.M., Stüdemann, W., Wilken, B., Kremser, G., 1988. Ring current development during the great geomagnetic storm of February 1986. J. Geophys. Res. 93 (A12), 14343–14355. https://doi.org/10.1029/JA093iA12p14343.

Harnett, E.M., Winglee, R.M., Stickle, A., Lu, G., 2008. Prompt ionospheric/magnetospheric responses 29 October 200 Halloween storm: outflow and energization. J. Geophys. Res. 113. A06209, https://doi.org/10.1029/2007JA012810.

He, H., Wang, H., 2008. Non-linear force-free coronal magnetic field extrapolation scheme based on the direct boundary integral formulation. J. Geophys. Res. 113. A05S90, https://doi.org/10.1029/2007JA012441.

Herbst, K., Kopp, A., Heber, B., 2013. Influence of the terrestrial magnetic field geometry on the cutoff rigidity of cosmic ray particles. Ann. Geophys. 31, 1637–1643. https://doi.org/10.5194/angeo-31-1637-2013.

Horne, R.B., Glauert, S.A., Meredith, N.P., Boscher, D., Maget, V., Heynderickx, D., Pitchford, D., 2013. Space weather impacts on satellites and forecasting the Earth's electron radiation belts with SPACECAST. Space Weather 11, 169–186. https://doi.org/10.1002/swe.20023.

Horvath, I., Lovell, B.C., 2010. Storm-enhanced plasma density features investigated during the Bastille Day Superstorm. J. Geophys. Res. 115. A06305, https://doi.org/10.1029/2009JA014674.

Huba, J.D., Warren, H.P., Joyce, G., Pi, X., Iijima, B., Coker, C., 2005. Global response of the low-latitude to midlatitude ionosphere due to the Bastille Day flare. Geophys. Res. Lett. 32. L15103, https://doi.org/10.1029/2005GL023291.

Hudson, H.S., 1995. Solar flares: no "myth" Eos. Trans. AGU 76 (41), 401. https://doi.org/10.1029/95EO00253.

Hudson, M.K., Kotelnikov, A.D., Li, X., Roth, I., Temerin, M., Wygant, J., Blake, J.B., Gussenhoven, M.S., 1995. Simulations of proton radiation belt formation during the March 24, 1991 SSC. Geophys. Res. Lett. 22, 291–294. https://doi.org/10.1029/95GL00009.

Hudson, H.K., Elkington, S.R., Lyon, J.G., Marchenko, V.A., Roth, I., Temerin, M., Blake, J.B., Gussenhoven, M.S., Wygant, J.R., 1997. Simulations of radiation belt formation during storm sudden commencements. J. Geophys. Res. 102, 14087–14102. https://doi.org/10.1029/97JA03995.

Hudson, M.K., Kress, B.T., Mazur, J.E., Perry, K.L., Slocum, P.L., 2004. 3D modeling of shock-induced trapping of solar energetic particles in the Earth's magnetosphere. J. Atmos. Terr. Phys. 66 (15–16), 1389–1397. https://doi.org/10.1016/j.jastp.2004.03.024.

Humble, J.E., Duldig, M.L., Smart, D.F., Shea, M.A., 1991. Detection of 0.5–15 GEV solar protons on 29 September 1989 at Australian stations. Geophys. Res. Lett. 18 (4), 737–740. https://doi.org/10.1029/91GL00017.

Hurford, G.J., Krucker, S., Lin, R.P., Schwartz, R.A., Share, G.H., Smith, D.M., 2006. Gamma-ray imaging of the 2003 October/November solar flares. Astrophys. J. 644, L93–L96.

Jackman, C.H., Marsh, D.R., Vitt, F.M., Garcia, R.R., Fleming, E.L., Labow, G.J., Randall, C.E., López-Puertas, M., Funke, B., von Clarmann, T., Stiller, G.P., 2008. Short- and medium-term atmospheric constituent effects of very large solar proton events. Atmos. Chem. Phys. 8, 765–785. https://doi.org/10.5194/acp-8-765-2008.

Jaeggli, S.A., Norton, A.A., 2016. The magnetic classification of solar active regions 1992–2015. Astrophys. J. Lett. 820, L11–L14.

Jamsen, T., Usoskin, I.G., Raiha, T., Sarkamo, J., Kovaltsov, G.A., 2007. Case study of Forbush decreases: energy dependence of the recovery. Adv. Space Res. 40, 342–347. https://doi.org/10.1016/j.asr.2007.02.025.

Kahler, S.W., 1993. Coronal mass ejections and long risetimes of solar energetic particle events. J. Geophys. Res. 98 (A4), 5607–5615. https://doi.org/10.1029/92JA02605.

Kahler, S.W., 1994. Injection profiles for solar energetic particles as functions of coronal mass ejection heights. Astrophys. J. 428, 837–842.

Kanbach, G., Bertsch, D.L., Fichtel, C.E., Hartman, R.C., Hunter, S.D., Kniffen, D.A., Kwok, P.W., Lin, Y.C., Mattox, J.R., Mayer-Hasselwander, H.A., Michelson, P.F., von Montigny, C., Nolan, P.L., Pinkau, K., Rothermel, H., Schneid, E., Sommer, M., Sreekumar, P., Thompson, D.J., 1993. Detection of a long-duration solar gamma-ray flare on June 11, 1991 with EGRET of COMPTON-GRO. Astron. Astrophys. 97, 349–353.

Kane, R.P., 2009. Cosmic ray ground level enhancements (GLEs) of October 28, 2003 and January 20, 2005: a simple comparison. Rev. Bras. Geogr. 27 (2). https://doi.org/10.1590/S0102-261X2009000200002.

Kil, H., Paxton, L.J., Pi, X., Hairston, M.R., Zhang, Y., 2003. Case study of the 15 July 2000 magnetic storm effects on the ionosphere-driver of the positive ionospheric storm in the winter hemisphere. J. Geophys. Res. 108 (A11), 1391. https://doi.org/10.1029/2002JA009782.

Kim, K.-C., Shprits, Y.Y., Blake, J.B., 2016. Fast injection of the relativistic electrons into the inner zone and the formation of the split-zone structure during the Bastille Day storm in July 2000. J. Geophys. Res. 121, 8329–8342. https://doi.org/10.1002/2015JA022072.

Klein, K.-L., Chupp, E.L., Trottet, G., Magun, A., Dunphy, P.P., Rieger, E., Urpo, S., 1999. Flare associated energetic particles in the corona at 1 AU. Astron. Astrophys. 348, 271–285.

Klimenko, M.V., Klimenko, V.V., Ratovsky, K.G., Goncharenko, L.P., 2011a. Ionospheric effects caused by the series of geomagnetic storms of September 9–14, 2005. Geomagn. Aeron. 51, 364–376. https://doi.org/10.1134/S0016793211030108.

Klimenko, M.V., Klimenko, V.V., Ratovsky, K.G., Goncharenko, L.P., Sahai, Y., Fagundes, P.R., de Jesus, R., de Abreu, A.J., Vesnin, A.M., 2011b. Numerical modeling of ionospheric effects in the middle- and low-latitude F region during geomagnetic storm sequence of 9–14 September 2005. Radio Sci. 46. RS0D03, https://doi.org/10.1029/2010RS004590.

Klobuchar, J.A., Kunches, J.M., Van Dierendonck, A.J., 1999. Eye on the ionosphere: potential solar radio burst effects on GPS signal to noise. GPS Solutions 3, 69–71.

Kockarov, L.G., Kovaltsov, G.A., Kocharov, G.E., Chuikin, E.I., Usoskin, I.G., Shea, M.A., Smart, D.F., Melnikov, V.F., Podstrigach, T.S., Armstrong, T.P., Zirin, H., 1994. Electromagnetic and corpuscular emission from the solar flare of 1991 June 15: Continuous acceleration of relativistic particles. Sol. Phys. 150, 267–283.

Koga, K., Matsumoto, H., Okudaira, O., Goka, T., Obara, T., Masuda, S., Muraki, Y., Shibata, S., Yamamoto, T., 2015. Results of a measurement of solar neutrons emitted on March 5, 2012 using a fiber-type neutron monitor onboard the SEDA-AP attached to the ISS. In: Proc. ICRC 34, The Hague, The Netherlands, 30-July–06-August 2015, p. 8.

Kohl, J.W., Bostrom, C.O., Williams, D.J., 1973. Particle observations of the August 1972 solar events by Explorers 41 and 43. In: Coffey, H.E. (Ed.), Collected Data Reports on August 1872 Solar-Terrestrial Events, NOAA National Geophysical Data Center, UAG-28, pp. 330–333. https://www.ngdc.noaa.gov/stp/space-weather/online-publications/stp_uag/. (Accessed 01.04.17).

Kosovichev, A.G., Zharkova, V.V., 2001. Magnetic energy release and transients in the solar flare of 2000 July 14. Astrophys. J. Lett. 550 (1), L105–L108.

Kovaltsov, G.A., Usoskin, I.G., Kocharov, L.G., Kananen, H., Tanskanen, P.T., 1995. Neutron monitor data on the 15 June 1991 flare: neutrons as a test for proton acceleration scenario. Sol. Phys. 158, 395–398. https://doi.org/10.1007/BF00795674.

Kozarev, K., Schwadron, N.A., Dayeh, M.A., Townsend, L.W., Desai, M.I., PourArsalan, M., 2010. Modeling the 2003 Halloween events with EMMREM: energetic particles, radial gradients and coupling to MHD. Space Weather. 8. S00E08, https://doi.org/10.1029/2009SW000550.

Kozyra, J.U., Liemohn, M.W., Clauer, C.R., Ridley, A.J., Thomsen, M.F., Borovsky, J.E., Roeder, J.L., Jordanova, V.K., Gonzalez, W.D., 2002. Multistep Dst development and ring current composition changes during the 4–6 June 1991 magnetic storm. J. Geophys. Res. 107(A8), https://doi.org/10.1029/2001JA00023.

Kozyra, J.U., Manchester IV, W.B., Escoubet, C.P., Lepri, S.T., Liemohn, M.W., Gonzalez, W.D., Thomsen, M.W., Tsurutani, B.T., 2013. Earth's collision with a solar filament on 21 January 2005: overview. J. Geophys. Res. 118, 5967–5978. https://doi.org/10.1002/jgra.50567.

Kozyra, J.U., Liemohn, M.W., Cattell, C., DeZeeuw, D., Escobet, C.P., Evans, D.S., Fang, X., Fok, M.-C., Frey, H.U., Gonzalez, W.D., Hairston, M., Heelis, R., Lu, G., Manchester IV, W.B., Mende, S., Paxton, L.J., Rastaetter, L., Ridley, A., Sandanger, M., Soraas, F., Sotirelis, T., Thomsen, M.W., Tsurutani, B.T., Verkhoglyadova, O., 2014. Solar filament impact on 21 January 2005: geospace consequences. J. Geophys. Res. 119, 5401–5448. https://doi.org/10.1002/2013JA019748.

Krall, J., Yurchyshyn, V.B., Slinker, S., Skoug, R.M., Chen, J., 2006. Flux rope model of the 2003 October 28–30 coronal mass ejection and interplanetary coronal mass ejection. Astrophys. J. 642, 541–553.

Kress, B.T., Hudson, M.K., Looper, M.D., Albert, J., Lyon, J.G., Goodrich, C.C., 2007. Global MHD test particle simulations of >10 MeV radiation belt electrons during storm sudden commencement. J. Geophys. Res. 112. A09215, https://doi.org/10.1029/2006JA012218.

Kress, B.T., Hudson, M.K., Looper, M.D., Lyon, J.G., Goodrich, C.C., 2008. Global MHD test particle simulations of solar energetic electron trapping in the Earth's radiation belts. J. Atmos. Terr. Phys. 70 (14), 1727–1737. https://doi.org/10.1016/j.jastp.2008.05.018.

Kumar, S., Kumar, A., Menk, F., Maurya, A.K., Singh, R., Veenadhari, B., 2015. Response of the low-latitude D region ionosphere to extreme space weather event of 14–16 December 2006. J. Geophys. Res. 120, 788–799. https://doi.org/10.1002/2014JA020751.

Kurt, V.G., Yushkov, B.Y., Kudela, K., Galkin, V.I., 2009. High-energy gamma-ray emission of solar flares as an indicator of acceleration of high-energy protons. In: Szabelski, J., Giller, M. (Eds.), Proc. ICRC 31, 07–15 July 2009, Lodz, Poland, p. 4.

Kuznetsov, S.N., Bogomolov, A.V., Denisov, Y.I., Kordylewski, Z., Kudela, K., Kurt, V.G., Lisin, D.V., Myagkova, I.N., Podorol'skii, A.N., Podosenova, T.B., Svertilov, S.I., Sylwester, J., Stepanov, A.I., Yushkov, B.Y., 2003a. The solar flare of November 4, 2001 and its manifestations in energetic particles from Coronas-F Data. Sol. Syst. Res. 37 (2), 121–127. https://doi.org/10.1023/A:1023384425209.

Kuznetsov, S.N., Kudela, K., Myagkova, I.M., Yushkov, B.Y., 2003b. Gamma and x-ray solar flare emissions: CORONAS-F Measurements. In: Kajita, T., Asaoka, Y., Kawachi, A., Matsubara, Y., Sasaki, M. (Eds.), Proc. ICRC 28, 31-July–07-August 2003, Trukuba, Japan, pp. 3183–3186.

Kuznetsov, S.N., Yushkova, B.Y., Kudelab, K., Myagkovaa, I.N., Starostina, L.I., Denisova, Y.I., 2005a. Dynamics of the earth's radiation belts during the magnetic storm of November 6th, 2001. Adv. Space Res. 36, 1997–2002. https://doi.org/10.1016/j.asr.2004.09.019.

Kuznetsov, S.N., Kurt, V.G., Yushkov, B.Yu., Myagkova, I.N., Kudela, K., Kassovicova, J.C., Slivka, M., 2005b. Proton acceleration during 20 January 2005 solar flare: CORONAS-F observations of high-energy gamma emission and GLE. In: Acharya, B.S., Gupta, S., Jagadeesan, P., Jain, A., Karthikeyan, S., Morris, S., Tonwar, S. (Eds.), Proc. ICRC 29, 03–10 August 2005, Pune, India, vol. 1. pp. 49–52.

Labrador, A.W., Leske, R.A., Mewaldt, R.A., Stone, E.C., von Rosenvinge, T.T., 2005. High energy ionic charge state composition in the October/November 2003 and January 20, 2005 SEP Events. In: Acharya, B.S.,

Gupta, S., Jagadeesan, P., Jain, A., Karthikeyan, S., Morris, S., Tonwar, S. (Eds.), Proc. ICRC 29, 03-10 August 2005, Pune, India, vol. 1. pp. 99–102.

Lario, D., Decker, R.B., Armstrong, T.P., 2001. Major solar proton events observed by IMP-8 (from November 1973 to May 2001). In: Kampert, K.-H., Heinzelmann, G., Spiering, C. (Eds.), Proc. ICRC 27, 05–15 Aug 2001, Hamburg, Germany, vol. 8. pp. 3254–3257.

Lario, D., Decker, R.B., 2002. The energetic storm particle event of October 20, 1989. Geophys. Res. Lett. 29, 1393–1396. https://doi.org/10.1029/2001GL014017.

Lario, D., Decker, R.B., Livi, S., Krimigis, S.M., Roelof, E.C., Russell, C.T., Fry, C.D., 2005. Heliospheric energetic particle observations during the October–November 2003 events. J. Geophys. Res. 110. A09S11, https://doi.org/10.1029/2004JA010940.

Le, G.M., Ye, Z.H., Gong, J.H., Tan, Y.H., Lu, H., Tang, Y.Q., 2003. Time determination of March 1991's CME hitting magnetosphere. In: Kajita, T., Asaoka, Y., Kawachi, A., Matsubara, Y., Sasaki, M. (Eds.), Proc. ICRC 28, 31-July–07-August 2003, Trukuba, Japan, pp. 3601–3604.

Lebed, O.M., Fedorenko, Y.V., Larchenko, V., Pil'gaev, S.V., 2015. Response of the auroral lower ionosphere to solar flares in March 2012 according to ELF observations. Geomagn. Aeron. 55 (6), 770–779. https://doi.org/10.1134/S0016793215060080.

Lebed, O.M., Larchenko, A.V., Pil'gaev, S.V., Fedorenko, Y.V., 2017. Reaction of the high-latitude lower ionosphere to solar proton events from observations in the ELF range. Geomagn. Aeron. 57 (1), 51–57. https://doi.org/10.1134/S0016793217010078.

Lei, J., Wang, W., Burns, A.G., Solomon, S.C., Richmond, A.D., Wiltberger, M., Goncharenko, L.P., Coster, A., Reinisch, B.W., 2008. Observations and simulations of the ionospheric and thermospheric response to the December 2006 geomagnetic storm: Initial phase. J. Geophys. Res. 113. A01314, https://doi.org/10.1029/2007JA012807.

Lei, J., Thayer, J.P., Lu, G., Burns, A.G., Wang, W., Sutton, E.K., Emery, B.A., 2011. Rapid recovery of thermosphere density during the October 2003 geomagnetic storms. J. Geophys. Res. 116. A03306, https://doi.org/10.1029/2010JA016164.

Lei, J., Burns, A.G., Thayer, J.P., Wang, W., Mlynczak, M.G., Hunt, L.A., Dou, X., Sutton, E., 2012. Overcooling in the upper thermosphere during the recovery phase of the 2003 October Storms. J. Geophys. Res. 117. A03314, https://doi.org/10.1029/2011JA016994.

Li, H., Schmieder, B., Song, M.T., Bommier, V., 2007. Interaction of magnetic field systems leading to an X1.7 flare due to large-scale flux tube emergence. Astron. Astrophys. 475, 1081–1091. https://doi.org/10.1051/0004-6361:20077500.

Li, W., Knipp, D., Lei, J., Raeder, J., 2011. The relation between dayside local poynting flux enhancement and Cusp reconnection. J. Geophys. Res. 116. A08301, https://doi.org/10.1029/2011JA016566.

Li, X., Roth, I., Temerin, M., Wygant, J.R., Hudson, M.K., Blake, J.B., 1993. Simulation of the prompt energization and transport of radiation belt particles during the March 24, 1991 SSC. Geophys. Res. Lett. 20, 2423–2426. https://doi.org/10.1029/93GL02701.

Liemohn, M.W., Kozyra, J.U., Clauer, C.R., Khazanov, G.V., Thomsen, M.F., 2002. Adiabatic energization in the ring current and its relation to other source and loss terms. J. Geophys. Res. 107 (A4), 1045. https://doi.org/10.1029/2001JA000243.

Lincoln, J.V., 1964. The listing of sudden ionospheric disturbances. Planet. Space Sci. 12, 419–434. https://doi.org/10.1016/0032-0633(64)90035-2.

Lingri, D., Papahliou, M., Mavromichalaki, H., Belov, A., Eroshenko, E., Yanke, V., 2013. Forbush decreases during the ascending phase of solar cycle 24. In: 11th Hellenic Astronomical Conference, Athens, Greece, 08–12 September 2013.

Lingri, D., Mavromichalaki, H., Belov, A., Eroshenko, E., Yanke, V., Abunin, A., Abunina, A., 2015. Solar activity parameters and associated Forbush decreases during the minimum between cycles 23 and 24 and the ascending phase of cycle 24. Sol. Phys. 291 (3), 1025–1041. https://doi.org/10.1007/s11207-016-0863-8.

Liu, C., Lee, J., Karlicky, M., Choudhary, D.P., Deng, N., Wang, H., 2009. Successive solar flares and coronal mass ejections on 2005 September 13 from NOAA AR 10808. Astrophys. J. 703, 757–768. https://doi.org/10.1088/0004-637X/703/1/757.

Liu, R., Luhr, H., Doornbos, E., Ma, S.-Y., 2010. Thermospheric mass density variations during geomagnetic storms and a prediction model based on the merging electric field. Ann. Geophys. 28, 1633–1645. https://doi.org/10.5194/angeo-28-1633-2010.

Liu, R., Ma, S.-Y., Luhr, H., 2011. Predicting storm-time thermospheric mass density variations at CHAMP and GRACE altitudes. Ann. Geophys. 29, 443–453. https://doi.org/10.5194/angeo-29-443-2011.

Liu, Y., Zhang, H., 2001. Relationship between magnetic field evolution and major flare event on July 14, 2000. Astron. Astrophys. 372, 1019–1029. https://doi.org/10.1051/0004-6361:20010550.

Liu, Y., Hayashi, K., 2006. The 2003 October November fast halo coronal mass ejections and the large-scale magnetic field structures. Astrophys. J. 640, 1135–1141.

Liu, J.Y., Lin, C.H., Chen, Y.I., Lin, Y.C., Fang, T.W., Chen, C.H., Chen, Y.C., Hwang, J.J., 2006. Solar flare signatures of the ionospheric GPS total electron content. J. Geophys. Res. 111. A05308, https://doi.org/10.1029/2005JA011306.

Liu, Y., Kurokawa, H., Liu, C., Brooks, D.H., Dun, J., Ishii, T.T., Zhang, H., 2007. The X10 flare on 29 October 2003: was it triggered by magnetic reconnection between counter-helical fluxes? Sol. Phys. 240, 253–262. https://doi.org/10.1007/s11207-006-0316-x.

Liu, Y., Luhmann, J.G., Muller-Mellin, R., Schroeder, P.C., Wang, L., Lin, R.P., Bale, S.D., Li, Y., Acuna, M.H., Sauvaud, J.-A., 2008. A comprehensive view of the 2006 December 13 CME: from the sun to interplanetary space. Astrophys. J. 689 (1), 563–571.

Liu, Y.D., Luhmann, J.G., Lugaz, N., Christian, M., Davies, J.A., Bale, S.D., Lin, R.P., 2013. On sun-earth propagation of coronal mass ejections. Astrophys. J. 769 (1), 15. https://doi.org/10.1088/0004-637X/769/1/45.

Liu, Y.D., Richardson, J.D., Wang, C., Luhmann, J.G., 2014. Propagation of the 2012 March coronal mass ejections from the sun to the heliopause. Astrophys. J. Lett. 788 (2), 6. https://doi.org/10.1088/2041-8205/788/2/L28.

Lockwood, J.A., Ryan, J.M., McConnell, M., Schonfelder, V., Aarts, H., Bennett, K., Winckler, C., 1993. Neutrons from the 15 June 1991 solar flare. In: Leahy, D.A., Hickws, R.B., VenkatesanIntern, D. (Eds.), Proc. ICRC 23, 19–30 July 1993, Alberta, Canada, vol. 3. pp. 115–118.

Looper, M.D., Blake, J.B., Mewaldt, R.A., Cummings, J.R., Baker, D.N., 1994. Observations of the remnants of the ultrarelativistic electrons injected by the strong SSC of 24 March 1991. Geophys. Res. Lett. 21, 2079–2082. https://doi.org/10.1029/94GL01586.

Looper, M.D., Blake, J.B., Mewaldt, R.A., 2005. Response of the inner radiation belt to the violent sun-earth connection events of October-November 2003. Geophys. Res. Lett. 32. L03S06, https://doi.org/10.1029/2004GL021502.

Lopez, R.E., Baker, D.N., Allen, J., 2004. Sun unleashes Halloween storm. Eos. Trans. AGU 85 (11), 105–108. https://doi.org/10.1029/2004EO110002.

Loto'aniu, T.M., Singer, H.J., Rodriguez, J.V., Green, J., Denig, W., Biesecker, D., Angelopoulos, V., 2015. Space weather conditions during the Galaxy 15 spacecraft anomaly. Space Weather 13, 484–502. https://doi.org/10.1002/2015SW001239.

Lovell, J.L., Duldig, M.L., Humble, J.E., 1998. An extended analysis of the September 1989 cosmic ray ground level enhancement. J. Geophys. Res. 103 (A10), 23733–23742. https://doi.org/10.1029/98JA02100.

Lu, G., Goncharenko, L.P., Richmond, A.D., Roble, R.G., Aponte, N., 2008. A dayside ionospheric positive storm phase driven by neutral winds. J. Geophys. Res. 113. A08304, https://doi.org/10.1029/2007JA012895.

Lu, G., Richmond, A.D., Luhr, H., Paxton, L., 2016. High-latitude energy input and its impact on the thermosphere. J. Geophys. Res. Space Phys. 121, 7108–7124. https://doi.org/10.1002/2015JA022294.

Malandraki, O.E., Marsden, R.G., Tranquille, C., Forsyth, R.J., Elliott, H.A., Geranios, A., 2008. Energetic particle measurements from the Ulysses/COSPIN/LET instrument obtained during the August/September 2005 events. Ann. Geophys. 26, 1029–1037. https://doi.org/10.5194/angeo-26-1029-2008.

Manchester IV, W.B., Kozyra, J.U., Lepri, S.T., Lavraud, B., 2014. Simulation of magnetic cloud erosion during propagation. J. Geophys. Res. 119, 5449–5464. https://doi.org/10.1002/2014JA019882.

Manju, G., Pant, T.K., Devasia, C.V., Ravindran, S., Sridharan, R., 2009. Electrodynamical response of the Indian low-mid latitude ionosphere to the very large solar flare of 28 October 2003—a case study. Ann. Geophys. 27, 3853–3860. https://doi.org/10.5194/angeo-27-3853-2009. www.ann-geophys.net/27/3853/2009/.

Mannucci, A.J., Tsurutani, B.T., Iijima, B.A., Komjathy, A., Saito, A., Gonzalez, W.D., Guarnieri, F.L., Kozyra, J.U., Skoug, R., 2005. Dayside global ionospheric response to the major interplanetary events of October 29–30, 2003 Halloween Storms. Geophys. Res. Lett. 32. L12S02, https://doi.org/10.1029/2004GL021467.

Marayama, T., Ma, G., 2004. TEC storm on November 6, 2001, derived from dense GPS receiver network and ionosonde chain over Japan Qingdao, China. In: Proc Asia-Pacific Radio Sci. Conf, pp. 359–362.

Marshall, R.A., Dalzell, M., Waters, C.L., Goldthorpe, P., Smith, E.A., 2012. Geomagnetically induced currents in the New Zealand power network. Space Weather 10. S08003, https://doi.org/10.1029/2012SW000806.

Martucci, M., 2014. Study of Forbush decrease during the 2012 March 7 solar particle event with the PAMELA experimen. In: Proc. 40th COSPAR Scientific Assembly, E1.6-53-14, Moscow, Russia, 02–10 August 2014.

Masson, S., Klein, K.-L., Buetikofer, R., Flueckiger, E., Kurt, V., Yushkov, B., Krucker, S., 2009. Acceleration of relativistic protons during the 20 January 2005 flare and CME. Sol. Phys. 257 (2), 305–322. https://doi.org/10.1007/s11207-009-9377-y.

Matandirotya, E., Cilliers, P.J., Van Zyl, R.R., 2015. Modelling geomagnetically induced currents in the South African power transmission network using the finite element method. Space Weather 13, 185–195. https://doi.org/10.1002/2014SW001135.

Matthia, D.H.B., Reitz, G., Sihver, L., Berger, T., Meier, M., 2009. The ground level event 70 on December 13th, 2006 and related effective doses at aviation altitudes. Radiat. Prot. Dosim. 136 (4), 304–310. https://doi.org/10.1093/rpd/ncp141.

Mayaud, P.N., 1980. Derivation, Meaning and Use of Geomagnetic Indices, Geophysical Monograph #22, IBSN 0-87590-022-4, American Geophysical Union, Washington, DC, 154 p.

McCracken, K.G., Moraal, H., Stoker, P.H., 2008. Investigation of the multiple-component structure of the 20 January 2005 cosmic ray ground level enhancement. J. Geophys. Res. 113. A12101, https://doi.org/10.1029/2007JA012829.

McIntosh, P., 1990. The classification of sunspot groups. Sol. Phys. 125, 251–267. https://doi.org/10.1007/BF00158405.

McKenna-Lawlor, S.M.P., Dryer, M., Fry, C.D., Smith, Z.K., Intriligator, D.S., Courtney, W.R., Deehr, C.S., Sun, W., Kecskemety, K., Kudela, K., Balaz, J., Barabash, S., Futaana, Y., Yamauchi, M., Lundin, R., 2008. Predicting interplanetary shock arrivals at Earth, Mars and Venus: a real-time modeling experiment following the solar flares of 5-14 December 2006. J. Geophys. Res. 113. A06101, https://doi.org/10.1029/2007JA012577.

McKenna-Lawlor, S., Li, L., Dandouras, I., Brandt, P.C., Zheng, Y., Barabash, S., Bucik, R., Kudela, K., Balaz, J., Strharsky, I., 2010. Moderate geomagnetic storm (21–22 January 2005) triggered by an outstanding coronal mass ejection viewed via energetic neutral atoms. J. Geophys. Res. 115. A08213, https://doi.org/10.1029/2009JA014663.

Medford, L.V., Lanzerotti, L.J., Kraus, J.S., Maclennan, C.G., 1989. Transatlantic Earth potential variations during the March 1989 magnetic storms. Geophys. Res. Lett. 16, 1145–1148. https://doi.org/10.1029/GL016i010p01145.

Meier, R.R., Warren, H.P., Nicholas, A.C., Bishop, J., Huba, J.D., Drob, D.P., Lean, J.L., Picone, J.M., Mariska, J.T., Joyce, G., Judge, D.L., Thonnard, S.E., Dymond, K.F., Budzien, S.A., 2002. Ionospheric and dayglow responses to the radiative phase of the Bastille Day flare. Geophys. Res. Lett. 29 (10), https://doi.org/10.1029/2001GL013956.

Mertens, C.J., Kress, B.T., Wiltberger, M., Blattnig, S.R., Slaba, T.S., Solomon, S.C., Engel, M., 2010. Geomagnetic influence on aircraft radiation exposure during a solar energetic particle event in October 2003. Space Weather 8. S03006, https://doi.org/10.1029/2009SW000487.

Mewaldt, R.A., Cohen, C.M.S., Labrador, A.W., Leske, R.A., Mason, G.M., Desai, M.I., Looper, M.D., Mazur, J.E., Selesnick, R.S., Haggert, D.K., 2005a. Proton, helium and electron spectra during the large solar particle events of October–November 2003. J. Geophys. Res. 110 (A9). A09S18, https://doi.org/10.1029/2005JA011038.

Mewaldt, R.A., Looper, M.D., Cohen, C.M.S., Mason, G.M., Haggerty, D.K., Desai, M.I., Labrador, A.W., Leske, R.A., Mazur, J.E., 2005b. Solar-particle energy spectra during the large events of October-November 2003 and January 2005. In: Acharya, B.S., Gupta, S., Jagadeesan, P., Jain, A., Karthikeyan, S., Morris, S., Tonwar, S. (Eds.), Proc. ICRC 29, 03–10 August 2005, Pune, India, vol. 1. pp. 111–114.

Meyer, P., Parker, E.N., Simpson, J.A., 1956. Solar cosmic rays of February, 1956 and their propagation through interplanetary space. Phys. Rev. 104, 768–783. https://doi.org/10.1103/PhysRev.104.768.

Mishin, E.V., Marcos, F.A., Burke, W.J., Cooke, D.L., Roth, C., Petrov, V.P., 2007. Prompt thermospheric response to the 6 November 2001 magnetic storm. J. Geophys. Res. 112. A05313, https://doi.org/10.1029/2006JA011783.

Miroshnichenko, L.I., DeKoning, C.A., Perez-Enriquez, R., 2000. Large solar event of September 29, 1989: ten years after. Space Sci. Rev. 91, 615–715.

Miroshnichenko, L.I., Klein, K.-L., Trottet, G., Lantos, P., Vashenyuk, E.V., Balabin, Y.V., Gvozdevsky, B.B., 2005a. Relativistic nucleon and electron production in the 2003 October 28 solar event. J. Geophys. Res. 110. A09S08, https://doi.org/10.1029/2004JA010936.

Miroshnichenko, L.I., Klein, K.-L., Trottet, G., Lantos, P., Vashenyuk, E.V., Balabin, Y.V., Gvozdevsky, B.B., 2005b. Correction to "Relativistic nucleon and electron production in the 28 October 2003 solar event" J. Geophys. Res. 110. A11S90, https://doi.org/10.1029/2005JA011441.

Mishev, A., Usoskin, I., 2016a. Analysis of the ground-level enhancements on 14 July 2000 and 13 December 2006 using neutron monitor data. Sol. Phys. 291, 1225–1239. https://doi.org/10.1007/s11207-016-0877-2.

Mishev, A., Usoskin, I., 2016b. Erratum to: analysis of the ground level enhancements on 14 July 2000 and on 13 December 2006 using neutron monitor data. Sol. Phys. 291 (5), 1579–1580. https://doi.org/10.1007/s11207-016-0877-2.

Miyasaka, H., Takahashi, E., Shimoda, S., Yamada, Y., Kondo, I., Tsuchiya, H., Makishima, K., Zhu, F.R., Tan, Y.H., Hu, H.B., Tang, Y.Q., Clem, J., YBJ NM collaboration, 2005. The solar event on 20 January 2005 observed with the Tibet YBJ neutron monitor observatory. In: Acharya, B.S., Gupta, S., Jagadeesan, P., Jain, A., Karthikeyan, S., Morris, S., Tonwar, S. (Eds.), Proc. ICRC 29, 03–10 August 2005, Pune, India, vol. 1. pp. 241–244.

Miyashita, Y., Miyoshi, Y., Matsumoto, Y., Ieda, A., Kamide, Y., Nose, M., Machida, S., Hayakawa, H., McEntire, R.W., Chrison, S.P., Evans, D.S., Troshichev, O.A., 2005. Geotail observations of signatures in the near-Earth magnetotail for the extremely intense substorms of the 30 October 2003 storm. J. Geophys. Res. 110. A09S25, https://doi.org/10.1029/2005JA011070.

Moraal, H., McCracken, K.G., Schoeman, C.C., Stoker, P.H., 2005. The ground level enhancements of 20 January 2005 and 28 October 2003. In: Acharya, B.S., Gupta, S., Jagadeesan, P., Jain, A., Karthikeyan, S., Morris, S., Tonwar, S. (Eds.), Proc. ICRC 29, 03–10 August 2005, Pune, India, vol. 1. pp. 221–223.

Moraal, H., Caballero-Lopez, R.A., McCracken, K.G., 2016. The cosmic-ray ground-level enhancements of 29 September 1989 and 20 January 2005. In: Proc. ICRC 34, The Hague, The Netherlands, 30-July–06-August 2015, p. 8.

Morton, Y.T., Mathews, J.D., 1992. Effects of the 13–14 March 1989 geomagnetic storm on the E region tidal ion layer structure at Arecibo during AIDA. J. Atmos. Terr. Phys. 55 (3), 467–485.

Muhr, N., Vrsnak, B., Temmer, M., Veronig, A.M., Magdalenic, J., 2010. Analysis of a global Moreton wave observed on 2003 October 28. Astrophys. J. 708, 1639–1649. https://doi.org/10.1088/0004-637X/708/2/1639.

Mullen, E.G., Gussenhoven, M.S., Ray, K., Violet, M., 1991. A double-peaked inner radiation belt: cause and effect as seen on CRRES. IEEE Trans. Nucl. Sci. 38, 1713–1718. https://doi.org/10.1109/23.124167.

Mulligan, T., Blake, J.B., Mewaldt, R.A., Leske, R.A., 2008. Unusual observations during the December 2006 solar energetic particle events within an interplanetary coronal mass ejection at 1 AU. In: Li, G., Hu, Q.,

Verkhoglyadova, O., Zank, G.P., Lin, R.P., Lulimann, J. (Eds.), Particle Acceleration and Transport in the Heliosphere and Beyond: 7th Annual International Astrophysics Conference.Proc AIP CP1039, Kauai, Hawaii, 07–13 March 2008, pp. 162–167. https://doi.org/10.1063/1.2982440.

Munakata, K., Kuwabara, T., Yasue, S., Kato, C., Akahane, S., Koyama, M., Ohashi, Y., Okada, A., Aoki, T., Mitsui, K., Kojima, H., Bieber, J.W., 2005. A "loss cone" precursor of an approaching shock observed by a cosmic ray muon hodoscope on October 28, 2003. Geophys. Res. Lett. 32. L03S04, https://doi.org/10.1029/2004GL021469.

Muraki, Y., Murakami, K., Shibata, S., Sakakibara, S., Yamada, T., Miyazaki, M., Takahashi, T., Mitsui, K., Sakai, Y., 1991. Solar neutrons associated with the large solar flare of June 1991. In: Cawley, M. et al., (Ed.), Proc. ICRC 22, 11–23 August 1991, Dublin, Ireland, vol. 3. pp. 49–52. paper, SH 2.2-5.

Myagkova, I.N., Panasyuk, M.I., Lazutin, L.L., Muravieva, E.A., Starostin, L.I., Ivanova, T.A., Pavlov, N.N., Rubinshtein, I.A., Vedenkin, N.N., Vlasova, N.A., 2009. December 2006 solar extreme events and their influence on the near-Earth space environment: "Universitetskiy-Tatiana" satellite observations. Adv. Space Res. 43 (4), 489–494. https://doi.org/10.1016/j.asr.2008.07.019.

Nagai, T., Yukimatu, A.S., Matsuoka, A., Asai, K.T., Green, J.C., Onsager, T.G., Singer, H.J., 2006. Timescales of relativistic electron enhancements in the slot region. J. Geophys. Res. 111. A11205, https://doi.org/10.1029/2006JA011837.

Nagashima, K., Isobe, H., Yokoyama, T., Ishii, T.T., Okamoto, T.J., Shibata, K., 2007. Triggering mechanism for the filament eruption on 2005 September 13 in NOAA Active Region 10808. Astrophys. J. 668 (1), 533–545.

Nemzek, R.J., Belian, R.D., Cayton, T.E., Reeves, G.D., 1994. The October 22, 1989 solar cosmic ray event measured at geosynchronous orbit. J. Geophys. Res. 99 (A3), 4221–4226. https://doi.org/10.1029/93JA03254.

Nishino, M., Makita, K., Yumoto, K., Miyoshi, Y., Schuch, N.J., Abdu, M.A., 2006. Energetic particle precipitation in the Brazilian geomagnetic anomaly during the "Bastille Day Storm" of July 2000. Earth Planets Space 58, 607–616. https://doi.org/10.1186/BF03351958.

Nitta, N.V., Cliver, E.W., Tylka, A.J., 2003. Low coronal signatures of large solar energetic particle events. Astrophys. J. 586, L103–L106.

Nozawa, S., Ogawa, Y., Oyama, S., Fujiwara, H., Tsuda, T., Brekke, A., Hall, C.M., Murayama, Y., Kawamura, S., Miyaoka, H., Fujii, R., 2010. Tidal waves in the polar lower thermosphere observed using the EISCAT long run data set obtained in September 2005. J. Geophys. Res. 115. A08312, https://doi.org/10.1029/2009JA015237.

Odenwald, S., 2015. Solar Storms: 2000 Years of Human Calamity!. CreateSpace Independent Publishing Platform, San Bernardino, CA, 172 p.

Olawepoa, A.O., Olanipo, O.A., Adeniyia, J.O., Doherty, P.H., 2015. TEC response at two equatorial stations in the African sector to geomagnetic storms. Adv. Space Res. 56 (1), 19–27. https://doi.org/10.1016/j.asr.2015.03.029.

Osepian, A., Kirkwood, S., Dalin, P., Tereschenko, V., 2009. D-region electron density and effective recombination coefficients during twilight – Experimental data and modelling during solar proton events. Ann. Geophys. 27, 3713–3724. https://doi.org/10.5194/angeo-27-3713-2009.

Pallamraju, D., Chakrabarti, S., 2005. First ground-based measurements of OI 6300 daytime aurora over Boston in response to the 30 October 2003 geomagnetic storm. Geophys. Res. Lett. 32. L03S10, https://doi.org/10.1029/2004GL021417.

Papailoiu, M., Mavromichalaki, H., Vassiliki, A., Kelesidis, K.M., Mertzanos, G.A., Petropoulos, B., 2009. Cosmic ray variations of solar origin in relation to human physiological state during the December 2006 solar extreme events. Adv. Space Res. 43 (4), 523–529. https://doi.org/10.1016/j.asr.2008.08.009.

Papaioannou, A., Mavromichalaki, H., Eroshenko, E., Belov, A., Oleneva, V., 2009. The burst of solar and geomagnetic activity in August–September 2005. Ann. Geophys. 27, 1019–1026. https://doi.org/10.5194/angeo-27-1019-2009.

Pavlov, A.V., Pavlova, N.M., 2011. Comparison of modeled electron densities and electron and ion temperatures with Arecibo observations during undisturbed and geomagnetic storm periods of 7–11 September 2005. J. Geophys. Res. 116. A03301, https://doi.org/10.1029/2010JA016067.

Pawlowski, D.J., Ridley, A.J., 2009. Modeling the ionospheric response to the 28 October 2003 solar flare due to coupling with the thermosphere. Radio Sci. 44. RS0A23, https://doi.org/10.1029/2008RS004081.

Pawlowski, D.J., Ridley, A.J., Kim, I., Bernstein, D.S., 2008. Global model comparison with Millstone Hill during September 2005. J. Geophys. Res. 113. A01312, https://doi.org/10.1029/2007JA012390.

Perez-Peraza, J.A., Vashenyuk, E.V., Gallegos-Cruz, A., Balabin, Y.V., Miroshnichenko, L.I., 2008. Relativistic proton production at the Sun in the 20 January 2005 solar event. Adv. Space Res. 41 (6), 947–954. https://doi.org/10.1016/j.asr.2007.04.054.

Plainaki, C., Belov, A., Eroshenko, E., Mavromichalaki, H., Yanke, V., 2007. Modeling ground level enhancements: event of 20 January 2005. J. Geophys. Res. 112. A04102, https://doi.org/10.1029/2006JA011926.

Plainaki, C., Mavromichalaki, H., Belov, A., Eroshenko, E., Yanke, V., 2009. Modeling the solar cosmic ray event of 13 December 2006 using ground level neutron monitor data. Adv. Space Res. 43, 474–479. https://doi.org/10.1016/j.asr.2008.07.011.

Plunkett, S.P., 2005. The extreme solar storms of October to November 2003, NRL Review, ADA523852, pp. 91–98.

PourArsalan, M., Townsend, L.W., Schwadron, N.A., Kozarev, K., Dayeh, M.A., Desai, M.I., 2010. Time-dependent estimates of organ dose and dose equivalent rates for human crews in deep space from the 26 October 2003 solar energetic particle event (Halloween event) using the Earth-Moon-Mars Radiation Environment Module. Space Weather 8. S00E05, https://doi.org/10.1029/2009SW000533.

Prikryl, P., Ghoddousi-Fard, R., Thomas, E.G., Ruohoniemi, J.M., Shepherd, S.G., Jayachandran, P.T., Danskin, D.W., Spanswick, E., Zhang, Y., Jiao, Y., Morton, Y.T., 2015a. GPS phase scintillation at high latitudes during geomagnetic storms of 7–17 March 2012—part 1: The North American sector. Ann. Geophys. 33, 637–656. https://doi.org/10.5194/angeo-33-637-2015.

Prikryl, P., Ghoddousi-Fard, R., Spogli, L., Mitchell, C.N., Li, G., Ning, B., Cilliers, P.J., Sreeja, V., Aquino, M., Terkildsen, M., Jayachandran, P.T., Jiao, Y., Morton, Y.T., Ruohoniemi, J.M., Thomas, E.G., Zhang, Y., Weatherwax, A.T., Alfonsi, L., De Franceschi, G., Romano, V., 2015b. GPS phase scintillation at high latitudes during geomagnetic storms of 7–17 March 2012—part 2: Interhemispheric comparison. Ann. Geophys. 33 (657–670), 2015. https://doi.org/10.5194/angeo-33-657-2015.

Prolss, G.W., 2004. Space weather effects in the upper atmosphere: Low and middle latitudes. In: Scherer, K., Fichtner, H., Heber, B., Mall, U. (Eds.), Space Weather: The Physics Behind the Slogan. ISBN:: 978-3-540-22907-0, pp. 195–213.

Qian, L., Burns, A.G., Chamberlin, P.C., Solomon, S.C., 2011. Variability of thermosphere and ionosphere responses to solar flares. J. Geophys. Res. 116. A10309, https://doi.org/10.1029/2011JA016777.

Rank, G., Debrunner, H., Kocharov, L., Kovaltsov, G., Lockwood, J., McConnell, M., 1997. The solar flare event on 15 June 1991. In: Potgieter, M.S., Raubenheimer, C., van der Walt, D.J. (Eds.), Proc. ICRC 25, 30 July–06 August 1997, Durban, South Africa, vol. 1. pp. 1–4.

Rank, G., Ryan, J., Debrunner, H., McConnell, M., Schonfelder, V., 2001. Extended gamma-ray emission of the solar flares in June 1991. Astron. Astrophys. 378, 1046–1066. https://doi.org/10.1051/0004-6361:20011060.

Rastatter, L., Hesse, M., Kuznetsova, M., Gombosi, T.I., DeZeeuw, D.L., 2002. Magnetic field topology during July 14–16 2000 (Bastille Day) solar CME event. Geophys. Res. Lett. 29 (15). https://doi.org/10.1029/2001GL014136.

Rawat, R., Alex, S., Lakhina, G.S., 2006. Low-latitude geomagnetic signatures during major solar energetic particle events of solar cycle-23. Ann. Geophys. 24, 3569–3583.

Reames, D.V., 1995. The dark side of the Solar Flare Myth. Eos. Trans. AGU 76 (41), 401. https://doi.org/10.1029/95EO00254.

Reames, 2009. Solar release times of energetic particles in ground-level events. Astrophys. J. 693 (1), 812–821.

Reid, J.H., 1963. Classification of solar flares. Ir. Astron. J. 6 (2), 44–50.

Remenets, G.F., Beloglazov, M.I., 1992. Dynamics of an auroral low ionospheric fringe at geophysical disturbances on 29 September 1989. Planet. Space Sci. 40 (8), 1101–1108. https://doi.org/10.1016/0032-0633(92)90039-Q.

Rich, F.J., Denig, W.F., 1992. The major magnetic storm of March 13–14, 1989 and associated ionosphere effects. Can. J. Phys. 70, 510–525. https://doi.org/10.1139/p92-086.

Richardson, J.D., Wang, C., Kasper, J.C., Liu, Y., 2005. Propagation of the October/November 2003 CMEs through the heliosphere. Geophys. Res. Lett. 32. L03S03, https://doi.org/10.1029/2004GL020679.

Riley, P., 2012. On the probability of occurrence of extreme space weather events. Space Weather 10. S02012, https://doi.org/10.1029/2011SW000734.

Rodger, C.J., Clilverd, M.A., Thomson, N.R., Gamble, R.J., Seppala, A., Turunen, E., Meredith, N.P., Parrot, M., Sauvaud, J.-A., Berthelier, J.-J., 2007. Radiation belt electron precipitation into the atmosphere: recovery from a geomagnetic storm. J. Geophys. Res. 112. A11307, https://doi.org/10.1029/2007JA012383.

Rodger, C.J., Clilverd, M.A., Seppala, A., Thomson, N.R., Gamble, R.J., Parrot, M., Sauvaud, J.-A., Ulich, T., 2010. Radiation belt electron precipitation due to geomagnetic storms: Significance to middle atmosphere ozone chemistry. J. Geophys. Res. 115. A11320, https://doi.org/10.1029/2010JA015599.

Rodriguez-Bilbao, I., Radicella, S.M., Rodríguez-Caderot, G., Herraiz, M., 2015. Precise point positioning performance in the presence of the 28 October 2003 sudden increase in total electron content. Space Weather 13, 698–708. https://doi.org/10.1002/2015SW001201.

Romanova, E., Kurkin, V., Zolotukhina, N., Polekh, N., 2014. Prompt and delayed effects of solar disturbances in magnetosphere-ionosphere system on March 4-7, 2012. In: 40th COSPAR Scientific Assembly, 02–10 Aug 2014, Moscow, Russia.

Ryan, J.M., 2000. Long-duration solar gamma-ray flares. Space Sci. Rev. 93, 581–610. https://doi.org/10.1023/A:1026547513730.

Saiz, A., Ruffolo, D., Rujiwarodom, M., Bieber, J.W., Clem, J., Evenson, P., Pyle, R., Duldig, M.L., Humble, J.E., 2005. Relativistic particle injection and interplanetary transport during the January 20, 2005 ground level enhancement. In: Acharya, B.S., Gupta, S., Jagadeesan, P., Jain, A., Karthikeyan, S., Morris, S., Tonwar, S. (Eds.), Proc. ICRC 29, 03–10 August 2005, Pune, India, vol. 1. pp. 229–232.

Sako, T., Watanabe, K., Muraki, Y., Matsubara, Y., Tsujihara, H., Yamashita, M., Sakai, T., Shibata, S., Valdes-Galicia, J.F., González, L.X., Hurtado, A., Musalem, O., Miranda, P., Martinic, N., Ticona, R., Velarde, A., Kakimoto, F., Ogio, S., Tsunesada, Y., Tokuno, H., Tanaka, Y.T., Yoshikawa, I., Terasawa, T., Saito, Y., Mukai, T., Gros, M., 2006. Long-lived solar neutron emission in comparison with electron-produced radiation in the 2005 September 7 solar flare. Astrophys. J. 651 (1), L69–L72.

Sako, T., Watanabe, K., Muraki, Y., Matsubara, Y., Sakai, T., Shibata, S., Kakimoto, F., Tsunesada, Y., Tokuno, H., Ogio, S., Valdes-Galicia, J.F., Gonzalez, L.X., Hurtado, A., Musalem, O., Miranda, P., Martinic, N., Ticona, R., Velarde, A., 2008. Emission profile of solar neutrons obtained from the ground-based observations for the 7 September 2005 event. In: Caballero, R., D'Olivo, J.C., Medina-Tanco, G., Nellen, L., Sánchez, F.A., Valdés-Galicia, J.F. (Eds.), Proc. ICRC 30, 03–11 July 2007, Merida, Yucatan, Mexico, vol. 1. pp. 53–56.

Sauer, H.H., 1993. GOES observations of energetic protons $E > 685$ MeV: Ground-level events from October 1983 to July 1992. In: Leahy, D.A., Hickws, R.B., VenkatesanIntern, D. (Eds.), Proc. ICRC 23, 19–30 July 1993, Calgary, Canada, vol. 3. pp. 205–253.

Schmidt, J.M., Cairns, I.H., 2014. Type II solar radio bursts predicted by 3-D MHD CME and kinetic radio emission simulations. J. Geophys. Res. Space Phys. 119, 69–87. https://doi.org/10.1002/2013JA019349.

Schmidt, J.M., Cairns, I.H., Lobzin, V.V., 2014. The solar type II radio bursts of 7 March 2012: detailed simulation analyses. J. Geophys. Res. 119, 6042–6061. https://doi.org/10.1002/2014JA019950.

Schwadron, N.A., Townsend, L., Kozarev, K., Dayeh, M.A., Cucinotta, F., Desai, M., Golightly, M., Hassler, D., Hatcher, R., Kim, M.-Y., Posner, A., PourArsalan, M., Spence, H.E., Suuier, R.K., 2010. Earth-Moon-Mars radiation environment module framework. Space Weather. 8. S00E02, https://doi.org/10.1029/2009SW000523.

Shea, M.A., Smart, D.F., 1992. Solar proton events: history, statistics and predictions. In: Hruska, J., Shea, M.A., Smart, D.F., Heckman, G. (Eds.), Proc. Solar-Terrestrial Predictions Workshop-IV, 18–22 May 1992, Ottawa, Canada, vol. 2. pp. 48–70.

Shea, M.A., Smart, D.F., 1993. March 1991 Solar-terrestrial phenomena and related technological consequences. In: Leahy, D.A., Hickws, R.B., VenkatesanIntern, D. (Eds.), Proc. ICRC 23, 19–30 July 1993, Alberta, Canada, vol. 3. pp. 739–742.

Shea, M.A., Smart, D.F., 1997. Dual acceleration and/or release of relativistic solar cosmic rays. In: Potgieter, M.S., Raubenheimer, C., van der Walt, D.J. (Eds.), Proc. ICRC 25, 30 July–06 August 1997, Durban, South Africa, vol. 1. pp. 129–132.

Shea, M.A., Smart, D.F., Wilson, M.D., Fluckiger, E.O., 1991. Possible ground-level measurements of solar neutron decay protons during the 19 October 1989 solar cosmic ray event. Geophys. Res. Lett. 18, 829–832. https://doi.org/10.1029/90GL02668.

Shea, M.A., Smart, D.F., Allen, J.H., Wilkinson, D.C., 1992. Spacecraft problems in association with episodes of intense solar activity and related terrestrial phenomena during March 1991. IEEE Trans. Nucl. Sci. 39, 1754–1760. https://doi.org/10.1109/23.211363.

Shen, C., Wang, Y., Pan, Z., Zhang, M., Ye, P., Wang, S., 2013. Full halo coronal mass ejections: do we need to correct the projection effect in terms of velocity? J. Geophys. Res. 118, 6858–6865. https://doi.org/10.1002/2013JA018872.

Shen, C., Wang, Y., Pan, Z., Miao, B., Ye, P., Wang, S., 2014. Full-halo coronal mass ejections: arrival at the Earth. J. Geophys. Res. 119, 5107–5116. https://doi.org/10.1002/2014JA020001.

Shim, J.S., et al., 2012. CEDAR Electrodynamics Thermosphere Ionosphere (ETI) Challenge for systematic assessment of ionosphere/thermosphere models: Electron density, neutral density, N_mF_2 and h_mF_2 using space based observations. Space Weather 10. S10004, https://doi.org/10.1029/2012SW000851.

Shim, J.S., Kuznetsova, M., Rastatter, L., Bilitza, D., Butala, M., Codrescu, M., Emery, B.A., Foster, B., Fuller-Rowell, T.J., Huba, J., Mannucci, A.J., Pi, X., Ridley, A., Scherliess, L., Schunk, R.W., Sojka, J.J., Stephens, P., Thompson, D.C., Weimer, D., Zhu, L., Anderson, D., Chau, J.L., Sutton, E., 2014. Systematic evaluation of ionosphere/thermosphere (IT) models. In: Huba, J., Schunk, R., Khazanov, G. (Eds.), Modeling the Ionosphere-Thermosphere System. John Wiley & Sons, Ltd, Chichester, United Kingdom. https://doi.org/10.1002/9781118704417.ch13.

Shiota, D., Kataoka, R., 2016. Magnetohydrodynamic simulation of interplanetary propagation of multiple coronal mass ejections with internal magnetic flux rope (SUSANOO-CME). Space Weather 14, 56–75. https://doi.org/10.1002/2015SW001308.

Shprits, Y.Y., Thorne, R.M., Horne, R.B., Glauert, S.A., Cartwright, M., Russell, C.T., Baker, D.N., Kanekal, S.G., 2006. Acceleration mechanism responsible for the formation of the new radiation belt during the 2003 Halloween solar storm. Geophys. Res. Lett. 33. L05104, https://doi.org/10.1029/2005GL024256.

Shprits, Y., Subbotin, D., Ni, B., Horne, R., Baker, D., Cruce, P., 2011. Profound change of the near-Earth radiation environment caused by solar superstorms. Space Weather 9. S08007, https://doi.org/10.1029/2011SW000662.

Shue, J.-H., et al., 1998. Magnetopause location under extreme solar wind conditions. J. Geophys. Res. 103 (A8), 17691–17700. https://doi.org/10.1029/98JA01103.

Simnett, G.M., 2003. Energetic particles and coronal mass ejections: a case study from ACE and Ulysses. Sol. Phys. 213, 387–412. https://doi.org/10.1023/A:1023903408107.

Simnett, G.M., Roelof, E.C., 2005. Timing of the relativistic proton acceleration responsible for the GLE on 20 January, 2005. In: Acharya, B.S., Gupta, S., Jagadeesan, P., Jain, A., Karthikeyan, S., Morris, S., Tonwar, S. (Eds.), Proc. ICRC 29, 03–10 August 2005, Pune, India, vol. 1. pp. 233–236.

Simnett, G.M., 2006. The timing of relativistic proton acceleration in the 20 January 2005 flare. Astron. Astrophys. 445, 715–724. https://doi.org/10.1051/0004-6361:20053503.

Simnett, G.M., 2007. Erratum—the timing of relativistic proton acceleration in the 20 January 2005 flare and other papers. Astron. Astrophys. 472, 309–310. https://doi.org/10.1051/0004-6361:20053503e.

Simpson, J.A., 2000. The cosmic ray nucleonic component: the invention and scientific uses of the neutron monitor. Space Sci. Rev. 93, 11–32. https://doi.org/10.1023/A:1026567706183.

Smart, D.F., Shea, M.A., 1991. A comparison of the magnitude of the 29 September 1989 high energy event with solar cycles 17, 18 and 19 events. In: Cawley, M. et al., (Ed.), Proc. ICRC 22, 11–23 August 1991, Dublin, Ireland, vol. 3. pp. 101–104.. Paper SH 3.1-3.

Smith, R.W., Hernandez, G., Price, K., Fraser, G., Clark, K.C., Schulz, W.J., Smith, S., Clark, M., 1994. The June 1991 thermospheric storm observed in the southern hemisphere. J. Geophys. Res. 99 (A9), 17609–17615. https://doi.org/10.1029/94JA01101.

Smith, S., Howard, R., 1968. Magnetic classification of active regions. In: Kiepenheuer, K.O. (Ed.), Structure and Development of Solat Active Regions. ISBN: 978-94-011-6817-5. pp. 33–42. IAU-35.

Sojka, J.J., Schunk, R.W., Denig, W.F., 1994. Ionospheric response to the sustained high geomagnetic activity during the March '89 Great Storm. J. Geophys. Res. 99 (A11), 21341–21352. https://doi.org/10.1029/94JA01765.

Somov, B.V., Kosugi, T., Hudson, H.S., Sakao, T., Masuda, S., 2002. Magnetic reconnection scenario of the Bastille Day 2000 flare. Astrophys. J. 579 (2), 863–873.

Sreeja, V., 2016. Impact and mitigation of space weather effects on GNSS receiver performance. Geosci. Lett. 3 (24), 11. https://doi.org/10.1186/s40562-016-0057-0.

Struminsky, A.B., 2005. Variations of solar proton spectrum during the ground level enhancement of 2005 January 20. In: Acharya, B.S., Gupta, S., Jagadeesan, P., Jain, A., Karthikeyan, S., Morris, S., Tonwar, S. (Eds.), Proc. ICRC 29, 03–10 August 2005, Pune, India, vol. 1. pp. 201–204.

Sutton, E.K., Forbes, J.M., Nerem, R.S., 2005. Global thermospheric neutral density and wind response to the severe 2003 geomagnetic storms from CHAMP accelerometer data. J. Geophys. Res. 110. A09S40, https://doi.org/10.1029/2004JA010985.

Svestka, Z., 1995. On 'The solar flare myth'. Sol. Phys. 160, 53–56. https://doi.org/10.1007/BF00679093.

Swinson, D.B., Shea, M.A., 1990. The September 29, 1989 ground-level event observed at high rigidity. Geophys. Res. Lett. 17, 1073–1975. https://doi.org/10.1029/GL017i008p01073.

Terekhov, O.V., Sunyaev, R.A., Tkachenko, A.Y., Denisenko, D.V., Kuznetsov, A.V., Barat, C., Dezalay, J.-P., Talon, R., 1996. Deuterium synthesis during the solar flare of March 22, 1991 (Granat Data). Astron. Lett. 22 (2), 143–147.

Tesema, T., Damtie, B., Nigussie, M., 2015. The response of the ionosphere to intense geomagnetic storms in 2012 using GPS-TEC data from East Africa longitudinal sector. J. Atmos. Sol. Terr. Phys. 135, 143–151. https://doi.org/10.1016/j.jastp.2015.10.021.

Timashkova, D.A., Balabinb, Y.V., Barbashinaa, N.S., Kokoulina, R.P., Kompanietsa, K.G., Mannocchic, G., Petrukhina, A.A., Saavedrad, O., Shutenkoa, V.V., Trincheroc, G., Vashenyukb, E.V., Yashina, I.I., 2008. Ground level enhancement of December 13, 2006 observed by means of muon hodoscope. Astropart. Phys. 30 (3), 117–123. https://doi.org/10.1016/j.astropartphys.2008.07.008.

Timofeev, V.E., Krivoshapkin, P.A., Grigoryev, V.G., Prihodko, A.N., Migunov, V.M., Filippov, A.T., 2005. Increase of the solar energetic particle flux on January 20, 2005. In: Acharya, B.S., Gupta, S., Jagadeesan, P., Jain, A., Karthikeyan, S., Morris, S., Tonwar, S. (Eds.), Proc. ICRC 29, 03–10 August 2005, Pune, India, vol. 1. pp. 205–208.

Torsti, J.J., Eronen, T., Mahonen, M., Riihonen, E., Schultz, C.G., udela, K.K., Kananen, H., 1991. Search of peculiarities in the flux profiles of GLE's in 1989. In: Cawley, M. et al., (Ed.), Proc. ICRC 22, 11–23 August 1991, Dublin, Ireland, vol. 3. pp. 141–144.

Torsti, J., Anttila, A., Vainio, R., Kocharov, L.G., 1995. Successive solar energetic particle events in the October 1989. In: Iucci, N., Lamanna, E. (Eds.), Proc. ICRC 24, 28 Aug–08 Sep 1995, Rome, Italy, vol. 4. pp. 139–142.

Toth, G., et al., 2005. Space weather modeling framework: a new tool for the space science community. J. Geophys. Res. 110. A12226, https://doi.org/10.1029/2005JA011126.

Toth, G., De Zeeuw, D.L., Gombosi, T.I., Manchester, W.B., Ridley, A.J., Sokolov, I.V., Roussev, I.I., 2007. Sun-to-thermosphere simulation of the 28–30 October 2003 storm with the Space Weather Modeling Framework. Space Weather. 5. S06003, https://doi.org/10.1029/2006SW000272.

Tsurutani, B.T., Judge, D.L., Guarnieri, F.L., Gangopadhyay, P., Jones, A.R., Nuttall, J., Zambon, G.A., Didkovsky, L., Mannucci, A.J., Iijima, B., Meier, R.R., Immel, T.J., Woods, T.N., Prasad, S., Floyd, L., Huba, J., Solomon, S.C., Straus, P., Viereck, R., 2005. The October 28, 2003 extreme EUV solar flare and resultant extreme ionospheric effects: comparison to other Halloween events and the Bastille Day event. Geophys. Res. Lett. 32. L03S09, https://doi.org/10.1029/2004GL021475.

Tsurutani, B.T., Verkhoglyadova, O.P., Mannucci, A.J., Lakhina, G.S., Li, G., Zank, G.P., 2009. A brief review of "solar flare effects" on the ionosphere. Radio Sci. 44. RS0A17, https://doi.org/10.1029/2008RS004029.

Tsurutani, B., Shibata, K., Echer, E., Mannucci, A., Verkhoglyadova, O.P., Mlynczak, M.G., 2012. The March 7-18 2012 CAWSES-SCOSTEP interplanetary events and their magnetospheric and ionospheric effects, Paper SH43C06. In: AGU Fall Meeting 2012.

Tsurutani, B.T., Echer, E., Shibata, K., Verkhoglyadova, O.P., Mannucci, A.J., Gonzalez, W.D., Kozyra, J.U., Patzold, M., 2014. The interplanetary causes of geomagnetic activity during the 7–17 March 2012 interval: a CAWSES II overview. Space Weather Space Clim. 4, https://doi.org/10.1051/swsc/2013056. A02-1–A02-8.

Urbarz, H., 1969. Weissenau solar radio astronomy observatory. Sol. Phys. 7 (1), 147–152. https://doi.org/10.1007/BF00148411.

Valdes-Galicia, J.F., Muraki, Y., Watanabe, K., Matsubara, Y., Sako, T., Gonzalez, L.X., Musalem, O., Hurado, A., 2009. Solar neutron events as a tool to study particle acceleration at the Sun. Adv. Space Res. 43 (4), 565–572. https://doi.org/10.1016/j.asr.2008.09.023.

Vashenyuk, E.V., Balabin, Y.V., Gvozdevsky, B.B., Miroshnichenko, L.I., Klein, K.L., Trottet, G., Lantos, P., 2005a. Energetic solar particle dynamics during 28 October, 2003 GLE. In: Acharya, B.S., Gupta, S., Jagadeesan, P., Jain, A., Karthikeyan, S., Morris, S., Tonwar, S. (Eds.), Proc. ICRC 29, 03–10 August 2005, Pune, India, vol. 1. pp. 217–220.

Vashenyuk, E.V., Balabin, Yu.V., Gvozdevsky, B.B., Karpov, S.N., Yanke, V.G., Eroshenk, E.A., Belov, A.V., 2005b. Relativistic solar cosmic rays in January 20, 2005 event on the ground based observations. In: Gushchina, R.T., Acharya, B.S., Gupta, S., Jagadeesan, P., Jain, A., Karthikeyan, S., Morris, S., Tonwar, S. (Eds.), Proc. ICRC 29, 03–10 August 2005, Pune, India, pp. 209–212.

Vashenyuk, E.V., Balabin, Y.V., Gvozdevskii, B.B., Karpov, S.N., 2006. Relativistic solar protons in the event of January 20, 2005: model studies. Geomagn. Aeron. 46 (4), 424–429. https://doi.org/10.1134/S0016793206040037.

Vashenyuk, E.V., Bazilevskaya, G.A., Balabin, Y.V., Gvozdevsky, B.B., Makhmutov, V.S., Stozhkov, Y.I., Svirzhevsky, N.S., Svirzhevskaya, A.K., Schur, L.I., 2008a. The GLE of December 13, 2006 according to the ground level and balloon observations. In: Caballero, R., D'Olivo, J.C., Medina-Tanco, G., Nellen, L., Sánchez, F.A., Valdés-Galicia, J.F. (Eds.), Proc. ICRC 30, 03–11 July 2007, Merida, Yucatan, Mexico, vol. 1. pp. 221–224.

Vashenyuk, E.V., Balabin, Y.V., Gvozdevsky, B.B., Shchur, L.I., 2008b. Characteristics of relativistic solar cosmic rays during the event of December 13, 2006. Geomagn. Aeron. 48, 149–153. https://doi.org/10.1134/S0016793208020035.

Veenadhari, B., Alex, S., Kikuchi, T., Shinbori, A., Singh, R., Chandrasekhar, E., 2010. Penetration of magnetospheric electric fields to the equator and their effects on the low-latitude ionosphere during intense geomagnetic storms. J. Geophys. Res. 115. A03305, https://doi.org/10.1029/2009JA014562.

Verkhoglyadova, O.P., Tsurutania, B.T., Mannuccia, A.J., Mlynczakc, M.G., Huntd, L.A., Paxton, L.J., 2014. Ionospheric TEC, thermospheric cooling and Σ[O/N2] compositional changes during the 6–17 March 2012 magnetic storm interval (CAWSES II). J. Atmos. Sol. Terr. Phys. 115–116, 42–51. https://doi.org/10.1016/j.jastp.2013.11.009.

Verkhoglyadova, O.P., Tsurutani, B.T., Mannucci, A.J., Mlynczak, M.G., Hunt, L.A., Paxton, L.J., Komjathy, A., 2016. Solar wind driving of ionosphere-thermosphere responses in three storms near St. Patrick's Day in 2012, 2013 and 2015. J. Geophys. Res. 121, 8900–8923. https://doi.org/10.1002/2016JA022883.

Verronen, P.T., Seppala, A., Clilverd, M.A., Rodger, C.J., Kyrola, E., Enell, C.-F., Ulich, T., Turunen, E., 2005. Diurnal variation of ozone depletion during the October–November 2003 solar proton events. J. Geophys. Res. 110. A09S32, https://doi.org/10.1029/2004JA010932.

Verronen, P.T., Rodger, C.J., Clilverd, M.A., Pickett, H.M., Turunen, E., 2007. Latitudinal extent of the January 2005 solar proton event in the northern hemisphere from satellite observations of hydroxyl. Ann. Geophys. 25, 2203–2215. http://www.ann-geophys.net/25/2203/2007/.

Verronen, P.T., Andersson, M.E., Kero, A., Enell, C.-F., Wissing, J.M., Talaat, E.R., Kauristie, K., Palmroth, M., Sarris, T.E., Armandillo, E., 2015. Contribution of proton and electron precipitation to the observed electron concentration in October–November 2003 and September 2005. Ann. Geophys. 33, 381–394. https://doi.org/10.5194/angeo-33-381-2015.

Veselovsky, I.S., Panasyuk, M.I., Avdyushin, S.I., Bazilevskaya, G.A., Belov, A.V., Bogachev, S.A., Bogod, V.M., Bogomolov, A.V., Bothmer, V., Boyarchuk, K.A., Vashenyuk, E.V., Vlasov, V.I., Gnezdilov, A.A., Gorgutsa, R.V., Grechnev, V.V., Denisov, Y.I., Dmitriev, A.V., Dryer, M., Yermolaev, Y.I., Eroshenko, E.A., Zherebtsov, G.A., Zhitnik, I.A., Zhukov, A.N., Zastenker, G.N., Zelenyi, L.M., Zeldovich, M.A., Ivanov-Kholodnyi, G.S., Ignat'ev, A.P., Ishkov, V.N., Kolomiytsev, O.P., Krasheninnikov, I.A., Kudela, K., Kuzhevsky, B.M., Kuzin, S.V., Kuznetsov, V.D., Kuznetsov, S.N., Kurt, V.G., Lazutin, L.L., Leshchenko, L.N., Litvak, M.L., Logachev, Y.I., Lawrence, G., Markeev, A.K., Makhmutov, V.S., Mitrofanov, A.V., Mitrofanov, I.G., Morozov, O.V., Myagkova, I.N., Nusinov, A.A., Oparin, S.N., Panasenco, O.A., Pertsov, A.A., Petrukovich, A.A., Podorol'sky, A.N., Romashets, E.P., Svertilov, S.I., Svidsky, P.M., Svirzhevskaya, A.K., Svirzhevsky, N.S., Slemzin, V.A., Smith, Z., Sobel'man, I.I., Sobolev, D.E., Stozhkov, Y.I., Suvorova, A.V., Sukhodrev, N.K., Tindo, I.P., Tokhchukova, S.K., Fomichev, V.V., Chashey, I.V., Chertok, I.M., Shishov, V.I., Yushkov, B.Y., Yakovchouk, O.S., Yanke, V.G., 2004. Solar and heliospheric phenomena in October-November 2003: Causes and effects. Cosm. Res. 42 (5), 453–508. https://doi.org/10.1023/B:COSM.0000046229.24716.02.

Vestrand, W.T., Forrest, D.J., 1993. Evidence for a spatially extended component of gamma rays from solar flares. Astrophys. J. 409, L69–L72. http://adsabs.harvard.edu/doi/10.1086/186862.

Villante, U., Regi, M., 2008. Solar flare effect preceding Halloween storm (28 October 2003): Results of a worldwide analysis. J. Geophys. Res. 113. A00A05, https://doi.org/10.1029/2008JA013132.

von Clarmann, T., Funke, B., López-Puertas, M., Kellmann, S., Linden, A., Stiller, G.P., Jackman, C.H., Harvey, V.L., 2013. The solar proton events in 2012 as observed by MIPAS. Geophys. Res. Lett. 40, 2339–2343. https://doi.org/10.1002/grl.50119.

von Rosenvinge, T.T., Richardson, I.G., Reames, D.V., Cohen, C.M.S., Cummings, A.C., Leske, R.A., Mewaldt, R.A., Stone, E.C., Wiedenbeck, M.E., 2009. The solar energetic particle event of 14 December 2006. Sol. Phys. 256 (1), 443–462. https://doi.org/10.1007/s11207-009-9353-6.

Wang, C., Richardson, J.D., Paularena, K.I., 2001. Predicted Voyager observations of the Bastille Day 2000 coronal mass ejection. J. Geophys. Res. 106 (A7), 13007–13013. https://doi.org/10.1029/2000JA000388.

Wang, H., Liu, C., 2010. Observational evidence of back reaction on the solar surface associated with coronal magnetic restructuring in solar eruptions. Astrophys. J. Lett. 716 (2), L195–L199.

Wang, R., 2009. Did the 2000 July 14 solar flare accelerate protons to ≥ 40 GeV? Astropart. Phys. 31, 149–155.

Wang, R., Liu, Y.D., Ying, Z. Yang, Hu, H., 2014. Magnetic field restructuring associated with two successive solar eruptions. Astrophys. J. 791(2). https://doi.org/10.1088/0004-637X/791/2/84.

Wang, W., Lei, J., Burns, A.G., Solomon, S.C., Wiltberger, M., Xu, J., Zhang, Y., Paxton, L., Coster, A., 2010. Ionospheric response to the initial phase of geomagnetic storms: Common features. J. Geophys. Res. 115. A07321, https://doi.org/10.1029/2009JA014461.

Washwell, E.R., 1996. GOES-8 and Beyond, SPIE-2812, 876 p.

Watanabe, K., Gros, M., Stoker, P.H., Kudela, K., Lopate, C., Valdés-Galicia, J.F., Hurtado, A., Musalem, O., Ogasawara, R., Mizumoto, Y., Nakagiri, M., Miyashita, A., Matsubara, Y., Sako, T., Muraki, Y., Sakai, T., Shibata, S., 2006. Solar neutron events of 2003 October-November. Astrophys. J. 635 (2), 1135–1144.

Watanabe, K., Sako, T., Muraki, Y., Matsubara, Y., Sakai, T., Shibata, S., Valdés-Galicia, J.F., González, L.X., Hurtado, A., Musalem, O., Miranda, P., Martinic, N., Ticona, R., Velarde, A., Kakimoto, F., Ogio, S., Tsunesada, Y., Tokuno, H., Tanaka, Y.T., Yoshikawa, I., Terasawa, T., Saito, Y., Mukai, T., Gros, M., 2007. Highly significant detection of solar neutrons on 2005 September 7. Adv. Space Res. 39 (9), 1462–1466. https://doi.org/10.1016/j.asr.2006.10.021.

Watanabe, K., Lin, R.P., Krucker, S., Murphy, R.J., Share, G.H., Harris, M.J., Gros, M., Muraki, Y., Sako, T., Matsubara, Y., Sakai, T., Shibata, S., Valdés-Galicia, J.F., González, L.X., Hurtado, A., Musalem, O., Miranda, P., Martinic, N., Ticona, R., Velarde, A., Kakimoto, F., Tsunesada, Y., Tokuno, H., Ogio, S., 2009. Physics of ion acceleration in the solar flare on 2005 September 7 determines gamma-ray and neutron production. Adv. Space Res. 44 (7), 789–793. https://doi.org/10.1016/j.asr.2009.06.002.

Wik, M., Pirjola, R., Lundstedt, H., Viljanen, A., Wintoft, P., Pulkkinen, A., 2009. Space weather events in July 1982 and October 2003 and the effects of geomagnetically induced currents on Swedish technical systems. Ann. Geophys. 27, 1775–1787. http://www.ann-geophys.net/27/1775/2009/.

Wittman, T.M., 2012. A quantitative analysis of solar flare characteristics as observed in the Solar Observing Optical Network and the Global Oscillation Network Group, Thesis, Air Force Institute of Technology, Wright-Patterson Air Force Base, 81 p.

Wu, C., Liou, K., Dryer, M., Wu, S., Plunkett, S.P., 2013. Evolution and interaction of six coronal mass ejections in March 2012, paper SH31B-2022. In: AGU Fall Mtg 2013.

Wu, C.-C., Liou, K., Lepping, R.P., Hutting, L., Plunkett, S., Howard, R.A., Socker, D., 2016. The first super geomagnetic storm of solar cycle 24: "The Saint Partick's day event (17 March 2015)". Earth Planets Space 68, 151. https://doi.org/10.1186/s40623-016-0525-y.

Xiong, B., et al., 2011. Ionospheric response to the X-class solar flare on 7 September 2005. J. Geophys. Res. 116. A11317, https://doi.org/10.1029/2011JA016961.

Xiong, B., Wan, W., Zhao, B., Yu, Y., Wei, Y., Ren, Z., Liu, J., 2015. Response of the American equatorial and low-latitude ionosphere to the X1.5 solar flare on 13 September 2005. J. Geophys. Res. 119, 10336–10347. https://doi.org/10.1002/2014JA020536.

Yamauchi, M., Iyemori, T., Frey, H., Henderson, M., 2006. Unusually quick development of a 4000 nT substorm during the initial 10 min of the 29 October 2003 magnetic storm. J. Geophys. Res. 111. A04217, https://doi.org/10.1029/2005JA011285.

Young, C.A., Arndt, M.B., Connors, A., McConnell, M., Rank, G., Ryan, J.M., Suleiman, R., Schonfelder, V., 2000. Energetic proton spectra in the 11 June 1991 solar flare, AIP CP510(1. In: McConnell, M.L., Ryan, J.M. (Eds.), The Fifth Compton Symposium, 15-17 September 1999, Portsmouth, New Hampshire, U.S.A, pp. 564–568.

Yue, X., et al., 2013. The effect of solar radio bursts on the GNSS radio occultation signals. J. Geophys. Res. Space Phys. 118, 5906–5918. https://doi.org/10.1002/jgra.50525.

Zank, G.P., Rice, W.K.M., Cairns, I.H., Bieber, J.W., Skoug, R.M., Smith, C.W., 2001. Predicted timing for the turn-on of radiation in the outer heliosphere due to the Bastille Day shock. J. Geophys. Res. 106 (A12), 29363–29372. https://doi.org/10.1029/2000JA000401.

Zhang, D.H., Xiao, Z., 2005. Study of ionospheric response to the 4B flare on 28 October 2003 using International GPS Service network data. J. Geophys. Res. 110. A03307, https://doi.org/10.1029/2004JA010738.

Zheng, Y., Pulkkinen, A., Taktakishvili, A., Mays, M.L., Kuznetsova, M.M., Hesse, M., 2013. The March 2012 solar storms and their impact: NASA/SWRC perspective, paper SH44A-04. In: AGU Spring Meeting 2013.

Zhu, F.R., Tang, Y.Q., Zhang, Y., Wang, Y.G., Lu, H., Zhang, J.L., Tan, Y.H., 2005. A possible GLE event in association with solar flare on January 20, 2005. In: Acharya, B.S., Gupta, S., Jagadeesan, P., Jain, A., Karthikeyan, S., Morris, S., Tonwar, S. (Eds.), Proc. ICRC 29, Pune, India, pp. 111–114.

Zhu, Q., Lei, J., Luan, X., Dou, X., 2016. Contribution of the topside and bottomside ionosphere to the total electron content during two strong geomagnetic storms. J. Geophys. Res. 121, 2475–2488. https://doi.org/10.1002/2015JA022111.

CHAPTER

NEAR-EARTH RADIATION ENVIRONMENT FOR EXTREME SOLAR AND GEOMAGNETIC CONDITIONS

13

Mikhail Panasyuk*[,†], Vladimir Kalegaev*, Leonty Miroshnichenko*[,‡], Nikolay Kuznetsov*, Rikho Nymmik*, Helen Popova*, Boris Yushkov*, Victor Benghin*[,§]

Skobeltsyn Institute of Nuclear Physics of Lomonosov Moscow State University, Moscow, Russia Physics Department of Lomonosov Moscow State University, Moscow Russia[†] Pushkov Institute of Terrestrial Magnetism, Ionosphere and Radio Wave Propagation of Russian Academy of sciences, Moscow, Troitsk, Russia[‡] Institute of Biomedical Problems of Russian Academy of Sciences, Moscow, Russia[§]

CHAPTER OUTLINE

1 Introduction	349
2 Solar Energetic Particles in Space	350
2.1 Concept of Extreme SEP Events	350
2.2 Largest SEP Events and Distribution Function	351
2.3 Solar Cosmic Rays: Penetration Boundary and Changes of Cutoff Rigidities	354
3 GCR Modulation Over the Solar Cycles	354
4 The Inner Proton Radiation Belt Variations Over Solar Cycles	359
5 Radiation Environment for Crewed Orbital Stations	361
6 Conclusions	367
Acknowledgments	368
References	368
Further Reading	372

1 INTRODUCTION

The near-Earth environment can be considered a unique place where different space radiation fields coexist; it can also play a significant role in the creation of radiation risks for both robotic and manned space missions. One cannot exclude that the cumulative effect for a particular spacecraft orbit would be the result of simultaneous impact of the different radiation fields. Among the types of space radiation, we consider: trapped radiation (inner Earth radiation belts, ERB) at low Earth orbit (LEO); solar energetic particles (SEP), including penetration of relativistic solar protons into the magnetosphere causing ground level enhancement (GLE) events; and variations of galactic cosmic rays (GCR) during periods of extreme solar and geomagnetic activity. SEP connected with GLE has sufficient energy and intensity to raise radiation levels, even on Earth's surface, to the degree that they are readily detected by ground-based neutron monitors. SEP can include other nuclei such as helium ions and heavy ions.

According to recent considerations, the near-Earth radiation environment is only one component of space weather (SW). This general term, SW, covers the "conditions on the Sun and in the solar wind, magnetosphere, ionosphere, and thermosphere that can influence the performance and reliability of space-borne and ground-based technological systems and can endanger human life or health" (U.S. National Space Weather Program, 2013). The SW field includes observational, modeling, and effect aspects. Special attention is paid to the studies of perturbation sources-solar flares and coronal mass ejections (CMEs).

Based on the requirements of SW users in the United States, the NOAA Space Environment Center (http://www.sec.noaa.gov) has identified three major SW threats: geomagnetic storms, solar radiation storms, and radio blackouts (Song, 2001). These three threats can be translated into the three physical phenomena in space physics: (1) geomagnetic field disturbances that induce the electric voltage or electric currents leading to damages in technological systems; (2) energetic particles which directly damage components of electronic systems of satellites and endanger human life or health in space; and (3) ionospheric electron density disturbances that vary the paths of radio signals leading to difficulties in radio transmissions and errors in the navigation systems that depend on radio signal propagation. These threats are considered in detail by Schrijver et al. (2015).

Below, we will concentrate on the solar radiation storms; that is, on the effects related to energetic particles only, including ERB particles, GCR, and SEP.

Both the results of empirical modeling of different radiation fields and some results of measurements of relevant physical parameters (energy spectra, space and time variations) for different solar (maxima and minima) and geomagnetic (major magnetic storms) conditions are analyzed. The models are based on the quantitative relationship between particle fluxes, taking into account the solar activity level (sunspot numbers).

We consider some examples of using models to estimate changes of radiation hazards in interplanetary space for the expected reduction of solar activity during the nearest solar cycles 25–26. The results of measurements of radiation doses and single-event effects (SEE) on LEO spacecraft and in comparison to interplanetary space conditions are presented as well.

2 SOLAR ENERGETIC PARTICLES IN SPACE
2.1 CONCEPT OF EXTREME SEP EVENTS

There are two different approaches to the definition of "extreme event" for solar cosmic rays. The first suggests that solar proton events (SPEs) may be considered as extreme events when the worldwide network of ground cosmic ray stations have detected the considerable flux of relativistic solar protons in a GLE. An alternative definition of extreme events is based on the detection of SEP flux enhancement at Earth's orbit in the energy range below the relativistic energies (Miroshnichenko, 2014). Note that the SEP energy spectrum covers several orders of magnitude on the energy scale, so relativistic particles are only part of the total SEP spectrum. Therefore, we use the term "SEP events" as the most common one to describe the particles of solar origin, including protons, electrons, and heavy ions.

In the context of our motivations and goals, one can suggest that we select the extreme SEP events by the integral fluence Φ of protons with the energy of $E \geq 30$ MeV. Up to now, the so-called "Carrington event" (September 1–2, 1859) with the integral fluence of Φ (≥ 30 MeV) = 1.9×10^{10} cm^{-2} (McCracken et al., 2001) was the most suitable for this purpose (see, Townsend et al., 2006; Miroshnichenko and Nymmik, 2014). Such events can be considered as the "worst-case" reference for radiation hazards in space.

Indeed, the two nearest candidates for the role of "worst-case" are the events of November 15, 1960, and August 4, 1972. They were characterized by lesser values of Φ (≥ 30 MeV), about 9×10^9 cm^{-2} and 5×10^9 cm^{-2}, respectively (Smart et al., 2006). Several years ago, some signatures of cosmic ray increase during AD 774–775 were found on the basis of tree ring analysis, which was carried out in Japan, Europe, Russia, and the United States (Miyake et al., 2012; Usoskin et al., 2013; Jull et al., 2014). Integral fluences Φ (≥ 30 MeV) were recently estimated as $(2.1–3.0) \times 10^{10}$ cm^{-2} (Miroshnichenko and Nymmik, 2014) and as $(4.5–8.0) \times 10^{10}$ cm^{-2} (Usoskin et al., 2013). Researchers from another group (Thomas et al., 2013) have examined possible sources of a substantial increase of ^{14}C content in AD 774–775. First of all, the authors rejected a CME as a possible cause of the effect. Further, they have modeled SEPs with three different fluences and two different spectra. Finally, they concluded that the data may be explained by an event with fluence about one order of magnitude (about 7 or more times greater) beyond the SEP of October 1989 (depending on the spectrum). Two hard spectrum cases considered by Thomas et al. (2013) may result in moderate ozone depletion, so no mass extinction is implied. At the same time, the authors do predict increases in erythema and damage to plants from enhanced solar UV. Also, they are able to rule out an event with a very soft spectrum that causes severe ozone depletion and subsequent biological impacts. Despite some doubts and discussions about the reliability of these estimates and even the nature of the event in AD 775 (e.g., Pavlov et al., 2013; Cliver et al., 2014; Miroshnichenko and Nymmik, 2014, Mekhaldi et al., 2015), we use them among all the available data to construct a new distribution function of SEP events by the integral proton fluence Φ (≥ 30 MeV).

2.2 LARGEST SEP EVENTS AND DISTRIBUTION FUNCTION

One of our previous studies (Miroshnichenko and Nymmik, 2014) was greatly inspired by the publication of new data on proton fluxes for large events from 1561 to 1994, identified by the so-called nitrate method (McCracken et al., 2001) as well as by the results of analysis and interpretation of these events (e.g., Townsend et al., 2006; Smart et al., 2006). A list of the largest SEP events (1561–1950) includes 11 events above Φ (≥ 30 MeV) = 6×10^9 cm^{-2} (all from Greenland ice core): 7.1×10^9 (1605); 8.0×10^9 (1619); 6.1×10^9 (1637); 7.4×10^9 (1719); 6.3×10^9 (1727); 6.4×10^9 (1813); 9.3×10^9 (1851); 1.9×10^{10} (1859); 7.0×10^9 (1864); 7.7×10^9 (1894); 1.1×10^{10} (1896). All these data have been used in (Miroshnichenko and Nymmik, 2014) to construct the modern distribution function presented below in Fig. 1.

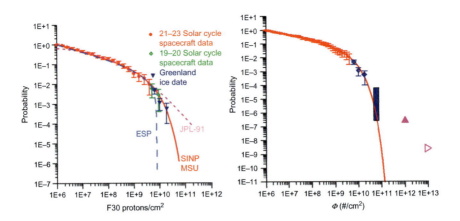

FIG. 1

Left panel: Comparison of several distribution functions for the fluences of Φ ($\geq$30 MeV): JPL91 (Feynman et al., 1993); emission of solar protons (ESP) (Xapsos et al., 2000); SINP MSU-modified model (Nymmik, 1999a,b; Miroshnichenko and Nymmik, 2014). *Right panel*: Space-era data *(red dots)*+Greenland ice data *(blue rhombus)* (Miroshnichenko and Nymmik, 2014), including Carrington event; event AD 775-*blue column*, upper and lower estimates: 4.5×10^{10}–8.0×10^{10} cm^{-2} (adapted *in this work* from the data by Usoskin, I.G., Kromer, B., Ludlow, F., Beer, J., M. Friedrich, M., et al., 2013. The AD775 cosmic event revisited: The Sun is to blame. Astron. Astrophys., 55, L3, doi:10.1051/0004-6361/201321080); *full and open triangles* demonstrate the extrapolation of integral fluences of Φ ($\geq$30 MeV) into the past for 1 and 100 My, respectively (Estimated in Miroshnichenko, L.I., Nymmik, R.A., 2014, Extreme fluxes in solar energetic particle events: methodological and physical limitations. Radiat. Meas. 61, 6–15; From the data for the protons $\geq$60 MeV by Kiraly, P., Wolfendale, A.W., 1999. Long-term particle fluence distributions and short-term observations. In: Proceedings of the 26th International Cosmic Ray Conference, Salt Lake City, USA, vol. 6, pp. 163–166).

All those values of the fluences Φ ($\geq$30 MeV) have been obtained by measurements of nitrate anomalies (excess) in thin layers of the Greenland and Antarctic ices (McCracken et al., 2001). These authors did not estimate the errors of the fluence values, but indicated a background (distinguishable) fluence of 5×10^8 cm^{-2} for the nitrate event of August 31, 1956 (GLE06). With this likely threshold for the detection of SEPs in ice cores and with the fluence Φ ($\geq$30 MeV)$=1.9 \times 10^{10}$ cm^{-2} for the Carrington event, we get a ratio 5×10^8 cm$^{-2}/1.9 \times 10^{10}$ cm$^{-2}=2.63 \times 10^{-2}$. It means that the estimated error for the fluence determination of the Carrington event is below 3%. By the way, the time of occurrence of the Carrington nitrate event is estimated with confidence to be 1859.75 ± 0.2 (McCracken et al., 2001). Those data underwent the analysis and interpretation, in particular, by Townsend et al. (2006) and Smart et al. (2006). Further, some doubts appeared (Wolff et al., 2012; Schrijver et al., 2012) on the validity of the "nitrate signal" itself for the Carrington event and, respectively, on the reliability of the fluence estimates $\Phi(\geq30$ MeV) by McCracken et al. (2001). These doubts, in our opinion, have some reasons, and a principal debate on methodological topics is still continuing (Smart et al., 2014; Wolff et al., 2016). But this debate does not close a possibility of the Carrington event data using (align with other events from the list by McCracken et al., 2001) for the construction of modern distribution function for the SEPs on their fluences of $\Phi(\geq 30$ MeV). Such an approach, in our opinion, will also help to resolve old and new doubts (Schrijver et al., 2012; Wolff et al., 2012, 2016). Moreover, it can be added that the studies by Kepko et al. (2008, 2009) give clear evidence in favor of the methodology and results obtained by McCracken et al. (2001).

As follows from Fig. 1, the model of SINP MSU seems to provide the best agreement with the data on "ancient" extreme SEP events. In this context, one important question arises: can our Sun produce a superflare with the fluence of $\Phi(\geq 30$ MeV) considerably larger, or even much larger, than the AD 775 event? With our current understanding of the problem, there is no direct answer to this question. Nevertheless, available data on the superflares on the Sun-like stars (Maehara et al., 2012), according to recent analysis and estimations (Shibata et al., 2013), can evidence indirectly that the Sun can produce superflares of 10^{34} erg once in 800 years.

It is important to note that the largest SEPs are produced, as a rule, by multiple sequenced solar flares and can be accompanied by a number of prominent geophysical disturbances (auroras, geomagnetic and ionospheric storms, etc.). Storm-time magnetic field depression changes the paths of the energetic particles and extends the region of energetic particle penetration inside the magnetosphere. As a result, strong geomagnetic storms can significantly enhance the effects of SEP. In particular, according to Tsurutani et al. (2003), the September 1–2, 1859, magnetic storm was the most intense geomagnetic disturbance in recorded history. These authors showed that the September 1, 1859, flare most likely had an associated extremely strong CME that led to the magnetic storm in the Earth's magnetosphere with $D_{st} \sim -1760$ nT. This is consistent with the response of $\Delta H = 1600 \pm 10$ nT, registered at the Colaba magnetic observatory near Mumbai, India.

It is generally accepted that the September 1859 solar-terrestrial disturbance was, in fact, the first recognized SW event (e.g., Cliver and Svalgaard, 2004). For example, the auroras associated with the 1859 magnetic storm occurred globally and have been reported by many researchers (see, e.g., Tsurutani et al., 2003). Moreover, the associated magnetic disturbances produced numerous disruptions to telegraph transmissions, which attracted much public attention and were widely reported (e.g., Boteler, 2006). The first magnetic disturbance started on the evening of August 28; telegraph operations were disrupted in North America and Europe through the next morning. The second disturbance started with a sudden commencement at 0440 UT on September 2 and a major disturbance followed immediately. Telegraph operations in Europe were severely affected by the initial magnetic disturbance. Instead, the North American telegraphs were affected by the second phase of the disturbance. Note that the telegraph was introduced in the 1840s and underwent rapid expansion, so by 1859 there were lines across North America, Europe, and parts of Australia and Asia. Both the August 28 and September 2 disturbances had a widespread impact on telegraph operation.

It was also found that both the September 1–2, 1859, flare energy and the speed of the associated CME were extremely high but not unique. Other events with more intense properties have been detected; therefore magnetic storms of this or even greater intensity may occur again. Because the data for the high-energy tails of solar flares and geomagnetic storms are extremely sparse, the tail distributions and therefore the probabilities of occurrence cannot be assigned with any reasonable accuracy. Finally, Tsurutani et al. (2003) concluded that a serious complication is a lack of knowledge of the saturation mechanisms of flares and magnetic storms.

Extreme magnetic storms can significantly extend the region of energetic particle penetration into the high-latitude magnetosphere. For example, changes in the radiation environment at LEO during a

SEP depend on the power of the solar event itself and the efficiency of the penetration of solar particles inside Earth's magnetic field. The stronger a magnetic storm, the larger the area affected by energetic particles will be observed. Sometimes such events can be detected on the ground by the network of neutron monitors. Such observations can be used to verify the magnetic field models (Desorgher et al., 2009). Below we describe the mechanism of penetration of particles at low altitudes.

2.3 SOLAR COSMIC RAYS: PENETRATION BOUNDARY AND CHANGES OF CUTOFF RIGIDITIES

An equatorward shift of the SEP penetration boundary reflects the storm-time magnetic field distortion. Penetration of SEP (electrons and protons) to the Earth's magnetosphere during solar flares and related geomagnetic disturbances in November 2001 and October–November 2003 has been analyzed using *CORONAS-F* energetic particle fluxes data by Kuznetsov et al. (2007). The authors obtained the relationships between the location of penetration boundary, the geomagnetic activity indices, and the local magnetic time. The correlation coefficient between the invariant latitude of the penetration boundary and the K_p and D_{st} indices for electrons with energies from 0.3 to 0.6 MeV in the dayside sector was demonstrated to be higher than that in the nightside sector. The correlation coefficient for protons with energies 1–5 MeV is higher in the nightside sector as compared to the dayside sector. For protons with energies 50–90 MeV, the correlation is higher in the nightside sector as compared to the dayside sector. For protons with energies 50–90 MeV, the correlation is high at all magnetic local times (MLTs). Fig. 2 illustrates variations in the position of the SEP penetration boundary and geomagnetic activity indices (Kuznetsov et al., 2007).

One of the causes of variations in the position of the SEP penetration boundary may be the considerable change of the particle cutoff rigidities. Nymmik et al. (2009, 2010) have developed a simple method for calculating the effective vertical cutoff rigidity of charged particles, taking into account the K_p-index and the local time, on the basis of generalization of the results of extensive trajectory calculations for trial particles moving in the geomagnetic field. The method is suitable for applications using cutoff calculations, such as evaluating particle penetration of spatial boundaries, calculating magnetospheric transmissions for low-orbital spacecraft flights, and interpreting the results of orbital experiments.

Fig. 3 illustrates the method applicability for the evaluation of cutoff rigidity changes during disturbed geomagnetic conditions. One can see that, with the increasing of the level of geomagnetic activity, the cutoff values are significantly reduced, especially for $R_c \leq 1.0$ GV. This decrease of a cutoff rigidity can lead to penetration of high-energy particles into low-latitude regions that are forbidden (not accessible) under quiet conditions. This penetration can occur during a series of solar flares when the CMEs from the initial flares cause a magnetic storm before the arrival of SEP, generated by subsequent flares. Such effect was observed in particular during the October–November 2003 and January 2005 events.

3 GCR MODULATION OVER THE SOLAR CYCLES

Fluxes of GCR are an important part of cosmic radiation that can affect the reliable operations of space equipment and the health of spacecraft crews. These fluxes depend on solar activity. The time dependence of the international sunspot number (version 1.0, http://www.sidc.be/silso/versionarchive) and

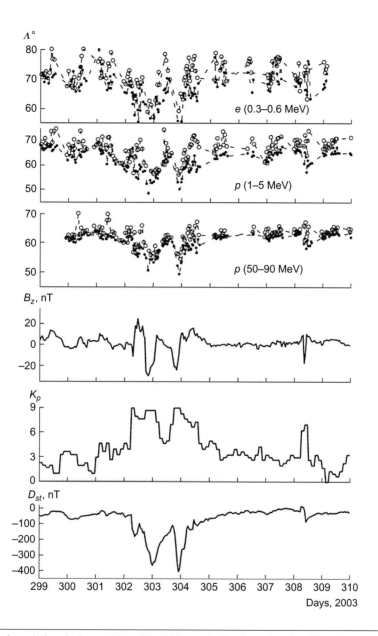

FIG. 2

Three upper panels: variations in the position of the SEP penetration boundary for different types of particles from October 26 (299th day of the year) to November 5, 2003, for MLT=06–09 h *(open circles)* and MLT=18–21 h *(filled dots)* by CORONAS-F data. *Three lower panels*: variations in B_z and geomagnetic activity indices.

From Kuznetsov, S.N., Yushkov, Y.B., Denisov, Y.I., Kudela, K., Myagkova, I.N., 2007, Dynamics of the boundary of the penetration of solar energetic particles to Earth's magnetosphere according to CORONAS-F data. Sol. Syst. Res. 41 (4), 348–353. ISSN 0038-0946.

© Pleiades Publishing, 2007. Original Russian Text © Kuznetsov, S.N., Yushkov, B.Y., Denisov, Y.I., Kudela, K., Myagkova, I.N., 2007, Astron. Vestnik 41 (4), 379–384.

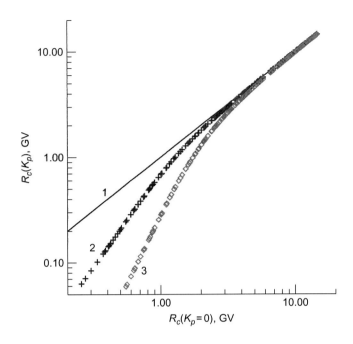

FIG. 3

Cutoff rigidities R_c calculated for the altitude of 450 km in the latitude range of ±52° (i.e., about the ISS orbit) for different K_p-values. At the X-axis we apply the rigidity values for $K_p=0$; at the Y-axis—for different K_p. A straight line 1 corresponds to $K_p=0$; the curves 2 and 3 have been calculated for $K_p=4$ and $K_p=8$, respectively.

an example of some experimental data for GCR fluxes modulated by solar activity during the last four cycles (21–24) of solar activity are shown in Fig. 4. The experimental data of proton and helium fluxes are obtained from detector channels of IMP-8, ACE, and SOHO spacecraft (http://www.srl.caltech.edu/ACE//data/imp/imp8; http://spdf.sci.gsfc.nasa.gov/pub; Kühl et al., 2016). One can see that solar activity generally decreased while GCR flux increased, so the solar forecast is important to estimate GCR flux for future cosmic missions.

Models describing GCR behavior during varying solar activity were constructed either by solving the transport equation describing the propagation of the charged particle fluxes in the heliosphere or by generalization of available experimental data. Models using the transport equations of charged particles (e.g., Potgieter, 2013; Bobik et al., 2012) are designed primarily to study the physical phenomena in the heliosphere. Such models cannot be used for the estimation of radiation hazards without further correction of their parameters by experimental data (O'Neill et al., 2015). The empirical models are more appropriate for these purposes. Such models take into account the particle flux's dependence on the global heliosphere characteristics: solar activity (11-year variations of sunspot numbers or modulation potential) and the spatial orientation of the large-scale solar magnetic field (22-year variations of tilt angle of the heliospheric current sheet). Currently available empirical models (Nymmik et al., 1994; Usoskin et al., 2005; Matthia et al., 2013) are designed to determine the GCR particle fluxes in interplanetary space at heliocentric distances about 1 AU (in the Earth's orbit).

3 GCR MODULATION OVER THE SOLAR CYCLES

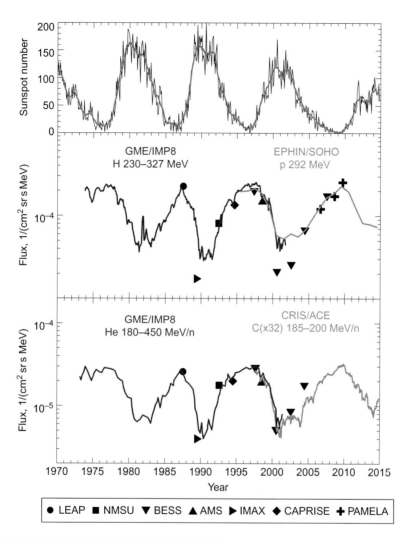

FIG. 4

Top panel: sunspot number during solar cycles 21–24. *Middle and bottom panels*: experimental fluxes of protons and nuclei. *Curves* present the detector data obtained from spacecraft (see text).

In the context of our above consideration, it seems timely to discuss here some prospects of the problem of radiation hazards in the light of available data on total weakening of solar activity in solar cycle 24 and in the nearest several decades. Now the Sun's arrhythmia in different parameters and displays is discussed at the level of observational data, empirical estimates, and theoretical expectations. Some of them are briefly given below. Miroshnichenko et al. (2012, 2013) have recently shown that: (a) the GCR oscillation distribution can be used to study a solar cycle, to predict SEPs, and in other practical applications in the SW problem; (b) oscillations in the SS (sunspots), CI (coronal index), and GCRs obey the hierarchy principle (e.g., GCR intensity is modulated by solar activity SS, etc.); and (c) meanwhile, oscillations in the GLE occurrence rate are apparently of an absolutely different nature, with the only statistically significant oscillation with a period of ~11 years. Only this period can be confirmed by two means in the wavelet analysis: (1) by oscillation spectrum of a pulse width modulation (PWM) series constructed for the registration dates of all 70 GLEs (Morlet scheme); and (2) by their coherence evolution analysis. All other oscillations (with shorter periods) are close to or below the red-noise level. Solar cycle 24 up to now displayed only one or two small GLEs.

Some empirical rules and methods can predict a maximum sunspot number of a future cycle or its amplitude (Petrovay, 2010). Norbury (2011) reviewed the solar cycles and the present level of solar activity and discussed their possible activities during the years 2020–40 with a perspective on space radiation and future space flights. In a number of works, it was assumed that the level of solar activity will be lower than usual in the coming decades (Obridko and Shelting, 2009; Nagovitsyn, 2008; Tlatov, 2015; Javaraiah, 2015). The last (23) and the present (24) solar cycles indicate a downward trend in solar activity in these cycles as compared to solar activity in previous cycles from the beginning of the satellite era. On the basis of statistical analysis of the sunspot numbers (http://www.gao.spb.ru/database/esai/), Nagovitsyn (2008) assumed that in the next two cycles of solar activity, the sunspot numbers will have an average value of 80–100 at maximum or 50–60 as in the Dalton minimum. The peak of the secular minimum is expected to be between the years 2025–35 (Tlatov, 2015). So the 25th cycle should be somewhat weaker than the current 24th and maximal Wolf number will be approximately equal to 68 ± 13. According to Javaraiah (2015) the amplitude of solar cycle 25 will be 50 ± 10; that is about 31% lower than the amplitude of cycle 24. Penn and Livingston (2010) suggested that cycle 25 will be almost without sunspots. However, these works predict only the maximum sunspot number during a cycle.

Several years ago, it was confidently established that a parameter of fluctuations of GCRs may be used as an indicator of the 11-year cycle activity growth phase (Kozlov and Kozlov, 2013). Due to long-term observations of GCR fluctuations, these authors, in fact, have predicted a phase upset in solar activity for cycle 24. Developing this approach, Kozlov and Kozlov (2016) do not exclude so-called "phase catastrophe" in cycle 25 (2020–30), similar to the epoch of Dalton's global minimum. This warning, in our opinion, deserves attention.

At last, some new ideas and suggestions appeared during the several last years based on the properties of solar background magnetic field (SBMF). We mention here especially a principal component analysis (PCA) of the SBMF (Zharkova et al., 2012; Shepherd et al., 2014). Also, it was probing latitudinal variations of the solar magnetic field in cycles 21–23 by Parker's two-layer dynamo model with meridional circulation (Popova et al., 2013). Finally, recent studies of the "heartbeat" of the Sun from PCA make it possible to predict solar activity on a millennial timescale (Zharkova et al., 2015).

In Fig. 5 the dependence of the smoothed monthly sunspot number $W(t)$ on time is shown, including well-known (until 2016) and predicted values on the basis of the ideas of Zharkova et al. (2015) for cycles 25 and 26 (upper panel).

We also can see in Fig. 5 that with the decrease of the amplitude of the sunspot number in solar cycles, the difference between the particle fluxes in solar minima and solar maxima decreases mainly due to a sharper increase of the particle flux in solar maxima (lower panel). Horizontal lines in Fig. 5 (lower panel) show a maximal flux level for heliocentric distance $r = 1$ AU and the given proton energy at the hypothetical heliospherical condition if sunspot number $W = 0$. However, we must admit that the present maximal level of GCR particle flux for the modern epoch cannot fully catch the maximal

4 THE INNER PROTON RADIATION BELT VARIATIONS OVER SOLAR CYCLES

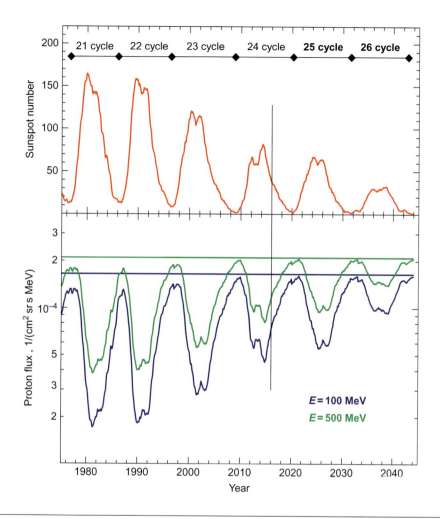

FIG. 5

Top panel: the sunspot numbers. *Bottom panel*: the GCR fluxes for proton energy $E=100$ and 500 MeV in the Earth's orbit. Predicted values are to the right of the *vertical line*.

From Kuznetsov, N.V., Popova, E.P., Panasyuk, M.I., 2017. Empirical model of long-time variations of galactic cosmic ray particle fluxes. J. Geophys. Res. Space Phys. 122, 1463–1472.

"ceiling" of the flux level when solar activity continues to decline steadily during many years (as in the Maunder minimum) (Owens and Lockwood, 2012; Owens et al., 2012; Potgieter et al., 2013).

4 THE INNER PROTON RADIATION BELT VARIATIONS OVER SOLAR CYCLES

Usually, when SEP fluxes are absent and do not affect the radiation conditions in near-Earth space, these conditions depend on ERB particle fluxes and GCR. A large amount of experimental and theoretical research established the basic processes in understanding the ERB formation mechanisms. Quantitative models of the spatial and energy distribution of the trapped particle fluxes in the stationary magnetic field of the Earth have been developed. However, temporal variations of the trapped particle fluxes due to solar and geomagnetic activity variations have been insufficiently studied.

The time dependence of trapped proton fluxes in the inner boundary of ERB was studied as a function of solar activity by Huston et al. (1998), Kuznetsov et al. (2010), and Qin et al. (2014). Fig. 6 from Kuznetsov et al. (2010) shows the experimental fluxes of protons with energies above 70 MeV at the geomagnetic equator for drift shells $L=1.14-1.2$. The same panel in Fig. 6 shows the calculated dependences of the particle fluxes taking into account the accumulation (from albedo neutrons) and removal (due to the interaction with atmosphere atoms) of trapped particles in different L-shells. It was shown that variations of trapped proton flux in the inner boundary of ERB are explained by variations of atmospheric density, which in turn varies due to changes of solar activity.

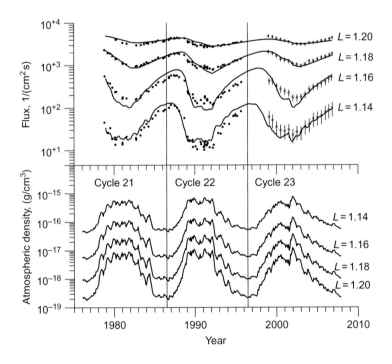

FIG. 6

Upper panel: the experimental *(points)* and calculated *(curves)* omnidirectional fluxes of trapped protons with the energy above 70 MeV in the geomagnetic equator. *Bottom panel*: the atmospheric density neutral and ionized atom density (MSIS-E-90 and IRI995 models) averaged over the geomagnetic equator for different L-shells.

From Kuznetsov, N.V., Nikolaeva, N.I, Panasyuk, M.I., 2010. Variation of the trapped proton flux in the inner radiation belt of the Earth as a function of solar activity. Cosm. Res. 48 (1), 80–85.

5 RADIATION ENVIRONMENT FOR CREWED ORBITAL STATIONS

The radiation environment on the International Space Station (ISS) is mainly due to GCR penetrating to the orbit through the geomagnetic field, and the protons of the inner radiation belt, extending to altitudes of the ISS flight in the area of the South Atlantic anomaly. These two sources (GCR and protons of radiation belts) give comparable contributions to the average absorbed dose rate onboard the station and can be considered as semipermanent exposure. Electrons that make a small contribution, in comparison with protons in the radiation load in this area of space, will not be considered here.

SEPs, as will be shown later, lead to additional radiation hardenings. Dose received from large SEP events in unprotected areas of the ISS for a couple of days of proton flux increasing can reach values that were accumulated under quiet radiation conditions for several months of flight.

Another source of additional exposure onboard is fluxes of high-energy electrons of the outer radiation belt, occasionally appearing in high-latitude parts of the orbit of the ISS. However, the dose they create inside the habitable compartments of the ISS does not exceed several percent from the average daily dose.

Very detailed information about the radiation environment on the ISS was obtained with onboard radiation detection equipment of the Russian, American, European, and Japanese partners.

The most important radiation instruments on-board ISS are TEPC, allowing to measure not only the absorbed, but also the equivalent dose (Zhou et al., 2007) and DOSTEL and Liulin (Burmeister et al., 2015; Dachev, 2009) allowing to measure neutron fluxes (Tretyakov et al., 2010). Important information about the nuclear composition of the radiation field inside the ISS was obtained with the help of ALTEA instrumentation (Narici et al., 2015; Chapter 17 of this volume). In this article, we will limit ourselves to data obtained through the Russian system for radiation monitoring onboard ISS.

The R-16 dosimeter onboard the MIR station and the Radiation Monitoring System (RMS) on the ISS just served for this purpose (Benghin et al., 1992; Lyagushin et al., 2002; Petrov et al., 2007). The sensitive elements of the R-16 dosimeter were two ionization chambers. The R-16 dosimeter and four DB-8 units are the sensitive units of the RMS. All the DB-8 units are similar. Each of them has two fully independent channels consisting of a semiconductor detector and electronic circuit. The sensitive component is a silicon semiconductor detector. The difference between the two channels is that one of the detectors has additional lead shielding. The shielding is a sphere surrounding the detector. The sphere wall thickness is 3 g/cm^2 of Pb. They have been installed inside the ISS "Zvezda" module in four places that are differently protected by the station equipment. The DB-8 No. 1 unit is the least protected, and the DB-8 No. 4 unit is the most protected (Kuznetsov et al., 2015b). The daily radiation doses measured on the ISS with DB-8 No. 1 and DB-8 No. 4 are presented in Fig. 7.

DB-8 units measured absorbed doses in silicon. In order to estimate the equivalent dose, we first need to calculate ratio of charged particle energy losses in tissue-equivalent material and silicon. The obtained ratio should be multiplied by the absorbed dose and by radiation quality factor.

The doses measured by other different detectors are in the interval between the values of these two detectors. The representative daily dose values are in the range of 0.15–0.4 mGy/days. One can see the noticeable increases in the daily doses caused by SEPs.

The separation of the contributions to daily doses from GCRs and the ERB was carried out. It was shown that the GCR contribution varies slightly and does not practically depend on the position of a measurement point (for four locations of the detectors). The GCR contribution increases slowly from

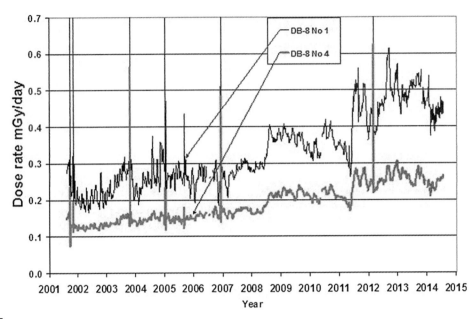

FIG. 7

Daily radiation doses measured onboard ISS with DB-8 No. 1 and DB-8 No. 4 during 2001–14. Dose values are smoothed with the period of average of 15 days for all time intervals (Benghin et al., 2014).

From Benghin, V.V., Petrov, V.M., Panasyuk, M.I., et al., 2013. Results of the radiation monitoring system measurements on service module of ISS during 2009–2013. Available from: http://www.wrmiss.org/workshops/eighteenth/Benghin.pdf.

0.08 to 0.11 mGy/days during the period of anomalously low solar activity since mid-2005 until mid-2009 (Lishnevskii et al., 2012). Furthermore, at an increase of solar activity, it began to decrease.

There were a lot of SEPs observed during the flights of MIR and ISS (up to mid-2014). However, most of them had no significant effect on the radiation environment onboard the stations. The SEP (more exactly, solar proton event, SPE) doses are listed in Table 1 for the events at which the dose was close to or exceeded the average daily value. Data were provided by the R-16 device or were recalculated to the R-16 device location on the basis of the DB-8 unit data.

The RMS data onboard ISS allowed the determination of the experimental difference of dose values in the living compartments of the station. In nine cases of SEPs, average daily values have been exceeded (Benghin et al., 2014) (Table 2).

Data of Table 2 illustrate considerable variety in SEP doses onboard the ISS within one station module. This variety is due to significant differences in the shielding conditions inside the ISS. Data for the events of October 2003 confirm the effectiveness of such protective measures as transitions to a more shielded area of the station in the periods of deterioration of radiation environment caused by SEPs.

A totally different situation occurred during two SEPs of January 17 and 20, 2005. The event of January 20 was characterized by the significant increase in count rate on neutron monitors, and one can assume that it will lead to essential deterioration of radiation conditions onboard the ISS. However,

5 RADIATION ENVIRONMENT FOR CREWED ORBITAL STATIONS

Table 1 The Large Solar Proton Events Dose Onboard the ISS and MIR Space Station

Event Date	MIR (mGy)	ISS (mGy)	Event Date	ISS (mGy)
March 7, 1989	0.35		April 18, 2001	0.16
March 23, 1989	0.20		September 25, 2001	2.78
September 28, 1989	4.65		April 11, 2001	4.08
October 19, 1989	27.20		November 22, 2001	0.46
October 23, 1989	3.00		December 26, 2001	3.22
October 25, 1989	1.70		April 21, 2002	1.32
May 24, 1990	0.20		October 28, 2003	6.37
March 23, 1991	2.45		October 29, 2003	7.90
April 6, 1991	0.60		November 4, 2003	0.56
June 11, 1991	2.05		November 10, 2004	0.26
June 15, 1991	0.75		January 16, 2005	2.67
April 20, 1998	0.35		January 20, 2005	1.86
July 14, 2000	7.80		September 8, 2005	0.83
November 8, 2000	2.80	1.40	December 6, 2006	6.65
April 2, 2001		0.17	December 13, 2006	8.02
April 15, 2001		0.50	March 7, 2012	1.56

Table 2 Doses, mGy, for Solar Proton Events, Measured With RMS Onboard the ISS

	DB-8 No. 1		DB-8 No. 2		DB-8 No. 3		DB-8 No. 4	
SPE Date	Unshield	Shielded	Unshield	Shielded	Unshield	Shielded	Unshield	Shielded
September 24, 2001	1.57	0.99	1.25	0.96	0.54	0.21	0.19	0.15
November 4, 2001	2.66	1.31	1.18	0.49	0.84	0.54	0.08	0.04
October 28, 2003	1.71	1.19	0.82	0.52	0.87	0.69	0.31	0.30
October 29, 2003	6.82	3.14	3.00	1.18	2.11	1.35	0.67	0.52
January 17, 2005	0.81	0.67	0.31	0.55	0.63	0.29	0.18	0.10
January 21, 2005	0.21	0.18	0.13	0.14	0.13	0.14	0.08	0.07
September 8, 2005	0.33	0.28	0.20	0.24	0.26	0.20	0.09	0.08
December 13, 2006	0.51	0.47	0.67	0.67	0.43	0.42	0.32	0.32
March 7, 2012	1.84	1.58	1.26	1.59	1.56	1.07	0.57	0.55

this did not occur because of the rather hard proton spectrum of the event. One can see from Table 2 that the extra dose on January 21 did not exceed 0.2 mGy that corresponds to the approximate daily dose under undisturbed conditions.

The difference of doses in the ISS "Zvezda" module compartments was small for this period. That was why any recommendations about transition to more shielded places within ISS were not viable. This illustrates that in each individual event, we need a strictly individual approach to the assessment of the radiation hazard and to the development of recommendations for the ISS crew on reducing the additional radiation doses during SEPs.

Fig. 8 illustrates the particle measurements and dose rate behavior inside ISS during one of the SEP events of the current solar cycle 24. The pair of left panels demonstrates SEP fluxes as measured by GOES-10 detectors during the event of March 7, 2012, along with data on the level of geomagnetic disturbance. Three right panels show corresponding dose data (Kuznetsov et al., 2015a).

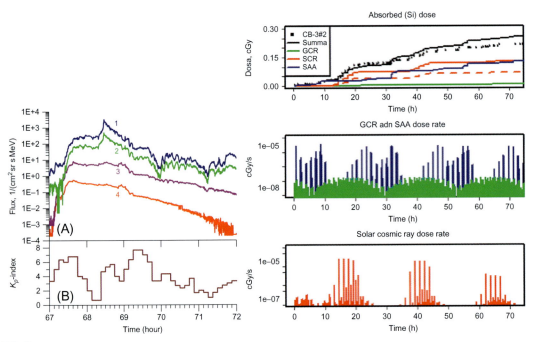

FIG. 8

Left panels: Energetic solar protons as measured by GOES-10 (A) between March 7–11, 2012, and K_p-index of geomagnetic disturbance (B) vs current time (h) measured from the 0000 UT March 7, 2012. Proton energies: (1) 4.2–8.7; (2) 8.7–14.5; (3) 38–82; (4) 84–200 MeV. *Right panels*: Absorbed dose *(upper panel)* and absorbed dose rate *(two bottom panels)* calculated for ISS using particle flux models *(curves)* and measured by dosimeter DB-8 № 2 *(points)* inside the ISS vs flight time during the SEP event. Curves show total dose *(black curves)* and the contribution to the dose from the ERB protons *(blue curves)*, GCR particles *(green curve)*, and SEP *(red curves)*. *Dashed red line* corresponds to a hypothetical case if the $K_p=2$.

Calculation results have satisfactorily coincided with the RMS data measured onboard ISS. This is good evidence of applicability of the proposed technique for the computing of radiation doses onboard the spacecraft during moderate SEPs (Kuznetsov et al., 2015a).

Comparison of data presented in Fig. 7 and in the last line of Table 2 shows that the additional dose due to this event approximately corresponds to the dose accumulated over the 4 days of flight in the undisturbed radiation conditions. If one can consider a presumptive two orders of magnitude more powerful than the SEP on March 7, 2012, the total dose may be close to the maximum permissible value for the annual flight, and even exceed it. However, the likelihood of such events, according to Fig. 1, is significantly less than 1%.

The temporal variations of the radiation environment in LEO are determined, in addition to the impact of the penetration of the SEP during magnetic storms on these altitudes, but also by changes in solar activity influence both on the spatial and energy structure of the ERB and GCR flux subjected to modulation during solar cycles.

Along with many persistent attempts to understand underlying mechanisms of the formation of the near-Earth radiation environment, much effort is directed at the estimation of the radiation hazard to spacecraft and crews of manned space stations (e.g., Durante and Cucinotta, 2011). The estimations are carried out using quantitative models of ERB, GCR, and SEP radiation that describe the spatial and temporal patterns of charged particle distribution in space. Computer codes of these models are part of a software packages (e.g., OMERE, http://www.trad.fr) or interactive information systems CREME (https://creme.isde.vanderbilt.edu), OLTARS (https://oltaris.larc.nasa.gov/), and SPENVIS (http://spenvis.oma.be). They allow calculating the characteristics of radiation hazards (absorbed or equivalent dose, rate of SEE, etc.) for space missions.

Let us consider the results of the calculations for the annual radiation dose that have been recently carried out using the programs and models developed in the SINP MSU. We calculate the characteristics of radiation hazards at the ISS orbit using the software package COSRAD (Kuznetsov and Malyshkin, 2011; Kuznetsov et al., 2015b). This software package calculates energy spectra of particle flux in the "open space" (for a given orbit) using the particle flux models AE8 and AP8 (Bilitza, 1996) for ERB, model ISO (ISO, 2004) for GCR, and the probabilistic model for the SEP event distribution function (Nymmik, 1999a,b). The model of SEP penetration (Nymmik et al., 2009, 2010) is also used to calculate the SEP and GCR fluxes inside the Earth's magnetosphere. Using the energy spectra of the particle fluxes (averaged over the orbit), the software package COSRAD calculates: (1) energy spectra and linear energy transfer (LET) spectra of particle fluxes behind spherical shielding, considering secondary radiation emerging from shielding (bremsstrahlung, protons, and neutrons); (2) absorbed or equivalent dose and SEU rate for integrated circuits.

Figs. 9 and 10 show the annual equivalent dose that can be obtained in the tissue-equivalent material at the ISS orbit with different thicknesses of spherical shielding. The SEP dose was calculated using particle fluence that can be exceeded during 1 year with 10% probability. Such a level of SEP fluence corresponds to a series of powerful events in October 1989. In the right panel of Fig. 9 we present similar estimates for a 1-year interplanetary mission.

As we can see in Fig. 9, the SEPs do not provide a significant contribution to the total dose (ERB +GCR+SEP) in the entire range of shielding thickness x. The total dose is accumulated due to ERB protons (at $x < \sim 15$ g/cm^2) and due to GCR particles (at $x > \sim 15$ g/cm^2).

Along with dose effects of radiation on humans and onboard systems of spacecraft, high-energy protons and heavier particles (HZE particles) of cosmic radiation can cause SEE in the electronic elements and a number of effects, including disturbances in the biostructures at the molecular level as well as neurophysiological disorders. In this respect, the modeling of fluxes of heavy particles present in all components of cosmic radiation is an urgent task. Let us consider the results of the model calculations of SEE for the radiation conditions typical for LEO.

Fig. 10 shows the single-event upset rate (SEU rate) expected in the ISS orbit in some commercial random-access memory (RAM) (cross-section from heavy charge particles = 1 cm^2/chip, threshold LET = 2 MeV/mg/cm^2). We consider the case when, during the day, the SEP peak flux corresponding to the SEP event in October 1989 exists along with the ERB and GCR particle fluxes. As one can see

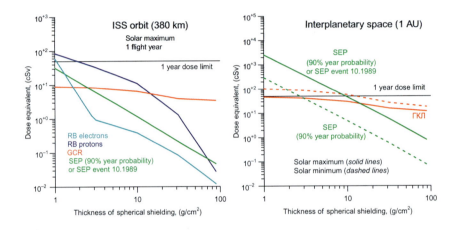

FIG. 9

Left panel: Annual equivalent dose that can be obtained in the tissue-equivalent material at LEO (ISS orbit) with different thicknesses of spherical shielding. *Right panel*: similar estimates for a 1-year interplanetary mission.

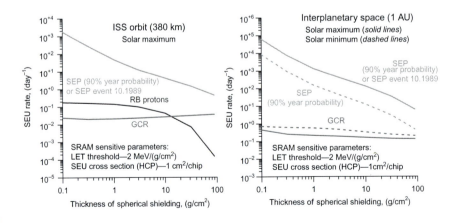

FIG. 10

Left panel: SEU rate expected in some commercial RAM at the ISS orbit. *Right panel*: Similar estimates for the case of free (interplanetary) space.

from Fig. 10, the ERB protons (at $x < \sim 15$ g/cm^2) or the GCR particles (at $x > \sim 15$ g/cm^2) make the main contribution to the SEU rate if SEP particles are absent. The appearance of SEP particles should considerably increase the SEU rate if the same SEP flux is present as in October 1989.

In conclusion, we emphasize that exemplified calculations of the dose (Fig. 9) and SEU rate (Fig. 10) qualitatively illustrate the relative contribution of individual components of cosmic radiation into the radiation hazard at LEOs. To determine the absolute values of the radiation hazard characteristics, specific information is needed about the conditions of space flight, the characteristics of the object, and the real distribution of the shielding thickness around this object.

6 CONCLUSIONS

In this paper, we presented the near-Earth radiation environment and its behavior under extreme solar and geomagnetic conditions. We have described the main sources of the high-energy particle fluxes in the near-Earth environment and analyzed the radiation conditions for space missions close to Earth at LEO orbit. Among them: GCR, SEP, and high-energy protons of the inner ERB.

It is shown that for short duration flights in the near-Earth region, the most dangerous are SEP fluxes that under extreme geomagnetic conditions can significantly increase the radiation load both on the spacecraft and on the crew at some parts of the orbits. For long-duration missions at LEO orbit, radiation hardening is mostly related to the presence of long-term variations of the proton radiation belts (mainly in the South Atlantic Anomaly) that are correlated with solar cycles. As to SEP events, the underestimation of SEP fluxes with an energy of more than ~ 100 MeV in the powerful events can lead to significant errors in the estimation of radiation hazards for spacecraft crews. However, in general, at the absence of SEP extreme events the radiation hazard for long-term interplanetary missions is associated with GCR flows, which vary with the change of solar activity and the polarity of the solar magnetic field. The maximum level of GCR flux during all the time of observations was reached in the modern era at the minimum of solar activity (2008–10), although such maximum level is not the limit for a longer period of absence of sunspots (as in the Maunder minimum).

For near-Earth orbital flights, the most significant factor for radiation hazards is the proton fluxes of ERB and their variations. The fluxes of trapped particles in the internal ERB are stable, but when evaluating the radiation hazard for spacecraft missions in LEO, it has to take into account long-term variations in proton fluxes. In the chapter this problem is illustrated by the experimental data on the absorbed dose, which are obtained by dosimetric sensors on the ISS. Model calculations allow us to explain these long-term solar-cyclical variations of trapped protons on low drift shells ($L = 1.14$–1.2) by a changing of atmospheric density with a solar activity variation.

Taking into account geomagnetic disturbances, the dynamics of SEP and GCR flows in near near-Earth space and the effects of their radiation effects become more complicated. The rate of accumulation of the absorbed dose in the orbit of the ISS increases during the appearance of the SEP event. Moreover, the increase level is growing if the arrival of the SEP event coincides with the perturbation of the magnetosphere. However, the level of fluxes and radiation doses of SEP and GCR depends not only on the amplitude of the geomagnetic disturbance, but also on the local time of localization of the satellite in orbit.

Model calculations of the dose and frequency of single faults in microcircuits performed using the Russian software complex COSRAD show, on the whole, the less significant role of the SEP and GCR fluxes in the formation of radiation hazards in the Earth's magnetosphere (especially LEO) than the fluxes of trapped particles in the radiation belts.

ACKNOWLEDGMENTS

This work was funded by the Russian Science Foundation (research of M. Panasyuk by Grant 16-17-00098). The work by L. Miroshnichenko was partially supported by the Russian Foundation for Basic Research (RFBR) (projects 13-02-00612, 13-02-91165, 14-02-90424). V. Kalegaev thanks support from RFBR Grant No 16-05-00760, N. Kuznetsov thanks support from RFBR Grant No 17-29-01022. H. Popova thanks support from RFBR (projects 17-29-01022, 16-05-00507).

REFERENCES

Benghin, V.V., Petrov, V.M., Teltsov, M.V., et al., 1992. Dosimetric control onboard the MIR space station during the solar proton events of September-October 1989. Nucl. Tracks Radiat. Meas. 20 (1), 21–23.

Benghin, V.V., Petrov, V.M., Panasyuk, M.I., et al., 2014. Nine years of the radiation monitoring system operation in service module of ISS. http://www.wrmiss.org/workshops/fifteenth/Benghin.pdf.

Bilitza, D., 1996. Models of Trapped Particle Fluxesae8 (Electrons) and AP8 (Protons) in Inner and Outer Radiation Belts. National Space Science Data Center. PT-11B, USA.

Bobik, P., Boella, G., Boschini, M.J., Consolandi, C., et al., 2012. Systematic investigation of solar modulation of galactic protons for solar cycle 23 using a Monte Carlo approach with particle drift effects and latitudinal dependence. Astrophys. J. 745, 132–152.

Boteler, D.H., 2006. The super storms of August/September 1859 and their effects on the telegraph system. Adv. Space Res. 38, 159–172.

Burmeister, S., Berger, T., Labrenz, J., Boehme, B., Haumann, L., Reitz, G., 2015. The DOSIS and DOSIS 3D experiments on-board the International Space Station—current status and latest data from the DOSTELs as active instruments. In: Workshop on Radiation Measurements on ISS, Cologne, Germany, September 8–10. Available from: http://wrmiss.org/workshops/twentieth/Burmeister.pdf.

Cliver, E.W., Svalgaard, L., 2004. The 1859 solar-terrestrial disturbance and the current limits of extreme space weather activity. Sol. Phys. 224, 407–422.

Cliver, E.W., Tylka, A.J., Dietrich, W.F., Ling, A.G., 2014. On a solar origin for the cosmogenic nuclide event of 775 A.D. Astrophys. J. 781 (32), 4. https://doi.org/10.1088/0004-637X/781/1/32.

Dachev, T.P., 2009. Characterization of near Earth radiation environment by Liulin type instruments. Adv. Space Res. 44, 1441–1449.

Desorgher, L., Kudela, K., Flückiger, E., Bütikofer, R., Storini, M., Kalegaev, V., 2009. Comparison of Earth's magnetospheric magnetic field models in the context of cosmic ray physics. Acta Geophys. 57 (1), 75–87. https://doi.org/10.2478/s11600-008-0065-3.

Durante, M., Cucinotta, F.A., 2011. Physical basis of radiation protection in space travel. Rev. Mod. Phys. 83, 1245–1281.

Feynman, J., Spitale, G., Wang, J., Gabriel, S., 1993. Interplanetary fluence model: JPL 1991. J. Geophys. Res. 98, 13281–13294.

Huston, S., Kuck, G., Pfitzer, K., 1998. Solar cycle variation of the low-altitude trapped proton flux. Adv. Space Res. 21 (12), 1625–1634.

International Standard ISO 15390, 2004. Space environment (natural and artificial)—galactic cosmic ray model. ISO.

Javaraiah, J., 2015. Long-term variations in the north-south asymmetry of solar activity and solar cycle prediction, III: prediction for the amplitude of solar cycle 25. New Astron. 34, 54–64.

Jull, A.J.T., Panyushkina, I.P., Lange, T.E., Kukarskih, V.V., et al., 2014. Excursions in the 14C record at A.D. 774–775 in tree rings from Russia and America. Geophys. Res. Lett. https://doi.org/10.1002/2014GL059874.

Kepko, L., Spence, H., Shea, M.A., Smart, D.F., Dreschhoff, G.A.M., 2008. Observations of impulsive nitrate enhancements associated with ground-level cosmic ray events 1–4 (1942–1949). In: Caballero, R., D'Olivo, J.C., Medina-Tanco, G., Nellen, L., Sánchez, F.A., Valdés-Galicia, J.F. (Eds.), Proc. 30th Int. Cosmic Ray Conf. Universidad Nacional Autónoma de México, Mexico City, Mexico, vol. 1, pp. 729–732.

Kepko, L., Spence, H., Smart, D.F., Shea, M.A., 2009. Interhemispheric observations of impulsive nitrate enhancements associated with the four large ground-level solar cosmic ray events (1940–1950). J. Atmos. Sol. Terr. Phys. 71, 1840–1845.

Kozlov, V.I., Kozlov, V.V., 2013. Cosmic ray fluctuation parameter as indicator of 11-year cycle activity growth phase. J. Phys. Conf. Ser. 409.012160. https://doi.org/10.1088/1742-6596/409/1/012160.

Kozlov, V.I., Kozlov, V.V., 2016. Arrhythmia of sun. In: Krymsky, G.F. (Ed.), Cosmic Rays. Institute of Cosmical Research and Aeronomy, Yakutsk.

Kühl, P., Gómez-Herrero, R., Heber, B., 2016. Annual cosmic ray spectra from 250 MeV up to 1.6 GeV from 1995–2014. Measured with the electron proton helium instrument onboard SOHO. Sol. Phys. 291, 965–974.

Kuznetsov, S.N., Yushkov, B.Y., Denisov, Y.I., Kudela, K., Myagkova, I.N., 2007. Dynamics of the boundary of the penetration of solar energetic particles to Earth's magnetosphere according to CORONAS-F data. Sol. Syst. Res. 41, 348–353.

Kuznetsov, N.V., Nikolaeva, N.I., Panasyuk, M.I., 2010. Variation of the trapped proton flux in the inner radiation belt of the Earth as a function of solar activity. Cosm. Res. 48 (1), 80–85.

Kuznetsov, N.V., Malyshkin, Yu.M., Nikolaeva, N.I., Nymmik, R.A., Panasyuk, M.I., et al., 2011. Software complex COSRAD for radiation environment forecasting onboard spacecrafts. In: Problems of Atomic Science and Technology. Physics of Radiation Effects on the Radio-Electronic Equipment, N 2. National Science Center, Kharkov, Ukraine, pp. 72–78 (in Russian).

Kuznetsov, N.V., Nymmik, R.A., Panasyuk, M.I., Yushkov, B.Yu., Benghin, V.V., Mitrikas, V.G., 2015a. Absorbed dose accumulation onboard the spacecraft at the near-Earth orbits under the influence of solar cosmic ray fluxes. In: Problems of Atomic Science and Technology. Physics of Radiation Effects on the Radio-Electronic Equipment, N 2. National Science Center, Kharkov, Ukraine, pp. 20–23 (in Russian).

Kuznetsov, N.V., Malyshkin, Yu.M., Nikolaeva, N.I., Nymmik, R.A., Panasyuk, M.I., et al., 2015b. Comparison of the models of charged particle fluxes in space. In: Proc. RADECS 2015pp. 178–181 IEEE catalog number: CFP15449-PRT.

Lishnevskii, A.E., Panasyuk, M.I., Nechayev, O.Y., Benghin, V.V., et al., 2012. Results of variations of the absorbed dose monitoring on the International Space Station during 2005-2011. Cosm. Res. 50, 419–424.

Lyagushin, V.I., Volkov, A.N., Aleksandrin, A.P., et al., 2002. Preliminary results of measuring absorbed dose rates with the radiation monitoring system of Russian segment of the International Space Station. In: Problems of Atomic Science and Technology. Physics of Radiation Effects on the Radio-Electronic Equipment, vol. 4. National Science Center, Kharkov, Ukraine, pp. 22–25 (in Russian).

Maehara, H., Shibayama, T., Notsu, S., et al., 2012. Superflares on solar-type stars. Nature 485, 478–485.

Matthia, D., Berger, T., Mrigakshi, A.I., Reitz, G., 2013. A ready-to-use galactic cosmic ray model. Adv. Space Res. 51, 329–338.

McCracken, K.G., Dreschhoff, G.A.M., Zeller, E.J., Smart, D.F., Shea, M.A., 2001. Solar cosmic ray events for the period 1561–1994. 1. Identification in polar ice, 1561–1950. J. Geophys. Res. 106 (A10), 21585–21598.

Mekhaldi, F., Muscheler, R., Adolphi, F., Aldahan, A., Beer, J., McConnell, J.R., Possnert, G., Sigl, M., Svenson, A., Synal, H.-A., Welten, K.C., Woodruff, T.F., 2015. Multiradionuclide evidence for the solar origin of the cosmic-ray events of AD 774/5 and 993/4. Nat. Commun. 6, 8611, https://doi.org/10.1038/ncomms9611.

Miroshnichenko, L.I., 2014. Solar Cosmic Rays: Fundamentals and Applications, second ed. Springer, Heidelberg/New York/Dordrecht/London.

Miroshnichenko, L.I., Nymmik, R.A., 2014. Extreme fluxes in solar energetic particle events: methodological and physical limitations. Radiat. Meas. 61, 6–15.

Miroshnichenko, L.I., Pérez-Peraza, J.A., Velasco-Herrera, V.M., Zapotitla, J., Vashenyuk, E.V., 2012. Oscillations of galactic cosmic rays and solar indices before the arrival of relativistic solar protons. Geomagn. Aeron. 52, 547–560.

Miroshnichenko, L.I., Vashenyuk, E.V., Perez-Peraza, J.A., 2013. Solar cosmic rays: 70 years of ground-based observations. Geomagn. Aeron. 53, 541–560.

Miyake, F., Nagaya, K., Masuda, K., Nakamura, T., 2012. A signature of cosmic-ray increase in AD 774–775 from tree rings in Japan. Nature 486, 240–242.

Nagovitsyn, Y.A., 2008. Global solar activity on long time scales. Astrophys. Bull. 63 (1), 45–55.

Narici, L., Casolino, M., Di Fino, L., Larosa, M., Picozza, P., Zaconte, V., 2015. Radiation survey in the International Space Station. J. Space Weather Space Clim. 5, A37–A57.

Norbury, J.W., 2011. Perspective on space radiation for space flights in 2020–2040. Adv. Space Res. 47, 611–621.

Nymmik, R.A., 1999a. SEP event distribution function as inferred from space born measurements and lunar rock isotopic data. In: Proceedings of the 26th International Cosmic Ray Conference, Salt Lake City, USA, vol. 6. pp. 268–271.

Nymmik, R.A., 1999b. Probabilistic model for fluences and peak fluxes of solar energetic particles. Radiat. Meas. 30, 287–296.

Nymmik, R.A., Panasyuk, M.I., Pervaya, T.I., Suslov, A.A., 1994. An analytical model describing dynamics of galactic cosmic ray heavy particles. Adv. Space Res. 14 (10), 750–763.

Nymmik, R.A., Panasyuk, M.I., Petrukhin, V.V., Yushkov, B.Y., 2009. A method of calculation of vertical cutoff rigidity in the geomagnetic field. Cosm. Res. 47, 191–197.

Nymmik, R.A., Yushkov, B.Y., Panasyuk, M.I., Petrukhin, V.V., 2010. A method for operative calculation of charged particle penetration into the LEO. Adv. Space Res. 46, 303–309.

Obridko, V., Shelting, B., 2009. Anomalies in the evolution of global and large-scale solar magnetic fields as the precursors of several upcoming low solar cycles. Astron. J. Lett. 35 (4), 247–252.

O'Neill, P.M., Golge, S., Slaba, T.C., 2015. Badhwar–O'Neill 2014 galactic cosmic ray flux model description. NASA/TP-2015-218569.

Owens, M.J., Lockwood, M., 2012. Cyclic loss of open solar flux since 1868: the link to heliospheric current sheet tilt and implications for the Maunder minimum. J. Geophys. Res. 117, https://doi.org/10.1029/2011JA017193. A04102.

Owens, M.J., Usoskin, I., Lockwood, M., 2012. Heliospheric modulation of galactic cosmic rays during grand solar minima: past and future variations. Geophys. Res. Lett. 39, L19102. https://doi.org/10.1029/2012GL053151.

Pavlov, A.K., Blinov, A.V., Konstantinov, A.N., Ostryakov, V.M., et al., 2013. AD 775 pulse of cosmogenic nuclide production as imprint of a galactic gamma-ray burst. Mon. Not. R. Astron. Soc. 435, 2878–2884. https://doi.org/10.1093/mnras/stt1468.

Penn, M.J., Livingston, W., 2010. Long-term evolution of sunspot magnetic fields. Proceedings of the International Astronomical Union, IAU Symposium. Phys. Sun Star Spots 273, 126–133.

Petrov, V.M., Tel'tsov, M.V., Mitrikas, V.G., Akatov, Y.A., Benghin, V.V., et al., 2007. Radiation dosimetry in space flight. In: Panasyuk, M.I. (Ed.), eighth ed. In: Space Model, vol. 1. SINP MSU, Moscow, pp. 642–667 (in Russian).

Petrovay, K., 2010. Solar cycle prediction. Living Rev. Sol. Phys. 7, 6.

Popova, E., Zharkova, V., Zharkov, S., 2013. Probing latitudinal variations of the solar magnetic field in cycles 21–23 by Parker's two-layer dynamo Model with meridional circulation. Ann. Geophys. 31, 2023–2038. https://doi.org/10.5194/angeo-31-2023-2013.

Potgieter, M.S., 2013. Solar modulation of cosmic rays. Living Rev. Sol. Phys. 10, 3.

Potgieter, M.S., Tout Strauss, D., De Simone, N., Boezio, M., 2013. The highest recorded proton spectrum at Earth since the beginning of the space age. In: Proceedings of the 33rd International Cosmic Ray Conference, Rio de Janeiro, Brazil. pp. 119–122.

Qin, M., Zhang, X., Ni, B., Song, H., Zou, H., Sun, Y., 2014. Solar cycle variations of trapped proton flux in the inner radiation belt. J. Geophys. Res. 119. https://doi.org/10.1002/2014JA020300.

Schrijver, C.J., Beer, J., Baltensperger, U., Cliver, E., Gudel, M., Hudson, H., McCracken, K.G., Osten, R., Peter, T., Soderblom, D., Usoskin, I., Wolff, E.W., 2012. Estimating the frequency of extremely energetic solar events, based on solar, stellar, lunar, and terrestrial records. J. Geophys. Res. 117, https://doi.org/10.1029/2012JA017706. A08103.

Schrijver, C.J., et al., 2015. Understanding space weather to shield society: a global road map for 2015–2025 commissioned by COSPAR and ILWS. Adv. Space Res. 55, 274–2807. https://doi.org/10.1016/j.asr.2015.03.023.

Shepherd, S.J., Zharkov, S.I., Zharkova, V.V., 2014. Prediction of solar activity from solar background magnetic field variations in cycles 21–23. Astrophys. J. 795 (46), 8. https://doi.org/10.1088/0004-637X/795/1/46.

Shibata, K., Isobe, H., Hillier, A., et al., 2013. Can superflares occur on our Sun? Publ. Astron. Soc. Jpn. 65, 49 (1)–49(8).

Smart, D.F., Shea, M.A., McCracken, K.G., 2006. The Carrington event: possible solar proton intensity–time profile. Adv. Space Res. 38, 215–225.

Smart, D.F., Shea, M.A., Melott, A.L., Laird, C.M., 2014. Low time resolution analysis of polar ice cores cannot detect impulsive nitrate events. J. Geophys. Res. Space Phys. 119, 9430–9440. https://doi.org/10.1002/2014JA020378.

Song, P., 2001. Foreword. J. Geophys. Res. 106 (A10), 20945–20946.

Thomas, B.C., Arkenberg, K.R., Brock, Snyder II, B.R., Melott, A.L., 2013. Terrestrial effects of possible astrophysical sources of an AD 774–775 increase in carbon-14 production. Geophys. Res. Lett. 40, 1237–1240.

Tlatov, A.G., 2015. The change of the solar cyclicity mode. Adv. Space Res. 55, 851–856.

Townsend, L.W., Stephens Jr., D.L., Hoff, J.L., Zapp, E.N., et al., 2006. The Carrington event: possible doses to crews in space from a comparable event. Adv. Space Res. 38, 226–231.

Tretyakov, V.I., Mitrofanov, I.G., Lyagushin, V.I., et al., 2010. The first stage of the «BTN-Neutron» space experiment onboard the Russian segment of the International Space Station. Kosm. Issled. 48, 293–307.

Tsurutani, B.T., Gonzalez, W.D., Lakhina, G.S., Alex, S., 2003. The extreme magnetic storm of 1–2 September 1859. J. Geophys. Res. 108 (A7), 1268. https://doi.org/10.1029/2002JA009504.

U.S. National Space Weather Program, 2013. http://www.swpc.noaa.gov/sites/default/files/images/u33/NationalSpaceWeatherProgram_MichaelBonadonna_NSWP.pdf.

Usoskin, I.G., Alanko-Huotari, K., Kovaltsov, G.A., Mursula, K., 2005. Heliospheric modulation of cosmic rays: monthly reconstruction for 1951–2004. J. Geophys. Res. 110.A12108.

Usoskin, I.G., Kromer, B., Ludlow, F., Beer, J., Friedrich, M., 2013. The AD775 cosmic event revisited: the sun is to blame. Astron. Astrophys. 55, L3. https://doi.org/10.1051/0004-6361/201321080.

Wolff, E.W., Bigler, M., Curran, M.A.J., Dibb, J.E., Frey, M.M., Legrand, M., McConnell, J.R., 2012. The Carrington event not observed in most ice core nitrate records. Geophys. Res. Lett. 39, https://doi.org/10.1029/2012GL051603. L08503.

Wolff, E.W., Bigler, M., Curran, M.A.J., Dibb, J.E., Frey, M.M., Legrand, M., McConnell, J.R., 2016. Comment on "Low time resolution analysis of polar ice cores cannot detect impulsive nitrate events" by D.F. Smart et al. J. Geophys. Res. Space Phys. 121, 1920–1924. https://doi.org/10.1002/2015JA021570.

Xapsos, M.A., Summers, G.P., Barth, J.L., Stassinopoulos, E.G., Burke, E.A., 2000. Probability model for cumulative solar proton event fluences. IEEE Trans. Nucl. Sci. 47, 486–490.

Zharkova, V.V., Shepherd, S.J., Zharkov, S.I., 2012. Principal component analysis of background and sunspot magnetic field variations during solar cycles 21–23. Mon. Not. R. Astron. Soc. 424, 2943–2953.

Zharkova, V.V., Shepherd, S.J., Popova, E.P., Zharkov, S.I., 2015. Heartbeat of the Sun from principal component analysis and prediction of solar activity on a millenium timescale. Nat. Sci. Rep. 5, 15689.

Zhou, D., Semones, E., Johnson, S., Weyland, M., 2007. Radiation measured with TEPC and CR-39 PNTDs in low Earth orbit. Adv. Space Res. 40, 1571–1574.

FURTHER READING

Benghin, V.V., Petrov, V.M., Panasyuk, M.I., et al., 2013. Results of the radiation monitoring system measurements on service module of ISS during 2009–2013. http://www.wrmiss.org/workshops/eighteenth/Benghin.pdf.

Kiraly, P., Wolfendale, A.W., 1999. Long-term particle fluence distributions and short-term observations. In: Proceedings of the 26th International Cosmic Ray Conference, Salt Lake City, USA, vol. 6. pp. 163–166.

Kuznetsov, N.V., Popova, H., Panasyuk, M.I., 2017. Empirical model of long-time variations of galactic cosmic ray particle fluxes. J. Geophys. Res. Space Phys. 122, 1463–1472.

Menn, W., Hof, M., Remer, O., et al., 2000. The absolute flux of protons and helium at the top of the atmosphere using IMAX. Astrophys. J. 533, 281–297.

Shikaze, Y., Abe, K., Anraku, K., et al., 2003. Solar modulation effect on the cosmic-ray proton spectra measured by BESS. In: Proceedings of the 28th International Cosmic Ray Conference. pp. 4027–4030.

CHAPTER 14

MAGNETOSPHERIC "KILLER" RELATIVISTIC ELECTRON DROPOUTS (REDs) AND REPOPULATION: A CYCLICAL PROCESS

Rajkumar Hajra[*], Bruce T. Tsurutani[†]

French National Center for Scientific Research (CNRS), Orléans, France California Institute of Technology, Pasadena, CA, United States[†]

CHAPTER OUTLINE

1 Introduction 374
2 Solar Wind/Interplanetary Driving and Geomagnetic Characteristics: A Schematic 375
3 Relativistic Electron Dropout and Acceleration: An Example 377
4 Solar Cycle Phase Dependence of Electron Acceleration 379
5 Maximum Energy-Level Dependence of Electron Acceleration 381
6 HILDCAA Duration Dependence of Electron Acceleration 384
7 Are CIR Storms Important? 385
8 Relativistic Electron Variation During ICME Magnetic Storms 387
    8.1 Fast Shock, Sheath, and First Magnetic Storm 387
    8.2 Magnetic Cloud (MC) and Second and Third Storms 389
    8.3 HSS and Storm Recovery Phase 389
    8.4 Relativistic Electron Flux Variability During the Complex Interplanetary Event: Shock Effects 389
    8.5 Electron Acceleration 390
9 Conclusions 390
References 391

1 INTRODUCTION

The Earth's magnetosphere is filled with energetic charged particles that gradient and curvature drift around the closed magnetic field region. These azimuthally drifting particles form part of the Earth's radiation belt known as the Van Allen belt (Van Allen and Frank, 1959; Vernov et al., 1960; Frank et al., 1963). The inner Van Allen belt ($1 \leq L < 3$) is composed of a combination of high-energy electrons ($\sim$100 keV) and very energetic protons ($\geq$10–100 MeV). The protons are produced by cosmic ray albedo neutron decay (CRAND: Singer, 1958; Dragt et al., 1966; Fennell et al., 2015; Selesnick, 2015; Selesnick et al., 2016; Su et al., 2016). The outer zone ($3 \leq L \leq 7$) is dominated by $\geq$100 keV electrons and $\sim$30–300 keV protons that are injected into the nightside magnetosphere by substorms and magnetic storms (e.g., Paulikas and Blake, 1979; Baker et al., 1979). The relativistic ($\geq$1 MeV) electrons are part of the outer belt, but their variability is not synchronous with those of the substorm and magnetic storm injected $\sim$10–100 keV electrons and $\sim$30–300 keV protons (Freeman, 1964; Paulikas and Blake, 1979; Baker et al., 1994; Friedel et al., 2002; Turner et al., 2014).

Understanding the cause and predicting the occurrence of these extremely energetic (relativistic) electrons in the Earth's outer belt are important because the particles cause hazards to Earth-orbiting spacecraft (Baker et al., 1994, 1998; Wrenn, 1995; Blake et al., 1997; Horne, 2003). The acceleration and decrease of the relativistic electrons are major aspects of extreme space weather effects. The electrons are known to be highly variable with orders of magnitude variations on time scales of a few minutes to several years. The variability depends on the solar cycle for long-term effects (Baker et al., 1986; Hajra et al., 2014c) and on solar wind and interplanetary variations for short-term effects (Tsurutani et al., 2006; Miyoshi and Kataoka, 2008, 2011; Kasahara et al., 2009; Baker et al., 2014; Hietala et al., 2014; Kilpua et al., 2015; Li et al., 2015).

The most important phase of the $\sim$11-year solar cycle for the acceleration of relativistic electrons is the declining phase (Paulikas and Blake, 1979; Baker et al., 1979, 1990; Li et al., 2001; Tsurutani et al., 2006). This is the interval where high-speed solar wind streams (HSSs) emanating from coronal holes (Sheeley et al., 1976) are dominant (see Tsurutani et al., 1995, 2006). In the HSSs, there are Alfvén wave trains (Belcher and Davis, 1971; Tsurutani and Gonzalez, 1987; Tsurutani et al., 1994, 1995) containing substantial and frequent southward interplanetary magnetic fields (IMFs) that lead to HILDCAA (high-intensity long-duration continuous auroral activity: Tsurutani and Gonzalez, 1987) events. HILDCAAs are intervals of continuous substorms and injection events (Tsurutani et al., 2004; Guarnieri, 2006; Hajra et al., 2013, 2014a; Souza et al., 2016; Mendes et al., 2017). There are only moderate intensity geomagnetic storms (-50 nT $\geq$ Dst >-100 nT: Gonzalez et al., 1994) associated with corotating interaction regions (CIRs) during this phase of the solar cycle (Tsurutani et al., 1995).

During the solar cycle maximum, intense geomagnetic magnetic storms (Dst < -100 nT) are induced by interplanetary coronal mass ejections (ICMEs) (see Tsurutani et al., 1988; Gonzalez et al., 1994; Chakraborty et al., 2008; Echer et al., 2008; Hajra et al., 2010; Hajra, 2011). Relativistic electron flux variability has been noted in and around these magnetic storm intervals (Baker et al., 1994; Li et al., 1997; Onsager et al., 2002; Horne et al., 2009). Although this phase of the solar cycle is less important from the overview of flux intensities, we will give a review of current hypotheses of loss and acceleration processes. We will also show one specific ICME storm interval for the reader.

Recently Tsurutani et al. (2016) proposed a new scenario for the relativistic electron dropout (RED) events that are not related to geomagnetic storms. These are caused by impingement of the interplanetary heliospheric plasma sheet (HPS) onto the magnetosphere. It should be noted that HPSs are located in slow solar wind streams. The HPSs and heliospheric current sheets (HCS: Smith et al., 1978) occur prior to the CIRs and HSSs (Tsurutani et al., 2006, 2016). Hajra et al. (2013, 2014c, 2015a, b) reported that the long and intense auroral activity intervals of HILDCAAs lead to relativistic electron acceleration irrespective of whether geomagnetic storms precede. We will discuss these new results in some detail. These studies indicate the predictability of the magnetospheric relativistic electrons well in advance of their occurrence.

In the present chapter we will review the outer zone Van Allen relativistic electron variation as a cyclical process. First HPSs cause the depopulation of the electrons and then later the HSSs lead to the repopulation of the electrons.

2 SOLAR WIND/INTERPLANETARY DRIVING AND GEOMAGNETIC CHARACTERISTICS: A SCHEMATIC

Fig. 1 illustrates schematically the slow solar wind and the fast solar wind interaction, associated interplanetary structures, and resulting geomagnetic effects. The interaction between the slow solar wind (on the left of the Vsw panel) and the fast solar wind (HSS, on the right) results in an interaction region characterized by the high plasma densities (Nsw), high IMF intensities (Bo), and high plasma temperatures (not shown), known as the CIR (Smith and Wolfe, 1976).

The *vertical dashed line* in Fig. 1 indicates the HCS. The high-density region adjacent to the HCS is the HPS (Winterhalter et al., 1994). The HCS is a region where the IMF reverses its polarity, that is, from an inward polarity to an outward one, or vice versa (Ness and Wilcox, 1964; Smith et al., 1978). An HCS crossing is identified by a reversal of both Bx and By components of the IMF (in either GSM or GSE coordinate systems). The HCS is accompanied by neighboring high-density cold plasma, typical of the slow solar wind. The cold plasma has been called the HPS. It should be noted that the HPS is typically part of the slow solar wind. The HPS occurs prior to the CIR and HSS as the HPS is typically "swept up" by the HSS (Tsurutani et al., 1995, 2006, 2016).

Fig. 1 shows an example of the causes of relativistic electron decreases and repopulation. The HPSs impact the magnetosphere, depleting it of the relativistic electrons (Tsurutani et al., 2016). The HPSs compress both the magnetosphere and the preexisting ~10–100 keV energetic particles within it. The betatron-accelerated protons generate coherent electromagnetic ion cyclotron (EMIC) waves in the dayside outer magnetosphere through a temperature anisotropy ($T_\perp/T_\parallel > 1$) instability. The waves in turn interact with relativistic electrons and cause the rapid loss to the atmosphere before they reach the magnetopause. By the time the CIR reached the magnetosphere, the relativistic electrons had already been lost. It is not until the HSS interval that the relativistic electrons repopulated the magnetosphere (Hajra et al., 2015a). The HSSs are accompanied by embedded Alfvén waves. The waves are convected to 1 AU and beyond by the solar wind (Belcher and Davis, 1971; Tsurutani et al., 1994). The southward component of the Alfvén waves causes magnetic reconnection at the Earth's dayside magnetopause (Dungey, 1961; Gonzalez and Mozer, 1974; Tsurutani et al., 1995), leading to substorms and convection events and energetic ~10–100 keV electron injections into the nightside sector of the magnetosphere (DeForest and McIlwain, 1971; Horne and Thorne, 1998). The temperature anisotropy of the heated electrons leads to plasma instability (Brice, 1964; Kennel and Petschek, 1966; Tsurutani and

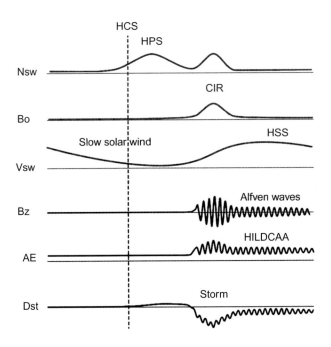

FIG. 1

A schematic of the slow stream-HSS interaction region. From top to bottom, the panels show: the solar wind density Nsw, the IMF magnitude Bo, the solar wind speed Vsw, the IMF Bz component, the geomagnetic AE, and Dst indices, respectively. The *dashed vertical line* indicates the HCS and the density associated with it (asymmetrically on the right side) is the HPS. A CIR and HSS HILDCAA are shown for context.

Modified from Tsurutani, B.T., Hajra, R., Tanimori, T., Takada, A., Bhanu, R., Mannucci, A.J., Lakhina, G.S., Kozyra, J.U., Shiokawa, K., Lee, L.C., Echer, E., Reddy, R.V., Gonzalez, W.D., 2016. Heliospheric plasma sheet (HPS) impingement onto the magnetosphere as a cause of relativistic electron dropouts (REDs) via coherent EMIC wave scattering with possible consequences for climate change mechanisms. J. Geophys. Res. 121, 10130–10156. doi: 10.1002/2016JA022499.

Lakhina, 1997), generating electromagnetic plasma waves called "chorus" (Tsurutani and Smith, 1977; Tsurutani et al., 1979, 2013; Inan et al., 1978; Meredith et al., 2001). Resonant interactions of the chorus waves with ~100 keV electrons lead to acceleration to relativistic energies (Inan et al., 1978; Horne and Thorne, 1998; Thorne et al., 2005, 2013; Summers et al., 2007; Reeves et al., 2013; Boyd et al., 2014).

The sporadic magnetic reconnection by the southward component of the interplanetary Alfvén wave train also results in the prolonged periods of intense auroral activity at Earth, which can last for days to weeks (Tsurutani et al., 1995, 2006; Gonzalez et al., 2006; Guarnieri, 2006; Kozyra et al., 2006; Turner et al., 2006; Hajra et al., 2013, 2014a,b, 2017). The auroral activity has been called HILDCAAs (Tsurutani and Gonzalez, 1987). By definition, HILDCAAs are characterized by peak AE intensity >1000 nT and a minimum duration of 2 days where AE never drops below 200 nT for >2 h at a time. Detailed studies of HILDCAA characteristics may be found in Hajra et al. (2013, 2014b). It was shown (Hajra et al., 2014c) that the lengthy and continuous intervals of AE activity are ideal for electron acceleration in the outer radiation belt.

3 RELATIVISTIC ELECTRON DROPOUT AND ACCELERATION: AN EXAMPLE

Fig. 2 shows an example of solar/interplanetary variations as well as geomagnetic and radiation belt effects during a HILDCAA event on September 15–20, 2003. This is taken and modified from Hajra et al. (2015b). The solar wind/interplanetary data were obtained from the OMNI database (http://omniweb.gsfc.nasa.gov/). These are time adjusted to take into account the solar wind convection time from the spacecraft to the bow shock. The IMFs are in the geocentric solar magnetospheric (GSM) coordinate system. The geomagnetic indices were obtained from the World Data Center for Geomagnetism, Kyoto, Japan (http://wdc.kugi.kyoto-u.ac.jp). The HILDCAA (denoted by the horizontal line in the AE panel) had a duration of ~5 days, from ~2102 UT on day 258 (September 15) to ~2203 UT on day 263 (September 20). The HILDCAA initiation was associated with a CIR, marked by the

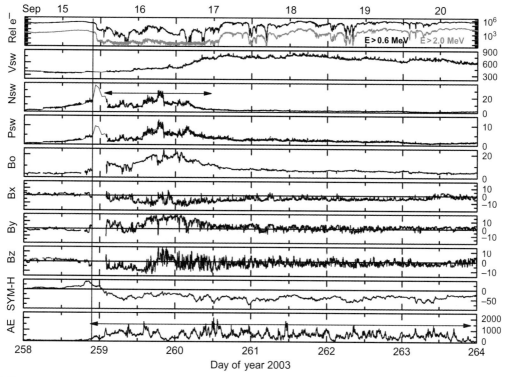

FIG. 2

HILDCAA event occurring on September 15–20, 2003. From top to bottom, the panels show the variations of $E>0.6$ (black curve) and $E>2.0$ MeV (gray curve) electron fluxes (FU) from GOES-8, solar wind speed (Vsw in km s^{-1}), plasma density (Nsw in cm^{-3}), ram pressure (Psw in nPa), IMF magnitude (Bo in nT), and Bx (nT), By (nT), and Bz (nT) components in GSM coordinate system, the SYM-H (nT) and AE (nT) indices, respectively. The data have 1 min resolution while thin lines in the Vsw, Nsw, and Psw panels show the 1 h average data. The horizontal arrows in the AE and Nsw panels indicate the HILDCAA event and the CIR interval, respectively. The vertical line shows the HCS.

compressed plasma density (Nsw) and IMF Bo from the start of day 259 to the middle of day 260. The CIR is followed by an HSS (peak Vsw of ~850 km s^{-1}). It is interesting to note that the HILDCAA event started with the CIR that led to only a weak geomagnetic activity for this case—the peak SYM-H during the CIR event was only −48 nT. This is not considered by Gonzalez et al. (1994) to be a magnetic storm.

Long-term moderate geomagnetic activity with peak SYM-H of −57 nT and peak AE of 2072 nT was recorded during the HSS interval. The onset of the HILDCAA event coincided with a north-to-southward turning of the IMF Bz component. The Bx and By components exhibited negative and positive polarity reversals, respectively, indicating an HCS crossing. This is denoted by a vertical line in the figure. In this case the HCS occurred in the slow-speed stream.

The top panel of Fig. 2 shows the variations of the integrated electron fluxes (in units of cm^{-2} s^{-1} sr^{-1}, hereafter called flux unit, FU) at two energy-levels: $E > 0.6$ and >2.0 MeV at geostationary orbit ($L=6.6$). The electron fluxes were obtained from the Geostationary Operational Environmental Satellite (GOES) 8 (http://www.ngdc.noaa.gov/stp/satellite/goes/dataaccess.html) (Onsager et al., 1996). The electron fluxes are measured by the solid-state detectors with pulse height discrimination in the energetic particle sensors onboard GOES. The data were corrected for secondary responses from other sources such as >32 MeV protons, and from directions outside the nominal detector entrance apertures. The running daily averages of the high-resolution (1 min) electron fluxes were estimated to remove diurnal variations, which are well-known features of geosynchronous flux data. This effectively removes the instrument background noise level as well (e.g., Turner and Li, 2008).

The HCS crossing at ~2136 UT on day 258 (September 15) coincides with the initiation of the electron flux "dropout" (RED). The $E > 0.6$ and >2.0 MeV electron fluxes exhibited decreases from ~19 × 10^4 to ~13 FU, and from ~23 × 10^2 to ~5 FU, respectively. The decreases took ~9.9 h and 8.2 h, respectively. The flux dropouts are time-coincident with the onset of an interplanetary high-density plasma feature. The plasma density Nsw increased from ~14 cm^{-3} at 2127 UT to the peak value of ~37 cm^{-3} at 2230 UT on day 258 (September 15). It decreased to ~9 cm^{-3} at 0240 UT on day 259 (September 16). The pressure pulse (Psw) started to rise at 2129 UT from a value of ~5 nPa until 2230 UT on day 258 when it reached the peak of ~10 nPa. The Psw slowly decreased to ~3 nPa at 0235 UT on day 259. It may be mentioned that Tsurutani et al. (2016) studied eight HPS pressure pulse events that were not followed by geomagnetic storms from solar cycle (SC) 23, and it was shown that all of them were associated with RED events.

Electron fluxes started to increase around the middle of day 260 (September 17) near the end of the CIR. The $E > 0.6$ and >2.0 MeV electron flux increases had time lags of >1 day and ~1.5 days, respectively, from the HILDCAA onset time. The entire HILDCAA interval thereafter was associated with enhanced fluxes of relativistic electrons.

An example of whistler-mode chorus wave generation during 1800–2100 UT on September 16, 2003, is shown in Fig. 3. This was obtained from the Cluster-4 satellite (Santolik et al., 2014). The red and yellow traces starting at ~1830 UT at ~4 kHz and descending to ~400 Hz by ~1945 UT are the chorus waves. The decrease in frequency was caused by the spacecraft moving to lower magnetic field strengths (chorus occurs from ~0.25 to ~0.75 the local equatorial electron cyclotron frequency: Tsurutani and Smith, 1974). These are marked by a red rectangle in the figure. The L value, expanding from 4.6 to ~13, corresponds to the entire outer zone magnetosphere between the nominal location of the plasmasphere to the magnetopause. The magnetic local time (MLT) for this

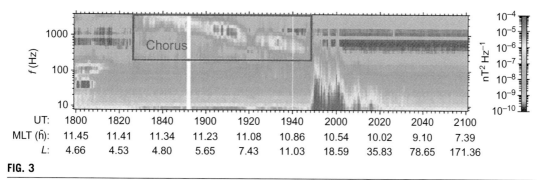

FIG. 3

The magnetic field spectrum measured by the Cluster-4 spacecraft on September 16, 2003, during the 1800–2100 UT period. The region marked by a *red rectangle* shows chorus signals.

Modified from Hajra, R., Tsurutani, B.T., Echer, E., Gonzalez, W.D., Santolík, O., 2015b. Relativistic (E> 0.6, >2.0, and >4.0 MeV) electron acceleration at geosynchronous orbit during high-intensity, long-duration, continuous ae activity (HILDCAA) events. Astrophys. J. 799, 39, doi: 10.1088/0004-637X/799/1/39.

interval varied from 11.4 to 10.5, a region where chorus is particularly intense. Hajra et al. (2015b) analysed Cluster-4 data in the $5 < L < 10$, $00 < MLT < 6$, and $6 < MLT < 12$ regions during 16 HILDCAA events occurring between 2001 and 2008. All of these 16 events were found to be associated with chorus waves in both local time sectors.

Hajra et al. (2015b) identified 35 HILDCAA events during SC 23 that had good quality solar wind/interplanetary and GOES 8 and 12 relativistic electron data available for a statistical study. It was observed that the peak fluxes of $E > 0.6$, >2.0, and >4.0 MeV electrons during HILDCAA intervals were always greater than those before HILDCAA initiation. The conclusion of Hajra et al. (2015b) was that HILDCAA events are always associated with flux enhancements of magnetospheric relativistic electrons.

4 SOLAR CYCLE PHASE DEPENDENCE OF ELECTRON ACCELERATION

To identify any possible dependence of relativistic electron acceleration at geostationary orbit ($L=6.6$) on SC phases, Hajra et al. (2014c) conducted a statistical study of 38 HILDCAA events during SC 23 (1995–2008). The events were sorted into different SC phases, namely, the ascending phase, solar maximum, descending phase, and solar minimum. The 11 events occurring during the ascending and solar maximum were combined and called the AMAX events. The remaining 27 events that occurred during the descending phase and solar minimum were combined and called the DMIN events.

Fig. 4 shows the superposed solar wind/interplanetary (Vsw, Nsw, IMF Bo, and Bz), geomagnetic parameters (Dst and AE) along with the fluxes of $E > 2$ MeV electrons during the HILDCAA intervals. The DMIN events are shown in red, and the AMAX events are shown in blue. The reference time ($t=0$) for the superposed epoch analysis is the time of HILDCAA onset. The typical CIR signatures as discussed in Figs. 1 and 2, that is, compressions in plasma and magnetic fields at the interface between the HSS and slow stream in the antisolar direction (upstream) of the HSS, may be noted. The HILDCAA

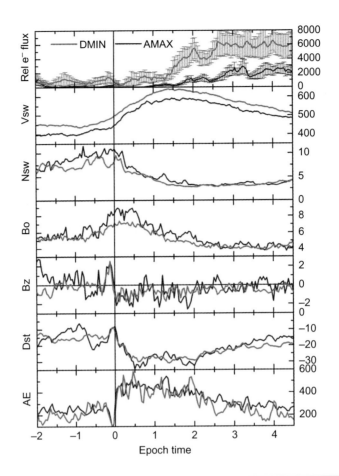

FIG. 4

Superposed time series of the relativistic electron fluxes (FU) from GOES, the solar wind speed (Vsw in km s^{-1}), the density (Nsw in cm^{-3}), the IMF magnitude (Bo in nT), the north-south component of the IMF (Bz in nT), the Dst (nT), and the AE (nT) indices during HILDCAA events. The HILDCAA onset time is taken as the zero epoch time *(vertical line)*. The time axis is in the unit of day. HILDCAA events occurring during DMIN phases and AMAX phases are shown by the *red and blue lines*, respectively.

event is found to start at the positive gradients of Vsw. This time clearly marks the sharp southward turning of the IMF Bz. The superposed HILDCAA events remarkably ordered the various interplanetary parameters.

From Fig. 4 it is clear that the solar wind/interplanetary parameters (Vsw, Nsw, IMF Bo, and Bz) exhibited more or less similar trends during both phases of the solar cycle. The Bo showed some systematic differences between the DMIN and AMAX phases, although the differences were less than the 1−σ levels. However, some significant differences were noted in the variation of Vsw. The peak HSS speed was higher for the DMIN events (∼650 km s^{-1}) than for the AMAX events (∼590 km s^{-1}). The HSSs also persisted for longer time in the DMIN phase than in the AMAX phase on average.

The Dst and AE indices are found to be likewise well ordered by the HILDCAA intervals during the DMIN and AMAX phase events. On the contrary, significant differences were noted in the variations of the relativistic electrons between the two phases. For the DMIN events, the flux decreased from the average initial flux level of >1000 FU to the lowest value of ~250 FU, ~4 h after the HILDCAA onset. The flux recovered to the preevent value within ~10 h of the dropout. For the AMAX events, the flux decreased from the average initial level of ~800 FU to the lowest value of ~32 FU at $t\sim 15$ h. It took longer (>1 day) than for the DMIN events to recover to the preevent flux value. The relativistic electron fluxes for both DMIN and AMAX phases start to increase at ~1.5 days after the HILDCAA onset. A maximum flux of ~6400 FU was recorded during the DMIN events. It is ~5 times the value of preevent fluxes. The maximum enhanced flux for the AMAX events was ~2900 FU, ~3.5 times the corresponding preevent value. The AMAX event peak flux was less than 50% of that during the DMIN events. During the HILDCAA intervals, the mean fluxes in the DMIN and AMAX phases were outside the $1-\sigma$ levels of one another. This implies that there may be statistically different values in the two phases.

Slightly higher and longer-lasting HSS during the DMIN phase than during the AMAX phase seems to be insignificant with respect to large flux differences between the phases. Higher solar wind speeds during the DMIN phases are expected to lead to higher solar wind electric fields and polar cap potentials. However, no such polar cap potential or electric field relative enhancements were noted in the analyses. Moreover, no other solar wind or magnetospheric parameter displayed any major difference between the solar cycle phases. To explain the solar cycle phase dependence of the relativistic electrons, Hajra et al. (2014c) proposed two possible explanations: either the relativistic electron loss rates are higher during the rising and maximum phases or the acceleration process is more efficient during the declining and minimum phases. Both the acceleration and loss of the electrons are known to be associated with the chorus (Horne and Thorne, 1998; Summers et al., 1998, 2004; Roth et al., 1999; Nakamura et al., 2000; Lorentzen et al., 2001; Meredith et al., 2002, 2003; Horne et al., 2003a, 2005a; Thorne et al., 2005; Tsurutani et al., 2006, 2013; Kasahara et al., 2009). Higher solar irradiance during solar maximum populates the dayside magnetosphere with higher thermal plasma densities (Jentsch, 1976). These higher plasma densities could act to reduce local wave phase speeds and enhance particle pitch angle scattering (Tsurutani and Smith, 1977), thus reducing the "seed" population of 10–100 keV electrons for relativistic electron growth. Another possibility is that the higher thermal plasma densities during solar maximum lead to the higher ratio of the electron plasma frequency to the electron gyrofrequency ω_{pe}/Ω_e. This might reduce the amount of acceleration, because the electron acceleration by whistler-mode waves is more efficient for the smaller ω_{pe}/Ω_e ratio (Summers et al., 1998; Horne et al., 2003b). It is possible that both mechanisms are contributing to the greater relativistic electron acceleration during the DMIN phase.

5 MAXIMUM ENERGY-LEVEL DEPENDENCE OF ELECTRON ACCELERATION

Hajra et al. (2015b) conducted a statistical analysis on the relativistic electrons at $L=6.6$ with $E>0.6$, >2.0, and >4.0 MeV to study the maximum energy-level dependence of the electron acceleration. Fig. 5 shows the superposed variations of the integrated electron fluxes for the SC 23 HILDCAA events. The Dst and AE panels are shown for reference. The HILDCAA onset time was taken as the zero epoch time of the superposed epoch analyses as before. The bold curves show the superposed mean values, and gray error bars show the standard $(1-\sigma)$ deviations.

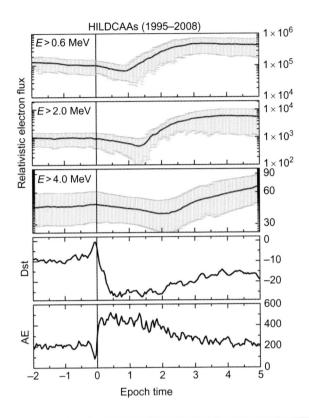

FIG. 5

Superposed time series of relativistic electron fluxes (FU) from GOES, Dst (nT), and AE (nT) indices. The energy-levels of the electrons are given in the electron panels. *Bold curves* show the mean values, and the error bars show the standard ($1-\sigma$) deviations. The zero epoch time corresponds to the initiation of HILDCAAs as in Fig. 4. The time axis is in the unit of day.

The HILDCAA interval is characterized by enhanced fluxes, and the flux enhancement is time-delayed from the HILDCAA onset. The enhancement of $E>0.6$ MeV electrons first started ~1.0 day after the statistical onset of HILDCAAs. The enhancements of $E>2.0$ and >4.0 MeV electrons occurred ~1.5 days and ~2.5 days after HILDCAA onset, respectively. Higher energy electrons took longer to respond to the HILDCAA events. After the start of flux enhancement, $E>0.6$ MeV electrons took ~2.3 days to reach the maximum level of 4.8×10^5 FU, $E>2.0$ MeV electrons took ~2.4 days to reach the maximum level of 5.8×10^3 FU, and $E>4.0$ MeV electrons reached the maximum flux level of ~7.2×10^1 FU after ~3.3 days. That is, acceleration timescales are longer for the higher energy electrons. The acceleration rates were ~1.8×10^5, 2.2×10^3, and 1.0×10^1 FU day^{-1} for $E>0.6$, >2.0, and >4.0 MeV electrons, respectively.

The estimated acceleration timescales are consistent with theoretical timescales of electron flux enhancement by whistler-mode chorus wave acceleration (e.g., Horne et al., 2003a). The delayed enhancement of higher energy electrons is a characteristic feature of the wave acceleration. It is proposed that the relativistic electrons are bootstrapped from high-energy ($\sim$100 keV) electrons. The $E>0.6$ MeV electrons are accelerated by chorus from HILDCAA-injected $E\sim$ 10–100 keV electrons, the $E>2.0$ MeV electrons are accelerated from the $E>0.6$ MeV electron population, and consequently the $E>4.0$ MeV electrons are accelerated from the $E>2.0$ MeV population. The enhancements of the energetic electron population followed by higher energy relativistic electron enhancements are also reported by other works as well (Baker et al., 1979, 1998; Li et al., 2005; Turner and Li, 2008; Boyd et al., 2014, 2016; Reeves et al., 2016).

After reaching their peak flux levels, the $E>0.6$, >2.0, and >4.0 MeV electron fluxes decayed at the rates of $\sim 0.5\times 10^5$, 0.9×10^3, and 0.6×10^1 FU day^{-1}, respectively. The decay rates are slower than the acceleration rates. From these latter values, we estimated the probable average decay timescales of $\sim$7.7, $\sim$5.5, and $\sim$4.0 days for $E>0.6$, >2.0, and >4.0 MeV electrons, respectively. These are the times taken to reach the preevent flux level from the peak flux. The decay timescales are found to be consistent with previously reported values (see Baker et al., 2004; Goldstein et al., 2005; Meredith et al., 2006 for post magnetic storm intervals).

The plausible mechanisms for the relativistic electron losses are cyclotron resonant interactions with EMIC waves (Thorne and Kennel, 1971; Horne and Thorne, 1998; Summers et al., 1998; Meredith et al., 2006; Tsurutani et al., 2016), diamagnetic influence of the partial ring current (Kim and Chan, 1997; Ukhorskiy et al., 2006), and magnetopause shadowing (West et al., 1981; Li et al., 1997; Desorgher et al., 2000; Lyons et al., 2005; Bortnik et al. 2006; Kim et al., 2008; Ohtani et al., 2009; Hietala et al., 2014; Hudson et al., 2014). EMIC waves are reported to lead to faster MeV electron losses compared to those due to chorus and hiss (e.g., Summers et al., 2007; Tsurutani et al., 2016). EMIC waves are shown to play an important role for >1 MeV electron losses, particularly at high L values ($L\geq 5$) (Albert, 2003; Meredith et al., 2006; Summers et al., 2007; Tsurutani et al., 2016).

Resonant interaction of relativistic electrons with chorus leading to pitch angle scattering and loss to the atmosphere as "microbursts" has been proposed by many authors (Abel and Thorne, 1998; Nakamura et al., 2000; Lorentzen et al., 2001; Horne and Thorne, 2003; Summers et al., 2005; Thorne et al., 2005). However, recently Tsurutani et al. (2013) have shown that typical $\sim$10–100 keV ionospheric microbursts with $\sim$0.3 s durations were associated with pitch angle scattering of the electrons in the equatorial plane where the chorus subelements are coherent. The electron interaction with coherent waves leads to a "pitch angle transport" (Tsurutani et al., 2009, 2011; Lakhina et al., 2010; Bellan, 2013) which is $\sim$3 orders of magnitude faster than that with incoherent waves as theoretically modeled by Kennel and Petschek (1966) and Tsurutani and Lakhina (1997). The problem with chorus interactions with relativistic electrons is that the interactions would have to take place away from the magnetic equator such that the local electron cyclotron frequency is considerably higher (than at the equator). Tsurutani et al. (2011) showed that off-axis chorus was only quasicoherent. Thus the cyclotron resonant interaction between relativistic electrons and quasicoherent chorus will not produce pitch angle diffusion fast enough to produce relativistic microbursts with timescales of $\sim$0.3 s (Tsurutani et al., 2013). The authors did leave open the possibility that if there were ducts guiding the chorus so that the waves remained coherent as they propagated outward away from the magnetic equator, relativistic microbursts would be possible. However, to the authors' knowledge relativistic microbursts have not been detected in the ionosphere to date.

6 HILDCAA DURATION DEPENDENCE OF ELECTRON ACCELERATION

During SC 23, the duration (D) of the HILDCAAs varied between ~2 and 5 days, with an average duration of ~2.9 days for all HILDCAAs. The events were separated into two groups: the short-duration HILDCAAs with $D \leq 3$ days and the longer-duration HILDCAAs with $D > 3$ days. Fig. 6 shows the comparison of electron flux enhancements at $L = 6.6$ during short-duration (black curves) and longer-duration (gray curves) HILDCAA events. Flux enhancements (with respect to preevent fluxes) are always larger during the longer-duration events compared to those during the short-duration ones. At the $E > 0.6$ MeV energy-level, the flux enhancements during the short- and long-duration events are ~250% and 290%, respectively. The same for the $E > 2.0$ MeV electrons are ~400% and 520%, respectively. The $E > 4.0$ MeV electron flux enhancements during the short- and long-duration events are ~27% and 82%, respectively. During the short events, the $E > 0.6$ and >2.0 MeV electrons reached slightly higher peak fluxes compared to the longer-duration ones and started to decrease after t ~3.0 and 4.0 days, respectively. During longer-duration HILDCAAs, the $E > 0.6$ and >2.0 MeV fluxes appeared to saturate around the peak values.

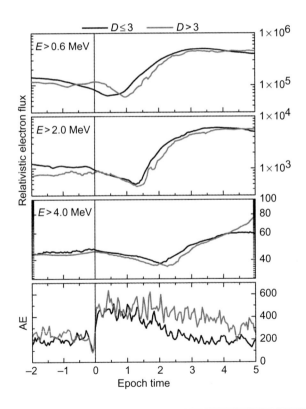

FIG. 6

Superposed time series of relativistic electron fluxes (FU) from GOES and AE (nT) index. HILDCAAs with durations $D \leq 3$ days and $D > 3$ days are shown by *black and gray curves*, respectively. The time axis is in the unit of day.

The maximum energy-level dependence on the HILDCAA duration is explained as follows. As discussed above, the intense substorm/convection events comprising the HILDCAA intervals are associated with injections of anisotropic ~10–100 keV electrons into the magnetosphere. These electrons are a source for the generation of chorus and the acceleration to even higher MeV energies. After the short-duration HILDCAAs end at ~3.0 days, the relativistic electrons undergo various loss processes, and the fluxes decay gradually. On the other hand, ~10–100 keV electrons are sporadically but continuously injected into the magnetosphere during the longer-duration HILDCAAs for longer time intervals. The electrons are accelerated to >0.6 MeV and consequently to higher ($E > 2.0$ and >4.0 MeV) energies, as proposed above. Different loss processes may occur simultaneously with the acceleration. The possible saturation-like effect observed in $E > 0.6$ MeV and $E > 2.0$ MeV electrons during the longer-HILDCAAs may be owing to a balance between acceleration and loss processes. The way to better understand the limits to this acceleration process is to make observations during extremely long HILDCAA intervals. Tsurutani et al. (1995) reported some events with ~12–25 day durations during 1973–75. It would be interesting to re-examine such older data or, better yet, see if events of this type occur during the modern Van Allen Probe epoch.

7 ARE CIR STORMS IMPORTANT?

Hajra et al. (2015a) studied the dependence of relativistic electron acceleration during HILDCAAs on the preceding geomagnetic storm main phase induced by CIRs. The HILDCAA events occurring during SC 23 were separated into: (1) events preceded by CIR-induced geomagnetic storms (SH-events) and (2) nonstorm/isolated events (H-events). The numbers of the SH- and H-events analysed in this case are 11 and 32, respectively.

Fig. 7 shows the superposed $E > 2.0$ MeV electron fluxes at $L = 6.6$, IMF Bz, Esw, SYM-H, and AE indices separately during the SH- and H-events (black and gray curves, respectively). The HILDCAA onset time was taken as the zero epoch time ($t = 0$) as before. The superposed geomagnetic indices (SYM-H and AE) display significantly different features before the HILDCAA onset and little differences during the HILDCAA interval between the SH- and H-events. The geomagnetic activity (SYM-H and AE indices) was enhanced before the SH-onset (zero epoch time) owing to the preceding geomagnetic storm main phases. A signature of the end of storm main phase ~6 h prior to the HILDCAA onset may be noted in the SYM-H variation. On the other hand, during the H-events, the SYM-H and AE indices indicate weak enhancements of ring current and auroral activity only during the HILDCAA interval. The latter interval occurred after geomagnetic calm (Tsurutani et al., 1995). The second panel of Fig. 7 shows the variation of IMF Bz. A small but significant southward Bz component may be noted before the SH onset time. This is responsible for the storm main phase. In the case of the H-events, Bz varied around 0 nT, consistent with the geomagnetic calm before event initiation. Interestingly, the initiation of the H-events was preceded by (~3 h) a prominent northward-to-southward turning of Bz. The Esw follows the variation of the IMF Bz.

The top panel of Fig. 7 shows the variation of the $E > 2.0$ MeV electron fluxes. There is no significant difference in electron fluxes between the SH- and H-events. In both cases, flux enhancements are noted to occur with time lags of ~1 day after the HILDCAA onset. The $E > 0.6$ and $E > 4.0$ MeV electron fluxes for the SH- and H-events are also similar to each other (not shown).

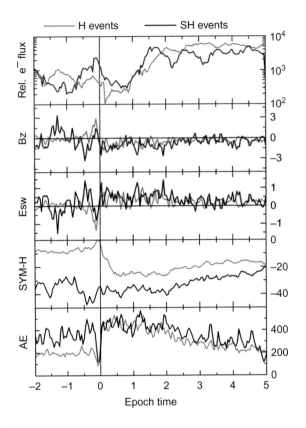

FIG. 7

Superposed time series of $E > 2.0$ MeV electron fluxes (cm^{-2} s^{-1} sr^{-1}), the IMF Bz (nT), Esw (mV m^{-1}), SYM-H (nT), and AE (nT) indices during HILDCAA events. The *gray and black curves* correspond to isolated HILDCAAs (H) and CIR storm-preceded HILDCAAs (SH), respectively. The time axis is in the unit of day.

Chorus generation is reported during geomagnetic storms (Horne and Thorne, 1998; Summers et al., 1998, 2002; Meredith et al., 2002; Horne et al., 2003a, 2005a,b, 2007). However, the particle injections during storms are deeper into the magnetosphere (stronger convection electric fields) and last for only hours. On the other hand, HSS/HILDCAAs can last for days to weeks (Tsurutani et al., 1995; Hajra et al., 2013), and presumably, the chorus lasts that long as well. The ~10–100 keV electron injection during HILDCAAs is somewhat shallow, involving only the outer portion of the magnetosphere, $L \sim 5$–10 (see Soraas et al., 2003) due to the relatively small convection electric fields (Tsurutani et al., 2006). Thus it is surmised that the electron acceleration is taking place at L values close to geosynchronous orbit. The NOAA GOES satellites might be in ideal locations to monitor events of this type. It may be mentioned that the results presented in the present review paper are based on observations by geostationary GOES satellites placed at the Earth's outer radiation belt ($L = 6.6$). Further study can be performed using measurements by Van Allen Probes that cover a wide range of L values to verify this hypothesis.

8 RELATIVISTIC ELECTRON VARIATION DURING ICME MAGNETIC STORMS

Relativistic electron losses have been noted in times during and around ICME-induced geomagnetic storms (e.g., Baker et al., 1994; Onsager et al., 2002; Reeves et al., 2003; Horne et al., 2009; Kim et al., 2010, 2011; Turner et al., 2013, 2014; Hietala et al., 2014; Hudson et al., 2014). A number of different processes have been identified: losses due to gradient drift in compressed magnetic fields such as due to shocks (Li et al., 1997; Desorgher et al., 2000; Lyons et al., 2005; Kim et al., 2008, 2010; Ohtani et al., 2009), losses due to magnetospheric inflation by the enhanced ring current (Kim and Chan, 1997; Ukhorskiy et al., 2006; Bortnik et al., 2006; Kim et al., 2010), and losses due to radial diffusion (Kellogg, 1959; Vernov et al., 1960; Fei et al., 2006; Shprits et al., 2006; Kim et al., 2011; Hietala et al., 2014). These processes are known as "magnetopause shadowing" that lead to particle losses at the magnetopause (Dessler and Karplus, 1961; West et al. 1972; Li et al., 1997; Desorgher et al., 2000; Lyons et al., 2005; Kim et al., 2008; Ohtani et al., 2009; Albert, 2014; Hietala et al., 2014; Hudson et al., 2014; Roederer and Zhang, 2014). There are also particle losses due to magnetic erosion or magnetic reconnection (Baker et al., 2013; Hudson et al., 2014; Ni et al., 2016). These latter particles are lost down tail. And finally, wave-particle interactions will lead to particle losses into the ionosphere (Thorne and Kennel, 1971; Meredith et al., 2003; Summers and Thorne, 2003; Thorne et al., 2005; Summers et al., 2007; Usanova et al., 2014; Tsurutani et al., 2016).

Particle acceleration during geomagnetic storms can be accomplished in several different ways. Particles can be energized by nightside storm time convection electric fields (Li et al., 2003; Hori et al., 2005; Xie et al., 2006; Wygant et al., 2013). PC5 oscillations can lead to radial diffusion (Schulz and Lanzerotti, 1974; Hudson et al., 2000). The electrons can be energized by substorm injection during the storms (e.g., Baker et al., 1998, 2014; Miyoshi et al., 2013; Thorne et al., 2013; Foster et al., 2014; Jaynes et al., 2015). Chorus generated during the storms and substorms can accelerate the electrons (Horne and Thorne, 1998; Summers et al., 1998; Roth et al., 1999; Meredith et al., 2003; Thorne et al., 2013). Other waves have been mentioned for electron acceleration as well (Selesnick and Blake, 2000; Meredith et al., 2002; Summers et al., 2002; Miyoshi et al., 2003; Horne et al., 2007; Thorne et al., 2013; Tsurutani et al., 2013; Boyd et al., 2014).

Fig. 8 shows an example of a geomagnetic storm that occurred during March 17–21, 2015. This is the biggest magnetic storm of SC 24 with peak SYM-H intensity of −234 nT. The associated interplanetary phenomena are characterized by an interplanetary shock and a magnetic cloud (MC) followed by a HSS.

8.1 FAST SHOCK, SHEATH, AND FIRST MAGNETIC STORM

The fast-forward shock at 0448 UT on March 17 (marked by the vertical dotted line in the figure) is identified by a jump in Vsw from $\sim$397 to 518 km s^{-1} (Fig. 8D), in Nsw from $\sim$11 to 36 cm^{-3} (Fig. 8E), in Psw from $\sim$4 to 19 nPa (Fig. 8E) and in IMF Bo from $\sim$8 to 28 nT (Fig. 8G). The shock is related to the C9-class solar flare erupted by the sunspot AR2297 on March 16. Higher resolution (3 s) interplanetary data from the WIND spacecraft (http://wind.gsfc.nasa.gov/mfi_swe_plot.php) were used to perform interplanetary shock analyses. The shock normal direction is calculated using the Abraham-Schrauner (1972) method, and then the Rankine-Hugoniot conservation equations are used to get the shock speed (Smith, 1985; Tsurutani and Lin, 1985; Tsurutani et al., 2011). The shock was found to have a magnetosonic Mach number of 2.9 and a shock normal angle of 63 degrees relative

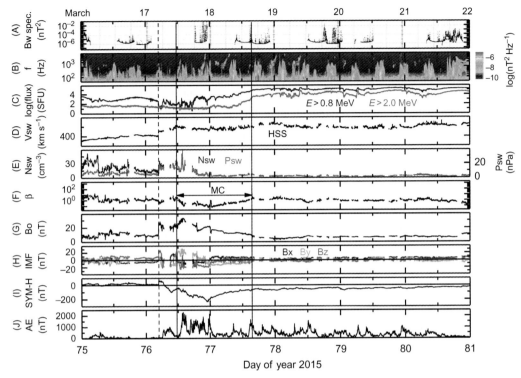

FIG. 8

Geomagnetic storm occurring on March 17–21, 2015. From top to bottom, the panels show the variations of (A) the lower-band chorus wave amplitudes integrated over 0.1–0.5fce measured by the EMFISIS instrument on VAP-A and VAP-B, (B) the frequency-time spectrogram of wave magnetic field spectral density in the WFR channel observed by VAP-A, (C) $E > 0.8$ *(blue)* and $E > 2.0$ MeV *(red)* electron fluxes measured by GOES-15, (D) solar wind speed Vsw, (E) plasma density Nsw (*blue*, scale on the left) and ram pressure Psw (*red*, scale on the right), (F) plasma β, (G) IMF magnitude Bo, (H) Bx *(blue)*, By *(green)*, and Bz *(red)* components of IMF in GSM coordinate system, (I) the SYM-H, and (J) AE indices, respectively. The *vertical dotted line* indicates an interplanetary fast-forward shock. The region between the *dotted line* and the first *solid line* is the sheath interval. This is followed by an MC, the region between the two *solid lines*. This is also marked by a *horizontal arrow* in the plasma β panel. An HSS follows the MC.

to the upstream magnetic field. The shock and the high plasma density sunward of it caused a sudden impulse (SI⁺) in SYM-H of ∼+67 nT (Fig. 8I). The shock triggered a nightside substorm with a peak AE of ∼1016 nT at 0852 UT (Fig. 8J). This is the storm initial phase where the high ram pressure behind the interplanetary shock compressed the magnetosphere.

Immediately after the shock is a magnetic sheath continuing until ∼1136 UT on March 17. The sheath is characterized by multiple IMF Bz changes with a peak value of ∼−22 nT (Fig. 8H), high plasma density Nsw ∼30 cm^{-3} (Fig. 8E), and high plasma β ∼7 (Fig. 8F). This is followed by an MC, from ∼1136 UT on March 17 to ∼1533 UT on March 18 (marked by a horizontal arrow in the plasma β panel). The MC is identified by the low plasma β ~0.05 (Fig. 8F), positive-negative rotation in Bx and negative-positive rotation in By (Fig. 8H). The IMF Bz turns southward from northward polarity during this interval (Fig. 8H). The MC had a peak southward IMF of ~−26 nT.

The first storm main phase started at ~0550 UT on March 17 following the interplanetary shock. SYM-H reached a local minimum of −101 nT at ~0937 UT. This storm main phase is caused by sheath southward IMF fields (peak ~−22 nT) immediately after the shock.

8.2 MAGNETIC CLOUD (MC) AND SECOND AND THIRD STORMS

Fig. 8I shows second storm main phase occurred with a deeper minimum of SYM-H = −177 nT at ~1728 UT. The third storm main phase with peak value of SYM-H = −234 nT was reached at ~2247 UT on March 17. Both storms are caused by southward IMFs within an MC. The second and third storm main phases correspond to peak IMF southward fields of ~−26 nT and −20 nT at ~1307 UT and 2153 UT on March 17, respectively.

The three storm main phases had a total duration of ~17 h. The peak AE intensity was ~2300 nT and occurred near the beginning of the second storm main phase. The final recovery phase of the storm starts with a northward turning of the IMF Bz to ~0 nT at ~2247 UT on March 17. It is believed that this is still part of the MC because of the low plasma β.

8.3 HSS AND STORM RECOVERY PHASE

The storm "recovery" corresponds to the HSS interval (Fig. 8D) following the MC. A long-duration and slow "recovery" in the SYM-H index is apparent in the figure from 2247 UT on March 17 to ~0033 UT on March 23. The "recovery phase" is characterized by a HILDCAA event.

An HSS follows immediately after the MC and extends from the end of the cloud at ~1533 UT on March 18 until the end of the interval shown (Fig. 8D). This is consistent with the detection of a solar coronal hole during March 15–19. The HSS had a peak speed of ~690 km s^{-1} at ~2301 UT on March 18. The magnetic field intensity was ~33 nT. The HSS carried Alfvén waves characterized by northward-southward fluctuations in IMF Bz with peak southward component of ~−7 nT.

The magnetic storm "recovery" is ~5 days, much longer than the typical ~10 h for the decay time scale of ring current particles by charge exchange, Coulomb collisions, wave-particle interactions, and plasma convection out of the dayside magnetopause (e.g., Kozyra et al., 1998). The southward component of the Alfvén waves leads to short bursts of magnetic reconnection (Tsurutani et al., 1995) causing the injection of plasma into the midnight sector of the magnetosphere, leading to a near steady state of the ring current energy (Soraas et al., 2003).

8.4 RELATIVISTIC ELECTRON FLUX VARIABILITY DURING THE COMPLEX INTERPLANETARY EVENT: SHOCK EFFECTS

Fig. 8C shows relativistic $E > 0.8$ and > 2.0 MeV electrons measured by GOES 15 at $L = 6.6$. The $E > 0.8$ MeV electron fluxes decrease by ~2 orders of magnitude and the $E > 2.0$ MeV electrons by ~1 order of magnitude during the main phase. The decrease in the $E > 0.8$ MeV electrons is quite abrupt and coincident with occurrence of the shock (shown by a vertical dotted line).

8.5 ELECTRON ACCELERATION

During the storm recovery phase, which is accompanied by the HSS, the relativistic electron fluxes are noted to be larger than the storm main phase as well as than the prestorm period (i.e., there is an overall flux enhancement). The increase is by ~4 orders of magnitude.

The wave activity during the geomagnetic storm was observed by NASA's twin Van Allen Probes: VAP-A and VAP-B (http://vanallenprobes.jhuapl.edu/index.php) (Mauk et al., 2012). Fig. 8A shows the lower-band chorus waves observed by the Electric and Magnetic Field Instrument Suit and Integrated Science (EMFISIS) (Kletzing et al., 2013; Wygant et al., 2013). The chorus wave amplitudes were calculated by integrating the magnetic wave power spectral density over 0.1–0.5 fce, where fce is the electron cyclotron frequency. The wave power spectral density obtained from the waveform receiver (WFR) on the EMFISIS instrument (Kletzing et al., 2013) is shown in Fig. 8B. Enhanced whistler-mode wave (hiss and chorus) activities may be noted during the geomagnetic storm recovery phase (Fig. 8B). The waves had the peak spectral density of $\sim 10^{-3}$ nT^2 Hz^{-1}. The waves are seen in all passes of the VAPs, indicating possibly continuous wave activity throughout the recovery phase.

9 CONCLUSIONS

The outer radiation belt relativistic electron variation is of physical interest in understanding the radiation belt dynamics and space weather in general. Also, these have practical importance for hazardous effects to orbiting spacecraft. These are known as "killer electrons" that can cause outages and failures of satellite electronics (Baker et al., 1994, 2004; Wrenn, 1995; Blake et al., 1997; Horne, 2003). The works by Hajra et al. (2013, 2014c, 2015a,b) and Tsurutani et al. (2016), reviewed above, clearly present the outer radiation belt relativistic electron variation as a cyclical process controlled by interplanetary phenomena.

The declining phase of the solar cycle is the most important period for large-scale variability in the relativistic electron loss and acceleration processes. Interplanetary space is dominated by HSSs during this phase. The interplanetary HPS impinging on the magnetosphere leads to the loss of the relativistic electrons through the interactions with EMIC waves and magnetopause shadowing. The HPS occurs in the slow solar wind followed by the CIR and HSS. The HSS, which carries the interplanetary Alfvén waves, causes substorms and convection events and energetic ~10–100 keV electron injections into the nightside sector of the magnetosphere. The electrons are accelerated to relativistic energies through resonant interactions with electromagnetic plasma waves called chorus.

The magnetic reconnection between the geomagnetic field and the southward component of the Alfvén waves results in intense auroral activity known as HILDCAA events. The HILDCAAs occurring in the geomagnetic storm recovery phase or in geomagnetic moderately quiet conditions are shown to be associated with magnetospheric electron acceleration to relativistic energies. Most interestingly, the electron acceleration starts ~1 day or more after the HILDCAA onset. This indicates the probability that magnetospheric relativistic electron acceleration may be predicted more than 1 day in advance using ground-based observations of auroral activity (HILDCAAs) during HSSs. Prediction models of geomagnetic indices based on solar wind parameters may be utilized for this purpose (e.g., Temerin and Li, 2006; Li et al., 2007; Luo et al., 2013).

Hajra et al. (2015b) noted that the acceleration of $E > 4.0$ MeV electrons is delayed from the acceleration of $E > 2.0$ MeV electrons and thus a "bootstrap" process was in effect. It will be interesting to find out what the upper limit of relativistic electron energy might be from the chorus-energetic electron interaction process. Tsurutani et al. (1995) noted extremely long-lasting HSSs during 1973–75, a feature that has never recurred since that epoch. Is it possible that even higher energies could be produced, and if so, what effects might that have on Earth-orbiting spacecraft?

The case study of an ICME storm shows clear evidence of relativistic electron losses by an interplanetary shock impingement. The loss process is most probably magnetopause shadowing and wave-particle interactions with EMIC waves (Tsurutani et al., 2016, and references therein). During the storm recovery, the one clear cause of relativistic electron repopulation is chorus associated with the HSS/HILDCAA event following the ICME. The flux levels in the storm recovery phase are larger than the preshock levels in this case. Other contributors to relativistic electron repopulation during magnetic storms would be (1) storm time convection electric fields, (2) substorms injections during and after the storm, and (3) chorus and other wave mode acceleration.

The reader can note some of the similarities between the ICME storm case and the low-speed high-speed solar declining phase case. For example, if it is shocks that cause the particle losses (instead of the HPS) and then injection and chorus or PC5 acceleration, then one notes a clear analogy.

REFERENCES

Abel, B., Thorne, R.M., 1998. Electron scattering loss in the Earth's inner magnetosphere: 1 Dominant physical processes. J. Geophys. Res. 103, 2385–2396. https://doi.org/10.1029/97JA02919.

Abraham-Schrauner, B., 1972. Determination of magnetohydrodynamic shock normal. J. Geophys. Res. 77, 736–739. https://doi.org/10.1029/JA077i004p00736.

Albert, J.M., 2003. Evaluation of quasi-linear diffusion coefficients for EMIC waves in a multispecies plasma. J. Geophys. Res. 108, 1249. https://doi.org/10.1029/2002JA009792.

Albert, J.M., 2014. Radiation diffusion simulations of the 20 September 2007 radiation belt dropout. Ann. Geophys. 32, 925–934. https://doi.org/10.5194/angeo-32-925-2014.

Baker, D.N., Higbie, P.R., Belian, R.D., Hones Jr., E.W., 1979. Do Jovian electrons influence the terrestrial outer radiation zone? Geophys. Res. Lett. 6, 531–534. https://doi.org/10.1029/GL006i006p00531.

Baker, D.N., Blake, J.B., Klebesadel, R.W., Higbie, P.R., 1986. Highly relativistic electrons in the Earth's outer magnetosphere: 1. Life-times and temporal history 1979–1984. J. Geophys. Res. 91, 4265–4276. https://doi.org/10.1029/JA091iA04p04265.

Baker, D.N., McPherron, R.L., Cayton, T.E., Klebesadel, R.W., 1990. Linear prediction filter analysis of relativistic electron properties at 6.6 RE. J. Geophys. Res. 95, 15133–15140.

Baker, D.N., Blake, J.B., Callis, L.B., Cummings, J.R., Hovestadt, D., Kanekal, S., Klecker, B., Mewaldt, R.A., Zwickl, R.D., 1994. Relativistic electron acceleration and decay time scales in the inner and outer radiation belts: SAMPEX. Geophys. Res. Lett. 21, 409–412.

Baker, D.N., Li, X., Blake, J.B., Kanekal, S., 1998. Strong electron acceleration in the Earth's magnetosphere. Adv. Space Res. 21, 609–613.

Baker, D.N., Kanekal, S.G., Li, X., Monk, S.P., Goldstein, J., Burch, J.L., 2004. An extreme distortion of the Van Allen belt arising from the 'Hallowe'en' solar storm in 2003. Nature 432, 878–881.

Baker, D.N., Kanekal, S.G., Hoxie, V.C., Henderson, M.G., Li, X., Spence, H.E., Elkington, S.R., Friedel, R.H.W., Goldstein, J., Hudson, M.K., Reeves, G.D., Thorne, R.M., Kletzing, C.A., Claudepierre, S.G., 2013.

A long-lived relativistic electron storage ring embedded within the Earth's outer Van Allen radiation zone. Science 340, 186–190. https://doi.org/10.1126/science.1233518.

Baker, D.N., Jaynes, A.N., Li, X., Henderson, M.G., Kanekal, S.G., Reeves, G.D., Spence, H.E., Claudepierre, S.G., Fennell, J.F., Hudson, M.K., Thorne, R.M., Foster, J.C., Erickson, P.J., Malaspina, D.M., Wygant, J.R., Boyd, A., Kletzing, C.A., Drozdov, A., Shprits, Y.Y., 2014. Gradual diffusion and punctuated phase space density enhancements of highly relativistic electrons: Van Allen Probes observations. Geophys. Res. Lett. 41, 1351–1358. https://doi.org/10.1002/2013GL058942.

Belcher, J.W., Davis Jr., L., 1971. Large-amplitude Alfvén waves in the interplanetary medium: 2. J. Geophys. Res. 76, 3534–3563. https://doi.org/10.1029/JA076i016p03534.

Bellan, P.M., 2013. Pitch angle scattering of an energetic magnetized particle by a circularly polarized electromagnetic wave. Phys. Plasmas 20. 042117. https://doi.org/10.1063/1.4801055.

Blake, J.B., Baker, D.N., Turner, N., Ogilvie, K.W., Lepping, R.P., 1997. Correlation of changes in the outer-zone relativistic electron population with upstream solar wind and magnetic field measurements. Geophys. Res. Lett. 24, 927–929.

Bortnik, J., Thorne, R.M., O'Brien, T.P., Green, J.C., Strangeway, R.J., Shprits, Y.Y., Baker, D.N., 2006. Observation of two distinct, rapid loss mechanisms during the 20 November 2003 radiation belt dropout event. J. Geophys. Res. 111. A12216. https://doi.org/10.1029/2006JA011802.

Boyd, A.J., Spence, H.E., Claudepierre, S.G., Fennell, J.F., Blake, J.B., Baker, D.N., Reeves, G.D., Turner, D.L., 2014. Quantifying the radiation belt seed population in the March 17, 2013 electron acceleration event. Geophys. Res. Lett. 41, 2275–2281. https://doi.org/10.1002/2014GL059626.

Boyd, A.J., Spence, H.E., Huang, C.L., Reeves, G.D., Baker, D.N., Turner, D.L., Claudepierre, S.G., Fennell, J.F., Blake, J.B., Shprits, Y.Y., 2016. Statistical properties of the radiation belt seed population. J. Geophys. Res. 121, 7636–7646. https://doi.org/10.1002/2016JA022652.

Brice, N., 1964. Fundamentals of very low frequency emission generation mechanisms. J. Geophys. Res. 69, 4515–4522. https://doi.org/10.1029/JZ069i021p04515.

Chakraborty, S.K., Hajra, R., Paul, A., 2008. Ionosphere near the anomaly crest in Indian zone during magnetic storm on 13–14 March 1989. Indian J. Radio Space Phys. 37, 396–407.

DeForest, S.E., McIlwain, C.E., 1971. Plasma clouds in the magnetosphere. J. Geophys. Res. 76, 3587–3611. https://doi.org/10.1029/JA076i016p03587.

Desorgher, L., Bühler, P., Zehnder, A., Flückiger, E.O., 2000. Simulation of the outer radiation belt electron flux decrease during the March 26, 1995, magnetic storm. J. Geophys. Res. 105, 21211–21223. https://doi.org/10.1029/2000JA900060.

Dessler, A.J., Karplus, R., 1961. Some effects of diamagnetic ring currents on Van Allen radiation. J. Geophys. Res. 66, 2289–2295. https://doi.org/10.1029/JZ066i008p02289.

Dragt, A.J., Austin, M.M., White, R.S., 1966. Cosmic ray and solar proton albedo neutron decay injection. J. Geophys. Res. 71, 1293–1304.

Dungey, J.W., 1961. Interplanetary magnetic field and the auroral zones. Phys. Rev. Lett. 6, 47–48.

Echer, E., Gonzalez, W.D., Tsurutani, B.T., 2008. Interplanetary conditions leading to superintense geomagnetic storms (Dst$\leq$-250 nT) during solar cycle 23. Geophys. Res. Lett. 35. L06S03. https://doi.org/10.1029/2007GL031755.

Fei, Y., Chan, A.A., Elkington, S.R., Wiltberger, M.J., 2006. Radial diffusion and MHD particle simulations of relativistic electron transport by ULF waves in the September 1998 storm. J. Geophys. Res. 111. A12209. https://doi.org/10.1029/2005JA011211.

Fennell, J.F., Claudepierre, S.G., Blake, J.B., O'Brien, T.P., Clemmons, J.H., Baker, D.N., Spence, H.E., Reeves, G.D., 2015. Van Allen Probes show that the inner radiation zone contains no MeV electrons: ECT/MagEIS data. Geophys. Res. Lett. 42, 1283–1289. https://doi.org/10.1002/2014GL062874.

Foster, J.C., Erickson, P.J., Baker, D.N., Claudepierre, S.G., Kletzing, C.A., Kurth, W., Reeves, G.D., Thaller, S.A., Spence, H.E., Shprits, Y.Y., Wygant, J.R., 2014. Prompt energization of relativistic and highly relativistic electrons during a substorm interval: Van Allen Probes observations. Geophys. Res. Lett. 41, 20–25. https://doi.org/10.1002/2013GL058438.

Frank, L., Van Allen, J.A., Whelpley, W.A., Craven, J.D., 1963. Absolute intensities of geomagnetically trapped particles with Explorer 14. J. Geophys. Res. 68, 1573–1579.

Freeman Jr., J.W., 1964. The morphology of the electron distribution in the outer radiation zone and near the magnetospheric boundary as observed by Explorer 12. J. Geophys. Res. 69, 1691–1723. https://doi.org/10.1029/JZ069i009p01691.

Friedel, R.H.W., Reeves, G.D., Obara, T., 2002. Relativistic electron dynamics in the inner magnetosphere—a review. J. Atmos. Sol. Terr. Phys. 64, 265–282. https://doi.org/10.1016/S1364-6826(01)00088-8.

Goldstein, J., Kanekal, S.G., Baker, D.N., Sandel, B.R., 2005. Dynamic relationship between the outer radiation belt and the plasmapause during March–May 2001. Geophys. Res. Lett. 32. L15104. https://doi.org/10.1029/2005GL023431.

Gonzalez, W.D., Mozer, F.S., 1974. A quantitative model for the potential resulting from reconnection with an arbitrary interplanetary magnetic field. J. Geophys. Res. 79, 4186–4194. https://doi.org/10.1029/JA079i028p04186.

Gonzalez, W.D., Joselyn, J.A., Kamide, Y., Kroehl, H.W., Rostoker, G., Tsurutani, B.T., Vasyliunas, V., 1994. What is a geomagnetic storm? J. Geophys. Res. 99, 5771–5792.

Gonzalez, W.D., Guarnieri, F.L., Clua-Gonzalez, A.L., Echer, E., Alves, M.V., Oginoo, T., Tsurutani, B.T., 2006. Magnetospheric energetics during HILDCAAs. In: Tsurutani, B.T. et al. (Ed.), Recurrent Magnetic Storms: Corotating Solar Wind Streams. Geophysical Monograph Series, vol. 167. AGU, Washington, DC, pp. 175–182. https://doi.org/10.1029/167GM15.

Guarnieri, F.L., 2006. The nature of auroras during high-intensity long-duration continuous AE activity (HILDCAA) events: 1998–2001. In: Tsurutani, B.T. et al. (Ed.), Recurrent Magnetic Storms: Corotating Solar Wind Streams. Geophysical Monograph Series, vol. 167. AGU, Washington, DC, pp. 235–243.

Hajra, R., 2011. A Study on the Variability of Total Electron Content Near the Crest of the Equatorial Anomaly in the Indian Zone (Ph.D. thesis). University of Calcutta, India.

Hajra, R., Chakraborty, S.K., DasGupta, A., 2010. Ionospheric effects near the magnetic equator and the anomaly crest of the Indian longitude zone during a large number of intense geomagnetic storms. J. Atmos. Sol. Terr. Phys. 72, 1299–1308.

Hajra, R., Echer, E., Tsurutani, B.T., Gonzalez, W.D., 2013. Solar cycle dependence of High-Intensity Long-Duration Continuous AE Activity (HILDCAA) events, relativistic electron predictors? J. Geophys. Res. 118, 5626–5638. https://doi.org/10.1002/jgra.50530.

Hajra, R., Echer, E., Tsurutani, B.T., Gonzalez, W.D., 2014a. Solar wind-magnetosphere energy coupling efficiency and partitioning: HILDCAAs and preceding CIR storms during solar cycle 23. J. Geophys. Res. 119, 2675–2690. https://doi.org/10.1002/2013JA019646.

Hajra, R., Echer, E., Tsurutani, B.T., Gonzalez, W.D., 2014b. Superposed epoch analyses of HILDCAAs and their interplanetary drivers: solar cycle and seasonal dependences. J. Atmos. Sol. Terr. Phys. 121, 24–31.

Hajra, R., Tsurutani, B.T., Echer, E., Gonzalez, W.D., 2014c. Relativistic electron acceleration during high-intensity, long-duration, continuous AE activity (HILDCAA) events: solar cycle phase dependences. Geophys. Res. Lett. 41, 1876–1881. https://doi.org/10.1002/2014GL059383.

Hajra, R., Tsurutani, B.T., Echer, E., Gonzalez, W.D., Brum, C.G.M., Vieira, L.E.A., Santolik, O., 2015a. Relativistic electron acceleration during HILDCAA events: are precursor CIR magnetic storms important? Earth Planets Space 67, 109. https://doi.org/10.1186/s40623-015-0280-5.

Hajra, R., Tsurutani, B.T., Echer, E., Gonzalez, W.D., Santolik, O., 2015b. Relativistic (E > 0.6, > 2.0, and > 4.0 MeV) electron acceleration at geosynchronous orbit during high-intensity, long-duration, continuous ae activity (HILDCAA) events. Astrophys. J. 799 (39). https://doi.org/10.1088/0004-637X/799/1/39.

Hajra, R., Tsurutani, B.T., Brum, C.G.M., Echer, E., 2017. High-speed solar wind stream effects on the topside ionosphere over Arecibo: a case study during solar minimum. Geophys. Res. Lett. 44, 7607–7617. https://doi.org/10.1002/2017GL073805.

Hietala, H., Kilpua, E.K.J., Turner, D.L., Angelopoulos, V., 2014. Depleting effects of ICME-driven sheath regions on the outer electron radiation belt. Geophys. Res. Lett. 41, 2258–2265. https://doi.org/10.1002/2014GL059551.

Hori, T., Lui, A.T.Y., Ohtani, S., Brandt, P.C., Mauk, B.H., McEntire, R.W., Maezawa, K., Mukai, T., Kasaba, Y., Hayakawa, H., 2005. Storm-time convection electric field in the near-Earth plasma sheet. J. Geophys. Res. 110. A04213. https://doi.org/10.1029/2004JA010449.

Horne, R.B., 2003. Rationale and requirements for a European space weather programme. In: Proceedings of ESA Workshop, ESTEC, Noordwijk, The Netherlands.

Horne, R.B., Thorne, R.M., 1998. Potential waves for relativistic electron scattering and stochastic acceleration during magnetic storms. Geophys. Res. Lett. 25, 3011–3014. https://doi.org/10.1029/98GL01002.

Horne, R.B., Thorne, R.M., 2003. Relativistic electron acceleration and precipitation during resonant interactions with whistler-mode chorus. Geophys. Res. Lett. 30, 1527. https://doi.org/10.1029/2003GL016973.

Horne, R.B., Meredith, N.P., Thorne, R.M., Heynderickx, D., Iles, R.H.A., Anderson, R.R., 2003a. Evolution of energetic electron pitch angle distributions during storm time electron acceleration to megaelectronvolt energies. J. Geophys. Res. 108, 1016. https://doi.org/10.1029/2001JA009165.

Horne, R.B., Glauert, S.A., Thorne, R.M., 2003b. Resonant diffusion of radiation belt electrons by whistler-mode chorus. Geophys. Res. Lett. 30, 1493. https://doi.org/10.1029/2003GL016963.

Horne, R.B., Thorne, R.M., Glauert, S.A., Albert, J.M., Meredith, N.P., Anderson, R.R., 2005a. Timescale for radiation belt electron acceleration by whistler mode chorus waves. J. Geophys. Res. 110. A03225. https://doi.org/10.1029/2004JA010811.

Horne, R.B., Thorne, R.M., Shprits, Y.Y., Meredith, N.P., Glauert, S.A., Smith, A.J., Kanekal, S.G., Baker, D.N., Engebretson, M.J., Posch, J.L., Spasojevic, M., Inan, U.S., Pickett, J.S., Decreau, P.M.E., 2005b. Wave acceleration of electrons in the Van Allen radiation belts. Nature 437 (7056), 227–230. https://doi.org/10.1038/nature03939.

Horne, R.B., Thorne, R.M., Glauert, S.A., Meredith, N.P., Pokhotelov, D., Santolik, O., 2007. Electron acceleration in the Van Allen radiation belts by fast magnetosonic waves. Geophys. Res. Lett. 34. L17107. https://doi.org/10.1029/2007GL030267.

Horne, R.B., Lam, M.M., Green, J.C., 2009. Energetic electron precipitation from the outer radiation belt during geomagnetic storms. Geophys. Res. Lett. 36. L19104. https://doi.org/10.1029/2009GL040236.

Hudson, M.K., Elkington, S.R., Lyon, J.G., Goodrich, C.C., 2000. Increase in relativistic electron flux in the inner magnetosphere: ULF wave mode structure. Adv. Space Res. 25, 2327–2337. https://doi.org/10.1016/S0273-1177(99)00518-9.

Hudson, M.K., Baker, D.N., Goldstein, J., Kress, B.T., Paral, J., Toffoletto, F.R., Wiltberger, M., 2014. Simulated magnetopause losses and Van Allen Probe flux dropouts. Geophys. Res. Lett. 41, 1113–1118. https://doi.org/10.1002/2014GL059222.

Inan, U.S., Bell, T.F., Helliwell, R.A., 1978. Nonlinear pitch angle scattering of energetic electrons by coherent VLF waves in the magnetosphere. J. Geophys. Res. 83, 3235–3253.

Jaynes, A.N., Baker, D.N., Singer, H.J., Rodriguez, J.V., Loto'aniu, T.M., Ali, A.F., Elkington, S.R., Li, X., Kanekal, S.G., Claudepierre, S.G., Fennell, J.F., Li, W., Thorne, R.M., Kletzing, C.A., Spence, H.E., Reeves, G.D., 2015. Source and seed populations for relativistic electrons: their roles in radiation belt changes. J. Geophys. Res. 120, 7240–7254. https://doi.org/10.1002/2015JA021234.

Jentsch, V., 1976. Electron precipitation in morning sector of the auroral zone. J. Geophys. Res. 81, 135–146.

Kasahara, Y., Miyoshi, Y., Omura, Y., Verkhoglyadova, O.P., Nagano, I., Kimura, I., Tsurutani, B.T., 2009. Simultaneous satellite observations of VLF chorus, hot and relativistic electrons in a magnetic storm "recovery" phase. Geophys. Res. Lett. 36. L01106. https://doi.org/10.1029/2008GL036454.

Kellogg, P.J., 1959. Van Allen radiation of solar origin. Nature 183, 1295–1297. https://doi.org/10.1038/1831295a0.

Kennel, C.F., Petschek, H.E., 1966. Limit on stable trapped particle fluxes. J. Geophys. Res. 71, 1–28. https://doi.org/10.1029/JZ071i001p00001.

Kilpua, E.K.J., Hietala, H., Turner, D.L., Koskinen, H.E.J., Pulkkinen, T.I., Rodriguez, J.V., Reeves, G.D., Claudepierre, S.G., Spence, H.E., 2015. Unraveling the drivers of the storm time radiation belt response. Geophys. Res. Lett. 42, 3076–3084. https://doi.org/10.1002/2015GL063542.

Kim, H.J., Chan, A.A., 1997. Fully adiabatic changes in storm time relativistic electron fluxes. J. Geophys. Res. 102, 22107–22116. https://doi.org/10.1029/97JA01814.

Kim, K.C., Lee, D.Y., Kim, H.J., Lyons, L.R., Lee, E.S., Öztürk, M.K., Choi, C.R., 2008. Numerical calculations of relativistic electron drift loss effect. J. Geophys. Res. 113. A09212. https://doi.org/10.1029/2007JA013011.

Kim, K.C., Lee, D.Y., Kim, H.J., Lee, E.S., Choi, C.R., 2010. Numerical estimates of drift loss and Dst effect for outer radiation belt relativistic electrons with arbitrary pitch angle. J. Geophys. Res. 115. A03208. https://doi.org/10.1029/2009JA014523.

Kim, K.C., Lee, D.Y., Shprits, Y., Kim, H.J., Lee, E., 2011. Electron flux changes in the outer radiation belt by radial diffusion during the storm recovery phase in comparison with the fully adiabatic evolution. J. Geophys. Res. 116. A09229. https://doi.org/10.1029/2011JA016642.

Kletzing, C.A., Kurth, W.S., Acuna, M., MacDowall, R.J., Torbert, R.B., Averkamp, T., Bodet, D., Bounds, S.R., Chutter, M., Connerney, J., Crawford, D., Dolan, J.S., Dvorsky, R., Hospodarsky, G.B., Howard, J., Jordanova, V., Johnson, R.A., Kirchner, D.L., Mokrzycki, B., Needell, G., Odom, J., Mark, D., Pfaff Jr., R., Phillips, J.R., Piker, C.W., Remington, S.L., Rowland, D., Santolik, O., Schnurr, R., Sheppard, D., Smith, C.W., Thorne, R.M., Tyler, J., 2013. The Electric and Magnetic Field Instrument Suite and Integrated Science (EMFISIS) on RBSP. Space Sci. Rev. 179, 127–181.

Kozyra, J.U., Jordanova, V.K., Borovsky, J.E., Thomsen, M.F., Knipp, D.J., Evans, D.S., McComas, D.J., Cayto, T.E., 1998. Effects of a high-density plasma sheet on ring current development during the November 2–6, 1993, magnetic storm. J. Geophys. Res. 103, 26285–26305.

Kozyra, J.U., Criwley, G., Emery, B.A., Fang, X., Maris, G., Mlynczak, M.G., Niciejewski, R.J., Palo, S.E., Paxton, L.J., Randal, C.E., Rong, P.P., Russell III, J.M., Skinner, W., Solomon, S.C., Talaat, E.R., Wu, Q., Yee, J.H., 2006. Response of the upper/middle atmosphere to coronal holes and powerful high-speed solar wind streams in 2003. In: Tsurutani, B.T., McPherron, R.L., Gonzalez, W.D., Lu, G., Sobral, J.H.A., Gopalswamy, N. (Eds.), Recurrent Magnetic Storms: Corotating Solar Wind Streams. Geophysical Monograph Series, vol. 167. AGU, Washington, DC, pp. 319–340.

Lakhina, G.S., Tsurutani, B.T., Verkhoglyadova, O.P., Pickett, J.S., 2010. Pitch angle transport of electrons due to cyclotron interactions with coherent chorus subelements. J. Geophys. Res. 115. A00F15. https://doi.org/10.1029/2009JA014885.

Li, X., Baker, D.N., Temerin, M.A., Cayton, T.E., Reeves, E.G.D., Christensen, R.A., Blake, J.B., Looper, M.D., Nakamura, R., Kanekal, S.G., 1997. Multi-satellite observations of the outer zone electron variation during the November 3–4, 1993, magnetic storm. J. Geophys. Res. 102, 14123–14140.

Li, X., Temerin, M., Baker, D.N., Reeves, G.D., Larson, G.D., 2001. Quantitative prediction of radiation belt electrons at geostationary orbit based on solar wind measurements. Geophys. Res. Lett. 28, 1887–1890. https://doi.org/10.1029/2000GL012681.

Li, X., Baker, D.N., Elkington, S., Temerin, M., Reeves, G.D., Belian, R.D., Blake, J.B., Singer, H.J., Peria, W., Parks, G., 2003. Energetic particle injections in the inner magnetosphere as a response to an interplanetary shock. J. Atmos. Sol. Terr. Phys. 65, 233–244. https://doi.org/10.1016/S1364-6826(02)00286-9.

Li, X., Baker, D.N., Temerin, M., Reeves, G., Friedel, R., Shen, C., 2005. Energetic electrons, 50 keV to 6 MeV, at geosynchronous orbit: their responses to solar wind variations. Space Weather 3. S04001. https://doi.org/10.1029/2004SW000105.

Li, X., Oh, K.S., Temerin, M., 2007. Prediction of the AL index using solar wind parameters. J. Geophys. Res. 112. A06224. https://doi.org/10.1029/2006JA011918.

Li, W., Thorne, R.M., Bortnik, J., Baker, D.N., Reeves, G.D., Kanekal, S.G., Spence, H.E., Green, J.C., 2015. Solar wind conditions leading to efficient radiation belt electron acceleration: a superposed epoch analysis. Geophys. Res. Lett. 42, 6906–6915. https://doi.org/10.1002/2015GL065342.

Lorentzen, K.R., Blake, J.B., Inan, U.S., Bortnik, J., 2001. Observations of relativistic electron microbursts in association with VLF chorus. J. Geophys. Res. 106, 6017–6027. https://doi.org/10.1029/2000JA003018.

Luo, B., Li, X., Temerin, M., Liu, S., 2013. Prediction of the AU, AL, and AE indices using solar wind parameters. J. Geophys. Res. 118. https://doi.org/10.1002/2013JA019188.

Lyons, L.R., Lee, D.Y., Thorne, R.M., Horne, R.B., Smith, A.J., 2005. Solar wind-magnetosphere coupling leading to relativistic electron energization during high-speed streams. J. Geophys. Res. 110. A11202. https://doi.org/10.1029/2005JA011254.

Mauk, B.H., Fox, N.J., Kanekal, S.G., Kessel, R.L., Sibeck, D.G., Ukhorskiy, A., 2012. Science objectives and rationale for the radiation belt storm probes mission. Space Sci. Rev. 179, 3–27. https://doi.org/10.1007/s11214-012-9908-y.

Mendes, O., Domingues, M.O., Echer, E., Hajra, R., Menconi, V.E., 2017. Characterization of high-intensity, long-duration continuous auroral activity (HILDCAA) events using recurrence quantification analysis. Nonlinear Process. Geophys. 24, 407–417.

Meredith, N.P., Horne, R.B., Anderson, R.R., 2001. Substorm dependence of chorus amplitudes: implications for the acceleration of electrons to relativistic energies. J. Geophys. Res. 106, 13165–13178.

Meredith, N.P., Horne, R.B., Iles, R.H.A., Thorne, R.M., Heynderickx, D., Anderson, R.R., 2002. Outer zone relativistic electron acceleration associated with substorm-enhanced whistler mode chorus. J. Geophys. Res. 107, 1144. https://doi.org/10.1029/2001JA900146.

Meredith, N.P., Cain, M., Horne, R.B., Thorne, R.M., Summers, D., Anderson, R.R., 2003. Evidence for chorus-driven electron acceleration to relativistic energies from a survey of geomagnetically disturbed periods. J. Geophys. Res. 108, 1248. https://doi.org/10.1029/2002JA009764.

Meredith, N.P., Horne, R.B., Glauert, S.A., Thorne, R.M., Summers, D., Albert, J.M., Anderson, R.R., 2006. Energetic outer zone electron loss timescales during low geomagnetic activity. J. Geophys. Res. 111. A05212. https://doi.org/10.1029/2005JA011516.

Miyoshi, Y., Kataoka, R., 2008. Flux enhancement of the outer radiation belt electrons after the arrival of stream interaction regions. J. Geophys. Res. 113. A03S09. https://doi.org/10.1029/2007JA012506.

Miyoshi, Y., Kataoka, R., 2011. Solar cycle variations of outer radiation belt and solar wind structures. J. Atmos. Sol. Terr. Phys. 73 (1), 77–87. https://doi.org/10.1016/j.jastp.2010.09.031.

Miyoshi, Y., Morioka, A., Obara, T., Misawa, T., Nagai, T., Kasahara, Y., 2003. Rebuilding process of the outer radiation belt during the 3 November 1993 magnetic storm: NOAA and Exos-D observations. J. Geophys. Res. 108, 1004. https://doi.org/10.1029/2001JA007542.

Miyoshi, Y., Kataoka, R., Kasahara, Y., Kumamoto, A., Nagai, T., Thomsen, M.F., 2013. High-speed solar wind with southward interplanetary magnetic field causes relativistic electron flux enhancement of the outer radiation belt via enhanced condition of whistler waves. Geophys. Res. Lett. 40, 4520–4525. https://doi.org/10.1002/grl.50916.

Nakamura, R., Isowa, M., Kamide, Y., Baker, D.N., Blake, J.B., Looper, M., 2000. SAMPEX observations of precipitating bursts in the outer radiation belt. J. Geophys. Res. 105, 15875–15885. https://doi.org/10.1029/2000JA900018.

Ness, N.F., Wilcox, J.M., 1964. Solar origin of the interplanetary magnetic field. Phys. Rev. Lett. 13, 461–464.

Ni, B., Xiang, Z., Gu, X., Shprits, Y.Y., Zhou, C., Zhao, Z., Zhang, X., Zuo, P., 2016. Dynamic responses of the Earth's radiation belts during periods of solar wind dynamic pressure pulse based on normalized superposed epoch analysis. J. Geophys. Res. 121, 8523–8536. https://doi.org/10.1002/2016JA023067.

Ohtani, S., Miyoshi, Y., Singer, H.J., Weygand, J.M., 2009. On the loss of relativistic electrons at geosynchronous altitude: its dependence on magnetic configurations and external conditions. J. Geophys. Res. 114. A01202. https://doi.org/10.1029/2008JA013391.

Onsager, T., Grubb, R., Kunches, J., Matheson, L., Speich, D., Zwickl, R.W., Sauer, H., 1996. Operational uses of the GOES energetic particle detectors. In: Proc SPIE 2812, GOES-8 and Beyond 281. https://doi.org/10.1117/12.254075.

Onsager, T.G., Rostoker, G., Kim, H.J., Reeves, G.D., Obara, T., Singer, H.J., Smithtro, C., 2002. Radiation belt electron flux dropouts: local time, radial, and particle-energy dependence. J. Geophys. Res. 107. (A11). https://doi.org/10.1029/2001JA000187.

Paulikas, G.A., Blake, J.B., 1979. Effects of the solar wind magnetospheric dynamics: energetic electrons at the synchronous orbit. In: Olson, W.P. (Ed.), Quantitative Modeling of Magnetospheric Processes. Geophysical Monograph Series, vol. 21. AGU, Washington, DC, p. 180.

Reeves, G.D., McAdams, K.L., Friedel, R.H.W., O'Brien, T.P., 2003. Acceleration and loss of relativistic electrons during geomagnetic storms. Geophys. Res. Lett. 30, 1529. https://doi.org/10.1029/2002GL016513.

Reeves, G.D., Spence, H.E., Henderson, M.G., Morley, S.K., Friedel, R.H.W., Funsten, H.O., Baker, D.N., Kanekal, S.G., Blake, J.B., Fennell, J.F., Claudepierre, S.G., Thorne, R.M., Turner, D.L., Kletzing, C.A., Kurth, W.S., Larsen, B.A., Niehof, J.T., 2013. Electron acceleration in the heart of the Van Allen radiation belts. Science 341, 991–994. https://doi.org/10.1126/science.1237743.

Reeves, G.D., Friedel, R.H.W., Larsen, B.A., Skoug, R.M., Funsten, H.O., Claudepierre, S.G., Fennell, J.F., Turner, D.L., Denton, M.H., Spence, H.E., Blake, J.B., Baker, D.N., 2016. Energy dependent dynamics of keV to MeV electrons in the inner zone, outer zone, and slot regions. J. Geophys. Res. 121, 397–412. https://doi.org/10.1002/2015JA021569.

Roederer, J.G., Zhang, H., 2014. Dynamics of Magnetically Trapped Particles—Foundations of the Physics of Radiation Belts and Space Plasmas. Springer, Heidelberg/New York/Dordrecht/London.

Roth, I., Temerin, M.A., Hudson, M.K., 1999. Resonant enhancement of relativistic electron fluxes during geomagnetically active periods. Ann. Geophys. 17, 631–638.

Santolik, O., Macusova, E., Kolmasova, I., Cornilleau-Wehrlin, N., de Conchy, Y., 2014. Propagation of lower-band whistler-mode waves in the outer Van Allen belt: systematic analysis of 11 years of multi-component data from the Cluster spacecraft. Geophys. Res. Lett. 41, 2729–2737.

Schulz, M., Lanzerotti, L., 1974. Particle Diffusion in the Radiation Belts. Springer, New York.

Selesnick, R.S., 2015. High-energy radiation belt electrons from CRAND. J. Geophys. Res. 120, 2912–2917. https://doi.org/10.1002/2014JA020963.

Selesnick, R.S., Blake, J.B., 2000. On the source location of radiation belt relativistic electrons. J. Geophys. Res. 105, 2607–2624. https://doi.org/10.1029/1999JA900445.

Selesnick, R.S., Su, Y.J., Blake, J.B., 2016. Control of the innermost electron radiation belt by large-scale electric fields. J. Geophys. Res. 121, 8417–8427. https://doi.org/10.1002/2016JA022973.

Sheeley Jr., N.R., Harvey, J.W., Feldman, W.C., 1976. Coronal holes, solar wind streams and recurrent geomagnetic disturbances: 1973–1976. Sol. Phys. 49, 271–278.

Shprits, Y.Y., Thorne, R.M., Friedel, R., Reeves, G.D., Fennell, J., Baker, D.N., Kanekal, S.G., 2006. Outward radial diffusion driven by losses at magnetopause. J. Geophys. Res. 111. A11214. https://doi.org/10.1029/2006JA011657.

Singer, S.F., 1958. Trapped albedo theory of the radiation belt. Phys. Rev. Lett. 1, 181–183.

Smith, E.J., 1985. Interplanetary shock phenomena beyond 1 AU. In: Tsurutani, B.T., Stone, R.G. (Eds.), Collisionless Shocks in the Heliosphere: Reviews of Current Research. Geophysical Monograph Series, vol. 35. AGU, Washington, DC, p. 69. https://doi.org/10.1029/GM035p0069.

Smith, E.J., Wolfe, J.H., 1976. Observations of interaction regions and corotating shocks between one and five AU: pioneers 10 and 11. Geophys. Res. Lett. 3, 137–140. https://doi.org/10.1029/GL003i003p00137.

Smith, E.J., Tsurutani, B.T., Rosenberg, R.L., 1978. Observations of the interplanetary sector structure up to helographic latitudes of 16°: pioneer 11. J. Geophys. Res. 83, 717–724.

Soraas, F., Oksavik, K., Aarsnes, K., Evans, D.S., Greer, M.S., 2003. Storm time equatorial belt—an "image" of RC behavior. Geophys. Res. Lett. 30. https://doi.org/10.1029/2002GL015636.

Souza, A.M., Echer, E., Bolzan, M.J.A., Hajra, R., 2016. A study on the main periodicities in interplanetary magnetic field Bz component and geomagnetic AE index during HILDCAA events using wavelet analysis. J. Atmos. Sol. Terr. Phys. 149, 81–86.

Su, Y.J., Selesnick, R.S., Blake, J.B., 2016. Formation of the inner electron radiation belt by enhanced large-scale electric fields. J. Geophys. Res. 121, 8508–8522. https://doi.org/10.1002/2016JA022881.

Summers, D., Thorne, R.M., 2003. Relativistic pitch angle scattering by electromagnetic ion cyclotron waves during geomagnetic storms. J. Geophys. Res. 108, 1143. https://doi.org/10.1029/2002JA009489.

Summers, D., Thorne, R.M., Xiao, F., 1998. Relativistic theory of wave-particle resonant diffusion with application to electron acceleration in the magnetosphere. J. Geophys. Res. 103, 20487–20500. https://doi.org/10.1029/98JA01740.

Summers, D., Ma, C., Meredith, N.P., Horne, R.B., Thorne, R.M., Heynderickx, D., Anderson, R.R., 2002. Model of the energization of outer-zone electrons by whistler-mode chorus during the October 9, 1990 geomagnetic storm. Geophys. Res. Lett. 29, 2174. https://doi.org/10.1029/2002GL016039.

Summers, D., Ma, C., Meredith, N.P., Horne, R.B., Thorne, R.M., Anderson, R.R., 2004. Modeling outer-zone relativistic electron response to whistler mode chorus activity during substorms. J. Atmos. Sol. Terr. Phys. 66, 133–146.

Summers, D., Mace, R.L., Hellberg, M.A., 2005. Pitch-angle scattering rates in planetary magnetospheres. J. Plasma Phys. 71, 237–250.

Summers, D., Ni, B., Meredith, N.P., 2007. Timescale for radiation belt electron acceleration and loss due to resonant wave-particle interactions: 2. Evaluation for VLF chorus, ELF hiss, and electromagnetic ion cyclotron waves. J. Geophys. Res. 112. A04207. https://doi.org/10.1029/2006JA011993.

Temerin, M., Li, X., 2006. Dst model for 1995–2002. J. Geophys. Res. 111. A04221. https://doi.org/10.1029/2005JA011257.

Thorne, R.M., Kennel, C.F., 1971. Relativistic electron precipitation during magnetic storm main phase. J. Geophys. Res. 76, 4446–4453. https://doi.org/10.1029/JA076i019p04446.

Thorne, R.M., O'Brien, T.P., Shprits, Y.Y., Summers, D., Horne, R.B., 2005. Timescale for MeV electron microburst loss during geomagnetic storms. J. Geophys. Res. 110. A09202. https://doi.org/10.1029/2004JA010882.

Thorne, R.M., Li, W., Ni, B., Ma, Q., Bortnik, J., Chen, L., Baker, D.N., Spence, H.E., Reeves, G.D., Henderson, M.G., Kletzing, C.A., Kurth, W.S., Hospodarsky, G.B., Blake, J.B., Fennell, J.F., Claudepierre, S.G., Kanekal, S.G., 2013. Rapid local acceleration of relativistic radiation-belt electrons by magnetospheric chorus. Nature 504, 411–414.

Tsurutani, B.T., Gonzalez, W.D., 1987. The cause of high-intensity long-duration continuous AE activity (HILDCAAs): interplanetary Alfvén wave trains. Planet. Space Sci. 35, 405–412.

Tsurutani, B.T., Lakhina, G.S., 1997. Some basic concepts of wave-particle interactions in collisionless plasmas. Rev. Geophys. 35, 491–501. https://doi.org/10.1029/97RG02200.

Tsurutani, B.T., Lin, R.P., 1985. Acceleration of >47 keV ions and >2 keV electrons by interplanetary shocks at 1 AU. J. Geophys. Res. 90, 1–11. https://doi.org/10.1029/JA090iA01p00001.

Tsurutani, B.T., Smith, E.J., 1974. Magnetospheric ELF (10-1000 Hz) electromagnetic-waves during storms and substorms. Trans. Am. Geophys. Union 55, 1019.

Tsurutani, B.T., Smith, E.J., 1977. Two types of magnetospheric ELF chorus and their substorm dependences. J. Geophys. Res. 82, 5112–5128. https://doi.org/10.1029/JA082i032p05112.

Tsurutani, B.T., Smith, E.J., West Jr., H.I., Buck, R.M., 1979. Chorus, energetic electrons and magnetospheric substorms. In: Palmadesso, P.J., Papadopoulos, K. (Eds.), Wave Instabilities in Space Plasmas. D. Reidel, Dordrecht, p. 55.

Tsurutani, B.T., Gonzalez, W.D., Tang, F., Akasofu, S.I., Smith, E.J., 1988. Origin of interplanetary southward magnetic fields responsible for major magnetic storms near solar maximum (1978-1979). J. Geophys. Res. 93, 8519–8531. https://doi.org/10.1029/JA093iA08p08519.

Tsurutani, B.T., Ho, C.M., Smith, E.J., Neugebauer, M., Goldstein, B.E., Mok, J.S., Arballo, J.K., Balogh, A., Southwood, D.J., Feldman, W.C., 1994. The relationship between interplanetary discontinuities and Alfvén waves: Ulysses observations. Geophys. Res. Lett. 21, 2267–2270. https://doi.org/10.1029/94GL02194.

Tsurutani, B.T., Gonzalez, W.D., Gonzalez, A.L.C., Tang, F., Arballo, J.K., Okada, M., 1995. Interplanetary origin of geomagnetic activity in the declining phase of the solar cycle. J. Geophys. Res. 110, 21717–21733.

Tsurutani, B.T., Gonzalez, W.D., Guarnieri, F., Kamide, Y., Zhou, X., Arballo, J.K., 2004. Are high-intensity long-duration continuous AE activity (HILDCAA) events substorm expansion events? J. Atmos. Sol. Terr. Phys. 66, 167–176.

Tsurutani, B.T., Gonzalez, W.D., Gonzalez, A.L.C., Guarnieri, F.L., Gopalswamy, N., Grande, M., Kamide, Y., Kasahara, Y., Lu, G., McPherron, R.L., Soraas, F., Vasyliunas, V., 2006. Corotating solar wind streams and recurrent geomagnetic activity: a review. J. Geophys. Res. 111. A07S01. https://doi.org/10.1029/2005JA011273.

Tsurutani, B.T., Verkhoglyadova, O.P., Lakhina, G.S., Yagitani, S., 2009. Properties of dayside outer zone chorus during HILDCAA events: loss of energetic electrons. J. Geophys. Res. 114. A03207. https://doi.org/10.1029/2008JA013353.

Tsurutani, B.T., Lakhina, G.S., Verkhoglyadova, O.P., Gonzalez, W.D., Echer, E., Guarnieri, F.L., 2011. A review of interplanetary discontinuities and their geomagnetic effects. J. Atmos. Sol. Terr. Phys. 73, 5–19. https://doi.org/10.1016/j.jastp.2010.04.001.

Tsurutani, B.T., Lakhina, G.S., Verkhoglyadova, O.P., 2013. Energetic electron (>10 keV) microburst precipitation, ~5-15 s X-ray pulsations, chorus, and wave-particle interactions: a review. J. Geophys. Res. 118, 2296–2312. https://doi.org/10.1002/jgra.50264.

Tsurutani, B.T., Hajra, R., Tanimori, T., Takada, A., Bhanu, R., Mannucci, A.J., Lakhina, G.S., Kozyra, J.U., Shiokawa, K., Lee, L.C., Echer, E., Reddy, R.V., Gonzalez, W.D., 2016. Heliospheric plasma sheet (HPS) impingement onto the magnetosphere as a cause of relativistic electron dropouts (REDs) via coherent EMIC wave scattering with possible consequences for climate change mechanisms. J. Geophys. Res. 121, 10130–10156. https://doi.org/10.1002/2016JA022499.

Turner, D.L., Li, X., 2008. Quantitative forecast of relativistic electron flux at geosynchronous orbit based on low-energy electron flux. Space Weather 6. S05005. https://doi.org/10.1029/2007SW000354.

Turner, N.E., Mitchell, E.J., Knipp, D.J., Emery, B.A., 2006. Energetics of magnetic storms driven by corotating interaction region: a study of geoeffectiveness. In: Tsurutani, B.T., McPherron, R.L., Gonzalez, W.D., Lu, G., Sobral, J.H.A., Gopalswamy, N. (Eds.), Recurrent Magnetic Storms: Corotating Solar Wind Streams. Geophysical Monograph Series, vol. 167. AGU, Washington, DC, pp. 113–124.

Turner, D.L., Angelopoulos, V., Li, W., Hartinger, M.D., Usanova, M., Mann, I.R., Bortnik, J., Shprits, Y., 2013. On the storm-time evolution of relativistic electron phase space density in Earth's outer radiation belt. J. Geophys. Res. 118, 2196–2212. https://doi.org/10.1002/jgra.50151.

Turner, D.L., Angelopoulos, V., Li, W., Bortnik, J., Ni, B., Ma, Q., Thorne, R.M., Morley, S.K., Henderson, M.G., Reeves, G.D., Usanova, M., Mann, I.R., Claudepierre, S.G., Blake, J.B., Baker, D.N., Huang, C.L., Spence, H., Kurth, W., Kletzing, C., Rodriguez, J.V., 2014. Competing source and loss mechanisms due to wave-particle interactions in Earth's outer radiation belt during the 30 September to 3 October 2012 geomagnetic storm. J. Geophys. Res. 119, 1960–1979. https://doi.org/10.1002/2014JA019770.

Ukhorskiy, A.Y., Anderson, B.J., Brandt, P.C., Tsyganenko, N.A., 2006. Storm time evolution of the outer radiation belt: transport and losses. J. Geophys. Res. 111. A11S03. https://doi.org/10.1029/2006JA011690.

Usanova, M.E., Drozdov, A., Orlova, K., Mann, I.R., Shprits, Y., Robertson, M.T., Turner, D.L., Milling, D.K., Kale, A., Baker, D.N., Thaller, S.A., Reeves, G.D., Spence, H.E., Kletzing, C., Wygant, J., 2014. Effect of EMIC waves on relativistic and ultrarelativistic electron populations: ground-based and Van Allen Probes observations. Geophys. Res. Lett. 41, 1375–1381. https://doi.org/10.1002/2013GL059024.

Van Allen, J.A., Frank, L.A., 1959. Radiation measurements to 658,300 km with Pioneer IV. Nature 184, 219–224.

Vernov, S.N., Chudakov, A.E., Vakulov, P.V., Logachev, Y.I., Nikolayev, A.G., 1960. Radiation measurements during the flight of the second Soviet space rocket. In: Kallmann-Bijl, H. (Ed.), Space Research: Proceedings of the First International Space Science Symposium. North-Holland Publications, Amsterdam, p. 845.

West Jr., H.I., Buck, R.M., Walton, J.R., 1972. Shadowing of electron azimuthal-drift motions near the noon magnetopause. Nature Phys. Sci. 240, 6–7. https://doi.org/10.1038/physci240006a0.

West Jr., H.I., Buck, R.M., Davidson, G.T., 1981. The dynamics of energetic electrons in the Earth's outer radiation belt during 1968 as observed by the Lawrence Livermore National Laboratory's Spectrometer on Ogo 5. J. Geophys. Res. 86, 2111–2142. https://doi.org/10.1029/JA086iA04p02111.

Winterhalter, D., Smith, E.J., Burton, M.E., Murphy, N., McComas, D.J., 1994. The heliospheric plasma sheet. J. Geophys. Res. 99, 6667–6680.

Wrenn, G.L., 1995. Conclusive evidence for internal dielectric charging anomalies on geosynchronous communications spacecraft. J. Spacecr. Rocket. 32, 514–520. https://doi.org/10.2514/3.26645.

Wygant, J.R., Bonnell, J.W., Goetz, K., Ergun, R.E., Mozer, F.S., Bale, S.D., Ludlam, M., Turin, P., Harvey, P.R., Hochmann, R., Harps, K., Dalton, G., McCauley, J., Rachelson, W., Gordon, D., Donakowski, B., Shultz, C., Smith, C., Diaz-Aguado, M., Fischer, J., Heavner, S., Berg, P., Malsapina, D.M., Bolton, M.K., Hudson, M., Strangeway, R.J., Baker, D.N., Li, X., Albert, J., Foster, J.C., Chaston, C.C., Mann, I., Donovan, E., Cully, C.M., Cattell, C.A., Krasnoselskikh, V., Kersten, K., Brenneman, A., Tao, J.B., 2013. The electric field and waves instruments on the radiation belt storm probes mission. Space Sci. Rev. 179, 183–220.

Xie, L., Pu, Z.Y., Zhou, X.Z., Fu, S.Y., Zong, Q.G., Hong, M.H., 2006. Energetic ion injection and formation of the storm-time symmetric ring current. Ann. Geophys. 24, 3547–3556.

CHAPTER

EXTREME SPACE WEATHER SPACECRAFT SURFACE CHARGING AND ARCING EFFECTS

15

Dale C. Ferguson
U.S. Air Force Research Laboratory, Albuquerque, NM, United States

CHAPTER OUTLINE

1 Limits of Discussion 402
2 Cause of Arcing 402
3 Effects of Arcing 402
4 Physics of Charging in Geosynchronous Earth Orbit 404
5 The Spacecraft Charging Equation 405
6 What Are the Worst Spacecraft Charging Events? 411
7 Limits on Spacecraft Charging Events 412
8 The Galaxy 15 Failure 415
9 Conclusion 416
References 416

We discuss the physics of charging, why spacecraft charge negatively, and the roles of photoemission and secondary electron emission in frame and differential charging. The spacecraft charging equation is explored. The discussion is limited to spacecraft surface charging events, excluding deep-dielectric discharge events. The cause of arcing is determined to be differential charging. The effects of arcing are described and illustrated. The role of spacecraft charging computer codes is delineated. We define extreme spacecraft charging events, and explain why the worst events for other effects may not be the worst for spacecraft charging. Examples are given from severe space weather events of the past. Natural limits on spacecraft charging are determined, and the maximum frame charging and differential charging are calculated. We give the reasons spacecraft charging is important for our modern civilization, and define the worst events as those with the longest duration, maximizing satellite exposures to charging conditions.

1 LIMITS OF DISCUSSION

Here, we will talk about traditional spacecraft charging, which involves charging the surfaces of the spacecraft. The other type of spacecraft charging, sometimes called deep-dielectric charging, bulk charging, or internal charging, involves electrons so energetic that they pass right through spacecraft surface materials and lodge themselves deep inside the spacecraft. Such high-energy electrons are often referred to as radiation, and deep-dielectric charging and arcing are then classified as radiation effects; they are treated in a separate chapter of this book (Chapter 16 of this volume). We shall see that the best definition of an extreme space weather event from a surface-charging standpoint is one where the propensity for arcing is maximized.

2 CAUSE OF ARCING

Of course, extreme spacecraft charging events are those that charge spacecraft to the highest potentials. Since the dawn of the space age, it has been known that spacecraft can suffer ESD events (arcs) because they charge in the plasma environment of outer space (DeForest, 1972). Indeed, the 1973 arc-related failure of the Defense Satellite Communications System satellite 9431 (Inouye, 1977), responsible for relaying the U.S. President's "red phone" orders (Odenwald, 2015), is what prompted NASA and the U.S. Air Force (USAF) to step up investigation of the causes and mitigation of charging in space. A well-instrumented satellite, SCATHA (Spacecraft Charging AT High Altitudes, P78-2), was built and launched in 1979 based on scientific theories of the causes of spacecraft charging; it proved to be a boon to the space-science and satellite-engineering communities. Data from the multiyear mission is still being analyzed to this day. Not only did SCATHA experience arcs, but it had exquisite sensors to measure both the space environment and the charging of the whole satellite and several of its surfaces in response to it (Mullen et al., 1986). Other well-instrumented satellites for surface charging are the Los Alamos National Laboratory (LANL) geosynchronous satellites (Thomsen et al., 2013), and the newest of the Geostationary Operational Environmental Satellites (GOES) (Dichter et al., 2015). Recently, the USAF Secretary mandated that future USAF satellites must be instrumented to measure the space environment to help determine whether upsets are due to the natural environment or man-made actions.

3 EFFECTS OF ARCING

Arcs on spacecraft are accompanied by the following:

(1) Contamination of adjacent surfaces
(2) Optical and radio interference
(3) Transient voltages and currents

Solar arrays are particularly prone to arcing, for these reasons:

(1) Dielectric surfaces (coverglasses) are closely adjacent to conductors (interconnects and cell edges), making for sites of high electric field.
(2) The charging electrons have practically free access to both dielectrics and conductors.
(3) The spacecraft frame (spacecraft ground) is connected to the array conductor at some point or points.

(4) Strings of solar cells require differing potentials along the string for power generation.
(5) String placement often places cells of differing potentials close together.
(6) Arcs between cells and coverglasses may produce a plasma that becomes a short circuit between adjacent strings, and the array can continue powering the arcs (producing "sustained" arcs).

Undoubtedly, sustained arcs are the most disastrous from a spacecraft viability standpoint. They invariably lead to complete loss of a string of solar cells (or two) when they occur on a solar array. When they occur elsewhere, such as in a power cable bundle, they can lead to a complete and permanent power loss as well as a loss of all satellite functions (for example, ADEOS-2, Maejima et al., 2004). Fig. 1 shows the result of sustained arcing on a solar array of a satellite recovered by the space shuttle. Sustained arcing led to the loss of the tethered satellite TSS-1R (Fig. 2, from NASA-HDBK-4006). Various standards have been produced that detail methods of preventing arcing and most of all, sustained arcing (NASA-HDBK-4006; NASA-HDBK-4002A; NASA-STD-4005; AIAA/ANSI Standard S-115; E-JAXA-JERG-2-211A; ECSS-E-ST-20-06C; ISO 11221).

FIG. 1
Sustained arcing on solar array of shuttle-returned satellite.

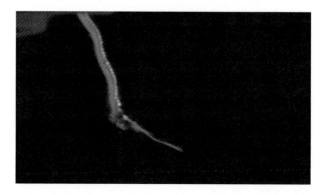

FIG. 2
Severed end of tether of TSS-1R after the satellite had already been lost. Sustained arc was to blame.

Even arcs that are not sustained can produce power degradation over a long period of time (Ferguson et al., 2016), damage or trip switches that are necessary for satellite operation (Ferguson et al, 2011), interfere with communications and satellite radio reception (IG Inside GNSS, 2009), and change satellite thermal properties (Carruth et al., 2001), etc. The higher the differential voltages on spacecraft, and the longer they are maintained, the more likely an arc, perhaps a disastrous one, will occur.

4 PHYSICS OF CHARGING IN GEOSYNCHRONOUS EARTH ORBIT

The following discussion is for spacecraft in geosynchronous Earth orbit (GEO). For low-Earth-orbiting (LEO) spacecraft in equatorial orbits, the low temperature ambient plasma usually keeps unbiased surfaces near the plasma potential; therefore, charging occurs only due to $v \times B$ effects and distributed potentials due to the power supply voltage. Thus, although extreme charging environments may exist in LEO, they may be those where electron or ion collection is restricted from reaching surfaces whose differential charging is due to the array string voltage, for instance. For spacecraft in LEO polar orbits, the auroral streams may charge spacecraft briefly to thousands of volts, and in those cases, the following discussion also holds for LEO.

Since the time of SCATHA, many satellites have incorporated charging monitors, environmental sensors, or both. Usually, the charging monitors may be surface potential monitors (essentially well-isolated surfaces with surface-to-spacecraft-ground voltage measurement) or ion spectrometers (to measure the "ion line," a spike in the energy spectrum of ion flux at the spacecraft ground voltage that in general varies with time). Spacecraft charging has been shown to be a function mainly of the electron density and temperature, usually measured with an electron spectrometer.

Spacecraft usually have a very low capacitance-to-space compared with the capacitances of their surfaces with respect to each other. Thus, absolute (spacecraft ground or frame) charging takes place on a much shorter timescale than differential charging (surface to surface). It can be shown that the equilibrium voltage (V_{max}) of a surface relative to an underlying conductor depends only on the *net* charge flux to the surface (J, A/cm^2), the bulk resistivity (ρ, ohm-m), and the thickness (d, m) [Ferguson et al, 2009].

$$V_{max} = J \cdot d \cdot \rho \qquad (1)$$

However, for a range of electron impact energies (between the so-called first and second crossover points), many dielectric materials have a secondary electron emission greater than one electron per incoming electron. Such electrons provide a discharging flux and make the surface more positive than before, as if the ion flux to the surface was greater. See Fig. 3 for a hypothetical material. Also, photoelectron emission in the presence of solar UV can help discharge surfaces. All these effects (capacitance, resistance, secondary electron emission, and photoemission), being different for different surface materials, can lead to differential charging with its concomitant electric field. It is the high electric fields that can build up from differential charging that can lead to arcing. In this sense, an extreme charging event is one where the propensity for arcing is maximized. Keeping track of all of these factors and the fluxes to different surfaces is the job of a three-dimensional charging code such as Nascap-2k (Mandell et al, 2006). Even without getting numerical, however, many of the spacecraft charging effects can be understood from simple physics.

For example, consider a spherical, one dielectric material GEO spacecraft in eclipse, where photoemission is zero. With electron temperatures being just a few times lower than ion temperatures, the

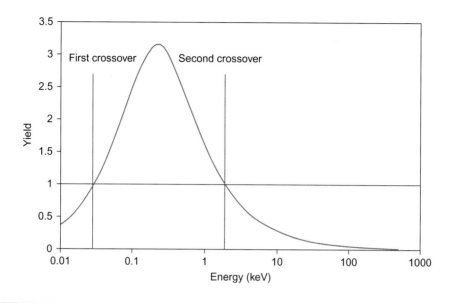

FIG. 3

Notional secondary electron emission characteristic showing first and second crossovers.

flux of the very light (1/1860 amu) electrons in the space plasma is much greater than the flux of the (≥ 1 amu) ions, and the spacecraft will charge negatively. To first order (ignoring secondary electron emission), it will charge until enough electrons are repelled that the remaining electron flux matches the incoming ion flux (for a full derivation, use Eq. 2). This means it will charge negatively to a potential about equal to the incoming electron energy in electron volts. So, a 5000 eV electron temperature will lead to a -5000 V potential on the spacecraft in eclipse. Now, let the spacecraft come into the full UV flux of sunlight, and about half the surface will undergo photoemission. For many conditions, this will fully discharge the sunlit surface, leading to a 5000 V differential voltage between sunlit and shaded surfaces. In the end, the spacecraft ground (frame) potential underneath the dielectric will readjust to a different potential somewhere between the two dielectric potentials. Of course, now the fluxes to the surfaces will be affected by their new potentials, which will alter the potentials, further affecting the fluxes, etc. To make things even more complicated, differential charging may build up, as the capacitance between surfaces and the spacecraft frame becomes charged. As one can imagine, the process is dynamic and complicated. Throw in different surface materials on different parts of the spacecraft and a more realistic spacecraft geometry, and the need to do numerical calculations to track all the surface potentials becomes obvious.

5 THE SPACECRAFT CHARGING EQUATION

$$\begin{aligned} &\text{Electron flux}(F_e) - \text{backscattered electrons}(F_{be}) - \text{secondary electron emission}(F_{se}) \\ &- \text{photoemission}(F_{pe})[-\text{conductivity term}(F_{cond})] \\ &= \text{fn}(\varphi) = \text{ion flux}(F_i) + \text{ion secondary electron emission}(F_{ise}) \end{aligned} \quad (2)$$

where φ is the local potential.

The entire spacecraft must obey the charging equation, and individual dielectric surfaces must also, with an added term [in brackets] describing the loss of charge through the dielectric to spacecraft ground. That is, because dielectrics have a finite conductivity, charge will gradually bleed off to spacecraft ground, changing both potentials in the process. Many of the dielectric surfaces are made of polymers, which typically have bleedoff times of many hours to days, but the solar arrays have glass coverglasses, with more conductivity and bleedoff times of a few hours. This has been confirmed by GPS arcing delays from peak fluxes and differential voltage delays from peak fluxes on Intelsat satellites. For differential charging, this means two things: first, the voltage on dielectrics depends not on the instantaneous fluxes, but rather on the fluence over the charge bleedoff time; and second, the differential charge will typically lag the flux in time by the charge bleedoff time (τ, Ferguson et al, 2009),

$$\tau = \varepsilon_0 \cdot \kappa \cdot \rho \tag{3}$$

where ε_0 is the permittivity of free space, κ is the material dielectric constant, and ρ is the bulk resistivity.

Illustrating this charge bleedoff time is Fig. 4 from Likar et al (2011), where the spacecraft charging on an Intelsat satellite lags the magnetic storm intensity (Kp) and the electron flux peaks by a few hours, and Fig. 5 (from Ferguson et al, 2016) showing GPS arcing event rates along with predicted electron fluxes. Typical spacecraft surface dielectrics have bleedoff times from about 3 h to several days. It has been recommended that certain materials (such as Teflon) not be used on spacecraft because of their extremely high resistivity and long bleedoff times (Purvis et al, 1984). Dielectric resistivities are usually a strong function of temperature, and are much greater under cold conditions (Ferguson et al, 2009).

Now, let's discuss the terms in the charging equation. First of all, all terms on the left side concern electron fluxes to or from the surface. The photoemission, secondary electron emission, and backscatter terms act to negate the electron flux to the surface. The terms on the right side correspond to the positive ion flux to the surface. Of interest here is the ion-secondary electron emission, which acts like an extra ion flux. The conductive flux term on the left accounts for the loss of electrons through dielectric materials to ground. Many of the terms depend on the surface potential, and through this dependence, the surface potential is determined. Of interest are the photoemission and secondary electron emission terms. These release only low energy electrons from the surface, and are easily overwhelmed if the surface is even 30 V or so positive of the surrounding plasma. However, it is clear that because the thermal electron flux overwhelms the thermal ion flux for any conceivable conditions, the higher the flux of electrons to the spacecraft, the more negative it should charge. The caveat is that electrons with energies between the first and second secondary electron emission crossover points act to discharge the spacecraft.

Thus, we can talk about a spacecraft-charging index as the net flux of electrons at energies above and below the second and first crossover points, respectively. Because the first crossover point is at very low energies, the best spacecraft-charging index is the electron flux above the second crossover point in secondary electron emission. Since this point is material-dependent, so will be the minimum energy of the "charging" electrons. Most modern spacecraft have solar array coverglasses (to prevent radiation damage to the cells) with antireflecting coatings made of magnesium fluoride (MgF_2), which has a second crossover point of perhaps 8 or 9 keV. Thus, it will be the total flux of electrons at energies higher than this that constitutes the charging flux. Detailed charging calculations done with Nascap-2k have confirmed this (Ferguson et al., 2015a).

A very important distinction between "frame" or "absolute" charging and "differential" charging must be made here. Because the capacitance of the entire spacecraft to space is very low, the spacecraft

5 THE SPACECRAFT CHARGING EQUATION

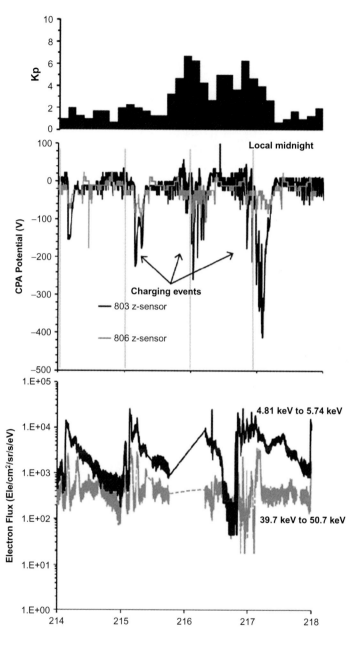

FIG. 4

Charging events on an Intelsat satellite (Likar et al, 2011).

"ground" or frame will charge very quickly (seconds) to a value given by the charging equation. After that time, the charging flux to the spacecraft will go into charging the surfaces of the spacecraft with respect to each other. These relative capacitances are usually much greater than the spacecraft capacitance, and may take minutes to hours to charge up. Thus, spacecraft surface charging is complicated by

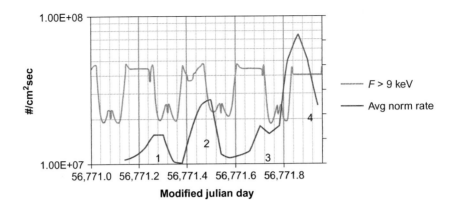

FIG. 5

Left axis, *blue*, estimated electron flux for $E \geq 10$ keV from the AE9/AP9/SPM model. *Red*, normalized USNDS (GPS) undispersed event rate (numbers not publicly available).

the relative capacitances of different surfaces to the spacecraft frame. It is the differences in charging that are important for arcing, as they lead to high electric fields. The fluxes to surfaces are determined, in turn, by the potentials of all the surfaces at any given time. It is this complex behavior that makes it necessary to use three-dimensional codes with surface material properties built in to fully understand spacecraft charging.

When in eclipse, spacecraft surfaces are not subject to discharge through photoemission. So, while absolute charging may be great in eclipse, differential charging usually is not, depending only on differences in secondary electron emission behavior between different surface materials. However, when the satellite comes out of eclipse, the photoemission often swamps the incident electron fluxes on sunlit surfaces, and differential potentials can build up quite rapidly between surface in sun and in shade. Thus, the most severe differential charging in GEO often occurs just after eclipse exit.

Nascap-2k runs (Ferguson et al., 2015a) have confirmed that the most severe charging conditions in the daytime in GEO are when the electron flux at energies higher than about 9 keV is highest. These fluxes are, however, constrained by an important energy density condition. The plasma conditions necessary for differential charging are that the fluences build up over some period of time (at least many minutes, to charge the relative capacitances of all surfaces), and so the electron densities and temperatures of the plasma must be stable over such time periods. This can only happen if the local energy density in the magnetic field is higher than the local energy density of all plasma particles perpendicular to the field (perpendicular "beta"). Thus, the local magnetic field must constrain the perpendicular particle motions so that they do not destabilize the magnetic field. For low beta (quiet times), particles are not constrained; they just undergo $E \times B$ drift and magnetic drift due to the magnetic field gradient. We quote from Gendrin (1980): "It is shown that whereas the beta of the plasma can reach very high values, the perpendicular kinetic energy density cannot…" The equation that must be obeyed for charging (Gendrin, 1980; Ferguson et al., 2015b, cgs units) is

$$B^2/8\pi \geq 3/2\left(N_i kT_{i\perp} + N_e kT_{e\perp}\right)/2 \tag{4}$$

5 THE SPACECRAFT CHARGING EQUATION

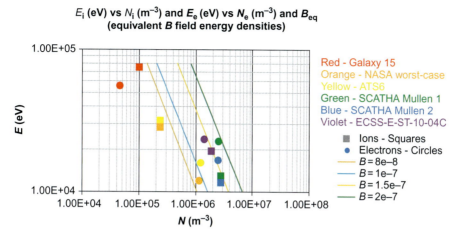

FIG. 6

Plasma temperature vs plasma density with equivalent magnetic field energy densities.

where $kT_{e,i \perp}$ is the energy of the electrons/ions perpendicular to the magnetic field, and the factor of 2 at the end is to average the electron and ion contributions. Thus, there is an upper limit to the electron flux related to the magnetic field strength. For quiet conditions, the upper limit is not important, as the fluxes are far below this limit. (Also, normally T_e is lower than T_i by a factor 3–8 in the plasmasheet. This should lower the max limit for electron flux below that given by T_e alone.)

So, the product of N (plasma density) and $T_\perp$ (plasma temperature perpendicular to the field) is constrained by the magnetic field strength. As can be seen from Fig. 6, the maximum ("worst-case") GEO charging condition that has been seen is at a magnetic field strength of about 2.5×10^{-7} T, whereas the typical GEO magnetic field is about 1.0×10^{-7} T, only a factor of six less in B^2!

The worst geomagnetic storms are those where the so-called Bz (north-south) component of the impinging field (IMF, for interplanetary magnetic field) is predominantly south, against the usual north-pointing direction of the Earth's equatorial magnetic field in GEO. The magnetopause is pushed inward until GEO satellites can enter the strongly Bz-negative magnetosheath region. By the same token, on the nightside, the field can become highly stretched and weakened under storm conditions as the partial ring current develops or prior to substorm injections. (Old-time GEO satellite attitudes were stabilized north-south by magnetic torques. Sometimes, when a geomagnetic storm hit, they would flip over when entering the magnetosheath!)

Disturbances in the Earth-wide magnetic field are often represented by a value called Kp. While Kp may indicate the severity of geomagnetic storm conditions on Earth due to the rapidly changing magnetic field, above a certain Kp value plasma surface charging conditions at GEO may become less severe. That is, the most extreme geomagnetic storm conditions on Earth's surface and in the ionosphere may not be the most extreme spacecraft-charging conditions for satellites in GEO, possibly because the magnetic field at the highest Kp may not be strong enough to constrain the highest charging fluxes. As evidence of this, Fig. 7 below shows measured charging (>20 V) for an LANL satellite (a "spinner") in GEO during the Halloween event of 2003 (Ferguson et al., 2015a,b). The most severe charging occurred for Kp values of about 7.7, much less than the Kp=9.0 at the peak

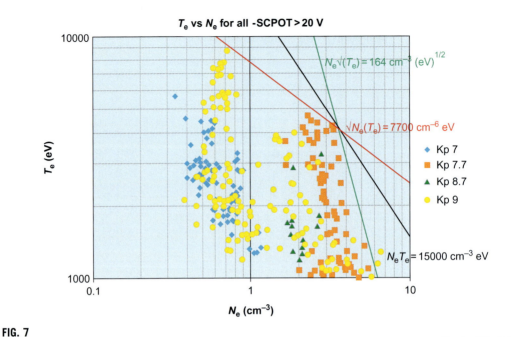

FIG. 7

Plasma moments for LANL satellite, charging >20 V, Halloween storm of 2003.

of the storm. And, no charging of more than 20 V negative was seen when the Kp value was less than 7.0. In fact, for three of the severe charging events given in Fig. 6 above, Kp never reached values above 8.0.

The results indicate that there is some broad peak value for charging flux at optimal Kp values (for more statistics, see Thomsen et al, 2013). This effect is incorporated into statistical models of charging in the magnetosphere (O'Brien, 2009). Eq. 4 (the Gendrin condition) might help to explain why. For low activity, the Gendrin condition is satisfied, but charging fluxes are low. For very disturbed conditions (Kp > 8), the magnetic field strength is weakened on the nightside, and the charging flux is limited by the weakened magnetic field. At any rate, Kp values of 7 and above can lead to severe charging if the magnetic field conditions are right.

Finally, even in severe geomagnetic storms, the GEO magnetic field very seldom gets to more than 250 nT either north or south, which places a limit on the maximum $N_e T_e$, and by implication the maximum $N_e T_e^{1/2}$, a measure of the total electron flux. Michael Bodeau (2015) has advocated the use of K at certain ground stations near the magnetic field line connecting a satellite to the Earth as a better indicator of charging for a given satellite at a given time, but the ground stations used are sparse enough that it is not always possible to use this indicator.

In addition to a disturbed magnetosphere, one other factor is important in spacecraft charging. Spacecraft exterior surfaces charge more on the morning side of the orbit than at any other time due to the increased electron density there. This is because high-energy electrons in the magnetosphere flow inward on the midnight meridian and move eastward in the magnetic field. Thus, the first few hours of satellite morning have a higher density of high-energy electrons. This is manifested in

6 WHAT ARE THE WORST SPACECRAFT CHARGING EVENTS?

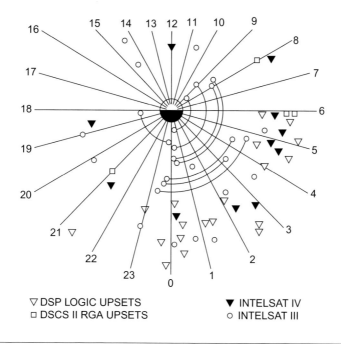

FIG. 8

Local times of satellite anomalies. Most anomalies occur on the morning side of the orbit.

Fig. 8 (Fennell et al, 2001; DeForest, 1972), showing the distribution of certain satellite upsets (anomalies) versus satellite local time. During eclipse seasons (near the equinoxes) the charging effects of local time are accentuated even more, as the interplanetary magnetic field interacts more strongly with the geomagnetic field (the Russell-McPherron effect, McPherron et al, 2009) and spacecraft charge more readily in eclipse, only to be partially discharged on eclipse exit.

6 WHAT ARE THE WORST SPACECRAFT CHARGING EVENTS?

So what are the most extreme spacecraft charging events? They are those where Kp gets above 7.3 and stays there for a day or more (there is no sharp cutoff of surface charging at high Kp). Then, GEO spacecraft at many different longitudes can experience differential charging as they reach the morning side of their orbits, and each satellite can experience severe charging for an extended period of time. Many of the extreme events reported in the literature fulfill these criteria. For example, the Halloween event of 2003 (already referred to in Fig. 7) is shown in Fig. 9 (NOAA, 2004).

Here, in the space of only 48 h, 27 h were at Kp=8 or greater. So, from October 23 to November 6, 2003, 47 satellites reported malfunctions and the ADEOS 2 scientific satellite (costing US$640 m) was a total loss due to surface charging (Webb and Allen, 2004). While many of these disruptions were due to deep dielectric discharges, others were due to surface charging. Other extreme events that happened before the space age could have produced even more satellite

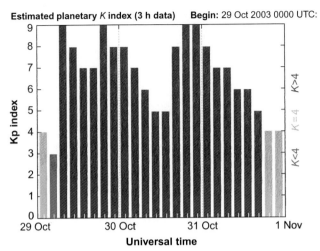

FIG. 9

Values of the planet-wide Kp magnetic field disturbance index for the 2003 Halloween storm of Fig. 7.

losses. For instance, the Carrington event on August 28, 1859, the March 25, 1940, (Easter Sunday) storm, and many others, such as the event of September 18–19, 1941, when Kp was at 8.7 or above for 24 straight hours, severely disrupted telegraph and radio communications and caused tremendous ground currents sufficient to blow up transformers, etc. Had satellites been in orbit during these events, it is likely that many of them would have experienced anomalies. For a good list of the most severe geomagnetic or auroral events in history, see http://www.solarstorms.org/SRefStorms.html.

7 LIMITS ON SPACECRAFT CHARGING EVENTS

Assume that the highest magnetic field in GEO is 250 nT, $N_e = N_i$ and $T_i = T_e$. This implies that $N_e T_e$ is $< 1.0 \times 10^5$ cm^{-3}eV. The total electron flux goes as $N_e \sqrt{T_e}$, so that $N_e \sqrt{T_e} < 1.0 \times 10^5 / \sqrt{T_e}$. This, then is the limit for total electron fluxes, which are closely related to spacecraft charging (Ferguson and Wimberly, 2013; Ferguson et al., 2015a), especially for $T_e > 9$ keV, which is true for all the "worst-case" environments. In Table 1 are the densities and temperatures for the extreme charging events in Fig. 6, along with derived quantities related to spacecraft charging.

Here it can be seen that the SCATHA Mullen 1 environment (NASA-HDBK-4002A) is almost 60% of theoretical maximum if the magnetic field is constrained to 250 nT. Fig. 10 shows the relationship between the electron flux proxy ($N_e \sqrt{T_e}$) and the energy density proxy ($N_e T_e$) for these six environments.

Here, it can be seen that there is a close relationship between the electron flux and the energy density for extreme charging events. Extrapolating to our maximum allowed energy density proxy

7 LIMITS ON SPACECRAFT CHARGING EVENTS

Table 1 Proposed Worst-Case Charging Environments

Environment	N_e (cm^{-3})	T_e (eV)	$N_e T_e$	$N_e \sqrt{T_e}$	$N_e T_e / 1.0 \times 10^5$
Galaxy 15	0.0458	5.56E+04	2.55E+03	1.080E+01	2.55E-02
NASA worst-case	1.12	12,000	1.34E+04	1.227E+02	1.34E-01
ATS-6	1.2	16,000	1.92E+04	1.518E+02	1.92E-01
SCATHA Mullen1	2.5	22,800	5.70E+04	3.775E+02	5.70E-01
SCATHA Mullen2	2.5	16,600	4.15E+04	3.221E+02	4.15E-01
ECSS	1.4	23,600	3.30E+04	2.151E+02	3.30E-01

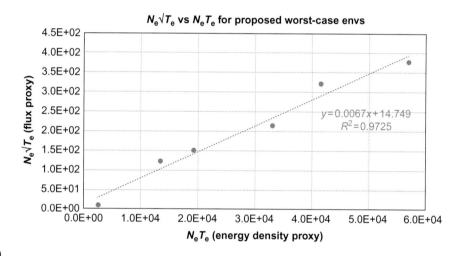

FIG. 10

Close relationship between total electron flux and particle energy density for many severe environments.

(1.0×10^5 eV cm^{-3}), we find that $N_e\sqrt{T_e}$ is 745 cm^{-3}eV$^{0.5}$. In Eq. 5, the total electron current flux is given in terms of electron densities and temperatures.

$$F_e = 2.68 \times 10^{-14} N_e (\text{m}^{-3}) T_e^{1/2} \text{ Amps/m}^2 \tag{5}$$

$N_e\sqrt{T_e} = 745$ cm^{-3}eV$^{0.5} = 7.45 \times 10^8$ m^{-3}eV$^{0.5}$, which corresponds to $F_e = 2 \times 10^{-5}$ Amps/m^2.

So what does this mean as far as frame and differential charging on typical spacecraft? To answer this, we refer in Figs. 11 and 12 to Nascap-2k results for a double Maxwellian electron energy distribution, where differential charging is shown on the vertical axis and the total electron flux (in Amps/m^2) is given on the horizontal axis (Ferguson and Wimberly, 2013). For these Nascap runs, the final time was chosen to be 2000 s, about the longest time extreme charging environments last on a given satellite due to its orbital motion and substorm durations, and a time long enough that potentials are approaching their asymptotic levels. Putting in our maximum flux from above, we find that the predicted theoretical maximum daytime frame charging (-Abs) = 400,000 V (!), and the predicted maximum differential charging (Max-Min) = 20,000 V. For comparison, the highest

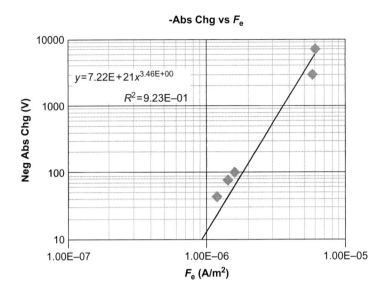

FIG. 11
Satellite frame (or absolute) charging vs total electron flux for double Maxwellian plasmas. Nascap-2k results for 2000 s of charging.

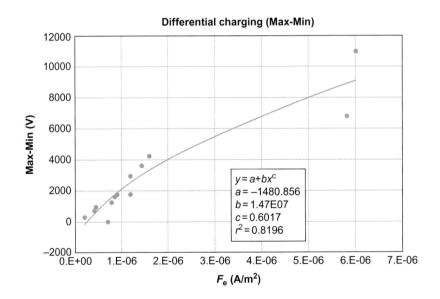

FIG. 12
Differential charging vs total electron flux. Nascap-2k results for 2000 s of charging.

electron temperature ever measured in the GEO magnetosphere was about 90,000 eV (Ferguson et al, 2011).

These extremely high charging values are beyond the design limits for any spacecraft, so we expect that the most extreme charging environment imaginable would cause any spacecraft exposed to that environment for a sufficient time to have surface arcs, regardless of whether our extrapolation is valid.

Using the commonly accepted value of a differential charging threshold for arcing (400 V, Cho et al, 2005) and Fig. 12 above, any even moderate geomagnetic substorm would be sufficient to cause arcing on satellites caught on the morning side of the orbit. And Mateo-Velez et al. (2016) have shown that conditions in medium Earth orbit (MEO), where GPS and other navigation satellites orbit, are more severe than in GEO. How extreme the environment is in terms of the number of anomalies would then depend on how long the charging conditions lasted. If more than a few hours, we could expect all satellites within a few hours longitude of the morning sector to undergo severe differential charging, and possibly damaging arcing. If the charging conditions last for 12 h or more, one might expect about half the orbiting satellites to undergo arcing with varying effects, from minor upsets to sustained arcs. See Fig. 17 in Ferguson et al., 2015b for the correlation of yearly spacecraft anomalies with the number of days with 24-h Kp sums greater than 35 (a Kp sum is the sum of all of the eight 3-h Kp values in the same day).

There has been much discussion in the literature (Lai, 1994, 2008; Lai and Tautz, 2006) that spacecraft charging is primarily determined by the plasma temperature. While it is true that below a certain temperature, the plasma discharges the spacecraft rather than charges it, and thus there is a "threshold" in temperature necessary for spacecraft charging, it is not true that a maximum plasma temperature implies a maximum in charging.

8 THE GALAXY 15 FAILURE

The ideal example of a surface-charging-related spacecraft failure is Galaxy 15 (Ferguson et al, 2011). The failure of this heavily used GEO telecommunication satellite (at 09:48 UT on April 5, 2010) happened shortly after an intense substorm hit at ∼09:10 UT (Connors et al, 2011) during the otherwise only moderate magnetic storm of April 5–6, 2010. In the event that led to the Galaxy 15 failure, the electron temperature abruptly increased for the last 20 min of eclipse up to the record maximum of 90 keV about 7 min before eclipse exit (leading to a possible 87 kV *frame charging*), then trailed off until the satellite underwent its failure about 28 min after eclipse exit. Even though the electron temperature at the time of the failure was still very high, Nascap-2k showed that the *differential potential* was only about 800 V because the minuscule electron density led to a low charging flux. And when considering the Galaxy 15 failure as one of the worst-case differential charging events, it is outranked by all other contenders (see Fig. 6). Ferguson et al (2011) argued that perhaps the Galaxy 15 failure was due to "deep" dielectric charging of wires just below the thermal blankets, because the fluence of >200 keV particles was within the range for such discharges (NASA-HDBK-4002A), but subsequent proprietary information definitively places the cause of the upset on surface charging, not interior charging.

9 CONCLUSION

Whether we like it or not, our modern civilization depends on functioning satellites (Ferguson et al., 2015a and also Chapter 23 of this volume):

(1) communications [TV, telephones (land and mobile)—GEO communications satellites];
(2) timekeeping (GPS, etc.);
(3) navigation (GPS, etc.);
(4) transportation (air traffic control—GPS, etc.; train and truck tracking—GPS, etc.);
(5) agriculture (planting and harvesting—GPS, etc.);
(6) wildlife management (GPS, etc.);
(7) earthquake, volcano, weather, and climate monitoring (GPS, etc.);
(8) defense (surveillance and other intelligence, weapons guidance).

Many of these require a functioning GPS system, which is dependent on at least 24 functioning satellites in the constellation. If only a few of them are knocked out by a severe charging event, the disruption to society would be incalculable.

We are led to the conclusion that as far as spacecraft charging goes, the most extreme events are those that last for the longest time, as more satellites will be affected by them and undergo arcing. Like Galaxy 15, whose severe substorm (AE > 2000 nT, Connors et al, 2011) hit while in eclipse and continued for at least a half-hour after eclipse exit, many satellites will find themselves in the wrong place at the wrong time.

REFERENCES

AIAA/ANSI Standard S-115, 2013. Low Earth Orbit Spacecraft Charging Design Standard Requirement and Associated Handbook. American Institute of Aeronautics and Astronautics.

Bodeau, M., 2015. Review of better space weather proxies for spacecraft surface charging. IEEE Trans. Plasma Sci. 43 (9), 3075–3085.

Carruth, M.R., et al., 2001. ISS and Space Environment Interactions Without Operating Plasma Contactor. AIAA Paper 2001-0401.

Cho, M., Kawakita, S., Nakamura, S., Takahashi, M., Sato, T., Nozaki, Y., 2005. Number of arcs estimated on solar array of a geostationary satellite. J. Spacecr. Rockets 42 (4), 740–748.

Connors, M., Russell, C.T., Angelopou, V., 2011. Magnetic flux transfer in the 5 April 2010 Galaxy 15 Substorm: an unprecedented observation. Ann. Geophys. 29, 619–622. https://doi.org/10.5194/angeo-29-619-201.

DeForest, S.E., 1972. Spacecraft charging at synchronous orbit. J. Geophys. Res. 77, 651–659.

Dichter, B.K., et al., 2015. Specification, design, and calibration of the space weather suite of instruments on the NOAA GOES-R Program spacecraft. IEEE Trans. Nucl. Sci. 62 (6), 2776–2783. https://doi.org/10.1109/TNS.2015.2477997.

ECSS-E-ST-20-06C, 2008. Space Engineering, Spacecraft Charging. ECSS Secretariat, ESA-ESTEC Requirements & Standards Division, Noordwijk. 31 July 2008.

E-JERG-2-211A, 2012. Design Standard, Spacecraft, Charging and Discharging. JAXA, Japan. May 10, 2012 Revision A.

Fennell, J.F., Koons, H.C., Roeder, J.R., Blake, J.B., 2001. Substorms and Magnetic Storms from the Satellite Charging Perspective, Aerospace Corp., El Segundo, CA, USA, Tech. Rep. TR-2000, 8570-2.

Ferguson, D.C., Wimberly, S.C., 2013. The best GEO daytime spacecraft charging index. In: 51st AIAA Aerospace Sciences Meeting including the New Horizons Forum and Aerospace Exposition, Grapevine, TX, Jan. 7–10.

Ferguson, D.C., Schneider, T.A., Vaughn, J.A., 2009. Effects of cryogenic temperatures on spacecraft internal dielectric discharges. In: 1st International Conference on Space Technology, Thessaloniki, Greece, August 24–26.

Ferguson, D.C., Denig, W.F., Rodriguez, J.V., 2011. Plasma Conditions During the Galaxy 15 Anomaly and the Possibility of ESD from Subsurface Charging, AFRL-RV-PS-TP-2011-0004, 15 September 2011, Air Force Research Laboratory, Space Vehicles Directorate, 3550 Aberdeen Ave SE, Air Force Materiel Command, Kirtland Air Force Base, NM 87117-5776.

Ferguson, D.C., Hilmer, R.V., Davis, V.A., 2015a. Best geosynchronous earth orbit daytime spacecraft charging index. J. Spacecr. Rocket. 52 (2), 526–543.

Ferguson, D.C., Worden, S.P., Hastings, D.E., 2015b. The space weather threat to situational awareness, communications, and positioning systems. IEEE Trans. Plasma Sci. 43 (9), 3086–3098.

Ferguson, D., Crabtree, P., White, S., Vayner, B., 2016. Anomalous GPS power degradation from Arc-induced contamination. J. Spacecr. Rocket. 53 (3), 464–470.

Gendrin, R., 1980. Kinetic energy density and diamagnetic effects during plasma injection events. Geophys. Res. Lett. 7 (12), 1105–1108.

IG Inside GNSS, 2009. http://www.insidegnss.com/node/1579.

Inouye, G.T., 1977. Spacecraft charging anomalies on the DSCS II, launch 2 satellites. In: Pike, C.P. (Ed.), Proc. of 1st Spacecraft Charging Technology Conference. AFGL-TR-77-0051, NASA TMX-73537, p. 829, 24 Feb.

ISO 11221, 2011. Space Systems—Space Solar Panels—Spacecraft Charging Induced Electrostatic Discharge Test Methods. International Organization for Standardization, Geneva, Switzerland, 01 Aug.

Lai, S.T., 1994. Recent Advances in Spacecraft Charging. AIAA 94-0329.

Lai, S.T., 2008. Critical temperature for the onset of spacecraft charging-the full story. In: 39th AIAA Plasmadynamics and Lasers Conference. AIAA-2008-3782.

Lai, S., Tautz, M., 2006. High-level spacecraft charging in eclipse at geosynchronous altitudes: a statistical study. J. Geophys. Res. 111, A09201. https://doi.org/10.1029/2004JA010733. AFRL-VS-HA-TR-2006-1089.

Likar, J.J., et al., 2011. Geosynchronous ESD environment characterization via in situ measurements on host spacecraft. In: Inst. of Electrical and Electronics Engineers International Symposium on Electromagnetic Compatibility, EMC, Inst. of Electrical and Electronics Engineers, 978-1-4577-0811-4/11, Aug. pp. 653–660.

Maejima, H., et al., 2004. Investigation of power system failure of a LEO satellite. In: 2nd International Energy Conversion Engineering Conference, International Energy Conversion Engineering Conference, IECEC.

Mandell, M.J., Davis, V.A., Cooke, D.L., Wheelock, A.T., Roth, C.J., 2006. Nascap-2k spacecraft charging code overview. IEEE Trans. Plasma Sci. 34 (5), 2084–2093.

Mateo-Velez, J.-C., et al., 2016. From GEO/LEO environment data to the numerical estimation of spacecraft surface charging at MEO. In: 14th Spacecraft Charging Technology Conference. ESA/ESTEC, Noordwijk, NL. 04–08 April.

McPherron, R.L., Baker, D.N., Crooker, N.U., 2009. Role of the Russell–McPherron effect in the acceleration of relativistic electrons. J. Atmos. Sol. Terr. Phys. 71, 1032–1044.

Mullen, E.G., Gussenhoven, M.S., Hardy, D.A., Aggson, T.A., Ledley, B.G., Whipple, E., 1986. SCATHA survey of high-level spacecraft charging in sunlight. J. Geophys. Res. 91 (A2), 1474–1490. https://doi.org/10.1029/JA091iA02p01474.

NASA-HDBK-4002A, 2011. Mitigating In-Space Charging Effects—A Guideline. NASA Standards, Washington, DC, p. 151.

NASA-HDBK-4006, 2007. Low earth orbit spacecraft charging design handbook. NASA Standards, June.
NASA-STD-4005, 2007. Low earth orbit spacecraft charging design standard. NASA Standards, June.
NOAA Technical Memorandum OAR SEC-88, 2004. Halloween Space Weather Storms of 2003. In: Weaver, M. et al., (Ed.), Space Environment Center, Boulder, CO. June.
O'Brien, P.T., 2009. SEAES-GEO: a spacecraft environmental anomalies expert system for geosynchronous orbit. Space Weather 7, S09003. https://doi.org/10.1029/2009SW000473.
Odenwald, S., 2015. http://www.solarstorms.org/Sscope.html.
Purvis, C.K., Garrett, H.B., Whittlesey, A.C., John Stevens, N., 1984. Design guidelines for assessing and controlling spacecraft charging effects. NASA TP-02361.
Thomsen, M.F., Henderson, M.G., Jordanova, V.K., 2013. Statistical properties of the surface charging environment at geosynchronous orbit. Space Weather 11, 237–244. https://doi.org/10.1002/swe.20049.
Webb, D., Allen, J., 2004. Spacecraft and ground anomalies related to the October–November 2003 solar activity. Space Weather 2. S03008. https://doi.org/10.1029/2004SW000075.

DEEP DIELECTRIC CHARGING AND SPACECRAFT ANOMALIES

CHAPTER 16

Shu T. Lai[*,†], Kerri Cahoy[‡], Whitney Lohmeyer[‡], Ashley Carlton[‡], Raichelle Aniceto[‡], Joseph Minow[§]

MIT Space Propulsion Laboratory, Cambridge, MA, United States[*] Boston College, Newton, MA, United States[†] MIT Space Systems Laboratory, Cambridge, MA, United States[‡] NASA Marshall Space Flight Center, Huntsville, AL, United States[§]

CHAPTER OUTLINE

1 Introduction 419
2 What Is Deep Dielectric Charging? 420
   2.1 The Roles of Ions 422
3 Space Environments 423
   3.1 Trapped Radiation 423
   3.2 Temporal Variation of the Radiation Belts 423
   3.3 Relative Role of Electrons and Ions 424
4 Deep Dielectric Charging and Discharging 426
5 Dependence from Geomagnetic Indices: Dst and Kp Index 428
6 Delay Time 428
7 Discharge Event Parameters 429
8 Spacecraft Design Guidelines 429
9 Conclusion 431
Acknowledgment 431
References 431

1 INTRODUCTION

The space radiation environment presents hazards to satellites and the services they provide by causing internal and surface charging, displacement damage, accumulated dose effects, etc., which have led to many documented satellite anomalies (e.g., Baker, 2000; Fennell et al., 2001; Frederickson, 1996). This chapter focuses on internal charging, which is one of the main causes of space environment-related anomalies (Koons et al., 1999). We present the physical mechanisms and environments that may lead to internal charging, an overview of some notable on-orbit spacecraft anomalies that are attributed to internal charging, and conclude with spacecraft design guidelines that can be used to mitigate internal charging threats.

2 WHAT IS DEEP DIELECTRIC CHARGING?

Deep dielectric charging (also called internal, or bulk charging) occurs when high-energy electrons and ions (MeV and higher) penetrate satellite shielding materials and deposit charge on internal spacecraft components. The terms "deep dielectric" or "bulk" charging refer to charge densities that accumulate within insulating (or dielectric) materials when exposed to penetration radiation. The more generic term "internal charging" includes charge densities that accumulate on the surfaces of conducting materials within the shielding afforded by the outer structure of a spacecraft. Both processes result in electric fields within spacecraft structures and materials and represent a threat to arcing which can damage spacecraft components.

The physics of internal charging for an insulator exposed to radiation environments is described by a set of two equations (Sessler, 1987; Sessler et al., 2004)

$$\nabla \cdot E = -\nabla^2 \Phi = \frac{\rho}{\kappa \varepsilon_0} \tag{1}$$

$$\frac{\partial \rho}{\partial t} = -\nabla \cdot (J_R + J_C) \tag{2}$$

where the Eq. (1) is Poisson's equation that relates the electric field, E, and electric potential, Φ, to the charge density, ρ, within the material. The continuity equation in Eq. (2) relates the charge density accumulating in the material to the incident radiation (electron) current density, J_R, and the conduction current density, J_C. The parameters κ and ε_0 are the dielectric constant for material exposed to the charging environment and permittivity of free space, respectively. Eqs. (1), (2) together describe the magnitude and direction of a time and spatially dependent electric field within a material due to the accumulation of charge from an external radiation source and loss of charge from the material through conduction. Typically, only electron currents are considered for internal charging because ions at similar energies as the electrons don't penetrate deeply into materials. Ions with sufficient energy to penetrate deep into the materials typically exhibit very low flux compared to the electrons with the same penetration depth. Thus, ions are not a significant contributor for internal charging currents in most internal charging environments. The role of ions will be considered in Section 2.1. Electron currents are given by

$$J_C = \sigma E = \left[\sigma_{dark} + k_p \left(\frac{d\gamma}{dt}\right)^\alpha\right] E \tag{3}$$

where the bulk conductivity of the insulator σ is divided into the σ_{dark} conductivity (in the absence of exposure to photons or charged particles) and a radiation induced conductivity that depends on the dose rate ($d\gamma/dt$) generated by interaction of the radiation field with the dielectric material. The parameter k_p is the coefficient of radiation-induced conductivity and the exponent $0.5 < \alpha < 1$ depends on the energy distribution of electron trapping states in the insulating material. Evaluation of internal charging requires solving both a radiation transport problem to compute the charge deposition and radiation dose rates within the material and the electrostatic problem to obtain the electrostatic fields and potentials due to the accumulating charge density.

The propagation of the high-energy particles is governed by the Bethe-Bloch equation (Garrett and Whittlesey, 2012; Ziegler, 1999):

$$S(E) = -\frac{dE}{dx} = \frac{4\pi Z^2 n_e}{mv^2}\alpha^2 \left[\log\left(\frac{2mv^2}{1-\beta^2}\right) - \beta^2 - \log I\right] \quad (4)$$

where $S(E)$ is the stopping power, which is the energy loss per unit distance of particle penetration into matter.

$$S(E) = -\frac{dE}{dx} \quad (5)$$

The range R, which means the depth, of penetration by a charged particle of initial energy E_p is given by

$$R = \int_0^{E_p} \frac{1}{S(E)} dE \quad (6)$$

Using Eqs.(4)–(6), one can plot the range, or penetration depth, for a given projectile and a given material.

Fig. 1 shows that the penetration depth of electrons is about two orders of magnitude as ions of about the same energy. For a typical spacecraft wall with a thickness of 100 mils (2.54 mm) of aluminum, electrons need energies in the range 0.5–5 MeV to penetrate, and protons need energies of 10–100 MeV. For a more in depth discussion, see (Ziegler, 1999).

One of the key characteristics of deep dielectric charging is the time it takes to build up electric fields within an insulating material or on a conductor compared to the time it takes for charge to leak

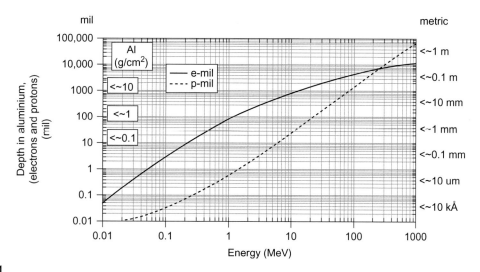

FIG. 1

Electron *(solid)* and proton *(dashed)* penetration depth in aluminum (in mils on the left y-axis and in meters on the right y-axis) for a range of energies (0.01–1000 MeV). For 100 mils (2.54 mm) of aluminum, protons need to have energies above about 20 MeV and electrons need to have energies above 1 MeV to pass through. *Reprinted with permission from Garrett, H., Whittlesey, A., 2012. Guide to mitigating spacecraft charging effects. In: JPL Space Science and Technology Series, Inc. John Wiley & Sons, Pasadena, CA.*

off the material. An estimate of this time constant can be obtained from a simple one dimensional, planar model for internal charging of an insulator where the electric field E at a depth X in a dielectric is given by Garrett and Whittlesey (2000).

$$\varepsilon(dE/dt) + \sigma E = J_R \tag{7}$$

The parameters E, J_R, and ε, and σ are the electric field, radiation current density, absolute permittivity of the material, and σ the volume conductivity of the insulating material described in Eq. (3). A solution to this equation assuming that J_R and σ are independent of time is

$$E = E_0 \exp(-\sigma t/\varepsilon) + (J_R/\sigma)(1 - \exp(-\sigma t/\varepsilon)) \tag{8}$$

where E_0 is the electric field at $t=0$. The absolute permittivity of the material is related to the permittivity in a vacuum ε_0 by the dielectric constant $\kappa = \varepsilon/\varepsilon_0$. Using this definition in Eq. (8), the time constant for charging of the material is estimated by $\tau = \varepsilon/\sigma$ or $\tau = \kappa\varepsilon_0/\sigma$. The charges deposited inside the dielectrics can stay inside for hours, days, or weeks (Rodgers and Sorenson, 2012; Bodeau, 2010). If the dielectric's resistivity is high, the rate of charge build up can overcome the leakage rate of the material and large charge densities can accumulate within a material. In addition to the properties of the dielectric, the time it takes for internal charging to occur depends on the duration of the elevated high-energy flux environment and the material thickness. Internal electric fields, material temperature, ionization properties, and history of discharges are also important factors (Lai, 2012).

The induced electric field may exceed the breakdown threshold for the material, causing electrostatic discharge (ESD) in the insulating material (Baker, 2000; Fennell et al., 2001; Frederickson, 1996; Garrett and Whittlesey, 2012). It is generally accepted that an internal electric field E^* of 10^6–10^8 V/m in dielectric materials may pose a risk for internal discharge. The resulting discharge is potentially hazardous directly to the material, or indirectly to other spacecraft components, causing spacecraft component anomalies.

2.1 THE ROLES OF IONS

The role of electrons penetrated deep inside dielectrics has been known for years. The electrons penetrated inside provide high internal electric fields which may be responsible for spacecraft anomalies in orbits exposed to MeV electron fluxes in the outer radiation belts. The high internal electric fields are necessary for deep dielectric charging. The roles of ions have not received much attention. We discuss two roles for ions in the internal charging process as follows.

The flux of electrons is about two orders of magnitude higher than that of ions because of the mass difference. During exposure to a high-energy electron environment, the electrons deposited deeply inside a material build up high electric fields. Days after the exposure, low energy ions are attracted to the surface of the dielectric by the electric fields generated by the internal charge density and deposit charge inside the material on the surface or at relatively shallower depths within the material. The double layer gradually formed inside the dielectric material enhances the internal electric fields.

Another role of high-energy ions is enhancing the probability of discharges (Fig. 2). It is generally believed that a high-energy heavy particle penetrating inside may cause a single event upset (SEU). A SEU may cause displacement damage or a sudden increase in ionization density in a local region, or track, enhance the conductivity there. In addition, if a high-electric field is already present in that

3 SPACE ENVIRONMENTS

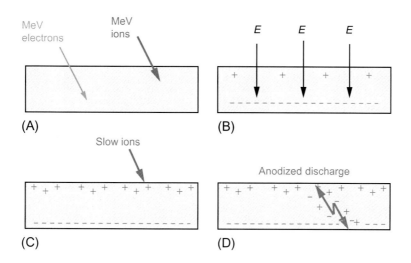

FIG. 2

(A) Penetration of high-energy electrons and ions into dielectrics. (B) Electric field build-up inside the dielectric material. (C) Ions coming in at shallower depths enhancing the internal electric field. (D) A spontaneous discharge may or may not occur if the electric field reaches a critical value. An external factor, such as a very high-energy particle, or a meteor, impact, may induce a discharge (Lai, 2012).

region, the increase of ionization density together with the high-electric fields, may enhance the mobility of the electrons there. As a result, a discharge may occur.

3 SPACE ENVIRONMENTS
In this section, we examine environments that may lead to deep charging.

3.1 TRAPPED RADIATION
High-energy particles can be found in the trapped energetic particle environment in Earth's magnetic field, a region known as the Van Allen radiation belts. The belts differ in distribution and energy for protons and electrons. They are generally located at about 1 R_E–4 R_E with energies of about 0.1–400 MeV and at about 2 R_E–8 R_E with energies of about 0.4–4.5 MeV for protons and electrons, respectively (Hastings and Garrett, 1996). The AE8 and AP8 models (and now AE9/AP9 models) are typically used to predict the variations in the belts (Sawyer and Vette, 1976; Vette, 1991; SPENVIS, n.d.).

3.2 TEMPORAL VARIATION OF THE RADIATION BELTS
Fig. 3 shows the modeled radiation belt fluxes during solar maximum. A similar figure can be obtained for solar minimum. The AE8 and AP8 models (now the updated AE9 and AP9 models) allow us to

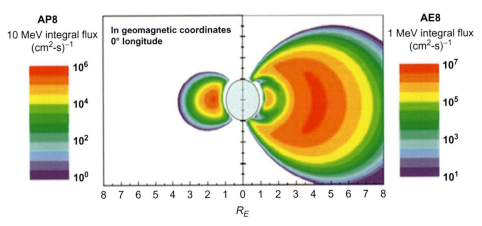

FIG. 3

Integral flux (cm^2-s)$^{-1}$ of protons (>10 MeV) and electrons (>1 MeV) as a function of altitude (where 1 R_E=6371 km) in Earth orbit at 0° longitude, using the AP8/AE8 models at solar maximum. Generated by using the AE8/AP8 models in SPENVIS (Sawyer and Vette, 1976; Vette, 1991; SPENVIS, n. d.). The mapping is done in magnetic coordinates given here in earth radii. There are up-to-date versions: AE9 and AP9. For illustration purpose, it is sufficient to use AE8/AP8.

estimate radiation belt fluxes at solar maximum and solar minimum, and hence the variation between these two cases. It should be noted that solar maximum and solar minimum do not necessarily reflect the solar cycle variation of outer belt electron fluxes. For example, we often see the highest fluxes in the declining phase of the cycle when high-speed solar wind streams are more stable and long-lived (Wrenn et al., 2002).

It is important to also note that the temporal variation of the electron fluxes in the belts often exhibits many periods of rapid intensification (minutes to hours) followed by slow decay (weeks)—as shown in Fig. 4. The temporal behavior is a significant issue in assessing deep dielectric charging threats to spacecraft in MEO and GEO orbits inside the radiation belts.

As shown in Fig. 4, the high-energy (2–6 MeV) electron fluxes in the radiation belts (especially the outer belt) do not vary smoothly with the solar cycle. The fluxes vary according to solar events, such as the arrival of a CME. In the solar cycle declining phase, very energetic particles from cosmic rays can sometimes come in. As will be discussed later, spacecraft anomalies related to deep dielectric charging and discharging occur more often during the declining phase of the solar cycle. Further details of radiation belt variations including solar cycle dependence and role of wave-particle interactions can be found in the Chapters 13 and 14 of this volume.

3.3 RELATIVE ROLE OF ELECTRONS AND IONS

Because of the mass difference between ions and electrons, the electron flux exceeds the ion flux in space by typically two orders of magnitude. When a severe geospace storm hits a spacecraft, the

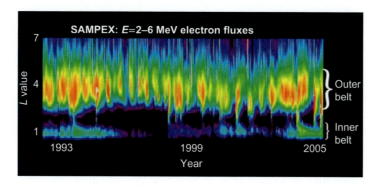

FIG. 4

Long-term (12-year) plot from SAMPEX showing temporal variation of the two radiation belts.

electrons and ions penetrate into the dielectric materials. For electrons and ions in the energy range of 1 MeV to several tens of MeV, the electrons penetrate deeper than the ions, as shown in Fig. 1. As a result, the electric fields building up inside depend very much on the electrons accumulating inside. The accumulation depends at the same time on the leakage rate, which is slow for dielectric materials. In other words, the electron accumulation plays the role of building up the internal electric fields. The accumulation takes time.

Fluence F is the integration of flux J over a period of time T

$$F(T) = \int_0^T J(t)\,dt \tag{9}$$

Relating to the information in Fig. 2, let us take the radiation-belt region in red, where the flux J is 10^7 electrons cm^2 s^{-1}. Such a flux J contributes a fluence $F = 10^{11}$ electrons cm^2 in 10^4 s or about 3 h. The actual accumulation depends also on the charge leakage, which is slow for dielectrics materials. Therefore, leakage slows the build-up of the internal electric field and lengthens the time needed for reaching the critical electric field E^*.

If the electric field built up inside reaches a critical value E^*, typically 10^6–10^8 V/m depending on the material, then a spontaneous discharge may occur. If a spontaneous discharge does not occur, the electric fields may continue to build up to higher values, forming an extremely hazardous situation. It takes time to build up the situation, which is unstable. Any disturbance applied to the situation may trigger a discharge. Because of the time needed for accumulation and for a probability of disturbance, such a discharge occurs with a delay.

When the peak of the severe geospace storm has passed, lower energy ions, being attracted by the electric fields, may come in and deposit at shallow depths. As a result, the internal electric fields are enhanced to even higher values. The probability of a discharge becomes higher than at the peak of the severe geospace storm. In other words, the protons play the role of enhancing the internal electric fields and potentially triggering the deep dielectric discharging. Again, the discharge occurs with a delay. Delay is a characteristic of deep dielectric charging and discharging.

A typical example of such a delay is hours, days, or weeks. Such a delay time is not a physical constant. It depends on the rate and duration of the incident energetic electron flux which varies in time at various locations in geospace, as well as the electrical properties of the material which control the rate at which charge may be lost, including electrical resistivity, temperature, and aging of the materials.

Extremely high energy protons of energies reaching many tens of MeV may penetrate well inside a material or spacecraft structure (Fig. 1). Suppose a beam of high-energy protons or extremely energetic cosmic ray ions (100 MeV, more or less) hit the dielectrics, which happen to be deeply charged a priori; the ions may penetrate deeply into the dielectrics generating a region, or trail, of ionization. Such a scenario may trigger a discharge. In other words, high-energy proton beams or cosmic-ray ions may play the role of triggering deep dielectric discharging. Although such a scenario requires a delay for charging accumulation followed by a rare probability of high-energy ion penetration and deposition. In this way, the ions play the roles of (1) enhancing the internal electric fields because of the differential ranges between electrons and ions and triggering is instantaneous, and (2) increasing the probability of triggering a discharge. Therefore, the delay time for an ion storm, or ion beam, to generate a discharge is comparably short, perhaps a day or a few hours. Again, it is not a physical constant but depends on the material properties, rate of depositions, etc.

4 DEEP DIELECTRIC CHARGING AND DISCHARGING

Deep dielectric charging is not directly related to spacecraft anomalies. In a hazardous environment, deep dielectric charging occurs and builds up the internal electric fields. A spontaneous discharge may, or may not, occur even when the electric field reaches sufficiently high values. A triggering impetus may bring forth the occurrence of a discharge. If very high energy cosmic ray particles, including protons or heavier elements, impacting on the dielectric impregnated with deep charging may trigger a discharge. A meteor impact may also trigger such a discharge.

In the laboratory, it has been repeatedly demonstrated with ease that a piece of plexiglass irradiated by MeV electrons for a long time may, or may not, undergo a spontaneous discharge (Fig. 5). This situation is a metastable equilibrium, which is unstable. Likewise, a supercharged dielectric material may, or may not, readily undergo spontaneous discharge. However, touching the deeply charged plexiglass with a grounded wire or impacting it with a hammer can readily induce a discharge (Fig. 5).

The solar cycle is an important factor in determining the likelihood of spacecraft anomalies related to deep dielectric charging. It has been found (Wrenn et al., 2002; Lohmeyer and Cahoy, 2013) that occurrence of these anomalies is higher during the declining/minimum phase of a solar cycle (Fig. 6). Two factors could contribute to the result. First, as noted above, outer radiation belts tend to intensify during the declining/solar minimum phase. It could explain the build-up of fluences required to produce a discharge. Second, during the declining phase of a solar cycle, the high-energy cosmic ray particles can penetrate into the Earth's geospace more readily (Mewaldt, 2013; Bazilevskaya et al., 2014). Therefore, there is higher probability of high-energy cosmic particles on the spacecraft surfaces during this period. The cosmic particles may initiate discharges in deeply charged dielectrics, but this factor is a hypothesis which needs further examination.

4 DEEP DIELECTRIC CHARGING AND DISCHARGING 427

FIG. 5

Triggering a discharge in a precharged plexiglass. *Reprinted with permission from Bert Hickman, http://capturedlightning.com.*

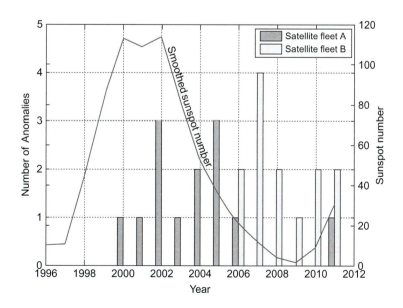

FIG. 6

Yearly total of SSPA anomalies. The sunspot curve is smoothed. Most of the anomalies occurred during the low period of the solar cycle (Lohmeyor and Cahoy, 2013).

5 DEPENDENCE FROM GEOMAGNETIC INDICES: DST AND KP INDEX

The disturbance storm time (Dst) is an index of magnetic activity giving information about the strength of the ring current and other magnetospheric currents and is derived from a network of near-equatorial geomagnetic observatories. During quiet time (Dst) typically takes on values that fluctuate around 0 nT. Negative Dst indicates a geomagnetic storm (e.g., Dst of −500 nT or less is considered an extreme storm). The Kp index is the averaged value of measurements by globally distributed sensors on the ground, which is used to characterize the magnitude of geomagnetic storms. The Kp index ranges from 0 (quiet) to 9 (extreme geomagnetic activity). It does not indicate the local flux rates at the location of a spacecraft. Unlike surface charging, discharges due to deep dielectric charging do not necessarily occur quickly in response to the storm level. It may take time to build up the internal electric field. Dst and Kp indices, and also Ae index, are therefore not usually considered when looking at discharge risks.

6 DELAY TIME

It has been observed that spacecraft anomalies often occur after, but not during, the peak energetic electron flux. With charge storage and decay time in dielectrics, the charge buildup peaks with a delay time. As a case in point, the GOES-5 satellite (Baker et al., 1987) experienced high fluence of high-energy electrons for days before an anomaly occurred (Fig. 7A) and both Equator-S and Galaxy 4 failures behaved similarly (Fig. 7B).

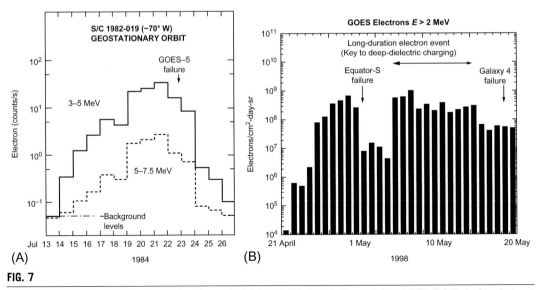

FIG. 7

(A) High-energy electron counts during the period July 13–26, 1984. The satellite GOES-5 failed after, but not during, the peak flux high energy electron flux (Baker et al., 1987).(B) Galaxy-4 and Equator-5 spacecraft failure events showing delays of failures occurring after the peak of the electron fluence (Baker et al., 1998).

7 DISCHARGE EVENT PARAMETERS

It is difficult to define an extreme deep dielectric charging event. One can define a space environment in which deep dielectric charging is likely to occur. One can also define a catastrophic discharge event as the result of deep dielectric charging.

When dielectric materials accumulate significant charge, they can become a hazard. However, there may be no impacts to the material or component until a discharge occurs. Discharge occurrences depend on several parameters:

1. The penetration depth of a charged particle from the environment depends on the *energy of the incoming particle* and the material properties. For 1 mm of aluminum, a proton must have approximately 10 MeV, and an electron must have 0.6 MeV.
2. The *particle energy distribution*. For incoming electrons and ions of approximately the same energy, the deposited charges form a double layer inside because of the difference in penetration depths of electrons and ions. The double layer not only enhances the internal electric field but also cancels the net electric field just outside the dielectric material. In this situation, any surface measurement would indicate a low electric field but the internal field may be building up to a high value.
3. The *fluence*—the integration of the incoming flux over a period of time—is also a key parameter (Han et al., 2005). The balance between the incoming and leakage flux results in a *delay time* τ for the internal field to build up to a critical value E^*. The delay may be hours, days, or longer (Ziegler, 1999). For example, (Lohmeyer and Cahoy, 2013; Lohmeyer et al., 2015) found that anomalies in the dielectric of high-power amplifiers were statistically more likely 14–21 days following high-energy electron flux events. A discharge may not occur during a high flux period, but may instead occur during a following lower flux period, due to the build-up of charge over time.
4. One major parameter is the *critical electric field for dielectric breakdown*, $E^* = 10^6$ to 10^8 V/m. It is possible, but unlikely, to have an extreme electric field $E \gg E^*$, inside the dielectric without any discharge. Such a situation is an unstable equilibrium. It requires another parameter to cause a discharge.
5. Another parameter is the *dielectric material age*. The E^* value decreases as the dielectric material ages. Small cracks may develop in the material, and multiple small local discharges may also lower E^* as tiny discharge paths may remain.
6. The *leakage flux* of the dielectric material is a critical parameter. Because the flux at high energies (0.6 MeV or higher) is lower than at lower energies, it can take time to build up high electric field inside. The time development of the charge build up is governed by the competition between the incoming flux and the leakage flux. It depends on the conductivity of the material and may change in space gradually as a result of material temperature, space radiation, charge density deposition, and other factors (Hanna et al., 2013).
7. Any *disturbance*, such as the impact of a high-energy proton, or heavy ion, penetration, a meteorite passing through, or a small discharge leading to avalanche ionization, may cause a significant discharge (Lai, 2012).

8 SPACECRAFT DESIGN GUIDELINES

Two sets of design guidelines for electrostatic discharge (ESD) hazards have been established. The first, and most commonly cited, is captured in NASA-HDBK-4002A (2011), which specifies a 10-h safe fluence level of $\leq 2 \times 10^{10}$ electrons/cm^2 (Violet and Frederickson, 1993). The safe-fluence level is derived from the Combined Release and Radiation Effects Satellite (CRRES), which was launched into an elliptical orbit in 1990 to study fields, plasmas, and energetic particles through the Van Allen radiation belts. CRRES was equipped with an Internal Discharge Monitor (IDM), which recorded deep dielectric charging events (Fig. 8) (Frederickson, 1996; Hanna et al., 2013). Using the CRRES data, (Frederickson et al., 1992) suggested a threshold for deep dielectric discharge risk at 2×10^{10} electrons/cm^2 for 10 h because CRRES experienced no ESD events when the accumulated fluence *inside the IDM* over a 10-h orbit was less than 2×10^{10} electrons/cm^2 (Frederickson et al., 1992).

Wrenn and Smith (1996) define two thresholds for space hazards. Threshold I states that significant probability of hazard exists when >2 MeV daily electron fluence outside the spacecraft exceeds 3.8×10^9 electrons/cm^2. Threshold II states that an extremely significant probability of hazard exists when >2 MeV daily electron fluence exceeds 3.8×10^{10} electrons/cm^2 (Wrenn and Smith, 1996). These thresholds are based upon empirical correlations of anomalies with the external electron flux data. However, keep in mind that the >2 MeV electron flux outside the spacecraft is only a proxy for the actual electron fluence reaching the source of ESD inside the spacecraft. The range of a 2 MeV electron is about 170 mils (4.3 mm) aluminum (Fig. 1). If the shielding between the ESD source and external environment is less than that, then lower energy electrons can penetrate the shielding and contribute to the charging of the ESD source. There are far more electrons at energies lower than 2 MeV, so the >2 MeV external flux level may understate the total flux of electrons reaching the ESD source if the shielding is substantially lower than 170 mils aluminum.

Recent findings by Bodeau (2010) and Lohmeyer et al. (2015) discuss statistical correlations between high-energy electron fluence over periods of 1–3 weeks prior to anomalies (Bodeau, 2010). Lohmeyer et al. (2015) confirm that a 26-component set of solid state power amplifier anomalies on eight geostationary communications satellites cluster at a higher than expected rate when >2 MeV electron fluence for 14 and 21 days prior to the anomalies was high compared to all >2 MeV electron fluence measurements for similar durations in the 1996–2012 interval. A Monte Carlo analysis conducted on daily >2 MeV electron fluence measurements show a higher rate of occurrence of anomalies after high fluence than expected by chance.

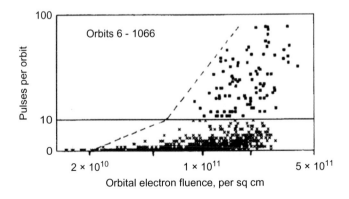

FIG. 8

Discharge pulses from orbits 6 to 1066 on the CRRES satellite as a function of the orbital electron fluence in electrons per cm^2. Reprinted with permission from Frederickson, A.R., Holeman E.G., Mullen, E.G., 1992. Characteristics of spontaneous electrical discharging of various insulators in space radiations. IEEE Trans. Nucl. Sci. 39(6), 1773–1782.

9 CONCLUSION

High-energy electrons and ions that have penetrated dielectric materials may remain inside for a long time because of low conductivity of the material. Electrons of 0.6 MeV can penetrate into 1 mm of aluminum, for example. The buildup of charges inside may be slow because of the low flux of high-energy electrons and ions. When high fluence of high-energy electrons arrives, they may not immediately generate a spacecraft anomaly but will continue to accumulate charges inside. When the charge accumulated reaches a certain level, the probability of a discharge becomes high. For typical dielectric materials, the critical electric field E^* for dielectric breakdown is 10^6–10^8 V/m. From the CRRES experience, the critical fluence of electrons of >2 MeV is 2×10^{10} electrons/cm^2 for 10 h. From experience of spacecraft failures, the delay time for spacecraft failure to occur after exposure to the peak of the high-energy electron flux is about one to several days.

These critical numbers are not physical constants. They depend on the satellite material properties such as conductivity, radiation induced conductivity, geometry, aging of material, history of partial discharges, etc. The dependency is also on the space environment properties such as the electron energy, the locality of the satellite, the duration of high-energy electron event, the duration of a satellite in a certain location, the solar activity, the phase of the solar cycle, etc.

ACKNOWLEDGMENT

We thank the reviewers for helpful comments.

REFERENCES

Baker, Daniel N., 2000. The occurrence of operational anomalies in spacecraft and their relationship to space weather. IEEE Trans. Plasma Sci. 28 (6), 2007–2016.

Baker, D.N., Belian, R.D., Higbie, P.R., Klebesadel, R.W., Blake, J.B., 1987. Deep dielectric charging effects due to high-energy electrons in earth's outer magnetosphere. J. Electrostat. 20, 3–19.

Baker, D.N., Allen, J.H., Kanekal, S.G., Reeves, G.D., 1998. Disturbed space environment may have been related to page satellite failure. EOS Trans. AGU 79, 477.

Bazilevskaya, G.A., Cliver, E.W., Kovaltsov, G.A., Ling, A.G., Shea, M.A., Smart, D.F., Usoskin, I.G., 2014. Solar cycle in the heliosphere and cosmic rays. Space Sci. Rev. 186, 409–435. https://doi.org/10.1007/s11214-014-0084-0.

Bodeau, Michael J., 2010. High energy electron climatology that supports deep charging risk assessment in GEO. In: 48th AIAA Aerospace Sciences Meeting Including the New Horizons Forum and Aerospace Exposition, Orlando.

Fennell, Joseph, Koons, H.C., Roeder, Jim, Blake, J.B., 2001. Spacecraft Charging: Observations and Relationships to Satellite Anomalies. The Aerospace Corporation, El Segundo, CA. (8579)-5 2001.

Frederickson, Arthur, 1996. Upsets related to spacecraft charging. IEEE Trans. Nucl. Sci. 23 (2), 426–441.

Frederickson, A.R., Holeman, E.G., Mullen, E.G., 1992. Characteristics of spontaneous electrical discharging of various insulators in space radiations. IEEE Trans. Nucl. Sci. 39 (6), 1773–1782.

Garrett, Henry, Whittlesey, Albert, 2012. Guide to mitigating spacecraft charging effects. JPL Space Science and Technology Series, Inc. John Wiley & Sons, Pasadena, CA.

Garrett, Henry B., Whittlesey, Albert C., 2000. Spacecraft charging, an update. IEEE Trans. Plasma Sci. 28, 2017–2028.

Han, J., Huang, J., Liu, Z., Wang, S., 2005. Correlation of Double Star anomalies with space environment. J. Spacecr. Rockets 42 (6), 1061–1065. https://doi.org/10.2514/1.14773.

Hanna, R., et al., 2013. Radiation induced conductivity in Teflon FEP irradiated with multi-energetic electron beam. IEEE Trans. Plasma Sci. 41 (12), 3520–3525.

Hastings, Daniel, Garrett, Henry, 1996. Spacecraft-environment interactions. Cambridge Atmospheric and Space Sciences Series, Cambridge University Press, Cambridge.

Koons, H.C., et al., 1999. The Impact of Space Environment on Space Systems. The Aerospace Corporation, El Segundo, CA. TR-99(1670)-1 TR-99(1670)-1.

Lai, Shu T., 2012. Fundamentals of Spacecraft Charging: Spacecraft Interactions with Space Plasmas. Princeton University Press, Princeton, NJ.

Lohmeyer, Whitney, Cahoy, Kerri, 2013. Space weather radiation effects on geostationary satellite solid-state power amplifiers. AGU Space Weather 11 (8), 476–488.

Lohmeyer, W.Q., et al., 2015. Response of geostationary communications satellite solid-state power amplifiers to high-energy electron fluence. AGU Space Weather 13 (5), 298–315.

Mewaldt, R.A., 2013. Cosmic rays in the heliosphere: requirements for future observations. Space Sci. Rev. 176, 365. https://doi.org/10.1007/s11214-012-9922-0.

Avoiding Problems Caused by Spacecraft On-Orbit: Internal Charging Effects. National Aeronautics and Space Administration, NASA-HDBK-4002A (2011).

Rodgers, D.J., Sorenson, J., 2012. Internal charging. In: Lai, S.T. (Ed.), Spacecraft Charging, Progress in Astronautics and Aeronautics. In: vol. 237. AIAA Press, Reston, VA, pp. 143–164.

Sawyer, D.M., Vette, J.I., 1976. AP-8 Trapped Proton Environment for Solar Maximum and Solar Minimum. NASA-GSFC, NSSDC-76-06.

Sessler, G.M. (Ed.), 1987. Electrets. second ed. Topicsin Applied Physics, vol. 33. Springer-Verlag, Berlin, Germany.

Sessler, G.M., et al., 2004. Models of charge transport in electron-beam irradiated insulators. IEEE Trans. Dielectr. Electr. Insul. 11, 192–202.

SPENVIS, https://www.spenvis.oma.be/.

Vette, J.I., 1991. The AE-8 Trapped Electron Model Environment. NASA-GSFC, NSSDC-91-24.

Violet, M.D., Frederickson, A.R., 1993. Spacecraft Anomalies on the CRRES satellite correlated with the environment and insulator samples. IEEE Trans. Nucl. Sci. 40 (6), 512–521.

Wrenn, G.L., Smith, R.K., 1996. Probability factors governing ESD effects in geosynchronous orbit. IEEE Trans. Nucl. Sci. 43 (6), 2783–2789.

Wrenn, G.L., Rodgers, D.J., Ryden, K.A., 2002. A solar cycle of spacecraft anomalies due to internal charging. Ann. Geophys. 20, 953–956.

Ziegler, J.F., 1999. The stopping of energetic light ions in elemental matter. J. Appl. Phys. 85 (1999), 1249–1272.

SOLAR PARTICLE EVENTS AND HUMAN DEEP SPACE EXPLORATION: MEASUREMENTS AND CONSIDERATIONS

17

Livio Narici*,†, Alessandro Rizzo*,†, Francesco Berrilli*,†,‡, Dario Del Moro*

University of Rome Tor Vergata, Rome, Italy *INFN Roma Tor Vergata, Rome, Italy† INAF - National Institute for Astrophysics, Rome, Italy‡*

CHAPTER OUTLINE

1 Radiation in Space and Health Risks for Astronauts	433
2 GCR vs SPEs: Different Approaches for Risk Mitigation	435
3 SPEs as Measured in a Space Habitat (International Space Station)	437
4 The ALTEA Detector Onboard the ISS	438
5 Results From SPE Measurements in the ISS	438
5.1 The December 13, 2006, SPE	438
5.2 The March 7, 2012, SPE	441
5.3 The May 17, 2012, SPE	444
6 Final Remarks	446
6.1 Forecasting at 1 AU	446
6.2 Forecasting at Space Habitat	447
6.3 Radiation on Mars	447
6.4 Countermeasures	448
7 Conclusions	449
Acknowledgements	449
References	449

1 RADIATION IN SPACE AND HEALTH RISKS FOR ASTRONAUTS

Radiation risks are probably the most important health hazard during long-duration missions in deep space. Other than the short voyages to the moon in the 1970s, the only long experience of the effect on humans of space radiation was in low Earth orbit (LEO) onboard space stations. In the last two decades, several astronauts and cosmonauts have lived in space for years, both on the MIR Station and more recently on the International Space Station (ISS). In LEO, radiation risks are significantly reduced due to the Earth's magnetic field and the thin residual atmosphere. Future exploration targets will take humans away from these "shelters" to explore the solar system. Therefore, they will require a significant effort in the improvement of the mitigation of radiation effects (Durante and Cucinotta, 2011).

The main radiation sources to consider for deep-space exploration are the galactic cosmic rays (GCRs) and the solar particle events. GCRs are characterized by a wide spectrum peaking at high energies (in the GeV/nucleon region), are mostly isotropic and constant in time, and feature a small modulation by the 11-year and 22-year solar cycles (for a review please see Potgieter, 2013). Their high energy makes them almost impossible to shield significantly. Conversely, solar particle events (SPEs) appear occasionally, are made mostly of low-energy protons (mainly below a few hundreds MeV), and feature unpredictable spectra peaked at energies much lower than GCRs. They are difficult to forecast due to the complexity of the associated physical processes, involving the dynamics of the emerging magnetic field in the solar photosphere, which is the visible layer of solar atmosphere; the acceleration associated with flares in the solar upper atmosphere, that is, the chromosphere and corona, or in interplanetary shocks; and finally the propagation in the interplanetary medium and along interplanetary magnetic fields (IMFs) (for a review Ryan et al., 2000). Considering their lower energies, SPEs are easier to shield than GCRs. Nevertheless, a few of the known historical events could have been lethal for unprotected crews (such as the August 4, 1972, solar event).

Astronauts/cosmonauts/taikonauts will never be directly exposed to space radiation. They will be living in a protected habitat, such as a spacecraft during the voyage or a base during their stay on an extraterrestrial body. In case an extra vehicular activity (EVA) is required, they will be protected by a space suit.

The space radiation (GCRs and SPEs) will therefore interact first with these protective materials through nuclear processes such as fragmentation (of the impinging radiation and of the nuclear target). From these types of nuclear interactions, neutrons can arise. These neutral massive particles may lead to significant risks due to their high penetration power. The flux and the spectrum of the radiation reaching the astronaut's skin is therefore significantly different from the one characterizing the radiation outside the space habitat.

Once the radiation reaches the astronaut's skin, it undergoes another series of transformations and interacts with underlying tissues and organs, releasing energy that can cause cellular damage due to ionization processes (direct or indirect) or free radicals formation. These interactions may be harmless, they may increase the probability of developing cancer, or they may lead to acute functional impairment. Extreme cases may even lead to death.

The quantification of space radiation risks, in terms of acute sickness probability but also long-term increase of death probability, relies on the understanding of the interactions between radiation and living matter. This is a major issue in space radiobiology research. The absorbed dose (D), the energy per unit of mass, is insufficient to describe such risks; the same absorbed doses of different kinds of radiation may lead to different biological effects. The dose equivalent (H) is defined as the absorbed dose weighted for the quality factor Q, which takes into account the radiobiological effectiveness of the radiation. The value of the Q factor is expressed as a function of the radiation linear energy transfer (LET). For a definition of these quantities in the frame of human space exploration, please see Durante and Cucinotta (2011). The evaluation of dose equivalent for astronauts in space is also found to be partly inadequate for a full description of the risk assessment in human space exploration. A more detailed characterization of the radiation might therefore be needed for a reliable risk estimate. However, for the scope of this chapter, we will use the dose equivalent as a parameter directly related to health risk.

The smaller contribution to the dose, and even more so to the dose equivalent given by protons, makes the detected particle flux the most sensitive parameter for their physical description. For this reason the solar events in this work will be characterized in terms of particle fluxes and for the largest event (March 2012), the dose and dose equivalent evaluation will also be discussed.

2 GCR VS SPEs: DIFFERENT APPROACHES FOR RISK MITIGATION

A typical dose from GCRs measured during a deep-space mission, such as the Earth to Mars transit by MSL Rad (Zeitlin et al., 2013), is 0.33 mGy/day, corresponding to a flux of about one particle/(cm^2 s). This value is mostly constant and inversely modulated by solar activity with 11-year and 22-year cycles (Potgieter, 2013). As mentioned, the GCR spectrum is peaked at a few GeV/nucleon. This flux is constituted mostly of protons and, to a much smaller extent, helium ions, with a very small contribution from higher Z ions. However, the high Z ions contribution to the relevant quantities for risk assessments (D and H) is high because of their larger amount of deposited energy during passage.

Conversely, SPEs are constituted mostly of protons with kinetic energies below 10^2 MeV. Their flux, however, can reach very high values (several orders of magnitude above GCRs, see for example Zeitlin et al. (2013)) and this may represent a high risk for astronaut health. Even though these lower energies make shielding feasible, the hull of a spacecraft, a space base, or a space suit is not sufficient to reduce the risk to a harmless level.

Space radiation countermeasures can intervene at different levels. The general idea is either to mitigate the damage at a biophysical level (pharmacological countermeasures) or to reduce the amount of radiation hitting an astronaut's body. While there is hope that in the future the first solution will provide significant protection, today the only real countermeasure is what is called "shielding" (Durante and Cucinotta, 2011), either "passive" (material interposed between the radiation source and the body) or "active" (magnetic fields able to deviate the charged radiation, mimicking the Earth's magnetic field).

The different characteristics of the two components of space radiation (GCRs and SPEs) suggest different approaches for risk mitigation.

To completely shield a spacecraft from GCRs (energetic and penetrating) in a deep-space voyage, a large-mass shielding system would be needed, which is unfeasible in space. This limit, set by the maximum amount of mass that can be sent into space, is currently valid for both passive and active solutions. It might be expected that technological advances would make an active shielding system more effective "per unit of mass." However, today a careful trade-off analysis, including a detailed analysis of the biological risks, must be carried out because only integrated countermeasure options appear feasible to obtain an acceptable risk mitigation for GCRs. Planetary bases would also benefit from materials found on the planet itself.

Protection from the high fluxes (but briefly extended in time) and low energies of the SPE events can follow a different approach. The idea is that, due to the lower energy, a sufficient amount of mass can be sent to space to shield a limited volume, not the entire spacecraft. The critical time lengths of harmful SPEs can be estimated to affect less than 1% of a mission. This leads to the concept of shelter where the crew might go for a limited amount of time to be protected from intense short-term critical events.

In principle, this shifts the problem from shielding to knowing, in due time, when the event will take place; that is, forecasting the SPE. In a first approximation, following the warning provided by forecasting, the crew would move into the shelter, completely solving the safety problem. Such a system relates precursor information and, for several years, this has been the goal of space weather services for support of space missions (e.g., the NOAA Space Environment Services Center in Boulder, Colorado).

A few important issues must be considered for risk mitigation during human exploration.

As highlighted in the following sections as well as elsewhere in this book, forecasting an SPE is the outcome of very complex solar and space physics related to particle acceleration during a solar flare associated with a CME, or by the shocks accompanying CMEs in interplanetary space, channeled by IMF lines. The massive effort produced over the years to provide reliable forecasting is driven by the significant damage these solar events may produce on Earth, including damage to satellites in LEO that causes severe problems on the ground (such as communication or GNSS satellites). The efficacy of a forecasting model (e.g., Papaioannou et al., 2015) as well as the evaluation of the effects of false negatives and false positives are therefore quantified in terms of money saved (for example from satellites as well as ground-based facility repairs). In the case of human exploration, this trade-off analysis is quite different: a false negative is life threatening and a false positive may have "only" mission-planning consequences (such as an unneeded move into the shelter). This consideration leads to the claim that SPE forecasting should have no (or almost no) false negatives. Unfortunately, with the presently available solar physics knowledge, this is not achievable.

A short-time-scale forecasting technique, often called *nowcasting*, is also under study in many laboratories in the world. The idea is to use photons coming from the SPE sources to warn about the arrival of the slowest component (charged particles). Photons are faster than ions not only because they travel at the speed of light, but also because their travel path is not affected by the solar and IMF, which may largely increase the path traveled by ions before reaching the spacecraft. Photons are also easily shielded by the spacecraft hull, thanks to their low energies. Therefore, the critical quantity is the delay Δt between the precursors (traveling at light speed) and the dangerous portion of the SPE, containing the fast ions. This delay is typically many minutes, but may approach zero when fast charged particles are emitted by the sun on almost straight trajectories to the spacecraft, in the most unfortunate topology of the magnetic field. A usable nowcasting for this situation is almost impossible, for the same reasons mentioned above.

A shelter solution is based on the warning in due time. Optimally this warning should arrive at least 30–60 minutes (this is the amount of time that is currently used for tests to move the crew into a shelter) before the arrival of dangerous solar particles. In the worst-case scenario (forecasting false negative plus fast ions, reaching the spacecraft with almost straight trajectory), this time can be much shorter. This implies that the crew will be exposed to the first part of the dangerous event. This introduces the need to define what could be the acceptable supplemental radiation dose for each individual astronaut.

In sum, there are three important elements of these SPE countermeasures: (i) forecasting, (ii) nowcasting, and (iii) a running definition of an acceptable supplemental radiation dose. This last item also requires an understanding of the risks due to a high radiation flux at low LET and of the SPE insurgence dynamics.

Mission plans will be developed on the basis of accurate forecasting, which might also be considered to set crew alertness levels. Nowcasting will drive the move into the shelter. This, however, shall also take into account the amount of the dose that each individual crewmember can still take. The described SPE countermeasure still has to prove a solid reliability in keeping the total exposure to radiation below the expected safety thresholds.

3 SPEs AS MEASURED IN A SPACE HABITAT (INTERNATIONAL SPACE STATION)

The major difference between the measurement of the SPE radiation in free space and the one measured inside a space habitat is the secondary radiation that arises from nuclear reactions such as target fragmentation and spallation reactions. The target material is not only the hull of the spacecraft, which is made mostly of aluminum, but also the other materials (such as racks and computers) inside the space habitat placed in between the astronauts and the radiation source. There is indeed an increasing awareness of the need for integrated solutions in the choice of materials to build a spacecraft in order to maximize radiation risk mitigation. Secondary radiation is of interest because it is conceivable that the risk due to this fragmentation might be comparable, or even higher, than the one that would have been caused by the original ion (in the SPE case, the original proton). Several active detectors were designed to measure with necessary detail the different characteristics of this radiation field (LET spectra, fluence, absorbed dose, dose equivalent, quality factor). Data obtained from these detectors onboard ISS (see for example Lee et al., 2007; Zhou et al., 2009,2010; Narici et al., 2015; Berger et al., 2016, 2017) allow us to obtain temporal and spatial variations of the radiation field at LEO, and support radiation risk assessment.

The only published detailed radiation data of an SPE in a space habitat are measurements performed on the ISS (Larosa et al., 2011; Di Fino et al., 2014; Berrilli et al., 2014; Semkova et al., 2013, 2014; Chapter 13 of this volume). A concurrent measurement inside and outside the ISS during an SPE would be highly desirable; however, no measurements of this kind have been performed or planned.

The radiation impinging on the ISS in LEO is composed of the deep space one (GCR and SPE) and the radiation trapped in the South Atlantic Anomaly (SAA), a portion of the lower Van Allen belts getting closer to Earth that is mostly composed of low-energy protons (mostly below a few hundreds MeV) and electrons. However, these are of minor importance for risk assessment and are not considered here. The SAA is over Brazil in the southern hemisphere. In order to study the deep-space radiation, the SAA contribution has to be disentangled from other contributions of interest. This result can be achieved using active detectors, provided the possibility of selecting the orbital tracts of interest. The data that will be considered in this chapter, therefore, do not contain measurements acquired during SAA passages.

Moreover, the protection offered by the Earth's magnetic field stops SPE ions from reaching the ISS in most of the orbital tracts. The SPE ions can reach the ISS only during the highest latitude passages (ISS orbits feature a 51.6-degree inclination). The effect of the SPE on the particle data is mostly additive to the normal radiation (GCR and SAA).

We will present here data from three events as measured by the ALTEA detector system on the ISS, with a focus on the March 2012 SPE.

4 THE ALTEA DETECTOR ONBOARD THE ISS

The ALTEA detector system is made of six identical silicon detector units (SDUs), which can be arranged in several configurations, read by a data acquisition unit (DAU), which also provides the power (Zaconte et al., 2008, 2010a,b; La Tessa et al., 2009a).

Each ALTEA SDU is a particle telescope with six silicon planes that can determine the energy loss and the trajectory of a passing-through particle. A single silicon plane (380 µm thick) presents two squared active areas of 8×8 cm^2, spaced by 5 mm; each square is segmented into 32 strips with a 2.5-mm pitch. Two consecutive planes inside an SDU are orthogonally segmented to reconstruct the x-y coordinate of the track, while the position of the paired planes inside the SDU gives the z coordinate. The interplanar space between a pair of silicon planes is 3.75 mm, while the distance between two pairs is 37.5 mm.

A single SDU is triggered by the passage of an ion through the whole stack of silicon planes, releasing an overthreshold amount of energy.

This threshold can be set via software. With the settings used during these measurements, the LET in silicon for the detected particles ranges from a threshold of 3–800 keV/µm. This corresponds to protons of about 25–45 MeV and He nuclei of about 25–250 MeV/n, all passing through ions with $Z > 2$ up to molybdenum.

Data is sent on ground and analyzed in real time (Di Fino et al., 2006, 2012).

ALTEA has been operating in ISS from August 2006 to November 2012, taking data for a total time of about 3.5 years in several configurations.

ALTEA participated in several measurement campaigns sponsored by different agencies (ASI, NASA, ESA). The first campaign (Narici, 2008) was aimed at studying ions traveling through the brains of astronauts (2006–09). For this experiment the six SDUs were arranged in a helmet configuration to be positioned on the astronaut's head. The same helmet, when not in use for these measurements, was positioned flat against a rack to continuously study the radiation environment. The first SPE presented here (Larosa et al., 2011) was measured with this configuration (see Fig. 1). The data shown are from SDU1, which is the telescope that was operative over the entire time frame considered for the SPE study. This telescope was oriented along the y axis of the USLab module. This direction has been evaluated to be the least shielded, and equivalent to 5 cm Al (La Tessa et al., 2009b; Di Fino et al., 2011; Narici et al., 2012)

From 2010 to June 2012, the ALTEA system was arranged in a Cartesian configuration (Narici et al., 2015): two SDUs were positioned with their field of view along each of the three axes of the ISS. With this configuration, ALTEA performed a survey of the radiation environment in the USLab, detecting the other two SPEs while on the USLab port side overhead (Fig. 2).

During this last period of operation (2010–12), ALTEA measured the effects of about 20 SPEs in the ISS. Reports on these measurements are in preparation. In this chapter we will show some of the results from the following SPEs: (1) December 13, 2006 (Larosa et al., 2011), (2) March 7, 2012 (Di Fino et al., 2014), and (3) May 17, 2012 (Berrilli et al., 2014).

5 RESULTS FROM SPE MEASUREMENTS IN THE ISS
5.1 THE DECEMBER 13, 2006, SPE

An X3.4 solar flare occurred on December 13, 2006, in the active region NOAA AR10930. As reported in Shibasaki (2012), it started at 0220 UT and lasted longer than 100 minutes. The author reports that

5 RESULTS FROM SPE MEASUREMENTS IN THE ISS

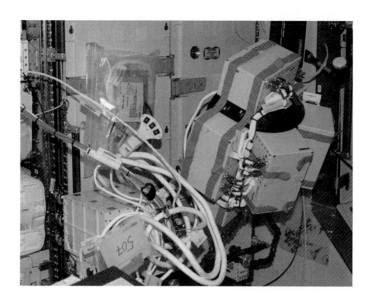

FIG. 1

The ALTEA configuration used when measuring, in the USLab, the SPE of December 13, 2006.

Photo NASA Courtesy.

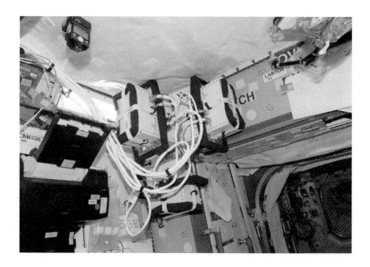

FIG. 2

The ALTEA configuration used when measuring, in the USLab (port side overhead), the SPE of March 7, March 13, and May 17, 2012.

Photo NASA Courtesy.

during the impulsive phase of the flare, two ribbon structures (appearing like filaments), parallel to the line separating two regions of opposite magnetic polarity (north from south) dividing the Active Region, were visible in HINODE-SOT Ca H images. Successively, a series of magnetic loops with two ribbons at their footpoints, and a cusp-like shape, were observed by the X-ray telescope HINODE-XRT. Consequently, this flare fits the standard solar flare model and represents a very good example of the magnetic reconnection, the physical process able to convert magnetic energy into electromagnetic radiation and kinetic energy. Such kinetic energy is released in the form of energetic particles generating an SPE. Given the aforementioned causal connection between flares and SPEs, it is important to monitor the flaring activity and the magnetic field of the Sun, although only a minority of solar flares produce SPEs detectable at LEO orbits.

In this first part of the presentation of the ALTEA radiation measurements in LEO during SPEs, we use these results to focus on the identification of the peculiar radiation modulations due to the Earth's magnetic field.

The particle count rate by ALTEA in 10 days around the flare is shown in Fig. 3 (Larosa et al., 2011), where the different modulations of the radiation flux can be readily appreciated. It should be noted that electromagnetic radiation from flares is not relevant for radiation risk assessment because it is easily shielded.

Focusing on the bottom part of the figure, we can see, on the left side, (i) a 1-day modulation (labeled "longitude effect"), (ii) periodically appearing peaks (with apparently random amplitude), labeled "SAA," and (iii) a clear increase of the (i) occurring on some specific days ("flare"). On the right side there is an explosion of a typical single-day flux, with the SAA cut out (as previously said the SAA contribution will be cut out in the following results as it is not relevant for deep-space exploration).

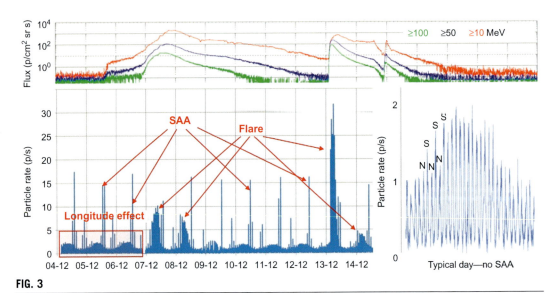

FIG. 3

GOES proton data (above) and ALTEA data below, during the December 13, 2006, SPE. The different components and modulation of the ALTEA measurements are marked. On the bottom right, a typical day in a quiet period to enlarge the south-north poles asymmetry.

This rightmost plot clearly shows the shielding effect of the Earth's magnetic field. During orbital tracts over the equator, the count rate is minimal, approaching zero. Than moving to the north pole (some of the north and south pole passages are marked respectively with N and S in the rightmost plot) the count rate increases, gets to a maximum value at maximum latitude (the ISS inclination is 51.6 degrees), then decreases again almost to zero over the equator to get to another maximum over the south pole. Rotational and magnetic Earth axes are not coincident, and the ISS orbit gets closer to the magnetic south pole than to the magnetic north pole. This can be seen from the higher peaks over the south pole. During the orbits the Earth rotates over its rotational axis, with the period of 1 day modulating, with the same period, the closest distance between the ISS orbit and the magnetic poles. This can be appreciated in the 1-day modulation ("longitudinal effect"). During passages over the SAA region, the protons trapped in this part of the Van Allen Belt add to the measured count rate and show up in those peaks, appearing in irregular bunches every day. The reason for this regularity is due to the daily transits over the SAA. The reason for the irregularity (in the number and heights of peaks) is the different track lengths over the SAA and the different distances of these tracks from the center of the SAA, featuring the highest radiation flux.

In sum, the effect of the SPE can be seen only when the magnetic rigidity is low enough to let the particles in. And to reach this situation, not only must the ISS be at the highest latitudes, but it must also be in an orbit close to the magnetic pole.

Because the satellite is in GEO, the GOES proton flux is not modulated by the Earth's magnetic field. In Fig. 3 it is possible to see that the December 7 flare began when the ISS was mostly protected and the first signal in the ISS could only be measured after several orbits. The activity as seen by GOES continued for a few days, reaching a maximum on December 8 and then slowly decaying. In the ISS it could be seen only when the geomagnetic shielding was weak enough during the orbital tracts closer to the magnetic poles. The same can be said for the following flare of December 13.

The effect inside the ISS is therefore due to external flux, magnetic field modulation/shielding, and the transformation of the radiation due to the interactions with the space vessel. A rough estimation of the difference between fluxes measured by GOES for the December 7 flare and the December 13 one suggests that: (i) the relevant energy window for the flux measured by GOES appears to be the ≥ 100 Mev (green curve); and (ii) the relation appears, at this level of approximation, more than linear because an increase of about 50% for the GOES proton flux corresponds in the ISS to an increase of about 200%.

The magnetic field modulation does not allow for an easy quantification of the effect inside the ISS. To provide information of the effect of an SPE during a deep-space voyage, the orbital tracts with the highest latitude should be selected (in Fig. 3 this would correspond to following the envelop of the longitude effect when the flux is highest). This selection would also get rid of the SAA passages.

In both cases, even the comparison between effects of two flares might not be straightforward. In this case it is easier, as during both flares' maxima (December 7 and 13) the ISS was at a minimum geomagnetic shielding; however this is not always so, as we will see later.

5.2 THE MARCH 7, 2012, SPE

Two bright X-class flares occurred on March 7, 2012, in the active region NOAA AR11429. The first flare (X5.4) started at 0002 UT and reached its maximum intensity 22 minutes later, while the second flare (X1.3) started at 0105 UT and reached its maximum 9 minutes later. The X-ray emission detected by the GOES satellite lasted several hours. The same satellite detected the protons associated with these

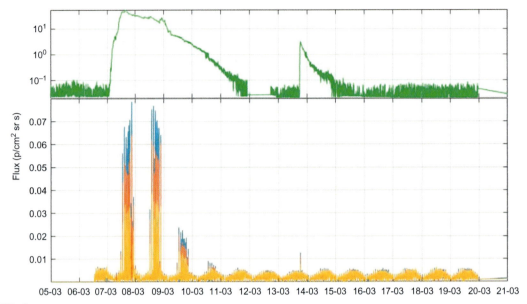

FIG. 4

GOES proton ($E > 100$ MeV) flux (above) and ALTEA flux below during the March 7 and March 13, 2012, SPEs. The three colors in the ALTEA data show the three different directions of the measuring detectors (*brown: X; red: Z; blue: Y*). The South Atlantic Anomaly passages are cut out.

flares in three energy bands. Fermi-LAT detected the brightest solar flare, showing that during most of the emission the gamma rays appeared to come from the AR11429 region responsible for the flare emission (Ajello et al., 2014).

Fig. 4 shows the GOES proton measurements ($E > 100$ MeV) as well as the ALTEA measurements (Di Fino et al., 2014) in the ISS. ALTEA during this time was in *XYZ* configuration, having two detectors pointing to each of three ISS axes (see Fig. 2). These results are presented, selecting out the SAA tracts.

The three colors in the ALTEA flux plot, Fig. 4, represent the measurements along the three ISS Cartesian axes. The brown line represents the flux along the *X* direction, which lies along the major ISS axis, looking through the largest portion of the ISS and therefore the most shielded direction. The blue line represents the flux along the *Y* direction (transversal) looking outside the USLab modulus, with the only shield from the modulus hull and the racks and apparatuses. The *Z* direction is pointing at the zenith (Earth's shadow has been taken into account), so the flux (red line) is also partially shielded by the big truss running over the port side of the USLab. These shielding differences can be appreciated more clearly during SPEs as these are mostly composed of protons with low LET and are therefore easier to be shielded against. Hidden behind this effect might also be a directionality effect of the SPE itself (which is not isotropic). Unfortunately, without an accurate estimation of the shielding (not available), these two effects cannot be separated.

The ISS is orbiting in regions where the Earth's magnetic shield is stronger during the beginning of the SPE and is hit by the SPE only after several orbits. The efficacy of the Earth's magnetic shielding is evident in these data even more than from the December 13, 2006, SPE, as can be seen from the intermittent flux measured inside, totally zeroed during orbits with high magnetic shielding.

The smaller SPE of March 13 produced a significant effect in the ISS only during one orbit. Note that apparently the flux measured by GOES remained high during successive orbits; however, no significant effect was appreciated after the first affected orbit.

Fig. 5 highlights the difference between southern hemisphere orbital tracts (in red) and northern hemisphere ones (in blue). The lower magnetic shield reached by the ISS while approaching the south pole is evident from the larger particle flux in the station.

As mentioned, an SPE contains mostly protons so the physical quantity most affected is the flux. Dose and dose equivalent, the quantities more relevant for risk assessment, are proportional to the energy transferred in the matter; the contribution brought by protons to such quantities is small, due to the low LET of the particles. This doesn't mean that the SPE contribution to the health risk for astronauts is negligible, because the particle flux during such an event is quite high. A small contribution in terms of dose and dose equivalent for each proton summed up on the proton multiplicity gives large values of doses, which may lead to health risks. In Fig. 6 the March 7 SPE data is plotted, averaging the three directions and integrating over a full day, for flux, dose, and dose equivalent. In this case only the measurements at high latitude are considered, being the most relevant for deep-space simulation.

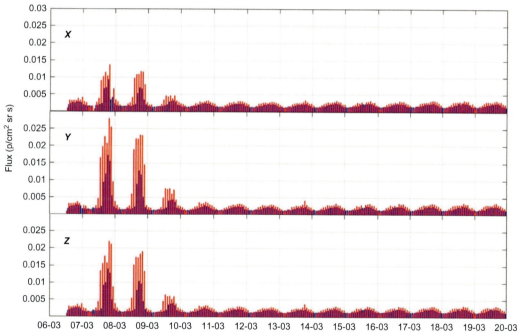

FIG. 5

ALTEA data during the March 7 and March 13, 2012, SPEs. From above: *X*, *Y*, and *Z* directions. In *red* are the measurements over the south hemisphere, in *blue* the measurements over the north hemisphere. The South Atlantic Anomaly passages are cut out.

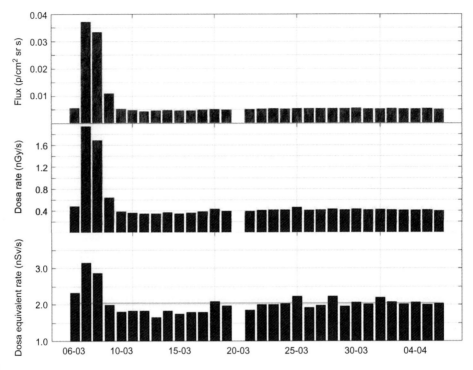

FIG. 6

ALTEA data during the March 7 and March 13, 2012, SPEs. The three directions (X, Y, and Z) are averaged (only measurements acquired at high latitude are considered). Data is integrated over 1 day. From above: flux, dose, dose equivalent. The *red line* in the dose equivalent plot serves as an eye guide to show the Forbush decrease.

This long integration time (1 day) evidence the decrease of the GCR radiation after the SPE. This is named "Forbush decrease," and describes the effect of the Sun magnetic field associated with the Flare, that swipes out the charged particles.

Dose equivalent (H) is the quantity that, in a simple approximation, best relates to the health risk. As expected, the reduction from quiet periods due to the Forbush effect is most evident in dose equivalent. This is due to the fact that the peak during the SPE is mostly produced by protons (with relatively low H) and the Forbush effect reduces all the GCRs, including high Z (and therefore high H) radiation. The radiation comes back to normal (quiet periods) after more than two weeks.

These measurements also suggest that the Forbush effect may partly cancel the increase in dose equivalent due to the SPE. This is partially true but we must also consider the ALTEA small energy acceptance window for protons. Furthermore, for a detailed risk assessment, even if the integrated decrease in dose equivalent due to the Forbush effect was comparable to the increase during the SPE, the radiation rate for the two effects is quite different and the acute risk due the SPE might be significant.

5.3 THE MAY 17, 2012, SPE

The M5.1 flare occurred on May 17, 2012, in the active region NOAA AR11476. The flare started at 0125 UT, reached its maximum at 0147 UT, and ended at 0214 UT. AR11476 was a complex AR, composed of two sections. The first section was bipolar with at least one spot having a penumbra as well as being open and asymmetric. The second section was a bipolar sunspot group with penumbra on both ends; it was elongated in longitude as well as large, symmetric, and open. The flare produced a halo CME, that is a CME launched toward the Earth and forming a bright halo around the Sun, recorded by LASCO aboard the SOHO satellite. The gamma ray burst monitor (GBM) on board the FERMI satellite, sensitive to X-rays and gamma rays with energies between 8 keV and 40 MeV, showed a peak in 6–12 and 12–25 keV channels simultaneous with the GOES maximum. Among those analyzed in this chapter, this flare had the unique characteristic of raising the radiation levels down to Earth's surface, as detected through the Neutron Monitor Network and the IceTop Cherenkov detectors (Balabin et al., 2013; Mishev et al., 2014) and to be therefore associated with the ground level enhancement (GLE) number 71. The proton intensity spectrum during the SPE event has been studied by Kühl et al. (2015).

The very fast increase of proton flux measured by GOES occurred while the ISS was flying well shielded by the Earth's magnetic field (Fig. 7). Only about 6 hours later the particle flux started reaching the ISS and at that time the SPE (as measured by the GOES proton flux) was already down by about 90% from the peak (Berrilli et al., 2014). The same directional difference seen for the March 7 SPE (as said, mostly due to the anisotropic shielding) can be appreciated during this SPE.

Note the SPE peaks in the ALTEA data: they all correspond to south hemisphere tracts (beside the fourth, the lowest peak, which is measured during a north hemisphere passage).

Even considering the partial proton acceptance of the ALTEA detector system (see ALTEA description), these data provide several insights, particularly when comparing different SPEs. This

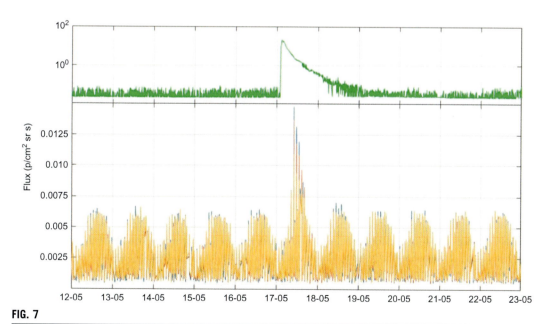

FIG. 7

GOES proton ($E > 100$ MeV) data (above) and ALTEA data below during the May 17, 2012, SPEs. The three colors in the ALTEA data show the three different directions of the measuring detectors (*brown*: X; *red*: Z; *blue*: Y). The South Atlantic Anomaly passages are cut out.

last-shown SPE is the only one associated with a GLE, producing effects measurable on the ground. Nevertheless, it caused a minor effect inside the ISS. This is because, at its maximum, it was shielded by the Earth's magnetic field; however, other causes, such as the emission of protons reaching 1 AU smaller than in the other shown cases, should be taken into account. As clear from what we've seen, the study of the effects of SPEs in the ISS aimed at gathering information for future deep-space voyages is somewhat complicated by the presence of the Earth's magnetic field. Today, the only radiation measurements in space habitat can be done in LEO (in the ISS). In this case it is mandatory to follow the SPE effects evolution only when the ISS is at low geomagnetic rigidity (limited magnetic shield). This corresponds to using these data as intermittent data, with the detector "on" a total of a few hours every day. In this sense a radiation measurement campaign in deep-space habitats (such as the Deep Space Habitat, DSH, to be launched in cis-Lunar space) is highly desirable. Furthermore, the observation of the presented results (see for example the measurements on March 10 and March 13, 2012) suggests that the proton flux at 1 AU cannot be the only parameter to describe the SPE if risks in the ISS are searched for. A complete proton acceptance would permit us to fully study the dose equivalent H, here shown only for the March 7, 2012, SPE (much smaller deviations of H can be shown for the other two SPEs). In this sense an upgrade of ALTEA is in preparation and should be launched in 2018. Finally, the possible partial balance of supplemental radiation from an SPE and a radiation Forbush decrease should also be the object of further investigations.

6 FINAL REMARKS
6.1 FORECASTING AT 1 AU

By the time SPE and X-ray flare events are detected, there is very little time to try to mitigate their effect. CMEs, on the other hand, provide up to 24–48 hours notice of arrival at 1 AU once they are launched from the solar corona (Pulkkinen, 2007; Schwenn, 2006). Once the CME start parameters are gathered, it is possible to estimate its speed and, to some extent, the density of the carried plasma and to determine the ram pressure of the arriving CME at the magnetosphere boundary. Also the effect of aerodynamic drag by the interplanetary medium and the acceleration/deceleration provided by the solar wind has been modeled (e.g., Vršnak and Žic, 2007). Present models provide significantly accurate determinations of the initial CME speed, its speed at 1 AU, and the transit time. These are extremely important parameters, but the geoeffectiveness of a CME is strongly determined by the magnetic field it carries. It is very well known that the most efficient energy transfer into the geomagnetic field occurs for "south-directed" CME field orientations. The orientation of the magnetic field of an Earth-impacting CME can only be determined by in situ measurements made by satellites at the Lagrangian point L1. As a consequence, the time span from the accurate assessment of the CME geoeffectiveness and its consequences on the near-Earth environments is extremely short.

Statistical results demonstrate that halo CMEs and X-class flares are not useful in predicting SPEs: only ∼25% of X-class flares have an associated SPE and ∼10% of halo CMEs have an associated SPE (Odenwald, 2010). On the other hand, a statistical correlation between radio bursts and SPE may prove to be very useful for early warning or short-term forecasting. For example, Li and Kang (2005) exploited a database of 28 SPE between 1997 and 2000 and correlated those events with solar radio bursts recorded between 245 and 15,400 MHz. They found a rather strong correlation between Type-III radio bursts at 245 MHz. This is interpreted as the acceleration of charged particles from the shock generated in the initial stages of CME ejection and, although this provides useful time only for a part of the SPEs, it is nevertheless of extreme relevance in actuating possible countermeasures, at least at distances of 1 AU or greater. This suggests that the way to a sensible form of space weather forecasting will likely need a multimessenger approach, by exploiting different types of instrumentation to apprehend the full complexity of the many physical processes that are occurring in what appears as a single space weather event. An interesting step in this direction is the technique proposed by Laurenza et al. (2009) to provide short-term SPE warnings. The method ingests flare location, flare size, evidences of particle acceleration/escape, time-integrated soft X-ray intensity, and time-integrated intensity of type III radio emissions at $\sim$1 MHz to predict the SPE event at Earth.

At present, there is also an effort to integrate the numerical codes for CME propagation such as the WSA-ENLIL+Cone model (Odstrcil et al., 2004) into the magnetic field transport to go beyond the limitation of only having information about the CME magnetic field direction at a few hours of the CME arrival on the magnetosphere. Such a model uses as input the CME parameters measured at $\sim$20 Rs, by coronagraph LASCO onboard SOHO, and supported when possible by data acquired by the SECCHI/COR2 instrument onboard STEREO-A and STEREO-B. This results in forecasting of CME effects at 1 AU with lead times of 20–30 h for the fastest CMEs. An efficient procedure could be to provide early warnings by running simulations with the CME parameters provided by the coronagraphs only, and then successively update and make more robust the results using the information provided by different instruments covering different part of the electromagnetic spectrum (such as radio).

6.2 FORECASTING AT SPACE HABITAT

If our capacity to provide a resilient forecast at Earth is poor, the capacity to provide a forecast for other points within the heliosphere is feeble. At present time, only timid attempts have been made to simulate the state of space weather beyond 1 AU, although this information is of interest for the management of the many different space missions that are and will be traveling the solar system.

Capital among these will be the forthcoming manned missions to return to the Moon and/or to reach Mars.

Those missions imply radiation exposure for astronauts that have to be carefully evaluated and that, according to the present estimates, can be higher than the thresholds currently set by NASA. On top of the concern for astronaut health, SPE and enhanced GCR periods may prevent an astronaut from performing his or her duties because of the requirement to return to a shielded shelter when the radiation flux exceeds the acceptable levels of risk. Because SPE fluencies are hardly predictable and may last for days, such conditions may persist for days as well. This could leave the astronauts unable to accomplish their mission objectives because they will be not permitted to leave the shelter.

6.3 RADIATION ON MARS

Let us briefly compare the radiation measured on Mars with what is measured on the ISS.

Onboard ISS, under quiet conditions (no SPEs), the dose equivalent rate was about 0.647 mSv/day at the end of May 2016 (Berger et al., 2016). During an SPE, this value can increase about one order of magnitude (Semkova et al., 2013). Of course, during extravehicular activity the exposure is several times that inside the spacecraft, and the radiation risk increases accordingly.

The average radiation environment on the Martian surface is measured to be about the same (0.64 ± 0.12 mSv/day) as that on the ISS (Hassler et al., 2014). This is due to the thin Martian atmosphere, to the shield provided by the planet itself, and to the presence of the radiation trapped in the SAA in the ISS measurements. The only SPE measured on the Martian surface, a mild one, provided an additional dose equivalent of 0.025 mSv. The situation is worst during the Mars transfer due to the lack of the protective actions of Mars and its thin atmosphere. During the trip to Mars (Hassler et al., 2014), 1.84 mSv/day have been measured for the GCRs (a factor three times that on ISS or the Martian surface), with the dose equivalent per SPE ranging from 1.2 to 19.5 mSv, corresponding to two-thirds of a day to more than 10 days of GCR radiation exposure, or about 2–30 years on the Earth's surface.

As a consequence of the values reported above, during an 860-day mission to Mars (180 days trip + 500 days permanence + 180 days trip, based on the NASA design reference mission (Drake et al., 2010)) an astronaut would be exposed to about 1 Sv.

To compare these values to what was measured with ALTEA during a quiet period (0.223 ± 0.003 mSv/day, Narici et al., 2015), which is about one third of what is mentioned above, all these must be taken into account: the solar cycle, the position of the detector, relative different shielding, and, most importantly, the limited energy window for protons.

6.4 COUNTERMEASURES

Given our limited long-term forecast capabilities, full avoidance of SPEs during a mission is not feasible. Much can be done in term of mitigation of the effects through shielding and mission rules, however. The radiation hazard can be greatly reduced with information on space weather, timely warnings, and access to appropriate shelter. As stated before, the duration of an SPE or a close series of SPEs can be up to a few days and the astronauts should be able, in principle, to spend that time inside the shelter. This would require access to food, water, and hygiene facilities for extended times. The size of the shelter would therefore become unfeasible. The definition of the acceptable supplemental radiation dose would become crucial to make it possible to leave the shelter for short periods (minutes to fractional hours) to perform necessary tasks.

EVAs should be scheduled for quiet periods when a severe solar storm is extremely unlikely. It is worth remembering that an astronaut performing an EVA on the lunar surface is exposed to a daily effective GCR dose of 0.85 mSv (and about twice this figure when in EVA in free space), while in a worst-case scenario of a massive SPE, it is foreseen that the astronaut would be exposed to 0.5 Sv on the Martian surface, 150 Sv on the lunar surface, and 300 Sv in free space (Bothmer and Daglis, 2007). In the unfortunate event of an SPE during an EVA far from a habitat with a radiation shelter, the "radiation officer" would note the hazard and direct the crewmember to seek further protection such as terrain that provides additional shielding (caves or cliffs) and employ portable shielding (personal or mini-habitat). Also, exploration vehicles should provide some shielding, which, given the characteristics of the mission, should focus more on SPE than GCR effect prevention and limit as much as possible the astronaut's exposure to prevent acute effects and allow a return to the habitat. In any case, astronauts on the Moon or Mars will need to limit the time they spend outside in their spacesuits and the distance they travel from available radiation shelters.

7 CONCLUSIONS

The mitigation of SPE effects on human space flights is based on a "shelter" approach. This would be truly effective if SPE forecasting (therefore CME and flare forecasting) were available at a reliability level much higher than what is expected over the next years. The cost of a false negative is, when humans are involved, too high to allow for the level of false negatives that are reached today and foreseen for tomorrow. Forecasting can therefore be used only to set a first level of alertness to guide successive mitigation actions.

The use of nowcasting is therefore mandatory. However, the possibility of connected SPEs arriving at the spacecraft very soon after (seconds, minutes?) the first precursors is small but not null. This implies that astronauts may be exposed to the first part of an SPE while still outside the shelter. The risks due to the exposure to the first part of fast-connected SPEs should therefore be assessed. The transformation of the radiation field due to the SPE when entering a space habitat, caused by the interaction with the many different materials between "outside" and "inside," should be addressed when studying such risk assessment. The data presented here are a first example of one of the sets of data needed to carry out this portion of such an endeavor.

Finally, nowcasting procedures should be optimized, identifying the best precursor and possibly suggesting ad-hoc detectors to measure the needed characteristics of such precursors with the appropriate detail and sensitivity.

ACKNOWLEDGMENTS

The authors acknowledge the support from ESA (program ALTEA-shield) and from ASI (that made the development of the ALTEA detector possible: Contracts ALTEA, MoMa-ALTEA, and ALTEA-support). The authors would also like to thank Dr. René Demets from ESA for his very valuable support during the ALTEA-shield project. Finally, the authors are thankful to the ISS crewmembers who made all this possible.

REFERENCES

Ajello, M., et al., 2014. Impulsive and long duration high-energy gamma-ray emission from the very bright 2012 March 7 solar flares. Astrophys. J. 789, 20. https://doi.org/10.1088/0004-637X/789/1/20.

Balabin, Yu.V., Vashenyuk, E.V., Gvozdevsky, B.B., Germanenko, A.V., 2013. Physics of auroral phenomena. In: Proc. XXXVI Annual Seminar, Apatity, Kola Science Centre, Russian Academy of Science, pp. 103–105.

Berger, T., Przybyla, B., Matthiä, D., Reitz, G., Burmeister, S., et al., 2016. DOSIS & DOSIS 3D: long-term dose monitoring onboard the Columbus Laboratory of the International Space Station (ISS). J. Space Weather Space Clim. 6. https://doi.org/10.1051/swsc/2016034. A39.

Berger, T., Burmeister, S., Matthiä, M., Przybyla, B., Reitz, G., et al., 2017. DOSIS & DOSIS 3D: radiation measurements with the DOSTEL instruments onboard the Columbus Laboratory of the ISS in the years 2009–2016. J. Space Weather Space Clim. 7. https://doi.org/10.1051/swsc/201700. A8.

Berrilli, F., Casolino, M., Del Moro, D., Di Fino, L., Larosa, M., Narici, L., Piazzesi, R., Picozza, P., Scardigli, S., Sparvoli, R., Stangalini, M., Zaconte, V., 2014. The relativistic solar particle event of May 17th, 2012 observed on board the International Space Station. J. Space Weather Space Clim. 4. https://doi.org/10.1051/swsc/2014014. A16.

Bothmer, V., Daglis, I.A., 2007. Space Weather: Physics and Effects. Springer, New York. Chapter 5, ISSN 1615-9454.

Di Fino, L., Belli, F., Bidoli, V., Casolino, M., Narici, L., Picozza, P., Rinaldi, A., Ruggieri, D., Zaconte, V., Carozzo, S., Sannita, W., Spillantini, P., Cotronei, V., Alippi, E., Gianelli, G., Galper, A., Korotkov, M., Popov, A., Petrov, V., Salnitskii, V., Avdeev, S., Bonvicini, V., Zampa, N., Zampa, G., Vittori, R., Fuglesang, C., Schardt, D., 2006. ALTEA data handling. Adv. Space Res. 1710–1715. https://doi.org/10.1016/j.asr.2005.01.105.

Di Fino, L., Casolino, M., De Santis, C., Larosa, M., La Tessa, C., Narici, L., Picozza, P., Zaconte, V., 2011. Heavy ions anisotropy measured by ALTEA in the International Space Station. Radiat. Res. 176 (3), 397–406. https://doi.org/10.1667/RR2179.1.

Di Fino, L., Zaconte, V., Ciccotelli, M., Larosa, M., Narici, L., 2012. Fast probabilistic particle identification algorithm using silicon strip detectors. Adv. Space Res. 50, 408–414. https://doi.org/10.1016/j.asr.2012.04.015.

Di Fino, L., Zaconte, V., Stangalini, M., Sparvoli, R., Picozza, P., Piazzesi, R., Narici, L., Larosa, M., Del Moro, D., Casolino, M., Berrilli, F., Scardigli, S., 2014. Solar Particle Event detected by ALTEA on board the International Space Station. The March 7th, 2012 X5.4 flare. J. Space Weather Space Clim. 4. https://doi.org/10.1051/swsc/201401. A19.

Drake, B.G., 2010. Human exploration of mars: challenges and design reference architecture 5.0. J. Cosmol. 12, 3578–3587.

Durante, M., Cucinotta, F.A., 2011. Physical basis of radiation protection in space travel. Rev. Mod. Phys. 83, 1245. https://doi.org/10.1103/RevModPhys.83.1245.

Hassler, D.M., Zeitlin, C., Wimmer-Schweingruber, R.F., Ehresmann, B., Rafkin, S., et al., 2014. Mars' surface radiation environment measured with the Mars Science Laboratory's Curiosity rover. Science. 343(6169) https://doi.org/10.1126/Science.1244797.

Kühl, P., Banjac, B., Dresing, N., Goméz-Herrero, R., Heber, B., Klassen, A., Terasa, C., 2015. Proton intensity spectra during the solar energetic particle events of May 17, 2012 and January 6, 2014. Astron. Astrophys. 576. https://doi.org/10.1051/0004-6361/201424874. A120.

Laurenza, M., Cliver, E.W., Hewitt, J., Storini, M., Ling, A.G., Balch, C.C., Kaiser, M.L., 2009. A technique for short-term warning of solar energetic particle events based on flare location, flare size, and evidence of particle escape. Space Weather 7, S04008. https://doi.org/10.1029/2007SW000379.

La Tessa, C., Di Fino, L., Larosa, M., Narici, L., Picozza, P., Zaconte, V., 2009a. Estimate of the space station thickness at a USLab site using ALTEA measurements and fragmentation cross sections. Nucl. Instrum. Methods Phys. Res. B 267, 3383–3387. https://doi.org/10.1016/j.nimb.2009.06.107.

La Tessa, C., Di Fino, L., Larosa, M., Lee, K., Mancusi, D., Matthiä, D., Narici, L., Zaconte, V., 2009b. Simulation of ALTEA calibration data with PHITS, FLUKA and GEANT4. Nucl. Instrum. Methods Phys. Res. B 267, 3549–3557. https://doi.org/10.1016/j.nimb.2009.06.086.

Larosa, M., Agostini, F., Casolino, M., De Santis, C., Di Fino, L., La Tessa, C., Narici, L., Picozza, P., Rinaldi, A., Zaconte, V., 2011. Ion rates in the International Space Station during the December 2006 Solar Particle Event. J. Phys. G Nucl. Part. Phys. 38, 095102. https://doi.org/10.1088/0954-3899/38/9/095102.

Lee, K., Flanders, J., Semones, E., Shelfer, T., Riman, F., 2007. Simultaneous observation of the radiation environment inside and outside the ISS. Adv. Space Res. 40, 1558–1561. https://doi.org/10.1016/j.asr.2007.02.083.

Li, X.C., Kang, L.S., 2005. Evidence for a strong correlation of solar proton events with solar radio bursts. Chin. J. Astron. Astrophys. 5 (1), 110. https://doi.org/10.1088/1009-9271/5/1/012.

Mishev, A.L., Kocharov, L.G., Usoskin, I.G., 2014. Analysis of the ground level enhancement on May 17, 2012 using data from the global neutron monitor network. J. Geophys. Res. Space Phys. 119, 670–679. https://doi.org/10.1002/2013JA019253.

Narici, L., 2008. Heavy ions light flashes and brain functions: recent observations at accelerators and in spaceflight. New J. Phys. 10, 075010. https://doi.org/10.1088/1367-2630/10/7/075010.

Narici, L., Casolino, M., Di Fino, L., Larosa, M., Larsson, O., Picozza, P., Zaconte, V., 2012. Iron flux inside the International Space Station is measured to be lower than predicted. Radiat. Meas. 47, 1030–1034. https://doi.org/10.1016/j.radmeas.2012.07.006.

Narici, L., Casolino, M., Di Fino, L., Larosa, M., Picozza, P., Zaconte, V., 2015. Radiation survey in the International Space Station. J. Space Weather Space Clim. 5. https://doi.org/10.1051/swsc/2015037. A37.

Odenwald, S., 2010. Space weather – impacts, mitigation and forecasting. In: Schrijver, C.J., Siscoe, G.L. (Eds.), Heliophysics: Space Storms and Radiation: Causes and Effects. Cambridge University Press, p. 15.

Odstrcil, D., Pizzo, V.J., Linker, A., Riley, P., Lionello, R., Mikic, Z., 2004. Initial coupling of coronal and heliospheric numerical magnetohydrodynamic codes. J. Atmos. Sol. Terr. Phys. 66 (15), 1311–1320. Towards an Integrated Model of the Space Weather System. https://doi.org/10.1016/j.jastp.2004.04.007.

Papaioannou, A., Anastasiadis, A., Sandberg, I., Georgoulis, M.K., Tsiropoula, G., Tziotziou, K., Jiggens, P., Hilgers, A., 2015. A novel forecasting system for solar particle events and flares (FORSPEF). J. Phys. Conf. Ser. 632, 012075. https://doi.org/10.1088/1742-6596/632/1/012075.

Potgieter, M.S., 2013. Solar modulation of cosmic rays. Living Rev. Sol. Phys. 10, 3. https://doi.org/10.12942/lrsp-2013-3.

Pulkkinen, T., 2007. Space weather: terrestrial perspective. Living Rev. Solar Phys. 4, 1. https://doi.org/10.12942/lrsp-2007-1.

Ryan, J.M., Lockwood, J.A., Debrunner, H., 2000. Solar energetic particles. Space Sci. Rev. 93, 35. https://doi.org/10.1023/A:1026580008909.

Schwenn, R., 2006. Space weather: the solar perspective. Living Rev. Solar Phys. 3. https://doi.org/10.12942/lrsp-2006-2.

Semkova, J., Dachev, T., Koleva, R., Maltchev, S., Bankov, N., Benghin, V., Shurshakov, V., Petrov, V., Drobyshev, S., 2013. Radiation environment on the International Space Station during the Solar Particle Events in March 2012. J. Astrobiol. Outreach 1, 102. https://doi.org/10.4172/2332-2519.1000102.

Semkova, J., Dachev, T., Koleva, R., Bankov, N., Maltchev, S., Benghin, V., Shurshakov, V., Petrov, V., 2014. Observation of radiation environment in the International Space Station in 2012–March 2013 by Liulin-5 particle telescope. J. Space Weather Space Clim. 4. https://doi.org/10.1051/swsc/2014029. A32.

Shibasaki, K., 2012. The Flare on December 13, 2006 and the Standard Solar Flare Model. In: ASP Conference Series, vol. 454, p. 315.

Vršnak, B., Žic, T., 2007. Transit times of interplanetary coronal mass ejections and the solar wind speed. Astron. Astrophys. 472 (3), 937–943. https://doi.org/10.1051/0004-6361:20077499.

Zaconte, V., Belli, F., Bidoli, V., Casolino, M., Di Fino, L., Narici, L., Picozza, P., Rinaldi, A., Sannita, W.G., Finetti, N., Nurzia, G., Rantucci, E., Scrimaglio, R., Segreto, E., Schardt, D., 2008. ALTEA: the instrument calibration. Nucl. Instrum. Methods Phys. Res. B 266, 2070–2078. https://doi.org/10.1016/j.nimb.2008.02.072.

Zaconte, V., Casolino, M., De Santis, C., Di Fino, L., La Tessa, C., Larosa, M., Narici, L., Picozza, P., 2010a. The radiation environment in the ISS-USLab measured by ALTEA: spectra and relative nuclear abundances in the polar, equatorial and SAA regions. Adv. Space Res. 46, 797–799. https://doi.org/10.1016/j.asr.2010.02.032.

Zaconte, V., Casolino, M., Di Fino, L., La Tessa, C., Larosa, M., Narici, L., Picozza, P., 2010b. High energy radiation fluences in the ISS-USLab: ion discrimination and particle abundances. Radiat. Meas. 45, 168–172. https://doi.org/10.1016/j.radmeas.2010.01.020.

Zeitlin, C., Hassler, D.M., Cucinotta, F.A., Ehresmann, B., Wimmer-Schweingruber, R.F., et al., 2013. Measurements of energetic particle radiation in transit to Mars on the Mars Science Laboratory. Science 340, 1080–1084. https://doi.org/10.1126/science.1235989.

Zhou, D., Semones, E., Gaza, R., Johnson, S., Zapp, N., Lee, K., George, T., 2009. Radiation measured during ISS-Expedition 13 with different dosimeters. Adv. Space Res. 43, 1212–1219. https://doi.org/10.1016/j.asr.2009.02.003.

Zhou, D., Semones, E., O'Sullivan, D., Zapp, N., Weyland, M., Reitz, G., Berger, T., Benton, E.R., 2010. Radiation measured for MATROSHKA-1 experiment with passive dosimeters. Acta Astronaut. 66, 301–308. https://doi.org/10.1016/j.actaastro.2009.06.014.

CHAPTER 18

CHARACTERIZING THE VARIATION IN ATMOSPHERIC RADIATION AT AVIATION ALTITUDES

W. Kent Tobiska*,†, Matthias M. Meier‡, Daniel Matthiae‡, Kyle Copeland§

Space Environment Technologies, Pacific Palisades, CA, United States Utah State University, Logan, UT, United States† Deutsches Zentrum für Luft- und Raumfahrt e.V. (DLR), German Aerospace Center, Institute of Aerospace Medicine, Cologne, Germany‡ Federal Aviation Administration, Civil Aerospace Medical Institute, Oklahoma City, OK, United States§

CHAPTER OUTLINE

1. Radiation Sources and Their Effects on Aviation 453
2. Status of Models 456
3. Status of Measurements 457
4. Status of Monitoring for Extreme Conditions 458
5. Classification of Aviation-Relevant Extreme Space Weather Radiation Events 460
6. Example of an Extreme Event 463
7. Conclusion 466
Acknowledgments 466
References 467

1 RADIATION SOURCES AND THEIR EFFECTS ON AVIATION

The aerospace environment has several sources of ionizing radiation. Exposure to this radiation is one of the natural hazards faced by aircrew, high-altitude pilots, frequent flyers, and, eventually, commercial space travelers. Galactic cosmic rays (GCRs) and solar energetic particles (SEPs) (Fig. 1) almost always are the most important sources of ionizing radiation, particularly when traveling at or above commercial aviation altitudes (8 km or 26,000 ft; Friedberg & Copeland, 2003, 2011; Tobiska et al., 2016).

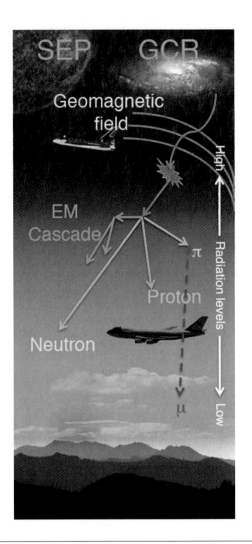

FIG. 1

Sources of primary and secondary cosmic radiation.

GCRs originate from outside the solar system and consist mostly of energetic protons (p^+), with some alpha particles (α) and a few heavier ions (HZE) such as iron (Fe^{26+}) (about 87 p^+:12 α:1 HZE from Simpson, 1983). SEPs originate on the Sun. They are similar in composition to GCRs, being predominantly protons, but with relatively fewer heavier ions. GCRs and SEPs are sometimes collectively referred to as *cosmic rays*.

Regardless of their source, some of these particles transit Earth's magnetosphere and interact with its atmosphere. Cosmic ray particle access to the neutral atmosphere depends upon rigidity (ratio of momentum to charge: particles of the same rigidity follow similar paths in a magnetic field). The Earth's magnetic field acts similar to a high-pass rigidity filter, and cosmic radiation particle access

1 RADIATION SOURCES AND THEIR EFFECTS ON AVIATION

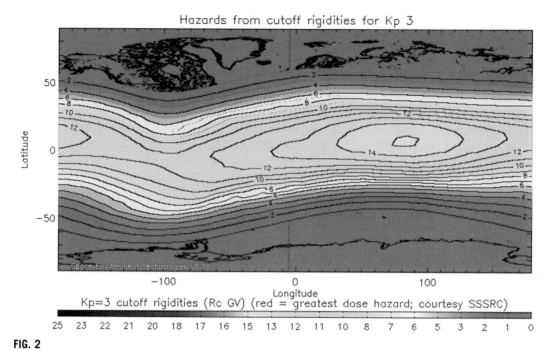

FIG. 2
Effective vertical magnetic cutoff rigidities for the 2010 epoch calculated by Smart and Shea using the IGRF 2010 internal reference field for $Kp=3$; the color bar indicates the notional hazard level based on the increased (lower rigidity) particle flux at higher latitudes (Shea & Smart, 2012).

is well described using a quantity called cutoff rigidity (Fig. 2). During normal geomagnetic conditions, cutoff rigidity varies approximately inversely with geographic latitude; only particles with relatively high rigidity can make it to the atmosphere at latitudes near the equator, while even the lowest rigidity particles can enter the atmosphere at the geomagnetic poles. As a result, the highest primary radiation fluxes enter at high latitudes, with maxima surrounding the geomagnetic poles.

Below the terrestrial atmosphere's mesopause near 85 km, the particles increasingly interact with neutral species, which are predominantly N_2 and O_2. The particle collisions with these target molecules create sprays of secondary and tertiary particles as well as photons with lower energies (collectively called *secondaries*), converting some of the initiating particle's energy into new particles (if the primary cosmic ray particle has enough energy, there will be many generations of secondary particles, called a *shower*) (Fig. 1). The secondaries in the showers produced by cosmic radiation include neutrons (n), p^+, e^-, e^+, α (and other nuclear fragments), pions (π), muons (μ), γ-rays, and x-rays (and to a much lesser degree more exotic particles). Under normal, GCR-dominated conditions (when there is no significant SEP contribution to atmospheric ionizing radiation) the primary particles lose energy, the secondary population increases, and the total ionization increases until this results in a maximum ionization rate between 15 and 20 km (49,000–65,000 ft) called the Regener-Pfotzer maximum (Regener & Pfotzer, 1935). During a time of increased SEP radiation, the relatively high flux and low average energy of the SEP particles compared to GCR can move the Regener-Pfotzer maximum to higher altitudes. Below the Regener-Pfotzer maximum down to the Earth's surface, the ionization rate continues to decrease because particles and photons are absorbed in an increasingly thick atmosphere. All of these particles are able to collide with an aircraft hull and its interior components, people, or fuel to further alter the radiation spectrum that affects tissue and avionics (IARC, 2000; UNSCEAR, 2000).

In human tissues, radiation can activate several injury pathways by causing atoms and molecules to become ionized, dissociated, or excited. These include (i) production of free radicals, (ii) breakage of chemical bonds, (iii) production of new chemical bonds and cross-linkage between macromolecules, and (iv) damage of molecules that regulate vital cell processes, such as deoxyribonucleic acid (DNA), ribonucleic acid (RNA), and proteins (UNSCEAR, 2000). Evidence indicates that high linear energy transfer (LET, a measure of energy lost by a radiation per unit track length) radiations are generally more harmful to living tissues per unit dose (energy deposited per unit of target mass) than low-LET radiations. Low-LET radiations include photons, muons, and electrons, while high-LET radiations are particles such as neutrons, alpha particles, and heavier ions. Protons and charged pions are often also considered low-LET radiation, but interact more like high-LET radiations often enough to be treated separately in dosimetry (ICRP, 2007). Although cells can usually repair damage from low doses of ionizing radiation, particularly if it is low-LET radiation such as that received daily from ambient radiation near the surface of the Earth, cell death is the most likely result from higher doses. At extremely high doses, the cell population in an organ can drop so rapidly that cells cannot be replaced quickly enough and the tissue fails to function normally (IARC, 2000; UNSCEAR, 2000). Even the mildest effects related to this mechanism are not observed until absorbed doses exceed 100 mGy; such doses or corresponding dose rates have not been observed as a result of cosmic radiation in the atmosphere so far, only from technological sources. Such extreme levels of radiation have only been theoretically calculated for hypothetical, extreme solar particle events (Fig. 3). Epidemiological studies in occupational groups have been conducted for several decades, usually with a focus on radiation-associated cancer, and there continues to be a broad discussion in this field of study.

In addition to potential health effects including an increased lifetime risk of cancer, damage to avionics is a matter of concern because it might endanger the safety of a flight (Dyer & Truscott, 1999; Dyer & Lei, 2001; Dyer et al., 2003). While this paper does not pursue a detailed discussion related to radiation effects on avionics, interested readers in this subject matter are pointed to a wide body of work. This includes studies and reports (Normand et al., 1994, 2006; Mutuel, 2016) as well as standards such as the International Commission on Radiation Units (ICRU) Joint Report (84), the International Electrotechnical Commission (IEC) SEE standard for avionics (International Electrotechnical Commission (IEC) 62396–1, 2012), the Joint Electron Device Engineering Council Solid State Technology Association (JEDEC) SEE standard for avionics (JESD89A), the World Meteorological Organization observing requirements (#709, #738), and the International Civil Aviation Organization (ICAO) regulatory guidelines (Standards and Recommended Practices 3.8.1).

2 STATUS OF MODELS

There have been many models developed that are capable of specifying the aviation radiation environment. They represent the breadth and depth of work done in this field for many years. Recently developed models include: AIR (Johnston, 2008), AVIDOS (Latocha et al., 2009; Latocha et al.,

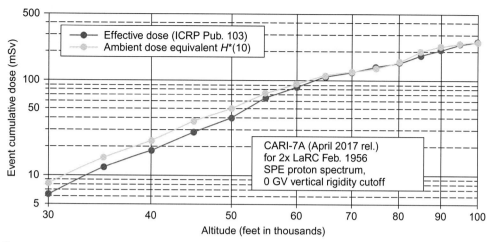

FIG. 3

Effect of altitude on the cumulative ambient equivalent $H^*(10)$ and effective doses for an extreme solar proton event at a polar location (0 GV vertical cutoff rigidity). For this example, twice the LaRC spectrum proton for the Feb. 1956 event was used, which remains the largest event to date of the neutron monitor era (Singleterry et al., 2010). CARI-7A was used for the calculations (Copeland, 2017).

2014), CARI-7 (Copeland, 2017), EPCARD.NET (Mares et al., 2009), FDOSCalc (Wissmann et al., 2010), FREE (Felsberger et al., 2009), KREAM (Hwang et al., 2014), NAIRAS (Mertens et al., 2013), PANDOCA (Matthiä et al., 2014), PARMA/EXPACS (Sato et al., 2008; Sato, 2015), and PC-AIRE (McCall et al., 2009). While all of these models are based on data, the data needed to drive an individual model varies from model to model. At one extreme are models like PC-AIRE and FDOSCalc, which are built from empirical functions fit to in-flight measurement databases. At the other extreme are models like CARI-7, EPCARD.NET, NAIRAS, and PANDOCA, which start from the particle spectrum (SEP or GCR local interstellar spectrum, based on measurements) and then model propagation of the particles through Earth's magnetosphere and atmosphere (and heliosphere for GCRs) using a collection of previously developed physics models. As an example of recent work, Joyce et al. (2014) utilized Monte Carlo simulations of showers coupled to CRaTER measurements (Spence et al., 2010; Schwadron et al., 2012) in deep space to estimate dose rates through the Earth's atmosphere at a range of different altitudes down to aviation heights. While most models could be used for nowcasting with proper data input and enough computing power, the purpose for their development was typically for retrospective evaluations.

3 STATUS OF MEASUREMENTS

Ground-based continuous monitoring of cosmic radiation-related particle data has been ongoing for more than half a century, using neutron monitors, ion chambers, muon telescopes, and other instruments. Space-based monitoring has been regularly performed since the start of the Geosynchronous Operational Environmental Satellite (GOES) program in the 1970s. But until recently, the typical method of measuring dose at commercial aviation altitudes was by in situ instruments that were returned after flight for analysis. A wealth of data related to the aviation radiation environment has made important contributions to model validations of the radiation field at altitude, especially for dose in human tissue. The vast majority of these measurements were made with the Tissue Equivalent Proportional Counters (TEPCs) under GCR background conditions, with very few solar events captured (perhaps fortunately, large SEP events are very rare) (Dyer et al., 1990; Beck et al., 1999; Kyllönen et al., 2001; ECRP 140, 2004; Getley et al., 2005; Beck et al., 2005; Latocha et al., 2007; Meier et al., 2009, 2016a,b; Beck et al., 2009; Dyer et al., 2009; Hands & Dyer, 2009; Getley et al., 2005; Gersey et al., 2012; and Tobiska et al., 2014a,b, 2015). Some solid-state detectors have been used (Dyer et al., 2009; Hands & Dyer, 2009; Ploc et al., 2013; Lee et al., 2015; Tobiska et al., 2016).

To date, however, the difficult task of continuous radiation environment monitoring, reporting, and modeling has not yet been achieved, putting extreme event monitoring at a disadvantage. Because monitoring does not exist, and because very few in-flight radiation measurements during significant SPEs have occurred, it remains an important task to fly calibrated instruments as widely and often as possible. This is needed to enable the accumulation of a data volume that can both validate models and potentially assist in creating data-assimilated "weather" of the radiation environment, similar to what has occurred in the tropospheric weather community during the past few decades.

4 STATUS OF MONITORING FOR EXTREME CONDITIONS

The most commonly used data in evaluation of past extreme space weather events has come from neutron monitors (Carmichael, 1964; Hatton, 1971; Simpson, 2000). In particular, ground level enhancements (GLEs) from solar cosmic rays have been used to identify extreme conditions for the aviation radiation environment with respect to dose rates above the background cosmic radiation levels. Neutron monitor data have been extremely useful for evaluation of the largest SEP events (O'Brien & Sauer, 2000; Copeland et al., 2008; Al Anid et al., 2009; Meier & Matthiä, 2014). Many works have studied this topic in detail (O'Brien et al., 1996; Gerontidou et al., 2002; Iles et al., 2004; Andriopoulou et al., 2011; Shea & Smart, 2012; McCracken et al., 2012; Mishev et al., 2015; Atwell et al., 2016) and a general conclusion from this body of work is that the energy spectra of GLEs are not identical to one another. Shea (private communication, 2017) has provided a list of 71 GLEs from 1942 through May 2012. The hardness, or particle energy distribution, is a feature that most distinguishes GLEs. For example, a hard spectra event of GLE 57 (May 6, 1998) showed a neutron monitor increase of 4% in the polar regions with a maximum integral >10 MeV proton flux increase of 239 protons cm^{-2} s^{-1} sr^{-1}. However, a soft spectra solar proton event on Nov. 8, 2000, had a maximum integral >10 MeV proton flux of 14,800 protons cm^{-2} s^{-1} sr^{-1} but also had no observable increase in neutron monitor datasets (Shea & Smart, 2012). Thus, not all proton events and GLEs are the same. Because a rigorous discussion of this topic is beyond the scope of this chapter, we will illustrate spectral differences with the example of a small number of GLE cases in the following discussion.

We note that coupling radiation transport models to neutron monitor data has been highly successful for monitoring variations in GCRs associated with changes in solar activity; this capability is incorporated into many of the models mentioned above. However, neutron monitor sensitivity to SEP radiation is limited by atmospheric and geomagnetic shielding, which are both much more effective for SEP radiation. While shielding has been identified, to some extent, by careful selection of monitoring sites with placement at different altitudes and latitudes and longitudes, the instruments are very massive, and thus poorly suited for in-flight use. Only the largest events can be observed, and often only at high latitudes. Another issue with neutron monitor use is that universities or science institutes operate many of these instruments with limited staff and budgets. These are often insufficient to support operational applications; thus, obtaining global data in near real time has been difficult or impractical. While older data are available from http://www.nmdb.eu/ and http://cr0.izmiran.ru/common/links.htm, some data providers have occasionally been reluctant to make their data available for operational users because corrections to generate the highest quality data may take a year or more to complete. These corrections account for local effects such as terrestrial weather as well as subtle changes that occur as the instruments age. Overall, the number of neutron monitor instruments supported has declined steadily for the past few decades even though there is broad interest in continuing these observations.

Continuous satellite-based monitoring of cosmic rays has augmented neutron monitor and other ground-based instrument data since the early 1970s. For aviation the most useful of these have been the particle detectors on GOES operated by the U.S. National Oceanic and Atmospheric Administration (NOAA). Unlike most satellite particle detectors, the detectors on GOES provide data for a wide energy range of protons and alpha particles from a few MeV to >1 GeV. The low-energy particles dominate the measurements and, thus, these satellites provide the SEP particle spectrum that does not generate a significant neutron monitor response because the particle energies are too low. The detectors have been limited by their upper energy threshold and directional sensitivity.

The early stages of a more practical monitoring capability are now being constructed. Among many possible models, one example of the current state of the art for an operational system is NASA Langley Research Center's (LaRC) *Nowcast of Atmospheric Ionizing Radiation for Aviation Safety* (NAIRAS) system (Mertens et al., 2012, 2013). NAIRAS is a data-driven, physics-based climatological model (Fig. 4) producing time-averaged weather conditions using the HZETRN radiation transport code that characterizes the global radiation environment from the surface to low Earth orbit (LEO) for dose rate and total dose hazards. Global, data-driven results are reported hourly at the NAIRAS public URL of http://sol.spacenvironment.net/~nairas/index.html. However, to produce the weather of the radiation environment, NAIRAS, as an example, needs assimilated real-time data. Consider the analogy of tropospheric weather models that need temperature, pressure, and humidity to make accurate weather reports. Similarly, for specifying radiation weather, models need data in near real time and from global locations. NAIRAS input data for assimilation could consist of total ionizing dose (TID), that is, absorbed dose in silicon or more complex dosimetric values where available.

The NASA *Automated Radiation Measurements for Aerospace Safety* (ARMAS) program creates these real-time TID data. For data assimilation into operational NAIRAS (Tobiska et al., 2015), TID can be used as an index (indicator of level of activity) for full energy spectrum measurements that is analogous to how total electron content (TEC) is used in ionospheric data assimilation models. ARMAS uses a TID commercial-off-the-shelf (COTS) microdosimeter combined with an Iridium data link to report the absorbed dose, $D(Si)$, from aircraft during flight. Between 2013 and 2017, ARMAS has obtained real-time radiation measurements from the ground to 17 km for 360 flights with 251,926 one-minute $\dot{D}(Si), \dot{D}(Ti), \dot{H}, \dot{E}$, and $\dot{H}^*(10)$ observed and derived data records (Tobiska et al., 2016). The data are available at the ARMAS URL http://sol.spacenvironment.net/armas_ops/Archive/ with an example shown in Fig. 5. The ARMAS data records are comparable in scope to the decade-long Liulin dataset (Ploc et al., 2013), which is also available online at http://hroch.ujf.cas.cz/~aircraft/ and contains 3699 flights with 133,438 $H^*(10)$ records having 5-minute resolution and covering one solar cycle from 2001 to 2011. The accuracy of all datasets continues to be assessed.

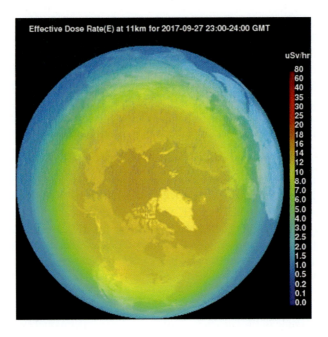

FIG. 4

Effective doses rates calculated for 11-km altitude by NAIRAS for the northern hemisphere on May 12, 2017 (Mertens et al., 2012, 2013).

A recent measurement project was the NASA Radiation Dosimetry Experiment (RaD-X) stratospheric balloon project (Mertens et al., 2016). RaD-X obtained dosimetric measurements from a balloon platform that was used to characterize cosmic ray primaries. In addition, radiation detectors were flown to assess their application to long-term, continuous monitoring of the aircraft radiation environment. The RaD-X balloon was launched from Fort Sumner, New Mexico, on Sept. 25, 2015. More than 18 h of flight data were obtained from each of the four different science instruments at altitudes above 20 km. The balloon data were supplemented by contemporaneous aircraft measurements. Flight-averaged dosimetric quantities were reported at seven altitudes to provide benchmark measurements for improving aviation radiation models. The altitude range of the flight data extends from commercial aircraft altitudes to above the Regener-Pfotzer maximum where the dosimetric quantities are influenced by cosmic ray primaries.

5 CLASSIFICATION OF AVIATION-RELEVANT EXTREME SPACE WEATHER RADIATION EVENTS

The assessment of space weather events in general and the identification of extreme space weather events in particular depend on the variation of observable physical parameters as well as upon their impacts and consequences. The challenge of quantifying space weather events related to aviation radiation, for practical purposes, consists in finding a relevant index that connects an observable physical

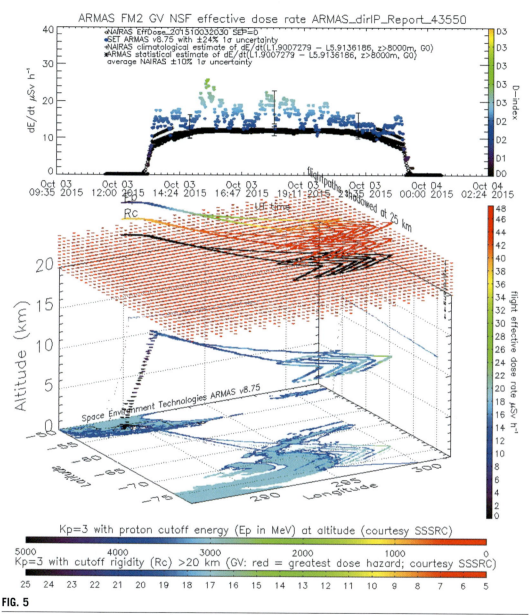

FIG. 5

Effective dose rates calculated as $\dot{E}$ from ARMAS (Tobiska et al., 2016).

quantity with the degree of impact. Furthermore, an index should communicate the relative severity of impacts and consequences of a particular space weather event to a nonexpert. NOAA introduced the Space Weather Scales, that is the G Index for geomagnetic storms, the S-Index for solar radiation storms, and the R-Index for radio blackouts in 1999 (Poppe, 2000; Poppe & Jorden, 2006). However, while these indices have proven useful for effects on power systems, radio communications,

GPS-based transportation, and geosynchronous (GEO) satellite single-event effects (SEEs), they are not applicable to the radiation environment at aviation altitudes.

An example of this inadequacy is for the well-known Halloween storms on aviation on Oct. 29, 2003, when several airlines reacted to information of an ongoing severe solar radiation storm with a level of S4 on the NOAA S-scale. As a consequence of this alert, some flights from the United States to Europe flew at lower altitudes because those airlines had established a radiation storm action level at a threshold of S3 (Lieber, 2003; U.S. DoC, 2004). A detailed analysis later showed that the response was generally ineffective in terms of mitigating radiation exposure on the corresponding flights (Meier & Matthiä, 2014). These mitigation measures resulted in higher flight costs in fuel consumption and time as well as contributed to additional atmospheric pollution.

Why is the NOAA S-scale not useful for aviation radiation alerts? It is based on the >10 MeV integral proton flux, which is detected by GOES in geosynchronous orbit, that is, outside the Earth's atmosphere. According to the S-scale, an extreme solar radiation storm is defined by particle fluxes above 10^5 proton flux units (pfu), that is, protons $cm^{-2} s^{-1} sr^{-1}$. It provides useful information for the prompt assessment of radiation impacts in the GEO space environment for the operation of satellites and manned spaceflight.

At aviation altitudes the situation is different from the GEO environment. The Earth's magnetosphere and atmosphere play important roles in modifying the radiation field as described above. Model calculations have shown that the threshold for primary cosmic particles to overcome the atmospheric shielding and contribute to the aviation radiation field at mid- to low-latitudes is about 600 MeV. The vast majority of the impinging particles during the Halloween storms had energies below this threshold (Matthiä et al., 2014). As a result, the radiation intensity in most of the atmosphere was only slightly increased. Direct measurements at aviation altitudes (Beck et al., 2005) showed an increase in dose rates of about 30%; the University of Oulu (Finland) neutron monitor measured a count rate variation of about 5% at sea level during the peak of the associated GLE 65. This event demonstrated the shielding capability of the magnetosphere and atmosphere, which reduced the increase in the primary flux of >10 MeV protons by about five orders of magnitude as observed onboard GOES. Oulu, at sea level, observed an increase of only about 5% resulting in only moderate dose rate increases at aviation altitudes. A subsequent Forbush decrease in the GCRs then reduced the atmospheric radiation intensity for several days (Meier & Matthiä, 2014).

The comparatively low radiation exposure at aviation altitudes during the Halloween storms was not represented by the S-index, but it did raise the awareness of the need for a relevant aviation industry index. The concept of the Dose index (D-index) was developed to provide warnings of elevated radiation levels. It is based on the radiation exposure, such as from solar particles, added to the background GCR levels and is formed from the effective dose rate $\dot{E}_{sol}$, which can be derived from either measurements or model calculations.

The D-index covers a wide range of radiation exposure at aviation altitudes using small natural numbers in a base 2 calculation using effective dose rates ($\mu Sv\ h^{-1}$). It is defined as the smallest natural number, including zero, to satisfy the inequality:

$$\dot{E}_{sol} < 5 \frac{\mu Sv}{h} 2^D \tag{1}$$

The indices from D0 to D8, their corresponding ranges of effective dose rates, and their comparison with other natural radiation sources are listed in Table 1. Quiet space weather for aviation is characterized by D0-, D1-, or D2-levels. The D3-level, where there is an additional dose rate of >20 $\mu Sv\ h^{-1}$, indicates an

Table 1 D Index Definitions and Comparisons With Other Exposure Scenarios

D Index	Dose Rate Interval (μSv h^{-1})	Exposure is Comparable With
D0	$\dot{E}_{sol} < 5$	Variation of the natural background at cruising altitudes
D1	$5 \leq \dot{E}_{sol} < 10$	Natural background at high latitudes up to FL400
D2	$10 \leq \dot{E}_{sol} < 20$	Natural background at high latitudes between FL400 and FL600
D3	$20 \leq \dot{E}_{sol} < 40$	Average dose rate inside the International Space Station (ISS)
D4	$40 \leq \dot{E}_{sol} < 80$	Average dose rate during extravehicular activity (EVA) on the ISS
D5	$80 \leq \dot{E}_{sol} < 160$	Dose of a north Atlantic return flight received or approximately one chest X-ray per hour
D6	$160 \leq \dot{E}_{sol} < 320$	Daily dose at aviation altitudes and high latitudes received in 1 h
D7	$320 \leq \dot{E}_{sol} < 640$	Daily average dose inside the ISS received in 1 h
D8	$640 \leq \dot{E}_{sol} < 1280$	Three-month dose for living on ground in most countries received in 1 h

elevated radiation intensity that can be used by air traffic management to trigger a radiation alert. The D-index can be used within the framework of already existing warning systems (Fig. 5, top panel). For example, this scale has already been used by the Federal Aviation Administration's (FAA) Solar Radiation Alert System (ESRAS); an alert is issued if D3 is exceeded at any altitude between 30,000 and 70,000 ft for each of three consecutive 5-min periods (Copeland et al., 2009; Copeland, 2016). The dose rate for triggering a D3 alert corresponds roughly to the average dose rate inside the International Space Station (ISS) that many astronauts and cosmonauts are exposed to for several months from GCR and trapped radiation. It is worth mentioning that the dose rates inside the ISS will generally be much higher during such a solar particle event due to the absence of atmospheric shielding. The D-index has also been used to provide space weather-induced radiation dose rate information for several European airlines since 2014.

An important feature of the D-index is its application within a particular volume cell of the atmosphere (latitude, longitude, altitude). In this context, it can be used to communicate a differentiated picture about the radiation field above specific geographic regions and at unique altitudes. This is similar to the communication of terrestrial weather parameters such as winds, temperature, air pressure, and humidity. For example, the increased radiation exposure for a particular region can be generalized for commercial aviation and a local warning index, D_L, can be derived from the maximum regional dose rate at a flight level of 41,000 ft (FL410). This would characterize the upper airspace as the worst-case scenario. Furthermore, the provision of individual indices for particular flights, as D_F-values, is possible as well.

6 EXAMPLE OF AN EXTREME EVENT

Although there have been no warning situations since the development of the D-index, as of May 2017 the application of the D-index can be demonstrated with GLE 70, which took place on Dec. 13, 2006. GLE 70 is an excellent example illustrating the D-index concept. This event showed a sharply peaked time profile and a spatial distribution leading to relatively large intensity increases at eastern latitudes

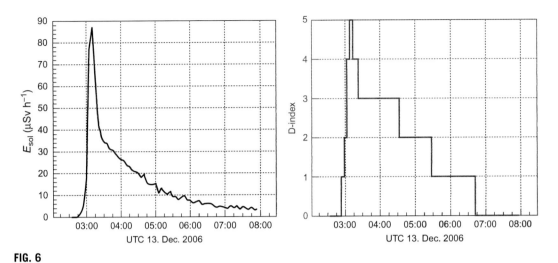

FIG. 6

Effective doses and corresponding D-indices at an altitude of 41,000 ft during GLE 70, a large solar proton event that occurred Dec. 13, 2006, as calculated by PANDOCA (Matthiä et al., 2014).

in Russia and Europe in the initial phase of the event (Fig. 6). In the later isotropic phase of the event a weaker response in cosmic ray intensities was recorded in North America.

The GLE was related to an X3 solar flare on the NOAA scale when it originated in the solar western hemisphere (5S23W) at 2:39 UTC. Neutron monitor stations in Europe, such as Kiel and Oulu, recorded count rate increases starting between 2:50 and 2:55 UTC marking the onset of GLE ~15 min after the peak in the X-ray flux. Maximum intensities were measured at 3:05 UTC. Stations in North America, such as Inuvik and Calgary, recorded the beginning of the event about 10 min later. Much weaker peak increases were measured by these stations about half an hour later between 3:30 and 3:40 UTC. Matthiä et al. (2009) performed a detailed analysis of this event using data from the complete neutron monitor network.

The left column of Fig. 7 illustrates the global distribution of dose rates at 41,000 ft when maximum values were reached (3:10 UTC) and about half an hour later (3:35 UTC) as calculated by the PANDOCA model (Matthiä et al., 2014). On the right of Fig. 7 are the corresponding D-indices. In the initial phase of the eventmostly eastern latitudes were affected as shown in the neutron monitor stations' measurements. Peak dose rates were calculated to be ~80–90 $\mu Sv\ h^{-1}$ corresponding to a D-index of 5 (Fig. 6). However, these relatively large values only occurred regionally. The Americas were minimally affected at that time. While the global D-index for the event was 5, a regionally derived index for North America would have been 0. About half an hour later at 3:35 UTC, however, the impact of the event on the radiation exposure at aviation altitudes was not limited to specific regions anymore and only showed a typical pattern of magnetic shielding (lower row of Fig. 7). At that time the maximum dose rate had decreased to about 30–40 $\mu Sv\ h^{-1}$ (D3), but previously unaffected regions showed an increase in exposure as well. All areas above 60°N and below 60°S were ultimately affected with a D-index of 2 or 3.

Matthiä et al. (2015) investigated the effectiveness of mitigation measures and their economic impacts related to delay and fuel consumption for a transatlantic flight during GLE 70. If the increase

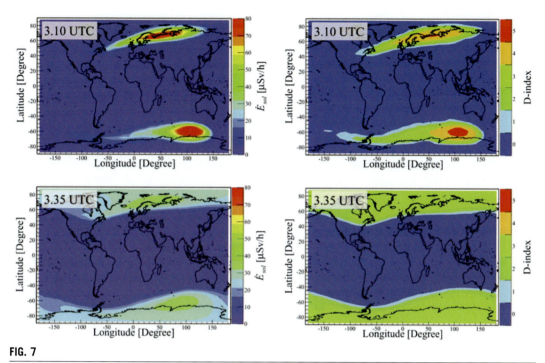

FIG. 7

Global distribution of effective dose rates and D-indices at an altitude of 41,000 ft during GLE 70, as calculated using PANDOCA.

in dose rate as expressed by the D-index could be communicated to the cockpit, then an appropriate action might be taken. The communication could typically be done through ACARS (Aircraft Communications Addressing and Reporting System). In Matthiä et al. (2015) the response to the event was (i) lower the flight altitude after the increase of the dose rates, (ii) adapt the flight velocity, and (iii) return to nominal flight altitudes after the additional dose rates had dropped below a threshold.

It was shown that in the ideal case, that is a prompt response to an increase in the radiation exposure caused by this GLE, the total effective dose on the flight could have been reduced up to 42% by lowering the flight altitude and using the contingency aircraft fuel (i.e., an additional fuel consumption of up to 5%). If the aircraft had returned to the most economical flight altitude after the dose rate had dropped below 10 μSv h^{-1} (D-index of 1 or 0), the total effective dose reduction would have been about 30%. For the ideal scenario this return to the original flight altitude occurred at 5:18 UTC. After this time the radiation exposure was at the GCR background or below. By comparison, the integral proton flux measured by GOES had not even reached its maximum at that time. The integral flux >10 MeV, which is the basis for the NOAA S-scale, reached its maximum not before 10:30 UTC. At that time, the event was essentially over in terms of radiation exposure at aviation altitudes. This example demonstrates that an ideal response to an event could be based on dose rates at aviation altitudes and could have been supported using the D-index. For the calculated flight, the D-index would have remained at D1 rather than D3 in case of no response.

7 CONCLUSION

The intensity of the radiation field due to cosmic radiation at aviation altitudes during quiet space weather conditions is, in terms of effective dose rate, $\dot{E}$, more than one order of magnitude higher than the average radiation environment from all natural sources on the ground in most countries. As a consequence, aircrew and frequent flyers are exposed to higher levels of ionizing radiation than the average population. This has led to legal regulations and the implementation of corresponding radiation protection measures in many countries, including in the European Union, Switzerland, and South Korea. In the United States, guidelines have been issued; the FAA funds research, provides software, operates a Solar Radiation Alert system, and maintains an issue-related website at http://www.faa.gov/data_research/research/med_humanfacs/aeromedical/radiobiology/ (Freidberg et al., 1999; Friedberg & Copeland, 2011).

Extreme space weather events are still not monitored during flight in real time and GLEs have often been used as a surrogate for extreme solar cosmic ray particle events. Their occurrence can bring about a further, albeit short-term, significant increase in radiation exposure, although not all GLEs have the same energy spectrum. The actual additional total exposure at cruising altitudes during *short-term* events is likely to be moderate in comparison to the ordinary *annual* radiation exposure from other natural and artificial sources, even though these events do present unknown consequences for an individual's health. Further research leading to effective and efficient real-time monitoring is necessary to better understand the effects of the aviation radiation environment.

An index for the assessment of the radiation field at aviation altitudes during extreme space weather events has merit and is discussed. Any index requires a close connection between an observable physical quantity and the degree of an impact. This is the basis of the D-index that is derived from the effective dose rate using either measurements or models. The effective dose rate, $\dot{E}$, is the fundamental physical quantity used for characterizing radiation fields in radiation protection practices. The D-index is a decision-aid tool that can provide timely and useful information to the aviation community about space weather effects related to radiation at aviation altitudes. The D-index is flexible in its global or regional application and is independent of the radiation model or measurement used for its assessment. In this respect, it also supports national level regional warning centers (RWCs), which have been successfully used in the field of terrestrial weather for many years. The responsible authorities for different countries or regions can select baseline measurements or models in consideration of their own needs. The feasibility of mitigating measures for aviation altitude radiation during a space weather event was described with the example of a study using GLE 70. Timely space weather information based on the D-index might have reduced the radiation exposure of crew and passengers during this event from D3 to D1, thus improving aviation crew and passenger health and safety.

ACKNOWLEDGMENTS

The authors thank the reviewers for their timely and insightful comments that have improved this paper. The authors acknowledge the financial support for ARMAS from the original NASA NAIRAS project contract NNL07AA00C, the NASA SBIR Phase I and Phase II program contracts NNX11CH03P and NNX12CA78C, the NASA AFRC Phase III contracts NND14SA64P and NND15SA55C, and the South Korean Space Weather

Center matching funds for SBIR Phase IIE. Gracious flight support for ARMAS instruments has been provided by the NASA Airborne Sciences Program and Armstrong Flight Research Center. The NOAA Space Weather Prediction Center facilitated use of the NOAA Gulfstream IV through their good offices as did the National Center for Atmospheric Research High Altitude Observatory for the use of the National Science Foundation Gulfstream V. In accordance with the AGU data policy, the ARMAS archival data used in this paper, as well as from all flights, is publically available from the ARMAS website at http://sol.spacenvironment.net/armas_ops/Archive/. Other support for this work was provided by the German Aerospace Center (DLR) and the Aerospace Medical Research Division of the FAA Civil Aerospace Medical Institute.

REFERENCES

Al Anid, H., Lewis, B.J., Bennett, L.G.I., Takada, M., 2009. Modelling of radiation exposure at high altitudes during solar storms. Radiat. Prot. Dosim. 136 (4), 311–316. https://doi.org/10.1093/rpd/ncp127.

Andriopoulou, M., Mavromichalaki, H., Preka-Papadema, P., Plainaki, C., Belov, A., Eroshenko, E., 2011. Solar activity and the associated ground level enhancements of solar cosmic rays during solar cycle 23. Astrophys. Space Sci. Trans. 7, 439–443.

Atwell, W., Tylka, A.J., Dietrich, W.F., Rojdev, K., Matzkind, C., 2016. Probability estimates of solar proton doses during periods of low sunspot number for short duration missions. In: 46th International Conference on Environmental Systems, ICES-2016-453, 10-14 July 2016, Vienna, Austria.

Beck, P., Ambrosi, P., Schrewe, U., O'Brien, K., 1999. ACREM, aircrew radiation exposure monitoring, Final report of European Commission contract F14P-CT960047, OEFZS, Rep. G-0008.

Beck, P., Latocha, M., Rollet, S., Stehno, G., 2005. TEPC reference measurements at aircraft altitudes during a solar storm. Adv. Space Res. 36 (9), 1627–1633.

Beck, P., Dyer, C., Fuller, N., Hands, A., Latocha, M., Rollet, S., Spurny, F., 2009. Overview of on-board measurements during solar storm periods. Radiat. Prot. Dosim. 136 (4), 297–303. https://doi.org/10.1093/rpd/ncp208.

Carmichael, H., 1964. Cosmic rays. In: IQSY Instruction Manual No. 7. IQSY Secretariat, London.

Copeland, K., 2016. ESRAS: an enhanced solar radiation alert system. Federal Aviation Administration, Civil Aerospace Medical Institute, Oklahoma City, OK. DOT Report No. DOT/FAA/AM-16/5.

Copeland, K., 2017. CARI-7A: development and validation. Radiat. Prot. Dosim. https://doi.org/10.1093/rpd/ncw369.

Copeland, K., Sauer, H.H., Duke, F.E., Friedberg, W., 2008. Cosmic radiation exposure of aircraft occupants on simulated high-latitude flights during solar proton events from 1 January 1986 through 1 January 2008. Adv. Space Res. 42, 1008–1029. https://doi.org/10.1016/j.asr.2008.03.001.

Copeland, K., Sauer, H., Friedberg, W., 2009. Solar radiation alert system. Federal Aviation Administration, Civil Aerospace Medical Institute, Oklahoma City, OK. DOT Report No. DOT/FAA/AM-09/6 (revised 30 May 2008).

Dyer, C., Lei, F., 2001. Monte-Carlo calculations of the influence on aircraft radiation environments of structures and solar particle events. IEEE Trans. Nucl. Sci. 48 (6), 1987–1995.

Dyer, C.S., Truscott, P., 1999. Cosmic radiation effects on avionics. Radiat. Prot. Dosim. 86 (4), 337–342.

Dyer, C.S., Sims, A.J., Farren, J., Stephen, J., 1990. Measurements of solar flare enhancements to the single event upset environment in the upper atmosphere. IEEE Trans. Nucl. Sci. 37, 1929–1937. https://doi.org/10.1109/23.101211.

Dyer, C.S., Lei, F., Clucas, S.N., Smart, D.F., Shea, M.A., 2003. Solar particle enhancements of single event effect rates at aircraft altitudes. IEEE Trans. Nucl. Sci. 50 (6), 2038–2045.

Dyer, C., Hands, A., Fan, L., Truscott, P., Ryden, K.A., Morris, P., Getley, I., Bennett, L., Bennett, B., Lewis, B., 2009. Advances in measuring and modeling the atmospheric radiation environment. IEEE Trans. Nucl. Sci. 6 (1), 3415–3422.

ECRP (European Commission Radiation Protection) 140, 2004. Cosmic Radiation Exposure of Aircraft Crew, Compilation of Measured and Calculated Data, European communities.

Felsberger, E., O'Brien, K., Kindl, P., 2009. IASON-FREE: theory and experimental comparisons. Radiat. Prot. Dosim. 136 (4), 267–273. https://doi.org/10.1093/rpd/ncp128.

Freidberg, W., Copeland, K., Duke, F.E., O'Brien, K., Darden, E.B., 1999. Guidelines and technical information provided by the U.S. Federal Aviation Administration to promote radiation safety for air carrier crew members. Radiat. Prot. Dosim 86, 323.

Friedberg, W., Copeland, K., 2003. What aircrews should know about their occupational exposure to ionizing radiation ionizing. Federal Aviation Administration, Civil Aerospace Medical Institute, Oklahoma City, OK. DOT Report No. DOT/FAA/AM-03/16.

Friedberg, W., Copeland, K., 2011. Ionizing radiation in earth's atmosphere and in space near earth. Federal Aviation Administration, Civil Aerospace Medical Institute, Oklahoma City, OK. DOT Report No. DOT/FAA/AM-11/9.

Gerontidou, M., Vassilaki, A., Mavromichalaki, H., Kurt, V., 2002. Frequency distributions of solar proton events. J. Atmos. Sol.-Terr. Phys. 64, 489–496.

Gersey, B., Wilkins, R., Atwell, W., Tobiska, W.K., Mertens, C., 2012. Tissue equivalent proportional counter microdosimetry measurements aboard high-altitude and commercial aircraft. AIAA 2012–3636In: AIAA 42nd International Conference on Environmental Systems, San Diego, California, 15–19 July.

Getley, I.L., Duldig, M.L., Smart, D.F., Shea, M.A., 2005. Radiation dose along North American transcontinental flight paths during quiescent and disturbed geomagnetic conditions. Space Weather 3. https://doi.org/10.1029/2004SW000110 S01004.

Hands, A., Dyer, C.S., 2009. A technique for measuring dose equivalent and neutron fluxes in radiation environments using silicon diodes. IEEE Trans. Nucl. Sci. 56 (1), 3442–3449.

Hatton, C.J., 1971. The neutron monitor. In: Progress in Elementary Particle and Cosmic Ray Physics, vol. 10. pp. 3–102.

Hwang, J., Dokgo, K., Choi, E., Kin, K.-C., Kim, H.-P., Cho, K.-S., 2014. Korean Radiation Exposure Assessment Model for aviation route dose. In: KREAM, KSS Fall meeting, Jeju, Korea, October 29-31.

IARC (International Agency for Research on Cancer), 2000. Ionizing radiation, Part 1, X- and γ-radiation and neutrons. In: IARC Monographs on the Evaluation of Carcinogenic Risks to Humans. vol. 75. IARC Press, Lyon, France. ISBN: 92 832 1275 4.

ICRP 2007, 2007. Recommendations of the ICRP, ICRP Pub. 103. Ann. ICRP 37 (2-4), 1–332.

Iles, R.H.A., Jones, J.B.L., Taylor, G.C., Blake, J.B., Bentley, R.D., Hunter, R., Harra, L.K., Coates, A.J., 2004. Effect of solar energetic particle (SEP) events on the radiation exposure levels to aircraft passengers and crew: case study of 14 July 2000 SEP event. J. Geophys. Res. 109. https://doi.org/10.1029/2003JA010343 A11103.

Johnston, C.O., 2008. A Comparison of EAST Shock-Tube Radiation Measurements With a New Radiation Model. AIAA Paper 2008-1245.

Joyce, C.J., Schwadron, N.A., Wilson, J.K., Spence, H.E., Kasper, J.C., Golightly, M., Blake, J.B., Townsend, L.W., Case, A.W., Semones, E., Smith, S., Zeitlin, C.J., 2014. Radiation modeling in the Earth and Mars atmospheres using LRO/CRaTER with the EMMREM Module. Space Weather 12, 112–119. https://doi.org/10.1002/2013SW000997.

Kyllönen, J.E., Lindborg, L., Samuelson, G., 2001. Cosmic radiation measurements on board aircraft with the variance method. Radiat. Prot. Dosim. 93, 197–205.

Latocha, M., Autischer, M., Beck, P., Bottolier–Depois, J.F., Rollet, S., Trompier, F., 2007. The results of cosmic radiation in-flight TEPC measurements during the CAATER flight campaign and comparison with simulation. Radiat. Prot. Dosim. 125 (1–4), 412–415. https://doi.org/10.1093/rpd/ncl123.

Latocha, M., Beck, P., Rollet, S., 2009. AVIDOS—a software package for European accredited aviation dosimetry. Radiat. Prot. Dosim. 136 (4), 286. https://doi.org/10.1093/rpd/ncp126.

Latocha, M., Beck, P., Bütikofer, P., Thommesen, H., 2014. AVIDOS 2.0—Current Developments for the Assessment of Radiation Exposure at Aircraft Altitudes Caused by Solar Cosmic Radiation Exposure, European Space Weather Week, Liege, 17-21 November. http://stce.be/esww11.

Lee, J., Nam, U.-W., Pyo, J., Kim, S., Kwon, Y.-J., Lee, J., Park, I., Kim, M.-H.Y., Dachev, T.P., 2015. Short-term variation of cosmic radiation measured by aircraft under constant flight conditions. Space Weather 13, 797–806. https://doi.org/10.1002/2015SW001288.

Lieber, R., 2003. Solar storm rekindles concern over whether radiation hurts fliers. Wall Street J. 30th October, http://online.wsj.com/news/articles/SB106744850896766200.

Mares, V., Maczka, T., Leuthold, G., Ruhm, M., 2009. Air crew dosimetry with a new version of EPCARD. Radiat. Prot. Dosim. 136 (4), 262–266. https://doi.org/10.1093/rpd/ncp129.

Matthiä, D., Heber, B., Reitz, G., Sihver, L., Berger, T., Meier, M., 2009. The ground level event 70 on December 13th, 2006 and related effective doses at aviation altitudes. Radiat. Prot. Dosim. 136 (4), 304–310. https://doi.org/10.1093/Rpd/Ncp141.

Matthiä, D., Meier, M.M., Reitz, G., 2014. Numerical calculation of the radiation exposure from galactic cosmic rays at aviation altitudes with the PANDOCA core model. Space Weather 12, 161. https://doi.org/10.1002/2013SW001022.

Matthiä, D., Schaefer, M., Meier, M.M., 2015. Economic impact and effectiveness of radiation protection measures in aviation during a ground level enhancement. J. Space Weather Space Clim. 5, A17. https://doi.org/10.1051/swsc/2015014.

McCall, M.J., Lemay, F., Bean, M.R., Lewis, B.J., Bennett, L.G., 2009. Development of a pre-dictive code for aircrew radiation exposure. Radiat. Prot. Dosim. 136 (4), 274–281. https://doi.org/10.1093/rpd/ncp130.

McCracken, K.G., Moraal, H., Shea, M.A., 2012. The high-energy impulsive ground-level enhancement. Astrophys. J. 761, 101 2012 December 20.

Meier, M.M., Matthiä, D.D., 2014. A space weather index for the radiation field at aviation altitudes. J. Space Weather Space Clim. 4, A13.

Meier, M.M., Hubiak, M., Matthiä, D., Wirtz, M., Reitz, G., 2009. Dosimetry at aviation altitudes (2006–2008). Radiat. Prot. Dosim. 136 (4), 251–255.

Meier, M.M., Trompier, F., Ambrozova, I., Kubancak, J., Matthiä, D., Ploc, O., Santen, N., Wirtz, M., 2016a. CONCORD: comparison of cosmic radiation detectors in the radiation field at aviation altitudes. J. Space Weather Space Clim. 6, A24. https://doi.org/10.1051/swsc/2016017.

Meier, M.M., Matthiä, D., Forkert, T., Wirtz, M., Scheibinger, M., Hübel, R., Mertens, C.J., 2016b. RaD-X: complementary measurements of dose rates at aviation altitudes. Space Weather 14. https://doi.org/10.1002/2016SW001418.

Mertens, C.J., Kress, B.T., Wiltberger, M., Tobiska, W.K., Grajewski, B., Xu, X., 2012. Atmospheric ionizing radiation from galactic and solar cosmic rays. In: Nenoi, M. (Ed.), Current Topics in Ionizing Radiation Research. InTech.

Mertens, C.J., Meier, M.M., Brown, S., Norman, R.B., Xu, X., 2013. NAIRAS aircraft radiation model development, dose climatology, and initial validation. Space Weather 11, 603. https://doi.org/10.1002/swe.20100.

Mertens, C.J., Gronoff, G.P., Norman, R.B., Hayes, B.M., Lusby, T.C., Straume, T., Tobiska, W.K., Hands, A., Ryden, K., Benton, E., Wiley, S., Gersey, B., Wilkins, R., Xu, X., 2016. Cosmic radiation dose measurements from the RaD-X flight campaign. Space Weather 14. https://doi.org/10.1002/2016SW001407.

Mishev, A.L., Adibpour, F., Usoskin, I.G., Felsberger, E., 2015. Computation of dose rate at flight altitudes during ground level enhancements no. 69, 70 and 71. Adv. Space Res. 55, 354–362.

Mutuel, L.H., 2016. Single Event Effects Mitigation Techniques Report, Department of Transportation/Federal Aviation Administration, TC-15/62, February 2016.

Normand, E., Oberg, D.L., Wert, J.L., Ness, J.D., Majewski, P.P., Wender, S., Gavron, A., 1994. Single event upset and charge collection measurements using high energy protons and neutrons. IEEE Trans. Nucl. Sci. 41 (6), 2203–2209.

Normand, E., Vranish, K., Sheets, A., Stitt, M., Kim, R., 2006. Quantifying the double-sided neutron SEU threat, from low energy (thermal) and high energy (>10 MeV) neutrons. IEEE Trans. Nucl. Sci. 53 (6), 3587–3595.

O'Brien, K., Sauer, H.H., 2000. An adjoint method of calculations of solar-particle-event dose rates. Technology 7 (2-4), 449–456.

O'Brien, K., Friedberg, W., Sauer, H.H., Smart, D.F., 1996. Atmospheric cosmic rays and solar energetic particles at aircraft altitudes. Environ. Int. 22 (Suppl. 1), S9–S44.

Ploc, O., Ambrozova, I., Kubancak, J., Kovar, I., Dachev, T.P., 2013. Publicly available database of measurements with the silicon spectrometer Liulin onboard aircraft. Radiat. Meas. 58, 107–112.

Poppe, B., 2000. New scales help public, technicians understand space weather. Eos Trans. Am. Geophys. Union 81 (29), 322–328.

Poppe, B., Jorden, K., 2006. Sentinels of the Sun. Johnson Books, Boulder, CO. ISBN 1-55566-379-6.

Regener, E., Pfotzer, G., 1935. Vertical intensity of cosmic rays by threefold coincidences in the stratosphere. Nature 136, 718.

Sato, T., 2015. Analytical model for estimating terrestrial cosmic ray fluxes nearly anytime and anywhere in the world: extension of PARMA/EXPACS. PLoS One. 10 (12), e0144679.

Sato, T., Yasuda, H., Niita, K., Endo, A., Sihver, L., 2008. Development of PARMAPHITS-based analytical radiation model in the atmosphere. Radiat. Res. 170, 244.

Schwadron, N.A., Baker, T., Blake, B., Case, A.W., Cooper, J.F., Golightly, M., Jordan, A., Joyce, C., Kasper, J., Kozarev, K., Mislinski, J., Mazur, J., Posner, A., Rother, O., Smith, S., Spence, H.E., Townsend, L.W., Wilson, J., Zeitlin, C., 2012. Lunar radiation environment and space weathering from the Cosmic Ray Telescope for the Effects of Radiation (CRaTER). J. Geophys. Res. Planets 117. https://doi.org/10.1029/2011JE003978. E00H13.

Shea, M.A., Smart, D.F., 2012. Space weather and the ground-level solar proton events of the 23rd solar cycle. Space Sci. Rev. 171, 161–188.

Simpson, J.A., 1983. Elemental and isotopic composition of the galactic cosmic rays. Annu. Rev. Nucl. Part. Sci. 33, 323–382. https://doi.org/10.1146/annurev.ns.33.120183.001543.

Simpson, J.A., 2000. The cosmic ray nucleonic component: the invention and scientific uses of the neutron monitor—(Keynote lecture). Space Sci. Rev. 93, 11–32. https://doi.org/10.1023/A:1026567706183.

Singleterry, R.C., Blattnig, S.R., Clowdsley, M.S., Qualls, G.D., Sandridge, C.A., Simonsen, L.C., Norbury, J.W., Slaba, T.C., Walker, S.A., Badavi, F.F., Spangler, J.L., Aumann, A.R., Zapp, E.N., Rutledge, R.D., Lee, K.T., Norman, R.B., 2010. OLTARIS: On-Line Tool for the Assessment of Radiation in Space, NASA/TP–2010-216722. NASA Langley Research Center, Hampton, Virginia.

Spence, H.E., Case, A., Golightly, M.J., Heine, T., Larsen, B.A., Blake, J.B., Caranza, P., Crain, W.R., George, J., Lalic, M., Lin, A., Looper, M.D., Mazur, J.E., Salvaggio, D., Kasper, J.C., Stubbs, T.J., Doucette, M., Ford, P., Foster, R., Goeke, R., Gordon, D., Klatt, B., O'connor, J., Smith, M., Onsager, T., Zeitlin, C., Townsend, L., Charara, Y., 2010. CRaTER: The Cosmic Ray Telescope for the Effects of Radiation Experiment on the Lunar Reconnaissance Orbiter Mission. Space Sci. Rev. 150 (1-4), 243–284.

Tobiska, W.K., Gersey, B., Wilkins, R., Mertens, C., Atwell, W., Bailey, J., 2014a. U.S. Government shutdown degrades aviation radiation monitoring during solar radiation storm. Space Weather 12. https://doi.org/10.1002/2013SW001015.

Tobiska, W.K., Gersey, B., Wilkins, R., Mertens, C., Atwell, W., Bailey, J., 2014b. Reply to comment by Rainer Facius et al. on "U.S. Government shutdown degrades aviation radiation monitoring during solar radiation storm," Space Weather 12, 320–321. https://doi.org/10.1002/2014SW001074.

Tobiska, W.K., Atwell, W., Beck, P., Benton, E., Copeland, K., Dyer, C., Gersey, B., Getley, I., Hands, A., Holland, M., Hong, S., Hwang, J., Jones, B., Malone, K., Meier, M.M., Mertens, C., Phillips, T., Ryden, K., Schwadron, N., Wender, S.A., Wilkins, R., Xapsos, M.A., 2015. Advances in atmospheric radiation measurements and modeling needed to improve air safety. Space Weather 13, 202–210. https://doi.org/10.1002/2015SW001169.

Tobiska, W.K., Bouwer, D., Smart, D., Shea, M., Bailey, J., Didkovsky, L., Judge, K., Garrett, H., Atwell, W., Gersey, B., Wilkins, R., Rice, D., Schunk, R., Bell, D., Mertens, C., Xu, X., Wiltberger, M., Wiley, S., Teets, E., Jones, B., Hong, S., Yoon, K., 2016. Global real-time dose measurements using the Automated Radiation Measurements for Aerospace Safety (ARMAS) system. Space Weather 14, 1053–1080.

U.S. DOC (Department of Commerce), 2004. Service Assessment—Intense Space Weather Storms October 19–November 07, 2003, 17-18. Available from http://www.swpc.noaa.gov/Services/SWstorms_assessment.pdf.

UNSCEAR (United Nations Scientific Committee on the Effect of Atomic Radiation)., 2000. Sources and effect of ionizing radiation, United Nations Scientific Committee on the Effect of Atomic Radiation UNSCEAR 2000 Report to the General Assembly, with Scientific Annexes, vol. II, Annex G.

Wissmann, F., Reginatto, M., Möller, T., 2010. Ambient dose equivalent at flight altitudes: a fit to a large set of data using a Bayesian approach. J. Radiol. Prot. 30, 513–524.

CHAPTER 19

HIGH-ENERGY TRANSIENT LUMINOUS ATMOSPHERIC PHENOMENA: THE POTENTIAL DANGER FOR SUBORBITAL FLIGHTS

Gali Garipov, Alexander Grigoriev, Boris Khrenov, Pavel Klimov, Mikhail Panasyuk

Skobeltsyn Institute of Nuclear Physics of Lomonosov Moscow State University, Moscow, Russia

CHAPTER OUTLINE

1 Introduction 473
2 Phenomenology of TLEs 474
3 Experimental Data on TLE from UVRIR Detector on Board Moscow State University Satellites 477
   3.1 TLE Types Measured by UVRIR Detector 478
   3.2 TLE Distribution Over Photon Numbers 480
   3.3 Series of TLE 480
4 Discussion 483
   4.1 Overview of TLE Models and Relation to Other Space Weather Phenomena 483
   4.2 TLE Energy Deposition 485
   4.3 TLEs as a Radiation Hazard 486
5 Results and Conclusions 487
References 488
Further Reading 490

1 INTRODUCTION

Extreme events in geospace are mostly tied to solar activity, geomagnetic storms, and variations of the near-Earth radiation environment; conventionally these are the main factors of space weather. But it is important to know and take into account that there are energetic events in the Earth's atmosphere of other origins, related to the atmospheric dynamics followed by perturbations of the electric field in the atmosphere and ionosphere at different altitudes.

Occasionally in the upper atmosphere (heights from tens to a hundred km) one can observe very short-time energetic flashes (from ms and more) of light in the ultraviolet (UV), red (R), and infrared (IR) ranges (and probably in other ranges of the electromagnetic spectrum), which are usually interpreted as effects accompanying electrical discharges in the upper atmosphere and called transient luminous (more than 1000 Raleigh) events (TLEs). Estimates of TLE energy show that the power of these events can reach values up to 10^{-2} Wt m^{-2}, which is large compared to the value of about 10^{-3} Wt m^{-2} for some space weather events (such as auroras in polar regions or precipitation of electrons from radiation belts). The typical duration of a TLE is a few milliseconds. This is much less than the duration of other space weather-related phenomena such as auroras, geomagnetic storms, etc. However, the timescales are comparable with so-called microburst precipitations of energetic electrons observed in different regions of the radiation belts (see Millan, 2011), which may indicate the community of their nature with some TLE types.

The first observation of TLEs in the upper atmosphere was done in 1990 (Franz et al., 1990) and until now, numerous experiments were performed with the help of various instruments based on the ground, mountains, airplanes, and satellites (see reviews by Wescott et al., 1995; Yair et al., 2004; Chen et al., 2008; Neubert et al., 2008; Surkov and Hayakawa, 2012). Different types of TLEs have been measured and classified according to their shape, duration, and spectral features (see classification in Pasco, 2003).

It was shown that luminous transients are related to lightning (detected by worldwide networks) but some of them can occur far from thunderstorm regions (Morozenko et al., 2016). There are indications of anthropogenic origin of some types of quasipermanent transients (Garipov et al., 2016), which are evidently connected with the modulation of UV radiation by powerful radio transmitters operating in very low frequency (VLF) range (about 20 kHz).

The phenomena of lightning and upper atmosphere TLEs are related to the global electric circuit and have a global distribution over the Earth's atmosphere that makes their role in the nearby geospace effects very important. A theoretical interpretation of these atmospheric events is far from a comprehensive description and much more experimental data are needed to fill in the gaps in our understanding of the origins of TLEs. Experimental and theoretical studies of TLEs and associated phenomena are also important for the development of space weather applications, because TLEs are very powerful and can significantly modify local properties of the ionosphere and radiation environments. This chapter will review these aspects of the upper atmosphere studies.

2 PHENOMENOLOGY OF TLEs

At least six types of upper atmospheric phenomena are among TLEs: elves, sprites, halos, blue gets, blue starters, and gigantic jets (Pasco, 2003). Elves are a ring expanding across the lower ionosphere (altitudes of about 80–100 km) in <1 ms and reaching the horizontal size of 200–300 km. Sprites appear at 60–90 km altitude and have a complex spatial structure with tens of kilometers of vertical extension. Usually sprites occur simultaneously with halos—a brief descending diffusive glow above the sprite. Blue jets and blue starters are the lowest transients propagating upward from the cloud tops and having vertical extension of 10–20 km. Gigantic jets are the most rare phenomena, propagating all the way from cloud to ionosphere.

The ISUAL experimental array (Chern et al., 2003) on board the FORMOSAT-2 satellite provided the most informative data regarding TLE. It consists of three instruments: the imager, open in the field of view (FOV) 20 degrees × 5 degrees with filters in a wheel for five wavelength ranges; the spectrophotometer of six channels serving as a trigger for the imager; and the array photometer in UV (360–450 nm) and red (540–650 nm) ranges. All of them look from a satellite at orbit height 890 km to the Earth limb.

The shape of lightning observed from space is a round spot determined by the scattering of lightning radiation in clouds. Duration of a lightning strike is <1 ms. Several types of TLEs above lightning were in agreement with the data of previous nonspace experiments. The elves were the most presented among all TLEs (80%). Elves develop at atmospheric heights of about 70–90 km, looking like a rapidly (duration of <1 ms) expanding ring with a diameter up to 300 km (Kuo et al., 2007). The time delay from lightning for elves is ~0.1 ms, as expected for the event initiated by the lightning electromagnetic pulse (EMP) expanding from clouds to the ionosphere.

A sprite is another type of TLE, starting at the bottom of the ionization D-layer ($H=70$ km). They were observed in detailed experiments (Cummer et al., 2006; Stenbaek-Nielsen and McHarg, 2008) with fast cameras having 0.2-ms frame intervals. It is possible to observe several stages of sprite development with such a camera. The first 10 frames of the sprite development are presented in Fig. 1 (Cummer et al., 2006). At the first stage the discharge starts in the D-layer of the ionosphere (see frames 1–3 in Fig. 1). This stage is short, <1 ms. In the second stage the electric field between the ionosphere and the ground (the electric potential between the cloud and ground is shorted by the lightning strike) becomes higher and avalanches (streamers) of the discharge start developing downward. In this stage the electric field increases due to the space charge generated by streamers in the head of the avalanche

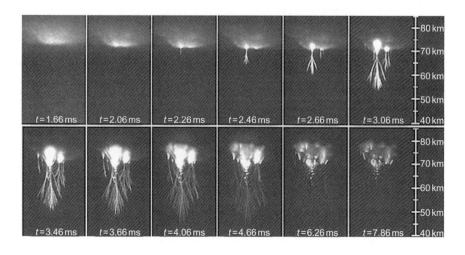

FIG. 1

Development of sprites in the upper atmosphere. Frame time interval is 0.2 ms.

Modified from Cummer, S.A., Jaugey, N., Li, J., Lyons, W.A., et al., 2006. Submillisecond imaging of sprite development and structure. Geophys. Res. Lett. 33, L04104.

(frames 4–6 in Fig. 1). The most luminous TLE stage (frames 7–9 in Fig. 1) is developed due to the maximal electric field according to the model (Raizer et al., 2010). Later on, when all electron energy is expired in a considered thunderstorm area, the discharge is finished. It is evident from Fig. 1 that all photons have been emitted in the upper atmosphere $H > 50$ km, where avalanche ionization energy losses could be estimated by measured luminosity. The last conclusion is based on the results of the calculation (Belyaev and Chudakov, 1966), which shows that UV photon flux at various atmosphere heights $H > 50$ km is due to the loss of excitation energy of nitrogen molecules by fluorescence, but not by the molecule collisions. At large heights in the atmosphere, correlation of discharge UV photon number and electron ionization losses does not depend on H and the simplest measure of TLE energy is the number of UV photons.

One of the most energetic sprites was presented in Stenbaek-Nielsen and McHarg (2008) where the luminosity of sprite stages is integrated over the whole event (Fig. 2), but several stages of sprite development are evident. An absolute luminosity for this sprite was estimated by comparison with a known star luminosity. The star was in the same camera frames (astronomical "star magnitude" units is −6). From this comparison the sprite luminosity in Fig. 2 was determined to be 10 Mega Raleigh (MR). Sprites measured in other experiments (ISUAL included) show much less luminosity of 0.1–1 MR (Kuo et al., 2008).

An important feature of sprite events is simultaneous development of several avalanches. In Figs. 1 and 2, they are seen as downward discharges neighboring the main discharge. The downward discharges make the width of TLE larger than expected for a single avalanche. Comparatively large transverse sizes of TLEs are important to estimate the radiation effects caused by TLEs when spacecraft are crossing the TLE space. In a calculation of sprite size, two main effects have to be accounted: scattering

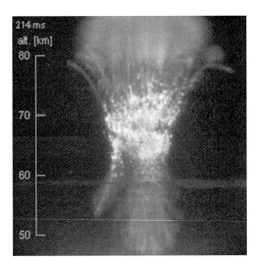

FIG. 2

Image of the most energetic sprite observed in experiment.

From Stenbaek-Nielsen, H.C., McHarg, M.G., 2008. High time-resolution sprite imaging observation and implication. J. Phys. D. Appl. Phys. 41, 234009.

of electrons in the atmosphere and their deflection by the geomagnetic field. In fractal model calculations (Pasco et al., 2000), the transverse size of sprites was estimated, which is in agreement with the observed size in Fig. 2. In the narrower part, it is of the order of ∼10–20 km, and in the wider part it is about ∼30–40 km. The longitudinal size of sprites is about 60 km (the atmosphere heights $H = 40–100$ km). Those estimates will be used to estimate radiation effects of TLE presented below, in Section 4.

The other two types of TLEs start from the top of thunderclouds: blue jets developing to an altitude of 40 km (Wescott et al., 1995) and "gigantic jets" moving higher upward and providing electrical connections between thunderclouds and the lower ionosphere (Su et al., 2003). Jets emerge above the thunderstorm regions, but their coincidence with a single lightning was not found.

The first observations of TLEs were performed using fast and sensitive cameras from the Earth's surface and airplanes, providing important information about their spatial structure and temporal development. Satellite measurements are favorable for studies of global TLE properties as well as their role in the electrical circuit and near Earth environment. They give more information about occurrence rate, geographical distribution, and energy spectrum. Spectral measurements are possible due to the transparency of upper layers of the atmosphere; they are used to distinguish TLEs from lightning (Adachi et al., 2016). Moreover, this approach allows searching for luminous events far away from thunderstorm regions, above oceans and at high latitudes.

3 EXPERIMENTAL DATA ON TLE FROM UVRIR DETECTOR ON BOARD MOSCOW STATE UNIVERSITY SATELLITES

In recent years a series of Moscow State University (MSU) satellites were launched: Universitetsky-Tatiana (for short—Tatiana-1, Garipov et al., 2005; Sadovnichii et al., 2007; Garipov et al., 2011); Universitetsky-Tatiana-2 (Tatiana-2, Sadovnichii et al., 2011; Vedenkin et al., 2011; Garipov et al., 2013), and Vernov (Panasyuk et al., 2016a,b). The study of TLEs in the atmosphere was among topics selected for satellite scientific programs. A program of the TLE study was prepared, taking into account already-measured TLE properties so it could be performed using simpler detectors on small satellites. The weight of the UV and red infrared (UVRIR) detector for the TLE study was <2 kg. It was designed (Garipov et al., 2006) with photomultipliers (PMs) as sensors.

The UVRIR detector measured the temporal profile of photon flux in the field of view limited by the collimator in its entrance so that the aperture of one channel was 0.024 cm^2 sr. In the Tatiana-2 and Vernov experiments, two channels served as two different spectral bands. In the first channel the entrance filter was transparent in the band of 240–400 nm (UV band), where quantum efficiency was 20%. In the second channel the filter cut wavelengths below 600 nm and the RIR band (600–800 nm) were determined as a band with average quantum efficiency of 2%. In the Tatiana-1 experiment one channel operated as a UV band 240–400 nm; the second one was closed by a black cover measuring noise from charged particles crossing the PMT. The number of time samples in the digital oscilloscope was 128 at both Tatiana satellites and 256 at the Vernov satellite. The duration of time samples varied: In the Tatiana-1 experiment it was 16 μs, in Tatiana-2 it was 1 ms, and in

Vernov 0.5 ms. In all MSU experiments, both channels were triggered by the UV signal. In the Tatiana-2 and Vernov satellites, the second channel served as an RIR detector, making it possible to measure the ratio of UV to RIR photon fluxes. The UV trigger was organized as a search for the largest one-time sample signal in time period T. In the Tatiana-1 experiment, events of UV flashes were recorded once per the night part of the orbit. In the Tatiana-2 experiment, time T was 1 min and in the Vernov experiment, time $T=4.5$ s. The measurements at those satellites gave new data concerning the global distribution of atmospheric UV and RIR flashes in a wide range of TLE energy. It also provided data on their temporal structure in millisecond and tens-millisecond time scales. The detailed discussion of this data follows.

3.1 TLE TYPES MEASURED BY UVRIR DETECTOR

Every channel of the UVRIR detector had the geometrical factor $\Omega S = 0.024$ cm^2 sr with the field of view $\Omega = 0.06$ sr and area $S = 0.4$ cm^2. TLEs were observed in the nadir direction from orbits of heights of 700–900 km. Triggering the UVRIR detector was done by signals from the UV channel. In this mode of triggering, the selection of TLE favors lightning selection, as nitrogen molecules—the main source of the UV signal—are dissociated in lightning due to its high temperature. It was shown in the calculation of the TLE spectrum (Milikh et al., 1998) that the percentage of UV signal in TLE is much higher than in the lightning spectrum (measured by Orville and Henderson, 1984). Experimental MSU results on the ratio of photon numbers in the UV band to photon numbers in the RIR band confirmed results of the calculation for TLE (Vedenkin et al., 2011; Garipov et al., 2013).

Event information is recorded as two variables (codes) M and N. M denotes the voltage of the PMT divider (PMTs gain) and N the analog digital converter (ADC) code for every time sample. The level of the M code is controlled by the automatic gain control system, which supports a given gain of the PMT. The algorithm of luminosity calculation from codes M and N is described in Garipov et al. (2011) and Vedenkin et al. (2011). For comparison with other experiments, we present here the final result of this calculation: the number of UV photons in the atmosphere as measured in one time step Q_a (N,M)

$$Q_a(N,M) = \frac{i(N,M) \times 4\pi R^2}{s} \tag{1}$$

Here $i(N,M)$ is the photon number in the FOV of the detector, which is transferred to the photon number $Q_a(N,M)$ in the point of TLE in the atmosphere with distance R from the satellite. As orbit height is larger in comparison with TLE altitudes, distance R varies in the range of 10%. For this reason the constant altitude (800 km) was used for TLE energy estimations. Photon radiation from TLE is supposed to be mainly fluorescence of molecular nitrogen and isotropic. The full photon numbers from TLE are a sum of photons over all time steps of the oscilloscope trace.

In space experiments, various temporal profiles were measured. The character examples of temporal profiles are presented in Fig. 3. Event A is a short single pulse in the UV and RIR bands. Event B contains two and more pulses A in one oscilloscope trace. In event C the main pulse is longer in time, while short pulses are also seen above the first one. Events A and B are the most frequent (60% of all selected events). Their short duration indicates a relation to elves events and lightning itself. Longer pulses in event C were related to sprites. Event D, without signals in C RIR band, is especially

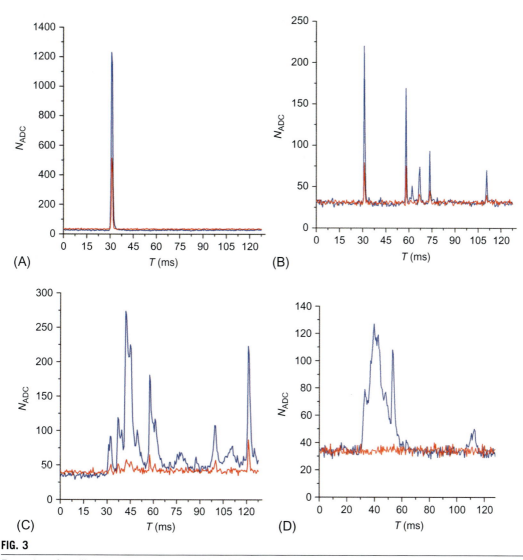

FIG. 3

Examples of oscilloscope traces measured by UVRIR detector at the Vernov satellite. In Y-axis code N. In X-axis—time in ms. Blue lines—UV signals. Red—RIR signals. In Y-axis ADC code N, related to photon numbers as explained in the text, is used. (A) A short single pulse in the UV and RIR bands, (B) two and more pulses in one oscilloscope trace, (C) longer pulses, and (D) event without signal in RIR band.

interesting. These events are rather dim in UV, but the maximum RIR number of photons doesn't exceed 3σ over the mean background value. Event D is a candidate to discharges occurring deep in the atmosphere where RIR fluorescence is restricted due to frequent molecular collisions, probably blue jets or blue starters.

3.2 TLE DISTRIBUTION OVER PHOTON NUMBERS

The data on the number of UV and RIR photons-Q_a distribution for all events selected in the Tatiana-2 experiment-are presented in Fig. 4. The left panel shows the differential fluxes and the right panel shows the corresponding integral fluxes. The rate of events with threshold photon number $Q_a \approx 10^{23}$ is $\approx 10^{-4}$ km^{-2} h^{-1} for all Earth regions and $\approx 10^{-3}$ km^{-2} h^{-1} for the active thunderstorm regions. The transient event distribution in this experiment has an interesting feature: the differential index of power low function approximating experimental data changes from -1 at photon numbers $Q_a < 10^{23}$ to -2 at $Q_a > 10^{23}$. The distribution of measured events on the Earth map confirms a difference between "luminous" ($Q_a > 10^{23}$) and "dim" ($Q_a < 10^{22}$) events. Faint transients show less concentration to the equator than luminous events and have more uniform geographical distribution (Garipov et al., 2013).

The rate of events with $Q_a > 10^{23}$ in MSU experiments is in agreement with the energetic TLE (sprite) rate measured by the ISUAL imager (Kuo et al., 2008). The global sprite rate mentioned in this work is 0.5 events per minute; this is compatible with Vernov data, taking into account the detector FOV.

It is important to underline that the events in the range of $Q_a > 10^{23}$ sometimes have extremely high photon numbers like in Fig. 2, while typical TLE started at Q_a of about 10^{22}–10^{23} for elves and sprites.

3.3 SERIES OF TLE

The selection of TLE by the maximum UV signal in one time step for a relatively long period T used in the triggering system made it possible to select a series of TLEs (Fig. 5). In the Tatiana-2

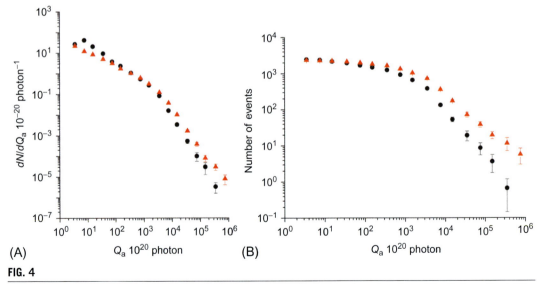

FIG. 4

Photon number distribution of TLEs selected by temporal profiles in experiments onboard the Tatiana-2 satellite. (A) Differential distribution and (B) integral distribution. Circles—UV photons, triangles—RIR photons.

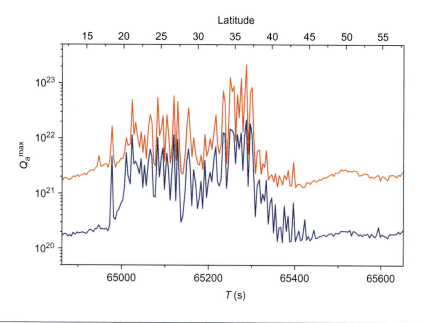

FIG. 5

A series of TLEs observed by the UVRIR detector in the Vernov mission. Every TLE is presented by the maximal number of photons from the atmosphere Q, detected in one time sample of 0.5 ms. Time interval between events $T=4.5$ s. Blue lines—UV signals, red lines—RIR signals. A series of TLEs are a mixture of different types (see text for the description of TLE types).

experiment, the interval T was 60 s and specific one-by-one TLEs were discovered (Vedenkin et al., 2011; Garipov et al., 2013). Events with a number of neighboring TLEs of ≥ 3 were called a series of TLE because the probability of random sequential events was very low. By definition, a series of TLEs could be a mixture of all four different types described in the section above. In contrary, single TLEs mostly consist of type A and are usually short and dim. At the same time, the correlation between TLEs separated by long distance L km is not trivial. For satellite velocity $V=8$ km s^{-1} and time $3T=180$ s, distance $L=3T \times V \approx 1500$ km. In the next Vernov experiment, the interval T was changed to $T=4.5$ s to obtain more information on the TLE series at smaller separation distances $L \approx 30$ km. In the Vernov experiment, the series of TLEs separated by those small distances were observed regularly (Fig. 5).

Fig. 6 presents the results of the Vernov and Tatiana-2 experiments for both TLEs in series (panel A, Tatiana-2 satellite results) and single TLEs (panel B, Vernov satellite results). The measurements from the Tatiana satellite were made from September 2009 to January 2010. The Vernov satellite was launched on July 8, 2014, and was operating until December 5, 2014. The latitudinal cutoff of measurements is explained by the orbit parameters, the period of year, and the possibility of measuring only

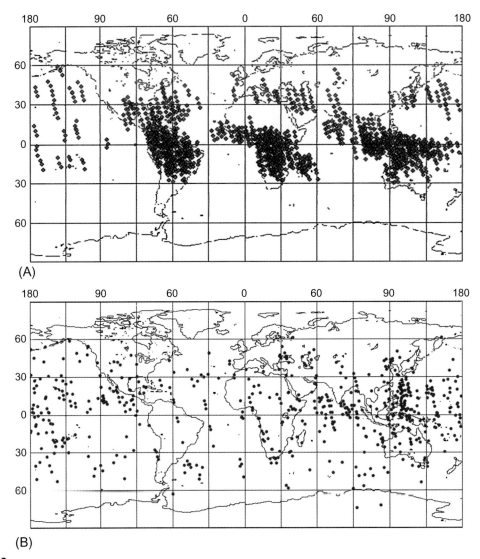

FIG. 6

Geography of TLEs in a series (data of Tatiana-2 experiment, panel A) and of single TLEs (data of Vernov experiment, panel B). Data from the Tatiana-2 experiment were collected during September 2009–January 2010. Data from the Vernov experiment were collected during July–December of 2014. Nightside of the orbit has borders determined by terminator position.

during nightside. The night period of observation has borders: latitude 60°N in the Northern hemisphere and latitude 30°S in the Southern hemisphere.

The coordinates of TLE in a series are correlated with the active thunderstorm zones while single TLEs are not (note the active thunderstorm areas on Fig. 6A coincide with continental land at low latitudes).

4 DISCUSSION

4.1 OVERVIEW OF TLE MODELS AND RELATION TO OTHER SPACE WEATHER PHENOMENA

Theoretical models of TLE are far from complete and their detailed description is not a task of this paper. Readers may see published reviews on the topic (e.g., Surkov and Hayakawa, 2012; Mareev et al., 2006). Here only qualitative conclusions from such models are considered.

Most of the theoretical models are based on the electric origin of events directly related to thunderstorm activity in the lower atmosphere. Existing experimental data on TLE in the upper atmosphere give evidence for generic correlation of events in the upper atmosphere with thunderstorm activity in the lower atmosphere. For example, the blue jets mechanism in the model (Surkov and Hayakawa, 2012) is related to the collection of extra positive charge at the top of the thunderstorm cloud and the increase of the vertical quasielectrostatic field (QE), which leads to electric breakdown at high atmosphere altitudes and, as a result, to the generation of the blue-jet type of TLE.

Taking into account that cloud formation and thunderstorm activity can be described as a consequence of cosmic factors such as ionization produced deep in the atmosphere at altitudes 5–15 km (Stozhkov, 2002; Svensmark et al., 2016), therefore one would understand that atmospheric events like at least some of TLEs are one of the space weather phenomena—the result of a chain of processes started by space factors.

Electric fields of the QE type play the key role in the generation of TLE at high altitudes—the sprite type of TLEs. Different dependence on the altitude of the QE field and of the value of air electric breakdown (both values decrease with altitudes but by different laws) determines altitudes where sprites are generated—in the mesosphere at altitudes of 70–80 km. However, one should also take into account QE dependence on the atmosphere conductivity, which exponentially decreases with altitude. In its turn the atmosphere conductivity depends on many outside factors: ionization by cosmic rays and ionization by particles (electrons) precipitated from radiation belts, which could be the key processes for the generation of certain TLE types.

An alternate model (or, maybe a complimentary one) of TLE generation at high altitudes is the model of "runaway electrons," which is a specific electric discharge (Gurevich et al., 1992; Gurevich and Zybin, 2001; Lehtinen et al., 1997; Fullekrug et al., 2011). This model is based on the assumption that discharge electrons are the product of seed electrons of EAS generated by galactic cosmic rays of energy around of $\sim 10^{15}$ eV. A runaway electron avalanche in interaction with air generates secondary electrons that are accelerated in the QE field, making lower the discharge breakdown value in the upper atmosphere (Colman et al., 2010).

The direct result of this kind of discharge is another high-altitude physical event: terrestrial gamma flashes (TGFs). These are transient processes in the X-ray and gamma ray bands of electromagnetic radiation (with energies up to tens of MeV and duration of 0.2–1 ms) generated in the process of bremsstrahlung emissions of high-energy electrons (Fishman et al., 1994; Briggs et al., 2010). One of the interesting phenomena in the runaway electron discharge is the production of electron-positron pairs, the observation of which proves the existence of gamma rays of energies above MeV in this process (Briggs et al., 2011). The efficiency of optical emissions in TGF events is not checked experimentally. Moreover, in the above review, it was shown that some of the UV optical events were observed far from thunderstorm lightning regions (Garipov et al., 2013; Morozenko et al., 2016). It could demonstrate that TGFs and TLEs are events of different natures.

So, looking for other models of TLE generation, one may assume one more model, the "top-down" type, to compare with the above "bottom-up" models. An example of the "top-down" model is the lightning-induced electron precipitation (LEP) process (Helliwell et al., 1973). This model is based on the precipitation of electrons from radiation belts induced by low-frequency electromagnetic waves of the "whistler" type generated by lightning and spread out along magnetic field lines. Resonance interaction of those waves with radiation belt particles (interaction of the "wave-particle" type) leads to the changing of the pitch-angle distribution of "trapped" particles and their coming to the loss cone, that is to precipitation.

The precipitation of electrons of hundreds keV energy could lead to the increase of the ionosphere D-layer conductivity directly above the active thunderstorm region and to the changing of the discharge breakdown threshold, respectively. Experimental proof of the existence of such a process is the EuroSprite data on pulses of electromagnetic VLF radiation (3–30kHz) (Inan et al., 1995). It was established that before sprite TLEs, there are early pulses of radiation directly related to the increase of the conductivity of the D-layer in a size of 100 km, exactly in place of the sprite (Neubert et al., 2008; Inan et al., 2010).

The investigation of electromagnetic radiation dynamics in VLF and in extremely low frequency (ELF, 0.3–3 kHz) bands accompanying the generation of sprites allowed the estimation of the value of the current momentum of sprites—of several hundreds of kA × km (Cummer et al., 2010), large enough to consider as a factor of space weather.

Direct spectrometric measurements of upper atmosphere TLEs on the FORMOSAT-2 satellite allowed the estimation of the electric field deviations from average values in place of TLE for altitudes 40–60 km. Deviations were found that were 2.1–3.7 times larger than the discharge threshold (Kuo et al., 2005). These electric field deviation ns were two times larger than in ground-based measurement. An analysis of sprite development in the frame of the plasma chemistry model of the sprite streamers (Sentman et al., 2008) model has shown hundred milliseconds delay time of the sprite optical emission to the beginning of the electric current in the lightning channel. At the maximum current of 110 kA, it is accompanied by a large deviation of the electric field from the average value E/N normalized to the density of neutral gas N. In a direct experiment on the satellite FORMOSAT-2, this value was larger than 440 Td (1 Townsend $= 10^{-17}$ V cm^2). In the results of model calculations (Evtushenko and Mareev, 2009) it is up to 200 Td during hundreds of ms. So, the results of the present experimental study of TLEs and the results of the modeling of physical and plasma-chemical processes responsible for TLE generation show that the upper atmosphere discharges are accompanied not only by intensive emission of photons in the UV band but also by bursts of radio emission in the VLF and ELF bands as well as increases of quasistable fields.

Following are some comments we have to make concerning the series of TLEs (multiple events) that we observed on MSU satellites (see Section 3.3).

From our point of view, the origin of TLEs in series is an open question. Their presence shows that there is a large volume of atmosphere that has long-duration flashing activity. One can suggest at least three hypotheses to explain these observations. First, it is the result of measurements above very large thunderstorm areas with ordinary lightning and TLEs. Second, the lightning can produce electromagnetic waves, which are propagating along magnetic field lines and interacting with magnetospheric (radiation belts) electrons (multiple "wave-particle" processes). Then particles can precipitate from trapped regions and therefore can modify the atmosphere conductivity. As a result, this creates conditions for further discharging processes at higher altitudes over the large area. And third, the lightning produces runaway relativistic electrons that escape atmosphere and precipitate in the conjugate point (Lehtinen et al., 1998). However, efficiency of this mechanism was checked in model calculations (Lehtinen et al., 1998) and was found to be low (10%).

Such multiple TLEs were predicted in the model calculation (Pasco et al., 2000), but the observed TLE series within a few minutes (within thousands of km!) show the rather global nature of registered processes and more broad separation in distance then predicted by the Pasco model. Therefore, their origins require further examination. For example, there is a discussion of the development of TLEs in regions with large differences in the atmosphere conductivity. Then, the development of TLEs in the region with high conductivity will cancel the electric field in this area, but it will not in the next area, separated from the first one by a region of small conductivity. Such features of "resistive" electrodes are used in the construction of particle detectors when efficiency of detection in large areas is the aim.

4.2 TLE ENERGY DEPOSITION

Evidently, energy loss for ionization of the atmosphere in its electric discharges, related to the intensity of electromagnetic emission, is a direct agent of ionosphere modification at altitudes of TLE generation, changing its conductivity and parameters of optical emission of TLE (Inan et al., 2010).

Modification of physical parameters of the ionosphere—its additional ionization in TLE development—may be the reason for changes in the amplitude and phase of radio stations operating in the VLF band (Evtushenko et al., 2013). Also, the generation of TLE at high altitudes may influence the accuracy of geophysical measurements by methods of radio enlightening (Evtushenko, 2007). Estimations (Cummer and Fullekrug, 2010) have shown that upper atmosphere sprites could transfer the charge of tens of Coulombs (C) and be accompanied by current up to 3 kA. For voltage between the top of the clouds and the mesosphere of several millions volts, this current leads to energy dissipation of hundreds of megajoules (MJ) (Mareev et al., 2006). Taking into account the estimate of average sprite rate—several events per minute from observations at the FORMSAT-2 satellite (Chen et al., 2008)—one finds the average dissipated energy in the global electric circuit only by the sprite type of TLE, of the order of tens MW.

The quantitative data on energy radiated by TLEs were obtained from direct measurements of the photon number Q_a emitted in the atmosphere by each TLE in the MSU space experiments. In the near-UV band (300–400 nm), the detectors have the advantage of lower background from the atmospheric glow. It has been shown in calculations (Belyaev and Chudakov, 1966) that the photon number Q_a emitted in TLE by the discharge electrons is proportional to their ionization energy loss E_{ion} in the atmosphere:

$$Q_a = E_{ion}/K(H) \times \varepsilon \qquad (2)$$

where ε is the mean energy of the emitted photons ($\varepsilon = 3.5$ eV for the UV band) and $K(H)$ is the coefficient depending on the height H in the atmosphere. Dependence on H comes from the variable ratio of radiative/nonradiative collisions of electrons with nitrogen molecules of the atmosphere. At heights $H > 50$ km this coefficient is constant $K \approx 10^3$ (Belyaev and Chudakov, 1966), and energy loss E_{ion} is characterized by the photon number Q_a:

$$E_{ion} = 5.6 \times 10^7 \times Q_a/10^{23} \qquad (3)$$

where energy E_{ion} is in Joules (J).

In the TLE range of $Q_a = 10^{23} - 10^{25}$ estimate (3) for Q_a distribution with exponent -2 (Section 3) gives the average TLE energy $E \sim 100$ MJ and, assuming rate of TLE with threshold energy ~ 50 MJ as ~ 0.1 s^{-1} (Chen et al., 2008), one found a yield of TLE power in the global electric current as 10 MW—in agreement with estimates (Mareev et al., 2006).

In MSU space experiments, the same UV detector that measured the intensity of the TLE radiation has also measured the intensity of polar (aurora) lights. It allowed us to make comparisons of aurora and TLE components of the global current. In both measurements, scattered UV radiation was measured in the same wavelength band but active zones are different. Aurora lights are active at latitudes of about 50°–70° North and South, but most TLEs are observed at latitudes 30°N–30°S. Also different is the temporal distribution of dissipated power. Aurora light permanently dissipates energy while transient events do it in pulses. A constant radiation of aurora of 1 kR corresponds to power dissipation of $6 \cdot 10^{-3}$ Wt m^{-2}. TLEs dissipate energy of about 0.5 Wt m^{-2} in pulses of 1–100 ms (i.e., two orders of magnitude more than aurora lights in the same short time). In longer time (minutes and hours), the average power dissipation of TLE becomes comparable with aurora. This comparison shows that transient events might be an important factor of the upper atmosphere at short time intervals.

Presented estimates of energy deposition of transient events to the upper atmosphere show that TLEs may be considered as principal agents of space weather.

4.3 TLES AS A RADIATION HAZARD

Consideration of TLEs as a possible hazardous factor for a vehicle crossing the upper atmosphere (sub-orbital flights) showed that UV and longer wavelength radiation is not a difficult problem. It could be protected by a thin satellite cover (several mm of aluminum). Potentially, it could be dangerous for open windows of optical detectors, and protection of them may be a specific problem.

However the transient electron flux of energy of electrons emitted from a thunderstorm area ("runaway electrons") has relativistic energies, reaching the MeV energy range. Effects of radiation doses from these electrons can be a potential danger factor for technological systems in the upper atmosphere. Estimations of effective radiation doses have been done (Dwyer et al., 2010) but are applicable just for aircraft altitudes. The dose of runaway electrons for human tissue has also been studied. Results of this work, in which the effective dose is calculated according to the relation $D = h \cdot \Phi$, where $h = 7 \cdot 10^{-16}$ Sv m^2, were used to make estimates for the upper atmosphere. In this way, the assessment of the instantaneous value of the effective dose is given by $D \approx 0.5 \cdot 10^{-4}$ Sv. Note that the value obtained is not a dangerous single-event dose for humans, for which a limit of 0.001 Sv is set. Evidently, radiation effects for higher altitudes will be much smaller than for an aircraft's altitude. On the other hand, the dose estimate may be different if one considers it in more detail. The spatial structure of the discharge, especially in the early stages when the region is strongly localized, may be important. Also, we have to take into account that radiation effect of a series of TLEs can be multiplied. This is, however, a special challenge for study using computer simulations of the discharge in the future.

Another source for the radiation hazard for spacecraft and even aircraft can be neutrons, generated as a secondary component of TLEs (Tavani et al., 2013; Drozdov et al., 2013). Neutrons are generated in photonuclear interactions when photon energy is above the photonuclear disintegration threshold (of about several MeV). Photons (gamma quanta) of such high energy proved to be existent in the atmosphere discharges TGF, discussed above in Sections 1 and 2. Data from the literature give the rate of TGF with total energy in gamma quanta of >1 kJ as 10^{-5} event km^{-2} h^{-1} in active thunderstorm zones—two orders of magnitude lower than the TLE rate. The expected neutron number in those TGFs is estimated as 10^{12} or the neutron fluency in TGF of size 10×10 km is $\approx 10^4$ neutrons m^{-2}. The fluency of thermal neutrons, which actually produce "radiation effects," is much larger. The above estimate of the neutron number in single TGF is an order of magnitude higher than the average neutron flux in the atmosphere produced by cosmic rays. The radiation of the neutron component of the TGF radiation was calculated for altitudes up to 50 km (Drozdov et al., 2013). The single-event value decreases fast with altitude. Assuming that the TGF source is located at an altitude of 10 km, the dose for human tissue near this point is about 10^{-4} Sv, and decreases almost exponentially down to 10^{-10} Sv by altitude 50 km. It should be noted that the aircraft sheathing is not so effective in this case as compared to the electron radiation discussed above.

5 RESULTS AND CONCLUSIONS

Transient luminous events (TLEs) observed in the upper atmosphere at altitudes from several tens up to 100 km are an important physical process for study, both for fundamental physics and for practical aspects. In this chapter, we performed an overview of recent measurements of TLEs onboard the Universitetsky-Tatiana, Universitetsky-Tatiana-2, and Vernov satellites. Satellite measurements demonstrate a large variety of temporal structures and energy of TLEs. Geographical distribution of some of them has shown a certain correlation with active thunderstorm regions above continents. Long series (several minutes of continuous flashes in the atmosphere) indicate a global nature of such TLEs. Instead, single TLEs are usually less bright and spread more uniformly over the globe. To summarize, the theoretical description of TLEs, both single and in a series, is far from complete and requires future examination.

The results also demonstrate the significant contribution of transient events to the energy deposition of electric discharges in the upper atmosphere. The most intense luminous events (a series of TLEs) dissipate energy up to hundreds MJ, indicating an important role of TLEs in the energetic balance of the upper atmosphere in comparison with other space weather effects, especially at low latitudes near the equator.

All these measurements show the global nature and high energy of such processes in the upper atmosphere, which reveal themselves in UV, X-rays, gamma rays, electron/positron beams, and neutron flux. These atmospheric events are connected with strong perturbations of ambient QS electric fields and the generation of low-frequency electromagnetic waves. They are possibly connected with the enhancement of ionization radiation caused by runaway electrons emitted from thunderstorm areas and/or precipitation of electrons from the near-Earth space environment. All these effects should be added to the list of space weather factors, as they are important in their large energy dissipation in the upper atmosphere. We can conclude that among the numerous physical effects, corresponding TLE generation, strong variations of radiation fields, and modification of conductivity of the D-layer of ionosphere are the most potentially dangerous factors in terms of space weather effects. If the first one has direct impact on the safety of suborbital flights, the second one can be responsible for the stability of radio communications.

REFERENCES

Adachi, T., Sato, M., Ushio, T., et al., 2016. Identifying the occurrence of lightning and transient luminous events by nadir spectrophotometric observation. JASTP 145, 85–97. https://doi.org/10.1016/j.jastp.2016.04.010.

Belyaev, V.A., Chudakov, A.E., 1966. Ionization radiation of the atmosphere and its possible application in measuring extensive air showers. Bull. Russ. Acad. Sci. Phys. 30 (10), 1700–1707.

Briggs, M.S., Fishman, G.J., Connaughton, V., et al., 2010. First results on terrestrial gamma-flashes from the Fermi gamma-ray burst monitor. J. Geophys. Res. 115. https://doi.org/10.1029/2009JA014853, A00E49.

Briggs, M.S., Connaughton, V., Wilson-Hodge, C., et al., 2011. Electron-positron beams from terrestrial lightning observed with Fermi GBM. Geophys. Res. Lett. 38. https://doi.org/10.1029/2010GL046259, L02808.

Chen, A.B., Kuo, C.-L., Lee, Y.-J., et al., 2008. Global distribution and occurrence rates of transient luminous events. J. Geophys. Res. 113. https://doi.org/10.1029/2008JA013101, A08306.

Chern, J.L., Hsu, R.R., Su, H.T., et al., 2003. Global survey of upper atmospheric transient luminous events on the FORMOSAT-2 satellite J. J. Atmos. Sol. Terr. Phys. 65 (5), 647–659.

Colman, J.J., Roussel-Dupre, R.A., Triplett, L., 2010. Temporally self-similar electron distribution functions in atmospheric breakdown: the thermal runaway regime. J. Geophys. Res. 115. https://doi.org/10.1029/2009JA014509, A00E16.

Cummer, S.A., Fullekrug, M., 2010. Unusually intense continuing current in lightning produces delayed mesospheric breakdown. Geophys. Res. Lett. 28, 495–498.

Cummer, S.A., Jaugey, N., Li, J., Lyons, W.A., et al., 2006. Submillisecond imaging of sprite development and structure. Geophys. Res. Lett. 33, L04104.

Drozdov, A., Grigoriev, A., Malyshkin, Y., 2013. Assessment of thunderstorm neutron radiation environment at altitudes of aviation flights. J. Geophys. Res. 118 (2), 947–955.

Dwyer, J.R., Smith, D.M., Uman, M.A., et al., 2010. Estimation of the fluency of high-energy electron bursts produced by thunderclouds and the resulting radiation doses received in aircraft. J. Geophys. Res. 115, D09206.

Evtushenko, A.A., 2007. Modeling the influence of a high altitude discharge on the chemical balance of the mesosphere. In: Earth: Our Changing Planet. Proceedings of the International Union of Geodesy and Geophysics (IUGG) XXIV General Assembly. Perugia, Italy (IUGG, Perugia, 2007), p. 151.

Evtushenko, A.A., Mareev, E.A., 2009. On the generation of charge layers in MCS stratiform regions. Atmos. Res. 45, 272–282.

Evtushenko, A.A., Mareev, E.A., Marshall, T.C., Stolzenburg, M., 2013. A model of sprite influence on the chemical balance of mesosphere. J. Atmos. Sol. Terr. Phys. 102, 298–310.

Fishman, G.J., Bhat, P.N., Mallozzi, R., et al., 1994. Discovery of intense gamma-ray flashes of atmospheric origin. Science 264 (5163), 1313–1316. https://doi.org/10.1126.

Franz, R.C., Nemzek, R.J., Winckler, J.R., 1990. Television image of a large upward electric discharge above a thunderstorm system. Science 249, 48–51.

Fullekrug, M., Roussel-Dupre, R.A., Symbalisty, E.M.D., et al., 2011. Relativistic electron beams above thunderclouds. Atmos. Chem. Phys. 11, 7747–7754.

Garipov, G.K., Khrenov, B.A., Panasyuk, M.I., et al., 2005. UV radiation from the atmosphere: Results of the MSU "Tatiana" satellite measurements. Astropart. Phys. 24, 400–408.

Garipov, G.K., Panasuyk, M.I., Rubinshtein, I.A., et al., 2006. UV detector of MSU scientific- educational micro satellite Universitetsky-Tatiana. ITE (1), 135–141.

Garipov, G.K., Klimov, P.A., Morozenko, V.S., et al., 2011. Time-energy characteristics of UV flashes in the atmosphere from data of Universitetsky-Tatiana satellite. Space Res. 49 (5), 391–398.

Garipov, G.K., Khrenov, B.A., Klimov, P.A., et al., 2013. Global transients in ultraviolet and red-infrared ranges from data of Universitetsky-Tatiana-2 satellite. J. Geophys. Res. 118 (2), 370–379.

REFERENCES

Garipov, G.K., Panasyuk, M.I., Svertilov, S.I., et al., 2016. Vernov satellite data on global techno genic events of UV and IR radiation in the night atmosphere. J. Exp. Theor. Phys. 123 (3), 470–479.

Gurevich, A.V., Zybin, K.P., 2001. Runaway breakdown and electric discharges in thunderstorms. Physics-Uspekhi 44 (11), 1119–1140.

Gurevich, A.V., Milikh, G.M., Roussel-Dupre, R.A., 1992. Runaway electron mechanism of air breakdown and preconditioning during a thunderstorm. Phys. Lett. A. 165, 463–468.

Helliwell, R., Katsufrakis, J.P., Trimpi, M., 1973. Whistler-induced amplitude perturbation in VLF propogation. J. Geophys. Res. 78, 4679–4688.

Inan, U.S., Bell, T.F., Pasko, V.P., et al., 1995. VLF signatures of ionospheric disturbances associated with sprites. J. Geophys. Res. Lett. 22. 3461–3464. https://doi.org/10.1029/95GL03507.

Inan, U.S., Cummer, S.A., Marshall, R.A., 2010. A survey of ELF and VLF research on lightning-ionosphere interaction and causative discharges. J. Geophys. Res. 115. https://doi.org/10.1029/2009JA014775, A00E36.

Kuo, C.-L., Hsu, R.R., Chen, A.B., et al., 2005. Electric fields and electron energies inferred from the ISUAL recorded sprites. Geophys. Res. Lett. 32. https://doi.org/10.1029/2005GL023389, L19103.

Kuo, C.-L., Chen, A.B., Lee, Y.J., et al., 2007. Modeling elves observed by FORMOSAT-2 satellite. J. Geophys. Res. 112. https://doi.org/10.1029/2007JA012407, A11312.

Kuo, C.-L., Chen, A.B., Chou, J.K., et al., 2008. Radiative emission and energy deposition in transient luminous events. J. Phys. D. Appl. Phys. 41, 234014 (14 pp.).

Lehtinen, N., Bell, T., Pasko, V., Inan, U., 1997. A two- dimensional model of runaway electron beams driven by quasi- electrostatic thunderclouds fields. Geophys. Res. Lett. 24, 2639–2642.

Lehtinen, N.G., Bell, T.F., Inan, U.S., 1998. Monte Carlo simulation of runaway MeV electron breakdown with application to red sprites and terrestrial gamma ray flashes. J. Geophys. Res. Space Phys. 104 (A11), 24699–24712.

Mareev, E.A., Evtushenko, A.A., Yashunin, S.A., 2006. On the modelling of sprites and sprite-producing clouds in the global electric circuit. In: Fullekrug, M. et al., (Ed.), Sprites, Elves and Intense Lightning Discharges. Springer, Netherland, pp. 313–340.

Milikh, G.M., Papadopoulos, K., Valdivia, J.A., 1998. Spectrum of Red Sprites. J. Atmos. Sol. Terr. Phys. 69, 907–915.

Millan, R.M., 2011. Understanding relativistic electron losses with BARREL. J. Atmos. Sol. Terr. Phys. 73. (11-12), 1425–1434. https://doi.org/10.1016/j.jastp.2011.01.006.

Morozenko, V.S., Klimov, P.A., Khrenov, B.A., et al., 2016. Far from thunderstorm UV transient events in the atmosphere measured by Vernov satellite. In: EGU General Assembly Conference. Abstracts,18, 496.

Neubert, T., Rycroft, M., Farges, T., et al., 2008. Recent results from studies of electric discharges in the mesosphere. Surv. Geophys. 29. 71–137. https://doi.org/10.1007/s10712-008-9043-1.

Orville, R.E., Henderson, R.W., 1984. Absolute spectral irradiance measurements of lightning from 375 to 880 nm. J. Atmos. Sci. 41 (21), 3180.

Panasyuk, M.I., Svertilov, S.I., Bogomolov, V.V., et al., 2016a. Vernov satellite experiment: transient energetic processes in the atmosphere and magnetosphere. Description of the experiment. Space Res. 54 (4), 277–285.

Panasyuk, M.I., Svertilov, S.I., Bogomolov, V.V., et al., 2016b. Vernov satellite experiment: transient energetic processes in the atmosphere and magnetosphere. First results. 5. Space Res. 54, 369–376.

Pasco, V.P., 2003. Electric jets. Nature 423, 927–929.

Pasco, V.P., Inan, U.S., Bell, T.F., 2000. Fractal structure of sprites. Geophys. Res. Lett. 27, 497–500.

Raizer, Y.P., Milikh, G.M., Shneider, M.N., 2010. Streamer- and leader-like processes in the upper atmosphere: models of red sprites and blue jets. J. Geophys. Res. 115. https://doi.org/10.102v/9/2009JA014645, A00E42.

Sadovnichii, V.A., Panasuyk, M.I., Boborovnikov, S.Yu., et al., 2007. The first results of space environment study on Universitetsky Tatiana satellite. Space Res. 45, 273–286.

Sadovnichii, V.A., Panasuyk, M.I., Yashin, I.V., 2011. Study of space environment on micro satellites Universitetsky-Tatiana and Universitetsky-Tatiana-2. Astronomy Bull. 45 (1), 1–27.

Sentman, D.D., Stenbaek-Nielsen, H.C., McHarg, M.G., Morrill, J.S., 2008. Plasma chemistry of sprite streamers. J. Geophys. Res. 113. https://doi.org/10.1029/2007JD008941, D11112.

Stenbaek-Nielsen, H.C., McHarg, M.G., 2008. High time-resolution sprite imaging observation and implication. J. Phys. D. Appl. Phys. 41, 234009.

Stozhkov, Y.I., 2002. The role of cosmic rays in the atmospheric process. J. Phys. G Nucl. Part. Phys. 29, 913–923.

Su, H.T., Hsu, R.R., Chen, A.B., et al., 2003. Gigantic jets between a thundercloud and the ionosphere. Nature 423, 974–976.

Surkov, V.V., Hayakawa, M., 2012. Underlying mechanisms of transient luminous events: a review. Ann. Geophys. 30. 1185–2012. https://doi.org/10.5194/angeo-30-1185-2012.

Svensmark, J., Enghoff, M.B., Shaviv, N., Svensmark, H., 2016. The response of clouds and aerosols to cosmic ray decreases. J. Geophys. Res. Space Phys. 121, 8152–8181. https://doi.org/10.1002/2016JA022689.

Tavani, M., Argan, A., Paccagnella, A., et al., 2013. Possible effects on avionics induced by terrestrial gamma-ray flashes. Nat. Hazards Earth Syst. Sci. 13, 1127–1133.

Vedenkin, N.N., Garipov, G.K., Klimov, P.A., et al., 2011. Atmospheric flashes in ultraviolet and red-infrared bands from data of Universitetsky-Tatiana-2 satellite. J. Exp. Theor. Phys. 140 (3(9)), 1–11.

Wescott, E.M., Sentman, D.D., Osborne, D.L., et al., 1995. Preliminary results from the Sprite-94 aircraft campaign: 2. Blue jets. Geophys. Rev. Lett. 22, 1209–1212.

Yair, Y., Israelevich, P., Devir, A.D., et al., 2004. New observations of sprites from space shuttle. J. Geophys. Res. 109. https://doi.org/10.1029/2003JD004497, D15201.

FURTHER READING

Babich, L.P., Donskoy, E.N., Il'kaev, R.I., Kutsyk, I.M., Roussel-Dupré, R.A., 2004. Fundamental parameters of a relativistic runaway electron avalanche in air. Plasma Phys. Rep. 30, 616–624.

Ostgaard, N., Stadsnes, J., Bjordal, J., et al., 1999. Global scale electron precipitation features seen in UV and X-rays during substorms. J. Geophys. Res. 104, 10191–10204.

Ostgaard, N., Stadsnes, J., Bjordal, J., et al., 2001. Auroral electron distributions derived from combined UV and x-ray emissions. J. Geophys. Res. 106, 26081–26089.

Rodger, C.J., Hendry, A.T., Clilverd, M.A., et al., 2015. High-resolution in situ observations of electron precipitation causing EMIC waves. Geophys. Res. Lett. 42. 9633–9641. https://doi.org/10.1002/2015GL066581.

Thorne, R.M., 2010. Radiation belt dynamics: the importance of wave-particle interactions. Geophys. Res. Lett. 37. https://doi.org/10.1029/2010GL044990, L22107.

Voss, H.D., Imhof, W.L., Walt, M., et al., 1984. Lightning-induced electron precipitation. Nature 312 (5996), 740–742.

Yair, Y., Price, C., Ziv, B., et al., 2005. Space shuttle observation of an unusual transient atmospheric emission. Geophys. Res. Lett. 32. https://doi.org/10.1029/2004GL021551, L02801.

PART 5

IONOSPHERIC/ THERMOSPHERIC EFFECTS AND IMPACTS

CHAPTER 20

IONOSPHERE AND THERMOSPHERE RESPONSES TO EXTREME GEOMAGNETIC STORMS*

Anthony J. Mannucci, Bruce T. Tsurutani
California Institute of Technology, Pasadena, CA, United States

CHAPTER OUTLINE

1 Historical Background	493
2 Electric Fields and the Creation of Large TEC Increases	495
2.1 Equatorial Plasma Irregularities and Scintillation	499
3 The Role of Ion-Neutral Coupling	501
3.1 Buoyancy (Gravity) Waves	503
4 Extreme Nighttime Responses Following the Storm Main Phase (Florida Effect)	504
5 Conclusions and Future Outlook	506
Acknowledgments	508
References	508
Further Reading	511

1 HISTORICAL BACKGROUND

Decades of research into ionospheric behavior have produced a consensus understanding of major perturbations that can occur. This understanding is far from complete, particularly considering variations at the mesoscale and finer. However, this understanding forms a basis for how to extrapolate ionospheric behavior to extreme events. Our focus is on those aspects of ionospheric extremes that cause large total electron content (TEC) values to appear during daytime. Affected technological systems include navigation (e.g., global navigation satellite systems), HF communications systems, and radar.

---

*Copyright 2017 California Institute of Technology, U.S. Government sponsorship acknowledged.

Ionospheric extremes that are "societally relevant" can fall into these two categories when considering TEC: large TEC values, in particular, that arise suddenly, and spatial TEC gradients. The former can cause navigation errors for a variety of systems and will also degrade radar range values. Spatial TEC gradients are of a particular concern to wide-area satellite-based navigation augmentation systems, such as the U.S. Wide Area Augmentation System (WAAS) used in commercial and private aviation (Sparks et al., 2011a,b; Enge et al., 1996). WAAS is designed to protect its aircraft navigation users from *hazardously misleading information* (HMI) that can arise from inadequate characterization of ionospheric spatial structure. Vulnerability to HMI increases in the presence of localized (~100 s of km or smaller) TEC enhancements or depletions. We focus in this chapter on mid-to-low latitudes (e.g., below 50° latitude) where many technology systems are affected.

When discussing ionospheric impacts, it is important to clearly define what is meant by "storm" in this chapter. Other chapters refer extensively to geomagnetic storms, which are of course associated with space weather impacts. "Ionospheric storms" refers to the ionospheric effects associated with geomagnetic activity. The root cause of ionospheric storms is generally accepted to be the same root cause of geomagnetic storms: enhanced energy input from the solar wind due to increased interplanetary magnetic field magnitudes, increased plasma speed, densities, and temperatures. Significant ionospheric variability may also be associated with other drivers, such as the lower atmosphere, although the largest effects likely arise from solar activity. It is important to understand the distinction between these geomagnetic and ionospheric storm types, despite their common cause, because geomagnetic storms are often associated with specific variations of geomagnetic indices. Ionospheric storms are general perturbations of the upper atmosphere that may or may not correlate well with variations in geomagnetic indices. Ionospheric storms have their own character and evolution, but it is widely believed that extreme geomagnetic storms will correlate with extreme ionospheric behavior.

To understand and plan for how technological systems may be affected by extremes, it is useful to have approximate quantitative estimates of key geophysical variables. One encounters the problem that there are limited numbers of observations from which to derive such estimates, as the observational record of extremes is likely incomplete. This limitation can often be partially overcome by focusing on the physical ideas, possibly in simplified form, from which reasonable extrapolations can be made. Extrapolations should not be made solely from the data, without this understanding. Theoretical simplifications can provide a form of confirmation: if approximate simple calculations cannot justify the expectation of extreme behavior, then confidence in more complex calculations is limited, particularly when physical justification is absent. Ultimately, a realistic and complex calculation could in theory provide the best answer, but with very limited observations, this can be difficult to justify. Thus, an emphasis on physical ideas is the focus of this chapter. We indicate where more complex calculations are needed and useful. Our discussion occurs with reference to the Carrington storm of 1859, where the measure of geomagnetic storm intensity given by the Dst index has been estimated to be as large as -1760 nT (see Chapter 7 of this volume), compared to -451 nT for the October 30, 2003, storm discussed in detail here (see Mannucci et al., 2005a, 2008 for discussions of superstorms from Solar Cycle 23).

It is well known that the largest storm-time TEC values generally occur during daytime (Mendillo and Klobuchar, 1975). This is in accord with the well-known diurnal cycle of ionospheric electron density whereby electron-ion pairs are created by photoionization, a process of plasma *production* that increases plasma density in a volume of the atmosphere (Schunk and Nagy, 2009). During disturbed periods, increased plasma density compared to quiet conditions is known as the *positive* ionospheric storm phase (Mendillo and Klobuchar, 1975). Initially, it was not understood how plasma density could increase during geomagnetic storms given that plasma production rates should not be higher during the storm; increased ionizing radiation due to solar extreme ultraviolet (EUV) fluxes are not associated with the storms themselves, but are rather associated with the possible solar flares that may have occurred several days earlier. Eventually, it was realized that geomagnetic storms did not increase EUV production significantly at middle and lower latitudes, where the largest increases occur. (At auroral to middle latitudes, production due to particle precipitation does increase, but this increase is not the focus of our report). Rather, increased TEC occurs despite constant solar ionizing radiation, because of decreased plasma recombination rate (loss). We will show how this basic physical idea permits us to set reasonable "benchmarks" on what can occur during extreme conditions.

For decreased loss to be a cause of the storm-time TEC increase requires that the "nominal" production rate due to daytime photoionization is sufficiently large to account for the TEC increase over time. If plasma is produced in the absence of recombination, can TEC increase as rapidly as is observed? In the next section, we quantitatively explore this idea.

2 ELECTRIC FIELDS AND THE CREATION OF LARGE TEC INCREASES

An example of a large column-integrated electron density increase is shown in Fig. 1 (Mannucci et al., 2005b). This figure shows the integrated electron content (IEC) above the CHAMP satellite orbiting at 400 km altitude as measured by the dual-frequency GPS receiver onboard the satellite. The dual-frequency measurements of delay are related to the IEC via the following formula (Stephens et al., 2011):

$$\Delta_{12} = 40.3\, \text{IEC} \left(1/f_2^2 - 1/f_1^2\right) \tag{1}$$

where Δ_{12} is the measured delay difference between the two GPS frequencies expressed in meters, f_1 is the L1 GPS frequency (1.5754 GHz), f_2 is the L2 GPS frequency (1.2276 GHz), and IEC is expressed in number of electrons per square meter (1 TEC unit is 10^{16} el/m^2). A constant interfrequency bias (IFB) for the receiver and each of the GPS satellites must generally be estimated for these types of measurements (Mannucci et al., 1998, 1999).

The blue trace in Fig. 1 is the prestorm IEC estimate above the satellite. There is often more than a single trace at a given time indicating tracking of more than one GPS satellite by the receiver, above the elevation angle cutoff of 40° used in this analysis. The next pass of the satellite, shown in red, is the first daytime pass following storm commencement, which is near to the time when a southward solar wind magnetic field B_z component reached the magnetopause (see Mannucci et al., 2005b for details). We note that this October 30, 2003, storm occurs during the recovery period of an intense storm occurring on the preceding day. The next and last pass of the satellite shown in this figure is the black trace, where the largest IEC values are observed at magnetic dipole latitudes near −30°. To summarize, in the span of ∼200 min, the global structure of electron density above 400 km altitude undergoes very large changes, including a factor of 13 IEC increase (from 26 to 335 TECU) at −28° dipole latitude.

Because the pre-storm (Fig. 1, blue) and post-storm (black and red) satellite passes traverse North America, we can find GPS receivers near the satellite ground tracks to validate and compare with the IEC. We expect that well-calibrated IEC/TEC measurements measured from ground receivers near to the satellite overpass will exceed IEC measurements from CHAMP. To the mid-bottom right of the

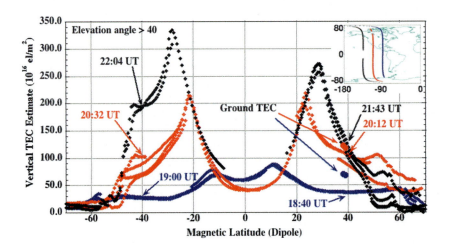

FIG. 1

Vertical integrated electron content (IEC) measured above the CHAMP satellite at 400 km altitude for daytime passes on October 20, 2003. Equatorial solar local time is 1300 LT.

From Mannucci, A.J., et al., 2005. Hemispheric daytime ionospheric response to intense solar wind forcing. In: Burch, J.L., Schulz, M., Spence, H. (Eds.), Inner Magnetosphere Interactions: New Perspectives From Imaging. In: Geophysical Monograph, vol. 159. American Geophysical Union, New York, pp. 261–275. doi:10.1029/159GM20; Mannucci, A.J., Tsurutani, B.T., Iijima, B.A., Komjathy, A., Saito, A., Gonzalez, W.D., Guarnieri, F.L. Kozyra, J.U., Skoug, R., 2005, Dayside global ionospheric response to the major interplanetary events of October 29–30, 2003 "Halloween Storms." Geophys. Res. Lett. 32(1), L12S02. https://doi.org/10.1029/2004GL021467.

figure, near 40° latitude, are indicated two vertical TEC estimates from ground-based receivers in North America near the time of the CHAMP overflights. The ground-based TEC is calibrated using the Global Ionospheric Mapping (GIM) technique (Mannucci et al., 1998). Pre-storm, we find that TEC is approximately twice the IEC from CHAMP. During the storm, TEC and IEC are approximately equal. The pre-storm ratio suggests that approximately half the TEC comes from electrons above 400 km altitude, which is reasonable based on "typical" mid-latitude electron density profiles, as can be verified using the International Reference Ionosphere (IRI) model for example (Bilitza et al., 2014).

During the storm (red trace), the IEC has grown to be ∼90% as large as the ground-based TEC (red circle near 40° latitude versus the closest CHAMP value). Thus, most of the plasma is above the altitude of the satellite after storm commencement. Considering pre- to poststorm conditions, comparison with ground-based TEC is strongly suggestive of plasma upward motion. The Halloween storm is an intense example of a *positive storm effect* that occurs during the main phase of a geomagnetic storm (Mendillo, 2006; Tsurutani et al., 2004). This daytime positive effect is generally caused by dawn-to-dusk (eastward during daytime) electric fields of magnetospheric origin, so-called "prompt penetration electric fields" (PPEF). The resulting $\mathbf{E} \times \mathbf{B}$ drift convects the plasma vertically upward. For fixed eastward field, such drift will be maximum near the geomagnetic dip equator where $\mathbf{B}$ is close to horizontal (north-south direction).

$\mathbf{E} \times \mathbf{B}$ drift speeds v are given by: $v = \mathbf{E} \times \mathbf{B}/B^2$ (Kelley, 2009). Storm-time values of electric field often reach ∼1 mV/m (Huang et al., 2005) or higher. The magnetic field magnitude $\mathbf{B}$ is ∼30,000 nT

2 ELECTRIC FIELDS AND THE CREATION OF LARGE TEC INCREASES

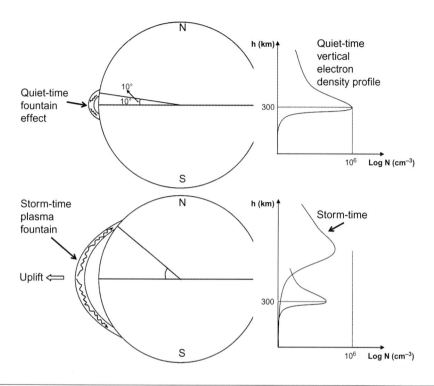

FIG. 2

A schematic description of the ionospheric superfountain created by large prompt penetration electric fields. Left: increased vertical transport broadens in latitude the region of the equatorial fountain effect. Right: schematic representation of vertical plasma uplift during storm conditions.
Based on Tsurutani, B., et al., 2004. Global dayside ionospheric uplift and enhancement associated with interplanetary electric fields. J. Geophys. Res. Space Phys. 109(A), A08302. doi:10.1029/2003JA010342.

near the dip equator at altitudes near 300 km. This leads to velocities in the range ~41.7 m/s, or ~150 km/h, sufficiently large to rapidly move plasma from altitudes where production rates are largest (~130–150 km) to altitudes where production rates are factors of 10 lower (~250 km). These vertical velocities suggest the uplift mechanism is amply capable of producing the observed TEC increases. The modified TEC structure has been referred to as an ionospheric "superfountain" effect by analogy to the "fountain effect" that causes the nominal low latitude TEC structure (equatorial ionization anomaly). The superfountain is schematically depicted in Fig. 2. Increased plasma vertical transport speeds and subsequent plasma diffusion along magnetic field lines contribute to increased ionization at latitudes more poleward than nominal conditions.

To gain further insight into positive phase storms, we use published production rates to calculate the column density of electrons generated per hour to compare with the measurements of Fig. 1. The result, based on production rates reported by Richards (2014) representative of solar maximum conditions on March 17, 1990, (188 F10.7 Solar Flux Units) is that 383 TECU per hour can be generated considering the three dominant ion species O^+, NO^+, and O_2^+.

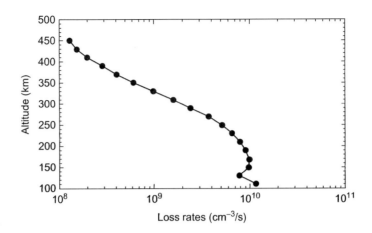

FIG. 3

Ion recombination rates for the three major ion species O^+, NO^+, and O_2^+, characteristic of solar maximum conditions on March 17, 1990.

Based on the model of Richards, P.G., 2014. Solar cycle changes in the photochemistry of the ionosphere and thermosphere. In: Modeling the Ionosphere-Thermosphere System, vol. 89. John Wiley & Sons, Ltd, Chichester, pp. 29–37.

These ion production rates are generally consistent with the IEC increases from Fig. 1, which can be explained by ion production of ~100–200 TECU/h, depending on how many ions are transported above CHAMP altitude. Using ion production rates alone to understand the ionospheric response to the extreme Halloween 2003 storms is overly simplistic because it does not account for recombination chemistry that occurs as the ions move upward to regions of lower concentrations of neutral and ion species. The consistency between the production rates and the measured TEC increases is suggestive, however, that reduced plasma recombination (loss) is sufficient to explain the observed TEC increases.

Ion recombination rates versus altitude characteristic of solar maximum, also based on Richards (2014), are shown in Fig. 3. The steep fall-off of recombination rates with altitude is why TEC can increase dramatically from upward drifting plasma. We note that photochemical ion production rates have a similar altitude dependence—peaking near altitudes between 120 and 150 km. As is well known, vertical transport of photochemically produced ions counteracts the sharp fall-off of plasma density with altitude, producing a peak plasma density in the F-region at altitudes near ~350 km (Rishbeth and Garriott, 1970).

A notable feature of Fig. 1 is the *reduced* TEC between $-10°$ and $15°$ dipole latitude in the second pass shown (red trace). Vertical TEC reduction, particularly at such low latitudes, is not consistent *per se* with an uplift mechanism. However, the reduced TEC could be due to transport associated with the equatorial fountain effect, as first modeled by Moffett and Hanson (1965). (See also Hanson and Moffett, 1966). Evidence that vertical drift reduces TEC near the equator is found in the quiet-time blue trace that shows the "nominal" equatorial ionization anomaly. The modeling work by Moffett and Hanson (1965) and others subsequently (Anderson, 1973) reproduce this feature associated with vertical and horizontal plasma transport away from the geomagnetic equator (see Fig. 2).

Detailed modeling of the ionospheric superfountain using SAMI-2 (Huba et al., 2000) has been reported (Tsurutani et al., 2007). Such modeling confirms that transport can reduce equatorial TEC under the strong uplift conditions characteristics of intense geomagnetic storms. Impactful thermospheric responses are likely as well, as we discuss below.

Heelis et al. (2009) have discussed another mechanism for TEC increases at higher middle latitudes (40° geomagnetic and higher) that are caused by the expansion of the high latitude convection pattern. These increases are related to horizontal plasma flows resulting from sunward plasma convection from dusk towards noon, "piling up" plasma at afternoon local times. As discussed in Chapter 21, for the largest storms, the expanded convection pattern flows will likely merge with the lower latitude increases due to the superfountain, forming the dramatic increases observed in the 2003 Halloween storms (Fig. 1), affecting the entire low-to-middle latitude ionosphere. Large plasma density gradients are likely to result from this combination of plasma motions, as has been reported by radars and GPS receivers in the Eastern North American sector (see Heelis et al., 2009 and references therein).

The CHAMP data (Fig. 1) show somewhat divergent TEC values among the multiple storm-time traces (red and black) appearing between 40° and 60° latitude (north and south). Multiple vertical IEC values at a single time are due to the IEC being derived from different GPS satellites above CHAMP. These divergent values reveal that the electron density distribution above CHAMP is not azimuthally symmetric, and that slant IEC depends on the direction of the line of sight between CHAMP and the individual transmitting GPS satellites. Such azimuthal dependence is likely due to horizontal TEC gradients above the satellite. Thus, the extreme ionospheric response we study is not only characterized by large electron density and TEC increases, but also larger horizontal gradients of TEC, which has a major impact on GPS positioning (see Chapter 23).

The physical mechanism by which electric fields "penetrate" to low latitudes has been investigated in several theoretical studies (e.g., Nopper and Carovillano 1978; Rothwell and Jasperse, 2006). Not all aspects of this "penetration" appear to be understood (e.g., Kikuchi et al., 1978), but much of the theoretical work is based on achieving consistency between magnetospheric Region-1 field aligned currents, the cross polar cap potential, and ionospheric conductance globally, dominated by E-region conductance (altitudes 100–120 km). Shielding by Region-2 currents was thought to mitigate much of the effect within 1 h for typical storms (Fejer and Scherliess, 1995; Richmond, 1995), but studies have clearly shown long-duration penetration for superstorms (Mannucci et al., 2008; Huang et al., 2005). It is appropriate to suggest that extreme geomagnetic storms will expose the low latitudes to very large electric fields of multi-hour duration with minimal shielding. Lakhina and Tsurutani (Chapter 7) suggest that an extreme storm may have a main phase lasting up to 24 h. The implications on dayside uplift for such an extended duration has not been investigated in the ionosphere to our knowledge, but would be important to pursue with detailed modeling and theory.

2.1 EQUATORIAL PLASMA IRREGULARITIES AND SCINTILLATION

As discussed in Chapter 1, a major technological impact from the Earth's upper atmosphere is the formation of turbulence or irregularities in the ionosphere that results in scintillation and possibly signal loss of radio waves used for communications, navigation, and radar purposes. It is reasonable to ask whether the same PPEFs that cause large daytime TEC increases may also lead to the formation of irregularities and whether their intensity may increase during extreme storms. We focus here on equatorial scintillation because this is most directly related to PPEF. Although there is little existing literature on the subject of extreme effects, extrapolation based on existing observations and theory is a reasonable approach.

Equatorial scintillation during nominal conditions is primarily a dusk to premidnight phenomenon. It originates from the Rayleigh-Taylor instability, when lower-than-normal plasma densities are situated below larger plasma densities aloft (Abdu, 2012). The buoyancy of the lower layer tends to lift plasma into the denser layer and large plasma depletions (or "bubbles") form in the F-region (altitudes above 300–400 km). The walls of the plasma depletions whose walls consist of irregularly structured plasma. An association of equatorial scintillation with the TEC increases due to PPEF is indirect. In fact, the dawn-to-dusk directed electric field on the dayside will lead to downward plasma convection on the nightside, tending to suppress the instability at night during the storm main phase when PPEF is active. However, as discussed below, intense storms have been found to destabilize the midnight ionosphere after the main phase is over, due to either a disturbance dynamo electric field, which raises the plasma at night, or over-shielding (described below). Because irregularities are associated with plasma depletion, and depletion cannot continue past zero plasma, it is unclear how the intensities of specific equatorial depletions increase due to extreme events, even though based on analysis of intense storms, the extent of equatorial scintillation is likely to occupy a broad range of post-sunset local times, as discussed below.

Basu et al. (2001a) studied two intense storms in 1999 and found increased equatorial scintillation associated with PPEF, in a range of local times extending to the postmidnight sector. Yeh et al. (2001) also found that, for the intense geomagnetic storm of October of 1999, scintillation extended past midnight into the predawn hours. Such a broad local time region of increased scintillation during extreme storms is a reasonable extrapolation based on our current understanding of the mechanisms. Basu et al. (2001a) hypothesized that the post-midnight scintillation was related to an apparent "driver" of scintillation: rapid decrease in the Dst index, indicating rapid changes in stormtime magnetospheric convection. This rapid convection increase, likely leading to efficient electric field penetration to low latitudes, destabilized the ionosphere at dusk. The later postmidnight irregularities were due to this destabilized region rotating to later local times. The intriguing results of Basu et al. (2001a) were reaffirmed in a study of the October and November 2003 superstorms (Basu et al., 2007) suggesting the robustness of the result across storm intensities. These works find significant dusk-time depletions occuring during rapid decreases in the Dst index. This association of irregularities with Dst behavior is further corroborated for the "Bastille Day" July 14, 2000, superstorm (Basu et al., 2001b). At times when Dst is not decreasing rapidly, there is less enhancement of dusk-time equatorial scintillation, and less destabilization of the ionosphere. Understanding the physical processes underlying why the rate of Dst variation affects equatorial scintillation, and how this would translate to extreme events, is an important area of future research.

As mentioned above, suppression of equatorial scintillation has also been reported associated with geomagnetic storms. Suppression can be due to downward plasma motion from the PPEF, which is directed westward at night. Based on the physical mechanism, we would expect this main-phase suppression to continue into extreme events. The review by Abdu (2012) discusses both suppression and increased occurrence of equatorial scintillation associated with disturbance electric fields. Increased occurrence is caused by the transient phenomenon of *over-shielding* where shielding electric fields that developed during the main phase create nighttime dusk-to-dawn electric fields that are opposite to the PPEF direction (see Fig. 8 in Chapter 10). This results in upward nighttime plasma drift, which is conducive to irregularity formation. Extreme events would likely lead to similar over-shielding cases. Yeh et al. (2001) emphasize disturbance dynamo electric fields as the source of midnight uplift. Carter et al. (2014) discuss the more common suppression case using the geomagnetic K_p index as a predictive parameter. Extreme storms may bring enhancement of scintillation across a broad range of local times, as discussed above.

Scintillation will of course occur at other latitudes as well. High-latitude scintillation is quite common and can occur during relatively weak geomagnetic storm conditions. Middle-latitude scintillation (below ~60° geomagnetic latitude) is virtually absent except during geomagnetic storms, where its occurrence is closely tied to the expansion of the high latitude convection pattern. These topics are covered in Chapter 23.

3 THE ROLE OF ION-NEUTRAL COUPLING

Tsurutani et al. (2007) were the first to report a possible connection between the neutral atmosphere and strong plasma uplift. In that study, it was shown that the rapid upward motion of O^+ ions due to PPEF can cause neutral oxygen (O) to be uplifted also, due to ion-neutral drag. These authors suggested that above ~400 km altitude, the neutral thermospheric density could substantially increase such that it can have a major impact on satellite drag. Because we now believe that PPEF during superstorms is an expected phenomenon, satellite drag impacts due to ion uplift will be a feature of extreme storms, in addition to the thermospheric expansion caused by temperature increases that has been studied extensively (e.g., Rees, 1995. See also Chapter 21).

A first-order understanding of the expected neutral density increase can be made from simple momentum exchange considerations. Momentum is imparted to the neutrals via collisions according to the following equation (e.g., Baron and Wand, 1983; Killeen et al., 1984; Brekke and Rino, 1978):

$$\rho_n \frac{\partial U}{\partial t} = \rho_i \nu_{in}(V_d - U) \qquad (2)$$

where U is the vertical neutral velocity imparted by the uplifting ions, ρ_n is the neutral density, ν_{in} is the momentum transfer collision rate from the ions to the neutrals, and V_d is the vertical uplift velocity. We have numerically integrated Eq. (2) under conditions similar to what occurred on November 10, 1979, when strong EUV emission from the Sun was present near solar maximum, and the F10.7 solar flux value of 367.0 was the largest value recorded after 1960. Values for ion and neutral densities under these conditions were obtained from the International Reference Ionosphere (IRI) model, and the Mass Spectrometer and Incoherent Scatter (MSIS) model (Picone et al., 2002), respectively. The collision frequencies were obtained using the formula of Bailey and Balan (1996), with temperature obtained from MSIS and the further approximation that neutral and ion temperatures are the same. These calculations were run for November 10, 1979, at a low latitude location where vertical uplift is generally largest (0 latitude, −60 W longitude).

Results of the integration are shown in Fig. 4 for two low-latitude electric field values: 1 and 4 mV/m, corresponding to ion vertical drift velocities of 153 and 612 km/h, respectively. The smaller value is large but not extreme and has been observed for several strong storms (e.g., Huang et al., 2005). The larger value is a slight extrapolation from the record value observed near November 9, 2004 (Kelley et al., 2010). We assume neutral vertical velocities are initially zero, and the integration is applied to neutral density parcels starting at 350 km altitude. These calculations are not self-consistent in that densities are not adjusted as the vertical uplift occurs, but are indicative of the consequences to density in an approximate way.

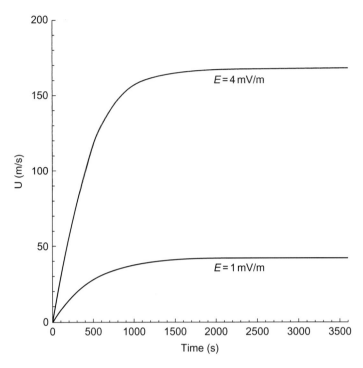

FIG. 4

Neutral velocity versus time according to Eq. (2) for two values of the ion upward velocity.

The calculations show that the "terminal" ion velocity is reached in a matter of tens of minutes for either electric field value. Neutral uplift distances of ~100–400 km are plausible in as little as 1–2 h. Table 1 shows densities, from IRI and MSIS runs, for dominant ion (O^+) and neutral (O) species, respectively. Starting at 350 km altitude, the ratio of the neutral density to its density higher up is calculated (200 km higher in Column 4, and 400 km higher in column 5). Above 350 km, O density is decreasing with altitude, so vertical uplift of O via ion-neutral collision has the potential to increase thermospheric density sufficiently to alter drag forces on a satellite. The ratios in Columns 4 and 5 of Table 1 are upper bounds for neutral density changes that can occur due to neutral uplift. Only a fraction of the neutrals would be uplifted by the ions, because the concentration of the neutral species is much larger than the ion concentration. Nevertheless, these large upper limits are a concern and suggest more detailed time-dependent modeling corresponding to extreme conditions is needed. A modeling study by Zhu et al. (2017) shows that under nominal conditions the thermospheric density at 400 km altitude can change due to vertical plasma drift by 5%–10% at low latitudes. (In this case, the density at 400 km is reduced). It is reasonable to expect much larger changes in neutral density during extreme storms. The vertical motion of the neutrals, in extreme cases, could change horizontal winds as well, due to mass conservation. This topic deserves further research. We note also that, using data from the Defense Meteorological Satellite Program satellites, Tsurutani et al. (2012) found oxygen ion densities could exceed neutral densities at altitudes above ~800 km. Therefore, at higher altitudes extreme storms can create satellite drag increases due to the uplifted ions that exceed neutral drag. This possibility is often over-looked when geomagnetic storms are analyzed.

Table 1 Ion and Neutral Oxygen Densities, From IRI and MSIS Models, Respectively

Altitude	O Density	O$^+$ Density	Ratio +200 km	Ratio +400 km
100	7.91E+17	0.00E+00	407.4	4270.0
150	2.93E+16	1.37E+11	28.0	274.6
200	8.63E+15	5.38E+11	14.9	139.7
250	3.82E+15	9.36E+11	11.7	105.7
300	1.94E+15	1.41E+12	10.5	91.2
350	1.04E+15	2.00E+12	9.8	82.6
400	5.78E+14	2.59E+12	9.4	76.4
450	3.25E+14	2.14E+12	9.0	71.4
500	1.85E+14	1.49E+12	8.7	
550	1.07E+14	9.97E+11	8.4	
600	6.18E+13	6.78E+11	8.2	
650	3.61E+13	4.75E+11	7.9	
700	2.13E+13	3.45E+11		
750	1.26E+13	2.58E+11		
800	7.56E+12	1.97E+11		
850	4.56E+12	1.53E+11		

Ratios refer to neutral densities between different altitudes, for 200 and 400 km vertical displacements, up to 850 km altitude.

3.1 BUOYANCY (GRAVITY) WAVES

We have neglected gravity and pressure forces in Eq. (2) (e.g., Killeen et al., 1984). Gravity is of most interest here because pressure decreases with altitude, so the pressure gradient force works in opposite direction to the upward collision-induced force. To quantify the effect of gravity, we compare the pressure gradient force due to gravity with the upward force from collisions. The pressure gradient force is estimated using MSIS densities and temperatures and the ideal gas law up to 600 km altitude. Fig. 5 contains the result of the calculation, which compares the acceleration of the neutrals due to collisions with ions with that from the gravity pressure gradient force, and their ratio. While gravity produces larger accelerations, the upward collisional force can reach 5% or more of the gravitational force, especially near altitudes of ~400 km. The net result is likely to be the development of buoyancy waves over a broad low latitude region where the uplift occurs (e.g. see Meng et al., 2015). As neutrals are transported upward, the disruption of hydrostatic equilibrium creates a downward force, decreasing the upward velocity of the parcel that eventually becomes downward. Overshooting produces oscillatory behavior that subsequently displaces neutral parcels aloft, and a vertical wave is launched. Such waves launched by PPEF have been analyzed using the nonhydrostatic GITM model under nominal low-latitude conditions (Zhu and Ridley, 2014).

Thus, significant modification of the neutral and ion densities is plausible for long-duration penetration electric fields. Detailed time dependent calculations are needed with a coupled thermosphere-ionosphere model (e.g., GITM, Ridley et al. 2006) under extreme conditions to understand these effects.

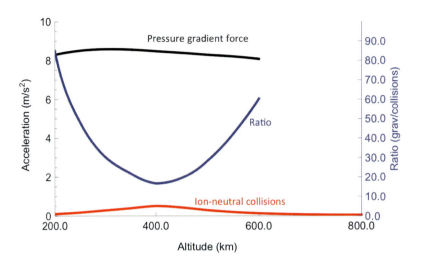

FIG. 5

Acceleration of the neutral oxygen species under the pressure gradient force *(black)* due to gravity and due to collisions with ions *(red)*. The ratio of the two accelerations *(blue)* is on the right axis.

4 EXTREME NIGHTTIME RESPONSES FOLLOWING THE STORM MAIN PHASE (FLORIDA EFFECT)

In this section, we review an unusual ionospheric feature that appeared over Florida in the United States during nighttime of October 31, 2003, in the aftermath of the large geomagnetic storm that occurred on the preceding day. We discuss the nighttime feature in this chapter because we believe it is associated with the extreme conditions associated with that storm, although it is currently unknown whether similar features may be associated with other intense geomagnetic storms.

A practical consequence of what occurred over Florida on October 31 is that it exposed the potential limitations of algorithms associated with the Wide Area Augmentation System (WAAS), designed to assist aviation users of GPS-based positioning systems. Such users require augmentation of the standard GPS signals to meet safety-of-life standards. Users in flight can only officially rely on a single GPS frequency (L1 at 1575.42 MHz) for navigation, until the modernized GPS constellation fully deploys another authorized civil frequency (L5, at 1176.45 MHz) designed for safety-of-life applications. Existing single-frequency WAAS users rely on a dual-frequency ground reference station network of GPS receivers to measure ionospheric TEC. An ionospheric correction is derived from this ground network and broadcast to WAAS users via geostationary satellites above the Continental United States (CONUS). The correction is encoded as a gridded data set, continuously updated, of vertical ionospheric TEC values over a $5° \times 5°$ grid covering the CONUS and extending to Hawaii and Alaska. Because the gridded values are interpolated from irregularly scattered measurements obtained by the ground network, localized TEC enhancements might not be adequately sampled by the reference measurements, leading to incorrect TEC values at the grid point locations. The ionospheric feature that appeared over Florida on October 31, 2003, was so unusual that it exposed potential limitations of algorithms designed to bound the uncertainties of the gridded TEC values. This led to new algorithms being implemented for WAAS, at significant expense, known as the *extreme storm detector* (Sparks, 2012).

In the remainder of this section, we review characteristics of the nighttime TEC enhancement that occurred on October 31, 2003, known as the *Florida Effect*. Its association with the intense Halloween storms of 2003 is postulated, but its physical origins are currently unknown, so it is not possible to extrapolate with confidence on how such a feature might behave during an extreme storm such as the Carrington event of 1859.

We review the spatial and temporal extent of the localized TEC enhancement based on the work of Datta-Barua et al. (2008). The purpose of that investigation was to estimate the vertical extent of the feature to help understand its physical causes. We can summarize the results as follows:

1. The duration of the enhancement was ~6 h, from 0000 UT to 0600 UT on October 31, 2003 (solar local times of 1900–0100 LT).
2. The horizontal dimensions of the feature reached ~700 × 700 km, persisting over a region centered at ~30 N latitude and 80 W longitude.
3. The altitude of peak electron density of the feature is ~160 km higher than the surrounding nighttime background.
4. The feature remained fixed in geographic coordinates until it dissipated at ~0600 UT.

The estimated magnitude of the enhancement based on sampling by the WAAS reference stations and other GPS receivers in its vicinity was ~60 vertical TECU above background vertical TEC values in the immediately surrounding area of ~15 TECU. The resulting slant range delays of up to 20 m are sufficient to cause positioning errors of 40 m or more, depending on details of the GPS satellites being tracked by a specific receiver. WAAS is designed to provide robust bounds for user positioning error, so an undetected positioning error of this magnitude is a significant concern (Sparks, 2012).

Fig. 6 (reproduced from Datta-Barua et al., 2008) shows the horizontal extent of the feature nears its peak at 0330 UT (2210 solar local time) on October 31, 2003. TEC measurements from nearby overflights of the JASON dual-frequency altimeter, capable of measuring TEC below altitudes of 1330 km, suggest that the enhancement was contained entirely below this altitude. TEC data from the SAC-C satellite orbiting at 700 km altitude, which also had a fortuitous nearby overflight, suggest some additional ionization above that altitude. Using typical electron density profiles derived from the IRI model, the peak density of the feature appeared to be near 500 km altitude, in the ionospheric F-region.

The relatively high altitude of the feature and where its peak density occurs argues against the cause being energetic particle precipitation. To achieve large density enhancements from precipitation would require large fluxes of very low-energy particles (<1 eV) that deposit their energy at relatively high altitudes (see Chapter 12 of Jursa, 1985). Such particle populations have never been observed precipitating at such low latitudes.

Of interest in considering extreme space weather is the cause of the Florida Effect feature. Associating this feature with the preceding ionospheric storm, where the largest TEC increases are due to PPEF, suggests that the ultimate source of ionization could have originated in the daytime and that this feature is a result of slow recombination of daytime plasma. For the ionospheric superfountain discussed in the previous section, we found that ion production rates below 200 km altitude are consistent with the large observed daytime TEC increases. We showed that ion recombination occurs

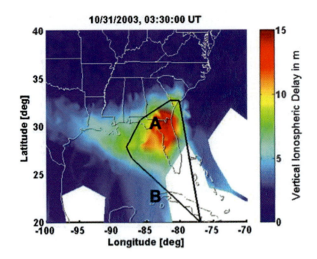

FIG. 6

The TEC enhancement known as the "Florida Effect" that occurred on October 31, 2003, as deduced from a ground network of GPS receivers. The region marked A is the expected shape of the enhanced region, based on GPS and ocean altimeter data. The region marked B is normal nighttime background.

relatively slowly for the uplifted plasma, consistent with the overall TEC increases (and with the observed increased altitude of the plasma, ascertained by comparing with ground-based TEC measurements, as discussed in Mannucci et al., 2005b). Daytime TEC increase is consistent with the Florida Effect if a sufficiently large quantity of plasma at high altitude might fail to recombine over a period of several hours. To assess plasma recombination rates, we use the IRI model in the vicinity of the Florida feature, comparing daytime and nighttime profiles. While the climatological IRI will not capture the storm-time features, it will capture the plasma recombination process versus altitude that occurs under typical conditions, which is of interest to us here.

The electron density profiles over Florida, day and night, and their ratio, are shown in Fig. 7. As expected, rapid recombination at altitudes lower than 225 km quickly reduces daytime plasma density by factors of 10 or greater. At altitudes of 500 km and above, the daytime plasma is reduced by only a factor of 3–4. Estimates of daytime TEC are 250.0 TECU near Florida (Fig. 1). Nighttime TEC reached ∼80 TECU in the Florida Effect feature, lower by a factor of ∼3 from the peak daytime value, which is consistent with slow upper altitude recombination. Of course, details of the processes that created the Florida Effect TEC feature are likely to be more complex than the picture offered here, including the causes of the structuring. These simplified calculations show it is plausible that the Florida Effect is a result of daytime plasma uplifted by long-duration PPEF.

5 CONCLUSIONS AND FUTURE OUTLOOK

The largest increases in ionospheric electron density and total electron content during intense geomagnetic storms are due to prompt penetration electric fields (PPEF), that tend to occur during the storm main phase, and can last for several hours. PPEF penetrating from high latitudes to the equatorial region

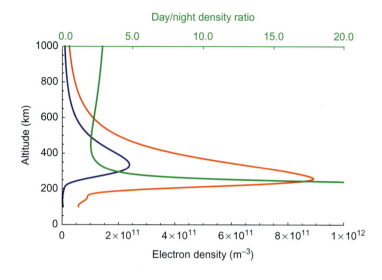

FIG. 7

Electron density profiles from the IRI-2016 model for nominal daytime conditions *(red trace)* on October 30, 2003, at 2120 UT (1600 LT solar), and nighttime conditions *(blue trace)* on October 31, 2003, at 0330 UT (2210 LT). The altitude-dependent ratio of the day to night density is plotted along the upper axis *(green)*.

create an ionospheric "superfountain" that vertically advects plasma from low to mid-latitudes via $\mathbf{E} \times \mathbf{B}$ drift. The large TEC increases that have been observed can be understood physically as a reduction of the plasma recombination process, whereby plasma created via photoionization at lower altitudes is rapidly raised to higher altitudes where recombination is much slower. The result is increased TEC and increased plasma densities at altitudes above 400 km. Combined with structured electric fields that may appear at subauroral latitudes, large electron density gradients are also a consequence of PPEF (see Chapter 23).

Extrapolating to the most extreme events, it is reasonable to expect increased electric field values and long-duration penetration (up to 24 h may be possible. An 18-h main phase is discussed in Chapter 7). Larger TEC values than observed during the October 2003 "Halloween" storms may be the result, perhaps spanning more than one day. We expect nonlinear effects (e.g., plasma heating) to play a role for the most extreme events, making simplistic extrapolation unreliable. Detailed modeling and theory are required to predict the consequences in more detail. Fortunately, the observations from 2003 show how rapidly TEC can grow, and it might get worse. Observations for this storm also show that at altitudes above 800 km, ion densities exceed neutral densities.

Applications that rely on predictable TEC spatial and temporal variations are likely to be affected by the aftermath of the large TEC increases, as occurred at night over Florida on October 31, 2003. The TEC magnitude and spatial variation of such structures are not currently understood, but it is likely, if they are indeed caused by remnant uplifted daytime plasma, that extreme geomagnetic storms will increase the severity of the effects resulting from such structures.

The prospect of catastrophic equatorial irregularities caused by extreme geomagnetic storms is not clear. Both suppression and enhancement of irregularities are associated with geomagnetic storms. Extreme events may bring over-shielding, PPEF, and fluctuating electric fields at low and middle latitudes, particularly if magnetospheric ring current plasma is enhanced in a way conducive to such effects (see Chapter 10). Dusktime scintillation enhancement associated with rapidly changing Dst is plausible, another cause of increased low-latitude irregularities affecting postmidnight local times. However, because these are transient and localized enhancements, the consequences for extreme events are hard to predict.

Further work is required to investigate the role of ion-neutral coupling of rapidly uplifted plasma. A consequence may be noticeably increased satellite drag at altitudes above 500 km. Increased satellite drag is a known consequence of extreme space weather events due to thermospheric heating and circulation changes (see Chapter 21). Rapidly increased drag can lead to loss of satellite positions for hours or days. The thermospheric consequences of PPEF should be better understood in preparation for a future extreme event.

ACKNOWLEDGMENTS

This research was carried out at the Jet Propulsion Laboratory, California Institute of Technology, under a contract with the National Aeronautics and Space Administration. Useful discussions with Xing Meng (NASA Jet Propulsion Laboratory, California Institute of Technology) are acknowledged.

REFERENCES

Abdu, M.A., 2012. Equatorial spread F/plasma bubble irregularities under storm time disturbance electric fields. J. Atmos. Sol. Terr. Phys. 75–76 (C), 44–56. https://doi.org/10.1016/j.jastp.2011.04.024.

Anderson, D.N., 1973. A theoretical study of the ionospheric F region equatorial anomaly—I. Theory. Planet. Space Sci. 21 (3), 409–419. https://doi.org/10.1016/0032-0633(73)90040-8.

Bailey, G.J., Balan, N., 1996. A low-latitude ionosphere-plasmasphere model. In: Schunk, R.W. (Ed.), Solar-Terrestrial Energy Program: Handbook of Ionospheric Models, pp. 173–206.

Baron, M.J., Wand, R.H., 1983. F region ion temperature enhancements resulting from Joule heating. J. Geophys. Res. 88 (A5), 4114. https://doi.org/10.1029/ja088ia05p04114.

Basu, S., et al., 2001a. Ionospheric effects of major magnetic storms during the international space weather period of September and October 1999: GPS observations, VHF/UHF scintillations, and in situ density structures at middle and equatorial latitudes. J. Geophys. Res. 106 (A), 30389–30414. https://doi.org/10.1029/2001JA001116.

Basu, S., Basu, S., Groves, K.M., Yeh, H.C., Su, S.Y., Rich, F.J., Sultan, P.J., Keskinen, M.J., 2001b. Response of the equatorial ionosphere in the South Atlantic Region to the Great Magnetic Storm of July 15, 2000. Geophys. Res. Lett. 28 (18), 3577–3580. https://doi.org/10.1029/2001GL013259.

Basu, S., Basu, S., Rich, F.J., Groves, K.M., MacKenzie, E., Coker, C., Sahai, Y., Fagundes, P.R., Becker-Guedes, F., 2007. Response of the equatorial ionosphere at dusk to penetration electric fields during intense magnetic storms. J. Geophys. Res. 112 (A8), A08308–A08314. https://doi.org/10.1029/2006JA012192.

Bilitza, D., Altadill, D., Zhang, Y., Mertens, C., Truhlik, V., Richards, P., McKinnell, L.-A., Reinisch, B., 2014. The International Reference Ionosphere 2012 ? a model of international collaboration. J. Space Weather Space Clim. 4, A07–A13. https://doi.org/10.1051/swsc/2014004.

Brekke, A., Rino, C.L., 1978. High-resolution altitude profiles of the auroral zone energy dissipation due to ionospheric currents. J. Geophys. Res. 83 (A6), 2517–2524. https://doi.org/10.1029/JA083iA06p02517.

Carter, B.A., et al., 2014. Geomagnetic control of equatorial plasma bubble activity modeled by the TIEGCM with Kp. Geophys. Res. Lett. 41 (1), 5331–5339. https://doi.org/10.1002/2014GL060953.

REFERENCES

Datta-Barua, S., Mannucci, A.J., Walter, T., Enge, P., 2008. Altitudinal variation of midlatitude localized TEC enhancement from ground- and space-based measurements. Space Weather 6 (10). https://doi.org/10.1029/2008SW000396.

Enge, P., Walter, T., Pullen, S., Kee, C., 1996. Wide area augmentation of the global positioning. Proceedings of the IEEE 84, 1063–1088.

Fejer, B.G., Scherliess, L., 1995. Time dependent response of equatorial ionospheric electric fields to magnetospheric disturbances. Geophys. Res. Lett. 22 (7), 851–854. https://doi.org/10.1029/95GL00390 (ISSN 0094–8276).

Hanson, W.B., Moffett, R.J., 1966. Ionization transport effects in the equatorial F region. J. Geophys. Res. Oceans 71 (23), 5559–5572. https://doi.org/10.1029/JZ071i023p05559.

Heelis, R.A., Sojka, J.J., David, M., Schunk, R.W., 2009. Storm time density enhancements in the middle-latitude dayside ionosphere. J. Geophys. Res. 114(A3). https://doi.org/10.1029/2008JA013690.

Huang, C.-S., Foster, J.C., Kelley, M.C., 2005. Long-duration penetration of the interplanetary electric field to the low-latitude ionosphere during the main phase of magnetic storms. J. Geophys. Res. 110. A11309, https://doi.org/10.1029/2005JA011202.

Huba, J.D., Joyce, G., Fedder, J.A., 2000. Sami2 is another model of the ionosphere (SAMI2): a new low-latitude ionosphere model. J. Geophys. Res. 105 (A), 23035–23054. https://doi.org/10.1029/2000JA000035.

Jursa, A.S. (Ed.), 1985. Handbook of Geophysics and the Space Environment. Air Force Geophysics Laboratory. National Technical Information Service, Springfield, VA, USA.

Kelley, M.C., 2009. The Earth's Ionosphere: Plasma Physics and Electrodynamics, second ed. Academic Press, New York.

Kelley, M.C., Ilma, R.R., Nicolls, M., Erickson, P., Goncharenko, L., Chau, J.L., Aponte, N., Kozyra, J.U., 2010. Spectacular low- and mid-latitude electrical fields and neutral winds during a superstorm. J. Atmos. Sol. Terr. Phys. 72 (4), 285–291. https://doi.org/10.1016/j.jastp.2008.12.006.

Kikuchi, T., Araki, T., Maeda, H., Maekawa, K., 1978. Transmission of polar electric fields to the equator. Nature 273 (5664), 650. https://doi.org/10.1038/273650a0.

Killeen, T.L., Hays, P.B., Carignan, G.R., Heelis, R.A., Hanson, W.B., Spencer, N.W., Brace, L.H., 1984. Ion-neutral coupling in the high-latitude F region evaluation of ion heating terms from dynamics explorer 2. J. Geophys. Res. 89 (A9), 7495–7508. https://doi.org/10.1029/JA089iA09p07495 (ISSN 0148–0227).

Mannucci, A.J., Wilson, B.D., Yuan, D.N., Ho, C.H., Lindqwister, U.J., Runge, T.F., 1998. A global mapping technique for GPS-derived ionospheric total electron content measurements. Radio Sci. 33 (3), 565–582. https://doi.org/10.1029/97rs02707.

Mannucci, A.J., Iijima, B.A., Lindqwister, U.J., Pi, X., Sparks, L., Wilson, B.D., 1999. GPS and ionosphere. In: Stone, W.R. (Ed.), Review of Radio Science 1996–1999. Review of Radio …, New York.

Mannucci, A.J., et al., 2005a. Hemispheric daytime ionospheric response to intense solar wind forcing. In: Burch, J.L., Schulz, M., Spence, H. (Eds.), Inner Magnetosphere Interactions: New Perspectives From Imaging. Geophysical-Monograph 159, American Geophysical Union, New York, pp. 261–275. https://doi.org/10.1029/159GM20.

Mannucci, A.J., Tsurutani, B.T., Iijima, B.A., Komjathy, A., Saito, A., Gonzalez, W.D., Guarnieri, F.L., Kozyra, J.U., Skoug, R., 2005b. Dayside global ionospheric response to the major interplanetary events of October 29–30, 2003 "Halloween Storms" Geophys. Res. Lett. 32 (1). L12S02, https://doi.org/10.1029/2004GL021467.

Mannucci, A.J., Tsurutani, B.T., Abdu, M.A., Gonzalez, W.D., Komjathy, A., Echer, E., Iijima, B.A., Crowley, G., Anderson, D., 2008. Superposed epoch analysis of the dayside ionospheric response to four intense geomagnetic storms. J. Geophys. Res. 113. https://doi.org/10.1029/2007ja012732.

Mendillo, M., 2006. Storms in the ionosphere: patterns and processes for total electron content. Rev. Geophys. 44 (4), RG4001–RG4047. https://doi.org/10.1029/2005RG000193.

Mendillo, M., Klobuchar, J.A., 1975. Investigations of the ionospheric F region using multistation total electron content observations. J. Geophys. Res. 80 (4), 643–650. https://doi.org/10.1029/JA080i004p00643.

Meng, X., Komjathy, A., Verkhoglyadova, O.P., Yang, Y.M., Deng, Y., Mannucci, A.J., 2015. A new physics-based modeling approach for tsunami-ionosphere coupling. Geophys. Res. Lett. 42 (1), 4736–4744. https://doi.org/10.1002/2015GL064610.

Moffett, R.J., Hanson, W.B., 1965. Effect of ionization transport on the equatorial F-region. Nature 206 (4), 705–706. https://doi.org/10.1038/206705a0.

Nopper, R.W.J., Carovillano, R.L., 1978. Polar-equatorial coupling during magnetically active periods. Geophys. Res. Lett. 5 (8), 699–702. https://doi.org/10.1029/GL005i008p00699.

Picone, J.M., Hedin, A.E., Drob, D.P., Aikin, A.C., 2002. NRLMSISE-00 empirical model of the atmosphere: statistical comparisons and scientific issues. J. Geophys. Res. 107 (A12), https://doi.org/10.1029/2002JA009430 SIA 15-1–SIA 15-16.

Rees, D., 1995. Observations and modelling of ionospheric and thermospheric disturbances during major geomagnetic storms: a review. J. Atmos. Sol. Terr. Phys. 57 (12), 1433–1457. https://doi.org/10.1016/0021-9169(94)00142-b.

Richards, P.G., 2014. Solar cycle changes in the photochemistry of the ionosphere and thermosphere. In: Modeling the Ionosphere-Thermosphere System, vol. 89. John Wiley & Sons, Ltd., Chichester, pp. 29–37

Richmond, A.D., 1995. Ionospheric electrodynamics. In: Volland, H. (Ed.), In: Handbook of Atmospheric Electrodynamics, vol. 2. pp. 249–290.

Ridley, A.J., Deng, Y., Tóth, G., 2006. The global ionosphere–thermosphere model. J. Atmos. Sol. Terr. Phys. 68 (8), 839–864. https://doi.org/10.1016/j.jastp.2006.01.008.

Rishbeth, H., Garriott, O.K., 1970. Introduction to ionospheric physics. International Geophysics Series, vol. 14. Academic Press, New York.

Rothwell, P.L., Jasperse, J.R., 2006. Modeling the connection of the global ionospheric electric fields to the solar wind. J. Geophys. Res. 111 (A3), A03211–A03216. https://doi.org/10.1029/2004JA010992.

Schunk, R.W., Nagy, A.F., 2009. Ionospheres: Physics, Plasma Physics, and Chemistry, second ed. Cambridge University Press, New York.

Sparks, L., 2012. Addressing the influence of space weather on Airline Navigation. In: Proceedings of the 6th Annual Guidance and Control Conference; 5 Feb. 2012; Breckenridge, CO; United States.

Sparks, L., Blanch, J., Pandya, N., 2011a. Estimating ionospheric delay using kriging: 1. Methodology. Radio Sci. 46 (6), 1–13. https://doi.org/10.1029/2011RS004667.

Sparks, L., Blanch, J., Pandya, N., 2011b. Estimating ionospheric delay using kriging: 2. Impact on satellite-based augmentation system availability. Radio Sci. 46 (6), 1–10. https://doi.org/10.1029/2011RS004781.

Stephens, P., Komjathy, A., Wilson, B., Mannucci, A., 2011. New leveling and bias estimation algorithms for processing COSMIC/FORMOSAT-3 data for slant total electron content measurements. Radio Sci. 46 (6). https://doi.org/10.1029/2010RS004588.

Tsurutani, B., et al., 2004. Global dayside ionospheric uplift and enhancement associated with interplanetary electric fields. J. Geophys. Res. 109 (A). A08302, https://doi.org/10.1029/2003JA010342.

Tsurutani, B.T., Verkhoglyadova, O.P., Mannucci, A.J., Araki, T., Sato, A., Tsuda, T., Yumoto, K., 2007. Oxygen ion uplift and satellite drag effects during the 30 October 2003 daytime superfountain event. Ann. Geophys. 25 (3), 569–574. https://doi.org/10.5194/angeo-25-569-2007.

Tsurutani, B.T., Verkhoglyadova, O.P., Mannucci, A.J., Lakhina, G.S., Huba, J.D., 2012. Extreme changes in the dayside ionosphere during a Carrington-type magnetic storm. J. Space Weather Space Clim. 2, A05–A07. https://doi.org/10.1051/swsc/2012004.

Yeh, H.C., Su, S.Y., Heelis, R.A., 2001. Storm-time plasma irregularities in the pre-dawn hours observed by the low-latitude ROCSAT-1 satellite at 600 km altitude. Geophys. Res. Lett. 28, 685.

Zhu, J., Ridley, A.J., 2014. Modeling subsolar thermospheric waves during a solar flare and penetration electric fields. J. Geophys. Res. 119 (1), 10. https://doi.org/10.1002/2014JA020473.

Zhu, Q., Deng, Y., Maute, A., Sheng, C., Lin, C.Y., 2017. Impact of the vertical dynamics on the thermosphere at low and middle latitudes: GITM simulations. J. Geophys. Res. Space Phys. 122 (6), 6882–6891. https://doi.org/10.1002/2017JA023939.

FURTHER READING

Heroux, L., Cohen, M., Higgins, J.E., 1974. Electron densities between 110 and 300 km derived from solar EUV fluxes of August 23, 1972. J. Geophys. Res. 79 (34), 5237–5244. https://doi.org/10.1029/JA079i034p05237.

Hinteregger, H.E., Hall, L.A., Schmidtke, G., 1965. Solar XUV radiation and neutral particle distribution in July 1963 themosphere. pp. 1175; Space Research V. In: King-Hele, D.G., Muller, P., Righini, G. (Eds.), Proceedings of the Fifth International Space Science Symposium, Florence, May 12–16. 1964; Organized by the Committee on Space Research—COSPAR and the Italian Space Research Committee. North-Holland Publishing Company, Amsterdam, New York, 1965.

Mannucci, A.J., Tsurutani, B.T., Verkhoglyadova, O., Komjathy, A., Pi, X., 2015. Use of radio occultation to probe the high-latitude ionosphere. Atmos. Meas. Tech. 8 (7), 2789–2800. https://doi.org/10.5194/amt-8-2789-2015.

CHAPTER 21

HOW MIGHT THE THERMOSPHERE AND IONOSPHERE REACT TO AN EXTREME SPACE WEATHER EVENT?

Tim Fuller-Rowell*, John Emmert†, Mariangel Fedrizzi*, Daniel Weimer‡, Mihail V. Codrescu§, Marcin Pilinski*, Eric Sutton¶, Rodney Viereck§, Joachim (Jimmy) Raeder∥, Eelco Doornbos#

University of Colorado at Boulder, CO, United States *Naval Research Laboratory, Washington, DC, United States*† *Virginia Tech, Blacksburg, VA, United States*‡ *NOAA Space Weather Prediction Center, Boulder, CO, United States*§ *Air Force Research Laboratory, Greene, OH, United States*¶ *University of New Hampshire, Durham, NH, United States*∥ *Delft University of Technology, Delft, Netherlands*#

CHAPTER OUTLINE

1 Introduction 513
2 Effects of Solar EUV and UV Radiation 517
3 Effect of an Extreme Solar Flare 519
4 Effects of an Extreme CME Driving a Geomagnetic Storm 520
   4.1 Defining the Drivers 521
   4.2 Neutral Atmosphere Response to an Extreme Geomagnetic Storm 525
   4.3 Ionospheric Response to an Extreme Geomagnetic Storm 529
5 Summary and Conclusions 532
Acknowledgments 533
References 534
Further Reading 539

1 INTRODUCTION

This chapter will focus on the response of the neutral atmosphere and ionosphere to extreme solar and geomagnetic conditions. An extreme event is categorized as a condition that is likely to occur once in a hundred years, or has about a one percent chance of occurring any given year. Of course, an extreme event could actually happen at any time. In light of the severe consequences, which will be discussed further, these odds are high enough that it is reasonable and prudent to invest in mitigation strategies. It is therefore appropriate to make an effort to understand the impact of extreme events on the neutral thermosphere and ionosphere. Predicting the response of the neutral atmosphere has less uncertainty, because the thermosphere has longer time constants, is somewhat easier to model, and the solar and magnetospheric drivers are easier to characterize. The ionosphere, on the other hand, has shorter time constants, responds to production and loss within minutes, and has more sources that drive significant increases in plasma density (e.g., solar radiation, auroral production, and energetic solar protons). In addition, plasma transport is also more complex. As well as the collisional interaction with the neutral atmosphere, the ionosphere also responds to electrodynamic transport, which can elevate plasma to greater altitudes, where the neutral atmosphere has less molecular nitrogen and recombination is slow.

Three possible solar sources of extreme conditions impacting the thermosphere and ionosphere will be considered: (1) an increase in ultraviolet and extreme ultraviolet (UV and EUV) radiation from the sun lasting several days, (2) impulsive changes in X-rays, UV, and EUV radiation from the sun lasting at most a few hours during a solar flare event, and (3) a coronal mass ejection (CME) arriving at Earth as an interplanetary CME (ICME) and driving an extreme geomagnetic storm. The first two of these sources will be considered briefly; the main focus of the rest of the chapter will be predicting the response to an extreme geomagnetic storm. The most extreme geomagnetic storms tend to be driven by solar wind conditions following the arrival at Earth of a CME. Storms driven by co-rotating interaction regions (CIRs) tend to be weaker, so are not considered separately in this chapter. CIRs, of course, can drive other extreme space weather conditions, for instance in the radiation belts (Chapter 14).

The response to solar and geomagnetic variability at all levels is referred to as *space weather*. The phrase encompasses variations in the space environment between the Sun and Earth that can affect technologies in space and on Earth. In the upper atmosphere, space weather disrupts satellite and airline operations, both military and commercial, hinders satellite/debris collision avoidance, and affects communications networks, navigation systems, and the electric power grid. As our society becomes ever more dependent upon these technologies, space weather poses an increasing risk to infrastructure and the economy. In addition to exploring how the thermosphere and ionosphere respond to extreme space weather events, this chapter will also examine if our current understanding of the salient physical process can be extrapolated to these more extreme conditions.

While space weather has been the subject of scientific study for several decades, it has only been recently that extreme space weather events have been recognized at the top levels of the U.S. government. For example, the *Strategic National Risk Assessment*[1] has identified space weather as a hazard that poses significant risk to the security of the nation. In November 2014, in recognition that extreme space weather could severely disrupt the U.S. economy and have the potential for social impacts and loss of life, the White House Office of Science and Technology Policy (OSTP) established the Space Weather Operations, Research, and Mitigation (SWORM) task force, an interagency group, chartered to develop a national strategy and a national action plan to enhance national preparedness for space-weather events. The resulting documents, the *National Space Weather*

---

[1]Department of Homeland Security, *The Strategic National Risk Assessment (SNRA) in Support of PPD 8: A Comprehensive Risk-Based Approach toward a Secure and Resilient Nation*, December 2011.

Strategy[2] and the accompanying *National Space Weather Action Plan*[3] (SWAP), identify risks associated with space weather and seek to improve resilience of essential facilities and systems. It is a challenge to definitively characterize the occurrence frequency and severity distribution of solar flares and CME eruptions, although we do have a historical record that gives us some indication as to the possible events that the Sun could have in store (see Chapters 2, 4, and 5).

The SWAP identified five benchmarks, including "Ionospheric Disturbances" and "Upper Atmosphere Expansion." Upper atmospheric expansion refers to changes in the thermosphere that can affect satellite drag at low-Earth-orbit (LEO) (Emmert, 2015a; NRC Reports, 1995, 2012). Ionospheric disturbances, refers to the plasma condition affecting radio communications and navigation.

The thermospheric parameters most relevant to space weather applications are: (1) *neutral density*, because of the impact on satellite drag, orbit prediction, and collision avoidance, (2) *neutral winds*, for satellite drag because the winds project onto the orbital velocity, changing the direction and magnitude of the resulting drag acceleration both in-track and cross-track, and winds also transport plasma and drive electrodynamics, (3) *neutral composition*, because it modulates thermal expansion (and hence density) through changes in scale height, impacts ionospheric production and loss processes, determines the nature of momentum exchange between thermospheric molecules and satellites impacting satellite drag, and corrodes satellite surfaces through reaction with atomic oxygen. In the ionosphere, we are mainly concerned about (4) *Plasma density*, and its distribution, because of the impact on radio wave propagation for communication and navigation.

Lack of knowledge of spatial structure or temporal variability of thermospheric neutral density is the main factor affecting the uncertainty in satellite drag at low-Earth-orbit (Emmert, 2015a,b; NRC Reports, 1995, 2012). The primary effect arises from an increase in temperature, which causes a thermal expansion of Earth's upper atmosphere and an increase in neutral density at a fixed altitude. Heating driving the thermal expansion can arise from solar or geomagnetic activity (UV/EUV, flares, CMEs). The neutral density response also depends on neutral composition. Composition varies as a function of altitude and solar cycle. At 250 km, molecular nitrogen (N_2) contributes a significant fraction, at 400 km, the thermosphere is dominated by atomic oxygen (O), and at 850 km, helium (He) begins to dominate. The relative composition of species at a fixed altitude is also dependent upon the level of solar activity, as the atmosphere expands and contracts (see Fig. 1). While assets at 850 km are not typically considered vulnerable to significant satellite drag, the variability in these forces may still constitute the largest contribution to overall orbital errors, and drag effects as high as 1000 km altitude have been reported. Moreover, during extreme events, the neutral density experienced at 850 km during storms would be comparable to the drag at 600 km altitude during quiet times.

At 250, 400, and 850 km altitudes, neutral species (e.g., N_2, O, He) tend to dominate. However, at 850 km altitude, the ionized atmosphere can begin to contribute a significant fraction of the total density. The ions impart acceleration to objects orbiting at these altitudes in a way similar to the neutral constituents. In addition, the ions can be transported to higher altitudes not just by thermal expansion, but also through transport by electric fields. During an extreme geomagnetic storm, vertical plasma drift at low and middle latitudes could be large, which if present for several hours, could raise the relative density concentration of the ionized atmosphere at 850 km to a significant fraction of the neutral density.

---

[2] https://www.whitehouse.gov/sites/default/files/microsites/ostp/final_nationalspaceweatherstrategy_20151028.pdf.
[3] https://www.whitehouse.gov/sites/default/files/microsites/ostp/final_nationalspaceweatheractionplan_20151028.pdf.

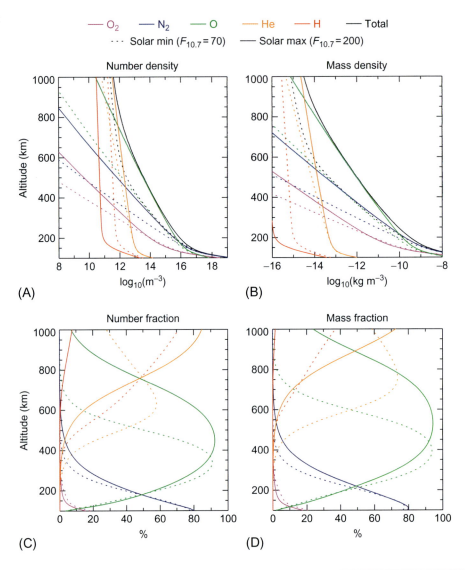

FIG. 1

Typical global average density and composition profiles from the NRLMSISE-00 empirical model: (A) Number density, (B) Mass density, (C) Number fraction (volume mixing ratio), and (D) Mass fraction (mass mixing ratio). Shown are profiles of O_2 *(purple)*, N_2 *(blue)*, O *(green)*, He *(orange)*, H *(red)*, and total density *(black)*, for solar maximum (solid lines, $F_{10.7} = 200$ solar flux units (1 sfu = 10^{-22} W m^{-2} Hz^{-1})) and solar minimum (dotted lines, $F_{10.7} = 70$ sfu) conditions.

Neutral density and winds at LEO altitudes pose two distinct risks to operational spacecraft:

- The direct effect of enhanced drag on the spacecraft, changing its orbit, increasing the uncertainty of its position, and reducing its orbital lifetime.

- The indirect effect that orbital change, caused by the atmospheric expansion, has on the ability to monitor the trajectories of debris, including objects with high area-to-mass ratios whose orbits are particularly sensitive to the state of the atmosphere, for collision avoidance.

The issue of predicting the orbits of debris is of particular concern for collision avoidance. Debris objects tend to be much smaller than operational satellites, have a larger area-to-mass ratio, have eccentric orbits that are affected by thermosphere horizontal and vertical structure, and consequently are more strongly and diversely affected by atmospheric drag. Collision avoidance has become of increasing concern due to the proliferation of space debris in LEO. The rapid increase in debris population is driven in part by collisions, such as the one between an Iridium satellite and the spent Russian satellite COSMOS 2251.

The issue of space debris has been brought to a new level of awareness and highlights the need for accurate orbit predictions and tracking of space objects. The Space Surveillance Network (SNN) currently tracks more than 20,000 objects greater than 10 cm. These objects consist of everything from active satellites, defunct satellites, and spent rocket stages, to smaller debris arising from erosion, explosion, and collision fragments. Because the orbits of these objects often overlap the trajectories of newer operational spacecraft, collision of the debris with active satellites is of serious concern.

As the number of objects in Earth orbit increases, there is a critical debris density at which the creation of new debris by collisions between resident space objects occurs faster than their removal from orbit by various natural forces. Beyond this point, a runaway chain reaction can occur that quickly reduces all objects in orbit to debris in a phenomenon known as the Kessler Syndrome (Kessler and Cour-Palais, 1978). A long-term trend of reduced density in the upper atmosphere due to CO_2 cooling (Emmert et al., 2004; Akmaev and Fomichev, 1998) will tend to increase the lifetime of debris and may exacerbate the challenge of collision avoidance, because atmospheric drag is currently the only mechanism by which debris are removed from orbit.

2 EFFECTS OF SOLAR EUV AND UV RADIATION

The likely response of neutral density to an increase in the solar UV and EUV lasting a few days can be estimated using empirical models, such as NRLMSISE-00 empirical neutral density model (see Fig. 1; Emmert, 2015b; Picone et al., 2002), JB2008 (Bowman et al., 2008), or DTM2013 (Bruinsma, 2015). These three empirical models are the ISO standard for Earth Atmosphere Density (ISO 14222, 2007). The empirical models have similar accuracy, all having been constructed with similar observed neutral density or tracking data. During quiet times, their uncertainty is ~15%–20% increasing to at least ~50% during geomagnetic storms, depending on storm severity. The empirical model uncertainty is expected to increase further when extrapolating for extreme events, to ~100%, for which no observational data are available. Empirical neutral density models of the effects of solar EUV and UV on upper atmosphere expansion are usually driven by proxies for EUV, such as the average of the daily solar 10.7-cm solar radio flux ($F_{10.7}$) and its 81-day mean. The $F_{10.7}$ radio emission originates high in the chromosphere and low in the corona of the solar atmosphere; it correlates well with the number of sunspots as well as EUV fluxes.

Over the past 68 years, the daily $F_{10.7}$ at 1 astronomical unit (AU) exceeded 300 solar flux units (sfu) (1 sfu = 10^{-22} W m^{-2} Hz^{-1}) on only 80 days (0.3% occurrence rate) (see Fig. 2). The peak value of the daily $F_{10.7}$ was 377 sfu at 1 AU and 390 sfu at the minimum Earth-Sun distance. These high

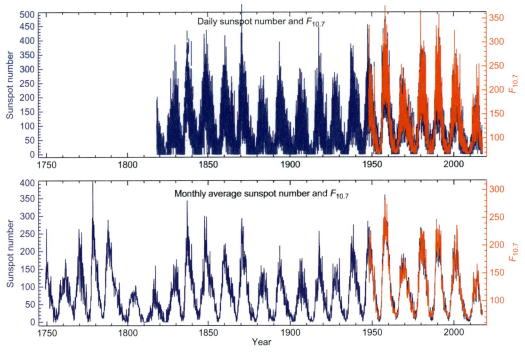

FIG. 2

Sunspot number (*blue*, referenced to the left axis) and $F_{10.7}$ (*red*, referenced to the right axis) time series. The top panel shows daily values, and the bottom panel shows monthly averages. The two vertical axes are scaled and offset based on a linear least-squares fit of the $F_{10.7}$ values to the sunspot number values. The source of the sunspot number data (Clette and Lefèvre, 2016 and references therein) is the World Data Center SILSO, Royal Observatory of Belgium, Brussels (http://www.sidc.be/silso/); version 2.0 data are shown. $F_{10.7}$ data (Tapping, 2013) were obtained from ftp://ftp.ngdc.noaa.gov/STP/GEOMAGNETIC_DATA/INDICES/KP_AP/.

values are usually associated with an active region rotating with the Sun, so could be elevated for a few days either side of these peak values, and possibly recur on the next solar rotation. Observations of the daily and 81-day mean $F_{10.7}$ radio flux were used to estimate a likely 100-year value of 390 sfu for the daily, and 280 sfu for the 81-day mean. These daily $F_{10.7}$ and 81-day mean values were used as input to the empirical NRLMSISE-00 neutral density model, and the values compared (as a percent change) to reference values using 240 and 200 sfu for the daily and 81-day mean, respectively. Relative to these background levels, at 250, 400, and 850 km, neutral density would increase by 50%, 100%, and 200%, respectively. Neutral density responds to EUV with an *e*-folding[4] time of about 1 day, and responds to the longer UV wavelengths that penetrate more deeply into the upper atmosphere, in 3–5 days. These

---

[4] *e*-folding time: the time for an exponential decay to decrease a factor of 1/*e*.

levels of events are not too far outside the parameter space used to develop the empirical model, so are likely to have uncertainties of around 20%–30%, not that much greater than the uncertainties at $F_{10.7}$ peak levels of ~300 sfu, which are included in the NRLMSISE-00 database.

Thermospheric neutral winds are not expected to increase substantially in response to increases in solar ultraviolet radiation, so their impact is likely to be small. The heating and thermal expansion will also cause a change in neutral composition at a fixed altitude. For instance, at 400 km, the thermosphere that was previously dominated by atomic oxygen would have a much higher fraction of molecular nitrogen, and at 850 km altitude, the thermosphere that was previously dominated by helium would develop a much higher fraction of atomic oxygen. These changes in composition may have some impact on the surface interactions with a spacecraft, degrade optics, organic films, advanced composites, and metallic surfaces (Tribble et al., 1999), but because their duration is short, are not likely to have significant long-term consequences.

The neutral atmosphere responds to an increase in solar UV and EUV for a few days, primarily as a result of the increased heating. If the neutral atmosphere rises, the levels of constant optical depth at a given wavelength will also rise, so at a particular wavelength the energy will be deposited (ionizing and heating the plasma and neutral gas) at a higher altitude. Existing plasma will also see the thermal expansion as upwelling, or a vertical wind (Fedrizzi et al., 2008). Away from the equator, the upward vertical neutral wind will have a component parallel to the magnetic field, which will also push the plasma along the magnetic field to greater altitude. In addition to the ionospheric response to the neutral atmosphere, the increased EUV radiation will also increase the ionization rate. At F-region altitudes, a doubling of the solar ionization will cause the plasma density to increase by about 100% (Philip Richards, private communication). Both the increased height of production and the upward transport of existing plasma will cause the height of the ionosphere to rise by 100 km or more.

3 EFFECT OF AN EXTREME SOLAR FLARE

Solar flares are categorized by the power in the 0.1–0.8 nm wavelength range in the X-ray part of the solar spectrum. The highest category of flare (X) has a power $>10^{-4}$ W m^{-2}. The largest solar flare on record is an X28 (2.8×10^{-3} W m^{-2} in the 0.1–0.8 nm wavelength range), recorded by the GOES satellite on November 6, 2003. However, the GOES instrument saturated for this event; the ionospheric response indicated the flare might have been closer to X45 (Tsurutani et al., 2005).

The X-rays themselves are deposited deep within the atmosphere (~80 to 100 km altitude) where they ionize the D region and cause absorption of HF radio waves. The X-ray source itself has very little impact on the ionosphere or neutral atmosphere above 100 km altitude. However, the EUV part of the solar spectrum can also increase during a flare, which does drive a response in the ionosphere and thermosphere, although the EUV flux is not always well correlated with the X-ray flux by which the flare is categorized. Several modeling studies have been conducted of the ionospheric response to solar flares including (Le et al., 2012, 2015; Huba et al., 2005; Fuller-Rowell et al., 2004). Smithtro et al. (2006) were able to explain the apparently anomalous response seen in the ionosonde at Bear Lake, Utah, where the peak F-region electron density (N_mF_2) decreased. They showed the increase in electron temperature caused the ionosphere to expand and store the plasma at higher altitudes, increasing h_mF_2, and decreasing N_mF_2 at the same time as increasing the total electron content. The CHAMP satellite (Lühr et al., 2004) has recorded the neutral density response to numerous solar flares greater than an X5 (Le et al., 2012). The peak neutral density response was typically 10%–13% at 400 km altitude. On October 28, 2003, during an X17 flare, CHAMP recorded a peak density increase of 50% on the dayside of the Earth for a few hours (Sutton et al., 2006; Tsurutani et al., 2005). Pawlowski and Ridley (2011) indicated that the density response at 400 km altitude is linearly dependent upon the total incident energy of the flare.

Based on an NRC study, Schrijver et al. (2012) estimated a 100-year flare event of ∼X30, which is actually not significantly greater than flares observed during the last few decades. The response of the D region may also not be significantly greater than previous large flares. Assuming a similar EUV spectrum and duration compared with the October 2003 event, the peak dayside neutral density response at 400 km at a median $F_{10.7}$ solar flux level of 150 sfu is expected to be about 50%–100%, relative to the background before the flare. The typical short duration of a flare of a few hours naturally constrains the temperature and density response, limiting the potential impact on satellite drag and orbit prediction. The impulsive nature of the energy injection is expected to launch gravity waves around the globe (Pawlowski and Ridley, 2011), but the increase in neutral winds is expected to be modest, and to have little impact on drag. The F-region plasma density changes are also expected to be fairly modest.

4 EFFECTS OF AN EXTREME CME DRIVING A GEOMAGNETIC STORM

Geomagnetic storms, which are driven by Coronal Mass Ejections (CMEs) emanating from the Sun, heat the thermosphere through auroral precipitation and Joule heating, and drive winds through ion drag. Observations of neutral density during large storms are available and reasonable estimates of the response can be predicted using empirical models (e.g., NRLMSISE-00), and simulated with physical models (e.g., CTIPe; Fuller-Rowell et al., 2008; Fedrizzi et al., 2008; Codrescu et al., 2012). Some of the largest storms in recent history stretch the limits of the empirical neutral atmosphere models, so they are likely to be uncertain in their estimates by at least 50% for significant storms, and ∼100% for the largest storms. Ionospheric empirical models (e.g., International Reference Ionosphere, IRI, Bilitza, 2001) tend to target the decreases in plasma density driven by changes in neutral composition during a geomagnetic storm (Fuller-Rowell et al., 1997, 2000, 2001). Capturing the increases in plasma density and total electron content in an empirical ionospheric model has been more challenging. The physical processes driving these "positive ionospheric phases" are beginning to be understood, with potential to capture in an empirical model. Empirical neutral and ionosphere models, together with observation of large storms and guidance from physical models, can be used to extrapolate the response to extreme events.

Although more likely to occur at the apex or descending phase of the solar cycle, large geomagnetic disturbances can theoretically occur during any level of background solar activity. For instance, the extreme CME event experienced by the STEREO-A spacecraft in July 2012 (Baker et al., 2013) occurred near solar maximum of Cycle 24, although at low solar activity. The relative thermosphere response to storms depends on solar activity; the same storm drives a larger relative density change at low solar activity, even though the absolute density is lower (see Fig. 3). Relative density enhancements and density structure are particularly important for orbit prediction of debris and collision avoidance, which is a hazard at all levels of solar activity.

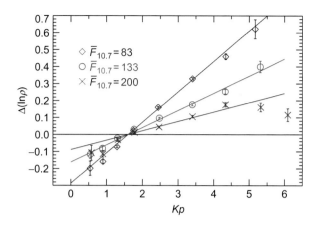

FIG. 3

Average variation of 1971–2007 orbit-derived global average log-density at altitude 400 km as a function of the 4-day K_p geomagnetic activity index (daily K_p averaged from 2 days prior to 1 day after the time of the density observation), in the following bins: {0.33–0.67, 0.67–1.00, 1.0–1.5, 1.5–2.0, 2–3, 3–4, 4–5, 5–6, 6–8}. Solar, geomagnetic activity, and long-term trend effects were filtered from the data using the global average mass density model (GAMDM2.1) (Emmert, 2015a) prior to computing the averages. Results are shown for low (*blue diamonds*; $F_{10.7} < 100$ sfu), moderate (*green circles*; $100 < F_{10.7} < 175$ sfu), and high (*red crosses*; $F_{10.7} > 175$ sfu) solar activity conditions; the average in each bin is given in the legend. Error bars denote the estimated 1σ uncertainty of the mean. Also shown are the corresponding variations from GAMDM2.1 (solid line).

Modified from Fig. 12 of Emmert, J.T., 2015b. Thermospheric mass density: a review. Adv. Space Res. 56, 773–824. https://doi.org/10.1016/j.asr.2015.05.038.

4.1 DEFINING THE DRIVERS

Estimates of the response to a 100-year geomagnetic storm are referenced to predictions of the solar wind condition of the important drivers: for example, the interplanetary magnetic field (IMF) magnitude and direction (particularly the magnitude of southward B_z), and solar wind velocity and density. Two events are often cited in the literature characterizing an extreme event: the Carrington storm of 1859 and an event observed by STEREO-A spacecraft in July 2012 (Baker et al., 2013). The impact on technology of the event in 1859, even at that time, was alarming as the induced currents in telegraph lines were sufficient to cause fires. The event is referred to as the "Carrington event," after the English Astronomer Richard Carrington, who viewed the white light flare on the Sun and associated it with the ensuing geomagnetic storm that followed a day later. Observations at the time were of course sparse. Based on the limited magnetometer record at the time, such as in Bombay, India, Li et al. (2006) were able to use the Temerin and Li (2002) model to estimate the expected solar wind conditions in the CME likely to produce the magnetic record of the time. Li et al. (2006) estimate a southward B_z between 60 and 70 nT, together with solar wind speed exceeding 2000 km s^{-1}. The predictions of solar wind density appeared to be unreasonably large, exceeding anything ever recorded by more than a factor of ten. The magnetic storm disturbance index, D_{st} (Sugiura, 1964), was predicted to be less than -1500 nT, although Siscoe et al. (2006) suggested a more modest value of -850 nT and Tsurutani et al. (2003; 2012) a value of -1760 nT.

The realism of the forcing estimates for the Carrington event can rightly be questioned because it is based on very little in the way of observations. However, on July 23, 2012, another coronal mass ejection similar in size to that predicted for the Carrington event was launched from the Sun. Fortunately, it was directed towards the STEREO-A spacecraft (Baker et al., 2013) and did not strike Earth. The CME speed was estimated to exceed 2000 km s^{-1}, and the measured B_z on STEREO-A reached −50 nT, values similar to the estimates for the Carrington event by Li et al. (2006), but with much more modest solar wind densities consistent with contemporary observations. Baker et al. (2013) used these observations to estimate that if the event had been directed at Earth, and had arrived at a time to maximize the magnetic dipole tilt angle with respect to the solar wind direction, the maximum D_{st} would have been −1182 nT, potentially greater than the Carrington event itself. Clearly a weak solar cycle like the last one does not preclude extreme solar events striking Earth. Note that the lowest measured D_{st} on record was recorded in March 1989 of −589 nT, during an event when the Hydro-Quebec and the U.S. East Coast power grid was severely disrupted.

Preliminary estimates of the possible magnetosphere/ionosphere plasma convection response to a Carrington-type event were obtained by using the solar wind values suggested by Li et al. (2006) or observed by STEREO-A as input to the Weimer (2005) empirical ion convection model. B_z was assumed to be equal to −60 nT, solar wind velocity of 1500 m s^{-1}, and a solar wind density of 50 cm^{-3}. Fig. 4 shows the polar region electric potential pattern and the Joule heating rate (or Poynting flux) from the Weimer empirical model. Because the solar wind conditions are at or beyond the limit of the data used to construct the empirical model, it would also be prudent to use physical magneto-hydrodynamic (MHD) global magnetosphere models to estimate the magnetospheric response.

The Weimer model predicts a cross-polar cap potential (CPCP) for the event exceeding 450 kV and a Joule heating power dissipation of ~7000 gigawatts (GW) in each hemisphere. The CPCP saturation in the Weimer empirical model could probably be extrapolated to more extreme cases. Based on expected magnetospheric saturation effects (Siscoe et al., 2004; Raeder and Lu, 2005), the estimates of Joule heating power are likely closer to 3000 GW in each hemisphere (see Fig. 5), or 6000 GW globally. These Joule heating rates are, about 1.5–2 times larger than events from the space era and 3 times the peak values observed in the last solar cycle. The convection pattern indicates auroral electrojet currents as far equatorward as Cuba (see Figs. 4 and 6), which is consistent with reports of auroral sightings during the 1859 Carrington event. The Weimer patterns appear consistent with reasonable physical expectations for the event; however, because the model is unconstrained for these parameters, nonlinearities in the system, such as known cross-polar cap potential saturation effects, make the empirical estimates uncertain. Again, it would be prudent to compare empirical estimates with those from a physical MHD magnetosphere modeling system, such as OpenGGCM (Raeder et al., 2001a).

The OpenGGCM magnetospheric model was used to simulate the more modest Bastille Day storm (Raeder et al., 2001a) and the January 10, 1997, storm, also known as the ISTP event (Raeder et al., 2001b). Individually, these storms reached IMF values of −60 nT and solar wind density in excess of 50 cm^{-3}, but not concurrently, as would be expected for a Carrington-like event. The Bastille Day modeling results compared quite well with observations. One result of these simulations was that the cross-polar potential saturates (Siscoe et al., 2004; Raeder and Lu, 2005). During the Bastille Day storm, OpenGGCM estimated the subsolar magnetopause distance to be as close as 4.5 R_E (Earth radii) (Raeder et al., 2001a), which was confirmed by the entry of geosynchronous satellites into the magnetosheath. During a Carrington-like event, the magnetopause would likely be much closer, possibly 3 R_E. This places the equatorward expansion of the convection pattern close to the

4 EFFECTS OF AN EXTREME CME DRIVING A GEOMAGNETIC STORM

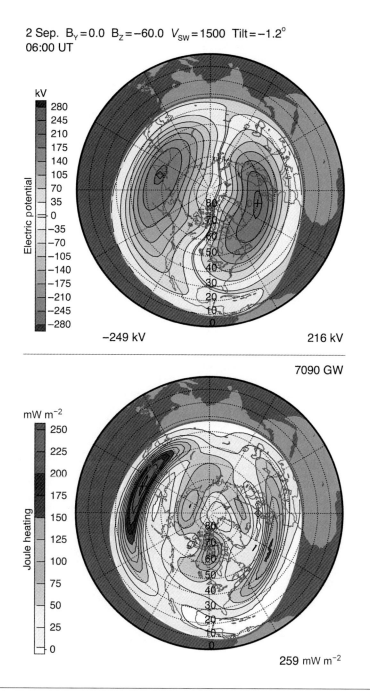

FIG. 4

Estimate of the polar region potential pattern (upper panel) and Joule heating rates (lower panel) for the Carrington event from the empirical model of Weimer (2005).

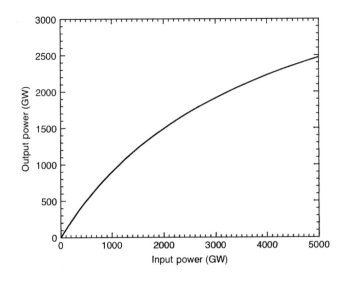

FIG. 5

Illustration of magnetospheric saturation function used to scale estimates of Joule heating from the Weimer empirical model.

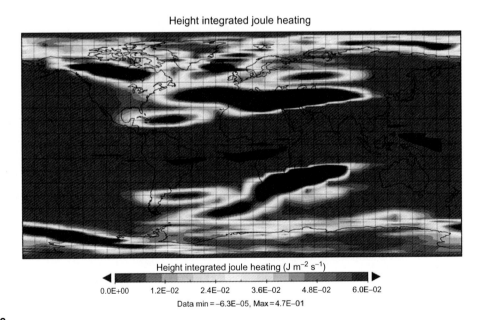

FIG. 6

CTIPe estimates of Joule heating rates illustrating the equatorward expansion of the magnetospheric source region to middle and low latitudes.

predictions from the Weimer empirical model. Ideally, simulations with coupled magnetosphere-ionosphere-thermosphere (M-I-T) models should be performed to check whether additional saturation of the cross-polar cap potential occurs for more extreme conditions or if new emergent behavior of the system unfolds. This has yet to be done.

4.2 NEUTRAL ATMOSPHERE RESPONSE TO AN EXTREME GEOMAGNETIC STORM

The Weimer estimates of magnetospheric potential were used to drive a coupled thermosphere-ionosphere-plasmasphere-electrodynamic model (CTIPe). CTIPe is a global, three-dimensional, time-dependent, nonlinear code that is a union of three physical components. The first is a global, non-linear, time-dependent, neutral thermospheric code developed by Fuller-Rowell and Rees (1980, 1983). The second is a mid- and high-latitude ionospheric convection model developed by Quegan et al. (1982). These first two components were initially coupled self-consistently and are known as the Coupled Thermosphere-Ionosphere Model (CTIM) (Fuller-Rowell et al., 1996), which is what is currently coupled to OpenGGCM. CTIM was further extended by including a third component, a plasmasphere and low-latitude ionosphere (Millward et al., 1996), to produce CTIP. Later the electrodynamics was solved self-consistently with the neutral dynamics and plasma components. The electrodynamic calculation was developed by Richmond and Roble (1987) and was included in the CTIP model by Millward et al. (2001), resulting in the creation of CTIPe. In principle, CTIPe simulates both the neutral, plasma, and electrodynamic response to the extreme event.

CTIPe was driven by the Weimer magnetospheric convection pattern, and the TIROS/NOAA statistical model of auroral precipitation (Fuller-Rowell and Evans, 1987). The auroral precipitation was scaled assuming an auroral power of 200 GW, where the most intense pattern on energy influx was scaled by a factor of two. Fig. 6 shows the pattern of Joule heating predicted by CTIPe driven by the Weimer estimates of magnetospheric convection, and illustrates the equatorward expansion of the sources to middle latitude. The neutral atmosphere density, composition, and wind response in CTIPe has been validated comprehensively for past storm responses (Fuller-Rowell et al., 2007, 2008; Fedrizzi et al., 2008, 2012; Codrescu et al., 2012; Negrea et al., 2012; Fang et al., 2014; Bruinsma and Fedrizzi, 2011; Fedrizzi and Bruinsma, 2012; Olsen et al., 2016). For example, Fig. 7A compares the CTIPe modeled neutral density with GOCE accelerometer observations (Gravity field and steady-state Ocean Circulation Explorer (GOCE), Bruinsma and Fedrizzi, 2012) of neutral density for the 2013 St. Patrick's Day storm. In Fig. 7B and C, the CTIPe neutral composition is compared to TIMED-GUVI observations of the height-integrated O/N_2 observations (Christensen et al., 2003). Fig. 7D and E show a comparison of the CTIPe neutral winds with GOCE cross-track meridional and zonal winds. The reasonable agreement of all the basic neutral parameters with observations provides at least some confidence that the model can be used to assess the possible response to a Carrington-level event, and to assess whether nonlinearities in the system start to impact the response.

Fig. 8A shows the neutral exospheric temperature at 12 UT in response to the estimated 6000 GW of Joule heating power imposed for about 12 h. The simulation produced exospheric temperatures in excess of 3000 K, about a factor two hotter than the 1500–2000 K observed for "super-storms" over the past 20 years. The temperature response was a fairly linear response to the larger Joule heating rates. This conclusion has to be treated with some caution because the cooling rates in the model were calculated based on the SNOE NO empirical model (NOEM) (Marsh et al., 2004). The cooling rates in CTIPe come firstly from vertical heat conduction from the upper to lower thermosphere, and then from

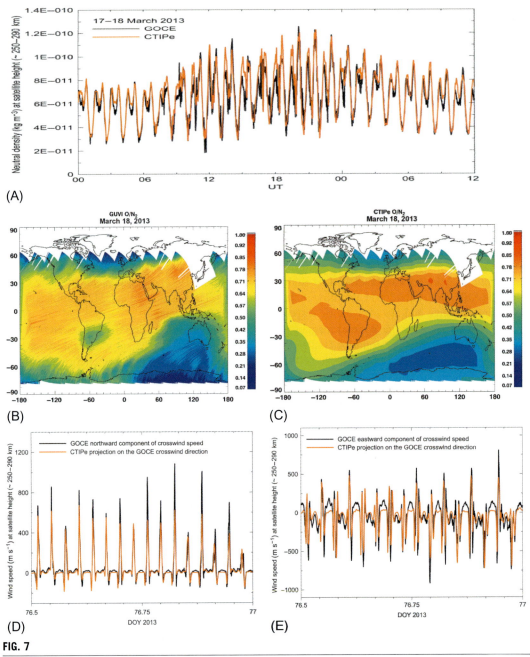

FIG. 7

Comparison of CTIPe during the March 18, 2013, St. Patrick's Day geomagnetic storm with observed neutral density and winds from GOCE satellite, and neutral composition O/N$_2$ ratio from the GUVI instrument on the TIMED satellite. (A) shows a comparison of CTIPe *(orange)* neutral density with GOCE *(black)*, (B) and (C) show a comparison of neutral composition O/N$_2$ ratio from GUVI and CTIPe, respectively, and (D) and (E) compare CTIPe with GOCE horizontal cross track winds in the meridional and zonal direction, respectively.

4 EFFECTS OF AN EXTREME CME DRIVING A GEOMAGNETIC STORM

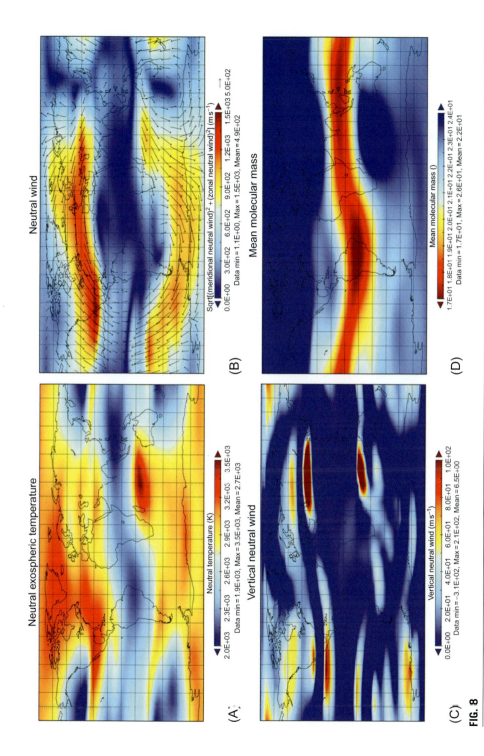

FIG. 8

CTIPe simulation of (A) exospheric temperature, (B) horizontal neutral wind, (C) vertical neutral wind, and (D) neutral composition, in response to a Carrington-level geomagnetic storm.

infrared radiative cooling, primarily from the 5.3-μm nitric oxide (NO) band, from the Kockarts (1980) formulism. NO cooling has been referred to as a "thermostat" for the system by Mlynczak et al. (2003); auroral precipitation, as well as serving as a source of heat and ionization, also produces NO and increases the cooling rate. Because the Marsh NOEM code has to be extrapolated from K_p 5 to K_p 9, the cooling rate estimates during these types of extreme events have significant uncertainty. NO is produced during a geomagnetic storm by the dissociation of N_2 by auroral electron precipitation, producing the excited state of atomic nitrogen $N(^2D)$, which is the precursor for NO (Fuller-Rowell, 1993). The auroral production of NO has significant uncertainty in its own right. The K_p scaling of NO and NO cooling rates used in CTIPe work well for following the neutral density recovery in simulations of previously observed storms, but it must be recognized that there is significant uncertainty in these cooling rates for this Carrington-level simulation.

Fig. 8B shows the neutral winds in the upper thermosphere in response to the ion drag associated with the Weimer magnetosphere-ionosphere convection. The peak neutral winds are in excess of 1500 m s^{-1}, which is about 50% larger than simulations of previous super-storms. Although the ion drag force is about twice that of previous simulations the winds are only about 50% increased, which is an illustration of a nonlinear response. The large winds are concentrated at middle to high latitudes in response to the predominately two-cell magnetospheric convection pattern, as illustrated in Fig. 4. The winds are typically sunward in the auroral oval, and antisunward over the polar cap. The winds are only accelerated to these high velocities while they remain in the convection channels. As the winds become larger, the neutral gas can be thrown out of the convection channels by inertia, through nonlinear advective transport and Coriolis forces (Fuller-Rowell and Schrijver, 2009; Fuller-Rowell et al., 2008; Fuller-Rowell, 2013), which naturally limit the response to the ion drag forces during this extreme event.

Fig. 8C shows a snapshot of the CTIPe vertical winds during the event. Vertical winds exceed 150 m s^{-1} upward in response to the intense Joule heating, temperature increase, and thermal expansion. Although the neutral atmosphere in CTIPe assumes hydrostatic balance, it does not exclude the possibility of large vertical winds (Fuller-Rowell, 2013). The assumption simply demands that the rate of heating is such that the atmosphere adjusts at a comparable rate. The term "quasi-hydrostatic balance" is the more correct expression in the case of accommodating vertical winds in the system. One component of the vertical wind is defined as the rate of change in the height of a pressure surface in the column of gas, $(\partial h/\partial t)_p$, the so called "barometric wind." Vertical winds in Earth's upper atmosphere in excess of 100 m s^{-1} can be accommodated within the quasi-hydrostatic assumption, and this is clearly seen in Fig. 8C. The assumption of hydrostatic balance has enabled the wide use of pressure as the vertical coordinate in atmospheric models. In fact, it is only fairly recently that Earth upper-atmosphere models are beginning to relax this assumption, and explicitly include a realistic adjustment process by acoustic waves (see e.g., Deng and Ridley, 2014; Ridley et al., 2006; Deng et al., 2008). The other component of vertical wind is the "divergence wind," where horizontal divergence or convergence must be balanced by a vertical flow. A local heat source will cause the local column of gas to expand thermally (the barometric wind). The horizontal pressure gradients so induced will drive a divergent wind that must be balanced by a vertical flow across the pressure surfaces. For extreme events like the one simulated here, the large vertical winds tend to arise from the barometric wind component.

Finally, in Fig. 8D, the CTIPe neutral composition mean molecular mass response to the Carrington event is shown. The neutral composition response to the event was unexpected in that it was not a simple extrapolation of previous smaller events. The expectation was that more extreme energy input would produce a stronger global pole-to-equator circulation and upwelling, leading to a large neutral composition perturbation. Instead, the composition change saturated. The explanation appears to be that the Joule heating source was so expanded in latitude (rather than more typically being concentrated at high latitude) that the energy spread globally very quickly. Instead of the strong global circulation developing, the energy was channeled into a more spatially uniform temperature increase and thermal expansion. However, this may also be an artifact of using the Weimer potential patterns, which predict rather uniform and smooth patterns of Joule heating. In reality, recent MHD simulations (Li et al., 2011) and observations (Knipp et al., 2011) suggest that the Joule heating pattern can be very localized, which could lead to a different thermospheric neutral composition response.

As expected, the neutral density responded to the Joule heating energy injection, and subsequent thermal expansion. To illustrate the neutral density response to an extreme event, Fig. 9 compares the time series of the neutral density response as seen from LEO at 400 km for a storm in December 2006, with an equivalent event where the peak energy input was tuned to match that expected for a Carrington-level event, but using the same UT history as the December event. The same time history is used in both cases. On the left side are the results for the December 2006 storm compared with the CHAMP accelerometer neutral density observations, which match reasonably well. On the right side, the same storm response is shown on a different scale and compared with the event scaled to match the Carrington peak storm power. The figure also shows the auroral power, Joule heating in the North and South Hemispheres, the global kinetic energy produced from ion drag, and the nitric oxide (NO) cooling rate. The peak neutral density response for the tuned Carrington event at the equivalent CHAMP altitude is about five times larger than the December 2006 storm, and would be about twice that for a Halloween or Bastille Day-level event.

The CTIPe model used to create Fig. 8 was initially unable to simulate the Carrington event because of numerical instabilities. The extreme vertical winds of 150 m s^{-1} and horizontal neutral winds of 1.5 km s^{-1} required improvement in the robustness of the solver. After some modification, CTIPe was able to complete the simulation. The increase in temperature, thermal expansion, and neutral density responded reasonably linearly. The winds and neutral composition changes displayed a more nonlinear response. The heating and thermal expansion of the neutral atmosphere caused the pressure level of the model upper boundary to rise well above 1000 km altitude, a pressure level that is normally at 500 km. The implication is that satellite drag would be significantly affected with degraded orbit prediction. If sustained for long periods (days), it would severely shorten satellite lifetimes and may cause premature re-entry of LEO satellites. However, a sustained period of increased density would cause some of the orbital debris to re-enter, thus reducing the hazards of collisions.

4.3 IONOSPHERIC RESPONSE TO AN EXTREME GEOMAGNETIC STORM

The CTIPe simulation of the Carrington storm has provided some insight into how the neutral atmosphere might respond to an extreme event. Simulating the ionospheric response was more challenging because the large increases in total electron content seen at middle and low latitude during some storms (e.g., Mannucci et al., 2005) are not well modeled.

Although CTIPe did not simulate the response well, it is useful to review some of the ionospheric physical processes that are likely to ensue during a Carrington like event (see also Chapters 20, 23, and 24 on the response to more modest storms), and speculate on the likely outcome for changes in

530 CHAPTER 21 THERMOSPHERE-IONOSPHERE RESPONSE TO EXTREME SPACE WEATHER EVENTS

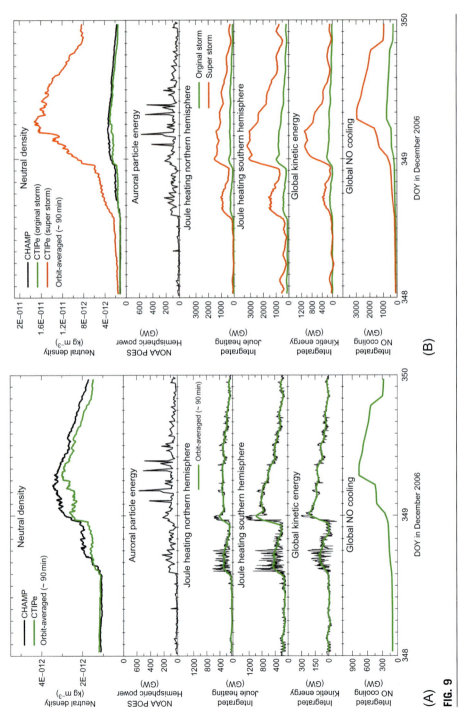

FIG. 9

Comparison of neutral density response during the real storm in December 2006 (A) on the left, with a simulated Carrington event with a similar time history (B) on the right with a different scale. The figure also shows the auroral power, Joule heating in the North and South Hemispheres, the global kinetic energy produced from ion drag, and the nitric oxide (NO) cooling rate.

ionospheric plasma content. The ionosphere is expected to undergo huge changes on the Earth's dayside during geomagnetic storms. Mannucci et al. (2005) used upward-looking GPS data from the CHAMP satellite at about 400 km altitude during the October 2003 Halloween storm to demonstrate that the plasma content above the satellite increased by more than a factor of 10. One of the main drivers of the increased total electron content (TEC) at middle latitudes is thought to be the intensification and expansion of the magnetospheric convection into middle latitudes (Heelis et al., 2009), and the development of storm-enhanced density (SED; Foster and Rideout, 2005, 2007; Heelis et al., 2009). The plasma caught in the sunward flow in the noon to dusk magnetic sector would stagnate and continue to increase due to solar production. The plasma in this sector can also start to move poleward and upward because the magnetic field is inclined. With a very expanded convection pattern, the poleward convection with significant magnetic inclination would very effectively raise the plasma to higher altitude, out of the neutral atmosphere to a region where recombination is slow. The combined effect would be a very significant storm-time increase in total electron content on the dayside, as observed by Mannucci et al. (2005).

In addition, the expansion of the high-latitude magnetospheric convection pattern does not cut off at middle latitudes, but is felt at low latitudes and at the magnetic equator. Here it is usually referred to as a "prompt penetration" electric field, because of its rapid onset in response to the solar wind drivers. The increase in the interplanetary electric field drives an eastward electric field on Earth's dayside that reaches equatorial latitudes. The near-horizontal magnetic field at low latitude combined with the eastward electric field renders the vertical plasma transport ($\mathbf{E} \times \mathbf{B}$) very effective in lifting the plasma to higher altitudes. The equatorial ionization anomalies (EIAs) tend to grow in magnitude and move to higher latitudes during these times, and can dominate the latitudinal plasma structure (Lin et al., 2005; Mannucci et al., 2005; Lei et al., 2014). Using the solar wind estimates for the Carrington event produced by Li et al. (2006), the equatorial vertical plasma drift driven by the penetration electric field using the model of Manoj and Maus (2012) was estimated to be 240 m s^{-1}, which is about twice the peak values ever observed at the Jicamarca incoherent scatter radar station on the magnetic equator (Kelley et al., 2010).

It is very likely the two mechanisms driving the buildup of plasma at middle latitude would interact during a Carrington-level event, in fact they could be indistinguishable. The expansion of the magnetospheric convection to middle latitudes would drive a plasma build-up at middle latitudes by the storm-enhanced density down to ~40° or 45° magnetic latitude. At the same time, the very strong vertical plasma drifts of 240 m s^{-1} predicted at low latitudes would move the EIA peaks to a similar latitude on the dayside, the so-called super-fountain effect (Tsurutani et al., 2008). The interaction between the two mechanisms could well cause TEC at middle latitudes to exceed 1000 TEC units.

Electrodynamics in the presence of solar EUV production is clearly a major driver of the TEC change. The second driver of the large changes in plasma density is the neutral atmosphere. High-latitude energy input drives enhanced equatorward winds, which also push the plasma to still greater altitudes at middle latitudes. The global wind circulation subsequently impacts neutral composition and drives the disturbance wind dynamo, all of which will impact the ionosphere. During geomagnetic storms, the neutral upper atmosphere can change dramatically (Rishbeth et al., 1987a,b; Burns et al., 1991; Prölss, 1997; Buonsanto, 1999; Fuller-Rowell et al., 1994, 1997, 2002; Fejer et al., 2002; Emmert et al., 2001, 2002, 2008).

The expansion of the high-latitude convection to middle latitudes will also drive plasma erosion, and the movement of the plasmapause equatorward (see for example, Goldstein et al., 2017). Fig. 10

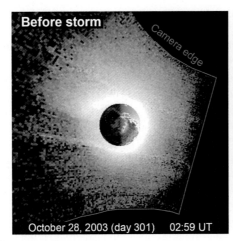 

FIG. 10

Satellite observations of the erosion of the plasmasphere during a storm, from observations by the IMAGE satellite before and after the Halloween storm of October 28, 2003.

Reproduced with permission from J. Goldstein.

shows satellite observations of the erosion of the plasmasphere during a storm, from observations by the IMAGE satellite before and after the Halloween storm of October 28, 2003. The size of the plasmasphere would appear to shrink considerably. The polar cap size, the boundary between open and closed field lines, could expand considerably, allowing plasma to escape, severely depleting the ionosphere. Poleward of ~40° to 50° latitude during an extreme geomagnetic storm, the ionospheric plasma would be wiped out by both the erosion of the plasmapause and by loss from recombination from the neutral composition change (see Fig. 8D). At the same time, the equatorial and low-latitude plasma would also be lost due to the migration of the EIA poleward by the eastward penetration electric field. All the plasma would be concentrated at middle latitudes in heavily populated areas, where impacts on communication and navigation would have maximum impact.

5 SUMMARY AND CONCLUSIONS

This chapter has attempted to estimate the impact on the thermosphere and ionosphere of three different flavors of extreme solar driven events: increases in solar UV and EUV radiation, an extreme solar flare, and an extreme geomagnetic storm. We have not explored the most extreme events likely, but chosen to target a level of event that is likely to occur once in a hundred years, or has a one percent chance of occurring any given year. This pragmatic approach we believe is more practical, and enables a mitigation strategy to be evaluated and prepared for. The consequences are potentially severe enough without having to consider a very unlikely doomsday scenario.

The main consequences of the increase in solar UV and EUV flux are clearly the increase in neutral density and the impact on satellite drag. Although the increase in drag would certainly be significant, the changes are not expected to be sudden, but spread out over a few days. This more gradual response enables the impact on orbit prediction and collision avoidance to be potentially planned for. If operators are aware of the possibility of a factor two to three increase in neutral density, then it is reasonable to plan a mitigation strategy, with fewer tracking failures and objects lost. A prolonged increase in solar UV and EUV radiation could actually have the added benefit of clearing out some of the debris in low-Earth-orbit. Plasma density is also anticipated to increase from the increased production, but again the changes are gradual, so is unlikely to have disastrous consequences on communication and navigation systems. The most likely impact could well be that scintillations of satellite radio signals will have more impact as background plasma density levels rise.

For the extreme flare scenario considered for the hundred-year event, the flare magnitude is not predicted to be significantly greater than flares in the last 40 years in the space era. Obviously D-region plasma density will impact absorption HF radio signals, but these effects are short lived, and operators have been dealing with these types of scenarios for many years. Orbit prediction operators will also need to be aware of the response to the impulsive nature of the flare, and the estimated 50%–100% increase in neutral density on the dayside.

Which brings us to, perhaps for more important solar-driven event, which is likely to have more severe consequences on operations, that of an extreme CME event driving a severe geomagnetic storm. Both neutral density changes and plasma density increases are potentially hazardous. Both the magnitude of the impulsive neutral-density changes and structure that a severe geomagnetic storm would likely create are potentially a serious concern for orbit tracking and collision avoidance. Satellite drag and collision avoidance are also dependent upon the motion of the atmosphere in the form of neutral winds. In-track winds change apparent density (proportional to velocity squared, V^2) along an orbit, and cross-track winds influence orbit trajectory. The neutral winds of 1500 m s^{-1} predicted introduces an apparent density change of 35%–40%, which are not huge changes in themselves, but are localized and make tracking debris a challenge.

The ionospheric changes predicted during a severe geomagnetic storm are also potentially of serious concern. Without guidance from physical model simulations, the likely response is very uncertain. We rely heavily on our understanding of the how the physical system responds to less extreme events such as the Halloween (October 2003) and Bastille Day (July 2000) geomagnetic storms. Extrapolating to the level of a Carrington event, we might expect a very large build up in plasma at middle latitudes, and severe depletions at high and low latitudes. Increases in TEC delay GNSS signals, which will reduce satellite navigation positioning accuracy. Steep ionospheric gradients and irregularities will cause scintillations in radio waves disrupting communication (see also Chapter 23).

In summary, the important consequences during a geomagnetic storm are upper atmosphere heating and thermal expansion increasing neutral density and the drag on satellites, impacting collision avoidance, and debris in low-Earth orbit, and the plasma density increase affecting critical communication and navigation systems.

ACKNOWLEDGMENTS

The GUVI data used in Fig. 7 are provided through support from the NASA MO&DA program. The GUVI instrument was designed and built by The Aerospace Corporation and The Johns Hopkins University. The Principal Investigator is Dr. Andrew B. Christensen, and the Chief Scientist and Co-PI is Dr. Larry J. Paxton.
The authors thank Karen O'Loughlin for careful editing of the manuscript.

REFERENCES

Akmaev, R.A., Fomichev, V.I., 1998. Cooling of the mesosphere and lower thermosphere due to doubling of CO_2. Ann. Geophys. 16, 1501–1512. https://doi.org/10.1007/s00585-998-1501-z.

Baker, D.N., Li, X., Pulkkinen, A., Ngwira, C.M., Mays, M.L., Galvin, A.B., Simunac, K.D.C., 2013. A major solar eruptive event in July 2012: defining extreme space weather scenarios. Space Weather 11, 585–591. https://doi.org/10.1002/swe.20097.

Bilitza, D., 2001. International reference ionosphere 2000. Radio Sci. 36 (2), 261–275. https://doi.org/10.1029/2000RS002432.

Bowman, B.R., Tobiska, W.K., Marcos, F.A., Huang, C.Y., Lin, C.S., Burke, W.J., 2008. A New Empirical Thermospheric Density Model JB2008 Using New Solar and Geomagnetic Indices. In: AIAA/AAS Astrodynamics Specialist Conference, AIAA 2008-6438.

Bruinsma, S.L., 2015. The DTM-2013 thermosphere model. J. Space Weather Space Clim. 5 (A1). https://doi.org/10.1051/swsc/2015001 (8 pp.).

Bruinsma, S.L., Fedrizzi, M., 2012. Simultaneous observations of TADs in GOCE, CHAMP and GRACE density data compared with CTIPe. In: AGU Fall Meeting, San Francisco, CA, December 3–7.

Bruinsma, S.L., Fedrizzi, M., 2011. Thermosphere variability inferred from CHAMP and GRACE accelerometer observations. In: AGU Fall Meeting, San Francisco, CA (presented by Bruinsma).

Buonsanto, M.J., 1999. Ionospheric storms—a review. Space Sci. Rev. 88, 563–601. https://doi.org/10.1023/A:1005107532631.

Burns, A.G., Killeen, T.L., Roble, R.G., 1991. A theoretical study of thermospheric composition perturbations during an impulsive geomagnetic storm. J. Geophys. Res. 96, 14153–14167. https://doi.org/10.1029/91JA00678.

Christensen, A.B., Paxton, L.J., Avery, S., et al., 2003. Initial observations with the Global Ultraviolet Imager (GUVI) in the NASA TIMED satellite mission. J. Geophys. Res. 108 (A12), 1451. https://doi.org/10.1029/2003JA009918.

Clette, F., Lefèvre, L., 2016. The new sunspot number: assembling all corrections. Sol. Phys. 291 (9–10), 2629–2651. https://doi.org/10.1007/s11207-016-1014-y.

Codrescu, M.V., Negrea, C., Fedrizzi, M., Fuller-Rowell, T.J., Dobin, A., Jakowsky, N., Khalsa, H., Matsuo, T., Maruyama, N., 2012. A real-time run of the coupled thermosphere ionosphere plasmasphere electrodynamics (CTIPe) model. Space Weather 10. S02001, https://doi.org/10.1029/2011SW000736.

Deng, Y., Ridley, A.J., 2014. The global ionosphere-thermosphere model and the nonhydrostatic processes. In: Huba, J., Schunk, R., Khazanov, G. (Eds.), Modeling the Ionosphere-Thermosphere. In: Geophysical Monograph Series, vol. 201. John Wiley & Sons, Ltd, Chichester, pp. 85–100. https://doi.org/10.1002/9781118704417.

Deng, Y., Richmond, A.D., Ridley, A.J., Liu, H.-L., 2008. Assessment of the non-hydrostatic effect on the upper atmosphere using a general circulation model (GCM). Geophys. Res. Lett. 35. L01104, https://doi.org/10.1029/2007GL032182.

Emmert, J.T., Fejer, B.G., Fesen, C.G., Shepherd, G.G., Solheim, B.H., 2001. Climatology of middle- and low latitude F region disturbance neutral winds measured by wind imaging interferometer (WINDII). J. Geophys. Res. 106, 24701–24712. https://doi.org/10.1029/2000JA000372.

Emmert, J.T., Fejer, B.G., Shepherd, G.G., Solheim, B.H., 2002. Altitude dependence of mid and low latitude daytime thermospheric disturbance winds measured by WINDII. J. Geophys. Res. 107 (A12), 1483. https://doi.org/10.1029/2002JA009646.

Emmert, J., Picone, J., Lean, S., Knowles, S., 2004. Global change in the thermosphere: compelling evidence of a secular decrease in density. J. Geophys. Res. 109. https://doi.org/10.1029/2003JA010176.

Emmert, J.T., Drob, D.P., Shepherd, G.G., Hernandez, G., Jarvis, M.J., Meriwether, J.W., Niciejewski, R.J., Sipler, D.P., Tepley, C.A., 2008. DWM07 global empirical model of upper thermospheric storm-induced disturbance winds. J. Geophys. Res. 113. A11319, https://doi.org/10.1029/2008JA013541.

Emmert, J.T., 2015a. Altitude and solar activity dependence of 1967–2005 thermospheric density trends derived from orbital drag. J. Geophys Res. Space Phys. 120. https://doi.org/10.1002/2015JA021047.

Emmert, J.T., 2015b. Thermospheric mass density: a review. Adv. Space Res. 56, 773–824. https://doi.org/10.1016/j.asr.2015.05.038.

Fang, T.-W., Fuller-Rowell, T., Wang, H., Akmaev, R., Wu, F., 2014. Ionospheric response to sudden stratospheric warming events at low and high solar activity. J. Geophys. Res. Space Phys. 119 (9), 7858–7869. https://doi.org/10.1002/2014JA020142.

Fedrizzi, M., Bruinsma, S., 2012. Modeling space weather in the thermosphere. In: EGU General Assembly 2012, Vienna, Austria, April, 22-27. (Presented by S. Bruinsma).

Fedrizzi, M., Fuller-Rowell, T.J., Codrescu, M., Maruyama, N., Khalsa, H., 2008. Sources of F-region height changes during geomagnetic storms at mid latitudes. In: Kintner, P.M. et al., (Ed.), Midlatitude Ionospheric Dynamics and Disturbances. Geophysical Monograph Series, vol. 181. AGU, Washington, DC, pp. 247–258.

Fedrizzi, M., Fuller-Rowell, T.J., Codrescu, M.V., 2012. Global Joule heating index derived from thermospheric density physics-based modeling and observations. Space Weather 10. S03001. https://doi.org/10.1029/2011SW000724.

Fejer, B.G., Emmert, J.T., Sipler, D.P., 2002. Climatology and storm-time dependence of nighttime thermospheric neutral winds over Millstone Hill. J. Geophys. Res. 107 (A5), 1052. https://doi.org/10.1029/2001JA000300.

Foster, J.C., Rideout, W., 2005. Midlatitude TEC enhancements during the October 2003 superstorm. Geophys. Res. Lett. 32. L12S04, https://doi.org/10.1029/2004GL021719.

Foster, J.C., Rideout, W., 2007. Storm enhanced density: magnetic conjugacy effects. Ann. Geophys. 25, 1791–1799. https://doi.org/10.5194/angeo-25-1791-2007.

Fuller-Rowell, T.J., 1993. Modeling the solar cycle change in nitric oxide in the thermosphere and upper mesosphere. J. Geophys. Res. 98 (A2), 1559–1570. https://doi.org/10.1029/92JA02201.

Fuller-Rowell, T.J., Evans, D.S., 1987. Height-integrated Pedersen and Hall conductivity patterns inferred from the TIROS-NOAA satellite data. J. Geophys. Res. 92 (A7), 7606–7618. https://doi.org/10.1029/JA092iA07p07606.

Fuller-Rowell, T.J., Rees, D., 1980. A three-dimensional time-dependent, global model of the thermosphere. J. Atmos. Sci. 37 (11), 2545–2567. https://doi.org/10.1175/1520-0469(1980)037<2545:ATDTDG>2.0.CO;2.

Fuller-Rowell, T.J., Rees, D., 1983. Derivation of a conservation equation for mean molecular weight for a two-constituent gas within a three-dimensional, time-dependent model of the thermosphere. Planet. Space Sci. 31 (10), 1209–1222. https://doi.org/10.1016/0032-0633(83)90112-5.

Fuller-Rowell, T.J., Rees, D., Quegan, S., Moffett, R.J., Codrescu, M.V., Millward, G.H., 1996. A coupled thermosphere-ionosphere model (CTIM). In: Schunk, R.W. (Ed.), STEP Handbook of Ionospheric Models. SCOSTEP, NOAA/NGDC, Boulder, CO, pp. 217–238.

Fuller-Rowell, T.J., Codrescu, M.V., Moffett, R.J., Quegan, S., 1994. Response of the thermosphere and ionosphere to geomagnetic storms. J. Geophys. Res. 99 (A3), 3893–3914. https://doi.org/10.1029/93JA02015.

Fuller-Rowell, T.J., Millward, G.H., Richmond, A.D., Codrescu, M.V., 2002. Storm-time changes in the upper atmosphere at low latitudes. J. Atmos. Sol. Terr. Phys. 64 (12–14), 1383–1391. https://doi.org/10.1016/S1364-6826(02)00101-3.

Fuller-Rowell, T.J., Codrescu, M.V., Roble, R.G., Richmond, A.D., 1997. How does the thermosphere and ionosphere react to a geomagnetic storm? In: Tsurutani, B.T., Gonzalez, W.D., Kamide, Y., Arballo, J.K. (Eds.), Magnetic Storms. Geophysical Monograph Series, vol. 98. AGU, Washington, DC, pp. 203–225.

Fuller-Rowell, T.J., Araujo-Pradere, E., Codrescu, M.V., 2000. An empirical ionospheric storm-time correction model. Adv. Space Res. 25 (1), 139–146. https://doi.org/10.1016/S0273-1177(99)00911-4.

Fuller-Rowell, T.J., Codrescu, M.V., Araujo-Pradere, E., 2001. Capturing the storm-time *F*-region ionospheric response in an empirical model. In: Song, P., Singer, H.J., Siscoe, G.L. (Eds.), Space Weather, Progress and Challenges in Research and Applications. Geophysical Monograph Series, vol. 125. AGU, Washington, DC, pp. 393–401.

Fuller-Rowell, T.J., Solomon, S.C., Viereck, R., Roble, R.G., 2004. Impact of solar EUV and X-ray variation on Earth's atmosphere. In: Pap, J.M., Fox, P.M. (Eds.), Solar Variability and its Effects on Climate. Geophysical Monograph Series, vol. 141. AGU, Washington, DC, pp. 341–354.

Fuller-Rowell, T., Codrescu, M., Maruyama, N., Fedrizzi, M., Araujo-Pradere, E., Sazykin, S., Bust, G., 2007. Observed and modeled thermosphere and ionosphere response to superstorms. Radio Sci. 42(4). RS4S90. https://doi.org/10.1029/2005RS003392.

Fuller-Rowell, T.J., Richmond, A., Maruyama, N., 2008. Global modeling of storm time thermospheric dynamics and electrodynamic. In: Kintner Jr., P.M., Coster, A.J., Fuller-Rowell, T.J., Mannucci, A.J., Mendillo, M., Heelis, R. (Eds.), Midlatitude Ionospheric Dynamics and Disturbances. Geophysical Monograph Series, vol. 181. AGU, Washington, DC, pp. 187–200. https://doi.org/10.1029/181GM18.

Fuller-Rowell, T.J., Schrijver, C.J., 2009. On the ionosphere and chromosphere. In: Schrijver, C.J., Siscoe, G.L. (Eds.), Heliophysics: Plasma Physics of the Local Cosmos. Cambridge University Press, Cambridge, pp. 324–359 (Chapter 12).

Fuller-Rowell, T., 2013. Physical characteristics and modeling of Earth's thermosphere. In: Huba, J.D., Schunk, R.W., Khazanov, G.V. (Eds.), Modeling the Ionosphere-Thermosphere. In: Geophysical Monograph Series, vol. 201. John Wiley & Sons, Ltd., Chichester, pp. 13–27. https://doi.org/10.1002/9781118704417.ch2.

Goldstein, J., Angelopoulos, V., De Pascuale, S., Funsten, H.O., Kurth, W.S., Lera, K., McComas, D.J., Perez, J.D., Reeves, G.D., Spence, H.E., Thaller, S.A., Valek, P.W., Wygant, J.R., 2017. Cross-scale observations of the 2015 St. Patrick's Day storm: THEMIS, Van Allen Probes, and TWINS. J. Geophys. Res. Space Phys. 122, 368–392. https://doi.org/10.1002/2016JA023173.

Heelis, R.A., Sojka, J.J., David, M., Schunk, R.W., 2009. Storm-time density enhancements in the middle latitude dayside ionosphere. J. Geophys. Res. 114. A03315. https://doi.org/10.1029/2008JA013690.

Huba, J.D., Warren, H.P., Joyce, G., Pi, X., Iijima, B., Coker, C., 2005. Global response of the low-latitude to midlatitude ionosphere due to the Bastille Day flare. Geophys. Res. Lett. 32. L15103. https://doi.org/10.1029/2005GL023291.

ISO 14222, 2007. Space environment (natural and artificial)—Process for determining solar irradiances. International Standards Organization, Geneva.

Kelley, M.C., Ilma, R.R., Nicolls, M., Erickson, P., Goncharenko, L., Chau, J.L., Aponte, N., Kozyra, J.U., 2010. Spectacular low- and mid-latitude electrical fields and neutral winds during a superstorm. J. Atmos. Sol. Terr. Phys. 72 (4), 285–291. https://doi.org/10.1016/j.jastp.2008.12.006.

Kessler, D.J., Cour-Palais, B.G., 1978. Collision frequency of artificial satellites: the creation of a debris belt. J. Geophys. Res. 83 (A6), 2637–2646. https://doi.org/10.1029/JA083iA06p02637.

Knipp, D.J., Kilcommons, L., Li, W., Raeder, J., Deng, Y., 2011. Data and Model Views of Energy Input to the Dayside Thermosphere, When the East-West Interplanetary Magnetic Field Is Large. In: AGU Fall Meeting, San Francisco, CA.

Kockarts, G., 1980. Nitric oxide cooling in the terrestrial thermosphere. Geophys. Res. Lett. 7, 137–140. https://doi.org/10.1029/GL007i002p00137.

Le, H., Liu, L., Wan, W., 2012. An analysis of thermospheric density response to solar flares during 2001–2006. J. Geophys. Res. 117. A03307. https://doi.org/10.1029/2011JA017214.

Le, H., Ren, Z., Liu, L., et al., 2015. Global thermospheric disturbances induced by a solar flare: a modeling study. Earth Planet Space 67, 3. https://doi.org/10.1186/s40623-014-0166-y.

Lei, J., Wang, W., Burns, A.G., Yue, X., Dou, X., Luan, X., Solomon, S.C., Liu, Y.C.-M., 2014. New aspects of the ionospheric response to the October 2003 superstorms from multiple-satellite observations. J. Geophys. Res. Space Phys. 119. https://doi.org/10.1002/2013JA019575.

Li, W., Raeder, J., Knipp, D., 2011. The relationship between dayside local Poynting flux enhancement and cusp reconnection. J. Geophys. Res. 116. A08301. https://doi.org/10.1029/2011JA016566.

Li, X., Temerin, M., Tsurutani, B.T., Alex, S., 2006. Modeling of 1–2 September 1859 super magnetic storm. Adv. Space Res. 38, 273–279. https://doi.org/10.1016/j.asr.2005.06/070.

Lin, C.H., Richmond, A.D., Heelis, R.A., Bailey, G.J., Lu, G., Liu, J.Y., Yeh, H.C., Su, S.-Y., 2005. Theoretical study of the low- and midlatitude ionospheric electron density enhancement during the October 2003 superstorm: relative importance of the neutral wind and the electric field. J. Geophys. Res. 110. A12312. https://doi.org/10.1029/2005JA011304.

Lühr, H., Rother, M., Köhler, W., Ritter, P., Grunwaldt, L., 2004. Thermospheric up-welling in the cusp region: evidence from CHAMP observations. Geophys. Res. Lett. 31. L06805. https://doi.org/10.1029/2003GL019314.

Mannucci, A.J., Tsurutani, B.T., Iijima, B.A., Komjathy, A., Saito, A., Gonzalez, W.D., Guarnieri, F.L., Kozyra, J.U., Skoug, R., 2005. Dayside global ionospheric response to the major interplanetary events of October 29–30, 2003 "Halloween Storms" Geophys. Res. Lett. 32. L12S02. https://doi.org/10.1029/2004GL021467.

Manoj, C., Maus, S., 2012. A real-time forecast service for the ionospheric equatorial zonal electric field. Space Weather 10. S09002. https://doi.org/10.1029/2012SW000825.

Marsh, D.R., Solomon, S.C., Reynolds, A.E., 2004. Empirical model of nitric oxide in the lower thermosphere. J. Geophys. Res. 109A07301. https://doi.org/10.1029/2003JA010199.

Millward, G.H., Moffett, R.J., Quegan, S., Fuller-Rowell, T.J., 1996. A coupled thermosphere–ionosphere–plasmasphere model (CTIP). In: Schunk, R.W. (Ed.), STEP Handbook of Ionospheric Models. SCOSTEP, NOAA/NGDC, Boulder, CO, pp. 239–279.

Millward, G.H., Müller-Wodarg, I.C.F., Aylward, A.D., Fuller-Rowell, T.J., Richmond, A.D., Moffett, R.J., 2001. An investigation into the influence of tidal forcing on F region equatorial vertical ion drift using a global ionosphere-thermosphere model with coupled electrodynamics. J. Geophys. Res. 106 (A11), 24,733–24,744. https://doi.org/10.1029/2000JA000342.

Mlynczak, M., Martin-Torres, F.J., Russell, J., Beaumont, K., Jacobson, S., Kozyra, J., Lopez-Puertas, M., Funke, B., Mertens, C., Gordley, L., Picard, R., Winick, J., Wintersteiner, P., Paxton, L., 2003. The natural thermostat of nitric oxide emission at 5.3 μm in the thermosphere observed during the solar storms of April 2002. Geophys. Res. Lett. 30 (21). https://doi.org/10.1029/2003GL017693.

National Research Council (NRC), 1995. Orbital Debris: A Technical Assessment. The National Academies of Science, Washington, DC.

National Research Council, 2012. Continuing Kepler's Quest: Assessing Air Force Space Command's Astrodynamics Standards. The National Academies Press, Washington, DC.

Negrea, C., Codrescu, M.V., Fuller-Rowell, T.J., 2012. On the validation effort of the Coupled Thermosphere Ionosphere Plasmasphere Electrodynamics model. Space Weather 10. S08010. https://doi.org/10.1029/2012SW000818.

Olsen, J.R., Fedrizzi, N., Fuller-Rowell, T.J., Codrescu, M.V., 2016. Validation of CTIPe O/N_2 column density ratios using TIMED/GUVI observations, submitted to Earth Space Sci.

Pawlowski, D.J., Ridley, A.J., 2011. The effects of different solar flare characteristics on the global thermosphere. J. Atmos. Sol. Terr. Phys. 73, 1840–1848. https://doi.org/10.1016/j.jastp.2011.04.004.

Picone, J.M., Hedin, A.E., Drob, D.P., Aikin, A.C., 2002. NRLMSISE-00 empirical model of the atmosphere: statistical comparisons and scientific issues. J. Geophys. Res. 107 (A12), 1468. https://doi.org/10.1029/2002JA009430.

Prölss, G.W., 1997. Magnetic storm associated perturbations of the upper atmosphere. In: Tsurutani, B.T. et al., (Ed.), Magnetic Storms. Geophysical Monograph Series, vol. 98. AGU, Washington, DC, pp. 227–241.

Quegan, S., Bailey, G.J., Moffett, R.J., Heelis, R.A., Fuller-Rowell, T.J., Rees, D., Spiro, R.W., 1982. A theoretical study of the distribution of ionization in the high-latitude ionosphere and the plasmasphere: first results on the mid-latitude trough and the light-ion trough. J. Atmos. Terr. Phys. 44 (7), 619–640. https://doi.org/10.1016/0021-9169(82)90073-3.

Raeder, J., Lu, G., 2005. Polar cap potential saturation during large geomagnetic storms. Adv. Space Res. 36, 1804. https://doi.org/10.1016/j.asr.2004.05.010.

Raeder, J., Wang, Y.L., Fuller-Rowell, T.J., Singer, H.J., 2001a. Global simulations of magnetospheric space weather effects of the Bastille Day storm. Sol. Phys. 204, 325. https://doi.org/10.1023/A:1014228230714.

Raeder, J., Wang, Y.L., Fuller-Rowell, T.J., 2001b. Geomagnetic storm simulation with a coupled, magnetosphere-ionosphere-thermosphere model. In: Song, P., Siscoe, G., Singer, H.J. (Eds.), Space Weather. Geophysical Monograph Series, vol. 125. AGU, Washington, DC, pp. 377–384. https://doi.org/10.1029/GM125p0377.

Richmond, A.D., Roble, R.G., 1987. Electrodynamic effects of thermospheric winds from the NCAR thermospheric general circulation model. J. Geophys. Res. 92 (A11), 12365–12376. https://doi.org/10.1029/JA092iA11p12365.

Ridley, A.J., Deng, Y., Tóth, G., 2006. The global ionosphere thermosphere model. J. Atmos. Sol. Terr. Phys. 68 (8), 839–864. https://doi.org/10.1016/j.jastp.2006.01.008.

Rishbeth, H., Fuller-Rowell, T.J., Rees, D., 1987a. Diffusive equilibrium and vertical motion in the thermosphere during a severe magnetic storm: a computational study. Planet. Space Sci. 35 (9), 1157–1165. https://doi.org/10.1016/0032-0633(87)90022-5.

Rishbeth, H., Fuller-Rowell, T.J., Rodger, A.D., 1987b. F-layer storms and thermospheric composition. Phys. Scripta 36 (2), 327–336. https://doi.org/10.1088/0031-8949/36/2/024.

Schrijver, C.J., Beer, J., Baltensperger, U., et al., 2012. Estimating the frequency of extremely energetic solar events, based on solar, stellar, lunar, and terrestrial records. J. Geophys. Res. 117. A08103. https://doi.org/10.1029/2012JA017706.

Siscoe, G., Raeder, J., Ridley, A.J., 2004. Transpolar potential saturation models compared. J. Geophys. Res. 109. https://doi.org/10.1029/2003JA010318.

Siscoe, G., Crooker, N.U., Clauer, C.R., 2006. D_{st} of the Carrington storm of 1859. Adv. Space Res. 38, 173–179. https://doi.org/10.1016/j.asr.2005.02.102.

Smithtro, C.G., Sojka, J.J., Berkey, T., Thompson, D., Schunk, R.W., 2006. Anomalous F region response to moderate solar flares. Radio Sci. 41. RS5S03. https://doi.org/10.1029/2005RS003350.

Sugiura, M., 1964. Hourly values of equatorial D_{st} for the IGY. Annals of the International Geophysical Year, vol. 35. Pergamon Press, Oxford (24 pp.).

Sutton, E.K., Forbes, J.M., Nerem, R.S., Woods, T.N., 2006. Neutral density response to the solar flares of October and November, 2003. Geophys. Res. Lett. 33. L22101. https://doi.org/10.1029/2006GL027737.

Tapping, K.F., 2013. The 10.7 cm solar radio flux (F10.7). Space Weather. 11. https://doi.org/10.1002/swe.20064.

Temerin, M., Li, X., 2002. A new model for the prediction of Dst on the basis of the solar wind. J. Geophys. Res. 107 (A12), 1472. https://doi.org/10.1029/2001JA007532.

Tribble, A.C., et al., 1999. The space environment. In: Wertz, J.R., Larson, W.J. (Eds.), Space Mission Analysis and Design. third ed. Springer, Netherlands.

Tsurutani, B.T., Gonzalez, W.D., Lakhina, G.S., Alex, S., 2003. The extreme magnetic storm of 1–2 September 1859. J. Geophys. Res. 108 (A7), 1268. https://doi.org/10.1029/2002JA009504.

Tsurutani, B.T., Judge, D.L., Guarnieri, F.L., Gangopadhyay, P., Jones, A.R., Nuttall, J., Zambon, G.A., Didkovsky, L., Mannucci, A.J., Iijima, B., Meier, R.R., Immel, T.J., Woods, T.N., Prasad, S., Floyd, L., Huba, J., Solomon, S.C., Straus, P., Viereck, R., 2005. The October 28 2003 extreme EUV flare and resultant extreme ionosphere effects: comparison to other Halloween events and the Bastille Day event. Geophys. Res. Lett. 32. L03S09. https://doi.org/10.1029/2004GL021475.

Tsurutani, B.T., et al., 2008. Prompt penetration electric fields (PPEFs) and their ionospheric effects during the great magnetic storm of 30–31 October 2003. J. Geophys. Res. 113. A05311. https://doi.org/10.1029/2007JA012879.

Tsurutani, B.T., Verkhoglyadova, O.P., Mannucci, A.J., Lakhina, G.S., Huba, J.D., 2012. Extreme changes in the dayside ionosphere during a Carrington-type magnetic storm. J. Space Weather Space Clim. 2 (A05). https://doi.org/10.1051/swsc/2012004.

Weimer, D.R., 2005. Improved ionospheric electrodynamics models and application to calculating joule heating rates. J. Geophys. Res. 110. A05306. https://doi.org/10.1029/2004JA010884.

FURTHER READING

Heelis, R.A., 2017. Longitude and hemispheric dependences in storm-enhanced density. In: Fuller-Rowell, T., Yizengaw, E., Doherty, P.H., Basu, S. (Eds.), Ionospheric Space Weather—Longitude Dependence and Lower Atmosphere Forcing. Geophysical Monograph Series, vol. 220. AGU, Washington, DC, pp. 61–70.

National Research Council (NRC), 2011. Limiting Future Collision Risk to Spacecraft: An Assessment of NASA's Meteoroid and Orbital Debris Programs. The National Academies of Science, Washington, DC.

Tsurutani, B.T., Lakhina, G.S., Verkhoglyadova, O.P., 2013. Energetic electron (>10 keV) microburst precipitation, ~5–15 s X-ray pulsations, chorus, and wave-particle interactions: a review. J. Geophys. Res. Space Phys. 118, 2296–2312. https://doi.org/10.1002/jgra.50264.

THE EFFECT OF SOLAR RADIO BURSTS ON GNSS SIGNALS

22

Xinan Yue*, Weixing Wan*, Limei Yan*, Wenjie Sun*, Lianhuan Hu*, William S. Schreiner[†]

Chinese Academy of Sciences, Beijing, PR China *University Corporation for Atmospheric Research, Boulder, CO, United States[†]*

CHAPTER OUTLINE

1 Introduction 541
   1.1 The Solar Radio Burst (SRB) 541
   1.2 The Global Navigation Satellite System (GNSS) 543
2 Review the Effect of SRBs on GNSS Signals 544
   2.1 Reduction of Signal-to-Noise Ratio (SNR) 545
   2.2 Signal Loss of Lock (LOL) 545
   2.3 Decrease of Positioning Precision 546
   2.4 Effect on Space-Based GNSS 546
   2.5 Threshold Value of SRBs Affecting GNSS 547
3 Extreme SRB Case on December 6, 2006 547
4 Discussions 551
5 Conclusions 552
Acknowledgments 552
References 552

1 INTRODUCTION
1.1 THE SOLAR RADIO BURST (SRB)

The SRB is the intense solar radio emission from the Sun's atmosphere usually associated with solar flares, during which large amounts of high-energy electrons, ions, and atoms are ejected through the solar atmosphere into space (Kundu, 1965). The solar flare occurrence is a necessary condition for an SRB. However, solar flares are not absolutely accompanied with SRBs, and SRB occurrence does not really depend on the intensity of solar flares (Kundu and Vlahos, 1982). SRB strength is measured in solar flux units (SFU; 1 SFU = 10^{-22} W m^{-2} Hz^{-1}). During violent flares, the solar radio emissions in a wide wavelength range show wild fluctuations and the intensity can exceed 100,000 SFU. SRBs with

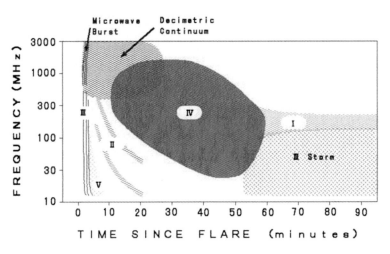

FIG. 1

A schematic diagram of different types of solar radio bursts.

From http://sunbase.nict.go.jp/solar/denpa/hiras/types.html.

an intensity of 1000 SFU can also occur during solar minimum (Yue et al., 2013). The SRBs can be observed in a broad wavelength range from millimeters and centimeters down to meters and decameters, originating from different altitudes of the solar atmosphere. SRBs are generally classified into five distinct types at the dynamical spectrum (Wild et al., 1963; see Fig. 1). A type I burst is the narrowband, short-term (0.1–2 s) burst appearing mainly in the meter region. The corresponding type I storm usually lasts a few hours to days. A type II burst is the plasma emission from the particles accelerated by the magneto hydrodynamics (MHD) shock waves at the levels of the local plasma frequency or its second harmonic. A Type II burst exhibits a slow drift from high to low frequencies, lasting for 10–30 min. Its frequency drift rate relies on the propagation speed of the shock front (with the order of 10^3 km/s). A type III burst is the most frequently occurring burst, with a short duration (~10 s to ~1 min) and a rapid drift from high to low frequencies in the decimeter and meter wavelength range. It is due to plasma emission generated by the beams of very fast (high-energy) electrons at the levels of the local plasma frequency or its second harmonic. A type IV burst is the long-period (from less than 1 h to several days) continuum radiation that follows a type II burst in the meter region with extension to decimeter and microwave ranges. Its radio sources could be either stationary or moving. A type V burst is a diffuse broadband quasicontinuous emission, following some type III bursts at frequencies less than ~200 MHz. It is assumed to be produced by the gyro-synchrotron radiation from high-energy electrons trapped in a magnetic loop; it can last 0.5–3 min. Besides these typical five types of SRBs mentioned above, there are also many fine structures superimposed on the solar radio emission, such as pulsations and spikes. The pulsations usually appear at periods from subseconds to several minutes and have been observed spectrographically at frequencies up to 2000 MHz (Chernov, 2011). Magnetic flux tube oscillations, periodic self-organizing systems of plasma instabilities, or the repetitive injection of particle streams from the source have been considered as basic mechanisms (Aschwanden, 1987). The spikes (as shown in Fig. 1) are isolated as a special kind of the shortest (less than

~100 ms) and narrow-band bursts, which are most prevailing in the decimeter wave bands. The peak flux observed in low time resolution is probably composed of a lot of spikes. Spikes are generally agreed to be a nonthermal, coherent emission closely connected with the particle acceleration and energy release in solar flares (Benz, 1986).

The SRBs are commonly circularly polarized. They can be either righthanded or lefthanded, depending on the magnetic field on the sun and the direction of wave propagation. However, random, linear, and elliptical polarizations may also occur. Type II bursts are even nonpolarized. Type I bursts, the stationary type IV bursts, pulsations, and spikes are strongly circularly polarized. The moving type IV bursts show weak circular polarization. Type III bursts have been found to be moderately elliptically polarized. The SRB is mainly monitored in solar observatories all over the world, frequently using techniques that include radio interference monitoring and the solar radio spectrometer, which could measure the radio wave flux from the Sun in either a specific frequency or continuous spectrum. Currently, the widely used SRB data come from the radio solar telescope network (RSTN), which is maintained and operated by the U.S. Air Force Weather Agency (USAFWA) and consists of ground-based observatories in Australia, Italy, Massachusetts, New Mexico, and Hawaii. Unless there is a special note, the SRB data used in this chapter always come from RSTN.

1.2 THE GLOBAL NAVIGATION SATELLITE SYSTEM (GNSS)

A GNSS consists of a space segment, a control segment, and a user segment. The space segment utilizes a constellation of 20–30+ satellites, mainly in medium Earth orbit (MEO) and spread between several orbital planes to transport L-band radio waves at multiple frequencies. This can then be used to derive the distance between the satellite and receiver (time of signal transfer) and therefore a satellite's location. The GNSS orbital data are transmitted in a data message that is superimposed on a code that serves as a timing reference. The satellite uses an atomic clock to maintain synchronization of all the satellites in the constellation. The GNSS signals are transmitted in righthand circular polarized (RHCP), but might change the polarization when passing through severe rain storms or reflection from ocean and ice surfaces. GNSS has been widely used in navigation, positioning, and timing (Leick et al., 2015). Since the signals pass through Earth's atmosphere and ionosphere, it could also be used to derive some environmental parameters such as water vapor and total electron content (TEC) along the ray. Furthermore, several new technologies based on GNSS signals have been developed recently and have shown great impact in related fields, such as ocean reflection and radio occultation (RO) (Anthes, 2011; Larson et al., 2014).

As of December 2016, only the U.S. Global Positioning System (GPS) and the Russian GLObal NAvigation Satellite System (GLONASS) are globally operational GNSSs (Leick et al., 2015; Urlichich et al., 2010). The European Union's Galileo is under development (http://www.esa.int). China is in the process of expanding its regional BeiDou Navigation Satellite System into the global Compass navigation system by 2020 (Yang et al., 2011). Several other regional navigation systems such as Japan's Quasi-Zenith Satellite System (QZSS) and India's Navigation with Indian Constellation (NAVIC) are also under development. There are large similarities in terms of mechanisms among those systems with the exceptions of carrier frequency and coding. GNSS technology is now widely used in industry, agriculture, transportation, aviation, maritime, land resources, disaster relief, etc. In other words, the GNSS is closely integrated into our modern economy. However, the GNSS signals could be oscillated and have less accuracy when passing through the ionosphere and atmosphere,

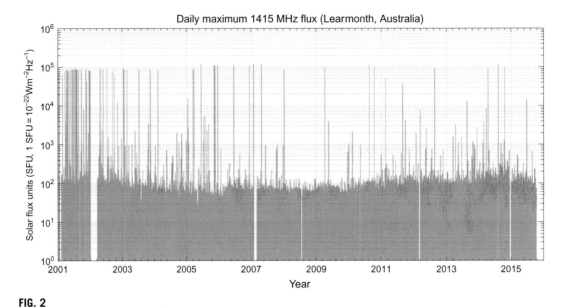

FIG. 2

Daily maximum radio wave flux in 1415 MHz in SFU (solar flux unit, 1 SFU = $10-22$ W m^{-2} Hz^{-1}) observed by Learmonth solar radio telescope during 2001–15. The gaps indicate no observations.

especially by ionospheric irregularities and tropospheric water vapor. In severe cases, the signal tracking can directly lose lock from minutes to hours (Blewitt, 1990; Yue et al., 2016).

In this chapter, we will review the effect of SRBs on GNSS signals. The occurrences of SRBs are far less than that of ionospheric irregularity. However, as will be noted below, the effect is more direct and more profound. For reference, we plotted the daily maximum radio wave flux in 1415 MHz observed by the Learmonth solar radio telescope from RSTN during 2001–15 in Fig. 2. We select the 1415 MHz frequency here because it is the most likely source of interference for the GNSS L-band signal. We can see many visible SRB events, with a higher occurrence rate during solar maximum. Specifically, there are a total of 122 events with a daily maximum 1415 MHz flux larger than 1000 SFU during the selected time period.

In Section 2 we will review previous investigations into the effects of SRBs on GNSS signals. Then we will show one extreme case in Section 3. We will make some necessary discussions in Section 4. Section 5 will summarize the whole chapter.

2 REVIEW THE EFFECT OF SRBs ON GNSS SIGNALS

To our knowledge, it is Klobuchar et al. (1999) who first pointed out that the SRB has the potential to affect GPS signals. After that, many researchers have studied SRB effects on GNSS in more detail, including receivers on both the ground and in low Earth orbit (LEO), in terms of theoretical estimation and real data analysis. Hereafter we will summarize these investigations in different categories.

2.1 REDUCTION OF SIGNAL-TO-NOISE RATIO (SNR)

In almost all the cited literature, researchers have reported the effects of SRBs on GNSS in terms of reduction of the signal-to-noise ratio (SNR) or the carrier-to-noise ratio (C/No) (Afraimovich et al., 2008, 2009; Carrano et al., 2009; Cerruti et al., 2006, 2008; Chen et al., 2005; Demyanov et al., 2012; Muhammad et al., 2015; Sreeja et al., 2013, 2014; Yue et al., 2013). Cerruti et al. (2006) first reported a 2.3 dB C/No decrease across all visible satellites in several American-region GPS receivers on the SRB event of September 7, 2005, when the peak value of 1600 MHz radio wave flux was ∼8700 SFU. They further studied a much larger SRB event that occurred on October 28, 2003, and found a deduction of ∼3 dB on GPS L1 and ∼10 dB on the semicodeless L2 signal. During several days in December—the 5th, 6th, 13th and 14th—a series of four X-class solar flares originated from the same active sunspot region 10,903 and resulted in three significant SRB events on December 6, December 13, and December 14, respectively. The maximum peak flux in the L band exceeded 1,000,000 SFU during the December 6, 2006, case. A decrease in SNR and/or C/No during these SRB events has been reported for globally distributed ground-based GNSS receivers (Afraimovich et al., 2008, 2009; Carrano et al., 2009; Cerruti et al., 2008; Demyanov et al., 2012) and even LEO-based RO receivers (Yue et al., 2013). By using high-rate GPS receivers from the Air Force Research Laboratory (AFRL) Scintillation Network Decision Aid (SCINDA) global network, Carrano et al. (2009) investigated the C/No reduction during these SRB events. They observed a C/No decrease exceeding 25 dB during the December 6 case. They also pointed out that the C/No reduction depends on the solar incidence angle due to the anisotropy of the antenna gain and derived an expression for the vertical equivalent C/No reduction due to an SRB. Huang et al. (2015) also studied the performance of GNSS receivers over the Asia-Pacific Ocean region during the December 13, 2006, case in terms of amplitude and the phase scintillation index rather than SNR reduction. They found that the SRB could result in amplitude scintillation up to 0.8 in this case due to the increased noise in the observations; no significant phase scintillation was observed. Using GPS RO observations from the Constellation Observing System for Meteorology, Ionosphere, and Climate (COSMIC), Yue et al. (2013) observed an average SNR reduction of 8% and 26% for L1 and L2 signals, respectively, during the December 6, 2006, SRB occurrence. Another frequently studied case is the SRB on September 24, 2011, during which a maximum 1415 MHz radio wave flux exceeded 100,000 SFU initiated by an M7.1 soft X-ray flare (Sreeja et al., 2013, 2014; Muhammad et al., 2015). Sreeja et al. (2013) evaluated the performance of GNSS receivers in the European and Latin American sectors during this case. Maximum reductions of 11, 22, and 10 dB in the C/No of the GPS L1C/A, L2P, and L2C signals, respectively, were observed. The C/No reduction is modulated by the local solar incidence angle for the GPS L1 and L2P signals, whereas such modulation was not observed for the GPS L2C signal. For the same case, Muhammad et al. (2015) observed a maximum reduction of 13 and 24 dB on L1 and L2P C/No, respectively, based on a slightly different data source. The average GPS RO SNR reduction was ∼4% and 7% for L1 and L2P signals during this case (Yue et al., 2013).

2.2 SIGNAL LOSS OF LOCK (LOL)

Significant reduction in SNR could result in instantaneous LOL (such as cycle slips in phase) or long-duration LOL from seconds to minutes. During the December 6, 2006, case, Cerruti et al. (2008) observed that the number of tracked GPS satellites decreased from more than 10 to less than 4 in several

Federal Aviation Administration (FAA) Wide Area Augmentation System (WAAS) stations. This lasted for more than 10 min, degrading the positioning performance significantly. Carrano et al. (2009) also found several minutes of LOL in AFRL-SCINDA observations for the same case. During December 13, Huang et al. (2015) observed up to ~6 min LOL over the Asia-Pacific Ocean region. Afraimovich et al. (2008, 2009) statistically analyzed the cycle slips and count omissions during the December 6 and December 13, 2006, SRB occurrences, using the U.S. Continuously Operating Reference Station (CORS) network and the Japanese GPS Earth Observation Network System (GEONET) data, respectively. A maximum slip rate of ~18.5% was found, which was 50 times larger than the reference background during December 6, 2006. During a 30 s observational period, the maximum LOL from the GPS PRN (Pseudo-Random Noise) 12 and 24 of CORS stations reached 82% and 69% on December 6. While on December 13, this value of GEONET stations was 50% and 39% for GPS PRN 28 and 8, respectively. During the September 24, 2011, SRB case, 14 out of 26 International GNSS Service (IGS) networks and European Geostationary Navigation Overlay System (EGNOS) Ranging Integrity Monitoring Stations (RIMS) network stations experienced less than four dual-frequency measurements, which resulted in large positioning errors in both vertical and horizontal directions (Muhammad et al., 2015).

2.3 DECREASE OF POSITIONING PRECISION

Several investigations have pointed out that the reduction in C/No and LOL could result in a larger positioning error or even positioning failure during an SRB occurrence (Afraimovich et al., 2008, 2009; Carrano et al., 2009; Demyanov et al., 2013; Huang et al., 2015; Muhammad et al., 2015; Sreeja et al., 2013, 2014). According to Carrano et al. (2009), peak positioning errors in the horizontal and vertical directions over the AFRL-SCINDA network reached 20 and 60 m, respectively during the December 6, 2006, SRB case. During the minor SRB case of September 24, 2011, Sreeja et al. (2014) analyzed the performance of 12 receivers from the Fugro network that had a reset in the position filter, and the maximum horizontal position error varied between 0.5 and 2.2 m. Out of these 12 locations, the position error is greater than 1.2 m for 5 locations. They also pointed out that although the position degradation due to the SRB lasted only for a few minutes, this had serious implications for high-accuracy (accuracies of the order of 10–20 cm) real-time applications that rely on the continuously available data of the specified quality. In addition, Sreeja et al. (2013) suggested that the degradation in the positioning accuracy at the peak of the radio burst is correlated with the local solar incidence angle. Based on IGS and RIMS network data and standard point positioning goGPS software, Muhammad et al. (2015) got a maximum positioning error of 303 m and 55 m in vertical and horizontal directions, respectively, during the September 24, 2011, SRB occurrence.

2.4 EFFECT ON SPACE-BASED GNSS

Yue et al. (2013) have evaluated the SRB effects on LEO-based GPS RO signals through both case study and statistical analysis based on RO observations from multiple missions during 2006–12, processed and archived in the University Corporation for Atmospheric Research (UCAR) COSMIC Data Analysis and Archive Center (CDAAC). Except SNR reduction and LOL mentioned above, they also evaluated the SRB effect on the RO data products. The LEO RO signals show frequent LOL, a simultaneous decrease on L1 and L2 SNR globally during daytime, small-scale perturbations of SNR, and decreased successful retrieval percentage (SRP) for both ionospheric and atmospheric occultations during SRB occurrence. Potential harmonic band interference was identified. They pointed out that either decreased data volume or data quality will influence weather prediction, climate study, and space weather monitoring by using RO data during SRB events. Statistically, the SRP of ionospheric and atmospheric occultation retrieval shows ~4% and ~13% decreases, respectively, while the SNR of L1 and L2 show ~5.7% and ~11.7% decreases, respectively.

2.5 THRESHOLD VALUE OF SRBs AFFECTING GNSS

Klobuchar et al. (1999) made a theoretical estimation and concluded that an SRB level of 40,000 SFU could result in an observable decrease in SNR and a SRB level of 200,000 SFU could be of serious concern to GNSS users. However, Chen et al. (2005) indicated that a lower threshold value of 4000–12,000 SFU should be adopted for codeless or semicodeless dual-frequency GPS receivers based on case study and correlation analysis. According to an analysis of the December 6, 2006, case, Afraimovich et al. (2009) derived a threshold value of less than 20,000 SFU at which GPS receiver failures occur at a high zenith angle. Through theoretical analysis, Demyanov et al. (2012) found that the solar radio emission power level of 1000 SFU or higher can cause GPS/GLONASS signal tracking failures, especially at the L2 frequency, which is much lower than the previous estimations. They then confirmed their estimation by analyzing the October 28, 2003, case and found that the selected GPS receiver began to degrade when 1420 MHz flux exceeding 1000 SFU. Demyanov et al. (2013) further made a much more detailed theoretical analysis of the threshold value of SRBs affecting GNSS, considering the tracking and filtering algorithms in the receiver. Based on their calculation, the encrypted signal tracking will fail when the solar radio noise powers are about 4000 SFU and 10,000 SFU at frequencies L2 and L1, respectively. The corresponding power levels that can cause the failure of the GLONASS high-precision signal tracking at frequencies L1 and L2 are 10,000 and 13,500 SFU, respectively. The situation is the worst when the "codeless" processing of encrypted signals is utilized. Yue et al. (2013) made a statistical analysis on multiple GPS RO data during 2006–12. They manually identified six SRB cases that have visible SNR reduction in both the L1 and L2 bands. Among these cases, the minimum SRB has maximum and average flux of 29,763 and 1807 SFU in 1415 MHz, initiated by an M7-class solar flare. So statistically, GPS RO signals could be degraded by SRBs when the average flux exceeds 1807 SFU. Furthermore, they also found that the GPS RO signal response is more sensitive to the time change of the SRB burst rather than the amplitude of the SRB flux when the flux exceeds a certain level. This might be related to the tracking algorithm in the receiver. There are several reasons that this value is lower than the investigations above. First, the LEO-based signal is not influenced by the lower atmosphere due to the attenuation and polarization loss (Demyanov et al., 2013), which is different from ground receivers used in the above studies. Second, the RO signal has a higher time resolution (1–50 Hz sampling rate) and the LEO receiver moves fast, which might result in much higher sensitivity of the receiver to the radio noise.

3 EXTREME SRB CASE ON DECEMBER 6, 2006

In all the literature mentioned above, there are eight SRB cases in total in which GNSS signals have shown visible decay, with the December 6, 2006, case being the biggest one. That SRB event was initiated by an X6.5-class solar flare; the peak value of 1415 MHz flux exceeded 1,000,000 SFU. It lasted for almost 1 h. Detailed solar radio flux evolution during this event in different frequencies

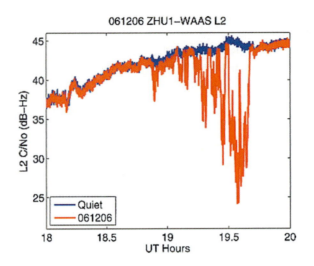

FIG. 3

Response of the WAAS ZHU1 receiver at L2 frequency to the December 6, 2006, solar radio burst. The *blue line* and *red line* represent the value for the reference quiet and SRB occurrence, respectively.

From Cerruti, A.P., Kintner Jr., P.M., Gary, D.E., Mannucci, A.J., Meyer, R.F., Doherty, P., Coster, A.J., 2008. Effect of intense December 2006 solar radio bursts on GPS receivers. Space Weather. 6, S10D07. doi: 10.1029/2007SW000375.

can be found in Yue et al. (2013). This case has been investigated widely by the community (Afraimovich et al., 2008, 2009; Carrano et al., 2009; Cerruti et al., 2008; Demyanov et al., 2013; Yue et al., 2013). Significant SNR reduction and LOL occurrence were found all over the world in both ground- and LEO-based GNSS receivers. Larger positioning errors and even positioning failure during this SRB case were also found. In addition, Yue et al. (2013) also indicated that this SRB event resulted in degraded GPS RO data products and therefore influenced the corresponding applications of numerical weather prediction (NWP) and space weather monitoring. As a typical SRB event, we will show several figures hereafter to demonstrate the SRB effect on GNSS signals.

Fig. 3 was adapted from Cerruti et al. (2008). It showed the GPS L2 C/No variations from one WAAS station during this SRB case and, for comparison, a reference quiet value. The reduction of L2 C/No was up to 20 dB. In addition, the corresponding L1 C/No decreased ~17 dB and similar fades were observed at all sunlit WAAS receivers (figures not shown, see Cerruti et al. (2008) for details).

In Fig. 4, we plotted the universal time (UT) variation of 1415 MHz radio flux and its time rate of change (A and B), and L1 and L2 SNR from precise orbit determination (POD) antennas for all the available LEO satellites (C–I). Note that the sampling rate of COSMIC POD data is ~1 Hz, while it is 0.1 Hz for CHAMP and GRACE-A POD data. The SNRs are plotted versus UT for all GPS PRNs and LEO-GPS elevation angles. Besides the SRB interference effect in this specific case, the LEO-observed SNR is mainly determined by the GPS transmitter gain pattern, receiver antenna gain pattern, and the influence of the ionosphere and atmosphere including absorption, multipath, and irregularities. Obvious LOL occurs frequently in time scales of seconds to minutes. During UT ~19.5–19.8, COSMIC FM6 lost most GPS signals. The COSMIC FM1 has a total LOL lasting ~1 min since UT 19.4. The CHAMP POD antenna only tracks 1 GPS around UT 19.3. A prominent feature illustrated in the figure

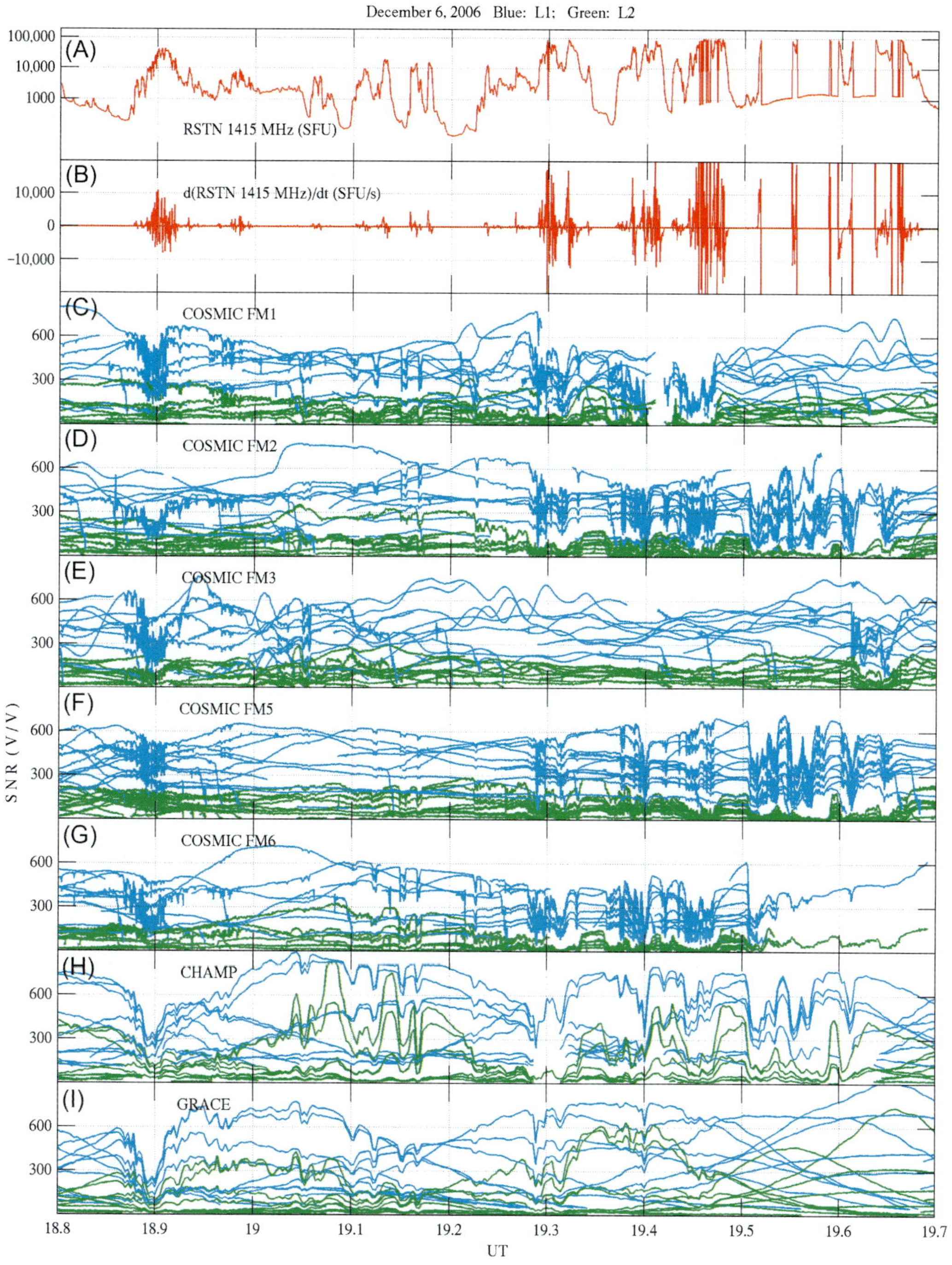

FIG. 4

Time variations around the December 6, 2006, SRB event (18.5–20 UT) of (A) The RSTN-observed solar radio flux at 1415 MHz in SFU; (B) Time change rates of RSTN 1415 MHz radio flux in unit of SFU/s; (C–I) signal-to-noise ratio (SNR) on L1 *(blue)* and L2 *(green)* channels observed by the precise orbit determination (POD) antennas of COSMIC (C) FM1, (D) FM2, (E) FM3, (F) FM5, (G) FM6, (H) CHAMP, and (I) GRACE-A. Except COSMIC FM5, all other LEOs fly from daytime to nighttime during this SRB interval.

From Yue, X., et al., 2013. The effect of solar radio bursts on the GNSS radio occultation signals. J. Geophys. Res. Space Physics. 118, 5906–5918. doi: 10.1002/jgra.50525.

is the oscillation of SNR in both L1 and L2 channels. In combination with the RSTN and its time change rates as shown in panels A and B, we can see that the SNR decreases simultaneously in both L1 and L2 channels and all antennas globally along with the RSTN 1415 MHz flux increase. When the satellites fly over the midnight time period, this oscillation is small or totally disappears. These cases include the FM1 during UT 19.5–19.7, FM3 during UT 19.0–19.6, CHAMP during 19.62–19.70, and GRACE-A during 19.42–19.70. During sunset or sunrise periods, the oscillations only occur in the signals from the daytime GPS PRNs. During the time interval of UT 18.5–20.0, the mean SNR of all LEO satellites decreased 8% and 26% in comparison with the daily mean value for L1 and L2, respectively.

Intense oscillations and sharp decreases in SNR will cause oscillations in both the pseudorange and carrier-phase observations. These observations' oscillations will therefore reduce the quality of the RO-derived data. As an example, in Fig. 5 we compared a quiet time (1272 UT) ionospheric RO event with one that occurred during an SRB event (1938 UT) by the same LEO (COSMIC FM2)-GPS (PRN 6) pair during December 6, 2006. Both the GPS L1 SNR (A) and the Abel-retrieved electron density profile (EDP, B) are shown here. Two cases are distinguished by the colors. As can be seen, both low-value and fluctuating SNR lead to a noisy EDP during the SRB time. By the way, we did not

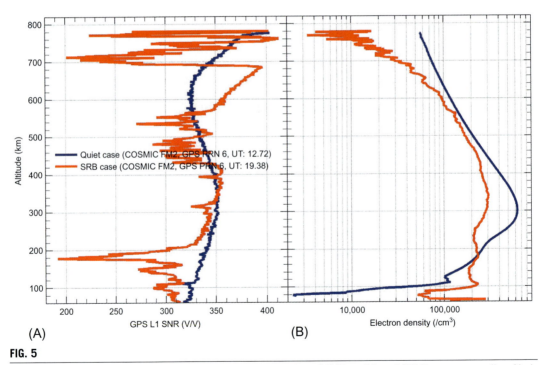

FIG. 5

(A) Altitude (impact height of *straight line* of GPS ray) variation of SNR on L1 and (B) the corresponding Abel-retrieved electron densities. A quiet (*blue line*, 12.72 UT) case and an SRB (*red line*, 19.38 UT) case observed by COSMIC FM2 from GPS PRN 6 during December 6, 2006, are shown.

From Yue, X., et al. 2013. The effect of solar radio bursts on the GNSS radio occultation signals. J. Geophys. Res. Space Physics. 118, 5906–5918. doi: 10.1002/jgra.50525.

see LOL over this time period. The SRB EDP deviates from the normal quiet time EDP very much, especially in the topside and bottomside regions. We have ruled out effects from other sources such as geomagnetic disturbance. The altitude variation of the EDP resulted from degraded SNR was unrealistic and therefore useless in this case. We also have shown an example where the RO event in the lower atmosphere was useless due to the SNR loss (Yue et al., 2013).

4 DISCUSSIONS

Although the monitoring of SRBs has more than 40 years of history, the potential threat of SRBs on GNSS operations was first identified by Chen et al. (2005) during the October 28, 2003, SRB case and then realized by the whole community by the intense series of SRBs in December 2006 (Afraimovich et al., 2008, 2009; Carrano et al., 2009; Cerruti et al., 2008; Demyanov et al., 2013; Yue et al., 2013). Kintner et al. (2009) tried to evaluate and estimate the effects of SRBs throughout the history of GNSS. However, they discovered inconsistencies and lapses within the RSTN data set, making it difficult to determine the true number of intense ($>$150,000 SFU) SRBs that may have occurred during the last 40 years. Nita et al. (2002) have statistically analyzed the SRB events during 1960–99 based on available observations all over the world with certain manual calibrations. According to their analysis, the SRB events with a maximum peak flux larger than 1000 SFU varies from 15–30 per solar cycle during solar cycles of 20, 21, and 22 for frequency range of 1–1.7 GHz. This estimation should be doubled considering that the available observations were lost for at least 50% of the SRB events. In addition, except for the radio frequency, they did not indicate any significant dependence of SRB occurrence on solar cycle and cycle phase. According to our plots in Fig. 2, there are 122 SRB events in total with daily maximum 1415 MHz radio flux larger than 1000 SFU during 2001–15, which means 84 per solar cycle. Our results generally agree with Nita et al. (2002) when taking the lost observations into account in their analysis and considering that our data might contain some wild points. During 2003–12, there were eight SRB events that showed degrading effects on GNSS signals in the literature. Roughly, the SRB occurrence ratio with negative effects is 8.8 per solar cycle. This occurrence rate is not significantly high in comparison with other space weather events such as solar proton events and geomagnetic storms (Riley, 2012). However, as indicated in Section 2, the SRBs could totally disturb the GNSS operations through direct signal interference immediately after occurrence. Given that there will be more GNSS satellites operating in the future, SRB effects should be given more attention by both the science and technology communities. In addition, GNSS receiver manufacturers should take into account the SRB effect in their design and production of receivers.

As summarized in Section 2, the SRBs could influence the GNSS signals in terms of SNR reduction and signal LOL. Degradation in SNR will increase the uncertainty and decrease the quality of the observed parameters such as pseudorange and phase. So theoretically, all applications based on these observations will be influenced, including RO as indicated by Yue et al. (2013) and precise GNSS positioning as demonstrated by Carrano et al. (2009). Other applications based on GNSS technology, such as NWP, space weather monitoring, geodesy, crustal deformation monitoring, agriculture, industry, military, transportation, traffic, telecommunication, etc., although not reported, can be influenced during severe SRB occurrences. The degraded performance of WAAS during SRB occurrences, as shown by Cerruti et al. (2008), will definitely cause economic loss for aviation. In modern society, human lives rely more and more on mobile intelligent terminals such as cell phones, which are usually equipped with GNSS receivers for positioning, navigation, and timing. Thus, we can expect that serious SRBs might also disorganize human lives and therefore cause economic loss. In addition to GNSS signals, other radio waves could also be disturbed by SRB occurrences, such as wireless cell sites as reported by Bala et al. (2002).

5 CONCLUSIONS

In this chapter, we first introduced the SRB phenomena and the GNSS system. Then we reviewed the previous investigations related to SRB effects on GNSS signals. Then we demonstrated the GNSS signal response to the SRB occurrence during an extreme SRB event that occurred on December 6, 2006. Several conclusions can be obtained:

1. SRBs could result in SNR reduction and instantaneous or long-period loss of lock on the GNSS signals and therefore decrease the observation quality, which subsequently will influence all the applications based on these observations such as RO technique and precise GNSS positioning. SRBs will mainly affect the stations located in the sunlit hemisphere during radio flux enhancement, while the influence strength depends on the solar incidence angle, antenna pattern, tracking algorithm, and some other factors.
2. The SRB occurrence does not really depend on the intensity of solar flares. The threshold value of SRB flux value that could result in visible effect on GNSS signals is believed to be between 1000 and 10,000 SFU in the L band. During 2003–12, there were eight SRB events that showed degrading effects on GNSS signals in the literature, which is 8.8 per solar cycle. Although the SRB occurrence ratio is not significantly high, we should pay sufficient attention to its side effects on modern society. Significant SRBs could occur during solar minimum.

ACKNOWLEDGMENTS

Xinan Yue acknowledges the support of the Thousand Young Talent Program of China. The USAF RSTN solar radio wave data were downloaded from NOAA/NGDC ftp server (ftp://ftp.ngdc.noaa.gov). The multiple GPS radio occultation data, including COSMIC, CHAMP, and GRACE, were processed and accessed through the University Corporation for Atmospheric Research (UCAR) COSMIC Data Analysis and Archive Center (CDAAC).

REFERENCES

Afraimovich, E.L., Demyanov, V.V., Ishin, A.B., Smolkov, G.Ya., 2008. Powerful solar radio bursts as a global and free tool for testing satellite broadband radio systems, including GPS–GLONASS–GALILEO. J. Atmos. Sol. Terr. Phys. 70 (15), 1985–1994. https://doi.org/10.1016/j.jastp.2008.09.008.

Afraimovich, E.L., Demyanov, V.V., Smolkov, G.Ya., 2009. The total failures of GPS functioning caused by the powerful solar radio burst on December 13, 2006. Earth Planets Space 61 (5), 637–641.

Anthes, R.A., 2011. Exploring Earth's atmosphere with radio occultation: contributions to weather, climate and space weather. Atmos. Meas. Tech. 4, 1077–1103. https://doi.org/10.5194/amt-4-1077-2011.

Aschwanden, M.J., 1987. Theory of radio pulsations in coronal loops. Sol. Phys. 111, 113–136.

REFERENCES

Bala, B., Lanzerotti, L.J., Gary, D.E., Thomson, D.J., 2002. Noise in wireless systems produced by solar radio bursts. Radio Sci. 37, 1018. https://doi.org/10.1029/2001RS002481.

Benz, A.O., 1986. Millisecond radio spikes. Sol. Phys. 104, 99–110. https://doi.org/10.1007/BF00159950.

Blewitt, G., 1990. An automatic editing algorithm for GPS data. Geophys. Res. Lett. 17, 199–202. https://doi.org/10.1029/GL017i003p00199.

Carrano, C.S., Bridgwood, C.T., Groves, K.M., 2009. Impacts of the December 2006 solar radio bursts on the performance of GPS. Radio Sci. 44. RS0A25, https://doi.org/10.1029/2008RS004071.

Cerruti, A.P., Kintner, P.M., Gary, D.E., Lanzerotti, L.J., de Paula, E.R., Vo, H.B., 2006. Observed solar radio burst effects on GPS/Wide Area Augmentation System carrier-to-noise ratio. Space Weather 4. S10006, https://doi.org/10.1029/2006SW000254.

Cerruti, A.P., Kintner Jr., P.M., Gary, D.E., Mannucci, A.J., Meyer, R.F., Doherty, P., Coster, A.J., 2008. Effect of intense December 2006 solar radio bursts on GPS receivers. Space Weather 6. S10D07, https://doi.org/10.1029/2007SW000375.

Chen, Z., Gao, Y., Liu, Z., 2005. Evaluation of solar radio bursts' effect on GPS receiver signal tracking within International GPS Service network. Radio Sci. 40. RS3012, https://doi.org/10.1029/2004RS003066.

Chernov, G.P., 2011. Fine Structures of Solar Radio Bursts. Springer, Berlin.

Demyanov, V.V., Afraimovich, E.L., Jin, S., 2012. An evaluation of potential solar radio emission power threat on GPS and GLONASS performance. GPS Solutions 16 (4), 411–424. https://doi.org/10.1007/s10291-011-0241-9.

Demyanov, V.V., Yasyukevich, Y.V., Jin, S., 2013. Effects of solar radio emission and ionospheric irregularities on GPS/GLONASS performance. In: Jin, S. (Ed.), Geodetic Sciences-Observations, Modeling and Applications. ISBN: 978-953-51-1144-3. (Chapter 5) Published: May 29, 2013 under CC BY 3.0 license. InTech, https://doi.org/10.5772/3439.

Huang, W., Ercha, A., Liu, S., Shen, H., Chen, Y., 2015. Effect of the 13 December 2006 solar radio burst on GPS observations. Chin. J. Space Sci. 35 (6), 679–686. https://doi.org/10.11728/cjss2015.06.679 (in Chinese).

Kintner Jr., P.M., O'Hanlon, B., Gary, D.E., Kintner, P.M.S., 2009. Global positioning system and solar radio burst forensics. Radio Sci. 44. RS0A08, https://doi.org/10.1029/2008RS004039.

Klobuchar, J.A., Kunches, J.M., Van Dierendonck, A.J., 1999. Eye on the ionosphere: potential solar radio burst effects on GPS signal to noise. GPS Solut. 3 (2), 69–71.

Kundu, M.R., 1965. Solar Radio Astronomy. Interscience Publishers, New York.

Kundu, M.R., Vlahos, L., 1982. Solar microwave bursts—a review. Space Sci. Rev. 32, 405.

Larson, K.M., Small, E.E., Braun, J.J., Zavorotny, V.U., 2014. Environmental sensing: a revolution in GNSS applications. Inside GNSS 9 (4), 36–46.

Leick, A., Lev Rapoport, L., Tatarnikov, D., 2015. GPS Satellite Surveying, fourth ed. Wiley.

Muhammad, B., Alberti, V., Marassi, A., Cianca, E., Messerotti, M., 2015. Performance assessment of GPS receivers during the September 24, 2011 solar radio burst event. J. Space Weather Space Clim. 5, A32. https://doi.org/10.1051/swsc/2015034.

Nita, G.M., Gary, D.E., Lanzerotti, L.J., Thomson, D.J., 2002. The peak flux distribution of solar radio bursts. Astrophys. J. 570, 423–438.

Riley, P., 2012. On the probability of occurrence of extreme space weather events. Space Weather 10. S02012, https://doi.org/10.1029/2011SW000734.

Sreeja, V., Aquino, M., de Jong, K., 2013. Impact of the 24 September 2011 solar radio burst on the performance of GNSS receivers. Space Weather 11, 306–312. https://doi.org/10.1002/swe.20057.

Sreeja, V., Aquino, M., de Jong, K., Visser, H., 2014. Effect of the 24 September 2011 solar radio burst on precise point positioning service. Space Weather 12, 143–147. https://doi.org/10.1002/2013SW001011.

Urlichich, Y., Subbotin, V., Stupak, G., Dvorkin, V., Povaliaev, A., Karutin, S., 2010. GLONASS developing strategy. In: Proceedings of the 23rd International Technical Meeting of The Satellite Division of the Institute of Navigation (ION GNSS 2010), Portland, OR, September 2010. pp. 1566–1571.

Wild, J.P., Smerd, S.F., Weiss, A.A., 1963. Solar bursts. Annu. Rev. Astron. Astrophys. 1, 291–366.
Yang, Y., Li, J., Xu, J., Tang, J., Guo, H., He, H., 2011. Contribution of the compass satellite navigation system to global PNT users. Chin. Sci. Bull. 56 (26), 2813–2819. https://doi.org/10.1007/s11434-001-4627-4.
Yue, X., et al., 2013. The effect of solar radio bursts on the GNSS radio occultation signals. J. Geophys. Res. Space Phys. 118, 5906–5918. https://doi.org/10.1002/jgra.50525.
Yue, X., Schreiner, W.S., Pedatella, N.M., Kuo, Y.-H., 2016. Characterizing GPS radio occultation loss of lock due to ionospheric weather. Space Weather 14, 285–299. https://doi.org/10.1002/2015SW001340.

CHAPTER 23

EXTREME IONOSPHERIC STORMS AND THEIR EFFECTS ON GPS SYSTEMS

Geoff Crowley, Irfan Azeem
Atmospheric & Space Technology Research Associates (ASTRA), Boulder, CO, United States

CHAPTER OUTLINE

1 Introduction 555
2 Global Positioning System 557
3 The Ionosphere 558
   3.1 Total Electron Content 559
   3.2 Low-Latitude Scintillation 560
   3.3 High-Latitude Scintillation 561
4 Ionospheric Structures Evident in TEC Data 565
   4.1 Storm Enhanced Densities, Patches and Blobs 565
   4.2 Traveling Ionospheric Disturbances 569
5 Event Studies for Large Ionospheric Storms 572
   5.1 October 2003 574
   5.2 November 2003 574
   5.3 November 2004 575
6 System Effects of Ionospheric Storms 575
7 Discussion 578
Acknowledgments 580
References 580
Further Reading 586

1 INTRODUCTION

The ionosphere is a region of the Earth's upper atmosphere from about 80 km to over 1000 km altitude that constitutes the boundary between Earth and interplanetary space. It is highly influenced by the sun and the plasma processes occurring within the Earth's magnetic field environment, as well as atmospheric waves and other disturbances propagating from the troposphere to high altitudes. The maximum electron density in the ionosphere is normally found in the F-region between approximately 150 and 450 km. However, under some circumstances, the maximum electron density can occur in the E-region between about 100 and 150 km, either as a result of energetic particle precipitation in the auroral region, or because the F-region has decayed. Because of its plasma properties, the ionosphere interacts with a broad range of electromagnetic waves at frequencies that are important to civilian and military activities. Moreover, since the processes that influence the ionosphere vary over time scales from seconds to years, it continues to be a challenge to adequately predict its behavior in many circumstances.

Ionospheric variability is an important aspect of space weather, which refers to conditions in space (the Sun, solar wind, magnetosphere, ionosphere, or thermosphere) that can influence the performance and reliability of space-borne and ground-based technological systems. It could be argued that the most important region of variability in space weather is the ionosphere because so many applications either depend on ionospheric space weather for their operation (HF communication, over-the-horizon or OTH radar), or can be deleteriously affected by ionospheric conditions (e.g., GNSS navigation and timing, UHF satellite communications, synthetic aperture radar, HF communications, the electric power grid). As the global community becomes ever more reliant on technology, the threats from space weather increase. What is not well known is the immense degree of inter-connectedness of these technological systems, as displayed for example in Fig. 1 for Global Positioning System (GPS) applications.

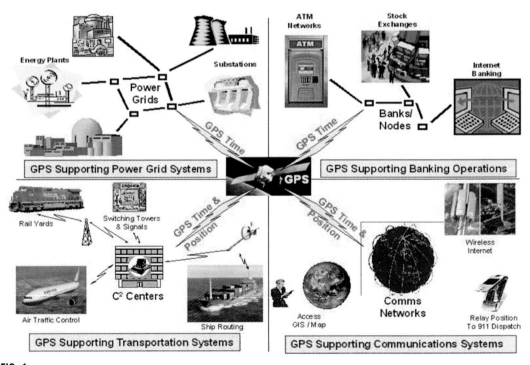

FIG. 1

Scope of interconnected critical systems dependent on availability of reliable GPS signals (after Department of Homeland Security).

The figure indicates that GPS outages can affect not only the surveyor (positioning) or farmer (precision agriculture), which are not part of the critical infrastructure, but also financial institutions, multimodal transportation, communications systems, and the internet, and perhaps most importantly, the power grid, which relies on timing signals provided by the GPS system.

Given the importance of GPS to our modern society, this chapter examines the effects of extreme ionospheric storms on GPS/GNSS signals and systems. In Section 2, we provide a brief description of the GPS satellite system and GPS signals. In Section 3, we describe ionospheric storms and some of the ionospheric structures that accompany them. In Section 4, we describe in more detail some of the system effects of extreme ionospheric storms – specifically on communications, navigation, and surveillance. In Section 5, we describe three of the largest ionospheric storms of the last 15 years, pointing out their features captured through data analysis and modeling. We also describe the impacts of the storms on GPS signals. In Section 6, we summarize some of the effects of the ionosphere on GPS systems. Finally, in Section 7, we discuss the implications for extreme ionospheric storms and their effects on operational systems, with a focus on GPS-reliant systems.

2 GLOBAL POSITIONING SYSTEM

The GPS satellite constellation consists of 32 satellites (31 operational satellites) in 6 equally-spaced orbital planes at 55 degrees inclination, separated by 60 degrees right ascension of the ascending node (angle along the equator from a reference point to the orbit's intersection), and at 20,200 km altitude (Misra and Enge, 2011). Each plane contains at least four "slots" occupied by GPS satellites. This orbital geometry and the distribution of GPS satellites ensures that users on the ground can view at least four satellites from virtually any point on Earth. Each of the legacy GPS satellites broadcasts its signals continuously at two radio frequencies in the L-band of the radio spectrum. The center frequencies of these two GPS signals are at 1575.42 MHz and 1227.60 MHz, commonly referred to as L1 and L2. The original GPS signal design contained two ranging codes: the Coarse/Acquisition code or C/A, which is freely available to the public, and the Precision (encrypted) code, or P(Y)-code, reserved for DoD-authorized users. The L1 GPS signals include C/A and P(Y) codes while the L2 signal is intended for DoD users only and thus includes the P(Y) code only. On more recent satellites, a second civilian signal (L2C) and a lower L5 frequency at 1176.45 MHz have been added.

As the signals from the GPS satellites propagate through the atmosphere they undergo path delays due to the refractive index. Variation of the refractive index along the path of the signal results in bending of the propagation path making it longer than the geometrical straight-line path expected in a vacuum. This bending of the signal path, changes the apparent range between the GPS receiver and the satellite. If the refraction of the GPS signal is not accounted for, it leads to atmosphere-induced position errors. The two largest sources of errors in GPS path delays are the ionosphere and troposphere. For a typical single-frequency application, the errors introduced in GPS measurements by the ionosphere and the troposphere are on the order of 5 m and 0.5 m, respectively (http://www.trimble.com/gps_tutorial/howgps-error2.aspx). These ionosphere errors are representative of periods when space weather is not a significant factor. It should be noted that unlike the neutral atmosphere, the ionosphere acts as a dispersive medium for RF waves, that is, the refractive index of the ionosphere is a function of frequency. As a result, refractive errors due to the ionosphere in GPS measurements can be corrected by using dual frequency GPS measurements (Spilker, 1978; Brunner and Gu, 1991; Ware et al., 1996).

The ionosphere can also degrade GPS receiver performance by producing diffraction of GPS signals from small-scale irregularities in the electron density distribution. These small-scale irregularities in the ionosphere, particularly those with spatial scales smaller than the Fresnel length at GPS frequencies, can produce rapid fluctuations in signal amplitude (C/No) and carrier phase, called amplitude scintillation and phase scintillation, respectively. Scintillation can cause a GPS receiver to suffer from cycle slips or even loss of phase lock in conditions of severe scintillation (Doherty et al., 2000; Skone et al., 2001; Aquino et al., 2005; Sreeja et al., 2012). Severe scintillations are often observed at high (auroral) and low latitudes (±20 degrees of the geomagnetic equator), and in general phase scintillation is dominant at high latitudes, whereas amplitude scintillation is dominant at low latitudes, as explained in more detail in subsequent sections. Thus, it is vitally important to understand the sources and characteristics of ionospheric irregularities in order to develop predictive capabilities for scintillation events.

In addition to the GPS constellation, there are other satellite navigation systems such as GLONASS (Russia), Beidou (China), and Galileo (Europe). Like GPS, these also operate at L-band, and are therefore subject to the same ionospheric effects as GPS.

3 THE IONOSPHERE

The ionosphere is formed by the stripping away of electrons from atmospheric atoms and molecules by solar UV, EUV, and X-rays and energetic particle precipitation. Enhanced levels of X-ray and EUV light from solar flares can rapidly increase the amount of ionization and can change the height at which the ionization is created in the ionosphere. Energetic particle precipitation (typically electrons and protons) can also play a role, especially in the auroral regions. The loss of ionospheric plasma is governed by atmospheric chemistry, and by transport of plasma both downwards and out to the plasmasphere. Within the ionosphere, the plasma can be transported by both neutral winds, and by electric fields. Increased solar wind speeds and densities in some Coronal Mass Ejections (CMEs) and High Speed Streams (HSSs), along with changes in the Interplanetary Magnetic Field (IMF), can increase the high-latitude electric fields, which also increase the deposition of Joule heating and momentum into the polar neutral atmosphere. These disturbances in turn change the global thermospheric composition, temperatures, winds, and electric fields which then cause changes in the global ionosphere (e.g., Crowley et al., 1989a,b). These variations in the ionosphere can alter the propagation of radio waves in general, including GNSS radio signals. During a geomagnetic storm the interactions of the solar wind, magnetosphere, thermosphere, and ionosphere, cause dynamical changes over different spatial and temporal scales. These disturbance conditions in the ionosphere can range from large scale (100–1000 km) electron density gradients to localized (1–100 km) regions of irregular plasma structure, with embedded irregularities of less than 1 km scale size.

As noted above, the ionosphere is not uniform globally. Therefore, location will have a major impact on whether the ionosphere has a significant effect on GPS signals. We noted above that the ionospheric plasma distribution is subject to production, loss and transport processes, therefore as these processes vary with latitude, the ionospheric plasma distribution can be expected to also vary with latitude, and it is important to distinguish the ionospheric phenomena that occur at high latitudes, mid-latitudes, and low latitudes. The mid-latitude ionosphere tends to be relatively benign. In contrast, both the low-latitude and high-latitude ionosphere include significant features and structures that must be taken into account when predicting radio propagation and GPS signal variability. Furthermore, these

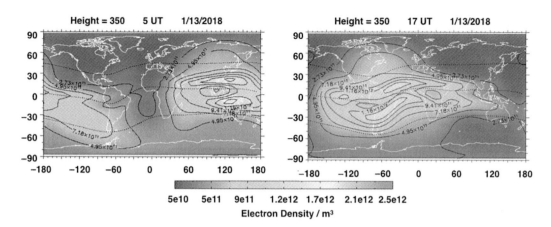

FIG. 2

Typical example of structure in the electron density at 350 km for 05 UT (*left*) and 17 UT (*right*) from the IRI climatological model predicted for January, 2018, under geomagnetically quiet conditions and low solar activity. The magnetic equator drives the location of major structures.

significant features are also very variable. Below, we describe major ionospheric features of the low-, mid-, and high-latitude ionosphere.

Fig. 2 shows the global ionospheric structure, including the Appleton Anomalies (peak values of electron density) at low latitudes, on either side of the magnetic equator. The two panels show the structure in the electron density at 350 km for 05 UT (*left*) and 17 UT (*right*) for January 2018 predicted from the IRI climatological model (Bilitza et al., 2014) under geomagnetically quiet conditions and low solar activity. These structures are also reproduced by full-physics models (e.g., Crowley et al., 2006). The Appleton Anomalies are the location of the largest electron densities and TEC. They are also the home of the most intense ionospheric irregularities, which we discuss in the next sub-section.

One way to monitor ionospheric conditions is to use specialized ground-based GPS receivers that are designed to measure ionospheric properties. These GPS receivers typically yield Total Electron Content (TEC), and some, like Atmospheric and Space Technology Research Associates (ASTRA's) CASES receivers can also measure amplitude and phase scintillation (Crowley et al., 2011; O'Hanlon et al., 2011). There are thousands of ground-based GPS receivers located across the United States and North America operated by federal, state, and local agencies, as well as private companies. Derived ionospheric data from many of these systems allow monitoring of the ionosphere, and these data could in turn be used for specification of the ionosphere-induced errors in the positioning and timing information that would be provided by the standard GNSS receivers in common use. Coster and Komjathy (2008) provided a brief history of GNSS development and its application to space weather measurements. In the sections below, we describe some of these GPS-derived TEC and scintillation measurements.

3.1 TOTAL ELECTRON CONTENT

The speed and propagation path of the GNSS signals in the ionosphere depend upon the number and distribution of free electrons in the path, or Total Electron Content (TEC). Changes in the TEC result in variable time delays in the signal propagation due to refraction. Space weather disturbances can cause the ionospheric electron density profile and the corresponding TEC to vary in a variety of different ways and on different time scales. Many GNSS receivers utilize the Klobuchar ionospheric model (Klobuchar, 1987) to account for the additional delay due to ionospheric TEC. However, during intense solar and geomagnetic disturbances, the Klobuchar ionospheric model is inappropriate and can result in increased Positioning Navigation and Timing (PNT) errors particularly for single frequency GPS receivers. The PNT corrections are discussed in more detail in Section 6 of this chapter.

To support the GPS user community, various groups provide maps of TEC in real-time (e.g., Fuller-Rowell, 2005). These products often use ionospheric data assimilation models and ingest ground-based GPS data from across the United States, Canada, and Mexico to produce 2D maps of TEC. These maps can, in principle, be used to estimate the GPS signal delay due to the ionospheric TEC between a receiver and a GPS satellite. This delay can then be translated into GPS positioning error. However, this processing is non-trivial. TEC maps can be characterized in terms of RMS error and provide quantitative information about the delays. This is similar for the WAAS system, which uses a similar approach (see detailed discussion in the accompanying chapter by Mannucci and Tsurutani).

3.2 LOW-LATITUDE SCINTILLATION

One of the often observed impacts of ionospheric irregularities on RF communication and navigation systems is scintillation. Ionospheric scintillation is a rapid temporal fluctuation in amplitude and/or phase of a GNSS signal due to the presence of ionospheric irregularities (Kintner et al., 2007). These rapid fluctuations can be represented by computing indices called S4 and sigma-phi (Van Dierendonck, 1999). S4 and sigma-phi are computed as the normalized standard deviation of the intensity and standard deviation of the phase, respectively, over some defined time interval (typically 60 or 100 s). The refraction and diffraction effects cause group delay, phase advance, and constructive/destructive interference of the radio signal as it interacts with electron density structures in the ionosphere. Scintillations are produced when a radio wave propagating through the ionosphere undergoes diffraction due to irregularities with scale sizes less than the Fresnel length ($L_F = (2\lambda z)^{1/2}$, where λ is the wavelength and z is the distance from the irregularity to the receiver). For the GPS L1 frequency (1.5 GHz) the Fresnel length is on the order of 350 m at 350 km altitude. Scintillations, when intense, can produce message errors in satellite communication systems and may result in outages or degradation of accuracy in GNSS navigation systems. The occurrence of scintillation depends on the local time, season, solar and magnetic activity and magnetic latitude. Scintillation is most intense at low latitudes (magnetic equator and Appleton Anomalies), and at high latitudes (Aarons, 1982, 1993; Basu and Basu, 1985; Kersley et al., 1988; MacDougall, 1990; Alfonsi et al., 2011).

In the equatorial regions, ionospheric irregularities can occur on a daily basis causing severe scintillations, and especially amplitude scintillation. Here, fades at L-band can exceed 30 dB and are more severe at lower frequencies. Fig. 3 compares the carrier-to-noise ratio of scintillating and non-scintillating signals recorded simultaneously in Brazil from two GPS satellites, PRN 21 (*blue* curve) and PRN 25 (*green* curve). The received power of the scintillating PRN 21 signal varies by more than 40 dB. These fades can cause loss of GPS receiver tracking lock, increase cycle slips, lengthen acquisition times, decrease PNT accuracy, and in severe cases, cause loss of navigation if multiple satellites are affected simultaneously. For typical GPS receivers, tracking normally fails below 26 dB-Hz. At UHF frequencies, scintillations are even more severe and can interfere with satellite communication networks. Thus, GPS scintillation can serve as an index for disruptions of UHF satellite communication links.

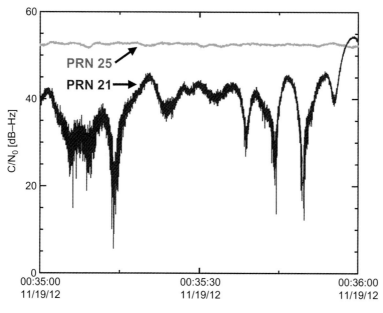

FIG. 3

Comparison of scintillating (PRN 21) and non-scintillating (PRN 25) signals using data taken with ASTRA CASES GPS receiver in Brazil

Low-latitude scintillation is associated with Rayleigh-Taylor instabilities that occur when the eastward electric field is enhanced and irregular plasma density depletions are generated on the bottom-side of the nighttime equatorial F region, which then rise to higher altitudes (e.g., Basu et al., 1978; Kelley, 1989; Fejer et al., 1999). The probability distribution of amplitude scintillations at low latitudes with values in excess of S4=0.5 is highest during the post-sunset to midnight local time sector (e.g., Rama Rao et al., 2009; Steenburgh et al., 2008). Using 7 years of data, Steenburgh et al. showed that low-latitude scintillation tended to maximize at the equinoxes, and showed a strong correlation with the solar cycle.

The impact of geomagnetic storms on low-latitude amplitude scintillations remains a topic of active research. Several studies have suggested that geomagnetic storm activity will inhibit the occurrence of ionospheric irregularities and corresponding scintillations during the storm periods (Aarons, 1991; Abdu et al., 1995, and the references therein). However, studies by Basu et al. (2001a,b); and Tulasi Ram et al. (2008) have shown that in the longitude sector where the main phase of the magnetic storm happens to coincide with local sunset period, there is a greater probability that the storm-induced electric fields will enhance the post-sunset vertical drifts at the equator, resulting in conditions conducive for the onset of irregularities and ionospheric scintillations.

3.3 HIGH-LATITUDE SCINTILLATION

High-latitude studies generally show that phase scintillation can be severe, but that the amplitude scintillation tends to be small. There are multiple physical causes of the ionospheric irregularities that produce high-latitude scintillation, and not all are well understood. Both high-latitude particle precipitation and convection dynamics can lead to plasma irregularities that cause scintillation of satellite signals. High-latitude GPS phase scintillation studies have shown that the variation in their occurrence rate is controlled by various factors including season and magnetic activity (Kinrade et al., 2013; Prikryl et al., 2011, 2013a,b). The main ionospheric regions that are affected by scintillation are the nightside auroral oval, the cusp/cleft on the dayside and the polar cap (Coker et al., 1995; Aarons, 1997; Aarons et al., 2000; Basu et al., 1995; Spogli et al., 2009). Using the Canadian High Arctic Ionospheric Network (CHAIN) of GPS receivers, Prikryl et al. (2010) has suggested that GPS phase scintillations are largest in the nightside auroral oval. The characteristics of the nightside phase scintillations have been shown to be significantly different from those of the dayside (cusp) ionosphere. Loucks et al. (2017) compared the climatology of phase scintillation measured in Alaska versus other sites, and pointed out the differences between auroral sites versus higher magnetic latitude polar cap sites.

To provide space weather services to GPS users and to initiate a detailed investigation of high-latitude scintillation ASTRA deployed a number of their CASES GPS receivers in Alaska, which have been operational since November 2012. These dual-frequency receivers form a longitudinal chain extending from Kaktovik to Anchorage in Alaska, and measure total electron content (TEC) and scintillation parameters, providing real-time data to users. The data and results are particularly valuable because they illustrate some of the challenges of using GPS systems for positioning and navigation in an auroral region such as Alaska.

Fig. 4 compares the GPS phase measured by a CORS receiver (*upper* panel) in Alaska, versus an ASTRA CASES receiver (*middle* panel). The CASES receivers are designed with special tracking

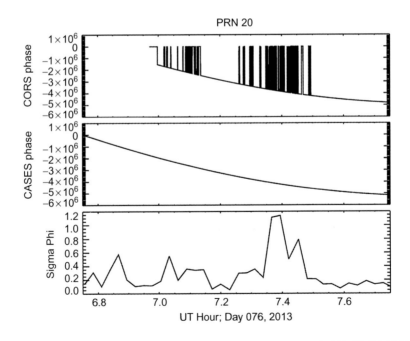

FIG. 4

Phase measurements from CORS receiver (*upper* panel) and CASES receiver (*middle* panel) for Day 76, 2013. Lower panel shows phase scintillation index measured by CASES receiver.

loops to provide resilience against loss of lock during strong scintillation events, and the middle panel indicates a gradual phase change with no loss of lock. In contrast, the CORS receiver (*upper* panel) exhibited frequent and sustained loss of lock at this time. The lower panel illustrates the scintillation level measured by the CASES receiver, and reveals the cause of the loss of lock on the CORS receiver. Loss of lock tends to occur for the CORS receiver when the scintillation index is elevated. The CASES GPS receiver is specifically designed for monitoring and measuring ionospheric scintillations. It utilizes specialized tracking loops designed for operation in both weak-signal and scintillating environments. The incorporation of these tracking loops allows the CASES GPS receiver to provide continuous phase tracking in a scintillating environment when other GPS receivers might suffer from loss of lock.

Data from the ASTRA CASES receivers include carrier-phase and pseudo-range measurements of the L1 ($f_1 = 1.57542$ GHz) and L2 ($f_2 = 1.22760$ GHz) GPS signals at 1 Hz. Fig. 5 shows the variability of the phase scintillation index, sigma-phi, measured by receivers in Alaska. Fig. 5A–D shows phase scintillation activity on November 13, 2012, from four stations in the CASES Alaskan chain. The four receivers shown here are located at sites in Kaktovik (70.1 degrees N, 143.6 degrees W), Fort Yukon (66.6 degrees N, 145.2 degrees W), Poker Flat (65.1 degrees N, −147.4 degrees W), and Gakona (62.4 degrees N, −145.2 degrees W). The *Kp* index is plotted in the lower panel (Fig. 5F) showing that for most of the day the geomagnetic activity was low. In general, moderate phase scintillation was present on November 13, 2012. The scintillation strength during this period is largest at Kaktovik and Fort Yukon, and shows a latitudinal dependence with the southernmost receiver recording the smallest phase scintillation. In Northern Alaska, scintillation will be a frequent problem due to the omnipresence of the aurora, versus southern Alaska where it will occur predominantly near midnight.

Poker Flat Research Range (PFRR) operates a meridian spectrograph to observe the location and intensity of key auroral emissions. Fig. 5 (panel-E) shows the corresponding auroral activity on November 13, 2012, measured by the PFRR meridian spectrograph. The imaging spectrograph maps a meridian slice of the sky aligned to the local magnetic meridian from horizon to horizon onto a reflection grating and collects the 2-D (spectral × spatial) images at a 15-s cadence. Auroral emission intensities at 630.0 nm, are shown in panel-e of Fig. 5. This was geomagnetically a moderately quiet day, and most of the auroral activity was observed poleward of Poker Flat at 65 degrees N. The Meridian Spectrograph data show that auroras were generally confined to higher latitudes early in the day (2–5 UT), corresponding to the larger scintillation levels observed at the more northerly GPS sites, versus the lower latitudes. At about 6–7 UT, the auroral brightness is concentrated near 67 degrees (Fort Yukon), and there is a corresponding spike in the GPS scintillation at Fort Yukon at that time. At about 15 UT, there is a brightening of the aurora across all latitudes in Alaska, and the GPS scintillation levels at all four sites increase. We did not have meridian spectrograph data after 17 UT because of the daylight, and therefore we could not ascertain whether the later scintillation increases were correlated with auroral intensity.

ASTRA uses the CASES GPS receiver measurements to create maps of phase-scintillation in real-time (http://cases.astraspace.net/alaska_sigphi.html). Fig. 6 shows the map of Alaska with phase scintillation superposed. Each panel represents 3 h of data from 9–12 UT. The arcs indicate motion of the ionospheric pierce-point during the 3 h, and the colors indicate scintillation intensity. The left panel is typical of quiet conditions, while the right panel is indicative of active conditions. We note that a relatively modest array of GPS receivers can map large regions of scintillation.

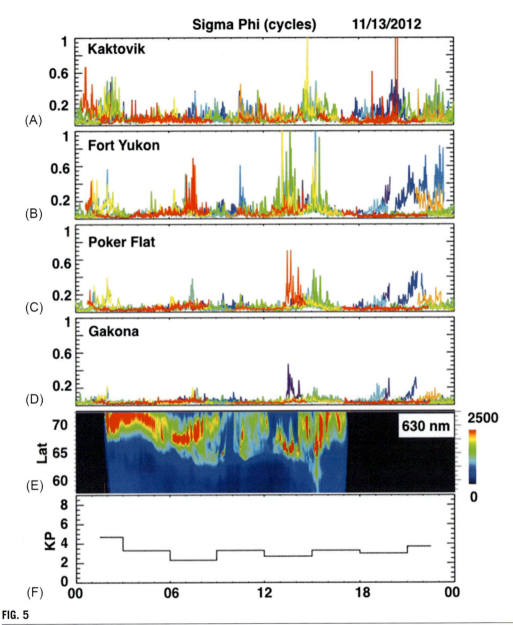

FIG. 5

Phase scintillation index measured by the CASES GPS receiver from (A) Kaktovik, (B) Fort Yukon, (C) Poker Flat, and (D) Gakona on November 12, 2012. (E) 630.0 nm (*red* line) auroral emissions from meridian scanning photometer; (F) 3-h Kp index. The keogram data is shown as a function of geographic latitude and UT. The emission intensity scale is in Rayleighs. Each color in the scintillation plots indicates a different PRN. Elevation mask of 20 degrees is used with GPS data to avoid spurious scintillations due to multipath effects.

After Azeem, I., Crowley, G., Reynolds, A., Santana, J., Hampton, D., 2013. First results of phase scintillation from a longitudinal chain of ASTRA's SM-211 GPS TEC and scintillation receivers in Alaska. Proceedings of the ION 2013 Pacific PNT Meeting, April 3–25, Honolulu, Hawaii.

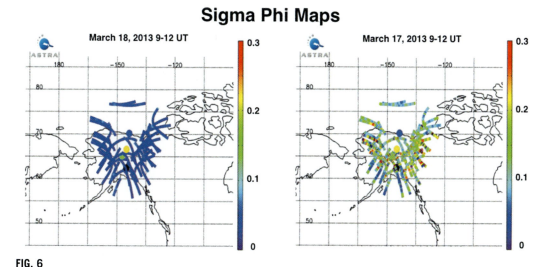

FIG. 6

Real-time maps of ionospheric scintillation over Alaska using ASTRA's CASES GPS receivers.

Fig. 7 shows the climatology of phase scintillations measured by three of the Alaskan CASES receivers as a function of solar local time (SLT) from November 9, 2012 to February 26, 2016. The climatology is computed by binning phase scintillation data into 30-min segments for each day of observations. To minimize spurious scintillations due to multipath effects we use an elevation mask of 20 degrees to compute the climatology. The most northerly receiver site, Kaktovik, AK shows the most intense phase scintillation while Poker Flat, which is the southernmost site in the figure, on average has the smallest phase scintillation values. Phase scintillations at Kaktovik show a preponderance of large values near midnight around 1000 UT (or 2300 MLT/0100 SLT). Comparison of phase scintillation measurements from the CASES receiver stations in the Alaska GPS chain reveals two main findings: (1) Severity of phase scintillation decreases with decreasing latitude and (2) largest phase scintillations occur near magnetic midnight (Azeem et al., 2013).

4 IONOSPHERIC STRUCTURES EVIDENT IN TEC DATA
4.1 STORM ENHANCED DENSITIES, PATCHES AND BLOBS

Although the ionosphere at mid-latitudes is generally more benign than in other regions, a number of significant perturbations can occur there. In the sub-auroral region, large-scale structures like Storm Enhanced Densities (SEDs) and tongues of ionization can be present. These are enhanced plumes of electron density typically 500–1000 km in width that are transported into the polar cap where they are broken up into patches and blobs. Polar cap patches are localized enhancements in ionospheric density which originate from solar EUV ionization on the dayside, enter the polar cap at the dayside cusp, drift anti-sunward with the plasma convection at up to 1–3 km/s velocities, and then exit the polar cap near midnight to merge with sunward returning flow patterns. There is evidence that sometimes particle precipitation in the cusp or aurora may also play a role in enhancing the electron density.

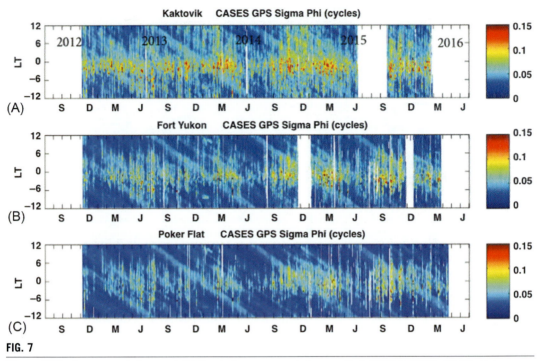

FIG. 7

Maps of 30-min averaged phase scintillation index over nearly 4 years from 2012 to 2016, shown as a function of Local Time (LT) and day for (A) Kaktovik, (B) Fort Yukon, and (C) Poker Flat. (On the time axis, S=September, D=December, etc.)

Crowley (1996) defined polar cap patches as plasma structures with a horizontal extent of at least 100 km and a plasma density of at least twice the density of the surrounding background plasma. Crowley et al. (2000) developed a plasma trajectory analysis code that used high-latitude convection patterns and plasma drift velocities obtained from the Assimilative Mapping of Ionospheric Electrodynamics (AMIE) algorithm. They showed that long thin strands of ionospheric F-region plasma are convected from the dayside, reorganize into quasi-circular patches over the polar cap due to the differential flows in the convection pattern, and then exit the polar cap as long thin strands known as "blobs." The trajectory code was later used with AMIE to analyze different events, and to identify and track measured patches of F-region plasma as they traversed the polar cap from the dayside to the nightside. These studies showed that the AMIE convection pattern has sufficient fidelity to specify the trajectory of plasma over the polar cap, while at the same time there is enough data for ionospheric electron density assimilation codes to specify the plasma distribution and its evolution with time.

Fig. 8 shows an example of SEDs, polar cap patches, and blobs derived from ASTRA's ionospheric electron density assimilation code. The figure shows AMIE potential patterns plotted over the southern hemisphere plasma densities (color-scale) in geographic coordinates for four selected UTs during the SED growth and decay. Each panel is labeled with its corresponding time and the color-scale appears at the bottom right. The AMIE electric potential contours are labeled in units of kiloVolts. The outer perimeter of each map corresponds to 37 degrees S geographic latitude, and a black line with a red "X"

4 IONOSPHERIC STRUCTURES EVIDENT IN TEC DATA

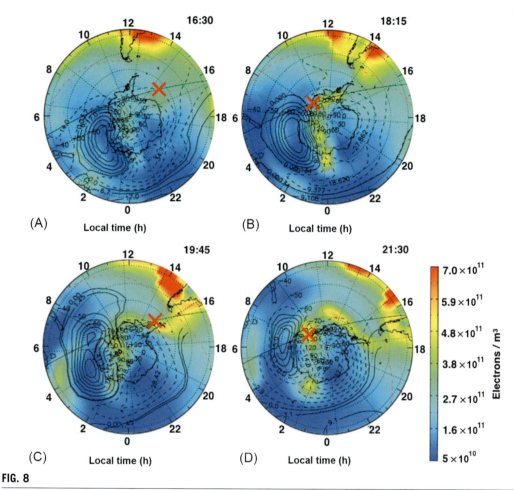

FIG. 8

Time evolution of the southern SED feature in electron density (color map) with AMIE potential patterns overlapped (line-contours). Results from ASTRA's assimilation algorithm are shown in geographic coordinates with the DICE orbit shown (*red* cross indicates precise DICE location at that UT). The assimilation includes data from 4559 ground-based TEC receivers, five radio occultation satellites, 3 beacon satellites combined with 32 beacon ground-stations, and 9 topside TEC instruments.

symbol indicates locations of ASTRA's DICE Cubesat (Fish et al., 2014) at this UT. By 18:15 UT (panel B) the SED plume was well developed, and a "tongue" of ionization extended from the Antarctic Peninsula near 14 LT, and between the two convection cells in the noon sector.

By this time, enhanced plasma densities appear to have been carried across the polar cap by convection, and are visible all the way into the midnight sector. A very similar structure was present at 19:45 UT (panel C), and the mid-latitude end of the SED has rotated towards a solar local time of 16 LT. By 21:30 UT the SED feature appears to originate closer to 18 LT, and wraps halfway around the dusk convection cell.

The SED in Fig. 8 occurred in the American longitude sector, and we note that SEDs do not seem to be as prevalent in other longitudes. SEDs regularly occur over the northern United States and Canada, but not in Europe, for example. Foster and Coster (2007) showed that SEDs can occur in conjugate hemispheres in the American sector. We do not understand exactly what causes SEDs, and we can only predict their occurrence and location on a statistical basis, therefore features like the SED and their effects on radio systems speak to the need for a better understanding of ionospheric variability and structure, and ultimately the need to predict its behavior more accurately.

SEDs seem to be associated with the occurrence of Sub-Auroral Plasma Streams (SAPS), which are regions of enhanced electric field. Observational data covering two solar cycles (1979–2000) from the Millstone Hill incoherent scatter radar have been used to study SAPS. The data show that there is a linear relationship between the MLT and the average magnetic latitudes (i.e., geomagnetic coordinates) in which SAPS are observed, indicating that SAPS are predominately an early evening/pre-midnight phenomenon (Foster and Vo, 2002).

Fig. 9 shows a map of North America with superposed the Foster and Vo (2002) magnetic latitude of the peak polarization stream as a function of magnetic local time (MLT) for several Kp values, from $Kp=4$ to $Kp=7$. $Kp=3$ corresponds to moderately disturbed geomagnetic conditions. The figure indicates that SAPS and therefore SEDs would not generally occur at latitudes south of Washington

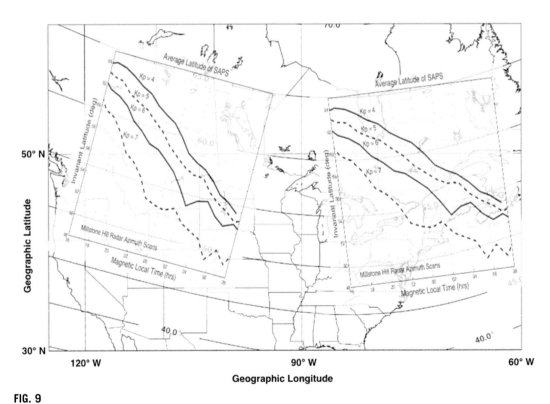

FIG. 9

Average SAPS characteristics applied to the U.S./Canadian region.

DC, although the dramatic low-latitude plasma enhancements associated with storms and SEDs occur at latitudes down to Florida and the Caribbean during Superstorm events, as indicated in Fig. 13. For completeness, we note that in the northern hemisphere Asian sector, SAPS and SEDs will occur at much higher geographic latitudes than in the United States because of the offset between the geographic and geomagnetic poles.

4.2 TRAVELING IONOSPHERIC DISTURBANCES

Again, although the mid-latitude ionosphere is generally considered benign, there is another type of disturbance called a Traveling Ionospheric Disturbance (TID) that can be present at almost all times and all latitudes. TIDs are propagating perturbations in the ionospheric electron density that can be detected by a number of different instruments, including ionosondes, GNSS receivers, incoherent scatter radars, and HF Doppler systems. Signatures of quasiperiodic structures detected by optical systems observing nighttime airglow emission were also linked to the occurrence of TIDs. The TID disturbance is driven by propagating acoustic gravity waves (AGWs) that perturb the ionospheric electron density. Based on their phase velocity and wave period, AGWs and classical TIDs are often classified into medium and large-scale waves. Some coarse guidelines on the properties of these two groups are summarized in Table 1. The large-scale gravity waves are generated by auroral sources. There are many sources of Medium Scale waves, including the aurora and several lower atmosphere sources including ocean waves, tsunamis, explosions, weather fronts, thunderstorms, and winds blowing over topography. Both LSTIDs and MSTIDs are thought to impact operational HF systems, and are generally considered to be the largest source of uncertainty in predicting the behavior of HF systems. Not shown in the Table are acoustic waves, below the Brunt-Vaisala period, as they are not thought to affect HF systems significantly.

TIDs and the underlying AGWs play a critical role in driving the day-to-day variability of the ionosphere. Theoretical understanding of upward AGW propagation in the atmosphere (Vadas and Crowley, 2010, 2017) suggests that while small spatial scale GWs are confined to below the stratopause, the medium and large scale AGWs have appropriate characteristic amplitudes and phase velocities to allow them to penetrate into the mesopause and ionosphere. The signature of an AGW in the ionosphere is manifested as oscillations of the ionospheric electron density resulting in a TID (Crowley and Williams, 1987; Crowley et al., 1987; Crowley and McCrea, 1988). The electron density perturbations are caused by AGWs via ion-neutral collisions (Hines, 1960; Hooke, 1968; Hocke and Schlegel, 1996).

Because of the variability of TIDs, we are unable to predict their occurrence even on a statistical basis. This variability arises from the multiple sources of gravity waves in both the troposphere and thermosphere, and the variability of the medium (background winds and temperature) through which they propagate. To add to the difficulty, both the gravity wave sources and the atmospheric medium vary with latitude, local time, and season.

Table 1 Typical properties of medium and large-scale AGWs/TIDs

	Period	V_H (m/s)	λ_H (km)
Medium scale	10–30 min	50–300	100–300
Large scale	0.5–5 h	300–1000	300–3000

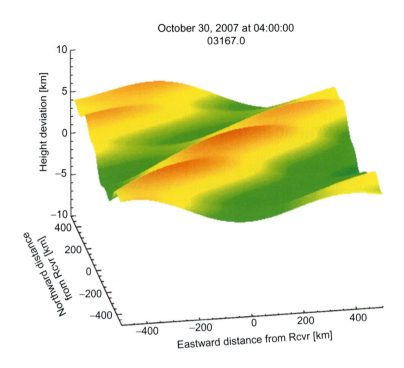

FIG. 10

Radio reflection surface perturbed by five TID components measured simultaneously by the TIDDBIT TID Mapping System, and extrapolated out to several hundred kilometers of horizontal distance to show the wave structure clearly. Color scale represents height perturbations from −10 km to +10 km, Consecutive frames can be viewed as a movie showing TIDs propagating with time.

TIDs and their effects on radio systems speak to the need for a better understanding of their morphology and climatology, and ultimately the need to predict their behavior more accurately. Before such a climatology can be constructed, more measurements of TIDs are needed, like those shown in Fig. 10 using systems like ASTRA's TIDDBIT Mapper, which provide continuous detailed TID information (Crowley and Rodrigues, 2012). The completeness of the wave information obtained from the TIDDBIT system (amplitudes, horizontal phase speeds, wavelengths, and propagation direction, vertical phase speeds all as a function of wave period) makes it possible to reconstruct the vertical displacement of radio reflection surfaces over the 200-km horizontal dimension of the sounder array. Fig. 10 illustrates a reconstruction of the TIDs/tilts for a given radio reflection surface, measured by ASTRA's TIDDBIT HF Doppler sounder array operating in Virginia, and reveals ionospheric corrugations caused by TIDs. In this figure, the wave field is comprised of five (5) wave components traveling in slightly different directions and with different speeds and wavelengths. The figure shows height perturbations of up to 10 km caused by the superposition of multiple TIDs. The resulting pattern resembles the surface of the ocean. Radio raytracing permits the effects of TIDs on operational systems to be explored. The amplitude of the TIDs depends on their source, propagation conditions, and damping factors, all of which can be difficult to determine. There is no useful F-region TID climatology at the present time that could be used to predict TID periods, amplitudes, speeds and directions at a given location.

Therefore, any operational requirement for TID information means the TID characteristics have to be measured at the required location.

While TIDDBIT provides the height perturbations of a given reflection surface, we often need to extend this to specify the ionospheric perturbations due to the TIDs at other altitudes. A number of approaches may be considered in order to extend the waves throughout the ionosphere. In practice, waves propagate with both horizontal and vertical velocity components that vary from one wave to another. Long period waves generally propagate more horizontally, while shorter period waves propagate more vertically. The TIDDBIT sounder, operating on two frequencies, can measure the vertical speed as a function of wave period. Therefore, the TIDDBIT system enables specification of both the vertical and horizontal components of wave velocity as a function of wave frequency.

Increasingly TEC measurements from networks of GPS receivers are being used to identify TIDs, and then to characterize the TID spectrum and propagation characteristics (Tsugawa et al., 2011; Nishioka et al, 2013; Crowley et al., 2016). Fig. 11 shows an example of medium scale TIDs seen in GPS TEC over the continental United States. When analyzed sequentially in time, these maps of TEC perturbations reveal that the TIDs were propagating towards the south-east for several hours. The right-hand panel in Fig. 11 shows a close-up of the TEC perturbations over Florida, where the phase-front alignment is clear.

In recent studies, ASTRA has used 5000 dual-frequency GPS receivers distributed throughout the United States to map TEC perturbations caused by TIDs (Azeem et al., 2015a,b; Crowley et al., 2016; Azeem et al., 2017), as shown in Fig. 11. Fig. 12 illustrates how the technique allows "imaging" of a wide range of TIDs including those caused by a tsunami, a thunderstorm, and auroral activity during a geomagnetic storm. Spectral analysis of these images can provide the TID frequency spectrum, horizontal phase velocities and corresponding wavelengths. We refer the reader to Azeem et al. (2017) for a detailed description of GPS TEC data processing used to create the TID maps shown in Figs. 11 and 12.

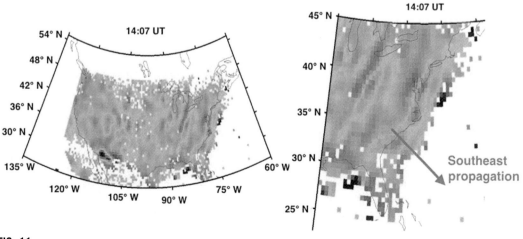

FIG. 11

Maps of GPS-TEC perturbations reveal the presence of propagating waves across the United States.

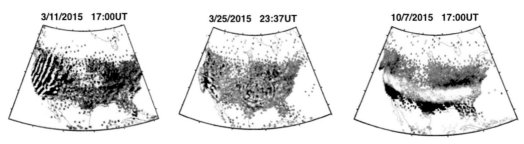

FIG. 12

GPS-TEC maps of TIDs generated by a tsunami (*left* panel), thunderstorm (*middle* panel), and auroral perturbations (*right* panel).

5 EVENT STUDIES FOR LARGE IONOSPHERIC STORMS

The most extreme storm on record was the Carrington event of 1859, which was described by Tsurutani et al. (2003). Unfortunately, there were few measurements available at the time, but newspapers reported fires in railway signals and telegraph wires due to the strong electrical currents induced in these long conductors. There was also an extreme event in 1921, and these two extreme events are described in more detail elsewhere in this book.

The Carrington event of 1859 was several times more intense than anything we have recorded in the space age (Ngwira et al., 2013). The impact of such an event today is difficult to imagine, but the most likely impact would be on the electrical grid due to the presence of long conductors. At the same time, the GPS satellites may find themselves outside of the protection of the magnetosphere, and subject to more radiation than normal, possibly causing problems with the satellite electronics. Finally, the ionospheric effects on GPS receivers could be significant. In the Carrington event, aurora sightings were reported as far south as Florida and the Caribbean, thus we would expect to see phase scintillation at most latitudes.

The largest ionospheric storms of the past 20 years occurred in October and November 2003, and in November 2004, when relatively few GPS receivers were deployed. We here describe TEC effects measured during the Halloween Storm of October 2003 to indicate the kind of effects likely to be experienced during extreme events. There have also been a number of papers published describing some of the GPS measurements obtained during these and other Super Storms, and we summarize some of those papers below.

The global thermospheric and ionospheric responses to the three superstorms mentioned above all had similarities. Specifically, the deposition of energy at high latitudes caused upwelling of the thermospheric gases, with nitrogen-rich air depleting the high-latitude ionospheric electron densities, and oxygen-rich air enhancing the low-latitude ionosphere. For the November 20, 2003, storm, these global effects were described by Meier et al. (2005), and by Crowley and Meier (2008).

Superposed on these global changes, in each superstorm were large SED events that created massively enhanced TEC values over the United States and Canada. Fig. 13 depicts the TEC distribution derived from ASTRA's ionospheric electron density assimilation code for two SEDs in October 2003.

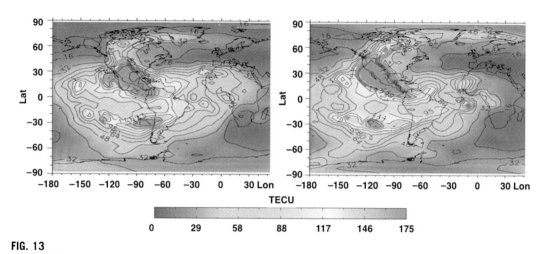

FIG. 13

TEC distribution for the SEDs that occurred at 21.15 UT on (*left*) Day 302, 2003, and (*right*) 21.15 UT on Day 303, 2003. These maps are generated by ASTRA's ionospheric electron density assimilation package. The assimilation includes data from 1200 ground-based TEC receivers, two radio occultation satellites, 5 beacon satellites combined with 41 beacon ground-stations, 5 topside TEC instruments, 23 ionosondes, and 3 DMSP satellites measuring in situ electron density.

These SEDs occurred on Day 302, 2003 between 20:00 to 24:00 UT (left panel), and on Day 303, 2003, between 20:00 to 24:00 UT (right panel). In each case, there is a region of enhanced mid-latitude plasma that reaches as far north as the U.S.-Canada border, and emanating from that location is a plume of plasma heading into the polar cap as a tongue of ionization that then is broken up into patches and blobs.

Fig. 14, shows the corresponding TEC values for PRN07 derived from eight (8) GPS receivers in the United States between 21:00 and 24:00 UT for each day. The receiver sites are more or less evenly spaced about 100 km apart every ~3 degrees longitude from −116.3 degrees to −94.7 degrees. The GPS receivers are at latitudes of 43 degrees to 47 degrees north. In the left-hand panel, the TEC reached values of about 160 TECU, and in the right panel, TEC reached about 230 TECU. However, what is remarkable is that even though they were all observing GPS signals from the same satellite (PRN07) at any given UT, the range of TEC values observed from these eight sites differed dramatically leading to gradients of about 70 TECU per hundred kilometers which existed for several hours. Similar effects were reported by Mannucci et al. (2008) for the November 20, 2003, storm, who showed an example of severe ionospheric gradients in the mid-Atlantic region of the United States, as demonstrated by rapidly changing ranging errors (as a function of time), corresponding to about 200 TECU.

An online search reveals dozens of scientific and engineering studies that have reported on the TEC and scintillation environments of the three Superstorms considered here, as well as many other smaller storms. It is not possible to list all of them here, but the studies mentioned below can be considered representative of the types of effects that might be expected from an extreme event like the Carrington event, although we are reminded that the Carrington event was much more extreme than the three Superstorms of the past 15 years.

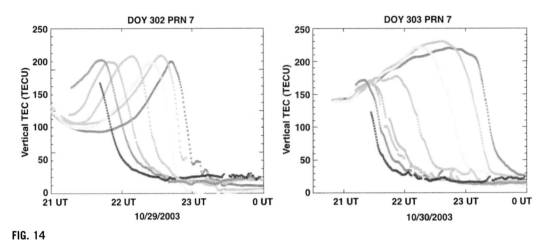

FIG. 14

TEC values from the 2003 Halloween Storm, measured from GPS receivers in the continental United States, located at longitudes of −116.28, −113.24, −111.53, −107.00, −103.27, −100.82, −97.96, −94.72 in a band of latitudes from 43 degrees to 47 degrees north. Only signals from PRN07 were used to construct this figure.

5.1 OCTOBER 2003

Mannucci et al. (2005) studied the dayside ionospheric TEC response to the storm. Mannucci et al. (2014) investigated the effect of the IMF B_Y component on the TEC response. Mannucci et al. (2008) compared the TEC response of all three superstorms listed here. Rao et al. (2010) compared the measured TEC against the TEC from the IRI model as preparation for development of the Indian SBAS system known as GPS Aided GEO Augmented Navigation (GAGAN). Bergeot et al. (2011) studied TEC variations and their impact on kinematic GPS positioning in Europe. Filjar et al. (2008) demonstrated significant vertical and horizontal positioning errors in Croatia during the storm. Mitchell et al. (2005) studied GPS TEC and scintillation measurements from the polar ionosphere. Bonelli (2008) studied scintillation in the Brazilian sector for several of the largest storms and found that scintillation is often inhibited during storms because of the westward electric field caused by the disturbance dynamo. Evans et al. (2004) described various impacts and operational challenges that occurred during the October 2003 time period.

5.2 NOVEMBER 2003

Meier et al. (2005), and Crowley and Meier (2008) described the development of the storm and used full physics models to explain the global changes occurring in the thermosphere and ionosphere. Kil et al. (2011) described the global TEC variations and their relationship to thermospheric composition. Astafyeva et al. (2014) described loss of lock and cycle slips from a global distribution of GPS receivers during four magnetic storms, including the November 2003 storm, and found that the loss of lock tends to peak during the maximum displacement of the geomagnetic Dst (or SYM-H) index.

5.3 NOVEMBER 2004

Mannucci et al. (2009) described the local time dependence of the prompt ionospheric response and TEC for the November 7, 9, and 10, 2004, superstorms. Rama Rao et al. (2009), demonstrated adverse effects on GPS, including phase slips, loss of lock, and range delay inferred from TEC in the Indian sector. Li et al. (2009) presented GPS TEC measurements of the November 2004 Superstorm in South East Asia. They found intense scintillations from low to middle latitudes in the Japanese sector during the second main phase (Nov. 10) of the storm, although the Chinese sector had no scintillation lower than 20 degrees latitude. Meggs et al. (2006) showed that the relationship between scintillation and loss of lock in northern Scandinavia depended on the receiver type.

Other large storms include the 2015 St. Patrick's Day Storm, which was the first storm of Solar Cycle 24 to reach a level of "Severe" on the NOAA geomagnetic storm scale (Wu et al., 2016). Astafyeva et al. (2015) and Carter et al. (2016) studied the global variation of plasma bubble occurrence during the storm. Jacobsen et al. (2016) provide an overview of scintillation in Norway, and its consequences for RTK and PPP positioning. Cherniak et al. (2015) and Prikryl et al. (2016) found that GPS phase scintillations were enhanced at the edge of the auroral region. Heine et al. (2017) identified small-scale irregularities along the poleward edge of the SED plume. Nayak et al. (2017) report that scintillation was suppressed in the Taiwanese sector on the storm day due to a reduced pre-reversal enhancement (PRE) in the eastward electric field, in contrast to the Indian sector, where both the PRE and scintillation were enhanced. A number of papers have described the ionospheric response in the Indian and Asian sectors (Ramsingh et al., 2015; Kakad et al., 2016; Nava et al., 2016; Tulasi Ram et al., 2016; Spogli et al., 2016). A special issue of Space Weather Journal was published in 2017 describing the responses to the St. Patrick's Day storms in 2013 and 2015 (Zhang et al., 2017, and references therein).

6 SYSTEM EFFECTS OF IONOSPHERIC STORMS

Because of its plasma properties, the ionosphere interacts with a broad range of electro-magnetic waves at frequencies that are important to civilian and military activities. Ionospheric "space weather" is one of the largest sources of error in Positioning, Navigation, and Timing (PNT) applications that use the Global Navigation Satellite System (GNSS) satellite constellations, including the U.S. government's GPS system. These space weather effects are important because GNSS signals are used by millions of people for positioning, navigation, and timing in a plethora of industries, including surveying, oil and mineral exploration, agriculture, construction, airlines, and the power industry (e.g., Hapgood, 2017).

In 2013, the U.K.'s Royal Academy of Engineering (Cannon et al., 2013) drafted a comprehensive report detailing the impacts of extreme space weather on engineered systems and infrastructure, including GNSS applications. As might be expected from the previous sections of this chapter, the main impacts will arise from TEC gradients and scintillation. Cannon et al. (2013) broke down the GPS-related infrastructure into different segments that will be impacted in different ways: single frequency receivers, dual frequency receivers, and augmented navigation systems.

- *Single-frequency* GPS receivers are now commonly used in automobile navigation systems and smart phones sold today. The single-frequency GPS receivers often operate on the L1 frequency, and their navigation accuracy is limited by ionospheric path delay. Earlier studies (Klobuchar,

1987) proposed a correction method to mitigate this delay using an empirical ionospheric model. Among the various models reported in the literature, the Klobuchar model is the one employed in most GPS receivers, and it makes use of eight broadcast coefficients (Klobuchar, 1987). As noted above, during intense solar and geomagnetic disturbances, the Klobuchar ionospheric model is inappropriate and will not be an accurate representation of the ionospheric TEC, resulting in increased PNT errors for single frequency GPS receivers. Cannon et al. point out that scintillation will add to the problem, because loss of signal from several satellites may result in greater dilution of precision, thus position errors of hundreds of meters may result, and there could even be a complete loss of positional and navigational solutions.

- *Dual-frequency* GPS receivers are an essential component of the precise position and timing services that are widely used in maritime and aviation navigation, surveying, agriculture, oil and mineral exploration, and banking industries. Dual-frequency GPS receivers require no modelling of the ionosphere because the availability of two signals, which have undergone the same ionospheric effects, is exploited to provide a direct measurement of TEC and a corresponding correction for ionospheric path delay. In the absence of measurement errors, the first-order ionospheric delay can be cancelled via a linear combination of the L1 and L2 pseudorange measurements (Misra and Enge, 2011). Thus, even in the presence of ionospheric electron density gradients, the dual-frequency receivers should remain operational. However, in an extreme event, it is likely that loss of signal from several satellites due to scintillation (e.g., Pi et al., 2017) will result in greater dilution of precision, large position errors and possibly complete loss of positional and navigational solutions. In an extreme event, Cannon et al suggest dual-frequency receivers may provide only a slight improvement over single frequency technology.

- *Augmented navigation systems* are designed to facilitate higher cadence aircraft landings by providing improved positioning and navigation capabilities. The FAA's Wide Area Augmentation System (WAAS) (Enge et al., 1996; Sparks et al., 2011a,b;) augments the GPS system with the goal to improve accuracy, integrity and availability. It enables aircraft to rely on GPS for all phases of flight, but especially landing. WAAS consists of a network of ground-based reference receivers that measure TEC and compute a "deviation correction" (Loh et al., 1995). These measurements are routed to Master Stations, and the deviations are transmitted up to dedicated WAAS satellites in geostationary orbit every 5 s. In turn, the satellites broadcast the corrections back to Earth, where WAAS-enabled receivers onboard aircraft can utilize corrections while computing their positions to improve accuracy. The accuracy goal of WAAS is 25 ft (vertical and horizontal positioning) or better for 95% of the time. Typically, accuracies are better than 3 ft both vertically and horizontally. Various countries are deploying their own WAAS-equivalent SBAS (Satellite Based Augmentation Systems) systems, including Brazil and India.

When strong gradients in the electron density occur, as during SED events, the WAAS system recognizes the situation and notifies users not to rely on it for precision navigation until the gradients have disappeared and the situation has returned to normal. Other differential systems, including those used for precision agriculture are similarly susceptible to ionospheric gradients, although many of them will warn users of the loss of integrity.

Cannon et al. (2013) point out that during the October 2003 Halloween Superstorm, while vertical navigation guidance from WAAS was unavailable for 30 h, horizontal guidance was continuously available, and the WAAS system worked exactly as planned by increasing the error bounds at the

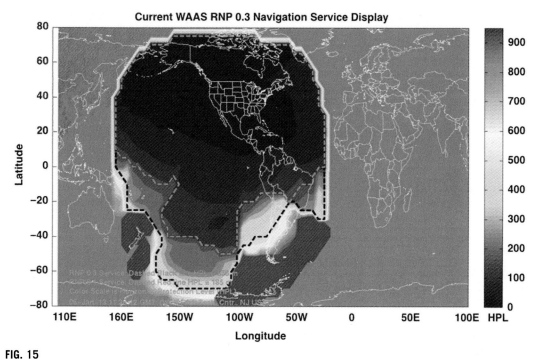

FIG. 15

Typical WAAS display showing a map of the Horizontal Protection Level (HPL). HPL is the radius of a circle in the horizontal plane (the plane tangent to the WGS-84 ellipsoid), with its center being at the true position, which describes the region that is assured to contain the indicated horizontal position. It is based upon the error estimates provided by WAAS.

affected locations. The magnitude of the October 2003 storm led to large TEC perturbations that came close to creating a hazardous situation for users, which necessitated the development of an "extreme storm detector" that has since been implemented in WAAS. This is discussed in more detail in the accompanying book chapter by Mannucci and Tsurutani (Fig. 15).

Figs. 11 and 12 depicted various TID structures seen in GPS TEC data over the Continental United States (CONUS). These TIDs in TEC data have spatial scales with a horizontal wavelength of 100 km to 3000 km, and periods of 10 min to several hours. These perturbations propagate at speeds of 50 to ~700 m/s. It is noteworthy that these TID speeds are similar to aircraft speeds, and the possibility exists that a TID and an airplane could be traveling in the same direction and at the same speed, so that the GPS receiver on the airplane "sees" a constant offset of TEC, and therefore height relative to the fixed ground systems surrounding an airport. If the TEC amplitude is large enough, the height difference between the airplane and ground-receiver could be significant enough to cause some danger. While this geometry is clear for a TID, the same problem arises for any phenomenon that causes the delay seen by the ground station to be significantly different from that seen by the airplane. Hence, SEDs and tongues of ionization that gradually move across the United States have the potential for steep TEC gradients that could similarly trigger the aforementioned situation. Clearly this is a potential problem for aircraft.

With sufficiently localized gradients, hazardous conditions could be created, as almost occurred after the storm of October 30, 2003. In the Chapter 20 of the same volume, Mannucci and Tsurutani discuss a different sort of feature that arose at night, but as yet has no adequate physical explanation.

7 DISCUSSION

This paper is intended to provide an understanding of how extreme ionospheric storms can affect GPS (or GNSS) systems. The GNSS systems are primarily affected by changes in TEC, and by the ionospheric irregularities that cause scintillation. We described a number of ionospheric phenomena that can lead to changes in the TEC, and we also presented GPS TEC measurements along with scintillation measurements. The largest ionospheric storms of the past 20 years occurred in October and November 2003, and in November 2004, when few GPS receivers were deployed. Since that time, the use of GPS/GNSS has exploded. Ionospheric space weather effects are now more important than ever, because GNSS signals are used by millions of people for positioning, navigation, and timing in a plethora of industries, including surveying, oil and mineral exploration, agriculture, construction, airlines and the power industry. Most users of these applications have never experienced the impact of a significant space weather event on their systems.

During a "geomagnetic storm" or ionospheric storm, the polar cap and aurora expand to lower latitudes, and enormous amounts of energy and momentum are deposited in the high-latitude region, resulting in global changes to the ionosphere and thermosphere as described in Section 3 above. The most extreme storm on record is the Carrington event of 1859, which was several times more intense than anything we have recorded in the space age (Ngwira et al., 2013). During large storms, the aurora, which is normally confined to the Arctic (or Antarctic) circle can be observed over the continental United States, United Kingdom and Europe, but in the Carrington event, auroras were reported as far south as Florida and the Caribbean. During such an event the ionospheric effects on GPS receivers would likely be significant. During large storms, the ionospheric electron densities at latitudes poleward of the aurora are significantly reduced due to upwelling molecular nitrogen gas, so the United States, Europe, Japan, and Southern Australia would experience dramatically reduced TEC compared to normal values. At other times, large variations in TEC could be expected due to dayside SEDs, polar cap patches, and nightside blobs. Because of the aurora and polar cap processes, phase scintillation could be expected throughout the United States, Europe, Japan, and Southern Australia for significant fractions of a day.

At the same time as the aurora expands equatorwards, large electric fields cause uplift of the low-latitude ionosphere, and the extent of the Appleton anomalies grows polewards. Thus, the mid-latitude ionosphere essentially disappears, squeezed by the high and low-latitude encroachment. Across Central America and Brazil, the Middle East, Southeast Asia, Northern Australia, and Africa, the Appleton anomalies will experience larger electron densities, but the embedded irregularities will generally cause scintillation within these regions.

Thus, at all latitudes, one can expect to see large changes in TEC, TEC gradients, and scintillation. At low latitudes, the amplitude scintillation will dominate, and at higher latitudes, the phase-scintillation will dominate. These changes will be modulated by variations of the Interplanetary Magnetic Field (IMF). During northward and southward IMF B_Z states, penetration electric fields will diminish or enhance, respectively, the conditions that produce ionospheric irregularities and cause low-latitude scintillation. Similarly, at high latitudes, a southward IMF B_Z strengthens and expands the aurora, whereas a northward IMF (Crowley et al., 1992) can result in transpolar arcs, both conditions causing phase scintillation.

Needless to say, the effects on GPS navigation and positioning systems will likely be severe. The large TEC variability will invalidate the Klobuchar model used in single frequency systems, and scintillation will likely result in dilution of precision for positioning users. Cannon et al. (2013) note that in an extreme storm, even dual frequency systems will likely be similarly affected. SBAS systems like WAAS will experience strong TEC gradients, so they will have reduced capability that will also likely be made worse by scintillation and loss of lock.

While the use of GNSS signals to provide precise position information for navigation, tracking, and mapping is well known, their use for precise timing, time intervals, and frequency measurements is much less well known. Precise timing is crucial for a variety of economic activities around the world. Communication systems, electrical power grids, and financial networks all rely on precision timing for synchronization and operational efficiency, as noted in the introduction to this chapter. The free availability of GPS time has enabled cost savings for companies that depend on precise time and has led to significant advances in capability. Significant timing errors can cause automated systems using GPS timing data to automatically shut down. GPS-disciplined timing servers are increasingly being utilized for providing precise timing information in a diverse array of settings including power grid operations. As our nation's dependence on reliable satellite navigation systems for precise position, navigation and timing applications increases, any errors/uncertainties or degradation of service will have significant life, safety, and economic impacts. The problem is exacerbated by the fact that many new applications of GPS have been deployed in the last 5 years, during which time there were few major space weather storms because the sun was at the lowest point in its eleven-year solar cycle. Thus, there are numerous customers for GPS correction products who do not yet know they are customers. Loss of GPS timing signals may be a greater threat than positioning and navigation problems during an extreme storm.

The Halloween 2003 and November 2004 storms lasted for 3–4 days due to the solar drivers. It seems likely that an extreme storm would develop under similar circumstances and would last for several days. During that time, any systems relying on GPS (or GNSS) signals will be unreliable. For this reason, GPS-reliant systems should consider mitigation techniques to withstand GPS outages lasting several days. In both the United States and Europe, extreme space weather has been recognized as a significant threat to critical infrastructure (e.g., Evans et al., 2004; Obama, 2016), and efforts are underway to understand and mitigate against this threat.

While engineering approaches to mitigation are likely to protect critical systems during quiet times and even during Superstorms like those in 2003 and 2004, there is a danger in relying on current knowledge and experience and assuming things will be the same during an Extreme Storm like the Carrington event. There is much that is still not understood about space weather, and several examples were mentioned above. It is important to continue the current strong thrust to study and understand space weather, and especially the ionospheric phenomena that have the greatest impact on operational systems and critical infrastructure. Areas of research that should be a high priority include the complex combination of physical processes involved in SAPS and SEDs, TIDs and their climatology, scintillation and its causes. It is critical to recognize the importance of fundamental research to understand the phenomenology and physical processes involved, as well as developing engineering solutions to mitigate the effects of space weather. Without understanding the physics and the phenomena, it seems likely that mitigation solutions will address perceived problems, but may lack the fundamental underpinning to avoid other phenomena that may be less common and just as dangerous or destructive.

ACKNOWLEDGMENTS

We are grateful for the opportunity to describe and explain ionospheric space weather and its potential impact on GPS systems and GPS-reliant technologies. We dedicate this chapter to the memory of our friend Dr. Paul Kintner of Cornell University who did so much to promote measurements of scintillation. His invention of the "Scintmon" receiver and his partnership with ASTRA led to the development and commercialization of the CASES receiver and later generations of ASTRA's scintillation monitors. GC's work at ASTRA was supported by Air Force contract FA9453-13-C-0035, ONR grant N00014-13-1-0343, NSF grants AGS-1144062 and AGS-1623962, and NASA grant NNX14AP88G. IA's participation in this study was supported by NASA contract NNH13CJ33C and NSF grant AGS-1552310 to ASTRA.

REFERENCES

Aarons, J., 1982. Global morphology of ionospheric scintillations. Proc. IEEE 70 (4), 360–378.

Aarons, J., 1991. The role of the ring current in the generation or inhibition of equatorial F layer irregularities during magnetic storms. Radio Sci. 26 (04), 1131–1149.

Aarons, J., 1993. The longitudinal morphology of equatorial F-layer irregularities relevant to their occurrence. Space Sci. Rev. 63 (3), 209–243.

Aarons, J., 1997. Global positioning system phase fluctuations at auroral latitudes. J. Geophys. Res. 102, 17219–17231. https://doi.org/10.1029/97JA01118.

Aarons, J., Lin, B., Mendillo, M., Liou, K., Codrescu, M., 2000. Global Positioning System phase fluctuations and ultraviolet images from the Polar satellite. J. Geophys. Res. 105, 5201–5213. https://doi.org/10.1029/1999JA900409.

Abdu, M.A., Batista, I.S., Walker, G.O., Sobral, J.H.A., Trivedi, N.B., De Paula, E.R., 1995. Equatorial ionospheric electric fields during magnetospheric disturbances: local time/longitude dependences from recent EITS campaigns. J. Atmos. Terres. Phys. 57 (10), 1065–1083.

Alfonsi, L., Spogli, L., De Franceschi, G., Romano, V., Aquino, M., Dodson, A., Mitchell, C.N., 2011. Bipolar climatology of GPS ionospheric scintillation at solar minimum. Radio Sci. 46, RS0D05. https://doi.org/10.1029/2010RS004571.

Aquino, M., Moore, T., Dodson, A., Waugh, S., Souter, J., Rodrigues, F.S., 2005. Implications of ionospheric scintillation for GNSS users in northern Europe. J. Navig. 58, 241–256. https://doi.org/10.1017/S0373463305003218.

Astafyeva, E., Yasyukevich, Y., Maksikov, A., Zhivetiev, I., 2014. Geomagnetic storms, super-storms, and their impacts on GPS-based navigation systems. Space Weather 12, 508–525. https://doi.org/10.1002/2014SW001072.

Astafyeva, E., Zakharenkova, I., Förster, M., 2015. Ionospheric response to the 2015 St. Patrick's Day storm: a global multi-instrumental overview. J. Geophys. Res. 120 (2015), 9023–9037. https://doi.org/10.1002/2015JA02162.

Azeem, I., Crowley, G., Reynolds, A., Santana, J., Hampton, D., 2013. First results of phase scintillation from a longitudinal chain of ASTRA's SM-211 GPS TEC and Scintillation Receivers in Alaska. In: Proceedings of the ION 2013 Pacific PNT Meeting, April 23–25, Honolulu, Hawaii, .

Azeem, I., Crowley, G., Honniball, C., 2015a. Global ionospheric response to the 2009 sudden stratospheric warming event using Ionospheric Data Assimilation Four-Dimensional (IDA4D) algorithm. J. Geophys. Res. Space Phys. 120. https://doi.org/10.1002/2015JA020993.

Azeem, I., Yue, J., Hoffmann, L., Miller, S.D., Straka III, W.C., Crowley, G., 2015b. Multisensor profiling of a concentric gravity wave event propagating from the troposphere to the ionosphere. Geophys. Res. Lett. 42. https://doi.org/10.1002/2015GL065903.

REFERENCES

Azeem, I., Vadas, S.L., Crowley, G., Makela, J.J., 2017. Traveling ionospheric disturbances over the United States induced by gravity waves from the 2011 Tohoku tsunami and comparison with gravity wave dissipative theory. J. Geophys. Res. 122. https://doi.org/10.1002/2016JA023659.

Basu, Su., Basu, S., 1985. Equatorial scintillations: advances since ISEA-6. J. Atmos. Terres. Phys. 47, 753–768. https://doi.org/10.1016/0021-9169(85) 90052-2.

Basu, S., Basu, S., Aarons, J., McClure, J., Cousins, M., 1978. On the coexistence of kilometer- and meter-scale irregularities in the Nighttime Equatorial F region. J. Geophys. Res. 83 (A9), 4219–4226.

Basu, S., Basu, S., Sojka, J.J., Schunk, R.W., MacKenzie, E., 1995. Macroscale modeling and mesoscale observations of plasma density structures in the polar cap. Geophys. Res. Lett. 22, 881–884. https://doi.org/10.1029/95GL00467.

Basu, S., et al., 2001a. Ionospheric effects of major magnetic storms during the international space weather period of September and October 1999: GPS observations, VHF/UHF scintillations, and in situ density structures at middle and equatorial latitudes. J. Geophys. Res. 106 (A), 30389–30414. https://doi.org/10.1029/2001JA001116.

Basu, S., Basu, S., Groves, K.M., Yeh, H.C., Su, S.Y., Rich, F.J., Sultan, P.J., Keskinen, M.J., 2001b. Response of the equatorial ionosphere in the South Atlantic Region to the Great Magnetic Storm of July 15, 2000. Geophys. Res. Lett. 28 (18), 3577–3580. https://doi.org/10.1029/2001GL013259.

Bergeot, N., Bruyninx, C., Defraigne, P., Pireaux, S., Legrand, J., Pottiaux, E., Baire, Q., 2011. Impact of the Halloween 2003 ionospheric storm on kinematic GPS positioning in Europe. GPS Solutions 15, 171–180.

Bilitza, D., Altadill, D., Zhang, Y., Mertens, C., Truhlik, V., Richards, P., McKinnell, L.-A., Reinisch, B., 2014. The International Reference Ionosphere 2012-A model of international collaboration. J. Space Weather Space Clim. 4, A07–A13. https://doi.org/10.1051/swsc/2014004.

Bonelli, E., 2008. Attenuation of GPS scintillation in Brazil due to magnetic storms. Space Weather. 6(9).

Brunner, F., Gu, M., 1991. An improved model for the dual frequency ionospheric correction of GPS observations. Manuscr Geodaet 16, 205–214.

Cannon, P.S., et al., 2013. Extreme Space Weather: Impacts on Engineered Systems. Royal Academy of Engineering, London, UK. p. 68. Available at http://www.raeng.org.uk/publications/reports/space-weather-full-report.

Carter, B.A., Yizengaw, E., Pradipta, R., Retterer, J.M., Groves, K., Valladares, C., Caton, R., Bridgwood, C., Norman, R., Zhang, K., 2016. Global equatorial plasma bubble occurrence during the 2015 St. Patrick's Day storm. J. Geophys. Res. 121 (2016), 894–905. https://doi.org/10.1002/2015JA02219.

Cherniak, L., Zakharenkova, I., Redmon, R.J., 2015. Dynamics of the high-latitude ionospheric irregularities during the 17 March 2015 St. Patrick's Day storm: ground-based GPS measurements. Space Weather 13 (2015), 585–597. https://doi.org/10.1002/2015SW00123.

Coker, C., Hunsucker, R., Lott, G., 1995. Detection of auroral activity using GPS satellites. Geophys. Res. Lett. 22 (23), 3259–3262. https://doi.org/10.1029/95GL03091.

Coster, A., Komjathy, A., 2008. Space weather and the Global Positioning System. Space Weather 6, S06D04. https://doi.org/10.1029/2008SW000400.

Crowley, G., 1996. Critical review on ionospheric patches and blobs. In: The Review of Radio Science 1992–1996. Oxford University Press.

Crowley, G., McCrea, I.W., 1988. A synoptic study of TIDs observed in the UK during the first WAGS campaign, October 10-18, 1985. Radio Sci. 23, 905–917.

Crowley, G., Meier, R.R., 2008. Disturbed O/N_2 ratios and their transport to middle and low latitudes, AGU Midlatitude Ionospheric Dynamics and Disturbances. AGU Geophys. Monograph Ser. 181, 221–234.

Crowley, G., Rodrigues, F., 2012. Characteristics of traveling ionospheric disturbances observed by the TIDDBIT Sounder. Radio Sci. 47, RS0L22. https://doi.org/10.1029/2011RS004959.

Crowley, G., Williams, P.J.S., 1987. Observation of the source and propagation of atmospheric gravity waves. Nature 328, 231–233.

Crowley, G., Jones, T.B., Dudeney, J.R., 1987. Comparison of short period TID morphologies in Antarctica during geomagnetically quiet and active intervals. J. Atmos. Terres. Phys. 49, 155.

Crowley, G., Emery, B.A., Roble, R.G., Carlson, H.C., Knipp, D.J., 1989a. Thermospheric dynamics during Sept. 18 and 19, 1984, Model Simulations. J. Geophys. Res. 94, 16925–16944.

Crowley, G., Emery, B.A., Roble, R.G., Carlson, H.C., Salah, J.E., Wickwar, V.B., Miller, K.L., Oliver, W.L., Burnside, R.G., Marcos, F.A., 1989b. Thermospheric dynamics during the Equinox Transition study of September 1984 II. Validation of the NCAR-TGCM. J. Geophys. Res. 94, 16945–16960.

Crowley, G., Cannon, P.S., Dozois, C.G., Reinisch, B.W., Buchau, J., 1992. Polar cap convection for B_z northward. Geophys. Res. Lett. 19, 657–660.

Crowley, G., Ridley, A.J., Deist, D., Wing, S., Knipp, D.J., Emery, B.A., Foster, J., Heelis, R., Hairston, M., Reinisch, B.W., 2000. The transformation of high-latitude ionospheric F-region patches into blobs during the March 21, 1990 storm. J. Geophys. Res. 105, 5215–5230.

Crowley, G., Hackert, C., Meier, R.R., Strickland, D.J., Paxton, L.J., Pi, X., Manucci, A., Christensen, A., Morrison, D., Bust, G., Roble, R.G., Curtis, N., Wene, G., 2006. Global thermosphere-ionosphere response to onset of November 20, 2003 Magnetic Storm. J. Geophys. Res. 111, A10S18. https://doi.org/10.1029/2005JA011518.

Crowley, G., Bust, G.S., Reynolds, A., Azeem, I., Wilder, R., O'Hanlon, B.W., Psiaki, M.L., Powell, S., Humphreys, T.E., Bhatti, J.A., 2011. CASES: a novel low-cost ground-based dual-frequency GPS software receiver and space weather monitor. In: Proceedings of the 24th International Technical Meeting of The Satellite Division of the Institute of Navigation, Portland, OR, pp. 1437–1446.

Crowley, G., Azeem, I., Reynolds, A., Duly, T.M., McBride, P., Winkler, C., Hunton, D., 2016. Analysis of traveling ionospheric disturbances (TIDs) in GPS TEC launched by the 2011 Tohoku earthquake. Radio Sci. 51, 507–514. https://doi.org/10.1002/2015RS005907.

Doherty, P.H., Delay, S.H., Valladares, C.E., Klobuchar, J.A., 2000. Ionospheric Scintillation Effects in the Equatorial and Auroral Regions. Paper presented at GPS 2000, Institute of Navigation, Salt Lake City, Utah.

Enge, P., Walter, T., Pullen, S., Kee, C., Chao, Y.-C., Tsai, Y.-J., 1996. Wide area augmentation of the global positioning system. Proc. IEEE 84, 1063–1088.

Evans, D.L., Lautenbacher, C.C., Rosen, R.D., Johnson, D.L., 2004. Intense Space Weather Storms October 19–Nov 7, 2003, Service assessment, U.S. Dept of Commerce, National Weather service, Silver Spring, Maryland.

Fejer, B.G., Scherliess, L., de Paula, E.R., 1999. Effects of the vertical plasma drift velocity on the generation and evolution of equatorial spread F. J. Geophys. Res. 104, 19859–19869.

Filjar, R., Kos, T., Cicin, V., 2008. GPS positioning performance in the wake of the Halloween 2003 geomagnetic storm. Published in proceedings of IEEE ELMAR 50th International Symposium, Zadar, Croatia, ISSN: 1334-2630.

Fish, C.S., Swenson, C.M., Crowley, G., Barjatya, A., Neilsen, T., Gunther, J., Azeem, I., Pilinski, M., Wilder, R., Cook, J., Nelsen, J., Burt, R., Whiteley, M., Bingham, B., Hansen, G., Wassom, S., Davis, K., Jensen, S., Patterson, P., Young, Q., Petersen, J., Schaire, S., Davis, C.R., Bokaie, M., Fullmer, R., Baktur, R., Sojka, J., Cousins, M., 2014. Design, development, implementation, and on orbit performance of the dynamic ionosphere CubeSat experiment mission. Space Sci. Rev. 181, 61–120. https://doi.org/10.1007/s11214-014-0034-x.

Foster, J.C., Coster, A.J., 2007. Conjugate localized enhancement of total electron content at low latitudes in the American sector. J. Atmos. Sol. Terr. Phys. 69, 1241–1252.

Foster, J.C., Vo, H.B., 2002. Average characteristics and activity dependence of the subauroral polarization stream. J. Geophys. Res.: Space Phys. 107(A12).

Fuller-Rowell, T., 2005. USTEC: a new product from the Space Environment Center characterizing the ionospheric total electron content. GPS Solutions 9 (3), 236–239.

Hapgood, M., 2017. Satellite navigation—amazing technology but insidious risk: why everyone needs to understand Space Weather. Space Weather 15, 545–548. https://doi.org/10.1002/2017SW001638.

Heine, T.R.P., Moldwin, M.B., Zou, S., 2017. Small-scale structure of the mid-latitude storm enhanced density plume during the March 17, 2015 St. Patrick's Day storm. J. Geophys. Res. 122, 3665–3677. https://doi.org/10.1002/2016JA022965.

Hines, C.O., 1960. Internal atmospheric gravity waves at ionospheric heights. Can. J. Phys. 38, 1441–1481.

Hocke, K., Schlegel, K., 1996. A review of atmospheric gravity waves and travelling ionospheric disturbances: 1982–1995. Ann. Geophys. 14 (9), 917.

Hooke, W.H., 1968. The response of the F-region ionosphere to internal atmospheric gravity waves. In: Acoustic-Gravity Waves in the Atmosphere. p. 367.

Jacobsen, K.S., et al., 2016. Overview of the 2015 St. Patrick's Day storm and its consequences for RTK and PPP positioning in Norway. J. Space Weather Space Climate 6, 12. https://doi.org/10.1051/swsc/2016004. id.A9.

Kakad, B., Gurram, P., Tripura Sundari, P.N.B., Bhattacharyya, A., 2016. Structuring of intermediate scale equatorial spread F irregularities during intense geomagnetic storm of solar cycle 24. J. Geophys. Res. 121 (2016), 7001–7012. https://doi.org/10.1002/2016JA022635.

Kelley, M.C., 1989. The Earth's Ionosphere, Plasma Physics and Electrodynamics. Academic, San Diego.

Kersley, L., Pryse, S.E., Wheadon, N.S., 1988. Amplitude and phase scintillation at high latitudes over northern Europe. Radio Sci. 23, 320–330. https://doi.org/10.1029/RS023i003p00320.

Kil, H., Kwak, Y.S., Paxton, L.J., Meier, R.R., Zhang, Y., 2011. O and N2 disturbances in the F region during the 20 November 2003 storm seen from TIMED/GUVI. J. Geophys. Res. 116, A02314. https://doi.org/10.1029/2010JA016227.

Kinrade, J., Mitchell, C.N., Smith, N.D., Ebihara, Y., Weatherwax, A.T., Bust, G.S., 2013. GPS phase scintillation associated with optical auroral emissions: first statistical results from the geographic South Pole. J. Geophys. Res. 118, 2490–2502. https://doi.org/10.1002/jgra.50214.

Kintner, P.M., Ledvina, B.M., De Paula, E.R., 2007. GPS and ionospheric scintillations. Space Weather. 5(9).

Klobuchar, J.A., 1987. Ionospheric time-delay algorithm for single-frequency GPS users. IEEE Trans. Aerosp. Electron. Syst. 3, 325–331.

Li, G., Ning, B., Zhao, B., Liu, L., Wan, W., Ding, F., Xu, J.S., Liu, J.Y., Yumoto, K., 2009. Characterizing the November 2004 storm-time middle latitude plasma bubble event in Southeast Asia using multi-instrument observations. J. Geophys. Res. 114, A07304. https://doi.org/10.1029/2009JA014057.

Loh, R., Wullschleger, V., Elrod, B., Lage, M., Haas, F., 1995. The U.S. Wide-Area Augmentation System (WAAS). Navigation 42, 435–465. https://doi.org/10.1002/j.2161-4296.1995.tb01900.x.

Loucks, D., Palo, S., Pilinski, M., Crowley, G., Azeem, I., Hampton, D., 2017. High-latitude GPS phase scintillation from E region electron density gradients during the 20–21 December 2015 geomagnetic storm. J. Geophys. Res. 122. https://doi.org/10.1002/2016JA023839.

MacDougall, J.W., 1990. The polar-cap scintillation zone. J. Geomag. Geoelectr. 42, 777–788.

Mannucci, A.J., Tsurutani, B.T., Iijima, B.A., Komjathy, A., Saito, A., Gonzalez, W.D., Guarnieri, F.L., Kozyra, J.U., Skoug, R., 2005. Dayside global ionospheric response to the major interplanetary events of October 29–30, 2003 "Halloween Storms" Geophys. Res. Lett. 32, L12S02. https://doi.org/10.1029/2004GL021467.

Mannucci, A.J., Tsurutani, B.T., Abdu, M.A., Gonzalez, W.D., Komjathy, A., Echer, E., Iijima, B.A., Crowley, G., Anderson, D., 2008. Superposed epoch analysis of the dayside ionospheric response to four intense geomagnetic storms. J. Geophys. Res. 113, A00A02. https://doi.org/10.1029/2007JA012732.

Mannucci, A.J., Tsurutani, B.T., Kelley, M.C., Iijima, B.A., Komjathy, A., 2009. Local time dependence of the prompt ionospheric response for the 7, 9, and 10 November 2004 superstorms. J. Geophys. Res. 114, A10308. https://doi.org/10.1029/2009JA014043.

Mannucci, A.J., Crowley, G., Tsurutani, B.T., Verkhoglyadova, O.P., Komjathy, A., Stephens, P., 2014. Interplanetary magnetic field B_Y control of prompt total electron content increases during superstorms. J. Atmos. Sol. Terr. Phys. 115. https://doi.org/10.1016/jastp.2014.01.001.

Meggs, R.W., Mitchell, C.N., Smith, A.M., 2006. An investigation into the relationship between ionospheric scintillation and loss of lock in GNSS receivers, in Characterising the Ionosphere, N. Atlantic Treaty Organization, Res. and Technol. Organ., Neuilly sur-Seine, France, pp. 5-1–5-10.

Meier, R.R., Crowley, G., Strickland, D.J., Christensen, A.B., Paxton, L.J., Morrison, D., 2005. First look at the November 20, 2003 super storm with TIMED/GUVI. J. Geophys. Res. 110, A09S41. https://doi.org/10.1029/2004JA010990.

Misra, P., Enge, P., 2011. Global Positioning System: Signals, Measurements and Performance: Revised, second ed. Ganga-Jamuna Press, Massachusetts.

Mitchell, C.N., Alfonsi, L., De Franceschi, G., Lester, M., Romano, V., Wernik, A.W., 2005. GPS TEC and scintillation measurements from the polar ionosphere during the October 2003 storm. Geophys. Res. Lett. 32, L12S03. https://doi.org/10.1029/2004GL021644.

Nava, B., Rodríguez-Zuluaga, J., Alazo-Cuartas, K., Kashcheyev, A., Migoya-Orue, Y., Radicella, S., Amory-Mazaudier, C., Fleury, R., 2016. Middle- and low-latitude ionosphere response to 2015 St. Patrick's Day geomagnetic storm. J. Geophys. Res. 121, 3421–3438. https://doi.org/10.1002/2015JA02229.

Nayak, C., Tsai, L.C., Su, S.Y., Galkin, I.A., Caton, R.G., Groves, K.M., 2017. Suppression of ionospheric scintillation during St. Patrick's Day geomagnetic super storm as observed over the anomaly crest region station Pingtung, Taiwan: a case study. Adv. Space Res. 60 (2), 396–405. https://doi.org/10.1016/j.asr.2016.11.036.

Ngwira, C.M., Pulkkinen, A., Wilder, F.D., Crowley, G., 2013. Extended study of extreme geoelectric field event scenarios for geomagnetically induced current applications. Space Weather 11, 121–131. https://doi.org/10.1002/swe.20021.

Nishioka, M., Tsugawa, T., Kubota, M., Ishii, M., 2013. Concentric waves and short-period oscillations observed in the ionosphere after the 2013 Moore EF5 tornado. Geophys. Res. Lett. 40, 5581–5586. https://doi.org/10.1002/2013GL057963.

O'Hanlon, B.W., Psiaki, M.L., Powell, S., Bhatti, J.A., Humphreys, T.E., Crowley, G., Bust, G.S., 2011. CASES: a smart, compact GPS software receiver for space weather monitoring. In: Proceedings of the 24th International Technical Meeting of the Satellite Division of the Institute of Navigation, Portland, OR, pp. 2745–2753.

Obama, B., 2016. Executive order—coordinating efforts to prepare the nation for space weather events. Available at https://obamawhitehouse.archives.gov/the-press-office/2016/10/13/executive-order-coordinating-efforts-prepare-nation-space-weather-events.

Pi, X., Iijima, B.A., Lu, W., 2017. Effects of ionospheric scintillation on GNSS-based positioning. NAVIGATION: J. Inst. Navigation 64 (1), 3–22.

Prikryl, P., Jayachandran, P.T., Mushini, S.C., Pokhotelov, D., MacDougall, J.W., Donovan, E., Spanswick, E., St. Maurice, J.-P., 2010. GPS TEC, scintillation and cycle slips observed at high latitudes during solar minimum. Ann. Geophys. 28, 1307–1316. https://doi.org/10.5194/angeo-28-1307-2010.

Prikryl, P., Jayachandran, P.T., Mushini, S.C., Chadwick, R., 2011. Climatology of GPS phase scintillation and HF radar backscatter for the high-latitude ionosphere under solar minimum conditions. Ann. Geophys. 29, 377–392. https://doi.org/10.5194/angeo-29-377-2011.

Prikryl, P., Zhang, Y., Ebihara, Y., Ghoddousi-Fard, R., Jayachandran, P.T., Kinrade, J., Mitchell, C.N., Weatherwax, A.T., Bust, G., Cilliers, P.J., Spogli, L., Alfonsi, L., De Franceschi, G., Romano, V., Ning, B., Li, G., Jarvis, M.J., Danskin, D.W., Spanswick, E., Donovan, E., Terkildsen, M., 2013a. An interhemispheric comparison of GPS phase scintillation with auroral emission observed at South Pole and from DMSP satellite. Special Issue Annals Geophys. 56 (2), R0216. https://doi.org/10.4401/ag-6227.

Prikryl, P., Ghoddousi-Fard, R., Kunduri, B.S.R., Thomas, E.G., Coster, A.J., Jayachandran, P.T., Spanswick, E., Danskin, D.W., 2013b. GPS phase scintillation and proxy index at high latitudes during a moderate geomagnetic storm. Ann. Geophys. 31, 805–816. https://doi.org/10.5194/angeo-31-805-2013.

Prikryl, P., et al., 2016. GPS phase scintillation at high latitudes during the geomagnetic storm of 17–18 March 2015. J. Geophys. Res. 121, 10448–10465. https://doi.org/10.1002/2016JA023171.

Rama Rao, P.V.S., Gopi Krishna, S., Vara Prasad, J., Prasad, S.N.V.S., Prasad, D.S.V.V.D., Niranjan, K., 2009. Geomagnetic storm effects on GPS based navigation. Ann. Geophys. 27, 2101–2110.

Ramsingh, S., Sripathi, S., Sreekumar, S., Banola, K., Emperumal, P., Tiwari, B.S., 2015. Kumar (2015), Low-latitude ionosphere response to super geomagnetic storm of 17/18 March 2015: results from a chain of ground-based observations over Indian sector. J. Geophys. Res. https://doi.org/10.1002/2015JA021509.

Rao, N.V., Madhu, T., Kishore, K.L., 2010. Geomagnetic storm effects on GPS aided navigation over low latitude South Indian region. Int. J. Comp. Sci. Network Sec. 10(3). http://paper.ijcsns.org/07_book/201003/20100306.pdf.

Skone, S., Kundsen, K., deJong, M., 2001. Limitation in GPS receiver tracking performance under ionospheric scintillation conditions. Phys. Chem. Earth 26, 613–621. https://doi.org/10.1016/S1464-1895(01)00110-7.

Sparks, L., Blanch, J., Pandya, N., 2011a. Estimating ionospheric delay using kriging: 1. Methodology. Radio Sci. 46 (6). https://doi.org/10.1029/2011RS004667. 41–13.

Sparks, L., Blanch, J., Pandya, N., 2011b. Estimating ionospheric delay using kriging: 2. Impact on satellite-based augmentation system availability. Rad. Sci. 46 (6). https://doi.org/10.1029/2011RS004781. 89–10.

Spilker, J., 1978. GPS signal structure and performance characteristics. J. Inst. Navigation 25, 121–146.

Spogli, L., Alfonsi, L., De Franceschi, G., Romano, V., Aquino, M.H.O., Dodson, A., 2009. Climatology of GPS ionospheric scintillations over high and mid-latitude European regions. Ann. Geophys. 27 (9), 3429–3437. https://doi.org/10.5194/angeo-27-3429-2009.

Spogli, L., et al., 2016. Formation of ionospheric irregularities over Southeast Asia during the 2015 St. Patrick's Day storm. J. Geophys. Res. 121, 12211–12233. https://doi.org/10.1002/2016JA023222.

Sreeja, V., Aquino, M., Elmas, Z.G., Forte, B., 2012. Correlation analysis between ionospheric scintillation levels and receiver tracking performance. Space Weather. 10, (6). S06005.

Steenburgh, R.A., Smithtro, C.G., Groves, K.M., 2008. Ionospheric scintillation effects on single frequency GPS. Space Weather 6, S04D02. https://doi.org/10.1029/2007SW000340.

Tsugawa, T., Saito, A., Otsuka, Y., Nishioka, M., Maruyama, T., Kato, H., Nagatsuma, T., Murata, K.T., 2011. Ionospheric disturbances detected by GPS total electron content observation after the 2011 off the Pacific coast of Tohoku Earthquake. Earth Planets Space 63, 875–879.

Tsurutani, B.T., Gonzalez, W.D., Lakhina, G.S., Alex, S., 2003. The extreme magnetic storm of 1–2 September 1859. J. Geophys. Res. 108 (A7), 1268. https://doi.org/10.1029/2002ja009504.

Tulasi Ram, S., Rama Rao, P.V.S., Prasad, D.S.V.V.D., Niranjan, K., Gopi Krishna, S., Sridharan, R., Ravindran, Sudha, 2008. Local time dependent response of postsunset ESF during geomagnetic storms. J. Geophys. Res. 113. https://doi.org/10.1029/2007JA01292.

Tulasi Ram, S., Yokoyama, T., Otsuka, Y., Shiokawa, K., Sripathi, S., Veenadhari, B., Heelis, R., Ajith, K.K., Gowtam, V.S., Gurubaran, S., et al., 2016. Duskside enhancement of equatorial zonal electric field response to convection electric fields during the St. Patrick's Day storm on 17 March 2015. J. Geophys. Res. 121, 538–548. https://doi.org/10.1002/2015JA021932.

Vadas, S.L., Crowley, G., 2010. Sources of the traveling ionospheric disturbances observed by the ionospheric TIDDBIT sounder near Wallops Island on 30 October 2007. J. Geophys. Res. 115, A07324. https://doi.org/10.1029/2009JA015053.

Vadas, S.L., Crowley, G., 2017. Neutral wind and density perturbations in the thermosphere created by gravity waves observed by the TIDDBIT sounder. J. Geophys. Res. 122, 6652–6678. https://doi.org/10.1002/2016JA023828.

Van Dierendonck, A.J., 1999. Eye on the ionosphere: measuring ionospheric scintillation events from GPS signals. GPS Solutions 2 (4), 60–63.

Ware, R., Exner, M., Feng, D., Gorbunov, M., Hardy, K., Herman, B., Kuo, H., Meehan, T., Melbourne, W., Rocken, C., Schreiner, W., Sokolovskiy, S., Solheim, F., Zou, X., Anthes, R., Businger, S., Trenberth, K., 1996. GPS sounding the atmosphere from low Earth orbit, preliminary results. Bull. Am. Meteorol. Soc. 77, 5–18.

Wu, C.C., Liou, K., Lepping, R.P., Hutting, L., Plunkett, S., Howard, R.A., Socker, D., 2016. The first super geomagnetic storm of solar cycle 24: "The St. Patrick's day event (17 March 2015)". Earth Planets Space 68, 151. https://doi.org/10.1186/s40623-016-0525-y.

Zhang, S.-R., Zhang, Y., Wang, W., Verkhoglyadova, O.P., 2017. Geospace system responses to the St. Patrick's Day storms in 2013 and 2015. J. Geophys. Res. 122. https://doi.org/10.1002/2017JA024232.

FURTHER READING

Trimble—GPS Tutorial—Error Correction, from http://www.trimble.com/gps_tutorial/howgps-error2.aspx (Accessed July 2017).

CHAPTER 24

RECENT GEOEFFECTIVE SPACE WEATHER EVENTS AND TECHNOLOGICAL SYSTEM IMPACTS

Robert J. Redmon*, William F. Denig*, Paul T.M. Loto'aniu*,[†], Dominic Fuller-Rowell[†,‡]

National Centers for Environmental Information, NOAA, Boulder, CO, United States *University of Colorado, Boulder, CO, United States*[†] *Space Weather Prediction Center, NOAA, Boulder, CO, United States*[‡]

CHAPTER OUTLINE

1 Introduction	588
2 Recent Events: Overview	588
3 Solar Origins of Activity	590
4 Geospace Response	591
4.1 Energetic Particles and Magnetic Field Observations at GEO	592
4.2 Geosynchronous Magnetopause Crossings	593
4.3 Radiation Environment at GEO	593
4.4 Radiation Environment at LEO	594
5 Ionospheric Effects	595
6 System Impacts	601
6.1. Technological System Impacts	601
6.2 Aviation Navigation System Impacts	602
7 Summary	605
Acknowledgments	605
References	605
Further Reading	609

CHAPTER 24 RECENT GEOEFFECTIVE SPACE WEATHER EVENTS AND TECHNOLOGY

KEY POINTS

1. NCEI maintains a broad, Sun-to-Earth space environment archive with a long history that allows analysts to follow disruptive space weather events from their Sun origins to technological system impacts;
2. We demonstrate geomagnetic storms with $Dst_{min} \sim -100/-200$ nT are able to partially or fully degrade the performance of U.S. and European aviation navigation systems WAAS and EGNOS. We speculate historic extreme events (Carrington-like events) will cause much worse adverse effects to those systems;
3. Operational parameters, beyond their usual real-time intentions, provide important future context for retrospective studies when used in concert with research quality products.

1 INTRODUCTION

The extremity of a space weather event is in the eye of the beholder. Moderate to extreme geomagnetic events can produce a range of adverse impacts from temporary degradation to permanent damage to a diverse array of costly technological systems (e.g., Oughton et al., 2017) and therefore, improving our understanding of the space weather susceptibility to critical infrastructure is an important activity (e.g., Baker et al., 2013; SWAP, 2015). We provide a Sun-to-Earth perspective, from mostly NOAA observations and model predictions, of the space environmental condition during three recent geoeffective events (February 2014, March 2015, and June 2015) to show the value of these products for general situational awareness and for activities attempting to draw a causal connection between technological anomalies and the space environment. We specifically demonstrate this point through a discussion of the availability of aviation navigational systems in the mid to high latitudes for the North American and European regions. The companion Chapter 12 of this volume focuses on the NOAA Geostationary Operational Environmental Satellite (GOES) constellation perspective of extreme space weather events which occurred over Solar Cycles 22 through 24. The remainder of this paper is laid out thusly: Section 2 provides an event overview, Section 3 the solar origins, Section 4 the geospace response, Section 5 the ionosphere response, Section 6 system impacts, and Section 7 a brief summary.

2 RECENT EVENTS: OVERVIEW

In this section, we provide a summary of the space environment and geospace response for all three events, with key details captured in Table 1. The content includes information regarding the occurrence of solar flares, coronal mass ejections (CMEs), solar energetic particles (SEPs), geostationary magnetopause crossings (GMC), global geomagnetic indices, spacecraft charging hazards, and aviation system impacts. Only the first and third events included an X class solar flare (Column 2). All three events saw geoeffective CME interactions with geospace, from a glancing blow (February event) to more direct impingement for the latter two events (Column 3). The first and third events saw moderate SEP levels (Column 4). All three events experienced significant compression/erosion of the magnetosphere, with the magnetopause moving inbound of geostationary orbit such that space assets with

Table 1 Event Summary

Event	Flares	CME	SEP	GMC	Indices and Scales (1–5)	Space Haz LEO	System Impacts
E1: February 2014	X4.9	Glancing blow	Moderate	Yes	Kp_{max} 5.3 Dst_{min} −94 nT Radio: R3 Radiation: S2 Geomag: G2	SC ↑ SEU ↑	WAAS and EGNOS LPV200 degraded
E2: March 2015 St. Patrick's	X2	1 Direct	Below SWPC threshold	Yes	Kp_{max} 7.7 Dst_{min} −223 nT Radio: R3 Radiation: N/A Geomag: G4	SC ↑↑	WAAS, EGNOS LPV200 degraded
E3: June 2015	No X 5 M class	3 Direct	Moderate	Yes	Kp_{max} 8.3 Dst_{min} −204 nT Radio: R2 Radiation: S3 Geomag: G4	SC ↑↑ SEU ↑	WAAS, EGNOS LPV200 degraded minorly

The eight columns are laid out thusly: (1) event information, (2) most significant solar flares, (3) coronal mass ejections, (4) solar energetic particles, (5) geostationary magnetopause crossings, (6) global indices and space weather scales, (7) space asset hazards, and (8) aviation system impacts.

orbital radii >6.6 geocentric found themselves in the magnetosheath or further out (Column 5, Section 4.2). These GMCs were observed by the GOES magnetometer and predicted by the Shue et al. (1998) model. Real-time estimates at the Earth's bow shock of the solar wind ram pressure (magnetospheric compression), and the interplanetary magnetic field southward component (erosion), when the model predicted a GMC event were on the order of (10 nPa, −16 nT), (22 nPa, −17 nT), and (30 nPa, −32 nT). The planetary geomagnetic activity index Kp peaked at 5.3, 7.7, and 8.3 for the three events, while the geomagnetic storm index Dst experienced the following minima: −94, −223, −204 nT (Column 6). By contrast, quiet geomagnetic periods (i.e., climatology) have $Kp \leq 1$ and $Dst \sim 0$ nT. The NOAA Space Weather Prediction Center (SWPC) issued alerts for "Strong R3" radio impacts for events E1 and E2, "Strong S3" radiation storm impacts for event E3, and "Severe G4" geomagnetic storm impacts for E2 and E3 (Column 6). The likelihood of Low Earth Orbit (LEO) spacecraft anomalies due to the radiation environment was elevated for all three events for spacecraft charging (SC) and for events E1 and E3 for single-event upsets (SEUs) (Column 7). All three events resulted in impacts to the WAAS (CONUS) and EGNOS navigation services due to the disturbed ionosphere (Column 8). These navigation aids allow aircraft pilots to make guided landing approaches even in poor weather conditions (see Section 6.2). As we focus on three recent events from the current Solar Cycle 24, we should reflect on the fact that much more extreme events can occur (as described in other chapters). The September 1859 (aka Carrington Event), March 1989, and Halloween 2003 storms of Solar Cycles 10, 22, and 23, respectively, resulted in a peak Dst geomagnetic activity index of approximately −850, −589, and −353 nT (Carrington, 1859, Riley and Love, 2017; Siscoe et al., 2006).

3 SOLAR ORIGINS OF ACTIVITY

Here we summarize solar observations from NOAA's GOES spacecraft. Observations of the solar corona in the X-ray band from the GOES-15 Solar X-ray imager (SXI) (top row) and X-ray sensor (XRS) (bottom row) are shown in Fig. 1. SXI images the Sun over a broad range of wavelengths from soft X-ray to extreme ultraviolet. The images shown here used the Thin Polymide filter (PTHN, 0.6–6.5 nm) with a 0.4 s integration time, which is particularly useful for imaging coronal holes (see below). Events one (E1), two (E2), and three (E3) are shown in the left, center, and middle columns, respectively. This three-column figure layout is repeated throughout the chapter. Common to all three events are the several bright, active regions in the equatorial and mid-latitudes, which are sources of increased X-ray flux, and potential flares, SEP, and CME events. It's worth noting that solar active regions aren't the only source of CMEs, and hence SEPs. CMEs are explosive events that may emanate from the same active region as solar flares or eruptive filament/prominence structures. SEPs are generated by the two primary acceleration mechanisms: flare, through magnetic reconnection, and shock, via CMEs propagating through the interplanetary medium (Gopalswamy et al., 2012). Darker high-latitude

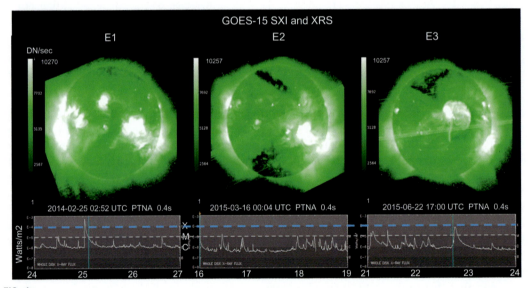

FIG. 1

GOES SXI solar X-ray (top) and XRS solar X-ray observations (bottom) for events E1 (left), E2 (middle) and E3 (right). SXI images were taken using the PTHN filter (0.6–6.5 nm), an exposure time of 0.4 s and are in uncalibrated units of (digital number (DN/s)). The images are for UTC date/times: E1 2014-02-25 02:52, E2 2015-03-16 00:04, E3 2015-06-22 17:00. The XRS time series is for the 0.1–0.8 band (are in watts/m^2) with all three events spanning 3 days (February 24–26, 2014; March 16–18, 2015; June 21–23, 2015). The vertical line *(cyan)* indicates the XRS observation at the time that the SXI image was generated. The dashed *blue* and *gray* lines demark the flux which needed to be classified as X-class and M-class events (10^{-4} and 10^{-5} watts/m^2). This three-column figure layout is repeated throughout the chapter.

From NOAA, https://ngdc.noaa.gov/stp.

patches are seen in events E2 and E3. These are representative of coronal holes, which are source regions for higher solar wind velocity, and in the downstream interplanetary medium create stream interaction regions (SIRs), which can produce roughly 27-day repeat frequency geomagnetic storms at geospace. In modern solar wind terminology, an SIR which persists for more than one solar rotation is known as a co-rotating interaction region (CIR). An SXI instrument has been included on GOES spacecraft since GOES-12 (first operational in January 22, 2003). For details on the SXI instrument design, modes of operation, and derived products, see Bornmann et al. (1996a), Hill et al. (2005), and references therein. Bands covered by the XRS instrument have been observed by NOAA satellites since 1974. For details on the XRS instrument design, modes of operation and derived products, see Bornmann et al. (1996b) and references therein.

Now we summarize the activities for our three events. For event E1, early on February 25, 2014, around 00:50 UTC, active region AR1990 (near equatorial, eastern solar limb) erupted, producing an X4.9 class flare, a moderate SEP event by midday, and a CME whose ejecta delivered a glancing blow 2.5 days later to geospace in the afternoon of February 27. For event E2 (aka St. Patrick's), on March 11, 2015, active region AR2297 (near center of solar disk) unleashed an X2 class flare and no significant SEP. On March 15, this active region (now near the western limb) erupted in concert with a nearby filament to propel a G4 producing CME toward Earth whose ejecta arrived at geospace in the early hours of March 17. While transiting the interplanetary medium, this CME interacted with a SIR which led to an enhanced period of IMF Bz south. For event E3, between June 18–25, 2015, active region AR2371 (near center of solar disk) yielded 5 M class flares, a moderate SEP event and a series of geoeffective CMEs which arrived between June 21–27. Noting the location on the Sun of eruptive events is indicative of the potential geoeffectiveness. For example, SEPs emanating from sources on the western solar hemisphere (right sided hemisphere as viewed by an Earth observer) are more likely to impact Earth as they nominally follow a Parker (Archimedean) spiral trajectory but their nominal trajectories can be deflected by the interaction with coronal holes emitting higher velocity solar wind (Gopalswamy et al., 2009).

Operational solar observing capabilities are in the midst of a dramatic enhancement with the dawn of the GOES-R series (first launched November 16, 2016), specifically provided by the new Extreme Ultraviolet and X-ray Irradiance Sensors (EXIS) and Solar Ultraviolet Imager (SUVI) instrument suites (e.g., automated flare location, thematic maps and dynamic feature tracking) (e.g., Eparvier et al., 2009; Chamberlin et al., 2009; Martínez-Galarce et al., 2010, 2013; Denig et al., 2014). In particular, advanced products being developed by NCEI aim to classify important regions of the solar disk such as coronal holes, filaments/prominences, flares, and bright regions. The GOES solar instruments provide critical operational observations for real-time nowcasting/forecasting, and valuable multidecadal contiguous data for retrospective studies alongside those from other sources such as the NASA Solar and Heliospheric Observatory.

4 GEOSPACE RESPONSE

In this section, we discuss the near-Earth geospace response at Geosynchronous Equatorial Orbit (GEO) and Low Earth Orbit (LEO) through the lens of charged particle and magnetic field observations from GOES and Polar Operational Environmental Satellite (POES) and Meteorological Operational (Metop) spacecraft.

4.1 ENERGETIC PARTICLES AND MAGNETIC FIELD OBSERVATIONS AT GEO

Observations of the GEO-charged particle and magnetic field environment from the GOES Space Environment Monitor (SEM) are shown in Fig. 2. Proton fluxes in 7 integral energy ranges from >1 to >100 MeV as 5-min time averages from the west viewing telescopes of the SEM Energetic Proton, Electron, and Alpha Detector (EPEAD) instrument are shown in the top row. The westward viewing telescopes for EPEAD are shown here as they observe larger solar proton fluxes than the eastward view due to the former seeing particles whose gyro centers lie outside geosynchronous orbit and are hence less filtered by the geomagnetic field (e.g., Rodriguez et al., 2010). Similarly, averaged

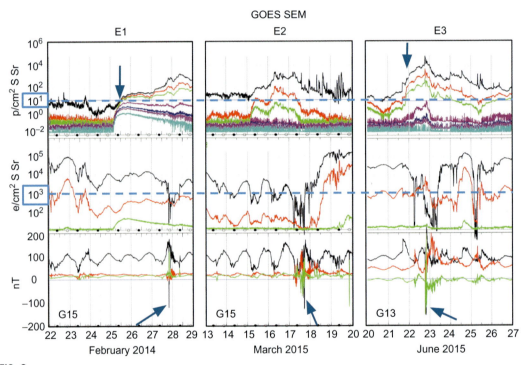

FIG. 2

GOES SEM measurements of charged particles and the magnetic field for events E1 (left), E2 (middle) and E3 (right) from the western (E1, E2) and eastern (E3) longitude observing location. Proton fluxes are shown in the top row for the 7 integral MeV energy ranges: >1 *(black)*, >5 *(red)*, >10 *(green)*, >30 *(magenta)*, >50 *(blue)*, >60 *(purple)*, and >100 *(cyan)*. SEP event onsets (E1, E3) are indicated by *blue* arrows where >10 MeV proton fluxes *(green)* exceed the dashed *blue* line *(green curve, top row)*. Electron fluxes are shown in the middle row for the 3 integral MeV ranges: >0.8 *(black)*, >2 *(red)*, >4 *(green)*. The dashed *blue* line is the alert level for >2 MeV electrons *(red curve)*. Vector magnetic measurements are shown in the bottom row with the EPN orientation: northward ("P," *black*), earthward ("E," *red*), and eastward ("N," *green*). Geostationary magnetopause crossings for all three events are annotated with *blue* arrows. The vertical trace in the MAG plot on the far right of E2 (middle row, bottom plot) is a calibration artifact. Events E1 and E2 show GOES-15, and E3 shows GOES-13, because GOES-15's magnetometer was in an anomalous state on June 25, 2015.

From NOAA, https://ngdc.noaa.gov/stp.

electron fluxes in the 3 integral energy ranges: >0.8, >2, and >4 MeV are shown in the middle row. The vector magnetic field averaged to 1 min from the MAG instrument is shown in the bottom row. An SEM package has been included on all NOAA geostationary satellites, starting with the Synchronous Meteorological Satellite (SMS)-1 (launched on May 17, 1974). For a summary of the EPEAD instrument design, and derived products, see Hanser (2011), Rodriguez et al. (2014, 2017), Rodriguez (2014) and references therein. For a similar summary of the GOES MAG instrument see Singer et al. (1996).

4.2 GEOSYNCHRONOUS MAGNETOPAUSE CROSSINGS

The impingement of solar wind on geospace has the effect of compressing (through dynamic pressure) and eroding (through reconnection) the dayside magnetosphere, with the solar wind dynamic pressure acting as the dominant parameter (Shue et al., 1998, their Fig. 13). As evident in the MAG observations negative Hp deflection (Fig. 2 bottom row, black trace), for all three events, the magnetosphere was compressed/eroded sufficiently to move the dayside magnetopause (nominally 10–12 Re geocentric) inbound of GEO orbit (6.6 Re), placing GOES and other GEO spacecraft in the magnetosheath. These GMCs were also predicted by the Shue et al. (1998) model (not shown here). Real-time estimates at the Earth's bow shock of the solar wind ram pressure (magnetospheric compression), and the interplanetary magnetic field southward component (erosion), when the model predicted a GMC event were on the order of (10 nPa, −16 nT), (22 nPa, −17 nT), and (30 nPa, −32 nT). Electron and ion density and temperature moments afforded by GOES-16's new MPS-LO will provide for improved detection of GMCs (i.e., Suvorova et al., 2005). The moments and magnetopause location products will be transitioned from NCEI and used operationally by SWPC (i.e., Petrinec et al., 2017; Redmon et al., 2014).

4.3 RADIATION ENVIRONMENT AT GEO

Solar energetic protons gained access to geostationary orbit for all three events (Fig. 2, top panel) but the >10 MeV particle flux (green) only exceeded the SWPC SEP alert threshold (>10 pfus, dashed blue line) for events E1 and E3 as indicated by the blue arrows. The energy spectrum is also much harder in E1. GOES >10 MeV observations are used for SWPC's solar radiation storm scale which reached alert levels of: S2 "moderate" and S3 "strong," for E1 and E3, respectively (Table 1, Column 6). For event E2, SWPC issued a warning that a S1 "minor" radiation storm might develop. It was predicted that Events E1 (S2) and E3 (S3) elevated the radiation risks to astronauts, the likelihood of space hardware Single Event Upsets (SEUs) (Table 1, Column 7), and transionospheric radio absorption (Fig. 4). It's important to note that these are modest, and thus by no means extreme, SEP events (e.g., He and Rodriguez, 2017; Shea and Smart, 2012).

Radiation belt electrons (middle row of Fig. 2) show signs of enhancement for all event periods, with the >2 MeV electrons (red) exceeding the SWPC alert threshold (>1000 pfus) for the latter two events and weakly before the E1 SEP onset. MeV electrons are known to penetrate typical spacecraft shielding, and this flux threshold has been chosen because spacecraft bathed in such an environment for prolonged periods risk degradation and irreversible damage to space hardware through internal electrostatic discharge (IESD) (Bodeau, 2010; Wrenn and Smith, 1996). For E2 and E3, >2 MeV electrons are elevated for about 11 days each beginning on March 18 and June 24, respectively. The slightly elevated flux in E1 is not a precursor of this event period; rather, it is more likely a remnant of previous magnetospheric activity. Additionally, GOES particle instruments do not observe solar origin, energetic electrons

594 CHAPTER 24 RECENT GEOEFFECTIVE SPACE WEATHER EVENTS AND TECHNOLOGY

(except perhaps briefly during GMCs). Finally, the dominant variation at these energies during quiescent and moderate times is diurnal, owing to the shape of the magnetosphere swept out by the GOES equatorial, circular orbit. Complementing the value of GOES electrons for IESD risk assessment, GOES-16's new Magnetospheric Particle Sensor-Low Energy Range (MPS-LO) extends the lower end of the electron energy range (covering 0.03–30 keV) (e.g., Dichter et al., 2015), a population known to contribute to frame (or bulk) charging at GEO (DeForest, 1972). Thus, a new frame charging product is being developed at NCEI for transition to SWPC.

4.4 RADIATION ENVIRONMENT AT LEO

Observations of the low-altitude, charged particle environment observed by the POES and Metop SEM-2 Medium Energy Proton and Electron Detector (MEPED) instrument for events E1–E3 are shown in Fig. 3. Highly inclined LEO spacecraft pass through the horns of the radiation belts, a regime

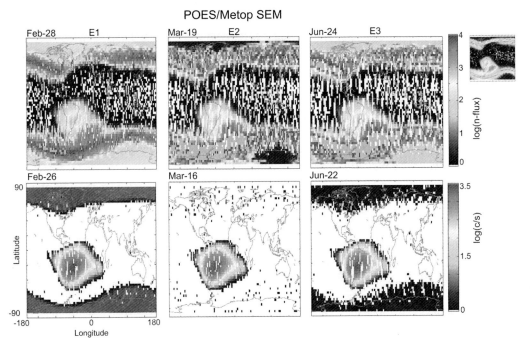

FIG. 3

POES/Metop SEM measurements of medium energy (>30 keV) electrons (top row) and high energy (>35 MeV aka "omni P7") protons (bottom row) during events E1 (left), E2 (middle) and E3 (right). The electron maps (top row) were created using the 90-degree telescope (aka spacecraft wake direction) and the units are directional number flux (log(#/cm^2 sr s)). The proton maps (bottom row) were created using the P7 omnidirectional dome detector and the units are log(counts/s). POES/Metop spacecraft M01, M02, N15, N16 (E1 only), N18, and N19 contributed to these event maps. The elevated feature covering the SAA in the electron maps (top row) is due to proton contamination of the electron telescope and should be ignored. The smaller inset map in the top right corner is representative of a quiescent period.

From NOAA, https://ngdc.noaa.gov/stp.

rich with energetic protons and electrons, several times per day. Depending on their precession rates and altitudes, LEO assets may also pass through the South Atlantic Anomaly (SAA) a region of inner radiation belt protons and often the source of Single Event Upsets (SEUs). The top row of Fig. 3 presents the average flux of medium energy (>30 keV) electrons from the telescope pointing in the satellite wake direction (i.e., 90-degree detector (90E1)). These electrons are known to increase the risk of surface charging (e.g., Anderson, 2012). The smaller inset map is representative of a quiescent period. The bottom row displays the average count rate for >35 MeV protons from the omni-directional (aka "omni") dome detector "P7." These ions are known to increase the likelihood of SEUs. While SWPC uses the GOES >10 MeV (Fig. 2, Table 1) ion measurements for their solar radiation storm scale, we discuss the >35 MeV ion measurements from POES/Metop here for two reasons: the lowest channel >16 MeV "omni" proton measurements are known to be contaminated by radiation belt electrons and the >35 MeV measurements are those used in the "human-in-the-loop" decision charts reported in O'Brien et al. (2012). A SEM-2 package (successor to SEM-1) has been included on all NOAA POES LEO satellites, starting with NOAA-15 (launched on May 13, 1998) and EUMETSAT Metop satellites (first launched on October 19, 2006). For a description of the availability and processing of the standard POES/Metop SEM-2 measurements see Evans and Greer (2004), Green (2013a,b), Redmon et al. (2015, 2017).

Evaluating the charged particle maps displayed in Fig. 3, a few observations are readily made. The specific days shown are noted on the figure and these were the days during our events which observed the most elevated electron fluxes (top row) or proton counts (bottom row). The >30 keV auroral electron fluxes (top row) were elevated for all three events with respect to climatology (see inset on the upper right). Additionally, the radiation belt indices which are derived from the same POES/Metop observations estimate that the daily average flux for our events was roughly: 100% (E1) and 800% (E2 and E3) above climatology (see Table 2 for access). These elevated observations imply an increased risk to surface charging of LEO spacecraft (as noted in Table 1) during high-latitude auroral crossings. The elevated feature covering the SAA in the electron maps (top row) is due mostly to proton contamination of the electron telescope and should be ignored. The elevated >35 MeV proton measurements at high latitudes (bottom row, left column) are due to the in-progress SEP events (see also top row of Fig. 2) and imply that spacecraft operators should consider SEU effects when assessing anomalies (i.e., O'Brien et al., 2012). The SAA feature in mid-to-low southern latitudes is a stable feature, generally independent of space weather activity, and provides a regular source of SEUs for LEO spacecraft. When assessing the space environmental state for potential attribution to satellite anomalies, it is necessary to assemble a multidisciplinary team that includes space environmental expertise early in the process and to consider the complete history of the space vehicle under similar environmental conditions (e.g., O'Brien et al., 2012; Mazur et al., 2015; Redmon et al., 2017).

5 IONOSPHERIC EFFECTS

In this section, we focus on the ionosphere response. Predictions of the degree of HF radio absorption predominantly in the D-region of the upper atmosphere, from the D-Region Absorption Prediction (DRAP) product for events E1–E3 are provided in Fig. 4. Details of the DRAP methodology and validation efforts are described in Sauer and Wilkinson (2008) and Akmaev et al. (2010) and references therein. Essentially, X-ray flares and solar particles during SEP events ionize the upper atmosphere at

Table 2 Data Source Locations

Domain	Platform	Provider	Access
Solar	GOES SXI	NCEI	https://sxi.ngdc.noaa.gov/
Magnetopause	GOES, Model	NCEI	https://www.ngdc.noaa.gov/stp/mag_pause/
Radiation Belts	GOES SEM	NCEI	https://www.ngdc.noaa.gov/stp/satellite/goes/
	POES/Metop SEM	NCEI	https://www.ngdc.noaa.gov/stp/satellite/poes/
	Belt Indices	NCEI	https://satdat.ngdc.noaa.gov/sem/poes/data/belt_indices/
Ionosphere	DRAP	NCEI	https://www.ngdc.noaa.gov/stp/drap/
	US-TEC	NCEI	https://www.ngdc.noaa.gov/stp/IONO/USTEC/
	GloTEC	SWPC	By request
	Madrigal	MIT Haystack	http://madrigal.haystack.mit.edu/madrigal/ Append "experiments/year/gps/ddmmmyy/images/"
	Aurora Forecast OVATION	SWPC	http://www.swpc.noaa.gov/products/aurora-30-minute-forecast
Indices	Kp, Dst	NASA	https://cdaweb.sci.gsfc.nasa.gov/cgi-bin/eval2.cgi The Dst index used for our 3 events is from the "provisional" OMNI database
Alerts	Radio, Radiation, Geomagnetic	SWPC	Scales: http://www.swpc.noaa.gov/noaa-scales-explanation Timeline: http://www.swpc.noaa.gov/products/notifications-timeline Archive: ftp://ftp.swpc.noaa.gov/pub/alerts/
Sun to Earth	Various	spaceweather.com	http://spaceweather.com/
Aviation	WAAS	FAA	Top: http://www.nstb.tc.faa.gov/DisplayDailyPlotArchive.htm Events: http://ftp.nstb.tc.faa.gov/pub/NSTB_data/24HOURPLOTS/ Append: W1781D4: 27Feb2014, W1836D2: 17Mar2015, W1850D1: 22Jun2015
	EGNOS	EDAS	Protection Level: https://egnos-user-support.essp-sas.eu/new_egnos_ops/protection_level LPV200: https://egnos-user-support.essp-sas.eu/new_egnos_ops/lpv200_availability Courtesy of ESSP and European GNSS Agency, produced under a program funded by the European Union

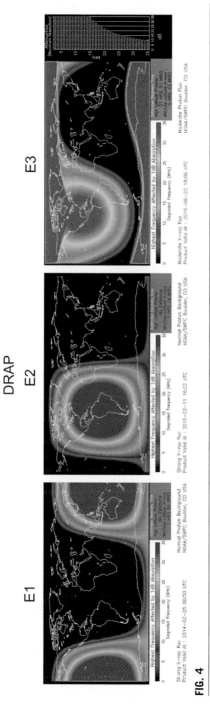

FIG. 4

DRAP estimates (in dB of attenuation) the most significant global ionospheric radio absorption effects due to X-rays (dayside) and to solar energetic protons (polar caps) for events E1 (left), E2 (middle), and E3 (right). Snapshots for E1, E2, and E3 are taken at 2014-02-25 00:50 UTC, 2015-03-11 16:22 UTC, and 2015-06-22 18:06 UTC.

From NOAA, https://ngdc.noaa.gov/stp.

equatorial to mid-latitudes, and high latitudes, respectively. These effects can significantly enhance ionospheric electron densities, potentially disrupting HF radio transmission, whose continuity is required for all commercial airline transpolar flights by the U.S. Federal Aviation Administration regulation (Section 121.99). DRAP operational real-time estimates are made available on a 1-min cadence. The specific times shown in Fig. 4 are chosen to be E1: 2014-02-25 00:50 UTC, E2: 2015-03-11 16:22 UTC, and E3: 2015-06-22 18:06 UTC. For event E1, early on February 25, 2014, around 00:50 UTC, active region AR1900 erupted, producing an X4.9 class flare (Fig. 1), and a moderate SEP event by midday (Fig. 2). Fig. 4, E1 (left) shows the intense increase in predicted radio absorption due to the X class flare's ionization of the D-region for all low-to-middle latitudes and dayside longitudes. Later in the day, the arriving solar protons entered the polar cap and DRAP estimated high-latitude radio absorption (not shown here). For event E2, on March 11, 2015, active region AR2297 unleashed an X2 class flare in the early afternoon, and no significant SEP. The predicted dayside absorption from this X2 flare is clearly seen in the E2 figure (middle). For event E3, between June 18–25, 2015, active region AR2371 yielded 5 M class flares, and a moderate SEP event. So, for the E3 DRAP figure (right), we have chosen a time that demonstrates the response to the M class flare in the afternoon of June 22 and the peak in the solar proton flux (see also Fig. 2, top-right panel). We should reinforce the idea that climatologically, the D-region is not sufficiently ionized to significantly affect HF propagation and thus the canonical quiet-time DRAP estimate would indicate no additional attenuation (a completely dark map). For these three events, we would be watchful for technological system impacts to: HF radio communications in the low to mid-latitudes (all) and the high latitudes (E1, E3) (e.g., Kurkin et al., 2015), and potentially aviation polar flight rerouting (E1, E3).

Estimates of the aurora latitudinal extent and the kinetic energy flux due to precipitating electrons and ions for events E1, E2, and E3 are presented in Fig. 5. These estimates are arrived at using the SWPC operational implementation of the Oval Variation, Assessment, Tracking, Intensity and Online Nowcasting (OVATION) Prime (OP) empirical model (Newell et al., 2014 and references therein). In Fig. 5, we are displaying the OP maps representing the peak hemispheric power estimated during our three events. For all three events, the auroral zone was expanded and the kinetic energy deposition was well above climatology (nominally 10 GW). In particular OP estimated these powers: 79 GW (E1), 152 GW (E2), and 124 GW (E3). Using these estimates of the location and intensity of the auroral zone, we would be watchful for technological system impacts to: spacecraft surface charging (e.g., Anderson, 2012), radar impacts due to auroral clutter (e.g., Ravan et al., 2012), and radio communication and mid- to-high latitude Global Navigation Satellite System (GNSS) impacts due to scintillation. An expanded auroral zone promotes the formation of storm enhanced density (SED) plumes (via plasmaspheric erosion), that then propagate into the polar cap (e.g., Coster et al., 2003; Foster et al., 2005; Mendillo, 2006; Heelis et al., 2009; Sojka et al., 2012) (see also Section 6.2). Radio waves transiting through steep gradients in the ionospheric and plasmaspheric TEC associated with SEDs are scintillated.

Estimates of the vertically integrated, total electron content from the US-TEC (bottom row), GloTEC (middle) and Madrigal (top) services for events E1–E3 are provided in Fig. 6. US-TEC is an operational real-time product created by SWPC. Details of its Global Positioning System (GPS) data assimilation scheme are described in Fuller-Rowell (2005) and Fuller-Rowell et al. (2006). Two independent efforts by Araujo-Pradere et al. (2007) and Minter et al. (2007) have estimated the slant path accuracy to be roughly 2.5 TEC units (or 2.5×10^{16} electrons/m^2). The US-TEC estimates shown herein were derived from slant TEC observations using several dozen GPS North American ground receivers isolated almost entirely to the Contiguous United States (CONUS) region

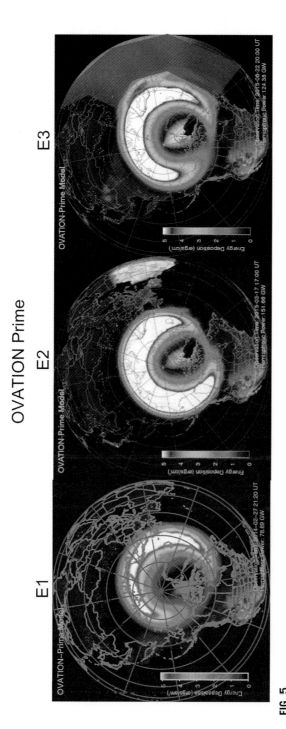

FIG. 5

OVATION Prime estimates of peak auroral power and latitudinal extent for events E1 (left), E2 (middle), and E3 (right). Snapshots for E1, E2, and E3 are taken at 2014-02-27 21:20UT (79 GW), 2015-03-17 17:00 UTC (152 GW), and 2015-06-22 20:00 UTC (124 GW). The units of the color bar are ergs/cm^2. The dusk to pre-midnight shading at mid-to-low latitudes for E3 is a model visualization artifact.

From NOAA, https://ngdc.noaa.gov/stp.

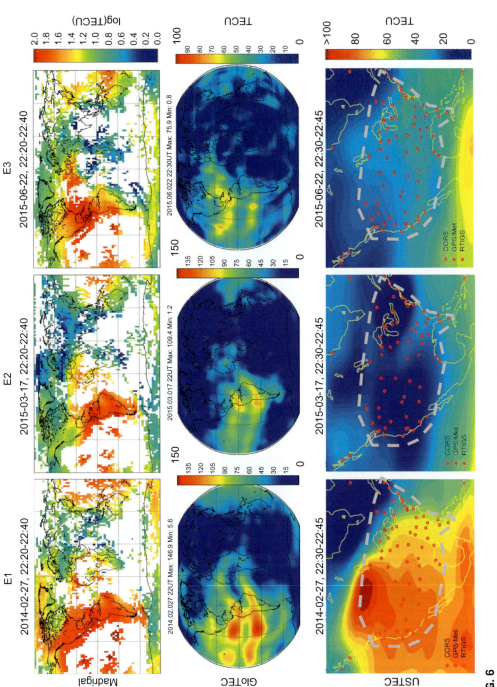

FIG. 6

Madrigal (top), GloTEC (middle), and US-TEC (bottom) estimates of impacts to the ionosphere for events E1 (left), E2 (middle), and E3 (right). Snapshots for E1, E2, and E3 are taken at roughly 22:30 UTC for all three events, placing the Eastern CONUS Coast predusk, at 17:30 LT. For Madrigal (top), the integration time is 20 min. For GloTEC (middle), the cadence is 20 min and the range is 0–100 TEC units, and up to 350 GPS ground stations and roughly 30–90 COSMIC occultations were used per hour. For US-TEC (bottom), the cadence is 15 min, and GPS ground receivers from the Continuously Operating Reference Station (CORS) (circles), GPS/Met (triangles) and Real-Time IGS (RTIGS) (stars) networks contributed to these maps and specific station locations are shown on each map. TEC estimates in regions not well covered by GPS ground stations are not reliable and the dashed overlay roughly encapsulates the US-TEC CONUS region of validity.

From NOAA and Madrigal, https://ngdc.noaa.gov/stp and http://madrigal.haystack.mit.edu/madrigal/.

6 SYSTEM IMPACTS   601

(gray dashed line, bottom row). Recently, SWPC has expanded the region of coverage for their operational US-TEC product to include all of North America (named NATEC), which is useful for assessing events occurring after those discussed herein. Another activity at SWPC is the Global TEC (GloTEC) project which assimilates both GPS ground receiver line of sight observations (like US-TEC) and space-based GPS radio occultation data from the COSMIC-1 constellation to provide optimal global estimates of TEC. We present preliminary TEC estimates from GloTEC herein. Efforts to validate the GloTEC method are largely based on Yue et al. (2014). Madrigal global TEC estimates (top row) are arrived at by assimilating GPS receiver data from >2000 ground stations and details of the processing scheme are discussed in Rideout and Coster (2006). All TEC maps shown in Fig. 6 were chosen to be around 22:30 UTC, which places the CONUS East Coast at 17:30 LT, and heading into dusk.

Several key features of the TEC maps in Fig. 6 are worth discussing. For all three events, the TEC, which has major contributions from both the ionosphere and plasmasphere, is elevated above climatology, and exhibits strong spatial gradients. Typical, climatological TEC values for central CONUS at predusk are ~30 TECU. Event E1 (left column) shows the most significant response with values as high as 100 TECU. All three events show signs of a SED plume propagating north-westward into a Tongue of Ionization (TOI) that convects into the polar cap. Event E1 shows the most clearly prominent TOI in this figure and its shape is quite classical (e.g., Foster et al., 2005, their Fig. 2). Other researchers have further leveraged TEC observations to create disturbance indices. Cherniak et al. (2015) used the Rate of TEC (ROT) and Rate of TEC index (ROTI) estimates to study the 2015 St. Patrick's storm (our event E2) and found the high-latitude, polar cap region to be highly structured due to the formation of SED and TOI contributions. Zhang et al. (2015, 2017) also reported on similar phenomenology during this storm. From the perspective of GPS, the ionosphere is a dispersive medium. The features described above attenuate, alter propagation delay (more or less than climatology) and scintillate GPS signals, and may result in the degradation of GNSS systems such as to Precise Point Positioning (PPP, e.g., surveying), Real Time Kinematic (RTK, e.g., farming) (Jacobsen and Andalsvik, 2016) and to aviation guidance applications (e.g., provided by WAAS and EGNOS, see also Section 6.2).

6 SYSTEM IMPACTS

Here we aggregate and summarize the risks posed to technological systems mentioned previously (Section 6.1) and discuss in detail impacts to the WAAS and EGNOS services (Section 6.2).

6.1 TECHNOLOGICAL SYSTEM IMPACTS

This subsection summarizes the impact risks introduced throughout previous sections, in a Sun-to-Earth order. System operators and analysts should be mindful of the potential for SEUs and/or increased radiation dose on interplanetary, GEO, and LEO assets and for astronauts and transpolar airline crew (E1, E3), complications to satellites using magnetic torquer attitude control systems (all); deep IESDs at GEO and MEO, and possibly LEO (E2, E3); increased risk of surface and subsurface discharges at LEO (all); transionospheric radio impacts to low to mid latitudes during daytime hours (all) and high latitudes (E1, E3); high latitude radar operations (all); electrical power distribution systems (E2, E3) through geomagnetically induced currents (GICs); and aviation navigation systems (all, see next section). It is worth noting that far more extreme events can and do occur, and thus pose far greater risks to technological systems than the three events we have reviewed from the current solar cycle. Collectively, the 1859 Carrington, March 1989, and Halloween 2003 super storms resulted in widespread brilliant and dynamic auroral light displays, and outages to critical contemporary systems. We'll summarize just a few impacts spanning diverse technologies. During the Carrington storm, telegraph systems, the modern form of telecommunications of the time, were nonfunctional over North America and Europe. During the 1989 storm, as the auroral zone expanded to lower latitudes, intense GICs developed, tripping several transmission lines from the James Bay hydroelectric power system and leaving inhabitants of the province of Quebec without power for around 9 h. The Halloween 2003 storm period was responsible for over half of the 70 reported anomalies experienced by spacecraft in 2003 (NRC Report, 2008).

6.2 AVIATION NAVIGATION SYSTEM IMPACTS

Now we'll employ a subset of the aforementioned space environment observations and model estimates presented in Section 5 to show their value in assessing the causality between space weather and impacts on GNSS and in particular on the U.S. Wide Area Augmentation System (WAAS) and the European Geostationary Navigation Overlay Service (EGNOS) aviation guidance systems for the three events discussed herein.

GNSS radio signals travel uniformly through space from an altitude of roughly 20,000 km until they encounter both the neutral and ionized plasma of Earth's atmosphere, where the signal is diffracted and refracted through the medium. This phenomenon can affect: aircraft operations, commercial farming equipment automation, geophysical surveying, and other processes that are reliant on GNSS technology. During significant space weather events, the propagation speed of the codes broadcast on the L1 and L2 signals are delayed, causing increased uncertainty in calculated positions of GNSS devices on the ground. In addition, the wave fronts of radio signals traveling through steep gradients in the ionospheric structure are distorted, causing scintillation. Rapid variations in the phase and amplitude of the scintillated radio signal reduces the accuracy of the arrival time measurements used to calculate receiver location with respect to each GNSS satellite. Thus, if a commercial or civilian user of GNSS technologies relies on a position correction and happens to be in an area affected by scintillation, they may experience degradation from reduced positioning accuracy to complete loss of positioning services ("loss of lock").

Commercial aviation uses various navigational aids and both the WAAS and EGNOS systems augment "GPS using the L1 (1575.42 MHz) Coarse/Acquisition (C/A) civilian signal function by providing correction data and integrity information for improving positioning, navigation and timing services over (North America and) Europe" (EGNOS, 2014; FAA, 2008). Both systems offer the Localizer Performance with Vertical guidance (LPV) and LPV200 (i.e., LPV down to 61 m (200 ft)) approach procedures which allows aircraft pilots to make guided approaches in even poor weather conditions to over 3000 runways in the North American and Hawaiian regions (WAAS) and a similar number of runways (across >100 airports) in the European region (EGNOS).

Geomagnetic storms can degrade mid-latitude trans-ionospheric radio waves through ionospheric absorption and scintillation. Very severe storms can degrade GPS capabilities to the point that it decreases WAAS and EGNOS service accuracy, continuity, integrity, and availability (e.g., Datta-Barua et al., 2014). Availability of the WAAS and EGNOS LPV200 services is defined as the percent of time the required protection level is below alert limits for the LPV200 service, which are Horizontal

Protection Level (HPL) < 40 m and Vertical Protection Level (VPL) < 35 m (FAA, 2008, their Table 3.2-1). The HPL "is the radius of a circle in the horizontal plane, with its center being at the true position, which describes the region assured to contain the indicated horizontal position" and the VPL "is the half length of a segment on the vertical axis with its center being at the true position, which describes the region assured to contain the indicated vertical position" (see Table 2 EGNOS EDAS row). The HPL and the VPL bound the horizontal and vertical position error, respectively. Regarding event E1, on February, 27, 2014, FAA personnel transmitted the following message to SWPC: "An Ionospheric Storm began on 2/27/14. The Satellite Operations Specialists were alerted at the WAAS O&M (Operations and Maintenance) by a Significant Event 757 at 2120 UTC. So far, LPV and LPV200 service has not been available in Eastern Alaska and Northeastern CONUS. At times, North Central CONUS and all of Alaska have lost (availability of) LPV and LPV200 Service" (Gordon, 2014; Rutledge, 2015; Viereck, 2014). Regarding event E2, "Iono activity affected WAAS (LP, LPV, 100 LPV-200) performance in Canada, Alaska, and CONUS on March 17 and March 18" (Wanner, 2015). Regarding event E3, WAAS LPV200 exhibited only minor degradation over CONUS as compared to nongeomagnetic storm conditions (described below). EGNOS VPL was degraded for all three events ordered from most to least: E1, E2, E3 (as described below).

The FAA provides public data access to LP, LPV, and LPV200 availability maps for the North American region and EGNOS provides public access to Vertical Protection level (VPL) maps for the European region for our three events (see also Table 2). Since late 2015 (after our three events), EGNOS now also provides LPV200 maps. In our subsequent discussion, we will focus on WAAS and EGNOS impacts through evaluation of LPV200 and VPL.

Fig. 7 shows WAAS LPV200 and EGNOS VPL maps for our three events. The specific dates shown are February 27, 2014, (left column) March 17, 2015 (middle), and June 22, 2015 (right). These are the days for each event experiencing the most significant impacts. The top two rows are dedicated to showing WAAS LPV200 performance degradation, and the bottom row is dedicated to showing EGNOS VPL degradation. The top row displays the percent of the CONUS area the LP, LPV, and LPV200 (red) WAAS services were available versus UTC time for the given day. All three events show loss in the percent of the CONUS area covered with E1 experiencing a significant (>50%), E2 a moderate (~25%), and E3 a minor (~5%) degree of degradation. In all three cases, the approximate onset (blue arrows) are at 20 UTC (E1, E3) and 22 UTC (E2), i.e., when the CONUS Eastern Coast local time is afternoon to dusk. The middle row displays the percent of the day that the WAAS LPV200 service was available from 100% (dark red) down to 85% (dark blue). Clearly, LPV200 was most impacted during event E1, with LPV200 (and LPV) services unavailable to Eastern Alaska and Northeastern CONUS for significant parts of the day (e.g., Viereck, 2014), while E2 was moderately impactful for the same regions (see also Wanner, 2015) and E3 only yielded a minor impact, mostly in eastern Canada. The bottom row displays maps of EGNOS 99% VPL contours with vertical protection ranging from <25 to >150 m. Similarly, position accuracy was degraded severely for E1 (most dark colored regions), moderately for E2 and minorly for E3 (most brightly colored regions) enough so that the vertical protection limit (35 m) for LPV200 would have been exceeded in all three cases, yielding unavailability of the LPV200 service for this region. For the first 2 events, southern Norway (~61° magnetic latitude) saw accuracy degradation to ~85 m (medium blue), a region which typically experiences vertical protection of better than 35 m (cyan contour). Referring back to Fig. 6, our TEC estimates imply the first event, E1 should experience the greatest ionosphere impact to GNSS systems, with moderate impacts to the latter two events, E2 and E3. These impacts were indeed realized by the WAAS and EGNOS LPV200 service.

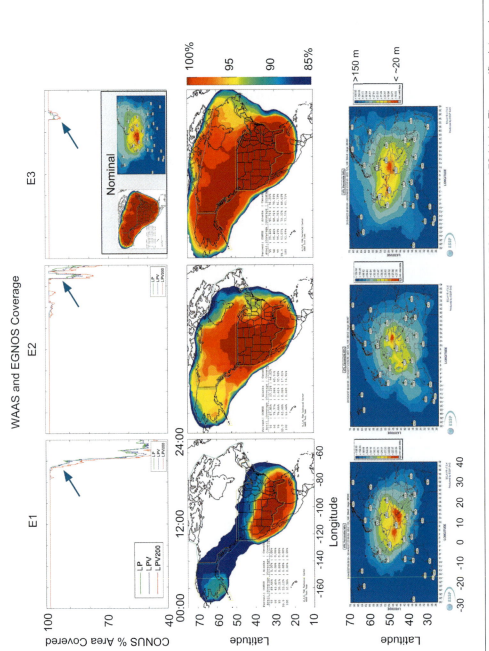

FIG. 7

North American WAAS and European EGNOS aviation system impacts for events E1 (left), E2 (middle), and E3 (right). The specific dates shown are February 27, 2014, March 17, 2015, and June 22, 2015. The top row displays the percent (40%–100%) of the CONUS area the LP (*green*), LPV (*blue*) and LPV200 (*red*) WAAS services were available versus UTC time for the given day. The middle row shows the percent of the day that the WAAS LPV200 service was available in North America from full coverage (100%, *dark red*) down to 85% (*blue*). The inset table displays the percent area covered for the CONUS, Alaska and Canada regions for a given color contour (percent of day). The bottom row displays the EGNOS 99% VPL contours ranging from approximately <25 to >150 m of vertical protection. The approximately 39 EGNOS Ranging Integrity Monitoring Stations (RIMS) ground GPS receivers are shown as *white* place marks. For comparison, the two inset maps in the upper right corner show climatological coverage contours for WAAS LPV200 (left) and EGNOS VPL (right). For both the WAAS and EGNOS maps, "*red*" is indicative of the best performance and "*blue*" the worst.

7 SUMMARY

We have presented a Sun-to-Earth perspective for three recent geoeffective space environment storms and have discussed the value in leveraging operational products for characterizing the space environmental state and for environmental attribution to technological system impacts including spacecraft charging, radio systems and aviation navigation tools. We then considered a subset of the observations and discussed their connection to impacts to the WAAS and EGNOS aviation services. Access to each data source used herein is described in Table 2. It's critical to remind the reader that the storms presented here were by most geophysical measures moderate yet they yielded real impacts to important technological infrastructure. Much more significant events will occur and with likely far greater costs to modern society. It is hoped that the space environment products and methodologies presented here will aid technology designers, builders, operators and scientists interested in maximizing the service life of current and future technology and addressing critical system anomalies.

ACKNOWLEDGMENTS

Two key NOAA organizations play roles that are critical to the U.S. and international space weather programs. SWPC is responsible for forecasting the state of the space environment and providing alerts during elevated space weather conditions helping to ensure the safety of important technological systems. The NCEI lab in Boulder, Colorado was known for decades as the National Geophysical Data Center (NGDC) and World Data Center A (WDC-A). NCEI's Solar Terrestrial Physics program works very closely with SWPC and is currently responsible for the calibration and validation of most of NOAA's space environmental instruments, the development of new products, the archival of key operational products used by SWPC, the creation and dissemination of reference environmental data records, and interacting with other governmental and space physics research communities to optimize the value and use of NOAA archives. The authors sincerely thank the many institutions and individuals responsible for the Sun-to-Earth observations and predictions presented herein. With respect to Madrigal, GPS TEC data products and access through the Madrigal distributed data system are provided to the community by the Massachusetts Institute of Technology under support from U.S. National Science Foundation grant AGS-1242204. Data for the TEC processing is provided from the following organizations: UNAVCO, Scripps Orbit and Permanent Array Center, Institut Geographique National, France, International GNSS Service, The Crustal Dynamics Data Information System (CDDIS), National Geodetic Survey, Instituto Brasileiro de Geografia e Estatística, RAMSAC CORS of Instituto Geográfico Nacional de la República Argentina, Arecibo Observatory, Low-Latitude Ionospheric Sensor Network (LISN), Topcon Positioning Systems, Inc., Canadian High Arctic Ionospheric Network, Institute of Geology and Geophysics, Chinese Academy of Sciences, China Meteorology Administration, Centro di Ricerche Sismologiche, Système d'Observation du Niveau des Eaux Littorales (SONEL), RENAG: REseau NAtional GPS permanent, GeoNet—the official source of geological hazard information for New Zealand, GNSS Reference Networks, Finnish Meteorological Institute, and SWEPOS—Sweden. The authors also wish to specifically thank these individuals for fruitful discussions during the authoring of this chapter: Rodney Viereck, Mihail Codrescu, Howard Singer, Dan Wilkinson, and Dan Autry. Data used in this chapter are available at the locations listed in Table 2.

REFERENCES

Akmaev, R.A., Newman, A., Codrescu, M., Schulz, C., Nerney, E., 2010. DRAP Model Validation: I. Scientific Report, pp. 10, January 13, 2010, https://www.ngdc.noaa.gov/stp/drap/DRAP-V-Report1.pdf Accessed 9 March 2017.

Anderson, P.C., 2012. Characteristics of spacecraft charging in low Earth orbit. J. Geophys. Res. 117 (A7), A07308. https://doi.org/10.1029/2011JA016875.

Araujo-Pradere, E.A., Fuller-Rowell, T.J., Spencer, P.S.J., Minter, C.F., 2007. Differential validation of the US-TEC model. Radio Sci. 42, RS3016. https://doi.org/10.1029/2006RS003459.

Baker, D.N., Li, X., Pulkkinen, A., Ngwira, C.M., Mays, M.L., Galvin, A.B., Simunac, K.D.C., 2013. A major solar eruptive event in July 2012: defining extreme space weather scenarios. Space Weather 11 (10), 585–591. https://doi.org/10.1002/swe.20097.

Bodeau, M., 2010. High Energy Electron Climatology that Supports Deep Charging Risk Assessment in GEO, 2010–1608.

Bornmann, P.L., Speich, D., Hirman, J., Pizzo, V.J., Grubb, R., et al., 1996a. GOES solar x-ray imager: overview and operational goals. In: Proc. SPIE 2812, GOES-8 and Beyond, 309 (October 18, 1996). https://doi.org/10.1117/12.254078.

Bornmann, P.L., Speich, D., Hirman, J., Matheson, L., Grubb, R., et al., 1996b. GOES x-ray sensor and its use in predicting solar-terrestrial disturbances. In: Proc. SPIE 2812, GOES-8 and Beyond, 291 (October 18, 1996). https://doi.org/10.1117/12.254076.

Carrington, R.C., 1859. Description of a singular appearance seen in the Sun on September 1, 1859. Mon. Not. R. Astron. Soc. 20, 13–15.

Chamberlin, P.C., Woods, T.N., Eparvier, F.G., Jones, A.R., 2009. Next generation x-ray sensor (XRS) for the NOAA GOES-R satellite series. In: Proc. SPIE 7438, Solar Physics and Space Weather Instrumentation III, 743802 (26 August 2009). https://doi.org/10.1117/12.826807.

Cherniak, I., Zakharenkova, I., Redmon, R.J., 2015. Dynamics of the high-latitude ionospheric irregularities during the March 17, 2015 St. Patrick's Day storm: Ground-based GPS measurements. Space Weather 13. https://doi.org/10.1002/2015SW001237.

Committee on the Societal and Economic Impacts of Severe Space Weather Events: A Workshop, National Research Council, 2008. Severe Space Weather Events—Understanding Societal and Economic Impacts: A Workshop Report, National Academies Press, p. 13, ISBN: 0-309-12769-6, Available from: https://www.nap.edu/read/12643/.

Coster, A.J., Foster, J., Erickson, P., 2003. Monitoring the ionosphere with GPS: space weather. GPS World 14 (5), 42.

Datta-Barua, S., Walter, T., Bust, G.S., Wanner, W., 2014. Effects of solar cycle 24 activity on WAAS navigation. Space Weather 12 (1), 46–63. https://doi.org/10.1002/2013SW000982.

DeForest, S.E., 1972. Spacecraft charging at synchronous orbit. J. Geophys. Res. 77 (4), 651–659. https://doi.org/10.1029/JA077i004p00651.

Denig, W.F., Redmon, R.J., Mulligan, P., 2014. NOAA operational space environmental monitoring—current capabilities and future directions. In: EGU General Assembly 2014, held 27 April–2 May, 2014 in Vienna, Austria. id.4525.

Dichter, B.K., Galica, G.E., McGarity, J.O., Tsui, S., Golightly, M.J., Lopate, C., Connell, J.J., 2015. Specification, design, and calibration of the space weather suite of instruments on the NOAA GOES-R Program Spacecraft. IEEE Trans. Nucl. Sci. 62 (6), 2776–2783. https://doi.org/10.1109/TNS.2015.2477997.

EGNOS, Open Service (OS) Service Definition Document, 2014. https://egnos-portal.gsa.europa.eu/sites/default/files/EGNOS_OS_SDD_2.1.pdf Accessed 2 March 2017.

Eparvier, F.G., Crotser, D., Jones, A.R., McClintock, W.E., Snow, M., Woods, T.N., 2009. The extreme ultraviolet sensor (EUVS) for GOES-R. In: Proc. SPIE 7438, Solar Physics and Space Weather Instrumentation III, 743804 (September 23, 2009). https://doi.org/10.1117/12.826445.

Evans, D.S., Greer, M.S., 2004. Polar Orbiting Environmental Satellite Space Environment Monitor-2: Instrument descriptions and archive data documentation. In: NOAA Tech. Mem. 93, Space Weather Prediction Center, Boulder, CO. version 1.4.

FAA, 2008. Global Positioning System Wide Area Augmentation System (WAAS) Performance Standard, first ed. U.S. Dept. of Trans., National Transportation Library. Available from: https://ntl.bts.gov/lib/31000/31300/31339/21_2008_GPS_WAAS_Performance_Standard.pdf. Accessed 8 March 2017.

Foster, J.C., Coster, A.J., Erickson, P.J., Holt, J.M., Lind, F.D., Rideout, W., et al., 2005. Multiradar observations of the polar tongue of ionization. J. Geophys. Res. Space Phys. 110 (A), A09S31. https://doi.org/10.1029/2004JA010928.

Fuller-Rowell, T., 2005. USTEC: a new product from the Space Environment Center characterizing the ionospheric total electron content. GPS Solutions 9 (3), 236–239. https://doi.org/10.1007/s10291-005-0005-5.

Fuller-Rowell, T., Araujo Pradere, E., Minter, C., Codrescu, M., Spencer, P., Robertson, D., Jacobson, A.R., 2006. US-TEC: A new data assimilation product from the Space Environment Center characterizing the ionospheric total electron content using real-time GPS data. Radio Sci. 41 (6). https://doi.org/10.1029/2005RS003393.

Gopalswamy, N., Mäkelä, P., Xie, H., Akiyama, S., Yashiro, S., 2009. CME interactions with coronal holes and their interplanetary consequences. J. Geophys. Res. 114, A00A22. https://doi.org/10.1029/2008JA013686.

Gopalswamy, N., Xie, H., Yashiro, S., Akiyama, S., Mäkelä, P., Usoskin, I.G., 2012. Properties of ground level enhancement events and the associated solar eruptions during solar cycle 23. Space Sci. Rev. 171 (1-4), 23–60. https://doi.org/10.1086/506903.

Gordon, B., 2014. Welcome to the 2014 space weather workshop. In: The 2014 Space Weather Workshop.

Green, J. C., 2013a. TED Data Processing Algorithm Theoretical Basis Document, version 1.0, 81 pp., NOAA National Geophysical Data Center. Available from: http://www.ngdc.noaa.gov/stp/satellite/poes/documentation.html.

Green, J.C., 2013b. MEPED Telescope Data Processing Algorithm Theoretical Basis Document. Version 1.0, 77 pp., NOAA National Centers for Environmental Information. Available from: http://www.ngdc.noaa.gov/stp/satellite/poes/documentation.html Accessed 29 February 2016.

Hanser, F.A., 2011. EPS/HEPAD Calibration and Data Handbook, Tech. Rep. GOESN-ENG-048D, Assurance Technology Corporation, Carlisle, Mass. Available from: http://www.ngdc.noaa.gov/stp/satellite/goes/documentation.html.

Heelis, R.A., Sojka, J.J., David, M., Schunk, R.W., 2009. Storm time density enhancements in the middle-latitude dayside ionosphere. J. Geophys. Res. 114 (A3), A03315. https://doi.org/10.1029/2008JA013690.

He, J., Rodriguez, J.V. 2017. Satellite and Ground-Based Observations of Solar Energetic Particle Event Onsets: A Comparison of Two Critical Observational Platforms, Abstract 729 Presented at 97th American Meteorological Society Annual Meeting, Seattle, WA.

Hill, S.M., Pizzo, V.J., Balch, C.C., et al., 2005. The NOAA Goes-12 Solar X-Ray Imager (SXI) 1. Instrument, operations, and data. Sol. Phys. 226, 255. https://doi.org/10.1007/s11207-005-7416-x.

Jacobsen, K.S., Andalsvik, Y.L., 2016. Overview of the 2015 St. Patrick's Day storm and its consequences for RTK and PPP positioning in Norway. J. Space Weather Space Clim. 6, A9. https://doi.org/10.1051/swsc/2016004.

Kurkin, V.I., Polekh, N.M., Ponomarchuk, S.N., Podlesny, A.V., Zolotukhina, N.A., Romanova, E.B., 2015. Characteristics of HF Radio Waves Propagation along Subauroral and Mid-Latitude Paths over Eastern Siberia during Magnetoactive Period in February 2014. In: Progress in Electromagnetics Research Symposium Proceedings, Prague. https://jpier.org/piersproceedings/piers2015PragueProc.php.

Martínez-Galarce, D., Harvey, J., Bruner, M., Lemen, J., Gullikson, E., et al., 2010. A novel forward-model technique for estimating EUV imaging performance: design and analysis of the SUVI telescope. In: Proc. SPIE 7732, Space Telescopes and Instrumentation 2010: Ultraviolet to Gamma Ray, 773237 (July 29, 2010). https://doi.org/10.1117/12.864577.

Martínez-Galarce, D., Soufli, R., Windt, D.L., Bruner, M., Gullikson, E., Khatri, S., Spiller, E., Robinson, J.C., Baker, S., Prast, E., 2013. Multisegmented, multilayer-coated mirrors for the Solar Ultraviolet Imager. Opt. Eng. 52(9), 095102. https://doi.org/10.1117/1.OE.52.9.095102.

Mazur, J., Guild, T., and O'Brien, P., 2015. The Anomaly Investigation Process, presentation at Spacecraft Anomalies and Failures Workshop, 06 October 2015, Chantilly, Virginia.

Mendillo, M., 2006. Storms in the ionosphere: patterns and processes for total electron content. Rev. Geophys. 44, RG4001. https://doi.org/10.1029/2005RG000193.

Minter, C.F., Robertson, D.S., Spencer, P.S.J., Jacobson, A.R., Fuller-Rowell, T.J., Araujo-Pradere, E.A., Moses, R.W., 2007. A comparison of Magic and FORTE ionosphere measurements. Radio Sci. 42, RS3026. https://doi.org/10.1029/2006RS003460.

Newell, P.T., Liou, K., Zhang, Y., Sotirelis, T., Paxton, L.J., Mitchell, E.J., 2014. OVATION Prime-2013: extension of auroral precipitation model to higher disturbance levels. Space Weather 12 (6), 368–379. https://doi.org/10.1002/2014SW001056.

O'Brien, P., Brinkman, D.G., Mazur, J.E., Fennell, J.F., Guild, T.B., 2012. A Human-in-the-Loop Decision Tool for Preliminary Assessment of the Relevance, of the Space Environment to a Satellite Anomaly, AEROSPACE NO. TOR-2011(8181)-2 Revision A.

Oughton, E.J., Skelton, A., Horne, R.B., Thomson, A.W.P., Gaunt, C.T., 2017. Quantifying the daily economic impact of extreme space weather due to failure in electricity transmission infrastructure. Space Weather 15, 65–83. https://doi.org/10.1002/2016SW001491.

Petrinec, S.M., Redmon, R.J., Rastaetter, L., 2017. Nowcasting and forecasting of the magnetopause and bow shock—a status update. Space Weather 15 (1), 36–43. https://doi.org/10.1002/2016SW001565.

Ravan, M., Riddolls, R.J., Adve, R.S., 2012. Ionospheric and auroral clutter models for HF surface wave and over-the-horizon radar systems. Radio Sci. 47, RS3010. https://doi.org/10.1029/2011RS004944.

Redmon, R.J., Loto'aniu, T.M., Berguson, M., Codrescu, S.M., Shue, J.H. Singer, H.J., Rowland, W.F., Denig, W.-F., 2014. Real-time monitoring of the dayside geosynchronous magnetopause location Abstract SM31A-4164 presented at 2014 Fall Meeting, AGU, San Francisco, CA, Dec.

Redmon, R.J., Rodriguez, J.V., Green, J.C., Ober, D., Wilson, G., Knipp, D., et al., 2015. Improved polar and geosynchronous satellite data sets available in common data format at the coordinated data analysis web. Space Weather 13 (5), 254–256. https://doi.org/10.1002/2015SW001176.

Redmon, R.J., Rodriguez, J.V., Gliniak, C., Denig, W.F., 2017. Internal charge estimates for satellites in low earth orbit and space environment attribution. IEEE Trans. Plasma Sci 45 (8), 1985–1997. https://doi.org/10.1109/TPS.2017.2656465.

Rideout, W., Coster, A., 2006. Automated GPS processing for global total electron content data. GPS Solutions 10, 219. https://doi.org/10.1007/s10291-006-0029-5.

Riley, P., Love, J.J., 2017. Extreme geomagnetic storms: probabilistic forecasts and their uncertainties. Space Weather 15, 53–64. https://doi.org/10.1002/2016SW001470.

Rodriguez, J.V., Onsager, T.G., Mazur, J.E., 2010. The east-west effect in solar proton flux measurements in geostationary orbit: a new GOES capability. Geophys. Res. Lett. 37, L07109. https://doi.org/10.1029/2010GL042531.

Rodriguez, J.V., Krosschell, J.C., Green, J.C., 2014. Intercalibration of GOES 8-15 solar proton detectors. Space Weather 12 (1), 92–109. https://doi.org/10.1002/2013SW000996.

Rodriguez, J.V., 2014b. GOES EPEAD Science-Quality Electron Fluxes Algorithm Theoretical Basis Document, version 1.0, 49 pp., NOAA National Centers for Environmental Information. Available from: http://www.ngdc.noaa.gov/stp/satellite/goes/documentation.html.

Rodriguez, J.V., Sandberg, I., Mewaldt, R.A., Daglis, I.A., Jiggens, P., 2017. Validation of the effect of cross-calibrated GOES solar proton effective energies on derived integral fluxes by comparison with STEREO observations. Space Weather 15, 290–309. https://doi.org/10.1002/2016SW001533.

Rutledge, B., 2015. "Space Weather and Impacts on GNSS Applications", National Geodetic Survey Webinar Series, October 8, 2015, 55 minutes, 42 slides. Available from: https://geodesy.noaa.gov/web/science_edu/webinar_series/archived-webinars.shtml Accessed 2 March 2017.

Sauer, H.H., Wilkinson, D.C., 2008. Global mapping of ionospheric HF/VHF radio wave absorption due to solar energetic protons. Space Weather 6, S12002. https://doi.org/10.1029/2008SW000399.

Sojka, J.J., David, M., Schunk, R.W., Heelis, R.A., 2012. A modeling study of the longitudinal dependence of storm time midlatitude dayside total electron content enhancements. J. Geophys. Res. 117(A2). https://doi.org/10.1029/JA083iA06p02695.

Shea, M.A., Smart, D.F., 2012. Space weather and the ground-level solar proton events of the 23rd solar cycle. Space Sci. Rev. 171, 161–188. https://doi.org/10.1007/s11214-012-9923-z.

Shue, J.-H., et al., 1998. Magnetopause location under extreme solar wind conditions. J. Geophys. Res. 103, 17,691–17,700. https://doi.org/10.1029/98JA01103.

Singer, H., Matheson, L., Grubb, R., Newman, A., Bouwer, D., 1996. Monitoring space weather with the GOES magnetometers. Proc. SPIE 2812 (2812), 299–308. https://doi.org/10.1117/12.254077.

Siscoe, G., Crooker, N.U., Clauer, C.R., 2006. Dst of the Carrington storm of 1859. Adv. Space Res. 38, 173–179. https://doi.org/10.1016/j.asr.2005.02.102.

"Space Weather Action Plan (SWAP)", United States Office of Science and Technology Policy, 2015. Available from: https://www.whitehouse.gov/sites/default/files/microsites/ostp/final_nationalspaceweatheractionplan_20151028.pdf Accessed 5 April 2016.

Suvorova, A., Dmitriev, A., Chao, J.-K., Thomsen, M., Yang, Y.-H., 2005. Necessary conditions for geosynchronous magnetopause crossings. J. Geophys. Res. 110, A01206. https://doi.org/10.1029/2003JA010079.

Yue, X., Schreiner, W.S., Ying-Hwa, K., Braun, J.J., Yu-Chen, L., Weixing, W., 2014. Observing system simulation experiment study on imaging the ionosphere by assimilating observations from ground GNSS, LEO-based radio occultation and ocean reflection, and cross link. IEEE Trans. Geosci. Remote Sens. 52 (7), 3759–3773. https://doi.org/10.1109/TGRS.2013.2275753.

Viereck, R., 2014. Supporting GNSS Users With Products to Help Identify and Mitigate the Impacts of Space Weather, 54th Meeting of the Civil GPS Service Interface Committee, September 8-9, 2014, Tampa, Florida, 29 slides. http://www.gps.gov/cgsic/meetings/2014/viereck.pdf. Accessed 7 March 2017.

Wanner, B., 2015. DR #127: Effect on WAAS from Iono Activity on March 17–18, 2015, WAAS Technical Report at the WAAS Test Team web-page, http://www.nstb.tc.faa.gov/DisplayDiscrepancyReport.htm Accessed 14 July 2015.

Wrenn, G.L., Smith, R.J.K., 1996. Probability factors governing ESD effects in geosynchronous orbit. IEEE Trans. Nucl. Sci. 43, 2783–2789.

Zhang, S.-R., et al., 2015. Thermospheric poleward wind surge at midlatitudes during great storm intervals. Geophys. Res. Lett. 42, 5132–5140. https://doi.org/10.1002/2015GL064836.

Zhang, S.-R., Erickson, P.J., Zhang, Y., Wang, W., Huang, C., Coster, A.J., Holt, J.M., Foster, J.F., Sulzer, M., Kerr, R., 2017. Observations of ion-neutral coupling associated with strong electrodynamic disturbances during the 2015 St. Patrick's Day storm. J. Geophys. Res. Space Phys. 122, 1314–1337. https://doi.org/10.1002/2016JA023307.

FURTHER READING

Fulbright, J.P., Kline, E., Pogorzala, D., MacKenzie, W., Williams, R., et al., 2016. Calibration/validation strategy for GOES-R L1b data products. In: Proc. SPIE 10000, Sensors, Systems, and Next-Generation Satellites XX, 100000 T. https://doi.org/10.1117/12.2242140.

Likar, J.J., Bogorad, A.L., Lombardi, R.E., Stone, S.E., Herschitz, R., 2012. On-orbit SEU rates of UC1864 PWM: comparison of ground based rate calculations and observed performance. IEEE Trans. Nucl. Sci. 59 (6), 3148–3153.

National Science and Technology Council, 2015. National Space Weather Action Plan, October 2015, https://obamawhitehouse.archives.gov/sites/default/files/microsites/ostp/final_nationalspaceweatheractionplan_20151028.pdf Accessed 9 March 2017.

CHAPTER 25

EXTREME SPACE WEATHER IN TIME: EFFECTS ON EARTH

Vladimir Airapetian

NASA/GSFC, Greenbelt, MD, United States; American University, Washington, DC, United States

CHAPTER OUTLINE

1 Introduction	611
2 Space Weather Events From the Current Sun	612
3 Space Weather Events From the Young Sun	613
3.1 Solar Superflares and CMEs	614
3.2 The Young Sun's Wind	616
4 3D MHD Model of Super-CME Interaction With the Early Earth	617
4.1 Effects of CMEs on the Magnetosphere of the Early Earth	618
4.2 Effects of XUV Flux on Atmospheric Escape From the Young Earth	622
5 Space Weather as a Factor of Habitability	624
6 Conclusions	628
Acknowledgments	629
References	629
Further Reading	632

1 INTRODUCTION

Our Sun is the closest star that serves as a major source of energy supporting life on Earth. The Sun was formed from a collapsing protostellar cloud 4.65 billion years ago. Our solar system formed from the protoplanetary disk left after the Sun's birth. It was bombarded by a vast amount of energy in the form of electromagnetic radiation, solar wind, magnetic clouds, shock waves, and energetic electrons and protons. Recent heliospheric missions including the Solar Dynamic Observatory (SDO) and Solar Terrestrial Observatory (STEREO) have provided a wealth of information about our star's activity. It has also helped to recover statistical information about spatial and temporal relationships between eruptive events occurring in the solar corona as well as provide clues to the physical mechanisms driving their underlying processes. In the meantime, recent X-ray and ultraviolet (UV) missions including CHANDRA, XMM-NEWTON, the Hubble Space Telescope, and the recent Kepler Space Telescope opened new windows on the lives of stars resembling our Sun at various phases of evolution. This provided a unique opportunity to trace the properties of the young Sun in its infancy by observing other young, solar-type stars.

In this chapter, I describe our recent progress in understanding space weather processes from the evolving Sun. Section 1 briefly reviews the major forms and properties of magnetic activity and associated space weather from the current Sun. This includes the properties of flares, coronal mass ejections (CMEs), and the solar wind. Section 2 is dedicated to the reconstruction models of magnetic activity and space weather from the young Sun at 0.7 Gyr, including properties of solar superflares and associated super-CMEs. In Section 3, I present magnetohydrodynamic models of interactions of CMEs with the Earth's magnetosphere and ionosphere. I also discuss the effects of atmospheric erosion introduced by solar flare-induced X-ray and UV emissions and its effect on habitability of the early Earth. Section 4 discusses the impact of the solar energetic particles from the young Sun on the atmospheric chemistry of the early Earth. I present the results of our recent photo-collisional atmospheric chemistry models that suggest that energetic particles could have been instrumental in setting environmental conditions required for the initiation of prebiotic chemistry as well as warming the early Earth to keep water in the liquid state under the faint young Sun. In the conclusion, I discuss the directions of future research in understanding the environmental conditions set by space weather from the young Sun on the atmosphere of the early Earth.

2 SPACE WEATHER EVENTS FROM THE CURRENT SUN

The Sun is a magnetically active star exhibiting the 11-year activity cycle. Its magnetic activity is driven by the interaction of solar magnetic fields with convective motions via magnetic dynamo (Parker, 1955; Brun et al., 2004). The solar convective zone constitutes the major power source for an energy flux, which contributes to the generation of UV, X-ray, and radio emissions from the solar chromosphere and corona and is also associated with the initiation of the solar wind. The emergent solar magnetic flux is also a major driver of the solar activity manifested in the form of solar eruptive events including solar flares, CMEs, and solar energetic particle events (SEPs). Solar flares are observed as a brightening of any emission across the electromagnetic spectrum (from hard X-rays to radio) occurring on a time scale of minutes to hours. The flare emission is driven by conversion of free magnetic energy of magnetic loop-like structures into the kinetic energy of flows, heat, and nonthermal energy of accelerated particles in the solar corona. Energy of solar flares spans the range between 10^{24} ergs to a few times 10^{32} ergs. During flare events, the magnetic loop structures undergo magnetic reconnection, forming a magnetic flux rope in the solar corona that expands to leave the solar corona as soon as it breaks up through the overlying magnetic field (Lepping et al., 1990; Antiochos et al., 1999; Longcope et al., 2007; Gopalswamy et al., 2017a,b; also see Chapter 2 of the same volume). This phenomenon is known as a CME event. The probability of a solar flare being associated with a CME increases with the GOES flare class reaching unity for X2 flares and greater (Yashiro and Gopalswamy, 2009). Flare energy distribution dN/dE is scaled with the flare energy, E, as a power law, $E^{-\alpha}$, where $\alpha = 1.732 \pm 0.008$ (Crosby et al., 1993). The CME expands outward by propagating in the interplanetary space over the background solar wind with speeds as high as 3000 km/s (e.g., Gopalswamy, 2009). The total CME energy may exceed the energy of the associated flare by a factor of 10 (Emslie et al., 2012). The fast (>1500 km/s) and massive CMEs produce strong interplanetary shocks that serve as sites for accelerated particles with energies from 0.1 to a few thousand

MeV/nucleon (Gopalswamy et al., 2016). As SEPs (electrons and protons) penetrate the Earth's atmosphere, the protons with energies >3 GeV reach the ground, causing ground level enhancement events (GLE) (Aschwanden, 2012). GLEs are mostly associated with large (>X2 class) white-light flares and fast and dense CMEs that form strong shocks, and as a result, SEP and GLE events.

Another form of dominant large-scale perturbation of density and magnetic field in the heliosphere associated with the formation of shocks is known as co-rotating interaction regions (CIRs, Smith and Wolfe, 1976). They are formed due to the interaction of fast solar wind with slower solar wind moving ahead due to the co-rotation with the Sun, and are the dominant large-scale perturbation of density and magnetic field in the heliosphere. CIRs usually occur on the declining and minimum phase of the solar activity cycle. Along with CMEs, CIRs have been known to produce powerful effects by depositing more energy in the ionosphere than would be expected from the electromagnetic energy input from the CIR (Chen et al., 2015). They appear to be more geoefficient, in the sense that the ratio of the measured energy deposited (ring current, Joule heating, and auroral precipitation) to energy input is greater than that for CMEs (Borovsky and Denton, 2013; Verkhoglyadova et al., 2016). Historic data suggest that the Sun produced much more catastrophic events in its recent past (Hayakawa et al., 2017). The Carrington magnetic storm that occurred on September 1–2, 1859, represents one of the most powerful storms in recorded history (Green and Boardsen, 2006; Siscoe et al., 2006). This storm was ignited by a powerful CME event propagating at 2500 km/s with the kinetic energy in the order of 2×10^{33} erg. Its impact on Earth's magnetosphere initiated extensive auroral events visible at low latitudes as low as 19° as well as fires and electrical shocks due to strong induced electric currents (Tsurutani and Lakhina, 2014). Recent study of 14C enhancement at ~775 AD suggests a strong proton storm possibly associated with a super-Carrington type CME event or series of CME events (Miyake et al., 2012). The total energy of these events is estimated as 2×10^{34} erg if one assumes the angular width of the CME is ~80 degrees (Airapetian et al., 2015, 2016). The frequency of occurrence of such energetic events in the Sun is estimated to be as 2×10^{-4}–1.8×10^{-3} events/year, or a relatively rare event (Shibata and Takasao, 2016).

3 SPACE WEATHER EVENTS FROM THE YOUNG SUN

Recent progress in our understanding of flare processes on the Sun and other solar-like stars, including young stars resembling our Sun in its infancy, provide a new perspective to reconstruct the properties of explosive events from the young Sun. Observations of young (a few Myr old) solar-like stars show that our Sun was about 30% less luminous in its past (4 billion years ago) due to a less dense core driven by thermonuclear fusion of hydrogen into helium (Gough, 1981). As the Sun evolves, its luminosity increases roughly 10% every billion years. Despite its lower bolometric luminosity, the young Sun represented a very magnetically active and rapidly rotating star (Güdel and Nazé, 2009 and references herein). Rapid rotation in combination with deep convection zones of these stars produced a strong surface magnetic field that emerged to the surface and formed a compact, dense (up to 10^{12} cm^{-3}), and hot corona (10 MK) of the young Sun (Güdel et al., 1997). Recent direct measurements of surface magnetic fields from young suns (*Bcool* project) shows that the surface magnetic flux of an 0.5-Myr-old star is by a factor of 30 greater than that measured from the current Sun as a star (Vidotto et al., 2014). Such strong surface magnetic fields may serve as the major energy source to produce frequent and energetic flares in their coronae, contributing to plasma heating and production of large X-ray luminosities that are 3–4 orders of magnitude greater than that observed today (Pevtsov et al., 2003; Güdel et al., 1997; Tu et al., 2015).

3.1 SOLAR SUPERFLARES AND CMEs

Recent detection of flare events on young solar-like stars based on observations performed by the *Kepler Space Telescope* opened a new avenue in understanding the properties of explosive events from the young Sun (Maehara et al., 2012). Photometric data obtained over the last 5 years show that young active solar-like stars as well as slowly rotating stars resembling our current Sun exhibit flares with the bolometric energy of 10^{33}–10^{35} ergs, referred to as *superflares* (Shibayama et al., 2013; Maehara et al., 2012). Most superflare stars show quasiperiodic light variations with the periods of 1–30 days and the amplitudes of 0.1%–10%. This suggests that the brightness variations are caused by rotationally modulated starspots that occupy a much larger area than sunspots observed on the Sun (e.g., Notsu et al., 2013). The statistics of the Kepler data suggests that the frequency of occurrence of the superflares observed on G-type dwarfs follow the power-law distribution with the spectral index in the range between 2.0 and 2.2, which is similar to those observed on dMe stars and the Sun (Gershberg, 2005; Shibayama et al., 2013). The occurrence rate of superflares with energies $>5 \times 10^{34}$ erg for young solar-like stars is higher than 0.1 events per day. Extrapolation of this value for a flare energy of $E > E_0 = 10^{33}$ erg provides a frequency of occurrence that is 2500 times greater, or 250 flare events per day! Recent SOHO/LASCO and STEREO observations of energetic and fast (>1000 km/s) CMEs show strong association with powerful solar flares (Yashiro and Gopalswamy, 2009; Aarnio et al., 2011; Tsurutani and Lakhina, 2014). This empirical correlation established for the events from the current Sun provides an estimate for CME occurrence frequencies. If every superflare with $E > E_0$ is associated with a CME event, then the early Sun was generating about 250 super-CMEs per day. Energy balance in observed flares and associated CME events suggests that the kinetic energy of CMEs carries at least 10 times more energy that the total energy in the associated flare event. Measurements of speeds and magnetic field strengths of propagating magnetic clouds in the interplanetary space (so called interplanetary CMEs or ICMEs) show that the magnetic energy density in an ICME event can be comparable or larger than its kinetic energy). Therefore, a super-CME possesses a total (kinetic+magnetic) energy of 5×10^{34} ergs. This energy is a factor of 3 greater than that suggested for the famous Carrington CME event (Tsurutani and Lakhina, 2014). Next, it is assumed that CMEs produce a random orientation of the magnetic field in the ejecting flux rope with respect to the Earth's dipole magnetic field. One can use these estimates to determine the rate of impacts of such super-CMEs with the early Earth's magnetosphere. To determine that, Airapetian et al. (2015) used the statistical correlations established for solar CMEs between the observed widths of a flare-associated solar CME and its mass and speed or its total kinetic energy (Belov et al., 2014). This correlation suggests that more energetic CMEs have larger widths. Next, the kinetic energy of a CME is scaled with associated flare X-ray fluence (Yashiro and Gopalswamy, 2009). This correlation suggest the width of 80 degrees for CMEs associated with the strongest, X-class solar flares. For stronger events like the one considered for a super-CME (SCME), the width should be proportionally higher. But if it is conservatively assumed that an SCME propagates within a cone of 80 degrees, then the probability of the magnetic cloud hitting the Earth's magnetosphere is about 5%. Geoeffective magnetic clouds introduce large disturbances of the magnetospheric field that cause generation of strong ionospheric currents. Joule dissipation of these currents heats the lower ionosphere and thermosphere causing increase in the temperature and density of the upper atmosphere. These effects are caused by adiabatic compression of the Earth's magnetosphere and magnetic reconnection that drives magnetic erosion at the front of the cloud with southward directed interplanetary magnetic field, IMF (see Lavraud et al., 2014 and discussion in Section 2 of this paper). For such a "perfect geomagnetic storm" when the IMF sheared with respect to the Earth's magnetic field, the low limit bound of the frequency of impacts of a super-CME cloud with Earth's magnetosphere is ∼1 event/day! Thus, Airapetian et al. (2015, 2016) concluded that the early Earth was constantly exposed to superstorms that were larger than the largest superstorm observed on September 1–2, 1959. What would be the impact of such a super-CME event on the Earth's magnetosphere and ionosphere? This question will be discussed in the next section.

The indication of the high frequency of flares on the young Sun comes from the direct comparison of the reconstructed energy distribution in the X-ray to UV band for the young Sun proxies, k^1 Cet and *EK* Dra, with the flux of an X-type solar flare (Airapetian et al., 2017a). Fig. 1 presents the reconstructed spectral energy distribution of the current Sun at the average level of activity (between solar minimum and maximum with the total flux, F_0 (5–1216 Å)=5.6 erg/cm^2/s; yellow dotted

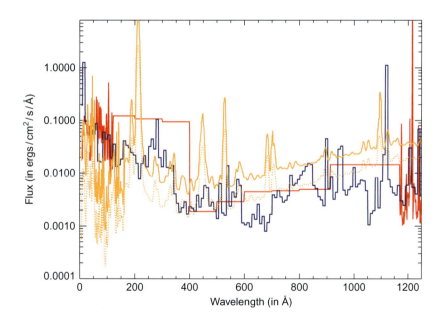

FIG. 1

The spectral energy distribution (SED); reconstructed for the solar X5.4 flare *(blue curve)* and the young Sun's SED *(orange curve)* and the quiet Sun at the average magnetic activity *(dotted orange curve)* scaled to 1 AU and GJ 832 SED *(red curve)* scaled to 0.16 AU.

From Airapetian, V.S., Glocer, A., Khazanov, G.V., Loyd, R.O.P., Kevin, F., et al., 2017a. Astrophys. J. Lett. 836, L3. http://adsabs.harvard.edu/abs/2017ApJ...836L...3A.

line), the X5.5 solar flare occurred on March 7, 2012 (blue line), the young Sun at 0.7 Gyr (yellow solid line), and an inactive M1.5 red dwarf, GJ 832 (red line). The spectra for the current Sun and the solar X5.5 flare in the XUV band (0.5–10 Å) are constructed from the Solar Dynamic Observatory (SDO)/EVE instrument data. Fig. 1 also shows the *XUV* spectrum of a moderately old and inactive M1.5 dwarf, GJ 832, that hosts a super-Earth planet at 0.16 AU using the Measurements of the Ultraviolet Spectral Characteristics of Low-mass Exoplanetary Systems (MUSCLES) Treasury Survey data (Loyd et al., 2016). Finally, to approximate the spectrum of the young Sun at 0.7 Gyr, Airapetian et al. (2017a) used the data obtained from the parameterization of the two young solar analogs of the Sun at around 0.7 Gyr, k^1 Cet, and *EK* Dra (Claire et al., 2012). The XUV flux from the young Sun and GJ 832 are comparable in magnitude and shape with the flux from the X5.5 solar flare at wavelengths shorter than (and including) the Ly-alpha emission line. *This suggests a dominant contribution of X-type flare activity flux to the "quiescent" fluxes from the young Sun and inactive M dwarfs. This suggests that frequent X-ray events contributed to the "quiescent" emission of the young Sun's proxy stars.*

3.2 THE YOUNG SUN'S WIND

Recent progress in direct measurements of stellar surface magnetic fields and the observational characterization of stellar chromospheric and coronal properties provides an opportunity to study the young Sun's properties. Recent "Sun as a star" studies suggest that our Sun's coronal XUV flux also has undergone the evolutionary phase analogous to young main-sequence stars and was also 2–3 orders of magnitude larger than that produced today (Ribas et al., 2005; Claire et al., 2012). Enhanced X-ray and extreme UV (XUV) flux and mass fluxes from the young Sun could have provided critical input to the erosion of exoplanetary atmospheres. Increased dynamic pressure from massive paleo solar wind should compress the magnetosphere and increase the polar cap area of the early Earth (Sterenborg et al., 2011; Tarduno et al., 2014; Airapetian et al., 2015). Photoionization of planetary atmospheres by solar coronal and flare XUV fluxes as high as 20 greater than that observed from the current Sun may be crucial for the evolution of the climate histories of the inner planets due to the escape of specific atmospheric species (Airapetian et al., 2017a). First simulations of the young solar wind and its interactions with the Earth's magnetosphere were performed by Sterenborg et al. (2011). Airapetian and Usmanov have recently simulated the fast and slow solar winds throughout the Sun's evolution using a three-fluid, three-dimensional magnetohydrodynamic (MHD) code, ALF3D (Airapetian and Usmanov, 2016). Unlike previous models, our model treats the wind thermal electrons, protons and pickup protons as separate fluids and incorporates turbulence transport, eddy viscosity, turbulent resistivity, and turbulent heating and models the entire heliosphere (Usmanov et al., 2016). In our model, the solar wind is driven by thermal pressure and the magnetic pressure supplied by the upward propagating Alfven waves excited by the photospheric convection. The winds from the young active sun can play a crucial role in the removal of angular momentum from them, resulting in spin-down in times as stars age. Our model treats the wind thermal electrons, protons, and pickup protons as separate fluids and incorporates turbulence transport, eddy viscosity, turbulent resistivity, and turbulent heating to properly describe proton and electron temperatures of the solar wind. Airapetian and Usmanov's model used three input model parameters, the plasma density, Alfvén wave amplitude, and the strength of the magnetic dipole field at the wind base for each of three solar wind evolution models. They concluded that the young (at 0.7 Gyr) solar

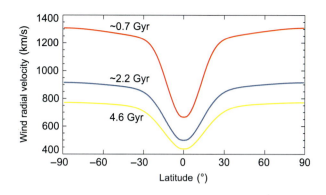

FIG. 2

The latitudinal distribution of the solar wind radial velocity at 1 AU. The *red, blue, and yellow curves* describe the wind's velocity profiles from the young Sun at 0.7 Gyr, the intermediate age Sun at 2 Gyr, and the current Sun respectively.

From Airapetian, V., Usmanov, A., 2016. Astrophys. J. Lett. 817, L24. http://adsabs.harvard.edu/abs/2016ApJ...817L..24A.

wind's speed at 1 AU was twice as fast, 100 times denser and five times hotter as compared to the current solar wind properties. Fig. 2 shows that latitudinal distribution of the young solar wind at 1 AU. The model suggests that the speed of the fast wind from the 0.7-Gyr-old Sun was ∼1400 km/s, while the speed of the slow wind was ∼700 km/s (Airapetian and Usmanov, 2016).

Fig. 3 shows the total mass loss rates from the Sun at the three evolutionary phases of the Sun, 0.7, 2, and 4.65 Gyr (the current epoch), superimposed on the range of empirically derived mass loss rates for solar-like stars at various phases of evolution (Wood et al., 2005). The evolution of the solar wind was driven mostly by the coronal magnetic field, the plasma density at the wind base, and the amplitude of Alfvén waves. Airapetian and Usmanov's young solar wind models suggest that the dynamic pressure from the young solar wind at 0.7 Gyr is expected to be up to 170 times greater than the wind pressure from the current Sun. The model of the intermediate age (∼2 Gyr) Sun produces a total mass loss rate that is smaller by a factor of ∼6 than that predicted from the young Sun. This model also appears to be consistent with the recent young Sun's three-dimensional MHD thermal conduction wind model that uses k^1 Cet's wind as a proxy (do Nascimento et al., 2016). However, more realistic constraints, including the surface magnetic field, density, and temperature imposed on the lower atmospheric boundary (upper atmosphere or lower), would be required to reproduce both X-ray luminosity of the young Sun and its wind properties.

4 3D MHD MODEL OF SUPER-CME INTERACTION WITH THE EARLY EARTH

The dynamic pressure from CMEs can significantly impact and modify the planetary magnetic field by inducing electric currents that can heat the upper atmosphere of Earth via resistive current dissipation in its polar regions (Airapetian et al., 2015; Verkhoglyadova et al., 2016). Also, magnetic reconnection between the geomagnetic field and the IMF is the primary mechanism whereby energy and momentum are coupled into the magnetosphere from the solar wind (Dungey, 1961). *What would be the effects of*

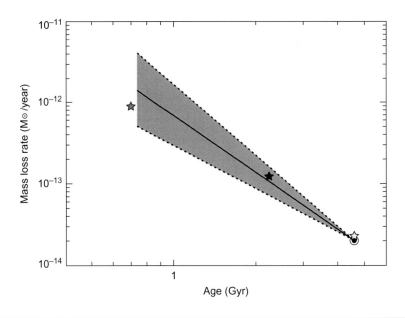

FIG. 3

The total mass loss rates from Alfvén-wave driven solar winds at 0.7 Gyr *(red star)*, 2.2 Gyr *(blue star)*, and 4.6 Gy *(yellow star)*, superimposed on the empirically derived values of mass loss rates *(gray area)* from a sample of solar-type stars of various ages (Wood et al., 2005).

From Airapetian, V., Usmanov, A., 2016. Astrophys. J. Lett. 817, L24. http://adsabs.harvard.edu/abs/2016ApJ...817L..24A.

the interaction of CME events with the magnetosphere of the young Earth? Fast and dense winds and CMEs can exert strong wind and CME dynamic pressures on the planetary magnetospheres, generating energy flux at the magnetopause that may cause the atmospheric erosion.

4.1 EFFECTS OF CMEs ON THE MAGNETOSPHERE OF THE EARLY EARTH

As ICMEs propagate toward Earth, they interact with the Earth's magnetosphere, compressing its dayside and nightsides. If the interplanetary magnetic field (IMF) is directed southward (or oppositely directed to the Earth's dipole field), then CMEs trigger geomagnetic storms due to the combined effects of magnetic reconnection on the dayside (as recently directly observed by MMS mission observations, Birch et al., 2016) and dynamic pressure effects (Birch et al., 2016). Also, CMEs perturb the nightside geomagnetic field, producing magnetic reconnection in the Earth's magnetotail (Zhao et al., 2016).

The St. Patrick's day CME event that occurred on March 17–18, 2015, was one of the strongest events that ignited the geomagnetic storm observed by a number of missions (Le et al., 2016). These observations suggest that the magnetopause driven by dayside magnetic reconnection between southward IMF and terrestrial fields moved inward at a subsolar standoff distance of 6.35 R_E. This distance is by 2 R_E closer to Earth than would be expected due to dynamic pressure effects only.

4 3D MHD MODEL OF SUPER-CME INTERACTION WITH THE EARLY EARTH

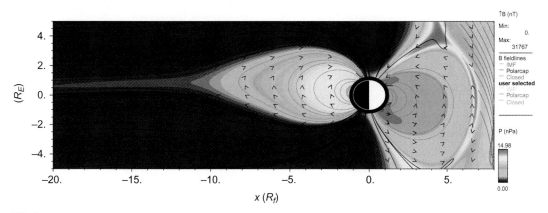

FIG. 4

The snapshot of the 3D MHD simulation of the plasma pressure as the St. Patrick's CME that occurred on March 17–18, 2015, transitions to the dayside of the Earth's magnetosphere. Plasma pressure (in nPa) and the magnetic field (in nT) at $t=1$ h 30 min. The combined effect of magnetic reconnection and dynamic pressure moves the Earth's magnetopause to 6.3 R_E.

From https://ccmc.gsfc.nasa.gov/results/viewrun.php?domain=GM&runnumber=Vladimir_Airapetian_20161026_GM-EXO_1.

To characterize this event, I performed global magnetospheric simulations of the interaction of this ICME with the Earth's magnetosphere at the parameters specified by observations of (Le et al., 2016). I utilized the Space Weather Modeling Framework (SWMF) available at the Community Coordinated Modeling Center (CCMC) at the NASA Goddard Space Flight Center (see at http://ccmc.gsfc.nasa.gov). A single-fluid, time-dependent fully nonlinear three-dimensional magnetohydrodynamic (MHD) code BATS-R-US (Block-Adaptive-Tree Solar-wind Roe-type Upwind Scheme) is a part of SWMF and was developed at the University of Michigan Center of Space Environment Modeling (CSEM). The spine of the SWMF is the BATS-R-US code (Powell et al., 1999), which is coupled to the Rice Convection Model (De Zeeuw et al., 2004) to model a propagation and interaction of the St. Patrick's day CME event with the magnetosphere and ionosphere of the young Earth. For the initial parameters of the CME event I used $N_{sw}=30$ cm^{-3}, $V_{x,sw}=-600$ km/s, IMF $B_z=-20$ nT, 1/16 R_E resolution with the inner boundary at 1.25 R_E. Fig. 4 shows the two-dimensional density with the background magnetic field in the x-z planet at $y=0$. Fig. 4 shows the formation of the wide bow shock extending to 10 R_E with the magnetopause at 6 R_E, which is consistent with observations (Le et al., 2016).

Fig. 5 shows the two-dimensional (x-y plane) distribution of the height integrated Joule heating due to resistive dissipation of ionospheric currents induced by the combined effects of reconnection and dynamic pressure. The Joule heating rate at the lower ionosphere (mid-thermosphere) reaches 0.04 W/m^2.

The energy deposition rate due to the dissipation of ionospheric currents at 110–150 km heats the Earth's thermosphere. The thermospheric temperature in the quiet state is 900 K. From the first law of thermodynamics

$$\frac{dQ}{dt}=\frac{7}{2}mk_D\frac{dT}{dt}$$

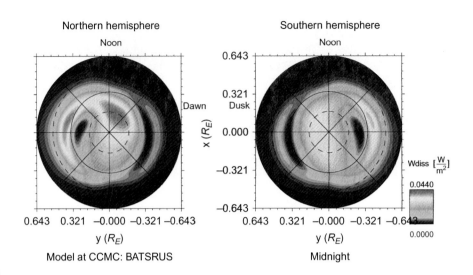

FIG. 5

The height-integrated Joule heating rate deposited in the northern *(left panel)* and southern hemispheres *(right panel)* due to resistive dissipation of ionospheric currents.

From https://ccmc.gsfc.nasa.gov/results/viewrun.php?domain=GM&runnumber=Vladimir_Airapetian_20161026_GM-EXO_1.

where $Q = 4 \times 10^{-6}$ erg/cm^3/s is the volumetric heating rate. Assuming the height of the thermosphere of 100 km, the temperature rate change is 13,600 K/day. The temperature increase over the course of the storm will drive the thermospheric expansion by a factor of 10 and increase thermospheric density. Thus, such energy deposition will induce the NO (at 5.3 μm) mediated radiative cooling, which is scaled linearly with the temperature increase (Weimer et al., 2015). Recent observations suggest that the strongest geomagnetic storms ignite thermospheric overcooling due to such effects (Knipp et al., 2017).

One of the strongest two CME successive events occurred on July 23–24, 2012. The first CME on July 23 had a peak shock speed over 2500 km/s with the peak southward magnetic field B_z exceeding 100 nT (Riley et al., 2016). This catastrophic event was comparable in its kinetic energy to the kinetic energy of the Carrington event of Sep. 1859. This rare energetic event missed the Earth. The modeling by Ngwira et al. (2014) using SWMF at CCMC/GSFC suggests that if these events had hit the Earth's magnetosphere, the stand-off distance would have been as low as 2 R_E. The height-integrated Joule heating rate deposited in the Earth's widened polar regions would have been as high as 2.5 W/m^2, or by a factor of 50 greater than in the St. Patrick's day event (Fig. 4). This would result in the temperature increase to at least 10,000 K and subsequent expansion of the thermosphere. Our estimates of the frequency of such events suggest that they would have hit the magnetosphere of the young Earth 4 billion years ago at a rate of a few events per day (Airapetian et al., 2015).

The results of the SWMF/CCMC-based simulations of an extreme CME event referred to as a super-Carrington event interacting with the magnetosphere of the young Earth were discussed by Airapetian et al. (2015, 2016). For a super-Carrington event, they implemented the model of the young solar wind discussed in Airapetian and Usmanov (2016). The magnetospheric cavity (outer and inner magnetosphere) in a computational box was defined by the following dimensions $-224\, R_E < x < 224$

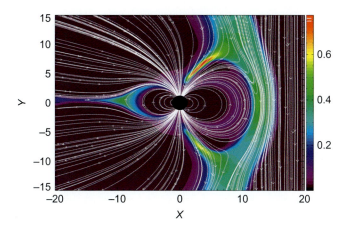

FIG. 6

3D MHD simulations of a super-Carrington CME event. Plasma pressure and the magnetic field lines at $t=0$. X and Y axes are given in the units of Earth radius.

From Airapetian, V.S., Glocer, A., Gronoff, G., Hébrard, E., Danchi, W., 2016. Nat. Geosci. 9, 459. doi:10.1038/NGEO2719. http://adsabs.harvard.edu/abs/2016NatGe...9..452A.

R_E, $-128\ R_E < y < 128\ R_E$, $-128\ R_E < z < 128\ R_E$. The simulations were carried out using a block-adaptive high-resolution grid with a minimum cell size of $1/16\ R_E$. The inner boundary is set at $1.25\ R_E$ with a density of 100 cm^{-3}. The young solar wind conditions are set at the upstream boundary and some period of local time stepping is used to get an initial steady state solution. The solar wind input parameters, including the three components of the interplanetary magnetic field, B_x, B_y, and B_z, the plasma density, and the wind velocity, V_x represent the physical conditions associated with a Carrington-type event as discussed by Tsurutani et al. (2003). Ngwira et al. (2014) and Airapetian et al. (2016). Fig. 6 presents a two-dimensional map of the steady-state plasma density superimposed by magnetic field lines for the magnetospheric configuration in the $Y=0$ plane corresponding to the initial 30 min of the simulations, when the Earth's magnetosphere was driven only by dynamic pressure from the solar wind assumed to have the properties of the current wind. The figure shows the formation of the bow shock at the standoff distance of $12\ R_E$, which is typical for the solar wind conditions. If we apply the values of the dynamic wind of the young Sun's wind (80 times the current wind's pressure) derived by Airapetian and Usmanov (2016), the magnetopause will be shifted by a factor of $\sim(80)^{1/6} \approx 2$ ot to 6 R_E.

A super-Carrington CME event characterized by the time profile of V_x as the CME approaches the Earth at the maximum velocity of 1800 km/s, was introduced at $t=30$ min. The CME magnetic field is directed southward or is sheared by 180 degrees with respect to the dipole field with the $B_z=-212$ nT.

As the CME propagates inward (from right to left in Figs. 6 and 7), its large dynamic pressure compresses and convects the magnetospheric field, inducing the convective electric field. It also compresses the nightside magnetosphere and ignites magnetic reconnection at the nightside of the Earth's magnetosphere, causing the magnetospheric storm as particles penetrate the polar regions of Earth. Another effect appears to be crucial in our simulations. The strong sheared magnetic field on the dayside (subsolar point) of Earth is also subject to reconnection, which dissipates the outer

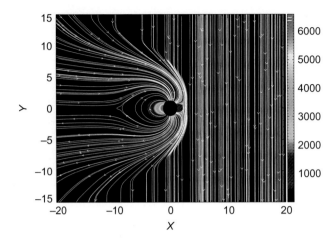

FIG. 7

3D MHD simulations of a super-Carrington CME event. Magnetospheric storm at $t=10$ h. Plasma pressure and the magnetic field lines. X and Y axes are given in the units of Earth radius.

From Airapetian, V.S., Glocer, A., Gronoff, G., Hébrard, E., Danchi, W., 2016. Nat. Geosci. 9, 459. doi:10.1038/NGEO2719. http://adsabs.harvard.edu/abs/2016NatGe...9..452A.

regions of the Earth's dipole field up until 1.5 R_E above the surface. The boundary of the open-closed field shifts to 36° in latitude. The ionospheric response is characterized by the ionospheric potential derived from the field-aligned current that was produced by the MHD solution and an ionospheric conductance. The CME drives large field-aligned currents that provide a Joule ionospheric heating at 110 km reaching ~4 W/m². Ionospheric heating produces large plasma pressure gradients in addition to the magnetic pressure gradients that drive mass outflow at velocities >20 km/s. Such a high-energy deposition rate will enhance the temperature, which will ignite rapid expansion of the thermospheric plasma. Thus, it will produce atmospheric evaporation of ionospheric plasma and adiabatic and radiative cooling. These processes are crucial factors that could contribute to habitability conditions on the early Earth and should be treated with a coupled ionosphere-thermosphere model.

4.2 EFFECTS OF XUV FLUX ON ATMOSPHERIC ESCAPE FROM THE YOUNG EARTH

Airapetian et al. (2017a) have recently modeled the effects of X-ray and UV (XUV) radiation from the young Sun on atmospheric escape from the 0.7-Gyr young Earth, when XUV fluxes were by a factor of 10 greater than that at the current epoch. *XUV* radiation induces nonthermal heating via photoabsorption and photoionization, raising the temperature of the exosphere, and, therefore, its pressure scale height. At high *XUV* fluxes, this process initiates hydrodynamic atmospheric escape of neutral atmospheric species, with the loss rate dependent on the molecular mass of atmospheric species. Hydrogen, as the lightest component, escapes more readily than any other species by this mechanism (Lammer et al., 2008; Tian et al., 2008). For the environments of active solar-type stars, much of the hydrogen likely escapes from a planet's atmosphere during the system's early evolution, leaving behind an atmosphere enriched in heavier elements such as N and O. Thus, processes of atmospheric ionization and loss via nonthermal mechanisms are crucial for modeling the evolution of oxygen and nitrogen-rich atmospheres as well as the efficiency of atmospheric loss of water as a critical factor of habitability of the young Earth. In the region above an Earth-sized planet's exobase, the layer where collisions are negligible, the incident *XUV* flux ionizes atmospheric atoms and molecules and produces photoelectrons. The polarization electric field is formed due to the different scale height between the thermal electrons and ions (Banks and Holzer, 1968). The photoelectrons can enhance the polarization field (increase Te comparing to Ti). The upward propagating photoelectrons outrun ions in the absence of a radially directed polarization electric field, forming the charge separation between electrons and atmospheric ions. Thus, a radially directed polarization electric field is established that enforces the quasineutrality and zero radial current. For ionospheric ions with energies over 10 eV, the polarization electric field cancels a substantial part of the Earth's gravitational potential barrier, greatly enhancing the flux of escaping ions and forming an ionospheric outflow.

The effects of XUV flux on the ionosphere were studied by coupling the ion hydrodynamics of the Polar Wind Outflow Model (PWOM) to the latest version of the SuperThermal Electron Transport (STET) code (Glocer et al., 2009; Glocer et al., 2012; Khazanov, 2011; Khazanov et al., 2015; Airapetian et al., 2017a). Full details of the model coupling will appear in a separate publication (Glocer et al., 2017). They have developed four models with the *XUV* flux from the evolving Sun expressed in terms of the total *XUV* flux, F_0, of the Sun at the average level of the magnetic cycle. The photoelectron flux increases approximately linearly with the input *XUV* flux. Then, the authors (Airapetian et al., 2017a) used *PWOM* to calculate the ionized atmospheric escape rates along an open single magnetic field line of the polar region at heights between 200 and 6000 km. The *XUV* flare flux at $10\,F_0$ corresponds to the associated super-Carrington type *CME* event discussed in Section 3.1. Thus, the steady state outflow rate of O^+ ions driven by the input *XUV* flux was calculated. In order to evaluate the effect of the base temperature on the O^+ outflow rate, two escape models for the XUV flux of $10\,F_0$ for these two exobase temperatures were calculated. As the base temperature is increased by a factor of 2 (from 1000 to 2000 K), the resulting O^+ outflow rates increase by a factor of 10. The total loss rate of O^+ at $h = 1000$ km is found from the integration of this value over the whole area. Fig. 8 shows that the mass loss of oxygen ions increases roughly linearly with the solar flux and reaches ~400 kg/s for $F = 20\,F_{sun}$. This estimate does not account for a number of effects typically contributing to the ion escape during space weather events associated with large solar flares. This mass loss rate can also be affected by precipitated energetic electrons from the day and nightsides of the Earth's magnetosphere. This input efficiently produces secondary superthermal electrons due to collisional ionization of species in the ambient ionosphere (Strangeway et al., 2005) and needs further study.

The simulations of the atmospheric escape by Airapetian et al. (2017a) suggest that the total expected escape rate for nitrogen ions along with oxygen ions at $10\,F_0$, characteristic of the Sun's flux 3.8 billion years ago, is about 400 kg/s. This suggests that Earth could have lost half of its 1-bar atmosphere in 300 million years after the secondary atmosphere was formed on the early Earth. A recent study of fossilized raindrops imprinted on 2.7-Gyr-old rocks suggests that the atmosphere of the early Earth was at least 0.5 bars (Som et al., 2016)

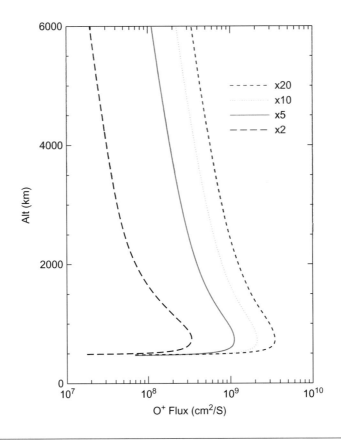

FIG. 8

The mass loss rate of oxygen ions from the Earth's atmosphere due to XUV and EUV irradiation from the young Sun at $F=2$ *(long dash)*, 5 *(dash-dot)*, 10 *(dot)*, and 20 *(short dash)*.

From Airapetian, V.S., Glocer, A., Khazanov, G.V., Loyd, R.O.P., Kevin, F., et al., 2017a. Astrophys. J. Lett. 836, L3. http://adsabs.harvard.edu/abs/2017ApJ...836L...3A.

5 SPACE WEATHER AS A FACTOR OF HABITABILITY

Our global magnetospheric simulations described in Section 3.1 suggest that a disturbance of the early Earth's magnetosphere by a super-Carrington CME event should shift the boundary of the open-closed field to 36° latitude, producing a polar cap opening ∼70% of the planet's dipole magnetic field. Thus, extended polar caps may provide a pathway for energetic electrons and protons accelerated in CME-driven shocks to penetrate the Earth's atmosphere along the open field lines (Airapetian et al., 2016).

The secondary atmosphere of the early Earth at ∼0.5 Gyr was nitrogen rich (80%–90%) and CO_2 rich (10%–20%) with traces of methane, CH_4, and water vapor, H_2O. Molecular nitrogen was mostly supplied by tectonic activity from the highly oxidized mantle wedges driven by subduction processes, while carbon dioxide, methane, and water vapor were released by intensive volcanic activity (Mikhail and Sverjensky, 2014). The Aeroplanets model (Gronoff et al., 2014) was recently applied by Airapetian et al. (2016) to simulate the atmospheric chemistry of such highly reduced, nitrogen-dominated (79% N_2, 20% CO_2, 0.4% CH_4, and 1% H_2O) rebiotic Earth atmosphere at a surface pressure of 1 bar with the photochemistry controlled by the XUV flux from the young Sun and a proton energy fluence that is comparable to the January 20, 2005, SEP event at 0.1 Mev, which was associated with an X7.1 solar flare (Mewaldt et al., 2005; Airapetian et al., 2016).

The Aeroplanets model calculates photoabsorption of the XUV flux from the early Sun and electron and proton fluxes to compute the corresponding energetic fluxes at all altitudes between 200 km to the surface (Gronoff et al., 2014). These fluxes are then used to calculate the photo- and particle-impact ionization/dissociation rates of the atmospheric species producing secondary electrons due to ionization processes. Then, using the XUV flux and the photoionization-excitation-dissociation cross-sections, the production of ionized and excited state species and the resulting photoelectrons are calculated.

In this steady-state model of the early Earth's atmosphere, energetic protons from an SEP event precipitate into the middle and lower atmosphere (stratosphere, mesosphere, and troposphere) and produce ionization, dissociation, dissociative ionization, and excitation of atmospheric species. The destruction of N2 into reactive nitrogen, $N(^2D)$ and $N(^4S)$, and the subsequent destruction of CO_2 and CH_4 produces NOx, CO, and NH in the polar regions of the atmosphere, as shown in Fig. 9. NOx molecules then get converted in the stratosphere to NO_2, HNO_2, and HNO_3.

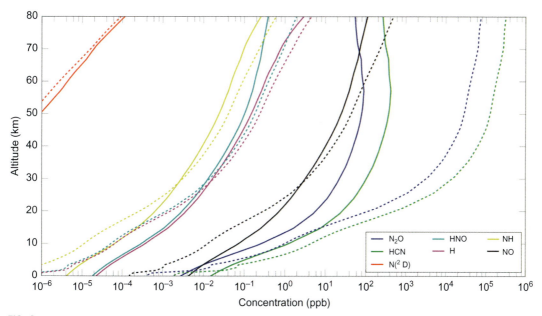

FIG. 9

Aeroplanets model predictions of production and mixing ratios of species under chemical equilibrium driven by frequent energetic SEPs on the early Earth.

From Airapetian, V.S., Glocer, A., Gronoff, G., Hébrard, E., Danchi, W., 2016. Nat. Geosci. 9, 459. doi:10.1038/NGEO2719. http://adsabs.harvard.edu/abs/2016NatGe...9..452A.

One of the major predictions of the atmospheric model by Airapetian et al. (2016) is the efficient production of nitrous oxide, N_2O, which is a potent greenhouse gas. This could represent a pathway to the resolution of the faint young Sun (FYS) paradox that suggests that the energy from the faint young Sun would be insufficient to support liquid water on the early Earth, contrary to geological evidence of its presence at that time (Sagan and Mullen, 1972; Ramirez, 2016). The proposed models of the atmospheric warming due to the large atmospheric concentration of CO_2, H_2O, and/or CH_4 cannot resolve the FYS paradox (Kasting, 2010; Rosing et al., 2010). This problem becomes even worse for the Martian atmosphere that would require up to 4 bars of the atmospheric abundance of CO (Ramirez et al., 2014). Our model proposes a resolution of the FYS paradox due to collisional dissociation of the atmospheric N_2, CO_2, CH_4, and NH_3 producing abundant NOx and NH molecules and efficient formation of N_2O through $NO + NH \rightarrow N_2O + H$ (Airapetian et al., 2016). Atmospheric N_2O density reaches a concentration with the mixing ratio of 0.3–1 ppmv in the lower atmosphere, depending on availability of gases shielding nitrous oxide from photodestruction. The sources and sinks for N_2O depend strongly on the chemical composition of the initial atmosphere and the energy flux in accelerated protons. Specifically, our simulations show that N_2O's abundance increases with an increasing CO_2/CH_4 ratio in the initial atmosphere. Moreover, the derived value should be considered as a lower bound because our model does not account for a number of factors—eddy diffusion and convection effects in the stratosphere-troposphere system, concentration of hazes, inclusion of SO_2 and H_2S volcanic outgassing sources, and Rayleigh scattering of solar EUV radiation—that significantly reduce photodestruction of N_2O, and therefore increase its production. Also, energetic protons associated with SEP events significantly enhance atmospheric ion production rates, which in turn drive the increased rate of formation/nucleation of newly formed and/or existing production of stratospheric aerosol particles by up to 1 order of magnitude in the polar regions at 10–25 km, which provides an efficient shield from UV emission around 240 nm (Mironova and Usoskin, 2014). Greater typical energies of SEP events from the young Sun could be another factor that contributes to the increased production rate by a factor of 5–10 greater than that conservatively assumed in our model. This is due to the fact that the young Sun's corona that represents the source of CMEs was at least by a factor of 10 denser as compared to the current Sun (Güdel et al., 1997; Güdel and Nazé, 2009). Thus, the denser corona provided correspondingly larger concentrations of seed particles that participated in the acceleration processes in CME-driven shockwaves closer to the solar surface. Also, recent kinetic simulations based on Particle Acceleration and Transport in the Heliosphere model (PATH) suggest that the particle acceleration via diffusive shock acceleration mechanisms on quasiparallel shocks produce mostly SEPs with harder spectrum (Airapetian et al., 2017b) similar to the February 1956 SEP event (see Fig. 10). This is consistent with the results of a statistical study by Gopalswamy et al. (2016, 2017a,b), who found a relationship between the initial acceleration of CME events and the highest starting frequency of associated Type II events with the hardness of proton spectra and fluence. CMEs represented by the magnetic flux ropes with a stronger initial magnetic field expand faster and form shocks closer to the solar surface, where the plasma density is greater. It is known (see Gopalswamy et al., 2017a,b) that CME events are associated with interplanetary Type II events initiated by shock-produced metric wavelength emissions at the fundamental mode (plasma frequency) or harmonic mode (double plasma frequency). Thus, the higher plasma density is associated with the higher Type II radio frequency events as would be expected, because the plasma frequency is proportional to the square root of plasma density. This would also suggest that the strongly magnetized CMEs from the young Sun should produce higher fluence and harder spectra SEP events.

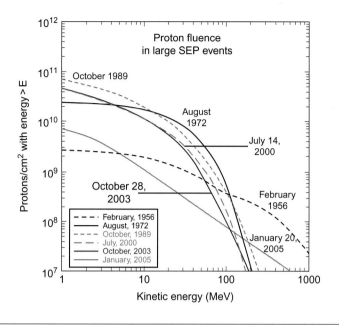

FIG. 10

Protons fluence spectra from the largest SEPs observed over the last 50 years.

From Mewaldt, R.A., et al., 2005. 29th International Cosmic Ray Conference Pune, 101–104. http://www.srl.caltech.edu/ACE/ASC/DATA/bibliography/ICRC2005/usa-mewaldt-RA-abs3-sh12-poster.pdf.

This empirical picture is consistent with our recent simulations of SEP-accelerated events from the young Sun suggesting that they are formed from the strong CME-driven magnetized shocks via diffusive shock acceleration mechanisms. This acceleration model produces hard particle energy spectra (Airapetian et al., 2017b) similar or harder to the February 1956 SEP event. Fig. 10 shows that the particle fluence of the February 1956 event at 500 Mev is by a factor of 10 greater than that of the January 20, 2005, SEP (see Fig. 10) event taken in our recent prebiotic chemistry model of the early Earth as a reference particle source (Airapetian et al., 2016). With the difference in the slopes of these two events, one can extrapolate that the flux of particles at 1 Gev is a factor of 30 greater than was assumed in this study. Protons with the energy ≥ 1 Gev deposit most of their energy in the low troposphere at ~3 km. This energy is deposited into the ionization of atmospheric molecules with production of cascades of electrons and dissociation and excitation of molecular species. As discussed in Airapetian et al. (2016), the abiotic fixation of nitrogen, carbon dioxide, methane, and water vapor produces NO_x (NO, NO_2), CH, and OH_x. The abundant production of the "hub" molecule, nitric oxide (NO), and its recombination with NH is the major source of nitrous oxide, the powerful greenhouse gas, N_2O, in the troposphere. Airapetian et al. (2016) proposed that the production of nitrous oxide in the atmosphere of the early Earth could potentially resolve the FYS paradox. Another important molecule, hydrogen cyanide (HCN), is produced via the reaction of NO with CH, the product of methane dissociation. Because the production rates of NO molecules and associated N_2O and HCN products are linearly scaled with the number of incoming protons, the resultant concentration of these molecules in the lower stratosphere (at 20 km) is by a factor of 30 larger than reported in Airapetian et al. (2016)

for a 1-bar atmosphere. Thus, the concentration of produced N_2O and HCN should be boosted by a factor of 100 at 10–15 km. However, recent studies of atmospheric loss via O^+ and N^+ ion escape by Airapetian et al. al. (2017a) suggest that the early Earth's atmospheric pressure should be at ~0.25 bar. At such a thin atmosphere, protons with the energy >200 Mev (or 5 times less energy) will reach the altitude of 3 km. This increases the flux of protons at the low troposphere by another factor of 30. This combination of two factors boosts the concentration of N_2O and HCN by a factor of 1000 to the level of a few hundreds of ppmv. The recent application of the three-dimensional Global Climate Model with the updated profiles of atmospheric nitrous oxide and 10% of CO_2 suggests that the long-standing FYS paradox could be resolved for the early Earth and Mars (Airapetian et al., 2017b).

6 CONCLUSIONS

As I discussed in this paper, space weather effects from the young Sun can contribute to the habitability conditions on the early Earth in a variety of ways. First, I have shown that the high magnitude southward IMF and large dynamic pressures from super Carrington-type CME events can restructure the Earth's magnetosphere due to reconnection events as well as widen its polar caps. Then, the dissipation of large induced geomagnetic currents can heat the thermospheric plasma to high temperatures that can support its escape from the planet in its earliest phase of evolution when CME events were frequent and energetic. Second, high XUV fluxes from associated superflare events can support the escape of oxygen and possibly nitrogen ions, due to production of photoelectrons. However, in our models the escape process was uncoupled with the thermal effects of inflated exosphere. They provide the lower bound of escape rates of oxygen ions. In these ways, CMEs can provide negative conditions for habitability on the early Earth and Mars, especially in the first 500 million years of the Sun's magnetically active phase of evolution. The conditions on early Mars were more severe because of a much lower surface gravity and the efficiency of photochemical escape via dissociative recombination of O_2 and possibly N_2 producing hot atomic oxygen and nitrogen. However, frequent SEP events from the young Sun probably played a positive role in setting the conditions for the formation of hydrogen cyanide and nitrous oxide in the lower stratosphere and upper troposphere of the early Earth and Mars. Organic molecules may subsequently rain out into surface reservoirs and ignite higher-order chemistry, producing more complex organics. For example, the hydrolysis of HCN produces formamide, $HCONH_2$. When irradiated with energetic protons, formamide can serve as a precursor of complex biomolecules that are capable of producing amino acids, the building blocks of proteins and nucleobases, sugars, and nucleotides, the constituents of RNA and DNA molecules (Saladino et al., 2015). In our recent experiments, the irradiation of a gas mixture resembling the young Earth's atmosphere with high-energy protons (~2.5 MeV) produced amino acids, including glycine and alanine (Kobayashi et al., 2001, 2017). The irradiation of the same mixture by the spark discharge (accelerated electrons) or UV irradiation (2500 Å), produced no amino acids at the CH_4 mixing ratio (<15%). Thus, considering fluxes of various energies on the primitive Earth, energetic protons appear to be a more efficient factor to produce N-containing organics than any other conventional energy sources like thundering or solar UV emission irradiating the early Earth atmosphere. Also, abiotic production of nitrous oxide in the lower troposphere at ~few hundred ppmv driven by energetic protons can provide an efficient way to resolve a long standing FYS paradox to explain the warming of our young planet to keep water in the liquid state in its early history (Airapetian et al., 2016; Airapetian et al., 2017b). In conclusion, I want to point out that the reconstruction of space weather events from the young Sun and their effects on the environments of the early Earth is crucial for understanding the properties of space weather from active G, K, and M dwarf stars that harbor rocky exoplanets and their effects on habitability. The recent discovery of the first Earth-sized planet in the habitable zone (HZ) of another star other than our Sun with the Kepler Space Telescope opens a new era in the study of exoplanets (Quantana et al., 2014). With a flotilla of upcoming missions capable of detecting potential Earth-like planets around G, K, and M dwarf stars, including TESS, K2, and PLATO (Ricker et al., 2010; Howell et al., 2014; Rauer et al., 2014), it becomes imperative that we understand the physical processes and environmental factors contributing to a habitable planetary system and its evolution in time.

This suggests that the question of habitability should be expanded beyond the scope of a habitable planet to the level of a habitable astrospheric system, where habitability phenomenon is treated as a result of the interaction between energy fluxes from the host star with magnetospheres, ionosphere, and thermospheres all the way to the surface. To understand whether an exoplanet is habitable, not only do we need to understand the changes in the chemistry of its atmosphere due to the penetration of ionizing radiation in the form of X-ray, UV, energetic particles and their interaction with constituent molecules, but also the loss of neutral and ionic species and the addition of molecules due to outgassing from volcanic and tectonic activity. These effects will produce a net gain or loss to the surface pressure, which will affect the surface temperature, as well as a net change in the molecular chemistry. Thus, due to the complexity of the problem, we have developed an interdisciplinary program, where the methodologies of heliophysics, astrophysics and planetary and Earth sciences are combined in a meaningful way to be applied for simulating the chemistry of exoplanetary atmospheres and deriving the strongest atmospheric biosignatures. This program should enlarge our understanding of exoplanet habitability and the potential for life in exoplanets around nearby stellar systems, especially for the most common low-mass active late M dwarf stars that are of great interest due to their potential for near-term observational studies of their atmospheres with transit and eclipse spectroscopy with the James Webb Space Telescope (JWST).

ACKNOWLEDGMENTS

Vladimir Airapetian's global magnetospheric simulations have been provided by the Community Coordinated Modeling Center at Goddard Space Flight Center through their public runs on request system (http://ccmc.gsfc.nasa.gov). The BATS-R-US model was developed by the University of Michigan. Vladimir Airapetian's work was supported by funding from NASA GSFC's Sellers Exoplanetary Environments Collaboration (SEEC) and by support from NASA's Exobiology program.

REFERENCES

Aarnio, A.N., Stassun, K.G., Hughes, W.J., McGregor, S.L., 2011. Sol. Phys. 268, 195.
Airapetian, V., Usmanov, A., 2016. Astrophys. J. Lett. 817, L24.
Airapetian, V., Glocer, A., Danchi, W., 2015. In: van Belle, G., Harris, H. (Eds.), Proceedings of the 18th Cambridge Workshop on Cool Stars, Stellar Systems, and the Sun, Lowell Observatory (9–13 June 2014). (eprint arXiv:1409.3833).

Airapetian, V.S., Glocer, A., Gronoff, G., Hébrard, E., Danchi, W., 2016. Nat. Geosci.. https://doi.org/10.1038/NGEO2719.
Airapetian, V.S., Glocer, A., Khazanov, G.V., Loyd, R.O.P., France, K., et al., 2017a. Astrophys. J. Lett. 836, L3.
Airapetian, V.S., Zank, G., Verkhodlyadova, O., Li, G., Gronoff, G., Del Genio, A., 2017a, to be submitted to Nat. Geosci.
Antiochos, S.K., DeVore, C.R., Klimchuk, J.A., 1999. Astrophys. J. 510, 485.
Aschwanden, M.J., 2012. Space Sci. Rev. 171, 3.
Banks, P.M., Holzer, T.E., 1968. J. Geophys. Res. 73, 6846.
Belov, A., Abunin, A., Abunina, M., Eroshenko, E., Oleneva, V., Yanke, V., Papaioannou, A., Mavromichalaki, H., Gopalswamy, N., Yashiro, S., 2014. Sol. Phys. 289, 3949.
Birch, J.L., et al., 2016. Science 352. aaf2939.
Borovsky, J.E., Denton, M.H., 2013. J. Geophys. Res. Space Phys. 118, 5506.
Brun, et al., 2004. Astrophys. J. Lett. 614 (2), 1073.
Chen, Y., Wang, W., Burns, A.G., Liu, S., Gong, J., Yue, X., Jiang, G., Coster, A., 2015. J. Geophys. Res. Space Phys. 120, 1394.
Claire, M.W., et al., 2012. Astrophys. J. 757, 95.
Crosby, N., Aschwanden, M., Dennis, B., 1993. Adv. Space Res. 13 (9), 179.
de Zeeuw, D.L., Sazykin, S., Wolf, R.A., Gombosi, T.I., Ridley, A.J., Toth, G., 2004. J. Geophys. Res. Space Phys. 109, A12.
do Nascimento Jr., J.-D., Vidotto, A.A., Petit, P., 2016. Astrophys. J. Lett. 820, L15.
Dungey, J.W., 1961. Phys. Rev. Lett. 6, 47.
Emslie, A.G., Dennis, B.R., Shih, A.Y., Chamberlin, P.C., Mewaldt, R.A., Moore, C.S., Share, G.H., Vourlidas, A., Welsch, B.T., 2012. Astrophys. J. Lett. 759, 71.
Gershberg, R.E., 2005. In: Solar-Type Activity in Main-Sequence Stars: Astronomy and Astrophysics Library. Springer, Berlin, Heidelberg. ISBN 978-3-540-21244-7.
Glocer, A., Toth, G., Gombosi, T., Welling, D., 2009. J. Geophys. Res. Space Phys 114 (A5), A05216.
Glocer, A., Kitamura, N., Tóth, G., Gombosi, T., 2012. J. Geophys. Res. Space Phys. 117, A04318.
Glocer, A., Khazanov, G., Liemohn, M., 2017. J. Geophys. Res. Space Phys. 122 (6), 6708.
Gopalswamy, N., 2009. Climate and Weather of the Sun-Earth System (CAWSES): Selected Papers from the 2007 Kyoto Symposium, Tsuda, T., Fujii, R., Shibata, K., Geller, M.A. (Eds.), 77.
Gopalswamy, N., Yashiro, S., Thakur, N., Mäkelä, P., Xie, H., Akiyama, S., 2016. Astrophys. J. Lett. 833(2), 216.
Gopalswamy, N., Mäkelä, P., Yashiro, S., Thakur, N., Akiyama, S., Xie, H., 2017a. J. Phys. Conf. Ser. 900 (1), 012009.
Gopalswamy, N., Mäkelä, P., Yashiro, S., Thakur, N., Akiyama, S., Xie, H., 2017b. In: Journal of Physics: Conference Series (JPCS), Proceedings of the 16th Annual International Astrophysics Conference held in Santa Fe, NM.
Gough, D.O., 1981. Sol. Phys. 74, 21–34.
Green, J., Boardsen, S., 2006. Adv. Space Res. 38 (2), 130–135.
Gronoff, G., Rahmati, A., Wedlund, C.S., Mertens, C.J., Cravens, T.E., Kallio, E., 2014. Geophys. Res. Lett. 41, 4844.
Güdel, M., Nazé, Y., 2009. Astron. Astrophys. Rev. 17 (3), 309.
Güdel, M., Guinan, E.F., Skinner, S.L., 1997. Astrophys. J. Lett. 483, 947.
Hayakawa, H., Mitsuma, Y., Fujiwara, Y., Kawamura, A.D., Kataoka, R., Ebihara, Y., Kosaka, S., Iwahashi, K., Tamazawa, H., Isobe, H., 2017. Publ. Astron. Soc. Jpn. 69 (2), 17.
Howell, S.B., Sobeck, C., Haas, M., et al., 2014. Publ. Astron. Soc. Jpn. 126 (938), 398.
Kasting, J.F., 2010. Nature 464, 687–689.
Khazanov, G., 2011. Kinetic theory of the inner magnetospheric plasma. Astrophysics and Space Science Library, vol. 372. Springer, London, ISBN 978-1-4419-6796-1.

Khazanov, G.V., Tripathi, A.K., Sibeck, D., Himwich, E., Glocer, A., Singhal, R.P., 2015. J. Geophys. Res. Space Phys. 20 (11), 9891.
Knipp, D.J., Pette, D.V., Lilcommons, L.M., et al., 2017. Space Weather 15. https://doi.org/10.1002/20165W001567.
Kobayashi, K., Masuda, H., Ushi, K., Ohashi, A., Yamanashi, H., et al., 2001. Adv. Space Res. 27 (2), 207.
Kobayashi, K., Aoki, R., Abe, H., Kebukawa, Y., Shibata, H., Yoshida, S., Fukuda, H., Kondo, K., Oguri, Y., Airapetian, V.S., 2017. Astrobiology Science Conference 2017, Abstract #3259.
Lammer, H., Kasting, J.F., Chassefiere, E., Johnson, R.E., Kulikov, Y.N., Fian, F., 2008. Space Sci. Rev. 139, 399.
Lavraud, B., Ruffenach, A., Rouillard, A.P., Kajdic, P., Manchester, W.B., Lugaz, N., 2014. J. Geophys. Res. Space Phys. 119, 26.
Le, et al., 2016. Geophys. Res. Lett. 43, 2396.
Lepping, R.P., Burlaga, L.F., Jones, J.A., 1990. J. Geophys. Res. 95, 11957.
Longcope, D., Beveridge, C., Qiu, J., Ravindra, B., Barnes, G., Dasso, S., 2007. Sol. Phys. 244, 45.
Loyd, R.O.P., France, K., Youngblood, A., Schneider, C., Brown, A., Hu, R., Linsky, K., Froning, C.S., Redfield, S., Rugheimer, S., Tian, F., 2016. Astrophys. J. Lett. 824, 102.
Maehara, H., et al., 2012. Nature 485, 478.
Mewaldt, R.A., Cohen, C.M.S., Labrador, A.W., Leske, R.A., Mason, G.M., Desai, M.I., Looper, M.D., Mazur, J.E., Selesnick, R.S., Haggerty, D.K., 2005. J. Geophys. Res. Space Phys. 110, A09S18.
Mikhail, Sverjensky, 2014. Nat. Geosci. 7, 816.
Mironova, I.A., Usoskin, I.G., 2014. Environ. Res. Lett. 9.
Miyake, F., Nagaya, K., Masuda, K., Nakamura, T., 2012. Nature 486, 240.
Ngwira, C.M., Pulkkinen, A., Kuznetsova, M.M., Glocer, A., 2014. J. Geophys. Res. Space Phys. 119, 4456.
Notsu, Y., et al., 2013. Astrophys. J. Lett. 771, 127.
Parker, E.N., 1955. Astrophys. J. Lett. 122, 293.
Pevtsov, A.A., Fisher, G.H., Acton, L.W., Longcope, D.W., Johns-Krull, C.M., Kankelborg, C.C., Metcalf, T.R., 2003. Astrophys. J. Lett. 589, 1387.
Powell, K.G., Roe, P.L., Linde, T.J., Gombosi, T.I., De Zeeuw, D.L., 1999. J. Comput. Phys. 154, 284.
Quantana, E., Barclay, T., Raymond, S.N., et al., 2014. Science 344, 277.
Ramirez, R.M., 2016. Nat. Geosci. 9 (6), 413.
Ramirez, R.M., Kopparapu, R., Zugger, M.E., Robinson, T.D., Freedman, R., Kasting, J., 2014. Nat. Geosci. 7 (1), 59.
Rauer, H., Catala, C., Aerts, C., et al., 2014. Exp. Astron. 38 (1–2), 249.
Ribas, I., Jordi, C., Vilardell, F., Fitzpatrick, E.L., Hilditch, R.W., Guinan, E.F., 2005. Astrophys. J. 635, L37.
Ricker, G.R., Latham, D.W., Vanderspek, R.K. et al., 2010. AAS Meeting #215, id.450.06, Bulletin of the American Astronomical Society, vol. 42, p. 459.
Riley, P., Caplan, R.M., Giacalone, J., Lario, D., Liu, Y., 2016. Astrophys. J. Lett. 819, 57.
Rosing, M.T., Bird, D.K., Sleep, N.H., Bjerrum, C.J., 2010. Nature 464, 744.
Sagan, C., Mullen, G., 1972. Science 177, 52.
Saladino, R., Carota, E., Botta, G., Kapralov, M., Timoshenko, G.N., Rozanov, A.Y., Krasavin, E., Ernesto Di Mauro, E., 2015. Publ. Nat. Acad. Sci. 112 (21), E2746.
Shibata, K., Takasao, S., 2016. Magnetic reconnection. Astrophysics and Space Science Library, Springer, Switzerland, ISBN 978-3-319-26430-1 (p. 373).
Shibayama, T., Maehara, H., Notsu, S., Notsu, Y., Nagao, T., Honda, S., Ishii, T.T., Nogami, D., Shibata, K., 2013. Astrophys. J. Suppl. Ser. 209, 5.
Siscoe, G., Crooker, N.U., Clauer, C.R., 2006. Dst of the Carrington storm of 1859. Adv. Space Res. 38, 173–179.
Smith, E.J., Wolfe, J.H., 1976. GeoRL 3, 137.

Som, S.M., Buick, R., Hagadorn, J.W., Blake, T.S., Perreault, J.M., Harnmeijer, J.P., Catling, D.C., 2016. Nat. Geosci. 9, 448.
Sterenborg, M.G., Cohen, O., Drake, J.J., Gombosi, T.I., 2011. J. Geophys. Res. 116, A01217.
Strangeway, R.J., Ergun, R.E., Su, Y.-J., Carlson, C.W., Elphic, R.C., 2005. J. Geophys. Res. Space Phys. 110 (A3). A03221.
Tarduno, J.A., Blackman, E.G., Mamajek, E.E., 2014. Phys. Earth Planet. Inter. 233, 68.
Tian, F., Kasting, J.F., Liu, H.L., Roble, R.G., 2008. J. Geophys. Res. Space Phys. 113 (E5), E05008.
Tsurutani, B.T., Lakhina, G.S., 2014. Geoph. Res. Let 41, 287.
Tsurutani, B.T., Gonzales, W.D., Lakhina, G.S., Alex, S., 2003. J. Geophys. Res. 108, 1268.
Tu, L., Johnstone, C., Gudel, M., Lammer, H., 2015. Astron. Astrophys. 577, L3.
Usmanov, A.V., Goldstein, M.L., Matthaeus, W.H., 2016. Astrophys. J. Lett. 820, 17.
Verkhoglyadova, O., Meng, X., Mannucci, A.J., Tsurutani, B.T., Hunt, L.A., Mlynczak, M.G., Hajra, R., Emery, B.A., 2016. J. Space Weather Space Clim. 6, id.A20.
Vidotto, A.A., Gregory, S.G., Jardine, M., Donati, J.F., Petit, P., et al., 2014. NMRAS 441, 2361.
Weimer, D.R., Mlynczak, M.G., Hunt, L.A., Tobiska, W.K., 2015. J. Geophys. Res. Space Phys. 120, 5998.
Wood, B.E., Müller, H.-R., Zank, G.P., Linsky, J.L., Redfield, S., 2005. Astrophys. J. Lett. 628, L143.
Yashiro, S., Gopalswamy, N., 2009. In: Gopalswamy, N., Webb, D.F. (Eds.), IAU Symp. 257, Universal Geophysical Processes. Cambridge University Press, Cambridge, p. 233.
Zhao, Y., Wang, R., Lu, Q., Du, A., Yao, Z., Wu, M., 2016. J. Geophys. Res. Space Phys. 121 (11), 10898.

FURTHER READING

Thakur, N., Gopalswamy, N., Xie, H., Makela, P., Akijama, P., Davila, J., Rickard, M., 2014. Astrophys. J. Lett. 790, L13.
Gopalswamy, N., 2011. In: Choudhuri, A.R., Banerjee, D. (Eds.), First Asia-Pacific Solar Physics Meeting ASI Conference Series. vol. 2. (pp. 2).
Konovalenko, A. A. and 17 co-authors, 2012. European Planetary Science Congress, vol. 7.
Kramar, M., Airapetian, V., Lin, H., 2016. Front. Astron. Space Sci. 3, id.25.
Love, J.J., 2012. Geophys. Res. Lett. 39, L10301.
Nitta, N.V., Aschwanden, M.J., Freeland, S.L., Lemen, J.R., Wülser, J.-P., Zarro, D.M., 2014. Sol. Phys. 289, 1257.
Ridley, A.J., Hansen, K.C., Tóth, G., de Zeeuw, D.L., Gombosi, T.I., Powell, K.G., 2002. J. Geophys. Res. Space Phys. 107, 1290.
Schrijver, C.J., Kauristie, K.A., Alan, D., Denardini, C.M., Gibson, S.E., Glover, A., Gopalswamy, N., Grande, M., Hapgood, M., Heynderickx, D., et al., 2015. Adv. Space Res. 55, 2745.
Schrijver, C.J., Title, A.M., 2013. Astrophys. J. Lett. 619, 1077.
Thomas, B.C., Melott, A.L., Arkenberg, K.R., Snyder, B.R., 2013. Geophys. Res. Lett. 40, 1237.
Usmanov, A.V., Goldstein, M.L., Matthaeus, W.H., 2014. Astrophys. J. Lett. 788, 43.

PART 6

DEALING WITH THE SPACE WEATHER

CHAPTER 26

DEALING WITH SPACE WEATHER: THE CANADIAN EXPERIENCE

David H. Boteler
Natural Resources Canada, Ottawa, ON, Canada

CHAPTER OUTLINE

1 Introduction 635
2 High-Frequency Radio Communications 637
3 Satellites 639
4 Ground Systems 642
5 Surveying and Navigation 646
6 Concluding Remarks 648
Acknowledgments 652
References 652
Further Reading 655

1 INTRODUCTION

Space weather has always been a feature of Canadian life. Its visual manifestation, the aurora—called "Arsaniit" in Inuktitut—has been observed by the indigenous people of Canada for millennia. The aurora occurrence was even recorded in the diaries of early European explorers and settlers (see review by Broughton, 2002). The first magnetic observatory was established in Canada, at Toronto in 1839 (Lam, 2016). Special magnetic recordings were made at Fort Rae during the International Polar Years in 1882–83 and 1932–33 (Newitt and Dawson, 1984). The Meanook Observatory was established in 1916 and was used for auroral observations as well as magnetic recordings during the second polar year (Vestine, 1943). However, it was the International Geophysical Year (IGY) in 1957–58 that saw an expansion of observing capabilities and scientific interest in space weather phenomena that has continued to the present day. This interest now includes widespread ground networks, scientific satellites, and solar radio telescopes.

Because of its northern location, Canada has had more experience dealing with space weather than probably any other country in the world. The term "space weather" is of fairly recent origin, but effects that we now attribute to space weather date from the mid 1800s (Prescott, 1866). The telegraph system suffered widespread effects in Canada and around the world during the Carrington storm of 1859. Telegraph problems continued during major magnetic disturbances for the next half-century (Boteler et al., 1998). In the first half of the 20th century, high-frequency (HF) radio communications became the dominant means for long-distance communications, something that was particularly important for a country the size of Canada. In the 1950s, with the start of the space age coinciding with the IGY, Canada became the third country in space, first with satellites studying the ionosphere to help with HF radio communications and then through embracing the opportunities for satellite communications and Earth observations. The 1950s and 1960s saw major developments and expansion of ground-based technology such as pipelines and long power lines to meet the energy needs of the increasing Canadian population (Fig. 1).

To describe the Canadian experience in dealing with space weather, we have chosen to start in the 1950s. This represents a pivotal time in both the expansion of space weather science and in the development of the technology that we rely on today that is affected by space weather. The 60 years since then also represents a time that is still within living memory. Earlier phenomena and effects, such as the Carrington storm, are brought into the story as they are "rediscovered" and used to provide information about extremes that is relevant to modern systems. The following sections describe the space weather

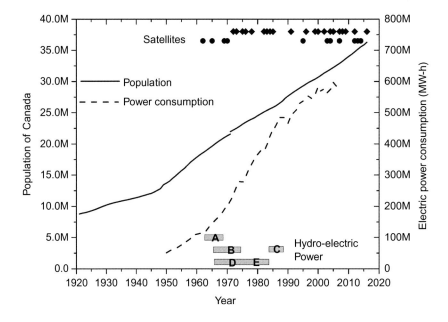

FIG. 1

Growth in population and electric power consumption in Canada. Also shown are the launch dates of Canadian science and Earth observation satellites ● and communication satellites ◆ and the dates of construction of major hydroelectric projects on northern rivers: Peace River, BC (A), Nelson River, Manitoba (B, C), Churchill Falls, Labrador (D), and James Bay, Québec (E).

Sources: Population growth 1921–71: Statistics Canada Table 051–0026; Population growth 1971–2016: Statistics Canada Table 051–001; Electric Power Statistics, annual: Statistics Canada Table 127–0001.

effects on different types of technology in Canada and the work done in dealing with those effects. Finally, some comments are made about how the continuing expansion of knowledge about space weather is being applied to protect critical infrastructure in Canada from future extremes of space weather.

2 HIGH-FREQUENCY RADIO COMMUNICATIONS

Canada has been involved with long-distance high-frequency radio communications since its beginning, with a transmission from the Marconi station in Glace Bay, Nova Scotia, in December 1902 becoming the world's first radio message to cross the Atlantic from North America (Parks Canada, 2017). The reflection of radio signals by the ionosphere that allowed radio transmissions to be received at remote stations beyond the curvature of the Earth not only allowed transoceanic communications but also benefitted communications over the vast distances within Canada. The communication needs in southern Canada could be met by telephone and radio relay networks, but the population in northern Canada was too small and scattered for the use of such systems to be economically justified. Instead, communication with outposts in northern Canada had only been provided by annual supply ships. Beginning in the late 1920s, HF radio communications started to be used in the north. In the 1930s, Hudson Bay outposts across the Arctic were equipped with shortwave radios. HF radio communications soon became an essential tool for business in the north as well as for medical emergencies (Roach, 2014).

The economic importance of HF radio communications for Canada led to considerable work on what we now refer to as space weather effects on the ionosphere. The Telecommunications Division of the Department of Transport, in collaboration with the Radio Physics Laboratory of the Defence Research Board (DRB), set up eight ionospheric monitoring stations (Davies, 1954). These observations showed how the ionosphere changed with the time of day, the season, and the sunspot cycle. This enabled DRB to provide radio operators with advice on the best frequencies to be used at different times and for different radio paths. This was crucial to establishing radio contact with a remote station: use of too low a frequency and the signal would be absorbed by the lower D region of the ionosphere while frequencies that were too high to be reflected by the E or F regions would just pass on into space. Radio operators became experienced in adjusting their frequencies to cope with these "regular" changes in ionospheric conditions. However, during times of ionospheric disturbances, radio communications in large parts of Canada would fail completely.

It was recognized that the ionosphere over northern and central Canada is often more disturbed than other regions of the Earth (Davies, 1954). Studies of ionospheric disturbances showed they occurred more frequently in the equinoctial months and in years with high sunspot activity. Radio disturbances were associated with aurora in a band across the country; farther north the aurora diminishes but ionospheric disturbances that affected radio communications still occurred. Davies (1954) commented that one of the main problems of the ionosphere over northern Canada, as elsewhere, is that of forecasting storm conditions in the ionosphere to improve radio communications. This work was undertaken by the Defence Research Telecommunications Establishment (DRTE). The DRTE contributed to the extensive ionospheric program that was conducted during the International Geophysical Year in 1957–58. In Canada, the ionospheric sounding network was expanded with nine more stations as well as the installation of other instruments, including riometer measurements of absorption at

Churchill and Ottawa (Meek, 1959). The DRTE also acted as the central warning agency for Canada and was responsible for distributing alerts about radio conditions.

Ionospheric research continued as a major topic of research into the 1960s. It was no coincidence that the first Canadian satellites put into orbit were designed to study the ionosphere (see next section). However, as satellite communications systems were introduced, the use of HF radio declined in many parts of the world. In northern Canada, it continued to have a role with geophysics field parties who used magnetic forecasts as a guide to when radio problems could be expected. HF radio communications also continued to be used by the aviation industry. Airline flights are managed by a limited number of air traffic control centers and, for safety reasons, the aircraft need to be able to communicate with these centers at all times. Over populated regions the aircraft-to-control-center communication is via VHF radio links from the aircraft directly to transponders on the ground that would relay the signals to the control center. However, over remote regions such as northern Canada and Greenland and over the oceans, the communication is via HF radio. During the space weather disturbance of October–November 2003, ionospheric disturbances caused considerable problems for HF radio communications between aircraft and air traffic control centers; many flights were diverted to lower latitude routes (NOAA, 2004).

Starting in the early 2000s, changes in airline operations also caused a revival of interest in HF radio communications in the Arctic. This change was the inauguration of commercial transpolar flights. Initially this was just a few flights a year by one airline but since then there has been a steady growth in transpolar operations so that there are now thousands of flights a year by aircraft from many airlines. These flights are managed by the NavCanada Arctic Air Traffic Control Center in Edmonton (Fig. 2) while in Canadian airspace. They approached the Canadian Space Weather Forecast Center (CSWFC) for forecasts of radio conditions at these high latitudes. This prompted a renewal of research on

FIG. 2

Operator at the Arctic Air Traffic Control Center, Edmonton, Canada.

Figure courtesy of NavCanada.

ionospheric conditions that affect HF radio communications and led to the installation of riometers at the Canadian magnetic observatories and other sites. This complemented riometers already installed by the University of Calgary for auroral research and has resulted in a combined Canadian network of 22 riometers (Danskin et al., 2008). These riometers and HF radio transmitters and receivers installed to monitor the operation of certain radio paths are now being used for research to improve forecasts for Arctic radio communications (Fiori and Danskin, 2016; Warrington et al., 2016).

An example of the difficulty in dealing with space weather effects on aircraft communications comes from December 2008. United Airlines had their pilots on transpolar flights report on the communication conditions they experienced. Of two flights a few days apart, one experienced good communications while the other had poor communications. On both days the space weather activity was low. United Airlines contacted the CSWFC to ask why poor communications had been experienced on the second day. Fortunately, we were able to obtain ionosonde and riometer data for the high-latitude regions, which provided information on the ability of the ionosphere to reflect or absorb radio signals relayed from Air Traffic Control to the aircraft. Understanding the physics behind what is happening is important because the conventional wisdom was that problems were always due to D region absorption; therefore to get above the absorption threshold operators need to move to higher frequencies. However, in this case it was a limited reflecting layer, meaning the better solution would be to go to a lower frequency that would be reflected.

3 SATELLITES

Launched on September 29, 1962, the Alouette-I scientific satellite marked Canada's entry into the space age; Canada became the third country after the Soviet Union and the United States to design and build its own artificial Earth satellite (Hartz and Paghis, 1982; Shepherd and Kruchio, 2008). Although the Alouette-I is now viewed as a purely scientific satellite, the documents from the time show that the ionospheric focus of its studies was strongly motivated by the need to improve understanding of ionospheric disturbances as a way to improve HF radio communications. The success of Alouette-I prompted the launch of further satellites. First the Alouette backup model was refurbished and flown in 1965 as "Alouette-II." Then two new satellites for the International Satellites for Ionospheric Studies (ISIS) program, named ISIS I and ISIS II, were launched in 1969 and 1971, respectively.

The Alouette and ISIS satellites were designed to study the ionosphere and the aurora borealis. Previously, the ionosphere had been studied from the ground, primarily by using ionosondes—a swept-frequency radar pointed upward that would measure the height of the ionospheric layers by the travel time of radio pulses transmitted from the ground and reflected by the ionosphere. The maximum frequency reflected by a layer gave a measure of its electron density, after which higher frequencies passed through that layer to perhaps be reflected by a higher layer until the highest electron density was reached and higher frequencies then went out into space. This provided a profile of the lower structure of the ionosphere but gave no information about the electron density profile above the electron density peak. To probe this upper part of the ionosphere, the Alouette and ISIS satellites employed "topside sounders" to obtain radio reflections of the ionosphere from above (Chapman and Warren, 1968). To say that these satellite-borne sounders were inverted versions of the ground-based ionosondes vastly understates the technical challenges of building equipment to work in space and working out how to deploy transmitting and receiving antennas after launch, but that is the type of measurements

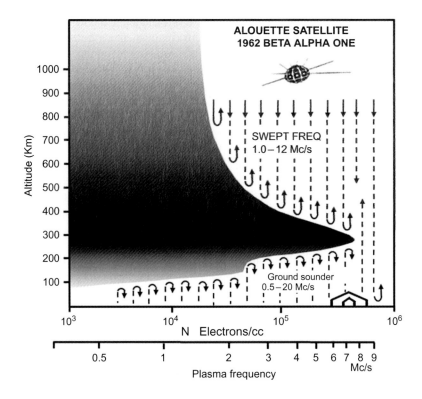

FIG. 3

Ionospheric sounding from ground ionosonde and topside sounder showing method of measurement of electron density distribution in the ionosphere.

Redrawn from Molozzi, A.R., 1964. Instrumentation of the Ionospheric Sounder Contained in the Satellite 1962 Beta Alpha (Alouette), Space Research IV. North-Holland Publishing Company, pp. 413–436.

they provided (Fig. 3). To complement the topside sounders, the satellites also carried other instruments, including auroral imagers from the University of Calgary. Measurements were taken over an entire 11-year solar cycle to determine how the ionosphere reacts to changes in the Sun's radiation. These measurements fed into many research publications at the time (e.g., Schmerling and Langille, 1969) and subsequent efforts to make these data more available have prompted further research efforts (Restoration, 2017).

In the late 1960s, the Canadian government's interest in space shifted to the possibility of using space for domestic communications (Chapman et al., 1967). In 1972, Telesat Canada, established as a government-owned company in 1969, launched the first in a series of geostationary communication satellites named Anik, meaning "little brother" in Inuktitut. The Anik A1, A2, and A3 satellites operated satisfactorily, with service availability exceeding 99.9%. Some operational issues were noted and spacecraft charging due to space weather activity was considered as a possible cause, although this was never conclusively determined (Wadham, 1980). Other satellites in the Anik series followed. Experience with Anik-B demonstrated another impact of space weather on satellite operation. Anik-B, along with many satellites, used "magnetic torquers" to control the attitude of the spacecraft; however, during a large magnetic field disturbance the torquers can actually increase the attitude error rather than reduce it. This occurred during a disturbance on February 8, 1986, and one other time during the 7 years of Anik-B service (Bedingfield et al., 1996). The Anik-Cs experienced some phantom commands due to space weather, as did Anik-D1 (Gubby and Evans, 2002), but these were dealt with by satellite operators. However, a more serious problem occurred with Anik-D2 on March 8, 1985, when a malfunction of the despin control system occurred after an interval of unusually high space weather activity. The satellite was eventually brought under control but at the expense of a year's worth of fuel for station keeping (Leach and Alexander, 1995).

Through their experience with the multiple Anik As, Bs, Cs, and Ds, Telesat satellite operators had become adept at managing space environment effects on their satellites. However, the Anik E satellites were going to present new challenges. Anik E2 was launched on April 4, 1991, and immediately problems occurred with the deployment of its C-band antenna. After extensive ground testing and simulations, Telesat performed a series of complex maneuvers that were never envisioned for this satellite and rarely, if ever, carried out before and the C-band antenna finally deployed at 02:30 on July 3, 1991 (Wang and Martens, 1993). Anik E1 was modified to avoid this problem and was launched in September 1991. For the next couple of years both satellites worked satisfactorily. However, in January 1994, spacecraft charging and subsequent discharge damaged the momentum wheel control circuits on both Anik E1 and Anik E2 (Harland and Lorenz, 2005). Anik E1 was recovered by switching over to the backup unit, but when this was tried on Anik E2, it was found that the backup had also failed, either during the same event or earlier. These effects followed a sustained period of elevated energetic electron fluxes in the outer fringe of the outer radiation belt (i.e., at geostationary orbit) that were produced by high-speed solar wind streams that could be traced back to a large coronal hole, elongated in longitude, on the Sun (Lam et al., 2012). The Anik E satellites were used to relay television programs across the country and this event produced one of the most serious consequences for Canadians–interrupting the television broadcast of a hockey game. The Anik E2 outage disrupted communications to northern Canada and sent technicians scurrying to realign satellite dishes to point at other satellites.

The loss of Anik E2 was potentially a major threat to the financial stability of Telesat. Industry commentators were predicting that the satellite was irrecoverable, as indeed had been the case with other satellites. However, after months of work the Telesat engineers developed a novel control system using ground-based measurements to determine the attitude of the spacecraft and then using the thrusters, in previously unheard of small bursts, to adjust the satellite attitude (Burlton, 1995). Thus in August 1994, Anik E2 was restored to service and continued operating until beyond its design lifetime (Lam et al., 2012). The space environment was a continuing concern for Telesat. The recognition that high fluxes of energetic electrons were responsible for the Anik-E failures led to energetic electron fluence forecasts being developed based on the relationship between energetic electrons and magnetic activity (Lam, 2004). Telesat's experience in dealing with the space environment was not all bad. Their experience with the Anik E2 problems and their heroic development of a radically new attitude control system meant that other satellite operators came to them for advice and consulting about satellite matters became a successful adjunct to Telesat's business.

Geostationary satellites provide good communications links for southern Canada, but as one goes farther north those satellites become lower in the sky and provide weaker signals. This can be compensated for by the use of larger, more expensive satellite dishes on the ground and with the use of spot-beam satellite transmissions, as on Anik F2. However, at very high latitudes such as in the Canadian Arctic, the geostationary satellites are not visible from the ground because of the curvature of the Earth. To provide communications and weather observations at these high latitudes, a different satellite configuration was proposed (Garand et al., 2014). This would use two satellites in highly elliptical orbits (HEO) that are not in a fixed position above the Earth, but, because of orbit dynamics, have a long "dwell" time over the pole to provide similar capabilities as geostationary satellites. However, satellites on such a highly elliptical orbit would pass through the active regions of the "radiation belts" where energetic particles are trapped in the Earth's magnetic field. To support the development of the HEO satellite proposal, considerable work was done to investigate the space environment that would be experienced by such satellites (Trichtchenko, 2012). This even included the identification of other orbits that would reduce the exposure to the worst of the space environment (Trichtchenko et al., 2014).

Space weather is taken into account, not only because of risks during extreme events but also in the routine operation of the satellite. The power for satellites is derived from solar panels, which are subject to damage by solar particles. An allowance for this degradation was built into the design of the solar panels to take account of the background particle flux as well as some allowance for solar flares. After each major flare, the engineers would check the solar array capability to see how much degradation had occurred. For satellites in low Earth orbit (LEO), such as Radarsat I and II, the satellite altitude is continually dropping because of atmospheric drag; thrusters are used periodically to boost the satellite so that it maintains orbit. Orbit maintenance is a particular concern for a modern Earth-observation satellite like Radarsat II where the operators are required to maintain the spacecraft's orbit to keep the ground track within a strict tolerance range (better than ± 5 km, with a goal of ± 1 km) during its operational lifetime. To maintain this tolerance required orbit maneuvers once every 1.5 months during periods of low solar activity, but this could increase to once every 3 weeks or more at a time of solar maximum. The sizing of solar panels and the fuel for thrusters have to be budgeted to last the projected life of the satellite, taking account of the variations in solar activity over that time. At the start of the satellite design process, this requires predictions of solar activity 30 years ahead: 10 years for design, construction, and launch of the satellite plus 20 years operational lifetime in orbit.

4 GROUND SYSTEMS

The first geomagnetic effects on power systems occurred on March 24, 1940, including tripping of transformers in Ontario (Davidson, 1940). In the following years there were no effects reported in the literature. However, the archives at the Geomagnetic Laboratory of Natural Resources Canada contain correspondence between the Ontario Power Commission (the forerunner of Ontario Hydro, now split into Hydro One and Ontario Power Generation) and the geomagnetism group in the Dominion Observatory saying that disturbance currents were observed in the power network on certain days in 1949 and requesting the magnetic charts for those days. The next notable effects on Canadian (and U.S.) power systems occurred during the magnetic storm of February 10–11, 1958. An article in the New Yorker (Brooks, 1959) claimed that the Toronto area in Ontario was plunged into temporary darkness. While two transformers tripped in the west of Ontario (Acres, 1975), this was too far away to impact the Toronto area, so the New Yorker article would appear to be incorrect. Canadian power utilities were periodically reminded of geomagnetic effects on their systems, but these effects were infrequent enough that they did not become part of the corporate memory and these effects had to be "rediscovered" by successive generations of engineers.

The first comprehensive study of geomagnetic effects on power systems, started in the late 1960s, was organized by the Edison Electric Institute and involved many power utilities in both the United States and Canada. This study included analytical work as well as recordings of geomagnetically induced currents (GICs) at many locations. These recordings highlighted that much larger GICs were recorded at Cornerbrook in Newfoundland than anywhere else (Albertson et al., 1974). Close analysis of this location showed that the power line constituted a low resistance path that bridged a resistive strip of land between the Atlantic on one side and the Gulf of St. Lawrence on the other. On August 4, 1972, a major geomagnetic disturbance occurred that produced effects on power systems and other systems. In Canada, power systems in Newfoundland, Quebec, Ontario, Manitoba, and Saskatchewan experienced a range of system effects, from tripping of equipment to voltage dips and frequency fluctuations (Acres, 1975). These would have kept system operators busy but produced no effects for consumers. The same cannot be said for the Bowater paper mill at Cornerbrook in Newfoundland, where the power fluctuations caused numerous paper breaks, disrupting the paper production process (Acres, 1975).

In the following years, geomagnetic activity declined and the power industry focused its attention on other issues. Some studies were made in collaboration with local universities but the industry view was that geomagnetic effects on power systems were of purely academic interest and that they would have no real effects on power system operation. That view continued until March 1989.

On March 6, 1989, a complex sunspot region came into view around the edge of the Sun's disk. Over the next 2 weeks, as it traversed the face of the Sun, it unleashed numerous solar flares, some of which were accompanied by the eruptions of plasma, now called coronal mass ejections (CMEs). The first CME impact on the Earth's magnetic field occurred at 00:30 UT on March 13, initiating a series of magnetic substorms that produced large magnetic field fluctuations across North America. The first two substorms produced relay trips and power surges on a number of power systems. But the third substorm had a particularly sudden onset and produced large electric fields across Canada, impacting many systems. The most notable effect occurred in Québec where the high resistance rock of the Canadian shield resulted in larger electric fields. The low resistance transmission lines of the 735-kV Hydro-Québec power system allowed these electric fields to drive large GICs across the network. Where these GICs flowed to and from ground through the power transformers, they pushed the transformers into saturation, resulting in the generation of harmonics of the AC frequency and increased (reactive) power use by the transformers (Kappenman, 2007). These effects had a double impact on the power system. The increased power demand by the transformers caused the voltage to drop. The extra power required by the transformers would normally be supplied by Static VAR Compensators (SVCs), but, just when they were needed most, they were tripped out of operation by the high levels of harmonics. Without the SVC support the voltage declined—automatic load shedding was not enough to compensate—and the system collapsed within 90 s (Guillon et al., 2016).

The people of Québec were treated to a magnificent auroral display, without their view being obstructed by streetlights. Within 9 h, 83% of the load had been restored and the remaining customers recovered electrical power that evening. Although this event was a surprise to the people of Québec, it represented more of a shock to the power industry. The unthinkable had happened: a geomagnetic disturbance had caused a power system collapse. Many power systems in North America, the United Kingdom, and Scandinavia had experienced problems, including overheating of transformers in the United States and the United Kingdom, and the industry was quick to organize workshops to find out exactly what happened (EPRI, 1989; IEEE, 1990). The Canadian Electricity Association funded a comprehensive study of the geomagnetic hazard to power systems (Boteler, 2001). This included involvement from many Canadian power utilities, which undertook modeling of the GICs in their systems. This laid the foundation of knowledge for studies that have continued to the present day.

Hydro-Quebec did a lot of work related to geomagnetic effects in the aftermath of the March 1989 power blackout (Bolduc, 2002; Guillon et al., 2016). However, the most significant change was motivated by increasing the efficiency and reliability of power transmission rather than by geomagnetic concerns. This was the introduction of "series compensation" on the major transmission lines bringing power from James Bay and Churchill Falls. All transmission lines have a certain impedance, comprised of the resistance and inductance of the lines and some power is used to overcome this impedance in the transmission of electrical power from the generators to the load centers. Capacitors introduced in series with the transmission lines can compensate for the line inductance (hence the name "series compensation"), thereby reducing the overall impedance of the line and the amount of power lost in the lines. The happy coincidence for the Hydro-Québec system is that this series compensation also blocks the flow of GICs in the lines where it is installed.

Many other Canadian power utilities have been involved in studies, often in collaboration with government scientists, to understand how magnetic disturbance characteristics, Earth conductivity structure, and system characteristics affect GICs in different systems across Canada (Zheng et al., 2013, 2014). This has led to developments in the modeling of GICs (Boteler and Pirjola, 2017) that are leading to the development of analysis tools for use by industry to assess the possible impacts on their systems (Jacobson et al., 2014). Processes have also been developed for real-time GIC simulation to provide information for power system controllers (see Fig. 4) to monitor conditions on their network (Marti et al., 2013).

Pipelines are another type of long conductor at the Earth's surface that are subjected to induced electric fields produced by geomagnetic disturbances. The geomagnetic field variations penetrate tens to hundreds of kilometers into the Earth, so the burial of pipelines at depths of a few meters does not shield them from the geomagnetic disturbance. Pipelines are fitted with "cathodic protection" systems that maintain the pipeline at approximately 1 volt negative with respect to the surrounding soil. This potential is normally sufficient to alter the electrochemical environment at the interface between the pipe and soil in such a way as to inhibit the corrosion of the pipeline. However, during geomagnetic disturbances, the induced telluric currents can take the pipeline potential outside the optimum range for cathodic protection.

Short pipelines had been constructed in Quebec and Ontario in the mid-1800s, but it was the oil and gas discoveries in Alberta in the late 1940s and early 1950s that prompted the construction of long pipelines to other parts of Canada and down into the United States. As early as 1954, the Canadian pipeline operators noted voltage fluctuations during geomagnetic disturbances (Russell and Nelson, 1954). However, it was the construction of the pipeline from Norman Wells, NWT, to Zama, Alberta, from 1983 to 1985 in the auroral zone that really prompted Canadian industry to confront the effects of telluric currents. The consulting corrosion engineers, Commonwealth Seager Group, conducted a series of special telluric studies on the Norman Wells–Zama pipeline. One study involved leapfrogging two potential recorders by helicopter to provide partially overlapping recordings of the telluric variations on the pipeline. By adjusting these variations for the level of geomagnetic activity measured at the Yellowknife magnetic observatory, they managed to obtain a picture of the phase relationship of the telluric variations on the pipeline. This showed that the variations in the northern half were in phase but out of phase with the variations in the southern half. This knowledge enabled cathodic protection systems to be designed that automatically adjusted their output to reduce the telluric potential variations (Seager, 1991).

Modeling of telluric effects on pipelines was introduced in the 1990s; the first test of the modeling was provided by comparison with the observations made on the Norman Wells–Zama pipeline (Boteler and Seager, 1998). Subsequent recordings of telluric effects were made in research collaborations between the Geomagnetic Laboratory in Ottawa and Canadian pipeline companies. The test area on the TransCanada pipeline was a section of pipeline where annual surveys could not be satisfactorily completed because of telluric fluctuations. This identified a region of unusually high telluric fluctuations where the pipeline crossed the boundary between different geologic zones. Increased understanding of geomagnetic induction in pipelines (Trichtchenko and Boteler, 2001, 2002) enabled telluric effects to be taken into account in the design of new pipelines (Rix and Boteler, 2001). The pipeline industry also developed methods for adjusting pipeline survey results to compensate for telluric currents (Place and Sneath, 2001; Carlson et al., 2004) and online services have been developed to provide pipeline operators with information about telluric activity that can be expected (Trichtchenko et al., 2008).

FIG. 4

Top, Power system control center. *Bottom*, Display of geomagnetically induced currents for system operators.

Figures courtesy of Hydro One.

Now, as pipelines are being proposed to bring hydrocarbons from fields farther north in the Arctic, space weather effects have become an integral part of the pipeline design (e.g. Fernberg et al., 2007).

5 SURVEYING AND NAVIGATION

Canada is recognized for its abundance of mineral reserves. In the early days, prospectors found such minerals on the ground but their activities were restricted to areas of rock outcrops. In the 1950s, the development of airborne magnetometers and electromagnetic (EM) systems allowed detection of ore bodies at depths of 100 m or more (see Cranstone, 2002, for history of mineral exploration in Canada). The availability of the airborne geophysical methods transformed the nature of mineral exploration and led to a great increase in mineral exploration that continues to the present day. In the early 1970s, the staff at the Ottawa Magnetic Observatory received a phone call from a surveyor asking if there had been a magnetic storm on the previous day. When they replied "Yes," there was an immediate sigh of relief at the other end of the phone that surprised the observatory staff. The surveyor went on to explain that they had been flying an aeromagnetic survey but had obtained very poor results with the magnetic field recordings varying all the time—not the flat recordings with the occasional "bump" to indicate the magnetic signature of an ore body that they usually saw. The surveyor was just about to tear apart his equipment to find out what was wrong with it when he remembered the problems that could be caused by magnetic storms and phoned the observatory to check. After his initial relief that the problem was a magnetic storm, the surveyor's next thought was that he still had unusable recordings from the previous day's flight so he would have to refly that route the following week. Thus his next question was whether there would be a magnetic storm then. Responding to such requests was one of the early drivers for the development of geomagnetic forecasting in Canada (Lam, 2011). Such forecasts were used to schedule aeromagnetic surveys and have been progressively refined from the initial national long-term forecasts (up to 27 days) to national short-term forecasts (up to 3 days in advance). However, magnetic forecasting in Canada is complicated by the existence of three distinct activity zones: the polar cap, the auroral zone, and the subauroral zone. In the mid-1980s, the short-term and long-term magnetic forecasts were divided into separate forecasts for each of these zones. More recently, regional short-term forecasts have been added to provide better services for users in different parts of the country (Lam, 2016). The spatial and temporal characteristics of specific types of geomagnetic activity are also being used to help plan aeromagnetic surveys in Canada (Vallée et al., 2007).

The Earth's magnetic field, as detected by a compass, has long been used for navigation. This works because a compass needle always points north, providing a portable reference for directing a course. The compass actually aligns with the geomagnetic field rather than pointing to the geographic pole. At lower latitudes this distinction is not so important, but at high latitudes the difference (called declination or variation) needs to be taken into account. In southern Canada the declination values range from 18 degrees E at Victoria on the west coast to 17 degrees W at Halifax on the east coast, while in the Arctic the values range from 22 degrees E at Inuvik to 27 degrees W at Iqaluit. Further north, as the magnetic field becomes more vertical, the horizontal component of the magnetic field becomes smaller, resulting in the magnetic field's influence on a compass needle becoming weaker. Accordingly, maps are now produced showing the region of compass unreliability, which includes large parts of the Northwest Passage. During space weather events the magnetic disturbances cause changes in declination. This is not so much of a concern for users of traditional compasses. However, for modern applications using electronic compasses this can be a concern. Directional drilling uses an electronic compass to provide measurements while drilling (MWD) to steer the drill bit. For example, multiple boreholes are commonly drilled in different directions from the same wellhead. Changes in the magnetic field during a space weather disturbance can cause errors in the direction of the drilling, which has led to drilling companies using real-time data from nearby magnetic observatories to compensate for the magnetic field changes. As the Canadian oil fields have matured, "fill-in" wells are being drilled between existing boreholes. So now errors in drilling could result in the new well intersecting with an existing borehole with possibly disastrous consequences, placing more stringent accuracy requirements on the magnetic information used to guide the drilling.

A radical change in navigation has occurred in the last 20 years with the development of global navigation satellite systems (GNSS): first the U.S. GPS, then the Russian GLONASS, and now the European GALILEO and the Chinese BeiDou systems. To obtain its position, a GNSS receiver measures the transmission times of radio signals from multiple satellites to perform a sophisticated triangulation. Radio signals from a satellite to the ground pass through the ionosphere, which delays the signals, introducing an error into the receiver position. Corrections for the delays due to the regular ionospheric variations are built into the receivers, but the irregular ionospheric changes produced by space weather disturbances are a major source of error in GNSS positioning in Canada. To compensate for this, measurements can be made at a reference station whose position is known to measure the ionospheric error and transmit this information to GNSS users in the area. These "differential GPS" (DGPS) services are provided by a variety of suppliers, both government and industry, in Canada as well as by the Canadian Coast Guard for maritime users in Canadian waters (Fig. 5). The use of differential GPS services assumes that the position errors seen by the reference station are the same as those experienced by the GPS users. However, during space weather disturbances this may not be the case because of sharp electron density gradients associated with such ionospheric features as plumes of storm enhance density (SED) that can reduce DGPS position accuracies by factors of 10–30 (Skone et al., 2004). This has led to further work to improve position accuracies.

Another concern for GNSS users is ionospheric scintillation, where rapid fluctuations in the amplitude and phase of the radio signals are caused by small-scale irregularities in the ionosphere. When these fluctuations are severe, the signal dips cause the receiver to lose lock on the satellite signal with detrimental effects on the position accuracy. Ionospheric scintillation is well known in equatorial regions where it has regular seasonal and local time variations. Scintillation also occurs in Canada in association with auroral activity and polar patches. Occurrence of auroral scintillation is closely related to the level of geomagnetic activity and is another aspect of navigation affected by space weather.

The use of GNSS positioning is expanding into many areas. For example, Canadian farmers are using GNSS systems for navigating their tractors. When they were a few meters off after traversing a field they were knowledgeable about possible causes and phoned the forecast center to inquire about the level of space weather activity. In aviation, GNSS is being used for runway approaches and en-route navigation. There are plans for the next generation of air traffic control systems to place even greater reliability on GNSS with the intention of halving the separation between aircraft. The reduction of ice in the Arctic will see an increase in marine traffic, even cruise ships, also making use of GNSS for navigation. While GNSS provides a great navigation tool for aircraft, ships, and other users 99% of the time; when these systems suddenly become unavailable because of a space weather event there is potential for problems (Kunches and Fritz, 2016).

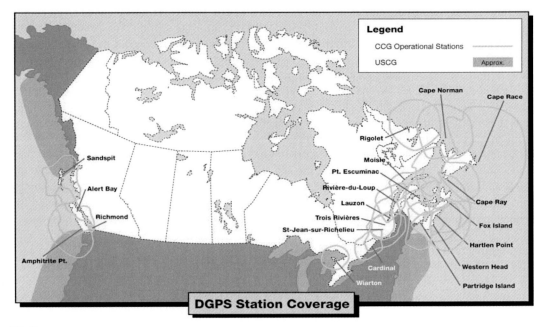

FIG. 5

Differential GPS coverage provided by the Canadian Coast Guard (CCG) to aid navigation on the east and west coasts of Canada and in the St. Lawrence River and Lake Huron.

Figure courtesy of the Canadian Coast Guard.

6 CONCLUDING REMARKS

The previous sections have examined the effects of space weather on various types of technology and how these effects have been dealt with in Canada. In parallel with the evolution of the technology that has been affected by space weather, there has been a great expansion in the knowledge about solar-terrestrial physics, the science behind space weather. The same high-latitude location that makes Canada the country most affected by space weather also puts it in the best position for observing it. Thus Canadian universities, with support from the Canadian Space Agency and federal and provincial funding agencies, have been at the forefront of solar-terrestrial physics. They have deployed instruments to use Canada as a giant space weather monitoring platform that now includes magnetometers, ionosondes, GPS TEC monitors, scintillation receivers, and auroral radars (see Fig. 6). There have also been extensive developments of rocket and satellite instruments that have played a key role in international missions and are now being launched on Canadian satellites such as CASSIOPE/ePOP (Yau and James, 2015).

Even where geographical location was not a particular advantage, other factors have contributed to producing a Canadian instrument that is now one of the key long-term monitors of solar activity. In 1946, using components from surplus radar equipment, scientists at the National Research Council (NRC) in Ottawa constructed Canada's first radio telescope (Tapping, 2013). Due to its limited sensitivity, the only radio emissions it could detect were those from the Sun. However, these showed

6 CONCLUDING REMARKS

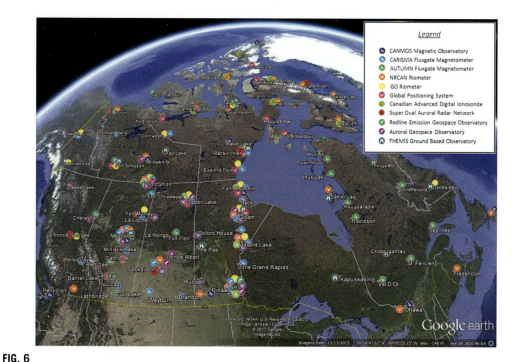

FIG. 6

Ground-based space weather research instruments in Canada.

Figure courtesy of P. Langlois, Canadian Space Agency.

interesting day-to-day variations and it was subsequently found that the operating frequency of the system (2800 MHz), corresponding to a wavelength of 10.7 cm, was very good for detecting radio emissions from the Sun's chromosphere and corona. Regular measurements of the radio flux at 10.7 cm (now referred to as F10.7) started in 1947 and continue to the present day, providing one of the longest running records of solar activity. The F10.7 Index has proven very valuable in specifying and forecasting space weather. It provides a good proxy for the solar extreme ultraviolet (EUV) emissions that impact the ionosphere and modify the upper atmosphere. Thus it has become a key input in modeling space weather effects on the ionosphere that affect radio communications as well as for modeling atmospheric drag on satellites.

Considerable knowledge about space weather phenomena has been accumulated and is being applied in forecasting and in understanding the effects on different systems. However, there is a lack of awareness of the extent of space weather effects on critical infrastructure in Canada. [There is also a debate about what constitutes critical infrastructure: with society's reliance on electricity, the power system is obviously regarded as critical, but there are also calls to classify satellites as critical infrastructure because of their importance for communications, Earth observation, and navigation (Lang, 2017).] The science and forecasting community has also not done a good job of explaining to policymakers the multifaceted nature of space weather. There is sometimes a perception that when a space weather event occurs, everything is affected at once. In fact, different space weather phenomena exhibit differences in occurrence across a wide range of temporal scales.

At the longest time scale, many phenomena approximately follow the 11-year solar cycle and the risk to power systems and HF radio communications is generally considered greatest during the maxima of these cycles. However, what is not often recognized is that the high fluxes of energetic electrons that are a risk for geostationary satellites have peak occurrences that are anticorrelated with the solar cycle (Fig. 7). Even when considering individual events, it is worth noting that 1989 was a peak year for both magnetic activity and solar energetic particles, but the peak magnetic activity occurred in March 1989 while the peak solar energetic particles occurred in August–September 1989. These differences are important, but sometimes difficult for users to appreciate. For example, some ionospheric disturbances that affect cross-polar flights are correlated with magnetic activity. It is possible to use magnetic activity forecasts as a proxy for this type of ionospheric disturbance. However, this is not the complete story: ionospheric disturbances can also be caused by solar energetic protons that are accelerated by the same CMEs that cause magnetic storms but often arrive 2 days earlier and produce ionospheric disturbances while the magnetic field remains quiet. Failure to recognize this could mean aircraft flying up into a zone with no communications on the misunderstanding that everything is okay, because of the low magnetic activity.

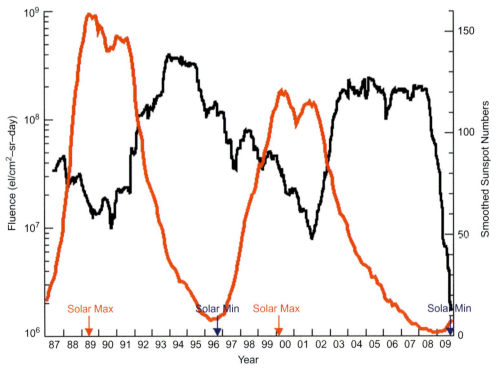

FIG. 7

Variation of energetic electron fluence (black) over solar cycles 22 and 23 showing its relation to the variation in sunspot number (red).

Figure courtesy of H.-L. Lam, Natural Resources Canada; extracted and redrawn from Fig. 1 of Lam, H.-L., 2017. On the predictive potential of Pc5 ULF waves to forecast relativistic electrons based on their relationships over two solar cycles. Space Weather 15, https://doi.org/10.1002/2016SW001492.

Even when talking about one type of phenomena, magnetic disturbances, it is often forgotten that not all magnetic disturbances are the same and different parts of the world are affected by different types of disturbances. The main phase of a magnetic storm is characterized by a worldwide depression of the magnetic field that lasts a day or so, measured by the Dst index, and is caused by a ring of electric current around the Earth in the magnetosphere. In contrast, the magnetic substorms and other rapid disturbances seen in Canada, measured by the AE and Kp indices, are produced by electric currents in the ionosphere at a height of only 100 km. Much has been written about the 1859 Carrington event (Cliver, 2006) that caused widespread effects on the technology of the time—the telegraph system (Boteler, 2006). This event is claimed to be the largest magnetic storm ever recorded, but this is based on an estimation of the Dst value, representing only the main phase of the storm. Values of the AE and Kp indices are poorly correlated with the Dst index so it is unclear how much reliance can be placed on the Carrington event as a guide to extreme space weather events in Canada. The magnetic storm of March 13–14, 1989, is second on the list of the largest recorded magnetic storms. However, the blackout of the Hydro-Quebec power system and other power-system problems in Canada were produced by strong ionospheric currents at a time when the Dst index was comparatively small. It was only later in the storm that notable Dst values occurred at a time when magnetic disturbances occurred at lower latitudes and power system effects had more of an impact in the United States, the United Kingdom, and Scandinavia.

Space weather is now being recognized as a new natural hazard of the technological age. A common factor, seen in different systems, is that as they become more technologically sophisticated they become more vulnerable to space weather. The experience with satellites is described by Gubby and Evans (2002): "… as design complexity increases, so does the number of space weather anomalies. [ ] many improvements in immunity have been incorporated into designs over the years. Even so, the improvements in immunity have not apparently kept pace with the increase in sensitivity." Increasing effects are being seen on different types of ground systems for different reasons. On pipelines the modern use of higher resistance coatings to reduce CP current requirements has led to larger telluric potential variations than on older pipelines with lower resistance coatings. On power systems, the use of higher-voltage transmission lines which, for engineering reasons, have lower resistance has led to larger GICs in power systems. The greater interconnectedness of systems is also leading to greater impacts. Society's growing reliance on electricity is increasing the consequences if that electricity supply is not available. Loss of electricity in a major Canadian city in winter could mean that apartment buildings, heated by electricity, would have to be vacated. Within a few days, the lack of water and sewage services–all driven by electric pumps–could mean that whole cities would need to be evacuated. This nearly happened during the ice storm in Montreal in January 1998 (Murphy, 2013).

Space weather phenomena are, of course, not restricted by national borders, but space weather information has traditionally been tailored to users within a particular country. However, many affected systems—for example, satellites or aviation services—transcend borders and the response to space weather events needs to be international in scope. Canada has long been an active member of the International Space Environment Service (ISES) that facilitates collaboration between space weather forecast centers around the world, and is actively participating in new international space weather initiatives being developed by the World Meteorological Organization (WMO) and the United Nations Committee on the Peaceful Use of Outer Space (UN-COPUOS). Even ground systems such as power systems and pipelines in Canada are interconnected with those in the United States and experience with the 2003 blackout has shown how disturbances can ripple across the border (Task Force, 2004). Thus, many power system issues are managed on a North American-wide basis by the North American Electric Reliability Corporation (NERC) with participation from both U.S. and Canadian power utilities.

The thinking about extreme events has been strongly influenced by several recent nonspace weather events. In 2005, Hurricane Katrina caused extensive flooding in the New Orleans area. This type of event, classified as "high impact, low frequency (HILF)," showed that extreme events may be rare but if they occur we better be prepared for them. The experience of Hurricane Katrina prompted emergency organizations in the United States to think about other HILF events that could cause problems and this concern spread into Canada. [A similar change in attitude to HILF events occurred in Europe, prompted by their experience with the Icelandic volcano eruptions in 2010.] HILF events include extreme space weather disturbances, prompting NERC to set up a Geomagnetic Disturbance Task Force (GMDTF) that has been developing guidelines for how power utilities should plan for a 1-in-100 year geomagnetic storm (NERC, 2013). Data from the Canadian magnetic observatory network have been analyzed to determine the occurrence of such extreme events in Canada (Nikitina et al., 2016).

In the last 150 years, Canadian technological systems have experienced the whole range of space weather phenomena. Although there have been some notable impacts during extreme space weather events, Canadian engineers have built up considerable experience in dealing with space weather effects. However, experience has shown that every time a new technology is introduced, hitherto unforeseen space weather effects are later observed. By chance, the new technology always seems to be introduced in years of low activity and works fine for a few years: an example being the CANTAT-3 submarine cable between Canada and Europe that was installed toward the end of solar cycle 22 and operated without incident during the following solar minimum. However, during the first major magnetic storm of the next cycle, large fluctuations occurring in the cable driving voltages were seen by the cable operators, revealing another potential impact of space weather. Now, the increasing reliance on GNSS is creating new possible vulnerabilities to extreme space weather, not just for navigation as described above, but for the increasing use of GNSS signals for timing of financial and control systems (Klatt, 2016). The need in the future is to be vigilant about how space weather can affect different systems and the greater reliance placed on these systems that work most of the time but could fail during extreme space weather events.

ACKNOWLEDGMENTS

Many people contributed information that has been used in this chapter. I especially appreciate the help from Gordon James, Fokke Creutzberg, Hing-Lan Lam, and Lorne McKee. I am also grateful for comments on the manuscript by Pierre Langlois and Ken Tapping.

REFERENCES

Acres Consulting Company, 1975. Study of the Disruption of Electric Power Systems by Magnetic Storms. Report to Department of Energy, Mines and Resources, Ottawa.

Albertson, V.D., Thorson, J.N., Miske, S.A., 1974. The effects of geomagnetic storms on electrical power systems. IEEE Trans. Power Appar. Syst. PAS-94, 1031.

Bedingfield, K.L., Leach, R.D., Alexander, M.B., 1996. Spacecraft System Failures and Anomalies Attributed to the Natural Space Environment, NASA Reference Publication 1390, NASA.

Bolduc, L., 2002. GIC observations and studies in the Hydro-Québec power system. J. Atmos. Sol. Terr. Phys. 64 (16), 1793–1802.

Boteler, D.H., 2001. Assessment of geomagnetic hazard to Canadian power systems. Nat. Hazards 23, 101–120.

Boteler, D.H., 2006. The super storms of August/September 1859 and their effects on the telegraph system. Adv. Space Res. 38, 159–172.

Boteler, D.H., Pirjola, R.J., 2017. Modelling geomagnetically induced currents. Space Weather 15, 258–276. https://doi.org/10.1002/2016SW001499.

Boteler, D.H., Seager, W.H., 1998. Telluric currents: a meeting of theory and observation. Corrosion 54, 751–755.

Boteler, D.H., Pirjola, R.J., Nevanlinna, H., 1998. The effects of geomagnetic disturbances on electrical systems at the Earth's surface. Adv. Space Res. 22, 17–27.

Brooks, J., 1959. A Reporter at Large: the subtle storm. New Yorker, 39–77.

Broughton, P., 2002. Auroral records from Canada 1769–1821. J. Geophys. Res. 107, A8. https://doi.org/10.1029/2001JA000241.

Burlton, B., 1995. The rescue of Anik E2. Can. Aeronaut. Space J. 41, 57–61.

Carlson, L., Dorman, B., Place, T., 2004. Telluric compensation for pipeline test station survey on the Alliance pipeline system. In: Proceedings of IPC 2004 International Pipeline Conference, Calgary, October 4–8. pp. 231–241. Paper IPC04-0762.

Chapman, J.H., Warren, E.S., 1968. Topside sounding of the Earth's ionosphere. Space Sci. Rev. 8 (5–6), 846–865. ISSN: 0038–6308; 1572–9672.

Chapman, J.H., Forsyth, P.A., Lapp, P.A., Patterson, G.N., 1967. Atmosphere and Space Programs in Canada, Special Study No 1, Science Secretariat, Government of Canada, 258 p.

Cliver, E.W., 2006. The 1859 space weather event: then and now. Adv. Space Res. 38, 119–129.

Cranstone, D.A., 2002. History of Mining and Mineral Exploration in Canada and Outlook for the Future. Natural Resources Canada, Catalogue No. M37-51/2002E, ISBN 0-662-32680-6.

Danskin, D.W., Boteler, D., Donovan, E., Spanswick, E., 2008. The Canadian riometer array. In: Proceedings of the 12th International Ionospheric Effects Symposium (IES 2008), Alexandria, VA, USA, May 13–15. National Technical Information Services (NTIS), U.S. Department of Commerce, Springfield, VA, pp. 80–86. Available at http://www.ntis.gov.

Davidson, W.F., 1940. The magnetic storm of March 24, 1940—effects in the power system. Edison Electric Inst. Bull. 8, 365.

Davies, F.T., 1954. The ionosphere over northern Canada. Arctic 7 (3 and 4), 188–190.

EPRI, 1989. Conference on Geomagnetically Induced Current, Burlingame, California, November 8–10.

Fernberg, P.A., Trichtchenko, L., Boteler, D.H., McKee, L., 2007. Telluric hazard assessment for northern pipelines. In: Proceedings of Corrosion/2007. NACE, Houston. Paper No. 07654.

Fiori, R.A.D., Danskin, D.W., 2016. Examination of the relationship between riometer-derived absorption and the integral proton flux in the context of modeling polar cap absorption. Space Weather 14 (11), 1032–1052. https://doi.org/10.1002/2016SW001461.

Garand, L., Trishchenko, A.P., Trichtchenko, L., Nassar, R., 2014. The polar communications and weather mission: addressing remaining gaps in the Earth observing system. Phys. Can. 70 (4), 247–254.

Gubby, R., Evans, J., 2002. Space environment effects and satellite design. J. Atmos. Solar Terr. Phys. 64, 1723–1733.

Guillon, S., Toner, P., Gibson, L., Boteler, D., 2016. A colorful blackout. IEEE Power Energy Mag. 14(6).

Harland, D.M., Lorenz, R.D., 2005. Space System Failures. Springer-Praxis, New York. 368 p.

Hartz, T.R., Paghis, I., 1982. Spacebound. Canadian Government Publishing Centre, Ottawa. 188p.

IEEE, 1990. Panel session on geomagnetic storm effects. In: IEEE Summer Power Meeting, Minneapolis, MN.

Jacobson, D.A.N., Shelemy, S., Chandrasena, W., Boteler, D., Pirjola, R., 2014. Development of advanced GIC analysis tools for the manitoba power grid. Paper C4-302, Proc. CIGRE Session 45, Paris.

Kappenman, J.G., 2007. Geomagnetic disturbances and impacts upon power system operation. In: Grigsby, L.L. (Ed.), The Electric Power Engineering Handbook, second ed. CRC Press/IEEE Press, Boca Raton, FL, pp. 16-1–16-22. Chapter 16.

Klatt, C., 2016. Precise timing from global navigation satellite systems & implications for critical infrastructure. IR3 Infrastruct. Resil. Risk Report. 1 (5), 3–9.

Kunches, J., Fritz, A., 2016. Space weather is a hidden risk to Arctic cruises, and it could be a significant one. The Washington Post, May 24.

Lam, H.-L., 2004. On the prediction of relativistic electron fluence based on its relationship with geomagnetic activity over a solar cycle. J. Atmos. Sol. Terr. Phys. 66, 1703–1714.

Lam, H.-L., 2011. From early exploration to space weather forecasts: Canada's geomagnetic odyssey. Space Weather 9. S05004. https://doi.org/10.1029/2011SW000664.

Lam, H.-L., 2016. The genesis and development of space weather forecast in Canada. IR3 Infrastruct. Resil. Risk Report. 1 (5), 10–25.

Lam, H.-L., Boteler, D.H., Burlton, B., Evans, J., 2012. Anik-E1 and E2 Satellite failures of January 1994 revisited. Space Weather 10. S10003. https://doi.org/10.1029/2012SW000811.

Lang, D. (chair), 2017. Military Underfunded: The Walk Must Match the Talk. Report of the Standing Senate Committee on National Security and Defence, 56 p.

Leach, R.D., Alexander, M.B., 1995. Failures and Anomalies Attributed to Spacecraft Charging, NASA Reference Publication 1375, NASA.

Marti, L., Rezaei-Zare, A., Yan, A., 2013. Modelling considerations for the Hydro One real-time GMD management system. Proceedings of IEEE Power & Energy Society General Meeting, Vancouver, BC, July 21–25, doi: 10.1109/PESMG.2013.6673069.

Meek, J.H., (Ed.), 1959. Report on the Canadian program for the International Geophysical Year. Canadian National Committee for the IGY, National Research Council, Ottawa, 153 p.

Murphy, R., 2013. Opinion: lessons of the 1998 ice storm, Special to the Gazette. http://www.montrealgazette.com/technology/Opinion+Lessons+1998+storm/7822490/story.html.

NERC, 2013. Geomagnetic Disturbance Planning Guide. North American Electric Reliability Corporation, Atlanta. 20 p.

Newitt, L.R., Dawson, E., 1984. Magnetic observations at International Polar Year stations in Canada. Arctic 37 (3), 255–262.

Nikitina, L., Trichtchenko, L., Boteler, D.H., 2016. Assessment of extreme values in geomagnetic and geoelectric field variation for Canada. Space Weather J. 14. https://doi.org/10.1002/2016SW001386.

NOAA, 2004. Intense Space Weather Storms October 19-November 07, 2003. NOAA National Weather Service, Silver Spring, MD, 50 p.

Parks Canada, 2017. Marconi national historic site. http://www.pc.gc.ca/en/lhn-nhs/ns/marconi/decouvrir-discover.

Place, T.D., Sneath, T.O., 2001. Practical telluric compensation for pipeline close-interval surveys. Mater. Perform. 40, 22–27.

Prescott, G.B., 1866. History, Theory and Practice of the Electric Telegraph. Tichnor and Fields, Boston.

Restoration Project: ISIS/Alouette Topside Sounder Data Restoration Project https://nssdc.gsfc.nasa.gov/space/isis/isis_read_analysis.html (Accessed March 2017).

Rix, B.C., Boteler, D.H., 2001. Telluric current considerations in the CP design for the Maritimes and Northeast Pipeline. In: Proceedings of Corrosion 2001. NACE, Houston. Paper 01317.

Roach, T.R., 2014. The saga of northern radio, Canada's History, Canada's History Society, Winnipeg. http://www.canadashistory.ca/Explore/Arts-Culture-Society/The-Saga-of-Northern-Radio.

Russell, G.I., Nelson, L.B., 1954. Extrinsic line current fluctuations seriously restrict progress of coating conductance surveys on large trunk line. Corrosion 10, 400.

Schmerling, E.R., Langille, R.C., 1969. Guest editors, special issue on topside sounding and the ionosphere. Proc. IEEE 57(6).

Seager, W.H., 1991. Adverse telluric effects on northern pipelines. In: Proceedings of the International Arctic Technology Conference, Anchorage, Alaska, May 29–31. Society of Petroleum Engineers (SPE 22178), pp. 771–783.

Shepherd, G., Kruchio, A., 2008. Canada's Fifty Years in Space. Apogee Books, Burlington, ON. 280 p.

Skone, S., Yousuf, R., Coster, A., 2004. Combating the perfect storm: improving marine differential GPS accuracy with a wide-area network. GPS World, pp. 31–38.

Tapping, K.F., 2013. The 10.7 cm solar radio flux ($F_{10.7}$). Space Weather 11, 394–406. https://doi.org/10.1002/swe.20064.

Task Force, 2004. Final Report on the August 14, 2003 Blackout in the United States and Canada: Causes and Recommendations. U.S.-Canada Power System Outage Task Force, https://reports.energy.gov/.

Trichtchenko, L.D., 2012. Radiation environment analysis for Canadian polar communication and weather mission. Geol. Surv. Canada, 139. https://doi.org/10.4095/291813. Open File 7252.

Trichtchenko, L., Boteler, D.H., 2001. Specification of geomagnetically induced electric fields and currents in pipelines. J. Geophys. Res. 106 (A10), 21039–21048.

Trichtchenko, L., Boteler, D.H., 2002. Modeling of geomagnetic induction in pipelines. Ann. Geophys. 20, 1063–1072.

Trichtchenko, L., Boteler, D.H., Fernberg, P., 2008. Space weather services for pipeline operation. In: Proceedings of ASTRO 2008, Montreal, April 29–May 1.

Trichtchenko, L.D., Nikitina, L.V., Trishchenko, A.P., Garand, L., 2014. Highly elliptical orbits for arctic observations: assessment of ionizing radiation. Adv. Space Res. 54, 2398–2414.

Vallée, M.A., Newitt, L., Mann, I.R., Moussaoui, M., Dumont, R., Keating, P., 2007. The spatial and temporal characteristics of Pc3 geomagnetic activity over Canada in 2000, as a guide to planning the times of aeromagnetic surveys. Pure Appl. Geophys. 164, 161–176.

Vestine, E.H., 1943. Remarkable auroral forms, Meanook Observatory, Polar Year, 1932-33. J. Geophys. Res. 48 (4), 233–236.

Wadham, P.N., 1980. Operational experience with Anik A. In: 8th Communications Satellite Systems Conference, Orlando, Florida, April 20–24. American Institute of Aeronautics and Astronautics, Inc, New York, pp. 445–449. Technical Papers. (A80-29526 11-32).

Wang, D., Martens, N., 1993. Anik-E2 recovery mission. Acta Astronaut. 29 (10/11), 811–815.

Warrington, E.M., Stocker, A.J., Siddle, D.R., Hallam, J., Zaalov, N.Y., Honary, F., Rogers, N.C., Boteler, D.H., Danskin, D.W., 2016. Near real-time input to an HF propagation model for nowcasting of HF communications with aircraft on polar routes. Radio Sci. 51 (7), 1048–1059.

Yau, A.W., James, H.G., 2015. CASSIOPE enhanced Polar Outflow Probe (e-POP) mission overview. Space Sci. Rev. 189 (1), 3–14. https://doi.org/10.1007/s11214-015-0135-1.

Zheng, K., Trichtchenko, L., Pirjola, R., Liu, L.-G., 2013. Effects of geophysical paramaters on GIC illustrated by benchmark network modelling. IEEE Trans. Power Delivery 28 (2), 1183–1191.

Zheng, K., Boteler, D.H., Pirjola, R., Liu, L.G., Becker, R., Marti, L., Boutilier, S., Guillon, S., 2014. Effects of system characteristics on geomagnetically induced currents. IEEE Trans. Power Delivery 29 (2), 890–898.

FURTHER READING

Albertson, V.D., Thorson Jr., J.M., 1974. Power system disturbances during a K-8 geomagnetic storm: August 4, 1972. IEEE Trans. Power Appar. Syst. PAS-93 (4), 1025–1030.

Basham, P.W., Newitt, L.R., 1993. A historical summary of geological Survey of Canada studies of earthquake seismology and geomagnetism. Can. J. Earth Sci. 30, 372–390.

Cerruti, A.P., Kintner Jr., P.M., Gary, D.E., Mannucci, A.J., Meyer, R.F., Doherty, P., Coster, A.J., 2008. Effect of intense December 2006 solar radio bursts on GPS receivers. Space Weather 6(10). https://doi.org/10.1029/2007SW000375.

Lam, H.-L., 2017. On the predictive potential of Pc5 ULF waves to forecast relativistic electrons based on their relationships over two solar cycles. Space Weather 15. https://doi.org/10.1002/2016SW001492.

Molozzi, A.R., 1964. Instrumentation of the Ionospheric Sounder Contained in the Satellite 1962 Beta Alpha (Alouette), Space Research IV. North-Holland Publishing Company. pp. 413–436.

Prikryl, P., Ghoddousi-Fard, R., Weygand, J.M., Viljanen, A., Connors, M., Danskin, D.W., Jayachandran, P.T., Jacobsen, K.S., Andalsvik, Y.L., Thomas, E.G., Ruohoniemi, J.M., Durgonics, T., Oksavik, K., Zhang, Y., Spanswick, E., Aquino, M., Sreeja, V., 2016. GPS phase scintillation at high latitudes during the geomagnetic storm of March 17–18, 2015. J. Geophys. Res. Space Phys. 121. https://doi.org/10.1002/2016JA023171.

RADARSAT-2, 2017. https://directory.eoportal.org/web/eoportal/satellite-missions/r/radarsat-2.

Skone, S., Yousuf, R., 2007. Performance of satellite-based navigation for marine users during ionospheric disturbances. Space Weather 5.S01006. https://doi.org/10.1029/2006SW000246.

Skone, S., Coster, A., Hoyle, V., Laurin, C., 2003. WAAS availability and performance at high latitudes. In: Proceedings of the 16th International Technical Meeting of the Satellite Division of The Institute of Navigation (ION GPS/GNSS 2003), Portland, OR, pp. 1279–1287.

Trichtchenko, L., Lam, H.-L., Boteler, D.H., Coles, R.L., Parmelee, J., 2009. Canadian space weather forecast services. Can. Aeronaut. Space J. 55 (2), 107–113.

Yousuf, R., Skone, S., 2005. WAAS performance evaluation under increased ionospheric activity. In: Proceedings of the 18th International Technical Meeting of the Satellite Division of The Institute of Navigation (ION GNSS 2005), Long Beach, CA, pp. 2316–2327.

CHAPTER 27

SPACE WEATHER: WHAT ARE POLICYMAKERS SEEKING?

Mike Hapgood

RAL Space, STFC Rutherford Appleton Laboratory, Didcot, United Kingdom;
Space and Planetary Physics Group, Lancaster University, Lancaster, United Kingdom

CHAPTER OUTLINE

1 Introduction	657
2 Key Concepts	659
2.1 Space Weather as a Natural Hazard	659
2.2 Policy Responses to Natural Hazards	662
2.3 The Importance of Science	664
3 How to Assess Extreme Risks	666
4 What Knowledge is Needed in the Future	671
4.1 The Need for Data	671
4.2 The Need to Learn From Meteorology	672
4.3 The Need for Better Science and Better Models	674
References	678
Further Reading	682

1 INTRODUCTION

Space weather is now widely recognized as a set of natural phenomena that can disrupt a range of technological infrastructures that are critical to everyday life in modern societies and their economies around the world, most obviously in that we all need continuous reliable access to electricity. As a result, space weather has gained much attention as a risk factor that must be addressed by the policy community, that is, the national and international networks of people in government, industry, and academia that develop ideas that influence the organization and governance of our societies. In this chapter we discuss how that community considers the issue of risk, and hence what knowledge and capabilities they seek from the scientific community.

The ownership and management of risk is an important issue for the policy community, especially as modern economies are highly interconnected such that disruption in one sector leads to disruption in other sectors (as illustrated by the linkages in Fig. 1).

658   CHAPTER 27 SPACE WEATHER: WHAT ARE POLICYMAKERS SEEKING?

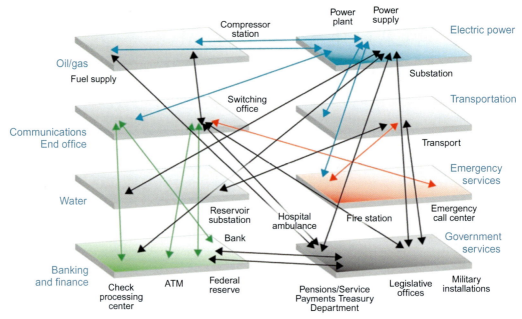

FIG. 1

Outline of linkages between different sectors. Space weather can directly impact sectors such as electric power, transportation, communications and finance—but linkages then lead to secondary impacts (e.g., as discussed by Oughton et al. (2017) in respect to impacts on electric power).

From U.S. Department of Homeland Security, National Infrastructure Protection Plan, circa 2008. https://www.dhs.gov/xprevprot/programs/editorial_0827.shtm.

It is important to consider how responsibility for managing risks is shared in this interconnected environment. The key word here is "shared," as it is utterly impractical for any single entity to take sole responsibility for managing risks in such a complex environment. Thus policymakers have to develop procedures to deliver effective risk sharing: engaging relevant expertise across different sectors, establishing a common understanding of the risks, and also establishing processes through which they can coordinate their risk management. Much of this work must be done well before the risk materializes, such as through analyses of potential risks and exercises performed against possible risk scenarios.

How does space weather fit into such policy activities? The primary concern is the impact of extreme space weather events: what are the expected economic and societal impacts of an extreme event, and what is the likelihood of such events? Policymakers generally focus on extreme events that are likely to occur with probabilities of 1 in 100 years or higher—what is often termed the "reasonable worst case." They also keep an eye on risks with lower probabilities (e.g., 1 in 1000 years), and the potential for catastrophic impacts. Thus they particularly look to the scientific community to describe the extreme space weather environments that we expect to occur at Earth—their scale, and their temporal and spatial extent. They also encourage the scientific community to work with other experts (engineers, economists, social scientists) to provide both (a) advice on how these environments physically interact with technological systems, thereby disrupting their operation; and (b) insights into the economic and societal consequences of that disruption.

While policymakers are not directly concerned with the chain of physics linking the Sun to the Earth to produce extreme environments, that chain is an important underpinning element in policy discussions. It is crucial evidence as to why extreme space weather events occur, and it is the fundamental basis for forecasting such events and developing products that will enable industry and government to prepare for, and respond to, such events.

In the rest of this chapter we will explore in more detail how policymakers engage with scientists as one of a range of stakeholders (engineers, economists, system operators, emergency managers and the general public) concerned with a particular policy area and give examples how that has been done in the case of space weather. We present a number of key concepts that underpin space weather engagement with the policy community: first, why it is important to present space weather as a natural hazard, then how policymakers approach natural hazards, and finally why science has a vital role, one that transcends practical experience. We then discuss how policymakers assess extreme risks and how such assessments can be applied to extreme space weather. Then in the final part of the chapter we discuss what future scientific knowledge is required to support policymakers dealing with extreme space weather, looking broadly at knowledge in terms of scientific data, of research to determine new facts, and of the transitioning of those facts into scientific models of space weather, and specifically into models that can inform policymaking, planning, and operational activities.

The ideas presented in this chapter draw heavily on the author's experience in assisting policymakers responsible for establishing concepts and procedures to deal with space weather, especially in its extreme forms. Much of this experience is based on work in the United Kingdom, as will be seen in many of the examples and references presented later in the chapter. But it also includes many fruitful interactions with experts from many other countries as well as experiences from working with international organizations. This international approach is important in dealing with such a global problem and I hope that becomes clear in the following sections.

2 KEY CONCEPTS
2.1 SPACE WEATHER AS A NATURAL HAZARD

Extreme space weather is a natural hazard that sits alongside other natural hazards of concern to the policy community, including extreme tropospheric weather, pandemic flu, and major volcanic eruptions. Such hazards are an unavoidable part of life—they arise because we live in environments that are usually good for life, but that occasionally exhibit dangerous conditions. For example, many human settlements have grown up close to rivers because such locations give access to food from nearby fertile lands, but, as a result, are also vulnerable to flooding. One can construct similar scenarios for human settlements in coastal regions (access to seafood, but vulnerable to storm surges and tsunamis), close to volcanoes (fertile volcanic soils, but vulnerable to eruptions), even those in rich pastoral regions (farming of animals for food can stimulate evolution of and exposure to viruses). Space weather fits perfectly in this scheme in that life flourishes on Earth because our planet lies at the right distance from the Sun to give conditions that sustain life on Earth, most obviously that water can exist in the liquid state on Earth's surface. But that distance also means that Earth is at risk when solar activity sends strong bursts of matter and radiation into interplanetary space.

Because natural hazards are unavoidable, successful human societies have always had to devise ways to be resilient to those hazards. For example, in centuries past many settlements vulnerable to flooding learned to deal with this by placing homes and public buildings some distance back from rivers, while using land close to the river for activities less vulnerable to flood damage. This practice has now evolved into a practical recognition of the value of land close to rivers being available to store excess water during floods. Today we face a similar challenge in that we need to devise ways to be resilient to extreme space weather. There are many ways to do this as we will discuss later in this section. One example that is analogous to the flood protection example above is how flights between North America and Asia make good use of short routes over the high Arctic when space weather conditions are calm, but divert to lower latitudes when adverse space weather blocks vital communications links between aircraft and control centers (see Box 1).

In developing resilience to natural hazards, scientists have to work alongside other professionals: engineers, economists, policymakers, emergency managers, and many others. We have to understand what these colleagues bring to the process of developing resilience, and thus how our scientific research can dovetail with their work. This is just as important for space weather as it is for other natural hazards, so we should consider how space weather affects the work of these other professional groups.

We first consider the engineering community. Here, we need to understand the physical interactions between the space weather environment and engineered systems, and, most importantly, what engineers can do to mitigate those interactions. These points are crucial to directing scientific knowledge toward effective mitigation. Space weather risk mitigation is ultimately the task of the engineers who design vulnerable systems and the teams who operate them. The engineers' first approach to mitigation will always be to try to design out vulnerabilities, such as in modern professional global navigation satellite systems (GNSS) that can automatically correct for most space-weather driven variations in the ionospheric delay of GNSS signals. But there will always be cases where mitigation by engineering design is not feasible, for example due to high costs, and thus it is better for operators of vulnerable systems to take actions to temporarily increase resilience or to fix problems as they occur. For example, power grids can increase resilience via "all-on" procedures, such as canceling planned maintenance (Cannon et al., 2013), while satellite operators can strengthen their operations teams ready to deal with enhanced anomaly rates (e.g., see pages 30–31 of Odenwald, 2010).

Thus the engineers need a range of space weather information from the scientific community. For example, they need reasonable worst-case scenarios (sometimes also called "benchmarks") to set engineering design requirements; these are not the absolute worst cases that scientists can conceive, but rather the worst cases at the 1-in-100-years level that is widely used by governments and insurers to assess major risks. In addition, engineers need situational awareness of space weather to guide their actions when adverse conditions occur. This obviously includes forecasts to plan those actions, but also includes information on current and recent space weather conditions to help operators determine if

---

BOX 1

Solar energetic particles are focused into the polar regions, where they generate ionization at altitudes around 90 km. At these altitudes the electron-neutral collision frequency is higher than that of the high-frequency radio transmissions that are the only long-range communications system available in the High Arctic, that is above 82° latitude where normal satcom is not available. Thus those radio transmissions are absorbed, rather than being reflected by ionization at higher altitudes.

anomalies in their systems are caused by space weather and hence take appropriate corrective measures. This range of information is broadly similar to that needed to mitigate other natural hazards, certainly as regards possible scenarios and forecasts. What is perhaps special about space weather is the need for information on current and recent conditions. For many hazards, operators can obtain the equivalent information from their own observations of what is happening and has happened. But space weather is a natural hazard with little direct impact on human senses (other than aurora in the sky if an event occurs during clear nighttime conditions). Thus engineers need good access to scientific information on current and recent conditions and advice on how best to use that information. Engineers should be also be encouraged to record, and where feasible to share, the results of their assessments so as to build up a base of experience on the impacts of extreme space weather.

We now turn to the role of economists. The study of the socioeconomic impact of space weather is an emerging discipline that we need to understand much better. Many past studies have simply highlighted the total impact of space weather, often based on very extreme scenarios. This produces scare stories that grip the public imagination but do little or nothing to address the real challenge of what we do about space weather. What is crucial here is to understand how various mitigation measures will reduce the economic impact of space weather and, thus, whether that mitigation is worthwhile. For example, how do we estimate the reduction in economic impact due to the forecasts that can be generated following deployment of a particular instrument, such as solar wind/IMF instruments at L1, a coronagraph at L1, or a heliospheric imager at L5? This kind of economic analysis is typical of what is done to justify monitoring of other environments and is essential to assess if it is worth investing in particular instruments. But to do that it is essential that scientists produce scenarios of the range of space weather conditions here and work with engineers to determine the technological impacts that may arise. Then economists can assess how much money may be saved using services based on data from that instrument, not forgetting to quantify all the uncertainties arising from the chain of scenarios, impacts, and economics.

And, finally, what of policymakers? They are the glue that brings all this together to deliver real benefits to society as a whole: encouraging others to work together to build the big picture of what we all need to do to protect our technological civilization from space weather, and creating the policy environments in which that work can be done well (and to do so at both national and international levels). The most important role science can play here is to build and sustain the interest of policymakers in space weather, to show that it is a natural hazard that needs attention. This must take space weather beyond being fascinating science (that's the easy bit); it must show policymakers that severe events will occur in their area of responsibility on timescales between a few decades and a couple of centuries. This is the classic timescale on which severe events are taken seriously by government and by key industries such as insurance. Severe space weather easily fits in this timescale, not just with the well-studied Carrington event of 1859, but also many other severe events in the intervening period, such as the great geomagnetic storm of May 1921 (Kappenman, 2006; Lundstedt et al., 2015). The wealth of evidence from 1859 through the first decade of this century has been fundamental to building the profile of space weather among policymakers.

But serendipity can also play a role in raising awareness of natural hazards. The eruption of Eyjafjallajökull in 2010 and the subsequent disruption of civil aviation by volcanic ash created economic impacts centered in Europe but also reaching far beyond Europe, such as to flower growers in Kenya who were unable to deliver their products to markets in Europe. This was a wake-up call for policymakers—it showed that they needed to work with the scientific community to identify and assess unusual natural hazards that could produce severe societal and economic effects on decadal to centennial timescales. For example, policy work in the United Kingdom following the Eyjafjallajökull eruption picked up not only on the future risk of other volcanic ash events, but also dangerous volcanic gas events (a well-documented example from the 18th century—see Schmidt, 2015) and space weather as risks that need policy attention.

However, it is important to recognize that there is a frequent turnover of working-level staff involved in different areas of policymaking. Thus it is important that scientists continue to raise the awareness of policymakers about space weather, both through formal processes and chance opportunities, to show the real base of evidence that we have available and to feed that evidence into discussions with policymakers and other experts as discussed above.

2.2 POLICY RESPONSES TO NATURAL HAZARDS

Once policymakers have identified that any phenomenon constitutes a significant natural hazard, the onus is on them to establish policies that will reduce its adverse impacts. The precise form of that response will depend on the political, legal, and administrative culture in which the policymakers operate, for example, a generic national framework for emergency management such as that developed in the United Kingdom under the Civil Contingencies Act of 2004 (The National Archives, 2016; see also background discussion in Wikipedia, 2017). But the general form of the policy response is likely to include:

a. measures to raise awareness of the hazard, and the consequent risks, across national, regional, and local government;
b. engagement with industry and academia to better understand the risks and the options for mitigation (including research to improve understanding and mitigation);
c. establishment of high-level plans to deal with the risk, most importantly including the capability to adapt plans to the actual impacts (reflecting that risks are extremely unlikely to occur in the exact same form as used in planning scenarios);
d. identification and development of capabilities to assist in mitigation of the risk, including both capabilities such as forecasting to reduce risk impacts and capabilities such as equipment and trained staff to speed recovery following the risk impact;

A key issue for policymakers is to assess how much the risk can be mitigated by generic capabilities designed to handle a range of different risks, and how much it requires specialist capabilities. An example of the former is the development of capabilities to handle large-scale failures in the distribution of electric power, whether from space weather or other causes. These include procedures to restart the power grid (a so-called "blackstart") and to provide emergency local power generation in areas where grid infrastructure has been damaged. An example of specialist capabilities is the provision of situational awareness of space weather to grid operators through measurements and forecasts, combined with prior training and discussion on the use of these capabilities.

Another key issue for policymakers is to communicate that the responsibility for implementing risk-management polices to manage risks is not just a matter for government; it needs to be shared by everyone. Government can and should provide overall leadership and accountability, but there needs to be a hierarchy of responses including:

- the responsibility of individuals to protect themselves and their families, namely that individuals should be able to cope with short power cuts (durations ranging from a few hours to a few days), such as by having alternative sources of lighting, cooking, and heating as well as reserve supplies of food.
- another generic and valuable capability is the willingness of communities (i.e., individuals working together informally for better effect) to protect the members of their community, such as checking and helping vulnerable neighbors.
- the responsibilities of public authorities at the local level to support individual and community efforts, such as by disseminating information about the risk, especially when the risk is likely to occur or has just occurred; also by providing help to deal with local problems where these go beyond the capabilities of individual and community efforts.
- the responsibilities of businesses to sustain commercial services critical to everyday life. The obvious case is the supply of electrical power and other forms of energy, but others include the distribution and sale of food and medicines and the provision of communications networks.
- the responsibility of local, regional, and national governments for the overall security of their populations, in particular to mobilize and coordinate resources needed to deal with major problems.

The space weather science community has an important role here. It can contribute to all these levels by helping to raise awareness and preparedness ahead of a major space weather event. While this task includes classic outreach activities, it must go well beyond that and include engagement in which scientists seek a dialogue with people working at all levels of risk management from the general public to government officials. Such dialogue is important to demonstrate that space weather is a significant and credible risk—and not just science fiction. An example of what is needed is the recent U.K. Space Weather Public Dialogue (Sciencewise/STFC, 2015; Hapgood, 2015). This was part of a U.K. national program in which members of the public interact with scientists, other stakeholders (e.g., research funders and businesses) and policymakers to deliberate on a wide range of technical issues relevant to future policy decisions. The dialogue outcomes help policymakers to gain a rich understanding of public aspirations and concerns, and thus make better, more robust decisions that reflect public values and societal implications. In the case of space weather, it was very clear that the adverse impact that catches the attention of the public is the loss of electric power. Expert concerns about impacts on civil aviation and GNSS occasionally got some attention but interest quickly returned to loss of electricity. This was no surprise—it simply highlights that the public recognizes that electricity is the fundamental infrastructure of modern societies. Loss of supply will become a nuisance within minutes, a serious inconvenience within hours, and a total disruption of everyday life within days.

The public dialogue showed that it is straightforward for scientists to communicate space weather risks to the public, that this is much appreciated by the public, and that it will be a valuable contribution to society's resilience against extreme space weather. A key phrase that came out of the dialogue is that "knowledge is power," meaning that when people have awareness of space weather (indeed, of any natural hazard), they are empowered to deal with the adverse impacts. They are then aware of the likely scale of the impacts, of the work being done to mitigate those impacts, and that governmental authorities are planning to deal with any ensuing crisis. This all acts to reassure people and to counter scaremongering about natural hazards. One very interesting point made by the public participants was that work on space weather models and forecasts is a great help in raising general awareness about space weather (e.g., getting people used to the idea of warnings and following up with further information about what actually happened).

2.3 THE IMPORTANCE OF SCIENCE

One of the greatest challenges in raising awareness of low-frequency risks, such as space weather, is overcoming personal and organizational experience that has not been exposed to a severe event (Lee et al., 2012). The author's own experience from many discussions on natural hazards is that personal and organizational experience typically has a time span of one or two decades and can be resistant to reports of severe events on longer timescales, unless challenged by good science.

A classic example from the wider field of natural hazards comes from the 1980 eruption of Mount St. Helens in the United States. In the weeks before the catastrophic eruption on May 18 of that year, it was clear that magma was building up under the volcano and likely to lead to a major eruption, so the authorities advised people to evacuate from areas at risk. But some people who had lived long and peaceful lives close by the volcano refused this advice and thus lost their lives during the eruption. This shows the importance of communicating how science gives insight into risks outside direct experience.

Turning to space weather, one example is the limited awareness of many satellite operators of what makes a truly severe radiation event, such as those that occurred in autumn 1989. During a few weeks in September and October of that year, the Sun produced a series of strong bursts of energetic particles such that the radiation environment in geosynchronous orbit was above the NOAA S4 radiation storm level for more than 7 days, and above the S3 level for a further 7 days. As shown in Fig. 2, the fluence accumulated in those few weeks is notably larger than any other events since that time. Thus satellite

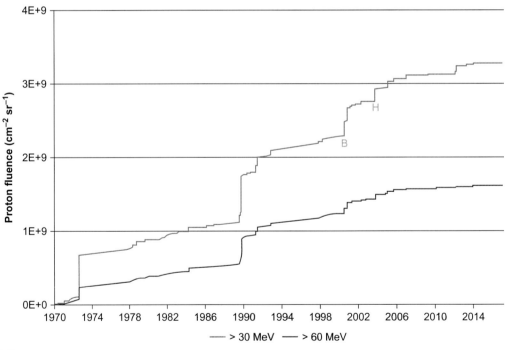

FIG. 2

The accumulation of high-energy proton fluence in the near-Earth environment during 1970–2016. Fluence estimated from proton flux data from the NASA Omniweb (https://omniweb.gsfc.nasa.gov/). The huge step near the center of the plot marks the severe radiation events of autumn 1989.

operators with experience since the 1990s have no direct experience of such a severe event. They may cite their experience during smaller events such as the Bastille Day radiation storm in 2000 and the radiation storms associated with the Halloween events of 2003 (marked by B and H, respectively, in Fig. 2) but, while those later events provided a challenge to satellite operators, they must not be regarded as a reasonable worst case.

This issue with experience is further complicated by the highly correlated nature of space weather events at Earth. Engineers managing critical infrastructures are well used to analyzing subsystem failures and using lessons learned to improve infrastructure resilience. But this is most often done in terms of independent failures of subsystems arising, for example, from manufacturing faults or localized hazards. Thus, using again the example of satellites, operators of large fleets of satellites can, within a decade or so, build up hundreds of in-orbit years of experience of such failures and thus gain good statistical insights into the worst cases for which they need to plan. But this approach does not work for extreme space weather. The effects of any extreme event on any set of vulnerable systems (such as satellites in geosynchronous orbit) will very likely be well correlated across that set of systems. Thus all anomalies arising in those systems during that event should be considered as a single event for statistical purposes. Furthermore, parallel periods of quiet space weather on different systems, not least satellites, can be counted only once for statistical purposes; that is, 10 satellites flying without space weather problems for the same 10 years provide only 10 in-orbit years of relevant experience, not 100 in-orbit years (see Box 2). This is a critical consideration in the analysis of space weather risks and one that is sometimes confused in wider discussions about experience with space weather risks.

To assess the reasonable worst case for space weather impacts, we need to adopt the 100- and 200-year perspectives generally used by governments and the insurance industry to assess major risks. Personal and organizational experience is ill suited for this perspective. Instead we need to study the extreme events over timescales going back several hundred years, not just the events of the last 20 years for which we have a wealth of experience and high-quality data. This is where science provides vital insights. We have a great deal of information on past space weather events, including:

- direct measurements such as the extensive ground magnetometer data that are available back to the 1840s, following the first development of a method to measure magnetic field strength by Gauss (1833);
- proxy data, such as cosmogenic isotopes, allow us to assess long-term solar activity back for many thousands of years (Usoskin, 2017);
- scientific reports of aurora and sunspots before the 19th century, such as Halley (1714) describes a then-recent auroral display over southern England comparable to that seen from the same region during the great geomagnetic storm of March 1989. He also cites reports of earlier displays back to 1560.

BOX 2

In general it is not possible to accumulate experience of space weather except in real time as, at any time, we have a single instance of space weather affecting the Earth and indeed much of the solar system. The only way to accumulate data from different instances of space weather is to study space weather around other Sun-like stars—and that is a topic that we are beginning to address. For example, data from NASA's Kepler mission has recently been used to study large flares on Sun-like stars (Maehara et al., 2012). Also it is anticipated that radio astronomers will soon be able to detect auroral radio emissions from the magnetospheres of exoplanets (Nichols, 2011).

- engineering reports of space weather impacts such as a damaging fire at a telephone exchange at Karlstad in Sweden during May 1921 (report by D. Stenquist, cited in Karsberg et al., 1959);
- more general reports of sky phenomena that were probably aurora, such as red aurora being misinterpreted as a distant fire in ancient Rome and in the early 20th century (Paton, 1946).
- growing information on space weather phenomena around other Sun-like stars—see Box 2

Thus the challenge is, how we can use and interpret this disparate information to produce a credible assessment of what are the worst cases for space weather. This is very much a challenge that science, and only science, can address because it allows us to transcend the natural limitations of personal and organizational experiences. It points to the fundamental importance of science in assessing space weather risks–and that this is a point that the space weather community must never tire of making. It requires dialogue and partnership between researchers, engineers, policymakers, and operators.

3 HOW TO ASSESS EXTREME RISKS

One important tool used in policy development is the risk matrix. This is a two-dimensional plot with one axis representing the probability of a particular risk and the other the societal and economic impact of the risk. Thus the risks from a range of hazards can be plotted in a single diagram, allowing policymakers to assess the relative importance of those risks. As an example, Fig. 3 shows the matrix for nonmalicious risks as presented in the U.K. 2015 National Risk Register (Cabinet Office, 2015). The horizontal axis shows the estimated probability of the risk affecting the United Kingdom in the next

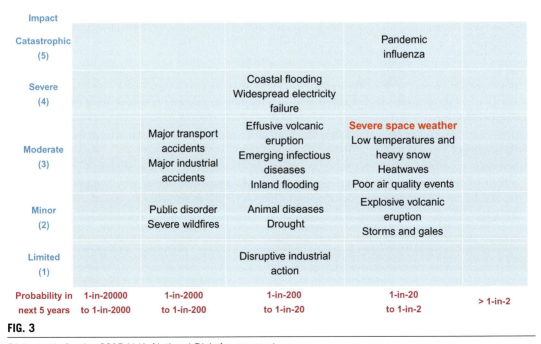

FIG. 3

Risk matrix for the 2015 U.K. National Risk Assessment.

Data from Cabinet Office, 2015. National Risk Register of Civil Emergencies, https://www.gov.uk/government/collections/national-risk-register-of-civil-emergencies. Accessed 8 March 2017.

5 years plotted on a logarithmic scale, while the vertical axis shows a logarithmic score that reflects the estimated impact of risk. The position of space weather in the matrix reflects:

a. the probability of severe space weather on the scale of the Carrington event at slightly greater than 1 in 100 years, such as the 79-year recurrence period used in the U.K. Royal Academy of Engineering report on extreme space weather impacts (Cannon et al., 2013).
b. the government's assessment that severe space weather would have moderate impact in terms of disruption to people's daily lives, of psychological impact, of harm to the overall economy, and of the number of fatalities and level of illness or injury.

It is important to recognize that the risk matrix is a guide, not a definitive ranking of the risks. There are considerable uncertainties in both the probability and the impact of these risks. Nonetheless, risks located toward the upper right of the matrix clearly require priority attention from policymakers. Others must not be ignored but will be given a priority that reflects their position in the matrix.

Unsurprisingly, the most important risk in the matrix is pandemic flu, but as you can see space weather is clearly also regarded as a significant risk, comparable to extreme temperatures. Thus space weather has been the subject of substantial study, in particular to assess how much it can be mitigated by generic capabilities designed to handle a range of different risks, and how much it requires specialist capabilities as discussed above.

It is this positioning in the risk matrix that drives much of the policy interest in space weather. This positioning is, in turn, driven by our understanding of the reasonable worst case. As noted above, the Carrington event provides a good basis for a space weather reasonable worst case, in that it was a severe event for which there is significant scientific and documentary evidence of major impacts on Earth. On the scientific side this includes geomagnetic records from a number of observatories including Greenwich, Kew, Mumbai, Helsinki, St. Petersburg, and Rome (Stewart, 1861; Nevanlinna, 2008; Cliver and Dietrich, 2013; Ptitsyna et al., 2012) as well as observations of the solar origin of the event (Carrington, 1859; Hodgson, 1859). The documentary evidence includes many reports of GICs in telegraph systems (Boteler, 2006) as well as auroral reports from around the world (Green and Boardsen, 2006). This weight of historical evidence is a powerful tool for engaging with policymakers because it demonstrates that the risk really can happen. Such historical evidence, sometimes from well before 1859, is an important driver of wider risk policies; for example, the positioning of effusive volcanic eruptions in Fig. 3 is influenced by evidence that noxious gases from the Laki eruption in Iceland in 1783–84 led to crop failures and thousands of excess deaths across Northern Europe.

But for space weather, we do not have to rely solely on the evidence from the Carrington event. Scientific and historical records since 1859 strongly suggest that the Carrington event is just the largest event in a spectrum of major geomagnetic storms, as shown in Fig. 4. Another outstanding event in this period was the "railroad storm" of May 1921, when geomagnetically induced currents (GIC) in the copper wires then used for communications systems led to fires that damaged rail-control systems in New York (New York Times, 1921) and that destroyed a telephone exchange in Sweden (Karsberg et al., 1959). This event is particularly significant as it arose from a short burst of intense solar activity in an otherwise fairly quiet period, and this warrants deeper study to see what other records are available. Fig. 4 also shows many other events, including the recent and well-studied events of March 1989 and October 2003. These other events also warrant deeper study, especially for those from the 1940s onward as the range of available scientific data steadily increases, first with the spread of ground-based sensors, especially in the period of the International Geophysical Year in 1957–58, and then with the growing deployment of space-based sensors.

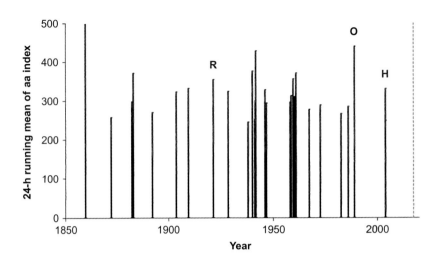

FIG. 4

The top 30 geomagnetic storms since 1868 as assessed using a 24-h running mean of the aa index. The events marked R, O, and H are the storms of May 1921, March 1989, and October 2003. The red dashed line indicates the time of writing in 2017. Based on other multiday storms in this dataset, the Carrington storm strength is taken as 1.25 times the daily aa value of 400 reported by Nevanlinna (2006).

This shows that there is a spectrum of known space weather events, ranging from the moderate to the severe, that adds to the base of evidence on what is a reasonable worst case for space weather risks. In particular, measurements from this spectrum of events can be analyzed and extrapolated to extreme conditions using a variety of statistical techniques such as fits to power law distributions (often observed in space weather environments) and extreme value methods. This approach has been applied to many parameters that characterize space weather environments, including geomagnetic indices (Riley and Love, 2017), rates of change of geomagnetic fields (Thomson et al., 2011), geoelectric fields (Love et al., 2016), solar flare X-ray intensity (Schrijver et al., 2012), and high-energy electron fluxes in geosynchronous orbit (Meredith et al., 2015) and low Earth orbit (Meredith et al., 2016). As a result, we have estimates of extreme values of many parameters, sometimes specified as the extreme expected on centennial timescales and sometimes explicitly linked to the value thought to have occurred during the Carrington event and thereby giving an estimate of the probability of a repeat of the Carrington event.

But it is not sufficient for a worst-case scenario to focus solely on the specification of the environment. To gain real traction with policymakers, these environmental effects need to be mapped into impacts on the technologies that underpin critical national and international infrastructures, and then of the risks that these pose to society. Thus scenarios need engineering input to show how, for example:

- extreme geoelectric fields can induce GIC in power grids and thereby disrupt their operation, possibly leading to power blackouts over wide areas;
- intense solar energetic particle fluxes can generate high levels of single-event effects in satellites, disrupting the smooth operation of many now-vital satellite applications including communications, precision timing and navigation, meteorology, tracking of ships and aircraft, and many others;

- if solar energetic particle fluxes include a strong high-energy (>400 MeV) component, they can penetrate the atmosphere down to aircraft flight altitudes and generate single-event effects in avionics, potentially disrupting the smooth operation of aircraft systems and increasing pilot workload to what may be dangerous levels.
- strong ionospheric scintillation can render L-band radio signals from satellites unusable, thereby blocking use of key services, in particular satellite navigation and timing (GPS, Galileo) and also satellite-based mobile communications.

It is important to present space weather risks to policymakers in these practical terms, showing how they arise from extreme environments together with a good understanding of how engineered systems interact with those systems. This builds a coherent picture linking space weather science to the responsibility of policymakers to assess and prepare for important risks.

An important element in this picture is understanding the spatial and temporal extent of extreme space weather. How will an extreme space weather event impact different parts of the world and how long will the impacts last? This is an important issue in planning how to handle space weather risks and one where scientific knowledge can be deployed to good effect. This is particularly true in respect to the headline space weather risk, namely the geomagnetic threat to power grids, where the critical role of substorms has sometimes been ignored. Substorms are the fundamental dynamical cycle of the magnetosphere, and are the primary factor modulating the large geomagnetic variations observed at high- and mid-latitudes (Viljanen et al., 2006; Pulkkinen et al., 2015; Chapter 8 of this volume). Thus, during an extreme space weather event, intense substorms will create a series of hotspots in which power grids in different high- and mid-latitude regions are disrupted at different times. In general we would expect to see a series of substorm impact footprints, each exhibiting activity over an hour or two, with new impact footprints appearing to the west as the Earth rotates eastward, and with impacts returning to already impacted regions if the space weather event lasts more than a day. Other space weather impacts exhibit different spatial and temporal patterns. For example, the geomagnetic focusing of solar energetic particles (SEPs) into the polar upper atmosphere will produce significant ionization at altitudes around 90 km in polar regions, leading to absorption of high-frequency radio waves in polar regions (see Box 1). The resulting blackout of aircraft communications will disrupt heavily used air routes that cross the Arctic and will continue for the duration of the SEP event, which may be several days.

However, it is vital to recognize that, like other major hazards, the societal impact of space weather is not confined to the regions of physical impact such as closed air space and blacked-out power grids. Economic linkages will create secondary disruptions in other regions, some many thousands of kilometers from the region of physical impact. This is a consequence of the modern interconnected global economy and is increasingly recognized as an important element in global risk management. Economists have developed techniques to model these linkages, in particular what they term input-output (IO) models. These models establish an extensive series of coupling factors that reflect the links between different economic elements, which may be different sectors (energy, transport, manufacturing, financial services, etc.) of a national economy and/or different national or regional economies. Thus, if we can use the physical impact of space weather to quantify the direct economic impacts on particular economic sectors, economists can use their IO models to show how this propagates to generate indirect impacts in other sectors and other regions and countries. This approach is now being applied by several workers; it is becoming clear that these indirect impacts are extensive and of importance both to the national economies of major nations and to the global economy (e.g., Schulte in den Bäumen et al., 2014; Eastwood et al., 2017; Oughton et al., 2017).

From this discussion it should be clear that extreme space weather is a significant natural hazard on the 1-in-100-year timescale and one that is likely to have global consequences. An extreme event will have direct impacts on a wide range of locations across the globe, such as in the Arctic where there is an increasing number of economically important activities, and in hotspots across mid-latitudes where power grids are disrupted by intense substorms. Global trade links will then generate indirect impacts in many other regions of the world.

However, it is important to recognize that this picture of the global threat from space weather is largely based on evidence accumulated over the past 160 years, and especially the last 100 years. There is now very strong evidence that solar activity has varied markedly over the past 9000 years and has been at a "grand maximum" for much of the past century—and that the weak solar and geomagnetic activity of current solar cycle 24 may indicate a decline towards a grand "solar minimum" (Lockwood, 2010). Thus there is today great scientific interest in understanding if and how this long-term modulation changes the likelihood of extreme space weather. This is very much a hot topic at the time of writing and so no conclusion will be offered here. It is important, though, to recognize that the assessment of longer-term trends in solar activity largely focuses on average conditions over periods of years or decades, such as through use of smoothed sunspot numbers or annual measures of cosmogenic isotope deposition. Thus it is hard to detect an isolated extreme event, on the scale of the Carrington event, occurring in an otherwise quiet period of solar activity. There are several examples of such events in the recent record, notably the solar minimum geomagnetic storm of February 1986. Furthermore, the recent solar cycle 24, even though very quiet at Earth, includes the spectacular coronal mass ejection (CME) of July 23, 2012. This would undoubtedly have caused an extreme event had it hit the Earth, but fortunately it launched from the west side of the Sun and traveled well away from the Earth. But, equally fortunately, it passed over the STEREO-A spacecraft, giving us an excellent opportunity for risk-free scientific observations of a CME that could drive a Carrington-class event (Baker et al., 2013). This opportunity has been widely exploited to carry out modeling work that reinforces the worst-case scenarios discussed above. The July 2012 event also provides a further warning that extreme space weather may appear in otherwise quiet space weather conditions.

This long-term view of space weather also raises the question of whether events more severe than Carrington can occur on millennial timescales. There is currently interest in the risks from all natural hazards on these timescales, because of the devastating example provided by the 2011 tsunami in Japan. This demonstrated that 1-in-1000 years' risks with severe impacts cannot be neglected. The recent identification (from radiocarbon data) of a very extreme atmospheric radiation event in AD 774 suggests that very extreme space weather events can occur (Miyake et al., 2012). This is reinforced by recent studies of large flares on Sun-like stars, where there is evidence that flares 10 times more energetic than the flare associated with the Carrington event can occur on millennial timescales (Maehara et al., 2012). We need to understand the potential consequences of such events in order to assess whether they should be included in risk management plans. The AD 774 event is particularly important here because it shows that we need to understand the practical consequences arising from a major enhancement of the surface radiation environment on Earth and the risk that this would pose to digital control systems in critical infrastructures.

In summary, the assessment of extreme space weather as a major risk to modern societies is based on a combination of two key factors: (1) that intense space weather events such as the Carrington event are likely to occur on centennial timescales, and (2) that such events will have major adverse impacts on modern societies because of our dependence on a range of technologies vulnerable to space weather. There is compelling scientific evidence for both factors but much can be done to extend and refine that evidence, leading to better management of the space weather risk. There is also emerging evidence that even more adverse impacts may occur on millennial timescales and that these need more study by both scientists and risk managers.

4 WHAT KNOWLEDGE IS NEEDED IN THE FUTURE

In the previous sections we have explored how policymakers assess and manage major risks, especially those from natural hazards. We have also outlined how space weather fits into this framework and how there is compelling evidence for extreme space weather to be considered as a major risk, one requiring a substantive policy response. In this final section we explore how scientific knowledge, including data and models, should be further developed to support continuing work by policymakers.

4.1 THE NEED FOR DATA

Probably the most important future development is to ensure the supply of space weather data needed to recognize and manage an extreme event. These include a wide range of observations of conditions on the Sun, in the solar wind, in Earth's magnetosphere and atmosphere, and on the surface of the Earth. The required observations have been the subject of extensive studies since the turn of this century (e.g., Hapgood, 2001). The observational requirements for operational use are now formally consolidated at the global level under the auspices of the International Space Environment Service (ISES) and the World Meteorological Organization (WMO-ISES, 2017), while the equivalent requirements for scientific research have been extensively discussed as part of the COSPAR Space Weather Roadmap (Schrijver et al., 2015).

The space weather community has already developed scenarios to show how extreme events may manifest at Earth (e.g., Cannon et al., 2013; Hapgood et al., 2016; National Science and Technology Council, 2017) and these are being used in several countries to drive work on the long-term planning needed to manage such events (BIS, 2015; OSTP, 2015a,b; Chapter 28 of this volume). While much further work is required to extend and refine these scenarios, it is very clear that those plans should include actions to ensure the supply of the data identified by WMO-ISES (2017). All countries at risk from space weather (which is essentially all countries on Earth) need the ability (1) to forecast when extreme space weather is imminent, so that contingency plans can be activated, and (2) to monitor the progress of an extreme event so that those contingency plans can be customized to the actual event. The latter is vital: the policy community has a very good understanding that no extreme event, from any natural hazard, follows the pattern set out in planning scenarios—and thus that their response must be customized to the actual manifestation of the event. They use planning scenarios to develop generic ideas and capabilities, but must customize to actual circumstances. Scientific input, both in the form of data and of expert advice, is an essential part of that customization and must be provided in ways that can respond quickly to a fast-moving situation.

The delivery of both abilities, forecasting and monitoring, critically requires a supply of space-based data at least comparable to what we have today. In particular, this means the long term maintenance of capabilities such as we have with SOHO and DSCOVR at L1, and, even better, the addition of a long-term commitment to an off-Sun-Earth line monitoring capability (ideally in stable orbit around L5) that builds on what we have learned from STEREO. Without these capabilities, particularly the ability to track Earthward-traveling CMEs, it will be difficult to detect the onset of extreme space weather at Earth before the aurora appears in the night sky. That will be too late.

Thus policy is a key driver to maintaining and expanding our existing satellite capabilities to monitor space weather. In particular, it shows the need for a program of operational satellites performing space weather measurements, akin to the meteorological satellite programs that underpin weather forecasting around the world. Space weather services should not continue to rely on opportunities that arise as a byproduct of measurements made by instruments on scientific satellites. The outstanding success of several scientific satellites launched in the 1990s under the International Solar-Terrestrial Physics program (notably ACE and SOHO) has given us almost 20 years' experience of 24/7 near-real-time space weather monitoring and stimulated huge growth in (and reliance on) space weather services. But the world has been slow to appreciate that this is a time-limited asset. The recent launch of DSCOVR has provided a temporary solution to sustain solar wind measurements at L1. But this was an innovative short-term solution—taking the opportunity to repurpose existing satellite hardware (see Fig. 5) to meet urgent space weather needs (NOAA, 2015).

Beyond DSCOVR there needs to be a commitment to continuing space weather observations optimized for purposes of forecasting and nowcasting, part of what is sometimes termed space situational awareness. This is now slowly taking shape, with good prospects for new instruments in orbit ahead of the next solar maximum in the mid-2020s.

4.2 THE NEED TO LEARN FROM METEOROLOGY

The analogy with meteorology is important here as it brings several important strengths that will enable an operational space weather program to address policy objectives:

- First, and perhaps most important, is to encourage international collaboration. Space weather is a global threat so it makes great sense to work together both for a better-quality outcome and to share costs. International collaboration also facilitates the delivery of a complementary set of space weather missions, such as an L5 mission as well as new L1 missions to replace SOHO and DSCOVR.
- Almost as important, this analogy will encourage a distinction between research and operational missions, a recognition that we need both kinds of missions, and with individual missions focused on requirements for research or for operations, but not trying to do both. It is very important to establish robust requirements for operational space weather measurements. It ensures that the design of the instruments, and of the host spacecraft, is targeted at what is really needed to deliver space weather services. For example, when observing a newly launched CME, one needs a wide-field coronagraph so that the observations track the CME long enough to determine an accurate speed and direction. It is not appropriate to simply reuse designs of existing scientific instruments. Those designs must be reviewed against, and where needed updated to address, operational requirements. It is important to understand that those requirements go beyond instrument performance and include critical issues not normally applied to science missions:

FIG. 5

The DSCOVR spacecraft under refurbishment at NASA Goddard Space Flight Center in February 2013.

- ○ the need for near-real-time delivery of data to space weather service providers (modern science missions usually store data in a large onboard memory and downlink later when a good ground station link becomes available)
- ○ the need for instruments that can operate well for many years, at least 10, maybe 15–20, in the harsh environment of space (science missions usually have a shorter design lifetime, typically 2 years)
- ○ the likely evolution of space weather services over the mission lifetime. Thus consideration should be given to what data will be required in, say, 10 years time.
- This analogy will also promote work to understand the socioeconomic value of each space weather measurement or set of measurements—by tracing how those measurements support space weather services and thence how those services can provide value to the wider community

at risk from space weather. This economic aspect is crucial to justify investment in operational measurements—and helps to focus investment on the most appropriate measurements. For example, an increasing number of studies show large economic impacts arising from the adverse effects of geomagnetic storms on power grids (Forbes and St. Cyr, 2008, 2012; Schulte in den Bäumen et al., 2014; Eastwood et al., 2017; Oughton et al., 2017); this provides a strong basis to justify investment in future satellite missions to monitor CMEs, as those are the space weather driver of the strongest geomagnetic storms.
- Finally the analogy with meteorology will encourage training of space weather scientists in the needs of forecasting, to understand how to apply the science of space weather in ways that improve forecasting skills, a point strongly advocated by Siscoe (2007).

4.3 THE NEED FOR BETTER SCIENCE AND BETTER MODELS

Thus space weather can learn from meteorology in using our scientific knowledge to build an operational monitoring program—one where the instruments are optimized to meet policy goals. Such instruments will still be of great value to science, so the data must be made accessible to promote advances in our scientific understanding of space weather. But, as with any satellite project, it is vital to maintain a sharp focus on core requirements and avoid the mission creep that could arise from adding other requirements (e.g., see Coughlin et al., 1999). For example, the resolution of imaging instruments should be matched to operational requirements set by policy goals and not increased to match additional scientific requirements. Such increases can significantly increase the costs of operating the host satellite by requiring more powerful or more complex downlink capabilities, well beyond what can be justified by policy goals.

In addition to this use of science to build up operational capabilities, there is also a strong need for continuing research to improve both our understanding of extreme space weather and our ability to forecast its impacts on Earth (Schrijver et al., 2015). Simulations enable more focused preparations for extreme events, providing valuable data for policy development and for exercising procedures to be followed by governments, emergency managers, and industry when an extreme event occurs (Chapter 28 of this volume). These better models will almost certainly be physics-based because we need models that can scale up from common to extreme space weather conditions. In general, parametric models cannot do this because they are based on fits to large datasets; this works well for common space weather conditions, but is almost certain to fail for the very limited data available on extreme conditions. As we discuss in the four examples below, models of extreme conditions need to capture some understanding of how the physics of space weather changes when we move toward those conditions.

Our first example is the need to mitigate space weather impacts on power grids. This is the headline impact that gets attention from senior policymakers for the very good reason that electric power is the fundamental infrastructure of modern society. Sustained loss of electricity for more than a few hours is extremely disruptive to everyday life and economic activities and would become a major emergency if it continued for even a few days. Thus we need good models of how the solar wind (especially major disturbances caused by CMEs) creates intense geoelectric fields on the surface of the Earth. This necessitates a better understanding of magnetospheric and ionospheric dynamics, and of how magnetic induction operates within the solid body of the Earth. We are today seeing good progress in these areas, such as with the recent introduction in the United States of the first physics-based magnetospheric model supporting operational services (NOAA, 2016; Tóth et al., 2005), and, also in the United States, the first magnetotelluric survey aiming to deliver a comprehensive mapping of the three-dimensional impedance matrices that allow us to derive geoelectric fields from geomagnetic variations (Schultz, 2009; Bonner and Schultz, 2017; Chapter 9 of this volume). Nonetheless, further work is undoubtedly needed, particularly to better incorporate the nonlinear effects and feedback loops that characterize coupling within the magnetosphere-ionosphere system, and its links with the thermosphere and plasmasphere. There are many open questions on how this complexity will influence the behavior of geomagnetic storms in extreme conditions. For example, how fast will the auroral oval expand and contract during an extreme event? This is crucial to assessing how intense substorms will impact on the dense infrastructures in the many advanced countries at mid-latitudes. Even this seemingly simple question probably needs significant advances in magnetospheric modeling as it will require self-consistent modeling of reconnection and hence a move beyond models based on magneto hydrodynamics (von Alfthan et al., 2014).

Our second example is the interest of policymakers in the effects of atmospheric radiation—the fluxes of neutrons and muons produced when high-energy ions collide with the nuclei inside the molecules and atoms that make up Earth's atmosphere. Galactic cosmic rays (GCRs) provide a low flux background of this radiation. But when solar radiation storms have significant ion fluxes at energies above 400 MeV/nucleon, they can cause major increases in the natural radiation environment on Earth's surface and in the atmosphere. More than 70 such events (ground level enhancements) have been observed since 1942 and in the worst case led to a 50-fold increase of radiation at ground level (Gold and Palmer, 1956; Marsden et al., 1956). It has been estimated that there was a 300-fold increase of radiation at aircraft flight altitudes during that same event (Dyer et al., 2007). The definitive database of ground-based neutron monitor measurements (University of Oulu, 2014) shows that these events produce markedly different radiation fluxes in different regions of the Earth. Thus, to assess radiation exposure during such events, we need good models of how atmospheric radiation fluxes vary in response to a number of factors including: (a) very obviously, the intensity and energy spectrum of the radiation storm impinging on Earth; (b) the transport of high-energy ions through the magnetosphere, taking account of the current state of the magnetosphere; (c) the interaction of these ions with atmospheric species; and (d) the subsequent atmospheric transport of particles. But we also need measurements of atmospheric radiation during radiation storms in order to validate these models. The latter is probably the critical element, especially in respect to aircraft altitudes. There are very limited measurements of radiation at aircraft altitudes—a few radiation monitors have been flown in recent years, but the datasets are fairly limited. This is particularly the case in respect to measurements during radiation storms, due to the recent dearth of such storms. Thus, while there are now a number of atmospheric radiation models, their development lacks good data to stimulate progress, especially in respect of extreme conditions. As a result there are a number of plans being developed for the rapid deployment of atmospheric radiation monitors during major events, such as through monitors on balloons (Phillips et al., 2016) and on atmospheric research aircraft such as the Met Office Civil Contingencies aircraft (WMO, 2013). These are important developments that will, at some point, give us data that can drive the development of better models of the atmospheric consequences of extreme radiation storms. Such models are an essential tool for managing the consequences of a radiation storm that leads to high radiation levels at aircraft altitudes, such as one similar to the February 1956 event analyzed by Dyer et al. (2007). When such an event next occurs, many thousands of passengers and aircrew will acquire a significant radiation dose, one far below the level that poses any immediate risk to their health but nonetheless one that will warrant expert assessment and advice on managing future exposure. Better measurements and models are critical to providing that assessment.

A third example is the growing appreciation that the services provided by GNSS are becoming deeply embedded in modern societies. Hence policymakers need to be alert to anything that can significantly disrupt GNSS, and to encourage the development of resilience against GNSS loss. Space weather is a major source of disruption to GNSS (Curry, 2014) through its impact on the propagation of radio signals from GNSS satellites through the ionosphere to receivers on Earth; this can lead to increased position errors and in the worst case to loss of the GNSS signals (Chapters 1 and 23 of this volume). Both effects may last up to several days in extreme cases (Cannon et al., 2013). In addition, solar radio bursts (Chapter 22 of this volume) can effectively jam GNSS on the dayside of the Earth, but only for short periods of around 5–10 min. Thus there are ongoing efforts to improve the resilience of GNSS-based services. Governments around the world have invested billions of dollars in systems that correct for position errors (Loh et al., 1995; Gauthier et al., 2001; Rao, 2007; Trinity House, 2016; Avanti, 2017). These systems can mitigate many space weather effects on position accuracy, most notably those caused by solar flares, so that those effects are no longer considered a serious risk for GNSS. There are also many efforts to develop more resilient GNSS receivers, such as devising software radio technologies that can keep a better lock on the GNSS signal (see for example, Kintner et al., 2007; van den IJssel et al., 2016). However, an extreme geomagnetic storm will likely produce conditions that overwhelm these technological solutions, and, in these conditions, it will be essential to switch to backup systems that use a different (non-GNSS) technology. The civil aviation and financial trading sectors are generally very well prepared for this, but other sectors such as shipping and road transport logistics may be seriously impacted and require forecasting and nowcasting services to enable them to recognize and wait out the storm (Hapgood, 2017).

Our final example is the need to assess how long-term change in the level of geomagnetic activity should affect policy on extreme space weather events. The past decade has seen astonishingly low activity compared to the previous 140 years, as shown in Fig. 6. This is widely attributed to the well-established decline in solar activity since 2005 though there are also suggestions that there may be a contribution from weaker coupling of solar wind energy into the magnetosphere (Yamauchi, 2015). It is also timely to consider whether other long-term terrestrial trends have any influence on the level of geomagnetic activity. Is this affected by the secular change of the internal magnetic field of the Earth (both the decline of the dipole moment and the rapid motion of the magnetic poles since 1900—see Figs. 7 and 8)? What is the impact of climate change on the electrical conductivity of the upper atmosphere, and hence on geomagnetic activity? We need a better understanding of which of these factors influence long-term changes in the level of geomagnetic activity. In particular, for policy purposes we need to understand how these factors affect the probability of extreme events, not just the average level of activity. As previously noted, an isolated extreme event, such as the solar minimum storm of February 1986, can occur during a period of otherwise low activity and thus have only a small influence on average activity. While some progress can be made here using statistical methods, it is essential to gain a physical understanding of how all the factors above combine to produce extreme geomagnetic storms. In the long run, we need to develop models that describe long-term change in the Sun, in the solar wind, and in the magnetosphere-ionosphere system. Given such models we will be much more confident in assessing how the risk from extreme geomagnetic storms is evolving and thus be able to provide better advice to policymakers.

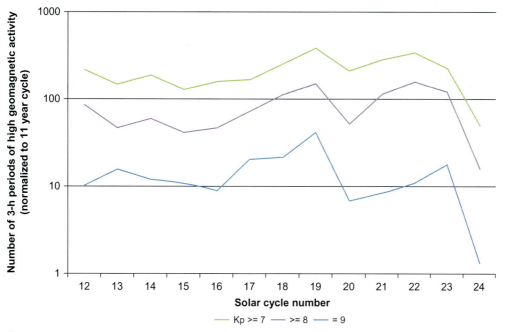

FIG. 6

Geomagnetic activity 1868–present, using an effective Kp derived from the aa index.

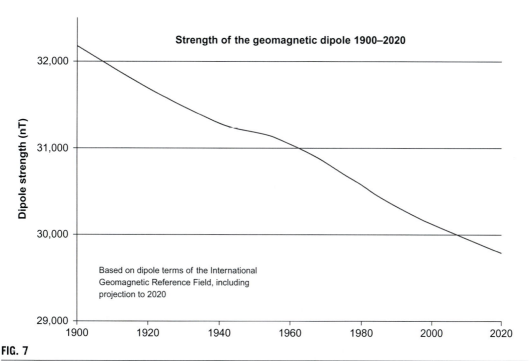

FIG. 7

Strength of the geomagnetic dipole 1900–2020. Based on the dipole terms of the IGRF spherical harmonic terms as follows $\sqrt{(g_1^0)^2 + (g_1^1)^2 + (h_1^1)^2}$.

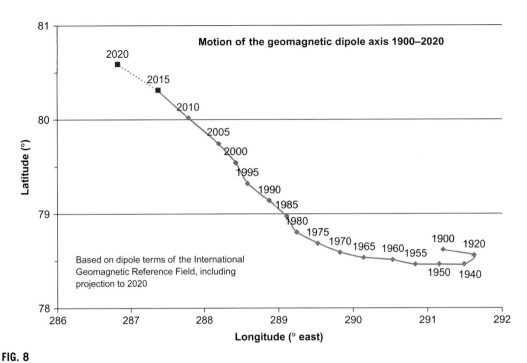

FIG. 8

Motion of the northern pole of geomagnetic dipole axis since 1900.

REFERENCES

Avanti, 2017. SBAS-AFRICA. http://sbas-africa.avantiplc.com/ (Accessed 1 April 2017).

Baker, D.N., Li, X., Pulkkinen, A., Ngwira, C.M., Mays, M.L., Galvin, A.B., Simunac, K.D.C., 2013. A major solar eruptive event in July 2012: defining extreme space weather scenarios. Space Weather 11, 585–591. https://doi.org/10.1002/swe.20097.

BIS, 2015. Space Weather Preparedness Strategy. https://www.gov.uk/government/publications/space-weather-preparedness-strategy (Accessed 8 March 2017).

Bonner IV, L.R., Schultz, A., 2017. Rapid prediction of electric fields associated with geomagnetically induced currents in the presence of three-dimensional ground structure: projection of remote magnetic observatory data through magnetotelluric impedance tensors. Space Weather 15, 204–227. https://doi.org/10.1002/2016SW001535.

Boteler, D.H., 2006. The super storms of August/September 1859 and their effects on the telegraph system. Adv. Space Res. 38, 159–172. https://doi.org/10.1016/j.asr.2006.01.013.

Cabinet Office, 2015. National Risk Register of Civil Emergencies. https://www.gov.uk/government/collections/national-risk-register-of-civil-emergencies (Accessed 8 March 2017).

Cannon, P., Angling, M., Barclay, L., Curry, C., Dyer, C., Edwards, R., et al., 2013. Extreme Space Weather: Impacts on Engineered Systems and Infrastructure. U.K. Royal Academy of Engineering, London. ISBN 1-903496-95-0.

Carrington, R.C., 1859. Description of a singular appearance seen in the Sun on September 1. Mon. Not. R. Astron. Soc. 20, 13–15. https://doi.org/10.1093/mnras/20.1.13.

REFERENCES

Cliver, E.W., Dietrich, W.F., 2013. The 1859 space weather event revisited: limits of extreme activity. J. Space Weather Space Clim. 3, https://doi.org/10.1051/swsc/2013053. A31.

Coughlin, T.B., Chiu, M.C., Dassoulas, J., 1999. Forty years of space mission management. Johns Hopkins APL Tech. Dig. 20, 507–510.

Curry, C., 2014. SENTINEL Project: Report on GNSS Vulnerabilities. Chronos Technology Ltd. http://www.chronos.co.uk/files/pdfs/gps/SENTINEL_Project_Report.pdf. (Accessed 28 April 2017).

Dyer, C.S., Lei, F., Hands, A., Truscott, P., 2007. Solar particle events in the QinetiQ atmospheric radiation model. IEEE Trans. Nucl. Sci. 54 (4), 1071–1075 10/1109/TNS.2007.893537.

Eastwood, J.P., Biffis, E., Hapgood, M.A., Green, L., Bisi, M.M., Bentley, R.D., Wicks, R., McKinnell, L.-A., Gibbs, M., Burnett, C., 2017. The economic impact of space weather: where do we stand? Risk Anal. 37, 206–218. https://doi.org/10.1111/risa.12765.

Forbes, K.F., St. Cyr, O.C., 2008. Solar activity and economic fundamentals: evidence from 12 geographically disparate power grids. Space Weather 6. S10003, https://doi.org/10.1029/2007SW000350.

Forbes, K.F., St. Cyr, O.C., 2012. Did geomagnetic activity challenge electric power reliability during solar cycle 23? Evidence from the PJM regional transmission organization in North America. Space Weather 10. S05001, https://doi.org/10.1029/2011SW000752.

Gauss, C.F., 1833. Die Intensität der erdmagnetischen Kraft zurückgeführt auf absolutes Maass. Ann. Phys. 104, 591–615. https://doi.org/10.1002/andp.18331040808.

Gauthier, L., et al., 2001. EGNOS: the first step in Europe's contribution to the global navigation satellite system. ESA Bull. 105, 35–42. http://esamultimedia.esa.int/multimedia/publications/ESA-Bulletin-105/.

Gold, T., Palmer, D.R., 1956. The solar outburst, 23 February 1956—observations by the Royal Greenwich Observatory. J. Atmos. Sol. Terr. Phys. 8, 287–290.

Green, J.L., Boardsen, S., 2006. Duration and extent of the great auroral storm of 1859. Adv. Space Res. 38, 130–135.

Halley, E., 1714. An account of the late surprizing appearance of the lights seen in the air, on the sixth of March last; with an attempt to explain the principal phaenomena thereof. Phil. Trans. 29 (338-350), 406–428. https://doi.org/10.1098/rstl.1714.0050.

Hapgood, M., 2001. ESA Space Weather Study: Project Implementation Plan and Final Report. http://purl.org/net/epubs/work/34262 (Accessed 11 May 2017).

Hapgood, M., 2015. Space weather: the public and policy, room. Space J. (March 2015 issue). https://room.eu.com/article/Space_weather_the_public__policy. (Accessed 6 March 2015).

Hapgood, M., 2017. Satellite navigation—amazing technology but insidious risk: Why everyone needs to understand space weather. Space Weather 15. https://doi.org/10.1002/2017SW001638.

Hapgood, M., Angling, M., Attrill, G., Burnett, C., Cannon, P., Gibbs, M., et al., 2016. Summary of space weather worst-case environments. Revised edition. RAL Technical Report RAL-TR-2016-06. http://purl.org/net/epubs/work/25015281. (Accessed 6 March 2017).

Hodgson, R., 1859. On a curious appearance seen in the Sun. Mon. Not. Roy. Astron. Soc 20, 15–16. https://doi.org/10.1093/mnras/20.1.15.

Kappenman, J.G., 2006. Great geomagnetic storms and extreme impulsive geomagnetic field disturbance events— an analysis of observational evidence including the great storm of May 1921. Adv. Space Res. 38, 188–199. https://doi.org/10.1016/j.asr.2005.08.055.

Karsberg, A., Swedenborg, G., Wyke, K., 1959. The influences of earth magnetic currents on telecommunication lines, tele (English ed.). Televerket, Stockholm. pp. 1–21.

Kintner, P.M., Ledvina, B.M., de Paula, E.R., 2007. GPS and ionospheric scintillations. Space Weather 5. S09003, https://doi.org/10.1029/2006SW000260.

Lee, B., Felix, P., Gemma, G., 2012. Preparing for high-impact, low-probability events: lessons from Eyjafjallajökull. Chatham House. https://www.chathamhouse.org/publications/papers/view/181179. (Accessed 11 May 2017).

Lockwood, M., 2010. Solar change and climate: an update in the light of the current exceptional solar minimum. Proc. R. Soc. A 466, 303–329. https://doi.org/10.1098/rspa.2009.0519.

Loh, R., Wullschleger, V., Elrod, B., Lage, M., Haas, F., 1995. The U.S. wide-area augmentation system (WAAS). Navigation 42, 435–465. https://doi.org/10.1002/j.2161-4296.1995.tb01900.x.

Love, J.J., et al., 2016. Geoelectric hazard maps for the continental United States. Geophys. Res. Lett. 43, 9415–9424. https://doi.org/10.1002/2016GL070469.

Lundstedt, H., Persson, T., Andersson, V., 2015. The extreme solar storm of May 1921: observations and a complex topological model. Ann. Geophys. 33, 109–116. https://doi.org/10.5194/angeo-33-109-2015.

Maehara, H., Shibayama, T., Notsu, S., Notsu, Y., Nagao, T., Kusaba, S., Honda, S., Nogami, D., Shibata, K., 2012. Superflares on solar-type stars. Nature 485, 478–481. https://doi.org/10.1038/nature11063.

Marsden, P.L., Berry, J.W., Fieldhouse, P., Wilson, J.G., 1956. Variation of cosmic-ray nucleon intensity during the disturbance of 23 February 1956. J. Atmos. Sol. Terr. Phys. 8, 278–281.

Meredith, N.P., Horne, R.B., Isles, J.D., Rodriguez, J.V., 2015. Extreme relativistic electron fluxes at geosynchronous orbit: Analysis of GOES $E > 2$ MeV electrons. Space Weather 13, 170–184. https://doi.org/10.1002/2014SW001143.

Meredith, N.P., Horne, R.B., Isles, J.D., Green, J.C., 2016. Extreme energetic electron fluxes in low Earth orbit: Analysis of POES $E > 30$, $E > 100$, and $E > 300$ keV electrons. Space Weather 14, 136–150. https://doi.org/10.1002/2015SW001348.

Miyake, F., Nagaya, K., Masuda, K., Nakamura, T., 2012. A signature of cosmic-ray increase in AD 774–775 from tree rings in Japan. Nature 486, 240–242. https://doi.org/10.1038/nature11123.

National Science and Technology Council, 2017. Space Weather Phase 1 Benchmarks. http://www.ofcm.gov/publications/spacewx/DRAFT_SWx_Phase_1_Benchmarks.pdf Accessed 8 March 2017.

Nevanlinna, H., 2006. A study on the great geomagnetic storm of 1859: comparisons with other storms in the 19th century. Adv. Space Res. 38, 180–187. https://doi.org/10.1016/j.asr.2005.07.076.

Nevanlinna, H., 2008. On geomagnetic variations during the August–September storms of 1859. Adv. Space Res. 42, 171–180. https://doi.org/10.1016/j.asr.2008.01.002.

New York Times, 1921. Cables Damaged by Sunspot Aurora, The New York Times, 17 May 1921. New York. http://query.nytimes.com/mem/archive-free/pdf?res=9407E2D61E3FEE3ABC4F52DFB366838A639EDE. (Accessed 12 May 2017).

Nichols, J.D., 2011. Magnetosphere–ionosphere coupling at Jupiter-like exoplanets with internal plasma sources: implications for detectability of auroral radio emissions. Mon. Not. Roy. Astro. Soc. 414, 2125–2138. https://doi.org/10.1111/j.1365-2966.2011.18528.x.

NOAA, 2015. DSCOVR: Deep Space Climate Observatory. https://www.nesdis.noaa.gov/sites/default/files/asset/document/dscovr_program_overview_info_sheet.pdf (Accessed 8 March 2017).

NOAA, 2016. News Item: New Experimental Regional Geomagnetic Products Available. http://www.swpc.noaa.gov/news/new-experimental-regional-geomagnetic-products-available (Accessed 8 March 2017).

Odenwald, S., 2010. Introduction to space storms and radiation. In: Schrijver, C., Siscoe, G. (Eds.), Heliophysics: Space Storms and Radiation: Causes and Effects. Cambridge University Press, Cambridge, pp. 15–42. https://doi.org/10.1017/CBO9781139194532.003.

OSTP, 2015a. National Space Weather Strategy. https://obamawhitehouse.archives.gov/sites/default/files/microsites/ostp/final_nationalspaceweatherstrategy_20151028.pdf (Accessed 8 March 2017).

OSTP, 2015b. National Space Weather Action Plan. https://obamawhitehouse.archives.gov/sites/default/files/microsites/ostp/final_nationalspaceweatheractionplan_20151028.pdf (Accessed 8 March 2017).

Oughton, E.J., Skelton, A., Horne, R.B., Thomson, A.W.P., Gaunt, C.T., 2017. Quantifying the daily economic impact of extreme space weather due to failure in electricity transmission infrastructure. Space Weather 15, 65–83. https://doi.org/10.1002/2016SW001491.

Paton, J., 1946. Aurora Borealis. Weather 1, 6–11. https://doi.org/10.1002/j.1477-8696.1946.tb00012.x.

Phillips, T., et al., 2016. Space weather ballooning. Space Weather 14, 697–703. https://doi.org/10.1002/2016SW001410.

Ptitsyna, N.G., Tyasto, M.I., Altamore, A., Ptitsyna, N.G., Tyasto, M.I., Altamore, A., 2012. New data on the giant September 1859 magnetic storm: an analysis of Italian and Russian historic observations. In: Proc. 9th Intl. Conf. "Problems of Geocosmos", Oct 8-12, 2012, St. Petersburg, Russia.

Pulkkinen, A., Bernabeu, E., Eichner, J., Viljanen, A., Ngwira, C., 2015. Regional-scale high-latitude extreme geoelectric fields pertaining to geomagnetically induced currents. Earth Planets Space 67, 93. https://doi.org/10.1186/s40623-015-0255-6.

Rao, K.N., 2007. GAGAN-The Indian satellite based augmentation system. Indian J. Radio Space Phys. 36, 293–302. http://nopr.niscair.res.in/handle/123456789/4707.

Riley, P., Love, J.J., 2017. Extreme geomagnetic storms: probabilistic forecasts and their uncertainties. Space Weather 15, 53–64. https://doi.org/10.1002/2016SW001470.

Schmidt, A., 2015. Volcanic gas and aerosol hazards from a future Laki-Type eruption in Iceland. In: Shroder, J.F., Papale, P. (Eds.), Volcanic Hazards, Risks and Disasters. Elsevier, Boston, pp. 377–397 http://doi.org/10.1016/B978-0-12-396453-3.00015-0.

Schrijver, C.J., et al., 2012. Estimating the frequency of extremely energetic solar events, based on solar, stellar, lunar, and terrestrial records. J. Geophys. Res.. 117A08103https://doi.org/10.1029/2012JA017706.

Schrijver, C.J., Kauristie, K., Aylward, A.D., Denardini, C.M., Gibson, S.E., Glover, A., Gopalswamy, N., Grande, M., Hapgood, M., Heynderickx, D., Jakowski, N., et al., 2015. Understanding space weather to shield society: a global road map for 2015–2025 commissioned by COSPAR and ILWS. Adv. Space Res. 55 (12), 2745–2807. https://doi.org/10.1016/j.asr.2015.03.023.

Schulte in den Bäumen, H., Moran, D., Lenzen, M., Cairns, I., Steenge, A., 2014. How severe space weather can disrupt global supply chains. Nat. Hazard. Earth Sys. Sci. 14 (10), 2749–2759.

Schultz, A., 2009. EMSCOPE: a continental scale magnetotelluric observatory and data discovery resource. Data Sci. J. 8, IGY6–IGY20. https://doi.org/10.2481/dsj.GG_IGY-009.

Sciencewise/STFC, 2015. Space Weather Public Dialogue: Final Report. http://www.sciencewise-erc.org.uk/cms/space-weather-dialogue (Accessed 8 March 2017).

Siscoe, G., 2007. Space weather forecasting historically viewed through the lens of meteorology. In: Bothmer, V., Daglis, I.A. (Eds.), Space Weather—Physics and Effects. Springer, Berlin.

Stewart, B., 1861. On the great magnetic disturbance which extended from August 28 to September 7, 1859, as recorded by photography at the Kew Observatory. Phil. Trans. R. Soc. 151, 423–430. https://doi.org/10.1098/rstl.1861.0023.

The National Archives, 2016. The Civil Contingencies Act 2004. http://www.legislation.gov.uk/ukpga/2004/36/contents. (Accessed 26 April 2017).

Thomson, A.W.P., Dawson, E.B., Reay, S.J., 2011. Quantifying extreme behavior in geomagnetic activity. Space Weather 9. S10001, https://doi.org/10.1029/2011SW000696.

Tóth, G., et al., 2005. Space weather modeling framework: a new tool for the space science community. J. Geophys. Res. 110, A12226, https://doi.org/10.1029/2005JA011126.

Trinity House, 2016. Satellite Navigation Ground Based Augmentations, https://www.trinityhouse.co.uk/dgps. (Accessed 1 April 2017).

University of Oulu, 2014. Database of Neutron Monitor count rates during Ground Level Enhancements. http://gle.oulu.fi/. (Accessed 8 March 2017).

Usoskin, I.G., 2017. A history of solar activity over millennia. Living Rev. Sol. Phys. 14, 3. https://doi.org/10.1007/s41116-017-0006-9.

van den IJssel, J., Forte, B., Montenbruck, O., 2016. Impact of Swarm GPS receiver updates on POD performance. Earth Planet Space 68, 85. https://doi.org/10.1186/s40623-016-0459-4.

Viljanen, A., Tanskanen, E.I., Pulkkinen, A., 2006. Relation between substorm characteristics and rapid temporal variations of the ground magnetic field. Ann. Geophys. 24, 725–733. https://doi.org/10.5194/angeo-24-725-2006.

von Alfthan, S., Pokhotelov, D., Kempf, Y., Hoilijoki, S., Honkonen, I., Sandroos, A., Palmroth, M., 2014. Vlasiator: first global hybrid-Vlasov simulations of earth's foreshock and magnetosheath. J. Atmos. Sol. Terr. Phys. 120, 24–35. https://doi.org/10.1016/j.jastp.2014.08.012.
Wikipedia, 2017. https://en.wikipedia.org/wiki/Civil_Contingencies_Act_2004 (Accessed 8 March 2017).
WMO (World Meteorlogical Organisation), 2013. Statement of Guidance on Space Weather Observations, see section on Radiation Dose Rate Observations. https://www.wmo.int/pages/prog/sat/meetings/documents/ICTSW-4_Doc_04-02_SoG-update.pdf (Accessed 8 March 2017).
WMO-ISES, 2017. Observing Systems Capability Analysis and Review Tool - Application: Space Weather. https://www.wmo-sat.info/oscar/applicationareas/view/25 (Accessed 11 May 2017).
Yamauchi, M., 2015. Decreased Sun-Earth energy-coupling efficiency starting from 2006. Earth Planets Space 67, 44. https://doi.org/10.1186/s40623-015-0211-5.

FURTHER READING

Krausmann, E., Andersson, E., Murtagh, W., Mitchison, N., 2013. Space Weather and Power Grids: Findings and Outlook. Report JRC86658European Commission Joint Research Centre, Ispra, Italy. ISBN 978-92-79-34812-9.
Tsurutani, B.T., Gonzales, W.D., Lakhina, G.S., Alex, S., 2003. The extreme magnetic storm of 1–2 September 1859. J. Geophys. Res. 108. https://doi.org/10.1029/2002JA009504.

CHAPTER 28

EXTREME SPACE WEATHER AND EMERGENCY MANAGEMENT

Mark H. MacAlester
Federal Emergency Management Agency, Washington, DC, United States

CHAPTER OUTLINE

1 Why Emergency Managers Care 684
2 Understanding the Risks of Extreme Solar Events 684
  2.1 The Hazard 684
3 What Emergency Managers Need From Researchers and Engineers 688
  3.1 Plain Language 688
  3.2 Expert Analysis Available on Demand 688
  3.3 Sharable Information 689
  3.4 Improved Forecast Products 690
4 Conclusion 699
References 699
Further Reading 700

In 2011, William "Bill" Murtagh of the National Oceanic and Atmospheric Administration (NOAA) Space Weather Prediction Center (SWPC) asked me to present the Department of Homeland Security (DHS) Federal Emergency Management Agency's (FEMA) initial work on extreme space weather's impact on communication systems at the annual Space Weather Workshop in Boulder, Colorado. The first 2 days of the workshop saw presentations from leading experts in space weather research filled with research-specific jargon, calculus equations, and advanced statistical analysis. As an emergency manager, I was often lost and, frankly, intimidated. When my turn to speak arrived, my presentation contained no equations, no statistics, and the simple language of operations. By the end, I felt embarrassed. After the presentation, two researchers came up to me and, to my surprise, thanked me. Never before, they said, had they understood the practical implications of their work. It was a revelation and the blossoming of a partnership with researchers and engineers that has helped the emergency management community understand both the promise and the limitations of space weather science, even as emergency managers helped them to understand the nature and needs of operations: research to operations, operations to research.

This chapter will focus on the operational needs of emergency managers and the organizations and people we serve. It will provide a brief overview of why we care about extreme space weather, our current understanding of the risk, and what we still need from researchers and engineers to improve our ability to prepare for, protect against, respond to, recover from, and mitigate the impacts of extreme space weather. This chapter is written in plain language and attempts, to the extent possible, to avoid the scientific and engineering jargon often used to describe this hazard.

1 WHY EMERGENCY MANAGERS CARE

Emergency managers are the people who work to make communities, businesses, and organizations more resilient to natural and man-made hazards. They advise leaders, managers, employees, and communities on the threats and vulnerabilities they face, and the actions everyone should take before, during, and after an emergency or disaster. Governmental emergency managers must also advise elected officials and the public on the nature of a hazard and the risks and vulnerabilities associated with that hazard as well as recommend courses of action. Clear, timely, and actionable guidance can—and often is—the difference between life and death. This is not an exaggeration.

Emergency managers rely on strong partnerships with the research and engineering communities to both understand and advise on hazard events. One well-known example in the United States is the partnership between NOAA's National Hurricane Center (NHC) and federal, state, territorial, tribal, local, and private-sector emergency managers and government officials. The NHC provides multiple forecast products—combined with expert advice—to inform the preparation for, response to, and recovery from tropical cyclones (e.g., hurricanes, typhoons, etc.). Emergency managers use this information to advise senior and elected officials and the public on preparedness, immediate protective measures, evacuations, activation of mutual aid agreements, and the prepositioning of emergency personnel, equipment, and commodities such as food, water, and blankets. The NOAA SWPC serves a similar role for space weather hazards.

2 UNDERSTANDING THE RISKS OF EXTREME SOLAR EVENTS

While there are many definitions of risk, emergency managers, in general, assess risk as (1) the likelihood that a hazard will occur in a geographic area, (2) the potential magnitude of the hazard in that geographic area, (3) the vulnerabilities of the community or organization to that hazard and magnitude, and (4) the consequences arising from those vulnerabilities.

The overall risk for an extreme space weather event is not well understood as of this writing. Although progress is being made, neither the science nor the engineering has matured to the point where likelihood, magnitude, vulnerabilities, and consequences can be predicted with the levels of confidence currently available for terrestrial severe weather hazards (e.g., tropical cyclones, etc.).

2.1 THE HAZARD

The Sun produces three types of space weather that emergency managers care about: radio blackouts (R), solar radiation storms (S), and geomagnetic storms (G). Other types of space weather are generally outside the scope of our concern. Current understanding suggests that extreme storms, in particular extreme geomagnetic storms, are low-occurrence, potentially high-consequence events.

2 UNDERSTANDING THE RISKS OF EXTREME SOLAR EVENTS

The risk assessments below are based on an internal FEMA impact assessment derived from current knowledge and in partnership with the SWPC, the National Aeronautics and Space Administration (NASA) Heliophysics Division at the Goddard Space Flight Center, the U.S. Department of Energy, the North American Electric Reliability Corporation (NERC), the Information Sharing and Analysis Center for Telecommunications (COMM-ISAC), and many others. The risk assessments do not incorporate the frequency of occurrence due to insufficient data. They do assess risk based on what the expected impact would be if the event did occur. A public version of this internal assessment is available from MacAlester and Murtagh (2014). Fig. 1 shows how FEMA has incorporated its internal risk assessments into an alert and notification strategy.

DHS and FEMA will periodically reassess space weather risk based on new research, engineering, and related efforts such as the implementation of the National Space Weather Action Plan (NSTC, 2015) and the work of the Federal Energy Regulatory Commission (FERC) in partnership with NERC, among others.

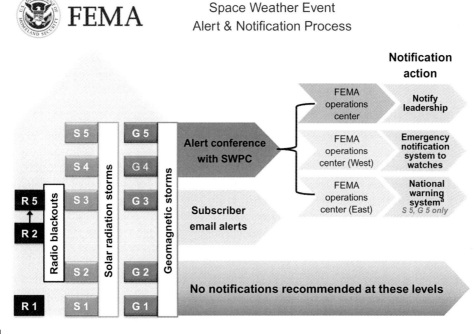

FIG. 1

FEMA Space Weather Event Alert and Notification Process for radio blackout, solar radiation storm, and geomagnetic storm events based on the NOAA Space Weather Scales. [a]This notification action includes the Washington Metropolitan Area Warning System.

Courtesy Federal Emergency Management Agency.

2.1.1 Radio Blackouts

This aspect of the space weather hazard is the best understood, though the name of this hazard is misleading. A radio blackout event *will not* black out all radio communications. Radio blackouts can degrade or prevent over-the-horizon communications in the 5–35 MHz range that rely on reflection from the ionosphere for hours up to 3 days on the daylight side of the Earth for a single event (SWPC, 2017). Multiple events could extend this duration. High frequency (HF) groundwave transmissions should be possible 10–60 miles from the transmitter (MacAlester and Murtagh, 2014). Solar radio emissions at microwave frequencies, if present, may degrade satellite communications for up to an hour (e.g., Cerruti et al., 2008). However, even in an extreme event, line-of-sight radio communications, including cellular, should see little or no impact from this hazard with the following possible exceptions: Solar radio bursts (SRBs) concurrent with this hazard may introduce sufficient noise in sun-facing antennas at sunrise or sunset to degrade or prevent communications for seconds up to 20 min (RAE, 2013), and air-to-ground communications with direct line-of-sight to the sun may experience increased noise.

In emergency management, HF radio (3–30 MHz) is generally a tertiary capability after landline, cellular, and public safety line-of-sight radio communications. In jurisdictions that can afford them, portable satellite units are often used before HF radio. Specific to the continental United States (excluding Alaska, Hawaii, and U.S. territories and possessions overseas), FEMA's internal assessment concluded that even an extreme event (R5) represents a LOW risk to emergency operations, unless a radio blackout coincides with the response phase of a major disaster where HF operations are more likely. Outside the continental United States, the risk would increase only if HF is a primary means of communications (e.g., isolated communities in Alaska).

Other users, such as commercial airlines that rely on HF for operational communications with aircraft over oceans and remote areas, may assess radio blackout risk differently.

2.1.2 Solar Radiation Storm

Energetic solar particles can damage spacecraft electronics, degrade solar panels, and increase the number of anomalies that can impact satellite operations. An extreme event (S5) could result in the temporary or permanent loss of between 10% and 15% of the satellite fleet and significantly increase operational anomalies that would create a challenging environment for ground controllers (Odenwald et al., 2005; RAE, 2013). In discussions with FEMA in 2011–12, satellite industry representatives to the COMM-ISAC disputed the severity of the Odenwald et al. predictions. Atmospheric drag on low-Earth orbit (LEO) satellite constellations used for communications should not present an immediate threat to emergency and disaster operations. HF radio communications in polar regions and satellite navigation may be degraded (SWPC, 2005).

Emergency responders rely on LEO and geosynchronous orbit (GEO) satellite communications as a secondary capability where landline, cellular, and line-of-sight radio are not available. Responders and emergency managers also rely on global navigation satellite systems (GNSS) (e.g., Global Positioning System, or GPS) that operate at medium-Earth orbit (MEO). Larger organizations and government jurisdictions generally have access to more satellite resources, which mitigates some of the risk. FEMA's internal assessment concluded that the risk in the United States to emergency and disaster operations from a minor (S1) through strong (S3) event is LOW with little or no impact to operations anticipated or observed. The risk of a severe (S4) or extreme (S5) event is assessed as MODERATE. The moderate risk assessment derives from the availability of alternate means of terrestrial communications and navigation, the popularity of resilient LEO satellite services (e.g., Iridium, Globalstar, etc.) among emergency responders and government organizations, and the ability of users and satellite service providers to access multiple satellites at GEO to support operations. An extreme event that coincides with the response phase of a major disaster would raise the risk to HIGH.

Users and organizations with heavy reliance on a single satellite, GNSS navigation and timing signals, or polar operations may assess solar radiation storm risk differently.

2.1.3 Geomagnetic Storms

Geomagnetic storms can effect power systems and pipelines as well as disrupt communications passing through or reflecting from the ionosphere due to scintillation (SWPC, 2005). Extreme geomagnetic storms have the potential to cause significant disruption to technological systems.

All 16 critical infrastructure sectors (e.g., communications, financial services, chemical, etc.) rely on electric power. In 2008, the National Research Council brought the space weather hazard to public attention with the release of "Severe Space Weather Events—Understanding Societal and Economic Impacts: A Workshop Report." The report predicted that an extreme geomagnetic storm would cause catastrophic damage to the energy sector—and by extension to the other critical infrastructure sectors—resulting in massive societal impact in the United States that would last for years. Recent work (e.g., NERC, 2012) disputes that conclusion, suggesting that permanent damage to electric power infrastructure on a large scale is unlikely, though widespread power blackouts lasting hours—not months or years—may still occur. However, the scientific and engineering communities have yet to reach consensus on impacts and much research remains.

Similarly, extreme geomagnetic storm impact on the ionosphere is not well understood. NOAA provides general guidance but cannot provide operational-level specifics. For instance, in an extreme event (G5), HF radio communications on the daylight side of the planet may initially be possible due to enhancement of the ionosphere F layer, but would quickly become unusable for several days (MacAlester and Murtagh, 2014). However, this assessment does not account for the dynamic nature of geomagnetic storm intensity and ionospheric response as a function of time over the duration of the storm. The same is true for satellite communications. Lower frequencies will be more affected than higher frequencies. Many emergency responders and all GNSS users rely on the L-band at the lower end of the satellite spectrum. This band is highly susceptible to scintillation and the disruption could be so severe as to block L-band signals for 7–12 h at a given geographic location. Once again, though, the dynamic nature of the storm event and ionospheric response is largely unknown. When an operator asks how bad it will be at their location, when will it be bad, and for how long, the science is short on specifics.

FEMA's internal assessment concluded that the risk in the United States to emergency and disaster operations of a minor (G1) though strong (G3) event is LOW with little or no impact to operations anticipated or observed. A severe (G4) event is assessed as a MODERATE risk because of the potential impact to communications and possible power disruptions, while an extreme (G5) event is assessed as a HIGH-risk. The high-risk assessment is partly due to the fact that the actual level of impact from an extreme event is still poorly understood. FEMA must assume a worst-case scenario until such time as the research and engineering matures sufficiently to provide high-confidence-level, high-fidelity impact forecasts. A worst-case scenario is currently defined as multiple states and regions without power for periods exceeding 2 days, with the potential for some areas to remain without power for weeks or even months. The uncertainty surrounding the impact of an extreme event presents a significant planning and operational challenge for emergency managers.

3 WHAT EMERGENCY MANAGERS NEED FROM RESEARCHERS AND ENGINEERS

The vast majority of emergency managers, elected officials, and members of the public do not speak the language nor understand the intricacies of geophysics, heliophysics, and space meteorology. They desire (1) forecast and impact products written in plain language, (2) expert analysis available on demand, (3) the ability to share information with senior government and private-sector decision makers and the public, as appropriate, and (4) greatly improved forecast and impact products.

3.1 PLAIN LANGUAGE

Emergency managers must not only understand the advisories, alerts, warnings, briefings, and other products generated by researchers and engineers, they must explain them to senior leaders, elected officials, and the public. As a general rule of thumb, products intended for a lay audience should be written at a fourth-grade level. They should avoid scientific and engineering jargon (e.g., horizontal geoelectric field, etc.) and units of measure if they are not widely known (e.g., MeV, nT, etc.). Instead, explain what the information means. The following example is from the Saffir-Simpson Hurricane Wind Scale for a category 1 hurricane: "**Very dangerous winds will produce some damage**: Well-constructed frame homes could have damage to roof, shingles, vinyl siding, and gutters. Large branches of trees will snap and shallowly rooted trees may be toppled. Extensive damage to power lines and poles likely will result in power outages that could last a few to several days" (NHC, 2012). This example is written in plain language that clearly explains what is expected in a category 1 storm in a manner that is understandable to the public. In an actual event, this description would be followed by recommended protective actions that people, businesses, and communities should take.

The final aspect of plain language is to avoid acronyms. Researchers, engineers, and government officials love them, but their meaning is often lost or confused when providing information to a broader audience.

3.2 EXPERT ANALYSIS AVAILABLE ON DEMAND

While forecast products are essential, they don't tell the whole story. Factors that are not part of forecast products can alter the preparation for a potential event. For instance, a tropical cyclone may appear to be a long- or medium-range threat to a coastal community, but factors such as air moisture, steering currents, sea surface temperatures, and storm measurements from hurricane-hunter aircraft can influence the forecast. The National Hurricane Center supports emergency management organizations like FEMA through audio and video teleconferences where an expert(s) on tropical cyclone science will discuss the forecast products and other factors and provide experienced guidance. This format is also valuable because it allows emergency managers and elected officials to ask questions and clarify aspects of the forecast. Representatives from the NHC also support public messaging through interviews and informational products provided to the news media, social networks, and other outlets.

Decision makers need to know the:

- Probable location(s) of impact
- Estimated time of impact at those locations

- Severity of impact at those locations
- Duration of impact at those locations
- Confidence level of the model(s)
- Other factors that may affect the forecast

It cannot be stressed enough how important expert analysis is to decision-making. As an example, prior to a major hurricane landfall in the United States, the director of the NHC will brief the storm forecast and analysis via a video teleconference hosted by FEMA that includes all federal departments and agencies with disaster support roles, the White House, all states that may be impacted by the storm, the American Red Cross (as the designated representative of the National Voluntary Organizations Active In Disaster), and critical stakeholders. The director's briefing and analysis directly affects decisions made about whether to evacuate coastal communities, emergency protective measures, and the movement and staging locations of disaster commodities (e.g., food, water, blankets, etc.) and emergency equipment. These are literally life-and-death decisions. In an extreme space weather event, an official of the Space Weather Prediction Center would fill the same role.

Finally, NOAA studies have shown that the public will attempt to verify a forecast through multiple sources (NOAA, 2016). If they find a consistent message across those multiple sources, they are more likely to act to prepare themselves and take protective measures. It is important that all messaging intended for public consumption be coordinated through a common clearing point, which for a space weather event in the United States is the SWPC. If a federal response is required, a Joint Information Center will be established to coordinate messaging to the press and the public.

3.3 SHARABLE INFORMATION

While the results of scientific observations, measurements, and modeling are generally available to the public, private-sector information is frequently considered proprietary and confidential. Private-sector entities, when not compelled by regulation, are often reluctant to share information that may expose operational vulnerabilities to competitors and customers. In some instances, the private-sector will share proprietary information with the government but only on the condition that the information not be shared with the public or even other governmental jurisdictions. Where regulations do compel the release of information, such information may not be timely or may be of limited value for operational decision-making.

Informed decision-making requires access to relevant and timely hazard-specific information. Information withheld is opportunity lost to affect a better outcome. Lives, property, and the environment are at risk. Emergency managers rarely need to know the proprietary vulnerabilities of a private-sector network. Instead, they need the expert assessment from the service provider about physical locations and services that are or could be impacted, the actual or potential effects of those impacts, the anticipated duration of those effects, and what assistance, if any, private-sector entities need to restore services. This information, at a minimum, needs to be shareable with impacted critical infrastructure owners, government decision makers, and key stakeholders. Service provider information that can be shared with the public is also desired. Before a space weather event, critical infrastructure owners and operators should work with relevant government agencies, regulators, customers, and key stakeholders to determine what information they are able to share and the limitations on the use and further sharing of such information.

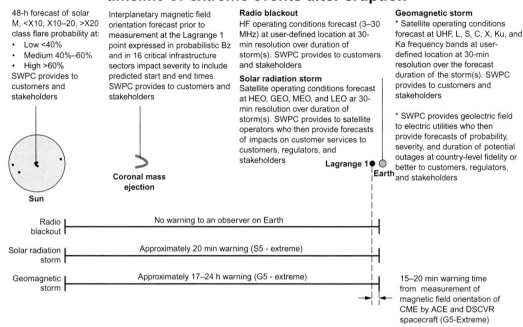

FIG. 2

"Sun to Mud" desired forecast products (United States) and timeline of extreme events after eruption.

Courtesy Federal Emergency Management Agency.

3.4 IMPROVED FORECAST PRODUCTS

NOAA often describes an extreme (G5) geomagnetic storm as a "Solar Katrina," a reference to Hurricane Katrina in 2005. At present, space weather forecasting and products lag those available for tropical cyclones. As such, it is instructive to look at the available NOAA forecast products for tropical cyclones, why they are so important for emergency and disaster planning, and discuss what emergency managers would like to see from "Sun to mud" for space weather. The requested forecast products below are aspirational and it is understood that some are years or even decades away (Fig. 2).

3.4.1 Solar Flare Forecast Product

Ideally, operators need a forecast of solar flares and their potential magnitudes before they occur. NOAA has two comparable forecast products for tropical cyclones: the Two-Day and Five-Day Graphical Tropical Weather Outlooks, available from the NHC (Fig. 3). The graphic uses shaded areas and yellow, orange, and red colors to provide a quick visual reference as to the area and likelihood of tropical cyclone formation within the forecast time period. Mousing over a shaded area displays a text-based pop-up that provides plain language details on each disturbance. For emergency managers

FIG. 3

The Five-Day Graphical Tropical Weather Outlook shows areas of disturbed weather that may become tropical cyclones. It uses shaded areas and a simple color scheme to highlight the likelihood and potential area of formation. Currently active cyclones are not shown. When a user mouses over the shaded area, a text box pops up providing detailed plain language analysis. This graphic shows the area *(red)* that eventually produced Hurricane Matthew in 2016.

Courtesy National Oceanic and Atmospheric Administration, National Weather Service, National Hurricane Center. http://www.nhc.noaa.gov/archive/xgtwo/gtwo_archive.php?current_issuance=201609281458&basin=atl&fdays=5.

and the public, the Two-Day and Five-Day Graphical Tropical Weather Outlook products are often the first indication that a threat may be looming. This raises awareness and begins the process of monitoring these and other forecast products for further developments. It can also motivate early preparedness and protective actions.

In emergency management, time matters. Given 48 h advanced notice of a potentially dangerous flare (and the potential for an associated solar radiation storm and coronal mass ejection (CME)), awareness could be raised, plans implemented, lines of communications opened and tested, and messaging to educate the public on the hazard and advise preparedness and protective actions could begin. This would require new research, models, and updated forecast products.

First, the requested forecast product(s) should be localized to specific sunspot groups. Flares from a sunspot group near the center of the solar disk, roughly defined as $\pm 20°$ solar latitude and between 20° east and 40° west solar longitude, are more likely to have CMEs that are Earth-directed. Flares occurring outside this "launch window" are less likely to be a threat, even if they are strong. Second, solar flares are classified on a logarithmic scale based on their intensity such that an X-class flare is 10 times as energetic as a comparable M-class flare (SWPC, 2017). However, the X scale is open ended. An X21-class flare is 10 times as powerful as an X11 flare but 100 times as powerful as an X1 flare. The Carrington Event of 1859 is estimated to have been X45$\pm$5 (Cliver and Dietrich, 2013). Generally, M-class flares are not of interest to the emergency management user community. $X < 10$ class flares are of marginal interest for HF operations should they occur during an emergency or disaster response. X10–X20 class flares are of interest to HF operations and other communications for emergency or disaster response. The primary interest of the emergency management user community is $X > 20$ class flares for their potential for wider impacts to communications and possible impacts to electrical power from an associated geomagnetic storm. While flare class cannot predict geomagnetic storm impact, a forecast product must use a recognized benchmark scale. If researchers determine a better benchmark scale, forecast products could be revised.

The desired solar eruptive forecast product should:

- Be graphical and easily accessible via the Internet, and available as subscription email, text, app, and social media product, hosted by the official space weather forecast office of the host nation;
- Be updated at least every 6 h;
- Give at least 12 h advanced notice of a potential flare with 48 h advanced notice desired (deterministic preferred, probabilistic acceptable as shown below);
- Provide separate forecasts for M-, $X < 10$-, X10–X20-, and $X > 20$-class flares;
- Give probabilities of each class of event and rate as low $<40\%$, medium $40\%-60\%$, and high $>60\%$;
- Use a simple color scheme localized to individual sunspot groups to indicate likelihood and potential magnitude of a solar flare (e.g., $X > 20$ with High probability = Red);
- Indicate the predicted movement of the sunspot group across the solar disk within the forecast period;
- Provide a text pop-up with more detailed analysis when a user selects a highlighted sunspot group;
- Have a comparable product(s) accessible for individuals with disabilities.

3.4.2 Interplanetary Magnetic Field Forecast Products

CMES and the geomagnetic storms they cause are of greatest interest to the emergency management community. The NOAA tropical cyclone product that is analogous to this is also the most familiar: the 3-day and 5-day Track Forecast Cone (Fig. 4). The Track Forecast Cone is rarely exact but it is good enough

3 WHAT EMERGENCY MANAGERS NEED

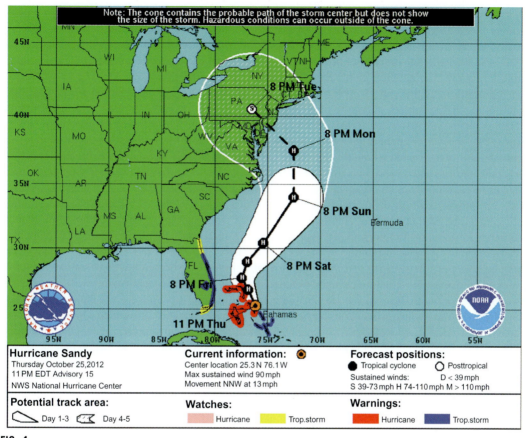

FIG. 4

The NHC 5-day Track Forecast Cone is formed by a set of circles along the forecast track at 12, 24, 36, 48, 72, 96, and 120 h and shows the cone within which the tropical cyclone track is expected to remain roughly 60%–70% of the time based on forecasts over the previous 5 years (NHC, 2016a). The product also provides a rough estimate of strength where S=tropical storm force winds (39–73 mph), H=category 1–2 hurricane (74–110 mph), M=category 3–5 major hurricane (>110 mph), and a *blank circle*=a posttropical system. This graphic shows the 5-day track forecast for Hurricane Sandy as of October 25, 2012.

Courtesy National Oceanic and Atmospheric Administration, National Weather Service, National Hurricane Center. http://www.nhc.noaa.gov/archive/2012/graphics/al18/loop_5W.shtml.

for emergency managers, elected officials, and the public to know that a threat may be approaching their area, when it is expected to arrive, and a rough understanding of its magnitude. This product informs precoordination between government jurisdictions and organizations, preparedness and protective actions, evacuation planning and public messaging. It may include the initial movement of commodities, equipment, and advance personnel to staging areas, which requires approximately 12–24 h to execute.

In space weather, the WSA-Enlil prediction model currently in operation at the SWPC can provide the approximate arrival time at Earth of a CME, similar to the Track Forecast Cone. However, it cannot provide the most important piece of information—the orientation of the magnetic field within the CME, which determines impact at Earth. A northward magnetic orientation may result in little impact while a southward orientation could drive substantial impact (geoeffective). An extreme CME may arrive at Earth in as little as 15–24 h (see Gopalswamy et al., 2005; Koskinen and Huttunen, 2006; Baker et al., 2013), yet the magnetic orientation will not be known until measured at the Lagrange 1 (L1) point by the Advanced Composition Explorer (ACE) and Deep Space Climate Observatory (DSCOVR) spacecraft approximately 15–20 min before arrival at Earth. From an operational standpoint, the ability to respond within this timeframe will depend heavily on whether an organization took early preparedness and protective actions when it received initial warnings of the CME from their space weather forecast office. If an organization took a wait-and-see approach, 15–20 min is almost no time at all. Customers receiving the warning products will need to analyze the information before beginning notification procedures to their employees, customers, and partners. Senior leaders will require additional information before making decisions, which then must be implemented. These human processes could take tens of minutes or longer depending on the time of day and the day of the week. Complicating this, the strength and magnetic orientation of the CME is not a constant. As the CME passes and the magnetic structure changes, potential impacts to technological systems will also change. As extreme geoeffective storms are rare in a human lifetime, most people are not familiar with their effects or dynamics, which will create a very challenging environment for informing senior and elected leaders and the public. As with other severe weather hazards, advanced warning time—particularly if a CME will be geoeffective—will allow emergency managers to better advise on preparedness and protective actions.

Emergency managers desire forecast products that can reliably predict CME magnetic field orientation prior to measurement at L1. Schrijver et al. (2015) set as a goal in their Pathway I recommendations to forecast the magnetic structure of a CME 12+h prior to arrival at Earth. Such warning time could significantly improve the ability to warn and advise organizations, critical infrastructure owners, and the public on preparedness and protective actions, and lead to a better outcome. Two forecast products are desired. The first is a technical product that would provide predicted values needed as inputs to Earth and near-Earth models (e.g., $-Bz$). The second is a text product with the following characteristics:

- Plain language;
- No jargon or scientific units of measure;
- Estimated arrival time and duration;
- Descriptions of possible impacts to critical infrastructures;
- Available from the official space weather forecast office website and via subscription email, text, app, and social media;
- Comparable product(s) accessible for individuals with disabilities.

3.4.3 Earth Environment Forecast and Impact Products

The Earth environment requires multiple products that speak directly to likely impacts from radio blackouts, solar radiation storms, and geomagnetic storms, tailored to specific user communities. Tropical cyclones also have three main damage mechanisms—winds, inland flooding, and storm surge—and emergency managers rely on multiple forecast products to predict likely impacts (Fig. 5).

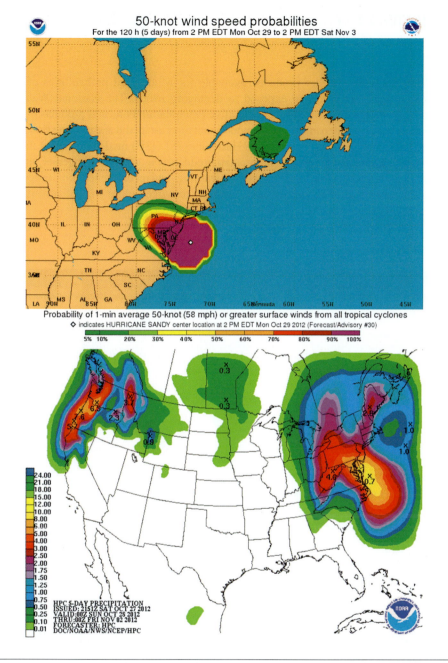

FIG. 5

(Top) 50-knot Wind Speed Probability product showing probability of 1-min average 50-knot (58 mph) or greater surface winds from all tropical cyclones. This product shows predicted 50-knot wind speed values prior to landfall of Hurricane Sandy in 2012 (Courtesy National Oceanic and Atmospheric Administration, National Weather Service, National Hurricane Center. http://www.nhc.noaa.gov/archive/2012/graphics/al18/loop_PROB50.shtml). (Bottom) The Weather Prediction Center offers Quantitative Precipitation Forecast products from 1 to 7 days showing predicted precipitation (e.g., rainfall) totals. This graphic shows the predicted rainfall totals prior to landfall of Hurricane Sandy in 2012 (Courtesy National Oceanic and Atmospheric Administration, National Weather Service, Weather Prediction Center. http://www.wpc.ncep.noaa.gov/archives/qpf/display_maps.php?prodtype=issued&proddate=10/28/2012&prodtime=00&allsent=no&imagetype=color&actualprods=d15).

The NHC Wind Speed Probability products, available for wind speed probabilities of 34, 50, and 64 knots, predict areas that could be exposed to damaging winds. The National Weather Service (NWS) National Prediction Center (NPC) Quantitative Precipitation Forecasts products predict rainfall totals for 1–7 days and are used to inform other models that forecast inland and riverine flooding. Finally, the NWS Sea, Lake, and Overland Surges from Hurricanes (SLOSH) model, hosted by the NHC, estimates storm surge heights and feeds inundation map products (Fig. 6). Taken together, these products allow emergency managers to issue location-specific warnings and recommendations, evacuation orders (if appropriate), and move commodities and response resources close to, but not in, anticipated impact zones.

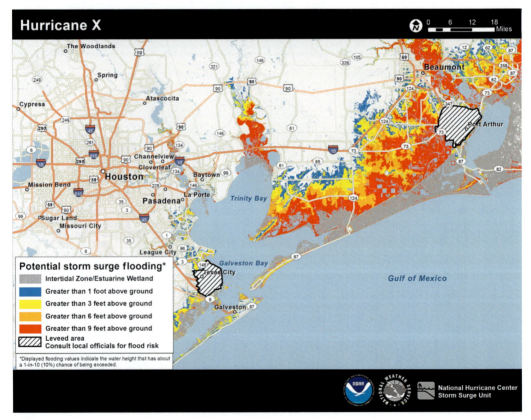

FIG. 6

This is a static example of a Potential Storm Surge Inundation Map, based on the NWS Sea, Lake, and Overland Surges from Hurricanes (SLOSH) model and hosted by the NHC (NHC, 2016b). It shows graphical areas where inundation could occur and the predicted heights above ground that water could reach.

Courtesy National Oceanic and Atmospheric Administration, National Weather Service, National Hurricane Center.

http://www.nhc.noaa.gov/surge/inundation/.

Radio blackouts and geomagnetic storms impact the ionosphere, disrupting communications. While multiple forecast models would be required to determine individual impacts, these would preferably feed a single end-user forecast product with the following characteristics:

- User defined location: web page and app that allows a user to either input their geographic location and receive a location-specific forecast or zoom into a specific location from a global map, or both, hosted by the official space weather forecast office;
- Provide forecast operating conditions for:
  - HF radio (3–30 MHz) and very high frequency (VHF) skywave up to 50 MHz
  - Satellite ultra-high frequency (UHF) (240–318 MHz), L-band (1–2 GHz), S-band (2–4 GHz), C-band (4–8 GHz), X-band (8–12 GHz), Ku-band (12–18 GHz), and Ka-band (26–40 GHz) communications
  - GNSS (reported separately because of its importance);
- 30-min forecast resolution for entire forecast duration of the storm(s);
- At least 2 h advanced forecast before arrival of a geomagnetic storm(s);
- Comparable product(s) accessible for individuals with disabilities.

Forecasts products for solar radiation storms should come from both government and private-sector sources. Government would produce at least two products, hosted by the official space weather forecast office:

- A technical forecast of the energetic particle environment at all orbits—defined, at a minimum, as high-Earth orbit (HEO), GEO, MEO, and LEO—with information needed by satellite and spacecraft operators to undertake preparedness and protective actions;
- A general forecast for nontechnical users and the public, provided in plain language with minimal scientific units of measure or jargon, that would describe potential operational impacts on satellite services at all orbits, including GNSS networks;
- Both products would provide 30-min forecast resolution for the entire forecast duration of the storm(s);
- Both products available via website, subscription email, text, app, and social media;
- Comparable public-facing product(s) for general forecast accessible for individuals with disabilities.

Satellite operators would also produce a forecast/impact product, after operational analysis, provided to their customers, regulators, and key stakeholders describing possible service impacts over the forecast duration of the storm. Preferably, these products would be available via website, subscription email, text, app, and social media. However, recognizing that some operators may prefer to limit distribution to actual customers, regulators, and key stakeholders, they could require user authentication prior to receiving the forecast product.

The final desired forecast products for geomagnetic storms are the most difficult—and the most important. Ideally, predictive solar flare models would feed CME interplanetary magnetic field models, which would feed Earth environment models, which would feed engineering models at electric power utilities, which would forecast potential short-, medium-, and long-term power outages with geographic resolution of at least county or major metropolitan level. Power outages that last for a few hours or even a few days are common enough each year due to terrestrial weather events and other faults that communities rarely need outside government assistance. Even if a geomagnetic storm knocked out power to a large section of a country, if that power were restored within hours, impacts would be minimal. Emergency managers, therefore, would prefer to know where power outages could extend to days or even weeks so they can move finite response assets to impacted areas to conduct lifesaving and life-sustaining operations, evacuations (if required), and restoration of critical services. Response operations on this scale require at least 24–72 h to implement. Tropical cyclone forecast products allow organizations to begin the movement of resources before landfall, significantly reducing response time to impacted communities. Emergency managers desire the same ability for geomagnetic storms.

At present, neither the science nor the engineering communities can provide specific locations of potential impacts at a geographic resolution sufficient to inform the movement of resources. Government emergency managers must wait until after impact to gauge where such impact has occurred and its severity. Response could be delayed by at least 24 h after the storm has passed as assessments of electric utilities are made and communicated. Meanwhile, critical infrastructure (e.g., communications, etc.) without power or whose backup power becomes exhausted will fail. As responders also depend on critical infrastructure (e.g., transportation, fuel, etc.), these failures will make responses more difficult. If this occurs during periods of excessive heat or cold, the situation could be much worse.

As with solar radiation storms, geomagnetically induced currents (GIC) forecast products from both government and private-sector sources are desired. Government would produce at least two products hosted by the official space weather forecast office:

- A technical forecast of the horizontal geoelectric field (see Pulkkinen et al., 2017) at county-level resolution or better, provided to individual electric utilities, regional power organizations, national power organizations, and appropriate government regulators and agencies. In the United States, these would be the North American Electric Reliability Corporation (NERC), the NERC regions and interconnections, the Federal Energy Regulatory Commission (FERC), and the Department of Energy (DOE). Similar products would be provided to telecommunications companies operating transoceanic cables as well as pipeline and railway operators. Operators should be able to input forecast values into engineering models that would forecast impacts to physical infrastructure (e.g., extra-high-voltage transformers, relays, generators, transmission lines, etc.);
- A general forecast for nontechnical users and the public, provided in plain language with minimal scientific units of measure or jargon, which would describe potential impacts for specific geographic areas;
- Comparable public-facing product(s) for general forecast accessible for individuals with disabilities.

Following engineering analysis, private-sector operators would provide to regulators and key stakeholders possible service impacts in their areas of responsibility with county-level resolution or better. Customers should also be informed but messaging to the public should be consistent at all levels of reporting as customers will try to confirm the information from other sources. In the United States, the DOE would consolidate the engineering impact forecasts from individual electric utilities and NERC into a national-level common operating picture with county-level fidelity, at a minimum, to include both probabilities of power outages and anticipated durations before restoration. This would permit emergency managers to advise senior leaders, elected officials, and the public on protection and preparedness actions, and position resources to speed response to areas that may be hardest hit.

4 CONCLUSION

Reliable physical impact prediction is the holy grail of space weather forecasting. At present, emergency managers cannot know what physical impacts will occur, where specifically they will occur, how bad they will be, or how long they will last. The public will know this event is coming. If history is any guide, the news media will speculate where governments and senior leaders cannot provide answers. The current national strategy within the United States is to raise awareness and advise general preparedness before an event (see https://www.ready.gov/severe-weather), provide consistent and authoritative information should an extreme event occur, advise and take protective and preparedness actions to the extent possible, and wait until the event passes to assess what impact occurred, where it occurred, how severe it is, and how long it may take to recover—a process that could take days.

The desired forecast products presented in this chapter may not be easy to achieve. FEMA's former administrator, Craig Fugate, said it best in his farewell speech to the Agency (FEMA, 2017): "... be unreasonable in your expectations, because reasonable people always fail in disaster response. Perfection is your enemy. Getting 100% answers to everything that can go wrong will keep you from ever making a decision ... In disasters, speed to action is what changes the outcome."

Current tropical cyclones and other severe weather forecast products are not perfect, but they are good enough for emergency managers, senior leaders, elected officials, and the public to act. Emergency managers need space weather forecast products that are good enough, as soon as possible, with enough information to act upon. As with other natural hazards, emergency managers will rely heavily on researchers and engineers for these products. This relationship needs to exist before an extreme event occurs: research to operations, operations to research.

We look forward to working with you.

REFERENCES

Baker, D.N., Li, X., Pulkkinen, A., Ngwira, C.M., Mays, M.L., Galvin, A.B., Simunac, K.D.C., 2013. A major solar eruptive event in July 2012: defining extreme space weather scenarios. Space Weather 11, 585–591. https://doi.org/10.1002/swe.20097.

Cerruti, A.P., Kintner Jr., P.M., Gary, D.E., Mannucci, A.J., Meyer, R.F., Doherty, P., Coster, A.J., 2008. Effect of intense December 2006 solar radio bursts on GPS receivers. Space Weather 6. S10D07. https://doi.org/10.1029/2007SW000375.

Cliver, E.W., Dietrich, W.F., 2013. The 1859 space weather event revisited: limits of extreme activity. Space Weather Space Clim. 3. https://doi.org/10.1051/swsc/2013053.

Federal Emergency Management Agency, 2017. Administrator Fugate farewell speech to FEMA employees. Personal communication, January 17, 2017.

Gopalswamy, N., Yashiro, S., Liu, Y., Michalek, G., Vourlidas, A., Kaiser, M.L., Howard, R.A., 2005. Coronal mass ejections and other extreme characteristics of the 2003 October–November solar eruptions. J. Geophys. Res. 110. A09S15. https://doi.org/10.1029/2004JA010958.

Koskinen, H.E., Huttunen, K.E., 2006. Geoeffectivity of coronal mass ejections. Space Sci. Rev. 124, 169–181. https://doi.org/10.1007/s11214-006-9103-0.

MacAlester, M.H., Murtagh, W., 2014. Extreme space weather impact: an emergency management perspective. Space Weather 12. https://doi.org/10.1002/2014SW001095.

National Hurricane Center, 2012. Saffir-Simpson Hurricane Wind Scale. National Hurricane Center (NHC), Miami, FL. Available from: http://www.nhc.noaa.gov/aboutsshws.php.

National Hurricane Center, 2016a. Definition of the NHC Track Forecast Cone. National Hurricane Center (NHC), Miami, FL. Available from: http://www.nhc.noaa.gov/aboutcone.shtml.

National Hurricane Center, 2016b. Potential Storm Surge Flooding Map. National Hurricane Center (NHC), Miami, FL Available from: http://www.nhc.noaa.gov/surge/inundation/.

National Oceanic and Atmospheric Administration, 2016. Risk Communication and Behavior: Best Practices and Research Findings. National Oceanic and Atmospheric Administration (NOAA) Social Science Committee, Silver Spring, MD. Available from: ftp://ftp.oar.noaa.gov/SAB/sab/Meetings/2016/Risk%20Communication%20and%20Behavior-%20Best%20Practices%20and%20Research%20Findings%20NOAASSC%20April%202016%20(1).pdf.

National Science and Technology Council, 2015. National Space Weather Action Plan. Office of Science and Technology Policy, Space Weather Operations, Research, and Mitigation (SWORM) Task Force. Available from: https://www.whitehouse.gov/sites/default/files/microsites/ostp/final_nationalspaceweatheractionplan_20151028.pdf.

North American Electric Reliability Corporation (NERC), 2012. 2012 Special Reliability Assessment Interim Report: Effects of a Geomagnetic Disturbance on the Bulk Power System. North American Electric Reliability Corporation (NERC), Washington, DC. Available from: http://www.nerc.com/pa/RAPA/ra/Reliability%20Assessments%20DL/2012GMD.pdf.

Odenwald, S., Green, J., Taylor, W., 2005. Forecasting the impact of an 1859-Calibre Superstorm on Satellite Resources. Adv. Space Res. 38, 280–297. https://doi.org/10.1016/j.asr.2005.10.046.

Pulkkinen, A., Bernabeu, E., Thomson, A., Viljanen, A., Pirjola, R., Boteler, D., 2017. Geomagnetically induced currents: science, engineering and applications readiness. Space Weather. https://doi.org/10.1002/2016SW001501.

Royal Academy of Engineering (RAE), 2013. Extreme Space Weather: Impacts on Engineered Systems and Infrastructure. Royal Academy of Engineering (RAE), London. Available from: http://www.raeng.org.uk/news/publications/list/reports/space_weather_full_report_final.pdf.

Schrijver, C.J., Kauristie, K., Aylward, A.D., Denardini, C.M., Gibson, S.E., Glover, A., et al., 2015. Understanding space weather to shield society: a global road map for 2015–2025 commissioned by COSPAR and ILWS. Adv. Space Res. 55, 2745–2807. https://doi.org/10.1016/j.asr.2015.03.023.

Space Weather Prediction Center, 2017. Solar Flares (Radio Blackouts). Space Weather Prediction Center (SWPC), Boulder, CO Available from: http://www.swpc.noaa.gov/phenomena/solar-flares-radio-blackouts.

Space Weather Prediction Center (SWPC), 2005. NOAA Space Weather Scales. Space Weather Prediction Center (SWPC), Boulder, CO. Available from: http://www.swpc.noaa.gov/NOAAscales/index.html.

FURTHER READING

National Hurricane Center, 2012. SANDY Graphics Archive. National Hurricane Center (NHC), Miami, FL. Available from: http://www.nhc.noaa.gov/archive/2012/graphics/al18/loop_5W.shtml.

National Hurricane Center, 2014. Five-Day Graphical Tropical Weather Outlook. National Hurricane Center (NHC), Miami, FL. Available from: http://www.nhc.noaa.gov/archive/xgtwo/gtwo_archive.php?current_issuance=201609281458&basin=atl&fdays=5.

National Research Council, 2008. Severe Space Weather Events—Understanding Societal and Economic Impacts: A Workshop Report. National Research Council (NRC), Space Studies Board, National Academies Press, Washington, DC. Available from: http://www.nap.edu/catalog.php?record_id=1250725toc.

Weather Prediction Center, 2012. WPC QPF Archive for Products Issued 00Z October 28, 2012. Weather Prediction Center (WPC), College Park, MD. Available from: http://www.wpc.ncep.noaa.gov/archives/qpf/display_maps.php?prodtype=issued&proddate=10/28/2012&prodtime=00&allsent=no&imagetype=color&actualprods=d15.

CHAPTER 29

THE SOCIAL AND ECONOMIC IMPACTS OF MODERATE AND SEVERE SPACE WEATHER

Stacey Worman*, Susan Taylor*, Terrance Onsager[†], Jeffery Adkins[‡],
Daniel N. Baker[§], Kevin F. Forbes[¶]

Abt Associates Inc., Bethesda, MD, United States *NOAA Space Weather Prediction Center, Boulder, CO, United States[†]
Integrated Systems Solutions, Dunn Loring, VA, United States[‡] University of Colorado, Boulder, CO, United States[§]
The Catholic University of America, Washington, DC, United States[¶]*

CHAPTER OUTLINE

1 Introduction 701
2 Approach 703
3 Results 705
   3.1 Electric Power 705
   3.2 Aviation 707
   3.3 Satellites 708
   3.4 GNSS Users 708
4 Next Steps and Concluding Remarks 709
References 709
Further Reading 710

1 INTRODUCTION

Severe space weather is a low-probability, high-consequence natural hazard that can interrupt and damage technologies critical to modern society such as electric grids, airplanes, pipelines, trains, satellites, and Global Navigation Satellite Systems (GNSS) (National Research Council, 2008; Baker and Lanzerotti, 2016). More moderate space weather that is less intense but occurs more frequently is also known to cause adverse impacts (Allen et al., 1989; Lopez et al., 2004). Despite such risks and our growing dependencies on these technologies for almost all aspects of daily life, there have been a limited number of studies on space weather's social and economic impacts (Eastwood et al., 2017).

The few impact studies that do exist estimate potential costs of historical moderate storms to be in the millions of dollars and of theoretical severe events to be in the billions to trillions of dollars. The

1989 Québec storm, for example, where power equipment was damaged and more than six million people lost electricity for 9 h, was estimated to cost $13.2 million (Bolduc, 2002). A recent study of the potential impacts of a severe storm on the power grid [defined by a geomagnetic storm size of Dst ≤ -1182 nT, reflecting an event that would be significantly larger than the 1859 Carrington event, Dst ≤ -880 nT, and that could have occurred if the July 2012 coronal mass ejection (CME) was Earth-directed (Baker et al., 2013)] estimates direct costs of $220 billion–$1.2 trillion with a total impact, including downstream supply-chain interruptions, ranging from $500 billion to $2.7 trillion (Oughton et al., 2016). Another study focused on the satellite industry estimates the costs of a worst-case storm (defined as a solar proton event (SPE) with a fluence of $\sim 5.6 \times 10^{10}$ particles/cm^2, reflecting an event ~ 3 times larger than the 1859 Carrington event) at $84 billion, without accounting for indirect impacts from service interruptions (Odenwald et al., 2006). For context, the costs of other major natural disasters are typically measured in tens of billions of dollars (Smith and Matthews, 2015) with the most devastating disasters to date, such as Hurricane Katrina in 2005 and the Japan Tsunami in 2011, estimated at $\sim$\$125 billion (Knabb et al., 2005) and $235 billion (World Bank, 2011), respectively.

Better understanding of the social and economic impacts of space weather therefore represents an important but formidable problem at the intersection of science, engineering, economics, and the social sciences (Baker, 2009). Modern society's fortunate lack of direct experience of a severe event poses theoretical and analytical challenges that require many assumptions, simplifications, extrapolations, and inferences. For example, impacts are not simply directly proportional to storm magnitude but are affected by a number of other factors such as geography, the time of day, or the season of the year that affect demand on the electric power grid (e.g., Forbes and Cyr, 2012). In addition, the definition of a "severe" event is debated and scientifically uncertain, and when studying impacts, is furthermore dependent and relative to the specific technology and impact pathway of interest (Cannon et al., 2013; Hapgood et al., 2016). The response of various technologies (e.g., power grids, satellites, airplanes, and GNSS) to severe events is furthermore largely unknown because they fall beyond engineering standards and operational experiences. Further complications arise from economic and social processes that are complex, interconnected, and rapidly evolving. Developing tools and strategies for overcoming these inherent challenges is essential for better understanding the potential socioeconomic impacts so that we can enhance preparedness and strategically reduce risks.

This chapter provides a brief overview of an ongoing effort to advance research on space weather impacts by identifying, describing, and quantifying the social and economic costs of space weather to the United States. A handful of other socioeconomic impact studies are concurrently underway or recently finished in Europe, but they differ somewhat in scope and are not yet published (Biffis and Burnett, 2017; Oughton, 2017; Luntama et al., 2017). Our study was in part initiated in response to the 2015 *Space Weather Action Plan* (SWAP), a much larger U.S. effort for addressing potential vulnerabilities and increasing resilience. One of our goals is to capture and synthesize what is known about the socioeconomic impacts of space weather across four technological sectors of prime concern: Electric power, commercial aviation, satellites, and GNSS users. Another is to estimate the potential costs of different-sized space weather events on a sector- and threat-specific basis. Our cost estimate of a severe event will be based on different working definitions, such as the 1-in-100-year storm parameters set by the U.S. SWAP Phase I benchmarks (Viereck, 2017) and the reasonable worst-case event according to Hapgood et al., 2016. We will also develop parameters and estimate costs for more moderate scenarios to assess the impacts of lower intensity, more frequently occurring events. Examining severe events allows us to explore upper boundaries on potential costs. Considering more moderate events will help us establish key thresholds above which significant impacts are expected based on empirical insight from sector and stakeholder experiences, including past events where notable impacts occurred and impact thresholds were therefore exceeded [e.g., the March 1989 event (Allen et al., 1989) and the 2003 Halloween Storm (Lopez et al., 2004)]. Because impact thresholds may increase or decrease as our technological infrastructure changes, analyzing the impacts of moderate space weather is essential for anticipating future trends. This chapter provides a summary of our research to date to construct impact mechanism diagrams that distil essential relationships from the literature, to identify the physical effects and socioeconomic impacts of space weather on each technological sector along with validation and findings from stakeholder discussions, and to develop methods and models to quantify these impacts.

2 APPROACH

This study began with an extensive literature review; however, similar to Eastwood et al., 2017, we also found a limited number of previous studies on the social and economic impacts of space weather. We therefore focused efforts on dissecting the far more extensive technical literature on the scientific and engineering effects (e.g., Baker and Lanzerotti, 2016) to map space weather events and physical effects by sector to the downstream social and economic impacts. Social and economic impacts, for example, are widely documented in historic accounts of notable space weather events (Allen et al., 1989; Lopez et al., 2004) or else are highlighted in technical work as a research motivation. To synthesize our findings for each sector, we constructed an impact mechanism diagram outlining the major causal pathways from solar events to physical effects to societal impacts (Fig. 1). These diagrams help simplify many of the complexities surrounding this interdisciplinary problem, allowing for a clear illustration of key processes and relationships.

After culling a wide range of socioeconomic impacts from the literature, we organized our findings into five broad but interrelated impact categories: defensive investments, mitigating actions, asset damages, service interruptions, and health effects (Fig. 2). Our first category, defensive investments, captures expenditures that help protect technologies against potential vulnerabilities such as on engineering designs or on situational awareness. Mitigating actions covers real-time decisions made by system operators to reduce the consequences of an event that is anticipated or underway. Asset damages refers to any physical damage to sector equipment that may result either suddenly ("acute") or slowly over time ("chronic"). Service interruptions addresses impacts seen by end users of technologies such as changes in provision, quality, and/or pricing. The final impact category is health effects, and it covers any direct potential hazard to human well-being or life such as from elevated radiation. Note that this category does not include any of the health effects that would result from other impact areas such as extended power outages. These health effects are accounted for in the downstream consequences of the impact category responsible for the primary disruption (e.g., service interruptions). Fig. 2 also shows the different types of space weather products and services most relevant to each impact category such as those provided by NOAA's Space Weather Prediction Center (SWPC).

We then used a Delphi-like approach (Helmer, 1967) to solicit feedback on our literature review findings from more than 30 stakeholders of diverse expertise across each sector from engineering to

704   CHAPTER 29 THE SOCIAL AND ECONOMIC IMPACTS

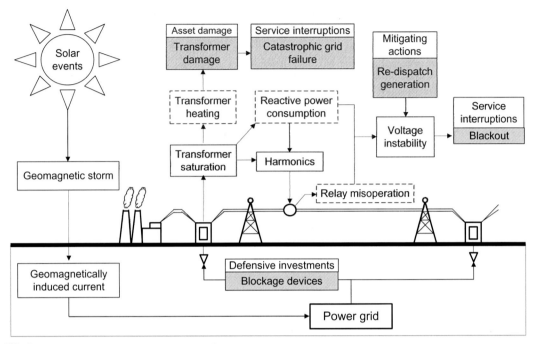

FIG. 1

Impact mechanism diagram developed for the power sector outlining primary causal pathways from a solar event (e.g., geomagnetic storm) to physical effects *(dashed boxes)* to outcomes leading to socioeconomic impacts *(gray boxes)*. Socioeconomic impacts identified represent specific outcome examples that can be organized according to larger impact categories (Fig. 2). Impact mechanism diagrams were developed for each of the four sectors and should be refined with additional outreach to sector experts.

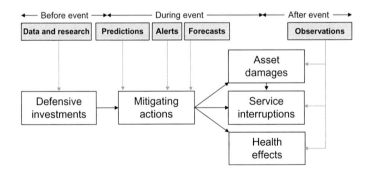

FIG. 2

The full range of potential economic and societal impacts of space weather to different technological sectors can be organized into five interrelated impact categories *(white boxes)*. A range of products and services offered by the SWPC *(gray boxes)* is valuable at different points in time in relationship to an event and can therefore be used by different sector stakeholders (engineers or operators) to reduce the many types of potential impacts.

operations to end users. To facilitate each conversation, a simple matrix was constructed for each sector to connect the identified physical effects with the five impact categories (see Table 1). The matrix was initially populated with specific examples culled from the literature, and then expanded or revised based on information provided by the stakeholder. An updated matrix was iteratively shown to subsequent stakeholders until further input yielded few substantive changes. Iterating this process allowing us to: (i) refine our list and descriptions of the physical effects; (ii) gauge the relative frequency and severity of their occurrence, including information on impact thresholds; (iii) identify likely future trends due to changing technologies, policies, and/or economics; (iv) recognize consensus or disagreement between stakeholders; (v) contextualize space weather with other sector hazards; (vi) gain insights on how the impacts may change as storm size escalates from moderate to severe; and (vii) understand how current applications of space weather information reduce sector vulnerabilities.

3 RESULTS

Stakeholders were very responsive to our inquiry. They emphasized the importance and challenges involved and generously provided their time, insights, and feedback to help guide and strengthen our endeavor. Across the 4 sectors, we identified a total of 17 different physical effects of space weather representing a considerable number of potential impact pathways (Table 1). Stakeholders were also able to readily discuss and connect the physical effects identified for their sector to our five different impact categories. Many emphasized, however, that almost all of the items in each cell of our impact matrix (Table 1) also apply to other nonspace weather hazards. For example, many of the defensive investments (e.g., conservative margins of safety, redundancies) we identified are essential for building generally robust and resilient systems, and many of the service interruptions caused by space weather (e.g., outages, degradation in quality) we identified also result from other natural- and man-made phenomena (e.g., terrestrial weather, natural disasters, terrorists, etc.). Most of the impacts in our matrix are empirically established with only a few identified as possible but unsubstantiated by evidence or beyond current experience (Table 1).

Stakeholders also noted that preventative expenditures associated with defensive investments and mitigating actions tend to be small relative to the much larger potential costs that would occur without them from the other impact categories, including asset damages, service interruptions, and health effects. They discussed the central importance of space weather products and services as trigger points for mitigating actions such as the forecasts and alerts provided by the SWPC. Many stakeholders also noted that a severe event might cause disproportionate impacts in regions that do not experience more moderate space weather events because they lack the familiarity, training, and preparedness that accompany more frequent exposure to potentially hazardous conditions. Next we provide a high-level summary and a few key findings for each sector.

3.1 ELECTRIC POWER

Industry stakeholders agree that reactive power consumption is the most concerning physical effect of space weather on the power grid. Other known and important effects include transformer heating, protective relay misoperation, generator tripping, real power imbalances, and the loss of precision timing. The consumption of reactive power represents a network threat that can occur when geomagnetically

Table 1 Literature Review and Stakeholder Outreach Findings Are Organized and Summarized in an Impact Matrix

Sector	Physical Effects	Social and Economic Impact Categories				
		Defensive Investments	Mitigating Actions	Asset Damages	Service Interruptions	Health Effects
Power grid	Reactive power loss	●	●		●	
	Transformer heating	●	●	●	●	
	Relay misoperation	●	●		●	
	Power imbalances		●		●	
	Generator tripping	●	●		●	
	Loss of precision timing	●	●		○	
Aviation	Communication	●	●	○	●	
	Navigation	●	●	○	●	○
	Human exposure		●		○	○
	Avionic upsets	●	○	○	○	○
Satellites	Cumulative dose	●		●	●	
	Anomalies	●	●	●	●	
	Link disruptions	●	●		●	
	Loss of orientation	●	●	●	●	
	Loss of altitude	●	●	●	●	
GNSS users	Loss of lock	●	○		●	
	Ranging errors	●	○		●	

Within each of the five sectors, the physical effects are listed in approximate order of relative concern as ranked by stakeholders (most concerning first but note the many ambiguities regarding relative importance discussed in the main text). These physical effects can cause different types of social and economic impacts that can be organized into five different impact categories (see Fig. 2 and definitions in text). The presence (or absence) of a circle denotes where we were able (or unable) to connect a given physical effect to a particular impact category based on detailed discussions with experts (see Fig. 1 for specific examples of physical effects and socioeconomic impacts for the power grid). Black circles denote that a physical effect is known, from direct past experience, to cause a particular category of societal impact. Open circles indicate that a given physical effect may cause a particular category of societal impact but lacks empirical support. For example, aviation stakeholders noted that avionic upsets and navigation disruptions may increase the potential risk of accidents and collisions, but this possibility is largely unknown and not supported by data.

induced currents (GICs) enter and flow along high-voltage transmission lines. It is particularly concerning because it has the potential, within less than one minute, to lead to voltage collapse and system instability. U.S. grid operators are now required by law to have contingency procedures for mitigating this threat (Federal Energy Regulatory Commission (FERC) TPL-001). Effective procedures include mitigating actions such as reducing transmission flows, redispatching generation, and/or reconfiguring network topology. It was further noted, however, that these mitigating actions are enacted more frequently for other nonspace weather reasons (e.g., terrestrial weather) so the costs of such measures must be compared to routine operational costs when trying to assess their potential significance. If such procedures are not enacted or are unable to boost reactive power to a sufficient level, far more costly (one stakeholder estimates potential costs of $\sim$$1 billion per event) wide-area blackouts could potentially result. Industry experts also stressed that transformer heating is only worrisome if it is substantial and sustained and is furthermore most likely to cause problematic thermal damage in older transformers at the end of their $\sim$40–50 year operational lives. This impact pathway (transformer heating) and related assumptions about the number and extent of transformer damage underpin previous, catastrophic grid-failure scenarios for extreme storms with costs to the U.S. and global economies that are upwards of $\sim$$1 trillion. Actual transformer vulnerability to a severe GIC (a 1-in-100-year storm) is currently unknown but is being actively assessed under FERC TPL-002 and will support decisions about what defensive investments (e.g., blocking capacitors, replacement transformers, etc.) are necessary for enhancing power reliability.

3.2 AVIATION

Depending on roles and responsibilities, industry stakeholders emphasized different concerns about the known effects of space weather on aviation including avionic upsets, interruptions to onboard communications and navigation systems, and human exposure to radiation. Airplane manufacturers, for example, make substantial defensive investments to safeguard airplane electronics ("avionics") against single-event upsets (SEUs). Although critical avionic equipment and circuits are rigorously designed and tested to minimize the potential problems that SEUs can cause, engineers noted that this hazard continuously evolves in complexity alongside industry trends such as the increasing dependence on aircraft computers, shrinking chip sizes, and increasing flight altitudes. Airplane operators, on the other hand, were primarily concerned about space weather's potential to disrupt different communications technologies, especially high frequency (HF) radio, which to date remains important to airline communications especially in certain locations around the world (e.g., polar and oceanic regions). Space weather disruptions to HF radio may prevent airlines from flying planned schedules or routes because the inability of an aircraft to be in continuous contact with air traffic control (ATC) violates Federal Aviation Administration (FAA) law. Delaying or rerouting airplanes around regions where HF communications are unavailable (e.g., polar regions for a moderate event and perhaps lower latitudes for a more severe event) are effective mitigating actions but can be costly to airlines, for example, in terms of additional fuel, reduced cargo, landing fees, and employee workloads. These service interruptions can also impact and delay airline passengers. Other stakeholders focused more directly on aircraft safety expressed unknown but potentially growing risks from the increasing role of GNSS in airplane navigation combined with the decommissioning of ground-based navigation aids around the world. And lastly pilots, flight attendants, and representatives throughout the industry also expressed concerns about potential health effects from radiation dosages to which those flying at altitude during a space weather event can be exposed. Although a connection between radiation exposure and health outcomes (e.g., cancer) remains uncertain, it is possible for airlines to take simple precautionary actions such as temporarily lowering cruising altitude if it does not jeopardize aircraft safety for other reasons (e.g., surrounding airspace is uncrowded and sufficient fuel is onboard to compensate for decreased fuel efficiency).

3.3 SATELLITES

The satellite industry is very diverse, requiring engineers and operators in different Earth orbits to overcome a range of environmental challenges in order to provide different types of satellite services (e.g., communications, Earth observations, etc.). Stakeholders across the industry emphasized different concerns about the effects of space weather on satellites that are directly associated with the various orbit types and missions. The physical effects of concern include cumulative dosage, anomalies, link disruptions, loss of altitude, and loss of orientation. Building communication satellites to operate in geostationary Earth orbit (GEO) and GNSS to operate in medium Earth orbit (MEO), where the radiation environment is relatively harsh, involves substantial defensive investments such as hardened components, shielding, or generous design margins to protect the integrity of satellites from high cumulative radiation dosages and from anomalous satellite behaviors that can be caused by surface- and deep-dielectric charging. Low Earth orbit (LEO) operators express an additional concern about an extreme space weather event and the large number of LEO spacecraft that could potentially lose altitude from excessive atmospheric heating: The growing orbital population is approaching a level where one accidental collision could trigger cascading collisions [e.g., "Kessler Syndrome" (Kessler et al., 2010)] to potentially devastate many critical societal services such as Earth observations and imagery. Despite these concerns, most industry stakeholders emphasized that the satellite population is rather robust and that it is more likely for space weather events to cause operational challenges, for example from link disruptions and anomalies, to potentially result in temporary service interruptions rather than any permanent damage that would compromise mission lifetimes and/or capabilities.

3.4 GNSS USERS

Space weather can interfere with GNSS signal transmission, causing ranging errors and loss of lock with potential impacts dependent on specific usages of the position, navigation, and timing (PNT) information broadcast by different GNSS satellite constellations (e.g., the U.S. Global Positioning System (GPS), the Russian GLONASS). Susceptibility and recovery to potential impacts from these different physical effects are most closely controlled by the design of a GNSS receiver, which varies as widely as GNSS uses (e.g., agriculture, resource exploration/extraction, transportation, surveying, financial trading, noncommercial, etc.). Ranging errors mostly affect single-frequency receivers so GNSS users needing higher accuracy, continuously available PNT information generally opt to mitigate this effect by purchasing dual-frequency receivers. Users can instead or in addition make other defensive investments to enhance the overall reliability of GNSS, for example, by building augmentation systems such as the FAA's Wide Area Augmentation System (WAAS). Space weather related ionospheric disturbances can also prevent GNSS receivers from being able to track, or lock onto, the signal sent from different GNSS satellites. In commercial applications, GNSS information typically increases efficiency and therefore reduces operational costs by saving companies labor, capital, and time. Therefore potential impacts from space weather are likely to result in decreased productivity and efficiency. In noncommercial applications, GNSS technology is of more nonmonetary value and contributes to daily life tasks and enjoyment (e.g., location-based services). Potential service interruptions to GNSS users may become more problematic in the future, as the PNT information provided by GNSS technology becomes further embedded into more complex, interdependent systems and processes.

4 NEXT STEPS AND CONCLUDING REMARKS

Our team is currently developing a set of methods and tools for estimating some of the socioeconomic impacts of space weather described above. This monetization is informed by and will complement the results of our literature review and stakeholder outreach. Given the numerous complexities and large uncertainties involved, as briefly discussed in our introduction, we are developing simple, tractable with estimates for each sector in close collaboration with stakeholders that focus on the social and economic costs that are apt to be largest and most plausible during space weather events of moderate and more severe sizes. Each estimate aims to capture essential science and engineering details and to address key social and economic relationships as determined by additional stakeholder outreach. This has involved, for example, iterative discussions about what is reasonable to expect under more "moderate" compared to more "severe" space weather scenarios. We are translating these stakeholder insights into tractable, quantitative statements that also incorporate standard socioeconomic tools to estimate specific key outcomes. For example, the costs of service interruptions such as power outages are being assessed using the value of lost load (e.g., London Economic International, 2013) and the costs of asset damages are being estimated using equipment replacement costs (e.g., Odenwald et al., 2006).

This effort advances our overall understanding of the potential impacts of space weather and, in addition to stimulating timely discussions, we hope it provides a foundation for future work. The systematic approach we developed for exploring and synthesizing the many qualitative and quantitative complexities of this problem can be readily applied to other sectors (e.g., rail, pipelines). Our initial findings and cost estimates should furthermore be iteratively refined with additional stakeholder input and collaboration. To the extent possible, our quantification framework should also be applied to historical events with existing data to estimate costs of actual storms in order to help ground-truth model outputs. To our knowledge, there has been little attempt to quantify the costs of historical events, perhaps because the appropriate tools are not readily available. It will also be pertinent to continuously modify our initial findings and estimates as new scientific information is acquired and as our technological infrastructure evolves. Such additional research is paramount, for example, if we want to design technologies and implement policies that will effectively and efficiently reduce our potential vulnerabilities to different space weather hazards. The study concludes in September 2017 and findings will be made available to the public.

REFERENCES

Allen, J., Sauer, H., Frank, L., Reiff, P., 1989. Effects of the March 1989 solar activity. Eos 70 (46), 1486–1488.
Baker, D.N., 2009. What does space weather cost modern societies? Space Weather. 7. S02003, https://doi.org/10.1029/2009SW000465.
Baker, D.N., Lanzerotti, L.J., 2016. Resource letter SW1: space weather. Am. J. Phys. 84 (3), 166–180.
Baker, D.N., Li, X., Pulkkinen, A., Nqwira, C.M., Mays, M.L., Galvin, A.B., Simunac, K.D.C., 2013. A major solar eruptive event in July 2012: defining extreme space weather scenarios. Space Weather 11, 585–591.
Biffis, E., Burnett, C., 2017. Report on the UKSA-funded IPSP SWx/SWe socio-economic study. Report on the ESA Space-Weather Socio-Economic Study L5 in Tandem with L1: Future Space-Weather Missions Workshop, March 6–9.
Bolduc, L., 2002. GIC observations and studies in the Hydro-Québec power system. J. Atmos. Sol. Terr. Phys. 64 (16), 1793–1802.

Cannon, P., Angling, M., Barclay, L., Curry, C., Dyer, C., Edwards, R., et al., 2013. Extreme Space Weather: Impacts on Engineered Systems and Infrastructure. U.K. Royal Academy of Engineering, London, ISBN: 1-903496-95-0.

Eastwood, J.P., Biffis, E., Hapgood, M.A., Green, L., Bisi, M.M., Bently, R.D., Wicks, R., McKinnel, L.A., Gibbs, M., Burnett, C., 2017. The economic impact of space weather: where do we stand? Risk Anal. 37 (2), 206–218.

Forbes, K.F., Cyr, O.C.S., 2012. Did geomagnetic activity challenge electric power reliability during solar cycle 23? Evidence from the PJM regional transmission organization in North America. Space Weather. 10. S05001, https://doi.org/10.1029/2011SW000752.

Hapgood, M., Angling, M., Attrill, G., Burnett, C., Cannon, P., Gibbs, M., et al., 2016. Summary of space weather worst-case environments. Revised ed., RAL Technical Report RAL-TR-2016-06.

Helmer, O., 1967. Analysis of the Future: The Delphi Method. The RAND Corporation, Santa Monica, CA.

Kessler, D.J., Johnson, N.L., Liou, J.-C., Matney, M., 2010. The Kessler syndrome: implications to future space operations. In: 33rd Annual AAS Guidance and Control Conference, 6–10 February. American Astronautical Society, Breckenridge, Colorado.

Knabb, R., Rhome, J.R. Brown, D.P., 2005. Tropical Cyclone Report, Hurricane Katrina, 23–30 August. http://www.nhc.noaa.gov/pdf/TCR-AL122005_Katrina.pdf.

London Economic International LLC, 2013. Estimating the value of lost load: briefing paper prepared for the Electric Reliability Council of Texas, Inc.

Lopez, R.E., Baker, D.N., Allen, J.H., 2004. Sun unleashes Halloween Storm. Eos 85 (11), 105–108.

Luntama, J.P, Bobrinsky, N., Kfraft, S., 2017. Report on the ESA Space-Weather Socio-Economic Study L5 in Tandem with L1: Future Space-Weather Missions Workshop. March 6–9.

National Research Council, 2008. Severe Space Weather Events—Understanding Societal and Economic Impacts: A Workshop Report. National Academies Press, Washington, DC.

Odenwald, S., Green, J., Taylor, W., 2006. Forecasting the impact of an 1859-calibre superstorm on satellite resources. Adv. Space Res. 38, 280–297.

Oughton, E., 2017. Report on the Cambridge SWx/SWe Socio-Economic Study, Report on the ESA Space-Weather Socio-Economic Study L5 in Tandem with L1: Future Space-Weather Missions Workshop. March 6–9.

Oughton, E., Copic, J., Skelton, A., Kesaite, V., Yeo, Z.Y., Ruffle, S.J., Tuveson, M., Coburn, A.W., Ralph, D., 2016. Helios solar storm scenario. Cambridge Risk Framework Series, Centre for Risk Studies, University of Cambridge, Cambridge, United Kingdom.

Smith, A., Matthews, J., 2015. Quantifying uncertainty and variable sensitivity within the U.S. billion-dollar weather and climate disaster cost estimates. Nat. Hazards 77 (3), 1829–1851.

Viereck, R., 2017. SWAP benchmarks on extreme event. In: NOAA/NWS Space Weather Workshop, May 1–5.

World bank, 2011. The recent earthquake and tsunami in Japan: implications for East Asia. East Asia and Pacific Economic Update, vol. 1. World bank, Washington, DC.

FURTHER READING

NOAA, 2004. Intense Space Weather Storms October 19–November 7, 2003. Government Printing Office, Washington, DC.

CHAPTER

SEVERE SPACE WEATHER EVENTS IN THE AUSTRALIAN CONTEXT

30

David Neudegg, Richard Marshall, Michael Terkildsen, Graham Steward

Space Weather Services, Bureau of Meteorology, Haymarket, NSW, Australia

CHAPTER OUTLINE

1 Introduction and Concept Development 711
2 The Nature of Severe Events and the Regional Context 712
3 The Severe Event Service 715
4 Policy Background 716
5 Stakeholder Technology Groups 716
6 Conclusion 717
References 718

1 INTRODUCTION AND CONCEPT DEVELOPMENT

The precursor to Space Weather Services (SWS), the Ionospheric Prediction Service (IPS), provided services in high-frequency (HF) radio (3–30 MHz, also known as shortwave) as well as monitored ionospheric and solar conditions from its inception in 1947, many decades prior to the advent of space weather as a concept. These activities were closely related to the development of solar-terrestrial physics in Australia (Fraser, 2016). HF radio refracts in the various layers in the ionosphere from 100–300 km altitude to propagate very long distances over the horizon, and has been pursued in detail by Australia for many years for communications, direction finding, and radar. However, HF radio systems cannot be damaged by geophysical activity, only impeded in their operation, albeit sometimes severely. Examples would be shortwave fadeouts (SWF), also known as sudden ionospheric disturbances (SID), when M- or X-class solar X-rays cause the lowest ionospheric D-region, near 90 km altitude, to absorb the lower end of the HF radio spectrum. For a large X-class flare, an SWF may block most or all of the frequencies in the HF band, which may be construed as a severe event. If there is a large geomagnetic storm, causing a large ionospheric storm, then usually the upper F2 region of the ionosphere at ~300 km altitude is strongly disturbed, resulting in suppression of the frequencies available in the HF band, which again may be construed as a severe event. However, ionospheric and HF conditions are continuously varying, even if solar and geomagnetic conditions are mild. Strong SWF and ionospheric storm conditions were seen for many decades as just the far end of a continuous spectrum of conditions, rather than discrete severe events. Services were developed to provide alerts and warnings for them but not in the context of an actual severe space weather event service.

In the 1990s, geomagnetic observations and geophysical services were added by IPS and a wider range of potential technology stakeholders (aeromagnetic surveys, power grids, pipeline corrosion) were added as the term "space weather" gained currency worldwide for these types of activities. In the mid-2000s, GPS observations and services (e.g., aviation precision GNSS) were added, again expanding the stakeholder technology base. The lead up to the maximum of solar cycle 23 also led to a small amount of spacecraft environment services being added on request from operators. Hence, by the time of the downslope after cycle 23 in the early 2000s, a far wider suite of space weather affected technologies was being considered for services than for prior decades of IPS operations. It also became apparent that some technologies had a high response threshold for space weather events rather than gradual degradation, with minimal response below the threshold but possibly responses potentially damaging to the technology above the threshold. These technologies are power grids, aviation GNSS precision approach, and geostationary spacecraft. The recognition of these "special" technologies and the absorption of IPS into the Australian Bureau of Meteorology in 2007, with its focus on special procedures for severe weather events, led to consideration of an Extreme/Severe Space Weather Event service soon after, to be prepared for the maximum in solar cycle 24.

2 THE NATURE OF SEVERE EVENTS AND THE REGIONAL CONTEXT

The nature of severe or extreme events needed to be defined in order to construct a service for stakeholders, but also what constituted this type of event for the Australasian region, if it were indeed different from a global event. It is assumed that severe space weather events will have origins in a solar disturbance, likely a coronal mass ejection (CME) striking the geomagnetic field in a favorable configuration with high speed, a southward interplanetary magnetic field (IMF) component (Bz = IMF north-south component), probably high mass density, and the center of the CME striking the Earth rather than the edge. This event would cause multiple large disturbances in the geomagnetic and ionospheric environments affecting key technologies. Other events such as X-ray flares, solar energetic particle events, high-speed solar wind streams, prolonged IMF Bz southwards, local ionospheric disturbances from the upper atmospheric (i.e., thermospheric) waves (e.g., traveling ionospheric disturbances, or TIDs), were therefore excluded from the extreme/severe category. Large CMEs striking the geomagnetic field are indeed a global event and no geographic region would escape the impact, and it may be questioned why a region would require a unique severe event service distinct from a global service. The answer lies in the fact that many of the disturbances affecting key technologies are in the ionosphere and geomagnetic field, which are at a low enough altitude to retain unique characteristics in large geographic regions.

Australia is in somewhat of a unique geographic position as a large island continent that has no land borders with adjacent countries. It also has responsibilities extending from the equatorial to the polar regions, encompassing approximately one eighth of the world's surface across a wide variety of geomagnetic and ionospheric conditions. To give an indication of the scale involved, a typical hourly map of the regional ionospheric F2 layer critical radio frequency foF2, useful for HF radio predictions, is seen in Fig. 1. The region is over 10,000 km per side, spanning 70 degrees of latitude and 90 degrees of longitude. Australia also did not have neighboring countries with approved RWCs until the Indonesian Space Agency LAPAN was approved as a Regional Warning Center (RWC) by the International Space Environment Service (ISES) body of ICSU in 2016, and so had to exhibit a fair degree of self-reliance in interpreting regional events.

It was decided that a measure of regional geomagnetic disturbance would best suit the requirements of a metric to determine the threshold of a severe event from a CME impact. There are numerous

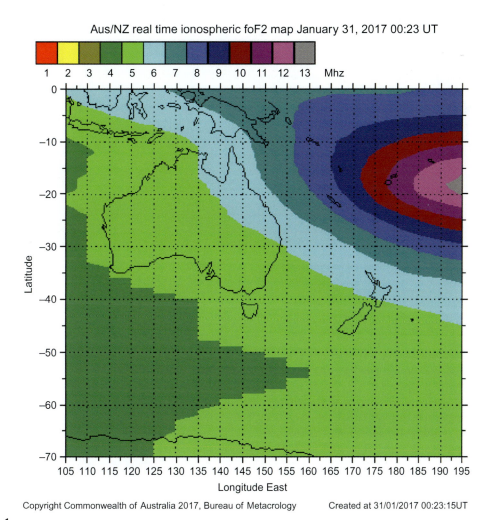

FIG. 1

A real-time map of the ionospheric layer critical radio frequency foF2 in the Australasian region on the SWS website (http://www.sws.bom.gov.au/HF_Systems/1/4).

714 CHAPTER 30 SEVERE SPACE WEATHER EVENTS IN THE AUSTRALIAN CONTEXT

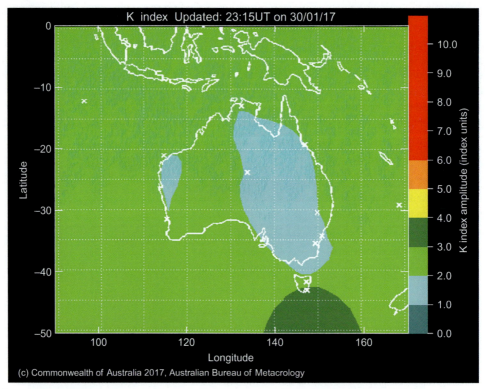

FIG. 2

A real-time map of the geomagnetic K index in the Australasian region on the SWS website (http://www.sws.bom.gov.au/Geophysical/1/3/1).

geomagnetic indices (e.g., Akasofu and Chapman, 1972) and many severe event services use the planetary K or A indices, Kp and Ap. The 3-h K indices for Australasian magnetometer sites leading to a regional K index, shown in Fig. 2, had been produced for several years and regional-K was initially considered as a metric. However, it was felt that the K-index did not accurately reflect true geomagnetic storm development, leading to relevant effects such as geomagnetically induced currents (GICs) affecting power grids (Marshall et al., 2013), ionospheric storms affecting aviation precision-GNSS (Terkildsen, 2013b), and enhanced geostationary orbit radiation levels affecting spacecraft. Also, for most severe events the regional K-index would likely be at the highest level of nine, reducing discrimination between levels of different events.

The geomagnetic storm index Dst was seen as a more discriminating measure of a proper storm with a developed ring current, rather than just a large geomagnetic disturbance. The global Dst index uses data from equatorial but predominantly northern hemisphere sites, and may not accurately reflect regional conditions. However, a regional index AusDst, shown in Fig. 3, had been developed from local data and was chosen as best representing regional geomagnetic storm impact.

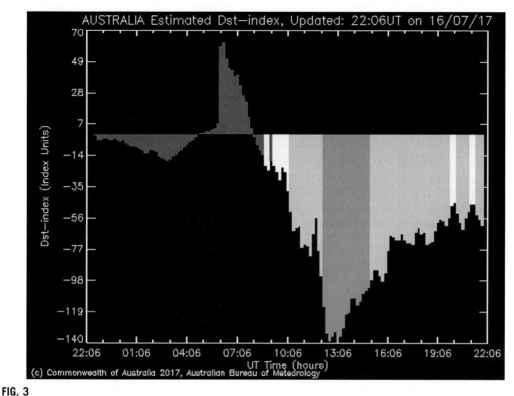

FIG. 3

Dst geomagnetic storm index on the SWS website (http://www.sws.bom.gov.au/Geophysical/1/2/4).

3 THE SEVERE EVENT SERVICE

An analysis was performed of a statistically significant number of geoeffective CMEs associated with X-ray flares from solar active regions (sunspot groups) and the resultant Dst. Parameters taken into consideration included the location of the active region on the solar disc, X-ray flare magnitude and duration, presence of halo CME and the CME width, and CME speed. Other parameters were considered but either Dst was found not to be dependent on them or they were not available for a wide enough range of events. A general linear statistical model was developed, providing both binary and probabilistic forecasts for events above a given threshold AusDst. After some consideration, a severe event threshold of $Dst < -250$ nT was set. An extreme event exact Dst threshold has not been set and is likely several hundred of nT for events with occurrence rates of one every few hundred years. Hence extreme events are treated as severe events for the purposes of prediction. A Severe Space Weather software suite and procedures for the forecasters in the Australian Space Forecast Center were developed, and protocols established for stakeholder interactions (Terkildsen, 2013a; Maher et al., 2016). A subscription email service is in operation at http://listserver.ips.gov.au/mailman/listinfo/ips-esws-general.

Solar cycle 24 was weak in terms of sunspot numbers and the size of geomagnetic disturbances compared with the previous seven cycles, going back to the mid-20th century and the advent of much of the technological systems vulnerable to space weather. Only three sunspot groups exceeded 1000 millionths of the solar disc. One of these, AR1429 in March 2012, was very active while on the Earthward-facing portion of the disc, emitting five substantial CMEs but none resulting in severe events. The largest geomagnetic storm of solar cycle 24 occurred on March 17, 2015, reaching a global Dst value of −223 nT. The second largest storm occurred on June 23, 2015, reaching a global Dst value of −204 nT (WDC Kyoto, 2017). A Severe Space Weather Warning was issued by SWS for the period of June 21–25, 2015; AR 2371 was responsible for a Dst of −236 nT at 14:25 UT on June 23, 2015. It is worth noting that the large CME issued on the solar far-side by AR1520 on July 23, 2012, would probably have constituted a severe event if aimed at Earth, based on the high speed and strong IMF Bz southward recorded on the STEREO spacecraft located on the solar far-side (Ngwira et al., 2013). For perspective on the storm magnitudes in cycle 24, the famous Carrington event of 1859 has been estimated to have had a Dst of −1760 nT (Tsurutani et al., 2003).

4 POLICY BACKGROUND

The policy driver for a regional Severe Space Weather Event Service was a "ground up" approach from within IPS/SWS and interested stakeholder groups, particularly aviation and power grids, rather than a "top down" approach from a national hazards risk register from which some countries (e.g., the United Kingdom) operate.

There are advantages and disadvantages of this approach and awareness of the threat would probably benefit from policy "top cover." The lack of severe events in the weak solar cycle 24 maximum and the subsequent deep minima being entered are, however, unlikely to bring this issue to the fore until the upswing in activity in the lead up to cycle 25 around 2020–21. However there is national coordination for infrastructure protection by the federal attorney general's Trusted Information Sharing Network (TISN) that has included severe space weather for many years in communications and energy infrastructure information sharing, and this has been invaluable in communicating the threat. Historically the detailed exchange of information between researchers in IPS/SWS and universities and industry stakeholders (e.g., Australian Energy Market Operator (AEMO) and AirServices Australia) has been conducted in bilateral exchanges established over many years.

5 STAKEHOLDER TECHNOLOGY GROUPS

Electricity power grids in Australia were developed and operated for many decades by each state, rather than federally. This was done originally by state government agencies and, in the last two decades, increasingly by private industry. As there are few states in Australia and they are large, the state capital cities where much of the grids are concentrated are several hundreds or thousands of kilometers apart. Hence the state grids were mostly independent until the last 25 years when interconnections to trade power between states became viable and profitable. A National Electricity Market (NEM) commenced in 1998, encompassing five states (SA, VIC, TAS, NSW, QLD) and one territory (ACT) and is coordinated by AEMO (2017). The long lengths of alternating-current (AC) high-voltage (HV) transmission lines are susceptible, as are many large grids around the world, to GICs, which are large quasi-DC electric currents that may be hazardous to HV transformers.

Traditionally only high-latitude power grids were seen as vulnerable to GICs, particularly after the 1989 Quebec event. However, GICs observed on the Australian grid (Marshall et al., 2013) and GIC-induced transformer damage in other mid-latitude countries (e.g., New Zealand 2001, Marshall et al., 2012) prompted IPS/SWS and AEMO to consider the effect of a severe event on the NEM. A set of standard operating procedures (SOPs) was implemented to best prepare the grid for events and limit any impact (AEMO, 2014).

Amongst GPS/GNSS-dependent systems used in Australia, it was realized that one susceptible to the greatest impact was the Ground Based Augmentation System (GBAS), a GPS-based precision landing system under evaluation as a future replacement for current instrument landing systems (ILS). Large ionospheric gradients in total electron content (TEC) caused by large geomagnetic storms (Terkildsen, 2013b) may cause erroneous correction signals to be broadcast from the GBAS if the gradients exceed the bounds of the inbuilt ionospheric threat model (412 mm/km). A study found that severe gradients were less prevalent in the Australian than U.S. ionosphere where GBAS was already operating. However, because they are associated with severe geomagnetic storms, the severe event service is used to identify space weather conditions with the potential to disrupt GBAS performance.

Australia has a relatively small space industry for an OECD country, with one major company operating geostationary communications satellites (Optus, 2017). However, the nation is an intensive user of overseas satellite data for weather, remote sensing, communications, defense, and satellite navigation/timing. Hence it has a vested interest in being aware if those spacecraft-dependent services are likely to be compromised. Modern spacecraft are designed to be very robust against the space radiation environment at geostationary and medium earth orbits, and likely have a very high threshold before significant damage occurs. So this technology is seen as an appropriate stakeholder in a severe event service, such that if a large geomagnetic storm developed, high-energy particles trapped in the radiation belts may become accelerated to a level that would make spacecraft damage possible. At this stage, direct high-energy protons from the Sun, in the largest of solar energetic particle (SEP) events, have not been considered, as SEP prediction is rudimentary. However advances in large X-ray flare statistical prediction, to which SEP events' likelihood may be related, and particle propagation from Sun to Earth are being advanced by collaborative research between SWS and the University of Sydney (Li et al., 2016; Steward et al., 2016). The national economic loss from having access to spacecraft-related services being strongly diminished by a severe space weather event has not been quantified, but is likely to be substantial.

6 CONCLUSION

The Severe Space Weather Event Service operated by SWS for the Australasian region is a baseline function that has been tested and operated for several years by the Australian Space Forecast Center, including across the maximum of solar cycle 24. There were no severe events across this maximum and the analysis algorithm provided minimal false positives, so with forecaster interpretation, no false warnings were issued. However, obviously the system has not been operated during an actual severe event, which may have to wait several years until the maximum of cycle 25 or even later. The efficiency of the system is high on the analysis and forecasting side, as it is operated regularly by forecasters during any Earth-directed CME events associated with M- or X-class flares.

Improvements are likely to be related to stakeholder interaction procedures and perhaps in national risk mitigation policy for such natural hazards to electromagnetic infrastructure.

REFERENCES

AEMO, 2014. Power system security guidelines—Section 12, Management of solar storms—geomagnetic disturbances (GMD), SO_OP_3715. https://www.aemo.com.au/media/Files/Other/SystemOperatingProcedures/SO_OP_3715_Power_System_Security_Guidelines_v65.pdf.

AEMO, 2017. National Electricity Market, Australian Energy Market Operator, https://www.aemo.com.au/Electricity/National-Electricity-Market-NEM.

AirServices Australia, 2017. GBAS, http://www.airservicesaustralia.com/projects/ground-based-augmentation-system-gbas/.

Akasofu, Chapman, 1972. Solar Terrestrial Physics. Oxford University Press, Oxford.

Fraser, 2016. A brief history of solar-terrestrial physics in Australia. Geosci. Let. 3, 23. https://doi.org/10.1186/s40562-016-0050-7 http://geoscienceletters.springeropen.com/articles/10.1186/s40562-016-0050-7.

Li, Cairns, Gosling, Steward, Francis, Neudegg, Schulte in den Bäumen, Player, Milne, 2016. Mapping magnetic field lines between the Sun and Earth. J. Geophys. Res. https://doi.org/10.1002/2015JA021853. http://onlinelibrary.wiley.com/doi/10.1002/2015JA021853/full.

Maher, Marshall, Terkildsen, Bouya, Kumar, Steward, Lobzin, 2016. Australian bureau of meteorology space weather services: recent initiatives. In: SWPC Space Weather Workshop 28 April 2016, and 4th AOSWA Workshop 24–27 October. http://www.swpc.noaa.gov/sites/default/files/images/u33/Maher%20SWW2016_AUS_BOM_SWS_P-Maher_v1.4.pdf, http://aoswa4.spaceweather.org/presentationfiles/20161024/S1-4.pdf.

Marshall, R.A., Dalzell, M., Waters, C.L., Goldthorpe, P., Smith, E.A., 2012. Geomagnetically induced currents in the New Zealand power network. Space Weather. 10. S08003, https://doi.org/10.1029/2012SW000806http://onlinelibrary.wiley.com/doi/10.1029/2012SW000806/full.

Marshall, Gorniak, Van Der Walt, Waters, Sciffer, Miller, Dalzell, Daly, Pouferis, Hesse, Wilkinson, 2013. Observations of geomagnetically induced currents in the Australian power network. Space Weather. https://doi.org/10.1029/2012SW000849 http://onlinelibrary.wiley.com/doi/10.1029/2012SW000849/full.

Ngwira, Pulkkinen, Mays, Kuznetsova, Galvin, Simunac, Baker, Li, Zheng, Glover, 2013. Simulation of the 23 July 2012 extreme space weather event: what if this extremely rare CME was Earth directed? Space Weather 11, 1–9. https://doi.org/10.1002/2013SW000990.

Optus, 2017. Satellite Network. https://www.optus.com.au/about/network/satellite.

Steward, Lobzin, Cairns, Li, Neudegg, 2016. Automatic recognition of complex magnetic regions on the Sun in SDO magnetogram images and prediction of flares: techniques for the revised Flarecast. In: Australian Space Research Conference.

Terkildsen, 2013a. International communication and coordination related to extreme space weather events. In: SWPC Space Weather Workshop, U.S.A. http://www.swpc.noaa.gov/sites/default/files/images/u33/ESW_Australia.pdf.

Terkildsen, 2013b. GNSS vulnerability to space weather. In: Australian Centre for Space Engineering Research UNSW Workshop. http://www.acser-archive.unsw.edu.au/events/GNSS_Vulnerability/02-MichaelTerkildsen-GNSS_vulnerability_to%20_space_weather.pdf.

Tsurutani, B.T., Gonzalez, W.D., Lakhina, G.S., Alex, S., 2003. The extreme magnetic storm of 1–2 September 1859. J. Geophys. Res. 108 (A7), 1268. https://doi.org/10.1029/2002JA009504.

WDC (World Data Center) for Geomagnetism, Kyoto Dst index service, 2017. http://wdc.kugi.kyoto-u.ac.jp/dstdir/index.html.

CHAPTER 31

EXTREME SPACE WEATHER RESEARCH IN JAPAN

Mamoru Ishii
National Institute of Information and Communications Technology, Tokyo, Japan

CHAPTER OUTLINE

1 Overview and History of Operational Space Weather Forecast 719
2 Action to Telecommunications and Satellite Positioning 720
3 Action to Aviation 721
4 Action to Satellite Saving 723
5 Action to GIC 724
6 Introduction of PSTEP 725
References 725

1 OVERVIEW AND HISTORY OF OPERATIONAL SPACE WEATHER FORECAST

Japan started its operational space weather observations more than 100 years ago. Since the Japanese government realized the importance of telecommunications with high-frequency (HF) radio waves in 1904, during the Japanese sea battle against Russia, they established the observatory in Hiraiso in 1915. The purpose of the observatory was monitoring the condition of HF telecommunications over the Pacific Ocean.

Around 1935, they began to monitor the ionosphere, which significantly influences HF propagation. They operated several observatories in Japan and Southeast Asia during the Pacific war.

After World War II, the Radio Research Laboratory (RRL; now known as the National Institute of Information and Communication Technology, or NICT) was established in 1952. RRL succeeded the operational ionosphere monitor, and started the solar observations at the time. RRL showed some marvelous results in space weather research; for example, it provided the worldwide distribution of critical ionospheric frequencies from the observational results of the ISS-b satellite, launched on February 16, 1978. (Fig. 1)

RRL was one of the seven original members of the International Ursigram and World Days Service (IUWDS) established in 1957 (International Geophysical Year, (IGY)). IUWDS was a consortium of space weather operations, sharing the information with early information networks such as TELEX.

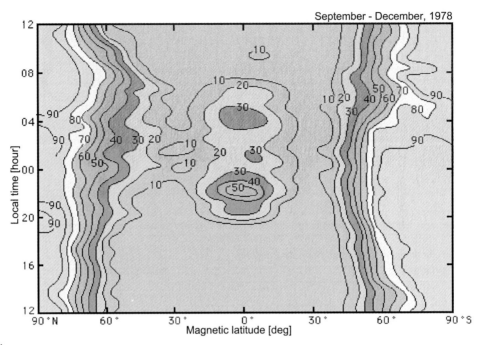

FIG. 1

Global distribution of the probability of occurrence of topside spread F (unit: %). This figure shows the global distribution of the probability of occurrence of spread F estimated from the observation with Ionosphere Sounding Satellite-b (ISS-b). Vertical axis shows local time [hour] and horizontal axis shows magnetic latitude [deg]. The observing period is August–December 1978.

From Matuura, N., Kotaki, M., Miyazaki, S., Sagawa, E., Iwamoto, I., 1981. IBB-B experimental results on global distributions of ionospheric parameters and thunderstorm activity. Acta Astronaut. 8(5–6), 527–548.

Now the function of IUWDS has been transferred to the International Space Environment Service (ISES), which has 17 Regional Warning Centers, 4 Associate Warning Centers and 2 Collaborative Expert Centers as of Sep. 2017.

2 ACTION TO TELECOMMUNICATIONS AND SATELLITE POSITIONING

As Japan is located in the mid-latitudes, it is rare that it is influenced by an ionospheric disturbance with an aurora. Instead, Japanese operators need to take care of the Dellinger effect, positive/negative ionospheric storms, day-to-day variabilities of equatorial anomalies, and the equatorial plasma bubble (EPB). In addition, it is well known that the sporadic-E is often observed in Japan from the spring to summer seasons. As extreme space weather, the sudden ionospheric disturbance (SID) is a rare but important phenomenon.

NICT has undertaken ionospheric observations in four sites in Japan (Wakkanai/Sarobetsu, 45.16N, 141.75E; Kokubunji, 35.71N, 139.49E; Yamagawa, 31.20N, 130.62E; and Okinawa, 26.68N, 128.15E)

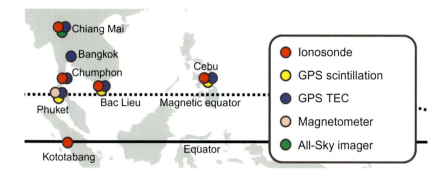

FIG. 2

The location and installed instruments of observatories of Southeast Asia Low-latitude Ionospheric Network (SEALION).

and one in Antarctica (Syowa, 69.00S, 39.58E) operationally. In addition, as a research project mainly studying the dynamics and characteristics of plasma bubbles, NICT has network observations of the ionosphere in Southeast Asia in cooperation with universities and academic institutes. At present, there are ionospheric observations with ionosondes in Chiang Mai (18.76N, 98.93E), Chumphon (10.72N, 99.37E) in Thailand, Bac Lieu (9.30N, 105.96E) in Vietnam, Cebu (10.35N, 123.91E) in the Philippines and Kototabang (0.2S, 100.32E) in Indonesia. In addition to GPS receivers, scintillation monitors and magnetometers are operated. The observation network named SEALION (Southeast Asia Low-latitude Ionospheric Network, Fig. 2) makes it possible to perform magnetic conjugate observation, which is ideal for studying plasma bubbles. Some fruitful results have already been published in scientific papers (e.g., Uemoto et al., 2007; Maruyama et al., 2014).

GEONET (GNSS Earth Observation Network System) is a GNSS receiver network deployed by the Geospatial Information Authority of Japan. It contains more than 1200 receivers all over Japan and monitors crustal movements and earthquakes. These receivers can receive multifrequency radio waves from the GNSS satellites, so we can calculate the total electron content (TEC) between the satellite and receiver. These dense networks enable us to plot a TEC map over Japan (Fig. 3). These results show the ionospheric variations, for example, the middle scale/large scale traveling ionospheric disturbances.

3 ACTION TO AVIATION

The International Civil Aviation Organization (ICAO) is interested in space weather information, which is important in three categories: HF communications, satellite positioning, and human exposure. Some descriptions about HF communications have already been shown in the previous section, so here the remaining two will be introduced.

The Global Navigation Satellite System (GNSS) is now important as a part of the aviation system. ICAO decided the international standards of the Satellite-Based Augmentation System (SBAS) and the Ground-Based Augmentation System (GBAS). The Japanese government launched the Multifunctional Transport SATellite (MTSAT) in 2005 and 2006 for preparing MSAS in the west Pacific area.

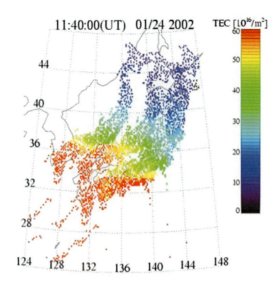

FIG. 3

An example of a Total Electron Content (TEC) map over Japan obtained from GEONET.

The significant ionospheric anomaly from the assumed shape in the model of GNSS is one of the largest factors in GNSS error. It is necessary to study the quantitative influence of ionospheric disturbances on MSAS for using the system in operational aviation.

The EPB is a type of ionospheric disturbance. EPB generates after sunset symmetrically with respect to the magnetic equator, with the size of several thousand km in North-South and several tens of km in East-West. Once generated, the EPB moves westward at about 30 km/h by the Earth's electric field [Fukao et al., 2006]. The southern area of Japan can be affected by EPB when its activity is high.

Recently, a three-dimensional high-resolution EPB model was developed that enables us to study the generation process and key parameters for EPB growth. Yokoyama et al. (2014) show the results of EPB growth with the high-resolution bubble models that have spatial resolution as fine as 1 km. Fig. 4 is an example of the growth of EPB.

The Electric Navigation Research Institute is responsible for research and development in the field of electronic navigation in Japan; many studies about the influence of ionospheric disturbances on MSAS are conducted. Saito et al. (2015) tested GAST-D (GBAS Approach Service Type D) at the real airport environment at New Ishigaki Airport. Their results showed that the ionospheric spatial gradient monitor (ISGM) worked satisfactory under nominal conditions, but some events of enhanced ISGM outputs, which may potentially cause an ISGM false alarm, were observed.

Saito et al. (2017) built an ionospheric delay scintillation model for GBAS specifically for the Asia-Pacific region to collect the observed data in various areas in that region.

When solar energetic particles (SEPs) are incident to the atmosphere, they can induce air showers by generating varieties of secondary particles. Such secondary particles can reach deep into the atmosphere and enhance the level of radiation doses, which can be a hazard to aircrews.

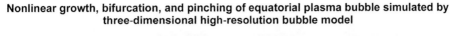

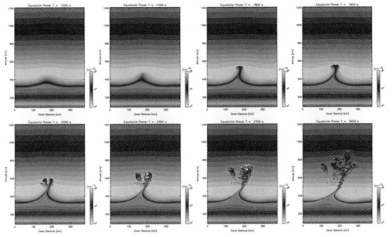

FIG. 4

An example of EPB growth calculated from the high-resolution bubble models.

From Yokoyama, T., Shinagawa, H., Jin, H., 2014. Nonlinear growth, bifurcation, and pinching of equatorial plasma bubble simulated by three-dimensional high-resolution bubble model. J. Geophys. Res. doi: 10.1002/2014JA020708.

To estimate the aviation exposure from SEP, several institutes in Japan and the United States developed WASAVIES, a Warning System for AVIation Exposure to Solar energetic particles (Sato et al., 2013; Kubo et al., 2015). WASAVIES has been tested and verified by making a comparison between the measured and calculated count rates of several neutron monitors during past ground level enhancement (GLE) events. The final goal of this project is to predict the enhancement of radiation doses due to SEP exposure within 6 h from the GLE onset.

4 ACTION TO SATELLITE SAVING

Before the Halloween 2003 event, Japanese satellites have often been affected by space weather disturbances. The Japanese Earth measurement satellite ADEOS-2 (Midori2) had fatal trouble on the solar panel on October 24, 2003; one of the most plausible reasons for this was an electric charge with auroral particles. Another example was shown on July 17, 2000, during the Bastille storm event. A Japanese astrophysics satellite Astro-D (ASCA) could not control its attitude and fell into the atmosphere. The reason is thought to be the significant increase of air drag with thermal expansion by auroral activities.

At present, the Japan Aerospace Exploration Agency (JAXA) prepared a simulation sequence for evaluating the robustness of the satellite against a severe space environment in the satellite development process. The model named MUSCAT (MUltility Spacecraft Charging Analysis Tool) can estimate the electric charge intensity and distribution with taking the shape and material of each satellite into consideration (Cho et al., 2012). Fig. 5 is an example of satellite simulation with MUSCAT.

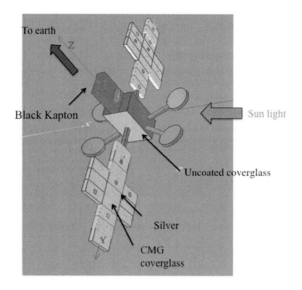

FIG. 5

An example of a satellite model with Multiutility Spacecraft Charging Analysis Tool (MUSCAT).
From Cho, M., Sumida, T., Masui, H., Toyoda, K., Kim, J.H., Hatta, S., Wong, F. K., Hoang, B., 2012. Spacecraft charging analysis of large GEO satellites using MUSCAT. IEEE Trans. Plasma Sci. 40(4).

Now a new attempt is under way to build a real-time or prediction warning system of satellite charge with connections between MUSCAT and the space environment monitoring system. The real-time space environment will be estimated from housekeeping data of geosynchronous satellites and predicted with simulation code. This project, named "Taylor-made space weather," is conducted on the Project for Solar-Terrestrial Environment Prediction (PSTEP), as shown in the last part of this section.

5 ACTION TO GIC

As Japan is located in mid- and low-latitudes in the geomagnetic coordinates, most people—including scientists—used to believe that the influence of geomagnetically induced currents (GICs) on the electric power grid was negligible.

However, some studies (e.g., Fujita et al., 2016) show that an inhomogeneous geological structure can enhance the effect of GIC even in Japan. Goto (2015) used a simulation code and determined that the amplitude of a geomagnetically induced electric field (GIE) on land can be estimated at more than twice as large as that of the homogeneous earth without the sea layer. This result is critical for Japan because most of the electric power plants are located in coastal areas.

In addition, the Japanese government began to consider rare but severe disasters in terms of keeping social infrastructures. In this stream, the Ministry of Economy, Trade and Industry (METI) presented a report about the influence of solar flares on the electric power grid. The document concluded that the scientists and electric companies needed to keep close communications for protecting electric infrastructures against future extreme space weather events.

6 INTRODUCTION OF PSTEP

PSTEP, launched in 2015, involved more than 100 scientists in Japan, supported by a Grant-in-Aid for Scientific Research on Innovative Areas from the Ministry of Education, Culture, Sports, and Technology (MEXT), Japan. The aim of this project is to develop a synergistic interaction between the space weather user needs and the cutting-edge scientific results through predictive operations. To realize this aim, this project put a large emphasis on the discussion between scientists and industry people (e.g., airline, electric power, telecommunications and satellite operators) One of the aims of PSTEP is to publish a Japanese version of a hazardous map against space weather disasters.

REFERENCES

Cho, M., Sumida, T., Masui, H., Toyoda, K., Kim, J.H., Hatta, S., Wong, F.K., Hoang, B., 2012. Spacecraft charging analysis of large GEO satellites using MUSCAT. IEEE Trans. Plasma Sci. 40 (4), 1248–1256.

Fujita, S., Kataoka, R., Fujii, I., Pulkkinen, A., Watari, S., 2016. Extremely severe space weather and geomagnetically induced currents in regions with locally heterogeneous ground resistivity. Earth Planets Space. https://doi.org/10.1186/s40623-016-0428-y.

Fukao, S., Yokoyama, T., Tayama, T., Yamamoto, M., Maruyama, T., Saito, S., 2006. Eastward traverse of equatorial plasma plumes observed with the Equatorial Atmosphere Radar in Indonesia. Ann. Geophys. 24, 1411–1418.

Goto, T., 2015. Numerical studies of geomagnetically induced electric field on seafloor and near coastal zones incorporated with heterogeneous conductivity distributions. Earth Planets Space. https://doi.org/10.1186/s40625-015-0356-2.

Kubo, Y., Kataoka, R., Sato, T., 2015. Interplanetary particle transpoert simulation for warning system for aviation exposure to solar energetic particles. Earth Planets Space. https://doi.org/10.1186/s40623-015-0260-9.

Maruyama, T., Uemoto, J., Ishii, M., Tsugawa, T., Supnithi, P., Komolmis, T., 2014. Low-latitude ionospheric height variation as observed by meridional ionosonde chain: Formation of ionospheric ceiling over the magnetic equator. J. Geophys. Res. https://doi.org/10.1002/2014JA020215.

Saito, S., Yoshihara, T., Nakahara, H., 2015. Performance of GAST-D ionospheric gradient monitor studied with low latitude ionospheric disturbance data obtained in a real airport environment. In: Proceedings of the ION 2015 Pacific PNT Meeting, Honolulu, Hawaii, April, pp. 815–820.

Saito, S., Sunda, S., Lee, J., Pullen, S., Supriadi, S., Yoshihara, T., Terkildsen, M., Lecat, F., ICAO APANPIRG ionospheric studies task force, 2017. Ionospheric delay gradient model for GBAS in the Asia-Pacific region, Submitted to GPS Solutions.

Sato, T., Kataoka, R., Yasuda, H., Yashiro, S., Kuwabara, t., Shiota, D., Kubo, Y., 2013. Air shower simulation for WASAVIES: warning system for aviation exposure to solar energetic particles. Radiat. Prot. Dosim. 1–5. https://doi.org/10.1093/rpd/nct332.

Uemoto, J., Ono, T., Maruyama, T., Saito, S., Iizima, M., Kumamoto, A., 2007. Magnetic conjugate observation of the F3 layer using the SEALION ionosonde network. Geophys. Res. Lett. https://doi.org/10.1029/2006GL028783.

Yokoyama, T., Shinagawa, H., Jin, H., 2014. Nonlinear growth, bifurcation, and pinching of equatorial plasma bubble simulated by three-dimensional high-resolution bubble model. J. Geophys. Res. https://doi.org/10.1002/2014JA020708.

Index

Note: Page numbers followed by *f* indicate figures, *t* indicate tables, and *b* indicate boxes.

A

Acoustic gravity waves (AGWs), 569, 569*t*
Active regions, magnetic fields and, 46–50, 47*f*, 49*f*, 51*f*
AD 774 event, 670
ADEOS-2 (Midori 2), 723
Advanced composition explorer (ACE), 143, 173–175, 234, 694
AE8/AP8 radiation belt model, 423–424, 424*f*
Aeroplanets model, 624–625, 625*f*
AFRL-SCINDA. *See* Air Force Research Laboratory Scintillation Network Decision Aid (AFRL-SCINDA)
AGWs. *See* Acoustic gravity waves (AGWs)
Airapetian and Usmanov's model, 616–617
Aircraft Communications Addressing and Reporting System (ACARS), 464–465
Aircraft fly, 25
Air Force Research Laboratory Scintillation Network Decision Aid (AFRL-SCINDA), 545–546
Alaska, CASES receiver, 562–563, 564*f*, 565
Alfven waves, 616–617, 618*f*
Alouette satellites, 639–642, 640*f*
ALTEA. *See* Anomalous long term effects in astronauts (ALTEA)
Aluminum, electron/proton penetration depth, 421, 421*f*
AMAX event, 381
Amplitude scintillation, 545, 558, 560–561, 578–579
Anik satellites, 640–642
Anomalous long term effects in astronauts (ALTEA), 437, 448
  solar proton events, 441–444, 442–444*f*
  USLab module, 439*f*
Appleton anomalies, 559
Arcing
  cause of, 402
  effects of, 402–404, 403*f*
Arctic Air Traffic Control Center, Edmonton, 638–639, 638*f*
Assimilative Mapping of Ionospheric Electrodynamics (AMIE) algorithm, 565–566
ASTRA. *See* Atmospheric & Space Technology Research Associates (ASTRA)
Astronauts, health risk, 433–435
Astronomical unit (AU) forecasting, 446–447
Atmospheric escape, XUV flux effects on, 622–623, 624*f*
Atmospheric radiation
  environment, 6
    space weather impacts, 25–27
  policymakers, 675–676
Atmospheric & Space Technology Research Associates (ASTRA)

CASES GPS receivers, 559, 561*f*, 562–563, 565*f*
DICE Cubesat, 566–567
dual-frequency GPS receivers, 495, 547, 571, 576
ionospheric electron density assimilation, 572–573, 573*f*
TIDDBIT Mapper, 570–571
AU. *See* Astronomical unit (AU)
Augmented navigation system, 576
Auroral electrojet index, distribution function of, 147, 147*f*
Aurora lights, 486
Australia
  K index, geomagnetic, 713–714, 714*f*
  OECD, 717
  power grids, 716–717
  severe events, 712–716
  solar-terrestrial physics, 711–712
  Space Weather Services, 711–712, 714–715*f*, 716
  stakeholder technology groups, 716–717
Australian Bureau of Meteorology, 712
Australian Space Forecast Center, 715
Automated Radiation Measurements for Aerospace Safety (ARMAS) program, 459, 461*f*
Aviation
  GNSS, 647
  Japan, 721–723
  navigation system impacts, 602–604, 604*f*
  radiation, 453–456
    events example, 463–465
    Halloween storm, 462
    measurements, 457–458
    models, 456–457
    monitoring for extreme conditions, 458–460
    space weather events, 460–463
  space weather effect on, 707–708
Avionics, radiation effect, 456

B

Barometric wind, 528
Bastille Day storm (July 2000), 16, 261, 270–274, 271–274*f*, 310–312, 500, 664–665
  OpenGGCM, MHD model, 522–525
BATS-R-US. *See* Block-Adaptive-Tree Solar-wind Roe-type Upwind Scheme (BATS-R-US)
Bear Lake, Utah, 519–520
BeiDou Navigation Satellite System, 543–544
Best-fit model parameters, 127
Bethe-Bloch equation, 420–421

727

Block-Adaptive-Tree Solar-wind Roe-type Upwind Scheme (BATS-R-US), 619
  MHD model, 241–242
    CRCM, ring current model, 248–250, 249f
Blue jets, 474–475, 477
Bootstrapping, 121
Brazil, CASES receiver, 561f
Brier skill score (BSS), 77, 81–82
Bulk charging. *See* Deep dielectric charging
Buoyancy waves, 503, 504f

C

Canada
  magnetic forecasting, 646
  population and electric power consumption in, 636f
  scintillation in, 647
  space weather, 635–637
    Alouette satellites, 639–642, 640f
    ground systems, 642–646
    high-frequency radio communications, 635–639
    instruments, 648, 649f
    International Satellites for Ionospheric Studies, 639–642
    surveying and navigation, 646–647
    telegraph problems, 635–636
Canadian Coast Guard (CCG), 648f
Canadian High Arctic Ionospheric Network (CHAIN), 561–562
Canadian hydro-quebec power system, 164
Canadian Space Weather Forecast Center (CSWFC), 638–639
Carrier-to-noise ratio (C/No), 545
Carrington storm, 1859, 116, 118–119, 132–133, 171–172, 494, 505, 521–522, 572, 578, 601–602
  ionospheric response, 529–532
  magnetic storm, 613
  magnetopause, 522–525
  modeling, 238–239
  neutral atmosphere response, 525–529
  occurrence probability of, 172
  Weimer empirical model, 522–525, 523f
Cathodic protection system, 15
CCMC. *See* Community Coordinated Modeling Center (CCMC)
Center of Space Environment Modeling (CSEM), 619
CHAIN. *See* Canadian High Arctic Ionospheric Network (CHAIN)
CHAMP satellite
  integrated electron content, 495–496, 496f, 499
  solar flares, 519–520
Charged particle, penetration depth, 421, 429
China, BeiDou Navigation Satellite System, 543–544
Chorus waves, 390–391
  generation, 386–387
  waves, 375–376, 378–379, 382–383, 390

CIMI ring current model. *See* Comprehensive inner magnetosphere-ionosphere model (CIMI)
CIRs. *See* Corotating interaction region (CIR)
Civil Contingencies Act of 2004, 662
Climate and Weather of the Sun-Earth System (CAWSES-II), 323–324
Cluster-4 spacecraft on September 16, 2003, 378–379, 379f
CME. *See* Coronal mass ejection (CME)
C/No. *See* Carrier-to-noise ratio (C/No)
Coherent waves, electron interaction with, 383
Colaba (Bombay)
  intense and superintense magnetic storms from, 165–166t
  magnetogram, 167, 167f
Collision avoidance, 517
Combined Release and Radiation Effects Satellite (CRRES), 429–430, 430f
Commercial-off-the-shelf (COTS), 459
Community Coordinated Modeling Center (CCMC), 83, 251, 619–621
Complementary cumulative distribution function (CCDF), 120, 122
  for geomagnetic storms, 124f
Complexity science, 140–141, 149
Comprehensive inner magnetosphere-ionosphere model (CIMI), 242
Comprehensive ring current model (CRCM), 235, 237
Conduction currents, 420
Connected autonomous space environment sensor (CASES) GPS receiver, 559, 562
  in Alaska, 562–563, 564f, 565
  in Brazil, 561f
  phase measurements, 562f
Constellation Observing System for Meteorology, Ionosphere, and Climate (COSMIC), 545, 548–551, 549–550f
Contiguous United States (CONUS), 504–505, 577–578, 598–601
  aviation system impacts, 602–604, 604f
Convective zone, 612–613
Coordinated universal time (UTC), 602–603, 604f
Coronal mass ejection (CME), 6, 66, 315–316, 424, 514, 590–591, 612–613, 643
  Carrington event, 613
  causing intense geomagnetic storms, 41f
  dynamic pressure from, 616–617
  effects on magnetosphere, 618–622, 619f, 621–622f
  forecasting at 1 AU, 446–447
  geomagnetic storm, 520–532
  IMF forecast products, 692–694, 697–698
  impact on Earth's magnetic field, 643
  July 2012, 620, 701–702
  kinetic energy of, 49f, 50
  parameters, 173–174

primary consequences of, 40, 51
solar flare, 436
solar sources of, 41*f*
solar superflares, 614–616
speeds, 42–43
   cumulative distribution, 43–44, 43*f*, 45*f*
   and kinetic energies, 43–44
St. Patrick's day, 618–619, 619*f*
at Sun, 37–38
Corotating interaction region (CIR), 99–100, 374, 514, 590–591, 613
magnetic storms, 162–164, 385–386, 386*f*
COSMIC. *See* Constellation Observing System for Meteorology, Ionosphere, and Climate (COSMIC)
COSMIC Data Analysis and Archive Center (CDAAC), 546–547
Cosmic ray, 4
continuous satellite-based monitoring, 459
galactic, 4, 6, 434, 453–458, 454*f*, 462–463
ground-based continuous monitoring, 457–458
ground level enhancements, 458
neutrons, 25
radiation dosimetry experiment, 460
radiation from, 25
solar, 354
solar energetic particles, 453–454, 454*f*, 456, 458–459
source, 453–455, 454*f*
Cosmic ray albedo neutron decay (CRAND), 374
Cosmic Ray Telescope for the Effects of Radiation (CRaTER), 456–457
COSPAR Space Weather Roadmap, 671
COTS. *See* Commercial-off-the-shelf (COTS)
Countermeasure, solar particle events, 436, 448
Coupled Thermosphere-Ionosphere Model (CTIM), 525
Coupled thermosphere-ionosphere-plasmasphere-electrodynamic (CTIPe) model, 525
ionospheric response, 529–531
Joule heating rates, 524*f*, 525
neutral atmosphere response, 525–529, 526–527*f*
neutral composition, 525, 526–527*f*, 528–529
simulation, 527*f*, 529
CPCP. *See* Cross-polar cap potential (CPCP)
CRAND. *See* Cosmic ray albedo neutron decay (CRAND)
CRCM. *See* Comprehensive ring current model (CRCM)
Crewed orbital stations, radiation environment for, 361–367, 362*f*, 364*f*, 366*f*
Cross-polar cap potential (CPCP), 197–198, 234, 522
saturation effects, 522
CRRES. *See* Combined Release and Radiation Effects Satellite (CRRES)
CTIM. *See* Coupled Thermosphere-Ionosphere Model (CTIM)

CTIPe model. *See* Coupled thermosphere-ionosphere-plasmasphere-electrodynamic (CTIPe) model
Cutoff rigidity, 454–455, 455*f*

D

Data acquisition unit (DAU), 438
Data-driven modeling, of extreme space weather, 139–145, 142*f*, 144–145*f*
DAU. *See* Data acquisition unit (DAU)
Deep dielectric charging, 28, 402, 420–423, 426–427
characteristics, 421–422
delay time, 425–426, 428
ions role, 422–423, 423*f*, 426
space environments, 423–426
Deep dielectric discharge, 426–427
event parameters, 429
Deep Space Climate Observatory (DSCOVR) spacecraft, 671–672, 673*f*, 694
Deep-space exploration, 434
Defence Defense Research Telecommunications Establishment (DRTE), 637–638
Defense Meteorological Satellite Program (DMSP), 303, 502
Delay time, deep dielectric charging, 425–426, 428
Delphi-like approach, 703–705
Detrended fluctuation analysis (DFA), 147
Differential GPS (DGPS), 647, 648*f*
Discharge
deep dielectric, 426–427, 429
electrostatic, 422, 429–430
Distributions
Gaussian, 146
photon number, 480, 480*f*
plasma pressure distribution, 260, 268*f*, 270, 275
power-law, 100, 116
ring current plasma, 232
spectral energy, 615*f*
storm enhanced density total electron content distribution, 572–573, 573*f*
Disturbance storm-time (Dst) index, geomagnetic, 55–56, 56*f*, 99–100, 118, 232, 428, 494
Australasian region, 714, 715*f*
equatorial scintillation, 500
estimates and confidence intervals for, 124*t*, 126*t*
geomagnetic activity, 588–589
time variation of measured Dst and pressure-corrected, 103*f*
Vuong's test statistics for, 126*t*
DMSP. *See* Defense Meteorological Satellite Program (DMSP)
Dose equivalent, 434–435, 444
Dose index (D-index)
concept, 462–463
exposure scenarios, 463*t*

Dose index (D-index) *(Continued)*
  GLE, 463–464, 464*f*
  PANDOCA model, 464, 464–465*f*
D-Region Absorption Prediction (DRAP), 595–598, 597*f*
DSCOVR spacecraft. *See* Deep Space Climate Observatory (DSCOVR) spacecraft
Dst index, geomagnetic. *See* Disturbance storm-time (Dst) index
Dual-frequency GPS receivers, 576

E

Earth
  Early Earth, CMEs effects on magnetosphere, 618–622, 619*f*, 621–622*f*
  environment, forecast product, 694–698
  induction in conducting, 211–212
  magnetosphere, 19, 99–100, 159–160, 232, 374
    outer radiation belt, 374
    radiation belt, 374, 376–378, 386, 390
      electron, 593–594
      for extreme storms, 239–240
      proton, variations over solar cycles, 359–360, 360*f*
      temporal variation, 423–424, 425*f*
  societal impacts on human activities on, 3
  space weather environments at, 4–8
  upper atmosphere, 16
EarthScope impedance tensors, 217–218
Earth's crust, 8–9
  magnetic induction in, 9
  radioactive elements in, 25
Edison Electric Institute, 643
EIA. *See* Equatorial ionization anomaly (EIA)
Electrical power systems, 190–192
Electric and Magnetic Field Instrument Suit and Integrated Science (EMFISIS), 390
Electric fields, and total electron content, 495–501
Electric power, 705–707
  consumption, Canada, 636*f*
Electromagnetic emission, from solar flares, 40–41
Electromagnetic ion cyclotron (EMIC) waves, 375–376, 383
Electron acceleration, 390
  HILDCAAs temporal length dependence of, 384–385, 384*f*
  maximum energy-level dependence of, 381–383, 382*f*
Electronic device, single-event effects
  cause, 6, 25
  rates, 25–26
Electrons
  energetic, 8
  high-energy, 402
  interaction with coherent waves, 383
  magnetospheric relativistic, 374–375, 379, 390
  penetration depth, 421–422, 421*f*
  radiation belts of, 232–233, 593–595
Electrostatic discharge (ESD), 422, 429–430
  events, 402
Elves, upper atmosphere, 474–475
Emergency management, 692
  high frequency radio, 686
  NHC *see* National Hurricane Center (NHC)
  user community, 692
Emergency managers, 684
  expert analysis available on demand, 688–689
  forecast products, 690–698
  geomagnetic storms, 687
  government, 698
  plain language, 688
  radio blackouts, 686
  sharable information, 689
  solar radiation storm, 686–687
EMFISIS. *See* Electric and Magnetic Field Instrument Suit and Integrated Science (EMFISIS)
Emission, photoelectron, 404, 405*f*
ENA. *See* Energetic neutral atom (ENA)
Energetic neutral atom (ENA), 232, 236–237, 260
Energetic proton, electron and alpha detector (EPEAD), 592–593
Energetic protons, 129–130, 159–160
Energetic solar particles. *See* Solar energetic particle (SEP) events
Energy deposition, transient luminous event, 485–486
Enhanced Solar Radiation Alert System (ESRAS), 462–463
EPEAD. *See* Energetic proton, electron and alpha detector (EPEAD)
Equatorial electrojet (EEJ) current, 195
Equatorial ionization anomaly (EIA), 311–313, 531–532
Equatorial plasma bubble (EPB), 722, 723*f*
Equatorial plasma irregularities, 499–501
Equatorial scintillation, 499–501
ESD. *See* Electrostatic discharge (ESD)
ESP. *See* Energetic storm particle (ESP)
Europe, neutron monitor stations, 464
European Geostationary Navigation Overlay Service (EGNOS), 588–589, 602–604, 604*f*
European Union's Galileo, 543–544
EUV. *See* Extreme ultraviolet (EUV)
EVA. *See* Extra vehicular activity (EVA)
EXIS. *See* Extreme Ultraviolet and X-ray Irradiance Sensors (EXIS)
Exoplanets, 628–629
Extra vehicular activity (EVA), 434, 448
Extreme, 139
  charging environments, 404, 413–415
  definition, 38–40
    science challenges, 199–201

drivers, science challenges, 195–197
GIC effects on ground systems, 191t
magnetic storms, 353–354
modeling, science challenges, 197–199
science challenges, 194–201
SEP events concept, 350–351
solar flares, 84–86
spacecraft charging events, 402
storms modeling of radiation belt response for, 239–240
Extremely low frequency (ELF), TLE, 484
Extreme space weather events, 115, 187–188, 194, 259–260, 588–589, 589t. *See also* Solar energetic particle (SEP) events; Solar particle events; Transient luminous event (TLE)
AE index, auroral, 118, 127–128, 129f, 130t
analysis, 130–135
in Australia, 711–718
Bastille day storm (July 2000), 293f, 310b
Carrington event, 522–525, 523f, 613, 667
data-driven modeling of, 139–145, 142f, 144–145f
datasets, 117–119
definitions, 37, 71, 87–88
Dxt and Dcx analysis, 117–118, 127, 128f
electrostatic discharge (ESD) events, 402
and emergency management, 684–699
estimation of, 42–50
  CME speeds, 42–44, 43f, 45f
  flare size distribution, 44–46, 46f
  active regions and magnetic fields, 46–50, 47f, 49f, 51f
evaluation of past, 458
geoeffective events, 588–589, 589t
Geostationary Operational Environmental Satellite (GOES) events, 283–324, 284f, 287t
ground level enhancement (GLE) event, 612–613, 723
Halloween storm (October–November 2003), 295f, 313b, 314t
in heliosphere, 129–130
HILDCAAs event, 377–380, 377f, 380f, 382
Hydro-Quebec event (March 1989), 287–300, 288f, 301b
impacts on radio systems, 20–22t
impacts on upper atmosphere, 16–17
  atmospheric drag, 22–24, 24f
  atmospheric radiation environment, 25–27
  satellite plasma environments, 27–29
  trans-ionospheric radio propagation, 17–22, 20–22t
intense and superintense magnetic storms from Colaba and Alibag magnetic observatories, 165–166t
in ionosphere, 127–128
March 1989 storm, 287–300, 288f, 301b
modelling, geomagnetic field
  data binning, 263
  model database, 264
  model structure, 261–262
model structure, 261–262
as natural hazard, 659–662
overview, 38–42
past supermagnetic storms, 166–169
policy responses to natural hazards, 662–663
predictability of, 146–149, 147–148f
radiation events, 453–456, 460–465
relativistic electron dropout (RED) event, 374–375
risk assessment, 666–671, 666f
St. Patrick's Day storms CME event, 618–619, 619f
single-event effects (SEE), 6, 25–26, 29, 350
single event upset (SEU), 422–423, 588–589, 593, 595, 707–708
SEU rate, 366–367, 366f
solar eruptions, 51–58
solar flare events, 84–86
solar origins of activity, 590–591
spacecraft charging events
  extreme, 402
  limits, 412–415, 413–414f
  worst, 411–412, 413t
spectral energy distribution (SED) event, 568–569, 572–573
from sun, 612–617
Super-Carrington CME event, 614–616, 620–621, 623–624
statistical modeling, 119–135
  best-fit parameters estimation, 120
  datasets, 117–119
  model comparison, 121–122
  nonparametric bootstrapping, 121
  tail in distribution, identification, 120–121
statistics of events, 142–143, 144f, 146
technological development, 29–31
time stationarity assumption validity assessment, 126
Extreme storm detector, 504–505
Extreme ultraviolet (EUV)
  effects, 517–519
  electrodynamics, 531
  flux, 519–520
  solar, 19
  Young Earth, 622–623, 624f
Extreme Ultraviolet and X-ray Irradiance Sensors (EXIS), 591
Eyjafjallajökull eruptions 2010, 661–662

F

Faint young Sun (FYS) paradox, 626–628
Faraday's law, 8–9, 188, 233
FDs. *See* Forbush Decreases (FDs)
Federal Aviation Administration (FAA)
  law, 707–708
  WAAS, 545–546

Index

Federal Emergency Management Agency (FEMA), 685–689, 685f
Federal Energy Regulatory Commission (FERC), 194, 685
Field-aligned currents (FACs) system, 262
Five-Day Graphical Tropical Weather Outlooks, 690–692, 691f
Flare. *See* Solar flare
Florida Effect, 504–506, 506f
Fokker-Plank equation, 239
Forbush decreases (FDs), 287–300
Forecasting, 87–88
  at 1 AU, 446–447
  short-time scale, 174–175
  solar flare *see* Solar flare forecasting
  solar particle events, SPEs, 436
  space habitat, 447
  of space weather, 144
Forecast product, 688
  Earth environment, 694–698
  emergency managers, 690–698
  geomagnetically induced currents, 698
  geomagnetic storms, 697–698
  interplanetary magnetic field, 692–694
  National Hurricane Center, 684
  single end-user, characteristics, 697
  solar eruptive, 692
  solar flare, 690–692
  solar radiation storms, 697
  "sun to mud", 690f
  tropical cyclones, 690
FORMOSAT-2 satellite, 475, 484–485
FYS paradox. *See* Faint young Sun (FYS) paradox

G

Galactic cosmic ray (GCR), 4, 6, 132–133, 361–362, 434, 453–458, 454f, 462–463, 675–676
  behavior, 356
  extra vehicular activity, 448
  modulation over solar cycles, 354–359, 359f
  parameter of fluctuations of, 358
  radiation, 232
    exposure, 448
  risk mitigation, 435–437
Galaxy 15 satellite failure, 415
Gamma ray line (GRL), 304
Gas pipelines, impact on ground systems, 192–193
Gaussian distribution, 146
GBAS. *See* Ground-Based Augmentation System (GBAS)
GBAS Approach Service Type D (GAST-D), 722
GCR. *See* Galactic cosmic ray (GCR)
GEO. *See* Geosynchronous Earth orbit (GEO)
Geocentric solar magnetospheric (GSM) coordinate system, 377–378
Geoelectric exceedance amplitudes, 222
Geoelectric field, 4, 209–210
  November 20, 2003, 196f
  during storm event on October 7 2015, 190f
Geoelectric hazard maps, 221–224, 223–224f
Geoelectric monitoring, 211
Geological interpretations, 219–221, 219–220f
Geomagnetic activity, 10–12, 16, 23, 54–55, 146, 158, 199, 234, 237–238, 676, 677f
  indices, 354, 355f, 385
  prediction of, 142f
  spatial structure of, 144
Geomagnetically induced currents (GICs), 8–16, 162–164, 187–190, 193, 211, 232–233, 724
  Canada, 643–644, 651
  Carrington event, 667
  forecast products, 698
  impact on
    pipelines, 15
    rail systems, 15
    telecommunication cable system, 16
  primary effect of, 190–191
  technological system impacts, 601–602
  theory of, 188–189
  U.S. federal actions relating to, 193–194
Geomagnetically induced electric field (GIE), 724
Geomagnetic dipole, 677–678f
Geomagnetic disturbance (GMD), 189, 194
Geomagnetic effects, on power systems, 642–644
Geomagnetic field, 476–477, 592–593
  modeling
    data binning, 263
    model database, 264
    model structure, 261–262
Geomagnetic indices
  Ap, 24f, 286–287, 713–714
  Dst, 118, 141, 144, 149–150, 159, 161, 232, 234–235, 428, 574
  Dxt and *Dcx* analysis, 117–118, 127, 128f
  K, 713–714, 714f
  Kp, 199, 286–287, 354, 364f, 412f, 428, 563, 564f, 713–714
  SYM-H index, 159, 161, 200, 200f, 232, 234–235, 574
Geomagnetic storm, 159, 232–233, 286, 312, 602–603, 692–693. *See also* Magnetic storm
  categories and types, 161, 162–163f
  cause of, 494
  CCDF for, 124f
  characteristics, 161–164
  CMEs causing intense, 41f
  coronal mass ejection, 520–532
  cumulative distribution of intense, 57f
  equatorial scintillation, 500–501
  extreme and intense, modeling, 236–240

forecast products, 697–698
hazard, 687
interplanetary causes of intense, 159–161, 160f
ionospheric response, 303, 529–532
March 1989, 24f
neutral atmosphere response, 525–529
NOAA space weather scales for, 286–287
positive storm effect, 496
pressure and current distribution, 242–244, 243–244f
ring current plasma role, 248–250, 249f
solar cycle 24, 716
thermospheric effects, 315–316
Geomagnetic waveform time series, 212–214, 214f
Geomagnetism, 157–158
GEONET. *See* GNSS Earth Observation Network System (GEONET); GPS Earth Observation Network System (GEONET)
Geospace response, 591–595
    CMEs with, 588–589
    solar wind on, 593
Geospatial Information Authority of Japan, 721
Geostationary Earth orbit (GEO), 285
Geostationary magnetopausing crossing (GMC) detection, 593
Geostationary Operational Environmental Satellite (GOES), 44–45, 402, 443, 519, 588
    GOES-16 space weather sensors, 285
    internal electrostatic discharge, 593–594
    program, 283, 285
    requirements for, 285
    solar instruments, 590–591
    space environmental data, 284f, 286
    space environment monitor, 592–593, 592f
    space weather instruments, 285
Geosynchronous Earth Orbit (GEO), 234, 409–410, 460–462
    Earth equatorial magnetic field in, 409
    emergency responders, 686–687
    energetic particles/magnetic field observations, 592–593
    geosynchronous magnetopause crossings, 593
    physics of charging in, 404–405
    radiation environment, 593–594, 594f
    technological system impacts, 601–602
    telecommunication satellite, 415
Geosynchronous magnetopause crossings, 593
Geosynchronous Operational Environmental Satellite (GOES) program, 457–459, 462, 465
GICs. *See* Geomagnetically induced currents (GICs)
Gigantic jets, 474–475, 477
GIM technique. *See* Global Ionospheric Mapping (GIM) technique
GLE. *See* Ground level enhancement (GLE)
Global coupled model, different modules in, 233f
Global Ionosphere Thermosphere Model (GTIM), 503
Global Ionospheric Mapping (GIM) technique, 495–496

Global magnetic hazard functions, 216–217
Global navigation satellite system (GNSS), 66, 543–544, 660, 721
    applications, 547–548, 551–552
    Canada, 647
    degradation, 601
    impacts, 598, 602–604
    ionospheric storms effect, 575–578
    low-latitude scintillation, 560
    risk for, 676
    space-based, 546–547
    to space weather, vulnerability, 30
    space weather effects on, 19
    SRB effects on, 544–547
    technology, 30
    total electron content, 559–560
    users, 708
Global positioning system (GPS), 8, 174, 312–313, 557–558
    CASES receiver *see* Connected autonomous space environment sensor (CASES) GPS receiver
    differential, 647, 648f
    interconnected critical systems, 556–557, 556f
    ionospheric storms effect, 575–578
    low-latitude scintillation, 560, 561f
    total electron content, 560, 598–601, 600f
    Wide Area Augmentation System, 504–505
Global Total Electron Content (GloTEC), 598–601, 600f
GMD. *See* Geomagnetic disturbance (GMD)
GNSS. *See* Global navigation satellite system (GNSS)
GNSS Earth Observation Network System (GEONET), 721, 722f
GOCE. *See* Gravity field and steady-state Ocean Circulation Explorer (GOCE)
GOES. *See* Geostationary Operational Environmental Satellite (GOES)
GPS. *See* Global positioning system (GPS)
GPS Aided GEO Augmented Navigation (GAGAN), 574
GPS Earth Observation Network System (GEONET), 545–546
Gravity field and steady-state Ocean Circulation Explorer (GOCE), 525
Gravity waves, 503, 504f
Grid topology, 10f
GRL. *See* Gamma ray line (GRL)
Ground-based augmentation system (GBAS), 717, 721–722
Ground level enhancement (GLE), 38, 287–300, 303, 317, 465
    dose index, 463–464, 464f
    events, 612–613, 723
    solar cosmic rays, 458
Ground system, impact on, 189–193, 191t
    electrical power system, 190–192
    oil and gas pipelines, 192–193
GTIM. *See* Global Ionosphere Thermosphere Model (GTIM)

H

Halloween solar eruptions, 38
Halloween storms (October–November, 2003), 14–15, 37–38, 169, 192, 209–210, 315–316, 409–410, 410f, 412f, 496f, 499, 505, 529–532, 532f, 572, 588–589, 601–602
    aviation, 462
    IMAGE satellite, 531–532, 532f
    impinging particles, 462
    positive effect, 496
    TEC, 574f
Halos, upper atmosphere, 474–475
Hanssen & Kuiper Skill Statistic (H&KSS), 77–78, 82
Hazard, 684–685
    definition, 210–211
    geomagnetic storms, 687
    radio blackouts, 686
    solar radiation storm, 686–687
Hazardously misleading information (HMI), 494
HCS. *See* Heliospheric current sheet (HCS)
Health hazard, astronauts, 433–435
Helioseismology, 73–74
Heliosphere, extreme space weather events in, 129–130
Heliospheric current sheet (HCS), 375, 376f
Heliospheric plasma sheet (HPS), 374–376
Helium Oxygen Proton Electrons (HOPE) instrument, 267
High-energy electron, 28, 402
High frequency (HF) radio, 686–687
    Australia, 711–712
    Canada, 635–639
    emergency management, 686
    space weather disruptions, 707–708
    waves, 719
High-intensity long-duration continuous auroral activities (HILDCAAs), 374–376
    event, 379–380, 380f, 382
        September 15–20, 2003, 377–378, 377f
        temporal length dependence of electron acceleration, 384–385, 384f
Highly elliptical orbits (HEO), 641–642
High-speed solar wind streams (HSSs), 374–376
    and storm recovery phase, 389
HILDCAAs. *See* High-intensity long-duration continuous auroral activities (HILDCAAs)
Hill estimator, 149
HINODE-XRT, 438–440
HMI. *See* Hazardously misleading information (HMI)
Hodgson, Richard, 158–159
Horizontal protection level (HPL), 602–603
Hot electron and ion drift integrator (HEIDI) code, 235
HPS. *See* Heliospheric plasma sheet (HPS)
Hudson Bay, 637
Hurricane Katrina, 701–702
Hurst exponent, 149
Hydro-Quebec event (March 1989), 169, 170f
Hydro-Quebec power system, 191–192, 644, 651

I

ICMEs. *See* Interplanetary coronal mass ejections (ICMEs)
IDM. *See* Internal Discharge Monitor (IDM)
IEC. *See* Integrated electron content (IEC)
IESD. *See* Internal electrostatic discharge (IESD)
Imager of Sprites and Upper Atmospheric Lightning (ISUAL), 475, 480
IMAGE satellite, 236–237
    Halloween storm, 531–532, 532f
IMF. *See* Interplanetary magnetic field (IMF)
India's Navigation with Indian Constellation (NAVIC), 543–544
Indonesian Space Agency LAPAN, 712–713
Inner proton radiation belt variations, solar cycles, 359–360, 360f
Input-output (IO) models, 669
Instrument landing systems (ILS), 717
Integrated electron content (IEC), CHAMP satellite, 495–496, 496f, 499
Intelsat satellite, charging events on, 406, 407f
Interfrequency bias (IFB), 495
Internal charging. *See* Deep dielectric charging
Internal Discharge Monitor (IDM), 429–430
Internal electrostatic discharge (IESD), GOES, 593–594
International Civil Aviation Organization (ICAO), 721
International Geophysical Year, 635–636, 667
International GNSS Service (IGS), 545–546
International Reference Ionosphere (IRI) model, 495–496
    climatological model, 559, 559f
    electron density, 505–506, 507f
    ion and neutral densities, 501–502, 503t
International Satellites for Ionospheric Studies (ISIS), 639–642
International Space Environment Service (ISES), 671, 712–713, 719–720
International Space Station (ISS), 361–362, 363t, 364, 433–434, 437, 462–463
    ALTEA detector system, 438, 439f
    South Atlantic Anomaly, 448
    SPE measurements, 437–446
International Ursigram and World Days Service (IUWDS), 719–720
Interplanetary coronal mass ejections (ICMEs), 99–103, 160–161, 167–169, 241, 374, 514, 614–615, 618–619
    magnetic storms, 162–163, 387–390
Interplanetary magnetic field (IMF), 99–100, 159–160, 374, 377–378, 521–525, 614–615, 618
    coronal mass ejection, 692–694, 697–698
    forecast product, 692–694

Ion and neutral oxygen densities, 501–502, 503t
Ionizing radiation
 solar flare associated, 66, 67f
 sources, 453–454
Ion-neutral coupling, 501–503, 504f
Ionosphere, 4, 234–235, 241–242, 555–556, 558–559
 Carrington event, 529–532
 DRAP, 595–598, 597f
 equatorial scintillation, 499–501
 extreme space weather events in, 127–128
 high-latitude scintillation, 561–565, 561f
 low-latitude scintillation, 560–561, 561f
 OVATION Prime (OP) empirical model, 598, 599f
 positive phases, 520
 response, 513–514, 529–532, 595–601, 597f, 599–600f
 total electron content, 559–560, 598–601, 600f
  storm enhanced densities, 565–569, 567f
  traveling ionospheric disturbances, 569–571, 569t, 570–572f
Ionosphere Sounding Satellite-b (ISS-b), 719, 720f
Ionosphere-thermosphere response, 248–250
Ionosphere-thermosphere system, 232–233
Ionospheric convection pattern, 18
Ionospheric disturbances, 515
Ionospheric electrodynamics (IE) module, 241–242
Ionospheric group delay, 20–22t
Ionospheric plasma, 18
Ionospheric Prediction Service (IPS), 711–712
Ionospheric spatial gradient monitor (ISGM), 722
Ionospheric storm, 494
 event studies for, 572–575
 GPS system effects, 575–578
 in November 2003, 574
 in November 2004, 575
 in October 2003, 574
 positive phase, 494–495
 superfountain effect, 496–499, 497f
Ions
 deep dielectric charging, 420, 422–423, 423f, 426
 electrons vs, 424–426
IRI model. See International Reference Ionosphere (IRI) model
ISES. See International Space Environment Service (ISES)
ISIS. See International Satellites for Ionospheric Studies (ISIS)
ISS. See International Space Station (ISS)
ISUAL. See Imager of Sprites and Upper Atmospheric Lightning (ISUAL)

J

Japan
 aviation system, 721–723
 geomagnetically induced currents, 724
 GEONET, 721, 722f
 data, 545–546
 Radio Research Laboratory, 719–721
 satellite
  development process, 723–724, 724f
  positioning, 720–721
 sporadic-E layer, 720
 telecommunications, 720–721
 Tsunami, 701–702
Japan Aerospace Exploration Agency (JAXA), 723
Japan's Quasi-Zenith Satellite System (QZSS), 543–544
Joint Information Center, 689
Joule heating, 522, 523–524f
 height-integrated, 619, 620f

K

Kepler Space Telescope, 614–615
Kessler Syndrome, 517
K index, geomagnetic, 713–714, 714f
Kinetic energy (KE), of CME, 49f, 50
Kirchhoff's law, 10
Klobuchar model, 559–560, 575, 579
Kp index, geomagnetic, 428, 588–589
Kullback-Leibler divergence, 121

L

Laki eruption in Iceland, 667
Langmuir waves, 8
Large Angle and Spectrometric Coronagraph (LASCO) instrument, 38, 310–311
Learmonth solar radio telescope, 544, 544f
LEO. See Low Earth orbit (LEO)
Lightning-induced electron precipitation (LEP) process, 484
Linear energy transfer (LET) radiations, 456
Line-of-sight radio communications, 686
Localizer performance with vertical guidance (LPV), WAAS, 602–603
Lorenz attractor, 140
Los Alamos National Laboratory (LANL) geosynchronous satellites, 402
 plasma moments for, 410f
Loss of lock (LOL), 545–546
Low Earth orbit (LEO), 4, 515
 emergency responders, 686–687
 polar, 27
 radiation
  environment, 594–595, 596t
  risks, 433–434
 RO signals, 546–547
 satellites, 29, 404, 642
 single-event upsets, 595
 SNR, 548–551

Low Earth orbit (LEO) *(Continued)*
  spacecraft anomalies, 588–589
  space debris in, 517
  technological system impacts, 601–602

M

Magnetic and Doppler Imager (MDI), 38
Magnetic clouds (MC), 99–100
  geoeffective, 614–615
  in interplanetary space, 614–615
Magnetic declination perturbations, 157–158
Magnetic fields
  active regions and, 46–50, 47f, 49f, 51f
  GEO, 592–593
  interplanetary *see* Interplanetary magnetic field (IMF)
Magnetic flux emergence, 80f
Magnetic hazard functions
  global, 216–217, 216f
  observatory, 215–216, 216f
Magnetic induction
  in Earth's crust, 9
  Faraday's law of, 8–9
Magnetic local time (MLT), 354, 378–379, 568–569
Magnetic observatory data, 212, 213f
Magnetic potential energy (MPE), 48, 49f
Magnetic reconnection, 66, 80f
Magnetic storm, 99–100, 123f, 157–158. *See also* Geomagnetic storm
  extreme, 353–354
  fast shock, sheath, 387–389, 388f
  ICMEs, 387–390
  intense, 165–166t, 209–210
    interplanetary causes of, 159–161, 160f
  magnetic cloud and second and third, 389
  solar and interplanetary sources of, 100–109, 102–103f, 105t, 106f, 107t, 108f
  strong, 259–260, 353–354
  superintense, 165–166t
Magnetische Ungewitter phenomenon, 157–158
Magnetohydrodynamic (MHD) model, 79–81, 233f, 234–235, 260, 617–623, 619f
  of magnetosphere, 259–260
  scale-free approaches, 81
  shock waves, 541–543
  simulation, 143, 198f
Magnetometer, 144–145, 145f, 713–714
Magnetopause, 588–589
  Carrington event, 522–525
  shadowing, 387
Magnetosphere, 4, 140–143, 147, 195, 374
  CMEs effects on, 618–622, 619f, 621–622f
  Earth, 99–100, 159–160, 232
  global MHD models of, 259–260
  models of, 234–236
Magnetosphere-ionosphere currents, 188–189
Magnetospheric multiscale (MMS), 237–238
Magnetospheric Particle Sensor-Low Energy Range (MPS-LO), 593–594
Magnetospheric relativistic electrons, 374–375, 379, 390
Magnetospheric specification model (MSM), 235
Magnetospheric storm, 621–622, 622f. *See also* Geomagnetic storm
Magnetospheric substorms, 195
Magnetotelluric (MT)
  impedances, 217–218, 219–220f
  surveying, 9
Marconi station, 637
MARIE instrument, 37–38
Maritime European Communications Satellite (MARECS-A), 306–307
Mars, radiation on, 447–448
Marsh NOEM code, 528
Martian surface, SPE, 448
Mass Spectrometer and Incoherent Scatter (MSIS) model, 501–502, 503t
Maunder, Walter, 158–159
Maximum likelihood estimate (MLE), 120
McIntosh classification system, 84
MDI. *See* Magnetic and Doppler Imager (MDI)
Meanook Observatory, 635
Measurements of the Ultraviolet Spectral Characteristics of Low-mass Exoplanetary Systems (MUSCLES) Treasury Survey, 615–616
Measurements while drilling (MWD), 646–647
Meteorological operational (Metop) SEM, 594–595, 594f
Meteorology, space weather, 672–674
Meteorology-derived metrics, 77
Met Office Civil Contingencies aircraft, 675–676
MHD. *See* Magnetohydrodynamic (MHD)
Ministry of Economy, Trade and Industry (METI), Japan, 724
MIR space station, 363t
MLT. *See* Magnetic local time (MLT)
Monte Carlo analysis, 430
Moscow neutron monitor (NM), 286, 303
Moscow State University (MSU) satellites, 477–478
  photon number distribution, 480, 480f
  series of TLE, 480–482
  UVRIR detector, 478–479, 479f, 481f, 486
Mount St. Helens eruption, 1980, 664
MPE. *See* Magnetic potential energy (MPE)
MSIS model. *See* Mass Spectrometer and Incoherent Scatter (MSIS) model

Multifunctional Transport SATellite (MTSAT), 721
MUltility Spacecraft Charging Analysis Tool (MUSCAT), 723–724, 724f

N

NASA, 447–448
  ARMAS program, 459, 461f
  Community Coordinated Modeling Center (CCMC), 83, 251, 619–621
  Goddard Space Flight Center (GSFC), 83, 619, 673f
  Langley Research Center, 459
  Living With a Star (LWS) Institute program, 192
  OMNIWeb service, 664f
  Radiation Dosimetry Experiment, 460
  Solar and Heliospheric Observatory, 38–40, 591
National Centers for Environmental Information (NCEI), 591, 593–594
National Electricity Market (NEM), 716–717
National Hurricane Center (NHC), 684
  Five-Day Graphical Tropical Weather Outlooks, 690–692, 691f
  Potential Storm Surge Inundation Map, 696f
  SLOSH model, 694–696, 696f
  Track Forecast Cone, 692–694, 693f
  tropical cyclone, 690–692, 691f, 693f
  Wind Speed Probability products, 694–696, 695f
National Institute of Information and Communication Technology (NICT). See Radio Research Laboratory (RRL)
National Oceanic and Atmospheric Administration (NOAA), 283, 285, 459, 687
  Active Region No. 12192, 79
  flare forecast, 81
  GOES see Geostationary Operational Environmental Satellite (GOES)
  Solar Katrin, 690
  space weather scales, 286–287, 460–462
  SWPC see Space Weather Prediction Center (SWPC)
  tropical cyclone, 690–693, 691f, 693f
National Research Council (NRC), 520, 687
National Risk Register, 2015, 666–667
National Space Weather Action Plan, 514–515, 685
National Space Weather Strategy, 84, 514–515
National Weather Service (NWS) Sea, Lake, and Overland Surges from Hurricanes (SLOSH) model, 694–696
Natural geoelectric fields, 4, 8–16
Natural hazard
  developing resilience, 660
  policy responses, 662–663
  space weather see Space weather
Natural radiation, 6

NavCanada Arctic Air Traffic Control Center in Edmonton, 638–639, 638f
Navigation with Indian Constellation (NAVIC), 543–544
Nearest neighbor (NN) data-mining algorithm, 263
Neutral atmosphere response, 513–514, 519, 525–529, 526–527f
Neutral composition, 515, 516f
  change, 519
  CTIPe, 525, 526–527f, 528–529
Neutral density, 515
  CHAMP satellite, 519–520
  NRLMSISE-00 empirical model, 516f, 517–519
  real storm in December 2006, 530f
Neutral wind, 515
  CTIPe, 525, 527f
  density change, 533
  horizontal, 527f, 529
  thermospheric, 519
  upper thermosphere, 528
  vertical, 519, 527f
Neutron monitor (NM)
  Moscow, 286, 303
  solar energetic particles, 458–459
  station, North America, 464
  University of Oulu, 462
Neutrons, cosmic ray, 25
NHC. See National Hurricane Center (NHC)
NLFFF models, 79–81
Nonequilibrium system, 139–141
Nonparametric bootstrapping, 121
Nonparametric discriminant analysis (NPDA), 83–84
Norman Wells–Zama pipeline, 644–646
North America, neutron monitor station, 464
North American Electric Reliability Corporation (NERC), 685, 698
Nowcasting technique, 436
Nowcast of Atmospheric Ionizing Radiation for Aviation Safety (NAIRAS), 459, 460f
NPDA. See Nonparametric discriminant analysis (NPDA)
Nyquist's theorem, 212–213

O

Ohm's law, 10, 188
Oil pipelines, impact on ground systems, 192–193
Open Geospace General Circulation MHD Model (OpenGGCM), 522–525
Organisation Organization for Economic Co-operation and Development (OECD), 717
Ottawa (Canada), geomagnetic field component, 13f
Ottawa Magnetic Observatory, 646
Oval Variation, Assessment, Tracking, Intensity and Online Nowcasting (OVATION) Prime (OP) empirical model, 598, 599f

P

Particle Acceleration and Transport in the Heliosphere model (PATH), 626
Pasco model, 485
PC5 oscillations, 387
Penetration depth
　charged particle, 421, 429
　electrons, 421–422, 421f
PFRR. *See* Poker Flat Research Range (PFRR)
Phase scintillation, 558, 578–579
　CASES receiver, 562f, 563, 564f, 565
　high-latitude GPS, 561–565
　nightside characteristics, 561–562
Photoelectron emission, 404, 405f
Photoionization, 494–495, 616–617
Photomultipliers (PMs), 477–478
Photon number distribution, TLE, 480, 480f
PILs. *See* Polarity inversion lines (PILs)
Pipelines, 644–646, 651
　GIC impacts on, 15
　impact on ground system, 192–193
　TransCanada, 645–646
Plane wave method, 189
Plasma density, 513–515
　D-region, 533
　F-region, 519–520
Plasma flows, 18
Plasma pressure, 265–267
　distribution, 260, 268f, 270, 275
　during March 2015 storm, 269f
Plasmasphere, during storm, 531–532, 532f
Plasma temperature, 409, 409f
Poisson's Equation, deep dielectric charging, 420
Poker Flat Research Range (PFRR), 563, 565
Polar cap patches, 565–566
Polarity inversion lines (PILs), 70
Polar LEOs, 27
Polar Operational Environmental Satellite (POES), 591, 594–595, 594f
Polar Wind Outflow Model (PWOM), 623
Policymakers, 658–659
　atmospheric radiation, 675–676
　data, space weather, 671–672
　meteorology, 672–674
　natural hazards, 662–663
　risk matrix, 666–667, 666f
　wake-up call for, 661–662
Position, navigation, and timing (PNT) information, 575, 708
Positioning error, solar radio burst, 546–548
Potential Storm Surge Inundation Map, 696f
Power grids, 4, 12, 14–15, 705–707, 716–717

Power-law bootstrap fit, parameters for, 125f.
　See also Bootstrapping
Power-law distributions, 100, 116
Power systems
　control center, 645f
　geomagnetic effects on, 642–644
　higher-voltage transmission line, 651
Poynting flux, 73–74
PPEF. *See* Prompt penetration electric field (PPEF)
Prebiotic chemistry, 612, 627–628
Pressure gradient force, 503, 504f
Principal component analysis (PCA), 358
Private-sector network, 689
Professional Aviation Dose Calculator (PANDOCA) model, 464, 464–465f
Project for Solar-Terrestrial Environment Prediction (PSTEP), 725
Prompt penetration electric field (PPEF), 164, 496, 497f, 531
　buoyancy waves, 503
　equatorial scintillation, 499–501
　during superstorms, 501
Protons, 374
　energetic, 129–130, 159–160
　radiation belts, 359–360, 360f
　radiation belt variations over solar cycles, 359–360, 360f
　solar proton events *see* Solar proton events (SPEs)

Q

Quasielectrostatic field (QE), 483
Quasipower-law behavior, 116
Quasi-Zenith Satellite System (QZSS), 543–544
Québec storm 1989, 701–702

R

Radarsat I/II satellite, 642
Radar system, 8
Radiation
　atmospheric radiation environment, 6
　from cosmic rays, 25
　effects on avionics, 456
　environment for crewed orbital stations, 361–367, 362f, 364f, 366f
　in human tissues, 456
　linear energy transfer, 456
　on mars, 447–448
　natural, 6
　sources, 453–456
Radiation belt environment (RBE) model, 242
Radiation belts, 374, 376–378, 386, 390. *See also* Van Allen radiation belts
　electrons, 232–233, 593–595

for extreme storms, 239–240
proton, variations over solar cycles, 359–360, 360f
temporal variation, 423–424, 425f
Radiation Belt Storm Probe Ion Composition Experiment (RBSPICE) instrument, 267, 270
Radiation Dosimetry Experiment (RaD-X), 460
Radiation environment
　GEO, 593–594, 594f
　LEO, 594–595, 596t
Radiation hazards, 448
　in space, 350–351, 365, 367
　transient luminous event, 486–487
Radiation-induced conductivity, 420
Radiation monitoring system (RMS), 361, 363t
Radiation trapping, 423
Radioactive elements, in Earth's crust, 25
Radio blackouts, 686
Radio frequency noise, 8
Radio occultation (RO), 545–548, 550–551
Radio Physics Laboratory of the Defence Research Board (DRB), 637
Radio propagation, trans-ionospheric, 17–22
Radio Research Laboratory (RRL), 719–721
Radio solar telescope network (RSTN), 543–544, 544f
Radio system
　space weather impacts on, 20–22t
　trans-ionospheric, 19
　VHF, 26–27
Radio waves, 17–18, 285
RaD-X. See Radiation Dosimetry Experiment (RaD-X)
Rail systems, GIC impact on, 15
RAM. See Ring current-atmosphere interactions model (RAM)
Random forest classifier (RFC), 83–84
Ranging Integrity Monitoring Stations (RIMS), 545–546
Rate of TEC (ROT), 601
Rate of TEC index (ROTI), 601
Rayleigh-Taylor instability, 500, 561
Receiver operating characteristic (ROC) curve, 77–78, 78f
Regener-Pfotzer maximum, 455–456, 460
Regional Warning Center (RWC), 712–713
Relativistic electron dropout (RED), 423–424
　and acceleration, 377–379, 377f, 379f
　events, 374–375
Relativistic electron flux variability, 389
Relativistic electron variation, during ICME magnetic storms, 387–390
Rice convection model (RCM), 198–199, 235–236
Righthand circular polarized (RHCP), 543
RIMS. See Ranging Integrity Monitoring Stations (RIMS)
Ring current-atmosphere interactions model (RAM), 235
Ring current plasma, 232
　distribution of, 232

Risk matrix, policymakers, 666–667, 666f
RMS. See Radiation monitoring system (RMS)
RO. See Radio occultation (RO)
Runaway electron, 483, 486
Russian satellite COSMOS 2251, 517

S

SAA. See South Atlantic Anomaly (SAA)
Sabine, Edward, 158
Saffir-Simpson Hurricane Wind Scale, 688
SAID. See Subauroral ion drifts (SAID)
Sami is Another Model of the Ionosphere-2 (SAMI-2), 498–499
SAMPEX. See Solar Anomalous and Magnetospheric Particle Explorer (SAMPEX)
SAPS. See Subauroral polarization streams (SAPS)
Satellite. See also specific types of satellites
　anomalies, 411f
　charging, 27
　drag, 16
　industry, 708
　Japan, 720–721, 723–724, 724f
　operations specialists, 602–603
　plasma environment, 6, 27–29
Satellite based augmentation systems (SBAS), 576, 579, 721
Scaling exponents, 146–147, 148f, 149–150
SCATHA (Spacecraft Charging AT High Altitudes, P78-2), 402, 404
Schwabe, S. Heinrich, 158
Scientific knowledge, 664–666, 674–677
Scintillation, 17, 558
　amplitude, 545, 558, 560–561, 578–579
　in Canada, 647
　equatorial, 499–501
　high-latitude ionosphere, 561–565, 561f
　low-latitude ionosphere, 560–561, 561f
SDUs. See Silicon detector units (SDUs)
Sea, Lake, and Overland Surges from Hurricanes (SLOSH) model, 694–696, 696f
Seawater, ground impedance, 11
SED. See Storm enhanced density (SED)
SEEs. See Single-event effects (SEE)
Self-organized criticality (SOC), 70, 133
SEON. See Solar electro-optical network (SEON)
SEPs. See Solar energetic particle (SEP)
Severe Space Weather Warning, 716
Shock effects, 389
Short-time scale forecasting, 174–175
SID. See Sudden ionospheric disturbance (SID)
Sigma-phi, 560
Signal loss of lock (LOL), 545–546
Signal-to-noise ratio (SNR), SRB effect, 545, 550f

Silicon detector units (SDUs), 438
SILSO. *See* Sunspot Index and Long-term Solar Oscillations (SILSO)
Single-event effects (SEE), 6, 25–26, 29, 350
Single event upset (SEU), 422–423, 588–589, 593, 595, 707–708
Single-event upset rate (SEU rate), 366–367, 366f
Single-frequency GPS receivers, 575
Skill-score metrics, 77
SNOE NO empirical model (NOEM), 525–528
SOC. *See* Self-organized criticality (SOC)
Societal impacts, 11
  on human activities on Earth, 3
  solar flares and, 66–69
  of space weather, 3
Socioeconomic impacts, space weather, 661, 703–705, 704f, 706t
SOC-type models, 87
SOHO. *See* Solar and Heliospheric Observatory (SOHO)
Solar active regions (ARs), 69f, 70, 72, 85f, 171
Solar activity, research on, 158
Solar and Heliospheric Observatory (SOHO), 38–40, 591
Solar Anomalous and Magnetospheric Particle Explorer (SAMPEX), 425f
Solar arrays, 402–403, 403f
Solar cosmic rays, 354
Solar cycle 24, 16, 588–589, 670, 716
  geomagnetic storm, 716
  St. Patrick's Day Storm, 575
Solar cycles, 426
  declining phase, 424, 426
  GCR modulation over, 354–359, 359f
  inner proton radiation belt variations over, 359–360, 360f
  modulation, 132
  from 1957 through 2016, 126t
  phase dependence of electron acceleration, 379–381, 380f
  SSPA anomalies occur during, 427f
  temporal variation, radiation belts, 423–424
Solar Dynamic Observatory (SDO), 611–612
Solar electro-optical network (SEON), 302
Solar energetic particle (SEP) events, 19, 25, 29, 41, 51–52, 52f, 55f, 164, 232, 349–350, 453–454, 454f, 626–628, 686, 722–723
  acceleration mechanisms, 590–591
  extreme SEP events concept, 350–351
  fluences, 52–54, 53f
  GEO, 592–593
  largest SEP events and distribution function, 351–354, 352f
  lower energy, 28–29
  neutron monitor, 458–459
  penetration of, 354, 355f, 365
  into polar upper atmosphere, 669
  protons fluence spectra, 626, 627f
  Regener-Pfotzer maximum, 456
  solar eruption *see* Solar eruption
  Space Weather Prediction Center, 593–594
Solar eruption, 37–38, 40
  consequences of, 51
    large geomagnetic storms, 54–58
    SEP events, 51–52, 52f, 55f
    SEP fluences, 52–54, 53t, 53f
  Halloween, 38
  origin of, 46–47
Solar EUV. *See* Extreme ultraviolet (EUV)
Solar flare, 19, 37–38, 44–45, 84, 116, 612–613, 676
  active region characteristics, 70t
  associated ionizing radiation, 66, 67f
  CME, 436
  Dec. 5–17, 2006, 321t
  detection, 304
  effects, 519–520
  electromagnetic emission from, 40–41
  extreme, 84–86
  forecast product, 690–692
  impact of, on radio-wave communications, 285
  occurrence, 541–543
  size distribution, 44–46, 46f
  size nomenclature, 68t
  and societal impacts, 66–69
  thermal heating, 68–69
  trigger mechanism, 88
  X-class *see* X-class solar flares
  X-type, 615–616
Solar flare forecasting, 82–83, 86–87
  addressing correlated variables, 88
  basic paradigms, 70
  challenges of, 68
  evaluation, 76–78
  event definitions, 71
  flare trigger mechanism, 88
  history, 68–69, 69f
  numerical models role, 78–81
  operational *vs.* research, 76
  parametrizations, 72–74, 72f, 74f
  self-organized criticality and related, 75–76
  statistical approaches, 89
  statistical classifiers, 74–75
Solar flare trigger mechanism, 88
Solar particle events, 434, 456, 457f, 458
  ALTEA detector system, 438, 439f
  countermeasures, 436, 448
  December 13, 2006, 438–441, 440f
  effect of, 441
  estimation and confidence intervals for, 132t

extra vehicular activity, 448
February 1956 event, GLE05, 626–628
ISS, 437–446
large, 131f
Martian surface, 448
risk mitigation, 435–437
Solar proton events (SPEs), 131f, 285, 364f
  extreme SEP events concept, 350–351
  large, 131f
  largest SEP events and distribution function, 351–354, 352f
  solar cosmic rays, 354, 355–356f
Solar radiation storms
  forecast products, 697
  hazard, 686–687
Solar radio burst (SRB), 676, 686
  altitude variation, 550f
  effect on GNSS, 544–547
  electron density profile, 550–551
  impact on December 6, 2006, 547–551, 548–549f
  loss of lock, 545–546
  positioning error, 546–548
  SNR reduction, 545
  time variations, 549f
  types, 541–543, 542f
  WAAS, 548, 551–552
Solar radio emissions, 541–543, 547, 686
Solar sources, CMEs, 41f
Solar superflares, 614–616
Solar system, 4
Solar Terrestrial Relations Observatory (STEREO), 611–612, 671–672, 716
  STEREO-A spacecraft, 116, 520–522, 670
Solar Ultraviolet Imager (SUVI), 591
Solar wind (SW), 108f, 232
  evolution, 617
  on geospace, 593
  interplanetary driving, 375–376, 376f
  type, 99–100
Solar wind–magnetosphere system, 141–142
Solar X-ray imager (SXI), 590–591, 590f
Solar X-ray irradiance sensor (XRS), 285
South Atlantic Anomaly (SAA), 29, 437, 441, 448, 594–595
Southeast Asia Low-latitude Ionospheric Network (SEALION), 720–721, 721f
Space
  hazards, thresholds, 430
  plasma environments, 7–8, 7f
  radiation hazards in, 350–351, 365
  solar proton events in, 350–354
Space-based GNSS, effect, 546–547
Space-based monitoring, 457–458
Spacecraft, design guidelines, 429–430
Spacecraft anomalies
  deep dielectric charging, 422, 426, 428
  low Earth Orbit, 588–589
Spacecraft charging
  equation, 405–411
  events
    extreme, 402
    limits, 412–415, 413–414f
    worst, 411–412, 413t
  index, 406
  surface, 406–408
  traditional, 402
Space debris, 517
Space Environment Monitor (SEM), 592–593, 592f
Space environments
  deep dielectric charging, 423–426
  during geoeffective events, 588–589, 589t
  1983–2014, 284f
Space habitat
  forecasting, 447
  SPE measurement, 437
Space radiation, 434
  characteristics, 435
  risks, 433–435
Space suit, 434–435
Space Surveillance Network (SNN), 517
Space weather (SW), 3–4, 115, 159, 173, 350, 514, 657
  adverse impacts of, 4
  assessment of, 460–462
  aviation radiation, 460–463
  Canada see Canada, space weather
  community, 671
  concepts, 659–662
  data, 671–672
  data-driven modeling of, 141–145, 142f, 144–145f
  effect on aviation, 707–708
  effects on GNSS, 19
  emergency managers see Emergency managers
  environments at Earth, 4–8
  Florida Effect, 504–506, 506f
  forecasting of, 144
  GNSS, 30, 708
  habitability conditions, 624–628
  impacts on upper atmosphere, 16–17
    atmospheric drag, 22–24, 24f
    atmospheric radiation environment, 25–27
    satellite plasma environments, 27–29
    trans-ionospheric radio propagation, 17–22, 20–22t
  importance of science, 664–666
  ionospheric variability see Ionosphere
  Japan, 719–720
    aviation system, 721–723

Space weather (SW) *(Continued)*
    geomagnetically induced currents, 724
        satellite development process, 723–724, 724f
        satellite positioning, 720–721
        telecommunications, 720–721
    linkages between sectors, 658f
    meteorology, 672–674
    moderate *vs.* severe, 701–703
    power grid, 705–707
    predictability of, 146–149, 147–148f
    processes and impacts, 5f
    risks evolve, 29–31
    satellites and, 708
    social and economic impacts, 661, 703–705, 704f, 706t
    societal impacts of, 3
    thermospheric parameters, 515
    threats, 350
    transient luminous event, 483–485
    types, 684
    worst case, 665–666
    WSA-Enlil prediction model, 694
    from young Sun, 613–617
Space Weather Action Plan (SWAP), 193–194, 702–703
Space weather modeling framework (SWMF), 195, 199f, 619–621
Space Weather Operations, Research, and Mitigation (SWORM), 514–515
Space Weather Prediction Center (SWPC), 588–589, 684, 689, 694
    OP empirical model, 593–594
    solar energetic particle events, 593–594
    total electron content, 598–601, 600f
Space Weather Services (SWS), 711–712, 716
    Dst geomagnetic storm index, 714, 715f
    geomagnetic K index, 714f
    policy, 716
Spatiotemporal model, 145f
Spectral energy distribution (SED), 615f
SPEs. *See* Solar proton events (SPEs)
Sprites, 474–477, 475f
    energetic, 476f
    size calculation, 476–477
SRB. *See* Solar radio burst (SRB)
SSNs. *See* Sunspot numbers (SSNs)
St. Patrick's Day storms, 322b, 575, 588–589, 591
    CME event, 618–619, 619f
    total electron content, 601
Stakeholders, 705, 706t
    aviation, 707–708
    electric power and, 705–707
    satellites, 708
    technology groups, 716–717

Standard operating procedures (SOPs), 717
Static VAR Compensators (SVCs), 643
STEREO. *See* Solar Terrestrial Relations Observatory (STEREO)
STET code. *See* SuperThermal Electron Transport (STET) code
Storm. *See also* specific types of storm
    behavior of, 101
    CME-induced, 104
    March 2015 storm, 264–270, 265–266f, 268–269f
Storm enhanced density (SED), 529–531, 565–569, 567f, 598
    electron density, 567f
    plasma density, 311–312
    Superstorm, 568–569, 572–573
    TEC distribution, 572–573, 573f
Strategic National Risk Assessment, 514–515
Stream interaction regions (SIRs), 590–591
Strong magnetic storms, 259–260, 353–354
Subauroral ion drifts (SAID), 309
Subauroral polarization streams (SAPS), 315–316, 568–569, 568f
Substorm current wedge (SCW), 195
Successful retrieval percentage (SRP), 546–547
Sudden ionospheric disturbance (SID), 304, 711–712, 720
Sudden Storm Commencement (SSC), 287–300
Sun, 611–613
    coronal mass ejection, 37–38, 41–42, 614–615
        parameters, 173–174
        speed, 57, 171, 174
    Faint young Sun paradox, 626–628
    magnetic field, 158–159, 444
    radio frequency noise, 8
    space weather, 612–617, 684
    spectral energy distribution, 615–616, 615f
    superflares, 45–46, 353
    wind, 616–617
Sunspot Index and Long-term Solar Oscillations (SILSO), 286
Sunspot magnetic field, 48
Sunspot numbers (SSNs), 286
Sun-to-thermosphere model, 234
Super-Carrington CME event, 614–616, 620–621, 623–624
Superfountain effect, 531
    ionospheric, 496–499, 497f
Superintense magnetic storms, 165–166t
Supermagnetic storms, 164
    future, 171–172
    intensity of, 171–172
    nowcasting and short-term forecasting of, 173–174
    past, 166–169
    present, 169–170
Superstorm, 572–575. *See also* Supermagnetic storm
    Halloween, October 2003, 576–577
    November 2004, 575

SED events, 568–569, 572–573
TEC, 574–575
SuperThermal Electron Transport (STET) code, 623
Surface flow, 73–74
SVCs. *See* Static VAR Compensators (SVCs)
SWPC. *See* Space Weather Prediction Center (SWPC)
SWS. *See* Space Weather Services (SWS)
SYM-H index, geomagnetic, 232
Synchronous Meteorological Satellite (SMS)-1, 592–593

T

Taqqu's theorem, 149
TEC. *See* Total electron content (TEC)
Telecommunications, Japan, 720–721
Telegraph problems, Canada, 635–636
Telesat Canada, 640–642
Telluric currents, 644–646
Temporal variation, radiation belts, 423–424, 425f
TEPCs. *See* Tissue Equivalent Proportional Counters (TEPCs)
Terrestrial gamma flashes (TGFs), 483, 486–487
Thermal heating, flare-related, 68–69
Thermosphere
  model, 233
  neutral wind, 519
  parameters, 515
  response to storm, 520
Thermostat, 525–528
Thomson, William, 158–159
Three-dimensional magnetohydrodynamic (MHD) model, 617–623, 619f
TID. *See* Traveling ionospheric disturbances (TID)
Time-dependent MHD models, 79–81
Time History of Events and Macroscale Interactions during Substorms (THEMIS), 264
Time series, geomagnetic waveform, 212–214
TISN. *See* Trusted Information Sharing Network (TISN)
Tissue Equivalent Proportional Counters (TEPCs), 457–458
Tongue of Ionization (TOI), 601
Total electron content (TEC)
  electric fields, 495–501
  Florida Effect, 504–506, 506f
  ground-based, 495–496
  Halloween Storm, 574f
  ionosphere response, 494–495, 559–560, 598–601, 600f
    storm enhanced densities, 565–569, 567f
    traveling ionospheric disturbances, 569–571, 569t, 570–572f, 577–578
  Japan, 722f
  pre-storm, 495–496
  St. Patrick's storm, 601
  storm enhanced densities, 572–573, 573f
  SWPC, 598–601, 599–600f

Total ionizing dose (TID), 459
Track Forecast Cone, 692–694, 693f
TransCanada pipeline, 645–646
Transient luminous event (TLE), 473–474
  energy deposition, 485–486
  Moscow State University satellites, 477–482
  phenomenology, 474–477
  photon number distribution, 480, 480f
  radiation hazard, 486–487
  series, 480–482, 481–482f
  theoretical models, 483–485
  UVRIR detector, 478–479, 479f
Trans-ionospheric radio propagation, 17–22
Traveling ionospheric disturbance detector built in Texas (TIDDBIT) system, 570–571, 570f
Traveling ionospheric disturbances (TID), 569–571, 569t, 570–572f, 577–578
Tropical cyclone, 688
  damage mechanisms, 694–696
  NHC, 690–692, 691f, 693f
  NOAA, 690–693, 691f, 693f
Trusted Information Sharing Network (TISN), 716
TS07D model, 260–262, 264–265

U

UK National Risk Assessment, 666–667, 666f
UK Space Weather Public Dialogue, 663
UK's Royal Academy of Engineering, 575–576
Ultralow frequency (ULF) waves, 163–164
Ultraviolet (UV), effects, 517–519
Universitetsky-Tatiana-2 spacecraft, 477–482
  data, 481–482, 482f
  photon number distribution, 480, 480f
University of Oulu (Finland), neutron monitor, 462
Upper atmosphere
  Earth, 16
  expansion, 515
  space weather impacts on, 16–17
    atmospheric drag, 22–24, 24f
    atmospheric radiation environment, 25–27
    satellite plasma environments, 27–29
    trans-ionospheric radio propagation, 17–22, 20–22t
  sprites development, 474–477, 475–476f
  TLE in *see* Transient luminous event (TLE)
U.S. Air Force (USAF) space weather operators, 302
U.S. Air Force Weather Agency (USAFWA), 543
U.S. Continuously Operating Reference Station (CORS), 545–546
U.S. FEMA, 683, 685, 688–689, 699
  impact assessment, 685
  internal assessment, 686–687
  in satellite industry (2011–2012), 686

U.S. FEMA *(Continued)*
    space weather event alert and notification process, 685*f*
U.S. Geological Survey (USGS), 213–214, 217–218, 219–220*f*
US Total Electron Content (US-TEC), 598–601, 600*f*
UTC. *See* Coordinated universal time (UTC)
UV and red infrared (UVRIR) detector
    transient luminous event, 478–479, 479*f*, 481*f*, 486
    Vernov satellite, 479*f*, 481*f*

V

Van Allen radiation belts, 374, 423, 441. *See also* Radiation belts
Vernov satellite, 477–482
    data, 480–482, 482*f*
    UVRIR detector, 479*f*, 481*f*
Versatile electron radiation belt (VERB) 3D code, 239
Vertical protection level (VPL), 602–603
Very high frequency (VHF) radio system, 26–27
Very low frequency (VLF), TLE, 474, 484–485
von Lamont, Johann, 158
Vuong's test, 121–124, 126*t*

W

WAAS. *See* Wide Area Augmentation System (WAAS)
Warning System for AVIation Exposure to Solar (WASAVIES), 723
Waveform receiver (WFR), 390
Weibull function, 46–47
Weimer empirical model
    Carrington event, 522–525, 523*f*
    Joule heating from, 522, 524*f*
White House Office of Science and Technology Policy (OSTP), 514–515
Wide Area Augmentation System (WAAS), 494, 504–505, 708
    aviation system impacts, 602–604, 604*f*
    Federal Aviation Administration, 545–546
    horizontal protection level, 576–577, 577*f*
    solar radio burst, 548, 551–552
Wind Speed Probability products, NHC, 694–696, 695*f*
World Meteorological Organization, 671
WSA-Enlil prediction model, 694

X

X-class solar flares, 311, 545, 547–548, 692
X-ray and UV (XUV) radiation, 615–616, 622–623, 624*f*
X-ray flares, solar energetic particles, 595–598
X-ray sensor (XRS), 285, 590–591, 590*f*

Y

Young Earth, atmospheric escape, 622–623, 624*f*
Young Sun
    space weather from, 613–617
    wind, 616–617

Edwards Brothers Inc.
Ann Arbor MI. USA
December 14, 2017